# The Right Tools to Drive Your Success

The power to succeed is in your hands. Stay focused on what's important, get the support you need, and practice the concepts regularly. Strategically placed learning aids throughout the book give you guidance right when you need it.

**Strategies** and **Procedures** appear throughout the text to sharpen your problem-solving skills. **Strategies** contain general guidelines while **Procedures** provide a step-by-step approach for solving specific types of problems.

> **STRATEGY**
> **Solving Exponential and Logarithmic Equations**
>
> 1. If the equation involves a single logarithm or a single exponential expression, then use the definition of logarithm: $y = \log_a(x)$ if and only if $a^y = x$.
> 2. Use the one-to-one properties when applicable:
>    a) if $a^M = a^N$, then $M = N$.
>    b) if $\log_a(M) = \log_a(N)$, then $M = N$.

Page 317

> **PROCEDURE**
> **Graphing a Rational Function**
>
> To graph a rational function in lowest terms:
> 1. Determine the asymptotes and draw them as dashed lines.
> 2. Check for symmetry.
> 3. Find any intercepts.
> 4. Plot several selected points to determine how the graph approaches the asymptotes.
> 5. Draw curves through the selected points, approaching the asymptotes.

Page 253

**Foreshadowing Calculus** gives a brief indication of the connection between certain algebraic topics and calculus, and identifies topics that you will see again in the future.

> **Foreshadowing Calculus**
>
> There are several ways to define the number $e$ and the function $f(x) = e^x$ in calculus. Calculus can deepen your understanding of many of the concepts that you study in algebra.

Page 282

**Function Galleries** show entire families of functions and their graphs, helping you make the connection between their visual and mathematical properties.

> FUNCTION **gallery...** Exponential and Logarithmic Functions
>
> **Exponential:** $f(x) = a^x$, domain $(-\infty, \infty)$, range $(0, \infty)$

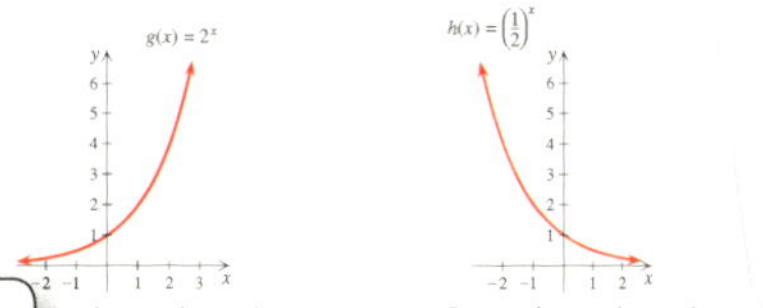


creasing on $(-\infty, \infty)$
intercept $(0, 1)$

Decreasing on $(-\infty, \infty)$
$y$-intercept $(0, 1)$

Increasing on $(-\infty, \infty)$
$y$-intercept $(0, 1)$

Page 295

**Tying It All Together** exercises cover the present and prior chapters so that you can integrate multiple concepts and skills.

AVAILABLE IN MyMathLab®

> TYING IT ALL **together...**
>
> **Chapters 1–4**
>
> *Solve each equation.*
>
> **1.** $(x - 3)^2 = 4$ **2.** $2 \cdot \log(x - 3) = \log(4)$ **3.** $\log_2(x - 3) = 4$
> **4.** $2^{x-3} = 4$ **5.** $\sqrt{x - 3} = 4$ **6.** $|x - 3| = 4$
> **7.** $x^2 - 4x = -2$ **8.** $2^{x-3} = 4^x$ **9.** $\sqrt[3]{x - 5} = 5$
> **10.** $2^x = 3$ **11.** $\log(x - 3) + \log(4) = \log(x)$ **12.** $x^3 - 4x^2 + x + 6 = 0$
>
> *Sketch the graph of each function.*
>
> **13.** $y = x^2$ **14.** $y = (x - 2)^2$ **15.** $y = 2^x$ **16.** $y = x^{-2}$
> **17.** $y = \log_2(x - 2)$ **18.** $y = x - 2$ **19.** $y = 2x$ **20.** $y = \log(2^x)$
> **21.** $y = e^2$ **22.** $y = 2 - x^2$ **23.** $y = \frac{2}{x}$ **24.** $y = \frac{1}{x - 2}$

Page 332

**Concepts of Calculus** discussions preview important topics of calculus, and may be used as a writing or collaborative learning assignment.

> CONCEPTS OF **calculus...**
>
> **The instantaneous rate of change of $f(x) = e^x$**
>
> *In the Concepts of Calculus for Chapter 2, we defined the average rate of change of a function f on the interval $[x, x + h]$ as*
>
> $$\frac{f(x + h) - f(x)}{h}.$$
>
> *We then defined the instantaneous rate of change of the function as the limit as h approaches zero of the average rate of change. The notation $f'(x)$, read "f prime of x," is used for the instantaneous rate of change. So*
>
> $$f'(x) = \lim_{h \to 0} \frac{f(x + h) - f(x)}{h}.$$
>
> *In the following exercises we will find the instantaneous rate of change of the base e exponential function. It will be necessary here to find a limit using a table of values as we did in the Concepts of Calculus for Chapter 1.*

Page 334

Continued next page

# The Right Tools to Drive Your Success

End-of-section exercises and review features help you check your progress and make sure you're mastering the material.

AVAILABLE IN MyMathLab®

**FOR thought...** True or False? Explain.

1. The first coordinate of an ordered pair in an exponential function is a logarithm.
2. $\log_{100}(10) = 2$
3. If $f(x) = \log_3(x)$, then $f^{-1}(x) = 3^x$.
4. $10^{\log(1000)} = 1000$
5. The domain of $f(x) = \ln(x)$ is $(-\infty, \infty)$.
6. $\ln(e^{2.451}) = 2.451$
7. For any positive real number $x$, $e^{\ln(x)} = x$.
8. For any base $a$, where $a > 0$ and $a \neq 1$, $\log_a(0) = 1$.
9. $\log(10^3) + \log(10^5) = \log(10^8)$
10. $\log_2(32) - \log_2(8) = \log_2(4)$

**For Thought** are true/false questions that review the basic concepts from the section, helping you check your understanding before starting on the exercises. Answers are available in the back of the book.

**EXERCISES 4.2**

*Fill in the blank.*

1. The inverse of an exponential function is a(n) ________ function.
2. A base 10 logarithm is a(n) ________ logarithm.
3. Base $e$ logarithm is a(n) ________ logarithm.
4. The function $f(x) = \log_a(x)$ is ________ if $a > 1$ and ________ if $0 < a < 1$.
5. The $y$-axis is a(n) ________ for the graph of $f(x) = \log_a(x)$.
6. The ________ of the function $f(x) = \log_a(x)$ is $(0, \infty)$.
7. The ________ of functions consists of all functions of the form $f(x) = b \cdot \log_a(x - h) + k$.
8. The ________ property of logarithms indicates that if $\log_a(m) = \log_a(n)$, then $m = n$.

*Determine the number that can be used in place of the question mark to make the equation true.*

9. $2^? = 64$
10. $2^? = 16$
11. $3^? = \frac{1}{81}$
12. $3^? = 1$
13. $16^? = 2$
14. $16^? = 16$
15. $\left(\frac{1}{5}\right)^? = 125$
16. $\left(\frac{1}{5}\right)^? = \frac{1}{125}$

*Find the indicated value of the logarithmic functions.*

17. $\log_2(64)$
18. $\log_2(16)$
19. $\log_3(1/81)$
20. $\log_3(1)$
21. $\log_{16}(2)$
22. $\log_{16}(16)$
23. $\log_{1/5}(125)$
24. $\log_{1/5}(1/125)$
25. $\log(0.1)$
26. $\log(10^6)$
27. $\log(1)$
28. $\log(10)$
29. $\ln(e)$
30. $\ln(0)$
31. $\ln(e^{-5})$
32. $\ln(e^9)$

*Sketch the graph of each function, and state the domain and range of each function.*

33. $y = \log_3(x)$
34. $y = \log_4(x)$
35. $f(x) = \log_5(x)$
36. $g(x) = \log_8(x)$
37. $y = \log_{1/2}(x)$
38. $y = \log_{1/4}(x)$
39. $h(x) = \log_{1/5}(x)$
40. $s(x) = \log_{1/10}(x)$
41. $f(x) = \ln(x - 1)$
42. $f(x) = \log_3(x + 2)$
43. $f(x) = -3 + \log(x + 2)$
44. $f(x) = 4 - \log(x + 6)$
45. $f(x) = -\frac{1}{2}\log(x - 1)$
46. $f(x) = -2 \cdot \log_2(x + 2)$

*Use a graph or a table to find each limit.*

47. $\lim_{x\to\infty} \log_3(x)$
48. $\lim_{x\to 0^+} \log_3(x)$
49. $\lim_{x\to 0^+} \log_{1/2}(x)$
50. $\lim_{x\to\infty} \log_{1/2}(x)$
51. $\lim_{x\to 0^+} \ln(x)$
52. $\lim_{x\to\infty} \ln(x)$
53. $\lim_{x\to\infty} \log(x)$
54. $\lim_{x\to 0^+} \log(x)$

*Write the equation of each graph in its final position.*

55. The graph of $y = \ln(x)$ is translated three units to the right and then four units downward.
56. The graph of $y = \log(x)$ is translated five units to the left and then seven units upward.
57. The graph of $y = \log_2(x)$ is translated five units to the right, reflected in the $x$-axis, and then translated one unit downward.
58. The graph of $y = \log_3(x)$ is translated four units upward, six units to the left, and then reflected in the $x$-axis.

Page 297

**NEW! Fill-in-the-blank exercises** now appear at the beginning of the section exercise sets to help you learn definitions, rules, and theorems.

**Rethinking** exercises appear at the end of the section exercise sets for a cumulative review of skills learned in previous sections and chapters.

AVAILABLE IN MyMathLab®

**RETHINKING**

149. Find the domain and range of the function
$$f(x) = -\frac{1}{2}3^{x-5} + 7.$$
150. Solve $2^{x-3} = 4^{5x-1}$.
151. Evaluate $(2 \times 10^{-9})^3(5 \times 10^3)^2$ without a calculator. Write the answer in scientific notation.
152. A pond contains 2000 fish of which 10% are bass. How many bass must be added so that 20% of the fish in the pond are bass?
153. Find all real and imaginary solutions to $x^3 + 13x = 4x^2$.
154. The cost of installing an oak floor varies jointly with the length and width of the room. If the cost is \$875.60 for a room that is 8 feet by 11 feet, what is the cost for a room that is 10 feet by 14 feet?

Page 300

AVAILABLE IN MyMathLab®

**POP QUIZ 4.1**

1. What is $f(4)$ if $f(x) = -2^x$?
2. Is $f(x) = 3^{-x}$ increasing or decreasing?
3. Find the domain and range for $y = e^{x-1} + 2$.
4. What is the horizontal asymptote for $y = 2^x - 1$?
5. Solve $(1/4)^x = 64$.
6. What is $a$, if $f(a) = 8$ and $f(x) = 2^{-x}$?
7. If \$1000 earns 4% annual interest compounded quarterly, then what is the amount after 20 years?
8. Find the amount in the last problem if the interest is compounded continuously.

Page 288

**Pop Quizzes** at the end of every section offer eight to ten questions to confirm that you understand the material before you continue reading.

**Linking Concepts** are multipart exercises that require the use of concepts from previous sections to illustrate the links among various concepts.

**LINKING concepts...** For Individual or Group Explorations

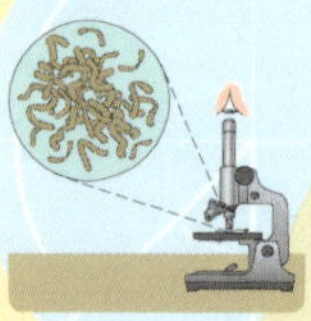

**Comparing Exponential and Linear Models**

*The function $f(t) = 300e^{0.5t}$ gives the number of bacteria present in a culture $t$ hours after the start of an experiment in which the bacteria are growing continuously at a rate of 50% per hour. Let $A[a, b]$ represent the average rate of change of $f$ on the time interval $[a, b]$.*

**a)** Fill in the following table.

| Interval $[a, b]$ | $f(a)$ | $A[a, b]$ | $A[a, b]/f(a)$ |
|---|---|---|---|
| $[3.00, 3.05]$ | | | |
| $[7.50, 7.51]$ | | | |
| $[8.623, 8.624]$ | | | |

**b)** What are the units for the quantity $A[a, b]$?

**c)** What can you conjecture about the ratio $A[a, b]/f(a)$?

**d)** Test your conjecture on a few more intervals of various lengths. Explain how the length of the interval affects the ratio.

**e)** State your conjecture in terms of variation (Section 2.6). What is the constant of proportionality?

**f)** Suppose the bacteria were growing in a linear manner, say $f(t) = 800t + 300$. Make a table like the given table and make a conjecture about the ratio $A[a, b]/f(a)$.

**g)** Explain how your conclusions about average rate of change can be used to justify an exponential model as better than a linear model for modeling growth of a bacteria (or human) population.

Page 288

PEARSON ALWAYS LEARNING

Mark Dugopolski

# Precalculus
## Functions and Graphs

Custom Edition for University of Southern Maine

Taken from:
*Precalculus: Functions and Graphs*, Fourth Edition
by Mark Dugopolski

Cover Art: Illustration by Nico Alba.

Taken from:

*Precalculus: Functions and Graphs*, Fourth Edition
by Mark Dugopolski

Published by Prentice Hall
Upper Saddle River, New Jersey 07458



This special edition published in cooperation with Pearson Learning Solutions.



Pearson Learning Solutions, 501 Boylston Street, Suite 900, Boston, MA 02116
A Pearson Education Company
www.pearsoned.com

Printed in the United States of America

000200010271661167

MC

PEARSON

ISBN 10: 1-256-76997-5
ISBN 13: 978-1-256-76997-2

18 2021

# Contents

# Preface

Making the transition to calculus is more than just finding a new classroom: it's about being prepared to grasp bigger and more complex mathematical concepts. *Precalculus: Functions and Graphs* is designed to make this transition seamless by focusing on all the skills that will be needed to succeed in calculus and beyond. This text contains the rigor essential for building a strong foundation of mathematical skills and concepts, and at the same time supports students' mathematical needs with a number of tools newly developed for this revision. With an eye toward future courses, this text provides students with an excellent opportunity to sharpen their critical thinking skills and introduces them to the usefulness and applicability of mathematics beyond the classroom. It is my goal that students will benefit from this approach and find that the mathematics presented here will enrich their classroom experience in this course as well as in future mathematics courses. Preparation is the foundation for success, and *Precalculus: Functions and Graphs* will help you succeed in this course and beyond.

## New to the Fourth Edition

For this edition of *Precalculus: Functions and Graphs*, I have updated explanations, examples, exercises, and art in response to comments from users of the last edition. In particular, I have written more than 900 new exercises that are specifically designed to increase student understanding and retention of the concepts that are taught in this text. Here are the major changes in this edition.

- **Updated real-data** in examples, exercises, and chapter openers make the text relevant for today's students.
- **Fill-in-the-blank exercises** are now used at the beginning of the section exercise sets to help students learn definitions, rules, and theorems.
- **Cumulative review exercises** are now used at the end of the section exercise sets to keep current the skills learned in previous sections and chapters. These exercises are under the heading "Rethinking."
- **Tying It All Together** exercises have been expanded to include fill-in-the-blank exercises that emphasize vocabulary.
- **Polynomial and rational inequalities** are now solved with one method, the test-point method. This change simplifies the procedure and makes it more consistent with techniques used in calculus.
- **Limit notation** is now introduced and used to describe the asymptotic behavior of exponential, logarithmic, rational, and trigonometric functions.
- **Try This** exercises have been included after every example in the text. These exercises are very similar to the corresponding examples and give the students the opportunity to immediately check their understanding of the example. Solutions to all of the Try This exercises are in the appendix of the *Student Edition*.
- **Suggested homework problems** are indicated in the Annotated Instructor's Edition. Each section exercise set contains 20 to 24 underlined exercise numbers. These exercises can be used as a set of homework exercises for the section.

## Continuing Features

With each new edition, all of the features are reviewed to make sure they are providing a positive impact on student success. The continuing features of the text are listed here.

### Strategies for Success

- **Chapter Opener** Each chapter begins with chapter opener text that discusses a real-world situation in which the mathematics of the chapter is used. Examples and exercises that relate back to the chapter opener are included in the chapter.
- **Foreshadowing Calculus** This feature gives a brief indication of the connection between certain algebraic topics and calculus, and identifies topics that will be continued in calculus.
- **Summaries** of important concepts are included to help students clarify ideas that have multiple parts.
- **Strategies** contain general guidelines for solving certain types of problems. They are designed to help students sharpen their problem-solving skills.
- **Procedures** are similar to *Strategies*, but are more specific and more algorithmic. *Procedures* are designed to give students a step-by-step approach for solving a specific type of problem.
- **Function Galleries** Located throughout the text, these function summaries are also gathered together at the beginning of the text. These graphical summaries are designed to help students link the visual aspects of various families of functions to the properties of the functions.
- **Historical Notes** Located in the margins throughout the text, these brief essays are designed to connect the topics of precalculus to the mathematicians who first studied them and to give precalculus a human face.
- **Hints** Selected applications include hints that are designed to encourage students to start thinking about the problem at hand. A *Hint* logo HINT is used where a hint is given.
- **Graphing Calculator Discussions** Optional graphing calculator discussions have been included in the text. They are clearly marked by graphing calculator icons so that they can be easily skipped if desired. Although the graphing calculator discussions are optional, all students will benefit from reading them. In this text, the graphing calculator is used as a tool to support and enhance algebraic conclusions, not to make conclusions.

### Section Exercises and Review

- **For Thought** Each exercise set is preceded by a set of ten true/false questions that review the basic concepts in the section, help check student understanding, and offer opportunities for writing and discussion. The answers to all *For Thought* exercises are included in the back of the *Student Edition*.
- **Exercise Sets** The exercise sets range from easy to challenging, and are arranged in order of increasing difficulty. Those exercises that require a graphing calculator are optional and are marked with an icon.
- **Writing/Discussion and Cooperative Learning Exercises** These exercises deepen students' understanding by giving them the opportunity to express mathematical ideas both in writing and to their classmates during small group or team discussions.
- **Thinking Outside the Box** Found throughout the text, these problems are designed to get students and instructors to do some mathematics just for fun. I enjoyed solving these problems and hope that you will too. The problems can be used for individual or group work. They may or may not have anything to do with the sections in which they are located and might not even require any techniques discussed in the text. So be creative and try *Thinking Outside the Box.* The answers are given in the *Annotated Instructor's Edition* only, and complete solutions can be found in the *Instructor's Solutions Manual.*
- **Pop Quizzes** Included at the end of every section of the text, the *Pop Quizzes* give instructors and students convenient quizzes of 8 to 10 questions that can be used in the classroom to check understanding of the basics. The answers appear in the *Annotated Instructor's Edition* only.

- **Linking Concepts** This feature is located at the end of nearly every exercise set. It is a multipart exercise or exploration that can be used for individual or group work. The idea of this feature is to use concepts from the current section along with concepts from preceding sections or chapters to solve problems that illustrate the links among various ideas. Some parts of these questions are open-ended, and require somewhat more thought than standard skill-building exercises. Answers are given in the *Annotated Instructor's Edition* only, and full solutions can be found in the *Instructor's Solutions Manual.*

**Chapter Review**

- **Highlights** This end-of-chapter feature contains an overview of all of the concepts presented in the chapter, along with brief examples to illustrate the concepts.
- **Chapter Review Exercises** These exercises are designed to give students a comprehensive review of the chapter without reference to individual sections and to prepare students for a chapter test.
- **Chapter Test** The problems in the *Chapter Test* are designed to measure the student's readiness for a typical one-hour classroom test. Instructors may also use them as a model for their own end-of-chapter tests. Students should be aware that their in-class test could vary from the *Chapter Test* due to different emphasis placed on the topics by individual instructors.
- **Tying It All Together** Found at the end of most chapters in the text, these exercises help students review selected concepts from the present and prior chapters, and require students to integrate multiple concepts and skills.
- **Concepts of Calculus** Most chapters end with a discussion of a particular concept of calculus, as well as exercises designed to illustrate that concept. This one-page feature will give students a preview of important topics of calculus, and may be used as a writing or collaborative learning assignment.
- **Index of Applications** The many applications contained within the text are listed in the *Index of Applications* that appears at the end of the text. The applications are page referenced and grouped by subject matter.

## Acknowledgments

Thanks to all of the professors and students who have used this text in previous editions. Special thanks to those who have taken the time to send me their comments and suggestions. I am always glad to hear from users of my texts. You can e-mail me at bookinit@charter.net. Thanks also to the professors who have reviewed all of the previous editions of this text.

Thanks to Edgar N. Reyes, Southeastern Louisiana University, for working all of the exercises and writing the *Solutions Manuals* and to Rebecca W. Muller, Southeastern Louisiana University, for writing the *Instructor's Testing Manual.* I wish to express my thanks to John Morin and Tom Wegleitner for accuracy checking. A special thanks to Nesbitt Graphics, the compositor, for the superb work they did on this book.

Finally, it has been another wonderful experience with the talented and dedicated Pearson team.

As always, thanks to my wife, Cheryl, whose love, encouragement, understanding, support, and patience are invaluable.

**Mark Dugopolski**
*Ponchatoula, Louisiana*

# Supplements List

## Student Supplements

### Student's Solutions Manual

- By Edgar N. Reyes, *Southeastern Louisiana University.*
- Provides detailed solutions to all odd-numbered text exercises.
- ISBN-13: 978-0-321-79118-4; ISBN-10: 0-321-79118-5

### A Review of Algebra

- By Heidi Howard, *Florida Community College at Jacksonville.*
- Provides additional support for those students needing further algebra review.
- ISBN-13: 978-0-201-77347-7; ISBN-10: 0-201-77347-3

## Instructor Supplements

### Annotated Instructor's Edition

- All answers included, with answers to most exercises on the page where they occur.
- Groups of exercises are keyed back to corresponding examples from the section.
- ISBN-13: 978-0-321-79063-7; ISBN-10: 0-321-79063-4

### Instructor's Solutions Manual

- By Edgar N. Reyes, *Southeastern Louisiana University.*
- Provides complete solutions to all text exercises, including the *For Thought* and *Linking Concepts* exercises.
- ISBN-13: 978-0-321-79117-7; ISBN-10: 0-321-79117-7

### Instructor's Testing Manual

- By Rebecca W. Muller, *Southeastern Louisiana University.*
- Contains six alternative forms of tests per chapter (two are multiple choice).
- Answer keys are included.
- Available for download from Pearson Education's online catalog.

### Graphing Calculator Manual

- By Darryl Nester, *Bluffton University.*
- Provides instructions and keystroke operations for the TI-83 Plus, TI-84 Plus, and TI-89.
- Also contains worked-out examples taken directly from the text.
- Available for download from Pearson Education's online catalog.

### PowerPoint Lecture Presentation

- Classroom presentation slides are geared specifically to sequence this textbook.
- Available within MyMathLab or from Pearson Education's online catalog.

## Technology Resources

### MathXL® Online Course (access code required)

**MathXL®** is the homework and assessment engine that runs MyMathLab. (MyMathLab is MathXL plus a learning management system.) With MathXL, instructors can:

- Create, edit, and assign online homework and tests using algorithmically generated exercises correlated at the objective level to the textbook.
- Create and assign their own online exercises and import TestGen tests for added flexibility.
- Maintain records of all student work tracked in MathXL's online gradebook.

With MathXL, students can:

- Take chapter tests in MathXL and receive personalized study plans and/or personalized homework assignments based on their test results.
- Use the study plan and/or the homework to link directly to tutorial exercises for the objectives they need to study.
- Access supplemental animations and video clips directly from selected exercises.

MathXL is available to qualified adopters. For more information, visit our website at www.mathxl.com, or contact your Pearson representative.

## MyMathLab® Online Course (access code required)

MyMathLab delivers **proven results** in helping individual students succeed. It provides **engaging experiences** that personalize, stimulate, and measure learning for each student. And, it comes from a **trusted partner** with educational expertise and an eye on the future.

- **NEW! Interactive Figures** enable you to manipulate figures to bring math concepts to life. They are assignable in MyMathLab and available in the Multimedia Library.
- **MyMathLab® Ready to Go Course (access code required)** These new Ready to Go courses provide students with all the same great MyMathLab features that you're used to, but make it easier for instructors to get started. Each course includes pre-assigned homework and quizzes to make creating your course even simpler. Ask your Pearson representative about the details for this particular course or to see a copy of this course.

To learn more about how MyMathLab combines proven learning applications with powerful assessment, visit **www.mymathlab.com** or contact your Pearson representative.

## MathXL® Tutorials on CD

This interactive tutorial CD-ROM provides algorithmically generated practice exercises that are correlated at the objective level to the exercises in the textbook. Every practice exercise is accompanied by an example and a guided solution designed to involve students in the solution process. Selected exercises may also include a video clip to help students visualize concepts. The software provides helpful feedback for incorrect answers and can generate printed summaries of students' progress.

## TestGen®

TestGen® (www.pearsonhighered.com/testgen) enables instructors to build, edit, print, and administer tests using a computerized bank of questions developed to cover all the objectives of the text. TestGen is algorithmically based, allowing instructors to create multiple but equivalent versions of the same question or test with the click of a button. Instructors can also modify test bank questions or add new questions. Tests can be printed or administered online. The software and testbank are available for download from Pearson Education's online catalog.

# FUNCTION gallery... Some Basic Functions and Their Properties

### Constant Function

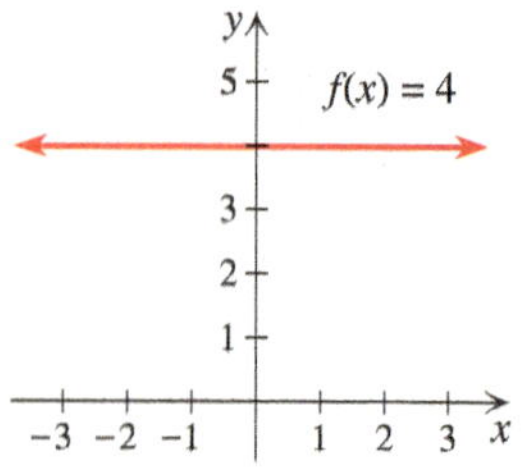


Domain $(-\infty, \infty)$
Range $\{4\}$
Constant on $(-\infty, \infty)$
Symmetric about $y$-axis

### Identity Function

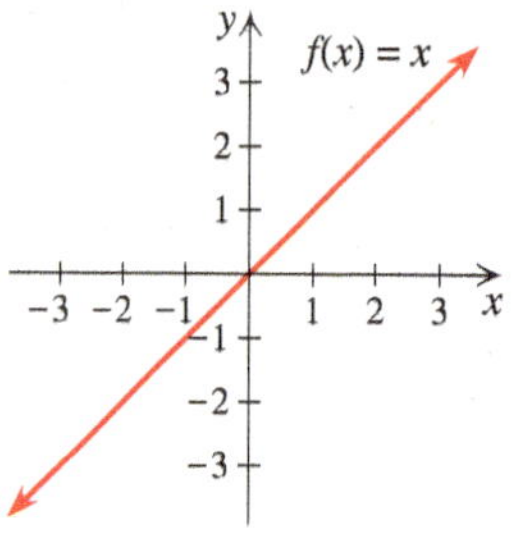


Domain $(-\infty, \infty)$
Range $(-\infty, \infty)$
Increasing on $(-\infty, \infty)$
Symmetric about origin

### Linear Function

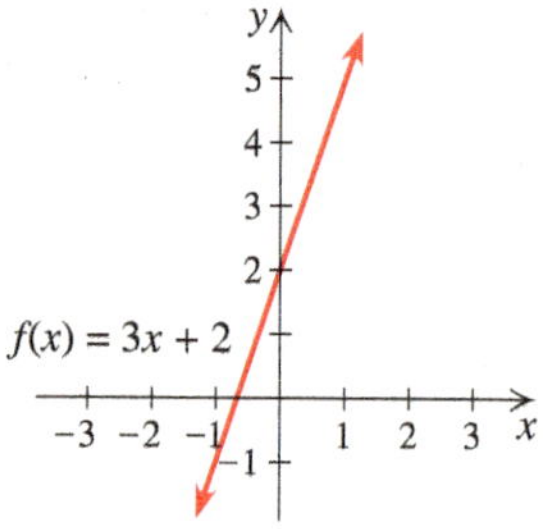


Domain $(-\infty, \infty)$
Range $(-\infty, \infty)$
Increasing on $(-\infty, \infty)$

### Absolute Value Function

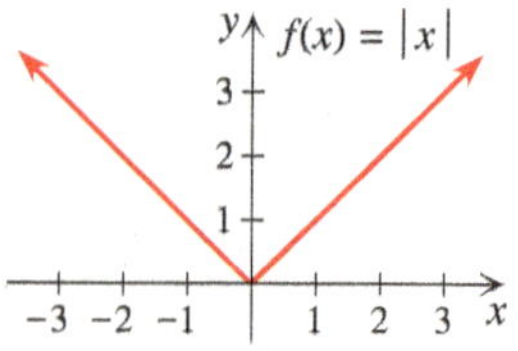


Domain $(-\infty, \infty)$
Range $[0, \infty)$
Increasing on $[0, \infty)$
Decreasing on $(-\infty, 0]$
Symmetric about $y$-axis

### Square Function

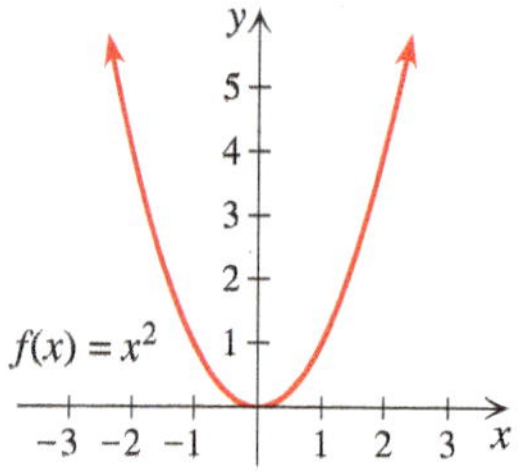


Domain $(-\infty, \infty)$
Range $[-0, \infty)$
Increasing on $[0, \infty)$
Decreasing on $(-\infty, 0]$
Symmetric about $y$-axis

### Square-Root Function

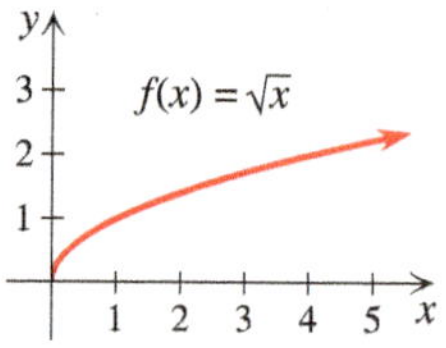


Domain $[0, \infty)$
Range $[0, \infty)$
Increasing on $[0, \infty)$

### Cube Function

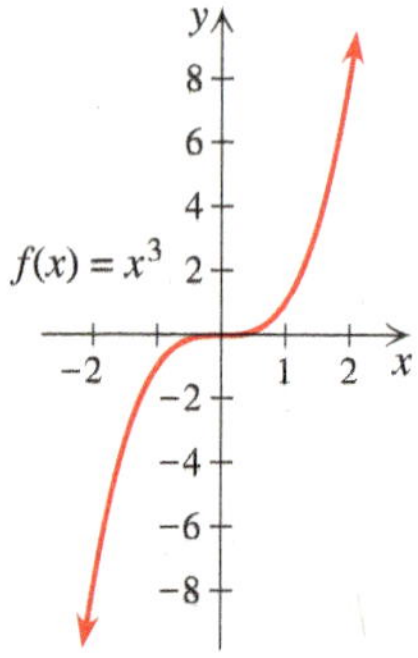


Domain $(-\infty, \infty)$
Range $(-\infty, \infty)$
Increasing on $(-\infty, \infty)$
Symmetric about origin

### Cube-Root Function

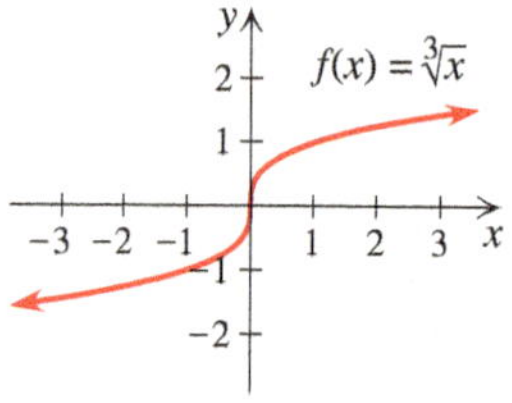


Domain $(-\infty, \infty)$
Range $(-\infty, \infty)$
Increasing on $(-\infty, \infty)$
Symmetric about origin

### Greatest Integer Function

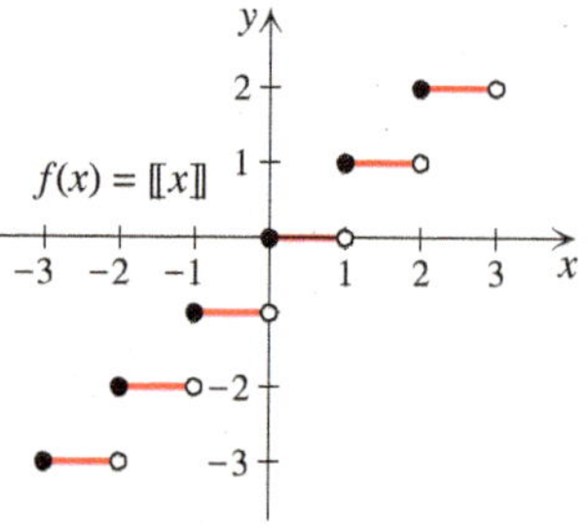


Domain $(-\infty, \infty)$
Range $\{n|n \text{ is an integer}\}$
Constant on $[n, n + 1)$
for every integer $n$

# FUNCTION gallery... Some Inverse Functions

## Linear

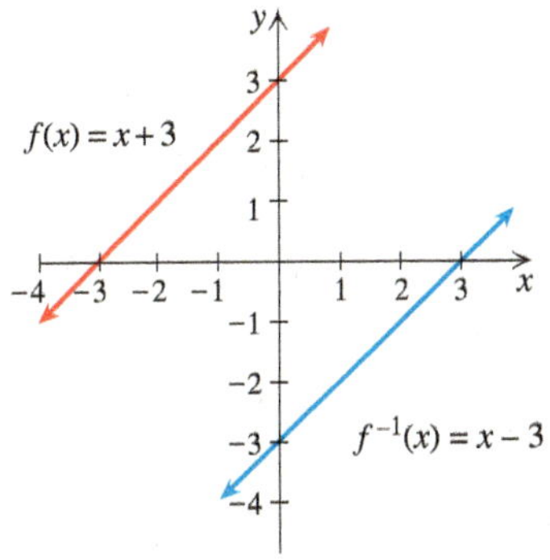


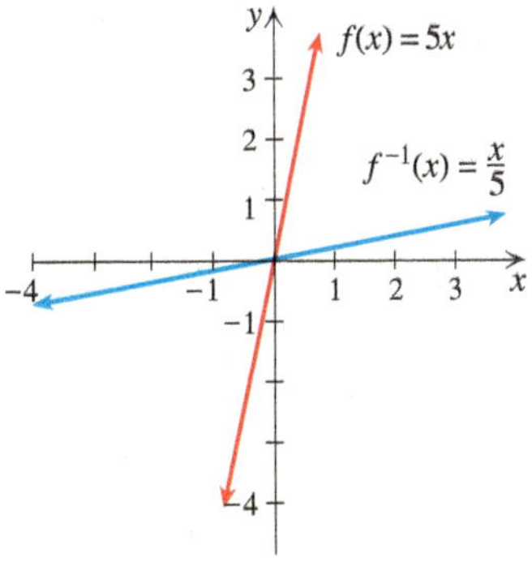


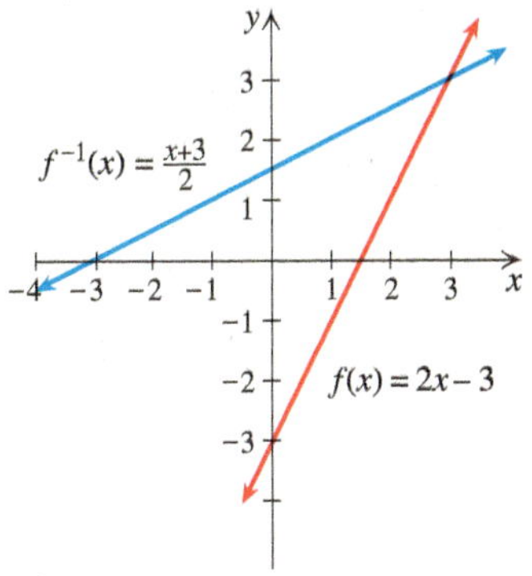


## Powers and Roots

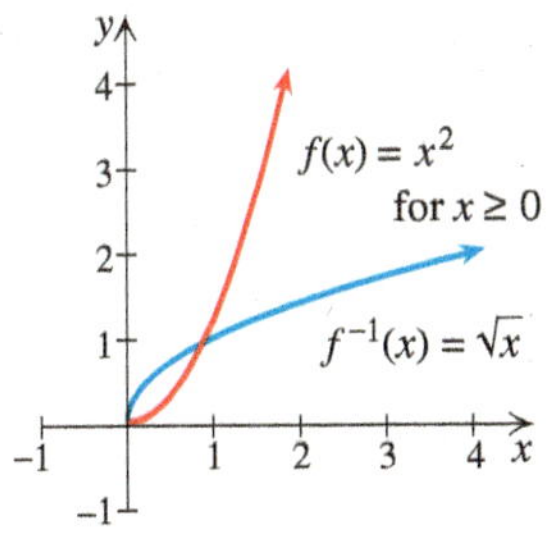


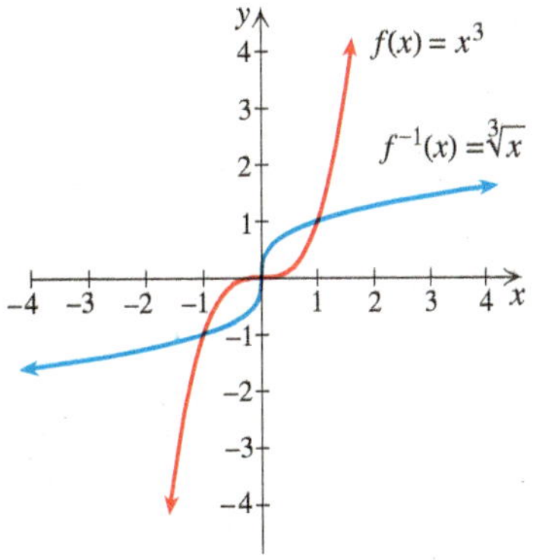


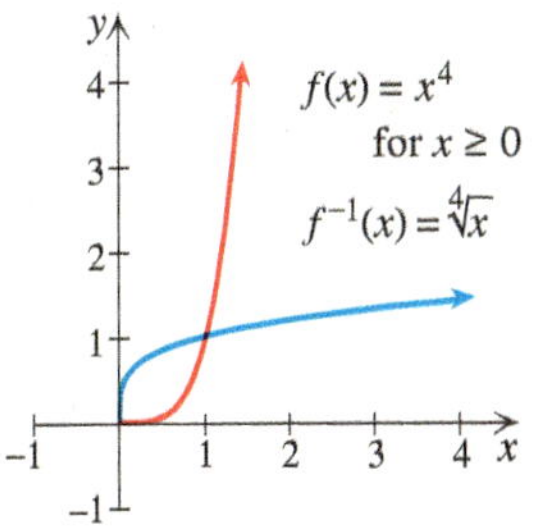

# FUNCTION gallery... Polynomial Functions

**Linear: $f(x) = mx + b$**

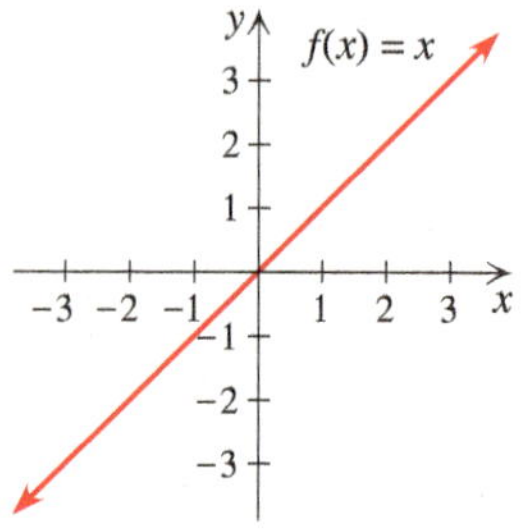


Slope 1, $y$-intercept (0, 0)

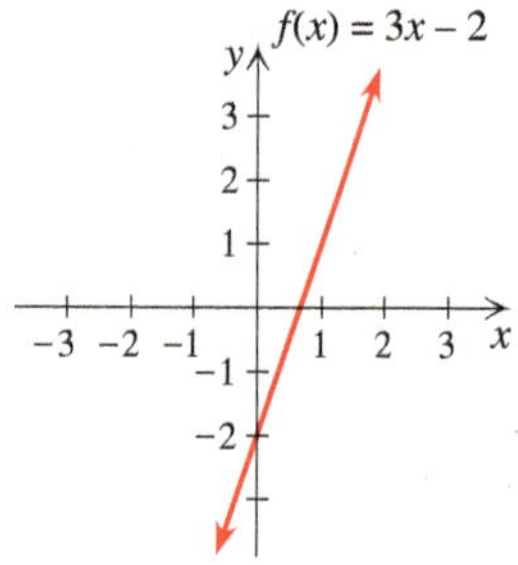


Slope 3, $y$-intercept (0, −2)

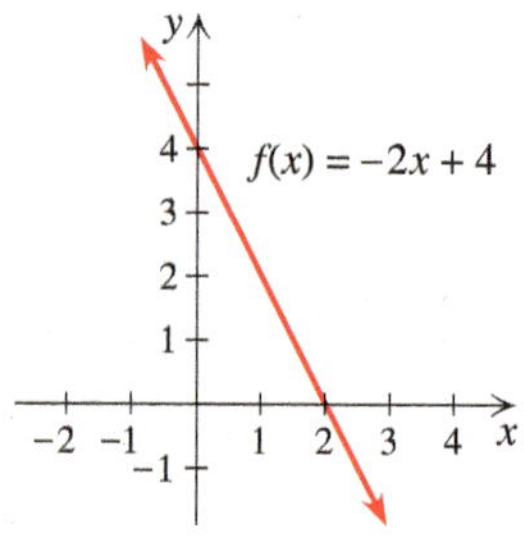


Slope −2, $y$-intercept (0, 4)

**Quadratic: $f(x) = ax^2 + bx + c$ or $f(x) = a(x - h)^2 + k$**

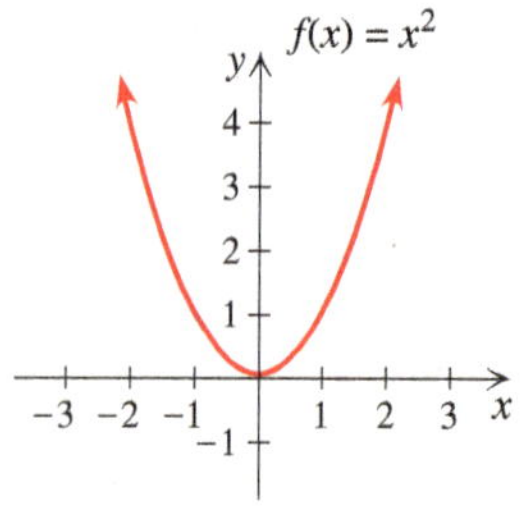


Vertex (0, 0)
Range $[0, \infty)$

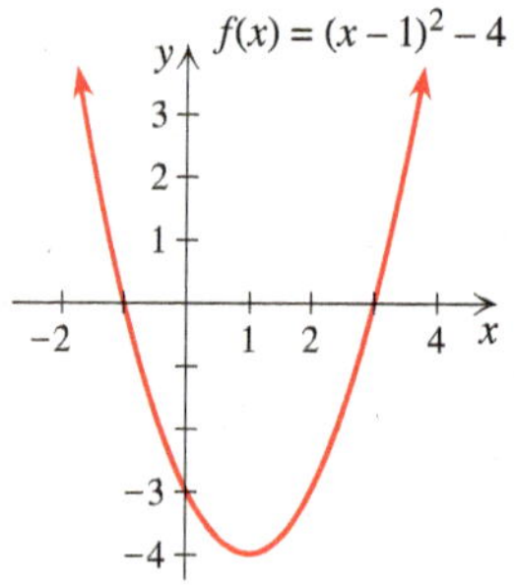


Vertex (1, −4)
Range $[-4, \infty)$

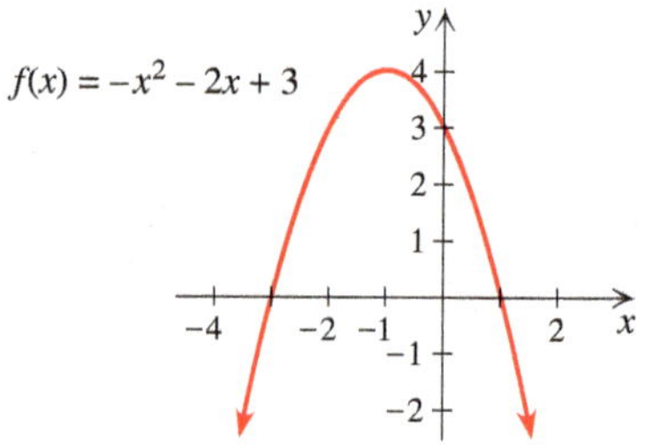


Vertex (−1, 4)
Range $(-\infty, 4]$

**Cubic: $f(x) = ax^3 + bx^2 + cx + d$**

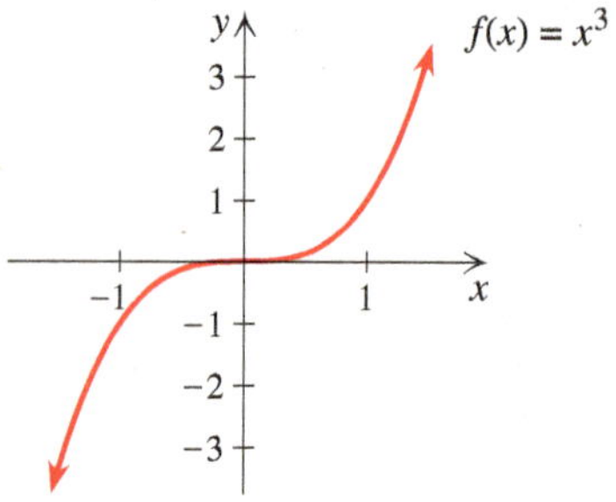


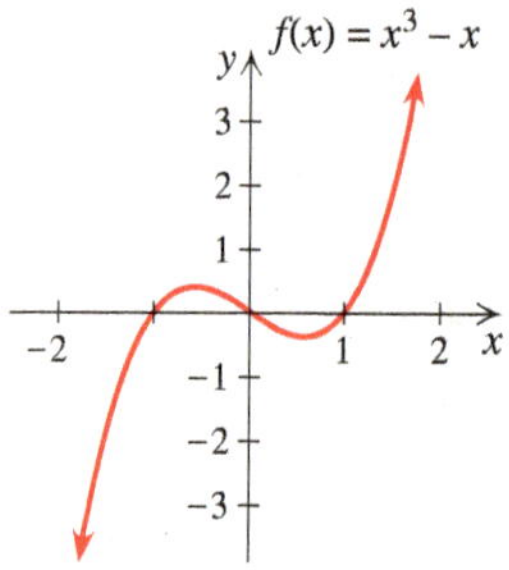


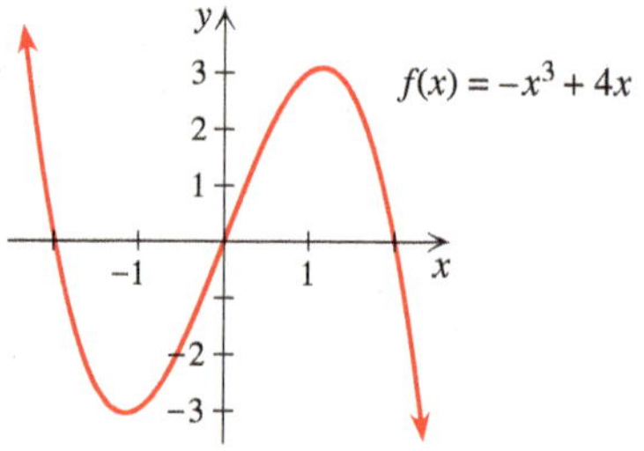


**Quartic or Fourth-Degree: $f(x) = ax^4 + bx^3 + cx^2 + dx + e$**

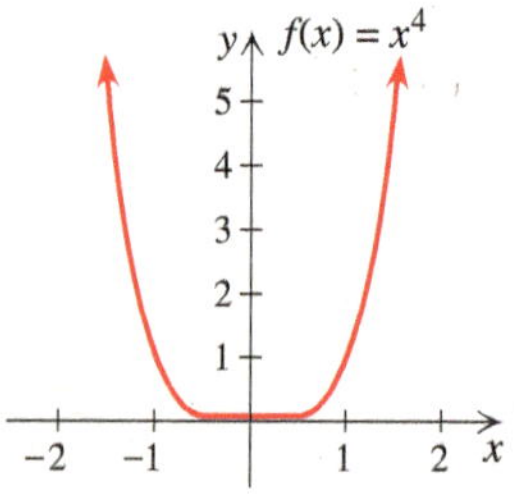


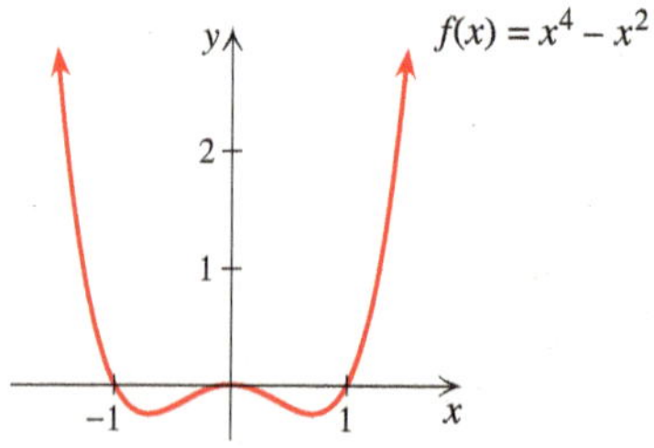


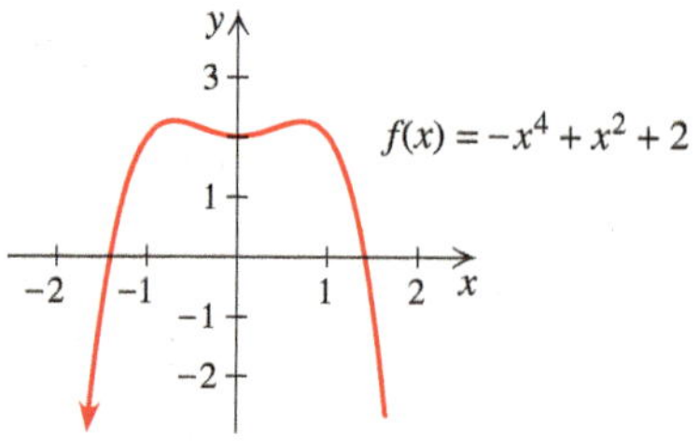

# FUNCTION gallery... Some Basic Rational Functions

### Horizontal Asymptote *x*-axis and Vertical Asymptote *y*-axis

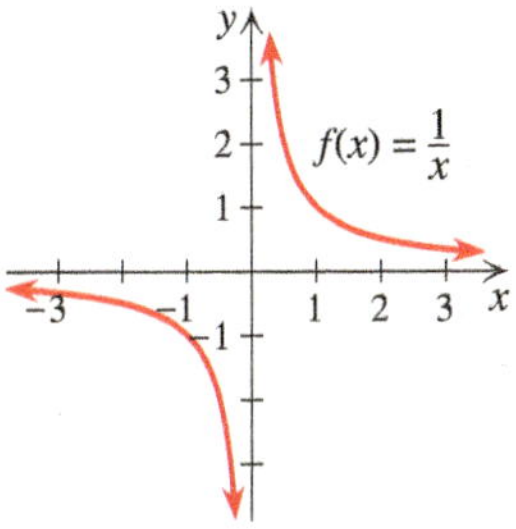


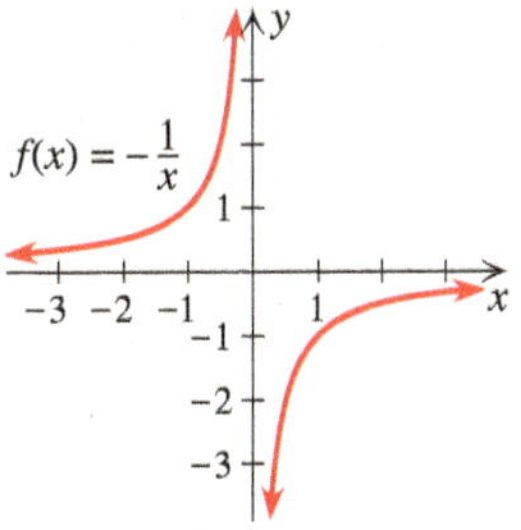


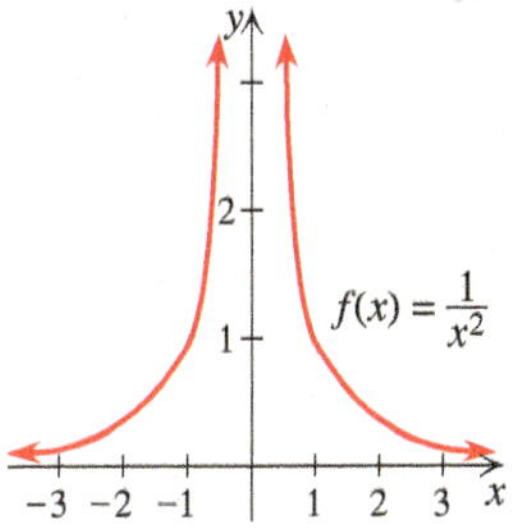


### Various Asymptotes

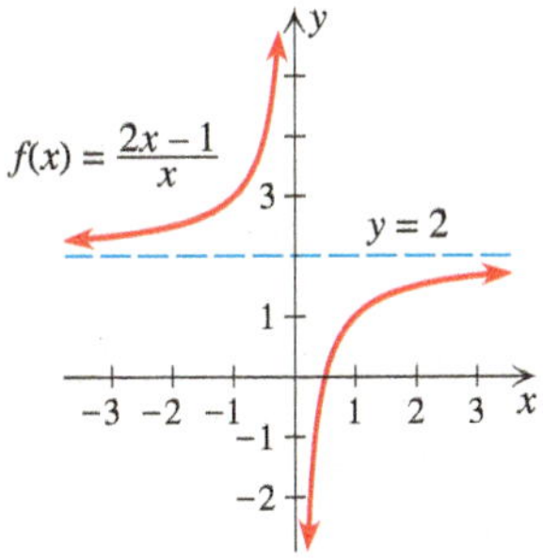


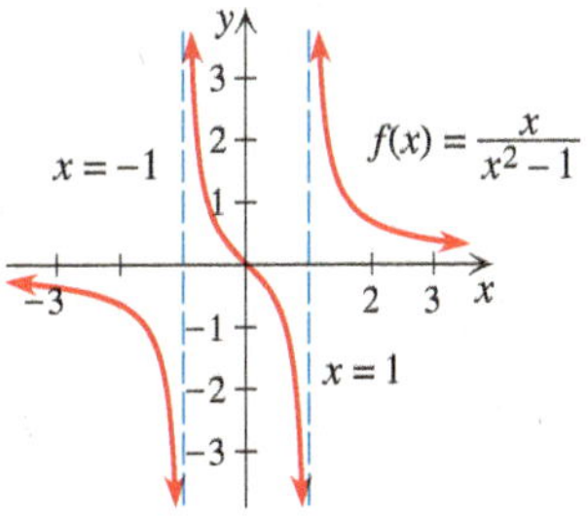


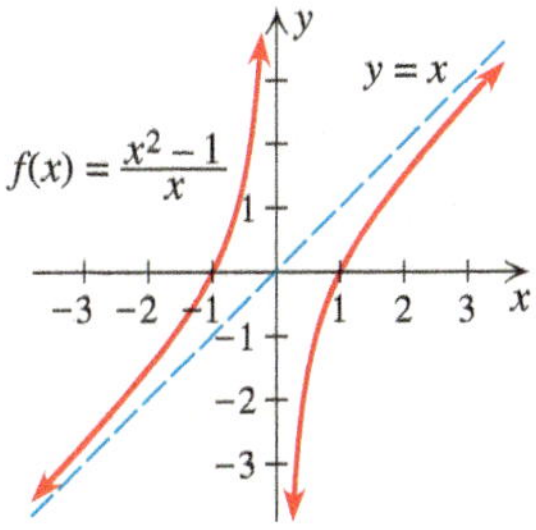

# FUNCTION gallery... Exponential and Logarithmic Functions

**Exponential: $f(x) = a^x$, domain $(-\infty, \infty)$, range $(0, \infty)$**

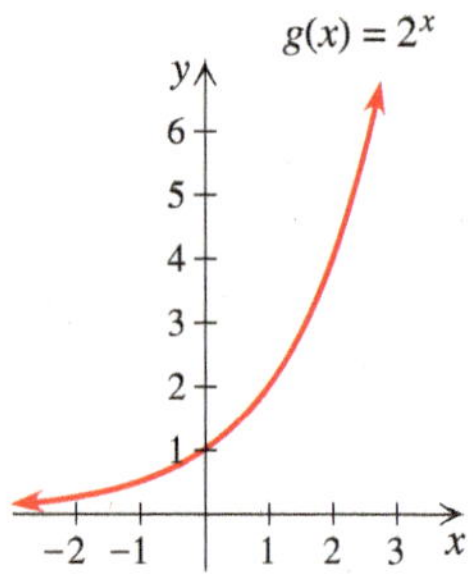


Increasing on $(-\infty, \infty)$
$y$-intercept $(0, 1)$

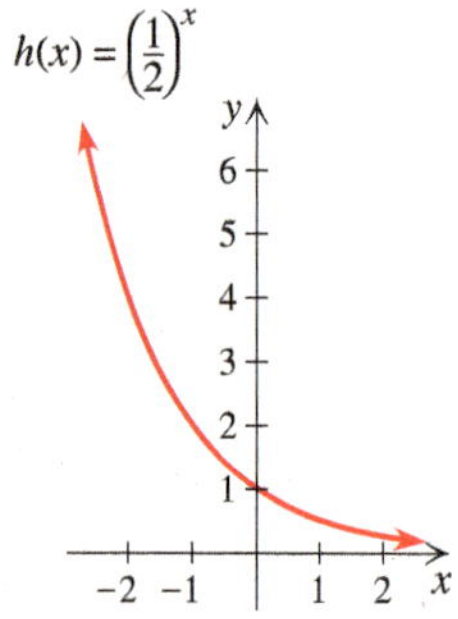


Decreasing on $(-\infty, \infty)$
$y$-intercept $(0, 1)$

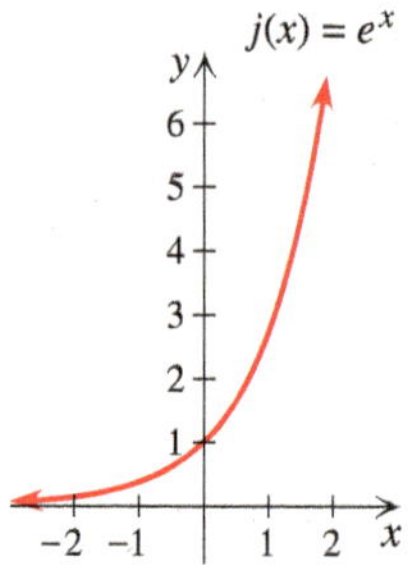


Increasing on $(-\infty, \infty)$
$y$-intercept $(0, 1)$

**Logarithmic: $f^{-1}(x) = \log_a(x)$, domain $(0, \infty)$, range $(-\infty, \infty)$**

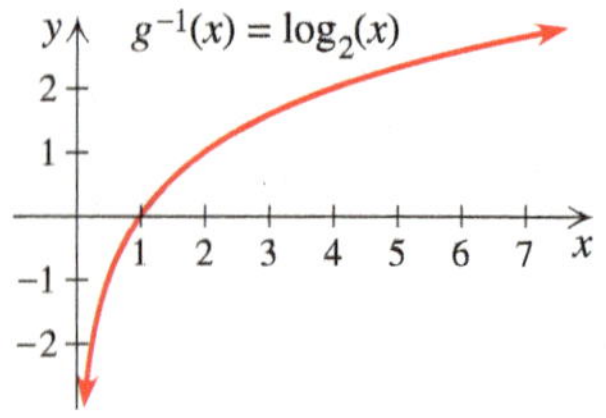


Increasing on $(0, \infty)$
$x$-intercept $(1, 0)$

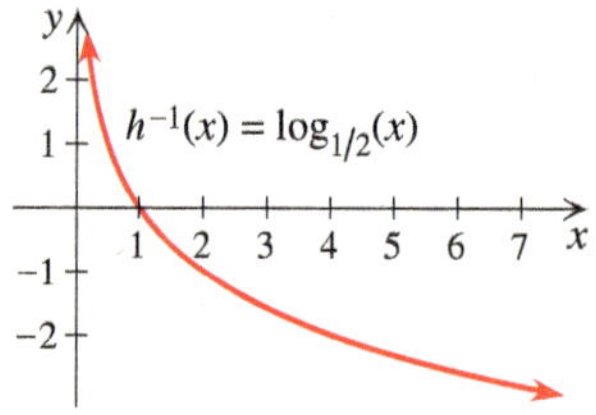


Decreasing on $(0, \infty)$
$x$-intercept $(1, 0)$

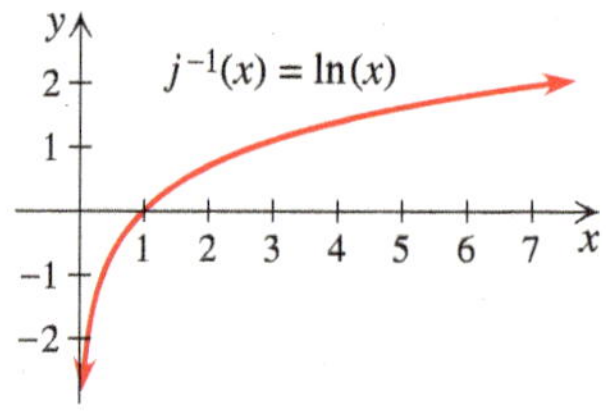


Increasing on $(0, \infty)$
$x$-intercept $(1, 0)$

# FUNCTION gallery... Some Basic Functions of Algebra with Transformations

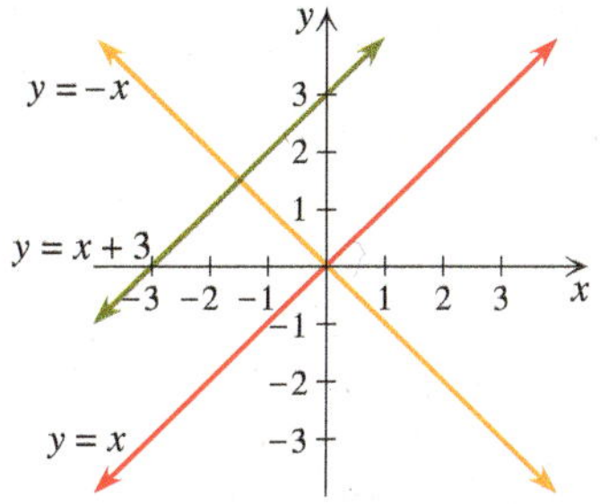

**Linear**

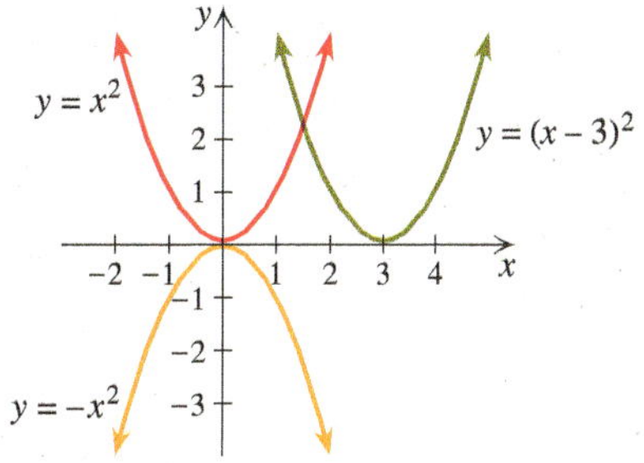

**Quadratic**

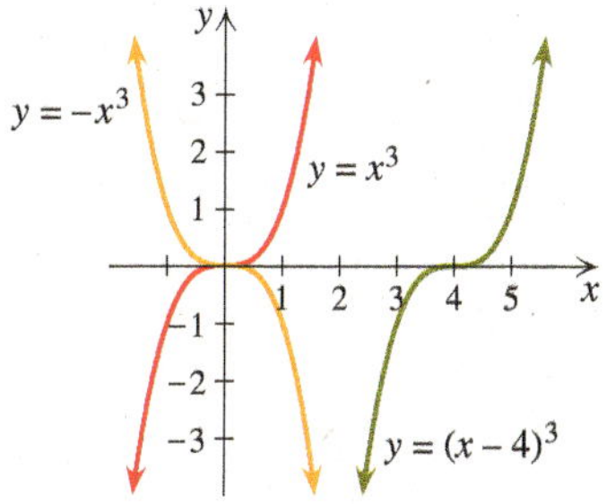

**Cubic**

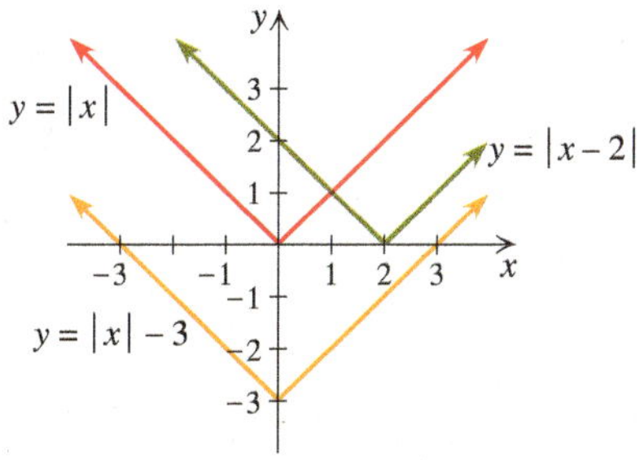

**Absolute value**

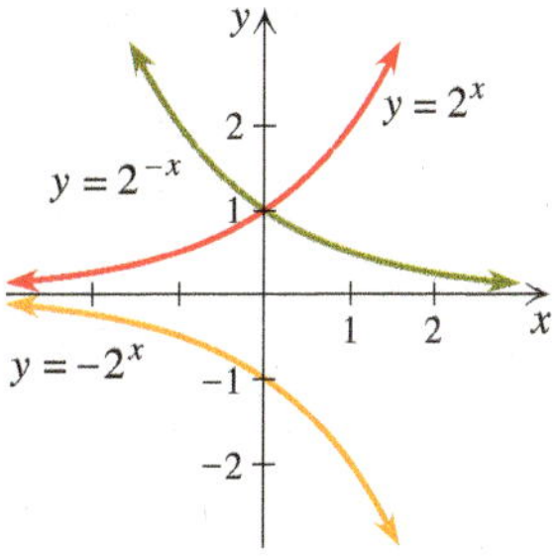

**Exponential**

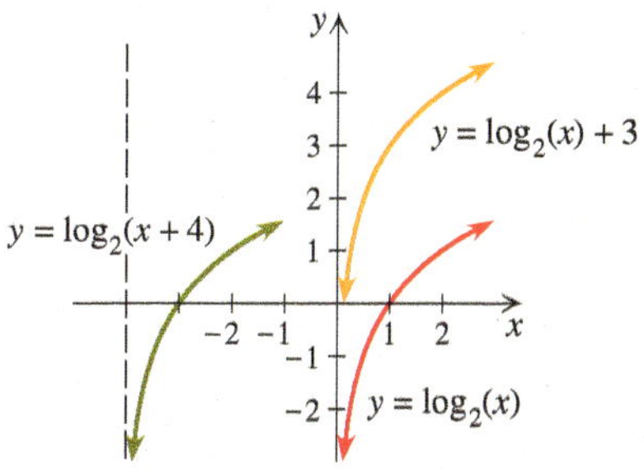

**Logarithmic**

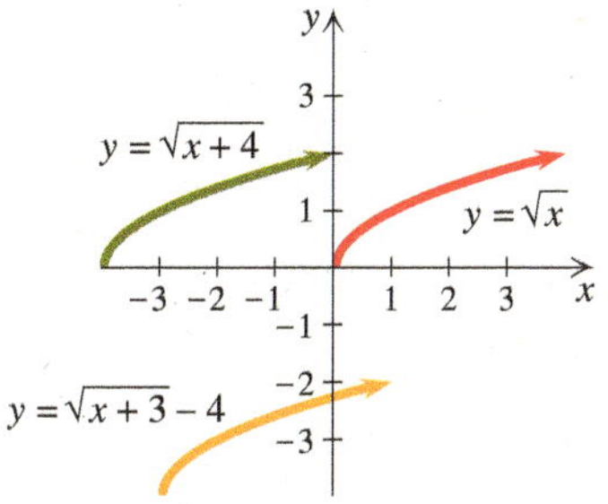

**Square root**

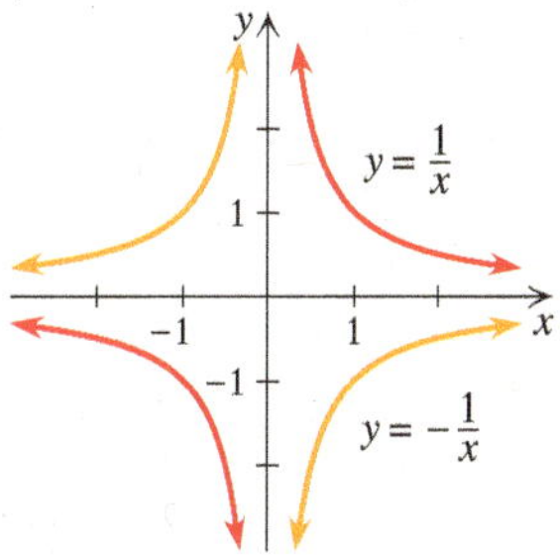

**Reciprocal**

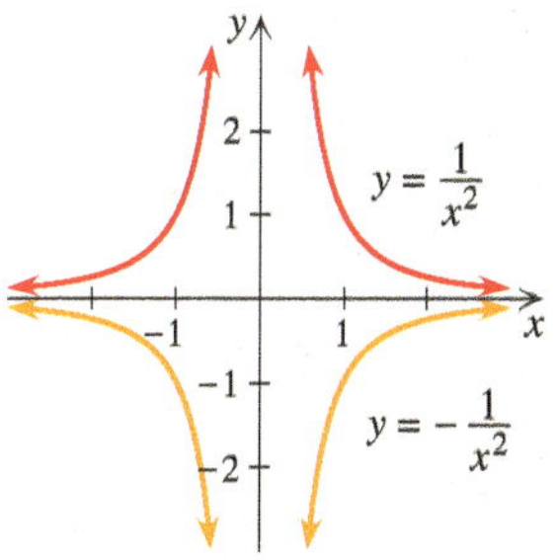

**Rational**

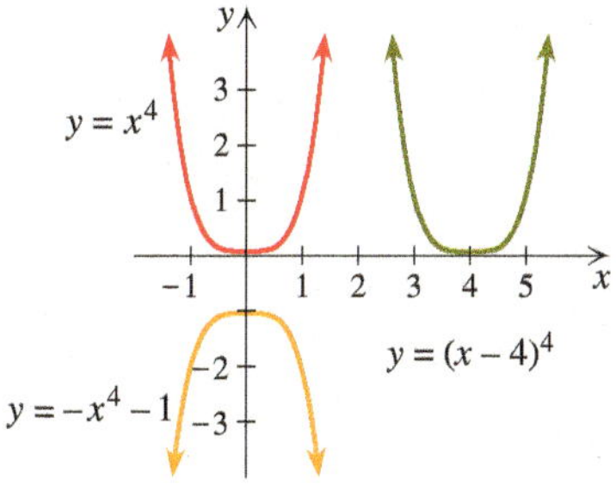

**Fourth degree**

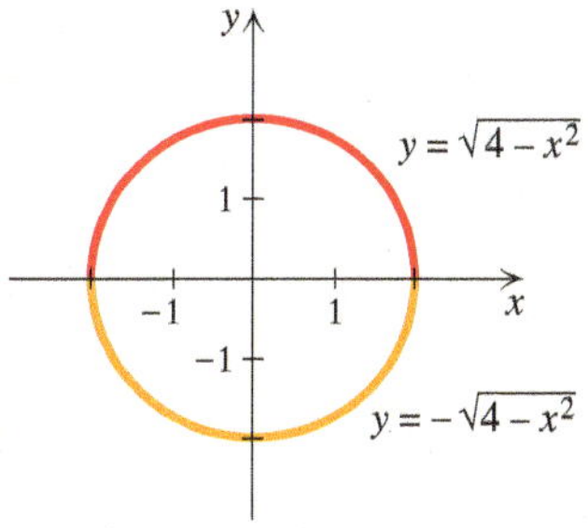

**Semicircle**

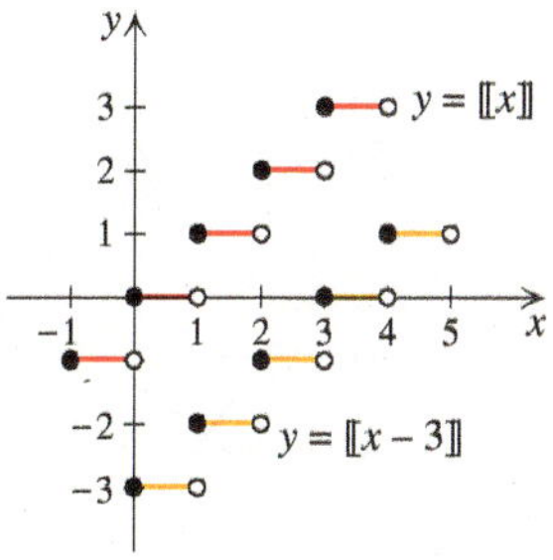

**Greatest integer**

## FUNCTION gallery... The Sine and Cosine Functions

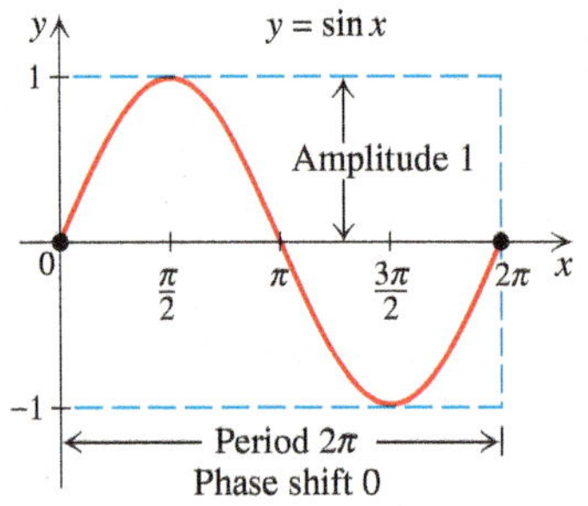


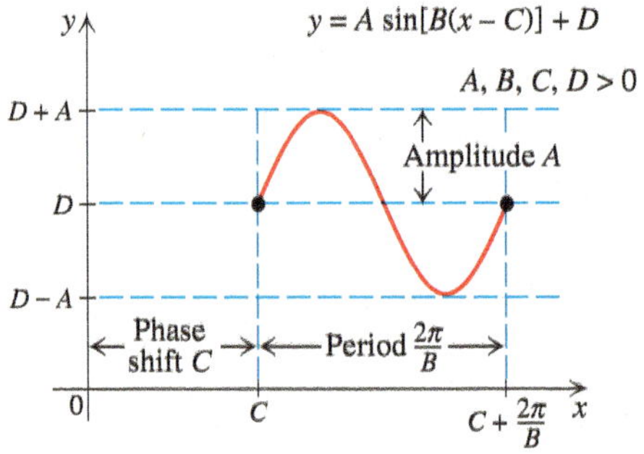


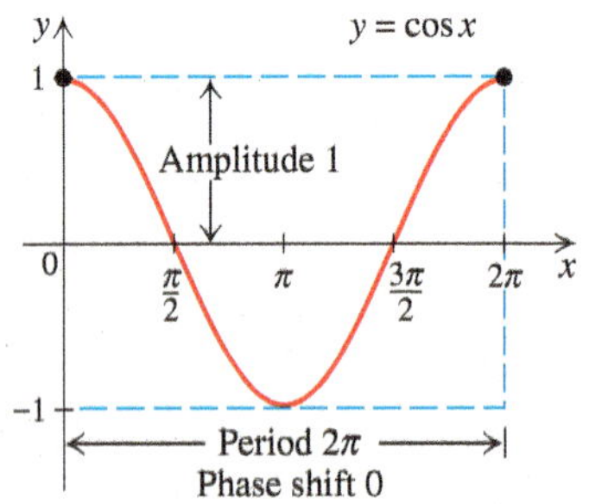


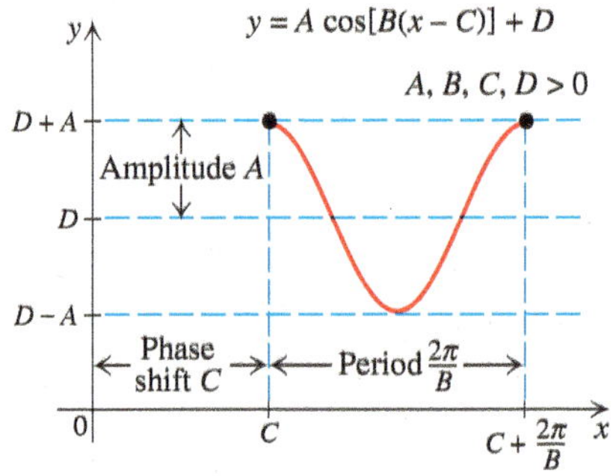


## FUNCTION gallery... Periods of Sine, Cosine, and Tangent ($B > 1$)

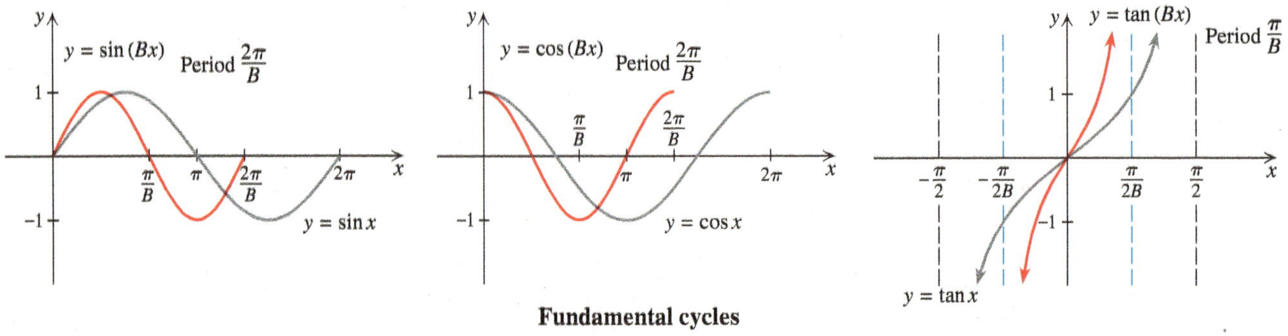


Fundamental cycles

# FUNCTION gallery... Trigonometric Functions

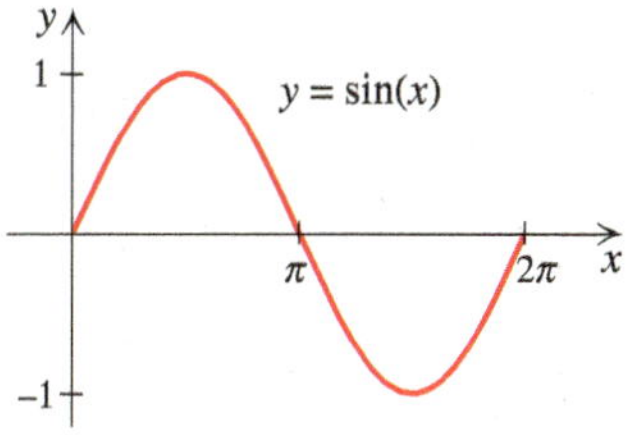


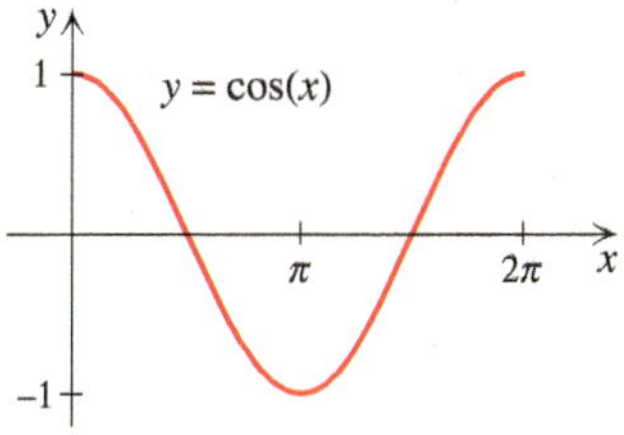


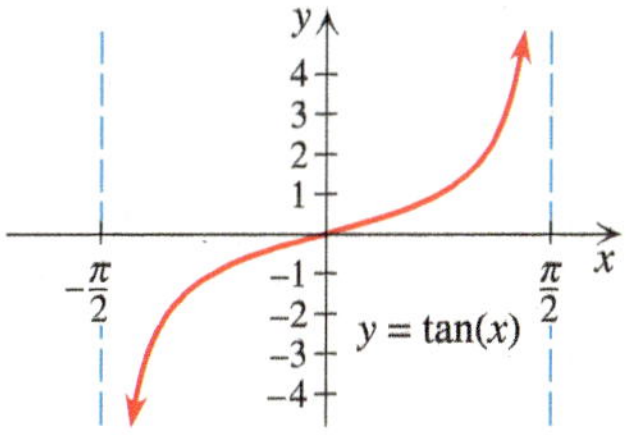


| | | | |
|---|---|---|---|
| Domain (*k* any integer) | $(-\infty, \infty)$ | $(-\infty, \infty)$ | $x \neq \frac{\pi}{2} + k\pi$ |
| Range | $[-1, 1]$ | $[-1, 1]$ | $(-\infty, \infty)$ |
| Period | $2\pi$ | $2\pi$ | $\pi$ |
| Fundamental cycle | $[0, 2\pi]$ | $[0, 2\pi]$ | $\left[-\frac{\pi}{2}, \frac{\pi}{2}\right]$ |

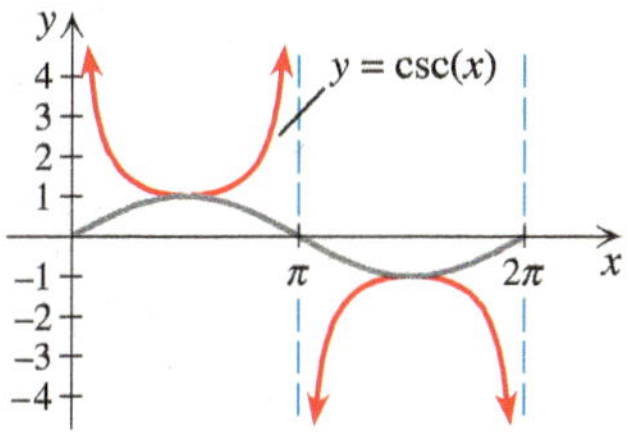


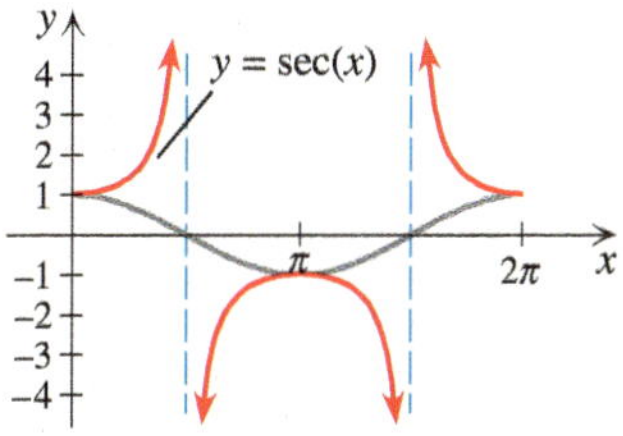


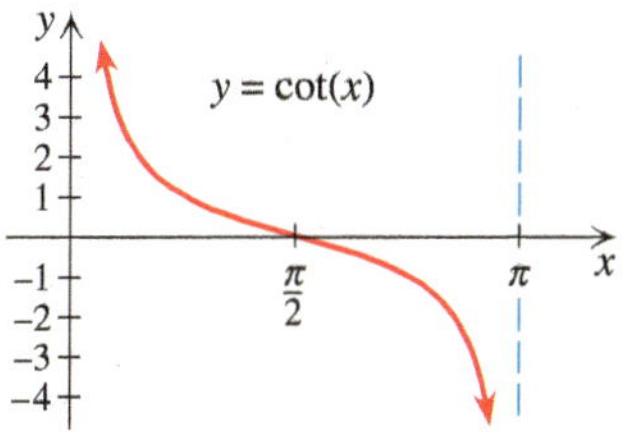


| | | | |
|---|---|---|---|
| Domain (*k* any integer) | $x \neq k\pi$ | $x \neq \frac{\pi}{2} + k\pi$ | $x \neq k\pi$ |
| Range | $(-\infty, -1] \cup [1, \infty)$ | $(-\infty, -1] \cup [1, \infty)$ | $(-\infty, \infty)$ |
| Period | $2\pi$ | $2\pi$ | $\pi$ |
| Fundamental cycle | $[0, 2\pi]$ | $[0, 2\pi]$ | $[0, \pi]$ |

# FUNCTION gallery... Inverse Trigonometric Functions

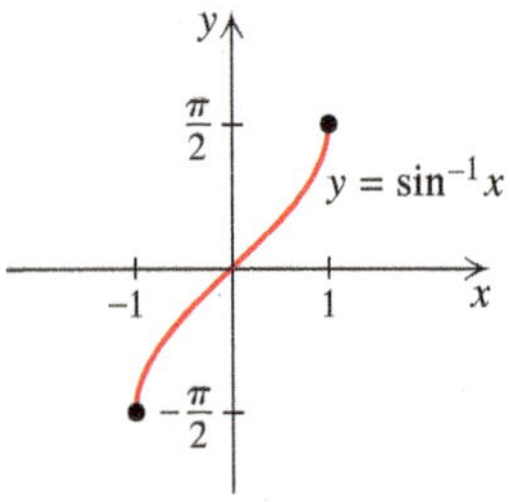


Domain $[-1, 1]$
Range $\left[-\frac{\pi}{2}, \frac{\pi}{2}\right]$

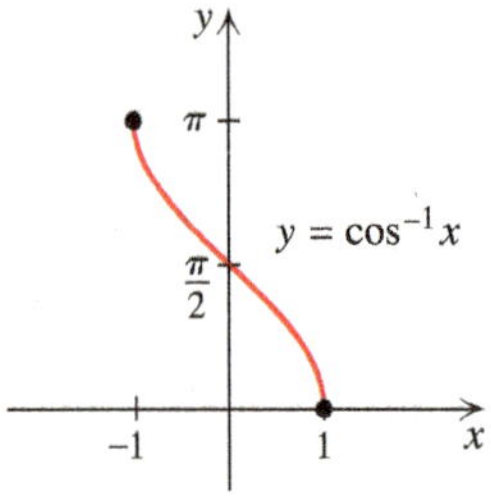


Domain $[-1, 1]$
Range $[0, \pi]$

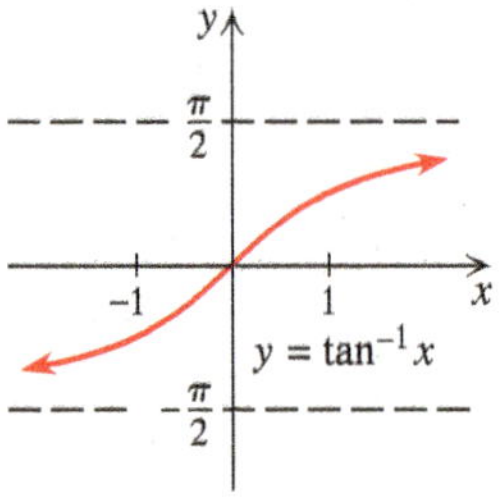


Domain $(-\infty, \infty)$
Range $\left(-\frac{\pi}{2}, \frac{\pi}{2}\right)$

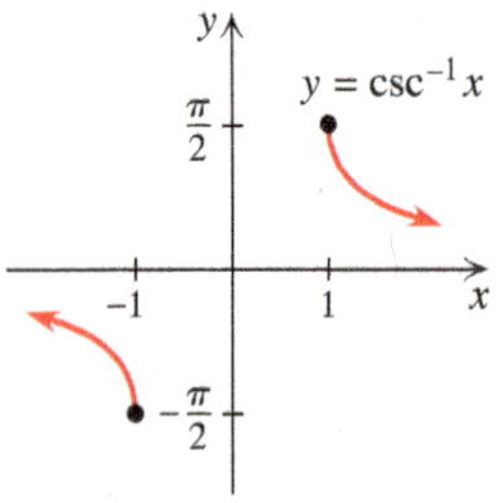


Domain $(-\infty, -1] \cup [1, \infty)$
Range $\left[-\frac{\pi}{2}, 0\right) \cup \left(0, \frac{\pi}{2}\right]$

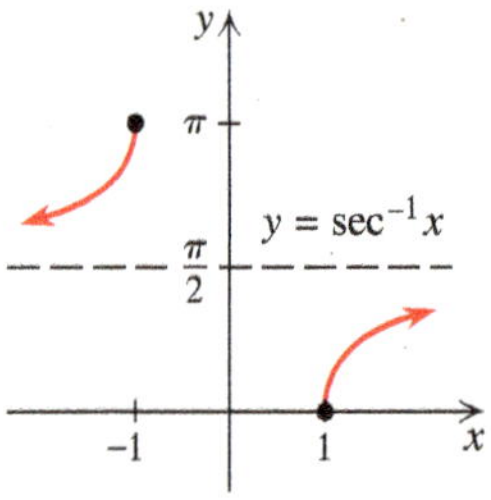


Domain $(-\infty, -1] \cup [1, \infty)$
Range $\left[0, \frac{\pi}{2}\right) \cup \left(\frac{\pi}{2}, \pi\right]$

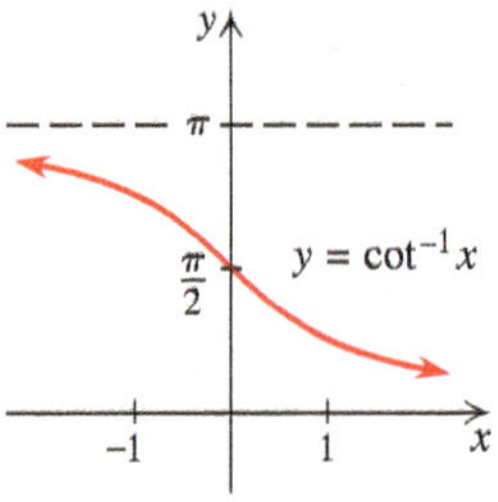


Domain $(-\infty, \infty)$
Range $(0, \pi)$

# FUNCTION gallery... Functions in Polar Coordinates

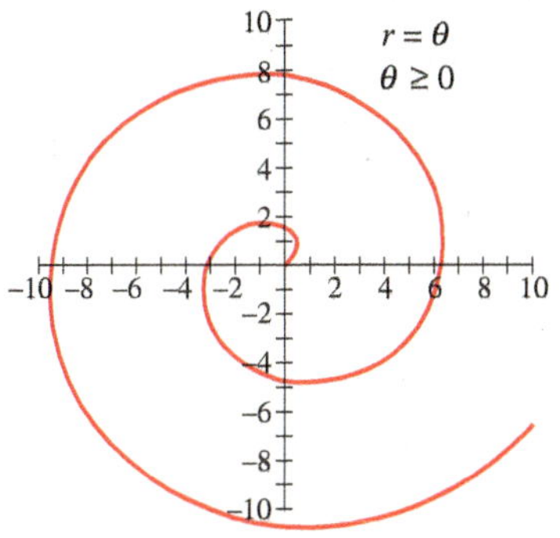


Spiral

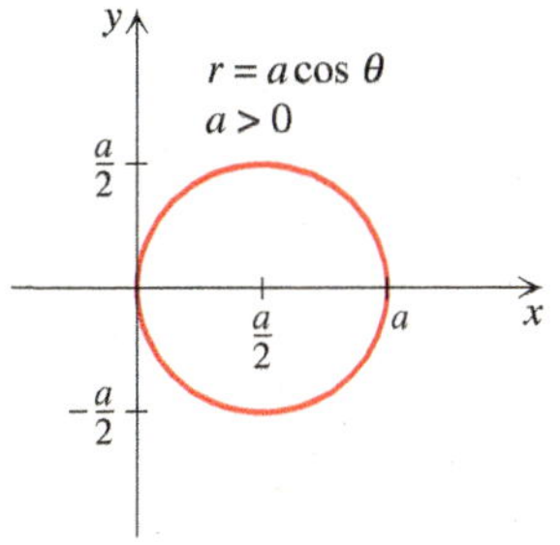


Circle

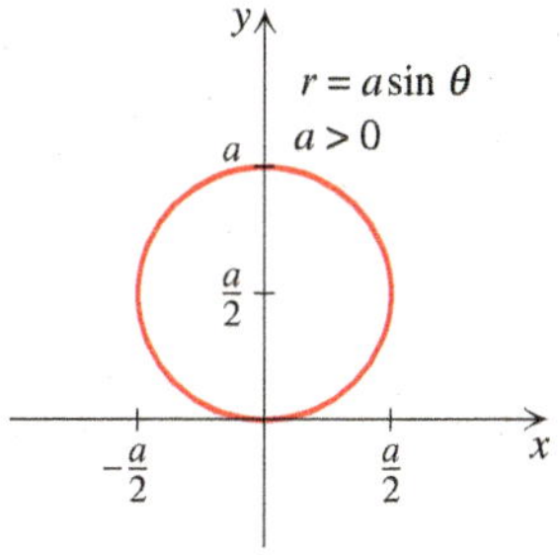


Circle

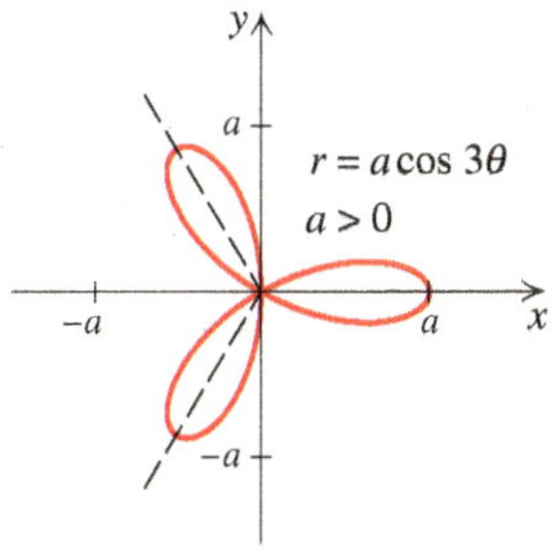


Three-Leaf Rose

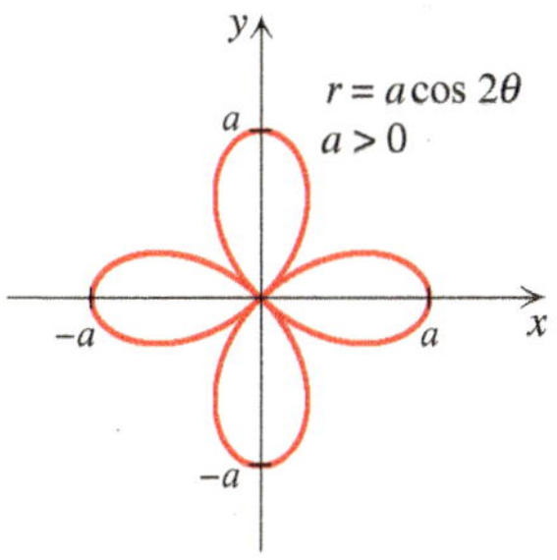


Four-Leaf Rose

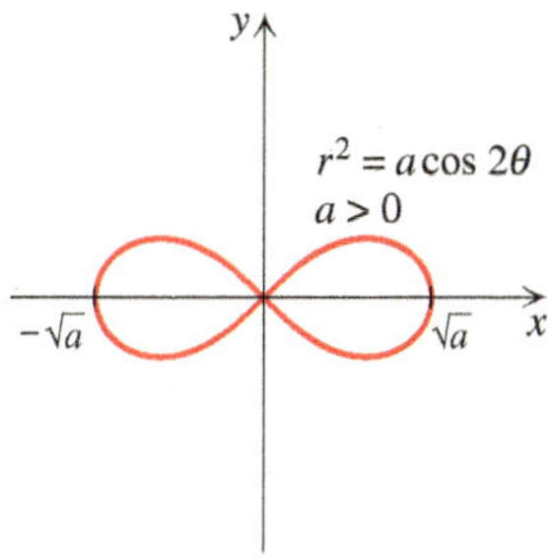


Lemniscate

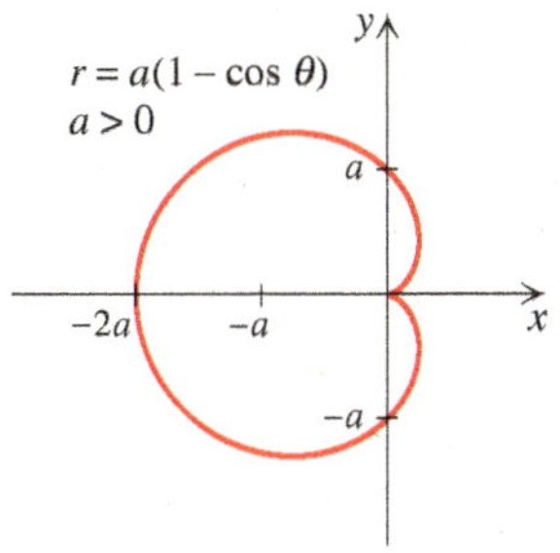


Cardioid

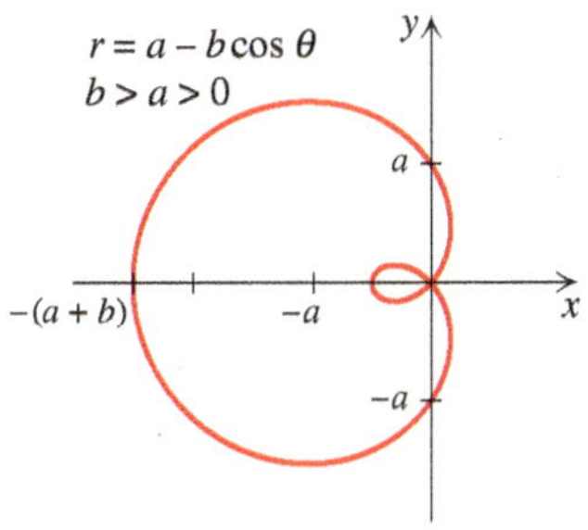


Limaçon

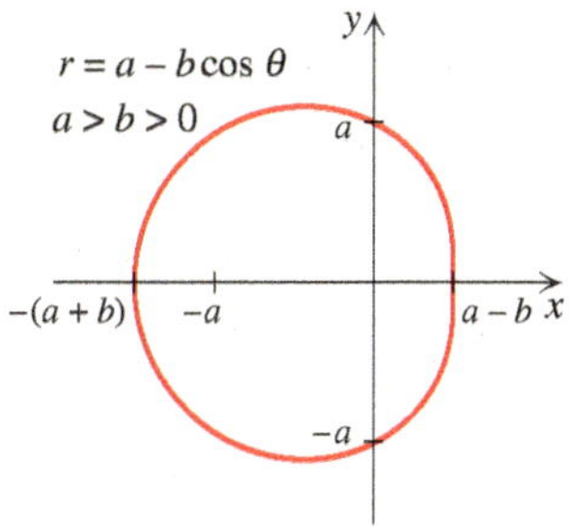


Limaçon

# 1 Equations, Inequalities, and Modeling

Even infamous Heartbreak Hill couldn't break the winning spirit of Ethiopian runner Dire Tune as she focused on first place. It was the 112th running of the Boston Marathon, a 26.2-mile ordeal that one runner called "14 miles of fun, 8 miles of sweat, and 4 miles of hell!"

Sporting events like the 2008 Marathon have come a long way since the first Olympic Games were held over 2500 years ago. Today, sports and science go hand in hand. Modern athletes often use mathematics to analyze the variables that help them increase aerobic capacity, reduce air resistance, or strengthen muscles.

**WHAT YOU WILL learn...** In this chapter you will see examples of sports applications while you learn to solve equations and inequalities. By the time you reach the finish line, you will be using algebra to model and solve problems.

# 1.1 Equations in One Variable

One of our main goals in algebra is to develop techniques for solving a wide variety of equations. In this section we will solve linear equations and other similar equations.

## Definitions

An **equation** is a statement (or sentence) indicating that two algebraic expressions are equal. The verb in an equation is the equality symbol. For example, $2x + 8 = 0$ is an equation. If we replace $x$ by $-4$, we get $2 \cdot (-4) + 8 = 0$, a true statement. So we say that $-4$ is a **solution** or **root** to the equation or $-4$ **satisfies** the equation. If we replace $x$ by 3, we get $2 \cdot 3 + 8 = 0$, a false statement. So 3 is not a solution.

Whether the equation $2x + 8 = 0$ is true or false depends on the value of $x$, and so it is called an **open sentence.** The equation is neither true nor false until we choose a value for $x$. The set of all solutions to an equation is called the **solution set** to the equation. To **solve** an equation means to find the solution set. The solution set for $2x + 8 = 0$ is $\{-4\}$. The equation $2x + 8 = 0$ is an example of a linear equation.

**Definition: Linear Equation in One Variable**

A **linear equation in one variable** is an equation of the form $ax + b = 0$, where $a$ and $b$ are real numbers, with $a \neq 0$.

Note that other letters can be used in place of $x$. For example, $3t + 5 = 0$, $2w - 6 = 0$, and $-2u + 7 = 0$ are linear equations.

## Solving Linear Equations

The equations $2x + 8 = 0$ and $2x = -8$ both have the solution set $\{-4\}$. Two equations with the same solution set are called **equivalent** equations. Adding the same real number to or subtracting the same real number from each side of an equation results in an equivalent equation. Multiplying or dividing each side of an equation by the same nonzero real number also results in an equivalent equation. These **properties of equality** are stated in symbols in the following box.

**Properties of Equality**

If $A$ and $B$ are algebraic expressions and $C$ is a real number, then the following equations are equivalent to $A = B$:

| | |
|---|---|
| $A + C = B + C$ | **Addition property of equality** |
| $A - C = B - C$ | **Subtraction property of equality** |
| $CA = CB\ (C \neq 0)$ | **Multiplication property of equality** |
| $\frac{A}{C} = \frac{B}{C}\ (C \neq 0)$ | **Division property of equality** |

We can use an algebraic expression for $C$ in the properties of equality, because the value of an algebraic expression is a real number. However, this can produce nonequivalent equations. For example,

$$x = 0 \quad \text{and} \quad x + \frac{1}{x} = 0 + \frac{1}{x}$$

appear to be equivalent by the addition property of equality. But the first is satisfied by 0 and the second is not. When an equation contains expressions that are undefined for some real number(s) then we must check all solutions carefully.

Any linear equation, $ax + b = 0$, can be solved in two steps. Subtract $b$ from each side and then divide each side by $a$ $(a \neq 0)$, to get $x = -b/a$. Although the equations in our first example are not exactly in the form $ax + b = 0$, they are often called linear equations because they are equivalent to linear equations.

## EXAMPLE 1 Using the properties of equality

Solve each equation.

**a.** $3x - 4 = 8$ **b.** $\frac{1}{2}x - 6 = \frac{3}{4}x - 9$ **c.** $3(4x - 1) = 4 - 6(x - 3)$

### Solution

**a.**

$$3x - 4 = 8$$

$$3x - 4 + 4 = 8 + 4 \qquad \text{Add 4 to each side.}$$

$$3x = 12 \qquad \text{Simplify.}$$

$$\frac{3x}{3} = \frac{12}{3} \qquad \text{Divide each side by 3.}$$

$$x = 4 \qquad \text{Simplify.}$$

Since the last equation is equivalent to the original, the solution set to the original equation is $\{4\}$. We can check by replacing $x$ by 4 in $3x - 4 = 8$. Since $3 \cdot 4 - 4 = 8$ is correct, we are confident that the solution set is $\{4\}$.

**b.** Multiplying each side of the equation by the least common denominator, LCD, will eliminate all of the fractions:

$$\frac{1}{2}x - 6 = \frac{3}{4}x - 9$$

$$4\left(\frac{1}{2}x - 6\right) = 4\left(\frac{3}{4}x - 9\right) \qquad \text{Multiply each side by 4, the LCD.}$$

$$2x - 24 = 3x - 36 \qquad \text{Distributive property}$$

$$2x - 24 - 3x = 3x - 36 - 3x \qquad \text{Subtract } 3x \text{ from each side.}$$

$$-x - 24 = -36 \qquad \text{Simplify.}$$

$$-x = -12 \qquad \text{Add 24 to each side.}$$

$$(-1)(-x) = (-1)(-12) \qquad \text{Multiply each side by } -1.$$

$$x = 12 \qquad \text{Simplify.}$$

Check 12 in the original equation. The solution set is $\{12\}$.

**c.** $3(4x - 1) = 4 - 6(x - 3)$

$$12x - 3 = 4 - 6x + 18 \qquad \text{Distributive property}$$

$$12x - 3 = 22 - 6x \qquad \text{Simplify.}$$

$$18x - 3 = 22 \qquad \text{Add } 6x \text{ to each side.}$$

$$18x = 25 \qquad \text{Add 3 to each side.}$$

$$x = \frac{25}{18} \qquad \text{Divide each side by 18.}$$

Check $25/18$ in the original equation. The solution set is $\left\{\frac{25}{18}\right\}$.

You can use a graphing calculator to calculate the value of each side of the equation when $x$ is $25/18$ as shown in Fig. 1.1.

Figure 1.1

**▸TRY THIS.** Solve $5(3x - 2) = 5 - 7(x - 1)$. ■

Note that checking the equation in Example 1(c) with a calculator did not prove that 25/18 is the correct solution. The properties of equality that were applied correctly in each step guarantee that we have the correct solution. The values of the two sides of the equation could agree for the 10 digits shown on the calculator and disagree for the digits not shown. Since that possibility is extremely unlikely, the calculator check does support our belief that we have the correct solution.

## Identities, Conditional Equations, and Inconsistent Equations

An equation that is satisfied by every real number for which both sides are defined is an **identity.** Some examples of identities are

$$3x - 1 = 3x - 1, \qquad 2x + 5x = 7x, \qquad \text{and} \qquad \frac{x}{x} = 1.$$

The solution set to the first two identities is the set of all real numbers, $R$. Since 0/0 is undefined, the solution set to $x/x = 1$ is the set of nonzero real numbers, $\{x \mid x \neq 0\}$.

A **conditional equation** is an equation that is satisfied by at least one real number but is not an identity. The equation $3x - 4 = 8$ is true only on condition that $x = 4$, and it is a conditional equation. The equations of Example 1 are conditional equations.

An **inconsistent equation** is an equation that has no solution. Some inconsistent equations are

$$0 \cdot x + 1 = 2, \qquad x + 3 = x + 5, \qquad \text{and} \qquad 9x - 9x = 8.$$

Note that each of these inconsistent equations is equivalent to a false statement: $1 = 2$, $3 = 5$, and $0 = 8$, respectively.

### EXAMPLE 2 Classifying an equation

Determine whether the equation $3(x - 1) - 2x(4 - x) = (2x + 1)(x - 3)$ is an identity, an inconsistent equation, or a conditional equation.

**Solution**

$$3(x - 1) - 2x(4 - x) = (2x + 1)(x - 3)$$

$$3x - 3 - 8x + 2x^2 = 2x^2 - 5x - 3 \qquad \text{Simplify each side.}$$

$$2x^2 - 5x - 3 = 2x^2 - 5x - 3$$

Since the last equation is equivalent to the original and the last equation is an identity, the original equation is an identity.

▶**TRY THIS.** Determine whether $x(x - 1) - 6 = (x - 3)(x + 2)$ is an identity, an inconsistent equation, or a conditional equation. ■

## Equations Involving Rational Expressions

Recall that division by zero is undefined and we can't have zero in the denominator of a fraction. Since the rational expressions in the next example have variables in their denominators, these variables can't be replaced by any numbers that would cause zero to appear in a denominator. Our first step in solving these equations is to multiply by the LCD and eliminate the denominators. But we must check our solutions in the original equations and discard any that cause undefined expressions.

## EXAMPLE 3 Equations involving rational expressions

Solve each equation and identify each as an identity, an inconsistent equation, or a conditional equation.

**a.** $\frac{y}{y-3} + 3 = \frac{3}{y-3}$ **b.** $\frac{1}{x-1} - \frac{1}{x+1} = \frac{2}{x^2-1}$ **c.** $\frac{1}{2} + \frac{1}{x-1} = 1$

### Solution

**a.** Since $y - 3$ is the denominator in each rational expression, $y - 3$ is the LCD. Note that using 3 in place of $y$ in the equation would cause 0 to appear in the denominators. So we know up front that 3 *is not a solution to this equation.*

$$(y-3)\left(\frac{y}{y-3} + 3\right) = (y-3)\frac{3}{y-3} \quad \text{Multiply each side by the LCD.}$$

$$(y-3)\frac{y}{y-3} + (y-3)3 = 3 \quad \text{Distributive property}$$

$$y + 3y - 9 = 3$$

$$4y - 9 = 3$$

$$4y = 12 \quad \text{Add 9 to each side.}$$

$$y = 3 \quad \text{Divide each side by 4.}$$

If we replace $y$ by 3 in the original equation, then we get two undefined expressions. So 3 is not a solution to the original equation. The original equation has no solution. The equation is inconsistent.

**b.** Since $x^2 - 1 = (x-1)(x+1)$, the LCD is $(x-1)(x+1)$. Note that using 1 or $-1$ for $x$ in the equation would cause 0 to appear in a denominator.

$$\frac{1}{x-1} - \frac{1}{x+1} = \frac{2}{x^2-1}$$

$$(x-1)(x+1)\left(\frac{1}{x-1} - \frac{1}{x+1}\right) = (x-1)(x+1)\frac{2}{x^2-1} \quad \text{Multiply by the LCD.}$$

$$(x-1)(x+1)\frac{1}{x-1} - (x-1)(x+1)\frac{1}{x+1} = 2 \quad \text{Distributive property}$$

$$x + 1 - (x - 1) = 2$$

$$2 = 2$$

Since the last equation is an identity, the original equation is also an identity. The solution set is $\{x \mid x \neq 1 \text{ and } x \neq -1\}$, because 1 and $-1$ cannot be used for $x$ in the original equation.

**c.** Note that we cannot use 1 for $x$ in the original equation. To solve the equation multiply each side by the LCD:

$$\frac{1}{2} + \frac{1}{x-1} = 1$$

$$2(x-1)\left(\frac{1}{2} + \frac{1}{x-1}\right) = 2(x-1)1 \quad \text{Multiply by the LCD.}$$

$$x - 1 + 2 = 2x - 2$$

$$x + 1 = 2x - 2$$

$$3 = x$$

Check 3 in the original equation. The solution set is $\{3\}$, and the equation is a conditional equation.

▶**TRY THIS.** Solve $\frac{2}{x-3} - \frac{3}{x+3} = \frac{4}{x^2-9}$. ■

In Example 3(a) the final equation had a root that did not satisfy the original equation, because the domain of the rational expression excluded the root. Such a root is called an **extraneous root.** If an equation has no solution, then its solution set is the **empty set** (the set with no members). The symbol $\varnothing$ is used to represent the empty set.

### EXAMPLE 4 Using a calculator in solving an equation

Solve

$$\frac{7}{2.4x} + \frac{3}{5.9} = \frac{1}{8.2}$$

with the aid of a calculator. Round the answer to three decimal places.

#### Solution

We could multiply each side by the LCD, but since we are using a calculator, we can subtract 3/5.9 from each side to isolate $x$.

$$\frac{7}{2.4x} + \frac{3}{5.9} = \frac{1}{8.2}$$

$$\frac{7}{2.4x} = \frac{1}{8.2} - \frac{3}{5.9}$$

$$\frac{7}{2.4x} \approx -0.38652336 \qquad \text{Use a calculator to simplify.}$$

$$\frac{7}{2.4} \approx -0.38652336x \qquad \text{Multiply each side by } x.$$

$$\frac{7}{2.4(-0.38652336)} \approx x \qquad \text{Divide each side by } -0.38652336.$$

$$x \approx -7.546 \qquad \text{Round to three decimal places.}$$

Figure 1.2

The solution set is $\{-7.546\}$. Since $-0.38652336$ is an approximate value, the sign $\approx$ (for "approximately equal to") is used instead of the equal sign. To get three-decimal-place accuracy in the final answer, use as many digits as your calculator allows until you get to the final computation.

Figure 1.2 shows the computations and the check. Note how the ANS key is used.

▶**TRY THIS.** Solve $\frac{2}{3.4x} - \frac{1}{8.9} = \frac{4}{4.7}$ with the aid of a calculator. ■

## Equations Involving Absolute Value

To solve equations involving absolute value, remember that $|x| = x$ if $x \geq 0$, and $|x| = -x$ if $x < 0$. The absolute value of $x$ is greater than or equal to 0 for any real number $x$. So an equation such as $|x| = -6$ has no solution. Since a number and its opposite have the same absolute value, $|x| = 4$ is equivalent to $x = 4$ or $x = -4$. The only number that has 0 absolute value is 0. These ideas are summarized as follows.

**SUMMARY**

**Basic Absolute Value Equations**

| Absolute value equation | Equivalent statement | Solution set |
|---|---|---|
| $\lvert x\rvert = k\ (k > 0)$ | $x = -k$ or $x = k$ | $\{-k, k\}$ |
| $\lvert x\rvert = 0$ | $x = 0$ | $\{0\}$ |
| $\lvert x\rvert = k\ (k < 0)$ | | $\varnothing$ |

## EXAMPLE 5 Equations involving absolute value

Solve each equation.

**a.** $\lvert x - 5\rvert = 4$ **b.** $2\lvert x + 8\rvert - 6 = 0$

### Solution

**a.** First write an equivalent statement without using absolute value symbols.

$$\lvert x - 5\rvert = 4$$

$$x - 5 = -4 \quad \text{or} \quad x - 5 = 4$$

$$x = 1 \quad \text{or} \quad x = 9$$

The solution set is $\{1, 9\}$. Check that $\lvert 1 - 5\rvert = 4$ and $\lvert 9 - 5\rvert = 4$.

**b.** First isolate $\lvert x + 8\rvert$.

$$2\lvert x + 8\rvert - 6 = 0$$

$$2\lvert x + 8\rvert = 6$$

$$\lvert x + 8\rvert = 3$$

Now write an equivalent statement without absolute value symbols.

$$x + 8 = -3 \quad \text{or} \quad x + 8 = 3$$

$$x = -11 \quad \text{or} \quad x = -5$$

The solution set is $\{-11, -5\}$. Check that $2\lvert -11 + 8\rvert - 6 = 0$ and $2\lvert -5 + 8\rvert - 6 = 0$.

▸**TRY THIS.** Solve $\lvert 2x - 3\rvert = 5$. ■

**STRATEGY**

**Equation-Solving Overview**

In solving equations, there are usually many different sequences of correct steps that lead to the correct solution. If the approach that you try first does not work, *try another approach*, but learn from your failures as well as your successes. Solving equations successfully takes patience and practice.

## Equations Used as Models

Equations are often used to describe or **model** real situations. In the coming sections we will learn how to write equations that model real situations. In the next example you are simply given an equation and asked to solve it.

### EXAMPLE 6 An equation used as a model

If $x$ is the number of years after 1990 and $y$ is the median income in dollars for working women in the United States, then the equation $y = 355.9x + 11{,}075.3$ models the real data (www.infoplease.com). In what year (to the nearest year) is the median income \$20,000?

**Solution**

Since $y$ is 20,000, we solve the following equation to find $x$:

$$20{,}000 = 355.9x + 11{,}075.3$$

$$8924.7 = 355.9x \qquad \text{Subtract 11,075.3 from each side.}$$

$$25.08 \approx x \qquad \text{Divide each side by 355.9.}$$

So to the nearest year the median income reaches \$20,000 in 25 years after 1990, or 2015.

▶**TRY THIS.** In what year (to the nearest year) will the median income be \$25,000? ■

## FOR thought... True or False? Explain.

1. The number 3 is in the solution set to $5(4 - x) = 2x - 1$.
2. The equation $3x - 1 = 8$ is equivalent to $3x - 2 = 7$.
3. The equation $x + \sqrt{x} = -2 + \sqrt{x}$ is equivalent to $x = -2$.
4. The solution set to $x - x = 7$ is the empty set.
5. The equation $12x = 0$ is an inconsistent equation.
6. The equation $x - 0.02x = 0.98x$ is an identity.
7. The equation $|x| = -8$ is equivalent to $x = 8$ or $x = -8$.
8. The equations $\frac{x}{x-5} = \frac{5}{x-5}$ and $x = 5$ are equivalent.
9. To solve $-\frac{2}{3}x = \frac{3}{4}$, we should multiply each side by $-\frac{2}{3}$.
10. If $a$ and $b$ are real numbers, then $ax + b = 0$ has a solution.

## EXERCISES 1.1

*Fill in the blank.*

1. A statement that two algebraic expressions are equal is an ________.
2. An equation of the form $ax + b = 0$ with $a \neq 0$ is a ________ equation.
3. Equations with the same solution set are ________ equations.
4. The set of all solutions to an equation is the ________.
5. An ________ is an equation that is satisfied by all real numbers for which both sides are defined.
6. An ________ is an equation that has no solution.
7. A ________ is an equation that has at least one solution but is not an identity.
8. A number that satisfies the final equation but not the original equation is an ________.

*Determine whether each given number is a solution to the equation following it.*

9. $3,\ 2x - 4 = 9$
10. $-2,\ \frac{1}{x} - \frac{1}{2} = -1$
11. $-3,\ (x - 1)^2 = 16$
12. $4,\ \sqrt{3x + 4} = -4$

*Solve each equation and check your answer.*

**13.** $3x - 5 = 0$

**14.** $-2x + 3 = 0$

**15.** $-3x + 6 = 12$

**16.** $5x - 3 = -13$

**17.** $8x - 6 = 1 - 6x$

**18.** $4x - 3 = 6x - 1$

**19.** $7 + 3x = 4(x - 1)$

**20.** $-3(x - 5) = 4 - 2x$

**21.** $-\frac{3}{4}x = 18$

**22.** $\frac{2}{3}x = -9$

**23.** $\frac{x}{2} - 5 = -12 - \frac{2x}{3}$

**24.** $\frac{x}{4} - 3 = \frac{x}{2} + 3$

**25.** $\frac{3}{2}x + \frac{1}{3} = \frac{1}{4}x - \frac{1}{6}$

**26.** $\frac{x}{2} + \frac{x}{5} = \frac{x}{6} - \frac{1}{3}$

*Solve each equation. Identify each equation as an identity, an inconsistent equation, or a conditional equation.*

**27.** $3(x - 6) = 3x - 18$

**28.** $2a + 3a = 6a$

**29.** $2x + 3x = 4x$

**30.** $4(y - 1) = 4y - 4$

**31.** $2(x + 3) = 3(x - 1)$

**32.** $2(x + 1) = 3x + 2$

**33.** $3(x - 6) = 3x + 18$

**34.** $2x + 3x = 5x + 1$

**35.** $\frac{3x}{x} = 3$

**36.** $\frac{x(x + 2)}{x + 2} = x$

*Solve each equation involving rational expressions. Identify each equation as an identity, an inconsistent equation, or a conditional equation.*

**37.** $\frac{1}{w - 1} - \frac{1}{2w - 2} = \frac{1}{2w - 2}$

**38.** $\frac{1}{x} + \frac{1}{x - 3} = \frac{9}{x^2 - 3x}$

**39.** $\frac{1}{x} - \frac{1}{3x} = \frac{1}{2x} + \frac{1}{6x}$

**40.** $\frac{1}{5x} - \frac{1}{4x} + \frac{1}{3x} = -\frac{17}{60}$

**41.** $\frac{z + 2}{z - 3} = \frac{5}{-3}$

**42.** $\frac{2x - 3}{x - 4} = \frac{5}{x - 4}$

**43.** $\frac{1}{x - 3} - \frac{1}{x + 3} = \frac{6}{x^2 - 9}$

**44.** $\frac{4}{x - 1} - \frac{9}{x + 1} = \frac{3}{x^2 - 1}$

**45.** $4 + \frac{6}{y - 3} = \frac{2y}{y - 3}$

**46.** $\frac{x}{x + 6} - 3 = 1 - \frac{6}{x + 6}$

**47.** $\frac{t}{t + 3} + 4 = \frac{2}{t + 3}$

**48.** $\frac{3x}{x + 1} - 5 = \frac{x - 11}{x + 1}$

*Use a calculator to help you solve each equation. Round each approximate answer to three decimal places.*

**49.** $0.27x - 3.9 = 0.48x + 0.29$

**50.** $x - 2.4 = 0.08x + 3.5$

**51.** $0.06(x - 3.78) = 1.95$

**52.** $0.86(3.7 - 2.3x) = 4.9$

**53.** $2a + 1 = -\sqrt{17}$

**54.** $3c + 4 = \sqrt{38}$

**55.** $\frac{0.001}{y - 0.333} = 3$

**56.** $1 + \frac{0.001}{t - 1} = 0$

**57.** $\frac{x}{0.376} + \frac{x}{0.135} = 2$

**58.** $\frac{1}{x} + \frac{5}{6.72} = 10.379$

**59.** $(x + 3.25)^2 = (x - 4.1)^2$

**60.** $0.25(2x - 1.6)^2 = (x - 0.9)^2$

**61.** $(2.3 \times 10^6)x + 8.9 \times 10^5 = 1.63 \times 10^4$

**62.** $(-3.4 \times 10^{-9})x + 3.45 \times 10^{-8} = 1.63 \times 10^4$

*Solve each absolute value equation. Use the summary on page 7.*

**63.** $|x| = 8$

**64.** $|x| = 2.6$

**65.** $|x - 4| = 8$

**66.** $|x - 5| = 3.6$

**67.** $|x - 6| = 0$

**68.** $|x - 7| = 0$

**69.** $|x + 8| = -3$

**70.** $|x + 9| = -6$

**71.** $|2x - 3| = 7$

**72.** $|3x + 4| = 12$

**73.** $\frac{1}{2}|x - 9| = 16$

**74.** $\frac{2}{3}|x + 4| = 8$

**75.** $2|x + 5| - 10 = 0$

**76.** $6 - 4|x + 3| = -2$

**77.** $8|3x - 2| = 0$

**78.** $5|6 - 3x| = 0$

**79.** $2|x| + 7 = 6$

**80.** $5 + 3|x - 4| = 0$

*Solve each equation.*

**81.** $x - 0.05x = 190$

**82.** $x + 0.1x = 121$

**83.** $0.1x - 0.05(x - 20) = 1.2$

**84.** $0.03x - 0.2 = 0.2(x + 0.03)$

**85.** $(x + 2)^2 = x^2 + 4$

**86.** $(x - 3)^2 = x^2 - 9$

**87.** $(2x - 3)^2 = (2x + 5)^2$

**88.** $(3x - 4)^2 + (4x + 1)^2 = (5x + 2)^2$

**89.** $\frac{x}{2} + 1 = \frac{1}{4}(x - 6)$

**90.** $-\frac{1}{6}(x + 3) = \frac{1}{4}(3 - x)$

**91.** $\frac{y - 3}{2} + \frac{y}{5} = 3 - \frac{y + 1}{6}$

**92.** $\frac{y - 3}{5} - \frac{y - 4}{2} = 5$

**93.** $5 + 7|x + 6| = 19$

**94.** $9 - |2x - 3| = 6$

**95.** $9 - 4|2x - 3| = 9$

**96.** $-7 - |3x + 1| = |3x + 1| - 7$

**97.** $8 - 5|5x + 1| = 12$

**98.** $5|7 - 3x| + 2 = 4|7 - 3x| - 1$

**99.** $\dfrac{3}{x - 2} + \dfrac{4}{x + 2} = \dfrac{7x - 2}{x^2 - 4}$

**100.** $\dfrac{2}{x - 1} - \dfrac{3}{x + 2} = \dfrac{8 - x}{x^2 + x - 2}$

**101.** $\dfrac{4}{x + 3} - \dfrac{3}{2 - x} = \dfrac{7x + 1}{x^2 + x - 6}$

**102.** $\dfrac{3}{x} - \dfrac{4}{1 - x} = \dfrac{7x - 3}{x^2 - x}$

**103.** $\dfrac{x - 2}{x - 3} = \dfrac{x - 3}{x - 4}$

**104.** $\dfrac{y - 1}{y + 4} = \dfrac{y + 1}{y - 2}$

*Solve each problem.*

**105.** *Working Mothers* The percentage of working mothers $y$ can be modeled by the equation

$$y = 0.0102x + 0.644$$

where $x$ is the number of years since 1990 (U.S. Census Bureau, www.census.gov).

**a.** Use the accompanying graph to estimate the year in which 70% of mothers were in the work force.

**b.** Is the percentage of mothers in the labor force increasing or decreasing?

**c.** Use the equation to find the year in which 90% of mothers will be in the labor force.

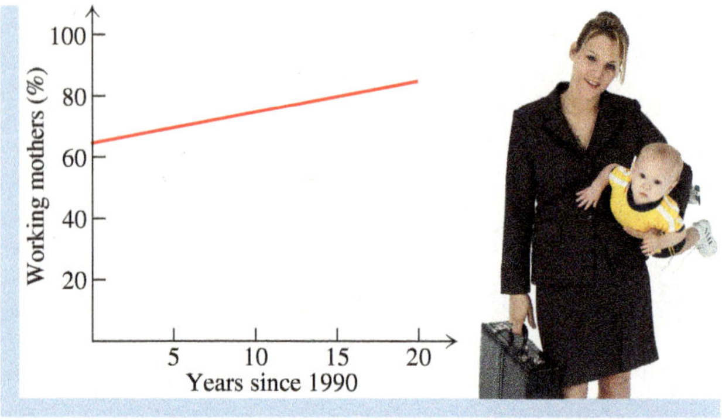


**Figure for Exercises 105 and 106**

**106.** *Working Mothers* Use the equation in the previous exercise to find the year in which 64.4% of mothers were in the labor force.

**107.** *Cost Accounting* An accountant has been told to distribute a bonus to the employees that is 15% of the company's net income. Since the bonus is an expense to the accountant, it must be subtracted from the income to determine the net income. If the company has an income of \$140,000 before the bonus, then the accountant must solve

$$B = 0.15(140{,}000 - B)$$

to find the bonus $B$. Find $B$.

**108.** *Corporate Taxes* For a class C corporation in Louisiana, the amount of state income tax $S$ is deductible on the federal return and the amount of federal income tax $F$ is deductible on the state return. With \$200,000 taxable income and a 30% federal tax rate, the federal tax is $0.30(200{,}000 - S)$. If the state tax rate is 6% then the state tax satisfies

$$S = 0.06(200{,}000 - 0.30(200{,}000 - S)).$$

Find the state tax $S$ and the federal tax $F$.

**109.** *Production Cost* An automobile manufacturer, who spent \$500 million to develop a new line of cars, wants the cost of development and production to be \$12,000 per vehicle. If the production costs are \$10,000 per vehicle, then the cost per vehicle for development and production of $x$ vehicles is $(10{,}000x + 500{,}000{,}000)/x$ dollars. Solve the equation

$$\frac{10{,}000x + 500{,}000{,}000}{x} = 12{,}000$$

to find the number of vehicles that must be sold so that the cost of development and production is \$12,000 per vehicle.

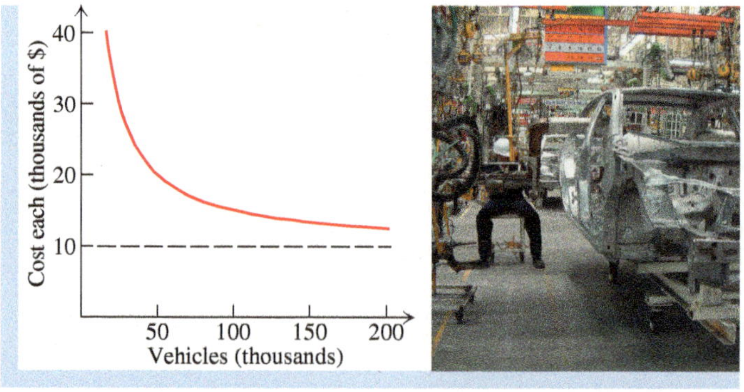


**Figure for Exercise 109**

**110.** *Harmonic Mean* The *harmonic mean* of the numbers $x_1, x_2, \ldots, x_n$ is defined as

$$HM = \frac{n}{\dfrac{1}{x_1} + \dfrac{1}{x_2} + \cdots + \dfrac{1}{x_n}}.$$

The accompanying table shows the gross domestic product (GDP) for the top five countries in trillions of U.S. dollars (www.worldbank.org).

**a.** What is the harmonic mean for the GDP for these five countries?

**b.** If France is included, the harmonic mean for the six countries is \$2.93 trillion. What is the GDP for France?

**Table for Exercise 110**

| Country | GDP (\$ trillions) |
|---|---|
| U.S. | 12.5 |
| Japan | 4.5 |
| Germany | 2.8 |
| China | 2.2 |
| U.K. | 2.2 |

**111.** *Inscribed Circle* A right triangle has sides of length 1 and 2 as shown in the accompanying figure. Find the exact radius of the inscribed circle.

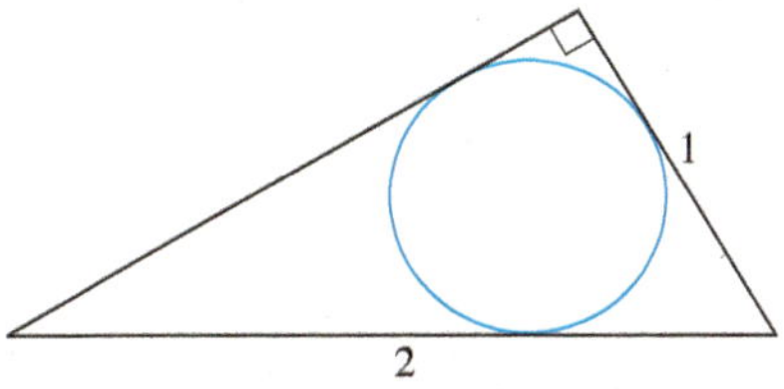


**Figure for Exercise 111**

**112.** *Inscribed Circle* A right triangle has sides of length 1 and 1 as shown in the accompanying figure. Find the exact radius of the inscribed circle.

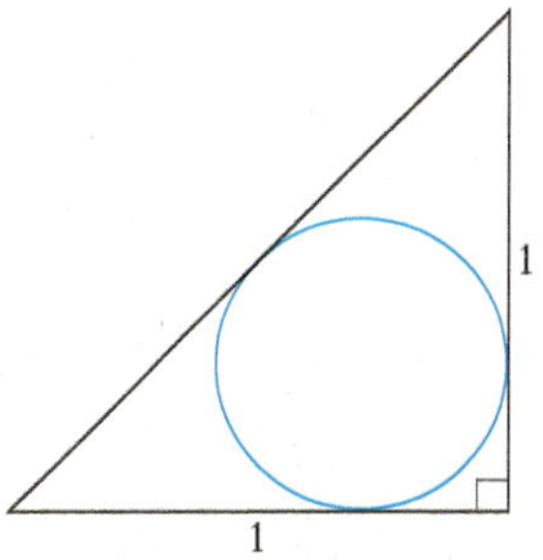


**Figure for Exercise 112**

## FOR WRITING/DISCUSSION

**113.** *Definitions* Without looking back in the text, write the definitions of linear equation, identity, and inconsistent equation. Use complete sentences.

**114.** *Cooperative Learning* Each student in a small group should write a linear equation, an identity, an inconsistent equation, and an equation that has an extraneous root. The group should solve each equation and determine whether each equation is of the required type.

## THINKING OUTSIDE THE BOX I

*Nines* Nine people applying for credit at the Highway 99 Loan Company listed nine different incomes each containing a different number of digits. Each of the nine incomes was a whole number of dollars and the maximum income was a nine-digit number. The loan officer found that the arithmetic mean of the nine incomes was \$123,456,789. What are the nine incomes?

## POP QUIZ 1.1

*Solve each equation and identify each as an identity, inconsistent equation, or a conditional equation.*

**1.** $7x - 6 = 0$

**2.** $\frac{1}{4}x - \frac{1}{6} = \frac{1}{3}$

**3.** $3(x - 9) = 3x - 27$

**4.** $|w - 1| = 6$

**5.** $2(x + 6) = 2x + 6$

**6.** $(x + 1)^2 = x^2 + 1$

LINKING **concepts...** For Individual or Group Explorations

### Modeling Oxygen Uptake in Runners

*In a study of oxygen uptake rate for marathon runners (Costill and Fox, Medicine and Science in Sports, Vol. 1) it was found that the power expended in kilocalories per minute is given by the formula $P = M(av - b)$ where M is the mass of the runner in kilograms and v is the speed in meters per minute. The constants a and b have values*

$$a = 1.02 \times 10^{-3}\ \text{kcal/kg m} \quad \textit{and} \quad b = 2.62 \times 10^{-2}\ \text{kcal/kg min.}$$

**a)** Runners with masses of 60 kg, 65 kg, and 70 kg are running together at 400 m/min. Find the power expenditure for each runner.

**b)** With a constant velocity, does the power expended increase or decrease as the mass of the runner increases?

**c)** Runners of 80 kg, 84 kg, and 90 kg are all expending power at the rate of 38.7 kcal/min. Find the velocity at which each is running.

**d)** With a constant power expenditure, does the velocity increase or decrease as the mass of the runner increases?

**e)** A 50-kg runner in training has a velocity of 480 m/min while carrying a weight of 2 kg. Assume that her power expenditure remains constant when the weight is removed, and find her velocity without the weight.

**f)** Why do runners in training carry weights?

**g)** Use a computer to create graphs showing power expenditure versus mass with a velocity fixed at 400 m/min and velocity versus mass with a fixed power expenditure of 40 kcal/min.

# 1.2 Constructing Models to Solve Problems

In Section 1.1 we solved equations. In this section we will use those skills in problem solving. To solve a problem we construct a mathematical model of the problem. Sometimes we use well-known formulas to model real situations, but we must often construct our own models.

## Formulas and Functions

A **formula** is an equation involving two or more variables. Consider the uniform-motion formula $D = RT$. If you know the value of $R$ (rate) and the value of $T$ (time), you can determine the value of $D$ (distance). This formula is a rule by which we can determine $D$: *Multiply the rate and time to get the distance.* We may also call this formula a *function*.

**Definition: Function**

A **function** is a rule for determining the value of one variable from the values of one or more other variables. We say that the first variable **is a function of** the other variable(s).

So $D$ is a function of $R$ and $T$. We also say that $D = RT$ expresses $D$ as a function of $R$ and $T$. We will see rules expressed in other ways besides formulas when we study functions in greater depth in Chapter 2.

Formulas and functions are used to **model** real-life situations. The function $D = RT$ models the relationship between distance, rate, and time in uniform motion. The formula $P = 2L + 2W$ models the relationship between the perimeter, length, and width of a rectangle. Since $P$ is determined by the values of $L$ and $W$, we say that $P$ is a function of $L$ and $W$. A list of common formulas that are commonly used as mathematical models is given inside the back cover of this text.

The formula $C = \frac{5}{9}(F - 32)$ is used to determine the Celsius temperature from the Fahrenheit temperature. So $C$ is a function of $F$. This formula is **solved for** $C$. When a formula is solved for a specified variable, that variable is isolated on one side of the equal sign and must not occur on the other side. The formula $F = \frac{9}{5}C + 32$ expresses $F$ as a function of $C$. It is solved for $F$.

## EXAMPLE 1 Solving a formula for a specified variable

Solve the formula $S = P + Prt$ for $P$.

### Solution

$$P + Prt = S \qquad \text{Write the formula with } P \text{ on the left.}$$

$$P(1 + rt) = S \qquad \text{Factor out } P.$$

$$P = \frac{S}{1 + rt} \qquad \text{Divide each side by } 1 + rt.$$

The formula $S = P + Prt$ expresses $S$ as a function of $P$, $r$, and $t$. The formula $P = \frac{S}{1 + rt}$ expresses $P$ as a function of $S$, $r$, and $t$.

▶**TRY THIS.** Solve $A = \frac{1}{2}hb_1 + \frac{1}{2}hb_2$ for $h$. ■

In some situations we know the values of all variables except one. After we substitute values for those variables, the formula is an equation in one variable. We can then solve the equation to find the value of the remaining variable.

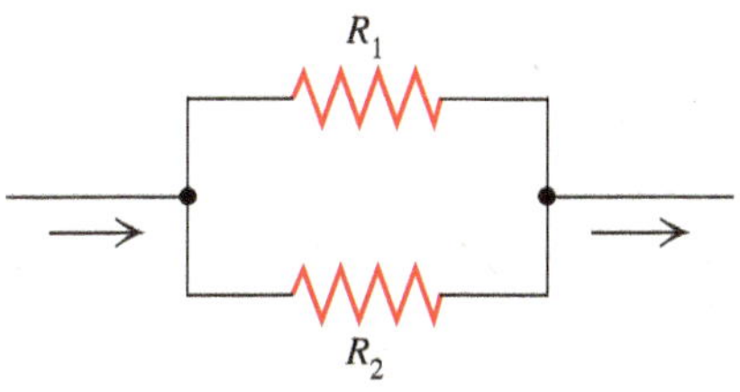

**Figure 1.3**

## EXAMPLE 2 Finding the value of a variable in a formula

The total resistance $R$ (in ohms) in the parallel circuit shown in Fig. 1.3 is modeled by the formula

$$\frac{1}{R} = \frac{1}{R_1} + \frac{1}{R_2}.$$

The subscripts 1 and 2 indicate that $R_1$ and $R_2$ represent the resistance for two different receivers. If $R = 7$ ohms and $R_1 = 10$ ohms, then what is the value of $R_2$?

### Solution

Substitute the values for $R$ and $R_1$ and solve for $R_2$.

$$\frac{1}{7} = \frac{1}{10} + \frac{1}{R_2} \qquad \text{The LCD of 7, 10, and } R_2 \text{ is } 70R_2.$$

$$70R_2 \cdot \frac{1}{7} = 70R_2\left(\frac{1}{10} + \frac{1}{R_2}\right) \qquad \text{Multiply each side by } 70R_2.$$

$$10R_2 = 7R_2 + 70$$

$$3R_2 = 70$$

$$R_2 = \frac{70}{3}$$

Check the solution in the original formula. The resistance $R_2$ is $70/3$ ohms, or about 23.3 ohms.

▸**TRY THIS.** Find $b_1$ if $A = 20$, $h = 2$, $b_2 = 3$, and $A = \frac{1}{2}h(b_1 + b_2)$. ■

## Constructing Your Own Models

Applied problems in mathematics often involve solving an equation. The equations for some problems come from known formulas, as in Example 2, while in other problems we must write an equation that models a particular problem situation.

The best way to learn to solve problems is to study a few examples and then solve lots of problems. We will first look at an example of problem solving, then give a problem-solving strategy.

### EXAMPLE 3 Solving a problem involving sales tax

Jeannie Fung bought a Ford Mustang GT for a total cost of \$18,966, including sales tax. If the sales tax rate is 9%, then what amount of tax did she pay?

#### Solution

There are two unknown quantities here, the price of the car and the amount of the tax. Represent the unknown quantities as follows:

$$x = \text{the price of the car}$$

$$0.09x = \text{the amount of sales tax}$$

The price of the car plus the amount of sales tax is the total cost of the car. We can model this relationship with an equation and solve it for $x$:

$$x + 0.09x = 18{,}966$$

$$1.09x = 18{,}966$$

$$x = \frac{18{,}966}{1.09} = 17{,}400$$

$$0.09x = 1566$$

You can check this answer by adding \$17,400 and \$1566 to get \$18,966, the total cost. The amount of tax that Jeannie Fung paid was \$1566.

▸**TRY THIS.** Joe bought a new computer for \$1506.75, including sales tax at 5%. What amount of tax did he pay? ■

No two problems are exactly alike, but there are similarities. The following strategy will assist you in solving problems on your own.

**STRATEGY**

**Problem Solving**

1. Read the problem as many times as necessary to get an understanding of the problem.
2. If possible, draw a diagram to illustrate the problem.
3. Choose a variable, write down what it represents, and if possible, represent any other unknown quantities in terms of that variable.

4. Write an equation that models the situation. You may be able to use a known formula, or you may have to write an equation that models only that particular problem.
5. Solve the equation.
6. Check your answer by using it to solve the original problem (not just the equation).
7. Answer the question posed in the original problem.

In the next example we have a geometric situation. Notice how we are following the strategy for problem solving.

### EXAMPLE 4 Solving a geometric problem

In 1974, Chinese workers found three pits that contained life-size sculptures of warriors, which were created to guard the tomb of an emperor (www.chinatour.com). The largest rectangular pit has a length that is 40 yards longer than three times the width. If its perimeter is 640 yards, what are the length and width?

**Solution**

First draw a diagram as shown in Fig. 1.4. Use the fact that the length is 40 yards longer than three times the width to represent the width and length as follows:

$$x = \text{the width in yards}$$
$$3x + 40 = \text{the length in yards}$$

**Figure 1.4**

The formula for the perimeter of a rectangle is $2L + 2W = P$. Replace $W$ by $x$, $L$ by $3x + 40$, and $P$ by 640.

$$\begin{aligned} 2L + 2W &= P && \text{Perimeter formula} \\ 2(3x + 40) + 2x &= 640 && \text{Substitution} \\ 6x + 80 + 2x &= 640 \\ 8x + 80 &= 640 \\ 8x &= 560 \\ x &= 70 \end{aligned}$$

If $x = 70$, then $3x + 40 = 250$. Check that the dimensions of 70 yards and 250 yards give a perimeter of 640 yards. We conclude that the length of the pit is 250 yards and its width is 70 yards.

▸**TRY THIS.** The length of a rectangle is 20 cm shorter than five times its width. If the perimeter is 800 cm, then what is the length? ■

The next problem is called a **uniform-motion** problem because it involves motion at a constant rate. Of course, people do not usually move at a constant rate, but their average speed can be assumed to be constant over some time interval. The problem also illustrates how a table can be used as an effective technique for organizing information.

## EXAMPLE 5 Solving a uniform-motion problem

A group of hikers from Tulsa hiked down into the Grand Canyon in 3 hours 30 minutes. Coming back up on a trail that was 4 miles shorter, they hiked 2 mph slower and it took them 1 hour longer. What was their rate going down?

### Solution

Let $x$ represent the rate going down into the canyon. Make a table to show the distance, rate, and time for both the trip down and the trip back up. Once we fill in any two entries in a row of the table, we can use $D = RT$ to obtain an expression for the third entry:

| | Rate | Time | Distance |
|---|---|---|---|
| **Down** | $x$ mi/hr | 3.5 hr | $3.5x$ mi |
| **Up** | $x - 2$ mi/hr | 4.5 hr | $4.5(x - 2)$ mi |

Using the fact that the distance up was 4 miles shorter, we can write the following equation:

$$4.5(x - 2) = 3.5x - 4$$

$$4.5x - 9 = 3.5x - 4$$

$$x - 9 = -4$$

$$x = 5$$

After checking that 5 mph satisfies the conditions given in the problem, we conclude that the hikers traveled at 5 mph going down into the canyon.

▸**TRY THIS.** Bea hiked uphill from her car to a waterfall in 6 hours. Hiking back to her car over the same route, she averaged 2 mph more and made the return trip in half the time. How far did she hike? ■

Average speed is not necessarily the average of your speeds. For example, if you drive 70 mph for 3 hours and then 30 mph for 1 hour, you will travel 240 miles in 4 hours. Your average speed for the trip is 60 mph, which is not the average of 70 and 30.

## EXAMPLE 6 Average speed

Shelly drove 40 miles from Peoria to Bloomington at 20 mph. She then drove back to Peoria at a higher rate of speed so that she averaged 30 mph for the whole trip. What was her speed on the return trip?

### Solution

Let $x$ represent the speed on the return trip. Make a table as follows:

| | Rate | Time | Distance |
|---|---|---|---|
| **Going** | 20 mph | 2 hr | 40 mi |
| **Returning** | $x$ mph | $40/x$ hr | 40 mi |
| **Round trip** | 30 mph | $2 + 40/x$ hr | 80 mi |

Use $D = RT$ to write an equation for the round trip and solve it.

$$80 = 30\left(2 + \frac{40}{x}\right)$$

$$80 = 60 + \frac{1200}{x}$$

$$20 = \frac{1200}{x}$$

$$20x = 1200$$

$$x = 60$$

So the speed on the return trip was 60 mph. It is interesting to note that the distance between the cities does not affect the solution. You should repeat this example using $d$ as the distance between the cities.

▶**TRY THIS.** Dee drove 20 miles to work at 60 mph. She then drove back home at a lower rate of speed, averaging 50 mph for the round trip. What was her speed on the return trip? ■

The next example involves mixing beverages with two different concentrations of orange juice. In other **mixture problems,** we may mix chemical solutions, candy, or even people.

## EXAMPLE 7 Solving a mixture problem

A beverage producer makes two products, Orange Drink, containing 10% orange juice, and Orange Delight, containing 50% orange juice. How many gallons of Orange Delight must be mixed with 300 gallons of Orange Drink to create a new product containing 40% orange juice?

Figure 1.5

### Solution

Let $x$ represent the number of gallons of Orange Delight and make a sketch as in Fig. 1.5. Next, we make a table that shows three pertinent expressions for each product: the amount of the product, the percent of orange juice in the product, and the actual amount of orange juice in that product.

| | Amount of Product | Percent Orange Juice | Amount Orange Juice |
|---|---|---|---|
| **Drink** | 300 gal | 10% | 0.10(300) gal |
| **Delight** | $x$ gal | 50% | $0.50x$ gal |
| **Mixture** | $x + 300$ gal | 40% | $0.40(x + 300)$ gal |

We can now write an equation expressing the fact that the actual amount of orange juice in the mixture is the sum of the amounts of orange juice in the Orange Drink and in the Orange Delight:

$$0.40(x + 300) = 0.10(300) + 0.50x$$

$$0.4x + 120 = 30 + 0.5x$$

$$90 = 0.1x$$

$$900 = x$$

Mix 900 gallons of Orange Delight with the 300 gallons of Orange Drink to obtain the proper mixture.

▶**TRY THIS.** How many gallons of a 40% acid solution must be mixed with 30 gallons of a 20% acid solution to obtain a mixture that is 35% acid? ■

**Work problems** are problems in which people or machines are working together to accomplish a task. A typical situation might have two people painting a house at different rates. Suppose Joe and Frank are painting a house together for 2 hours and Joe paints at the rate of one-sixth of the house per hour while Frank paints at the rate of one-third of the house per hour. Note that

$$\left(\frac{1}{6}\text{ of house per hour}\right)(2\text{ hr}) = \frac{1}{3}\text{ of house}$$

and

$$\left(\frac{1}{3}\text{ of house per hour}\right)(2\text{ hr}) = \frac{2}{3}\text{ of house.}$$

The product of the rate and the time gives the fraction of the house completed by each person, and these fractions have a sum of 1 because the entire job is completed. Note how similar this situation is to a uniform-motion problem where $RT = D$.

## EXAMPLE 8 Solving a work problem

Aboard the starship *Nostromo,* the human technician, Brett, can process the crew's medical history in 36 minutes. However, the android Science Officer, Ash, can process the same records in 24 minutes. After Brett worked on the records for 1 minute, Ash joined in and both crew members worked until the job was done. How long did Ash work on the records?

### Solution

Let $x$ represent the number of minutes that Ash worked and $x + 1$ represent the number of minutes that Brett worked. Ash works at the rate of $1/24$ of the job per minute, while Brett works at the rate of $1/36$ of the job per minute. The following table shows all of the pertinent quantities.

| | Rate | Time | Work Completed |
|---|---|---|---|
| **Ash** | $\frac{1}{24}$ job/min | $x$ min | $\frac{1}{24}x$ job |
| **Brett** | $\frac{1}{36}$ job/min | $x + 1$ min | $\frac{1}{36}(x + 1)$ job |

The following equation expresses the fact that the work completed together is the sum of the work completed by each worker alone.

$$\frac{1}{24}x + \frac{1}{36}(x + 1) = 1$$

$$72\left[\frac{1}{24}x + \frac{1}{36}(x + 1)\right] = 72 \cdot 1 \quad \text{Multiply by the LCD 72.}$$

$$3x + 2x + 2 = 72$$

$$5x = 70$$

$$x = 14$$

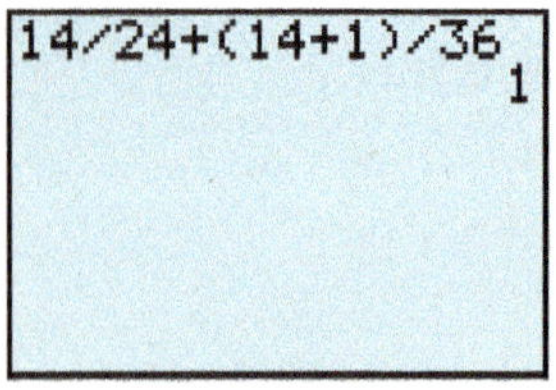

**Figure 1.6**

Ash worked for 14 minutes.

Check 14 in the original equation as shown in Fig. 1.6.

**▸TRY THIS.** A tank can be filled by a small pipe in 12 hours or a large pipe in 8 hours. How long will it take to fill the tank if the small pipe is used alone for 2 hours and then both pipes are used until the tank is full? ■

## FOR thought... True or False? Explain.

1. If we solve $P + Prt = S$ for $P$, we get $P = S - Prt$.
2. The perimeter of any rectangle is the product of its length and width.
3. If $n$ is an odd integer, then $n + 1$ and $n + 3$ represent odd integers.
4. The equation $y = x + 2$ expresses $y$ as a function of $x$.
5. Two numbers that have a sum of $-3$ can be represented by $x$ and $-3 - x$.
6. If $P$ is the number of professors and $S$ is the number of students at the play, and there are twice as many professors as students, then $2P = S$.
7. If you need \$100,000 for your house, and the agent gets 9% of the selling price, then the agent gets \$9000, and the house sells for \$109,000.
8. If John can mow the lawn in $x$ hours, then he mows the lawn at the rate of $1/x$ of the lawn per hour.
9. If George hiked $3x$ miles and Anita hiked $4(x - 2)$ miles, and George hiked 5 more miles than Anita, then $3x + 5 = 4(x - 2)$.
10. Two numbers that differ by 9 can be represented as 9 and $x + 9$.

## EXERCISES 1.2

*Fill in the blank.*

1. A ________ is an equation that involves two or more variables.
2. If the value of $y$ can be determined from the value of $x$, then $y$ is a ________ of $x$.
3. Motion at a constant rate is ________ motion.
4. In uniform motion, distance is a function of ________ and ________.

*Solve each formula for the specified variable. The use of the formula is indicated in parentheses.*

5. $I = Prt$ for $r$ (simple interest)
6. $D = RT$ for $R$ (uniform motion)
7. $F = \frac{9}{5}C + 32$ for $C$ (temperature)
8. $C = \frac{5}{9}(F - 32)$ for $F$ (temperature)
9. $A = \frac{1}{2}bh$ for $b$ (area of a triangle)
10. $A = \frac{1}{2}bh$ for $h$ (area of a triangle)
11. $Ax + By = C$ for $y$ (equation of a line)
12. $Ax + By = C$ for $x$ (equation of a line)
13. $\frac{1}{R} = \frac{1}{R_1} + \frac{1}{R_2} + \frac{1}{R_3}$ for $R_1$ (resistance)
14. $\frac{1}{R} = \frac{1}{R_1} + \frac{1}{R_2} + \frac{1}{R_3}$ for $R_2$ (resistance)
15. $a_n = a_1 + (n - 1)d$ for $n$ (arithmetic sequence)
16. $S_n = \frac{n}{2}(a_1 + a_n)$ for $a_1$ (arithmetic series)
17. $S = \frac{a_1 - a_1 r^n}{1 - r}$ for $a_1$ (geometric series)
18. $S = 2LW + 2LH + 2HW$ for $H$ (surface area)
19. *The 2.4-Meter Rule* A 2.4-meter sailboat is a one-person boat that is about 13 ft in length, has a displacement of about 550 lb, and a sail area of about 81 $\text{ft}^2$. To compete in the 2.4-meter class, a boat must satisfy the formula

$$2.4 = \frac{L + 2D - F\sqrt{S}}{2.37},$$

where $L$ = length, $F$ = freeboard, $D$ = girth, and $S$ = sail area. Solve the formula for $D$.

20. *Finding the Freeboard* Solve the formula in the previous exercise for $F$.

*Find the requested formula.*

21. Write a formula that expresses rate $R$ as a function of distance $D$ and time $T$ in uniform motion.
22. Write a formula that expresses time $T$ as a function of distance $D$ and rate $R$ in uniform motion.
23. Write a formula that expresses the width $W$ of a rectangle as a function of its area $A$ and length $L$.

**24.** Write a formula that expresses the width $W$ of a rectangle as a function of its perimeter $P$ and length $L$.

**25.** Write a formula that expresses the radius $r$ of a circle as a function of its diameter $d$.

**26.** Write a formula that expresses the diameter $d$ of a circle as a function of its radius $r$.

*Use the appropriate formula to solve each problem.*

**27.** *Simple Interest* If \$51.30 in interest is earned on a deposit of \$950 in one year, then what is the simple interest rate?

**28.** *Simple Interest* If you borrow \$100 and pay back \$105 at the end of one month, then what is the simple annual interest rate?

**29.** *Uniform Motion* How long does it take an SR-71 Blackbird, one of the fastest U.S. jets, to make a surveillance run of 5570 mi if it travels at an average speed of Mach 3 (2228 mph)?

**30.** *Circumference of a Circle* If the circumference of a circular sign is $72\pi$ in., then what is the radius?

**31.** *Fahrenheit Temperature* If the temperature at 11 P.M. on New Year's Eve in Times Square was 23°F, then what was the temperature in degrees Celsius?

**32.** *Celsius Temperature* If the temperature at 1 P.M. on July 9 in Toronto was 30°C, then what was the temperature in degrees Fahrenheit?

*Solve each problem. See the strategy for problem solving on pages 14–15.*

**33.** *Cost of a Car* Jeff knows that his neighbor Sarah paid \$40,230, including sales tax, for a new Buick Park Avenue. If the sales tax rate is 8%, then what is the cost of the car before the tax?

**34.** *Real Estate Commission* To be able to afford the house of their dreams, Dave and Leslie must clear \$128,000 from the sale of their first house. If they must pay \$780 in closing costs and 6% of the selling price for the sales commission, then what is the minimum selling price for which they will get \$128,000?

**35.** *Adjusting the Saddle* The saddle height on a bicycle should be 109% of the inside leg measurement of the rider (www.harriscyclery.com). See the figure. If the saddle height is 37 in., then what is the inside leg measurement?

**Figure for Exercise 35**

**36.** *Target Heart Rate* For a cardiovascular workout, fitness experts recommend that you reach your target heart rate and stay at that rate for at least 20 minutes (www.healthstatus.com). To find your target heart rate find the sum of your age and your resting heart rate, then subtract that sum from 220. Find 60% of that result and add it to your resting heart rate. If the target heart rate for a 30-year-old person is 144, then what is that person's resting heart rate?

**37.** *Classic Car Auction* A buyer paid \$50,600, including the buyer's premium, for a 1963 Corvette at a Mecum auction. If Mecum adds a 10% buyer's premium to the sale price of the car, then what was the sale price of the car?

**38.** *Art Auction* A buyer paid 2.835 million pounds, including a buyer's premium, for a Salvador Dali painting at a Sotheby's auction. If Sotheby's adds a 5% buyer's premium to the winning bid, then what was the winning bid?

**39.** *Investment Income* Tara paid one-half of her game-show winnings to the government for taxes. She invested one-third of her winnings in Jeff's copy shop at 14% interest and one-sixth of her winnings in Kaiser's German Bakery at 12% interest. If she earned a total of \$4000 on the investments in one year, then how much did she win on the game show?

**40.** *Construction Penalties* Gonzales Construction contracted Kentwood High and Memorial Stadium for a total cost of \$4.7 million. Because the construction was not completed on time, Gonzales paid 5% of the amount of the high school contract in penalties and 4% of the amount of the stadium contract in penalties. If the total penalty was \$223,000, then what was the amount of each contract?

**41.** *Trimming a Garage Door* A carpenter used 30 ft of molding in three pieces to trim a garage door. If the long piece was 2 ft longer than twice the length of each shorter piece, then how long was each piece?

**Figure for Exercise 41**

**42.** *Increasing Area of a Field* Julia's soybean field is 3 m longer than it is wide. To increase her production, she plans to increase both the length and width by 2 m. If the new field is 46 $m^2$ larger than the old field, then what are the dimensions of the old field?

**43.** *Fencing a Feed Lot* Peter plans to fence off a square feed lot and then cross-fence to divide the feed lot into four smaller square feed lots. If he uses 480 ft of fencing, then how much area will be fenced in?

**Figure for Exercise 43**

**44.** *Fencing Dog Pens* Clint is constructing two adjacent rectangular dog pens. Each pen will be three times as long as it is wide, and the pens will share a common long side. If Clint has 65 ft of fencing, what are the dimensions of each pen?

**Figure for Exercise 44**

**45.** *Racing Speed* Bobby and Rick are in a 10-lap race on a one-mile oval track. Bobby, averaging 90 mph, has completed two laps just as Rick is getting his car onto the track. What speed does Rick have to average to be even with Bobby at the end of the tenth lap?

HINT Bobby does 8 miles in the same time as Rick does 10 miles.

**46.** *Rowing a Boat* Boudreaux rowed his pirogue from his camp on the bayou to his crab traps. Going down the bayou, he caught a falling tide that increased his normal speed by 2 mph, but coming back it decreased his normal speed by 2 mph. Going with the tide, the trip took only 10 min; going against the tide, the trip took 30 min. How far is it from Boudreaux's camp to his crab traps?

HINT With the tide his rate is $x + 2$ mph and against the tide it is $x - 2$ mph.

**47.** *Average Speed* Junior drove his rig on Interstate 10 from San Antonio to El Paso. At the halfway point he noticed that he had been averaging 80 mph, while his company requires his average speed to be 60 mph. What must be his speed for the last half of the trip so that he will average 60 mph for the entire trip?

HINT The distance from San Antonio to El Paso is irrelevant. Use $d$ or simply make up a distance.

**48.** *Basketball Stats* Two-thirds of the way through the basketball season, Tina Thompson of the Houston Comets has an average of 18 points per game. What must her point average be for the remaining games to average 22 points per game for the season?

**49.** *Start-Up Capital* Norma invested the start-up capital for her Internet business in two hedge funds. After one year one of the funds returned 5% and the other returned 6%, for a total return of \$5880. If the amount on which she made 6% was \$10,000 larger than the amount on which she made 5%, then what was the original amount of her start-up capital?

**50.** *Combining Investments* Brent lent his brother Bob some money at 8% simple interest, and he lent his sister Betty half as much money at twice the interest rate. Both loans were for one year. If Brent made a total of 24 cents in interest, then how much did he lend to each one?

**51.** *Percentage of Minority Workers* At the Northside assembly plant, 5% of the workers were classified as minority, while at the Southside assembly plant, 80% of the workers were classified as minority. When Northside and Southside were closed, all workers transferred to the new Eastside plant to make up its entire work force. If 50% of the 1500 employees at Eastside are minority, then how many employees did Northside and Southside have originally?

**52.** *Mixing Alcohol Solutions* A pharmacist needs to obtain a 70% alcohol solution. How many ounces of a 30% alcohol solution must be mixed with 40 ounces of an 80% alcohol solution to obtain a 70% alcohol solution?

HINT Add $x$ ounces of 30% solution to 40 ounces of 80% solution to get $x + 40$ ounces of 70% solution.

**53.** *Harvesting Wheat* With the old combine, Nikita's entire wheat crop can be harvested in 72 hr, but a new combine can do the same job in 48 hr. How many hours would it take to harvest the crop with both combines operating?

HINT The rate for the old combine is $1/72$ crop/hr and for the new one it is $1/48$ crop/hr. Together the rate is $1/x$ crop/hr.

**54.** *Processing Forms* Rita can process a batch of insurance claims in 4 hr working alone. Eduardo can process a batch of insurance claims in 2 hr working alone. How long would it take them to process a batch of claims if they worked together?

**55.** *Batman and Robin* Batman can clean up all of the crime in Gotham City in 8 hr working alone. Robin can do the same job alone in 12 hr. If Robin starts crime fighting at 8 A.M. and Batman joins him at 10 A.M., then at what time will they have all of the crime cleaned up?

**56.** *Scraping Barnacles* Della can scrape the barnacles from a 70-ft yacht in 10 hr using an electric barnacle scraper. Don can do the same job in 15 hr using a manual barnacle scraper. If Don starts scraping at noon and Della joins him at 3 P.M., then at what time will they finish the job?

**57.** *Planning a Race Track* If Mario plans to develop a circular race track one mile in circumference on a square plot of land, then what is the minimum number of acres that he needs? (One acre is equal to 43,560 $ft^2$.)

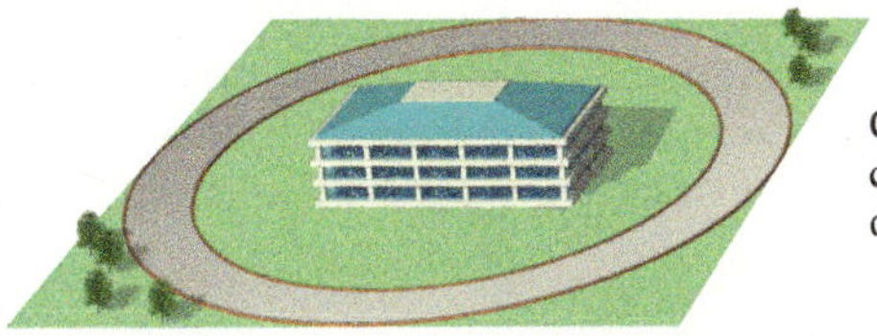

Circular track has a circumference of one mile.

**Figure for Exercise 57**

**58.** *Volume of a Can of Coke* If a can of Coke contains 12 fluid ounces and the diameter of the can is 2.375 in., then what is the height of the can? (One fluid ounce equals approximately 1.8 $\text{in.}^3$.)

**59.** *Area of a Lot* Julio owns a four-sided lot that lies between two parallel streets. If his 90,000-$\text{ft}^2$ lot has 500-ft frontage on one street and 300-ft frontage on the other, then how far apart are the streets?

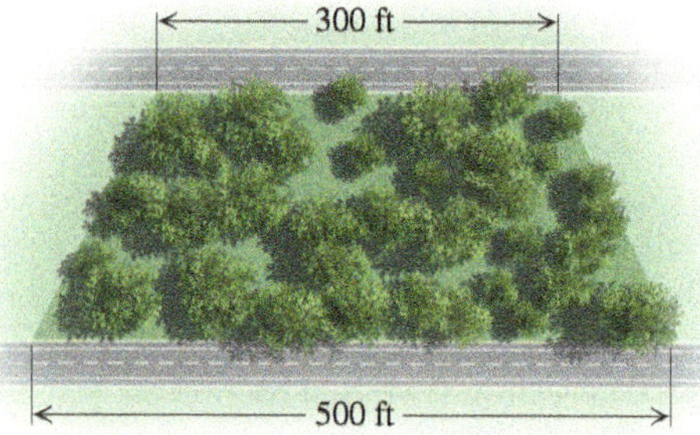


**Figure for Exercise 59**

**60.** *Width of a Football Field* If the perimeter of a football field in the NFL, including the end zones, is 1040 ft and the field is 120 yd long, then what is the width of the field in feet?

**61.** *Depth of a Swimming Pool* A circular swimming pool with a diameter of 30 ft and a horizontal bottom contains 22,000 gal of water. What is the depth of the water in the pool? (One cubic foot contains approximately 7.5 gal of water.)

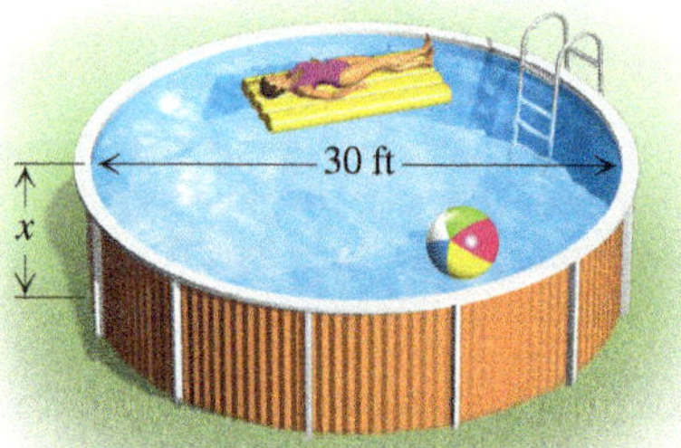


**Figure for Exercise 61**

**62.** *Depth of a Reflecting Pool* A rectangular reflecting pool with a horizontal bottom is 100 ft by 150 ft and contains 200,000 gal of water. How deep is the water in the pool?

**63.** *Olympic Track* To host the Summer Olympics, a city plans to build an eight-lane track. The track will consist of parallel 100-m straightaways with semicircular turns on either end as shown in the figure. The distance around the outside edge of the oval track is 514 m. If the track is built on a rectangular lot as shown in the drawing, then how many hectares (1 hectare $=$ 10,000 $\text{m}^2$) of land are needed in the rectangular lot?

**64.** *Green Space* If the inside radius of the turns is 30 m and grass is to be planted inside and outside the track of Exercise 63, then how many square meters of the rectangular lot will be planted in grass?

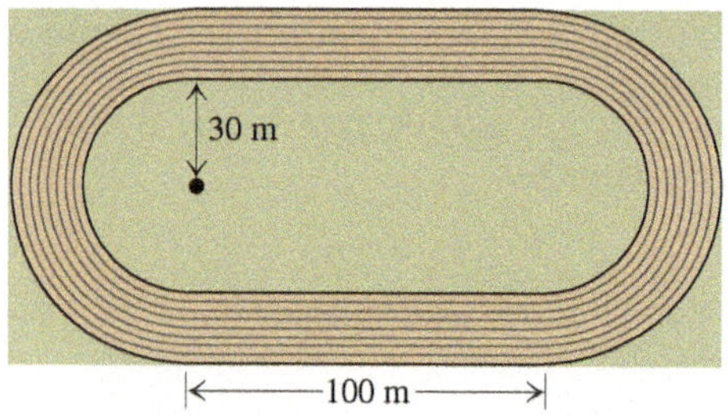


**Figure for Exercises 63 and 64**

**65.** *Taxable Income* According to the IRS (Internal Revenue Service, www.irs.gov), a single taxpayer with a taxable income between \$82,400 and 171,850 in 2010 paid \$16,781.25, plus 28% of the amount over \$82,400. If Lorinda paid \$22,280.17 in federal income tax for 2010, then what was her taxable income in 2010?

**66.** *Higher Bracket* According to the IRS (Internal Revenue Service, www.irs.gov), a single taxpayer with a taxable income over \$373,650 in 2010 paid \$108,421.25 plus 35% of the amount over \$373,650. If Glen paid \$284,539.85 in federal income tax for 2010, then what was his taxable income for 2010?

**67.** *Diluting Baneberry* How much water must Poison Ivy add to a 4-liter solution that contains 5% extract of baneberry to get a solution that contains 3% extract of baneberry?

**68.** *Strengthening Vinegar* A food chemist has 200 gallons of vinegar that is 4% acetic acid. How much acetic acid must be added to get the acetic acid up to 5%?

**69.** *Mixing Gasoline* A chemist has 500 gallons of gasoline that contain 5% ethanol. How many gallons of gasoline containing 12% ethanol should she add to get a mixture that contains 10% ethanol?

**70.** *Diluting Antifreeze* A mechanic is working on a car with a 20-qt radiator containing a 60% antifreeze solution. How much of the solution should he drain and replace with pure water to get a solution that is 50% antifreeze?

**71.** *Mixing Dried Fruit* The owner of a health-food store sells dried apples for \$1.20 per quarter-pound, and dried apricots for \$1.80 per quarter-pound. How many pounds of each must he mix together to get 20 lb of a mixture that sells for \$1.68 per quarter-pound?

**72.** *Mixing Breakfast Cereal* Raisins sell for \$4.50/lb, and bran flakes sell for \$2.80/lb. How many pounds of raisins should be mixed with 12 lb of bran flakes to get a mixture that sells for \$3.14/lb?

**73.** *Coins in a Vending Machine* Dana inserted eight coins, consisting of dimes and nickels, into a vending machine to purchase a Snickers bar for 55 cents. How many coins of each type did she use?

**74.** *Cost of a Newspaper* Ravi took eight coins from his pocket, which contained only dimes, nickels, and quarters, and bought the Sunday Edition of *The Daily Star* for 75 cents. If the number of nickels he used was one more than the number of dimes, then how many of each type of coin did he use?

**75.** *Active Ingredients* A pharmacist has 200 milliliters of a solution that is 40% active ingredient. How much pure water should she add to the solution to get a solution that is 25% active ingredient?

**76.** *Mixed Nuts* A manager bought 12 pounds of peanuts for \$30. He wants to mix \$5 per pound cashews with the peanuts to get a batch of mixed nuts that is worth \$4 per pound. How many pounds of cashews are needed?

**77.** *Salt Solution I* A chemist has 5 gallons of salt solution with a concentration of 0.2 pound per gallon and another solution with a concentration of 0.5 pound per gallon. How many gallons of the stronger solution must be added to the weaker solution to get a solution that contains 0.3 pound per gallon?

**78.** *Salt Solution II* Suppose the 5-gallon solution in the previous problem is contained in a 5-gallon container. The chemist plans to remove some amount of the 0.2 pound per gallon solution and replace it with 0.5 pound per gallon solution so that he ends up with 5 gallons of a 0.3 pound per gallon solution. What amount should he remove?

**79.** *Draining a Pool I* A small pump can drain a pool in 8 hours. A large pump could drain the same pool in 5 hours. How long (to the nearest minute) will it take to drain the pool if both pumps are used simultaneously?

**80.** *Draining a Pool II* Suppose that the pumps in the previous exercise could not be used simultaneously and that the pool was drained in exactly 6 hours. How long was each pump used?

**81.** *Alcohol Solution* How many gallons of a 15% alcohol solution and how many gallons of a 10% alcohol solution should be mixed together to obtain 20 gallons that contain 12% alcohol?

**82.** *Alcohol Solution* How many quarts of a 20% alcohol solution and how many quarts of pure alcohol should be mixed together to obtain 100 quarts that contain 30% alcohol?

**83.** *Dining In* Revenue for restaurants and supermarkets can be modeled by the equations $R = 13.5n + 190$ and $S = 7.5n + 225$, where $n$ is the number of years since 1986 (Forbes, www.forbes.com).

**a.** Use the accompanying graph to estimate the year in which restaurant and supermarket revenue were equal.

**b.** Use the equations to find the year in which restaurant and supermarket revenue were equal.

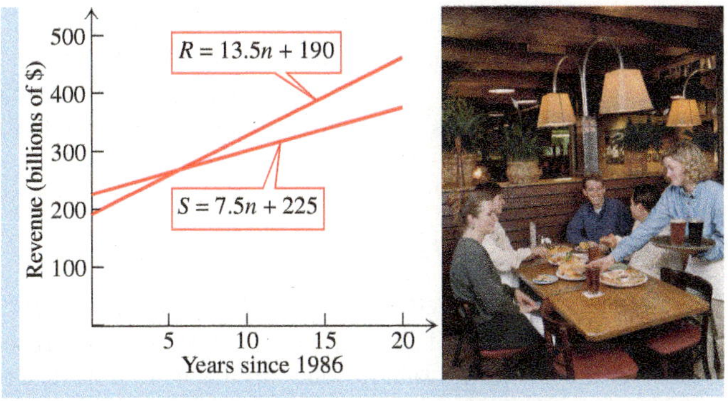


**Figure for Exercises 83 and 84**

**84.** *Dining Out* Use the equations in the previous exercise to find the year in which restaurant revenue will double supermarket revenue.

## FOR WRITING/DISCUSSION

**85.** *Working Together* If one lost hiker can pick a gallon of wild berries in 2 hr, then how long should it take two lost hikers working together to pick one gallon of wild berries? If one mechanic at Spee-Dee Oil Change can change the oil in a Saturn in 6 min, then how long should it take two mechanics working together to change the oil in a Saturn? How long would it take 60 mechanics? Are your answers reasonable? Explain.

**86.** *Average Speed* Imagine that you are cruising down I-75 headed to spring break in Florida.

**a.** Suppose that you cruise at 70 mph for 3 hours and then at 60 mph for 1 hour. What is your average speed for those 4 hours? When is the average speed over two time intervals actually the average of the two speeds? Explain your answer.

**b.** Suppose that you cruise at 60 mph for 180 miles and then cruise at 40 mph for 160 miles. What is your average speed? When is the average speed over two distance intervals actually the average of the two speeds? Explain your answer.

## ▶ RETHINKING

**87.** Solve $2x - 3 = 8$.

**88.** Solve $\frac{x}{2} + \frac{1}{3} = \frac{x}{9} + \frac{1}{6}$.

**89.** Solve $x - 0.001x = 9990$.

**90.** Solve $|2x - 3| = 8$.

**91.** Solve $|2x - 3| = -8$.

**92.** Solve $|2x - 3| = 0$.

## THINKING OUTSIDE THE BOX II

*Roughing It* Milo and Bernard are planning a three-day canoe trip on the Roaring Fork River. Their friend Vince will drop them off at the Highway 14 bridge. From there they will paddle upstream for 12 hours on the first day and 9 hours on the second day. They have been on this river before and know that their average paddling rate is twice the rate of the current in the river. At what time will they have to start heading downstream on the third day to meet Vince at the Highway 14 bridge at 5 P.M.?

## POP QUIZ 1.2

1. Solve $dx + dy = w$ for $y$.

2. If the length of a rectangle is 3 feet longer than the width and the perimeter is 62 feet, then what is the width?

3. How many liters of water should be added to 3 liters of a 70% alcohol solution to obtain a 50% alcohol solution?

## LINKING concepts... For Individual or Group Explorations

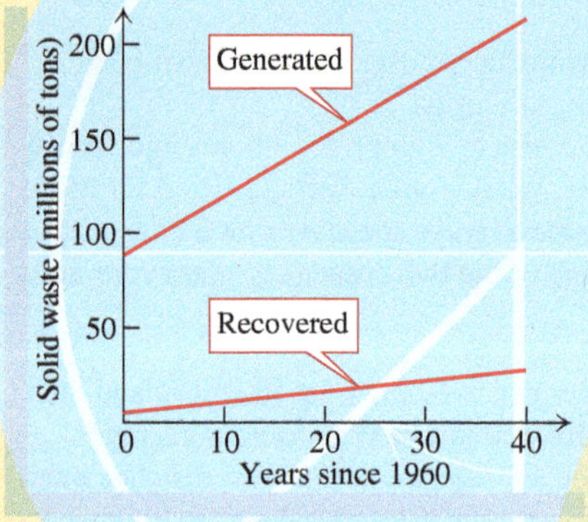


### A Recycling Model

*In* 1960 *the United States generated* 87.1 *million tons of municipal solid waste and recovered (or recycled) only* 4.3% *of it (U.S. EPA, www.epa.gov). The amount of municipal solid waste generated in the United States can be modeled by the formula* $w = 3.14n + 87.1$, *while the amount recovered can be modeled by the formula* $w = 0.576n + 3.78$, *where w is in millions of tons and n is the number of years since* 1960.

**a)** Use the graph to estimate the first year in which the United States generated over 100 million tons of municipal solid waste.

**b)** Use the formulas to determine the year in which the United States generated 150 million more tons than it recovered.

**c)** Find the year in which 13% of the municipal solid waste generated will be recovered.

**d)** Find the years in which the recovery rates will be 14%, 15%, and 16%.

**e)** Will the recovery rate ever reach 25%?

**f)** According to this model, what is the maximum percentage of solid waste that will ever be recovered?

# 1.3 Equations and Graphs in Two Variables

In Section 1.1 we studied equations in one variable. In this section we study pairs of variables. For example, $p$ might represent the price of gasoline and $n$ the number of gallons that you consume in a month; $x$ might be the number of toppings on a medium pizza and $y$ the cost of that pizza; or $h$ might be the height of a two-year-old child and $w$ the weight. To study relationships between pairs of variables we use a two-dimensional coordinate system.

**Table 1.1**

| Toppings $x$ | Cost $y$ |
|---|---|
| 0 | \$5 |
| 1 | 7 |
| 2 | 9 |
| 3 | 11 |
| 4 | 13 |

## The Cartesian Coordinate System

If $x$ and $y$ are real numbers then $(x, y)$ is called an **ordered pair** of real numbers. The numbers $x$ and $y$ are the **coordinates** of the ordered pair, with $x$ being the **first coordinate** or **abscissa,** and $y$ being the **second coordinate** or **ordinate.** For example, Table 1.1 shows the number of toppings on a medium pizza and the corresponding cost. The ordered pair (3, 11) indicates that a three-topping pizza costs \$11. The order of the numbers matters. In this context, (11, 3) would indicate that an 11-topping pizza costs \$3.

To picture ordered pairs of real numbers we use the **rectangular coordinate system** or **Cartesian coordinate system,** named after the French mathematician René Descartes (1596–1650). The Cartesian coordinate system consists of two number lines drawn perpendicular to one another, intersecting at zero on each number line as shown in Fig. 1.7. The point of intersection of the number lines is called the **origin.** The horizontal number line is the ***x*-axis** and its positive numbers are to the right of the origin. The vertical number line is the ***y*-axis** and its positive numbers are above the origin. The two number lines divide the plane into four regions called **quadrants,** numbered as shown in Fig. 1.7. The quadrants do not include any points on the axes. We call a plane with a rectangular coordinate system the **coordinate plane** or the ***xy*-plane.**

Just as every real number corresponds to a point on the number line, every ordered pair of real numbers $(a, b)$ corresponds to a point $P$ in the $xy$-plane. For this reason, ordered pairs of numbers are often called **points.** So $a$ and $b$ are the coordinates of $(a, b)$ or the coordinates of the point $P$. Locating the point $P$ that corresponds to $(a, b)$ in the $xy$-plane is referred to as **plotting** or **graphing** the point, and $P$ is called the *graph* of $(a, b)$. In general, a **graph** is a set of points in the rectangular coordinate system.

HISTORICAL NOTE

René Descartes (1596–1650) was a noted French philosopher, mathematician, and scientist. He has been called the "Founder of Modern Philosophy" and the "Father of Modern Mathematics." He ranks as one of the most important and influential thinkers of modern times.

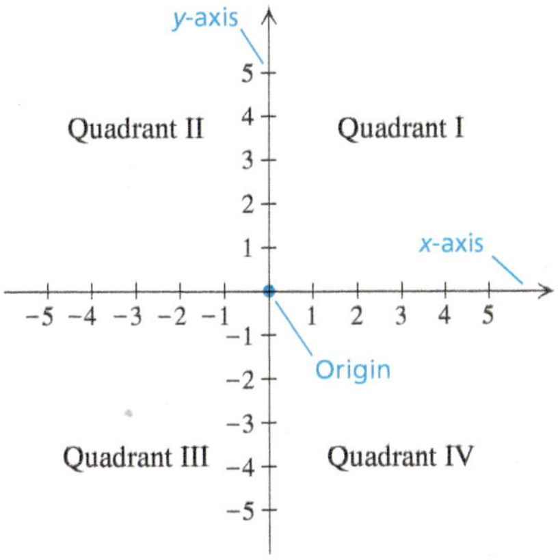


Figure 1.7

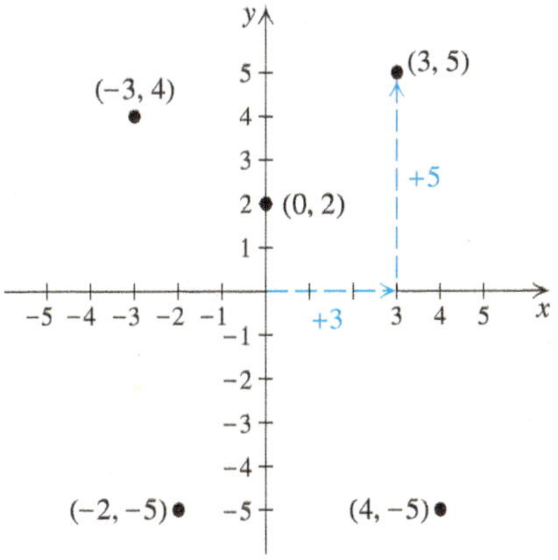


Figure 1.8

HISTORICAL NOTE

Pythagoras of Samos (582 B.C.–507 B.C.) was a Greek mathematician and philosopher. He founded the mystic, religious, and scientific society called Pythagoreans. He is best known for the Pythagorean theorem. Pythagoras made influential contributions to philosophy and religious teaching in the late 6th century B.C. Pythagoras and his students believed that everything was related to mathematics, and that through mathematics everything could be predicted and measured in rhythmic patterns or cycles.

## EXAMPLE 1 Plotting points

Plot the points $(3, 5)$, $(4, -5)$, $(-3, 4)$, $(-2, -5)$, and $(0, 2)$ in the $xy$-plane.

### Solution

The point $(3, 5)$ is located three units to the right of the origin and five units above the $x$-axis as shown in Fig. 1.8. The point $(4, -5)$ is located four units to the right of the origin and five units below the $x$-axis. The point $(-3, 4)$ is located three units to the left of the origin and four units above the $x$-axis. The point $(-2, -5)$ is located two units to the left of the origin and five units below the $x$-axis. The point $(0, 2)$ is on the $y$-axis because its first coordinate is zero.

▸**TRY THIS.** Plot $(-3, -2)$, $(-1, 3)$, $(5/2, 0)$, and $(2, -3)$. ■

Note that for points in quadrant I, both coordinates are positive. In quadrant II the first coordinate is negative and the second is positive, while in quadrant III, both coordinates are negative. In quadrant IV the first coordinate is positive and the second is negative.

## The Pythagorean Theorem and the Distance Formula

You have probably studied the Pythagorean theorem in an algebra or a trigonometry course that you have taken. This theorem states that the *sum of the squares of the legs of any right triangle is equal to the square of the hypotenuse.*

**The Pythagorean Theorem**

The triangle shown here is a right triangle if and only if $a^2 + b^2 = c^2$.

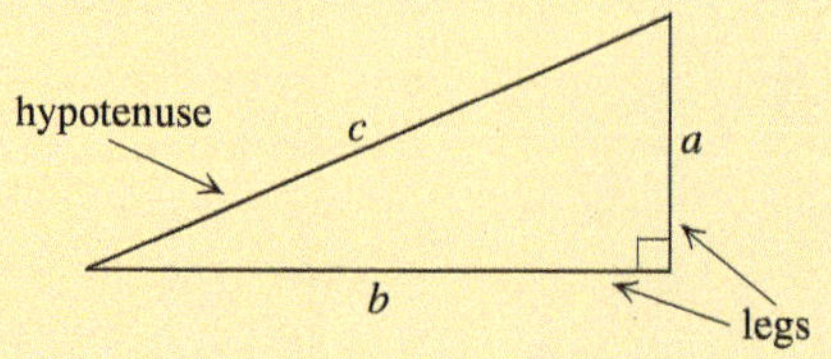


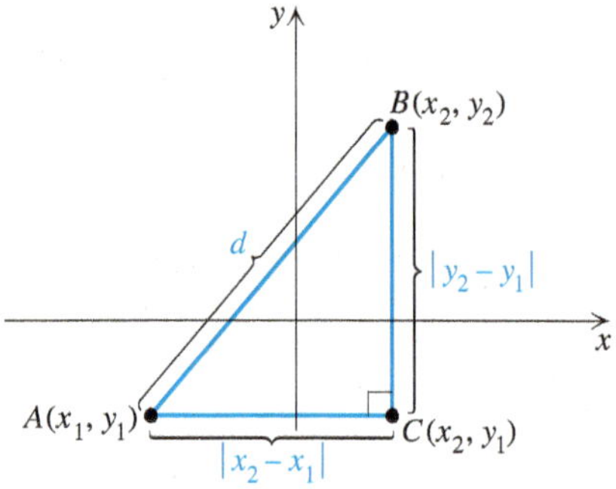


Figure 1.9

We will not prove the Pythagorean theorem here. If you want to see a proof, try an Internet search for the Pythagorean theorem. You will get many sites with numerous proofs. This author found one site with 79 proofs, including one by a former president of the United States. Here we will show how the Pythagorean theorem leads to a formula for the distance between two points in the plane and the equation for a circle.

If $a$ and $b$ are real numbers, then the distance between them on the number line is $|a - b|$. Now consider the points $A(x_1, y_1)$ and $B(x_2, y_2)$ shown in Fig. 1.9. Let $AB$ represent the length of line segment $\overline{AB}$. Now $\overline{AB}$ is the hypotenuse of the right triangle in Fig. 1.9. Since $A$ and $C$ lie on a horizontal line, the distance between them is $|x_2 - x_1|$. Likewise $CB = |y_2 - y_1|$. Since the sum of the squares of the legs of a right triangle is equal to the square of the hypotenuse (the Pythagorean theorem) we have

$$d^2 = |x_2 - x_1|^2 + |y_2 - y_1|^2.$$

Since the distance between two points is a nonnegative real number, we have $d = \sqrt{(x_2 - x_1)^2 + (y_2 - y_1)^2}$. The absolute value symbols are replaced with parentheses, because $|a - b|^2 = (a - b)^2$ for any real numbers $a$ and $b$.

**The Distance Formula**

The distance $d$ between the points $(x_1, y_1)$ and $(x_2, y_2)$ is given by the formula

$$d = \sqrt{(x_2 - x_1)^2 + (y_2 - y_1)^2}.$$

### EXAMPLE 2 Finding the distance between two points

Find the exact distance between each pair of points.

**a.** $(5, -3), (-1, -6)$ **b.** $(\pi/2, 0), (\pi/3, 1)$

**Solution**

**a.** Let $(x_1, y_1) = (5, -3)$ and $(x_2, y_2) = (-1, -6)$. These points are shown on the graph in Fig. 1.10. Substitute these values into the distance formula:

$$\begin{aligned} d &= \sqrt{(-1 - 5)^2 + (-6 - (-3))^2} \\ &= \sqrt{(-6)^2 + (-3)^2} \\ &= \sqrt{36 + 9} = \sqrt{45} = 3\sqrt{5} \end{aligned}$$

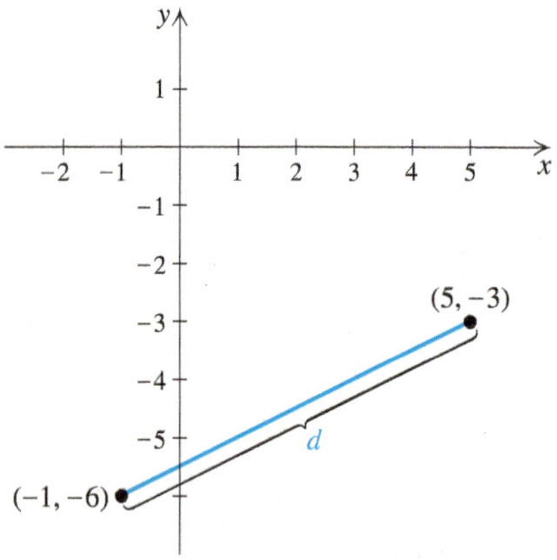


Figure 1.10

The exact distance between the points is $3\sqrt{5}$.

**b.** Use the distance formula as follows.

$$d = \sqrt{\left(\frac{\pi}{2} - \frac{\pi}{3}\right)^2 + (0 - 1)^2}$$

$$= \sqrt{\left(\frac{3\pi}{6} - \frac{2\pi}{6}\right)^2 + (-1)^2} \quad \text{Get common denominators.}$$

$$= \sqrt{\left(\frac{\pi}{6}\right)^2 + 1} = \sqrt{\frac{\pi^2}{36} + 1} = \sqrt{\frac{\pi^2 + 36}{36}} = \frac{\sqrt{\pi^2 + 36}}{6}$$

The exact distance between the points is $\sqrt{\pi^2 + 36}/6$.

▶**TRY THIS.** Find the distance between $(-3, -2)$ and $(-1, 4)$. ■

Note that the distance between two points is the same regardless of which point is chosen as $(x_1, y_1)$ or $(x_2, y_2)$.

## The Midpoint Formula

When you average two test scores (by finding their sum and dividing by 2), you are finding a number midway between the two scores. Likewise, the midpoint of the line segment with endpoints $-1$ and 7 in Fig. 1.11 is $(-1 + 7)/2$ or 3. In general, $(a + b)/2$ is the midpoint of the line segment with endpoints $a$ and $b$ shown in Fig. 1.12.

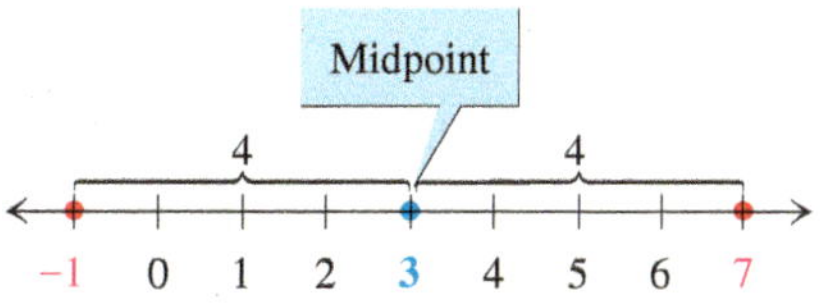


**Figure 1.11**

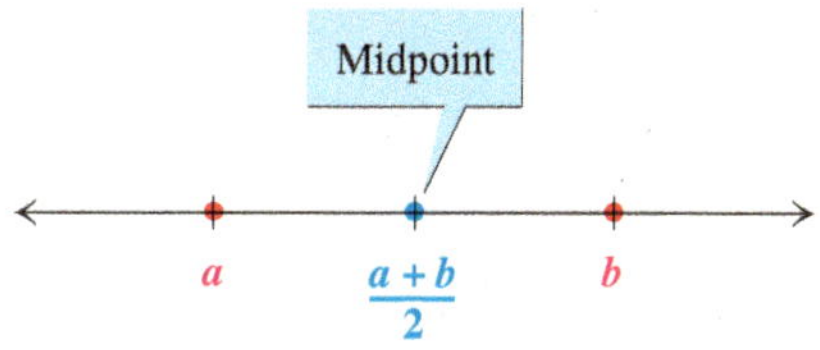


**Figure 1.12**

**Theorem: Midpoint on a Number Line**

If $a$ and $b$ are real numbers, then $\frac{a+b}{2}$ is midway between them on the number line.

**PROOF** The distance between two numbers on the number line is the absolute value of their difference. So

$$\left|a - \frac{a+b}{2}\right| = \left|\frac{2a}{2} - \frac{a+b}{2}\right| = \left|\frac{a-b}{2}\right| = \frac{|a-b|}{2}$$

and

$$\left|b - \frac{a+b}{2}\right| = \left|\frac{2b}{2} - \frac{a+b}{2}\right| = \left|\frac{b-a}{2}\right| = \frac{|b-a|}{2}.$$

Since $|a - b| = |b - a|$, the distances from $\frac{a+b}{2}$ to $a$ and from $\frac{a+b}{2}$ to $b$ are equal. Since $\frac{a+b}{2}$ is equidistant from $a$ and $b$, it must be between $a$ and $b$. ■

We can find the midpoint of a line segment in the $xy$-plane in the same manner.

**Theorem: The Midpoint Formula**

The midpoint of the line segment with endpoints $(x_1, y_1)$ and $(x_2, y_2)$ is

$$\left(\frac{x_1 + x_2}{2}, \frac{y_1 + y_2}{2}\right).$$

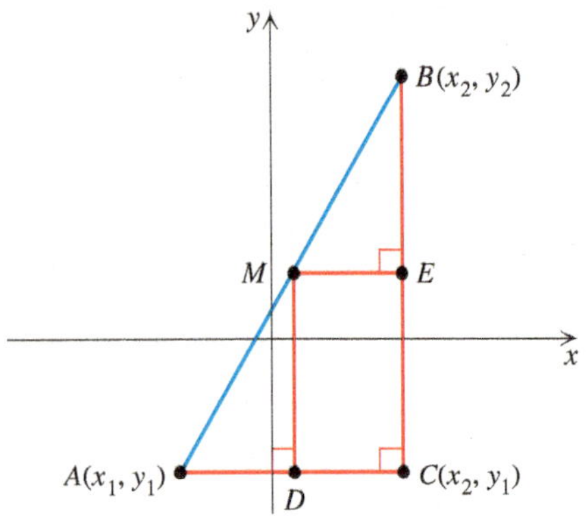


Figure 1.13

**PROOF** Start with a line segment with endpoints $A(x_1, y_1)$ and $B(x_2, y_2)$, as shown in Fig. 1.13. Let $M$ be the midpoint of $\overline{AB}$. Draw horizontal and vertical line segments that form three right triangles, as shown in Fig. 1.13. Since $M$ is the midpoint of $\overline{AB}$, the two small right triangles are congruent. So $D$ is the midpoint of $\overline{AC}$ and $E$ is the midpoint of $\overline{BC}$. Since the midpoint on a number line is found by adding and dividing by two, the $x$-coordinate of $D$ is $\frac{x_1 + x_2}{2}$ and the $y$-coordinate of $E$ is $\frac{y_1 + y_2}{2}$. So the midpoint $M$ is $\left(\frac{x_1 + x_2}{2}, \frac{y_1 + y_2}{2}\right)$. ■

## EXAMPLE 3 Finding the midpoint

Find the midpoint of the line segment with the given endpoints.

**a.** $(5, -3), (-1, -6)$ **b.** $(\pi/2, 0), (\pi/3, 1)$

### Solution

**a.** To get the midpoint, add the corresponding coordinates and divide by 2:

$$\left(\frac{5 + (-1)}{2}, \frac{-3 + (-6)}{2}\right) = \left(\frac{4}{2}, \frac{-9}{2}\right) = \left(2, -\frac{9}{2}\right)$$

The midpoint is $(2, -9/2)$.

**b.** To get the midpoint, add the corresponding coordinates and divide by 2:

$$\left(\frac{\frac{\pi}{2} + \frac{\pi}{3}}{2}, \frac{0 + 1}{2}\right) = \left(\frac{\frac{3\pi}{6} + \frac{2\pi}{6}}{2}, \frac{1}{2}\right) = \left(\frac{\frac{5\pi}{6}}{2}, \frac{1}{2}\right) = \left(\frac{5\pi}{12}, \frac{1}{2}\right)$$

The midpoint is $(5\pi/12, 1/2)$.

▶**TRY THIS.** Find the midpoint of the line segment with endpoints $(4, -1)$ and $(3, 1/2)$. ■

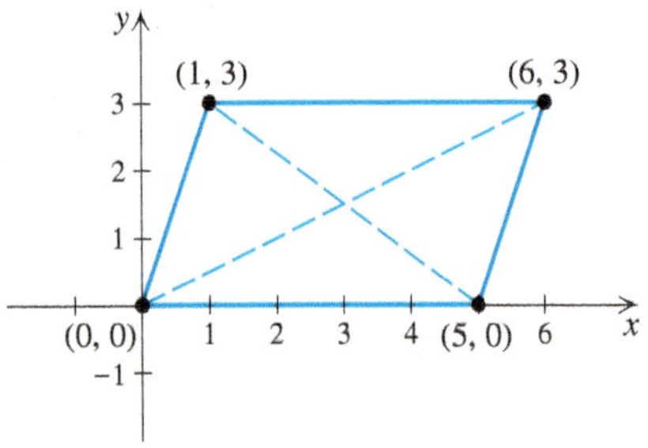


Figure 1.14

## EXAMPLE 4 Using the midpoint formula

Prove that the diagonals of the parallelogram with vertices $(0, 0)$, $(1, 3)$, $(5, 0)$, and $(6, 3)$ bisect each other.

### Solution

The parallelogram is shown in Fig. 1.14. The midpoint of the diagonal from $(0, 0)$ to $(6, 3)$ is

$$\left(\frac{0 + 6}{2}, \frac{0 + 3}{2}\right),$$

or $(3, 1.5)$. The midpoint of the diagonal from $(1, 3)$ to $(5, 0)$ is

$$\left(\frac{1 + 5}{2}, \frac{3 + 0}{2}\right),$$

or $(3, 1.5)$. Since the diagonals have the same midpoint, they bisect each other.

▶ **TRY THIS.** Prove that the diagonals of the square with vertices $(0, 0)$, $(4, 1)$, $(3, 5)$, and $(-1, 4)$ bisect each other. ■

## The Circle

An ordered pair is a **solution to** or **satisfies** an equation in two variables if the equation is correct when the variables are replaced by the coordinates of the ordered pair. For example, $(3, 11)$ satisfies $y = 2x + 5$ because $11 = 2(3) + 5$ is correct. The

**solution set** to an equation in two variables is the set of all ordered pairs that satisfy the equation. The graph of (the solution set to) an equation is a geometric object that gives us a visual image of an algebraic object. Circles provide a nice example of this relationship between algebra and geometry.

A **circle** is the set of all points in a plane that lie a fixed distance from a given point in the plane. The fixed distance is called the **radius,** and the given point is the **center.** The distance formula can be used to write an equation for the circle shown in Fig. 1.15 with center $(h, k)$ and radius $r$ for $r > 0$. A point $(x, y)$ is on the circle if and only if it satisfies the equation

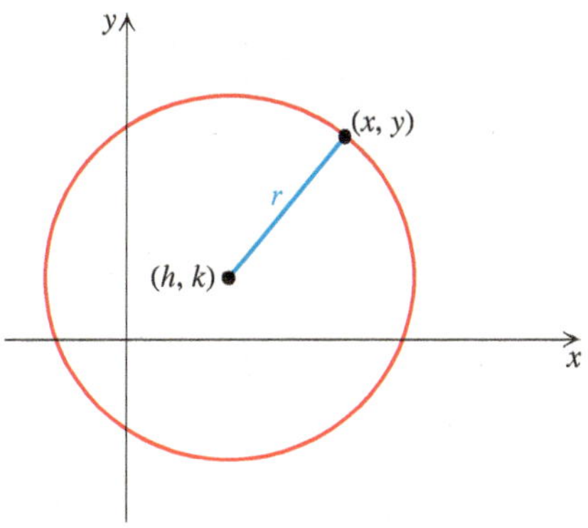

Figure 1.15

$$\sqrt{(x-h)^2+(y-k)^2}=r.$$

Since both sides of $\sqrt{(x-h)^2+(y-k)^2}=r$ are positive, we can square each side to get the following **standard form** for the equation of a circle.

### Theorem: Equation for a Circle in Standard Form

The equation for a circle with center $(h, k)$ and radius $r$ for $r > 0$ is

$$(x-h)^2+(y-k)^2=r^2.$$

A circle centered at the origin has equation $x^2+y^2=r^2$.

Note that squaring both sides of an equation produces an equivalent equation only when both sides are positive. If we square both sides of $\sqrt{x}=-3$, we get $x=9$. But $\sqrt{9}\neq-3$ since the square root symbol always represents the nonnegative square root. Raising both sides of an equation to a power is discussed further in Section 3.4.

## EXAMPLE 5 Graphing a circle

Sketch the graph of the equation $(x-1)^2+(y+2)^2=9$.

### Solution

To determine the center and radius of this circle, compare its equation to the standard form:

$$(x-h)^2+(y-k)^2=r^2$$
$$(x-1)^2+(y+2)^2=9$$
$$(x-1)^2+(y-(-2))^2=3^2$$

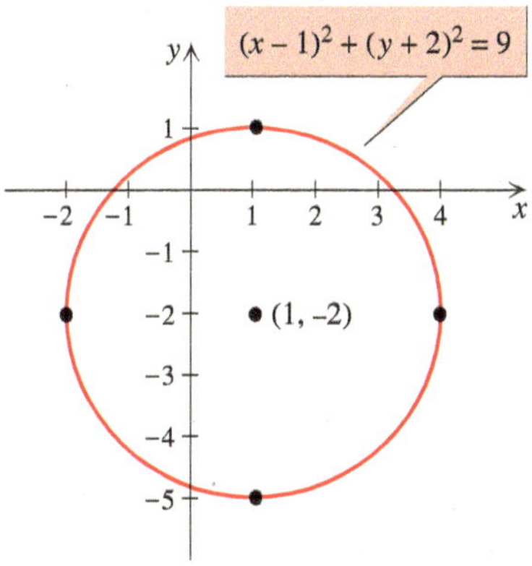

Figure 1.16

It is clear that $h=1$, but you must rewrite (or at least think of) $y+2$ as $y-(-2)$ to determine that $k=-2$. So the center of the circle is $(1,-2)$ and the radius is 3. You can draw the circle as in Fig. 1.16 with a compass or computer. To draw a circle by hand, locate the points that lie 3 units above, below, right, and left of the center, as shown in Fig. 1.16. Then sketch a circle through these points.

To support these results with a graphing calculator you must first solve the equation for $y$:

$$(x-1)^2+(y+2)^2=9$$
$$(y+2)^2=9-(x-1)^2$$
$$y+2=\pm\sqrt{9-(x-1)^2}$$
$$y=-2\pm\sqrt{9-(x-1)^2}$$

Now enter $y_1$ and $y_2$ as in Fig. 1.17(a) on the next page. Set the viewing window as in Fig. 1.17(b). The graph in Fig. 1.17(c) supports our previous conclusion. A circle looks round only if the same unit distance is used on both axes. Some calculators

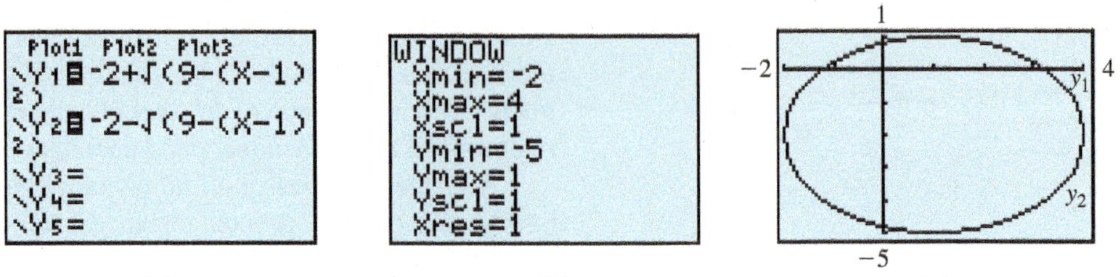


**Figure 1.17**

automatically draw the graph with the same unit distance on both axes given the correct command (Zsquare on a TI-83).

▸**TRY THIS.** Sketch the graph of $(x + 2)^2 + (y - 4)^2 = 25$. ■

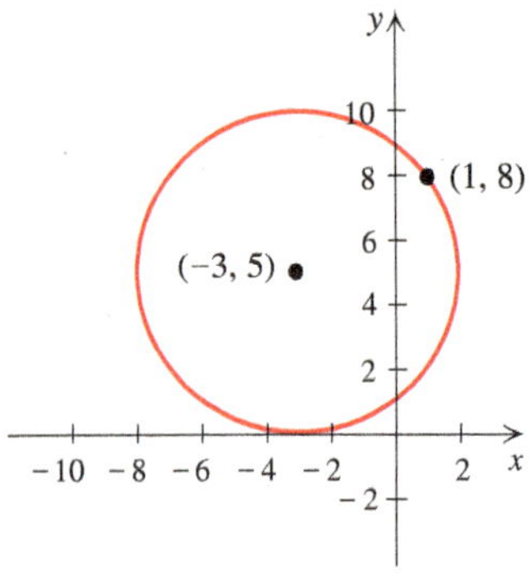


**Figure 1.18**

Note that an equation such as $(x - 1)^2 + (y + 2)^2 = -9$ is not satisfied by any pair of real numbers, because the left-hand side is a nonnegative real number, while the right-hand side is negative. The equation $(x - 1)^2 + (y + 2)^2 = 0$ is satisfied only by $(1, -2)$. Since only one point satisfies $(x - 1)^2 + (y + 2)^2 = 0$, its graph is sometimes called a degenerate circle with radius zero. We study circles again later in this text when we study the conic sections.

In the next example we start with a description of a circle and write its equation.

## EXAMPLE 6 Writing an equation of a circle

Write the standard equation for the circle with the center $(-3, 5)$ and passing through $(1, 8)$ as shown in Fig. 1.18.

### Solution

The radius of this circle is the distance between $(-3, 5)$ and $(1, 8)$:

$$r = \sqrt{(1 - (-3))^2 + (8 - 5)^2} = \sqrt{16 + 9} = 5$$

Now use $h = -3$, $k = 5$, and $r = 5$ in the standard equation of the circle $(x - h)^2 + (y - k)^2 = r^2$.

$$(x - (-3))^2 + (y - 5)^2 = 5^2$$

So the equation of the circle is $(x + 3)^2 + (y - 5)^2 = 25$.

▸**TRY THIS.** Write the standard equation for the circle with center $(2, -1)$ and passing through $(3, 6)$. ■

A circle in standard form could be rewritten as follows:

$$(x + 3)^2 + (y - 5)^2 = 4 \qquad \text{Standard form for a circle}$$

$$x^2 + 6x + 9 + y^2 - 10y + 25 = 4 \qquad \text{Square the binomials.}$$

$$x^2 + 6x + y^2 - 10y = -30$$

To find the center and radius of the circle that is given by the last equation, we go back to standard form by *completing the square.*

**Completing the square** means finding the third term of a perfect square trinomial when given the first two. That is, if we start with $x^2 + bx$, then what third term will make a perfect square trinomial? Since

$$\left(x + \frac{b}{2}\right)^2 = x^2 + 2\cdot\frac{b}{2}\cdot x + \left(\frac{b}{2}\right)^2 = x^2 + bx + \left(\frac{b}{2}\right)^2$$

adding $\left(\frac{b}{2}\right)^2$ to $x^2 + bx$ completes the square. For example, the perfect square trinomial that starts with $x^2 + 6x$ is $x^2 + 6x + 9$. Note that 9 can be found by taking one-half of 6 and squaring.

**Rule for Completing the Square of $x^2 + bx + ?$**

> The last term of a perfect square trinomial (with $a = 1$) is the square of one-half of the coefficient of the middle term. In symbols, the perfect square trinomial whose first two terms are $x^2 + bx$ is
>
> $$x^2 + bx + \left(\frac{b}{2}\right)^2.$$

### EXAMPLE 7 Changing an equation of a circle to standard form

Graph the equation $x^2 + 6x + y^2 - 5y = -\frac{1}{4}$.

**Solution**

Complete the square for both $x$ and $y$ to get the standard form.

$$x^2 + 6x + 9 + y^2 - 5y + \frac{25}{4} = -\frac{1}{4} + 9 + \frac{25}{4} \qquad \left(\tfrac{1}{2}\cdot 6\right)^2 = 9,\ \left(\tfrac{1}{2}\cdot 5\right)^2 = \tfrac{25}{4}$$

$$(x + 3)^2 + \left(y - \frac{5}{2}\right)^2 = 15 \qquad \text{Factor the trinomials on the left side.}$$

The graph is a circle with center $\left(-3, \frac{5}{2}\right)$ and radius $\sqrt{15}$. See Fig. 1.19.

**▶TRY THIS.** Graph the equation $x^2 + 3x + y^2 - 2y = 0$. ■

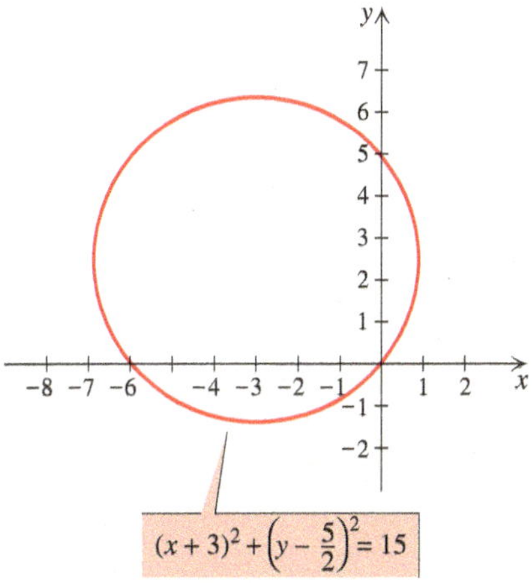


**Figure 1.19**

Whether an equation of the form $x^2 + y^2 + Ax + By = C$ is a circle depends on the values of $A$, $B$, and $C$. We can always complete the squares for $x^2 + Ax$ and $y^2 + By$ as in Example 7. Since a circle must have a positive radius, this equation is the equation of a circle only if the right side of the equation turns out to be positive *after* completing the squares. If the right side turns out to be zero, then only one point satisfies the equation. If the right side turns out to be negative, then no points satisfy the equation.

**HISTORICAL NOTE**

Euclid of Alexandria (325–265 B.C.) was a Greek mathematician who lived in Alexandria, Egypt, during the reign of Ptolemy I (323 B.C.–283 B.C.). He is considered to be the "father of geometry." His most popular work, *Elements*, is one of the most successful textbooks in the history of mathematics. Within it, the properties of geometrical objects are deduced from a small set of axioms, thereby founding the axiomatic method of mathematics.

## The Line

For the circle, we started with the geometric definition and developed the algebraic equation. We would like to do the same thing for lines, but geometric definitions of lines are rather vague. Euclid's definition was "length without breadth." A modern geometry textbook states that a line is "a straight set of points extending infinitely in both directions." Another modern textbook defines lines algebraically as the graph of an equation of the form $Ax + By = C$. So we will accept the following theorem without proof.

### Theorem: Equation of a Line in Standard Form

If $A$, $B$, and $C$ are real numbers, then the graph of the equation

$$Ax + By = C$$

is a straight line, provided $A$ and $B$ are not both zero. Every straight line in the coordinate plane has an equation in the form $Ax + By = C$, the **standard form** for the equation of a line.

An equation of the form $Ax + By = C$ is called a **linear equation in two variables.** The equations

$$2x + 3y = 5, \qquad x = 4, \qquad \text{and} \qquad y = 5$$

are linear equations in standard form. An equation such as $y = 3x - 1$ that can be rewritten in standard form is also called a linear equation.

There is only one line containing any two distinct points. So to graph a linear equation we simply find two points that satisfy the equation and draw a line through them. We often use the point where the line crosses the $x$-axis, the **$x$-intercept,** and the point where the line crosses the $y$-axis, the **$y$-intercept.** Since every point on the $x$-axis has $y$-coordinate 0, we find the $x$-intercept by replacing $y$ with 0 and then solving the equation for $x$. Since every point on the $y$-axis has $x$-coordinate 0, we find the $y$-intercept by replacing $x$ with 0 and then solving for $y$. If the $x$- and $y$-intercepts are both at the origin, then you must find another point that satisfies the equation.

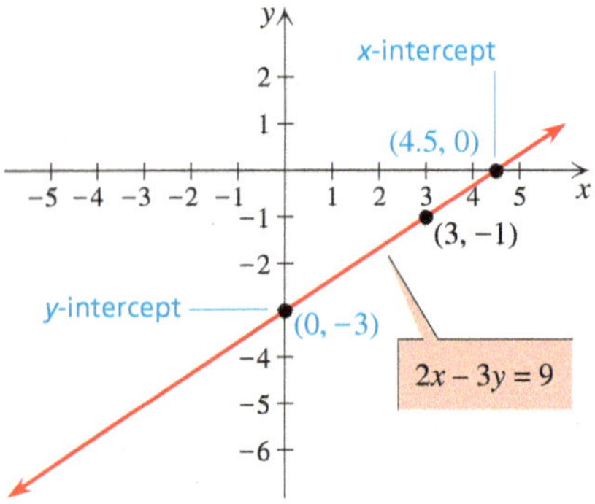


Figure 1.20

## EXAMPLE 8 Graphing lines and showing the intercepts

Graph each equation. Be sure to find and show the intercepts.

**a.** $2x - 3y = 9$ **b.** $y = 40 - x$

### Solution

**a.** Since the $y$-coordinate of the $x$-intercept is 0, we replace $y$ by 0 in the equation:

$$2x - 3(0) = 9$$
$$2x = 9$$
$$x = 4.5$$

To find the $y$-intercept, we replace $x$ by 0 in the equation:

$$2(0) - 3y = 9$$
$$-3y = 9$$
$$y = -3$$

The $x$-intercept is $(4.5, 0)$ and the $y$-intercept is $(0, -3)$. Locate the intercepts and draw the line as shown in Fig. 1.20. To check, locate a point such as $(3, -1)$, which also satisfies the equation, and see if the line goes through it.

**b.** If $x = 0$, then $y = 40 - 0 = 40$ and the $y$-intercept is $(0, 40)$. If $y = 0$, then $0 = 40 - x$ or $x = 40$. The $x$-intercept is $(40, 0)$. Draw a line through these points as shown in Fig. 1.21. Check that $(10, 30)$ and $(20, 20)$ also satisfy $y = 40 - x$ and the line goes through these points.

The calculator graph shown in Fig. 1.22 is consistent with the graph in Fig. 1.21. Note that the viewing window is set to show the intercepts.

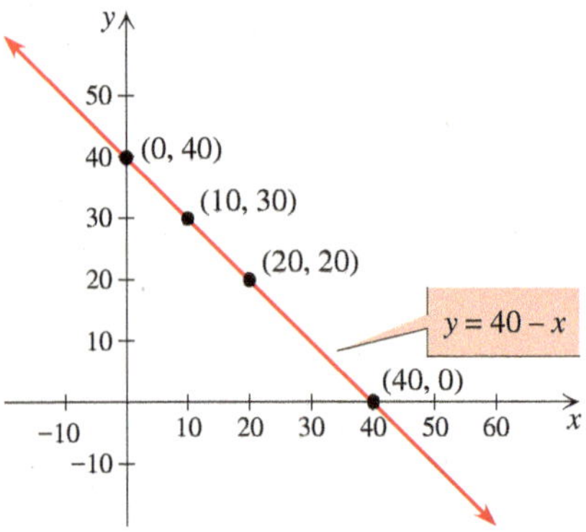


Figure 1.21

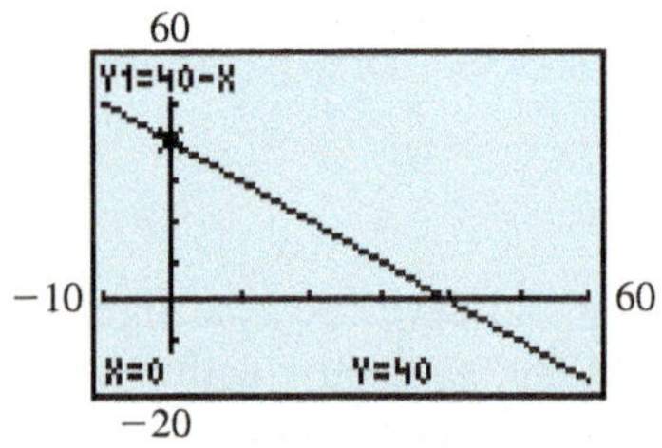


Figure 1.22

**▸TRY THIS.** Graph $2x + 5y = 10$ and determine the intercepts. ■

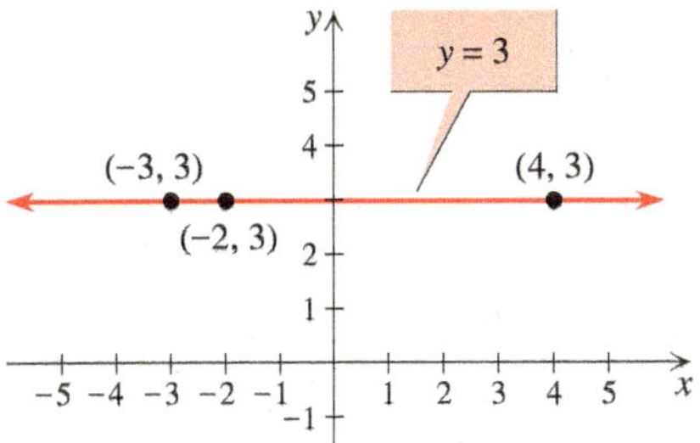


Figure 1.23

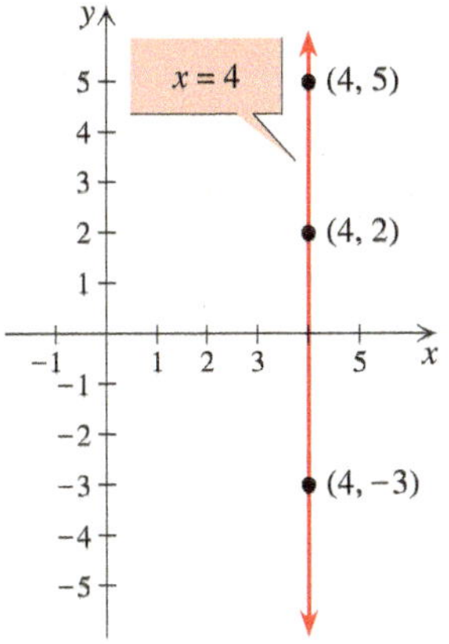


Figure 1.24

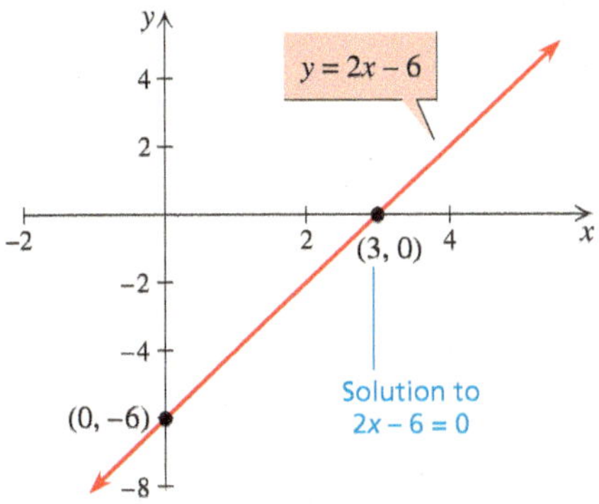


Figure 1.25

## EXAMPLE 9 Graphing horizontal and vertical lines

Sketch the graph of each equation in the rectangular coordinate system.

**a.** $y = 3$ **b.** $x = 4$

### Solution

**a.** The equation $y = 3$ is equivalent to $0 \cdot x + y = 3$. Because $x$ is multiplied by 0, we can choose any value for $x$ as long as we choose 3 for $y$. So ordered pairs such as $(-3, 3)$, $(-2, 3)$, and $(4, 3)$ satisfy the equation $y = 3$. The graph of $y = 3$ is the horizontal line shown in Fig. 1.23.

**b.** The equation $x = 4$ is equivalent to $x + 0 \cdot y = 4$. Because $y$ is multiplied by 0, we can choose any value for $y$ as long as we choose 4 for $x$. So ordered pairs such as $(4, -3)$, $(4, 2)$, and $(4, 5)$ satisfy the equation $x = 4$. The graph of $x = 4$ is the vertical line shown in Fig. 1.24.

Note that you cannot graph $x = 4$ using the Y= key on your calculator. You can graph it on a calculator using polar coordinates or parametric equations (discussed later in this text).

▶**TRY THIS.** Graph $y = 5$ in the rectangular coordinate system. ■

In the context of two variables the equation $x = 4$ has infinitely many solutions. Every ordered pair on the vertical line in Fig. 1.24 satisfies $x = 4$. In the context of one variable, $x = 4$ has only one solution, 4.

## Using a Graph to Solve an Equation

Graphing and solving equations go hand in hand. For example, the graph of $y = 2x - 6$ in Fig. 1.25 has $x$-intercept $(3, 0)$, because if $x = 3$ then $y = 0$. Of course, 3 is also the solution to the corresponding equation $2x - 6 = 0$ (where $y$ is replaced by 0). For this reason, the solution to an equation is also called a **zero** or **root** of the equation. Every $x$-intercept on a graph provides us a solution to the corresponding equation. However, an $x$-intercept on a graph may not be easy to identify. In the next example we see how a graphing calculator identifies an $x$-intercept and thus gives us the approximate solution to an equation.

## EXAMPLE 10 Using a graph to solve an equation

Use a graphing calculator to solve $0.55(x - 3.45) + 13.98 = 0$.

### Solution

First graph $y = 0.55(x - 3.45) + 13.98$ using a viewing window that shows the $x$-intercept as in Fig. 1.26(a). Next press ZERO or ROOT on the CALC menu. The calculator can find a zero between a *left bound* and a *right bound*, which you must

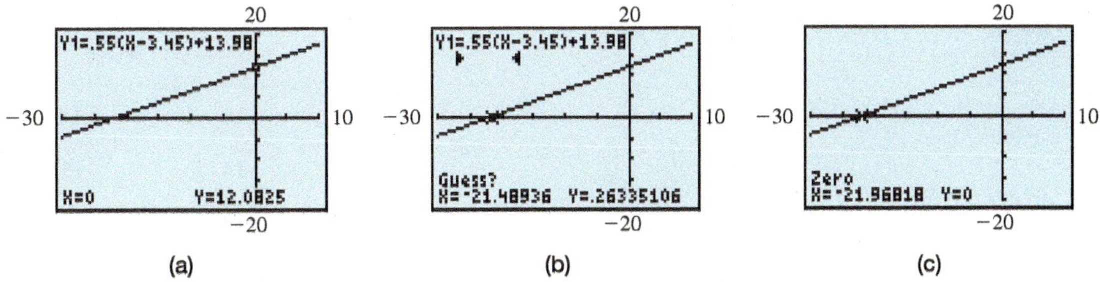


Figure 1.26

enter. The calculator also asks you to make a guess. See Fig. 1.26(b). The more accurate the guess, the faster the calculator will find the zero. The solution to the equation rounded to two decimal places is $-21.97$. See Fig. 1.26(c). As always, consult your calculator manual if you are having difficulty.

▶**TRY THIS.** Use a graphing calculator to solve $0.34(x - 2.3) + 4.5 = 0$. ■

## FOR thought... True or False? Explain.

1. The point $(2, -3)$ is in quadrant II.
2. The point $(4, 0)$ is in quadrant I.
3. The distance between $(a, b)$ and $(c, d)$ is $\sqrt{(a - b)^2 + (c - d)^2}$.
4. The equation $3x^2 + y = 5$ is a linear equation.
5. The solution to $7x - 9 = 0$ is the $x$-coordinate of the $x$-intercept of $y = 7x - 9$.
6. $\sqrt{7^2 + 9^2} = 7 + 9$
7. The origin lies midway between $(1, 3)$ and $(-1, -3)$.
8. The distance between $(3, -7)$ and $(3, 3)$ is 10.
9. The $x$-intercept for the graph of $3x - 2y = 7$ is $(7/3, 0)$.
10. The graph of $(x + 2)^2 + (y - 1)^2 = 5$ is a circle centered at $(-2, 1)$ with radius 5.

## EXERCISES 1.3

*Fill in the blank.*

1. If $x$ and $y$ are real numbers, then $(x, y)$ is an ________ pair of real numbers.
2. The first coordinate in $(x, y)$ is the ________ and the second coordinate is the ________.
3. Ordered pairs are graphed in the rectangular coordinate system or the ________ coordinate system.
4. The intersection of the $x$-axis and the $y$-axis is the ________.
5. The set of all points in a plane that lie a fixed distance from a given point in the plane is a ________.
6. Finding the third term of a perfect square trinomial when given the first two is ________.
7. An equation of the form $Ax + By = C$ is a ________ in two variables.
8. The point where a line crosses the $y$-axis is the ________.

*In Exercises 9–18, for each point shown in the xy-plane, write the corresponding ordered pair and name the quadrant in which it lies or the axis on which it lies.*

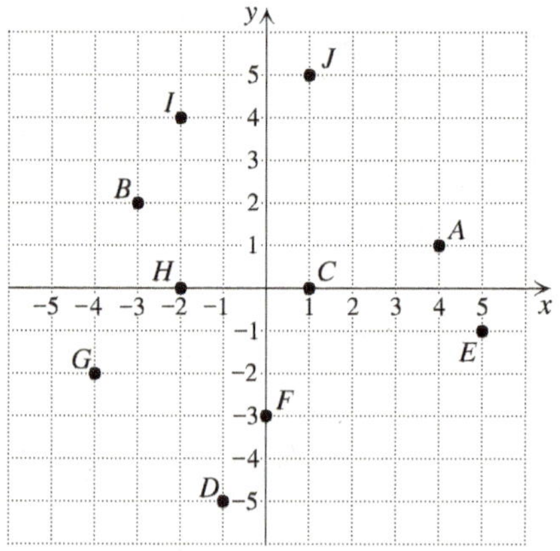


9. $A$
10. $B$
11. $C$
12. $D$
13. $E$
14. $F$
15. $G$
16. $H$
17. $I$
18. $J$

*For each pair of points find the distance between them and the midpoint of the line segment joining them.*

**19.** $(1, 3), (4, 7)$

**20.** $(-3, -2), (9, 3)$

**21.** $(-1, -2), (1, 0)$

**22.** $(-1, 0), (1, 2)$

**23.** $(12, -11), (5, 13)$

**24.** $(-4, -7), (4, 8)$

**25.** $(-1, 1), \left(-1 + 3\sqrt{3}, 4\right)$

**26.** $\left(1 + \sqrt{2}, -2\right), \left(1 - \sqrt{2}, 2\right)$

**27.** $(1.2, 4.8), (-3.8, -2.2)$

**28.** $(-2.3, 1.5), (4.7, -7.5)$

**29.** $(a, 0), (b, 0)$

**30.** $(a, 0), \left(\frac{a+b}{2}, 0\right)$

**31.** $(\pi, 0), (\pi/2, 1)$

**32.** $(0, 0), (\pi/2, 1)$

*Determine the center and radius of each circle and sketch the graph.*

**33.** $x^2 + y^2 = 16$

**34.** $x^2 + y^2 = 1$

**35.** $(x + 6)^2 + y^2 = 36$

**36.** $x^2 + (y - 2)^2 = 16$

**37.** $y^2 = 25 - (x + 1)^2$

**38.** $x^2 = 9 - (y - 3)^2$

**39.** $(x - 2)^2 = 8 - (y + 2)^2$

**40.** $(y + 2)^2 = 20 - (x - 4)^2$

*Write the standard equation for each circle.*

**41.** Center at $(0, 0)$ with radius 7

**42.** Center at $(0, 0)$ with radius 5

**43.** Center at $(-2, 5)$ with radius $1/2$

**44.** Center at $(-1, -6)$ with radius $1/3$

**45.** Center at $(3, 5)$ and passing through the origin

**46.** Center at $(-3, 9)$ and passing through the origin

**47.** Center at $(5, -1)$ and passing through $(1, 3)$

**48.** Center at $(-2, -3)$ and passing through $(2, 5)$

*Determine the center and radius of each circle and sketch the graph. See the rule for completing the square on page 31.*

**49.** $x^2 + y^2 = 9$

**50.** $x^2 + y^2 = 100$

**51.** $x^2 + y^2 + 6y = 0$

**52.** $x^2 + y^2 = 4x$

**53.** $x^2 + 6x + y^2 + 8y = 0$

**54.** $x^2 - 8x + y^2 - 10y = -5$

**55.** $x^2 - 3x + y^2 + 2y = \frac{3}{4}$

**56.** $x^2 + 5x + y^2 - y = \frac{5}{2}$

**57.** $x^2 - 6x + y^2 - 8y = 0$

**58.** $x^2 + 10x + y^2 - 8y = -40$

**59.** $x^2 + y^2 = 4x + 3y$

**60.** $x^2 + y^2 = 5x - 6y$

**61.** $x^2 + y^2 = \frac{x}{2} - \frac{y}{3} - \frac{1}{16}$

**62.** $x^2 + y^2 = x - y + \frac{1}{2}$

*Write the standard equation for each of the following circles.*

**63. a.**

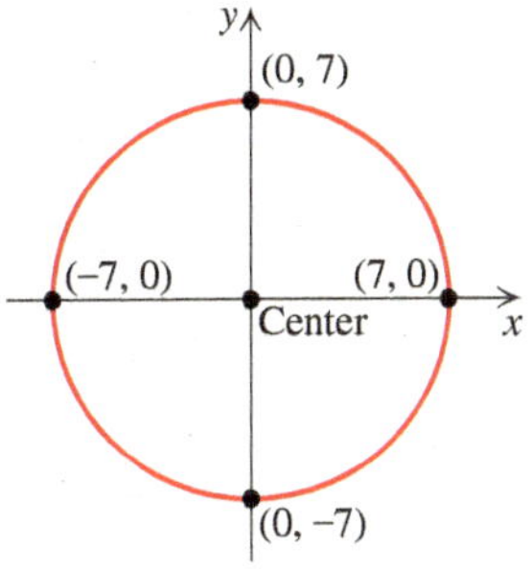


**b.**

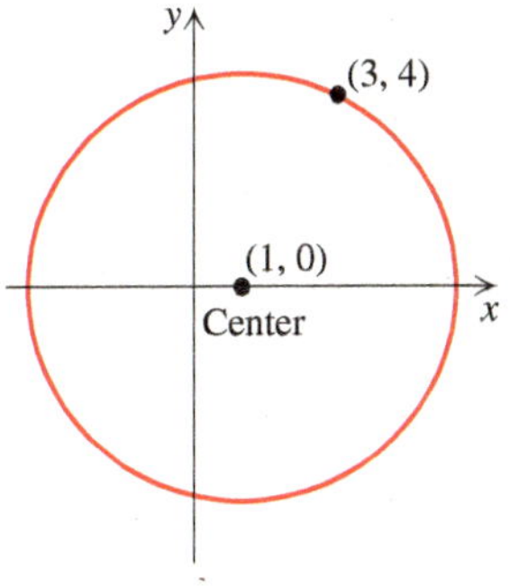


**c.**

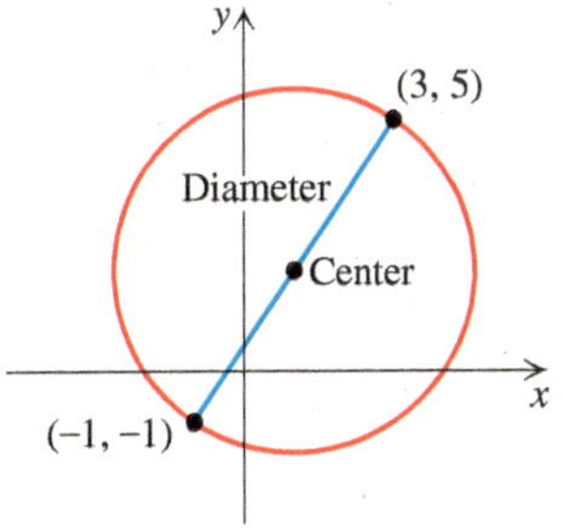


**64. a.**

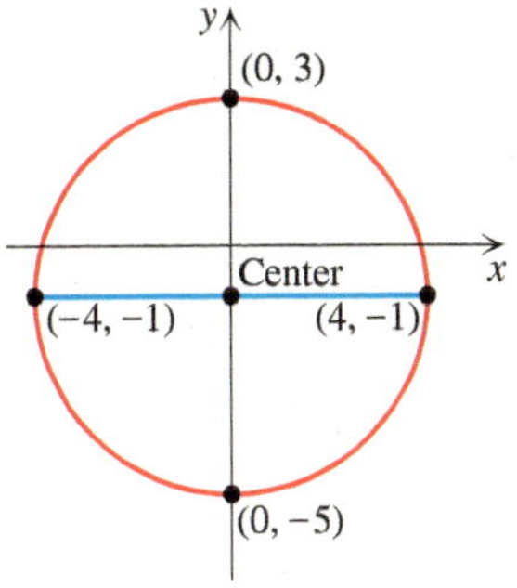


**b.**

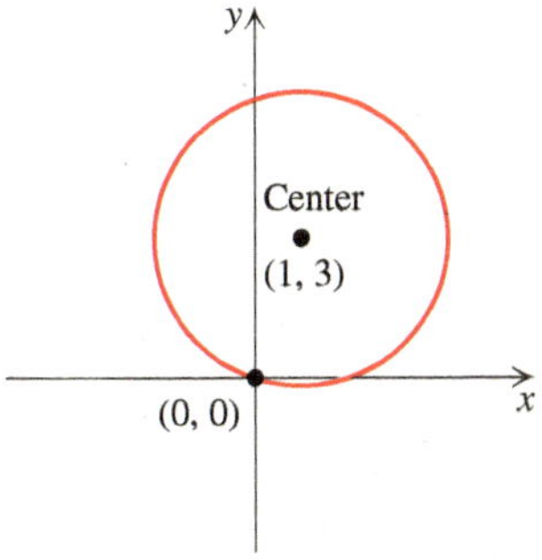


**c.**

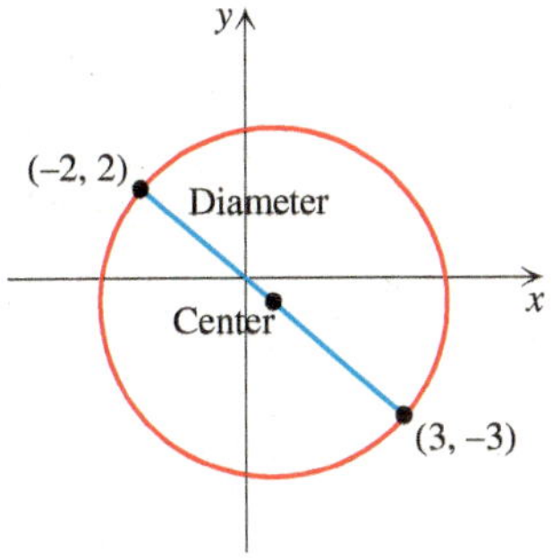

*Write the equation of each circle in standard form. The coordinates of the center and the radius for each circle are integers.*

**65. a.**

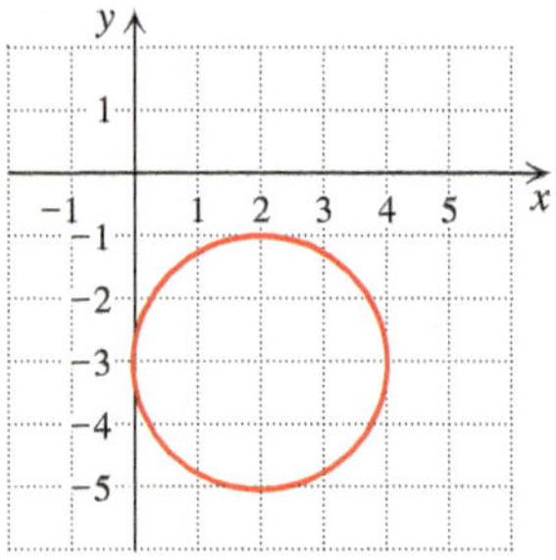


**b.**

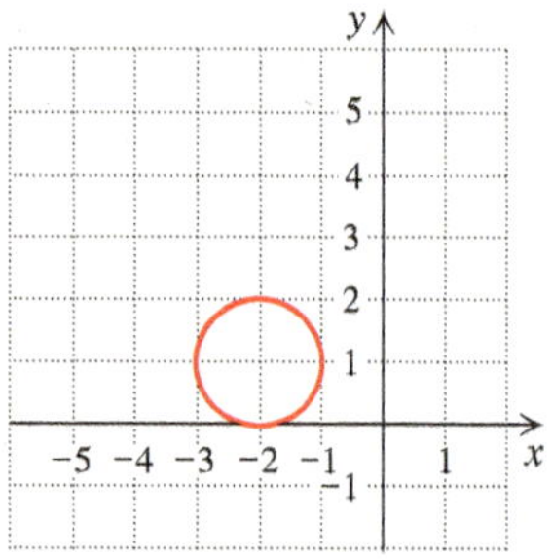


**c.**

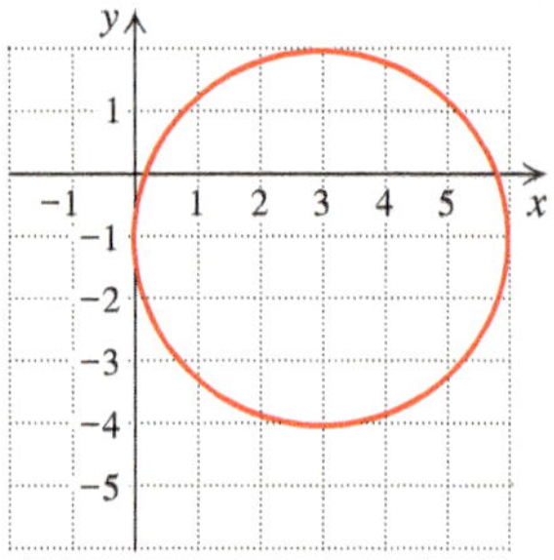


**d.**

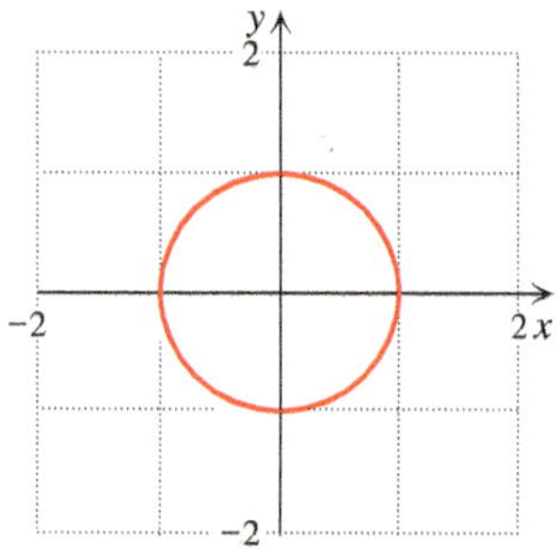


**66. a.**

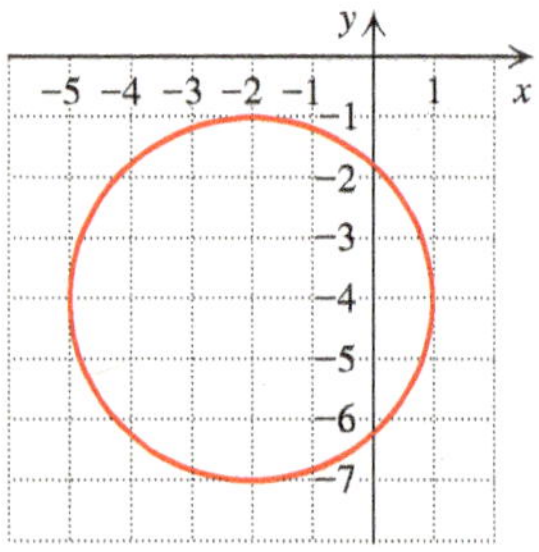


**b.**

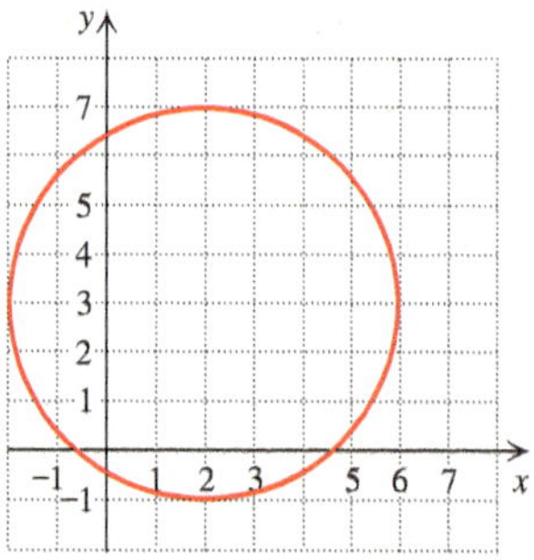


**c.**

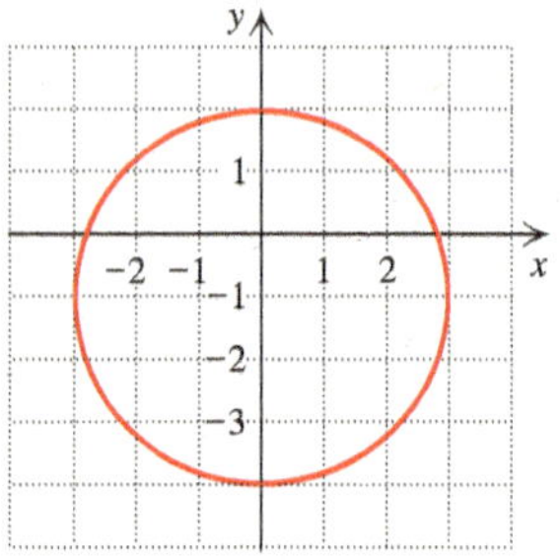


**d.**

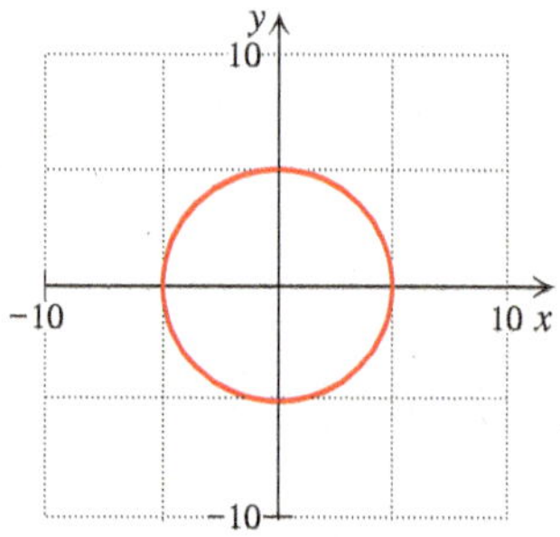


*Sketch the graph of each linear equation. Be sure to find and show the x- and y-intercepts.*

**67.** $y = 3x - 4$

**68.** $y = 5x - 5$

**69.** $3x - y = 6$

**70.** $5x - 2y = 10$

**71.** $x = 3y - 90$

**72.** $x = 80 - 2y$

**73.** $\frac{2}{3}y - \frac{1}{2}x = 400$

**74.** $\frac{1}{2}x - \frac{1}{3}y = 600$

**75.** $2x + 4y = 0.01$

**76.** $3x - 5y = 1.5$

**77.** $0.03x + 0.06y = 150$

**78.** $0.09x - 0.06y = 54$

*Graph each equation in the rectangular coordinate system.*

**79.** $x = 5$

**80.** $y = -2$

**81.** $y = 4$

**82.** $x = -3$

**83.** $x = -4$

**84.** $y = 5$

**85.** $y - 1 = 0$

**86.** $5 - x = 4$

*Find the solution to each equation by reading the accompanying graph.*

**87.** $2.4x - 8.64 = 0$

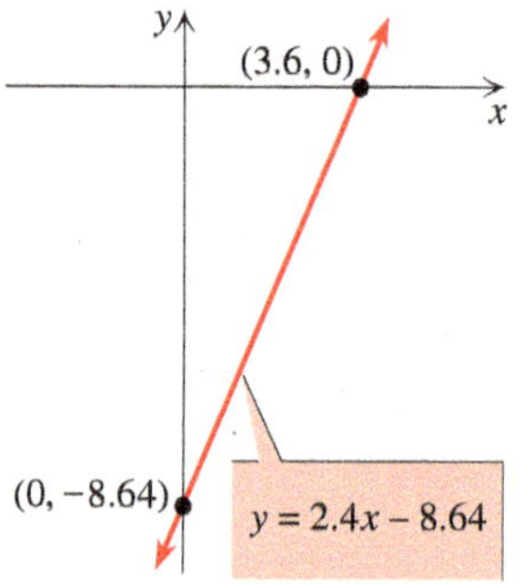


**88.** $8.84 - 1.3x = 0$

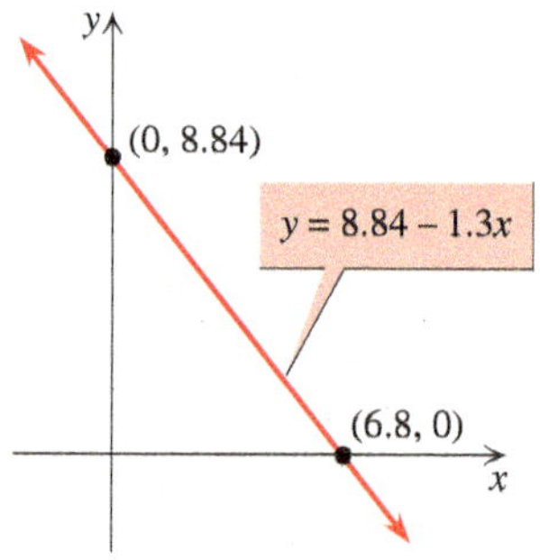


**89.** $-\frac{3}{7}x + 6 = 0$

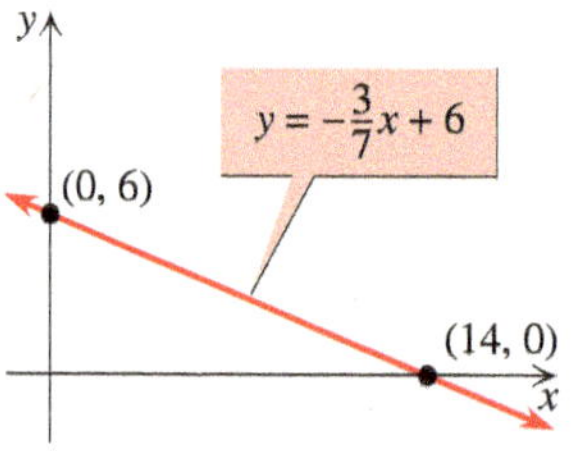


**90.** $\frac{5}{6}x + 30 = 0$

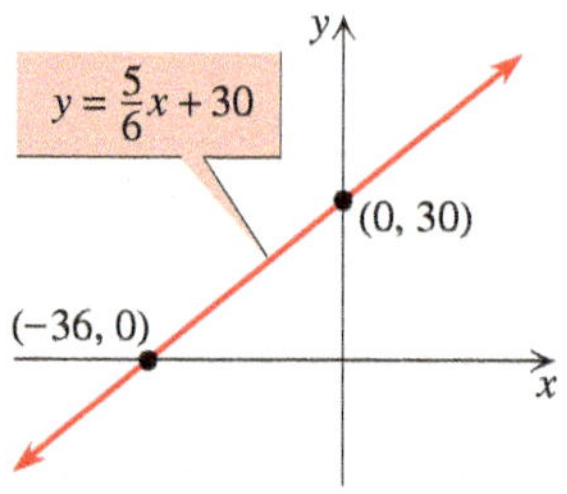


*Use a graphing calculator to estimate the solution to each equation to two decimal places. Then find the solution algebraically and compare it with your estimate.*

**91.** $1.2x + 3.4 = 0$

**92.** $3.2x - 4.5 = 0$

**93.** $1.23x - 687 = 0$

**94.** $-2.46x + 1500 = 0$

**95.** $0.03x - 3497 = 0$

**96.** $0.09x + 2000 = 0$

**97.** $4.3 - 3.1(2.3x - 9.9) = 0$

**98.** $9.4x - 4.37(3.5x - 9.76) = 0$

*Solve each problem.*

**99.** *First Marriage* The median age at first marriage for women went from 20.8 in 1970 to 25.1 in 2000 as shown in the accompanying figure (U.S. Census Bureau, www.census.gov).

**a.** Find the midpoint of the line segment in the figure and interpret your result.

**b.** Find the distance between the two points shown in the figure and interpret your result.

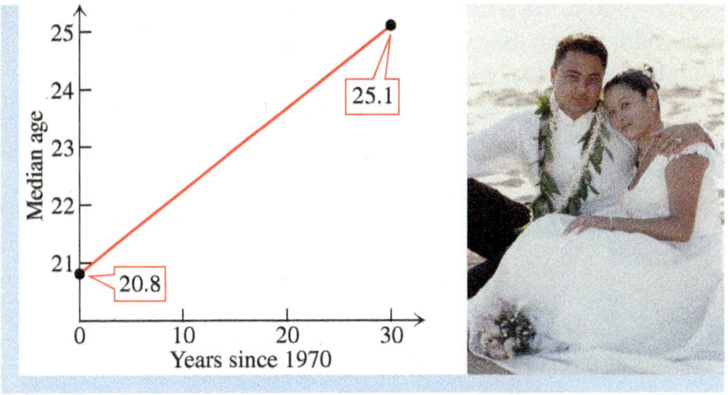


**Figure for Exercise 99**

**100.** *Unmarried Couples* The number of unmarried-couple households $h$ (in millions) can be modeled using the equation $h = 0.171n + 2.913$, where $n$ is the number of years since 1990 (U.S. Census Bureau, www.census.gov).

**a.** Find and interpret the $n$-intercept for the line. Does it make sense?

**b.** Find and interpret the $h$-intercept for the line.

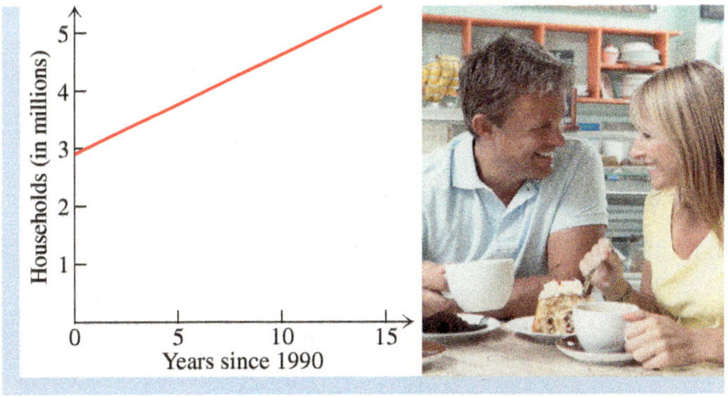


**Figure for Exercise 100**

**101.** *Capsize Control* The capsize screening value $C$ is an indicator of a sailboat's suitability for extended offshore sailing. $C$ is determined by the formula

$$C = 4D^{-1/3}B,$$

where $D$ is the displacement of the boat in pounds and $B$ is its beam (or width) in feet. Sketch the graph of this equation for $B$ ranging from 0 to 20 ft, assuming that $D$ is fixed at 22,800 lb. Find $C$ for the Island Packet 40, which has a displacement of 22,800 pounds and a beam of 12 ft 11 in. (Island Packet Yachts, www.ipy.com).

**102.** *Limiting the Beam* The International Offshore Rules require that the capsize screening value $C$ (from the previous exercise) be less than or equal to 2 for safety. What is the maximum allowable beam (to the nearest inch) for a boat with a displacement of 22,800 lb? For a fixed displacement, is a boat more or less likely to capsize as its beam gets larger?

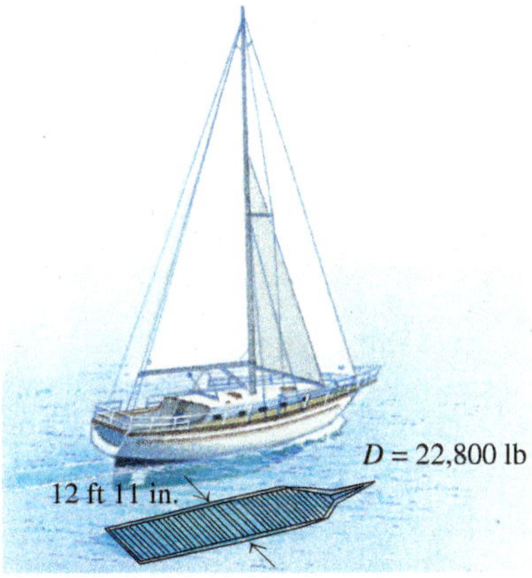


**Figure for Exercises 101 and 102**

**103.** Show that the points $A(-4, -5)$, $B(1, 1)$, and $C(6, 7)$ are colinear.

HINT Points $A$, $B$, and $C$ lie on a straight line if $AB + BC = AC$.

**104.** Show that the midpoint of the hypotenuse of any right triangle is equidistant from all three vertices.

## FOR WRITING/DISCUSSION

**105.** *Finding Points* Can you find two points such that their coordinates are integers and the distance between them is 10? $\sqrt{10}$? $\sqrt{19}$? Explain.

**106.** *Plotting Points* Plot at least five points in the $xy$-plane that satisfy the inequality $y > 2x$. Give a verbal description of the solution set to $y > 2x$.

**107.** *Cooperative Learning* Work in a small group to plot the points $(-1, 3)$ and $(4, 1)$ on graph paper. Assuming that these two points are adjacent vertices of a square, find the other two vertices. Now select your own pair of adjacent vertices and "complete the square." Now generalize your results. Start with the points $(x_1, y_1)$ and $(x_2, y_2)$ as adjacent vertices of a square and write expressions for the coordinates of the other two vertices. Repeat this exercise assuming that the first two points are opposite vertices of a square.

**108.** *Distance to the Origin* Let $m$ and $n$ be any real numbers. What is the distance between $(0, 0)$ and $(2m, m^2 - 1)$? What is the distance between $(0, 0)$ and $(2mn, m^2 - n^2)$?

## ▶ RETHINKING

**109.** Identify each equation as an identity, inconsistent equation, or conditional equation.

**a.** $2x + 4 = 5$

**b.** $2x + 4 = 2(x + 2)$

**c.** $2x + 4 = 2x$

**110.** Solve $\frac{4}{x - 3} + \frac{1}{x + 3} = \frac{x}{x^2 - 9}$.

**111.** Solve $\frac{x - 2}{x + 3} = \frac{x + 4}{x + 9}$.

**112.** The Wilsons got \$180,780 for their house after paying a sales commission that was 8% of the selling price. What was the selling price?

**113.** Solve $ax + b = cx + d$ for $x$.

**114.** Solve $\frac{1}{a} + \frac{1}{b} = \frac{1}{x}$ for $x$.

### THINKING OUTSIDE THE BOX III

*Methodical Mower* Eugene is mowing a rectangular lawn that is 300 ft by 400 ft. He starts at one corner and mows a swath of uniform width around the outside edge in a clockwise direction. He continues going clockwise, widening the swath that is mowed and shrinking the rectangular section that is yet to be mowed. When he is half done with the lawn, how wide is the swath?

## POP QUIZ 1.3

**1.** Find the distance between $(-1, 3)$ and $(3, 5)$.

**2.** Find the center and radius for the circle

$$(x - 3)^2 + (y + 5)^2 = 81.$$

**3.** Find the center and radius for the circle

$$x^2 + 4x + y^2 - 10y = -28.$$

**4.** Find the equation of the circle that passes through the origin and has center at $(3, 4)$.

**5.** Find all intercepts for $2x - 3y = 12$.

**6.** Which point is on both of the lines $x = 5$ and $y = -1$?

## LINKING concepts... For Individual or Group Explorations

### Modeling Energy Requirements

*Clinical dietitians must design diets for patients that meet their basic energy requirements and are suitable for the condition of their health. The basic energy requirement B (in calories) for a male is given by the formula*

$$B = 655.096 + 9.563W + 1.85H - 4.676A,$$

*where W is the patient's weight in kilograms, H is the height in centimeters, and A is the age in years (www.eatwell.com). For a female the formula is*

$$B = 66.473 + 13.752W + 5.003H - 6.755A.$$

**a)** Find your basic energy requirement.

**b)** Replace $H$ and $A$ with your actual height and age. Now draw a graph showing how $B$ depends on weight for suitable values of $W$.

**c)** For a fixed height and age, does the basic energy requirement increase or decrease as weight increases?

**d)** Replace $W$ and $H$ with your actual weight and height. Now draw a graph showing how $B$ depends on age for suitable values of $A$.

**e)** For a fixed weight and height, does the basic energy requirement increase or decrease as a person gets older?

**f)** Replace $W$ and $A$ with your actual weight and height. Now draw a graph showing how $B$ depends on height for suitable values of $H$.

**g)** For a fixed weight and age, does the basic energy requirement increase or decrease for taller persons?

**h)** For the three graphs that you have drawn, explain how to determine whether the graph is increasing or decreasing from the formula for the graph.

# 1.4 Linear Equations in Two Variables

In Section 1.3 we graphed lines, including horizontal and vertical lines. We learned that every line has an equation in standard form $Ax + By = C$. In this section we will continue to study lines.

## Slope of a Line

A road that rises 4 feet in a horizontal run of 100 feet has a grade of $\frac{4}{100}$ or 4%. A roof that rises 5 feet in a horizontal run of 12 feet has a 5-12 pitch. See Fig. 1.27. Grade and pitch are measurements of *steepness*.

**Figure 1.27**

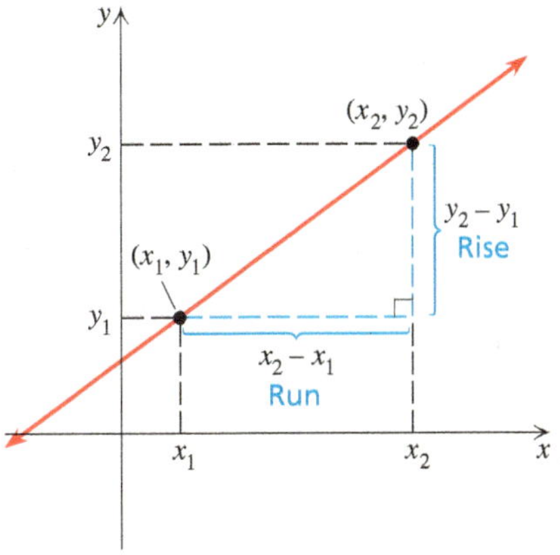


**Figure 1.28**

The steepness or *slope* of a line in the *xy*-coordinate system is the ratio of the **rise** (the change in *y*-coordinates) to the **run** (the change in *x*-coordinates) between two points on the line. The notation $\Delta x$ (read "delta *x*") is often used to represent the change in the *x*-coordinates. Likewise, $\Delta y$ is used to represent the change in *y*-coordinates. So

$$\text{slope} = \frac{\text{change in } y\text{-coordinates}}{\text{change in } x\text{-coordinates}} = \frac{\Delta y}{\Delta x} = \frac{\text{rise}}{\text{run}}.$$

If $(x_1, y_1)$ and $(x_2, y_2)$ are the coordinates of the two points in Fig. 1.28, then the rise is $y_2 - y_1$ and the run is $x_2 - x_1$:

**Definition: Slope**

> The **slope** of the line through $(x_1, y_1)$ and $(x_2, y_2)$ with $x_1 \neq x_2$ is
>
> $$\frac{y_2 - y_1}{x_2 - x_1}.$$

Note that if $(x_1, y_1)$ and $(x_2, y_2)$ are two points for which $x_1 = x_2$ then the line through them is a vertical line. Since this case is not included in the definition of slope, a vertical line does not have a slope. We also say that the slope of a vertical line is undefined. If we choose two points on a horizontal line then $y_1 = y_2$ and $y_2 - y_1 = 0$. For any horizontal line the rise between two points is 0 and the slope is 0.

### EXAMPLE 1 Finding the slope

In each case find the slope of the line that contains the two given points.

**a.** $(-3, 4), (-1, -2)$ **b.** $(-3, 7), (5, 7)$ **c.** $(-3, 5), (-3, 8)$

**Solution**

**a.** Use $(x_1, y_1) = (-3, 4)$ and $(x_2, y_2) = (-1, -2)$ in the formula:

$$\text{slope} = \frac{y_2 - y_1}{x_2 - x_1} = \frac{-2 - 4}{-1 - (-3)} = \frac{-6}{2} = -3$$

The slope of the line is $-3$.

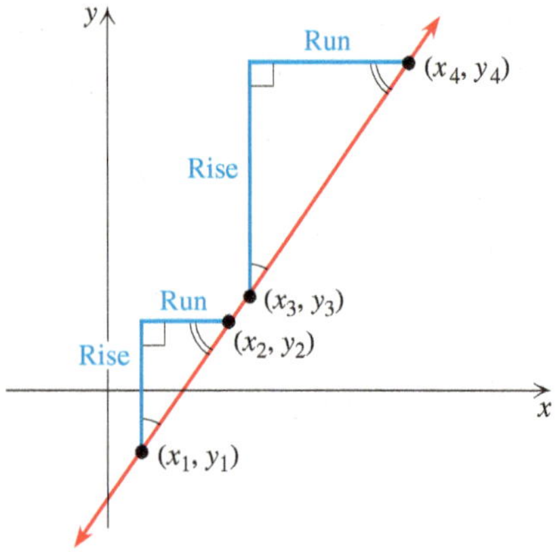


Figure 1.29

**b.** Use $(x_1, y_1) = (-3, 7)$ and $(x_2, y_2) = (5, 7)$ in the formula:

$$\text{slope} = \frac{y_2 - y_1}{x_2 - x_1} = \frac{7 - 7}{5 - (-3)} = \frac{0}{8} = 0$$

The slope of this horizontal line is 0.

**c.** The line through $(-3, 5)$ and $(-3, 8)$ is a vertical line and so it does not have a slope.

▶**TRY THIS.** Find the slope of the line that contains $(-2, 5)$ and $(-1, -3)$. ■

The slope of a line is the same number regardless of which two points on the line are used in the calculation of the slope. To understand why, consider the two triangles shown in Fig. 1.29. These triangles have the same shape and are called similar triangles. Because the ratios of corresponding sides of similar triangles are equal, the ratio of rise to run is the same for either triangle.

## Point-Slope Form

Suppose that a line through $(x_1, y_1)$ has slope $m$. Every other point $(x, y)$ on the line must satisfy the equation

$$\frac{y - y_1}{x - x_1} = m$$

because any two points can be used to find the slope. Multiply both sides by $x - x_1$ to get $y - y_1 = m(x - x_1)$, which is the **point-slope form** of the equation of a line.

### Theorem: Point-Slope Form

The equation of the line (in point-slope form) through $(x_1, y_1)$ with slope $m$ is

$$y - y_1 = m(x - x_1).$$

In Section 1.3 we started with the equation of a line and graphed the line. Using the point-slope form, we can start with a graph of a line or a description of the line and write the equation for the line.

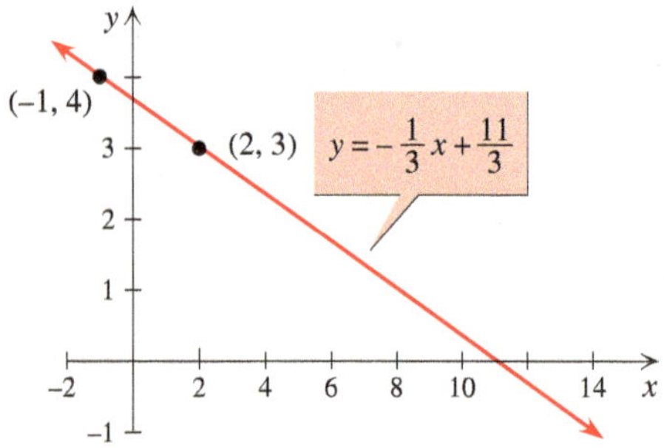


Figure 1.30

### EXAMPLE 2 The equation of a line given two points

In each case graph the line through the given pair of points. Then find the equation of the line and solve it for $y$ if possible.

**a.** $(-1, 4), (2, 3)$ **b.** $(2, 5), (-6, 5)$ **c.** $(3, -1), (3, 9)$

**Solution**

**a.** Find the slope of the line shown in Fig. 1.30 as follows:

$$m = \frac{y_2 - y_1}{x_2 - x_1} = \frac{3 - 4}{2 - (-1)} = \frac{-1}{3} = -\frac{1}{3}$$

Now use a point, say $(2, 3)$, and $m = -\frac{1}{3}$ in the point-slope form:

$$y - y_1 = m(x - x_1)$$

$$y - 3 = -\frac{1}{3}(x - 2) \quad \text{The equation in point-slope form}$$

$$y - 3 = -\frac{1}{3}x + \frac{2}{3}$$

$$y = -\frac{1}{3}x + \frac{11}{3} \quad \text{The equation solved for } y$$

**b.** The slope of the line through $(2, 5)$ and $(-6, 5)$ shown in Fig. 1.31 is 0. The equation of this horizontal line is $y = 5$.

**c.** The line through $(3, -1)$ and $(3, 9)$ shown in Fig. 1.32 is vertical and it does not have slope. Its equation is $x = 3$.

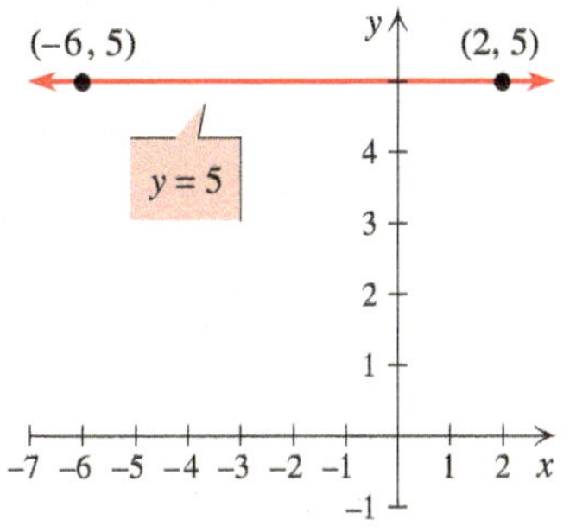


**Figure 1.31**

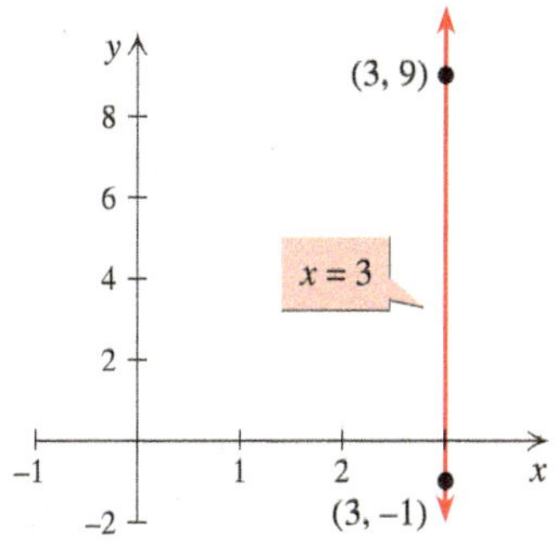


**Figure 1.32**

▸**TRY THIS.** Find the equation of the line that contains $(-2, 5)$ and $(-1, -3)$ and solve the equation for $y$. ■

## Slope-Intercept Form

The line $y = mx + b$ goes through $(0, b)$ and $(1, m + b)$. Between these two points the rise is $m$ and the run is 1. So the slope is $m$. Since $(0, b)$ is the $y$-intercept and $m$ is the slope, $y = mx + b$ is called slope-intercept form. Any equation in standard form $Ax + By = C$ can be rewritten in **slope-intercept form** by solving the equation for $y$ provided $B \neq 0$. If $B = 0$, then the line is vertical and has no slope.

**Theorem: Slope-Intercept Form**

The equation of a line (in slope-intercept form) with slope $m$ and $y$-intercept $(0, b)$ is

$$y = mx + b.$$

Every nonvertical line has an equation in slope-intercept form.

If you know the slope and $y$-intercept for a line then you can use slope-intercept form to write its equation. For example, the equation of the line through $(0, 9)$ with slope 4 is $y = 4x + 9$. In the next example we use the slope-intercept form to determine the slope and $y$-intercept for a line.

### EXAMPLE 3 Find the slope and $y$-intercept

Identify the slope and $y$-intercept for the line $2x - 3y = 6$.

**Solution**

First solve the equation for $y$ to get it in slope-intercept form:

$$2x - 3y = 6$$
$$-3y = -2x + 6$$
$$y = \frac{2}{3}x - 2$$

So the slope is $\frac{2}{3}$ and the $y$-intercept is $(0, -2)$.

▸**TRY THIS.** Identify the slope and $y$-intercept for $3x + 5y = 15$. ■

Given any point and a slope, we can find the equation of the line using the point-slope form. We can also use the slope-intercept form as shown in the next example.

### EXAMPLE 4 Using slope-intercept form with any point and a slope

Find the equation of the line in slope-intercept form through $(-2, 3)$ with slope $1/2$.

**Solution**

Since $(-2, 3)$ must satisfy the equation of the line, we can use $x = -2$, $y = 3$, and $m = 1/2$ in the slope-intercept form to find $b$:

$$y = mx + b \qquad \text{Slope-intercept form}$$

$$3 = \frac{1}{2}(-2) + b \qquad x = -2, y = 3, \text{ and } m = \tfrac{1}{2}.$$

$$3 = -1 + b$$

$$4 = b$$

So the equation is $y = \frac{1}{2}x + 4$.

**▶TRY THIS.** Find the equation of the line in slope-intercept form through $(4, -6)$ with slope $1/4$. ■

## Using Slope to Graph a Line

Slope is the ratio $\frac{\text{rise}}{\text{run}}$ that results from moving from one point to another on a line. A positive rise indicates a motion upward and a negative rise indicates a motion downward. A positive run indicates a motion to the right and a negative run indicates a motion to the left. If you start at any point on a line with slope $\frac{1}{2}$, then moving up 1 unit and 2 units to the right will bring you back to the line. On a line with slope $-3$ or $\frac{-3}{1}$, moving down 3 units and 1 unit to the right will bring you back to the line.

### EXAMPLE 5 Graphing a line using its slope and $y$-intercept

Graph each line.

**a.** $y = 3x - 1$ **b.** $y = -\frac{2}{3}x + 4$

**Solution**

**a.** The line $y = 3x - 1$ has $y$-intercept $(0, -1)$ and slope 3 or $\frac{3}{1}$. Starting at $(0, -1)$ we obtain a second point on the line by moving up 3 units and 1 unit to the right. So the line goes through $(0, -1)$ and $(1, 2)$ as shown in Fig. 1.33.

**b.** The line $y = -\frac{2}{3}x + 4$ has $y$-intercept $(0, 4)$ and slope $-\frac{2}{3}$ or $\frac{-2}{3}$. Starting at $(0, 4)$ we obtain a second point on the line by moving down 2 units and then 3 units to the right. So the line goes through $(0, 4)$ and $(3, 2)$ as shown in Fig. 1.34.

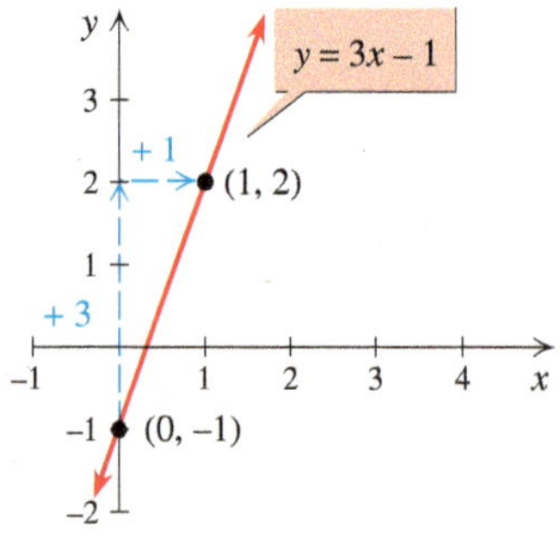


**Figure 1.33**

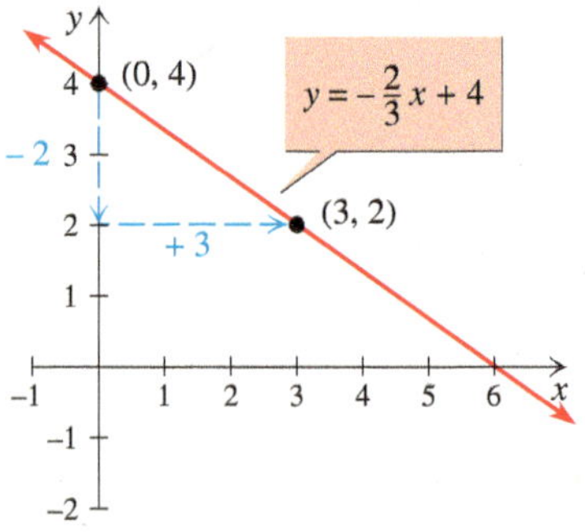


**Figure 1.34**

**▶TRY THIS.** Use the slope and $y$-intercept to graph $y = -\frac{3}{2}x + 1$. ■

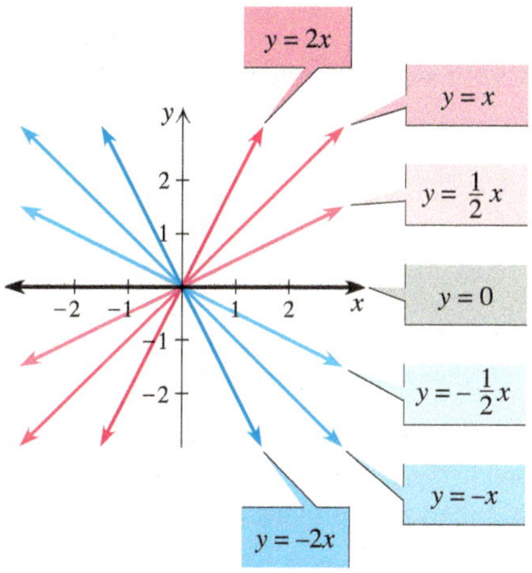


Graphs of $y = mx$

**Figure 1.35**

As the $x$-coordinate increases on a line with positive slope, the $y$-coordinate increases also. As the $x$-coordinate increases on a line with negative slope, the $y$-coordinate decreases. Figure 1.35 shows some lines of the form $y = mx$ with positive slopes and negative slopes. Observe the effect that the slope has on the position of the line. Note that lines with positive slope go up as you move from left to right and lines with negative slope go down as you move from left to right as shown in Fig. 1.36.

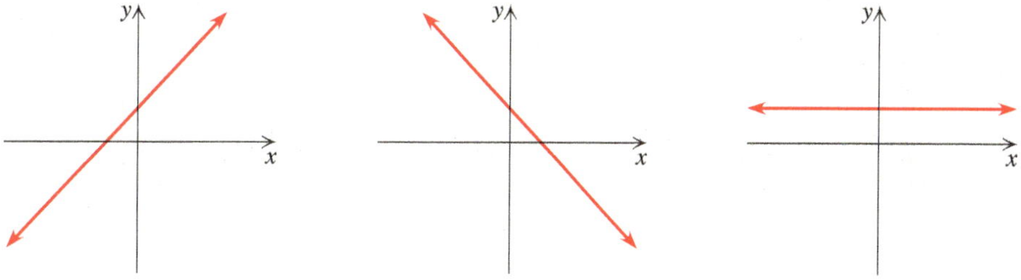


**Figure 1.36**

## The Three Forms for the Equation of a Line

There are three forms for the equation of a line. The following strategy will help you decide when and how to use these forms.

**STRATEGY**

### Finding the Equation of a Line

| | |
|---|---|
| Standard form | $Ax + By = C$ |
| Slope-intercept form | $y = mx + b$ |
| Point-slope form | $y - y_1 = m(x - x_1)$ |

1. Since vertical lines have no slope, they can't be written in slope-intercept or point-slope form.
2. All lines can be described with an equation in standard form.
3. For any constant $k$, $y = k$ is a horizontal line and $x = k$ is a vertical line.
4. If you know two points on a line, then find the slope.
5. If you know the slope and a point on the line, use point-slope form. If the point is the $y$-intercept, then use slope-intercept form.
6. Final answers are usually written in slope-intercept or standard form. Standard form is often simplified by using only integers for the coefficients.

### EXAMPLE 6 Standard form using integers

Find the equation of the line through $\left(0, \frac{1}{3}\right)$ with slope $\frac{1}{2}$. Write the equation in standard form using only integers.

### Solution

Since we know the slope and $y$-intercept, start with slope-intercept form:

$$y = \frac{1}{2}x + \frac{1}{3} \quad \text{Slope-intercept form}$$

$$-\frac{1}{2}x + y = \frac{1}{3}$$

$$-6\left(-\frac{1}{2}x + y\right) = -6 \cdot \frac{1}{3} \quad \text{Multiply by } -6 \text{ to get integers.}$$

$$3x - 6y = -2 \quad \text{Standard form with integers}$$

Any integral multiple of $3x - 6y = -2$ would also be standard form, but we usually use the smallest possible positive coefficient for $x$.

▸**TRY THIS.** Find the equation of the line through $\left(0, \frac{1}{2}\right)$ with slope $\frac{3}{4}$ and write the equation in standard form using only integers. ■

## Parallel Lines

Two lines in a plane are said to be **parallel** if they have no points in common. Any two vertical lines are parallel, and slope can be used to determine whether nonvertical lines are parallel. For example, the lines $y = 3x - 4$ and $y = 3x + 1$ are parallel because their slopes are equal and their $y$-intercepts are different.

**Theorem: Parallel Lines**

> Two nonvertical lines in the coordinate plane are parallel if and only if their slopes are equal.

A proof to this theorem is outlined in Exercises 113 and 114.

### EXAMPLE 7 Writing equations of parallel lines

Find the equation in slope-intercept form of the line through $(1, -4)$ that is parallel to $y = 3x + 2$.

### Solution

Since $y = 3x + 2$ has slope 3, any line parallel to it also has slope 3. Write the equation of the line through $(1, -4)$ with slope 3 in point-slope form:

$$y - (-4) = 3(x - 1) \quad \text{Point-slope form}$$

$$y + 4 = 3x - 3$$

$$y = 3x - 7 \quad \text{Slope-intercept form}$$

The line $y = 3x - 7$ goes through $(1, -4)$ and is parallel to $y = 3x + 2$.

The graphs of $y_1 = 3x - 7$ and $y_2 = 3x + 2$ in Fig. 1.37 support the answer.

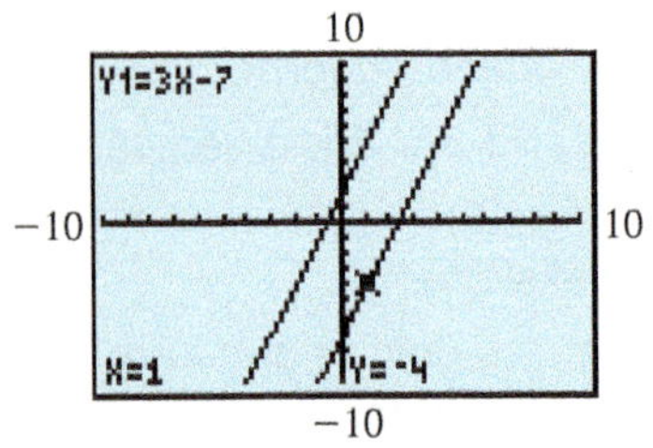


Figure 1.37

▸**TRY THIS.** Find the equation in slope-intercept form of the line through $(2, 4)$ that is parallel to $y = -\frac{1}{2}x + 9$. ■

## Perpendicular Lines

Two lines are **perpendicular** if they intersect at a right angle. Slope can be used to determine whether lines are perpendicular. For example, lines with slopes such as $2/3$ and $-3/2$ are perpendicular. The slope $-3/2$ is the opposite of the reciprocal of

2/3. In the following theorem we use the equivalent condition that the product of the slopes of two perpendicular lines is $-1$, provided they both have slopes.

**Theorem: Perpendicular Lines**

> Two lines with slopes $m_1$ and $m_2$ are perpendicular if and only if $m_1 m_2 = -1$.

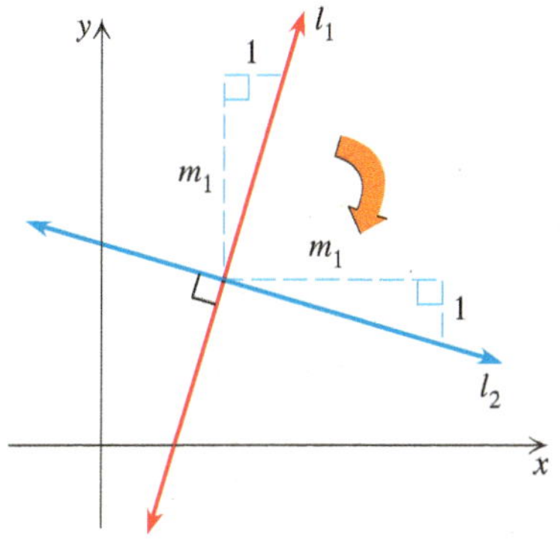


Figure 1.38

**PROOF** The phrase "if and only if" means that there are two statements to prove. First we prove that if $l_1$ with slope $m_1$ and $l_2$ with slope $m_2$ are perpendicular, then $m_1 m_2 = -1$. Assume that $m_1 > 0$. At the intersection of the lines draw a right triangle using a rise of $m_1$ and a run of 1, as shown in Fig. 1.38. Rotate $l_1$ (along with the right triangle) 90 degrees so that it coincides with $l_2$. Now use the triangle in its new position to determine that $m_2 = \frac{1}{-m_1}$ or $m_1 m_2 = -1$.

The second statement to prove is that $m_1 m_2 = -1$ or $m_2 = \frac{1}{-m_1}$ implies that the lines are perpendicular. Start with the two intersecting lines and the two congruent right triangles, as shown in Fig. 1.38. It takes a rotation of 90 degrees to get the vertical side marked $m_1$ to coincide with the horizontal side marked $m_1$. Since the triangles are congruent, rotating 90 degrees makes the triangles coincide and the lines coincide. So the lines are perpendicular. ■

## EXAMPLE 8 Writing equations of perpendicular lines

Find the equation of the line perpendicular to the line $3x - 4y = 8$ and containing the point $(-2, 1)$. Write the answer in slope-intercept form.

### Solution

Rewrite $3x - 4y = 8$ in slope-intercept form:

$$-4y = -3x + 8$$

$$y = \frac{3}{4}x - 2 \qquad \text{Slope of this line is } 3/4.$$

Since the product of the slopes of perpendicular lines is $-1$, the slope of the line that we seek is $-4/3$. Use the slope $-4/3$ and the point $(-2, 1)$ in the point-slope form:

$$y - 1 = -\frac{4}{3}(x - (-2))$$

$$y - 1 = -\frac{4}{3}x - \frac{8}{3}$$

$$y = -\frac{4}{3}x - \frac{5}{3}$$

The last equation is the required equation in slope-intercept form. The graphs of these two equations should look perpendicular.

If you graph $y_1 = \frac{3}{4}x - 2$ and $y_2 = -\frac{4}{3}x - \frac{5}{3}$ with a graphing calculator, the graphs will not appear perpendicular in the standard viewing window because each axis has a different unit length. The graphs appear perpendicular in Fig. 1.39 because the unit lengths were made equal with the ZSquare feature of the TI-83.

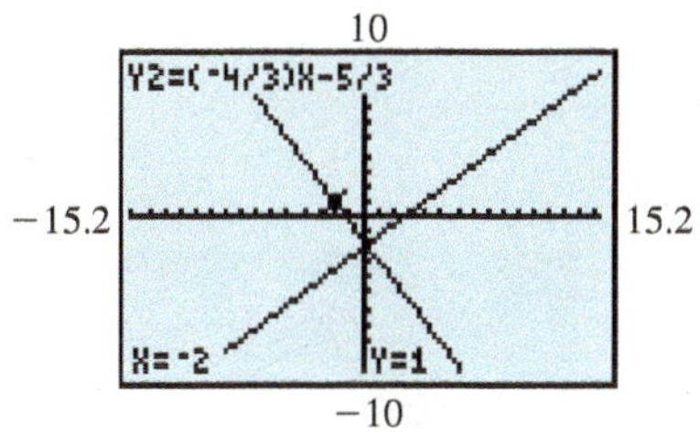


Figure 1.39

▸**TRY THIS.** Find the equation in slope-intercept form of the line through $(-2, 1)$ that is perpendicular to $2x - y = 8$. ■

## Applications

In Section 1.3 the distance formula was used to establish some facts about geometric figures in the coordinate plane. We can also use slope to prove that lines are parallel or perpendicular in geometric figures.

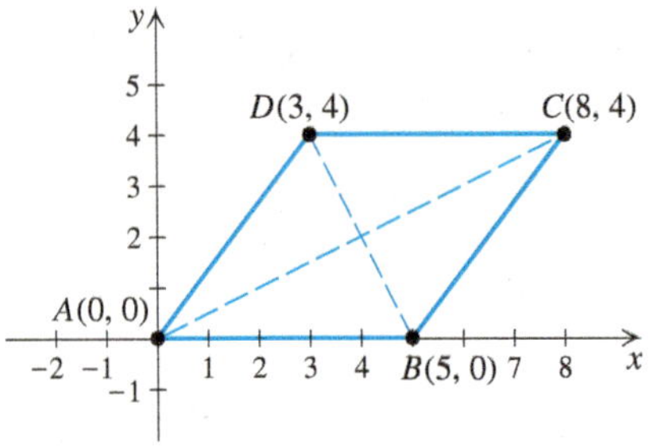


Figure 1.40

## EXAMPLE 9 The diagonals of a rhombus are perpendicular

Given a rhombus with vertices $(0, 0)$, $(5, 0)$, $(3, 4)$, and $(8, 4)$, prove that the diagonals of this rhombus are perpendicular. (A rhombus is a four-sided figure in which the sides have equal length.)

### Solution

Plot the four points $A$, $B$, $C$, and $D$ as shown in Fig. 1.40. You should show that each side of this figure has length 5, verifying that the figure is a rhombus. Find the slopes of the diagonals and their product:

$$m_{AC} = \frac{4-0}{8-0} = \frac{1}{2} \qquad m_{BD} = \frac{0-4}{5-3} = -2$$

$$m_{AC} \cdot m_{BD} = \frac{1}{2}(-2) = -1$$

Since the product of the slopes of the diagonals is $-1$, the diagonals are perpendicular.

▸**TRY THIS.** Prove that the triangle with vertices $(0, 0)$, $(5, 2)$, and $(1, 12)$ is a right triangle. ■

If the value of one variable can be determined from the value of another variable, then we say that the first variable is a function of the second variable. This idea was discussed in Section 1.2. Because the area of a circle can be determined from the radius by the formula $A = \pi r^2$, we say that $A$ is a function of $r$. If $y$ is determined from $x$ by using the slope-intercept form of the equation of a line $y = mx + b$, then **$y$ is a linear function of $x$.** The formula $F = \frac{9}{5}C + 32$ expresses $F$ as a linear function of $C$. We will discuss functions in depth in Chapter 2.

## EXAMPLE 10 A linear function

From ABC Wireless the monthly cost for a cell phone with 100 minutes per month is \$35, or 200 minutes per month for \$45. See Fig. 1.41. The cost in dollars is a linear function of the time in minutes.

**a.** Find the formula for $C$.

**b.** What is the cost for 400 minutes per month?

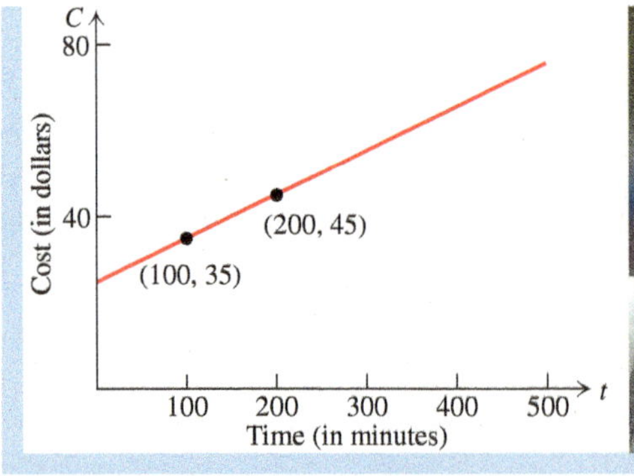

Figure 1.41

### Solution

**a.** First find the slope:

$$m = \frac{C_2 - C_1}{t_2 - t_1} = \frac{45 - 35}{200 - 100} = \frac{10}{100} = 0.10$$

The slope is \$0.10 per minute. Now find $b$ by using $C = 35$, $t = 100$, and $m = 0.10$ in the slope-intercept form $C = mt + b$:

$$35 = 0.10(100) + b$$
$$35 = 10 + b$$
$$25 = b$$

So the formula is $C = 0.10t + 25$.

**b.** Use $t = 400$ in the formula $C = 0.10t + 25$:

$$C = 0.10(400) + 25 = 65$$

The cost for 400 minutes per month is \$65.

▸**TRY THIS.** The cost of a truck rental is \$70 for 100 miles and \$90 for 200 miles. Write the cost as a linear function of the number of miles. ■

Note that in Example 10 we could have used the point-slope form $C - C_1 = m(t - t_1)$ to get the formula $C = 0.10t + 25$. Try it.

The situation in the next example leads naturally to an equation in standard form.

## EXAMPLE 11 Interpreting slope

A manager for a country market will spend a total of \$80 on apples at \$0.25 each and pears at \$0.50 each. Write the number of apples she can buy as a linear function of the number of pears. Find the slope and interpret your answer.

### Solution

Let $a$ represent the number of apples and $p$ represent the number of pears. Write an equation in standard form about the total amount spent:

$$0.25a + 0.50p = 80$$
$$0.25a = 80 - 0.50p$$
$$a = 320 - 2p \quad \text{Solve for } a.$$

The equation $a = 320 - 2p$ or $a = -2p + 320$ expresses the number of apples as a function of the number of pears. Since $p$ is the first coordinate and $a$ is the second, the slope is $-2$ apples per pear. So if the number of apples is decreased by 2, then the number of pears can be increased by 1 and the total is still \$80. This makes sense because the pears cost twice as much as the apples.

▸**TRY THIS.** A manager will spend \$3000 on file cabinets at \$100 each and bookshelves at \$150 each. Write the number of file cabinets as a function of the number of bookshelves and interpret the slope. ■

## FOR thought... True or False? Explain.

1. The slope of the line through (2, 2) and (3, 3) is 3/2.
2. The slope of the line through (−3, 1) and (−3, 5) is 0.
3. Any two distinct parallel lines have equal slopes.
4. The graph of $x = 3$ in the coordinate plane is the single point (3, 0).
5. Two lines with slopes $m_1$ and $m_2$ are perpendicular if $m_1 = -1/m_2$.
6. Every line in the coordinate plane has an equation in slope-intercept form.
7. The slope of the line $y = 3 - 2x$ is 3.
8. Every line in the coordinate plane has an equation in standard form.
9. The line $y = 3x$ is parallel to the line $y = -3x$.
10. The line $x - 3y = 4$ contains the point (1, −1) and has slope 1/3.

## EXERCISES 1.4

*Fill in the blank.*

1. The change in $y$-coordinate between two points on a line is the ________.
2. The change in $x$-coordinate between two points on a line is the ________.
3. The rise divided by the run is the ________ of a line.
4. The equation $y - y_1 = m(x - x_1)$ is the ________ form of the equation of a line.
5. The equation $y = mx + b$ is the ________ form of the equation of a line.
6. Two lines that have no points in common are ________.
7. Two lines with slopes $m_1$ and $m_2$ are ________ if and only if $m_1 m_2 = -1$.
8. If the value of one variable can be determined from the value of another variable, then the first variable ________ the second variable.

*Find the slope of the line containing each pair of points.*

9. (−2, 3), (4, 5)
10. (−1, 2), (3, 6)
11. (1, 3), (3, −5)
12. (2, −1), (5, −3)
13. (5, 2), (−3, 2)
14. (0, 0), (5, 0)
15. $\left(\frac{1}{8}, \frac{1}{4}\right), \left(\frac{1}{4}, \frac{1}{2}\right)$
16. $\left(-\frac{1}{3}, \frac{1}{2}\right), \left(\frac{1}{6}, \frac{1}{3}\right)$
17. (5, −1), (5, 3)
18. (−7, 2), (−7, −6)

*Find the equation of the line through the given pair of points. Solve it for y if possible.*

19. (−1, −1), (3, 4)
20. (−2, 1), (3, 5)
21. (−2, 6), (4, −1)
22. (−3, 5), (2, 1)
23. (3, 5), (−3, 5)
24. (−6, 4), (2, 4)
25. (4, −3), (4, 12)
26. (−5, 6), (−5, 4)

*Write an equation in slope-intercept form for each of the lines shown.*

**27.**

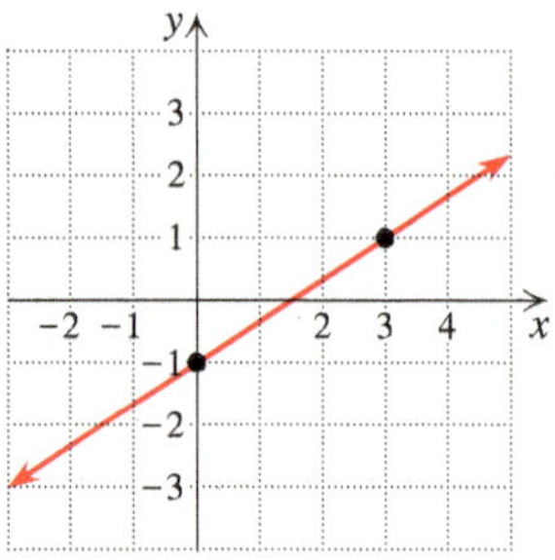


**28.**

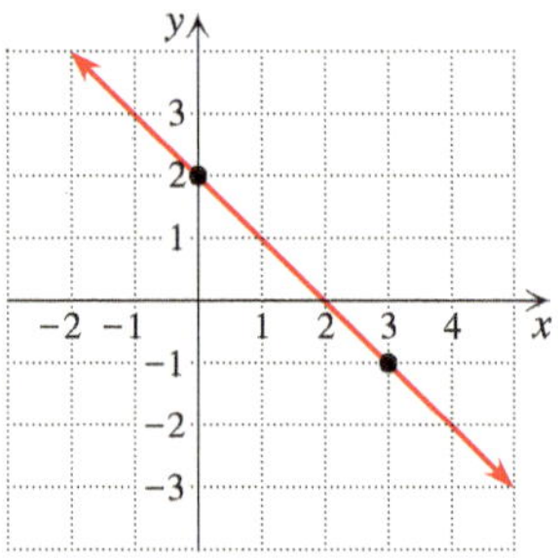


**29.**

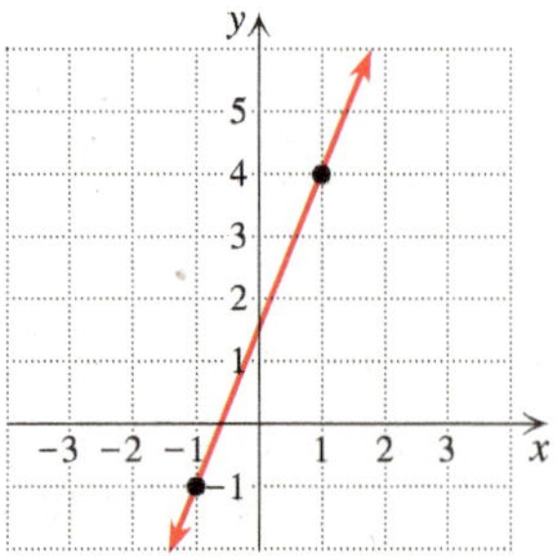


**30.**

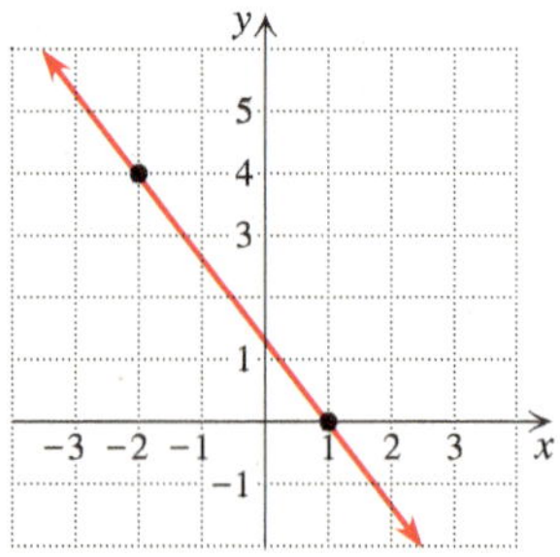

31.

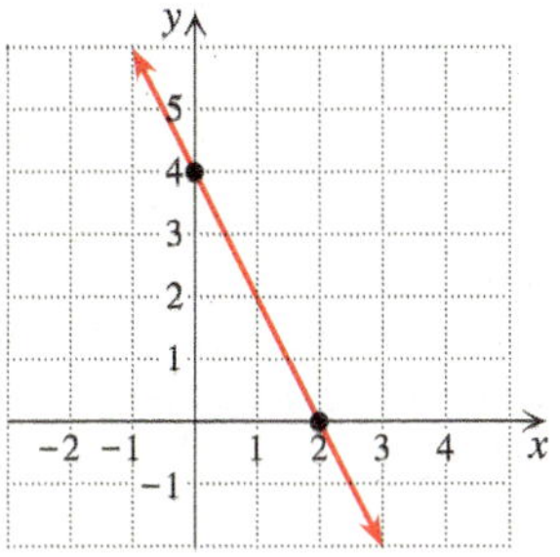


32.

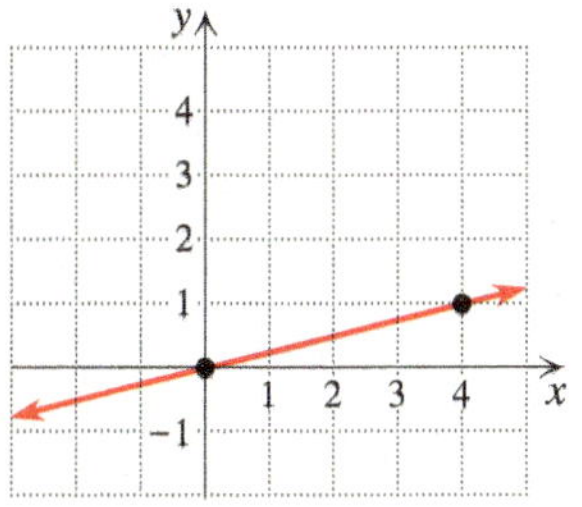


33.

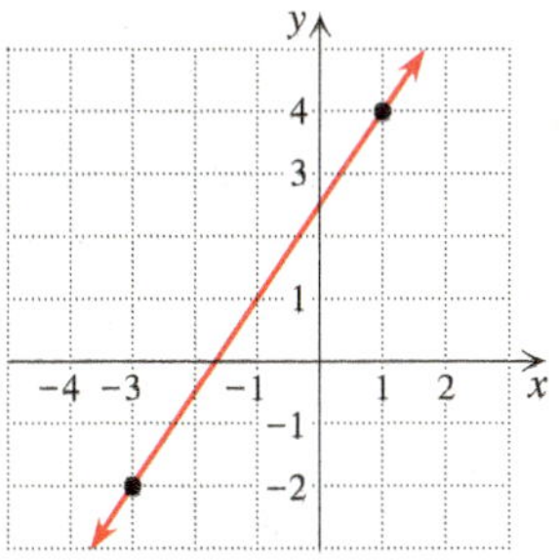


34.

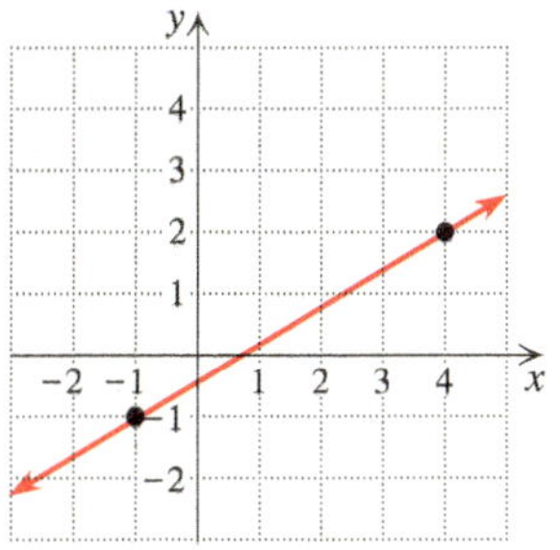


*Write each equation in slope-intercept form and identify the slope and y-intercept of the line.*

**35.** $3x - 5y = 10$

**36.** $2x - 2y = 1$

**37.** $y - 3 = 2(x - 4)$

**38.** $y + 5 = -3(x - (-1))$

**39.** $y + 1 = \frac{1}{2}(x - (-3))$

**40.** $y - 2 = -\frac{3}{2}(x + 5)$

**41.** $y - 4 = 0$

**42.** $-y + 5 = 0$

*Find the equation of each line in slope-intercept form.*

**43.** The line through $(-8, 5)$ with slope $1/4$

**44.** The line through $(6, 9)$ with slope $-1/3$

**45.** The line through $(-3, -2)$ with slope $-1/2$

**46.** The line through $(4, 3)$ with slope $2/3$

*Use the y-intercept and slope to sketch the graph of each equation.*

**47.** $y = \frac{1}{2}x - 2$

**48.** $y = \frac{2}{3}x + 1$

**49.** $y = -3x + 1$

**50.** $y = -x + 3$

**51.** $y = -\frac{3}{4}x - 1$

**52.** $y = -\frac{3}{2}x$

**53.** $x - y = 3$

**54.** $2x - 3y = 6$

**55.** $y - 5 = 0$

**56.** $6 - y = 0$

*Find the equation of the line through the given pair of points in standard form using only integers. See the strategy for finding the equation of a line on page 43.*

**57.** $(3, 0)$ and $(0, -4)$

**58.** $(-2, 0)$ and $(0, 3)$

**59.** $(2, 3)$ and $(-3, -1)$

**60.** $(4, -1)$ and $(-2, -6)$

**61.** $(-4, 2)$ and $(-4, 5)$

**62.** $(-3, 6)$ and $(9, 6)$

**63.** $\left(2, \frac{2}{3}\right), \left(-\frac{1}{2}, -2\right)$

**64.** $\left(\frac{3}{4}, -3\right), \left(-5, \frac{1}{8}\right)$

**65.** $\left(\frac{1}{2}, \frac{1}{4}\right), \left(-\frac{1}{3}, \frac{1}{5}\right)$

**66.** $\left(-\frac{3}{8}, \frac{1}{4}\right), \left(\frac{1}{2}, -\frac{1}{6}\right)$

*Find the slope of each line described.*

**67.** A line parallel to $y = 0.5x - 9$

**68.** A line parallel to $3x - 9y = 4$

**69.** A line perpendicular to $3y - 3x = 7$

**70.** A line perpendicular to $2x - 3y = 8$

**71.** A line perpendicular to the line $x = 4$

**72.** A line parallel to $y = 5$

*Write an equation in standard form using only integers for each of the lines described. In each case make a sketch.*

**73.** The line with slope 2, going through $(1, -2)$

**74.** The line with slope $-3$, going through $(-3, 4)$

**75.** The line through $(1, 4)$, parallel to $y = -3x$

**76.** The line through $(-2, 3)$ parallel to $y = \frac{1}{2}x + 6$

**77.** The line parallel to $5x - 7y = 35$ and containing $(6, 1)$

**78.** The line parallel to $4x + 9y = 5$ and containing $(-4, 2)$

**79.** The line perpendicular to $y = \frac{2}{3}x + 5$ and containing $(2, -3)$

**80.** The line perpendicular to $y = 9x + 5$ and containing $(5, 4)$

**81.** The line perpendicular to $x - 2y = 3$ and containing $(-3, 1)$

**82.** The line perpendicular to $3x - y = 9$ and containing $(0, 0)$

**83.** The line perpendicular to $x = 4$ and containing $(2, 5)$

**84.** The line perpendicular to $y = 9$ and containing $(-1, 3)$

*Find the value of a in each case.*

**85.** The line through $(-2, 3)$ and $(8, 5)$ is perpendicular to $y = ax + 2$.

**86.** The line through $(3, 4)$ and $(7, a)$ has slope $2/3$.

**87.** The line through $(-2, a)$ and $(a, 3)$ has slope $-1/2$.

**88.** The line through $(-1, a)$ and $(3, -4)$ is parallel to $y = ax$.

*Either prove or disprove each statement. Use a graph only as a guide. Your proof should rely on algebraic calculations.*

**89.** The points $(-1, 2)$, $(2, -1)$, $(3, 3)$, and $(-2, -2)$ are the vertices of a parallelogram.

**90.** The points $(-1, 1)$, $(-2, -5)$, $(2, -4)$, and $(3, 2)$ are the vertices of a parallelogram.

**91.** The points $(-5, -1)$, $(-3, -4)$, $(3, 0)$, and $(1, 3)$ are the vertices of a rectangle.

**92.** The points $(-5, -1)$, $(1, -4)$, $(4, 2)$, and $(-1, 5)$ are the vertices of a square.

**93.** The points $(-5, 1)$, $(-2, -3)$, and $(4, 2)$ are the vertices of a right triangle.

**94.** The points $(-4, -3)$, $(1, -2)$, $(2, 3)$, and $(-3, 2)$ are the vertices of a rhombus.

*Use a graphing calculator to solve each problem.*

**95.** Graph $y_1 = (x - 5)/3$ and $y_2 = x - 0.67(x + 4.2)$. Do the lines appear to be parallel? Are the lines parallel?

**96.** Graph $y_1 = 99x$ and $y_2 = -x/99$. Do the lines appear to be perpendicular? Should they appear perpendicular?

**97.** Graph $y = (x^3 - 8)/(x^2 + 2x + 4)$. Use TRACE to examine points on the graph. Write a linear function for the graph. Explain why this function is linear.

**98.** Graph $y = (x^3 + 2x^2 - 5x - 6)/(x^2 + x - 6)$. Use TRACE to examine points on the graph. Write a linear function for the graph. Factor $x^3 + 2x^2 - 5x - 6$ completely.

*Solve each problem.*

**99.** *Celsius to Fahrenheit Formula* Fahrenheit temperature $F$ is a linear function of Celsius temperature $C$. The ordered pair $(0, 32)$ is an ordered pair of this function because 0°C is equivalent to 32°F, the freezing point of water. The ordered pair $(100, 212)$ is also an ordered pair of this function because 100°C is equivalent to 212°F, the boiling point of water. Use the two given points and the point-slope formula to write $F$ as a function of $C$. Find the Fahrenheit temperature of an oven at 150°C.

**100.** *Cost of Business Cards* Speedy Printing charges $23 for 200 deluxe business cards and $35 for 500 deluxe business cards. Given that the cost is a linear function of the number of cards printed, find a formula for that function and find the cost of 700 business cards.

**101.** *Volume Discount* Mona Kalini gives a walking tour of Honolulu to one person for $49. To increase her business, she advertised at the National Orthodontist Convention that she would lower the price by $1 per person for each additional person. Write the cost per person $c$ as a function of the number of people $n$ on the tour. How much does she make for a tour with 40 people?

HINT Find the equation of the line through $(1, 49)$, $(2, 48)$, $(3, 47)$, etc.

**102.** *Ticket Pricing* At $10 per ticket, Willie Williams and the Wranglers will fill all 8000 seats in the Assembly Center. The manager knows that for every $1 increase in the price, 500 tickets will go unsold. Write the number of tickets sold $n$ as a function of the ticket price $p$. How much money will be taken in if the tickets are $20 each?

HINT Find the equation of the line through $(10, 8000)$, $(11, 7500)$, $(12, 7000)$, etc.

**103.** *Lindbergh's Air Speed* Charles Lindbergh estimated that at the start of his historic flight the practical economical air speed was 95 mph and at 4000 statute miles from the starting point it was 75 mph (www.charleslindbergh.com). Assume that the practical economical air speed $S$ is a linear function of the distance $D$ from the starting point as shown in the accompanying figure. Find a formula for that function.

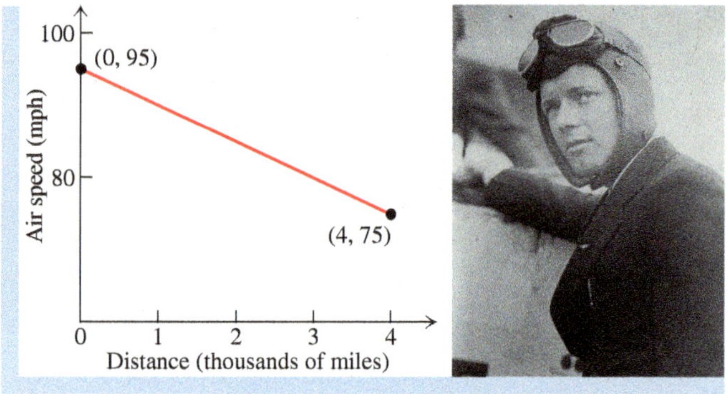


**Figure for Exercise 103**

**104.** *Speed over Newfoundland* In Lindbergh's flying log he recorded his air speed over Newfoundland as 98 mph, 1100 miles into his flight. According to the formula from Exercise 103, what should have been his air speed over Newfoundland?

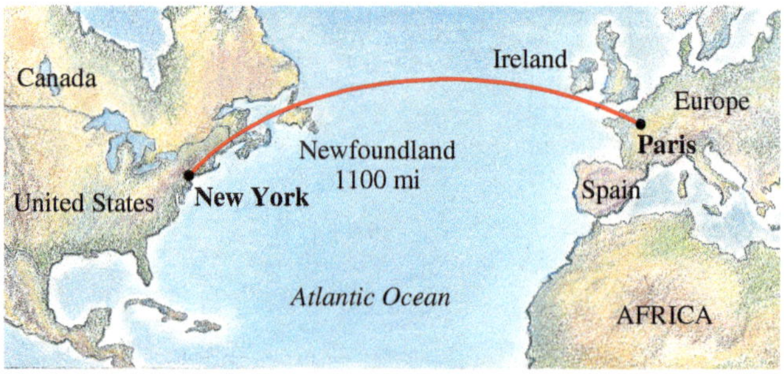


**Figure for Exercise 104**

**105.** *Computers and Printers* An office manager will spend a total of $60,000 on computers at $2000 each and printers at $1500 each. Write the number of computers purchased as a function of the number of printers purchased. Find and interpret the slope.

HINT Start with standard form.

**106.** *Carpenters and Helpers* Because a job was finished early, a contractor will distribute a total of \$2400 in bonuses to 9 carpenters and 3 helpers. The carpenters all get the same amount and the helpers all get the same amount. Write the amount of a helper's bonus as a function of the amount of a carpenter's bonus. Find and interpret the slope.

**107.** *Integral Coordinates* Find all points on the line through $(-9, 14)$ and $(21, -4)$ that lie between these two points and have integral coordinates.

**108.** *Parallelograms* The points $(-1, 2)$, $(4, 1)$, and $(2, -3)$ are the vertices of a parallelogram in the coordinate plane. Find all possibilities for the fourth vertex.

*The distance d from the point $(x_1, y_1)$ to the line $Ax + By = C$ is given by the formula*

$$d = \frac{|Ax_1 + By_1 - C|}{\sqrt{A^2 + B^2}}.$$

*Find the exact distance from each given point to the given line.*

**109.** $(3, -6)$, $5x - 12y = 2$

**110.** $(-4, 8)$, $3x + 4y = 9$

**111.** $(1, 3)$, $y = 5x - 4$

**112.** $(-2, 5)$, $y = -4x + 1$

## FOR WRITING/DISCUSSION

**113.** *Equal Slopes* Show that if $y = mx + b_1$ and $y = mx + b_2$ are equations of lines with equal slopes, but $b_1 \neq b_2$, then they have no point in common.

HINT Assume that they have a point in common and see that this assumption leads to a contradiction.

**114.** *Unequal Slopes* Show that the lines $y = m_1x + b_1$ and $y = m_2x + b_2$ intersect at a point with $x$-coordinate $(b_2 - b_1)/(m_1 - m_2)$ provided $m_1 \neq m_2$. Explain how this exercise and the previous exercise prove the theorem that two nonvertical lines are parallel if and only if they have equal slopes.

## ▶ RETHINKING

**115.** Solve $3 - 5|x - 4| = 0$.

**116.** Hall can shovel the snow from a long driveway in 24 minutes. If Shanna helps, they can do the job in 18 minutes. How long would it take Shanna to do the job alone?

**117.** The diameter of a circle has endpoints $(1, 3)$ and $(3, 9)$. Find the equation of the circle.

**118.** Find the center and radius of the circle $x^2 + 3x + y^2 - 4y = 0$.

**119.** Solve $-5|x - 4| = 0$.

**120.** Find the midpoint of the line segment with endpoints $(\pi/2, 1)$ and $(\pi, 1)$.

## THINKING OUTSIDE THE BOX IV AND V

*Army of Ants* An army of ants is marching across the kitchen floor. If they form columns with 10 ants in each column, then there are 6 ants left over. If they form columns with 7, 11, or 13 ants in each column, then there are 2 ants left over. What is the smallest number of ants that could be in this army?

*Summing Angles* Consider the angles shown in the accompanying figure. Show that the degree measure of angle *A plus* the degree measure of angle *B* is equal to the degree measure of angle *C*.

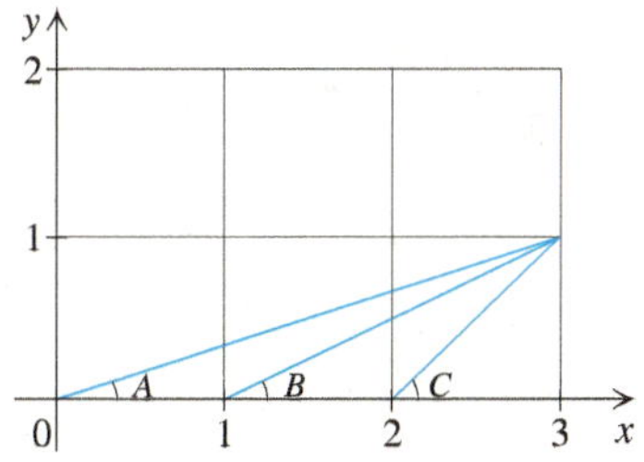


**Figure for Thinking Outside the Box V**

## POP QUIZ 1.4

**1.** Find the slope of the line through $(-4, 9)$ and $(5, 6)$.

**2.** Find the equation of the line through $(3, 4)$ and $(6, 8)$.

**3.** What is the slope of the line $2x - 7y = 1$?

**4.** Find the equation of the line with $y$-intercept $(0, 7)$ that is parallel to $y = 3x - 1$.

**5.** Find the equation of the line with $y$-intercept $(0, 8)$ that is perpendicular to $y = \frac{1}{2}x + 4$.

LINKING
# concepts... For Individual or Group Explorations

## The Negative Income Tax Model

*One idea for income tax reform is that of a negative income tax. Under this plan families with an income below a certain level receive a payment from the government in addition to their income while families above a certain level pay taxes. In one proposal a family with an earned income of \$6000 would receive a \$4000 payment, giving the family a disposable income of \$10,000. A family with an earned income of \$24,000 would pay \$2000 in taxes, giving the family a disposable income of \$22,000. A family's disposable income D is a linear function of the family's earned income E.*

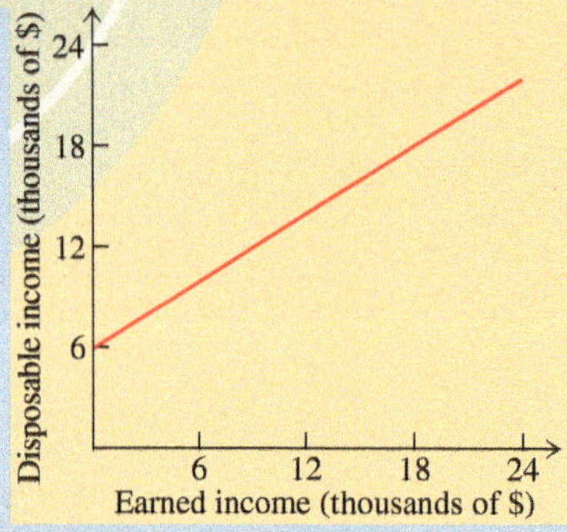

**a)** Use the given data to write a formula for $D$ as a linear function of $E$.

**b)** What is the disposable income for a family with an earned income of \$60,000?

**c)** Find the break-even point, the income at which a family pays no taxes and receives no payment.

**d)** What percentage of their earned income is paid in taxes for a family with an earned income of \$25,000? \$100,000? \$2,000,000?

**e)** What is the maximum percentage of earned income that anyone will pay in taxes?

**f)** Compare your answer to part (e) with current tax laws. Does this negative income tax favor the rich, poor, or middle class?

# 1.5 Scatter Diagrams and Curve Fitting

The Cartesian coordinate system is often used to illustrate real data and relationships between variables. In this section we will graph real data and see how to fit lines to that data.

## Scatter Diagrams and Types of Relationships

In statistics we often gather paired data, such as height and weight of an individual. We seek a relationship between the two variables. We can graph pairs of data as

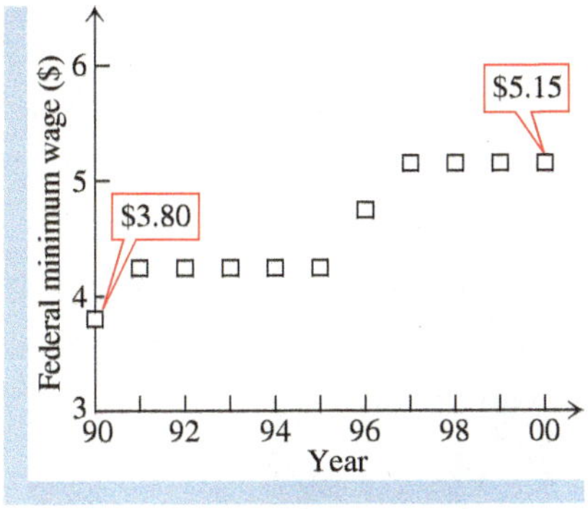


Figure 1.42

ordered pairs in the Cartesian coordinate system as shown in Fig. 1.42. The graph is called a **scatter diagram** because real data are more likely to be scattered about rather than perfectly lined up. If a pattern appears in the scatter diagram, then there is a relationship between the variables. If the points in the scatter diagram look like they are scattered about a line, then the relationship is **linear.** Otherwise the relationship is **nonlinear.**

### EXAMPLE 1 Classifying scatter diagrams

For each given scatter diagram determine whether there is a relationship between the variables. If there is a relationship then determine whether it is linear or nonlinear.

**a.**

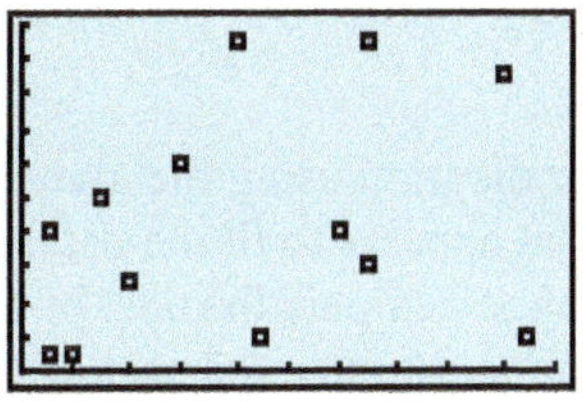

**b.**

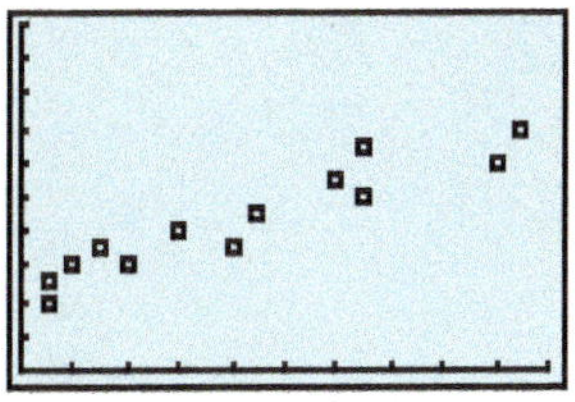

**c.**

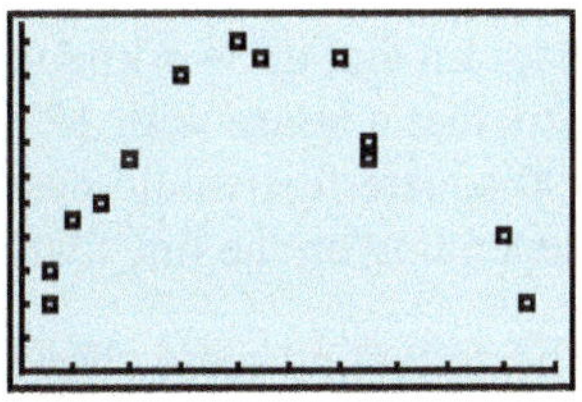

**d.**

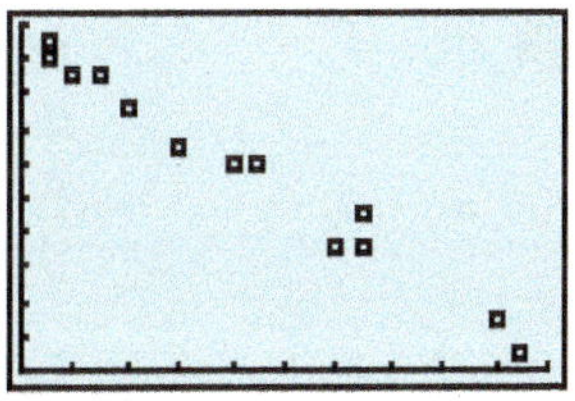

#### Solution

**a.** The points are just scattered about and it does not appear that there is a relationship between the variables.
**b.** It appears that there is a linear relationship between the variables.
**c.** It appears that there is a pattern to the data and that there is a nonlinear relationship between the variables.
**d.** It appears that there is a linear relationship between the variables.

**▸TRY THIS.** Determine whether the variables depicted in the following scatter diagrams have a linear or nonlinear relationship.

**a.**

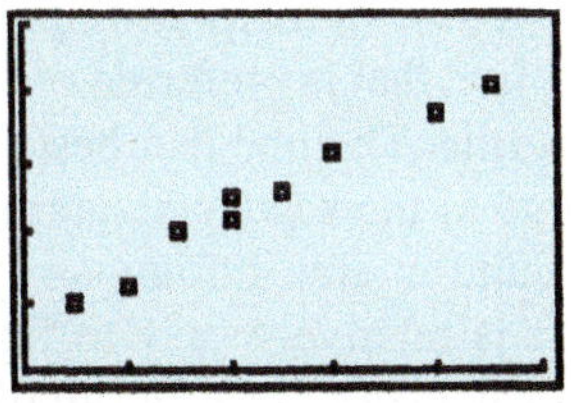

**b.**

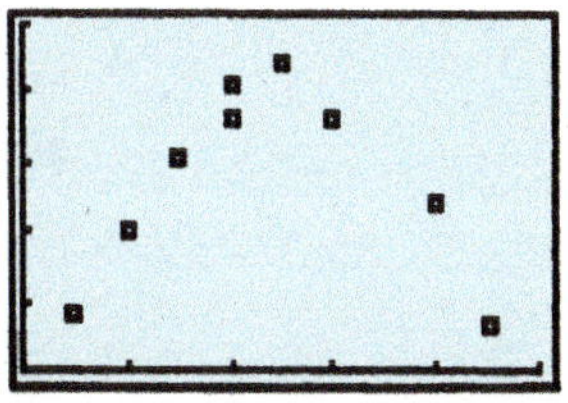

■

Using a scatter diagram to determine whether there is a relationship between variables is not precise. We will see a more precise method in Example 3.

## Finding a Line of Best Fit

If there appears to be a linear relationship in a scatter diagram, we can draw in a line and use it to make predictions.

## EXAMPLE 2 Fitting a line to data

The following table shows the relationship between the size of a house in square feet and the selling price of the house in dollars.

| Size (sq ft) | Price (thousands of dollars) |
|---|---|
| 1100 | 259 |
| 1200 | 280 |
| 1350 | 299 |
| 1625 | 335 |
| 2200 | 419 |

**a.** Make a scatter diagram using the given data.
**b.** Draw a line that appears to fit the data.
**c.** Use your graph to estimate the cost of a house with 1500 square feet.
**d.** Use your graph to estimate the cost of a house with 2400 square feet.

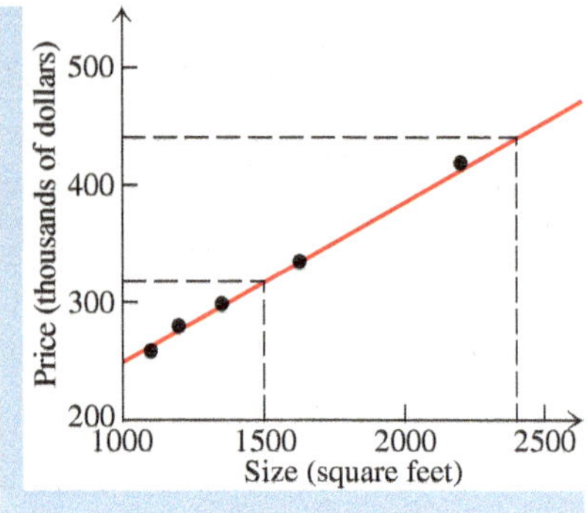

Figure 1.43

### Solution

Make a scatter diagram and draw a line to fit the data as shown in Fig. 1.43. From the graph it appears that a house with 1500 square feet would cost about \$320,000. A house with 2400 square feet would cost approximately \$450,000. Of course, answers will vary depending on the line that is drawn.

**▸TRY THIS.** The following table shows the average cost of a gallon of regular unleaded gas in selected years. Predict the cost of a gallon of regular gas in 2011 by using a scatter diagram for the given data.

| Year | 1999 | 2001 | 2005 | 2007 | 2008 |
|---|---|---|---|---|---|
| Price per gallon (dollars) | 1.17 | 1.64 | 2.48 | 2.93 | 4.15 |

■

Making a prediction within the range of the data (as in Example 2c) is called **interpolating** and making a prediction outside the range of the data (as in Example 2d) is called **extrapolating.** Remember that a prediction made by extrapolating or interpolating the data is a guess and it may be incorrect.

We could have drawn many different lines to fit the data in Example 2. However, there is one line that is preferred over all others. It is called the **line of best fit** or the **regression line.** The line that best fits the data minimizes the total "distance" between the points in the scatter diagram and the line. More details on the line of best fit can be found in any introductory statistics text and we will not give them here. However, it is relatively easy to find an equation for the regression line because a graphing calculator performs all of the computations for you. The calculator even indicates how well the equation fits the data by giving a number $r$ called the **correlation coefficient.** The correlation coefficient is between $-1$ and 1. If $r = 1$ (a **positive correlation**) the data are perfectly in line, and increasing values of one variable correspond to increasing values of the other. If $r = -1$ (a **negative correlation**) the data are also perfectly in line, but increasing values of one variable correspond to decreasing values of the other. If $r = 0$ (or $r$ is close to 0) the data are scattered about with no linear correlation. The closer $r$ is to 1 or $-1$, the more the scatter diagram will look linear. See Fig. 1.44.

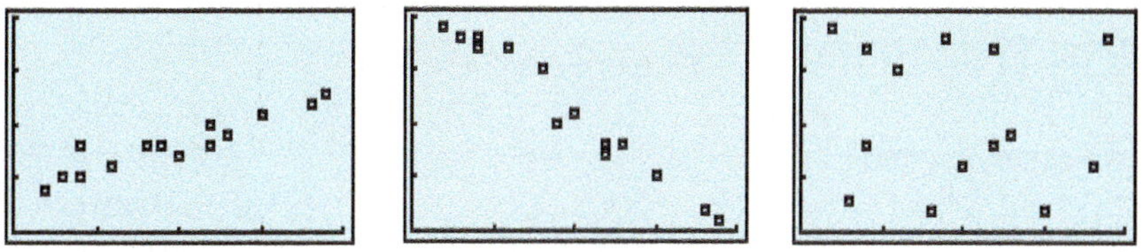

$r = 0.93$
Positive linear correlation

$r = -0.98$
Negative linear correlation

$r = -0.12$
No correlation

Figure 1.44

| Temp (°F) | % Injured |
|---|---|
| 46 | 4.7 |
| 72 | 12.3 |
| 59 | 6.5 |
| 56 | 5.9 |
| 54 | 6.6 |
| 70 | 10.3 |
| 68 | 8.4 |
| 48 | 4.0 |

Source: Larry Tye, "Ambulatory Laboratory Marathon Is a Unique Research Arena," *Boston Globe*, 20 April 1992, p. 28.

## EXAMPLE 3 Finding the line of best fit

The table shows the temperature at the start of the race and the percentage of runners injured in eight runnings of the Boston Marathon.

**a.** Use a graphing calculator to make a scatter diagram and find the equation of the regression line for the data.
**b.** Use the regression equation to estimate the percentage of runners that would be injured if the marathon was run on a day when the temperature was 82°F.
**c.** Use the regression equation to estimate the temperature at the start of the race for a race in which 9% of the runners were injured.

### Solution

**a.** First enter the data into the calculator using the STAT EDIT feature as shown in Fig. 1.45(a). Enter the temperature in the first list or $x$-list and the percent injured in the second list or $y$-list. Consult your calculator manual if you have trouble or your calculator does not accept data in this manner. Use the STAT CALC feature and the choice LinReg, to find the equation as shown in Fig. 1.45(b). The calculator gives the equation in the form $y = ax + b$ where $a \approx 0.27$ and $b \approx -8.46$. So the equation is $y = 0.27x - 8.46$. Since $r \approx 0.94$, the equation is a good model for the data. The scatter diagram and the regression line are shown in Fig. 1.45(c).

(a) (b) (c)

Figure 1.45

**b.** If $x = 82$, then $y = 0.27(82) - 8.46 = 13.68$. When the temperature is 82°F about 13.7% of the runners would be injured.
**c.** If 9% of the runners are injured, then $9 = 0.27x - 8.46$. Solve for $x$ to get $x \approx 64.7$. So the temperature should be about 65°F if 9% are injured.

▸**TRY THIS.** Predict the cost in 2010 to the nearest cent by using linear regression and the given data. Use the equation given by the calculator to make the prediction.

| Year | 1980 | 1985 | 1990 | 2000 | 2005 |
|---|---|---|---|---|---|
| Cost ($) | 12 | 21 | 33 | 49 | 61 |

■

The percentage of runners injured in the Boston Marathon is a function of many variables, one of which is temperature. Linear regression removes all of the variables except temperature. With linear regression we get a formula that determines the approximate percentage of runners who are injured as a function of temperature alone.

## FOR thought... True or False? Explain.

**1.** A scatter diagram is a graph.

**2.** If data are roughly in line in a scatter diagram, then there is a linear relationship between the variables.

**3.** If there is no pattern in a scatter diagram, then there is a nonlinear relationship between the variables.

**4.** The line of best fit is the regression line.

**5.** If $r = 0.999$, then there is a positive correlation between the variables.

**6.** If $r = -0.998$, then there is a negative correlation between the variables.

**7.** If $r = 0.002$, then there is a positive correlation between the variables.

**8.** If $r = -0.001$, then there is a negative correlation between the variables.

**9.** If we make a prediction outside the range of the data, then we are interpolating.

**10.** If we make a prediction within the range of the data, then we are extrapolating.

## EXERCISES 1.5

*Fill in the blank.*

**1.** The graph of pairs of data in the Cartesian coordinate system is a ________.

**2.** The line that best fits the data is the line of best fit or the ________ line.

*For each given scatter diagram determine whether there is a linear relationship, a nonlinear relationship, or no relationship between the variables.*

**3.**

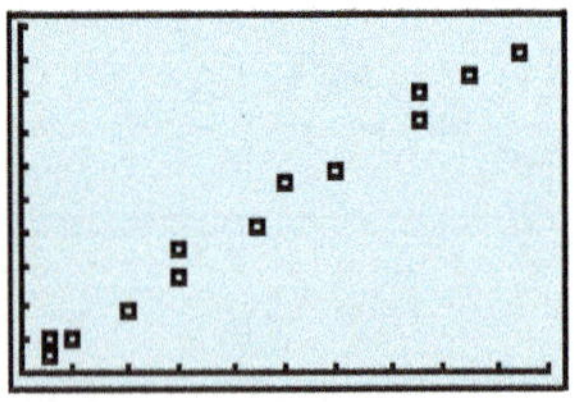

**4.**

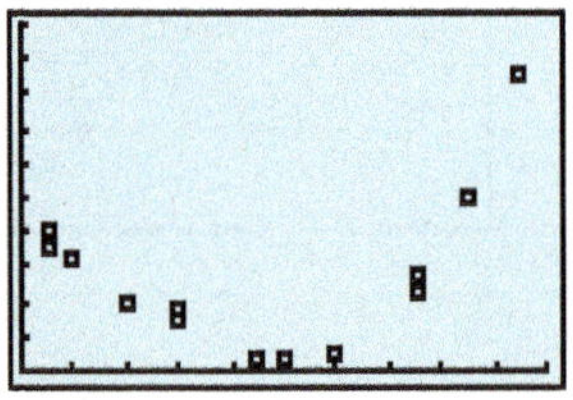

**5.**

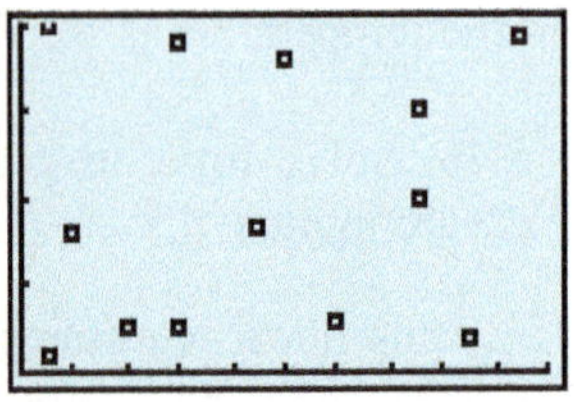

**6.**

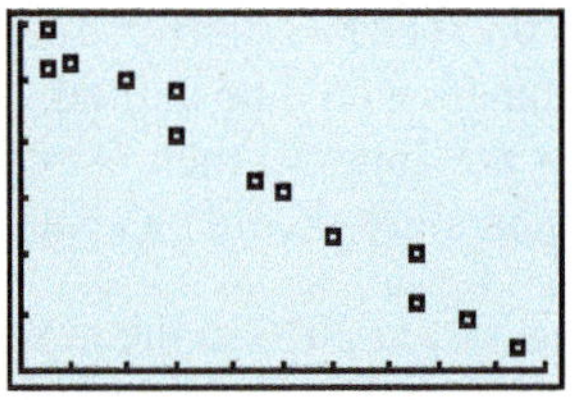

**7.**

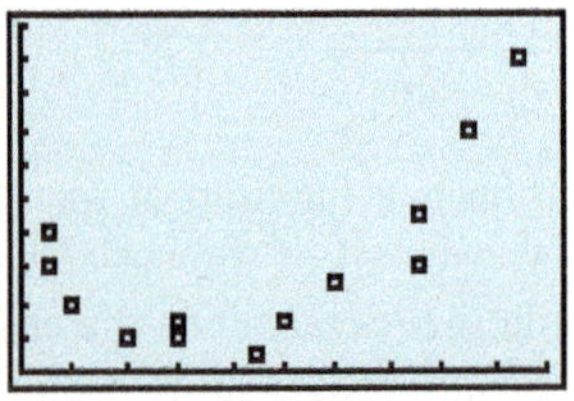

**8.**

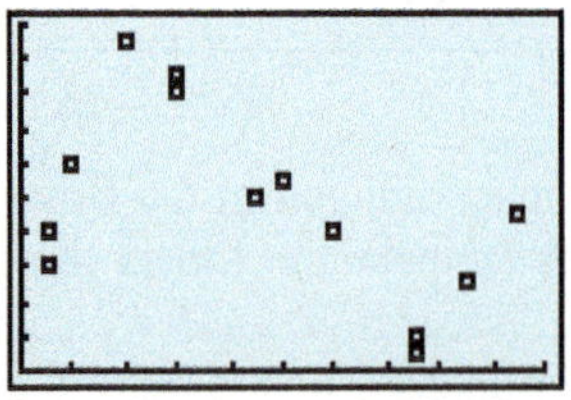

**9.**

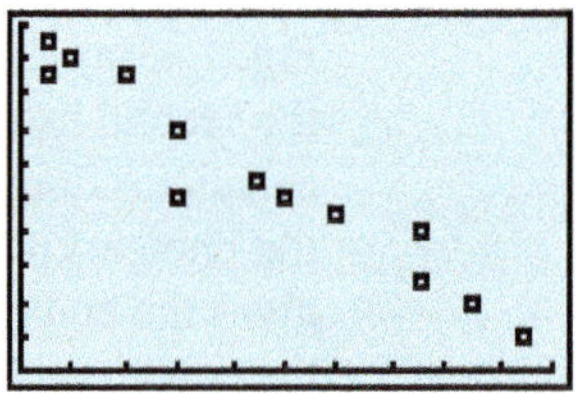

**10.**

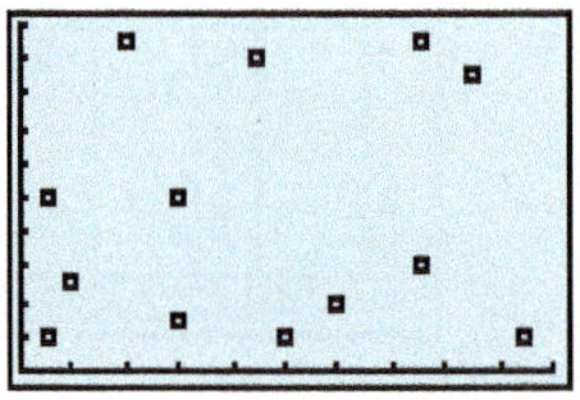

*Draw a scatter diagram for each given data set. Use graph paper. From your graph determine whether there is a linear relationship between the variables.*

**11.**

| Height (in.) | Weight (lb) |
|---|---|
| 24 | 52 |
| 30 | 62 |
| 32 | 66 |
| 18 | 40 |
| 22 | 50 |
| 36 | 68 |
| 40 | 78 |
| 28 | 60 |
| 22 | 42 |

**12.**

| Age (yr) | Income (thousands of $) |
|---|---|
| 33 | 24 |
| 30 | 23 |
| 32 | 26 |
| 30 | 22 |
| 38 | 36 |
| 35 | 34 |
| 32 | 30 |
| 40 | 44 |
| 39 | 42 |

**13.**

| ACT Score | Grade Point Average |
|---|---|
| 18 | 2.2 |
| 26 | 3.4 |
| 20 | 2.8 |
| 17 | 2.0 |
| 19 | 2.6 |
| 23 | 3.2 |
| 25 | 3.6 |
| 30 | 4.0 |

**14.**

| Distance (mi) | Time (min) |
|---|---|
| 8 | 34 |
| 5 | 20 |
| 6 | 26 |
| 16 | 70 |
| 12 | 50 |
| 6 | 32 |
| 20 | 90 |
| 18 | 70 |

**15.**

| Height (in.) | Grade Point Average |
|---|---|
| 61 | 4.0 |
| 59 | 1.9 |
| 66 | 3.8 |
| 72 | 1.6 |
| 62 | 2.5 |
| 74 | 3.8 |
| 60 | 1.2 |
| 68 | 2.6 |
| 65 | 2.0 |

**16.**

| Price (thousands of $) | Miles per gal |
|---|---|
| 16 | 40 |
| 18 | 20 |
| 22 | 16 |
| 24 | 33 |
| 29 | 8 |
| 32 | 30 |
| 35 | 40 |
| 38 | 12 |
| 40 | 66 |

*Draw a scatter diagram using the given ordered pairs and a line that you think fits the data. Use graph paper. Complete the missing entries in the table by reading them from your graph. Answers may vary.*

**17.**

| Weight (thousands of lb) | Stopping Distance (ft) |
|---|---|
| 2.0 | 96 |
| 2.4 | 110 |
| 2.7 | 130 |
| 2.8 | 130 |
| 2.8 | 142 |
| ? | 160 |
| 3.7 | 180 |
| 3.8 | 180 |
| 4.0 | ? |

**18.**

| Prime Rate (%) | Unemployment Rate (%) |
|---|---|
| 8.2 | 6.1 |
| 5.1 | 3.0 |
| ? | 4.6 |
| 9.7 | 7.7 |
| 5.5 | 3.2 |
| 7.3 | 5.2 |
| 5.5 | 3.7 |
| 7.8 | 5.6 |
| 9.9 | ? |

**19.**

| Pressure ($lb/in.^2$) | Volume ($in.^3$) |
|---|---|
| 122 | 40 |
| 110 | 50 |
| 125 | 38 |
| 132 | ? |
| 141 | 30 |
| 138 | 32 |
| 117 | 40 |
| 144 | 26 |
| ? | 25 |

**20.**

| Rate (ft/sec) | Time (sec) |
|---|---|
| 10.0 | 6 |
| 10.1 | 6 |
| 10.2 | 5.7 |
| 10.3 | ? |
| 10.4 | 5.6 |
| 10.5 | 5.4 |
| 10.6 | 5.5 |
| 10.7 | 5.4 |
| 10.8 | 5.2 |
| ? | 5.0 |

*Solve each problem.*

**21.** *Fossil Fuels* The amount of coal used to generate electricity in the United States is shown in the accompanying table (Energy Information Administration, www.eia.doe.gov).

**Table for Exercise 21**

| Year | Amount (thousand tons) |
|---|---|
| 2001 | 964,433 |
| 2002 | 977,507 |
| 2003 | 1,005,116 |
| 2004 | 1,016,268 |
| 2005 | 1,037,485 |
| 2006 | 1,026,636 |
| 2007 | 1,046,424 |

**a.** Use linear regression on your graphing calculator to find the regression equation. Let $x = 1$ correspond to 2001.

**b.** Use the regression equation to predict the amount of coal used for electricity in 2010.

**c.** Interpret the slope of the regression equation.

**22.** *Earthquakes* The accompanying table shows the number of earthquakes worldwide with magnitude between 5.0 and 9.9 (U.S. Geological Survey, www.usgs.gov).

**Table for Exercise 22**

| Year | Number of Earthquakes |
|---|---|
| 2000 | 1518 |
| 2001 | 1385 |
| 2002 | 1361 |
| 2003 | 1358 |
| 2004 | 1672 |
| 2005 | 1844 |
| 2006 | 1865 |
| 2007 | 2190 |

**a.** Use linear regression on your graphing calculator to find the regression equation. Let $x = 0$ correspond to 2000.

**b.** Use the regression equation to predict the number of earthquakes in 2015.

**c.** According to the regression equation, in what year were there no earthquakes?

**23.** *Money Supply* The M1 money supply is a measure of money in its most liquid forms, including currency and checking accounts. The accompanying table shows the M1 in billions of dollars in the United States (www.economagic.com).

**Table for Exercise 23**

| Year | M1 Money Supply ($ billions) |
|---|---|
| 2001 | 1121 |
| 2002 | 1096 |
| 2003 | 1221 |
| 2004 | 1290 |
| 2005 | 1354 |
| 2006 | 1375 |
| 2007 | 1369 |
| 2008 | 1388 |

**a.** Use linear regression on your graphing calculator to find the regression equation. Let $x = 1$ correspond to 2001.

**b.** Use the regression equation to predict the year in which the M1 money supply will reach $2 trillion.

**c.** According to the regression equation, what was the M1 money supply in 1995?

**24.** *Price of Oil* The accompanying table shows the price of West Texas Intermediate Crude at the beginning of each year for the years 2000 through 2008 (www.economagic.com).

**Table for Exercise 24**

| Year | Price ($/barrel) |
|---|---|
| 2000 | 27.18 |
| 2001 | 29.58 |
| 2002 | 19.67 |
| 2003 | 32.94 |
| 2004 | 34.27 |
| 2005 | 46.84 |
| 2006 | 65.51 |
| 2007 | 54.57 |
| 2008 | 92.95 |

**a.** Use linear regression on your graphing calculator to find the regression equation. Let $x = 0$ correspond to 2000.

**b.** Use the regression equation to predict the year in which the price will reach $200 per barrel.

**c.** According to the regression equation, what was the price in 1999?

**25.** *Injuries in the Boston Marathon* A study of the 4386 male runners in the Boston Marathon showed a high correlation between age and the percentage of injured runners for men aged 59 and below (*The Boston Globe,* www.boston.com/marathon). The bar graph shows the percentage of injured men age 59 and below grouped into five categories. There appears to be a relationship between the age group $A$ and the percentage $p$ of those injured. Use the linear regression feature of a graphing calculator to find an equation that expresses $p$ in terms of $A$. Find the percentage of runners that are predicted to be injured in age group 4 according to the equation and compare the answer to the actual percentage injured.

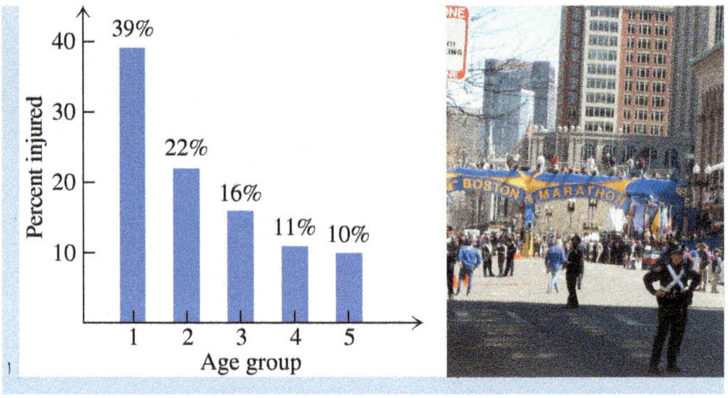


1: under 20, 2: 20–29, 3: 30–39, 4: 40–49, 5: 50–59

**Figure for Exercise 25**

**26.** *Injuries in the Boston Marathon* Use the equation of Exercise 25 to predict the percentage of runners injured in age group 6 (60-plus). Actually, 21% of the runners in the 60-plus age group were injured. Can you offer an explanation for the difference between the predicted and the actual figures?

## ▶ RETHINKING

**27.** Solve $(x - 5)^2 = (x + 7)^2$.

**28.** The length of a rectangle is 3 feet more than its width. If the perimeter is 46 feet, then what is the width?

**29.** Find the center and radius of the circle $x^2 + 8x + y^2 - 10y = 0$.

**30.** Write the equation of the line through $(-1, -2)$ and $(3, 1)$ in slope-intercept form.

**31.** Find the slope of the line $3x - 5y = -11$.

**32.** Write the equation of the line through $(5, -2)$ with slope $(-1/2)$ in standard form using only integers.

## THINKING OUTSIDE THE BOX VI

*Counting Coworkers* Chris and Pat work at Tokyo Telemarketing. One day Chris said to Pat, "19/40 of my coworkers are female." Pat replied, "That's strange, 12/25 of my coworkers are female." If both are correct, then how many workers are there at Tokyo Telemarketing and what are the genders of Chris and Pat?

## POP QUIZ 1.5

1. Use your calculator to find the regression equation for the data given in the table:

| Year | 01 | 02 | 03 | 04 | 05 |
|---|---|---|---|---|---|
| Cost ($) | 10 | 12 | 13 | 16 | 17 |

2. Use the regression equation to estimate the cost in 2010.

3. Use the regression equation to estimate the cost in 1998.

# 1.6 Complex Numbers

Our system of numbers developed as the need arose. Numbers were first used for counting. As society advanced, the rational numbers were formed to express fractional parts and ratios. Negative numbers were invented to express losses or debts. When it was discovered that the exact size of some very real objects could not be expressed with rational numbers, the irrational numbers were added to the system, forming the set of real numbers. The set of real numbers was later expanded to the set of complex numbers, which includes the real and imaginary numbers. The imaginary numbers are the most abstract numbers in our system. They will be used in Section 1.7 to solve quadratic equations and again in Chapter 3 when we study polynomial functions and equations.

## Definitions

There are no even roots of negative numbers in the set of real numbers. So, the real numbers are inadequate or incomplete in this regard. The imaginary numbers were invented to complete the set of real numbers. Using real and imaginary numbers, every nonzero real number has *two* square roots, *three* cube roots, *four* fourth roots, and so on. (Actually finding all of the roots of any real number is done in trigonometry.)

The imaginary numbers are based on the symbol $\sqrt{-1}$. Since there is no real number whose square is $-1$, a new number called $i$ is defined such that $i^2 = -1$.

**Definition: Imaginary Number $i$**

The **imaginary number** $i$ is defined by

$$i^2 = -1.$$

We may also write $i = \sqrt{-1}$.

A complex number is formed as a real number plus a real multiple of $i$.

**Definition: Complex Numbers**

The set of **complex numbers** is the set of all numbers of the form $a + bi$, where $a$ and $b$ are real numbers.

In $a + bi$, $a$ is called the **real part** and $b$ is called the **imaginary part.** Two complex numbers $a + bi$ and $c + di$ are **equal** if and only if their real parts are equal $(a = c)$ and their imaginary parts are equal $(b = d)$. If $b = 0$, then $a + bi$ is a **real number.** If $b \neq 0$, then $a + bi$ is an imaginary number.

The form $a + bi$ is the **standard form** of a complex number, but for convenience we use a few variations of that form. If either the real or imaginary part of a complex number is 0, then that part can be omitted. For example,

$$0 + 3i = 3i, \qquad 2 + 0i = 2, \qquad \text{and} \qquad 0 + 0i = 0.$$

If $b$ is a radical, then $i$ is usually written before $b$. For example, we write $2 + i\sqrt{3}$ rather than $2 + \sqrt{3}i$, which could be confused with $2 + \sqrt{3i}$. If $b$ is negative, a subtraction symbol can be used as in $3 + (-2)i = 3 - 2i$. A complex number with fractions, such as $\frac{1}{3} - \frac{2}{3}i$, may be written as $\frac{1 - 2i}{3}$.

### EXAMPLE 1 Standard form of a complex number

Determine whether each complex number is real or imaginary and write it in the standard form $a + bi$.

**a.** $3i$ **b.** $87$ **c.** $4 - 5i$ **d.** $0$ **e.** $\dfrac{1 + \pi i}{2}$

**Solution**

**a.** The complex number $3i$ is imaginary, and $3i = 0 + 3i$.
**b.** The complex number 87 is a real number, and $87 = 87 + 0i$.
**c.** The complex number $4 - 5i$ is imaginary, and $4 - 5i = 4 + (-5)i$.
**d.** The complex number 0 is real, and $0 = 0 + 0i$.
**e.** The complex number $\frac{1 + \pi i}{2}$ is imaginary, and $\frac{1 + \pi i}{2} = \frac{1}{2} + \frac{\pi}{2}i$.

▶**TRY THIS.** Determine whether $i - 5$ is real or imaginary and write it in standard form. ■

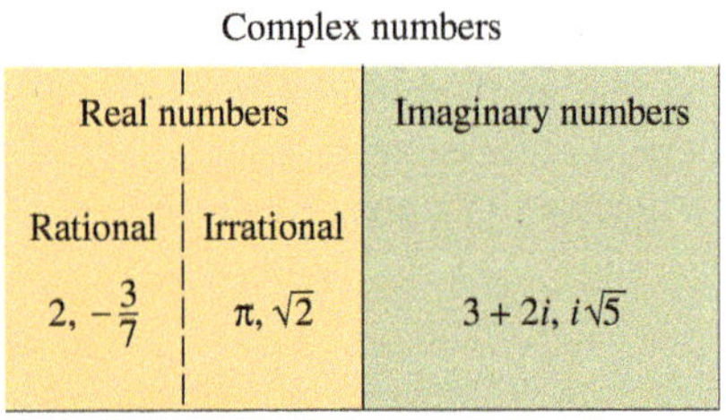


Figure 1.46

The real numbers can be classified as rational or irrational. The complex numbers can be classified as real or imaginary. The relationship between these sets of numbers is shown in Fig. 1.46.

## Addition, Subtraction, and Multiplication

Now that we have defined complex numbers, we define the operations of arithmetic with them.

**Definition: Addition, Subtraction, and Multiplication**

If $a + bi$ and $c + di$ are complex numbers, we define their sum, difference, and product as follows.

$$(a + bi) + (c + di) = (a + c) + (b + d)i$$
$$(a + bi) - (c + di) = (a - c) + (b - d)i$$
$$(a + bi)(c + di) = (ac - bd) + (bc + ad)i$$

It is not necessary to memorize these definitions, because the results can be obtained by performing the operations as if the complex numbers were binomials with $i$ being a variable, replacing $i^2$ with $-1$ wherever it occurs.

### EXAMPLE 2 Operations with complex numbers

Perform the indicated operations with the complex numbers.

**a.** $(-2 + 3i) + (-4 - 9i)$ **b.** $(-1 - 5i) - (3 - 2i)$ **c.** $2i(3 + i)$
**d.** $(3i)^2$ **e.** $(-3i)^2$ **f.** $(5 - 2i)(5 + 2i)$

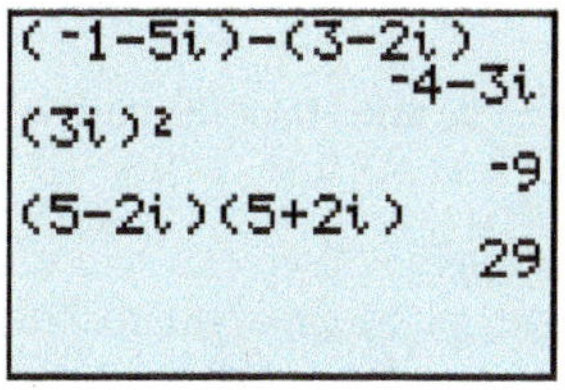


Figure 1.47

### Solution

**a.** $(-2 + 3i) + (-4 - 9i) = -2 + (-4) + 3i + (-9i) = -6 - 6i$
**b.** $(-1 - 5i) - (3 - 2i) = -1 - 5i - 3 + 2i = -4 - 3i$
**c.** $2i(3 + i) = 6i + 2i^2 = 6i + 2(-1) = -2 + 6i$
**d.** $(3i)^2 = 3^2i^2 = 9(-1) = -9$
**e.** $(-3i)^2 = (-3)^2i^2 = 9(-1) = -9$
**f.** $(5 - 2i)(5 + 2i) = 25 - 4i^2 = 25 - 4(-1) = 29$

Check these results with a calculator that does computations with complex numbers, as in Fig. 1.47.

▶**TRY THIS.** Find the product $(4 - 3i)(1 + 2i)$. ■

We can find whole-number powers of $i$ by using the definition of multiplication. Since $i^1 = i$ and $i^2 = -1$, we have

$$i^3 = i^1 \cdot i^2 = i(-1) = -i \quad \text{and} \quad i^4 = i^1 \cdot i^3 = i(-i) = -i^2 = -(-1) = 1.$$

The first eight powers of $i$ are listed here.

$$i^1 = i \qquad i^2 = -1 \qquad i^3 = -i \qquad i^4 = 1$$
$$i^5 = i \qquad i^6 = -1 \qquad i^7 = -i \qquad i^8 = 1$$

This list could be continued in this pattern, but any other whole-number power of $i$ can be obtained from knowing the first four powers. We can simplify a power of $i$ by using the fact that $i^4 = 1$ and $(i^4)^n = 1$ for any integer $n$.

## EXAMPLE 3 Simplifying a power of *i*

Simplify.

**a.** $i^{83}$ **b.** $i^{-46}$

### Solution

**a.** Divide 83 by 4 and write $83 = 4 \cdot 20 + 3$. So

$$i^{83} = (i^4)^{20} \cdot i^3 = 1^{20} \cdot i^3 = 1 \cdot i^3 = -i.$$

**b.** Since $-46 = 4(-12) + 2$, we have

$$i^{-46} = (i^4)^{-12} \cdot i^2 = 1^{-12} \cdot i^2 = 1(-1) = -1.$$

▶**TRY THIS.** Simplify $i^{35}$. ■

## Division of Complex Numbers

The complex numbers $a + bi$ and $a - bi$ are called **complex conjugates** of each other.

## EXAMPLE 4 Complex conjugates

Find the product of the given complex number and its conjugate.

**a.** $3 - i$ **b.** $4 + 2i$ **c.** $-i$

### Solution

**a.** The conjugate of $3 - i$ is $3 + i$, and $(3 - i)(3 + i) = 9 - i^2 = 10$.
**b.** The conjugate of $4 + 2i$ is $4 - 2i$, and $(4 + 2i)(4 - 2i) = 16 - 4i^2 = 20$.
**c.** The conjugate of $-i$ is $i$, and $-i \cdot i = -i^2 = 1$.

▶**TRY THIS.** Find the product of $3 - 5i$ and its conjugate. ■

In general we have the following theorem about complex conjugates.

**Theorem: Complex Conjugates**

If $a$ and $b$ are real numbers, then the product of $a + bi$ and its conjugate $a - bi$ is the real number $a^2 + b^2$. In symbols,

$$(a + bi)(a - bi) = a^2 + b^2.$$

We use the theorem about complex conjugates to divide imaginary numbers, in a process that is similar to rationalizing a denominator.

### EXAMPLE 5 Dividing imaginary numbers

Write each quotient in the form $a + bi$.

**a.** $\dfrac{8 - i}{2 + i}$ **b.** $\dfrac{1}{5 - 4i}$ **c.** $\dfrac{3 - 2i}{i}$

#### Solution

**a.** Multiply the numerator and denominator by $2 - i$, the conjugate of $2 + i$:

$$\frac{8 - i}{2 + i} = \frac{(8 - i)(2 - i)}{(2 + i)(2 - i)} = \frac{16 - 10i + i^2}{4 - i^2} = \frac{15 - 10i}{5} = 3 - 2i$$

Check division using multiplication: $(3 - 2i)(2 + i) = 8 - i$.

**b.** $$\frac{1}{5 - 4i} = \frac{1(5 + 4i)}{(5 - 4i)(5 + 4i)} = \frac{5 + 4i}{25 + 16} = \frac{5 + 4i}{41} = \frac{5}{41} + \frac{4}{41}i$$

Check: $$\left(\frac{5}{41} + \frac{4}{41}i\right)(5 - 4i) = \frac{25}{41} - \frac{20}{41}i + \frac{20}{41}i - \frac{16}{41}i^2$$

$$= \frac{25}{41} + \frac{16}{41} = 1.$$

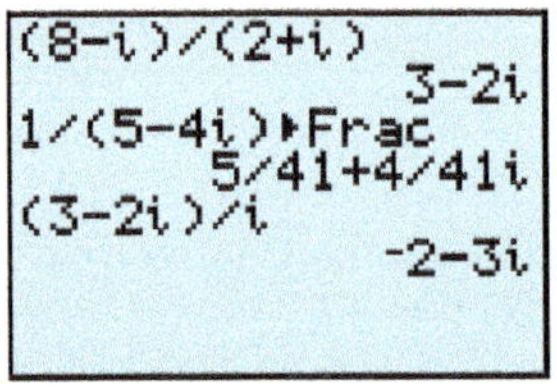


Figure 1.48

You can also check with a calculator that handles complex numbers, as in Fig. 1.48. □

**c.** $$\frac{3 - 2i}{i} = \frac{(3 - 2i)(-i)}{i(-i)} = \frac{-3i + 2i^2}{-i^2} = \frac{-2 - 3i}{1} = -2 - 3i$$

Check: $(-2 - 3i)(i) = -2i - 3i^2 = 3 - 2i$.

▶**TRY THIS.** Write $\dfrac{4}{1 + i}$ in the form $a + bi$. ■

## Roots of Negative Numbers

In Examples 2(d) and 2(e), we saw that both $(3i)^2 = -9$ and $(-3i)^2 = -9$. This means that in the complex number system there are two square roots of $-9$, $3i$ and $-3i$. For any positive real number $b$, we have $(i\sqrt{b})^2 = -b$ and $(-i\sqrt{b})^2 = -b$. So there are two square roots of $-b$. They are $i\sqrt{b}$ and $-i\sqrt{b}$. We call $i\sqrt{b}$ the **principal square root** of $-b$ and make the following definition.

**Definition: Square Root of a Negative Number**

For any positive real number $b$, $\sqrt{-b} = i\sqrt{b}$.

In the real number system, $\sqrt{-2}$ and $\sqrt{-8}$ are undefined, but in the complex number system they are defined as $\sqrt{-2} = i\sqrt{2}$ and $\sqrt{-8} = i\sqrt{8}$. Even though we now have meaning for a symbol such as $\sqrt{-2}$, *all operations with complex numbers must be performed after converting to the $a + bi$ form.* If we perform operations with roots of negative numbers using properties of the real numbers, we can get contradictory results:

$$\sqrt{-2} \cdot \sqrt{-8} = \sqrt{(-2)(-8)} = \sqrt{16} = 4 \quad \text{Incorrect}$$
$$i\sqrt{2} \cdot i\sqrt{8} = i^2 \cdot \sqrt{16} = -4 \quad \text{Correct}$$

The product rule $\sqrt{a} \cdot \sqrt{b} = \sqrt{ab}$ is used *only* for nonnegative numbers $a$ and $b$.

## EXAMPLE 6 Square roots of negative numbers

Write each expression in the form $a + bi$, where $a$ and $b$ are real numbers.

**a.** $\sqrt{-8} + \sqrt{-18}$

**b.** $\dfrac{-4 + \sqrt{-50}}{4}$

**c.** $\sqrt{-27}(\sqrt{9} - \sqrt{-2})$

### Solution

The first step in each case is to replace the square roots of negative numbers by expressions with $i$.

**a.** $\sqrt{-8} + \sqrt{-18} = i\sqrt{8} + i\sqrt{18} = 2i\sqrt{2} + 3i\sqrt{2}$
$= 5i\sqrt{2}$

**b.** $\dfrac{-4 + \sqrt{-50}}{4} = \dfrac{-4 + i\sqrt{50}}{4} = \dfrac{-4 + 5i\sqrt{2}}{4}$
$= -1 + \dfrac{5}{4}i\sqrt{2}$

**c.** $\sqrt{-27}(\sqrt{9} - \sqrt{-2}) = 3i\sqrt{3}(3 - i\sqrt{2}) = 9i\sqrt{3} - 3i^2\sqrt{6}$
$= 3\sqrt{6} + 9i\sqrt{3}$

▶**TRY THIS.** Write $\dfrac{2 - \sqrt{-12}}{2}$ in the form $a + bi$. ■

## FOR thought... True or False? Explain.

1. The multiplicative inverse of $i$ is $-i$.
2. The conjugate of $i$ is $-i$.
3. The set of complex numbers is a subset of the set of real numbers.
4. $(\sqrt{3} - i\sqrt{2})(\sqrt{3} + i\sqrt{2}) = 5$
5. $(2 + 5i)(2 + 5i) = 4 + 25$
6. $5 - \sqrt{-9} = 5 - 9i$
7. $(3i)^2 + 9 = 0$
8. $(-3i)^2 + 9 = 0$
9. $i^4 = 1$
10. $i^{18} = 1$

# EXERCISES 1.6

*Fill in the blank.*

**1.** Numbers of the form $a + bi$ where $a$ and $b$ are real numbers are ________.

**2.** In $a + bi$, $a$ is the ________ and $b$ is the ________.

**3.** A number of the form $a + bi$ with $b \neq 0$ is an ________.

**4.** The ________ square root of $-b$ where $b$ is a positive real number is $i\sqrt{b}$.

*Determine whether each complex number is real or imaginary and write it in the standard form a + bi.*

**5.** $6i$ **6.** $-3i + \sqrt{6}$ **7.** $\frac{1+i}{3}$

**8.** $-72$ **9.** $\sqrt{7}$ **10.** $-i\sqrt{5}$

**11.** $\frac{\pi}{2}$ **12.** $0$

*Perform the indicated operations and write your answers in the form a + bi, where a and b are real numbers.*

**13.** $(3 - 3i) + (4 + 5i)$ **14.** $(-3 + 2i) + (5 - 6i)$

**15.** $(1 - i) - (3 + 2i)$ **16.** $(6 - 7i) - (3 - 4i)$

**17.** $(1 - i\sqrt{2}) + (3 + 2i\sqrt{2})$

**18.** $(5 + 3i\sqrt{5}) + (-4 - 5i\sqrt{5})$

**19.** $\left(5 + \frac{1}{3}i\right) - \left(\frac{1}{2} - \frac{1}{2}i\right)$ **20.** $\left(\frac{1}{2} - \frac{2}{3}i\right) - \left(3 - \frac{1}{4}i\right)$

**21.** $-6i(3 - 2i)$ **22.** $-3i(5 + 2i)$

**23.** $(2 - 3i)(4 + 6i)$ **24.** $(3 - i)(5 - 2i)$

**25.** $(4 - 5i)(6 + 2i)$ **26.** $(3 + 7i)(2 + 5i)$

**27.** $(5 - 2i)(5 + 2i)$ **28.** $(4 + 3i)(4 - 3i)$

**29.** $(\sqrt{3} - i)(\sqrt{3} + i)$

**30.** $(\sqrt{2} + i\sqrt{3})(\sqrt{2} - i\sqrt{3})$

**31.** $(3 + 4i)^2$ **32.** $(-6 - 2i)^2$

**33.** $(\sqrt{5} - 2i)^2$ **34.** $(\sqrt{6} + i\sqrt{3})^2$

**35.** $i^{17}$ **36.** $i^{24}$ **37.** $i^{98}$ **38.** $i^{19}$

**39.** $i^{-1}$ **40.** $i^{-2}$ **41.** $i^{-3}$ **42.** $i^{-4}$

**43.** $i^{-13}$ **44.** $i^{-27}$ **45.** $i^{-38}$ **46.** $i^{-66}$

*Find the product of the given complex number and its conjugate.*

**47.** $3 - 9i$ **48.** $4 + 3i$ **49.** $\frac{1}{2} + 2i$

**50.** $\frac{1}{3} - i$ **51.** $i$ **52.** $-i\sqrt{5}$

**53.** $3 - i\sqrt{3}$ **54.** $\frac{5}{2} + i\frac{\sqrt{2}}{2}$

*Write each quotient in the form a + bi.*

**55.** $\frac{1}{2 - i}$ **56.** $\frac{1}{5 + 2i}$

**57.** $\frac{-3i}{1 - i}$ **58.** $\frac{3i}{-2 + i}$

**59.** $\frac{-3 + 3i}{i}$ **60.** $\frac{-2 - 4i}{-i}$

**61.** $\frac{1 - i}{3 + 2i}$ **62.** $\frac{4 + 2i}{2 - 3i}$

**63.** $\frac{2 - i}{3 + 5i}$ **64.** $\frac{4 + 2i}{5 - 3i}$

*Write each expression in the form a + bi, where a and b are real numbers.*

**65.** $\sqrt{-4} - \sqrt{-9}$ **66.** $\sqrt{-16} + \sqrt{-25}$

**67.** $\sqrt{-4} - \sqrt{16}$ **68.** $\sqrt{-3} \cdot \sqrt{-3}$

**69.** $(\sqrt{-6})^2$ **70.** $(\sqrt{-5})^3$

**71.** $\sqrt{-2} \cdot \sqrt{-50}$ **72.** $\frac{-6 + \sqrt{-3}}{3}$

**73.** $\frac{-2 + \sqrt{-20}}{2}$ **74.** $\frac{9 - \sqrt{-18}}{-6}$

**75.** $-3 + \sqrt{3^2 - 4(1)(5)}$ **76.** $1 - \sqrt{(-1)^2 - 4(1)(1)}$

**77.** $\sqrt{-8}(\sqrt{-2} + \sqrt{8})$ **78.** $\sqrt{-6}(\sqrt{2} - \sqrt{-3})$

*Evaluate the expression* $\frac{-b + \sqrt{b^2 - 4ac}}{2a}$ *for each choice of a, b, and c.*

**79.** $a = 1, b = 2, c = 5$ **80.** $a = 5, b = -4, c = 1$

**81.** $a = 2, b = 4, c = 3$ **82.** $a = 2, b = -4, c = 5$

*Evaluate the expression* $\frac{-b - \sqrt{b^2 - 4ac}}{2a}$ *for each choice of a, b, and c.*

**83.** $a = 1, b = 6, c = 17$ **84.** $a = 1, b = -12, c = 84$

**85.** $a = -2, b = 6, c = 6$ **86.** $a = 3, b = 6, c = 8$

*Perform the indicated operations. Write the answers in the form $a + bi$ where a and b are real numbers.*

**87.** $(3 - 5i)(3 + 5i)$

**88.** $(2 - 4i)(2 + 4i)$

**89.** $(3 - 5i) + (3 + 5i)$

**90.** $(2 - 4i) + (2 + 4i)$

**91.** $\frac{3 - 5i}{3 + 5i}$

**92.** $\frac{2 - 4i}{2 + 4i}$

**93.** $(6 - 2i) - (7 - 3i)$

**94.** $(5 - 6i) - (8 - 9i)$

**95.** $i^5(i^2 - 3i)$

**96.** $3i^7(i - 5i^3)$

## FOR WRITING/DISCUSSION

**97.** Explain in detail how to find $i^n$ for any positive integer $n$.

**98.** Find a number $a + bi$ such that $a^2 + b^2$ is irrational.

**99.** Let $w = a + bi$ and $\overline{w} = a - bi$, where $a$ and $b$ are real numbers. Show that $w + \overline{w}$ is real and that $w - \overline{w}$ is imaginary. Write sentences (containing no mathematical symbols) stating these results.

**100.** Is it true that the product of a complex number and its conjugate is a real number? Explain.

**101.** Prove that the reciprocal of $a + bi$, where $a$ and $b$ are not both zero, is $\frac{a}{a^2 + b^2} - \frac{b}{a^2 + b^2}i$.

**102.** Evaluate $i^{0!} + i^{1!} + i^{2!} + \cdots + i^{100!}$, where $n!$ (read "$n$ factorial") is the product of the integers from 1 through $n$ if $n \geq 1$ and $0! = 1$.

## ▶ RETHINKING

**103.** Solve $\frac{x}{2} + 4 = \frac{x}{6} - 3$.

**104.** Solve $\frac{1}{x - 1} + \frac{x}{x + 1} = \frac{x^2}{x^2 - 1}$.

**105.** Write a formula that expresses the width $W$ of a rectangle as a function of its perimeter $P$ and its length $L$.

**106.** Write a formula that expresses the diameter $d$ of a circle as a function of its circumference $C$.

**107.** E10 consists of 10% ethanol and 90% gasoline. E20 is 20% ethanol and 80% gasoline. How many gallons of ethanol must be added to 50 gallons of E10 to obtain E20?

**108.** How many gallons of gasoline must be added to 50 gallons of E20 to obtain E10? See the previous exercise.

## THINKING OUTSIDE THE BOX VII & VIII

*Reversing the Digits* Find a four-digit integer $x$ such that $4x$ is another four-digit integer whose digits are in the reverse order of the digits of $x$.

*Summing Reciprocals* There is only one way to write 1 as a sum of the reciprocals of three different positive integers:

$$\frac{1}{2} + \frac{1}{3} + \frac{1}{6} = 1$$

Find all possible ways to write 1 as a sum of the reciprocals of four different positive integers.

## ▶ POP QUIZ 1.6

1. Find the sum of $3 + 2i$ and $4 - i$.
2. Find the product of $4 - 3i$ and $2 + i$.
3. Find the product of $2 - 3i$ and its conjugate.
4. Write $\frac{5}{2 - 3i}$ in the form $a + bi$.
5. Find $i^{27}$.
6. What are the two square roots of $-16$?

# 1.7 Quadratic Equations

One of our main goals is to solve polynomial equations. In Section 1.1 we learned to solve equations of the form $ax + b = 0$, the linear equations. Linear equations are first-degree polynomial equations. In this section we will solve second-degree polynomial equations, the quadratic equations.

### Definition

Quadratic equations have second-degree terms that linear equations do not have.

**Definition: Quadratic Equation**

> A **quadratic equation** is an equation of the form
> $$ax^2 + bx + c = 0,$$
> where $a$, $b$, and $c$ are real numbers with $a \neq 0$.

The condition that $a \neq 0$ in the definition ensures that the equation actually does have an $x^2$-term. There are several methods for solving quadratic equations. Which method is most appropriate depends on the type of quadratic equation that we are solving. We first consider the factoring method.

## Solving Quadratic Equations by Factoring

Many second-degree polynomials can be factored as a product of first-degree binomials. When one side of an equation is a product and the other side is 0, we can write an equivalent statement by setting each factor equal to zero. This idea is called the **zero factor property.**

**The Zero Factor Property**

> If $A$ and $B$ are algebraic expressions, then the equation $AB = 0$ is equivalent to the compound statement $A = 0$ or $B = 0$.

### EXAMPLE 1 Quadratic equations solved by factoring

Solve each equation by factoring.

**a.** $x^2 - x - 12 = 0$
**b.** $(x + 3)(x - 4) = 8$

**Solution**

**a.**

$$x^2 - x - 12 = 0$$

$$(x - 4)(x + 3) = 0 \qquad \text{Factor the left-hand side.}$$

$$x - 4 = 0 \quad \text{or} \quad x + 3 = 0 \qquad \text{Zero factor property}$$

$$x = 4 \quad \text{or} \quad x = -3$$

Check: $(-3)^2 - (-3) - 12 = 0$ and $4^2 - 4 - 12 = 0$. The solution set is $\{-3, 4\}$.

The graph of $y = x^2 - x - 12$ intersects the $x$-axis at $(-3, 0)$ and $(4, 0)$ as shown in Fig. 1.49. □

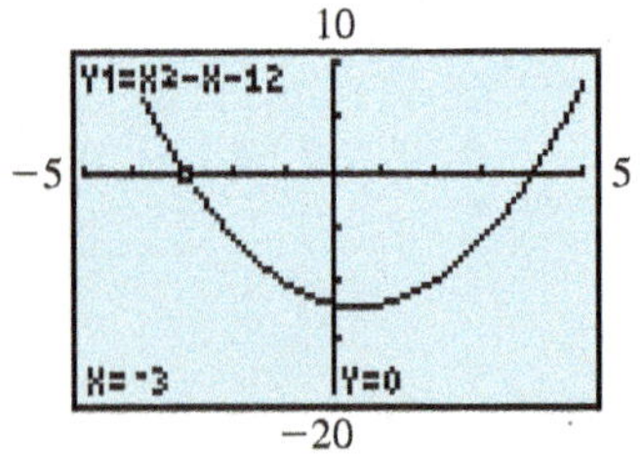


Figure 1.49

**b.** We must first rewrite the equation with 0 on one side, because the zero factor property applies only when the factors have a product of 0.

$$(x + 3)(x - 4) = 8$$

$$x^2 - x - 12 = 8 \qquad \text{Multiply on the left-hand side.}$$

$$x^2 - x - 20 = 0 \qquad \text{Get 0 on the right-hand side.}$$

$$(x - 5)(x + 4) = 0 \qquad \text{Factor.}$$

$$x - 5 = 0 \quad \text{or} \quad x + 4 = 0 \qquad \text{Zero factor property}$$

$$x = 5 \quad \text{or} \quad x = -4$$

Check in the original equation. The solution set is $\{-4, 5\}$.

The graph of $y = x^2 - x - 20$ intersects the $x$-axis at $(-4, 0)$ and $(5, 0)$ as shown in Fig. 1.50.

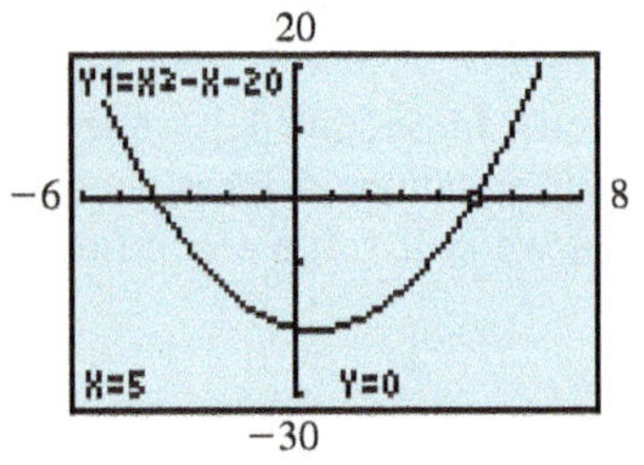


Figure 1.50

▸**TRY THIS.** Solve $x^2 - 7x - 18 = 0$ by factoring. ■

## The Square Root Property

An equation of the type $x^2 = k$ can be solved by factoring a difference of two squares:

$$x^2 - k = 0$$

$$(x + \sqrt{k})(x - \sqrt{k}) = 0 \quad \text{Factor.}$$

$$x + \sqrt{k} = 0 \quad \text{or} \quad x - \sqrt{k} = 0 \quad \text{Zero factor property}$$

$$x = -\sqrt{k} \quad \text{or} \quad x = \sqrt{k}$$

If $k > 0$, then $x^2 = k$ has two real solutions. If $k < 0$, then $\sqrt{k}$ is an imaginary number and $x^2 = k$ has two imaginary solutions. If $k = 0$, then 0 is the only solution to $x^2 = k$. Instead of solving equations of this type by factoring, it is worth remembering this result as the square root property.

**The Square Root Property**

For any real number $k$, the equation $x^2 = k$ is equivalent to $x = \pm\sqrt{k}$.

### EXAMPLE 2 Using the square root property

Find all complex solutions to each equation.

**a.** $x^2 - 9 = 0$
**b.** $2x^2 - 1 = 0$
**c.** $(2x - 1)^2 = 0$
**d.** $(x - 3)^2 + 8 = 0$

#### Solution

**a.** Before using the square root property, isolate $x^2$.

$$x^2 - 9 = 0$$

$$x^2 = 9$$

$$x = \pm\sqrt{9} = \pm 3 \quad \text{Square root property}$$

Check: $3^2 - 9 = 0$ and $(-3)^2 - 9 = 0$. The solution set is $\{-3, 3\}$.

**b.** $2x^2 - 1 = 0$

$$2x^2 = 1$$

$$x^2 = \frac{1}{2}$$

$$x = \pm\sqrt{\frac{1}{2}} = \pm\frac{1}{\sqrt{2}} = \pm\frac{1 \cdot \sqrt{2}}{\sqrt{2} \cdot \sqrt{2}} = \pm\frac{\sqrt{2}}{2}$$

Check: $2\left(\pm\frac{\sqrt{2}}{2}\right)^2 - 1 = 2\left(\frac{1}{2}\right) - 1 = 0$. The solution set is $\left\{\pm\frac{\sqrt{2}}{2}\right\}$.

**c.** $(2x - 1)^2 = 0$

$$2x - 1 = \pm\sqrt{0} \quad \text{Square root property}$$

$$2x - 1 = 0$$

$$2x = 1$$

$$x = \frac{1}{2}$$

The solution set is $\left\{\frac{1}{2}\right\}$.

**d.** $(x - 3)^2 + 8 = 0$

$$(x - 3)^2 = -8$$

$$x - 3 = \pm\sqrt{-8} \quad \text{Square root property}$$

$$x = 3 \pm 2i\sqrt{2} \quad \sqrt{-8} = i\sqrt{8} = i\sqrt{4}\sqrt{2} = 2i\sqrt{2}$$

The solution set is $\{3 - 2i\sqrt{2}, 3 + 2i\sqrt{2}\}$. Note that there are no real solutions to this equation, because the square of every real number is nonnegative.

**▶TRY THIS.** Solve $(x - 3)^2 = 16$ by the square root property. ■

In Section 1.3 we learned that the solution to an equation in one variable corresponds to an $x$-intercept on a graph in two variables. Note that the solutions to $x^2 - 9 = 0$ and $(2x - 1)^2 = 0$ in Example 2 correspond to the $x$-intercepts on the graphs of $y = x^2 - 9$ and $y = (2x - 1)^2$ in Figs. 1.51(a) and 1.51(b). Because there are no real solutions to $(x - 3)^2 + 8 = 0$ in Example 2(d), the graph of $y = (x - 3)^2 + 8$ in Fig. 1.51(c) has no $x$-intercepts.

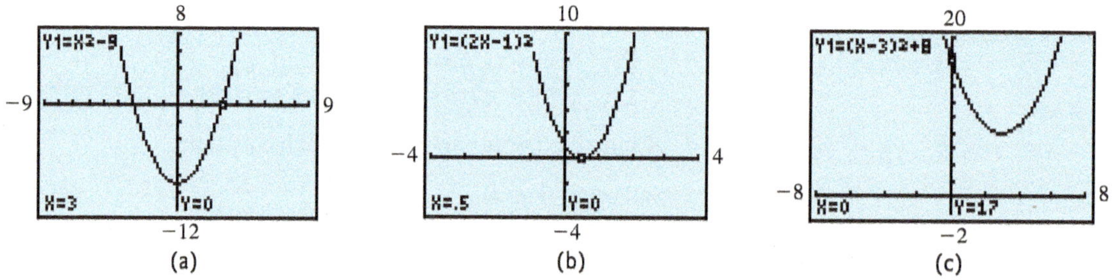


**Figure 1.51**

We will study graphs and their properties extensively in the coming chapters. In this chapter we will point out, using calculator graphs, how graphs in two variables can help us understand the solutions to equations and inequalities in one variable. So even if you are not using a graphing calculator, you should pay close attention to the calculator graphs shown in this chapter.

## Completing the Square

Solving quadratic equations by factoring is limited to equations that we can factor. However, we can solve any quadratic equation by completing the square and then using the square root property. We used completing the square in Section 1.3 to convert an equation of a circle into standard form.

Since it is simplest to complete the square when the leading coefficient of the polynomial is 1, we divide by the leading coefficient before completing the square. This is illustrated in part (b) of the next example.

### EXAMPLE 3 Solving a quadratic equation by completing the square

Solve each equation by completing the square.

**a.** $x^2 + 6x + 7 = 0$
**b.** $2x^2 - 3x - 4 = 0$

## Solution

**a.** Since $x^2 + 6x + 7$ is not a perfect square trinomial, we must find a perfect square trinomial that has $x^2 + 6x$ as its first two terms. Since one-half of 6 is 3 and $3^2 = 9$, our goal is to get $x^2 + 6x + 9$ on the left-hand side:

$$x^2 + 6x + 7 = 0$$

$$x^2 + 6x = -7 \qquad \text{Subtract 7 from each side.}$$

$$x^2 + 6x + 9 = -7 + 9 \qquad \text{Add 9 to each side.}$$

$$(x + 3)^2 = 2 \qquad \text{Factor the left-hand side.}$$

$$x + 3 = \pm\sqrt{2} \qquad \text{Square root property}$$

$$x = -3 \pm \sqrt{2}$$

Check in the original equation. The solution set is $\{-3 - \sqrt{2}, -3 + \sqrt{2}\}$.

**b.** $2x^2 - 3x - 4 = 0$

$$x^2 - \frac{3}{2}x - 2 = 0 \qquad \text{Divide each side by 2 to get } a = 1.$$

$$x^2 - \frac{3}{2}x = 2 \qquad \text{Add 2 to each side.}$$

$$x^2 - \frac{3}{2}x + \frac{9}{16} = 2 + \frac{9}{16} \qquad \frac{1}{2} \cdot \frac{3}{2} = \frac{3}{4} \text{ and } \left(\frac{3}{4}\right)^2 = \frac{9}{16}.$$

$$\left(x - \frac{3}{4}\right)^2 = \frac{41}{16} \qquad \text{Factor the left-hand side.}$$

$$x - \frac{3}{4} = \pm\frac{\sqrt{41}}{4} \qquad \text{Square root property}$$

$$x = \frac{3}{4} \pm \frac{\sqrt{41}}{4} = \frac{3 \pm \sqrt{41}}{4}$$

The solution set is $\left\{\frac{3 - \sqrt{41}}{4}, \frac{3 + \sqrt{41}}{4}\right\}$.

(a)

(b)

**Figure 1.52**

To check, enter $y_1 = 2x^2 - 3x - 4$ as in Fig. 1.52(a). On the home screen in Fig. 1.52(b), enter $y_1$ followed in parentheses by the $x$-value at which you want to calculate $y_1$.

▶**TRY THIS.** Solve $2x^2 - 4x - 1 = 0$ by completing the square. ■

**STRATEGY**

### Completing the Square

The goal is to get a perfect square trinomial on one side of the equation.

1. If the coefficient of $x$ is not 1, divide by it.
2. To complete $x^2 + bx$, add $(b/2)^2$ to both sides of the equation.
3. Factor the perfect square trinomial.
4. Apply the square root property.
5. Simplify the results.

## The Quadratic Formula

The method of completing the square can be applied to any quadratic equation

$$ax^2 + bx + c = 0, \qquad \text{where } a \neq 0.$$

Assume for now that $a > 0$, and divide each side by $a$.

$$x^2 + \frac{b}{a}x + \frac{c}{a} = 0$$

$$x^2 + \frac{b}{a}x = -\frac{c}{a} \qquad \text{Subtract } \tfrac{c}{a} \text{ from each side.}$$

$$x^2 + \frac{b}{a}x + \frac{b^2}{4a^2} = -\frac{c}{a} + \frac{b^2}{4a^2} \qquad \tfrac{1}{2}\cdot\tfrac{b}{a} = \tfrac{b}{2a} \text{ and } \left(\tfrac{b}{2a}\right)^2 = \tfrac{b^2}{4a^2}.$$

Now factor the perfect square trinomial on the left-hand side. On the right-hand side get a common denominator and add.

$$\left(x + \frac{b}{2a}\right)^2 = \frac{b^2 - 4ac}{4a^2} \qquad \tfrac{c}{a}\cdot\tfrac{4a}{4a} = \tfrac{4ac}{4a^2}$$

$$x + \frac{b}{2a} = \pm\sqrt{\frac{b^2 - 4ac}{4a^2}} \qquad \text{Square root property, assuming } b^2 - 4ac \geq 0$$

$$x = -\frac{b}{2a} \pm \frac{\sqrt{b^2 - 4ac}}{2a} \qquad \text{Because } a > 0,\ \sqrt{4a^2} = 2a.$$

$$x = \frac{-b \pm \sqrt{b^2 - 4ac}}{2a}$$

We assumed that $a > 0$ so that $\sqrt{4a^2} = 2a$ would be correct. If $a$ is negative, then $\sqrt{4a^2} = -2a$, and we get

$$x = -\frac{b}{2a} \pm \frac{\sqrt{b^2 - 4ac}}{-2a}.$$

However, the negative sign in $-2a$ can be deleted because of the $\pm$ symbol preceding it. For example, $5 \pm (-3)$ gives the same values as $5 \pm 3$. After deleting the negative sign on $-2a$, we get the same formula for the solution. It is called the **quadratic formula.** Its importance lies in its wide applicability. Any quadratic equation can be solved by using this formula.

## The Quadratic Formula

The solutions to $ax^2 + bx + c = 0$, with $a \neq 0$, are given by the formula

$$x = \frac{-b \pm \sqrt{b^2 - 4ac}}{2a}.$$

### EXAMPLE 4 Using the quadratic formula

Use the quadratic formula to find all real or imaginary solutions to each equation.

**a.** $x^2 + 8x + 6 = 0$ **b.** $x^2 - 6x + 11 = 0$ **c.** $4x^2 + 9 = 12x$

#### Solution

**a.** For $x^2 + 8x + 6 = 0$ we use $a = 1$, $b = 8$, and $c = 6$ in the formula:

$$x = \frac{-8 \pm \sqrt{8^2 - 4(1)(6)}}{2(1)} = \frac{-8 \pm \sqrt{40}}{2} = \frac{-8 \pm 2\sqrt{10}}{2}$$

$$= \frac{2(-4 \pm \sqrt{10})}{2} = -4 \pm \sqrt{10}$$

The solution set is $\{-4 - \sqrt{10}, -4 + \sqrt{10}\}$.

Check by using a calculator as in Fig. 1.53. □

Figure 1.53

**b.** Use $a = 1$, $b = -6$, and $c = 11$ in the quadratic formula:

$$x = \frac{-(-6) \pm \sqrt{(-6)^2 - 4(1)(11)}}{2(1)}$$

$$= \frac{6 \pm \sqrt{-8}}{2} = \frac{6 \pm 2i\sqrt{2}}{2} = 3 \pm i\sqrt{2}$$

Check by evaluating $x^2 - 6x + 11$ with $x = 3 + i\sqrt{2}$ as follows:

$$(3 + i\sqrt{2})^2 - 6(3 + i\sqrt{2}) + 11 = 9 + 6i\sqrt{2} + (i\sqrt{2})^2 - 18 - 6i\sqrt{2} + 11$$
$$= 9 + 6i\sqrt{2} - 2 - 18 - 6i\sqrt{2} + 11$$
$$= 0$$

Figure 1.54

The reader should check $3 - i\sqrt{2}$. The imaginary solutions are $3 - i\sqrt{2}$ and $3 + i\sqrt{2}$.

The solutions can be checked with a calculator as in Fig. 1.54. □

**c.** To use the quadratic formula, the equation must be in the form $ax^2 + bx + c = 0$. So rewrite the equation as $4x^2 - 12x + 9 = 0$. Now use $a = 4$, $b = -12$, and $c = 9$ in the formula:

$$x = \frac{-(-12) \pm \sqrt{(-12)^2 - 4(4)(9)}}{2(4)} = \frac{12 \pm \sqrt{0}}{8} = \frac{3}{2}$$

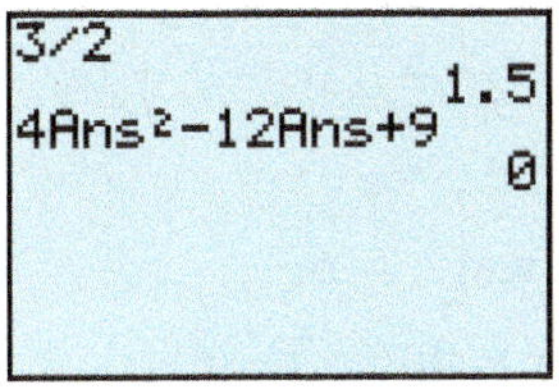


Figure 1.55

You should check that $4\left(\frac{3}{2}\right)^2 - 12\left(\frac{3}{2}\right) + 9 = 0$. The solution set is $\left\{\frac{3}{2}\right\}$.

You can check with a calculator as in Fig. 1.55.

▸**TRY THIS.** Solve $2x^2 - 3x - 2 = 0$ by the quadratic formula. ■

To decide which of the four methods to use for solving a given quadratic equation, use the following strategy.

**STRATEGY**

**Solving $ax^2 + bx + c = 0$**

1. If $b = 0$, solve $ax^2 + c = 0$ for $x^2$ and apply the square root property.
2. If $ax^2 + bx + c$ can be easily factored, then solve by factoring. If $a < 0$ multiply each side of the equation by $-1$ for easier factoring.
3. If the equation cannot be solved by the square root property or factoring, then use the quadratic formula.
4. Completing the square can be used on any quadratic equation, but the quadratic formula is usually easier.
5. Graph $y = ax^2 + bx + c$. The $x$-coordinates of the $x$-intercepts on the graph are the solution(s) to $ax^2 + bx + c = 0$. The graph may not give exact solutions, but you can easily check your solutions from the other methods with a graph.

## The Discriminant

The expression $b^2 - 4ac$ in the quadratic formula is called the **discriminant,** because its value determines the number and type of solutions to a quadratic equation. If $b^2 - 4ac > 0$, then $\sqrt{b^2 - 4ac}$ is real and the equation has two real solutions. If $b^2 - 4ac < 0$, then $\sqrt{b^2 - 4ac}$ is not a real number and there are no real solutions.

There are two imaginary solutions. If $b^2 - 4ac = 0$, then $-b/(2a)$ is the only solution. See Table 1.2.

**Table 1.2** Number of real solutions to a quadratic equation

| Value of $b^2 - 4ac$ | Number of Real Solutions |
|---|---|
| Positive | 2 |
| Zero | 1 |
| Negative | 0 |

## EXAMPLE 5 Using the discriminant

For each equation, state the value of the discriminant and the number of real solutions.

**a.** $x^2 + 8x + 6 = 0$ **b.** $5x^2 - 4x + 1 = 0$ **c.** $4x^2 + 12x + 9 = 0$

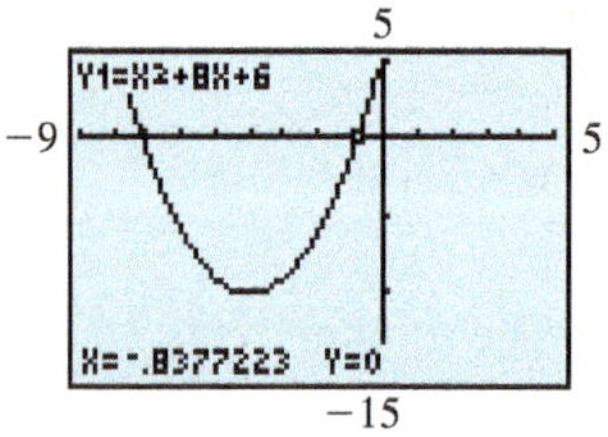


Figure 1.56

### Solution

**a.** Find the value of $b^2 - 4ac$ using $a = 1$, $b = 8$, and $c = 6$:

$$b^2 - 4ac = 8^2 - 4(1)(6) = 40$$

The value of the discriminant is 40 and the equation has two real solutions.
The graph of $y = x^2 + 8x + 6$ has two $x$-intercepts as shown in Fig. 1.56. □

**b.** Find the value of the discriminant for the equation $5x^2 - 4x + 1 = 0$:

$$b^2 - 4ac = (-4)^2 - 4(5)(1) = -4$$

Because the discriminant is negative, the equation has no real solutions.
The graph of $y = 5x^2 - 4x + 1$ has no $x$-intercepts as shown in Fig. 1.57. □

**c.** For $4x^2 + 12x + 9 = 0$, we have $b^2 - 4ac = 12^2 - 4(4)(9) = 0$. So the equation has one real solution.
The graph of $y = 4x^2 + 12x + 9$ has one $x$-intercept as shown in Fig. 1.58.

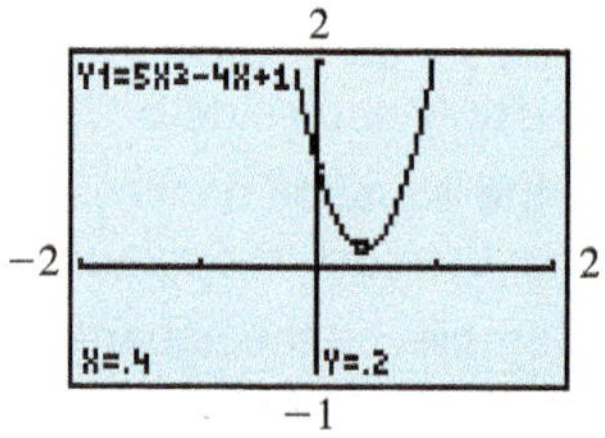


Figure 1.57

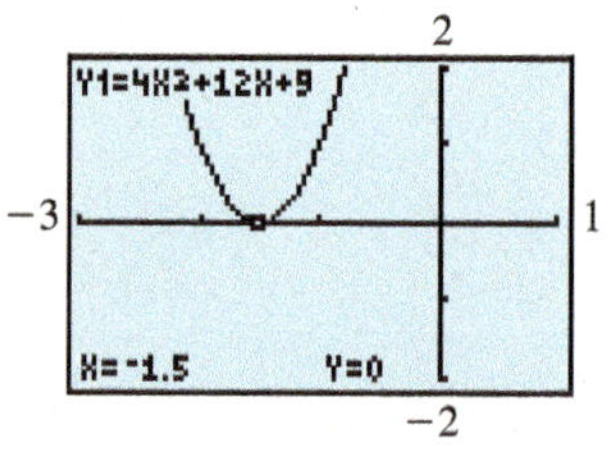


Figure 1.58

▸**TRY THIS.** State the value of the discriminant and the number of real solutions to $5x^2 - 7x + 9 = 0$. ■

The number of solutions to polynomial equations is a topic of considerable interest in algebra. In Section 3.2 we will learn that the number of solutions to any polynomial equation of degree $n$ is less than or equal to $n$ (the fundamental theorem of algebra). Notice how the number of solutions to linear and quadratic equations agrees with these results.

## Using Quadratic Models in Applications

The problems that we solve in this section are very similar to those in Section 1.2. However, in this section the mathematical model of the situation results in a quadratic equation.

### EXAMPLE 6 A problem solved with a quadratic equation

It took Susan 30 minutes longer to drive 275 miles on I-70 west of Green River, Utah, than it took her to drive 300 miles east of Green River. Because of a sand storm and a full load of cantaloupes, she averaged 10 mph less while traveling west of Green River. What was her average speed for each part of the trip?

#### Solution

Let $x$ represent Susan's average speed in mph east of Green River and $x - 10$ represent her average speed in mph west of Green River. We can organize all of the given information as in the following table. Since $D = RT$, the time is determined by $T = D/R$.

| | Distance | Rate | Time |
|---|---|---|---|
| **East** | 300 mi | $x$ mi/hr | $\frac{300}{x}$ hr |
| **West** | 275 mi | $x - 10$ mi/hr | $\frac{275}{x - 10}$ hr |

The following equation expresses the fact that her time west of Green River was $\frac{1}{2}$ hour greater than her time east of Green River.

$$\frac{275}{x - 10} = \frac{300}{x} + \frac{1}{2}$$

$$2x(x - 10) \cdot \frac{275}{x - 10} = 2x(x - 10)\left(\frac{300}{x} + \frac{1}{2}\right)$$

$$550x = 600(x - 10) + x(x - 10)$$

$$-x^2 - 40x + 6000 = 0$$

$$x^2 + 40x - 6000 = 0$$

$$(x + 100)(x - 60) = 0 \quad \text{Factor.}$$

$$x = -100 \quad \text{or} \quad x = 60$$

The solution $x = -100$ is a solution to the equation, but not a solution to the problem. The other solution, $x = 60$, means that $x - 10 = 50$. Check that these two average speeds are a solution to the problem. Susan's average speed east of Green River was 60 mph, and her average speed west of Green River was 50 mph.

▶**TRY THIS.** It took Josh 30 minutes longer to drive 100 miles than it took Bree to drive 90 miles. If Josh averaged 5 mph less than Bree, then what was the average speed of each driver? ■

In the next example we use the formula for the height of a projectile under the influence of gravity. Although we can use any letters we want for variables, it is customary to use $s$ (upper case or lower case) for height and $t$ for time.

## EXAMPLE 7 Applying the quadratic formula

In tennis a lob can be used to buy time to get into position. The approximate height $S$ in feet for a tennis ball that is hit straight upward at $v_0$ ft/sec from a height of $s_0$ ft is modeled by

$$S = -16t^2 + v_0 t + s_0,$$

where $t$ is time in seconds.

**a.** Use Fig. 1.59 to estimate the time that it takes a ball to return to the court when it is hit straight upward with velocity 60 ft/sec from a height of 5 ft.
**b.** Find the time for part (a) by using the quadratic formula.

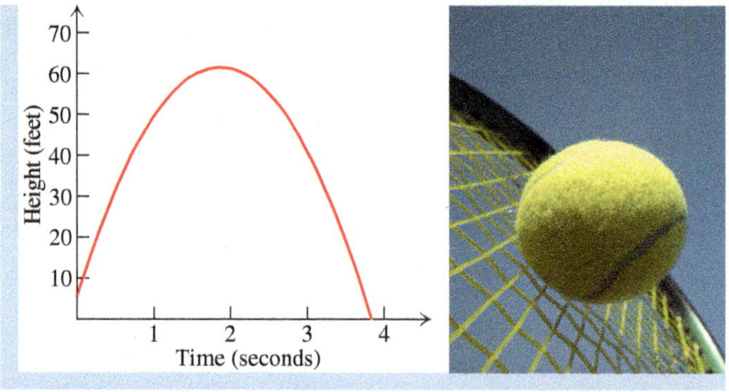


Figure 1.59

### Solution

**a.** At the $t$-intercept in Fig. 1.59, $S = 0$. Since the $t$-intercept is approximately $(3.7, 0)$, the ball is in the air for about 3.7 sec.
**b.** We are looking for the value of $t$ for which the height $S$ is zero. Use $v_0 = 60$, $s_0 = 5$, and $S = 0$ in the given formula:

$$0 = -16t^2 + 60t + 5$$

Use the quadratic formula to solve the equation:

$$t = \frac{-60 \pm \sqrt{60^2 - 4(-16)(5)}}{2(-16)} = \frac{-60 \pm \sqrt{3920}}{-32} = \frac{-60 \pm 28\sqrt{5}}{-32}$$

$$= \frac{15 \pm 7\sqrt{5}}{8}$$

Figure 1.60

Now $(15 - 7\sqrt{5})/8 \approx -0.082$ and $(15 + 7\sqrt{5})/8 \approx 3.83$. Since the time at which the ball returns to the earth must be positive, it will take exactly $(15 + 7\sqrt{5})/8$ seconds or approximately 3.83 seconds for the ball to hit the earth. Check as shown in Fig. 1.60. Due to round-off errors the calculator answer is not zero, but really close to zero $(-1 \times 10^{-11} = -0.00000000001)$.

**▸TRY THIS.** A ball is tossed straight upward. Its height $h$ in feet at time $t$ in seconds is given by $h = -16t^2 + 40t + 6$. For how long is the ball in the air? ■

Quadratic equations often arise from applications that involve the Pythagorean theorem from geometry: *A triangle is a right triangle if and only if the sum of the squares of the legs is equal to the square of the hypotenuse.*

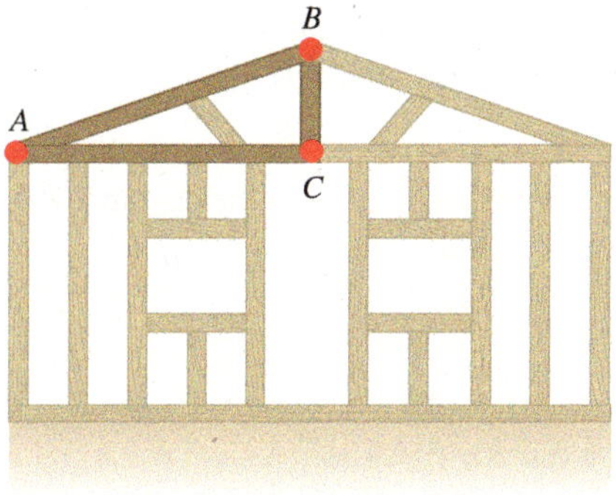


Figure 1.61

## EXAMPLE 8 Using the Pythagorean theorem

In the house shown in Fig. 1.61 the ridge of the roof at point $B$ is 6 feet above point $C$. If the distance from $A$ to $C$ is 18 feet, then what is the length of a rafter from $A$ to $B$?

### Solution

Let $x$ be the distance from $A$ to $B$. Use the Pythagorean theorem to write the following equation.

$$x^2 = 18^2 + 6^2$$
$$x^2 = 360$$
$$x = \pm\sqrt{360} = \pm 6\sqrt{10}$$

Since $x$ must be positive in this problem, $x = 6\sqrt{10}$ feet or $x \approx 18.97$ feet.

▸**TRY THIS.** One leg of a right triangle is 2 feet longer than the other leg. If the hypotenuse is 6 feet, then what are the lengths of the legs? ■

## Quadratic Regression

In Section 1.5 we used a graphing calculator to find the line of best fit. In the next example we will use the quadratic regression feature of a calculator to find an equation of the form $y = ax^2 + bx + c$ that fits a set of data points.

## EXAMPLE 9 Quadratic regression

The accompanying table shows the average monthly temperature in degrees Fahrenheit for Athens, Georgia, for the year 2008.

| Month | Temp. |
|---|---|
| Jan | 49.2 |
| Feb | 45.0 |
| Mar | 54.5 |
| Apr | 65.6 |
| Jun | 77.4 |
| Jul | 81.4 |
| Aug | 80.7 |
| Sep | 71.5 |
| Oct | 60.4 |
| Nov | 53.1 |

**a.** Use quadratic regression on your graphing calculator to find the regression equation.
**b.** Draw the scatter diagram and the regression curve on the calculator.
**c.** According to the regression equation, what should be the average temperature for May and for December?

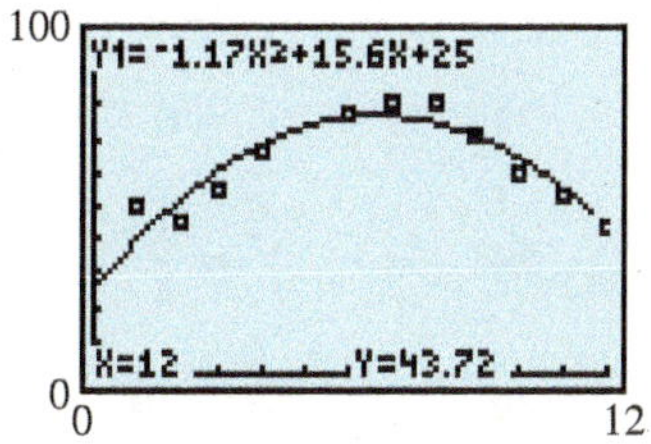


Figure 1.62

### Solution

**a.** Use the number of the month starting at 1 and enter the data using the STAT EDIT menu. Note that May is missing from the table. Use 6 for June. Choose quadratic regression from the STAT CALC menu to get the equation

$$y = -1.17x^2 + 15.6x + 25.0.$$

**b.** The regression curve and the scatter diagram are shown in Fig. 1.62.

**c.** Let $x = 5$ in the equation to get the average monthly temperature for May of 73.8°. Use $x = 12$ in the equation to get the average monthly temperature for December of 43.7°.

**▸TRY THIS.** Use quadratic regression and the given data to predict the cost in 2010.

| Year | 1980 | 1985 | 1990 | 2000 | 2005 |
|---|---|---|---|---|---|
| Cost ($) | 11.9 | 8.15 | 7.04 | 12.3 | 18.40 |

■

## FOR thought... True or False? Explain.

1. The equation $(x - 3)^2 = 4$ is equivalent to $x - 3 = 2$.
2. Every quadratic equation can be solved by factoring.
3. The trinomial $x^2 + \frac{2}{3}x + \frac{4}{9}$ is a perfect square trinomial.
4. The equation $(x - 3)(2x + 5) = 0$ is equivalent to $x = 3$ or $x = \frac{5}{2}$.
5. All quadratic equations have two distinct solutions.
6. If $x^2 - 3x + 1 = 0$, then $x = \frac{3 \pm \sqrt{5}}{2}$.
7. If $b = 0$, then $ax^2 + bx + c = 0$ cannot be solved by the quadratic formula.
8. All quadratic equations have at least one real solution.
9. For $4x^2 + 12x + 9 = 0$ the discriminant is 0.
10. A quadratic equation with real coefficients might have only one real solution.

## EXERCISES 1.7

*Fill in the blank.*

1. An equation of the form $ax^2 + bx + c = 0$ where $a$, $b$, and $c$ are real numbers with $a \neq 0$ is a ________ equation.
2. The equation $ax^2 + bx + c = 0$ can be solved by using the ________ formula.
3. The expression $b^2 - 4ac$ in the quadratic formula is the ________.
4. According to the ________ property, $x^2 = k$ is equivalent to $x = \pm\sqrt{k}$.

*Solve each equation by factoring.*

5. $x^2 - x - 20 = 0$
6. $x^2 + 2x - 8 = 0$
7. $a^2 + 3a = -2$
8. $b^2 - 4b = 12$
9. $2x^2 - 5x - 3 = 0$
10. $2x^2 - 5x + 2 = 0$
11. $6x^2 - 7x + 2 = 0$
12. $12x^2 - 17x + 6 = 0$
13. $(y - 3)(y + 4) = 30$
14. $(w - 1)(w - 2) = 6$

*Use the square root property to find all real or imaginary solutions to each equation.*

15. $x^2 - 5 = 0$
16. $x^2 - 8 = 0$
17. $3x^2 + 2 = 0$
18. $2x^2 + 16 = 0$
19. $(x - 3)^2 = 9$
20. $(x + 1)^2 = \frac{9}{4}$
21. $(3x - 1)^2 = 0$
22. $(5x + 2)^2 = 0$
23. $\left(x - \frac{1}{2}\right)^2 = \frac{25}{4}$
24. $(3x - 1)^2 = \frac{1}{4}$
25. $(x + 2)^2 = -4$
26. $(x - 3)^2 = -20$
27. $\left(x - \frac{2}{3}\right)^2 = \frac{4}{9}$
28. $\left(x + \frac{3}{2}\right)^2 = \frac{1}{2}$

*Find the perfect square trinomial whose first two terms are given.*

**29.** $x^2 - 12x$

**30.** $y^2 + 20y$

**31.** $r^2 + 3r$

**32.** $t^2 - 7t$

**33.** $w^2 + \frac{1}{2}w$

**34.** $p^2 - \frac{2}{3}p$

*Find the real or imaginary solutions by completing the square. See the strategy for completing the square on page 69.*

**35.** $x^2 + 6x + 1 = 0$

**36.** $x^2 - 10x + 5 = 0$

**37.** $n^2 - 2n - 1 = 0$

**38.** $m^2 - 12m + 33 = 0$

**39.** $h^2 + 3h - 1 = 0$

**40.** $t^2 - 5t + 2 = 0$

**41.** $2x^2 + 5x = 12$

**42.** $3x^2 + x = 2$

**43.** $3x^2 + 2x + 1 = 0$

**44.** $5x^2 + 4x + 3 = 0$

*Find the real or imaginary solutions to each equation by using the quadratic formula.*

**45.** $x^2 + 3x - 4 = 0$

**46.** $x^2 + 8x + 12 = 0$

**47.** $2x^2 - 5x - 3 = 0$

**48.** $2x^2 + 3x - 2 = 0$

**49.** $9x^2 + 6x + 1 = 0$

**50.** $16x^2 - 24x + 9 = 0$

**51.** $2x^2 - 3 = 0$

**52.** $-2x^2 + 5 = 0$

**53.** $x^2 + 5 = 4x$

**54.** $x^2 = 6x - 13$

**55.** $x^2 - 2x + 4 = 0$

**56.** $x^2 - 4x + 9 = 0$

**57.** $-2x^2 + 2x = 5$

**58.** $12x - 5 = 9x^2$

**59.** $4x^2 - 8x + 7 = 0$

**60.** $9x^2 - 6x + 4 = 0$

*Use a calculator and the quadratic formula to find all real solutions to each equation. Round answers to two decimal places.*

**61.** $3.2x^2 + 7.6x - 9 = 0$

**62.** $1.5x^2 - 6.3x - 10.1 = 0$

**63.** $3.25x^2 - 4.6x + 20 = 42$

**64.** $4.76x^2 + 6.12x = 55.3$

*For each equation, state the value of the discriminant and the number of real solutions.*

**65.** $9x^2 - 30x + 25 = 0$

**66.** $4x^2 + 28x + 49 = 0$

**67.** $5x^2 - 6x + 2 = 0$

**68.** $3x^2 + 5x + 5 = 0$

**69.** $7x^2 + 12x - 1 = 0$

**70.** $3x^2 - 7x + 3 = 0$

*Find the solutions to each equation by reading the accompanying graph.*

**71.** $6x^2 + x - 2 = 0$

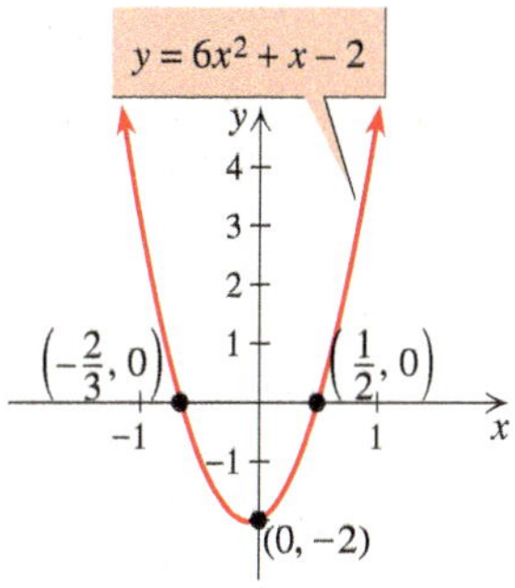


**72.** $-2x^2 - 2x + 12 = 0$

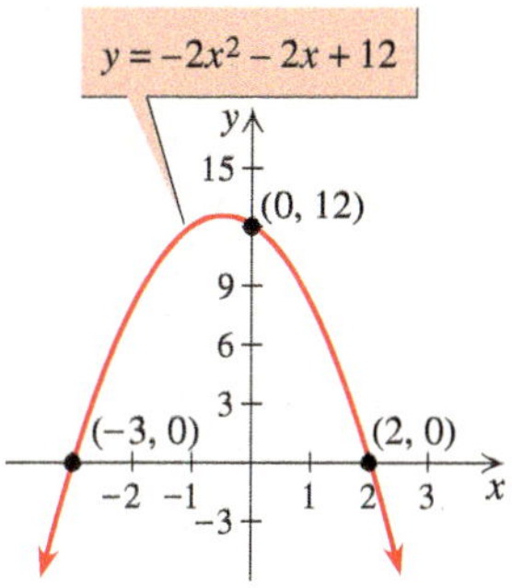


**73.** $-0.5x^2 + x + 7.5 = 0$

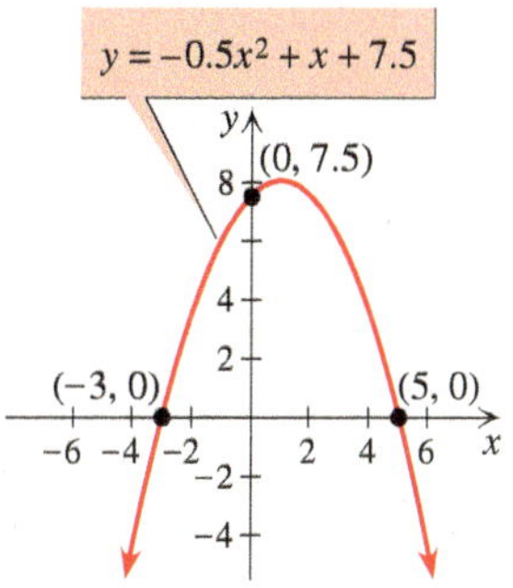


**74.** $0.5x^2 - 2.5x + 2 = 0$

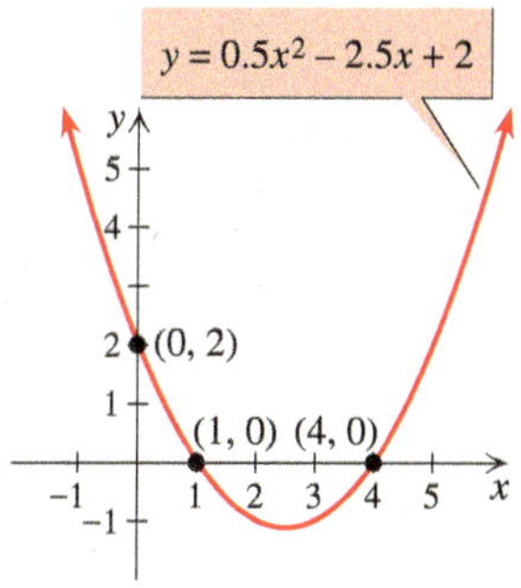


*For each equation of the form $ax^2 + bx + c = 0$, determine the number of real solutions by examining the graph of $y = ax^2 + bx + c$.*

**75.** $1.44x^2 - 8.4x + 12.25 = 0$

**76.** $4.41x^2 - 5.46x + 1.69 = 0$

**77.** $x^2 + 3x + 15 = 0$

**78.** $-2x^2 + 5x - 40 = 0$

**79.** $x^2 + 3x - 160 = 0$

**80.** $-x^2 + 5x - 6.1 = 0$

*Use the method of your choice to find all real solutions to each equation. See the strategy for solving $ax^2 + bx + c = 0$ on page 71.*

**81.** $x^2 = \frac{4}{3}x + \frac{5}{9}$

**82.** $x^2 = \frac{2}{7}x + \frac{2}{49}$

**83.** $x^2 - \sqrt{2} = 0$

**84.** $\sqrt{2}x^2 - 1 = 0$

**85.** $12x^2 + x\sqrt{6} - 1 = 0$

**86.** $-10x^2 - x\sqrt{5} + 1 = 0$

**87.** $x(x + 6) = 72$

**88.** $x = \frac{96}{x + 4}$

**89.** $x = 1 + \frac{1}{x}$

**90.** $x = \frac{1}{x}$

**91.** $\frac{28}{x} - \frac{7}{x^2} = 7$

**92.** $\frac{20}{x} - \frac{46}{x^2} = 2$

**93.** $\frac{x - 12}{3 - x} = \frac{x + 4}{x + 7}$

**94.** $\frac{x - 9}{x - 2} = -\frac{x + 3}{x + 1}$

**95.** $\frac{x - 8}{x + 2} = -\frac{1 - 2x}{x + 3}$

**96.** $\frac{x - 1}{x} = \frac{3x}{x + 1}$

**97.** $\frac{2x + 3}{2x + 1} = \frac{8}{2x + 3}$

**98.** $\frac{2x + 3}{6x + 5} = \frac{2}{2x + 3}$

*Use the methods for solving quadratic equations to solve each formula for the indicated variable.*

**99.** $A = \pi r^2$ for $r$

**100.** $S = 2\pi rh + 2\pi r^2$ for $r$

**101.** $x^2 + 2kx + 3 = 0$ for $x$

**102.** $hy^2 - ky = p$ for $y$

**103.** $2y^2 + 4xy = x^2$ for $y$

**104.** $\frac{\frac{1}{x + h} - \frac{1}{x}}{h} = 1$ for $x$

*Find an exact solution to each problem. If the solution is irrational, then find an approximate solution also.*

**105.** *Demand Equation* The demand equation for a certain product is $P = 40 - 0.001x$, where $x$ is the number of units sold per week and $P$ is the price in dollars at which each one is sold. The weekly revenue $R$ is given by $R = xP$. What number of units sold produces a weekly revenue of \$175,000?

**106.** *Average Cost* The total cost in dollars of producing $x$ items is given by $C = 0.02x^3 + 5x$. For what number of items is the average cost per item equal to \$5.50?

**HINT** Average cost is total cost divided by the number of items.

**107.** *Height of a Ball* A juggler tosses a ball into the air with an initial velocity of 40 ft/sec from an initial height of 4 ft. Find how long it takes for the ball to return to a height of 4 ft.

**HINT** The formula $S = -16t^2 + v_0t + s_0$ gives the height $S$ in feet at time $t$ in seconds where $v_0$ is the initial velocity and $s_0$ is the initial height.

**108.** *Height of a Sky Diver* A sky diver steps out of an airplane at 5000 ft. Use the formula $S = -16t^2 + v_0t + s_0$ to find how long it takes the sky diver to reach 4000 ft.

**109.** *Diagonal of a Football Field* A football field is 100 yd long from goal line to goal line and 160 ft wide. If a player ran diagonally across the field from one goal line to the other, then how far did he run?

**110.** *Dimensions of a Flag* If the perimeter of a rectangular flag is 34 in. and the diagonal is 13 in., then what are the length and width?

**111.** *Long Shot* To avoid hitting the ball out, a tennis player in one corner of the 312 $yd^2$ court hits the ball to the farthest corner of the opponent's court as shown in the diagram. If the length $L$ of the tennis court is 2 yd longer than twice the width $W$, then how far did the player hit the ball?

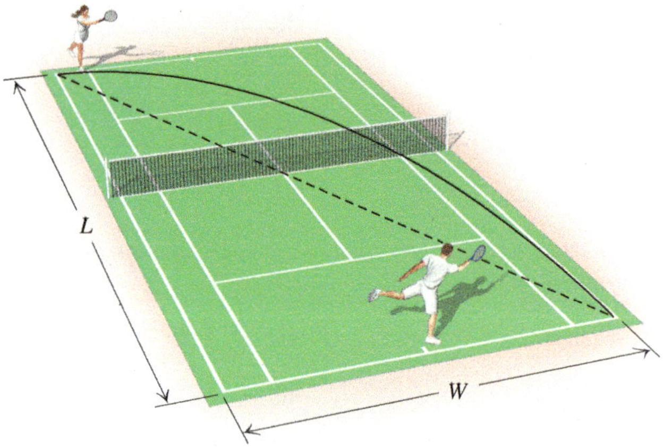


**Figure for Exercise 111**

**112.** *Open-Top Box* Imogene wants to make an open-top box for packing baked goods by cutting equal squares from each corner of an 11 in. by 14 in. piece of cardboard as shown in the diagram. She figures that for versatility the area of the bottom must be 80 $in.^2$. What size square should she cut from each corner?

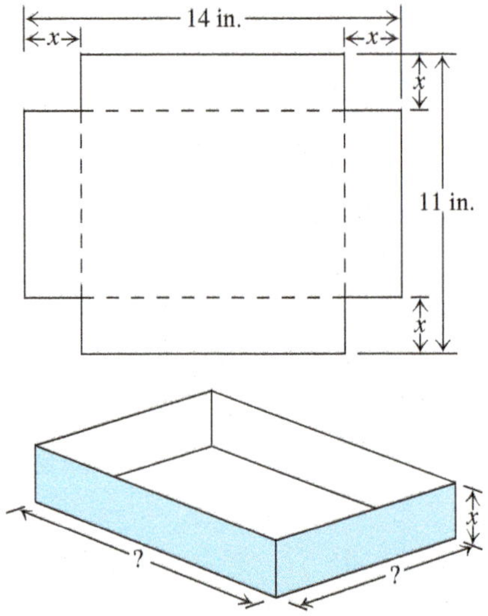


**Figure for Exercise 112**

**113.** *Finding the Displacement* The sail area-displacement ratio $S$ measures the amount of power available to drive a sailboat in moderate to heavy winds (Ted Brewer Yacht Design, www.tedbrewer.com). For the Sabre 402 shown in the figure, the sail area $A$ is 822 ft$^2$ and $S = 18.8$. The displacement $d$ (in pounds) satisfies $2^{-12}d^2S^3 - A^3 = 0$. Find $d$.

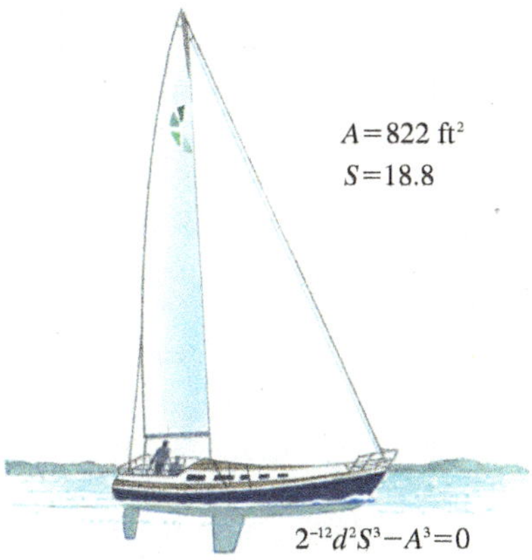


**Figure for Exercise 113**

**114.** *Charleston Earthquake* The Charleston, South Carolina, earthquake of 1886 registered 7.6 on the Richter scale and was felt over an area of 1.5 million square miles (U.S. Geological Survey, www.usgs.gov). If the area in which it was felt was circular and centered at Charleston, then how far away was it felt?

**115.** *Radius of a Pipe* A large pipe is placed next to a wall and a 1-ft-high block is placed 5 ft from the wall to keep the pipe in place as shown in the figure. What is the radius of the pipe? **HINT** Draw in the radius at several locations.

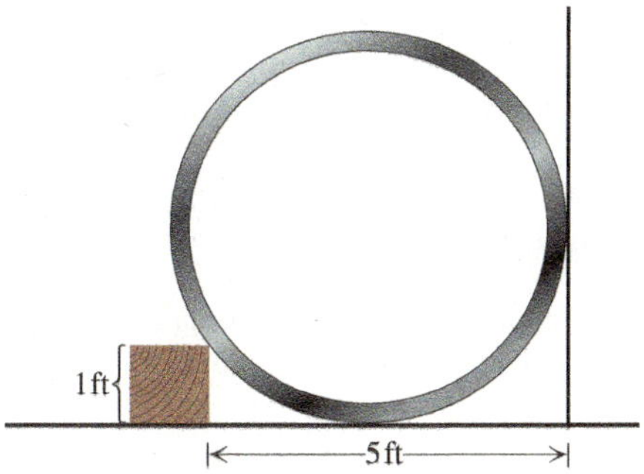


**Figure for Exercise 115**

**116.** *Another Pipe* A small pipe is placed against a wall as in the previous exercise, but no block is used to keep it in place. There is a point on the edge of this pipe that is both 5 in. from the ground and 10 in. from the wall. Find two possibilities for the radius of the pipe.

**117.** *Speed of a Tortoise* When a tortoise crosses a highway, his speed is 2 ft/hr faster than normal. If he can cross a 24-ft lane in 24 min less time than he can travel that same distance off the highway, then what is his normal speed?

**118.** *Speed of an Electric Car* An experimental electric-solar car completed a 1000-mi race in 35 hr. For the 600 mi traveled during daylight the car averaged 20 mph more than it did for the 400 mi traveled at night. What was the average speed of the car during the daytime?

**119.** *Initial Velocity of a Basketball Player* Vince Carter is famous for his high leaps and "hang time," especially while slam-dunking. If Carter leaped for a dunk and reached a peak height at the basket of 1.07 m, what was his upward velocity in meters per second at the moment his feet left the floor? The formula $v_1^2 = v_0^2 + 2gS$ gives the relationship between final velocity $v_1$, initial velocity $v_0$, acceleration of gravity $g$, and his height $S$. Use $g = -9.8$ m/sec$^2$ and the fact that his final velocity was zero at his peak height. Use $S = \frac{1}{2}gt^2 + v_0t$ to find the amount of time he was in the air.

**Figure for Exercise 119**

**120.** *Hazards of Altitude* As shown in the table (www.nasa.gov), atmospheric pressure $a$ (atm) decreases as height above sea level $h$ (in feet) increases. The equation

$$a = 3.89 \times 10^{-10}h^2 - 3.48 \times 10^{-5}h + 1$$

can be used to model this relationship.

**a.** Mountain climbers begin to deteriorate at 18,000 ft. Find $a$ at that height.

**b.** The atmospheric pressure at the highest human settlements is 0.52 atm. Find the height for $a = 0.52$.

**c.** Use the quadratic regression feature of your graphing calculator and the data given in the table to find an equation expressing atmospheric pressure $a$ in terms of $h$.

**Table for Exercise 120**

| Altitude (ft) | Atmospheric Pressure (atm) |
|---|---|
| 28,000 | 0.33 |
| 20,000 | 0.46 |
| 18,000 | |
| | 0.52 |
| 10,000 | 0.69 |
| 0 | 1 |

**121.** *Teen Birth Rate* The accompanying table gives the number of births per 1000 females ages 15–19 (www.infoplease.com).

**a.** Use your graphing calculator to find a quadratic regression equation that expresses the birth rate in terms of the year.

**b.** In what year will the birth rate reach zero according to the quadratic equation of part (a)?

**Table for Exercise 121**

| Year | Births per 1000 Females (ages 15–19) |
|---|---|
| 1980 | 53.0 |
| 1985 | 51.0 |
| 1990 | 59.9 |
| 1995 | 56.8 |
| 2000 | 47.7 |
| 2005 | 41.4 |

**122.** *Teen Birth Rate* Use the data in the previous exercise.

**a.** Find the equation of the regression line that expresses the birth rate in terms of the year.

**b.** In what year will the birth rate reach zero according to the regression line?

**123.** *Computer Design* Using a computer design package, Tina can write and design a direct-mail package in two days less time than it takes to create the same package using traditional design methods. If Tina uses the computer and her assistant Curt uses traditional methods, and together they complete the job in 3.5 days, then how long would it have taken Curt to do the job alone using traditional methods?

HINT Curt's rate is $1/x$ job/day and Tina's rate is $1/(x - 2)$ job/day.

**124.** *Making a Dress* Rafael designed a sequined dress to be worn at the Academy Awards. His top seamstress, Maria, could sew on all the sequins in 10 hr less time than his next-best seamstress, Stephanie. To save time, he gave the job to both women and got all of the sequins attached in 17 hr. How long would it have taken Stephanie working alone?

**125.** *Percentage of White Meat* The Kansas Fried Chicken store sells a Party Size bucket that weighs 10 lb more than the Big Family Size bucket. The Party size bucket contains 8 lb of white meat, while the Big Family Size bucket contains 3 lb of white meat. If the percentage of white meat in the Party Size is 10 percentage points greater than the percentage of white meat in the Big Family Size, then how much does the Party Size bucket weigh?

**126.** *Mixing Antifreeze in a Radiator* Steve's car had a large radiator that contained an unknown amount of pure water. He added two quarts of antifreeze to the radiator. After testing, he decided that the percentage of antifreeze in the radiator was not large enough. Not knowing how to solve mixture problems, Steve decided to add one quart of water and another quart of antifreeze to the radiator to see what he would get. After testing he found that the last addition increased the percentage of antifreeze by three percentage points. How much water did the radiator contain originally?

## ▶ RETHINKING

**127.** Find the equation (in standard form) of the line through $(-2, 6)$ that is parallel to $4x - 5y = 8$.

**128.** Find the product $(3 - 2i)(3 + 2i)$.

**129.** Chandra invested in a CD that returned 5% and a mutual fund that returned 6% after one year. The amount invested in the mutual fund was \$4000 more than the amount invested in the CD, and the total amount of interest was \$1230. How much did she invest in the CD?

**130.** Find the radius of the circle $x^2 + y^2 = 3y$.

**131.** Find the additive inverse of $2 + 3i$.

**132.** Use the linear regression feature of a graphing calculator to find the equation of the regression line for the points $(-2, 1)$, $(-1, 2)$, $(0, 5)$, $(1, 4)$, and $(2, 4)$.

## THINKING OUTSIDE THE BOX IX

*As the Crow Flies* In Perfect City the avenues run east and west, the streets run north and south, and all of the blocks are square. A crow flies from the corner of 1st Ave and 1st Street to the corner of $m$th Ave and $n$th Street, "as the crow flies." Assume that $m$ and $n$ are positive integers greater than 1 and the streets and avenues are simply lines on a map. If the crow flies over an intersection, then he flies over only two of the blocks that meet at the intersection.

**a.** If $m - 1$ and $n - 1$ are relatively prime (no common factors), then how many city blocks does the crow fly over?

**b.** If $d$ is the greatest common factor for $m$ and $n$, then how many city blocks does the crow fly over?

## POP QUIZ 1.7

**1.** Solve $x^2 = 2$.

**2.** Solve $x^2 - 2x = 48$ by factoring.

**3.** Solve $x^2 - 4x = 1$ by completing the square.

**4.** Solve $2x^2 - 4x = 3$ by the quadratic formula.

**5.** How many real solutions are there to $5x^2 - 9x + 5 = 0$?

## LINKING concepts... For Individual or Group Explorations

### Baseball Statistics

*Baseball fans keep up with their favorite teams through charts where the teams are ranked according to the percentage of games won. The chart usually has a column indicating the games behind (GB) for each team. If the win-loss record of the number one team is $(A, B)$, then the games behind of another team whose win-loss record is $(a, b)$ is calculated by the formula*

$$GB = \frac{(A - a) + (b - B)}{2}.$$

**a)** Find *GB* for Atlanta and Philadelphia in the following table.

| Team | Won | Lost | Pct. | GB |
|---|---|---|---|---|
| New York | 38 | 24 | .613 | — |
| Atlanta | 35 | 29 | .547 | ? |
| Philadelphia | 34 | 31 | .523 | ? |

**b)** In the following table Pittsburgh has a higher percentage of wins than Chicago, and so Pittsburgh is in first place. Find *GB* for Chicago.

| Team | Won | Lost | Pct. | GB |
|---|---|---|---|---|
| Pittsburgh | 18 | 13 | .581 | — |
| Chicago | 22 | 16 | .579 | ? |

**c)** Is Chicago actually behind Pittsburgh in terms of the statistic *GB*?

**d)** Another measure of how far a team is from first place, called the deficit *D*, is the number of games that the two teams would have to play against each other to get equal percentages of wins, with the higher-ranked team losing all of the games. Find the deficits for Atlanta, Philadelphia, and Chicago.

**e)** Is it possible for *GB* or *D* to be negative? Does it make any sense if they are?

**f)** Compare the values of *D* and *GB* for each of the three teams. Which is a better measure of how far a team is from first place?

# 1.8 Linear and Absolute Value Inequalities

An equation states that two algebraic expressions are equal, while an **inequality** or **simple inequality** is a statement that two algebraic expressions are not equal in a particular way. Inequalities are stated using less than $(<)$, less than or equal to $(\leq)$, greater than $(>)$, or greater than or equal to $(\geq)$. In this section we study some basic inequalities. We will see more inequalities in Chapter 3.

## Interval Notation

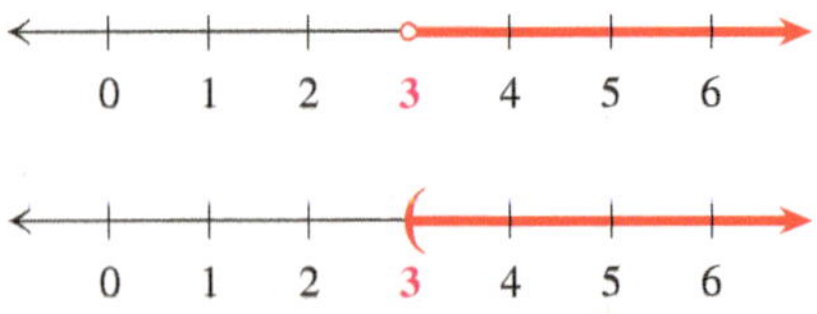


Figure 1.63

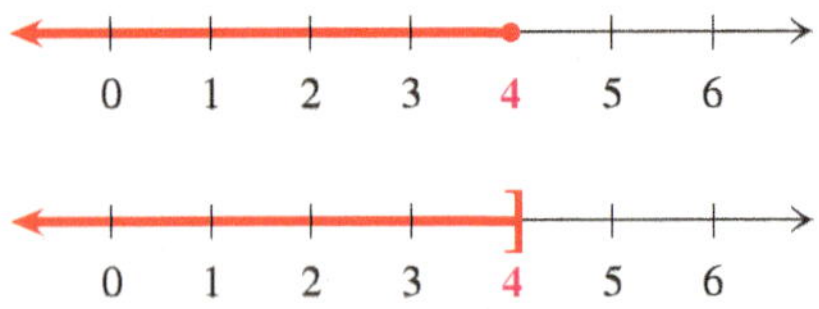


Figure 1.64

The solution set to an inequality is the set of all real numbers for which the inequality is true. The solution set to the inequality $x > 3$ is written as $\{x \mid x > 3\}$ and consists of all real numbers to the right of 3 (but not including 3) on the number line. This set is also called the **interval** of numbers greater than 3, and it is written in **interval notation** as $(3, \infty)$. To graphically show that 3 is not in the interval, you can use a parenthesis or an open circle as shown on the graphs in Fig. 1.63. In this text, a parenthesis is used because it matches the interval notation. The infinity symbol $(\infty)$ is not a number. It indicates that there is no bound on the numbers greater than 3.

The solution set to $x \leq 4$ is written in set notation as $\{x \mid x \leq 4\}$ and it consists of the number 4 along with all real numbers to the left of 4 on the number line. It is written in interval notation as $(-\infty, 4]$. The symbol $-\infty$ indicates that all numbers to the left of 4 on the number line are in the interval. To graphically show that 4 is in the interval, you can use a bracket or a solid circle as shown on the graphs in Fig. 1.64. In this text, a bracket is used because it matches the interval notation. Open circles and solid circles will be used occasionally on some two-dimensional graphs in this text.

Intervals that use the infinity symbol are **unbounded** intervals. The following summary lists the different types of unbounded intervals used in interval notation and the graphs of those intervals on a number line. An unbounded interval with an endpoint is **open** if the endpoint is not included in the interval and **closed** if the endpoint is included.

**SUMMARY**

### Interval Notation for Unbounded Intervals

| Set | Interval notation | Type | Graph |
|---|---|---|---|
| $\{x \mid x > a\}$ | $(a, \infty)$ | Open |  |
| $\{x \mid x < a\}$ | $(-\infty, a)$ | Open |  |
| $\{x \mid x \geq a\}$ | $[a, \infty)$ | Closed |  |
| $\{x \mid x \leq a\}$ | $(-\infty, a]$ | Closed |  |
| Real numbers | $(-\infty, \infty)$ | Open |  |

We use a parenthesis when an endpoint of an interval is not included in the solution set and a bracket when an endpoint is included. A bracket is never used next to $\infty$ because infinity is not a number. On the graphs above, the number lines are shaded, showing that the solutions include all real numbers in the given interval.

## EXAMPLE 1 Interval notation

Write an inequality whose solution set is the given interval.

**a.** $(-\infty, -9)$ **b.** $[0, \infty)$

### Solution

**a.** The interval $(-\infty, -9)$ represents all real numbers less than $-9$. It is the solution set to $x < -9$.

**b.** The interval $[0, \infty)$ represents all real numbers greater than or equal to 0. It is the solution set to $x \geq 0$.

▶**TRY THIS.** Write an inequality whose solution set is $(-\infty, 5]$. ■

## Linear Inequalities

Replacing the equal sign in the general linear equation $ax + b = 0$ by any of the symbols $<$, $\leq$, $>$, or $\geq$ gives a **linear inequality.** Two inequalities are **equivalent** if they have the same solution set. We solve linear inequalities like we solve linear equations by performing operations on each side to get equivalent inequalities. However, the rules for inequalities are slightly different from the rules for equations.

Adding any real number to both sides of an inequality results in an equivalent inequality. For example, adding 3 to both sides of $-4 < 5$ yields $-1 < 8$, which is true. Adding or subtracting the same number simply moves the original numbers to the right or left along the number line and does not change their order.

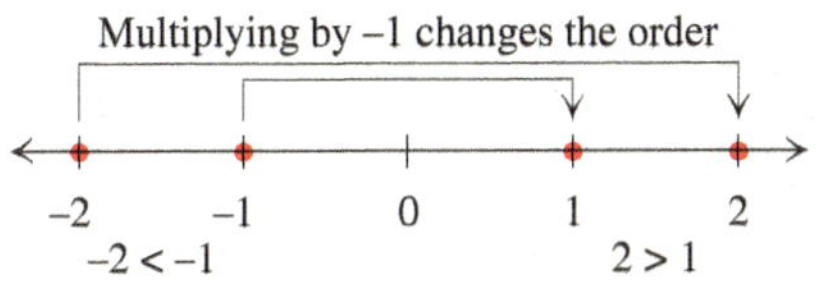


Figure 1.65

The order of two numbers will also be unchanged when they are multiplied or divided by the same positive real number. For example, $10 < 20$, and after dividing both numbers by 10 we have $1 < 2$. However, multiplying or dividing any two numbers by a negative number will change their order. For example, $-2 < -1$, but after multiplying both numbers by $-1$ we have $2 > 1$. See Fig. 1.65. Likewise, $-10 < 20$, but after dividing both numbers by $-10$ we have $1 > -2$. *When an inequality is multiplied or divided by a negative number, the direction of the inequality symbol is reversed.* These ideas are stated symbolically in the following box for $<$, but they also hold for $>$, $\leq$, and $\geq$.

**Properties of Inequality**

If $A$ and $B$ are algebraic expressions and $C$ is a nonzero real number, then the inequality $A < B$ is equivalent to

1. $A \pm C < B \pm C$,
2. $CA < CB$ (for $C$ positive), $CA > CB$ (for $C$ negative),
3. $\frac{A}{C} < \frac{B}{C}$ (for $C$ positive), $\frac{A}{C} > \frac{B}{C}$ (for $C$ negative).

## EXAMPLE 2 Solving a linear inequality

Solve $-3x - 9 < 0$. Write the solution set in interval notation and graph it.

### Solution

Isolate the variable as is done in solving equations.

$$-3x - 9 < 0$$

$$-3x - 9 + 9 < 0 + 9 \quad \text{Add 9 to each side.}$$

$$-3x < 9$$

$$x > -3 \quad \text{Divide each side by } -3 \text{, reversing the inequality.}$$

Figure 1.66

The solution set is the interval $(-3, \infty)$ and its graph is shown in Fig. 1.66. Checking the solution to an inequality is generally not as simple as checking an equation, because usually there are infinitely many solutions. We can do a "partial check" by checking one number in $(-3, \infty)$ and one number not in $(-3, \infty)$. For example, $0 > -3$ and $-3(0) - 9 < 0$ is correct, while $-6 < -3$ and $-3(-6) - 9 < 0$ is incorrect.

▸**TRY THIS.** Solve $2 - 5x \le 7$. Write the solution set in interval notation and graph it. ■

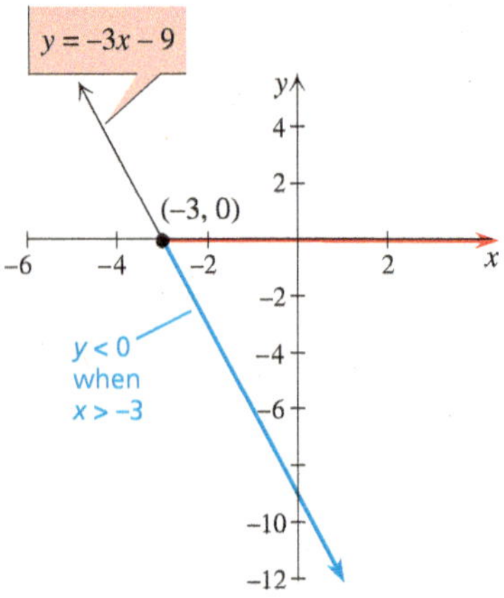


Figure 1.67

We can read the solution to an inequality in one variable from a graph in two variables in the same manner that we read the solution to an equation in one variable. Figure 1.67 shows the graph of $y = -3x - 9$. From this figure we see that the $y$-coordinates on this line are negative when the $x$-coordinates are greater than $-3$. In other words, $-3x - 9 < 0$ when $x > -3$.

We can also perform operations on each side of an inequality using a variable expression. Addition or subtraction with variable expressions will give equivalent inequalities. However, we must always watch for undefined expressions. *Multiplication and division with a variable expression are usually avoided because we do not know whether the expression is positive or negative.*

## EXAMPLE 3 Solving a linear inequality

Solve $\frac{1}{2}x - 3 \ge \frac{1}{4}x + 2$ and graph the solution set.

### Solution

Multiply each side by the LCD to eliminate the fractions.

$$\frac{1}{2}x - 3 \ge \frac{1}{4}x + 2$$

$$4\left(\frac{1}{2}x - 3\right) \ge 4\left(\frac{1}{4}x + 2\right) \quad \text{Multiply each side by 4.}$$

$$2x - 12 \ge x + 8$$

$$x - 12 \ge 8$$

$$x \ge 20$$

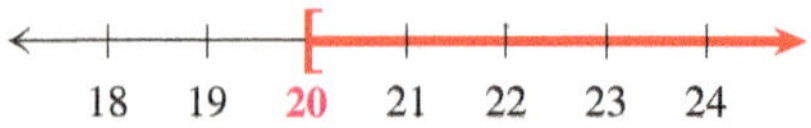


Figure 1.68

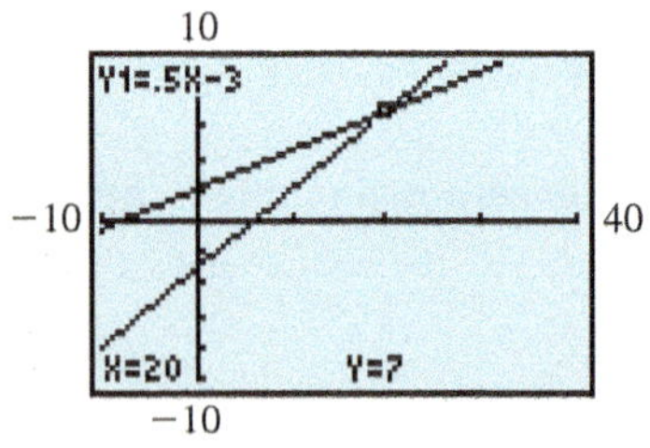


Figure 1.69

The solution set is the interval $[20, \infty)$. See Fig. 1.68 for its graph.

The algebraic solution to $\frac{1}{2}x - 3 \ge \frac{1}{4}x + 2$ proves that the graph of $y = \frac{1}{2}x - 3$ is at or above the graph of $y = \frac{1}{4}x + 2$ when $x \ge 20$, as it appears to be in Fig. 1.69. Conversely, the graphs in Fig. 1.69 support the conclusion that $x \ge 20$ causes the inequality to be true.

▸**TRY THIS.** Solve $\frac{1}{2}x + \frac{1}{3} \le \frac{1}{3}x + 1$ and graph the solution set. ■

## Compound Inequalities

A **compound inequality** is a sentence containing two simple inequalities connected with "and" or "or." The solution to a compound inequality can be an interval of real numbers that does not involve infinity, a **bounded** interval of real numbers. For example, the solution set to the compound inequality $x \ge 2$ and $x \le 5$ is the set of real numbers between 2 and 5, inclusive. This inequality is also written as $2 \le x \le 5$. Its solution set is $\{x \mid 2 \le x \le 5\}$, which is written in interval notation as $[2, 5]$. Because $[2, 5]$ contains both of its endpoints, the interval is **closed.** The following summary lists the different types of bounded intervals used in interval notation and the graphs of those intervals on a number line.

**SUMMARY**

**Interval Notation for Bounded Intervals**

| Set | Interval notation | Type | Graph |
|---|---|---|---|
| $\{x \mid a < x < b\}$ | $(a, b)$ | Open |   |
| $\{x \mid a \le x \le b\}$ | $[a, b]$ | Closed |   |
| $\{x \mid a \le x < b\}$ | $[a, b)$ | Half open or half closed |   |
| $\{x \mid a < x \le b\}$ | $(a, b]$ | Half open or half closed |   |

The notation $a < x < b$ is used only when $x$ is between $a$ and $b$, and $a$ is less than $b$. We do *not* write inequalities such as $5 < x < 3$, $4 > x < 9$, or $2 < x > 8$.

The **intersection** of sets $A$ and $B$ is the set $A \cap B$ (read "$A$ intersect $B$"), where $x \in A \cap B$ if and only if $x \in A$ and $x \in B$. (The symbol $\in$ means "belongs to.") The **union** of sets $A$ and $B$ is the set $A \cup B$ (read "$A$ union $B$"), where $x \in A \cup B$ if and only if $x \in A$ or $x \in B$. In solving compound inequalities it is often necessary to find intersections and unions of intervals.

## EXAMPLE 4 Intersections and unions of intervals

Let $A = (1, 5)$, $B = [3, 7)$, and $C = (6, \infty)$. Write each of the following sets in interval notation.

**a.** $A \cup B$ **b.** $A \cap B$ **c.** $A \cup C$ **d.** $A \cap C$

### Solution

**a.** Graph both intervals on the number line, as shown in Fig. 1.70(a). The union of two intervals is the set of points that are in one, the other, or both intervals. For a union, nothing is omitted. The union consists of all points shaded in the figure. So $A \cup B = (1, 7)$.

**b.** The intersection of $A$ and $B$ is the set of points that belong to both intervals. The intersection consists of the points that are shaded twice in Fig. 1.70(a). So $A \cap B = [3, 5)$.

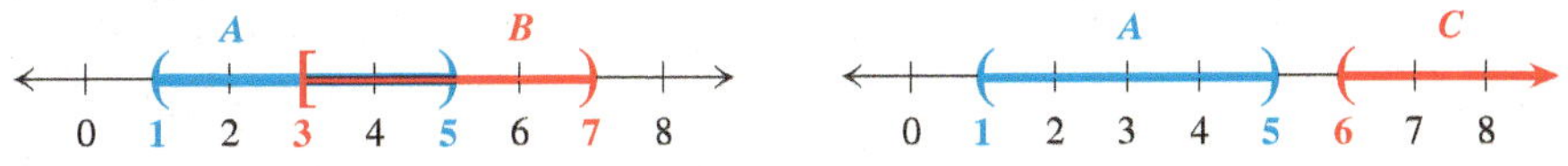


(a) (b)

**Figure 1.70**

**c.** Graph both intervals on the number line, as shown in Fig. 1.70(b). For a union, nothing is omitted. So $A \cup C = (1, 5) \cup (6, \infty)$. Note that $A \cup C$ cannot be written as single interval.

**d.** Since there are no points shaded twice in Fig. 1.70(b), $A \cap C = \varnothing$.

▶**TRY THIS.** Find $A \cup B$ and $A \cap B$ if $A = (1, 6)$ and $B = [4, 9)$. ■

The solution set to a compound inequality using the connector "or" is the union of the two solution sets, and the solution set to a compound inequality using "and" is the intersection of the two solution sets.

## EXAMPLE 5 Solving compound inequalities

Solve each compound inequality. Write the solution set using interval notation and graph it.

**a.** $2x - 3 > 5$ and $4 - x \le 3$ **b.** $4 - 3x < -2$ or $3(x - 2) \le -6$
**c.** $-4 \le 3x - 1 < 5$

### Solution

**a.** 

$$\begin{aligned} 2x - 3 &> 5 \quad \text{and} \quad & 4 - x &\le 3 \\ 2x &> 8 \quad \text{and} \quad & -x &\le -1 \\ x &> 4 \quad \text{and} \quad & x &\ge 1 \end{aligned}$$

Graph $(4, \infty)$ and $[1, \infty)$ on the number line, as shown in Fig. 1.71(a). The intersection of the intervals is the set of points that are shaded twice in Fig. 1.71(a). So the intersection is the interval $(4, \infty)$, and $(4, \infty)$ is the solution set to the compound inequality. Its graph is shown in Fig. 1.71(b).

(a) (b)

Figure 1.71

**b.** 

$$\begin{aligned} 4 - 3x &< -2 \quad \text{or} \quad & 3(x - 2) &\le -6 \\ -3x &< -6 \quad \text{or} \quad & x - 2 &\le -2 \\ x &> 2 \quad \text{or} \quad & x &\le 0 \end{aligned}$$

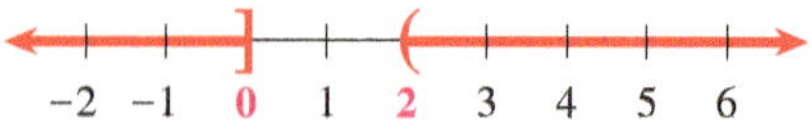

Figure 1.72

The union of the intervals $(2, \infty)$ and $(-\infty, 0]$ consists of all points that are shaded in Fig. 1.72. This set cannot be written as a single interval. So the solution set is $(-\infty, 0] \cup (2, \infty)$. Its graph is shown in Fig. 1.72.

**c.** We could write $-4 \le 3x - 1 < 5$ as the compound inequality $-4 \le 3x - 1$ and $3x - 1 < 5$, and then solve each simple inequality. Since each is solved using the same sequence of steps, we can solve the original inequality without separating it:

$$-4 \le 3x - 1 < 5$$

$$-4 + 1 \le 3x - 1 + 1 < 5 + 1 \qquad \text{Add 1 to each part of the inequality.}$$

$$-3 \le 3x < 6$$

$$\frac{-3}{3} \le \frac{3x}{3} < \frac{6}{3} \qquad \text{Divide each part by 3.}$$

$$-1 \le x < 2$$

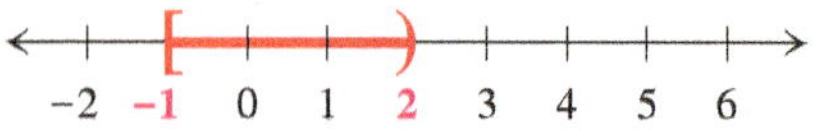

Figure 1.73

The solution set is the half-open interval $[-1, 2)$, graphed in Fig. 1.73.

▸**TRY THIS.** Solve $2x > -4$ and $4 - x \ge 0$ and graph the solution set. ■

It is possible that all real numbers satisfy a compound inequality or no real numbers satisfy a compound inequality.

## EXAMPLE 6 Solving compound inequalities

Solve each compound inequality.

**a.** $3x - 9 \le 9$ or $4 - x \le 3$

**b.** $-\frac{2}{3}x < 4$ and $\frac{3}{4}x < -6$

**Solution**

**a.** Solve each simple inequality and find the union of their solution sets:

$$3x - 9 \le 9 \quad \text{or} \quad 4 - x \le 3$$
$$3x \le 18 \quad \text{or} \quad -x \le -1$$
$$x \le 6 \quad \text{or} \quad x \ge 1$$

The union of $(-\infty, 6]$ and $[1, \infty)$ is the set of all real numbers, $(-\infty, \infty)$.

**b.** Solve each simple inequality and find the intersection of their solution sets:

$$-\frac{2}{3}x < 4 \quad \text{and} \quad \frac{3}{4}x < -6$$
$$\left(-\frac{3}{2}\right)\left(-\frac{2}{3}x\right) > \left(-\frac{3}{2}\right)4 \quad \text{and} \quad \left(\frac{4}{3}\right)\left(\frac{3}{4}x\right) < \left(\frac{4}{3}\right)(-6)$$
$$x > -6 \quad \text{and} \quad x < -8$$

Since $(-6, \infty) \cap (-\infty, -8) = \varnothing$, there is no solution to the compound inequality.

▶**TRY THIS.** Solve $3x + 2 > -1$ and $5 < -3 - 4x$. ■

## Absolute Value Inequalities

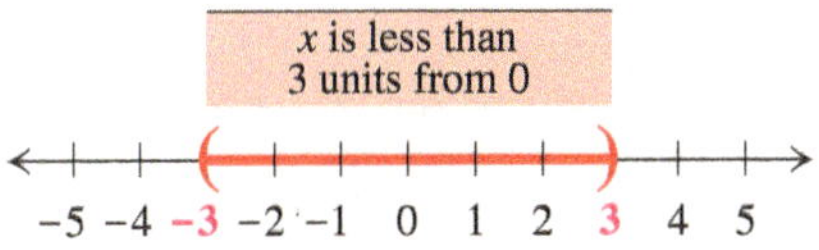


**Figure 1.74**

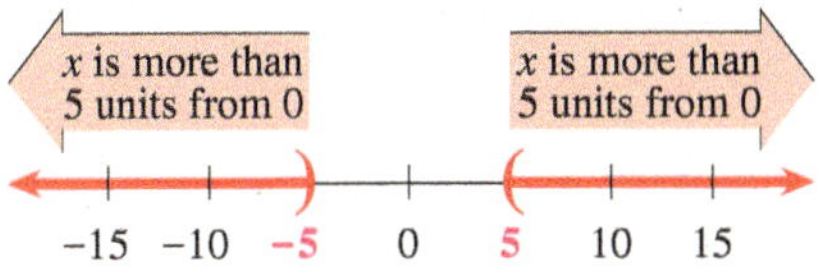


**Figure 1.75**

We studied absolute value equations in Section 1.1. Absolute value inequalities are closely related to absolute value equations. Remember that the absolute value of a number is its distance from 0 on the number line. The equation $|x| = 3$ means that $x$ is exactly three units from 0 on the number line. Since 3 and $-3$ are three units from 0, both 3 and $-3$ satisfy the equation. The inequality $|x| < 3$ means that $x$ is less than three units from 0. See Fig. 1.74. Any number between $-3$ and 3 is less than three units from 0. So $|x| < 3$ is equivalent to $-3 < x < 3$ and the solution set is the open interval $(-3, 3)$.

The inequality $|x| > 5$ means that $x$ is more than five units from 0 on the number line, which is equivalent to the compound inequality $x > 5$ or $x < -5$. See Fig. 1.75. So the solution to $|x| > 5$ is the union of two intervals, $(-\infty, -5) \cup (5, \infty)$. These ideas about absolute value inequalities are summarized as follows.

**SUMMARY**

**Basic Absolute Value Inequalities (for $k > 0$)**

| Absolute value inequality | Equivalent statement | Solution set in interval notation | Graph of solution set |
|---|---|---|---|
| $\|x\| > k$ | $x < -k$ or $x > k$ | $(-\infty, -k) \cup (k, \infty)$ |   |
| $\|x\| \ge k$ | $x \le -k$ or $x \ge k$ | $(-\infty, -k] \cup [k, \infty)$ |   |
| $\|x\| < k$ | $-k < x < k$ | $(-k, k)$ |   |
| $\|x\| \le k$ | $-k \le x \le k$ | $[-k, k]$ |   |

In the next example we use the rules for basic absolute value inequalities to solve more complicated absolute value inequalities.

## EXAMPLE 7 Absolute value inequalities

Solve each absolute value inequality and graph the solution set.

**a.** $|3x + 2| < 7$ **b.** $-2|4 - x| \le -4$ **c.** $|7x - 9| \ge -3$

### Solution

**a.**

$$|3x + 2| < 7$$

$$-7 < 3x + 2 < 7 \qquad \text{Write the equivalent compound inequality.}$$

$$-7 - 2 < 3x + 2 - 2 < 7 - 2 \qquad \text{Subtract 2 from each part.}$$

$$-9 < 3x < 5$$

$$-3 < x < \frac{5}{3} \qquad \text{Divide each part by 3.}$$

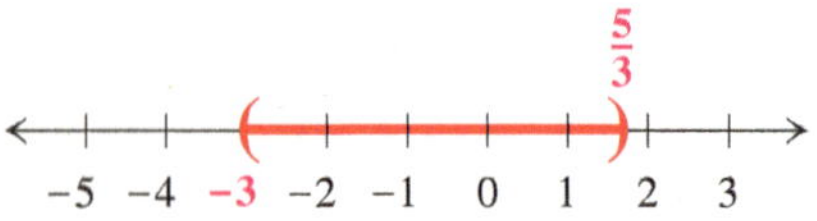

Figure 1.76

The solution set is the open interval $\left(-3, \frac{5}{3}\right)$. The graph is shown in Fig. 1.76.

**b.**

$$-2|4 - x| \le -4 \qquad \text{Divide each side by } -2\text{, reversing the inequality.}$$

$$|4 - x| \ge 2$$

$$4 - x \le -2 \quad \text{or} \quad 4 - x \ge 2 \qquad \text{Write the equivalent compound inequality.}$$

$$-x \le -6 \quad \text{or} \quad -x \ge -2$$

$$x \ge 6 \quad \text{or} \quad x \le 2 \qquad \text{Multiply each side by } -1.$$

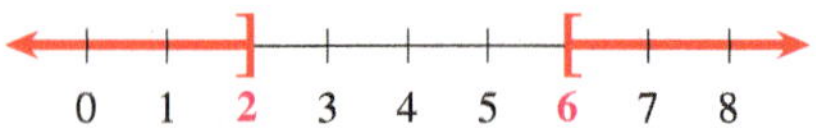

Figure 1.77

The solution set is $(-\infty, 2] \cup [6, \infty)$. Its graph is shown in Fig. 1.77. Check 8, 4, and 0 in the original inequality. If $x = 8$, we get $-2|4 - 8| \le -4$, which is true. If $x = 4$, we get $-2|4 - 4| \le -4$, which is false. If $x = 0$, we get $-2|4 - 0| \le -4$, which is true.

**c.** The expression $|7x - 9|$ has a nonnegative value for every real number $x$. So the inequality $|7x - 9| \ge -3$ is satisfied by every real number. The solution set is $(-\infty, \infty)$, and its graph is shown in Fig. 1.78.

Figure 1.78

▶**TRY THIS.** Solve $|x - 6| - 3 \le -2$ and graph the solution set. ■

## Modeling with Inequalities

Inequalities occur in applications just as equations do. In fact, in real life, equality in anything is usually the exception. The solution to a problem involving an inequality is generally an interval of real numbers. In this case we often ask for the range of values that solve the problem.

## EXAMPLE 8 An application involving inequality

Remington scored 74 on his midterm exam in history. If he is to get a B, the average of his midterm and final exam must be between 80 and 89 inclusive. In what range must his final exam score lie for him to get a B in the course? Both exams have a maximum of 100 points possible.

### Solution

Let $x$ represent Remington's final exam score. The average of his midterm and final must satisfy the following inequality.

$$80 \le \frac{74 + x}{2} \le 89$$

$$160 \le 74 + x \le 178$$

$$86 \le x \le 104$$

His final exam score must lie in the interval [86, 104]. Since 100 is the maximum possible score, we can shorten the interval to [86, 100].

▶**TRY THIS.** Kelly's commissions for her first two sales of the Whirlwind Vacuum Cleaner were \$80 and \$90. In what range must her third commission lie for her average to be between \$100 and \$110? ■

When discussing the error made in a measurement, we may refer to the *absolute error* or the *relative error*. For example, if $L$ is the actual length of an object and $x$ is the length determined by a measurement, then the absolute error is $|x - L|$ and the relative error is $|x - L|/L$.

### EXAMPLE 9 Application of absolute value inequality

A technician is testing a scale with a 50-lb block of steel. The scale passes this test if the relative error when weighing this block is less than 0.1%. If $x$ is the reading on the scale, then for what values of $x$ does the scale pass this test?

**Solution**

If the relative error must be less than 0.1%, then $x$ must satisfy the following inequality:

$$\frac{|x - 50|}{50} < 0.001$$

Solve the inequality for $x$:

$$|x - 50| < 0.05$$
$$-0.05 < x - 50 < 0.05$$
$$49.95 < x < 50.05$$

So the scale passes the test if it shows a weight in the interval (49.95, 50.05).

▶**TRY THIS.** A gas pump is certified as accurate if the relative error when dispensing 10 gallons of gas is less than 1%. If $x$ is the actual amount of gas dispensed, then for what values of $x$ is the pump certified as accurate? ■

## FOR thought... True or False? Explain.

1. The inequality $-3 < x + 6$ is equivalent to $x + 6 > -3$.
2. The inequality $-2x < -6$ is equivalent to $\frac{-2x}{-2} < \frac{-6}{-2}$.
3. The smallest real number that satisfies $x > 12$ is 13.
4. The number $-6$ satisfies $|x - 6| > -1$.
5. $(-\infty, -3) \cap (-\infty, -2) = (-\infty, -2)$
6. $(5, \infty) \cap (-\infty, -3) = (-3, 5)$
7. All negative numbers satisfy $|x - 2| < 0$.
8. The compound inequality $x < -3$ or $x > 3$ is equivalent to $|x| < -3$.
9. The inequality $|x| + 2 < 5$ is equivalent to $-5 < x + 2 < 5$.
10. The fact that the difference between your age $y$ and my age $m$ is at most 5 years is written $|y - m| \leq 5$.

# EXERCISES 1.8

*Fill in the blank.*

**1.** The set $\{x \mid x > a\}$ is written in ________ notation as $(a, \infty)$.

**2.** An ________ interval does not include its endpoints.

**3.** A ________ interval includes its endpoints.

**4.** An interval that involves the infinity symbol is an ________ interval.

**5.** Two simple inequalities connected with "and" or "or" is a ________ inequality.

**6.** The ________ of sets $A$ and $B$ is the set of all elements that are in both $A$ and $B$.

*For each interval write an inequality whose solution set is the interval, and for each inequality, write the solution set in interval notation. See the summary of interval notation for unbounded intervals on page 82.*

**7.** $(-\infty, 12)$ **8.** $(-\infty, -3]$ **9.** $[-7, \infty)$

**10.** $(1.2, \infty)$ **11.** $x \geq -8$ **12.** $x < 54$

**13.** $x < \pi/2$ **14.** $x \geq \sqrt{3}$

*Solve each inequality. Write the solution set using interval notation and graph it.*

**15.** $3x - 6 > 9$ **16.** $2x + 1 < 6$

**17.** $7 - 5x \leq -3$ **18.** $-1 - 4x \geq 7$

**19.** $\frac{1}{2}x - 4 < \frac{1}{3}x + 5$ **20.** $\frac{1}{2} - x > \frac{x}{3} + \frac{1}{4}$

**21.** $\frac{7 - 3x}{2} \geq -3$ **22.** $\frac{5 - x}{3} \leq -2$

**23.** $\frac{2x - 3}{-5} \geq 0$ **24.** $\frac{5 - 3x}{-7} \leq 0$

**25.** $-2(3x - 2) \geq 4 - x$ **26.** $-5x \leq 3(x - 9)$

*Solve each inequality by reading the accompanying graph.*

**27.** $1.8x + 6.3 < 0$ **28.** $1.2x - 3 \geq 0$

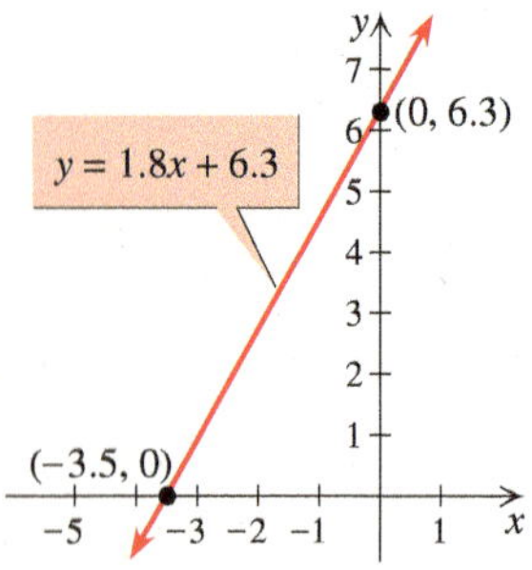


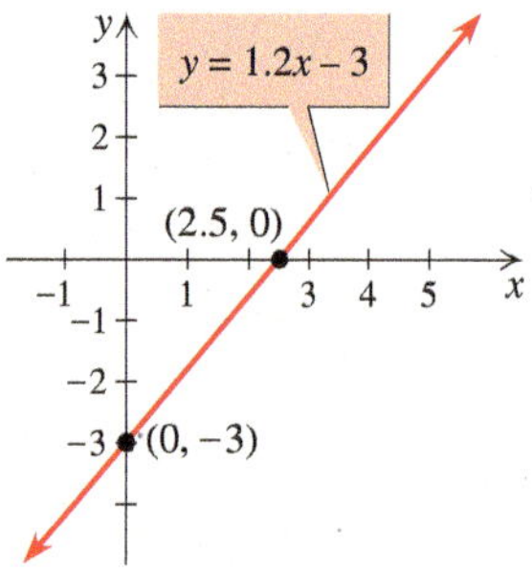


**29.** $6.3 - 4.5x \geq 0$ **30.** $-5.1 - 1.7x < 0$

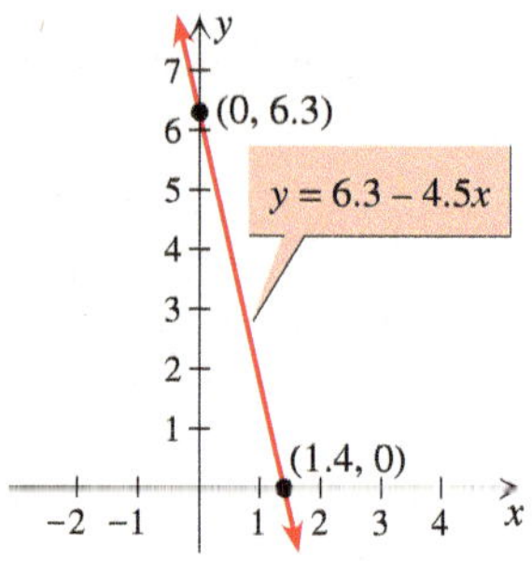


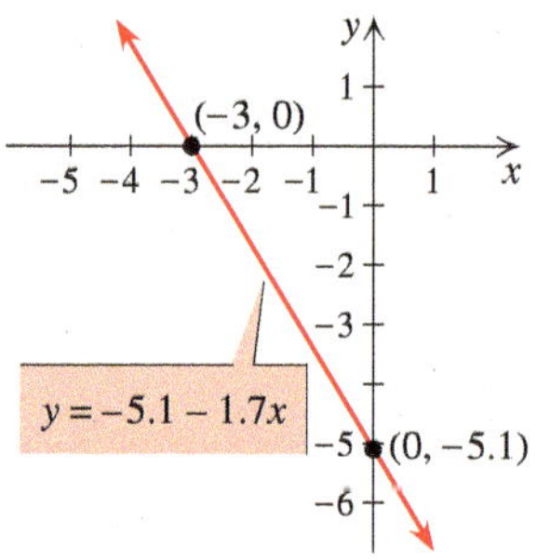


*Solve the inequalities in Exercises 31–34 by reading the following graphs.*

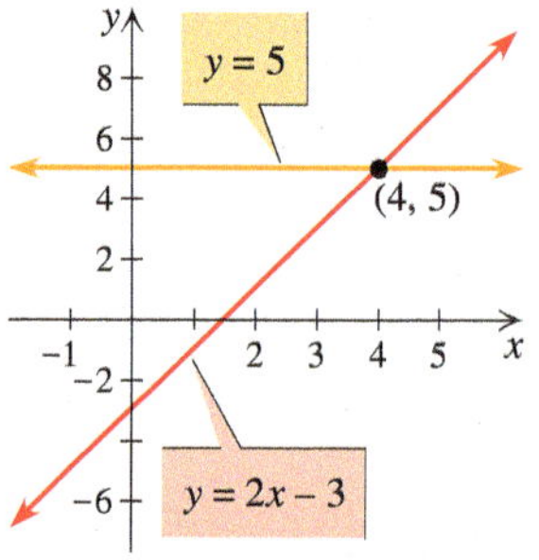


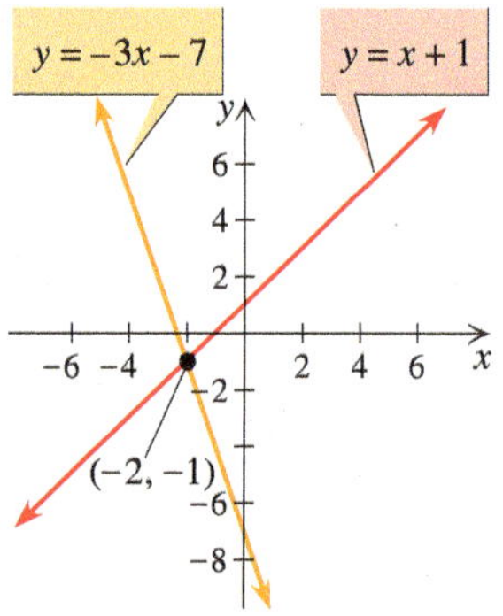


**31.** $2x - 3 > 5$ **32.** $5 \geq 2x - 3$

**33.** $x + 1 \leq -3x - 7$ **34.** $x + 1 > -3x - 7$

*Write as a single interval. See the summary of interval notation for bounded intervals on page 85.*

**35.** $(-3, \infty) \cup (5, \infty)$ **36.** $(-\infty, 0) \cup (-\infty, 6)$

**37.** $(3, 5) \cup (-3, \infty)$ **38.** $(4, 7) \cap (3, \infty)$

**39.** $(-\infty, -2) \cap (-5, \infty)$ **40.** $(-3, \infty) \cap (2, \infty)$

**41.** $(-\infty, -5) \cap (-2, \infty)$ **42.** $(-\infty, -3) \cup (-7, \infty)$

**43.** $(-\infty, 4) \cup [4, 5]$ **44.** $[3, 5] \cup [5, 7]$

*Solve each compound inequality. Write the solution set using interval notation and graph it.*

**45.** $5 > 8 - x$ and $1 + 0.5x < 4$

**46.** $5 - x < 4$ and $0.2x - 5 < 1$

**47.** $\frac{2x - 5}{-2} < 2$ and $\frac{2x + 1}{3} > 0$

**48.** $\frac{4 - x}{2} > 1$ and $\frac{2x - 7}{-3} < 1$

**49.** $1 - x < 7 + x$ or $4x + 3 > x$

**50.** $5 + x > 3 - x$ or $2x - 3 > x$

**51.** $\frac{1}{2}(x + 1) > 3$ or $0 < 7 - x$

**52.** $\frac{1}{2}(x + 6) > 3$ or $4(x - 1) < 3x - 4$

**53.** $1 - \frac{3}{2}x < 4$ and $\frac{1}{4}x - 2 \le -3$

**54.** $\frac{3}{5}x - 1 > 2$ and $5 - \frac{2}{5}x \ge 3$

**55.** $1 < 3x - 5 < 7$

**56.** $-3 \le 4x + 9 \le 17$

**57.** $-2 \le 4 - 6x < 22$

**58.** $-13 < 5 - 9x \le 41$

*Solve each absolute value inequality. Write the solution set using interval notation and graph it. See the summary of basic absolute value inequalities on page 87.*

**59.** $|3x - 1| < 2$

**60.** $|4x - 3| \le 5$

**61.** $|5 - 4x| \le 1$

**62.** $|6 - x| < 6$

**63.** $|x - 1| \ge 1$

**64.** $|x + 2| > 5$

**65.** $|5 - x| > 3$

**66.** $|3 - 2x| \ge 5$

**67.** $5 \ge |4 - x|$

**68.** $3 < |2x - 1|$

**69.** $|5 - 4x| < 0$

**70.** $|3x - 7| \ge -5$

**71.** $|4 - 5x| < -1$

**72.** $|2 - 9x| \ge 0$

**73.** $3|x - 2| + 6 > 9$

**74.** $3|x - 1| + 2 < 8$

**75.** $\left|\frac{x - 3}{2}\right| > 1$

**76.** $\left|\frac{9 - 4x}{2}\right| < 3$

*Write an inequality of the form $|x - a| < k$ or of the form $|x - a| > k$ so that the inequality has the given solution set.*

**HINT** $|x - a| < k$ means that $x$ is less than $k$ units from $a$ and $|x - a| > k$ means that $x$ is more than $k$ units from $a$ on the number line.

**77.** $(-5, 5)$

**78.** $(-2, 2)$

**79.** $(-\infty, -3) \cup (3, \infty)$

**80.** $(-\infty, -1) \cup (1, \infty)$

**81.** $(4, 8)$

**82.** $(-3, 9)$

**83.** $(-\infty, 3) \cup (5, \infty)$

**84.** $(-\infty, -1) \cup (5, \infty)$

*For each graph write an absolute value inequality that has the given solution set.*

**85.**
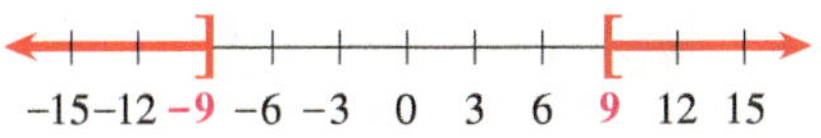


**86.**
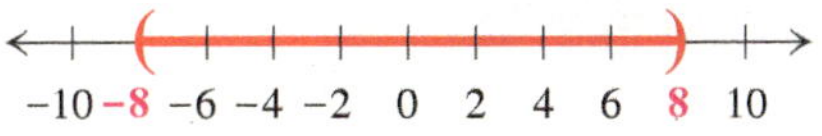


**87.**
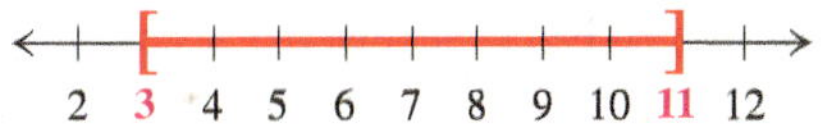


**88.**
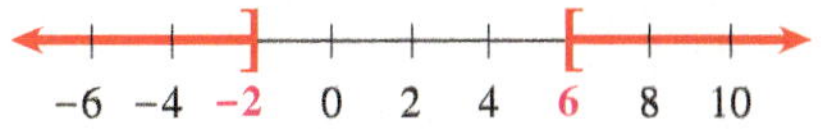


**89.**
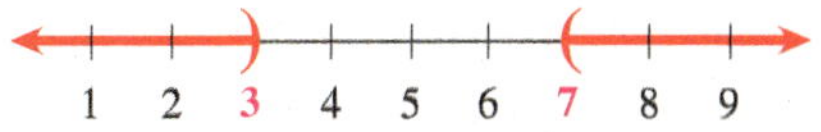


**90.**
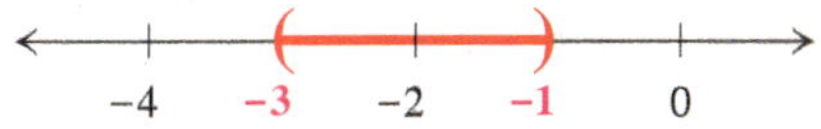


*Recall that $\sqrt{w}$ is a real number only if $w \ge 0$ and $1/w$ is a real number only if $w \ne 0$. For what values of $x$ is each of the following expressions a real number?*

**91.** $\sqrt{x - 2}$

**92.** $\sqrt{3x - 1}$

**93.** $\frac{1}{\sqrt{2 - x}}$

**94.** $\frac{5}{\sqrt{3 - 2x}}$

**95.** $\sqrt{|x| - 3}$

**96.** $\sqrt{5 - |x|}$

*Solve each problem.*

**97.** *Price Range for a Car* Yolanda is shopping for a used car in a city where the sales tax is 10% and the title and license fee is \$300. If the maximum that she can spend is \$8000, then she should look at cars in what price range?

**98.** *Price of a Burger* The price of Elaine's favorite Big Salad at the corner restaurant is 10 cents more than the price of Jerry's hamburger. After treating a group of friends to lunch, Jerry is certain that for 10 hamburgers and 5 salads he spent more than \$9.14, but not more than \$13.19, including tax at 8% and a 50 cent tip. In what price range is a hamburger?

**99.** *Final Exam Score* Lucky scored 65 points on his Psychology 101 midterm. If the average of his midterm and final must be between 79 and 90 inclusive for a B, then for what range of scores on the final exam would Lucky get a B? Both tests have a maximum of 100 points.

**HINT** Write a compound inequality with his average between 79 and 90 inclusive.

**100.** *Bringing Up Your Average* Felix scored 52 points and 64 points on his first two tests in Sociology 212. What must he get on the third test to get an average for the three tests of at least 70? All tests have a maximum of 100 points.

**101.** *Weight Average with Whole Numbers* Ingrid scored 65 points on her calculus midterm. If her final exam counts twice as much as her midterm exam, then for what range of scores on her final would she get an average between 79 and 90? Both tests have a maximum of 100 points.

**102.** *Weighted Average with Fractions* Elizabeth scored 64, 75, and 80 on three equally weighted tests in French. If the final exam score counts for two-thirds of the grade and the other tests count for one-third, then what range of scores on the final exam would give her a final average over 70? All tests have a maximum of 100 points.

**HINT** For this weighted average multiply the final exam score by 2/3 and the average of the other test scores by 1/3.

**103.** *Maximum Girth* United Parcel Service (UPS) defines girth of a box as the sum of the length, twice the width, and twice the height. The maximum length that can be shipped with UPS is 108 in. and the maximum girth is 130 in. If a box has a length of 40 in. and a width of 30 in. then in what range must the height fall?

**104.** *Raising a Batting Average* At one point during the 2008 season, a baseball player had 97 hits in 387 times at bat for an average of 0.251.

**a.** How many more times would he have to bat to get his average over 0.300, assuming he got a hit every time?

**b.** How many more times would he have to bat to get his average over 0.300, assuming he got a hit 50% of the time?

Figure for Exercise 104

**105.** *Bicycle Gear Ratio* The gear ratio $r$ for a bicycle is defined by the following formula

$$r = \frac{Nw}{n},$$

where $N$ is the number of teeth on the chainring (by the pedal), $n$ is the number of teeth on the cog (by the wheel), and $w$ is the wheel diameter in inches. The following chart gives uses for the various gear ratios.

| Ratio | Use |
|---|---|
| $r < 90$ | down hill |
| $70 < r \leq 90$ | level |
| $50 < r \leq 70$ | mild hill |
| $35 < r \leq 50$ | long hill |

A bicycle with a 27-in.-diameter wheel has 50 teeth on the chainring and 5 cogs with 14, 17, 20, 24, and 29 teeth. Find the gear ratio with each of the five cogs. Does this bicycle have a gear ratio for each of the four types of pedaling described in the table?

**106.** *Selecting the Cogs* Use the formula from the previous exercise to answer the following.

**a.** If a single-speed 27-in. bicycle has 17 teeth on the cog, then for what numbers of teeth on the chainring will the gear ratio be between 60 and 80?

**b.** If a 26-in. bicycle has 40 teeth on the chainring, then for what numbers of teeth on the cog will the gear ratio be between 60 and 75?

**107.** *Expensive Models* Two of the ten most expensive cars in the world are the BMW Z8 and the Mercedes Benz CL 600 (www.forbes.com). The prices of these two cars differ by more than \$10,000. The price of the Z8 is \$130,645.

**a.** Assuming that you do not know which model is more expensive, write an absolute value inequality that describes this situation. Use $x$ for the price of the CL 600.

**b.** What are the possibilities for the price of the CL 600?

**108.** *Differences in Prices* There is less than a \$5100 difference between the base price of a \$21,195 Subaru Forester 2.5X and a comparable Ford (www.subaru.com).

**a.** Write an absolute value inequality that describes this situation. Use $x$ for the price of the Ford.

**b.** What are the possibilities for the price of the Ford?

**109.** *Controlling Temperature* Michelle is trying to keep the water temperature in her chemistry experiment at 35°C. For the experiment to work, the relative error for the actual temperature must be less than 1%. Write an absolute value inequality for the actual temperature. Find the interval in which the actual temperature must lie.

**110.** *Laying Out a Track* Melvin is attempting to lay out a track for a 100-m race. According to the rules of competition, the relative error of the actual length must be less than 0.5%. Write an absolute value inequality for the actual length. Find the interval in which the actual length must fall.

**111.** *Acceptable Bearings* A spherical bearing is to have a circumference of 7.2 cm with an error of no more than 0.1 cm. Use an absolute value inequality to find the acceptable range of values for the diameter of the bearing.

**112.** *Acceptable Targets* A manufacturer makes circular targets that have an area of 15 $ft^2$. According to competition rules, the area can be in error by no more than 0.5 $ft^2$. Use an absolute value inequality to find the acceptable range of values for the radius.

Figure for Exercise 112

**113.** *Per Capita Income* The 2007 per capita income for the United States was \$38,611 (U.S. Census Bureau, www.census.gov). The accompanying table shows the per capita income for 2007 for selected states.

**Table for Exercise 113**

| State | Per Capita Income |
|---|---|
| Alabama | 32,404 |
| Colorado | 41,042 |
| Georgia | 33,457 |
| Iowa | 35,023 |
| Maryland | 46,021 |
| Missouri | 34,389 |
| New Jersey | 49,194 |
| Ohio | 34,874 |
| S. Carolina | 31,013 |
| Vermont | 36,670 |

**a.** If $a$ is the per capita income for a state, then for which states is $|a - 38{,}611| < 3000$?

**b.** For which states is $|a - 38{,}611| > 5000$?

**114.** *Making a Profit* A strawberry farmer paid \$4200 for planting and fertilizing her strawberry crop. She must also pay \$2.40 per flat (12 pints) for picking and packing the berries and \$300 rent for space in a farmers' market where she sells the berries for \$11 per flat. For what number of flats will her revenue exceed her costs?

## ▶ RETHINKING

**115.** Solve $x^2 + 2x = 0$.

**116.** Solve $x^2 + 2x = 9$.

**117.** Find the equation of the line (in standard form) through $(3, -4)$ that is perpendicular to $2x - y = 1$.

**118.** Find the distance between the points $(-3, 5)$ and $(2, 8)$ and the midpoint of the line segment joining the points.

**119.** Solve $3y - w = ay + 9$ for $y$.

**120.** Solve $5|2x - 9| = 0$.

## THINKING OUTSIDE THE BOX X

*One in a Million* If you write the integers from 1 through 1,000,000 inclusive, then how many ones will you write?

## POP QUIZ 1.8

*Solve each inequality. Write the solution set using interval notation.*

**1.** $x - \sqrt{2} \ge 0$

**2.** $6 - 2x < 0$

**3.** $x > 4$ or $x \ge -1$

**4.** $x - 1 > 5$ and $2x < 18$

**5.** $|x| > 6$

**6.** $|x - 1| \le 2$

## LINKING concepts... For Individual or Group Explorations

### Modeling the Cost of Copying

*A company can rent a copy machine for five years from American Business Machines for \$105 per month plus \$0.08 per copy. The same copy machine can be purchased for \$6500 with a per copy cost of \$0.04 plus \$25 per month for a maintenance contract. After five years the copier is worn out and worthless. To help the company make its choice, ABM sent the accompanying graph (see next page).*

**a)** Write a formula for the total cost of renting and using the copier for five years in terms of the number of copies made during five years.

*(continued on next page)*

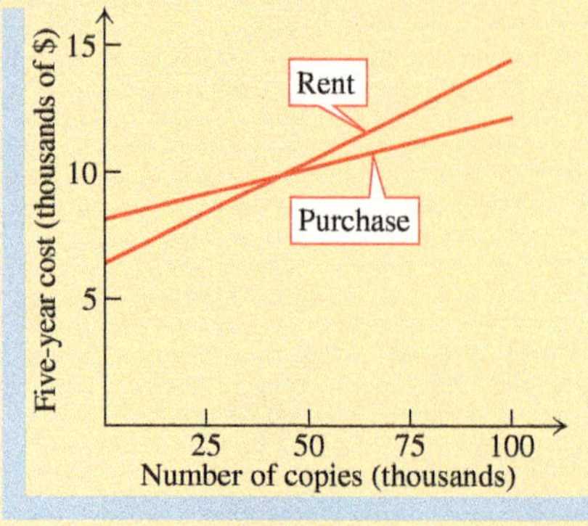

**b)** Write a formula for the total cost of buying and using the copier for five years in terms of the number of copies made during five years.

**c)** For what number of copies does the total cost of renting exceed \$10,000?

**d)** For what number of copies does the total cost of buying exceed \$10,000?

**e)** Use an absolute value inequality and solve it to find the number of copies for which the two plans differ by less than \$1000.

**f)** For what number of copies is the cost of renting equal to the cost of buying?

**g)** If the company estimates that it will make between 40,000 and 50,000 copies in five years, then which plan is better?

# Highlights

## 1.1 Equations in One Variable

| | | |
|---|---|---|
| **Linear Equation** | $ax + b = 0$ where $a$ and $b$ are real numbers with $a \neq 0$. | $5x - 10 = 0$ |
| **Properties of Equality** | An equivalent equation is obtained by adding, subtracting, multiplying, or dividing each side of an equation by the same nonzero real number. | $5x = 10$<br>$x = 2$ |
| **Solution Set** | The set of all real numbers that satisfy an equation | Solution set to $5x - 10 = 0$ is $\{2\}$. |
| **Identity** | An equation satisfied by all real numbers for which both sides are defined | $\frac{y}{y} = 1$, solution set: $(-\infty, 0) \cup (0, \infty)$ |
| **Conditional Equation** | An equation that has at least one solution but is not an identity | $2z - 1 = 0$<br>Solution set: $\{1/2\}$ |
| **Inconsistent Equation** | An equation that has no solutions | $w = w + 1$, solution set: $\varnothing$ |
| **Absolute Value Equations** | $\lvert x\rvert = k$ for $k > 0 \Leftrightarrow x = k$ or $x = -k$<br>$\lvert x\rvert = k$ for $k < 0$ is inconsistent<br>$\lvert x\rvert = 0 \Leftrightarrow x = 0$ | $\lvert t\rvert = 1 \Leftrightarrow t = 1$ or $t = -1$<br>$\lvert m\rvert = -2$, solution set: $\varnothing$<br>$\lvert q - 4\rvert = 0 \Leftrightarrow q = 4$ |

## 1.2 Constructing Models to Solve Problems

| | | |
|---|---|---|
| **Formula** | An equation in two or more variables that is usually of an applied nature | $A = \pi r^2$, $P = 2L + 2W$<br>$E = mc^2$, $F = \frac{9}{5}C + 32$ |
| **Function** | A rule for determining the value of one variable from the values of one or more other variables. The first variable is a function of the other variable(s). | $A = \pi r^2$<br>Area of a circle is a function of its radius. |
| **Solving for a Variable** | The properties of equality are used to solve a formula for one variable in terms of the others. | $P = 2L + 2W$<br>$2L = P - 2W$<br>$L = (P - 2W)/2$ |

## 1.3 Equations and Graphs in Two Variables

| | | |
|---|---|---|
| **Distance Formula** | Distance between $(x_1, y_1)$ and $(x_2, y_2)$ is $\sqrt{(x_2 - x_1)^2 + (y_2 - y_1)^2}$. | For $(1, 2)$ and $(4, -2)$,<br>$\sqrt{(4 - 1)^2 + (-2 - 2)^2} = 5$. |
| **Midpoint Formula** | The midpoint of the line segment with endpoints $(x_1, y_1)$ and $(x_2, y_2)$ is $\left(\frac{x_1 + x_2}{2}, \frac{y_1 + y_2}{2}\right)$. | For $(0, -4)$ and $(6, 2)$,<br>$\left(\frac{0 + 6}{2}, \frac{-4 + 2}{2}\right) = (3, -1)$. |
| **Equation of a Circle** | The graph of $(x - h)^2 + (y - k)^2 = r^2\ (r > 0)$ is a circle with center $(h, k)$ and radius $r$. | Circle: $(x - 1)^2 + (y + 2)^2 = 9$<br>Center $(1, -2)$, radius 3 |
| **Linear Equation: Standard Form** | $Ax + By = C$ where $A$ and $B$ are not both zero, $x = h$ is a vertical line, $y = k$ is a horizontal line. | $2x + 3y = 6$ is a line.<br>$x = 5$ is a vertical line.<br>$y = 7$ is a horizontal line. |

## 1.4 Linear Equations in Two Variables

| | | |
|---|---|---|
| **Slope Formula** | The slope of a line through $(x_1, y_1)$ and $(x_2, y_2)$ is $(y_2 - y_1)/(x_2 - x_1)$ provided $x_1 \neq x_2$. | $(1, 2), (3, -6)$<br>slope $\frac{-6 - 2}{3 - 1} = -4$ |
| **Slope-Intercept Form** | $y = mx + b$, slope $m$, $y$-intercept $(0, b)$, $y$ is a linear function of $x$. | $y = 2x + 5$, slope 2,<br>$y$-intercept $(0, 5)$ |
| **Point-Slope Form** | The line through $(x_1, y_1)$ with slope $m$ is $y - y_1 = m(x - x_1)$. | Point $(-3, 2)$, $m = 5$<br>$y - 2 = 5(x - (-3))$ |
| **Parallel Lines** | Two nonvertical lines are parallel if and only if their slopes are equal. | $y = 7x - 1$ and $y = 7x + 4$ are parallel. |
| **Perpendicular Lines** | Two lines with slopes $m_1$ and $m_2$ are perpendicular if and only if $m_1 m_2 = -1$. | $y = \frac{1}{2}x + 4$ and $y = -2x - 3$ are perpendicular. |

## 1.5 Scatter Diagrams and Curve Fitting

| | |
|---|---|
| **Scatter Diagram** | A graph of a set of ordered pairs |
| **Linear Relationship** | If the points in a scatter diagram appear to be scattered about a line, then there is a linear relationship between the two variables. |

## 1.6 Complex Numbers

| | | |
|---|---|---|
| **Standard Form** | Numbers of the form $a + bi$ where $a$ and $b$ are real numbers, $i = \sqrt{-1}$, and $i^2 = -1$ | $2 + 3i$, $-\pi + i\sqrt{2}$, 6, 0, $\frac{1}{2}i$ |
| **Add, Subtract, Multiply** | Add, subtract, and multiply like binomials with variable $i$, using $i^2 = -1$ to simplify. | $(3 - 2i)(4 + 5i)$<br>$= 12 + 7i - 10i^2$<br>$= 22 + 7i$ |

| | | |
|---|---|---|
| **Divide** | Divide by multiplying the numerator and denominator by the complex conjugate of the denominator. | $6/(1+i)$ $= \frac{6(1-i)}{(1+i)(1-i)} = 3 - 3i$ |
| **Square Roots of Negative Numbers** | Square roots of negative numbers must be converted to standard form using $\sqrt{-b} = i\sqrt{b}$, for $b > 0$, before doing computations. | $\sqrt{-4} \cdot \sqrt{-9} = 2i \cdot 3i = -6$ |

## 1.7 Quadratic Equations

| | | |
|---|---|---|
| **Quadratic Equation** | $ax^2 + bx + c = 0$ where $a$, $b$, and $c$ are real and $a \neq 0$ | $x^2 + 2x - 3 = 0$ |
| **Methods for Solving Quadratic Equations** | Factoring: factor the quadratic polynomial and set the factors equal to zero. | $(w+3)(w-1) = 0 \Rightarrow$ $w + 3 = 0$ or $w - 1 = 0$ |
| | Square root property: $x^2 = k$ is equivalent to $x = \pm\sqrt{k}$. | $m^2 = 5 \Rightarrow m = \pm\sqrt{5}$ |
| | Completing the square: complete the square and then apply the square root property. | $(w+1)^2 = 4 \Rightarrow$ $w + 1 = \pm 2$ |
| | Quadratic formula: solutions to $ax^2 + bx + c = 0$ are $x = \frac{-b \pm \sqrt{b^2 - 4ac}}{2a}$. | $t^2 + 2t - 3 = 0 \Rightarrow$ $t = \frac{-2 \pm \sqrt{2^2 - 4(1)(-3)}}{2(1)}$ |

## 1.8 Linear and Absolute Value Inequalities

| | | |
|---|---|---|
| **Linear Inequalities** | The inequality symbol is reversed if the inequality is multiplied or divided by a negative number. | $4 - 2x > 10$ <br> $-2x > 6$ <br> $x < -3$ |
| **Absolute Value Inequalities** | $\lvert x \rvert > k\ (k > 0) \Leftrightarrow x > k$ or $x < -k$ <br> $\lvert x \rvert < k\ (k > 0) \Leftrightarrow -k < x < k$ <br> $\lvert x \rvert \leq 0 \Leftrightarrow x = 0$ <br> $\lvert x \rvert \geq 0 \Leftrightarrow x$ is any real number | $\lvert y \rvert > 1 \Leftrightarrow y > 1$ or $y < -1$ <br> $\lvert z \rvert < 2 \Leftrightarrow -2 < z < 2$ <br> $\lvert 2b - 5 \rvert \leq 0 \Leftrightarrow 2b - 5 = 0$ <br> All real numbers satisfy $\lvert 3s - 7 \rvert \geq 0$. |

# Chapter 1 Review Exercises

*Find all real solutions to each equation.*

**1.** $3x - 2 = 0$

**2.** $3x - 5 = 5(x + 7)$

**3.** $\frac{1}{2}y - \frac{1}{3} = \frac{1}{4}y + \frac{1}{5}$

**4.** $\frac{1}{2} - \frac{w}{5} = \frac{w}{4} - \frac{1}{8}$

**5.** $\frac{2}{x} = \frac{3}{x - 1}$

**6.** $\frac{5}{x + 1} = \frac{2}{x - 3}$

**7.** $\frac{x + 1}{x - 1} = \frac{x + 2}{x - 3}$

**8.** $\frac{x + 3}{x - 8} = \frac{x + 7}{x - 4}$

*For each pair of points, find the distance between them and the midpoint of the line segment joining them.*

**9.** $(-3, 5), (2, -6)$

**10.** $(-1, 1), (-2, -3)$

**11.** $(1/2, 1/3), (1/4, 1)$

**12.** $(0.5, 0.2), (-1.2, 2.1)$

*Sketch the graph of each equation. For the circles, state the center and the radius. For the lines state the intercepts.*

**13.** $x^2 + y^2 = 25$

**14.** $(x - 2)^2 + y^2 = 1$

**15.** $x^2 + 4x + y^2 = 0$

**16.** $x^2 - 6x = 2y - y^2 - 1$

**17.** $x + y = 25$

**18.** $2x - y = 40$

**19.** $y = 3x - 4$

**20.** $y = -\frac{1}{2}x + 4$

**21.** $x = 5$

**22.** $y = 6$

*Solve each problem.*

**23.** Write in standard form the equation of the circle that has center $(-3, 5)$ and radius $\sqrt{3}$.

**24.** Find the center and radius for the circle $x^2 + y^2 = x - 2y + 1$.

**25.** Find the $x$- and $y$-intercepts for the graph of $3x - 4y = 12$.

**26.** What is the $y$-intercept for the graph of $y = 5$?

**27.** Find the slope of the line that goes through $(3, -6)$ and $(-1, 2)$.

**28.** Find the slope of the line $3x - 4y = 9$.

**29.** Find the equation (in slope-intercept form) for the line through $(-2, 3)$ and $(5, -1)$.

**30.** Find the equation (in standard form using only integers) for the line through $(-1, -3)$ and $(2, -1)$.

**31.** Find the equation (in standard form using only integers) for the line through $(2, -4)$ that is perpendicular to $3x + y = -5$.

**32.** Find the equation (in slope-intercept form) for the line through $(2, -5)$ that is parallel to $2x - 3y = 5$.

*Solve each equation for y.*

**33.** $2x - 3y = 6$

**34.** $x(y - 2) = 1$

**35.** $xy = 1 + 3y$

**36.** $x^2y = 1 + 9y$

**37.** $ax + by = c$

**38.** $\frac{1}{y} = \frac{1}{x} + \frac{1}{2}$

*Evaluate the discriminant for each equation, and use it to determine the number of real solutions to the equation.*

**39.** $x^2 + 2 = 4x$

**40.** $y^2 + 2 = 3y$

**41.** $4w^2 = 20w - 25$

**42.** $2x^2 - 3x + 10 = 0$

*Write each expression in the form a + bi, where a and b are real numbers.*

**43.** $(3 - 7i) + (-4 + 6i)$

**44.** $(-6 - 3i) - (3 - 2i)$

**45.** $(4 - 5i)^2$

**46.** $7 - i(2 - 3i)^2$

**47.** $(1 - 3i)(2 + 6i)$

**48.** $(0.3 + 2i)(0.3 - 2i)$

**49.** $(2 - 3i) \div i$

**50.** $(-2 + 4i) \div (-i)$

**51.** $(1 - i) \div (2 + i)$

**52.** $(3 + 6i) \div (4 - i)$

**53.** $\frac{1 + i}{2 - 3i}$

**54.** $\frac{3 - i}{4 - 3i}$

**55.** $\frac{6 + \sqrt{-8}}{2}$

**56.** $\frac{-2 - \sqrt{-18}}{2}$

**57.** $\frac{-6 + \sqrt{(-2)^2 - 4(-1)(-6)}}{-8}$

**58.** $\frac{-9 - \sqrt{(-9)^2 - 4(-3)(-9)}}{-6}$

**59.** $i^{34} + i^{19}$

**60.** $\sqrt{6} + \sqrt{-3}\sqrt{-2}$

*Find all real or imaginary solutions to each equation. Use the method of your choice on each problem.*

**61.** $x^2 - 5 = 0$

**62.** $3x^2 - 54 = 0$

**63.** $x^2 + 8 = 0$

**64.** $x^2 + 27 = 0$

**65.** $2x^2 + 1 = 0$

**66.** $3x^2 + 2 = 0$

**67.** $(x - 2)^2 = 17$

**68.** $(2x - 1)^2 = 9$

**69.** $x^2 - x - 12 = 0$

**70.** $2x^2 - 11x + 5 = 0$

**71.** $b^2 + 10 = 6b$

**72.** $4t^2 + 17 = 16t$

**73.** $s^2 - 4s + 1 = 0$

**74.** $3z^2 - 2z - 1 = 0$

**75.** $4x^2 - 4x - 5 = 0$

**76.** $9x^2 - 30x + 23 = 0$

**77.** $x^2 - 2x + 2 = 0$

**78.** $x^2 - 4x + 5 = 0$

**79.** $\frac{1}{x} + \frac{1}{x - 1} = \frac{3}{2}$

**80.** $\frac{2}{x - 2} - \frac{3}{x + 2} = \frac{1}{2}$

*Solve each equation.*

**81.** $|3q - 4| = 2$

**82.** $|2v - 1| = 3$

**83.** $|2h - 3| = 0$

**84.** $4|x - 3| = 0$

**85.** $|5 - x| = -1$

**86.** $|3y - 1| = -2$

*Solve each inequality. State the solution set using interval notation and graph the solution set.*

**87.** $4x - 1 > 3x + 2$

**88.** $6(x - 3) < 5(x + 4)$

**89.** $5 - 2x > -3$

**90.** $7 - x > -6$

**91.** $\frac{1}{2}x - \frac{1}{3} > x + 2$

**92.** $0.06x + 1000 > x + 60$

**93.** $-2 < \frac{x - 3}{2} \le 5$

**94.** $-1 \le \frac{3 - 2x}{4} < 3$

**95.** $3 - 4x < 1$ and $5 + 3x < 8$

**96.** $-3x < 6$ and $2x + 1 > -1$

**97.** $-2x < 8$ or $3x > -3$

**98.** $1 - x < 6$ or $-5 + x < 1$

**99.** $|x - 3| > 2$

**100.** $|4 - x| \le 3$

**101.** $|2x - 7| \le 0$

**102.** $|6 - 5x| < 0$

**103.** $|7 - 3x| > -4$

**104.** $|4 - 3x| \ge 1$

*Solve each equation or inequality by reading the accompanying graph.*

**105.** $x - 0.5(30 - x) = 0$

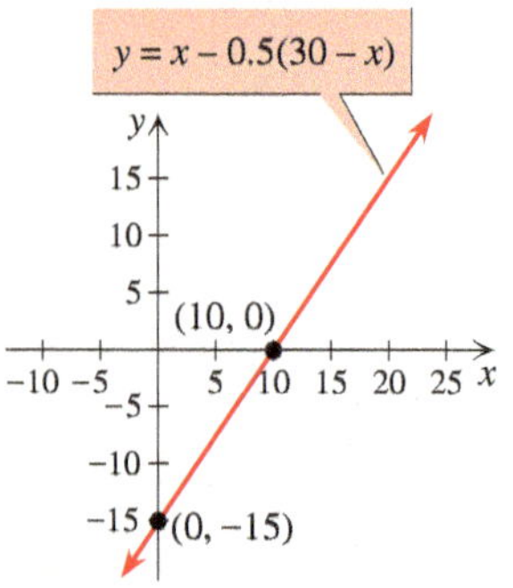


**106.** $x^2 + 4x - 780 = 0$

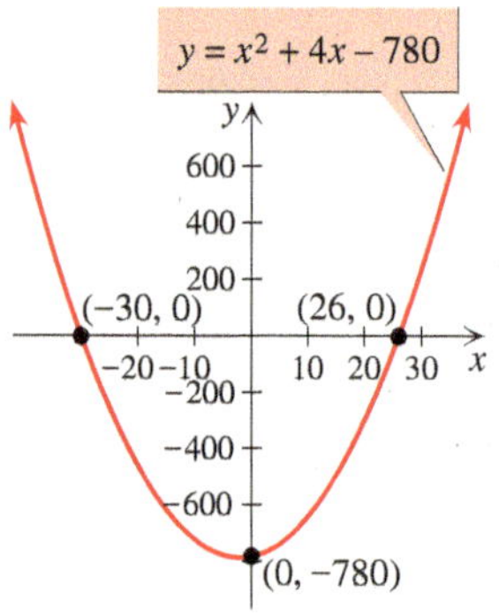


**107.** $0.6x - 4.8 < 0$

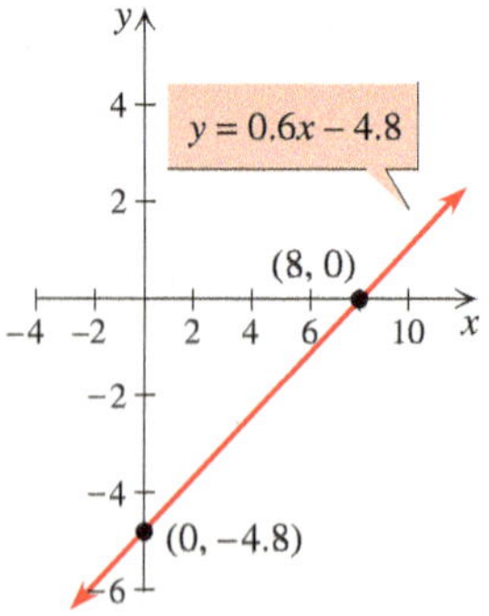


**108.** $36 - 1.2x \geq 0$

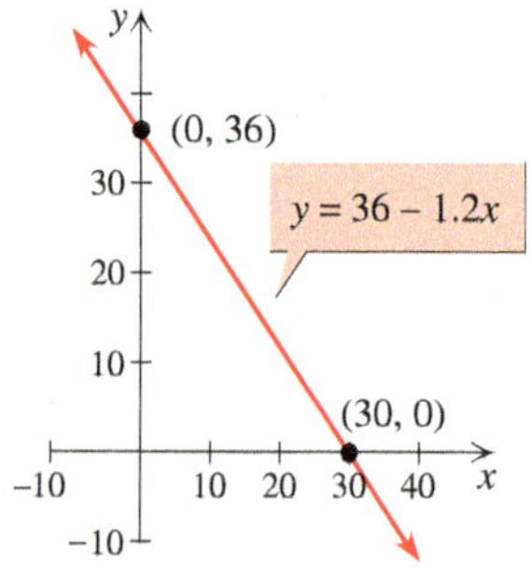


*Solve each problem. Use an equation or an inequality as appropriate. For problems in which the answer involves an irrational number, find the exact answer and an approximate answer.*

**109.** *Folding Sheet Metal* A square is to be cut from each corner of an 8-in. by 11-in. rectangular piece of copper and the sides are to be folded up to form a box as shown in the figure. If the area of the bottom is to be 50 in.$^2$, what is the length of the side of the square to be cut from the corner?

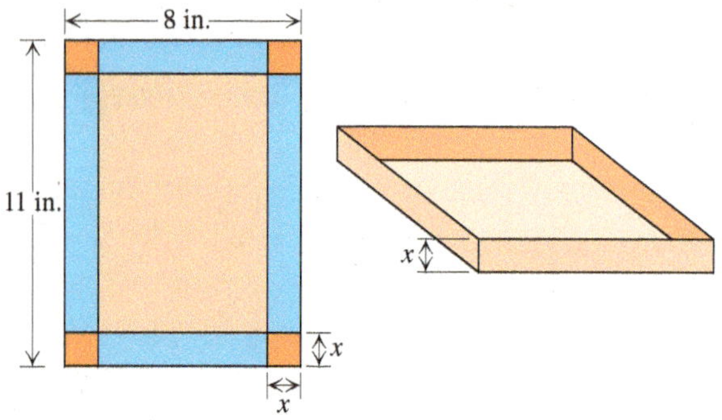


**Figure for Exercise 109**

## Foreshadowing Calculus

Making an open-top box as in Exercise 109 is a classic problem in calculus. However, in calculus we usually find the value of $x$ that maximizes the volume of the box.

**110.** *Peeling Apples* Bart can peel a batch of 3000 lb of apples in 12 hr with the old machine, while Mona can do the same job in 8 hr using the new machine. If Bart starts peeling at 8 A.M. and Mona joins him at 9 A.M., then at what time will they finish the batch working together?

**111.** *Meeting Between Two Cities* Lisa, an architect from Huntsville, has arranged a business luncheon with her client, Taro Numato from Norwood. They plan to meet at a restaurant off the 300-mi highway connecting their two cities. If they leave their offices simultaneously and arrive at the restaurant simultaneously, and Lisa averages 50 mph while Taro averages 60 mph, then how far from Huntsville is the restaurant?

**112.** *Driving Speed* Lisa and Taro agree to meet at the construction site located 100 mi from Norwood on the 300-mi highway connecting Huntsville and Norwood. Lisa leaves Huntsville at noon, while Taro departs from Norwood 1 hr later. If they arrive at the construction site simultaneously and Lisa's driving speed averaged 10 mph faster than Taro's, then how fast did she drive?

**113.** *Fish Population* A channel was dug to connect Homer and Mirror Lakes. Before the channel was dug, a biologist estimated that 20% of the fish in Homer Lake and 30% of the fish in Mirror Lake were bass. After the lakes were joined, the biologist made another estimate of the bass population in order to set fishing quotas. If she decided that 28% of the total fish population of 8000 were bass, then how many fish were in Homer Lake originally?

**114.** *Support for Gambling* Eighteen pro-gambling representatives in the state house of representatives bring up a gambling bill every year. After redistricting, four new representatives are added to the house, causing the percentage of pro-gambling representatives to increase by 5 percentage points. If all four of the new representatives are pro-gambling and they still do not constitute a majority of the house, then how many representatives are there in the house after redistricting?

**115.** *Hiking Distance* The distance between Marjorie and the dude ranch was $4\sqrt{34}$ mi straight across a rattlesnake-infested canyon. Instead of crossing the canyon, she hiked due north for a long time and then hiked due east for the shorter leg of the journey to the ranch. If she averaged 4 mph and it took her 8 hr to get to the ranch, then how far did she hike in a northerly direction?

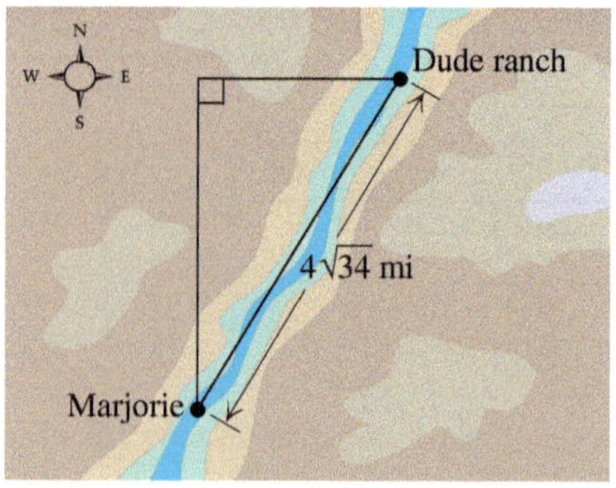


**Figure for Exercise 115**

**116.** *Golden Rectangle* The *golden rectangle* of the ancient Greeks was thought to be the rectangle with the shape that was most pleasing to the eye. The golden rectangle was defined as a rectangle that would retain its original shape after removal of a $W$ by $W$ square from one end, as shown in the figure. So the length and width of the original rectangle must satisfy

$$\frac{L}{W} = \frac{W}{L - W}.$$

If the length of a golden rectangle is 20 m, then what is its width? If the width of a golden rectangle is 8 m, then what is its length?

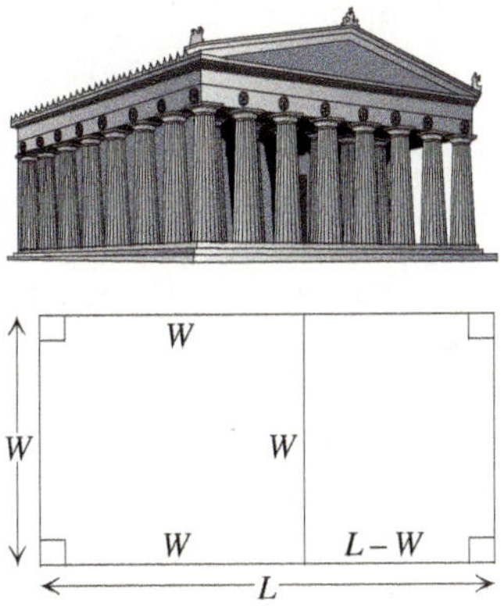


**Figure for Exercise 116**

**117.** *Price Range for a Haircut* A haircut at Joe's Barber Shop costs \$50 less than a haircut at Renee's French Salon. In fact, you can get five haircuts at Joe's for less than the cost of one haircut at Renee's. What is the price range for a haircut at Joe's?

**118.** *Selling Price of a Home* Elena must get at least \$120,000 for her house in order to break even with the original purchase price. If the real estate agent gets 6% of the selling price, then what should the selling price be?

**119.** *Dimensions of a Picture Frame* The length of a picture frame must be 2 in. longer than the width. If Reginald can use between 32 and 50 in. of frame molding for the frame, then what are the possibilities for the width?

**120.** *Saving Gasoline* If Americans drive $10^{12}$ mi per year and the average gas mileage is raised from 27.5 mpg to 29.5 mpg, then how many gallons of gasoline are saved?

**121.** *Increasing Gas Mileage* If Americans continue to drive $10^{12}$ mi per year and the average gas mileage is raised from 29.5 to 31.5 mpg, then how many gallons of gasoline are saved? If the average mpg is presently 29.5, then what should it be increased to in order to achieve the same savings in gallons as the increase from 27.5 to 29.5 mpg?

**122.** *Thickness of Concrete* Alfred used 40 yd$^3$ of concrete to pour a section of interstate highway that was 12 ft wide and 54 ft long. How many inches thick was the concrete?

**Figure for Exercise 122**

**123.** *U.S. Cell Phone Subscribers* The number of cell phone subscribers in the United States reached 256 million in 2008, up from 34 million in 1995, as seen in the accompanying graph.

**a.** Find the equation of the line that passes through the two given points.

**b.** Use the equation to estimate the number of cell phone subscribers in the United States in 2012.

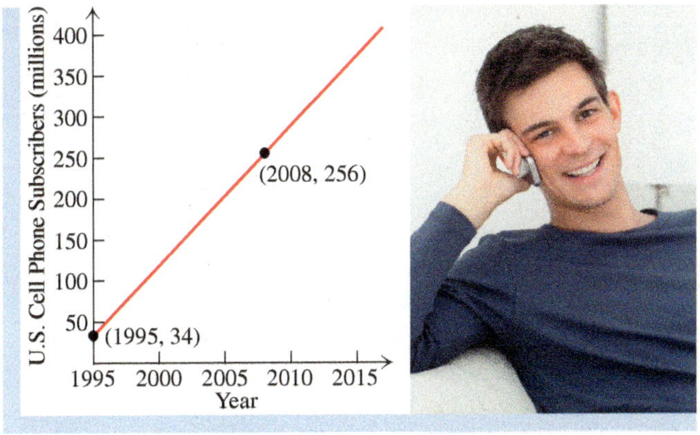


**Figure for Exercise 123**

**124.** *Worldwide Cell Phone Subscribers* The number of cell phone subscribers worldwide reached approximately 3300 million in 2007, up from 145 million in 1996, as seen in the accompanying graph.

**a.** Find the equation of the line that passes through the two given points.

**b.** Use the equation to estimate the number of worldwide cell phone subscribers in 2015.

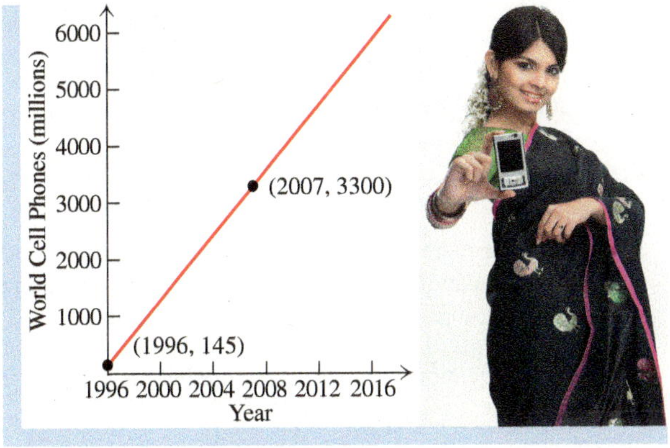


**Figure for Exercise 124**

**125.** *Percent of Body Fat* The average 20-year-old woman has 23% body fat and the average 50-year-old woman has 47% body fat (Infoplease, www.infoplease.com). Assuming that the percentage of body fat is a linear function of age, find the function. Use the function to determine the percentage of body fat in the average 65-year-old woman.

**126.** *Olympic Gold* In 2004 Shawn Crawford won the 200-meter race with a time of 19.79 seconds (www.infoplease.com). In 2008 Usain Bolt won the race with a time of 19.30 seconds. Assuming the winning time is decreasing linearly, as shown in the accompanying figure, express the winning time as a linear function of the year. Use the function to predict the winning time in the 2012 Olympics.

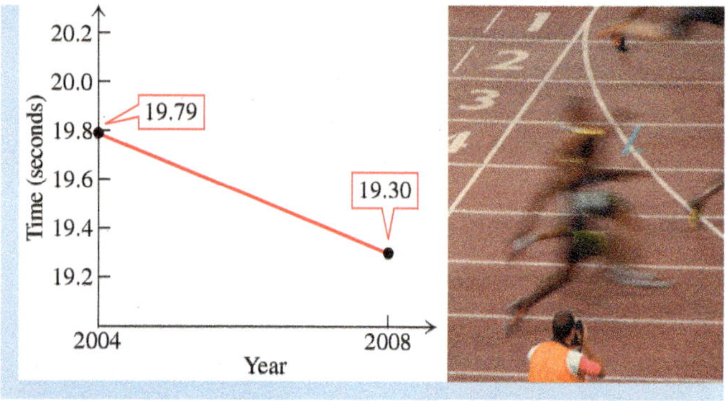


Figure for Exercise 126

**127.** *Price of Prescription Drugs* The accompanying table shows the average price of a prescription drug in the United States for the years 2000 through 2007 (U.S. Census Bureau, www.census.gov).

Table for Exercise 127

| Year | Average Price ($) |
|---|---|
| 2000 | 45.79 |
| 2001 | 50.06 |
| 2002 | 55.37 |
| 2003 | 59.52 |
| 2004 | 63.59 |
| 2005 | 64.86 |
| 2006 | 68.26 |
| 2007 | 72.16 |

**a.** Use linear regression with a graphing calculator to find a linear equation that expresses the price as a function of the year. Let $x = 0$ correspond to the year 2000.

**b.** Use the equation from part (a) to predict the average price in 2015.

**128.** *Number of Prescriptions* The accompanying table shows the number of prescriptions written in the United States for the years 2000 through 2007 (U.S. Census Bureau, www.census.gov).

Table for Exercise 128

| Year | Number (in millions) |
|---|---|
| 2000 | 2865 |
| 2001 | 3009 |
| 2002 | 3139 |
| 2003 | 3215 |
| 2004 | 3274 |
| 2005 | 3279 |
| 2006 | 3420 |
| 2007 | 3516 |

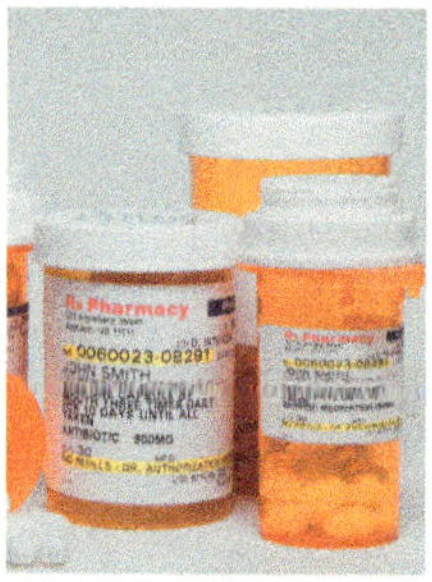

**a.** Use linear regression with a graphing calculator to find a linear equation that expresses the number of prescriptions as a function of the year. Let $x = 0$ correspond to the year 2000.

**b.** Use the equation from part (a) to predict the number of prescriptions of 2012.

**129.** *Tangent Circles* Circle $A$ with center $(1, 1)$ and radius 1 is drawn in the first quadrant as shown in the accompanying figure. Circle $B$ with radius 2 and circle $C$ are placed as shown so that each circle is tangent to the other two circles and the $x$-axis. Find the equations of circles $A$, $B$, and $C$.

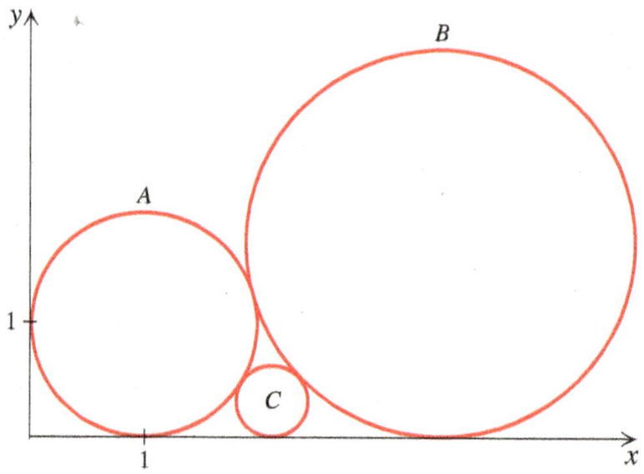


Figure for Exercise 129

## THINKING OUTSIDE THE BOX XI & XII

*Unlucky Number* What is the smallest positive integer that is 13 times the sum of its digits?

*Run for Your Life* A hiker is one-fourth of the way through a narrow train tunnel when he looks over his shoulder to see a train approaching the tunnel at 30 mph. He quickly figures that if he runs in either direction at his top speed he can just make it out of the tunnel. What is his top running speed?

# Chapter 1 Test

*Find all real or imaginary solutions to each equation.*

**1.** $2x + 1 = x - 6$

**2.** $\frac{1}{2}x - \frac{1}{6} = \frac{1}{3}x$

**3.** $3x^2 - 2 = 0$

**4.** $x^2 + 1 = 6x$

**5.** $x^2 + 14 = 9x$

**6.** $\frac{x-1}{x+3} = \frac{x+2}{x-6}$

**7.** $x^2 = 2x - 5$

**8.** $x^2 + 1 = 0$

*Sketch the graph of each equation in the xy-coordinate system.*

**9.** $3x - 4y = 120$

**10.** $x^2 + y^2 = 400$

**11.** $x^2 + y^2 + 4y = 0$

**12.** $y = -\frac{2}{3}x + 4$

**13.** $y = 4$

**14.** $x = -2$

*Perform the indicated operations, and write the answer in the form $a + bi$, where $a$ and $b$ are real.*

**15.** $(4 - 3i)^2$

**16.** $\frac{2-i}{3+i}$

**17.** $i^6 - i^{35}$

**18.** $\sqrt{-8}(\sqrt{-2} + \sqrt{6})$

*Solve each problem.*

**19.** Find the slope of the line $3x - 5y = 8$.

**20.** Find the slope of the line through $(-3, 6)$ and $(5, -4)$.

**21.** Find the slope-intercept form of the equation of the line that goes through $(1, -2)$ and is perpendicular to the line $2x - 3y = 6$.

**22.** Find the equation of the line in slope-intercept form that goes through $(3, -4)$ and is parallel to the line through $(0, 2)$ and $(3, -1)$.

**23.** Find the exact distance between $(-3, 1)$ and $(2, 4)$.

**24.** Find the midpoint of the line segment with endpoints $(-1, 1)$ and $(1, 0)$.

**25.** Find the value of the discriminant for $x^2 - 5x + 9 = 0$. How many real solutions are there for this equation?

**26.** Solve $5 - 2y = 4 + 3xy$ for $y$.

*Solve each inequality in one variable. State the solution set using interval notation and graph it on the number line.*

**27.** $3 - 2x > 7$

**28.** $\frac{x}{2} > 3$ and $5 < x$

**29.** $|2x - 1| \le 3$

**30.** $2|x - 3| + 1 > 5$

*Solve each problem.*

**31.** Terry had a square patio. After expanding the length by 20 ft and the width by 10 ft, the area was 999 $ft^2$. What was the original area?

**32.** How many gallons of 20% alcohol solution must be mixed with 10 gal of a 50% alcohol solution to obtain a 30% alcohol solution?

**33.** *Median Price* The median price of a home in San Diego was \$311,000 in 1997 and \$495,000 in 2007.

**a.** Assuming that the price is a linear function of the year, write a formula for that function.

**b.** Use the formula to predict the median price to the nearest thousand in 2015.

**34.** *Number of Farms* The accompanying table gives the number of farms in the United States for the years 2001 through 2007. (U.S. Department of Agriculture, www.usda.gov).

**Table for Exercise 34**

| Year | Farms (in thousands) |
|---|---|
| 2001 | 2149 |
| 2002 | 2135 |
| 2003 | 2127 |
| 2004 | 2113 |
| 2005 | 2099 |
| 2006 | 2090 |
| 2007 | 2082 |

**a.** Find the equations that express the number of farms as a function of the year by using linear regression and quadratic regression. Let $x = 1$ correspond to 2001.

**b.** Predict the number of farms in the United States in 2012 using linear regression and quadratic regression.

# CONCEPTS OF calculus...

## Limits

*In algebra we can evaluate algebraic expressions for any acceptable value of the variable. The following tables show values of $x^2 + 5$ for certain values of x.*

| $x$ | 2.9 | 2.99 | 2.999 |
|---|---|---|---|
| $x^2 + 5$ | 13.41 | 13.9401 | 13.994001 |

| $x$ | 3.1 | 3.01 | 3.001 |
|---|---|---|---|
| $x^2 + 5$ | 14.61 | 14.0601 | 14.006001 |

*In calculus we look for trends. What happens to the value of $x^2 + 5$ as x gets closer and closer to 3? From the tables we see that the closer x is to 3, the closer $x^2 + 5$ is to 14. We say that the limit of $x^2 + 5$ as x approaches 3 is 14, and abbreviate this statement as $\lim_{x\to 3}(x^2 + 5) = 14$ or $\lim_{x\to 3} x^2 + 5 = 14$.*

*Note that we get 14 if we evaluate $3^2 + 5$, but that is not the idea of limits. We are looking for the trend as x approaches but never actually reaches a number. In fact, we often let x approach a number for which the expression cannot be evaluated.*

## Exercises

**1. a.** Fill in the second row of each table.

| $x$ | 1.9 | 1.99 | 1.999 |
|---|---|---|---|
| $5x - 4$ | | | |

| $x$ | 2.1 | 2.01 | 2.001 |
|---|---|---|---|
| $5x - 4$ | | | |

**b.** Can you evaluate $5x - 4$ for $x = 2$?

**c.** What is $\lim_{x\to 2}(5x - 4)$?

**2. a.** Fill in the second row of each table.

| $x$ | 0.6 | 0.66 | 0.666 |
|---|---|---|---|
| $\dfrac{24x^2 - 25x + 6}{3x - 2}$ | | | |

| $x$ | 0.7 | 0.67 | 0.667 |
|---|---|---|---|
| $\dfrac{24x^2 - 25x + 6}{3x - 2}$ | | | |

**b.** Can you evaluate $\dfrac{24x^2 - 25x + 6}{3x - 2}$ for $x = 2/3$?

**c.** What is $\lim_{x\to 2/3}\dfrac{24x^2 - 25x + 6}{3x - 2}$?

**3. a.** Fill in the second row of each table.

| $x$ | 0.01 | 0.0001 | 0.000001 |
|---|---|---|---|
| $(1 + \lvert x\rvert)^{1/\lvert x\rvert}$ | | | |

| $x$ | $-0.01$ | $-0.0001$ | $-0.00001$ |
|---|---|---|---|
| $(1 + \lvert x\rvert)^{1/\lvert x\rvert}$ | | | |

**b.** Can you evaluate $(1 + |x|)^{1/|x|}$ for $x = 0$?

**c.** What is $\lim_{x\to 0}(1 + |x|)^{1/|x|}$?

**4. a.** Use a calculator in radian mode to fill in the second row of each table.

| $x$ | 0.1 | 0.001 | 0.0001 |
|---|---|---|---|
| $\dfrac{\sin(x)}{x}$ | | | |

| $x$ | $-0.1$ | $-0.001$ | $-0.0001$ |
|---|---|---|---|
| $\dfrac{\sin(x)}{x}$ | | | |

**b.** Can you evaluate $\dfrac{\sin(x)}{x}$ for $x = 0$?

**c.** What is $\lim_{x\to 0}\dfrac{\sin(x)}{x}$?

# 2 Functions and Graphs

The rainforests cover less than an eighth of a percent of the earth's surface, yet they are home to over half of our animal and plant species. Rainforests are also a source of foods such as chocolate, vanilla, pineapples, and cinnamon, and about twenty percent of all medicines.

However, the rainforests are disappearing at an alarming rate and along with them a thousand species of plants and animals per year. Destroying the rainforests will irrevocably change our planet's future.

Depletion of the rainforests is only one of many ecological issues that include the ozone layer, global warming, and hazardous waste. In assessing these problems, scientists often look for relationships between variables.

**WHAT YOU WILL learn...** In this chapter you will study relationships between variables, using graphic, numeric, and algebraic points of view.

# 2.1 Functions

We introduced the idea of a function in Section 1.2 when we discussed formulas, and we saw linear functions in Section 1.4. In this section we continue the study of functions.

## The Function Concept

If you spend \$10 on gasoline, then the price per gallon determines the number of gallons that you get. There is a rule: The number of gallons is \$10 divided by the price per gallon. The number of hours that you sleep before a test might be related to your grade on the test, but does not determine your grade. There is no rule that will determine your grade from the number of hours of sleep. If the value of a variable $y$ is determined by the value of another variable $x$, then $y$ is a function of $x$. The phrase "is a function of" means "is determined by." If there is more than one value for $y$ corresponding to a particular $x$-value, then $y$ is not determined by $x$ and $y$ is not a function of $x$.

### EXAMPLE 1 Using the phrase "is a function of"

Decide whether $a$ is a function of $b$, $b$ is a function of $a$, or neither.

**a.** Let $a$ represent a positive integer smaller than 100 and $b$ represent the number of divisors of $a$.
**b.** Let $a$ represent the age of a U.S. citizen and $b$ represent the number of days since his/her birth.
**c.** Let $a$ represent the age of a U.S. citizen and $b$ represent his/her annual income.

#### Solution

**a.** We can determine the number of divisors of any positive integer smaller than 100. So $b$ is a function of $a$. We cannot determine the integer knowing the number of its divisors, because different integers have the same number of divisors. So $a$ is not a function of $b$.
**b.** The number of days since a person's birth certainly determines the age of the person in the usual way. So $a$ is a function of $b$. However, you cannot determine the number of days since a person's birth from their age. You need more information. For example, the number of days since birth for two 1-year-olds could be 370 and 380 days. So $b$ is not a function of $a$.
**c.** We cannot determine the income from the age or the age from the income. We would need more information. Even though age and income are related, the relationship is not strong enough to say that either one is a function of the other.

▶**TRY THIS.** Let $p$ be the price of a grocery item and $t$ be the amount of sales tax at 5% on that item. Determine whether $p$ is a function of $t$, $t$ is a function of $p$, or neither. ■

In Section 1.2 we defined a function as a rule by which the value of one variable is determined from the value of one or more other variables. In this section we will continue studying functions of a single variable. The formula $A = \pi r^2$ is a function because it provides a rule for finding the value of $A$ from the value of the single variable $r$. We can also make an equivalent definition in terms of sets, as follows.

**Definition: Function**

> A **function** is a rule that assigns each element in one set to a unique element in a second set.

$A = \pi r^2$ is a function by this definition because it assigns each element in a set of radii to a unique element in a set of areas. Understanding this definition depends on knowing the meanings of the words *rule*, *assigns*, and *unique*. Using the language of ordered pairs we can make the following equivalent definition in which we don't use those words.

**Definition: Function**

A **function** is a set of ordered pairs in which no two ordered pairs have the same first coordinate and different second coordinates.

If we start with two related variables, we can identify one as the first variable and the other as the second variable and consider the set of ordered pairs containing their corresponding values. If the set of ordered pairs satisfies the function definition, then we say that the second variable is a function of the first. The variable corresponding to the first coordinate is the **independent variable,** and the variable corresponding to the second coordinate is the **dependent variable.**

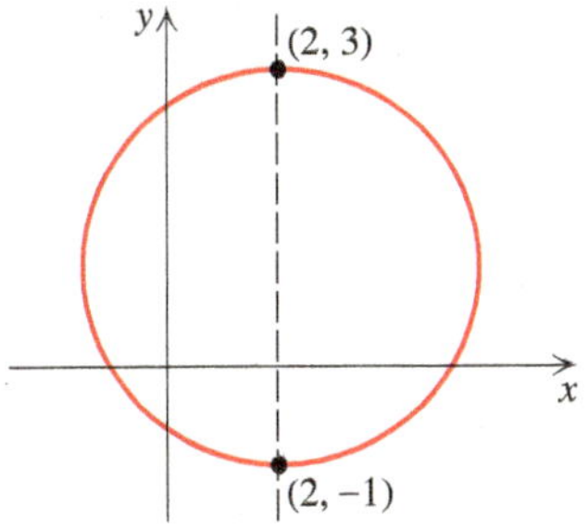

Figure 2.1

## Identifying Functions

Any set of ordered pairs is called a **relation.** A relation can be indicated by a verbal description, a graph, a formula or equation, or a table, but there is always an underlying set of ordered pairs. Not every relation is a function. A function is a special relation.

When a relation is given by a graph, we can visually check whether there are two ordered pairs with the same first coordinate and different second coordinates. For example, the circle shown in Fig. 2.1 is not the graph of a function, because there are two points on the circle with the same first coordinate. These points lie on a vertical line. In general, if there is a vertical line that crosses a graph more than once, the graph is not the graph of a function. This criterion is known as the **vertical line test.**

**Theorem: The Vertical Line Test**

A graph is the graph of a function if and only if there is no vertical line that crosses the graph more than once.

Every nonvertical line is the graph of a function, because every vertical line crosses a nonvertical line exactly once. Note that the vertical line test makes sense only because we always put the independent variable on the horizontal axis.

### EXAMPLE 2 Identifying a function from a graph

Determine which of the graphs shown in Fig. 2.2 are graphs of functions.

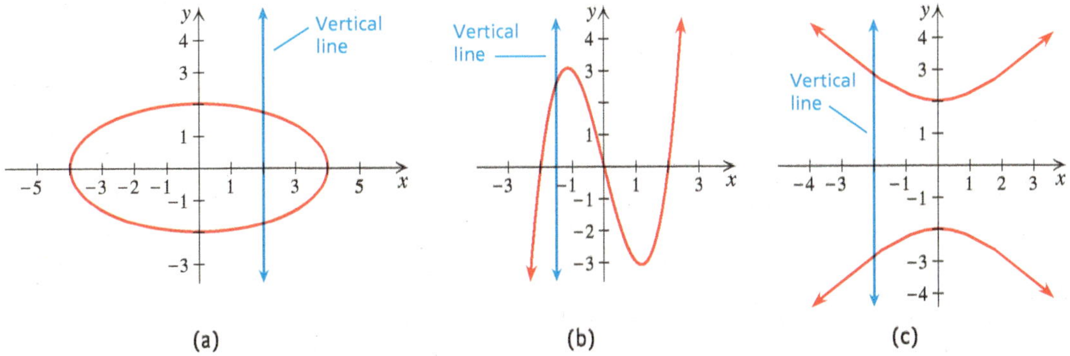

Figure 2.2

### Solution

For each graph we want to decide whether $y$ is a function of $x$. That is, can $y$ be uniquely determined from $x$? In parts (a) and (c) we can draw a vertical line that crosses the graph more than once. So in each case there is an $x$-coordinate that corresponds to two different $y$-coordinates. In parts (a) and (c) we cannot always determine a unique value of $y$ from a given $x$-coordinate. So $y$ is not a function of $x$ in parts (a) and (c). The graph in Fig. 2.2(b) is the graph of a function because every vertical line appears to cross the graph at most once. In this case we can determine $y$ for a given $x$-coordinate and $y$ is a function of $x$.

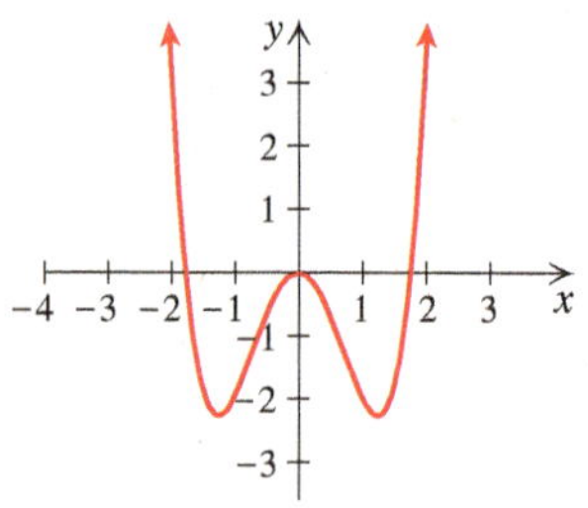


**Figure 2.3**

**▸TRY THIS.** Use the graph in Fig. 2.3 to determine whether $y$ is a function of $x$. ■

A calculator is a virtual function machine. Built-in functions on a calculator are marked with symbols such as $\sqrt{x}$, $x^2$, $x!$, $10^x$, $e^x$, $\ln(x)$, $\sin(x)$, $\cos(x)$, etc. When you provide an $x$-coordinate and use one of these symbols, the calculator finds the appropriate $y$-coordinate. The ordered pairs certainly satisfy the definition of function because the calculator will not produce two different second coordinates corresponding to one first coordinate.

In the next example we determine whether a relation is a function when the relation is given as a list of ordered pairs or as a table.

## EXAMPLE 3 Identifying a function from a list or table

Determine whether each relation is a function.

**a.** $\{(1, 3), (2, 3), (4, 9)\}$ **b.** $\{(9, -3), (9, 3), (4, 2), (0, 0)\}$

**c.**

| Quantity | Price Each |
|---|---|
| 1–5 | \$9.40 |
| 5–10 | \$8.75 |

### Solution

**a.** This set of ordered pairs is a function because no two ordered pairs have the same first coordinate and different second coordinates.

**b.** This set of ordered pairs is not a function because both $(9, 3)$ and $(9, -3)$ are in the set and they have the same first coordinate and different second coordinates.

**c.** The quantity 5 corresponds to a price of \$9.40 and also to a price of \$8.75. Assuming that quantity is the first coordinate, the ordered pairs $(5, \$9.40)$ and $(5, \$8.75)$ both belong to this relation. If you are purchasing items whose price was determined from this table, you would certainly say that something is wrong, the table has a mistake in it, or the price you should pay is not clear. The price is not a function of the quantity purchased.

**▸TRY THIS.** Determine whether each relation is a function.

**a.** $\{(4, 5), (5, 5), (5, 7)\}$ **b.**

| Time (minutes) | Cost (\$) |
|---|---|
| 0–30 | 40 |
| 31–60 | 75 |

■

We have seen and used many functions as formulas. For example, the formula $c = \pi d$ defines a set of ordered pairs in which the first coordinate is the diameter of a circle and the second coordinate is the circumference. Since each diameter corresponds to a unique circumference, the circumference is a function of the diameter. If the set of ordered pairs satisfying an equation is a function, then we say that the equation is a function or the equation defines a function. Other well-known formulas such as $C = \frac{5}{9}(F - 32)$, $A = \pi r^2$, and $V = \frac{4}{3}\pi r^3$ are also functions, but, as we will see in the next example, not every equation defines a function. The variables in the next example and all others in this text represent real numbers unless indicated otherwise.

### EXAMPLE 4 Identifying a function from an equation

Determine whether each equation defines $y$ as a function of $x$.

**a.** $|y| = x$ **b.** $y = x^2 - 3x + 2$ **c.** $x^2 + y^2 = 1$ **d.** $3x - 4y = 8$

**Solution**

**a.** We must determine whether there are any values of $x$ for which there is more than one $y$-value satisfying the equation. We arbitrarily select a number for $x$ and see. If we select $x = 2$, then the equation is $|y| = 2$, which is satisfied if $y = \pm 2$. Both $(2, 2)$ and $(2, -2)$ satisfy $|y| = x$. So this equation does *not* define $y$ as a function of $x$.

**b.** If we select any number for $x$, then $y$ is calculated by the equation $y = x^2 - 3x + 2$. Since there is only one result when $x^2 - 3x + 2$ is calculated, there is only one $y$ for any given $x$. So this equation *does* define $y$ as a function of $x$.

**c.** Is it possible to pick a number for $x$ for which there is more than one $y$-value? If we select $x = 0$, then the equation is $0^2 + y^2 = 1$ or $y = \pm 1$. So $(0, 1)$ and $(0, -1)$ both satisfy $x^2 + y^2 = 1$ and this equation does *not* define $y$ as a function of $x$. Note that $x^2 + y^2 = 1$ is equivalent to $y = \pm\sqrt{1 - x^2}$, which indicates that there are many values for $x$ that would produce two different $y$-coordinates.

**d.** The equation $3x - 4y = 8$ is equivalent to $y = \frac{3}{4}x - 2$. Since there is only one result when $\frac{3}{4}x - 2$ is calculated, there is only one $y$ corresponding to any given $x$. So $3x - 4y = 8$ *does* define $y$ as a function of $x$.

▶**TRY THIS.** Determine whether $x^3 + y^2 = 0$ defines $y$ as a function of $x$. ■

## Domain and Range

A relation is a set of ordered pairs. The **domain** of a relation is the set of all first coordinates of the ordered pairs. The **range** of a relation is the set of all second coordinates of the ordered pairs. The relation

$$\{(19, 2.4), (27, 3.0), (19, 3.6), (22, 2.4), (36, 3.8)\}$$

shows the ages and grade point averages of five randomly selected students. The domain of this relation is the set of ages, $\{19, 22, 27, 36\}$. The range is the set of grade point averages, $\{2.4, 3.0, 3.6, 3.8\}$. This relation matches elements of the domain (ages) with elements of the range (grade point averages), as shown in Fig. 2.4. This relation is not a function because two 19-year-old students have different grade point averages.

For some relations, all of the ordered pairs are listed, but for others, only an equation is given for determining the ordered pairs. When the domain of the relation is not stated, it is understood that the domain consists of only values of the independent

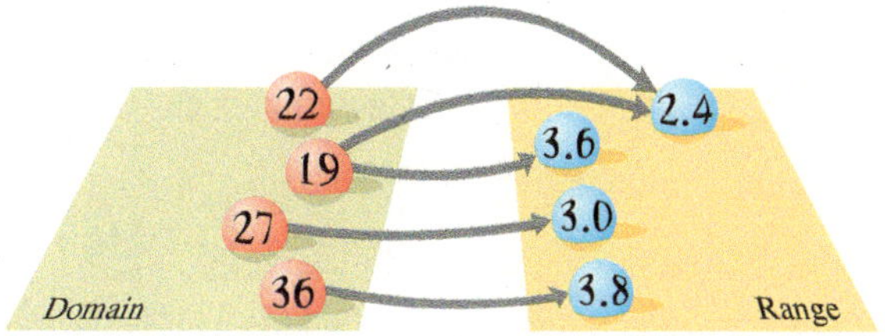


Figure 2.4

variable that can be used in the expression defining the relation. When we use $x$ and $y$ for the variables, we always assume that $x$ is the independent variable and $y$ is the dependent variable.

### EXAMPLE 5 Determining domain and range

State the domain and range of each relation and whether the relation is a function.

**a.** $\{(-1, 1), (3, 9), (3, -9)\}$ **b.** $y = \sqrt{2x - 1}$ **c.** $x = |y|$

#### Solution

**a.** The domain is the set of first coordinates $\{-1, 3\}$, and the range is the set of second coordinates $\{1, 9, -9\}$. Note that an element of a set is not listed more than once. Since $(3, 9)$ and $(3, -9)$ are in the relation, the relation is not a function.

**b.** Since $y$ is determined uniquely from $x$ by the formula $y = \sqrt{2x - 1}$, $y$ is a function of $x$. In our discussions of functions, $x$ and $y$ must be real numbers. Now $\sqrt{2x - 1}$ is a real number only if $2x - 1 \geq 0$, or $x \geq 1/2$. So the domain of the function is the set of all real numbers greater then or equal to $1/2$, which is written in set notation as $\{x \mid x \geq 1/2\}$ or in interval notation as $[1/2, \infty)$. Since interval notation is simpler, we will use it. If $2x - 1 \geq 0$ and $y = \sqrt{2x - 1}$, we have $y \geq 0$. So the range of the function is the interval $[0, \infty)$.

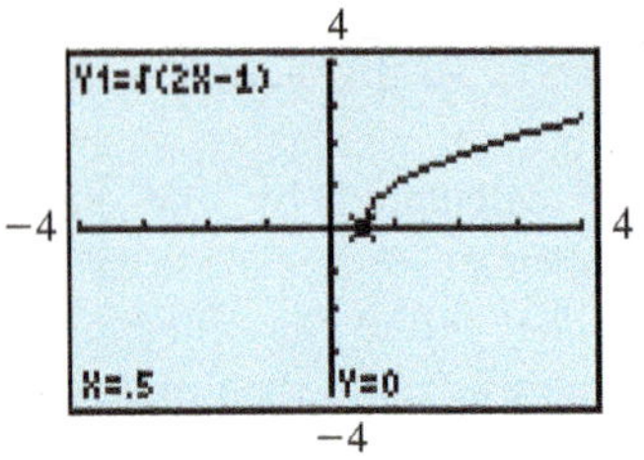


Figure 2.5

The graph of $y = \sqrt{2x - 1}$ shown in Fig. 2.5 supports these answers because the points that are plotted appear to have $x \geq 1/2$ and $y \geq 0$. □

**c.** The expression $|y|$ is defined for any real number $y$. So the range is the interval of all real numbers, $(-\infty, \infty)$. Since $|y|$ is nonnegative, the values of $x$ must be nonnegative. So the domain is $[0, \infty)$. Since ordered pairs such as $(2, 2)$ and $(2, -2)$ satisfy $x = |y|$, this equation does not give $y$ as a function of $x$.

▶**TRY THIS.** Determine whether $y = \sqrt{x + 3}$ is a function and find its domain and range. ■

In Example 5(b) and (c) we found the domain and range by examining an equation defining a relation. If the relation is a function as in Example 5(b), you can easily draw the graph with a graphing calculator and use it to support your answer. However, to choose an appropriate viewing window you must know the domain and range to begin with. So it is best to use a calculator graph to support your conclusions about domain and range rather than to make conclusions about domain and range. □

## Function Notation

A function defined by a set of ordered pairs can be named with a letter. For example,

$$f = \{(2, 5), (3, 8)\}.$$

Since the function $f$ pairs 2 with 5 we write $f(2) = 5$, which is read as "the value of $f$ at 2 is 5" or simply "$f$ of 2 is 5." We also have $f(3) = 8$.

A function defined by an equation can also be named with a letter. For example, the function $y = x^2$ could be named by a new letter, say $g$. We can then use $g(x)$, read "$g$ of $x$" as a symbol for the second coordinate when the first coordinate is $x$. Since $y$ and $g(x)$ are both symbols for the second coordinate we can write $y = g(x)$ and $g(x) = x^2$. Since $3^2 = 9$, the function $g$ pairs 3 with 9 and we write $g(3) = 9$. This notation is called **function notation.**

### EXAMPLE 6 Using function notation

Let $h = \{(1, 4), (6, 0), (7, 9)\}$ and $f(x) = \sqrt{x - 3}$. Find each of the following.

**a.** $h(7)$ **b.** $w$, if $h(w) = 0$ **c.** $f(7)$ **d.** $x$, if $f(x) = 5$

### Solution

**a.** The expression $h(7)$ is the second coordinate when the first coordinate is 7 in the function named $h$. So $h(7) = 9$.

**b.** We are looking for a number $w$ for which $h(w) = 0$. That is, the second coordinate is 0 for some unknown first coordinate $w$. By examining the function $h$ we see that $w = 6$.

**c.** To find $f(7)$ replace $x$ by 7 in $f(x) = \sqrt{x - 3}$:

$$f(7) = \sqrt{7 - 3} = \sqrt{4} = 2$$

**d.** To find $x$ for which $f(x) = 5$ we replace $f(x)$ by 5 in $f(x) = \sqrt{x - 3}$:

$$5 = \sqrt{x - 3}$$
$$25 = x - 3$$
$$28 = x$$

▶**TRY THIS.** Let $f(x) = x - 3$. **a.** Find $f(4)$. **b.** Find $x$ if $f(x) = 9$. ■

Function notation such as $f(x) = 3x + 1$ provides a rule for finding the second coordinate: Multiply the first coordinate (whatever it is) by 3 and then add 1. The $x$ in this notation is called a **dummy variable** because the letter used is unimportant. We could write $f(t) = 3t + 1$,

$$f(\text{first coordinate}) = 3(\text{first coordinate}) + 1,$$

or even $f(\ ) = 3(\ ) + 1$ to convey the same idea. Whatever appears in the parentheses following $f$ must be used in place of $x$ on the other side of the equation.

## EXAMPLE 7 Using function notation with variables

Given that $f(x) = x^2 - 2$ and $g(x) = 2x - 3$, find and simplify each of the following expressions.

**a.** $f(a)$ **b.** $f(a + 1)$ **c.** $f(x + h) - f(x)$ **d.** $g(x - 2)$ **e.** $g(x + h) - g(x)$

### Solution

**a.** Replace $x$ by $a$ in $f(x) = x^2 - 2$ to get $f(a) = a^2 - 2$.

**b.** 
$$\begin{aligned} f(a + 1) &= (a + 1)^2 - 2 && \text{Replace } x \text{ by } a + 1 \text{ in } f(x) = x^2 - 2. \\ &= a^2 + 2a + 1 - 2 \\ &= a^2 + 2a - 1 \end{aligned}$$

**c.** 
$$\begin{aligned} f(x + h) - f(x) &= (x + h)^2 - 2 - (x^2 - 2) && \text{Replace } x \text{ with } x + h \text{ to get } f(x + h). \\ &= (x + h)^2 - x^2 \\ &= x^2 + 2hx + h^2 - x^2 \\ &= 2hx + h^2 \end{aligned}$$

**d.** 
$$\begin{aligned} g(x - 2) &= 2(x - 2) - 3 && \text{Replace } x \text{ by } x - 2 \text{ in } g(x) = 2x - 3. \\ &= 2x - 7 \end{aligned}$$

**e.** 
$$\begin{aligned} g(x + h) - g(x) &= 2(x + h) - 3 - (2x - 3) \\ &= 2x + 2h - 3 - 2x + 3 \\ &= 2h \end{aligned}$$

▶**TRY THIS.** Let $f(x) = x^2 - 4$. Find and simplify $f(x + 2)$. ■

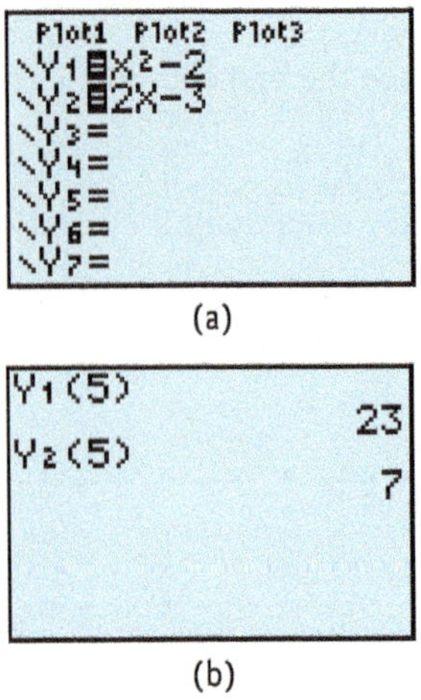


Figure 2.6

A graphing calculator uses subscripts to indicate different functions. For example, if $y_1 = x^2 - 2$ and $y_2 = 2x - 3$, then $y_1(5) = 23$ and $y_2(5) = 7$ as shown in Fig. 2.6(a) and (b). □

If a function describes some real application, then a letter that fits the situation is usually used. For example, if watermelons are \$3 each, then the cost of $x$ watermelons is given by the function $C(x) = 3x$. The cost of five watermelons is $C(5) = 3 \cdot 5 = \$15$. In trigonometry the abbreviations sin, cos, and tan are used rather than a single letter to name the trigonometric functions. The dependent variables are written as $\sin(x)$, $\cos(x)$, and $\tan(x)$.

## The Average Rate of Change of a Function

In Section 1.4 we defined the slope of the line through $(x_1, y_1)$ and $(x_2, y_2)$ as $\frac{y_2 - y_1}{x_2 - x_1}$. We now extend that idea to any function (linear or not).

### Definition: Average Rate of Change from $x_1$ to $x_2$

If $(x_1, y_1)$ and $(x_2, y_2)$ are two ordered pairs of a function, we defined the **average rate of change** of the function as $x$ varies from $x_1$ to $x_2$, as the change in $y$-coordinates divided by the change in $x$-coordinates, $\Delta y/\Delta x$, or

$$\frac{y_2 - y_1}{x_2 - x_1}.$$

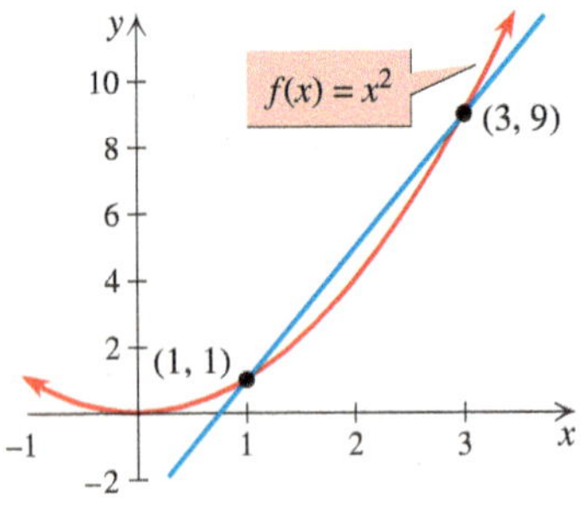


Figure 2.7

Note that the $x$-values can be specified with interval notation. For example, the average rate of change $f(x) = x^2$ on $[1, 3]$ is found as follows:

$$\frac{\Delta y}{\Delta x} = \frac{f(3) - f(1)}{3 - 1} = \frac{9 - 1}{3 - 1} = 4$$

The average rate of change is simply the slope of the line that passes through two points on the graph of the function as shown in Fig. 2.7.

It is not necessary to have a formula for a function to find an average rate of change, as is shown in the next example.

### EXAMPLE 8 Finding the average rate of change

The population of California was 29.8 million in 1990 and 33.9 million in 2000 (U.S. Census Bureau, www.census.gov). What was the average rate of change of the population over that time interval?

#### Solution

The population is a function of the year. The average rate of change of the population is the change in population divided by the change in time:

$$\frac{\Delta p}{\Delta t} = \frac{33.9 - 29.8}{2000 - 1990} = \frac{4.1}{10} = 0.41$$

The average rate of change of the population was 0.41 million people/year or 410,000 people/year. Note that 410,000 people/year is an average and that the population did not actually increase by 410,000 every year.

**▸TRY THIS.** A BMW was purchased for \$28,645 in 2003 and sold for \$13,837 in 2009. What was the average rate of change of the car's value for that time interval? ■

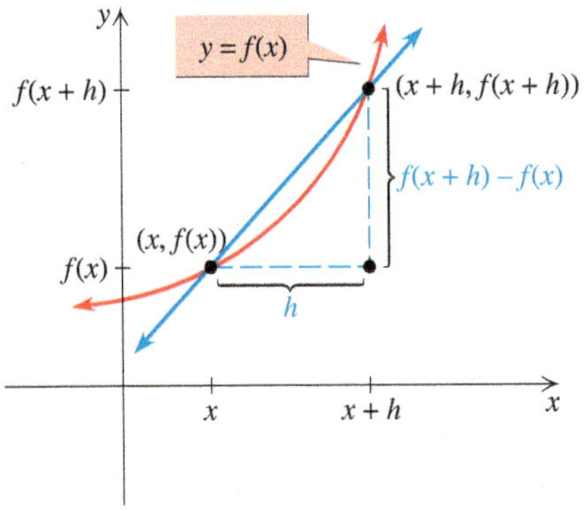


Figure 2.8

The average rate of change of a function between two points that are labeled as shown in Fig. 2.8 is called the *difference quotient.*

**Definition: Difference Quotient**

The **difference quotient** is the expression $\dfrac{f(x + h) - f(x)}{h}$.

Note that $h$ is the change in value of the $x$-coordinates. So using $\Delta x$ in place of $h$, the difference quotient is written as

$$\frac{f(x + \Delta x) - f(x)}{\Delta x}.$$

In calculus it is often necessary to find and simplify the difference quotient for a function.

## EXAMPLE 9 Finding a difference quotient

Find and simplify the difference quotient for each of the following functions.

**a.** $j(x) = 3x + 2$ **b.** $f(x) = x^2 - 2x$ **c.** $g(x) = \sqrt{x}$ **d.** $y = \dfrac{5}{x}$

### Solution

**a.**
$$\frac{j(x + h) - j(x)}{h} = \frac{3(x + h) + 2 - (3x + 2)}{h}$$
$$= \frac{3x + 3h + 2 - 3x - 2}{h}$$
$$= \frac{3h}{h} = 3$$

**b.**
$$\frac{f(x + h) - f(x)}{h} = \frac{[(x + h)^2 - 2(x + h)] - (x^2 - 2x)}{h}$$
$$= \frac{x^2 + 2xh + h^2 - 2x - 2h - x^2 + 2x}{h}$$
$$= \frac{2xh + h^2 - 2h}{h}$$
$$= 2x + h - 2$$

**c.**
$$\frac{g(x + h) - g(x)}{h} = \frac{\sqrt{x + h} - \sqrt{x}}{h}$$
$$= \frac{(\sqrt{x + h} - \sqrt{x})(\sqrt{x + h} + \sqrt{x})}{h(\sqrt{x + h} + \sqrt{x})} \quad \text{Rationalize the numerator}$$
$$= \frac{x + h - x}{h(\sqrt{x + h} + \sqrt{x})}$$
$$= \frac{1}{\sqrt{x + h} + \sqrt{x}}$$

**d.** Use the function notation $f(x) = \frac{5}{x}$ for the function $y = \frac{5}{x}$:

$$\frac{f(x + h) - f(x)}{h} = \frac{\dfrac{5}{x + h} - \dfrac{5}{x}}{h} = \frac{\left(\dfrac{5}{x + h} - \dfrac{5}{x}\right)x(x + h)}{h \cdot x(x + h)}$$
$$= \frac{5x - 5(x + h)}{hx(x + h)} = \frac{-5h}{hx(x + h)} = \frac{-5}{x(x + h)}$$

▶**TRY THIS.** Find and simplify the difference quotient for $f(x) = x^2 - x$. ■

Note that in Examples 9(b), 9(c), and 9(d) the average rate of change of the function depends on the values of $x$ and $h$, while in Example 9(a) the average rate of change of the function is constant. In Example 9(c) the expression does not look much different after rationalizing the numerator than it did before. However, we did remove $h$ as a factor of the denominator, and in calculus it is often necessary to perform this step.

## Constructing Functions

In the next example we find a formula for, or **construct,** a function relating two variables in a geometric figure.

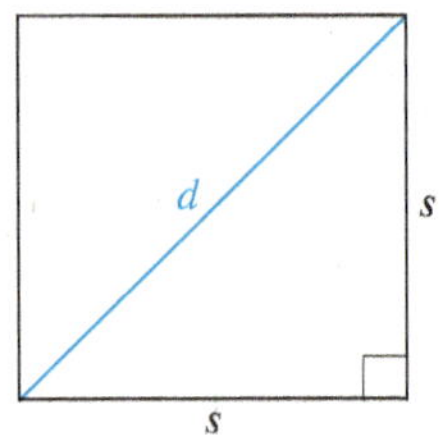


**Figure 2.9**

### EXAMPLE 10 Constructing a function

Given that a square has diagonal of length $d$ and side of length $s$, write the area $A$ as a function of the length of the diagonal.

**Solution**

The area of any square is given by $A = s^2$. The diagonal is the hypotenuse of a right triangle as shown in Fig. 2.9. By the Pythagorean theorem, $d^2 = s^2 + s^2$, $d^2 = 2s^2$, or $s^2 = d^2/2$. Since $A = s^2$ and $s^2 = d^2/2$, we get the formula

$$A = \frac{d^2}{2}$$

expressing the area of the square as a function of the length of the diagonal.

▶**TRY THIS.** A square has perimeter $P$ and sides of length $s$. Write the side as a function of the perimeter. ■

## FOR thought... True or False? Explain.

1. Any set of ordered pairs is a function.
2. If $f = \{(1, 1), (2, 4), (3, 9)\}$, then $f(5) = 25$.
3. The domain of $f(x) = 1/x$ is $(-\infty, 0) \cup (0, \infty)$.
4. Each student's exam grade is a function of the student's IQ.
5. If $f(x) = x^2$, then $f(x + h) = x^2 + h$.
6. The domain of $g(x) = |x - 3|$ is $[3, \infty)$.
7. The range of $y = 8 - x^2$ is $(-\infty, 8]$.
8. The equation $x = y^2$ does not define $y$ as a function of $x$.
9. If $f(t) = \frac{t - 2}{t + 2}$, then $f(0) = -1$.
10. The set $\{(\frac{3}{8}, 8), (\frac{4}{7}, 7), (0.16, 6), (\frac{3}{8}, 5)\}$ is a function.

## EXERCISES 2.1

*Fill in the blank.*

1. Any set of ordered pairs is a ________.
2. A set of ordered pairs in which no two have the same first coordinate and different second coordinates is a ________.
3. For a set of ordered pairs, the variable corresponding to the first coordinate is the ________ variable and the variable corresponding to the second coordinate is the ________ variable.
4. For a set of ordered pairs, the set of all first coordinates is the ________ and the set of all second coordinates is the ________.
5. The expression $\frac{f(x + h) - f(x)}{h}$ is the ________.

6. If $(x_1, y_1)$ and $(x_2, y_2)$ are two ordered pairs of a function, then $\dfrac{y_2 - y_1}{x_2 - x_1}$ is the ________ of the function on $[x_1, x_2]$.

*For each pair of variables determine whether a is a function of b, b is a function of a, or neither.*

**7.** $a$ is the radius of any U.S. coin and $b$ is its circumference.

**8.** $a$ is the length of any rectangle with a width of 5 in. and $b$ is its perimeter.

**9.** $a$ is the length of any piece of U.S. paper currency and $b$ is its denomination.

**10.** $a$ is the diameter of any U.S. coin and $b$ is its value.

**11.** $a$ is the universal product code for an item at Wal-Mart and $b$ is its price.

**12.** $a$ is the final exam score for a student in your class and $b$ is his/her semester grade.

**13.** $a$ is the time spent studying for the final exam for a student in your class and $b$ is the student's final exam score.

**14.** $a$ is the age of an adult male and $b$ is his shoe size.

**15.** $a$ is the height of a car in inches and $b$ is its height in centimeters.

**16.** $a$ is the cost for mailing a first-class letter and $b$ is its weight.

*Use the vertical line test on each graph in Exercises 17–22 to determine whether y is a function of x.*

**17.**

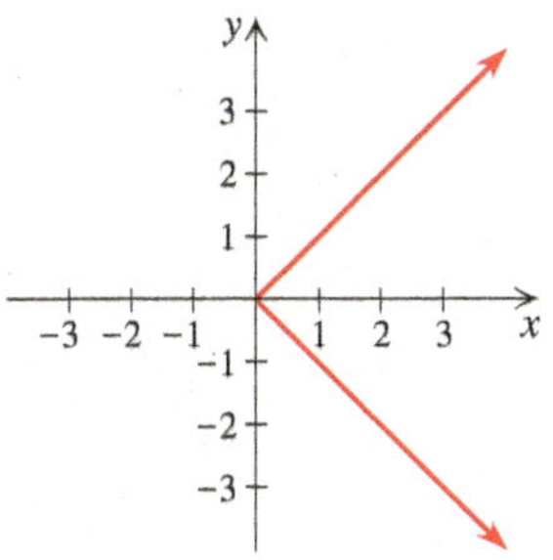


**18.**

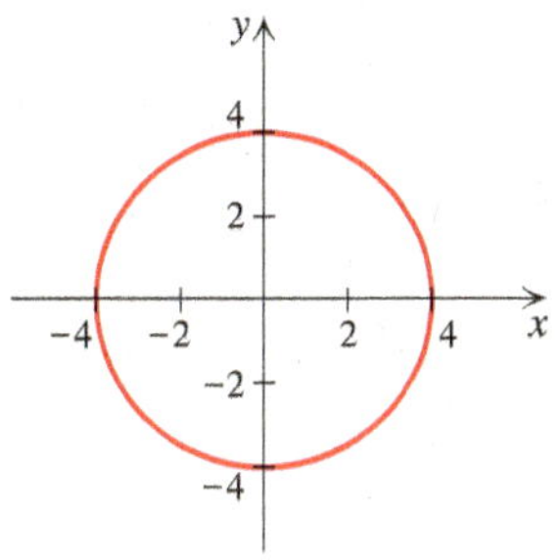


**19.**

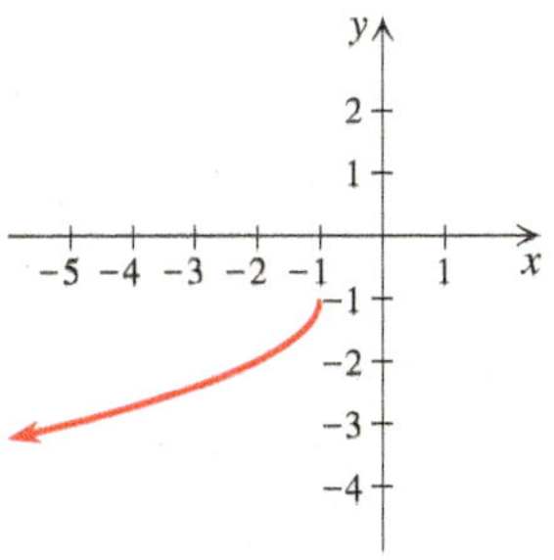


**20.**

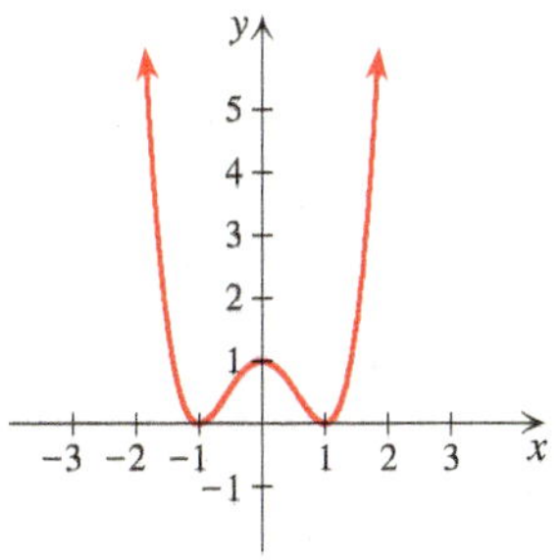


**21.**

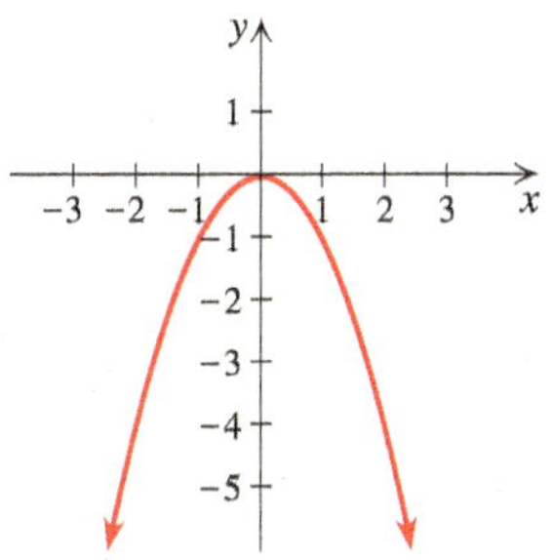


**22.**

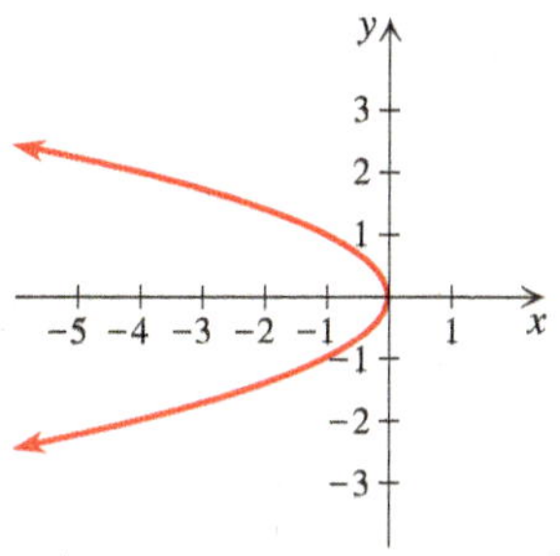


*Determine whether each relation is a function.*

**23.** $\{(-1, -1), (2, 2), (3, 3)\}$

**24.** $\{(0.5, 7), (0, 7), (1, 7), (9, 7)\}$

**25.** $\{(25, 5), (25, -5), (0, 0)\}$

**26.** $\{(1, \pi), (30, \pi/2), (60, \pi/4)\}$

**27.**

| $x$ | $y$ |
|---|---|
| 3 | 6 |
| 4 | 9 |
| 3 | 12 |

**28.**

| $x$ | $y$ |
|---|---|
| 1 | 6.98 |
| 5 | 5.98 |
| 9 | 6.98 |

**29.**

| $x$ | $y$ |
|---|---|
| −1 | 1 |
| 1 | 1 |
| −5 | 1 |
| 5 | 1 |

**30.**

| $x$ | $y$ |
|---|---|
| 1 | 1 |
| 2 | 4 |
| 3 | 9 |
| 4 | 16 |

*Determine whether each equation defines y as a function of x.*

**31.** $y = 3x - 8$

**32.** $y = x^2 - 3x + 7$

**33.** $x = 3y - 9$

**34.** $x = y^3$

**35.** $x^2 = y^2$

**36.** $y^2 - x^2 = 9$

**37.** $x = \sqrt{y}$

**38.** $x = \sqrt[3]{y}$

**39.** $y + 2 = |x|$

**40.** $y - 1 = x^2$

**41.** $x = |2y|$

**42.** $x = y^2 + 1$

*Determine the domain and range of each relation.*

**43.** $\{(-3, 1), (4, 2), (-3, 6), (5, 6)\}$

**44.** $\{(1, 2), (2, 4), (3, 8), (4, 16)\}$

**45.** $\{(x, y) \mid y = 4\}$

**46.** $\{(x, y) \mid x = 5\}$

**47.** $y = |x| + 5$

**48.** $y = x^2 + 8$

**49.** $x + 3 = |y|$

**50.** $x + 2 = \sqrt{y}$

**51.** $y = \sqrt{x - 4}$

**52.** $y = \sqrt{5 - x}$

**53.** $x = -y^2$

**54.** $x = -|y|$

*Let* $f = \{(2, 6), (3, 8), (4, 5)\}$ *and* $g(x) = 3x + 5$. *Find the following.*

**55.** $f(2)$

**56.** $f(4)$

**57.** $g(2)$

**58.** $g(4)$

**59.** $x$, if $f(x) = 8$

**60.** $x$, if $f(x) = 6$

**61.** $x$, if $g(x) = 26$

**62.** $x$, if $g(x) = -4$

**63.** $f(4) + g(4)$

**64.** $f(3) - g(3)$

*Let* $f(x) = 3x^2 - x$ *and* $g(x) = 4x - 2$. *Find the following.*

**65.** $f(a)$

**66.** $f(w)$

**67.** $g(a + 2)$

**68.** $g(a - 5)$

**69.** $f(x + 1)$

**70.** $f(x - 3)$

**71.** $g(x + h)$

**72.** $f(x + h)$

**73.** $f(x + 1) - f(x)$

**74.** $g(x + 2) - g(x)$

**75.** $f(x + h) - f(x)$

**76.** $g(x + h) - g(x)$

*The following problems involve average rate of change.*

**77.** *Depreciation of a Mustang* If a new Mustang is valued at \$20,000 and five years later it is valued at \$8000, then what is the average rate of change of its value during those five years?

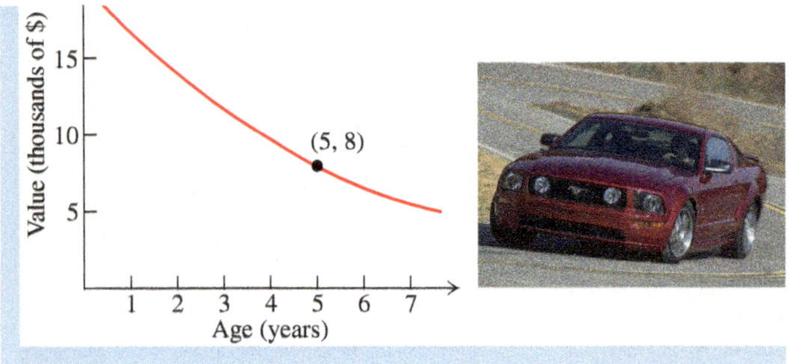


**Figure for Exercise 77**

**78.** *Cost of Gravel* Wilson's Sand and Gravel will deliver 12 $yd^3$ of gravel for \$240, 30 $yd^3$ for \$528, and 60 $yd^3$ for \$948. What is the average rate of change of the cost as the number of cubic yards varies from 12 to 30? What is the average rate of change as the number of cubic yards varies from 30 to 60?

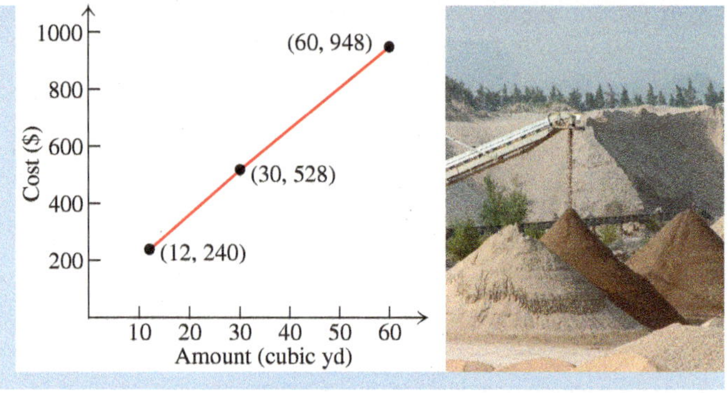


**Figure for Exercise 78**

**79.** *Dropping a Watermelon* If a comedian drops a watermelon from a height of 64 ft, then its height (in feet) above the ground is given by the function $h(t) = -16t^2 + 64$ where $t$ is time (in seconds). To get an idea of how fast the watermelon is traveling when it hits the ground find the average rate of change of the height on each of the time intervals $[0, 2]$, $[1, 2]$, $[1.9, 2]$, $[1.99, 2]$, and $[1.999, 2]$.

**80.** *Bungee Jumping* Billy Joe McCallister jumped off the Tallahatchie Bridge, 70 ft above the water, with a bungee cord tied to his legs. If he was 6 ft above the water 2 sec after jumping, then what was the average rate of change of his altitude as the time varied from 0 to 2 sec?

**81.** *Deforestation* In 1988 tropical forest covered 1970 million hectares. In 2008 tropical forest covered 1768 million hectares (1 hectare $= 10{,}000 \text{ m}^2$). What was the average rate of change in hectares per year of the area of tropical forest over those 20 years?

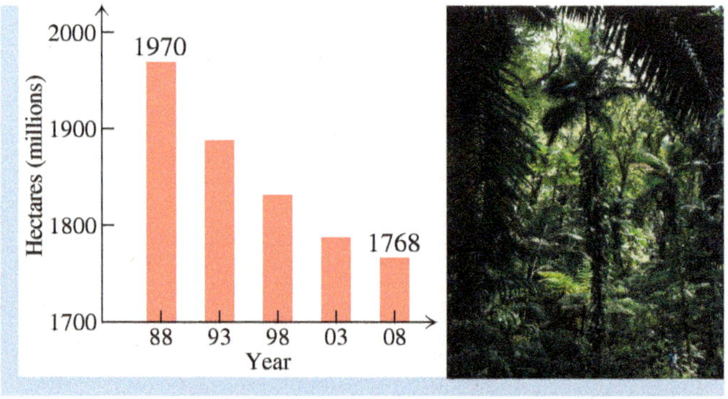


**Figure for Exercises 81 and 82**

**82.** *Elimination* If the deforestation described in the previous exercise continues at the same rate, then in which year will the tropical moist forest be totally eliminated?

*Find the difference quotient* $\dfrac{f(x + h) - f(x)}{h}$ *for each function and simplify it.*

**83.** $f(x) = 4x$

**84.** $f(x) = \dfrac{1}{2}x$

**85.** $f(x) = 3x + 5$

**86.** $f(x) = -2x + 3$

**87.** $y = x^2 + x$

**88.** $y = x^2 - 2x$

**89.** $y = -x^2 + x - 2$

**90.** $y = x^2 - x + 3$

**91.** $g(x) = 3\sqrt{x}$

**92.** $g(x) = -2\sqrt{x}$

**93.** $f(x) = \sqrt{x + 2}$

**94.** $f(x) = \sqrt{\dfrac{x}{2}}$

**95.** $g(x) = \dfrac{1}{x}$

**96.** $g(x) = \dfrac{3}{x}$

**97.** $g(x) = \dfrac{3}{x + 2}$

**98.** $g(x) = 3 + \dfrac{2}{x - 1}$

*Solve each problem.*

**99.** *Constructing Functions* Consider a square with side of length $s$, diagonal of length $d$, perimeter $P$, and area $A$.

**a.** Write $A$ as a function of $s$.

**b.** Write $s$ as a function of $A$.

**c.** Write $s$ as a function of $d$.

**d.** Write $d$ as a function of $s$.

**e.** Write $P$ as a function of $s$.

**f.** Write $s$ as a function of $P$.

**g.** Write $A$ as a function of $P$.

**h.** Write $d$ as a function of $A$.

**100.** *Constructing Functions* Consider a circle with area $A$, circumference $C$, radius $r$, and diameter $d$.

**a.** Write $A$ as a function of $r$.

**b.** Write $r$ as a function of $A$.

**c.** Write $C$ as a function of $r$.

**d.** Write $d$ as a function of $r$.

**e.** Write $d$ as a function of $C$.

**f.** Write $A$ as a function of $d$.

**g.** Write $d$ as a function of $A$.

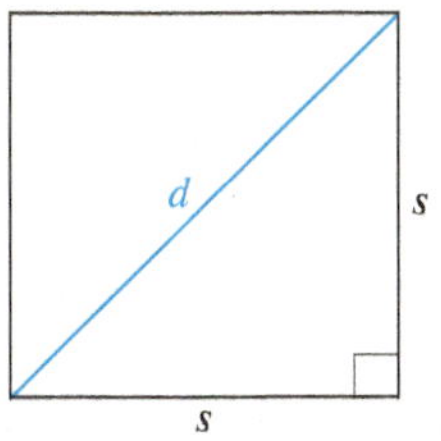


Figure for Exercise 99

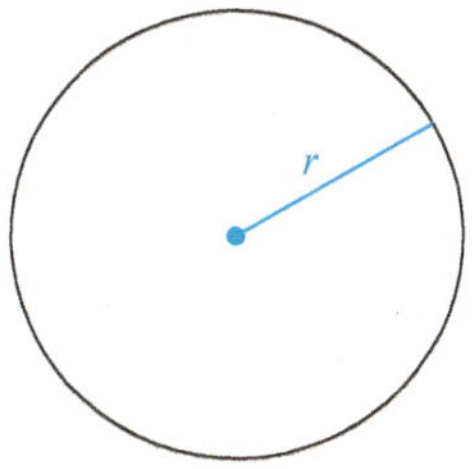


Figure for Exercise 100

**101.** *Cost of Window Cleaning* A window cleaner charges \$50 per visit plus \$35 per hour. Express the total charge as a function of the number of hours worked, $n$.

**102.** *Below Sea Level* The accompanying table shows the depth below sea level $d$ and the atmospheric pressure $A$ (www.sportsfigures.espn.com). The equation $A(d) = 0.03d + 1$ expresses $A$ as a function of $d$.

**a.** Find the atmospheric pressure for a depth of 100 ft, where nitrogen narcosis begins.

**b.** Find the depth at which the pressure is 4.9 atm, the maximum depth for intermediate divers.

Table for Exercise 102

| Depth (ft) | Atmospheric Pressure (atm) |
|---|---|
| 21 | 1.63 |
| 60 | 2.8 |
| 100 | |
| | 4.9 |
| 200 | 7.0 |
| 250 | 8.5 |

**103.** *Computer Spending* The amount spent online for computers in the year $2000 + n$ can be modeled by the function $C(n) = 0.95n + 5.8$ where $n$ is a whole number and $C(n)$ is billions of dollars.

**a.** What does $C(4)$ represent and what is it?

**b.** Find the year in which online spending for computers will reach \$15 billion?

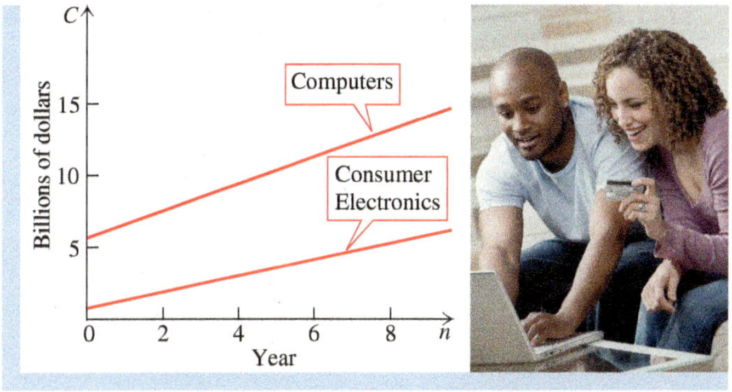


Figure for Exercises 103 and 104

**104.** *Electronics Spending* The amount spent online for consumer electronics (excluding computers) in the year $2000 + n$ can be modeled by the function $E(n) = 0.5n + 1$ where $n$ is a whole number and $E(n)$ is billions of dollars.

**a.** What does $E(4) + C(4)$ represent and what is it?

**b.** Find the year in which the total spending for computers and electronics will reach \$20 billion?

**c.** In which category is spending growing faster? See the accompanying figure.

**105.** *Pile of Pipes* Six pipes, each with radius $a$, are stacked as shown in the accompanying figure. Construct a function that gives the height $h$ of the pile in terms of $a$.

HINT Connect the centers of three circles to form an equilateral triangle.

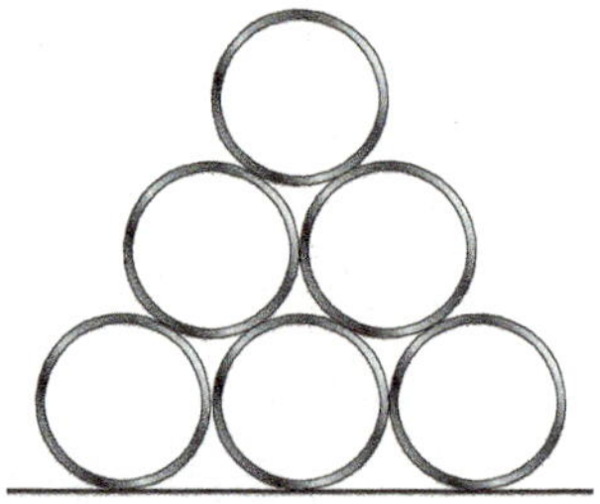

Figure for Exercise 105

**106.** *Angle Bisectors* The angle bisectors of any triangle meet at a single point. Let $a$ be the length of the hypotenuse of a 30-60-90 triangle and $d$ be the distance from the vertex of the right angle to the point where the angle bisectors meet. Write $d$ as function of $a$.

HINT Draw a diagram and label as many sides and angles as you can.

**107.** *Concert Revenue* The revenue in dollars from the sale of concert tickets at $x$ dollars each is given by the function

$$R(x) = 20{,}000x - 500x^2.$$

Find the difference quotient when $x = 18$ and $h = 0.1$ and when $x = 22$ and $h = 0.1$. Interpret your answers.

**108.** *Surface Area* The amount of tin $A$ (in square inches) needed to make a tin can with radius $r$ inches and volume 36 in.$^3$ can be found by the function

$$A(r) = \frac{72}{r} + 2\pi r^2.$$

Find the difference quotient when $r = 1.4$ and $h = 0.1$ and when $r = 2$ and $h = 0.1$. Interpret your answers.

## FOR WRITING/DISCUSSION

**109.** *Find a Function* Give an example of a function and an example of a relation that is not a function, from situations that you have encountered outside of this textbook.

**110.** *Cooperative Learning* Work in a small group to consider the equation $y^n = x^m$ for any integers $n$ and $m$. For which integers $n$ and $m$ does the equation define $y$ as a function of $x$?

## RETHINKING

**111.** Solve $\frac{3}{2}x + \frac{5}{6} = \frac{5}{9}x + \frac{1}{3}$.

**112.** There are twice as many males as females in a classroom. If the total number of people is 36, then how many males are in the classroom?

**113.** Find the distance between the points $(-6, 3)$ and $(-4, -3)$.

**114.** Find the equation of the line through the points $(-1, 2)$ and $(5, 3)$. Write the answer in slope-intercept form.

**115.** Solve $(x - 3)(x + 2) = 36$.

**116.** Solve $|2x - 9| < 13$. Write the solution set in interval notation.

## THINKING OUTSIDE THE BOX XIII

*Lucky Lucy* Lucy's teacher asked her to evaluate $(20 + 25)^2$. As she was trying to figure out what to do she mumbled, "twenty twenty-five." Her teacher said, "Good, 2025 is correct." Find another pair of two-digit whole numbers for which the square of their sum can be found by Lucy's method.

## POP QUIZ 2.1

**1.** Is the radius of a circle a function of its area?

**2.** Is $\{(2, 4), (1, 8), (2, -4)\}$ a function?

**3.** Does $x^2 + y^2 = 1$ define $y$ as a function of $x$?

**4.** What is the domain of $y = \sqrt{x - 1}$?

**5.** What is the range of $y = x^2 + 2$?

**6.** What is $f(2)$ if $f = \{(1, 8), (2, 9)\}$?

**7.** What is $a$ if $f(a) = 1$ and $f(x) = 2x$?

**8.** If the cost was \$20 in 1998 and \$40 in 2008, then what is the average rate of change of the cost for that time period?

**9.** Find and simplify the difference quotient for $f(x) = x^2 + 3$.

LINKING
**concepts...** For Individual or Group Explorations

### Modeling Debt and Population Growth

*The following table gives the U.S. federal debt in billions of dollars as a function of the year and the population in millions of people as a function of the year (U.S. Treasury Department, www.treas.gov). Do parts (a) through (e) for each function.*

| Year | Debt | Pop. |
|---|---|---|
| 1940 | 51 | 131.7 |
| 1950 | 257 | 150.7 |
| 1960 | 291 | 179.3 |
| 1970 | 381 | 203.3 |
| 1980 | 909 | 226.5 |
| 1990 | 3207 | 248.7 |
| 2000 | 5666 | 274.8 |

**a)** Draw an accurate graph of the function.

**b)** Find the average rate of change of the function over each ten-year period.

**c)** Take the average rate of change for each ten-year period (starting with 1950–1960) and subtract from it the average rate of change for the previous ten-year period.

**d)** Are the average rates of change for the function positive or negative?

**e)** Are the answers to part (c) mostly positive or mostly negative?

**f)** Judging from the graphs and the average rates of change, which is growing out of control, the federal debt or the population?

**g)** Explain the relationship between your answer to part (f) and your answer to part (e).

# 2.2 Graphs of Relations and Functions

When we graph the set of ordered pairs that satisfy an equation, we are combining algebra with geometry. We saw in Section 1.3 that the graph of any equation of the form $(x - h)^2 + (y - k)^2 = r^2$ is a circle and that the graph of any equation of the form $Ax + By = C$ is a line. In this section we will see that graphs of equations have many different geometric shapes.

## Graphing Equations

The circle and the line provide nice examples of how algebra and geometry are interrelated. When you see an equation that you recognize as the equation of a circle or a line, sketching a graph is easy to do. Other equations have graphs that are not such familiar shapes. Until we learn to recognize the kinds of graphs that other equations have, we graph other equations by calculating enough ordered pairs to determine the shape of the graph. When you graph equations, try to anticipate what the graph will look like, and after the graph is drawn, pause to reflect on the shape of the graph and the type of equation that produced it. You might wish to look ahead to the Function Gallery on pages 140 and 141, which shows the basic functions that we will be studying.

Of course, a graphing calculator can speed up this process. Remember that a graphing calculator shows only finitely many points and a graph usually consists of infinitely many points. After looking at the display of a graphing calculator, you must still decide what the entire graph looks like.

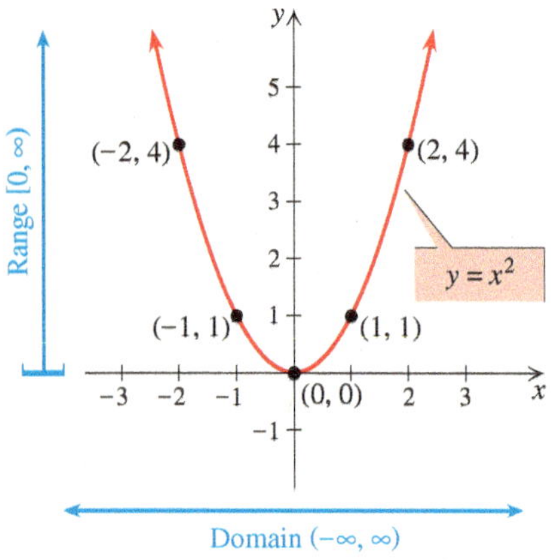


**Figure 2.10**

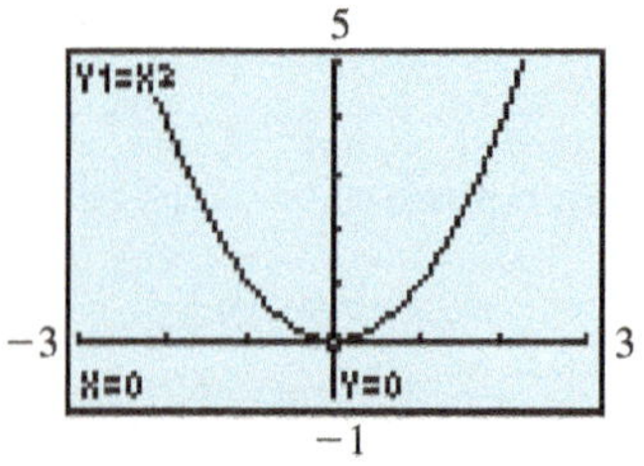


**Figure 2.11**

## EXAMPLE 1 The square function

Graph the equation $y = x^2$ and state the domain and range. Determine whether the relation is a function.

### Solution

Make a table of ordered pairs that satisfy $y = x^2$:

$x$ is any real number

| $x$ | 0 | 1 | −1 | 2 | −2 |
|---|---|---|---|---|---|
| $y = x^2$ | 0 | 1 | 1 | 4 | 4 |

$y \geq 0$

These ordered pairs indicate a graph in the shape shown in Fig. 2.10. The domain is $(-\infty, \infty)$ because any real number can be used for $x$ in $y = x^2$. Since all $y$-coordinates are nonnegative, the range is $[0, \infty)$. Because no vertical line crosses this curve more than once, $y = x^2$ is a function.

The calculator graph shown in Fig. 2.11 supports these conclusions.

▸**TRY THIS.** Determine whether $y = \frac{1}{2}x^2$ is a function, graph it, and state the domain and range. ■

The graph of $y = x^2$ is called a **parabola.** We study parabolas in detail in Chapter 3. The graph of the **square-root function** $y = \sqrt{x}$ is half of a parabola.

## EXAMPLE 2 The square-root function

Graph $y = \sqrt{x}$ and state the domain and range of the relation. Determine whether the relation is a function.

### Solution

Make a table listing ordered pairs that satisfy $y = \sqrt{x}$:

$x \geq 0$

| $x$ | 0 | 1 | 4 | 9 |
|---|---|---|---|---|
| $y = \sqrt{x}$ | 0 | 1 | 2 | 3 |

$y \geq 0$

Plotting these ordered pairs suggests the graph shown in Fig. 2.12. The domain of the relation is $[0, \infty)$ and the range is $[0, \infty)$. Because no vertical line can cross this graph more than once, $y = \sqrt{x}$ is a function.

The calculator graph in Fig. 2.13 supports these conclusions.

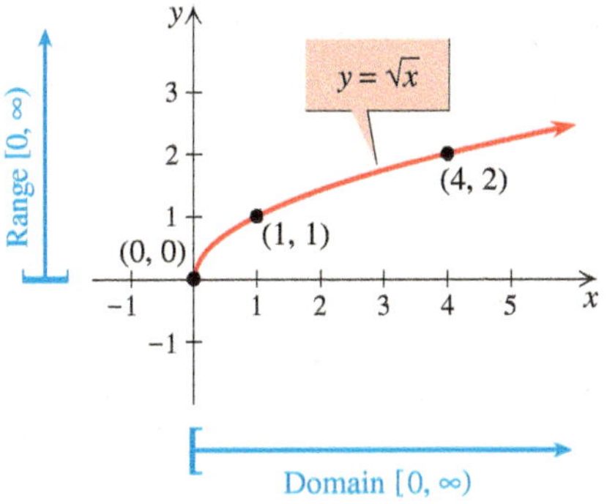


Figure 2.12

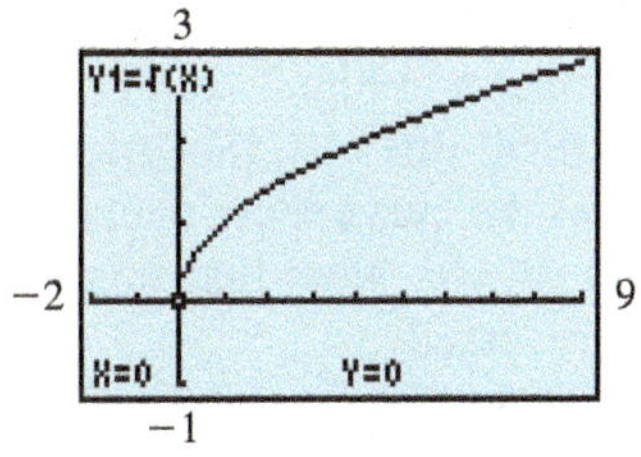


Figure 2.13

**▸TRY THIS.** Determine whether $y = \sqrt{1 - x}$ is a function, graph it, and state the domain and range. ■

In the next example we graph $x = y^2$ and see that its graph is also a parabola.

## EXAMPLE 3 A parabola opening to the right

Graph $x = y^2$ and state the domain and range of the relation. Determine whether the relation is a function.

### Solution

Make a table listing ordered pairs that satisfy $x = y^2$. In this case choose $y$ and calculate $x$:

$x \geq 0$

| $x = y^2$ | 0 | 1 | 1 | 4 | 4 |
|---|---|---|---|---|---|
| $y$ | 0 | 1 | $-1$ | 2 | $-2$ |

$y$ is any real number

Note that these ordered pairs are the same ones that satisfy $y = x^2$ except that the coordinates are reversed. For this reason the graph of $x = y^2$ in Fig. 2.14 has the same shape as the parabola in Fig. 2.10 and it is also a parabola. The domain of $x = y^2$ is $[0, \infty)$ and the range is $(-\infty, \infty)$. Because we can draw a vertical line that crosses this parabola twice, $x = y^2$ does not define $y$ as a function of $x$. Because $x = y^2$ is equivalent to $y = \pm\sqrt{x}$, the top half of the graph of $x = y^2$ is $y = \sqrt{x}$ and the bottom half is $y = -\sqrt{x}$.

To support these conclusions with a graphing calculator, graph $y_1 = \sqrt{x}$ and $y_2 = -\sqrt{x}$ as shown in Fig. 2.15.

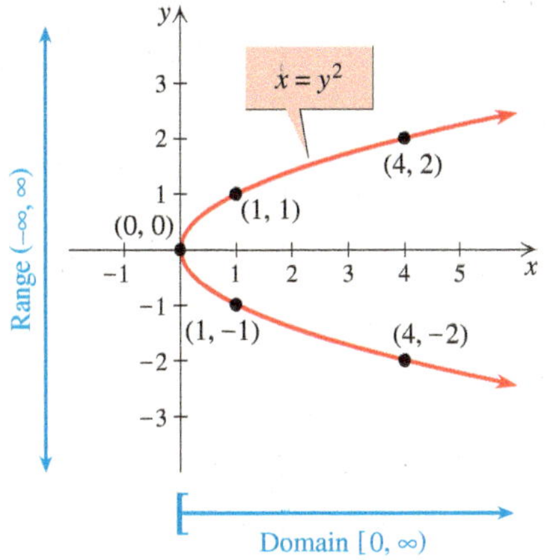


Figure 2.14

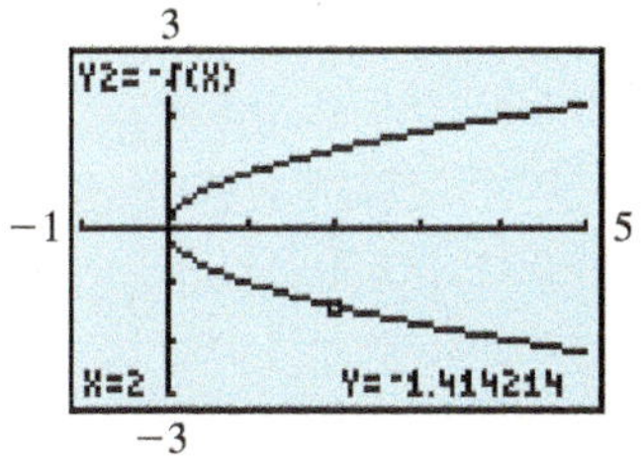


Figure 2.15

**▸TRY THIS.** Determine whether $x = -y^2$ is a function, graph it, and state the domain and range. ■

In the next example we graph the **cube function** $y = x^3$ and the **cube-root function** $y = \sqrt[3]{x}$.

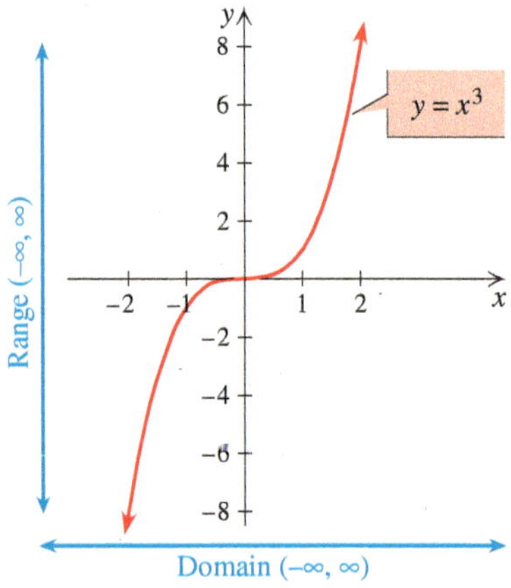


Figure 2.16

### EXAMPLE 4 Cube and cube-root functions

Graph each equation. State the domain and range.

**a.** $y = x^3$ **b.** $y = \sqrt[3]{x}$

**Solution**

**a.** Make a table listing ordered pairs that satisfy $y = x^3$:

*x* is any real number

| $x$ | $-2$ | $-1$ | $0$ | $1$ | $2$ |
|---|---|---|---|---|---|
| $y = x^3$ | $-8$ | $-1$ | $0$ | $1$ | $8$ |

*y* is any real number

These ordered pairs indicate a graph in the shape shown in Fig. 2.16. The domain is $(-\infty, \infty)$ and the range is $(-\infty, \infty)$. By the vertical line test, this graph is the graph of a function because no vertical line crosses the curve more than once.

**b.** Make a table listing ordered pairs that satisfy $y = \sqrt[3]{x}$. Note that this table is simply the table for $y = x^3$ with the coordinates reversed.

*x* is any real number

| $x$ | $-8$ | $-1$ | $0$ | $1$ | $8$ |
|---|---|---|---|---|---|
| $y = \sqrt[3]{x}$ | $-2$ | $-1$ | $0$ | $1$ | $2$ |

*y* is any real number

These ordered pairs indicate a graph in the shape shown in Fig. 2.17. The domain is $(-\infty, \infty)$ and the range is $(-\infty, \infty)$. Because no vertical line crosses this graph more than once, $y = \sqrt[3]{x}$ is a function.

The calculator graph in Fig. 2.18 supports these conclusions.

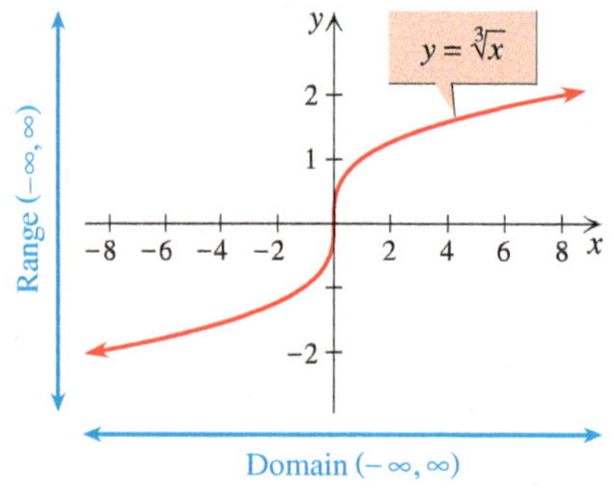


Figure 2.17

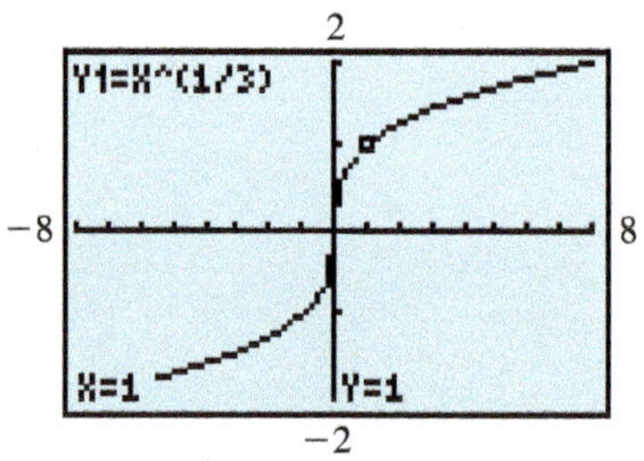


Figure 2.18

▸**TRY THIS.** Determine whether $y = -\sqrt[3]{x}$ is a function, graph it, and state the domain and range. ■

## Semicircles

The graph of $x^2 + y^2 = r^2 (r > 0)$ is a circle centered at the origin of radius $r$. A circle does not pass the vertical line test, and is not the graph of a function. We can find an equivalent equation by solving for $y$:

$$x^2 + y^2 = r^2$$

$$y^2 = r^2 - x^2$$

$$y = \pm\sqrt{r^2 - x^2}$$

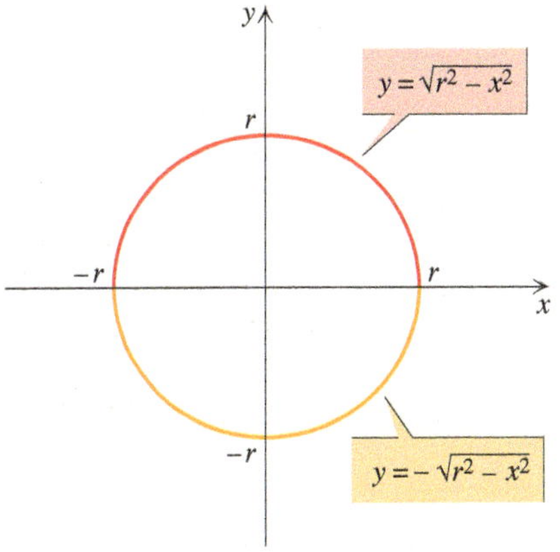


Figure 2.19

The equation $y = \sqrt{r^2 - x^2}$ does define $y$ as a function of $x$. Because $y$ is nonnegative in this equation, the graph is the top semicircle in Fig. 2.19. The top semicircle passes the vertical line test. Likewise, the equation $y = -\sqrt{r^2 - x^2}$ defines $y$ as a function of $x$, and its graph is the bottom semicircle in Fig. 2.19.

## EXAMPLE 5 Graphing a semicircle

Sketch the graph of each function and state the domain and range of the function.

**a.** $y = -\sqrt{4 - x^2}$ **b.** $y = \sqrt{9 - x^2}$

### Solution

**a.** Rewrite the equation in the standard form for a circle:

$$y = -\sqrt{4 - x^2}$$

$$y^2 = 4 - x^2 \quad \text{Square each side.}$$

$$x^2 + y^2 = 4 \quad \text{Standard form for the equation of a circle}$$

The graph of $x^2 + y^2 = 4$ is a circle of radius 2 centered at $(0, 0)$. Since $y$ must be negative in $y = -\sqrt{4 - x^2}$, the graph of $y = -\sqrt{4 - x^2}$ is the semicircle shown in Fig. 2.20. We can see from the graph that the domain is $[-2, 2]$ and the range is $[-2, 0]$.

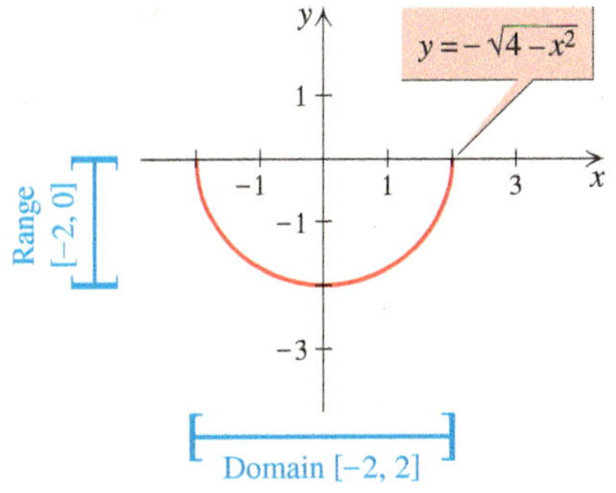


Figure 2.20

**b.** Rewrite the equation in the standard form for a circle:

$$y = \sqrt{9 - x^2}$$

$$y^2 = 9 - x^2 \quad \text{Square each side.}$$

$$x^2 + y^2 = 9 \quad \text{Standard form for the equation of a circle}$$

The graph of $x^2 + y^2 = 9$ is a circle with center $(0, 0)$ and radius 3. But this equation is not equivalent to the original. The value of $y$ in $y = \sqrt{9 - x^2}$ is nonnegative. So the graph of the original equation is the semicircle shown in Fig. 2.21. We can read the domain $[-3, 3]$ and the range $[0, 3]$ from the graph.

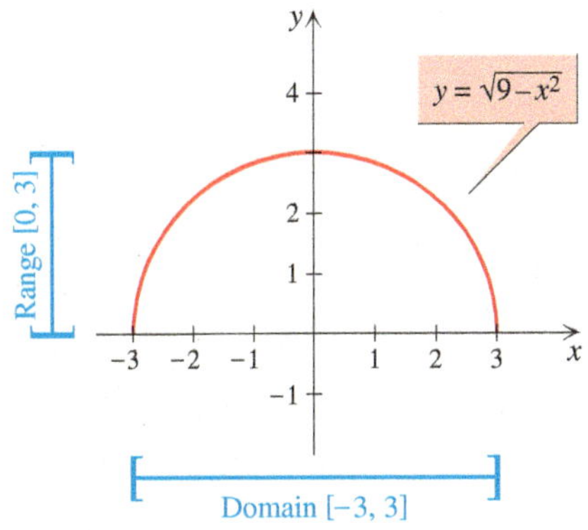


Figure 2.21

▸**TRY THIS.** Graph $y = -\sqrt{9 - x^2}$ and state the domain and range. ■

## Piecewise Functions

For some functions, different formulas are used in different regions of the domain. Since these functions are pieced together from two or more functions, they are called **piecewise functions.** The simplest example of such a function is the **absolute value function** $f(x) = |x|$, which can be written as

$$f(x) = \begin{cases} x & \text{for} \quad x \geq 0 \\ -x & \text{for} \quad x < 0. \end{cases}$$

For $x \geq 0$ the equation $f(x) = x$ is used to obtain the second coordinate, and for $x < 0$ the equation $f(x) = -x$ is used. The graph of the absolute value function is shown in the next example. Note how the graph is pieced together from the graphs of $y = x$ and $y = -x$.

## EXAMPLE 6 The absolute value function

Graph the equation $y = |x|$ and state the domain and range. Determine whether the relation is a function.

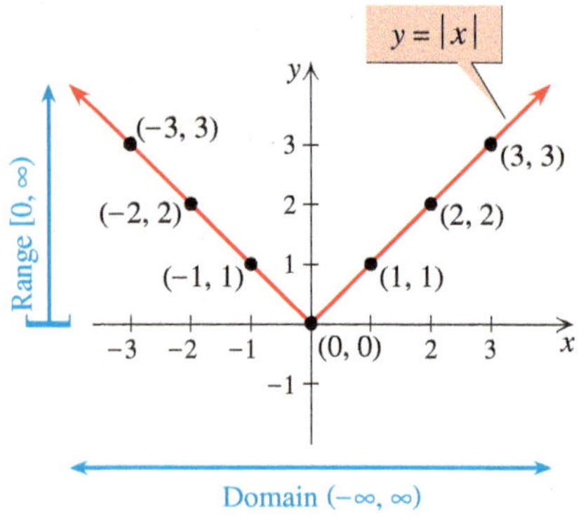


Figure 2.22

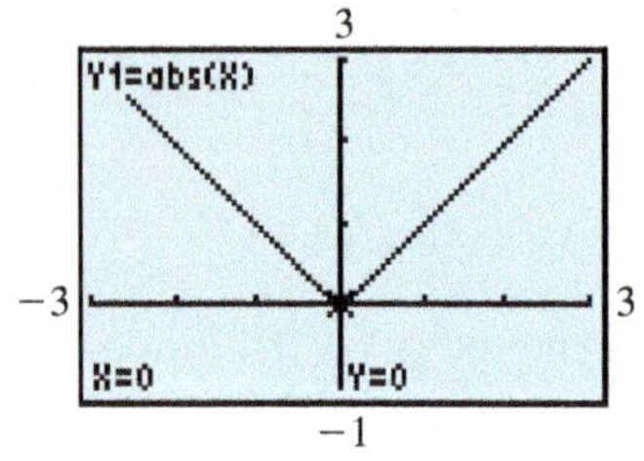


Figure 2.23

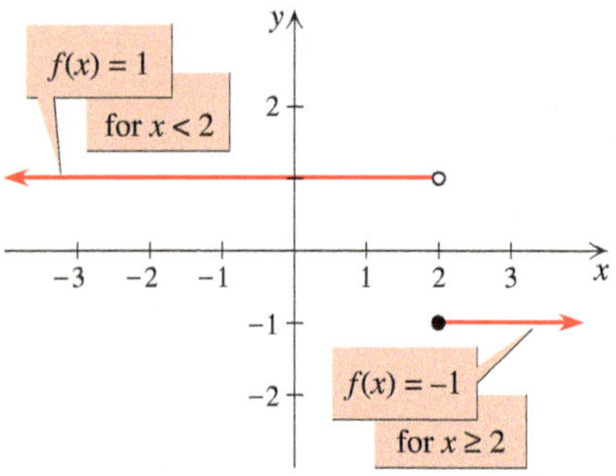


Figure 2.24

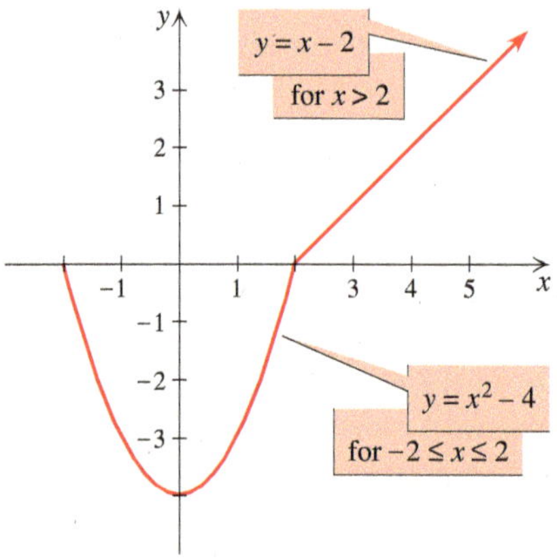


Figure 2.25

### Solution

If $x \geq 0$ use $y = x$ to determine ordered pairs that satisfy $y = |x|$. If $x < 0$ use $y = -x$:

| $x\ (x \geq 0)$ | 0 | 1 | 2 | 3 |
|---|---|---|---|---|
| $y = x$ | 0 | 1 | 2 | 3 |

| $x\ (x < 0)$ | −1 | −2 | −3 |
|---|---|---|---|
| $y = -x$ | 1 | 2 | 3 |

Plot these ordered pairs to get the V-shaped graph shown in Fig. 2.22. Because no vertical line crosses the graph more than once, $y = |x|$ is a function. Note that $x$ can be any real number, but $y$ is nonnegative. So the domain is $(-\infty, \infty)$ and the range is $[0, \infty)$.

To support these conclusions with a graphing calculator, graph $y_1 = \text{abs}(x)$ as shown in Fig. 2.23.

▶**TRY THIS.** Determine whether $y = |x| + 2$ is a function, graph it, and state the domain and range. ■

In the next example we graph two more piecewise functions.

## EXAMPLE 7 Graphing a piecewise function

Sketch the graph of each function and state the domain and range.

**a.** $f(x) = \begin{cases} 1 & \text{for} \quad x < 2 \\ -1 & \text{for} \quad x \geq 2 \end{cases}$ **b.** $f(x) = \begin{cases} x^2 - 4 & \text{for} \quad -2 \leq x \leq 2 \\ x - 2 & \text{for} \quad x > 2 \end{cases}$

### Solution

**a.** For $x < 2$ the graph is the horizontal line $y = 1$. For $x \geq 2$ the graph is the horizontal line $y = -1$. Note that $(2, -1)$ is on the graph shown in Fig. 2.24 but $(2, 1)$ is not, because when $x = 2$ we have $y = -1$. The domain is the interval $(-\infty, \infty)$ and the range consists of only two numbers, $-1$ and $1$. The range is not an interval. It is written in set notation as $\{-1, 1\}$. Note that the graph consists of two separate sections that do not touch each other.

**b.** Make a table of ordered pairs using $y = x^2 - 4$ for $x$ between $-2$ and $2$ and $y = x - 2$ for $x > 2$.

| $x$ | −2 | −1 | 0 | 1 | 2 |
|---|---|---|---|---|---|
| $y = x^2 - 4$ | 0 | −3 | −4 | −3 | 0 |

| $x$ | 2.1 | 3 | 4 | 5 |
|---|---|---|---|---|
| $y = x - 2$ | 0.1 | 1 | 2 | 3 |

For $x$ in the interval $[-2, 2]$ the graph is a portion of a parabola as shown in Fig. 2.25. For $x > 2$, the graph is a portion of a straight line through $(2.1, 0.1)$, $(3, 1)$, $(4, 2)$, and $(5, 3)$. The domain is $[-2, \infty)$, and the range is $[-4, \infty)$.

To graph a piecewise function on your calculator, use the inequality symbols from the TEST menu as shown in Fig. 2.26(a). On the calculator, an inequality or a compound inequality has a value of 1 when it is satisfied and 0 when it is not satisfied. So $y_1 = (x^2 - 4)/(x \geq -2 \text{ and } x \leq 2)$ will be graphed as the parabola $y = x^2 - 4$ only when both inequalities are satisfied. The word "and" is found in the TEST LOGIC menu. To graph $y = x - 2$ only for $x > 2$ enter $y_2 = (x - 2)/(x > 2)$. The calculator graph is shown in Fig. 2.26(b).

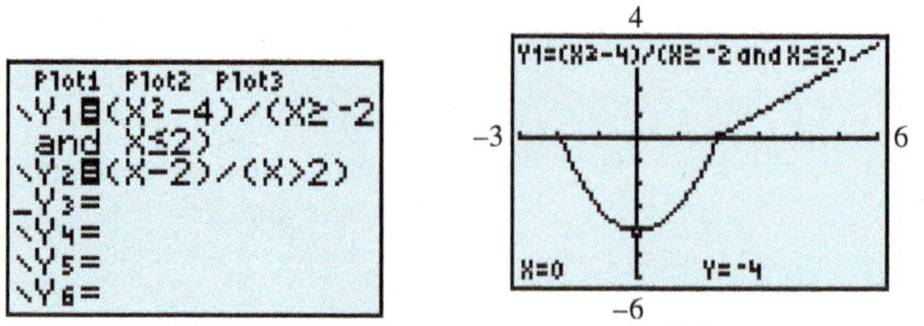


(a) (b)

Figure 2.26

▶**TRY THIS.** Graph $y = \begin{cases} x & \text{for} \quad x \geq 0 \\ -2x & \text{for} \quad x < 0 \end{cases}$ and state the domain and range. ■

Piecewise functions are often found in shipping charges. For example, if the weight in pounds of an order is in the interval $(0, 1]$, the shipping and handling charge is \$3. If the weight is in the interval $(1, 2]$, the shipping and handling charge is \$4, and so on. The next example is a function that is similar to a shipping and handling charge. This function is referred to as the **greatest integer function** and is written $f(x) = [\![x]\!]$ or $f(x) = \text{int}(x)$. The symbol $[\![x]\!]$ is defined to be the largest integer that is less than or equal to $x$. For example, $[\![5.01]\!] = 5$, because the greatest integer less than or equal to 5.01 is 5. Likewise, $[\![3.2]\!] = 3$, $[\![-2.2]\!] = -3$, and $[\![7]\!] = 7$.

## EXAMPLE 8 Graphing the greatest integer function

Sketch the graph of $f(x) = [\![x]\!]$ and state the domain and range.

### Solution

For any $x$ in the interval $[0, 1)$ the greatest integer less than or equal to $x$ is 0. For any $x$ in $[1, 2)$ the greatest integer less than or equal to $x$ is 1. For any $x$ in $[-1, 0)$ the greatest integer less than or equal to $x$ is $-1$. The definition of $[\![x]\!]$ causes the function to be constant between the integers and to "jump" at each integer. The graph of $f(x) = [\![x]\!]$ is shown in Fig. 2.27. The domain is $(-\infty, \infty)$, and the range is the set of integers.

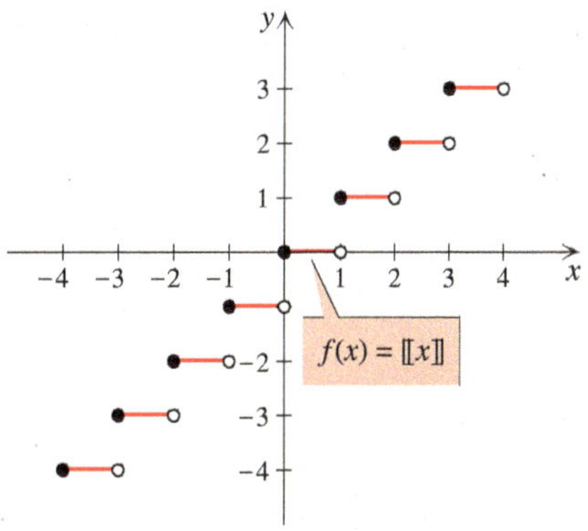


Figure 2.27

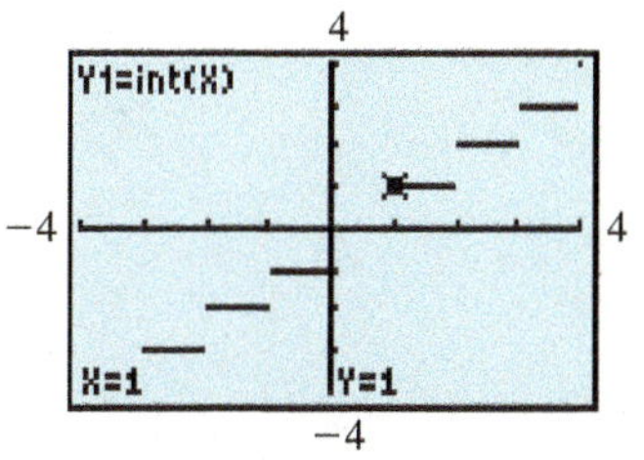


Figure 2.28

The calculator graph of $y_1 = \text{int}(x)$ looks best in dot mode as in Fig. 2.28, because in connected mode the calculator connects the disjoint pieces of the graph. The calculator graph in Fig. 2.28 supports our conclusion that the graph of this function looks like the one drawn in Fig. 2.27. Note that the calculator is incapable of showing whether the endpoints of the line segments are included.

▶**TRY THIS.** Graph $y = -[\![x]\!]$ and state the domain and range. ■

In the next example we vary the form of the greatest integer function, but the graph is still similar to the graph in Fig. 2.27.

## EXAMPLE 9 A variation of the greatest integer function

Sketch the graph of $f(x) = [\![x - 2]\!]$ for $0 \le x \le 5$.

### Solution

If $x = 0$, $f(0) = [\![-2]\!] = -2$. If $x = 0.5$, $f(0.5) = [\![-1.5]\!] = -2$. In fact, $f(x) = -2$ for any $x$ in the interval $[0, 1)$. Similarly, $f(x) = -1$ for any $x$ in the interval $[1, 2)$.

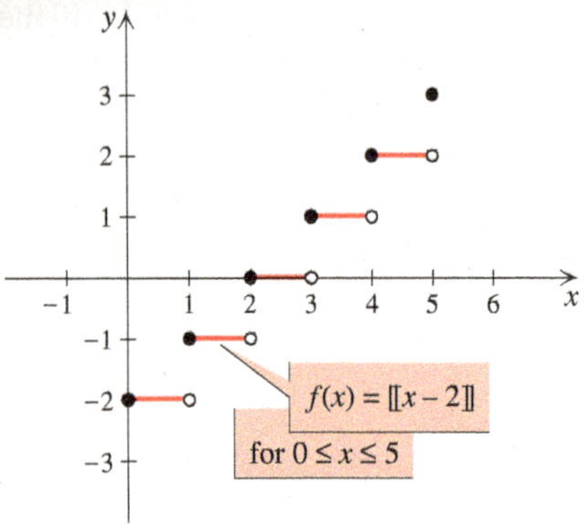


**Figure 2.29**

This pattern continues with $f(x) = 2$ for any $x$ in the interval $[4, 5)$, and $f(x) = 3$ for $x = 5$. The graph of $f(x) = [\![x - 2]\!]$ is shown in Fig. 2.29.

▶**TRY THIS.** Graph $y = [\![x + 2]\!]$ and state the domain and range. ■

## Increasing, Decreasing, and Constant

Imagine a point moving from left to right along the graph of a function. If the $y$-coordinate of the point is getting larger, smaller, or staying the same, then the function is said to be *increasing, decreasing,* or *constant,* respectively. More precisely we make the following definition.

**Definition: Increasing, Decreasing, Constant**

If $a < b$ implies $f(a) < f(b)$ for any $a$ and $b$ in the domain of $f$, then $f$ is an **increasing** function.
If $a < b$ implies $f(a) > f(b)$ for any $a$ and $b$ in the domain of $f$, then $f$ is a **decreasing** function.
If $a < b$ implies $f(a) = f(b)$ for any $a$ and $b$ in the domain of $f$, then $f$ is a **constant** function.

### EXAMPLE 10 Increasing, decreasing, or constant functions

Determine whether each function is increasing, decreasing, or constant by examining its graph.

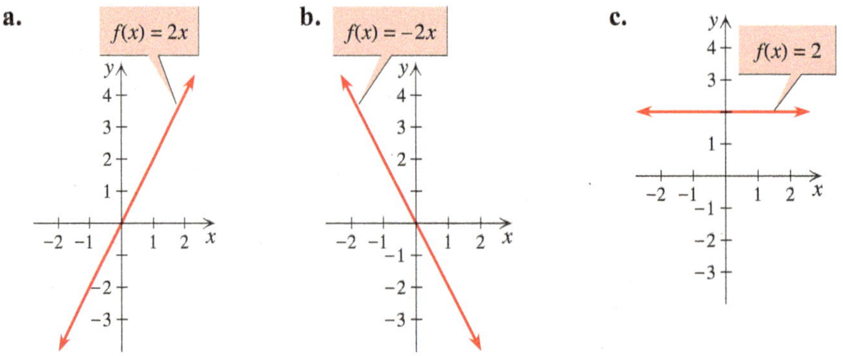


#### Solution

If a point is moving left to right along the graph of $f(x) = 2x$, then its $y$-coordinate is increasing. So $f(x) = 2x$ is an increasing function. If a point is moving left to right along the graph of $f(x) = -2x$, then its $y$-coordinate is decreasing. So $f(x) = -2x$ is a decreasing function. The function $f(x) = 2$ has a constant $y$-coordinate and it is a constant function.

▶**TRY THIS.** Determine whether $f(x) = -3x$ is increasing, decreasing, or constant by examining its graph. ■

We can also discuss whether a function is increasing, decreasing, or constant on a subset of its domain. For example, the absolute value function $f(x) = |x|$, which we graphed in Example 6, is decreasing on the interval $(-\infty, 0]$ and increasing on the interval $[0, \infty)$.

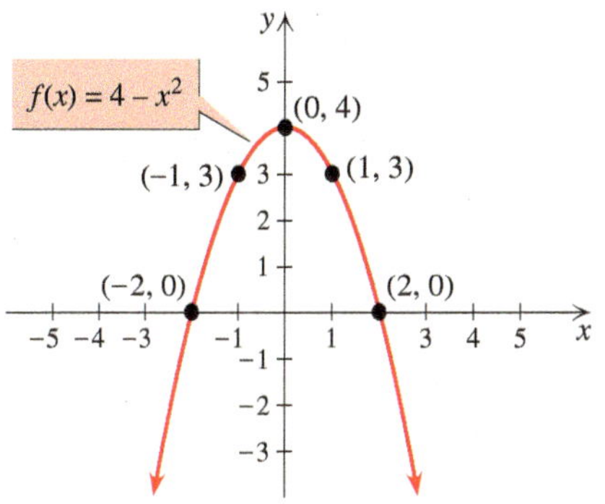

Figure 2.30

## EXAMPLE 11 Increasing, decreasing, or constant on an interval

Sketch the graph of each function and identify any intervals on which the function is increasing, decreasing, or constant.

**a.** $f(x) = 4 - x^2$ **b.** $g(x) = \begin{cases} 0 & \text{for} \quad x \le 0 \\ \sqrt{x} & \text{for} \quad 0 < x < 4 \\ 2 & \text{for} \quad x \ge 4 \end{cases}$

### Solution

**a.** The graph of $f(x) = 4 - x^2$ includes the points $(-2, 0)$, $(-1, 3)$, $(0, 4)$, $(1, 3)$, and $(2, 0)$. The graph is shown in Fig. 2.30. The function is increasing on the interval $(-\infty, 0]$ and decreasing on $[0, \infty)$.

**b.** The graph of $g$ is shown in Fig. 2.31. The function $g$ is constant on the intervals $(-\infty, 0]$ and $[4, \infty)$, and increasing on $[0, 4]$.

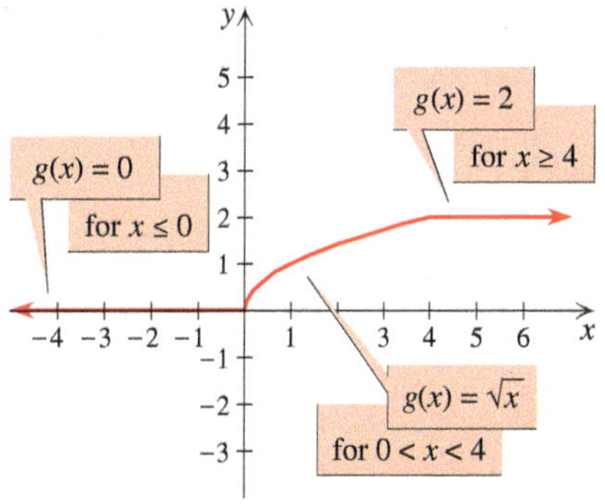

Figure 2.31

**▸TRY THIS.** Graph $f(x) = \begin{cases} x & \text{for} \quad x \ge 0 \\ -2x & \text{for} \quad x < 0 \end{cases}$ and identify any intervals on which the function is increasing, decreasing, or constant. ■

## FOR thought... True or False? Explain

1. The range of $y = -x^2$ is $(-\infty, 0]$.
2. The function $y = -\sqrt{x}$ is increasing on $[0, \infty)$.
3. The function $f(x) = \sqrt[3]{x}$ is increasing on $(-\infty, \infty)$.
4. If $f(x) = [\![x + 3]\!]$, then $f(-4.5) = -2$.
5. The range of the function $f(x) = \frac{|x|}{x}$ is the interval $(-1, 1)$.
6. The range of $f(x) = [\![x - 1]\!]$ is the set of integers.
7. The only ordered pair that satisfies $(x - 5)^2 + (y + 6)^2 = 0$ is $(5, -6)$.
8. The domain of the function $y = \sqrt{4 - x^2}$ is the interval $[-2, 2]$.
9. The range of the function $y = \sqrt{16 - x^2}$ is the interval $[0, \infty)$.
10. The function $y = \sqrt{4 - x^2}$ is increasing on $[-2, 0]$ and decreasing on $[0, 2]$.

# EXERCISES 2.2

*Fill in the blank.*

1. The graph of $y = x^2$ is a ________.

2. The absolute value function is an example of a ________ function.

*Make a table listing ordered pairs that satisfy each equation. Then graph the equation. Determine the domain and range, and whether y is a function of x.*

**3.** $y = 2x$ **4.** $x = 2y$ **5.** $x - y = 0$

**6.** $x - y = 2$ **7.** $y = 5$ **8.** $x = 3$

**9.** $y = 2x^2$ **10.** $y = x^2 - 1$ **11.** $y = 1 - x^2$

**12.** $y = -1 - x^2$ **13.** $y = 1 + \sqrt{x}$ **14.** $y = 2 - \sqrt{x}$

**15.** $x = y^2 + 1$ **16.** $x = 1 - y^2$ **17.** $x = \sqrt{y}$

**18.** $x - 1 = \sqrt{y}$ **19.** $y = \sqrt[3]{x} + 1$ **20.** $y = \sqrt[3]{x} - 2$

**21.** $x = \sqrt[3]{y}$ **22.** $x = \sqrt[3]{y} - 1$ **23.** $y^2 = 1 - x^2$

**24.** $x^2 + y^2 = 4$ **25.** $y = \sqrt{1 - x^2}$ **26.** $y = -\sqrt{25 - x^2}$

**27.** $y = x^3$ **28.** $y = -x^3$ **29.** $y = 2|x|$

**30.** $y = |x - 1|$ **31.** $y = -|x|$ **32.** $y = -|x + 1|$

**33.** $x = |y|$ **34.** $x = |y| + 1$

*Make a table listing ordered pairs for each function. Then sketch the graph and state the domain and range.*

**35.** $f(x) = \begin{cases} 2 & \text{for } x < -1 \\ -2 & \text{for } x \geq -1 \end{cases}$

**36.** $f(x) = \begin{cases} 3 & \text{for } x < 2 \\ 1 & \text{for } x \geq 2 \end{cases}$

**37.** $f(x) = \begin{cases} x + 1 & \text{for } x > 1 \\ x - 3 & \text{for } x \leq 1 \end{cases}$

**38.** $f(x) = \begin{cases} 5 - x & \text{for } x \leq 2 \\ x + 1 & \text{for } x > 2 \end{cases}$

**39.** $f(x) = \begin{cases} \sqrt{x + 2} & \text{for } -2 \leq x \leq 2 \\ 4 - x & \text{for } x > 2 \end{cases}$

**40.** $f(x) = \begin{cases} \sqrt{x} & \text{for } x \geq 1 \\ -x & \text{for } x < 1 \end{cases}$

**41.** $f(x) = \begin{cases} \sqrt{-x} & \text{for } x < 0 \\ \sqrt{x} & \text{for } x \geq 0 \end{cases}$

**42.** $f(x) = \begin{cases} 3 & \text{for } x < 0 \\ 3 + \sqrt{x} & \text{for } x \geq 0 \end{cases}$

**43.** $f(x) = \begin{cases} x^2 & \text{for } x < -1 \\ -x & \text{for } x \geq -1 \end{cases}$

**44.** $f(x) = \begin{cases} 4 - x^2 & \text{for } -2 \leq x \leq 2 \\ x - 2 & \text{for } x > 2 \end{cases}$

**45.** $f(x) = [\![x + 1]\!]$

**46.** $f(x) = 2[\![x]\!]$

**47.** $f(x) = [\![x]\!] + 2$ for $0 \leq x < 4$

**48.** $f(x) = [\![x - 3]\!]$ for $0 < x \leq 5$

*From the graph of each function in Exercises 49–56, state the domain, the range, and the intervals on which the function is increasing, decreasing, or constant.*

**49. a.**

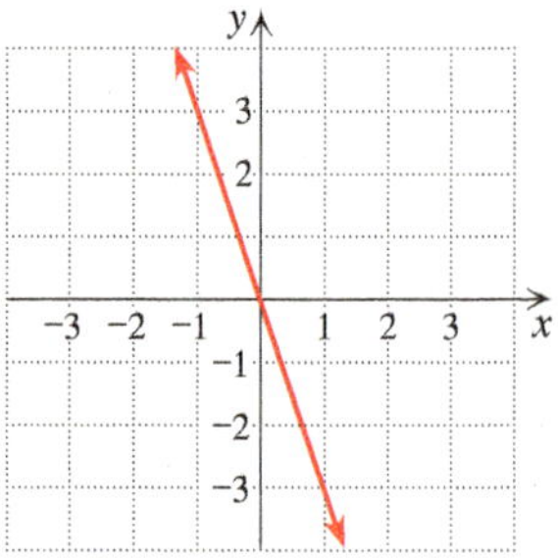


**b.**

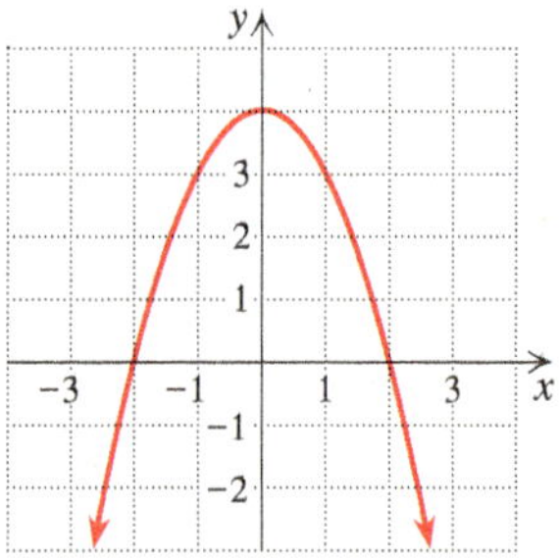


**50. a.**

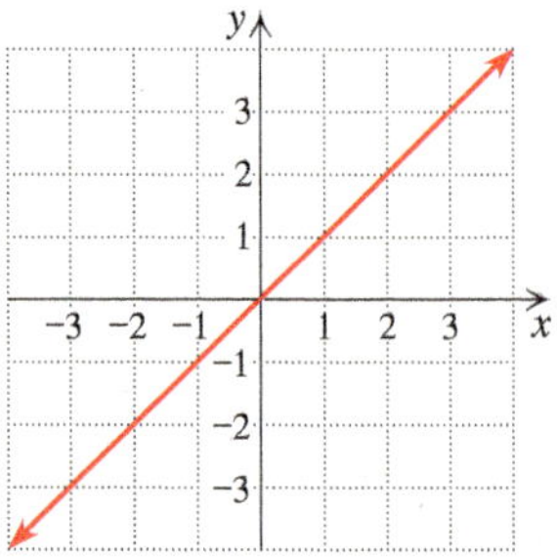


**b.**

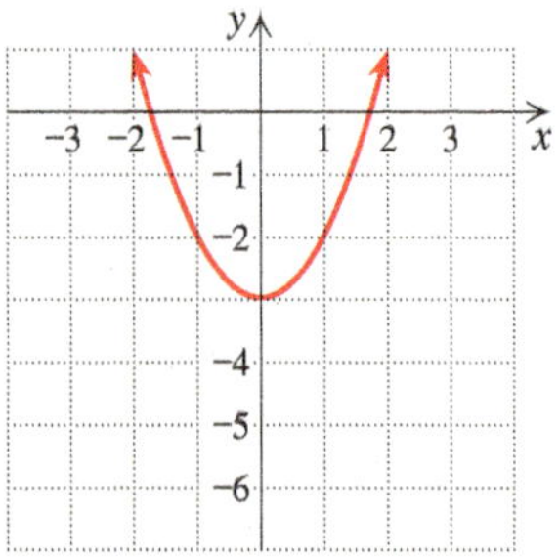


**51. a.**

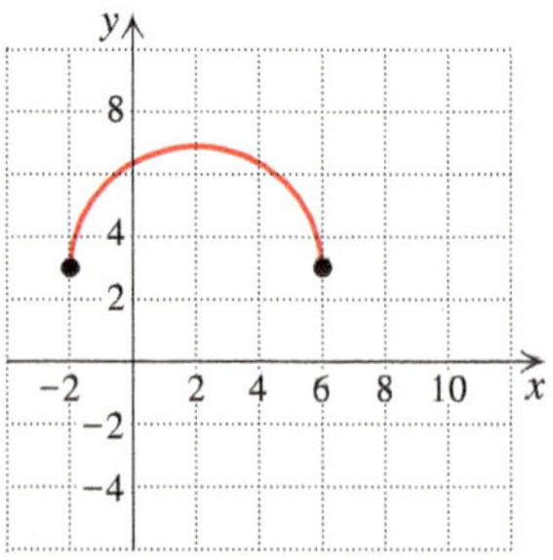


**b.**

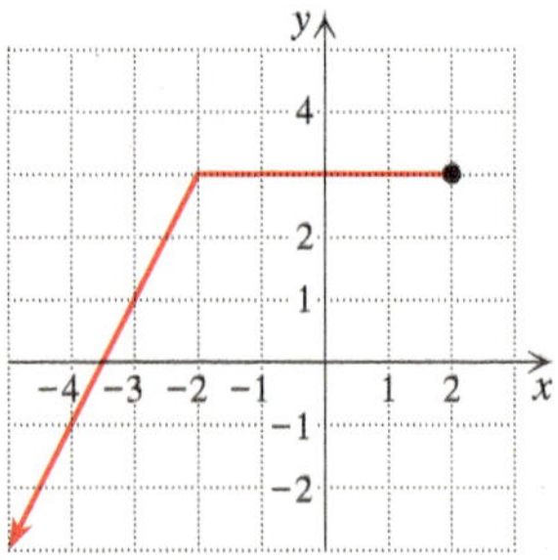

**52. a.**

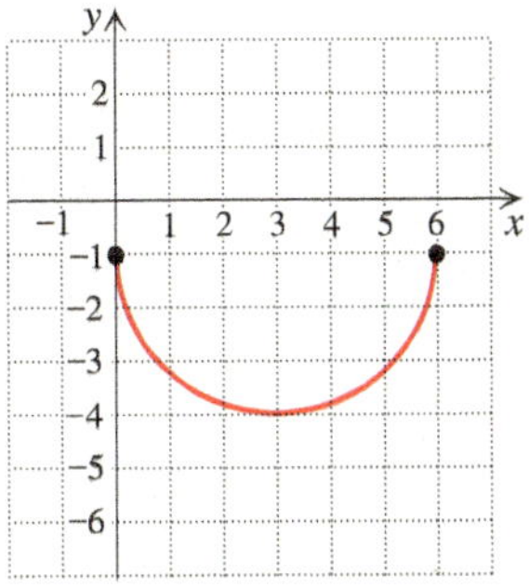

**b.**

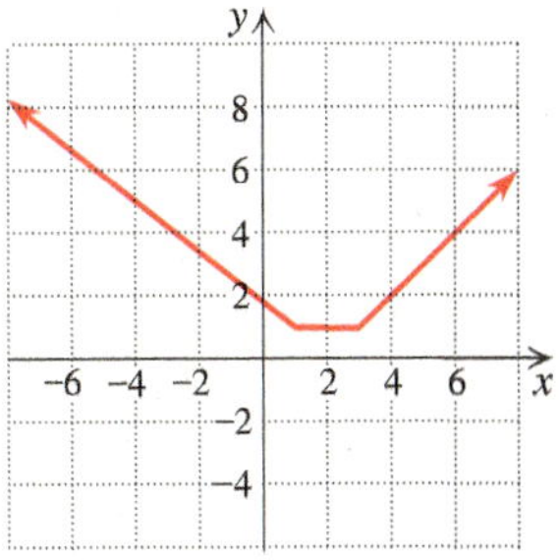

**53. a.**

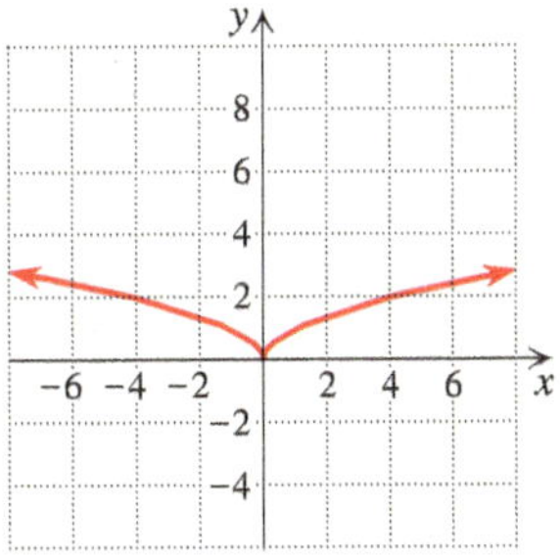

**b.**

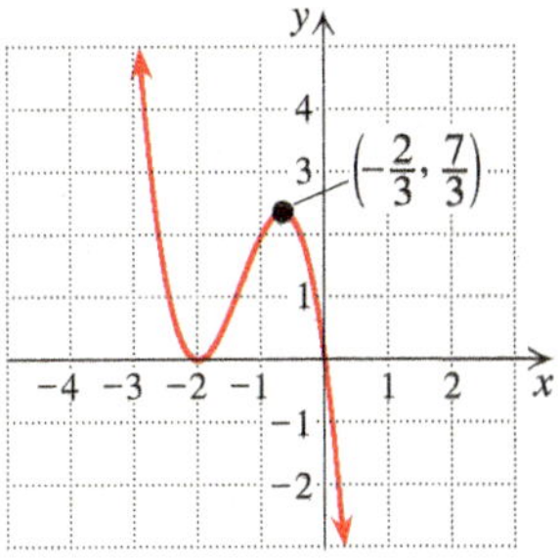


**54. a.**

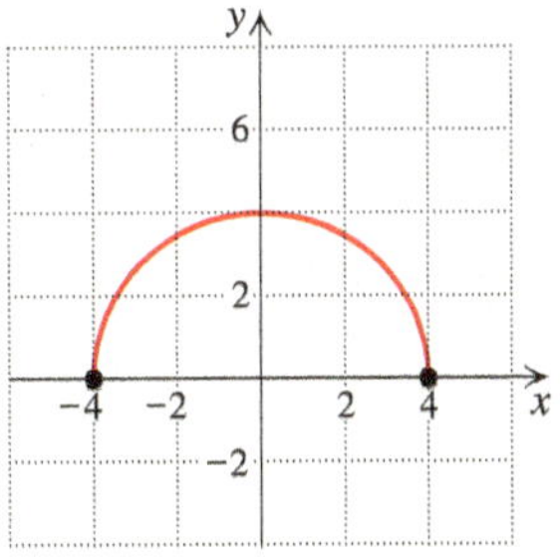

**b.**

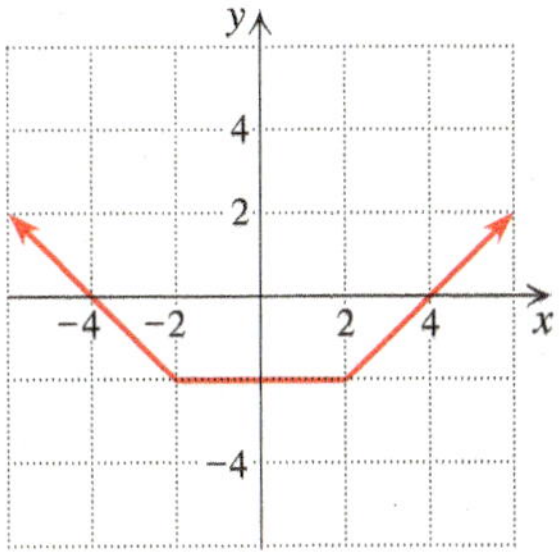

**55. a.**

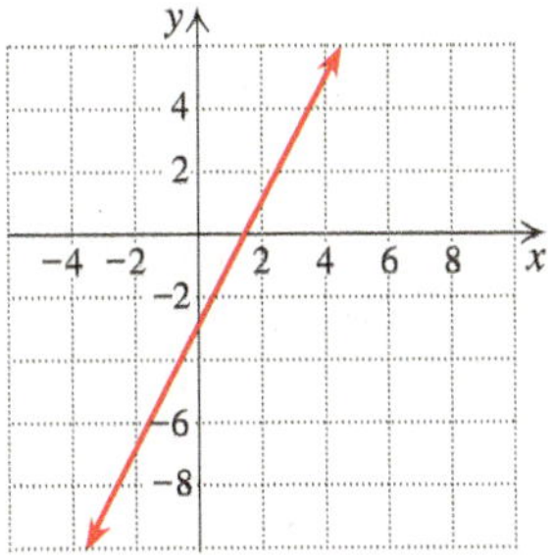

**b.**

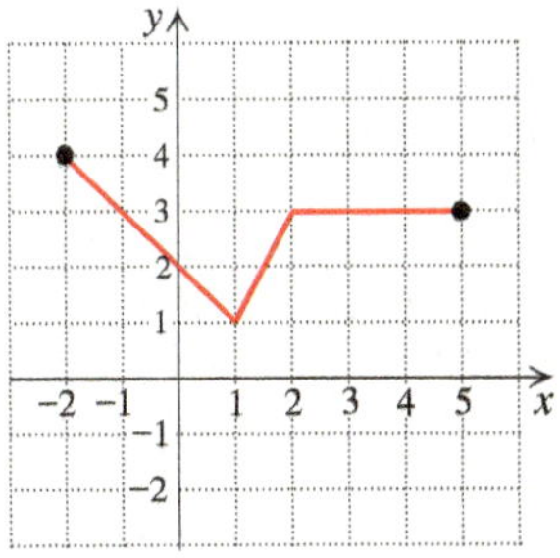

**56. a.**

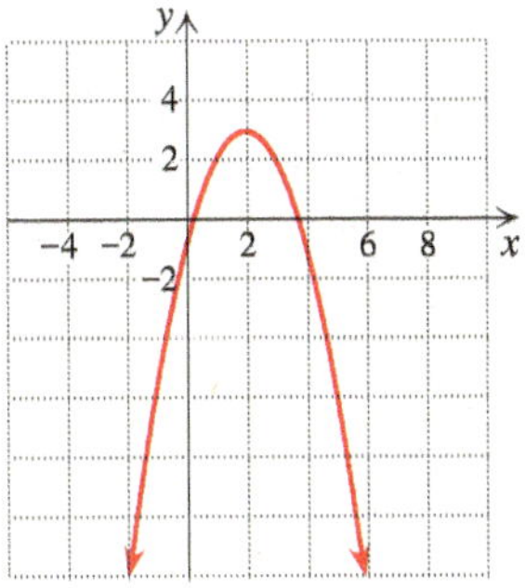

**b.**

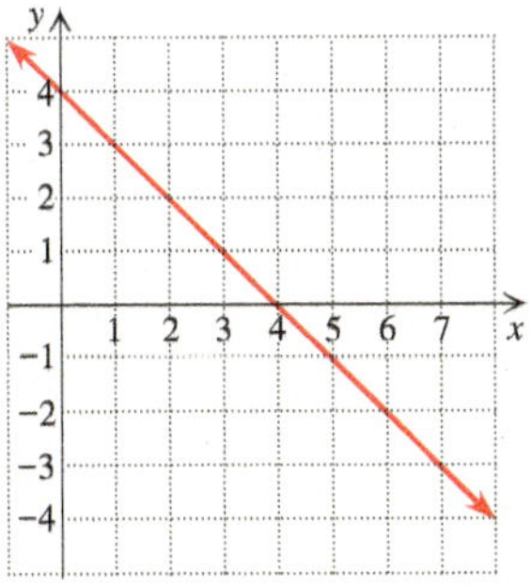

*Make a table listing ordered pairs for each function. Then sketch the graph and state the domain and range. Identify any intervals on which f is increasing, decreasing, or constant.*

**57.** $f(x) = 2x + 1$

**58.** $f(x) = -3x$

**59.** $f(x) = |x - 1|$

**60.** $f(x) = |x| + 1$

**61.** $f(x) = \dfrac{|x|}{x}$

**62.** $f(x) = \dfrac{2x}{|x|}$

**63.** $f(x) = \sqrt{9 - x^2}$

**64.** $f(x) = -\sqrt{1 - x^2}$

**65.** $f(x) = \begin{cases} x + 1 & \text{for } x \geq 3 \\ x + 2 & \text{for } x < 3 \end{cases}$

**66.** $f(x) = \begin{cases} \sqrt{-x} & \text{for } x < 0 \\ -\sqrt{x} & \text{for } x \geq 0 \end{cases}$

**67.** $f(x) = \begin{cases} x + 3 & \text{for } x \leq -2 \\ \sqrt{4 - x^2} & \text{for } -2 < x < 2 \\ -x + 3 & \text{for } x \geq 2 \end{cases}$

**68.** $f(x) = \begin{cases} 8 + 2x & \text{for } x \leq -2 \\ x^2 & \text{for } -2 < x < 2 \\ 8 - 2x & \text{for } x \geq 2 \end{cases}$

*Write a piecewise function for each given graph.*

**69.**

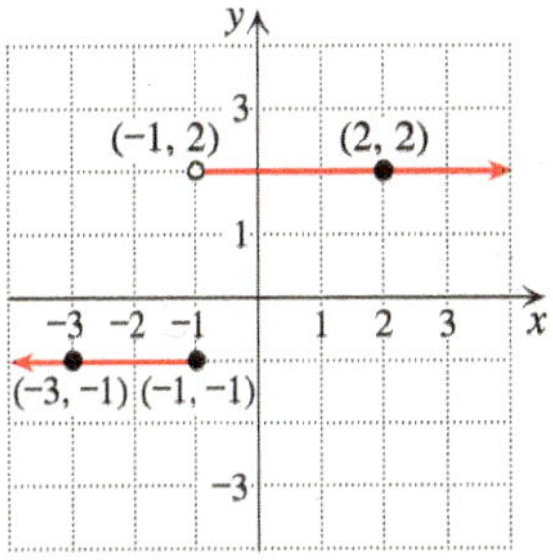


**70.**

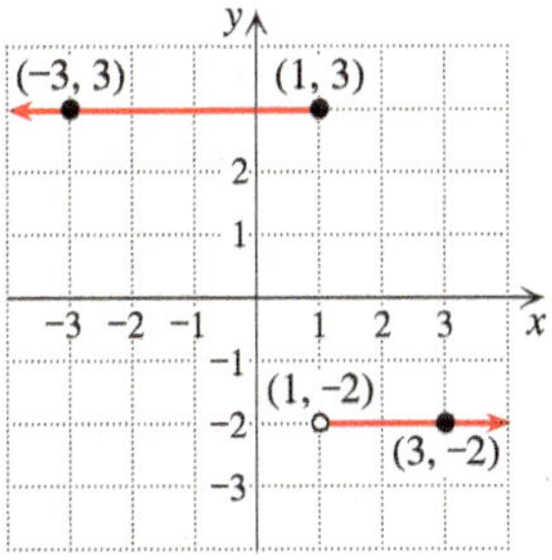


**71.**

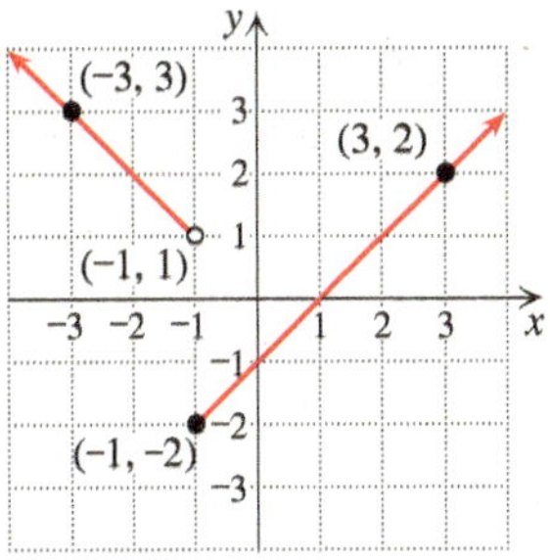


**72.**

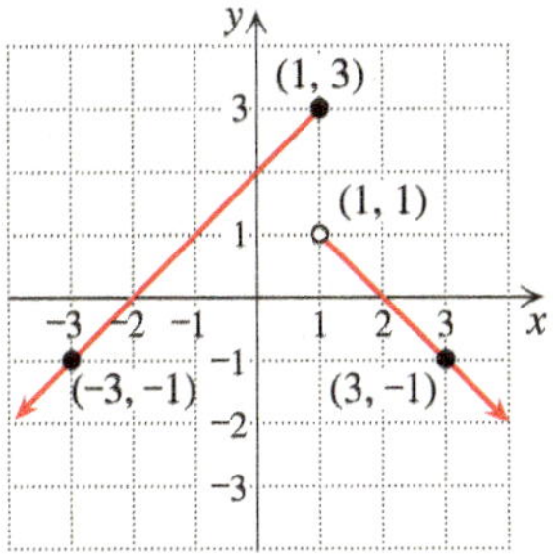

73.

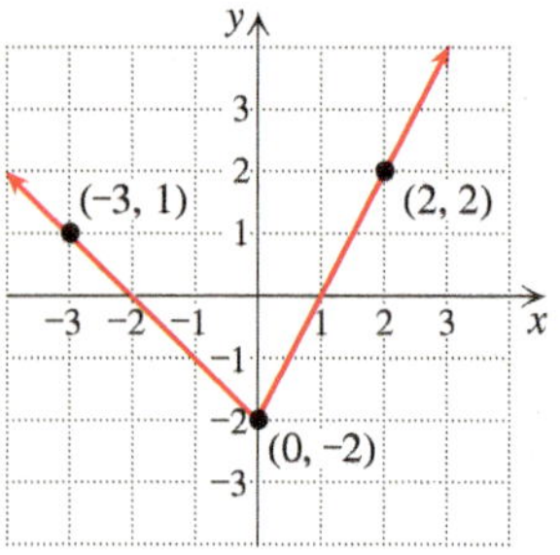


74.

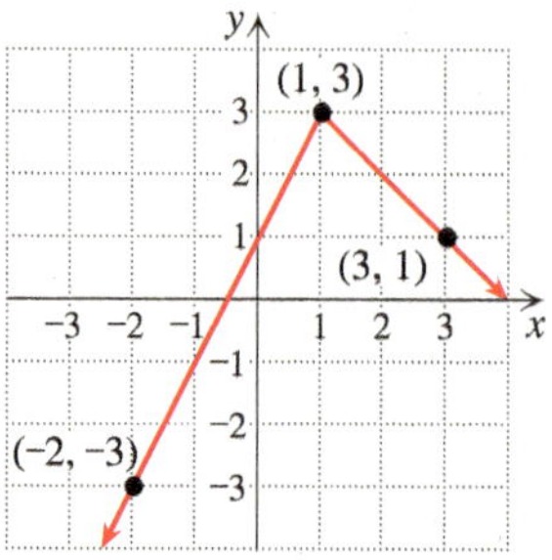


*Use the minimum and maximum features of a graphing calculator to find the intervals on which each function is increasing or decreasing. Round approximate answers to two decimal places.*

**75.** $y = 3x^2 - 5x - 4$

**76.** $y = -6x^2 + 2x - 9$

**77.** $y = x^3 - 3x$

**78.** $y = x^4 - 11x^2 + 18$

**79.** $y = 2x^4 - 12x^2 + 25$

**80.** $y = x^5 - 13x^3 + 36x$

**81.** $y = |20 - |x - 50||$

**82.** $y = x + |x + 30| - |x + 50|$

*Select the graph (a, b, c, or d) that best depicts each of the following scenarios.*

**83.** Profits for Cajun Drilling Supplies soared during the seventies, but went flat when the bottom fell out of the oil industry in the eighties. Cajun saw a period of moderate growth in the nineties.

**84.** To attain a minimum therapeutic level of Flexeril in her blood, Millie takes three Flexeril tablets, one every four hours. Because the half-life of Flexeril is four hours, one-half of the Flexeril in her blood is eliminated after four hours.

**85.** The bears dominated the market during the first quarter with massive sell-offs. The second quarter was an erratic period in which investors could not make up their minds. The bulls returned during the third quarter, sending the market to record highs.

**86.** Medicare spending soared during the eighties. Congress managed to slow the rate of growth of Medicare during the nineties and actually managed to decrease Medicare spending in the first decade of the twenty-first century.

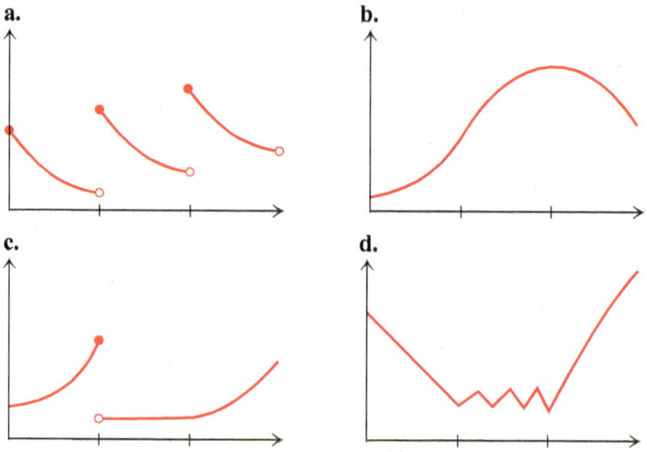


**Figure for Exercises 83 to 86**

*Draw a graph that pictures each situation. Explain any choices that you make. Identify the independent and dependent variables. Determine the intervals on which the dependent variable is increasing, decreasing, or constant. Answers may vary.*

**87.** Captain Janeway left the holodeck at 7:45 to meet Tuvok, her chief of security, on the Bridge. After walking for 3 minutes, she realized she had forgotten her tricorder and returned to get it. She picked up the tricorder and resumed her walk, arriving on the Bridge at 8:00. After 15 minutes the discussion was over, and Janeway returned to the holodeck. Graph Janeway's distance from the holodeck as a function of time.

**88.** Starting from the pit, Helen made three laps around a circular race track at 40 seconds per lap. She then made a 30-second pit stop and two and a half laps before running off the track and getting stuck in the mud for the remainder of the five-minute race. Graph Helen's distance from the pit as a function of time.

**89.** Winona deposited \$30 per week into her cookie jar. After 2 years she spent half of her savings on a stereo. After spending 6 months in the outback she proceeded to spend \$15 per week on CDs until all of the money in the jar was gone. Graph the amount in her cookie jar as a function of time.

**90.** Michael started buying Navajo crafts with \$8000 in his checking account. He spent \$200 per day for 10 days on pottery, then \$400 per day for the next 10 days on turquoise and silver jewelry. For the next 20 days, he spent \$50 per day on supplies while he set up his retail shop. He rested for 5 days, then took in \$800 per day for the next 20 days from the resale of his collection. Graph the amount in his checking account as a function of time.

*Solve each problem.*

**91.** *Motor Vehicle Ownership* World motor vehicle ownership in developed countries can be modeled by the function

$$M(t) = \begin{cases} 17.5t + 250 & \text{for} \quad 0 \le t \le 20 \\ 10t + 400 & \text{for} \quad 20 < t \le 40 \end{cases}$$

where $t$ is the number of years since 1970 and $M(t)$ is the number of motor vehicles in millions in the year $1970 + t$ (World Resources Institute, www.wri.org). See the accompanying figure. How many vehicles were there in developed countries in 1988? How many will there be in 2010? What was the average rate of change of motor vehicle ownership from 1984 to 1994?

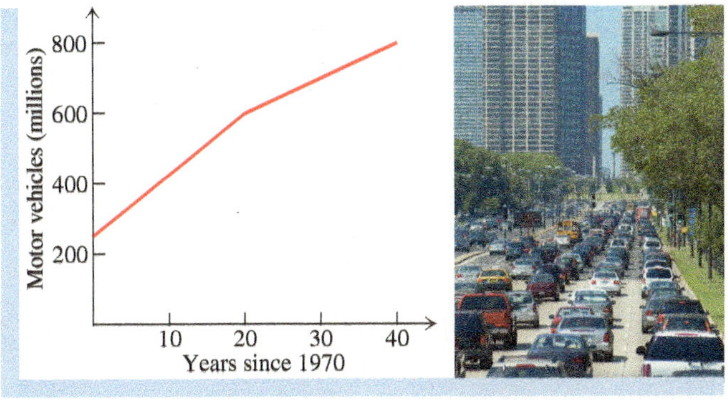


**Figure for Exercise 91**

**92.** *Motor Vehicle Ownership* World motor vehicle ownership in developing countries and Eastern Europe can be modeled by the function

$$M(t) = 6.25t + 50$$

where $t$ is the number of years since 1970 and $M(t)$ is in millions of vehicles (World Resources Institute, www.wri.org). Graph this function. What is the expected average rate of change of motor vehicle ownership from 1990 through 2010? Is motor vehicle ownership expected to grow faster in developed or developing countries in the period 1990 through 2010? (See the previous exercise.)

**93.** *Water Bill* The monthly water bill in Hammond is a function of the number of gallons used. The cost is \$10.30 for 10,000 gal or less. Over 10,000 gal, the cost is \$10.30 plus \$1 for each 1000 gal over 10,000 with any fraction of 1000 gal charged at a fraction of \$1. On what interval is the cost constant? On what interval is the cost increasing?

**94.** *Gas Mileage* The number of miles per gallon obtained with a new Toyota Prius is a function of the speed at which it is driven. Is this function increasing or decreasing on its domain? Explain.

**95.** *Filing a Tax Return* An accountant determines the charge for filing a tax return by using the function $C = 50 + 40[\![t]\!]$ for $t > 0$, where $C$ is in dollars and $t$ is in hours. Sketch the graph of this function. For what values of $t$ is the charge over \$235?

**96.** *Shipping Machinery* The cost in dollars of shipping a machine is given by the function $C = 200 + 37[\![w/100]\!]$ for $w > 0$, where $w$ is the weight of the machine in pounds. For which values of $w$ is the cost less than \$862?

**97.** *Delivering Concrete* A concrete company charges \$150 for delivering less than 3 $\text{yd}^3$ of concrete. For 3 $\text{yd}^3$ and more, the charge is \$50/$\text{yd}^3$ with a fraction of a yard charged as a fraction of \$50. Use function notation to write the charge as a function of the number $x$ of cubic yards delivered, where $0 < x \le 10$, and graph this function.

**98.** *Parking Charges* A garage charges \$4/hr up to 3 hr, with any fraction of an hour charged as a whole hour. Any time over 3 hr is charged at the all-day rate of \$15. Use function notation to write the charge as a function of the number of hours $x$, where $0 < x \le 8$, and graph this function.

## FOR WRITING/DISCUSSION

**99.** *Steps* Find an example of a function in real life whose graph has "steps" like that of the greatest integer function. Graph your function and find a formula for it if possible.

**100.** *Cooperative Learning* Select two numbers $a$ and $b$. Then define a piecewise function using different formulas on the intervals $(-\infty, a]$, $(a, b)$, and $[b, \infty)$ so that the graph does not "jump" at $a$ or $b$. Give your function to a classmate to graph and check.

## ▶ RETHINKING

**101.** Find the domain and range of the function $f(x) = \sqrt{x - 2} + 3$.

**102.** Solve $|13x - 55| = -9$.

**103.** Solve $|13x - 55| = 0$.

**104.** Solve $|13x - 55| = 9$.

**105.** Solve $4x^2 - 20x + 25 = 0$.

**106.** Solve $2w^2 - 5w - 9 = 0$.

## THINKING OUTSIDE THE BOX XIV

*Best-Fitting Pipe* A work crew is digging a pipeline through a frozen wilderness in Alaska. The cross section of the trench is in the shape of the parabola $y = x^2$. The pipe has a circular cross section. If the pipe is too large, then the pipe will not lie on the bottom of the trench.

**a.** What is the radius of the largest pipe that will lie on the bottom of the trench?

**b.** If the radius of the pipe is 3 and the trench is in the shape of $y = ax^2$, then what is the largest value of $a$ for which the pipe will lie in the bottom of the trench?

## POP QUIZ 2.2

1. Find the domain and range for $y = 1 - \sqrt{x}$.
2. Find the domain and range for $y = \sqrt{9 - x^2}$.
3. Find the range for $f(x) = \begin{cases} 2x & \text{for } x \geq 1 \\ 3 - x & \text{for } x < 1 \end{cases}$.
4. On what interval is $y = x^2$ increasing?
5. On what interval is $y = |x - 3|$ decreasing?

## LINKING concepts... For Individual or Group Explorations

### Social Security and Life Expectancy

*The average annual Social Security benefit in 2009 was about \$13,800. The amount of a retiree's benefit depends on the retiree's lifetime earnings, the person's full retirement age, and the age at the time of retirement. The following function gives the annual benefit in dollars for a person retiring in 2009 at ages 62 through 70 with a full retirement benefit of \$12,000 at full retirement age of 66.*

$$B = \begin{cases} 600a - 28{,}200 & 62 \leq a < 64 \\ 804a - 41{,}064 & 64 \leq a < 67 \\ 960a - 51{,}360 & 67 \leq a \leq 70 \end{cases}$$

**a)** Graph the benefit function.

**b)** What is the annual benefit for a person who retires at age 64?

**c)** At what age does a person receive an annual benefit of \$14,880?

**d)** What is the average rate of change of the benefit for the ages 62 through 63? Ages 64 through 66? Ages 66 through 70?

**e)** Do the answers to part (d) appear in the annual benefit formula?

*The life expectancy L for a U.S. white male with present age a can be modeled by the formula*

$$L = 67.0166(1.00308)^a.$$

**f)** Bob is a white male retiring in 2009 at age 62. What total amount can he be expected to draw from Social Security before he dies? Figure his Social Security benefit using the formula given above.

**g)** Bill is a white male retiring in 2009 at age 70. What total amount can he be expected to draw from Social Security before he dies? Figure his Social Security benefit using the formula given above.

# 2.3 Families of Functions, Transformations, and Symmetry

If $a$, $h$, and $k$ are real numbers with $a \neq 0$, then $y = af(x - h) + k$ is a **transformation** of the function $y = f(x)$. All of the transformations of a function form a **family of functions.** For example, any function of the form $y = a\sqrt{x - h} + k$ is in the square-root family, any function of the form $y = a|x - h| + k$ is in the absolute value family, and any function of the form $y = a(x - h)^2 + k$ is in the square or quadratic family. All of the functions in a family of functions have similar graphs. The graph of any function in the square or quadratic family is called a **parabola.** We will now see what effect each of the numbers $a$, $h$, and $k$ has on the graph of the original function $y = f(x)$.

## Horizontal Translation

According to the order of operations, the first operation to perform in the formula $y = af(x - h) + k$ is to subtract $h$ from $x$. Then $f(x - h)$ is multiplied by $a$, and finally $k$ is added on. The order is important here and we will study the effects of these numbers in the order $h$, $a$, and $k$. Subtracting $h$ from $x$ causes the graph of $y = f(x)$ to move horizontally.

**Definition: Translation to the Right or Left**

If $h > 0$, then the graph of $y = f(x - h)$ is a **translation of $h$ units to the right** of the graph of $y = f(x)$. If $h < 0$, then the graph of $y = f(x - h)$ is a **translation of $|h|$ units to the left** of the graph of $y = f(x)$.

### EXAMPLE 1 Translations to the right or left

Graph $f(x) = \sqrt{x}$, $g(x) = \sqrt{x - 3}$, and $h(x) = \sqrt{x + 5}$ on the same coordinate plane.

**Solution**

First sketch $f(x) = \sqrt{x}$ through $(0, 0)$, $(1, 1)$, and $(4, 2)$ as shown in Fig. 2.32. Since the first operation of $g$ is to subtract 3, we get the corresponding points by adding 3 to each $x$-coordinate. So $g$ goes through $(3, 0)$, $(4, 1)$, and $(7, 2)$. Since the first operation of $h$ is to add 5, we get corresponding points by subtracting 5 from the $x$-coordinates. So $h$ goes through $(-5, 0)$, $(-4, 1)$, and $(-1, 2)$.

The calculator graphs of $f$, $g$, and $h$ are shown in Fig. 2.33. On many calculators the radical does not contain the radicand. Be sure to note the difference between $y = \sqrt{\ }(x) - 3$ and $y = \sqrt{\ }(x - 3)$ on a calculator.

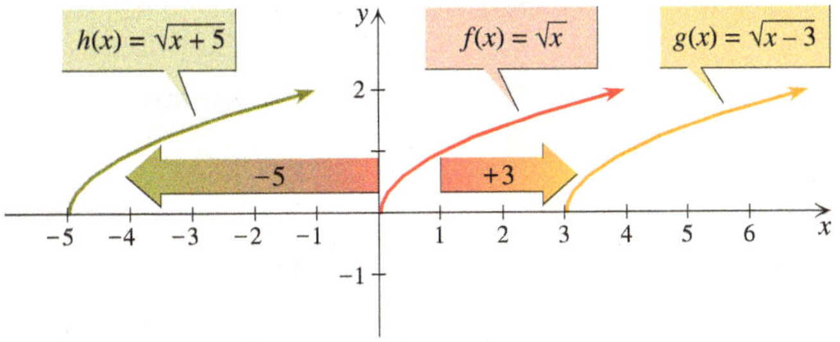


Figure 2.32

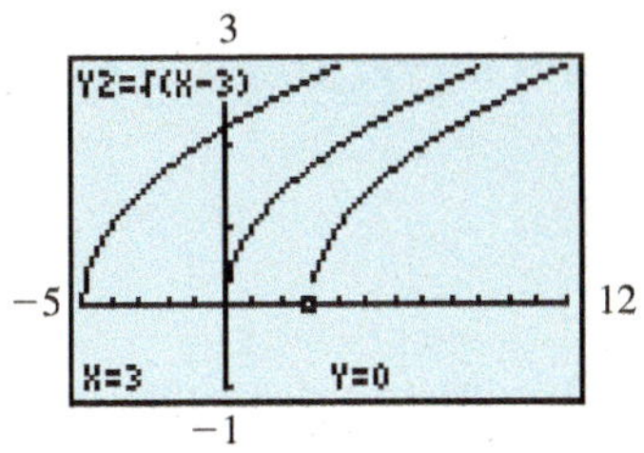


Figure 2.33

▶**TRY THIS.** Graph $f(x) = \sqrt{x}$, $g(x) = \sqrt{x - 2}$, and $h(x) = \sqrt{x + 1}$ on the same coordinate plane. ■

Notice that $y = \sqrt{x - 3}$ lies 3 units to the *right* and $y = \sqrt{x + 5}$ lies 5 units to the *left* of $y = \sqrt{x}$. The next example shows two more horizontal translations.

## EXAMPLE 2 Horizontal translations

Sketch the graph of each function.

**a.** $f(x) = |x - 1|$ **b.** $f(x) = (x + 3)^2$

### Solution

**a.** The function $f(x) = |x - 1|$ is in the absolute value family, and its graph is a translation one unit to the right of $g(x) = |x|$. Calculate a few ordered pairs to get an accurate graph. The points $(0, 1)$, $(1, 0)$, and $(2, 1)$ are on the graph of $f(x) = |x - 1|$ shown in Fig. 2.34.

**b.** The function $f(x) = (x + 3)^2$ is in the square family, and its graph is a translation three units to the left of the graph of $g(x) = x^2$. Calculate a few ordered pairs to get an accurate graph. The points $(-3, 0)$, $(-2, 1)$, and $(-4, 1)$ are on the graph shown in Fig. 2.35.

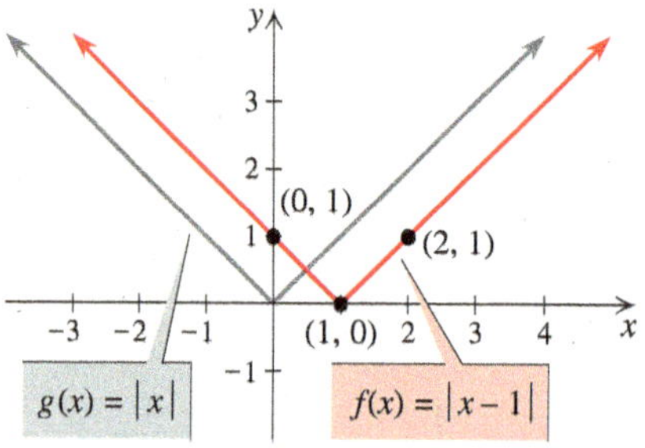


Figure 2.34

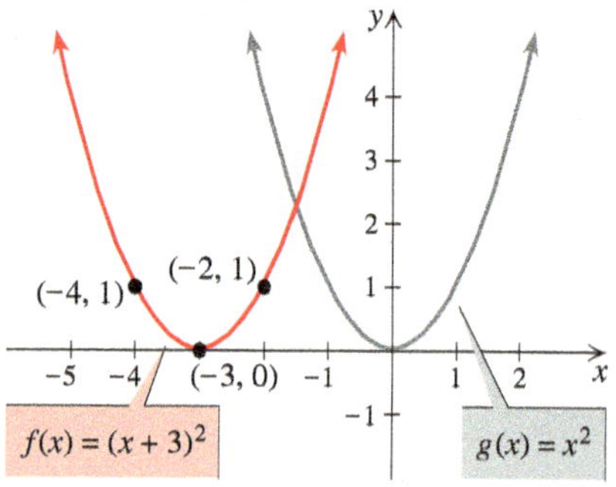


Figure 2.35

▸**TRY THIS.** Graph $f(x) = (x - 2)^2$. ■

## Reflection

In the formula $y = af(x - h) + k$, we multiply $f(x - h)$ by $a$ before adding $k$. So next we study the effect that $a$ has on the graph. We first consider the case where $a = -1$. Multiplying by $-1$ is the same as taking the opposite, so we usually just write $y = -f(x)$. All points of the graph of $y = -f(x)$ can be obtained by simply changing the signs of all of the $y$-coordinates of the points on the graph of $y = f(x)$. This causes the graphs to be mirror images of each other with respect to the $x$-axis.

**Definition: Reflection**

The graph of $y = -f(x)$ is a **reflection** in the $x$-axis of the graph of $y = f(x)$.

## EXAMPLE 3 Graphing using reflection

Graph each pair of functions on the same coordinate plane.

**a.** $f(x) = x^2$, $g(x) = -x^2$
**b.** $f(x) = x^3$, $g(x) = -x^3$
**c.** $f(x) = |x|$, $g(x) = -|x|$

### Solution

**a.** The graph of $f(x) = x^2$ goes through $(0, 0)$, $(\pm 1, 1)$, and $(\pm 2, 4)$. The graph of $g(x) = -x^2$ goes through $(0, 0)$, $(\pm 1, -1)$, and $(\pm 2, -4)$ as shown in Fig. 2.36.

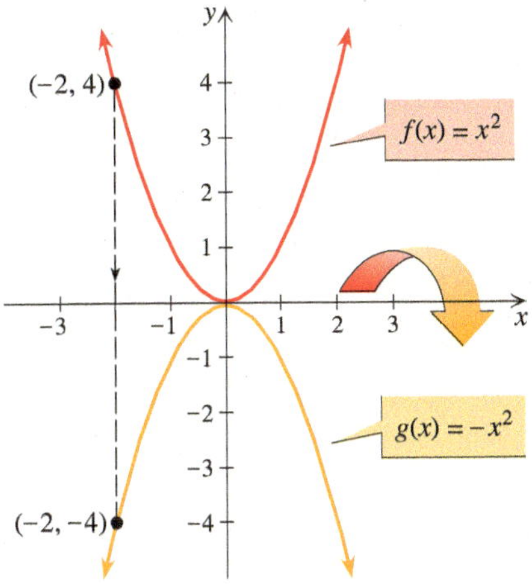


Figure 2.36

**b.** Make a table of ordered pairs for $f$ as follows:

| $x$ | $-2$ | $-1$ | $0$ | $1$ | $2$ |
|---|---|---|---|---|---|
| $f(x) = x^3$ | $-8$ | $-1$ | $0$ | $1$ | $8$ |

Sketch the graph of $f$ through these ordered pairs as shown in Fig. 2.37. Since $g(x) = -f(x)$, the graph of $g$ can be obtained by reflecting the graph of $f$ in the $x$-axis. Each point on the graph of $f$ corresponds to a point on the graph of $g$ with the opposite $y$-coordinate. For example, the point $(2, 8)$ on the graph of $f$ corresponds to the point $(2, -8)$ on the graph of $g$. Both graphs are shown in Fig. 2.37.

**c.** The graph of $f$ is the familiar V-shaped graph of the absolute value function as shown in Fig. 2.38. Since $g(x) = -f(x)$, the graph of $g$ can be obtained by reflecting the graph of $f$ in the $x$-axis. Each point on the graph of $f$ corresponds to a point on the graph of $g$ with the opposite $y$-coordinate. For example, $(2, 2)$ on $f$ corresponds to $(2, -2)$ on $g$. Both graphs are shown in Fig. 2.38.

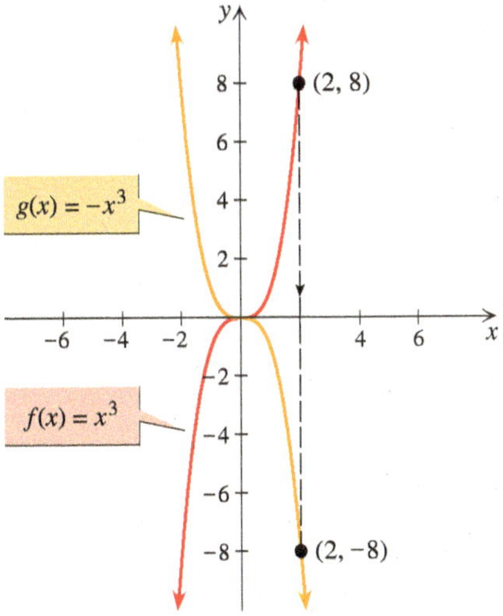


Figure 2.37

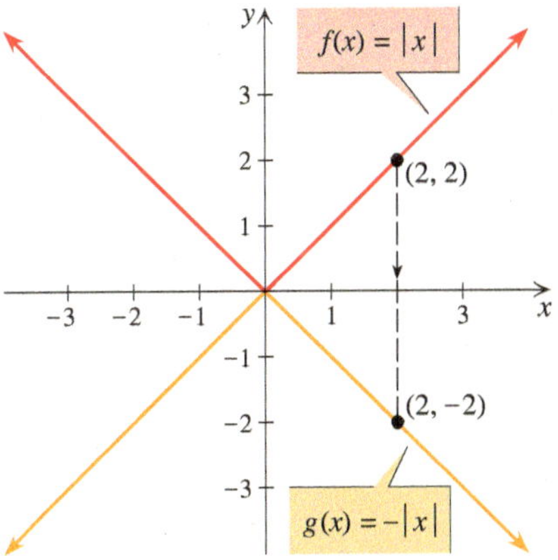


Figure 2.38

▸**TRY THIS.** Graph $f(x) = \sqrt{x}$ and $g(x) = -\sqrt{x}$ on the same coordinate plane. ■

We do not include reflections in the $y$-axis in our function families, but it is interesting to note that you get a reflection in the $y$-axis simply by replacing $x$ with $-x$ in the formula for the function. For example, graph $y_1 = \sqrt{x}$ and $y_2 = \sqrt{-x}$ on your graphing calculator.

## Stretching and Shrinking

If $a > 1$ then all of the $y$-coordinates on $y = af(x)$ are obtained by multiplying the $y$-coordinates on the graph of $y = f(x)$ by $a$. So the graph of $y = f(x)$ is *stretched* by a factor of $a$. If $0 < a < 1$ then the graph of $y = f(x)$ is *shrunk* by a factor of $a$. If $a$ is negative, then reflection occurs along with the stretching or shrinking.

**Definitions: Stretching and Shrinking**

The graph of $y = af(x)$ is obtained from the graph of $y = f(x)$ by

1. **stretching** the graph of $y = f(x)$ by $a$ when $a > 1$, or
2. **shrinking** the graph of $y = f(x)$ by $a$ when $0 < a < 1$.

## EXAMPLE 4 Graphing using stretching and shrinking

In each case graph the three functions on the same coordinate plane.

**a.** $f(x) = \sqrt{x}$, $g(x) = 2\sqrt{x}$, $h(x) = \frac{1}{2}\sqrt{x}$

**b.** $f(x) = x^2$, $g(x) = 2x^2$, $h(x) = \frac{1}{2}x^2$

### Solution

**a.** The graph of $f(x) = \sqrt{x}$ goes through $(0, 0)$, $(1, 1)$, and $(4, 2)$ as shown in Fig. 2.39. The graph of $g$ is obtained by stretching the graph of $f$ by a factor of 2. So $g$ goes through $(0, 0)$, $(1, 2)$, and $(4, 4)$. The graph of $h$ is obtained by shrinking the graph of $f$ by a factor of $\frac{1}{2}$. So $h$ goes through $(0, 0)$, $\left(1, \frac{1}{2}\right)$, and $(4, 1)$. The functions $f$, $g$, and $h$ are shown on a graphing calculator in Fig. 2.40. Note how the viewing window affects the shape of the graph. The curves do not appear as separated on the calculator as they do in Fig. 2.39. □

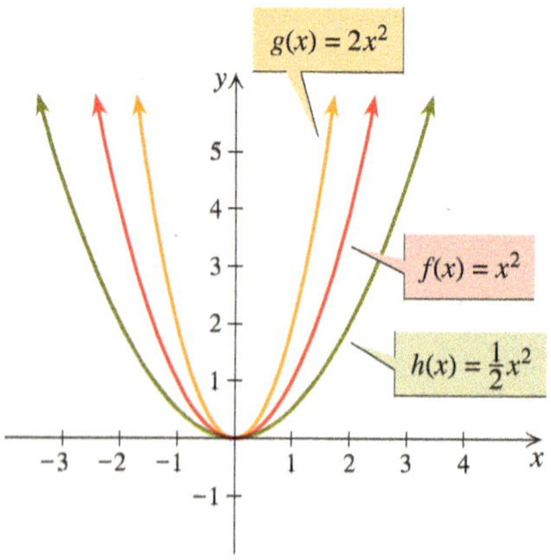


Figure 2.41

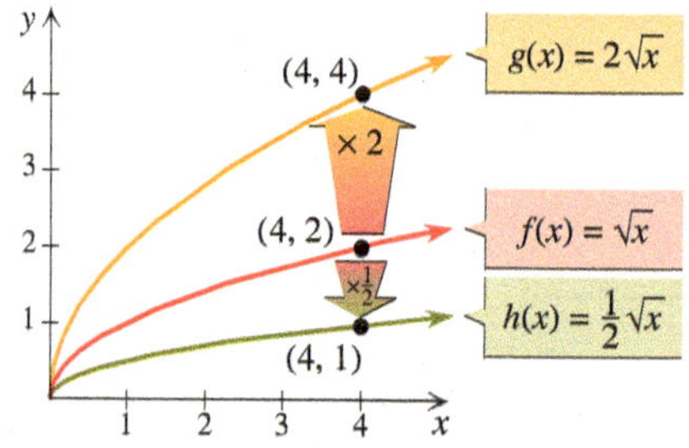


Figure 2.39

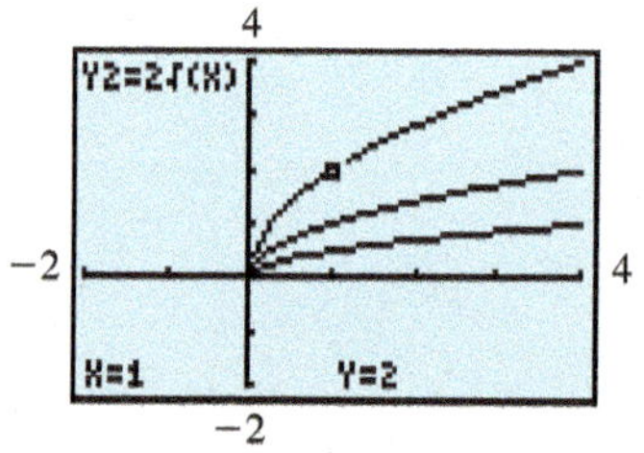


Figure 2.40

**b.** The graph of $f(x) = x^2$ is the familiar parabola shown in Fig. 2.41. We stretch it by a factor of 2 to get the graph of $g$ and shrink it by a factor of $\frac{1}{2}$ to get the graph of $h$.

►**TRY THIS.** Graph $f(x) = |x|$, $g(x) = 3|x|$, and $h(x) = \frac{1}{3}|x|$ on the same coordinate plane. ■

Note that the graph of $y = 2x^2$ has exactly the same shape as the graph of $y = x^2$ if we simply change the scale on the $y$-axis as in Fig. 2.42.

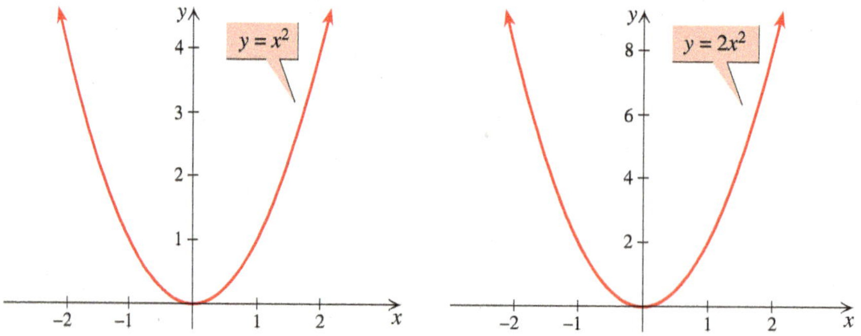


Figure 2.42

## Vertical Translation

In the formula $y = af(x - h) + k$, the last operation is addition of $k$. If we add the constant $k$ to all $y$-coordinates on the graph of $y = f(x)$, the graph will be translated up or down depending on whether $k$ is positive or negative.

**Definition: Translation Upward or Downward**

> If $k > 0$, then the graph of $y = f(x) + k$ is a **translation of $k$ units upward** of the graph of $y = f(x)$. If $k < 0$, then the graph of $y = f(x) + k$ is a **translation of $|k|$ units downward** of the graph of $y = f(x)$.

### EXAMPLE 5 Translations upward or downward

Graph the three given functions on the same coordinate plane.

**a.** $f(x) = \sqrt{x}$, $g(x) = \sqrt{x} + 3$, $h(x) = \sqrt{x} - 5$
**b.** $f(x) = x^2$, $g(x) = x^2 + 2$, $h(x) = x^2 - 3$

**Solution**

**a.** First sketch $f(x) = \sqrt{x}$ through $(0, 0)$, $(1, 1)$, and $(4, 2)$ as shown in Fig. 2.43. Since $g(x) = \sqrt{x} + 3$ we can add 3 to the $y$-coordinate of each point to get $(0, 3)$, $(1, 4)$, and $(4, 5)$. Sketch $g$ through these points. Every point on $f$ can be moved up 3 units to obtain a corresponding point on $g$. For $h(x) = \sqrt{x} - 5$, we subtract 5 from the $y$-coordinates on $f$ to obtain points on $h$. So $h$ goes through $(0, -5)$, $(1, -4)$, and $(4, -3)$.

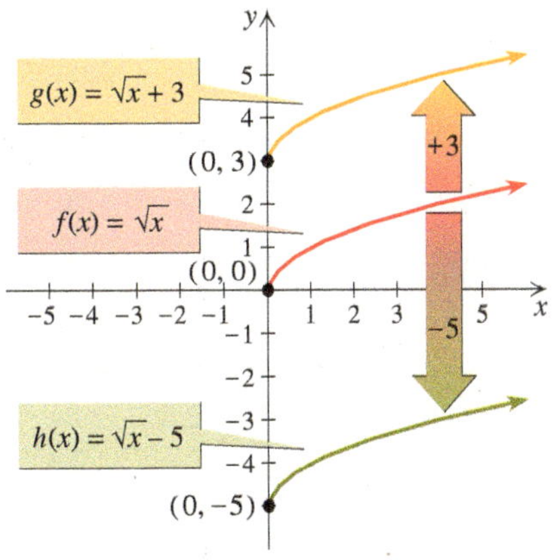


Figure 2.43

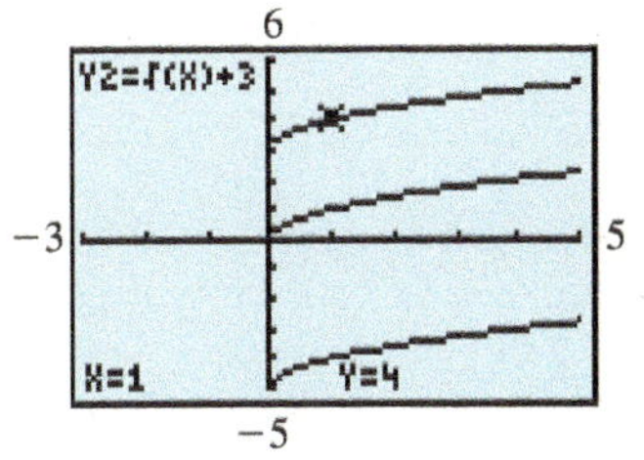


Figure 2.44

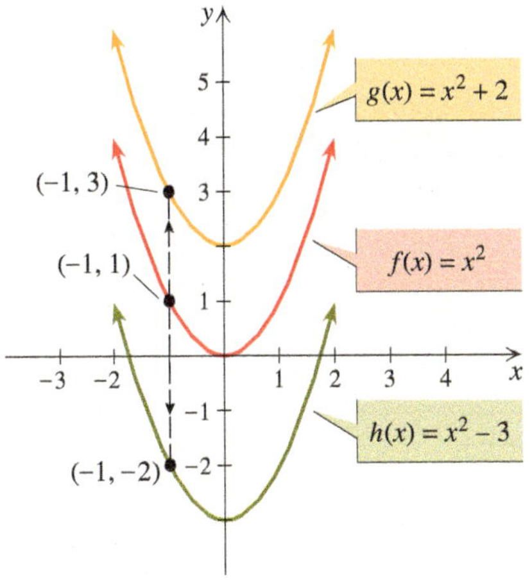


Figure 2.45

The relationship between $f$, $g$, and $h$ can be seen with a graphing calculator in Fig. 2.44. You should experiment with your graphing calculator to see how a change in the formula changes the graph. □

**b.** First sketch the familiar graph of $f(x) = x^2$ through $(\pm 2, 4)$, $(\pm 1, 1)$, and $(0, 0)$ as shown in Fig. 2.45. Since $g(x) = f(x) + 2$, the graph of $g$ can be obtained by translating the graph of $f$ upward two units. Since $h(x) = f(x) - 3$, the graph of $h$ can be obtained by translating the graph of $f$ downward three units. For example, the point $(-1, 1)$ on the graph of $f$ moves up to $(-1, 3)$ on the graph of $g$ and down to $(-1, -2)$ on the graph of $h$ as shown in Fig. 2.45.

**▶TRY THIS.** Graph $f(x) = \sqrt{x}$, $g(x) = \sqrt{x} + 1$, and $h(x) = \sqrt{x} - 2$ on the same coordinate plane. ■

## Multiple Transformation

Any combination of stretching, shrinking, reflecting, horizontal translation, and vertical translation transforms one function into a new function. If a transformation does not change the shape of a graph, then it is a **rigid** transformation. If the shape changes, then it is **nonrigid.** Stretching and shrinking are nonrigid transformations. Translating (horizontally or vertically) and reflection are rigid transformations.

When graphing a function involving more than one transformation, $y = af(x - h) + k$, apply the transformations in the order that we discussed them: *h-a-k*. Remember that this order is simply the order of operations.

**PROCEDURE**

**Multiple Transformations**

To graph $y = af(x - h) + k$ apply the transformations in the following order:

1. Horizontal translation ($h$)
2. Reflecting/stretching/shrinking ($a$)
3. Vertical translation ($k$).

Note that the order in which you reflect and stretch or reflect and shrink does not matter. It does matter that you do vertical translation last. For example, if $y = x^2$ is reflected in the $x$-axis and then translated up one unit, the equation for the graph is $y = -x^2 + 1$. If $y = x^2$ is translated up one unit and then reflected in the $x$-axis, the equation for the graph in the final position is $y = -(x^2 + 1)$ or $y = -x^2 - 1$. Changing the order has resulted in different functions.

### EXAMPLE 6 Graphing using several transformations

Use transformations to graph each function.

**a.** $y = -2(x - 3)^2 + 4$ **b.** $y = 4 - 2\sqrt{x + 1}$

**Solution**

**a.** This function is in the square family. So the graph of $y = x^2$ is translated 3 units to the right, reflected and stretched by a factor of 2, and finally translated 4 units upward. See Fig. 2.46.

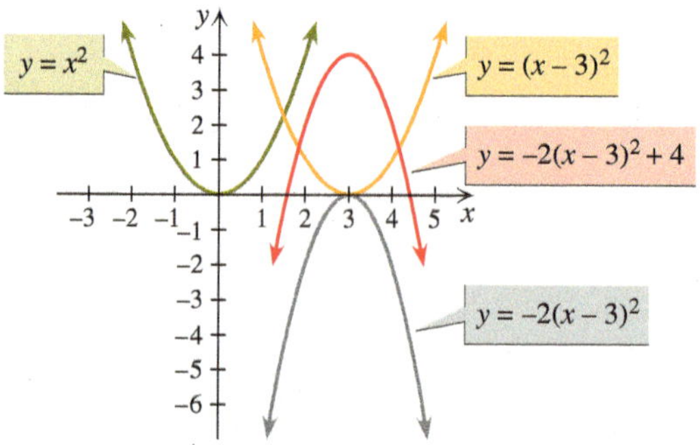


**Figure 2.46**

**b.** Rewrite the function as $y = -2\sqrt{x + 1} + 4$ and recognize that it is in the square root family. So we start with the graph of $y = \sqrt{x}$. The graph of $y = \sqrt{x + 1}$ is a horizontal translation one unit to the left of $y = \sqrt{x}$. Stretch by a factor of 2 to get $y = 2\sqrt{x + 1}$. Reflect in the $x$-axis to get $y = -2\sqrt{x + 1}$.

Finally, translate 4 units upward to get the graph of $y = -2\sqrt{x+1} + 4$. All of these graphs are shown in Fig. 2.47.

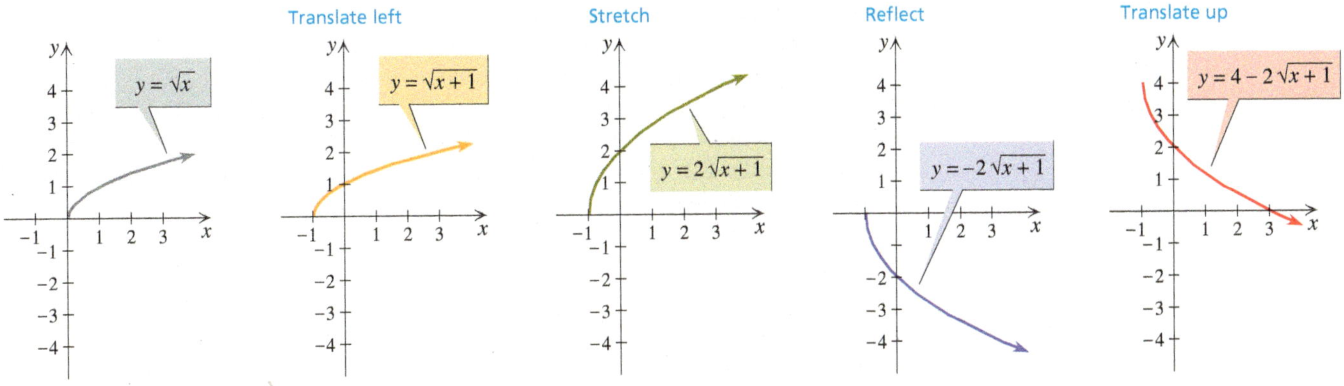


**Figure 2.47**

**▸TRY THIS.** Graph $y = 4 - 2|x + 1|$.

## The Linear Family of Functions

The function $f(x) = x$ is called the **identity function** because the coordinates in each ordered pair are identical. Its graph is a line through $(0, 0)$ with slope 1. A member of the **linear family,** a **linear function** is a transformation of the identity function: $f(x) = a(x - h) + k$ where $a \neq 0$. Since $a$, $h$, and $k$ are real numbers, we can rewrite this form as a multiple of $x$ plus a constant. So a linear function has the form $f(x) = mx + b$, with $m \neq 0$ (the slope-intercept form). If $m = 0$, then the function has the form $f(x) = b$ and it is a **constant function.**

### EXAMPLE 7 Graphing linear functions using transformations

Sketch the graphs of $y = x$, $y = 2x$, $y = -2x$, and $y = -2x - 3$.

### Solution

The graph of $y = x$ is a line through $(0, 0)$, $(1, 1)$, and $(2, 2)$. Stretch the graph of $y = x$ by a factor of 2 to get the graph of $y = 2x$. Reflect in the $x$-axis to get the graph of $y = -2x$. Translate downward three units to get the graph of $y = -2x - 3$. See Fig. 2.48.

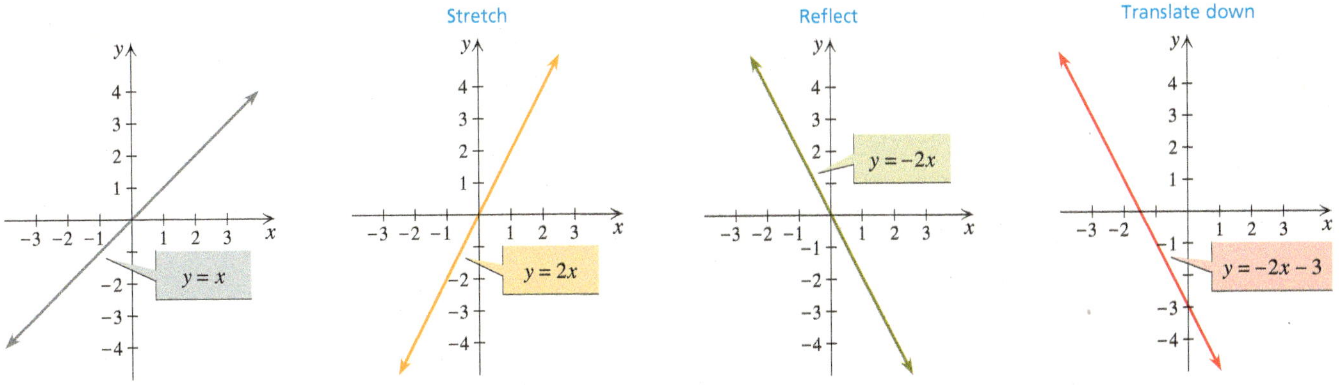


**Figure 2.48**

**▸TRY THIS.** Graph $y = -2x + 5$. ■

## Symmetry

The graph of $g(x) = -x^2$ is a reflection in the $x$-axis of the graph of $f(x) = x^2$. If the paper were folded along the $x$-axis, the graphs would coincide. See Fig. 2.49. The symmetry that we call reflection occurs between two functions, but the graph of $f(x) = x^2$ has a symmetry within itself. Points such as $(2, 4)$ and $(-2, 4)$ are on the graph and are equidistant from the $y$-axis. Folding the paper along the $y$-axis brings all such pairs of points together. See Fig. 2.50. The reason for this symmetry about the $y$-axis is the fact that $f(-x) = f(x)$ for any real number $x$. We get the same $y$-coordinate whether we evaluate the function at a number or at its opposite.

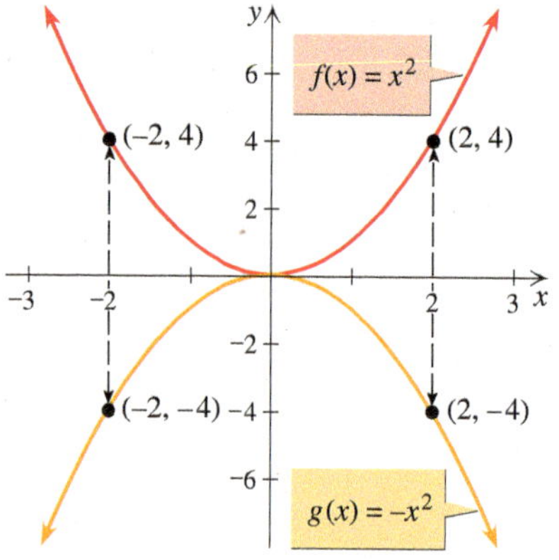

Figure 2.49

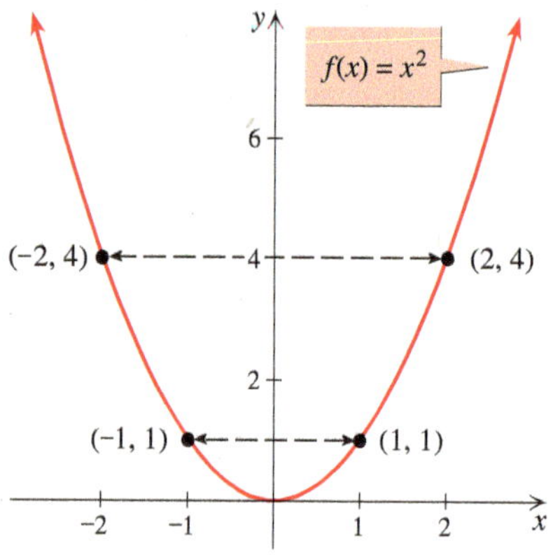

Figure 2.50

**Definition: Symmetric about the $y$-Axis**

If $f(-x) = f(x)$ for every value of $x$ in the domain of the function $f$, then $f$ is called an **even function** and its graph is **symmetric about the $y$-axis.**

Consider the graph of $f(x) = x^3$ shown in Fig. 2.51. On the graph of $f(x) = x^3$ we find pairs of points such as $(2, 8)$ and $(-2, -8)$. The odd exponent in $x^3$ causes the second coordinate to be negative when the sign of the first coordinate is changed from positive to negative. These points are equidistant from the origin and on opposite sides of the origin. So the symmetry of this graph is about the origin. In this case $f(x)$ and $f(-x)$ are not equal, but $f(-x) = -f(x)$.

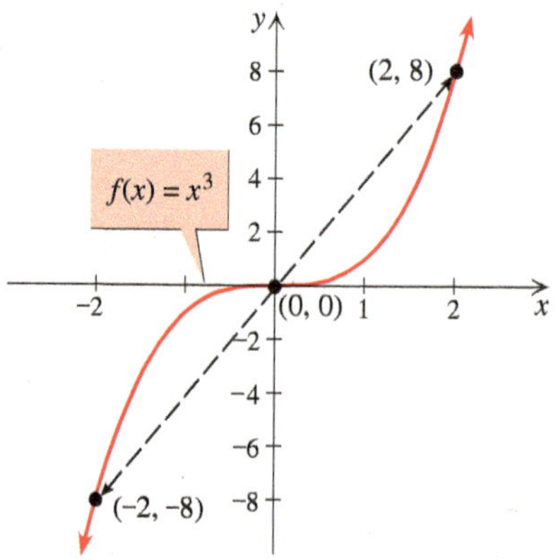

Figure 2.51

**Definition: Symmetric about the Origin**

If $f(-x) = -f(x)$ for every value of $x$ in the domain of the function $f$, then $f$ is called an **odd function** and its graph is **symmetric about the origin.**

A graph might look like it is symmetric about the $y$-axis or the origin, but the only way to be sure is to use the definitions of these terms as shown in the following example. Note that an odd power of a negative number is negative and an even power of a negative number is positive. So for any real number $x$ we have $(-x)^n = -x^n$ if $n$ is odd and $(-x)^n = x^n$ if $n$ is even.

## EXAMPLE 8 Determining symmetry in a graph

Discuss the symmetry of the graph of each function.

**a.** $f(x) = 5x^3 - x$ **b.** $f(x) = |x| + 3$ **c.** $f(x) = x^2 - 3x + 6$

### Solution

**a.** Replace $x$ by $-x$ in the formula for $f(x)$ and simplify:

$$f(-x) = 5(-x)^3 - (-x) = -5x^3 + x$$

Is $f(-x)$ equal to $f(x)$ or the opposite of $f(x)$? Since $-f(x) = -5x^3 + x$, we have $f(-x) = -f(x)$. So $f$ is an odd function and the graph is symmetric about the origin.

**b.** Since $|-x| = |x|$ for any $x$, we have $f(-x) = |-x| + 3 = |x| + 3$. Because $f(-x) = f(x)$, the function is even and the graph is symmetric about the $y$-axis.

**c.** In this case, $f(-x) = (-x)^2 - 3(-x) + 6 = x^2 + 3x + 6$. So $f(-x) \neq f(x)$, and $f(-x) \neq -f(x)$. This function is neither even nor odd and its graph has neither type of symmetry.

▸**TRY THIS.** Discuss the symmetry of the graph of $f(x) = -2x^2 + 5$. ■

Do you see why functions symmetric about the $y$-axis are called *even* and functions symmetric about the origin are called *odd?* In general, a function defined by a polynomial with even exponents only, such as $f(x) = x^2$ or $f(x) = x^6 - 5x^4 + 2x^2 + 3$, is symmetric about the $y$-axis. (The constant term 3 has even degree because $3 = 3x^0$.) A function with only odd exponents such as $f(x) = x^3$ or $f(x) = x^5 - 6x^3 + 4x$ is symmetric about the origin. A function containing both even- and odd-powered terms such as $f(x) = x^2 + 3x$ has neither symmetry. For other types of functions (such as absolute value) you must examine the function more carefully to determine symmetry.

Note that functions cannot have $x$-axis symmetry. A graph that is symmetric about the $x$-axis fails the vertical line test. For example, the graph of $x = y^2$ is symmetric about the $x$-axis, but the graph fails the vertical line test and $y$ is not a function of $x$.

## Reading Graphs to Solve Inequalities

In Chapter 1 we learned how to use a two-dimensional graph to help us solve an inequality in one variable. Using transformations we can graph more quickly and use those graphs to solve a wide variety of inequalities in one variable.

## EXAMPLE 9 Using a graph to solve an inequality

Solve the inequality $(x - 1)^2 - 2 < 0$ by graphing.

### Solution

The graph of $y = (x - 1)^2 - 2$ is obtained by translating the graph of $y = x^2$ one unit to the right and two units downward. See Fig. 2.52. To find the $x$-intercepts we solve $(x - 1)^2 - 2 = 0$:

$$(x - 1)^2 = 2$$
$$x - 1 = \pm\sqrt{2}$$
$$x = 1 \pm \sqrt{2}$$

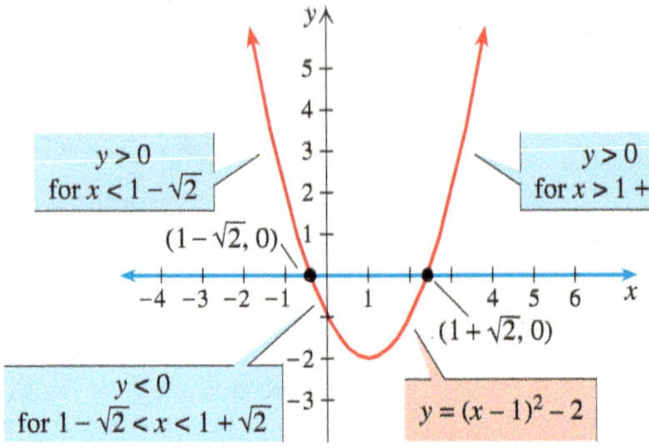


Figure 2.52

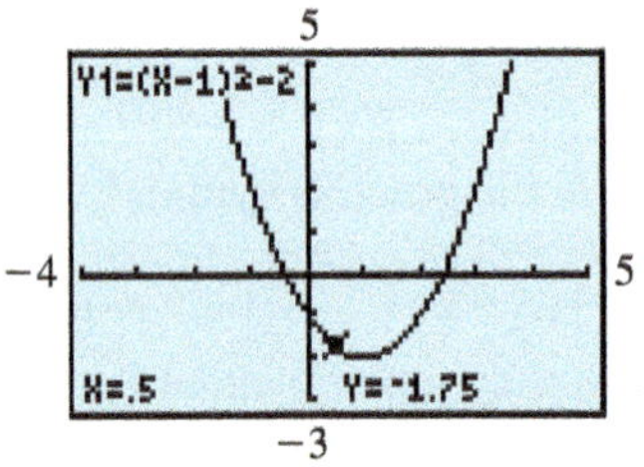


Figure 2.53

The $x$-intercepts are $(1 - \sqrt{2}, 0)$ and $(1 + \sqrt{2}, 0)$. If the $y$-coordinate of a point on the graph is negative, then the $x$-coordinate satisfies $(x - 1)^2 - 2 < 0$. So the solution set to $(x - 1)^2 - 2 < 0$ is the open interval $(1 - \sqrt{2}, 1 + \sqrt{2})$. Although a graphing calculator will not find the exact solution to this inequality, you can use TRACE to support the answer and see that $y$ is negative between the $x$-intercepts. See Fig. 2.53.

▶**TRY THIS.** Solve $2 - |x - 1| \leq 0$ by graphing. ■

Note that the solution set to $(x - 1)^2 - 2 \geq 0$ can also be obtained from the graph in Fig. 2.52. If the $y$-coordinate of a point on the graph is positive or zero, then the $x$-coordinate satisfies $(x - 1)^2 - 2 \geq 0$. So the solution set to $(x - 1)^2 - 2 \geq 0$ is $(-\infty, 1 - \sqrt{2}] \cup [1 + \sqrt{2}, \infty)$.

## FUNCTION gallery... Some Basic Functions and Their Properties

**Constant Function**

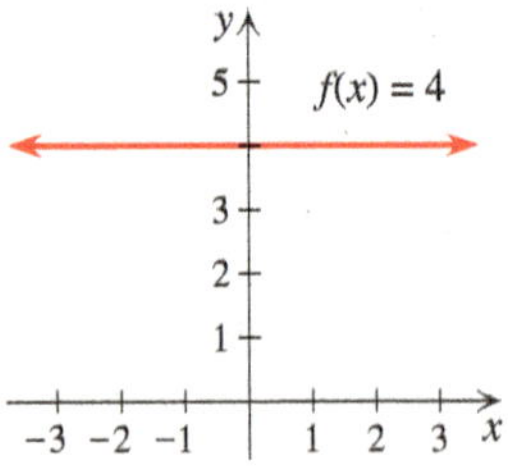


Domain $(-\infty, \infty)$
Range $\{4\}$
Constant on $(-\infty, \infty)$
Symmetric about $y$-axis

**Identity Function**

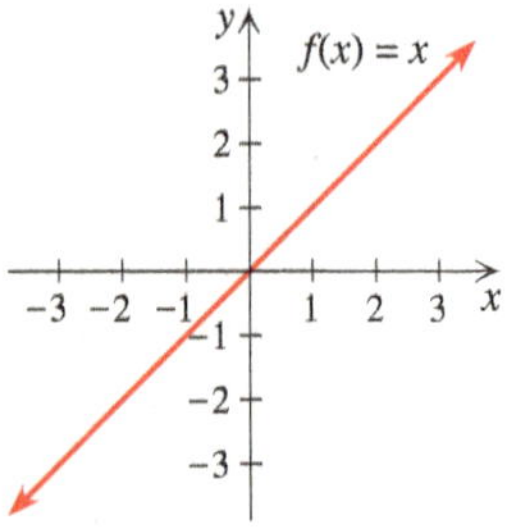


Domain $(-\infty, \infty)$
Range $(-\infty, \infty)$
Increasing on $(-\infty, \infty)$
Symmetric about origin

**Linear Function**

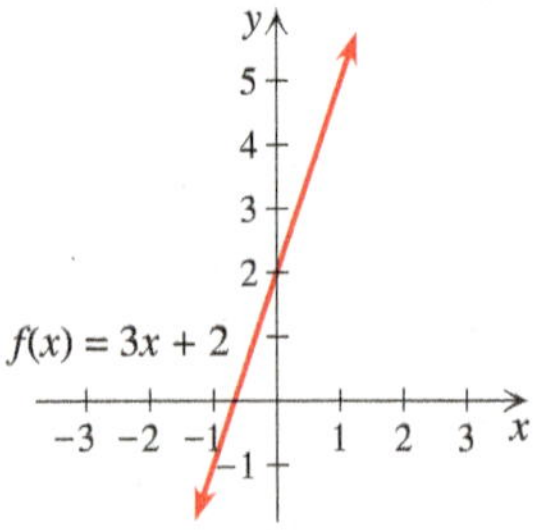


Domain $(-\infty, \infty)$
Range $(-\infty, \infty)$
Increasing on $(-\infty, \infty)$

### Absolute Value Function

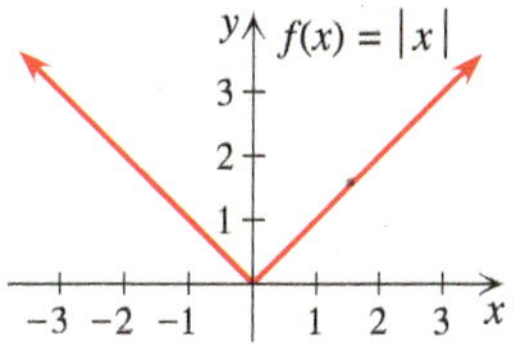


Domain $(-\infty, \infty)$
Range $[0, \infty)$
Increasing on $[0, \infty)$
Decreasing on $(-\infty, 0]$
Symmetric about $y$-axis

### Square Function

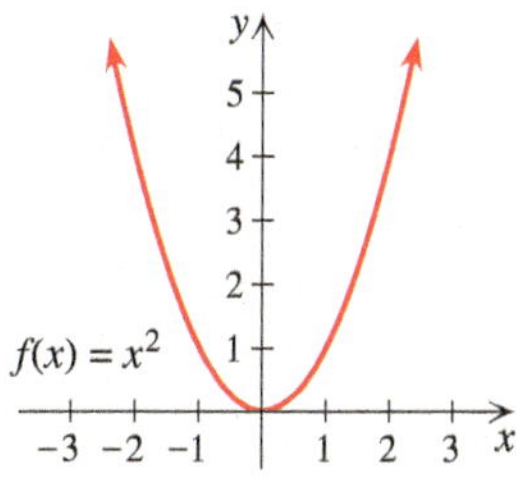


Domain $(-\infty, \infty)$
Range $[0, \infty)$
Increasing on $[0, \infty)$
Decreasing on $(-\infty, 0]$
Symmetric about $y$-axis

### Square-Root Function

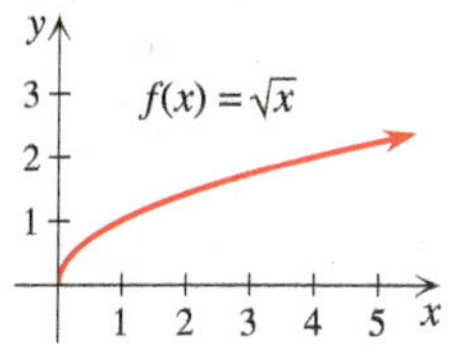


Domain $[0, \infty)$
Range $[0, \infty)$
Increasing on $[0, \infty)$

### Cube Function

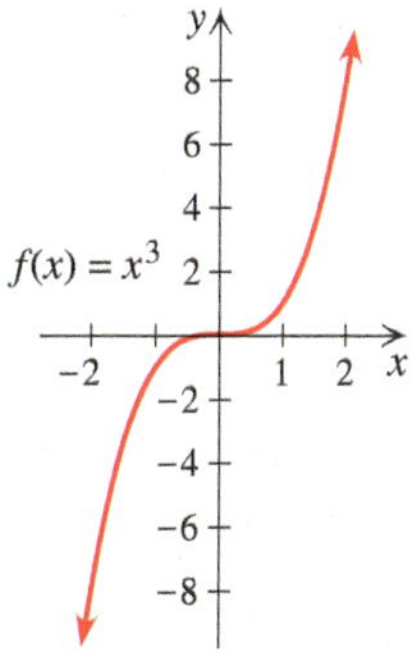


Domain $(-\infty, \infty)$
Range $(-\infty, \infty)$
Increasing on $(-\infty, \infty)$
Symmetric about origin

### Cube-Root Function

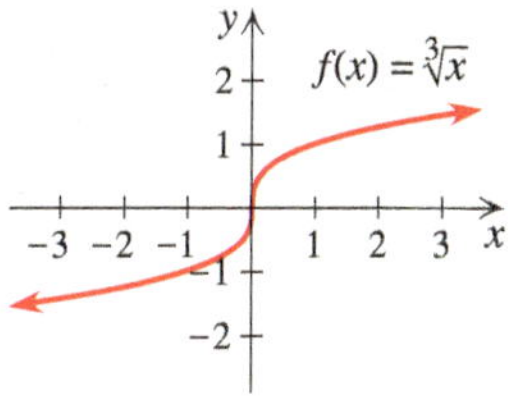


Domain $(-\infty, \infty)$
Range $(-\infty, \infty)$
Increasing on $(-\infty, \infty)$
Symmetric about origin

### Greatest Integer Function

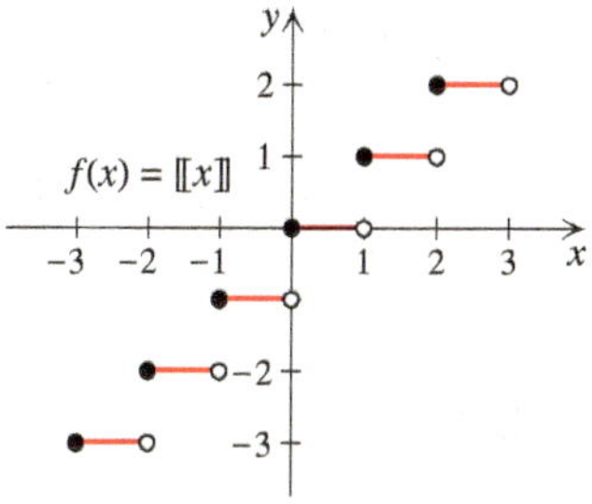


Domain $(-\infty, \infty)$
Range $\{n \mid n \text{ is an integer}\}$
Constant on $[n, n + 1)$
for every integer $n$

## FOR thought... True or False? Explain.

1. The graph of $f(x) = (-x)^4$ is a reflection in the $x$-axis of the graph of $g(x) = x^4$.

2. The graph of $f(x) = x - 4$ lies four units to the right of the graph of $f(x) = x$.

3. The graph of $y = |x + 2| + 3$ is a translation two units to the right and three units upward of the graph of $y = |x|$.

4. The graph of $f(x) = -3$ is a reflection in the $x$-axis of the graph of $g(x) = 3$.

5. The functions $y = x^2 + 4x + 1$ and $y = (x + 2)^2 - 3$ have the same graph.

6. The graph of $y = -(x - 3)^2 - 4$ can be obtained by moving $y = x^2$ three units to the right and down four units, and then reflecting in the $x$-axis.

7. If $f(x) = -x^3 + 2x^2 - 3x + 5$, then $f(-x) = x^3 + 2x^2 + 3x + 5$.

8. The graphs of $f(x) = -\sqrt{x}$ and $g(x) = \sqrt{-x}$ are identical.

9. If $f(x) = x^3 - x$, then $f(-x) = -f(x)$.

10. The solution set to $|x| - 1 \leq 0$ is $[-1, 1]$.

## EXERCISES 2.3

*Fill in the blank.*

**1.** Translating and reflecting are ________ transformations.

**2.** Stretching and shrinking are ________ transformations.

**3.** The graph of a function of the form $y = a(x - h)^2 + k$ where $a \neq 0$ is a ________.

**4.** The graph of $y = f(x - h)$ is a ________ of the graph of $y = f(x)$.

**5.** The graph of $y = -f(x)$ is a ________ of the graph of $y = f(x)$.

**6.** The function $f(x) = x$ is the ________ function.

**7.** The function $f(x) = mx + b$ with $m \neq 0$ is a ________ function.

**8.** The function $f(x) = b$ is a ________ function.

**9.** If $f(-x) = -f(x)$ for every $x$ in the domain of $f$, then $f$ is an ________ function.

**10.** If $f(-x) = f(x)$ for every $x$ in the domain of $f$, then $f$ is an ________ function.

*Sketch the graphs of each pair of functions on the same coordinate plane.*

**11.** $f(x) = |x|, g(x) = |x| - 4$

**12.** $f(x) = \sqrt{x}, g(x) = \sqrt{x} + 3$

**13.** $f(x) = x, g(x) = x + 3$

**14.** $f(x) = x^2, g(x) = x^2 - 5$

**15.** $y = x^2, y = (x - 3)^2$

**16.** $y = |x|, y = |x + 2|$

**17.** $y = \sqrt{x}, y = \sqrt{x + 9}$

**18.** $y = x^2, y = (x - 1)^2$

**19.** $f(x) = \sqrt{x}, g(x) = -\sqrt{x}$

**20.** $f(x) = x, g(x) = -x$

**21.** $y = \sqrt{x}, y = 3\sqrt{x}$

**22.** $y = \sqrt{1 - x^2}, y = 4\sqrt{1 - x^2}$

**23.** $y = x^2, y = \frac{1}{4}x^2$

**24.** $y = |x|, y = \frac{1}{3}|x|$

**25.** $y = \sqrt{4 - x^2}, y = -\sqrt{4 - x^2}$

**26.** $f(x) = x^2 + 1, g(x) = -(x^2 + 1)$

*Match each function in Exercises 27–34 with its graph (a)–(h). See the procedure for multiple transformations on page 136.*

**27.** $y = x^2$

**28.** $y = (x - 4)^2 + 2$

**29.** $y = (x + 4)^2 - 2$

**30.** $y = -2(x - 2)^2$

**31.** $y = -2(x + 2)^2$

**32.** $y = -\frac{1}{2}x^2 - 4$

**33.** $y = \frac{1}{2}(x + 4)^2 + 2$

**34.** $y = -2(x - 4)^2 - 2$

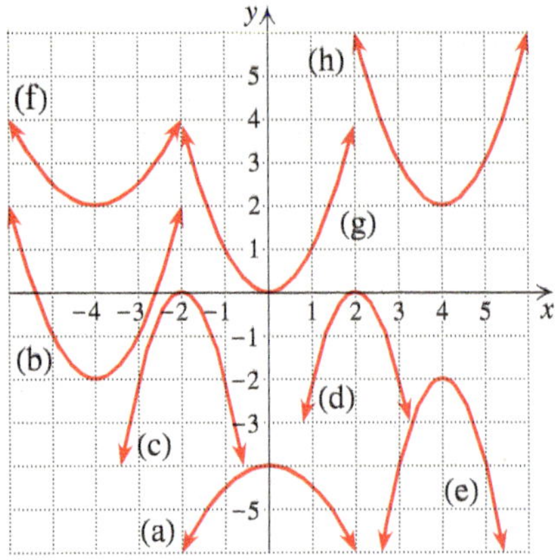


**Figure for Exercises 27 to 34**

*Write the equation of each graph after the indicated transformation(s).*

**35.** The graph of $y = \sqrt{x}$ is translated two units upward.

**36.** The graph of $y = \sqrt{x}$ is translated three units downward.

**37.** The graph of $y = x^2$ is translated five units to the right.

**38.** The graph of $y = x^2$ is translated seven units to the left.

**39.** The graph of $y = x^2$ is translated ten units to the right and four units upward.

**40.** The graph of $y = \sqrt{x}$ is translated five units to the left and twelve units downward.

**41.** The graph of $y = \sqrt{x}$ is stretched by a factor of 3, translated five units upward, then reflected in the $x$-axis.

**42.** The graph of $y = x^2$ is translated thirteen units to the right and six units downward, then reflected in the $x$-axis.

**43.** The graph of $y = |x|$ is reflected in the $x$-axis, stretched by a factor of 3, then translated seven units to the right and nine units upward.

**44.** The graph of $y = x$ is stretched by a factor of 2, reflected in the $x$-axis, then translated eight units downward and six units to the left.

*Use transformations to graph each function and state the domain and range.*

**45.** $y = (x - 1)^2 + 2$

**46.** $y = (x + 5)^2 - 4$

**47.** $y = |x - 1| + 3$

**48.** $y = |x + 3| - 4$

**49.** $y = 3x - 40$

**50.** $y = -4x + 200$

**51.** $y = \frac{1}{2}x - 20$

**52.** $y = -\frac{1}{2}x + 40$

**53.** $y = -\frac{1}{2}|x| + 40$

**54.** $y = 3|x| - 200$

**55.** $y = -\frac{1}{2}|x + 4|$

**56.** $y = 3|x - 2|$

**57.** $y = -\sqrt{x - 3} + 1$

**58.** $y = -\sqrt{x + 2} - 4$

**59.** $y = -2\sqrt{x + 3} + 2$

**60.** $y = -\frac{1}{2}\sqrt{x + 2} + 4$

*Determine algebraically whether the function is even, odd, or neither. Discuss the symmetry of each function.*

**61.** $f(x) = x^4$

**62.** $f(x) = x^4 - 2x^2$

**63.** $f(x) = x^4 - x^3$

**64.** $f(x) = x^3 - x$

**65.** $f(x) = (x + 3)^2$

**66.** $f(x) = (x - 1)^2$

**67.** $f(x) = |x - 2|$

**68.** $f(x) = |x| - 9$

**69.** $f(x) = x$

**70.** $f(x) = -x$

**71.** $f(x) = 3x + 2$

**72.** $f(x) = x - 3$

**73.** $f(x) = x^3 - 5x + 1$

**74.** $f(x) = x^6 - x^4 + x^2$

**75.** $f(x) = 1 + \frac{1}{x^2}$

**76.** $f(x) = (x^2 - 2)^3$

**77.** $f(x) = \sqrt{x}$

**78.** $f(x) = \sqrt{9 - x^2}$

**79.** $f(x) = |x^2 - 3|$

**80.** $f(x) = \sqrt{x^2 + 3}$

*Match each function with its graph (a)–(h).*

**81.** $y = 2 + \sqrt{x}$

**82.** $y = \sqrt{2 + x}$

**83.** $y = \sqrt{x^2}$

**84.** $y = \sqrt{\frac{x}{2}}$

**85.** $y = \frac{1}{2}\sqrt{x}$

**86.** $y = 2 - \sqrt{x - 2}$

**87.** $y = -2\sqrt{x}$

**88.** $y = -\sqrt{-x}$

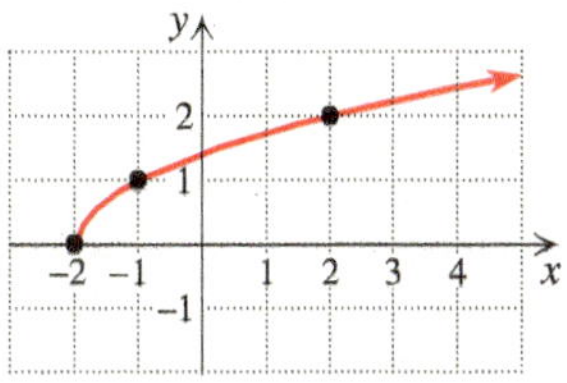

(a)

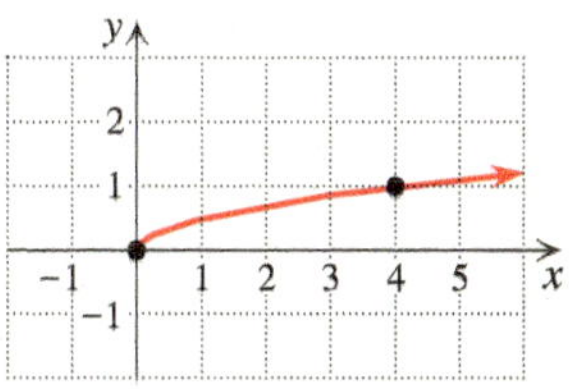

(b)

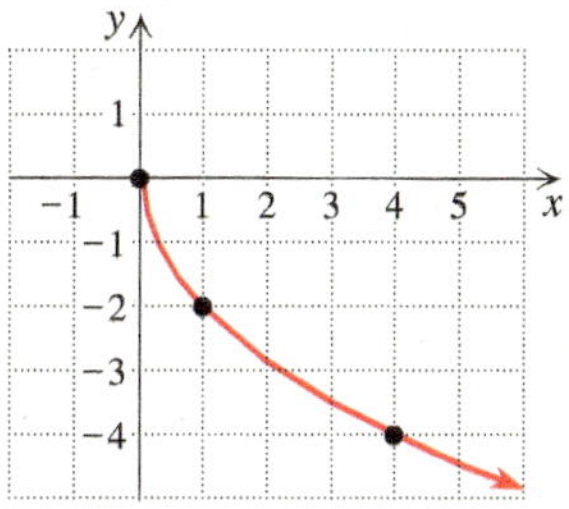

(c)

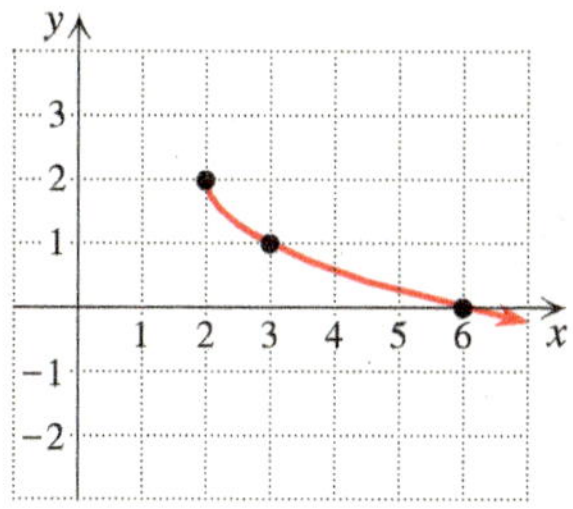

(d)

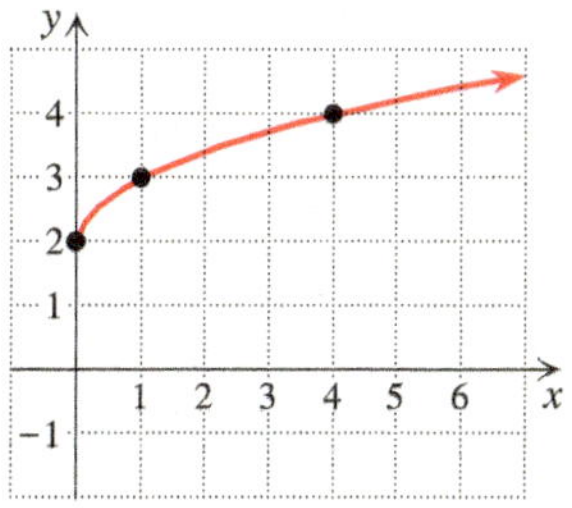

(e)

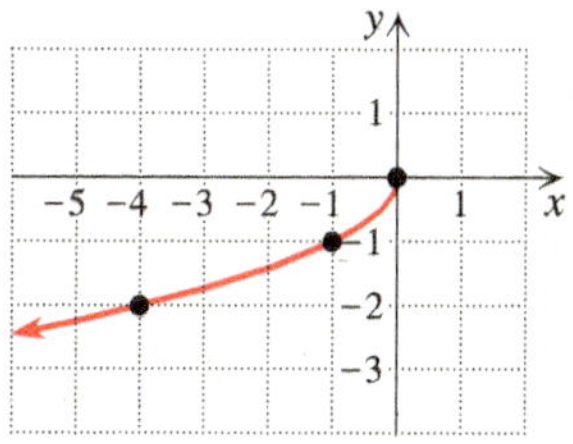

(f)

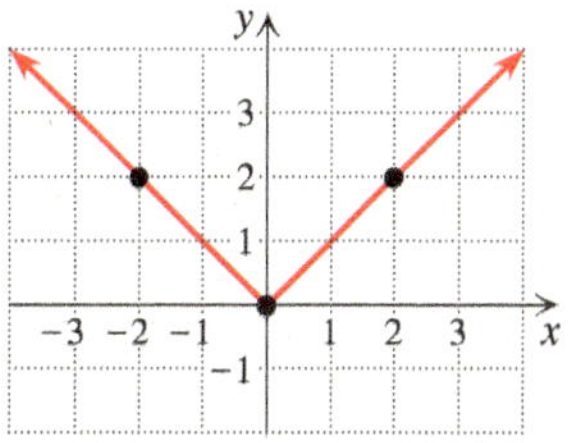

(g)

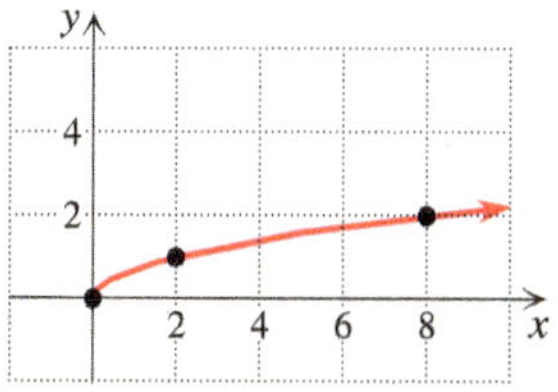

(h)

**Figure for Exercises 81 to 88**

*Solve each inequality by reading the corresponding graph.*

**89.** $x^2 - 1 \geq 0$

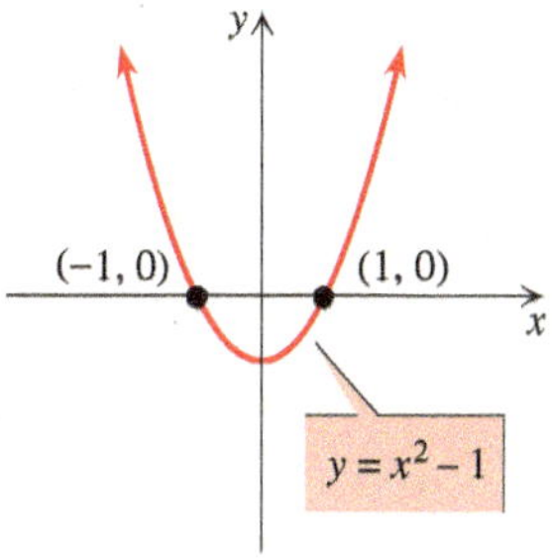


**90.** $2x^2 - 3 < 0$

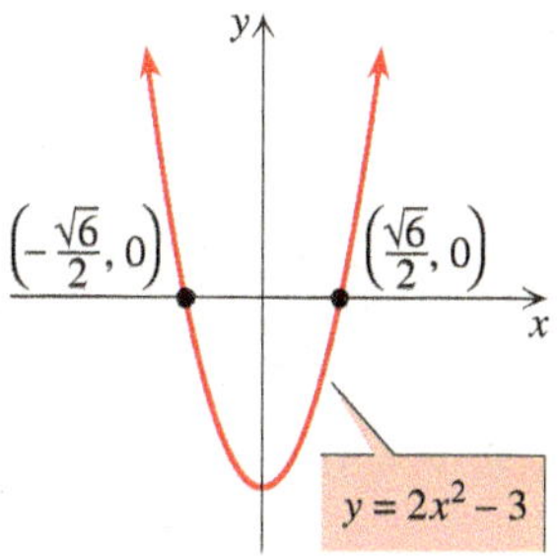


**91.** $|x - 2| - 3 > 0$

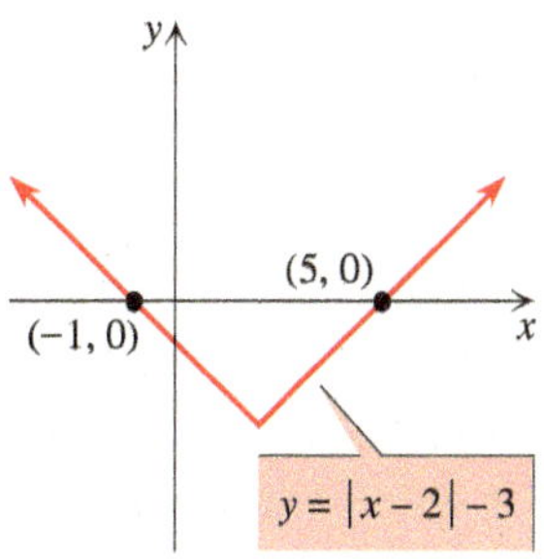


**92.** $2 - |x + 1| \geq 0$

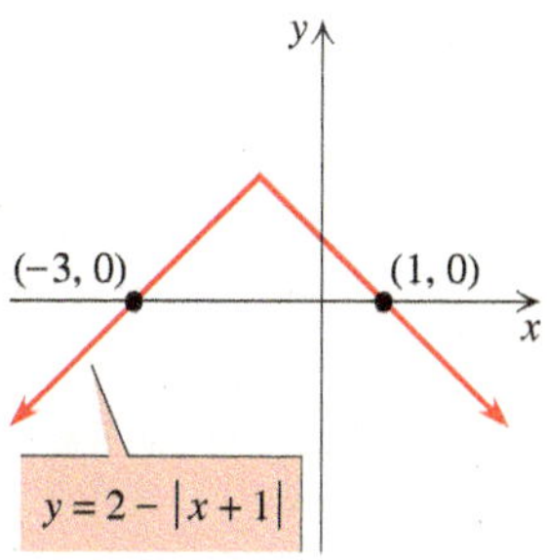

*Solve each inequality by graphing an appropriate function. State the solution set using interval notation.*

**93.** $(x - 1)^2 - 9 < 0$

**94.** $\left(x - \frac{1}{2}\right)^2 - \frac{9}{4} \geq 0$

**95.** $5 - \sqrt{x} \geq 0$

**96.** $\sqrt{x + 3} - 2 \geq 0$

**97.** $(x - 2)^2 > 3$

**98.** $(x - 1)^2 < 4$

**99.** $\sqrt{25 - x^2} > 0$

**100.** $\sqrt{4 - x^2} \geq 0$

*Use a graphing calculator to find an approximate solution to each inequality by reading the graph of an appropriate function. Round to two decimal places.*

**101.** $\sqrt{3x^2 + \pi x - 9} < 0$

**102.** $x^3 - 5x^2 + 6x - 1 > 0$

*Graph each of the following functions by transforming the given graph of $y = f(x)$.*

**103.** **a.** $y = 2f(x)$

**b.** $y = -f(x)$

**c.** $y = f(x + 1)$

**d.** $y = f(x - 3)$

**e.** $y = -3f(x)$

**f.** $y = f(x + 2) - 1$

**g.** $y = f(x - 1) + 3$

**h.** $y = 3f(x - 2) + 1$

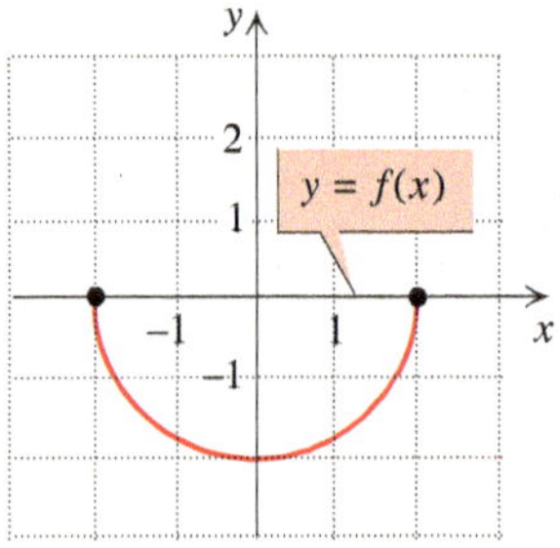


**Figure for Exercise 103**

**104.** **a.** $y = -f(x)$

**b.** $y = 2f(x)$

**c.** $y = -3f(x)$

**d.** $y = f(x + 2)$

**e.** $y = f(x - 1)$

**f.** $y = f(x - 2) + 1$

**g.** $y = -2f(x + 4)$

**h.** $y = 2f(x - 3) + 1$

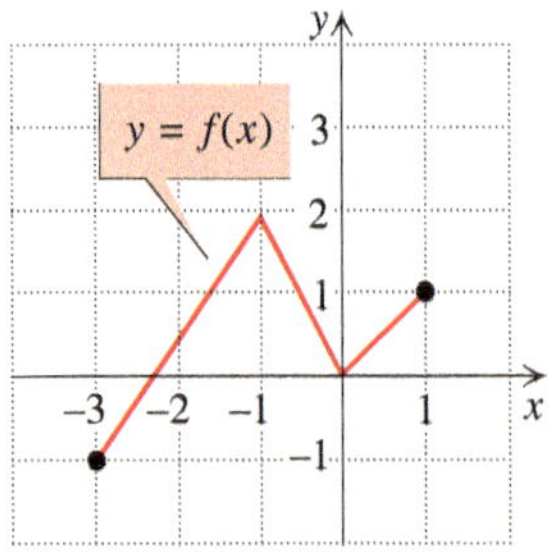


**Figure for Exercise 104**

*Solve each problem.*

**105.** *Across-the-Board Raise* Each teacher at C. F. Gauss Elementary School is given an across-the-board raise of \$2000. Write a function that *transforms* each old salary $x$ into a new salary $N(x)$.

**106.** *Cost-of-Living Raise* Each registered nurse at Blue Hills Memorial Hospital is first given a 5% cost-of-living raise and then a \$3000 merit raise. Write a function that *transforms* each old salary $x$ into a new salary $N(x)$. Does the order in which these raises are given make any difference? Explain.

**107.** *Unemployment Versus Inflation* The Phillips curve shows the relationship between the unemployment rate $x$ and the inflation rate $y$. If the equation of the curve is $y = 1 - \sqrt{x}$ for a certain Third World country, then for what values of $x$ is the inflation rate less than 50%?

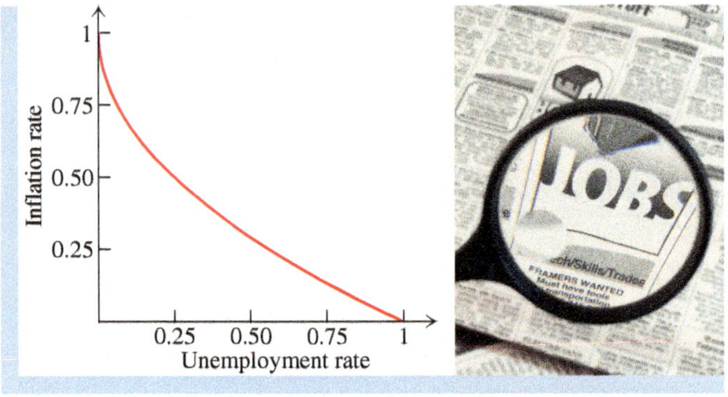


**Figure for Exercise 107**

**108.** *Production Function* The production function shows the relationship between inputs and outputs. A manufacturer of custom windows produces $y$ windows per week using $x$ hours of labor per week, where $y = 1.75\sqrt{x}$. How many hours of labor are required to keep production at or above 28 windows per week?

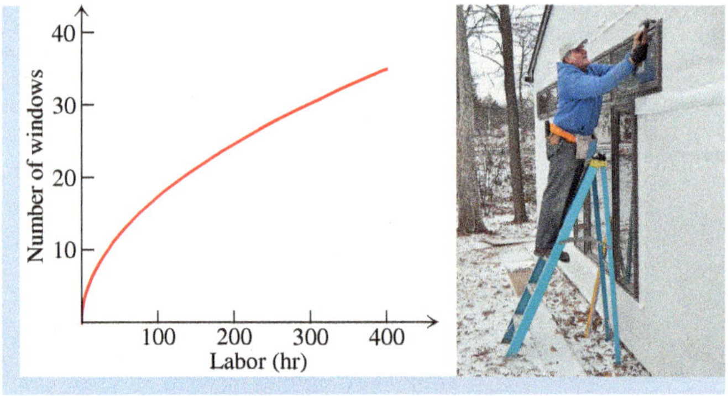


**Figure for Exercise 108**

## FOR WRITING/DISCUSSION

**109.** Graph each pair of functions (without simplifying the second function) on the same screen of a graphing calculator and explain what each exercise illustrates.

**a.** $y = x^4 - x^2$, $y = (-x)^4 - (-x)^2$

**b.** $y = x^3 - x$, $y = (-x)^3 - (-x)$

**c.** $y = x^4 - x^2$, $y = (x + 1)^4 - (x + 1)^2$

**d.** $y = x^3 - x$, $y = (x - 2)^3 - (x - 2) + 3$

**110.** Graph $y = x^3 + 6x^2 + 12x + 8$ on a graphing calculator. Explain how this graph could be obtained as a transformation of a simpler function.

## RETHINKING

**111.** Find and simplify $f(a + 2) - f(a)$ given that $f(x) = x^2 + 1$.

**112.** Find the domain and range of the function $y = -\sqrt{36 - x^2}$.

**113.** Solve $2 - 3|x| \leq 0$. Write the solution set in interval notation.

**114.** Simplify $i^{83}$.

**115.** Solve $ay - 3 + by = cy - 5$ for $y$.

**116.** Express $\frac{-2 + i}{4 + 2i}$ in the form $a + bi$.

## THINKING OUTSIDE THE BOX XV

*Lucky Lucy* Ms. Willis asked Lucy to come to the board to find the mean of a pair of one-digit positive integers. Lucy slowly wrote the numbers on the board. While trying to think of what to do next, she rested the chalk between the numbers to make a mark that looked like a decimal point to Ms. Willis. Ms. Willis said "correct" and asked her to find the mean for a pair of two-digit positive integers. Being a quick learner, Lucy again wrote the numbers on the board, rested the chalk between the numbers, and again Ms. Willis said "correct." Lucy had to demonstrate her ability to find the mean for a pair of three-digit and a pair of four-digit positive integers before Ms. Willis was satisfied that she understood the concept. What four pairs of integers did Ms. Willis give to Lucy? Explain why Lucy's method will not work for any other pairs of one-, two-, three-, or four-digit positive integers.

## POP QUIZ 2.3

1. What is the equation of the curve $y = \sqrt{x}$ after it is translated 8 units upward?

2. What is the equation of the curve $y = x^2$ after it is translated 9 units to the right?

3. What is the equation of the curve $y = x^3$ after it is reflected in the $x$-axis?

4. Find the domain and range for $y = -2\sqrt{x - 1} + 5$.

5. If the curve $y = x^2$ is translated 6 units to the right, stretched by a factor of 3, reflected in the $x$-axis, and translated 4 units upward, then what is the equation of the curve in its final position?

6. Is $y = \sqrt{4 - x^2}$ even, odd, or neither?

## LINKING concepts... For Individual or Group Explorations

### Designing a Racing Boat

*According to International America's Cup Class rules, the basic dimensions of any yacht competing for the silver trophy must satisfy the inequality*

$$16.96 + 9.8D^{1/3} - L - 1.25S^{1/2} \geq 0,$$

*where L is the length in meters, S is the sail area in square meters, and D is the displacement in cubic meters (Americas Cup, www.americascup.org). Use a graphing utility to get approximate answers to the following questions.*

**a)** A team of British designers wants its new boat to have a length of 20.85 m and a displacement of 17.67 m$^3$. Write the inequality that must be satisfied by the sail area $S$. Use the graphing technique described in this section to find the interval in which $S$ must lie.

**b)** A team of Australian designers wants its new boat to have a sail area of 312.54 m$^2$ and length of 21.45 m. Write the inequality that must be satisfied by the displacement $D$. Use the graphing technique to find the interval in which $D$ must lie.

**c)** A team of American designers wants its new boat to have a sail area of 310.28 m$^2$ and a displacement of 17.26 m$^3$. Write the inequality that must be satisfied by the length $L$. Use the graphing technique to find the interval in which $L$ must lie.

**d)** When two of the three variables are fixed, the third variable either has a maximum or minimum value. Explain how you can determine whether it is a maximum or minimum by looking at the original inequality.

# 2.4 Operations with Functions

In Sections 2.2 and 2.3 we studied the graphs of functions to see the relationships between types of functions and their graphs. In this section we will study various ways in which two or more functions can be combined to make new functions. The emphasis here will be on formulas that define functions.

## Basic Operations with Functions

A college student is hired to deliver new telephone books and collect the old ones for recycling. She is paid \$6 per hour plus \$0.30 for each old phone book she collects. Her salary for a 40-hour week is a function of the number of phone books collected. If $x$ represents the number of phone books collected in one week, then the function $S(x) = 0.30x + 240$ gives her salary in dollars. However, she must use her own car for this job. She figures that her car expenses average \$0.20 per phone book collected plus a fixed cost of \$20 per week for insurance. We can write her expenses as a function of the number of phone books collected, $E(x) = 0.20x + 20$. Her profit for one week is her salary minus her expenses:

$$\begin{aligned} P(x) &= S(x) - E(x) \\ &= 0.30x + 240 - (0.20x + 20) \\ &= 0.10x + 220 \end{aligned}$$

By subtracting, we get $P(x) = 0.10x + 220$. Her weekly profit is written as a function of the number of phone books collected. In this example we obtained a new function by subtracting two functions. In general, there are four basic arithmetic operations defined for functions.

**Definition: Sum, Difference, Product, and Quotient Functions**

For two functions $f$ and $g$, the **sum, difference, product,** and **quotient functions,** functions $f + g$, $f - g$, $f \cdot g$, and $f/g$, respectively, are defined as follows:

$$(f + g)(x) = f(x) + g(x)$$
$$(f - g)(x) = f(x) - g(x)$$
$$(f \cdot g)(x) = f(x) \cdot g(x)$$
$$(f/g)(x) = f(x)/g(x) \quad \text{provided that } g(x) \neq 0.$$

### EXAMPLE 1 Evaluating functions

Let $f(x) = 3\sqrt{x} - 2$ and $g(x) = x^2 + 5$. Find and simplify each expression.

**a.** $(f + g)(4)$ **b.** $(f - g)(x)$ **c.** $(f \cdot g)(0)$ **d.** $\left(\frac{f}{g}\right)(9)$

**Solution**

**a.** $(f + g)(4) = f(4) + g(4) = 3\sqrt{4} - 2 + 4^2 + 5 = 25$

**b.** $(f - g)(x) = f(x) - g(x) = 3\sqrt{x} - 2 - (x^2 + 5) = 3\sqrt{x} - x^2 - 7$

**c.** $(f \cdot g)(0) = f(0) \cdot g(0) = (3\sqrt{0} - 2)(0^2 + 5) = (-2)(5) = -10$

**d.** $\left(\frac{f}{g}\right)(9) = \frac{f(9)}{g(9)} = \frac{3\sqrt{9} - 2}{9^2 + 5} = \frac{7}{86}$

Parts (a), (c), and (d) can be checked with a graphing calculator as shown in Figs. 2.54(a) and (b).

(a)

(b)

**Figure 2.54**

▶**TRY THIS.** Let $h(x) = x^2$ and $j(x) = 3x$. Find $(h + j)(5)$, $(h \cdot j)(2)$, and $(h/j)(a)$. ■

Think of $f + g$, $f - g$, $f \cdot g$, and $f/g$ as generic names for the sum, difference, product, and quotient of the functions $f$ and $g$. If any of these functions has a particular meaning, as in the phone book example, we can use a new letter to identify it. The domain of $f + g$, $f - g$, $f \cdot g$, or $f/g$ is the intersection of the domain of $f$ with the domain of $g$. Of course, we exclude from the domain of $f/g$ any number for which $g(x) = 0$.

## EXAMPLE 2 The sum, product, and quotient functions

Let $f = \{(1, 3), (2, 8), (3, 6), (5, 9)\}$ and $g = \{(1, 6), (2, 11), (3, 0), (4, 1)\}$. Find $f + g$, $f \cdot g$, and $f/g$. State the domain of each function.

### Solution

The domain of $f + g$ and $f \cdot g$ is $\{1, 2, 3\}$ because that is the intersection of the domains of $f$ and $g$. The ordered pair $(1, 9)$ belongs to $f + g$ because

$$(f + g)(1) = f(1) + g(1) = 3 + 6 = 9.$$

The ordered pair $(2, 19)$ belongs to $f + g$ because $(f + g)(2) = 19$. The pair $(3, 6)$ also belongs to $f + g$. So

$$f + g = \{(1, 9), (2, 19), (3, 6)\}.$$

Since $(f \cdot g)(1) = f(1) \cdot g(1) = 3 \cdot 6 = 18$, the pair $(1, 18)$ belongs to $f \cdot g$. Likewise, $(2, 88)$ and $(3, 0)$ also belong to $f \cdot g$. So

$$f \cdot g = \{(1, 18), (2, 88), (3, 0)\}.$$

The domain of $f/g$ is $\{1, 2\}$ because $g(3) = 0$. So

$$\frac{f}{g} = \left\{\left(1, \frac{1}{2}\right), \left(2, \frac{8}{11}\right)\right\}.$$

▸**TRY THIS.** Let $h = \{(2, 10), (4, 0), (6, 8)\}$ and $j = \{(2, 5), (6, 0)\}$. Find $h + j$ and $h/j$ and state the domain of each function. ■

In Example 2 the functions are given as sets of ordered pairs, and the results of performing operations with these functions are sets of ordered pairs. In the next example the sets of ordered pairs are defined by means of equations, so the result of performing operations with these functions will be new equations that determine the ordered pairs of the function.

## EXAMPLE 3 The sum, quotient, product, and difference functions

Let $f(x) = \sqrt{x}$, $g(x) = 3x + 1$, and $h(x) = x - 1$. Find each function and state its domain.

**a.** $f + g$ **b.** $\dfrac{g}{f}$ **c.** $g \cdot h$ **d.** $g - h$

### Solution

**a.** Since the domain of $f$ is $[0, \infty)$ and the domain of $g$ is $(-\infty, \infty)$, the domain of $f + g$ is $[0, \infty)$. Since $(f + g)(x) = f(x) + g(x) = \sqrt{x} + 3x + 1$, the equation defining the function $f + g$ is

$$(f + g)(x) = \sqrt{x} + 3x + 1.$$

**b.** The number 0 is not in the domain of $g/f$ because $f(0) = 0$. So the domain of $g/f$ is $(0, \infty)$. The equation defining $g/f$ is

$$\left(\frac{g}{f}\right)(x) = \frac{3x + 1}{\sqrt{x}}.$$

**c.** The domain of both $g$ and $h$ is $(-\infty, \infty)$. So the domain of $g \cdot h$ is $(-\infty, \infty)$. Since $(3x + 1)(x - 1) = 3x^2 - 2x - 1$, the equation defining the function $g \cdot h$ is

$$(g \cdot h)(x) = 3x^2 - 2x - 1.$$

**d.** The domain of both $g$ and $h$ is $(-\infty, \infty)$. So the domain of $g - h$ is $(-\infty, \infty)$. Since $(3x + 1) - (x - 1) = 2x + 2$, the equation defining $g - h$ is

$$(g - h)(x) = 2x + 2.$$

▶**TRY THIS.** Let $h(x) = \sqrt{x}$ and $j(x) = x$. Find $h + j$ and $h/j$ and state the domain of each function. ■

## Composition of Functions

It is often the case that the output of one function is the input for another function. For example, the number of hamburgers purchased at \$1.49 each determines the subtotal. The subtotal is then used to determine the total (including sales tax). So the number of hamburgers actually determines the total, and that function is called the *composition* of the other two functions. The composition of functions is defined using function notation as follows.

**Definition: Composition of Functions**

If $f$ and $g$ are two functions, the **composition** of $f$ and $g$, written $f \circ g$, is defined by the equation

$$(f \circ g)(x) = f(g(x)),$$

provided that $g(x)$ is in the domain of $f$. The composition of $g$ and $f$, written $g \circ f$, is defined by

$$(g \circ f)(x) = g(f(x)),$$

provided that $f(x)$ is in the domain of $g$.

Note that $f \circ g$ is not the same as the function $f \cdot g$, the product of $f$ and $g$. It might be helpful for you to read the composition symbol $\circ$ as "after." This reading makes the order clear. So $(f \circ g)(x)$ is read "$f$ after $g$ of $x$." So $f$ is applied after $g$ is applied to $x$.

For the composition $f \circ g$ to be defined at $x$, $g(x)$ must be in the domain of $f$. So the domain of $f \circ g$ is the set of all values of $x$ in the domain of $g$ for which $g(x)$ is in the domain of $f$. The diagram shown in Fig. 2.55 will help you to understand the composition of functions.

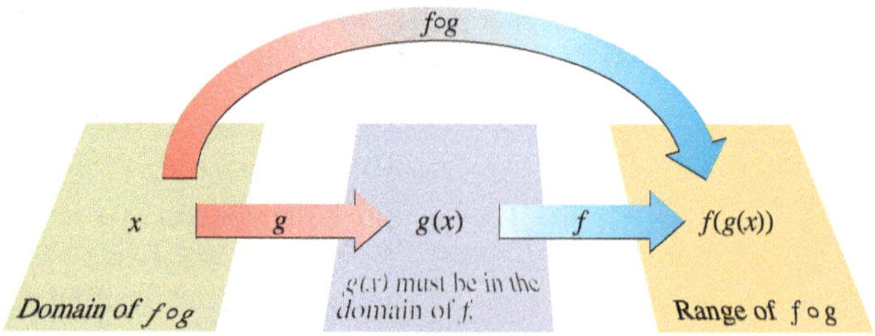


**Figure 2.55**

## EXAMPLE 4 Composition of functions defined by sets

Let $g = \{(1, 4), (2, 5), (3, 6)\}$ and $f = \{(3, 8), (4, 9), (5, 10)\}$. Find $f \circ g$.

### Solution

Since $g(1) = 4$ and $f(4) = 9$, $(f \circ g)(1) = 9$. So the ordered pair $(1, 9)$ is in $f \circ g$. Since $g(2) = 5$ and $f(5) = 10$, $(f \circ g)(2) = 10$. So $(2, 10)$ is in $f \circ g$. Now $g(3) = 6$, but 6 is not in the domain of $f$. So there are only two ordered pairs in $f \circ g$:

$$f \circ g = \{(1, 9), (2, 10)\}$$

▶**TRY THIS.** Let $h = \{(2, 0), (4, 0), (6, 8)\}$ and $j = \{(0, 7), (6, 0)\}$. Find $j \circ h$ and state the domain. ■

In the next example we find specific values of compositions that are defined by equations.

## EXAMPLE 5 Evaluating compositions defined by equations

Let $f(x) = \sqrt{x}$, $g(x) = 2x - 1$, and $h(x) = x^2$. Find the value of each expression.

**a.** $(f \circ g)(5)$ **b.** $(g \circ f)(5)$ **c.** $(h \circ g \circ f)(9)$

### Solution

**a.** $(f \circ g)(5) = f(g(5))$ Definition of composition

$= f(9)$ $g(5) = 2 \cdot 5 - 1 = 9$

$= \sqrt{9}$

$= 3$

**b.** $(g \circ f)(5) = g(f(5)) = g(\sqrt{5}) = 2\sqrt{5} - 1$

**c.** $(h \circ g \circ f)(9) = h(g(f(9))) = h(g(3)) = h(5) = 5^2 = 25$

You can check these answers with a graphing calculator as shown in Figs. 2.56(a) and (b). Enter the functions using the Y = key, then go back to the home screen to evaluate. The symbols $Y_1$, $Y_2$, and $Y_3$ are found in the variables menu (VARS) on a TI-83.

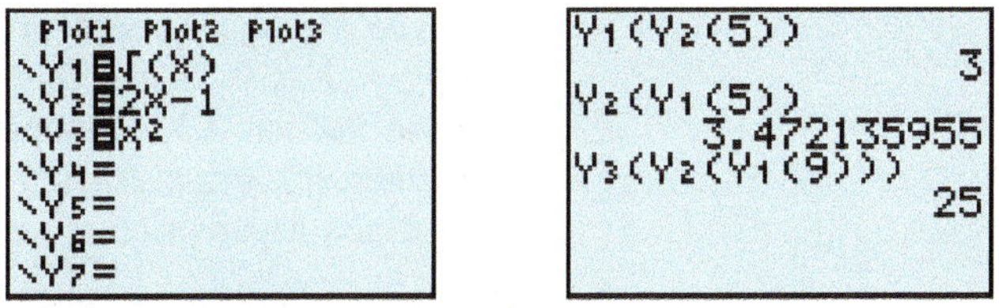


(a) (b)

**Figure 2.56**

▶**TRY THIS.** Let $h(x) = \sqrt{x - 1}$ and $j(x) = 2x$. Find $(h \circ j)(5)$ and $(j \circ h)(5)$. ■

In Example 4, the domain of $g$ is $\{1, 2, 3\}$ while the domain of $f \circ g$ is $\{1, 2\}$. To find the domain of $f \circ g$, we remove from the domain of $g$ any number $x$ such that $g(x)$ is not in the domain of $f$. In the next example we construct compositions of functions defined by equations and determine their domains.

## EXAMPLE 6 Composition of functions defined by equations

Let $f(x) = \sqrt{x}$, $g(x) = 2x - 1$, and $h(x) = x^2$. Find each composition function and state its domain.

**a.** $f \circ g$ **b.** $g \circ f$ **c.** $h \circ g$ **d.** $f \circ h$ **e.** $f \circ g \circ h$

### Solution

**a.** In the composition $f \circ g$, $g(x)$ must be in the domain of $f$. The domain of $f$ is $[0, \infty)$. If $g(x)$ is in $[0, \infty)$, then $2x - 1 \geq 0$, or $x \geq \frac{1}{2}$. So the domain of $f \circ g$ is $\left[\frac{1}{2}, \infty\right)$. Since

$$(f \circ g)(x) = f(g(x)) = f(2x - 1) = \sqrt{2x - 1},$$

the function $f \circ g$ is defined by the equation $(f \circ g)(x) = \sqrt{2x - 1}$, for $x \geq \frac{1}{2}$.

**b.** Since the domain of $g$ is $(-\infty, \infty)$, $f(x)$ is certainly in the domain of $g$. So the domain of $g \circ f$ is the same as the domain of $f$, $[0, \infty)$. Since

$$(g \circ f)(x) = g(f(x)) = g(\sqrt{x}) = 2\sqrt{x} - 1,$$

the function $g \circ f$ is defined by the equation $(g \circ f)(x) = 2\sqrt{x} - 1$. Note that $g \circ f$ is generally not equal to $f \circ g$, but in Section 2.5 we will study special types of functions for which they are equal.

**c.** Since the domain of $h$ is $(-\infty, \infty)$, $g(x)$ is certainly in the domain of $h$. So the domain of $h \circ g$ is the same as the domain of $g$, $(-\infty, \infty)$. Since

$$(h \circ g)(x) = h(g(x)) = h(2x - 1) = (2x - 1)^2 = 4x^2 - 4x + 1,$$

the function $h \circ g$ is defined by $(h \circ g)(x) = 4x^2 - 4x + 1$.

**d.** The domain of $f(x) = \sqrt{x}$ is the set of nonnegative real numbers. Since $x^2$ is nonnegative, $h(x)$ is in the domain of $f$. So the domain of $f \circ h$ is the same as the domain of $h$, the set of all real numbers or $(-\infty, \infty)$. Since

$$(f \circ h)(x) = f(h(x)) = f(x^2) = \sqrt{x^2} = |x|,$$

the function $f \circ h$ is defined by $(f \circ h)(x) = |x|$.

**e.** Apply the functions in the order $h$, $g$, and then $f$:

$$(f \circ g \circ h)(x) = (f \circ g)(x^2) = f(2x^2 - 1) = \sqrt{2x^2 - 1}$$

Since the square root of a negative number is not allowed, we must have $2x^2 - 1 \geq 0$. Solve this inequality by reading the graph of $y = 2x^2 - 1$. (See Section 2.3, Example 9.) The graph of $y = 2x^2 - 1$ is a parabola with $x$-intercepts $(\pm\sqrt{2}/2, 0)$. The $y$-coordinates are negative between the intercepts and non-negative otherwise. So $(f \circ g \circ h)(x) = \sqrt{2x^2 - 1}$ and its domain is $(-\infty, -\sqrt{2}/2] \cup [\sqrt{2}/2, \infty)$.

►**TRY THIS.** Let $h(x) = \sqrt{x + 3}$ and $j(x) = 3x$. Find $h \circ j$ and state its domain.

In Example 6 we composed some simple functions to get more complicated functions. Now we will reverse that process. We start with a function that involves several operations and express it as a composition of simpler functions. *The order of operations is the key to this process.* For example, consider $H(x) = (x + 3)^2$. By the order of operations, we add 3 to $x$ first, then square the result. So $H$ is a composition of a function that adds 3 $(f(x) = x + 3)$ and a function that squares $(g(x) = x^2)$.

Since we write compositions from right to left, we have $H = g \circ f$. To check, we find $(g \circ f)(x)$:

$$(g \circ f)(x) = g(f(x)) = g(x + 3) = (x + 3)^2$$

Note that $(f \circ g)(x) = x^2 + 3$ and it is not the same as $H(x)$. The order is important.

### EXAMPLE 7 Writing a function as a composition

Let $f(x) = \sqrt{x}$, $g(x) = x - 3$, and $h(x) = 2x$. Write each given function as a composition of appropriate functions chosen from $f$, $g$, and $h$.

**a.** $F(x) = \sqrt{x - 3}$ **b.** $G(x) = x - 6$ **c.** $H(x) = 2\sqrt{x} - 3$

#### Solution

**a.** By the order of operations, to evaluate $F(x) = \sqrt{x - 3}$ we first subtract 3 from $x$ $(g(x) = x - 3)$ and then find the square root of the result $(f(x) = \sqrt{x})$. Writing these functions from right to left, we have $F = f \circ g$. Check as follows:

$$(f \circ g)(x) = f(g(x)) = f(x - 3) = \sqrt{x - 3} = F(x)$$

**b.** This one is a bit tricky, because $G(x) = x - 6$ involves only one operation and none of the functions given involve a 6. However, you can subtract 6 by subtracting 3 $(g(x) = x - 3)$ and then again subtracting 3. So $G = g \circ g$. Check as follows:

$$(g \circ g)(x) = g(g(x)) = g(x - 3) = x - 3 - 3 = x - 6 = G(x)$$

**c.** By the order of operations, to evaluate $H(x) = 2\sqrt{x} - 3$, we first find the square root $(f(x) = \sqrt{x})$, next we multiply by 2 $(h(x) = 2x)$, and finally we subtract 3 $(g(x) = x - 3)$. Writing these functions from right to left, we have $H = g \circ h \circ f$. Check as follows:

$$(g \circ h \circ f)(x) = g(h(f(x))) = g(h(\sqrt{x})) = g(2\sqrt{x}) = 2\sqrt{x} - 3$$

▶**TRY THIS.** Write $K(x) = 2\sqrt{x - 3}$ as a composition of $f$, $g$, and $h$. ■

**Foreshadowing Calculus**

One of the big topics in calculus is the instantaneous rate of change of a function. To find the instantaneous rate of change, we often view a complicated function as a composition of simpler functions as is done in Example 7.

## Applications

In applied situations, functions are often defined with formulas rather than function notation. In this case, composition can be simply a matter of substitution.

### EXAMPLE 8 Composition with formulas

The radius of a circle is a function of the diameter $(r = d/2)$ and the area is a function of the radius $(A = \pi r^2)$. Construct a formula that expresses the area as a function of the diameter.

#### Solution

The formula for $A$ as a function of $d$ is obtained by substituting $d/2$ for $r$:

$$A = \pi r^2 = \pi\left(\frac{d}{2}\right)^2 = \pi\frac{d^2}{4}$$

The function $A = \pi d^2/4$ is the composition of $r = d/2$ and $A = \pi r^2$.

▶**TRY THIS.** The diameter of a circle is a function of the radius $(d = 2r)$ and the radius is a function of the circumference $(r = C/(2\pi))$. Construct a formula that expresses $d$ as a function of $C$. ■

In the next example we find a composition using function notation for the salary of the phone book collector mentioned at the beginning of this section.

### EXAMPLE 9 Composition with function notation

A student's salary (in dollars) for collecting $x$ phone books is given by $S(x) = 0.30x + 240$. The amount of withholding (for taxes) is given by $W(x) = 0.20x$, where $x$ is the salary. Express the withholding as a function of the number of phone books collected.

#### Solution

Note that $x$ represents the number of phone books in $S(x) = 0.30x + 240$ and $x$ represents salary in $W(x) = 0.20x$. So we can replace the salary $x$ in $W(x)$ with $S(x)$ or $0.30x + 240$, which also represents salary:

$$W(S(x)) = W(0.30x + 240) = 0.20(0.30x + 240) = 0.06x + 48$$

Use a new letter to name this function, say $T$. Then $T(x) = 0.06x + 48$ gives the amount of withholding (for taxes) as a function of $x$, where $x$ is the number of phone books.

▸**TRY THIS.** The revenue for a sale of $x$ books at \$80 each is given by $R(x) = 80x$. The commission for a revenue of $x$ dollars is given by $C(x) = 0.10x$. Express the commission as a function of the number of books sold. ■

## FOR thought... True or False? Explain.

1. If $f = \{(2, 4)\}$ and $g = \{(1, 5)\}$, then $f + g = \{(3, 9)\}$.
2. If $f = \{(1, 6), (9, 5)\}$ and $g = \{(1, 3), (9, 0)\}$, then $f/g = \{(1, 2)\}$.
3. If $f = \{(1, 6), (9, 5)\}$ and $g = \{(1, 3), (9, 0)\}$, then $f \cdot g = \{(1, 18), (9, 0)\}$.
4. If $f(x) = x + 2$ and $g(x) = x - 3$, then $(f \cdot g)(5) = 14$.
5. If $s = P/4$ and $A = s^2$, then $A$ is a function of $P$.
6. If $f(3) = 19$ and $g(19) = 99$, then $(g \circ f)(3) = 99$.
7. If $f(x) = \sqrt{x}$ and $g(x) = x - 2$, then $(f \circ g)(x) = \sqrt{x} - 2$.
8. If $f(x) = 5x$ and $g(x) = x/5$, then $(f \circ g)(x) = (g \circ f)(x) = x$.
9. If $F(x) = (x - 9)^2$, $g(x) = x^2$, and $h(x) = x - 9$, then $F = h \circ g$.
10. If $f(x) = \sqrt{x}$ and $g(x) = x - 2$, then the domain of $f \circ g$ is $[2, \infty)$.

## EXERCISES 2.4

*Fill in the blank.*

1. For two functions $f$ and $g$, the function $f + g$ is the ________ function.
2. For two functions $f$ and $g$, the function $f \circ g$ is the ________ of $f$ and $g$.

*Let $f(x) = x - 3$ and $g(x) = x^2 - x$. Find and simplify each expression.*

3. $(f + g)(2)$
4. $(g + f)(3)$
5. $(f - g)(-2)$
6. $(g - f)(-6)$
7. $(f \cdot g)(-1)$
8. $(g \cdot f)(0)$

**9.** $(f/g)(4)$ **10.** $(g/f)(4)$ **11.** $(f+g)(a)$

**12.** $(f-g)(b)$ **13.** $(f \cdot g)(a)$ **14.** $(f/g)(b)$

*Let $f = \{(-3, 1), (0, 4), (2, 0)\}$, $g = \{(-3, 2), (1, 2), (2, 6), (4, 0)\}$, and $h = \{(2, 4), (1, 0)\}$. Find each function and state the domain of each function.*

**15.** $f+g$ **16.** $f+h$ **17.** $f-g$ **18.** $f-h$

**19.** $f \cdot g$ **20.** $f \cdot h$ **21.** $g/f$ **22.** $f/g$

*Let $f(x) = \sqrt{x}$, $g(x) = x - 4$, and $h(x) = \frac{1}{x-2}$. Find an equation defining each function and state the domain of the function.*

**23.** $f+g$ **24.** $f+h$ **25.** $f-h$ **26.** $h-g$

**27.** $g \cdot h$ **28.** $f \cdot h$ **29.** $g/f$ **30.** $f/g$

*Let $f = \{(-3, 1), (0, 4), (2, 0)\}$, $g = \{(-3, 2), (1, 2), (2, 6), (4, 0)\}$, and $h = \{(2, 4), (1, 0)\}$. Find each function.*

**31.** $f \circ g$ **32.** $g \circ f$ **33.** $f \circ h$

**34.** $h \circ f$ **35.** $h \circ g$ **36.** $g \circ h$

*Let $f(x) = 3x - 1$, $g(x) = x^2 + 1$, and $h(x) = \frac{x+1}{3}$. Evaluate each expression. Round approximate answers to three decimal places.*

**37.** $f(g(-1))$ **38.** $g(f(-1))$ **39.** $(f \circ h)(5)$

**40.** $(h \circ f)(-7)$ **41.** $(f \circ g)(4.39)$ **42.** $(g \circ h)(-9.87)$

**43.** $(g \circ h \circ f)(2)$ **44.** $(h \circ f \circ g)(3)$

**45.** $(f \circ g \circ h)(2)$ **46.** $(h \circ g \circ f)(0)$

**47.** $(f \circ h)(a)$ **48.** $(h \circ f)(w)$

**49.** $(f \circ g)(t)$ **50.** $(g \circ f)(m)$

*Let $f(x) = x - 2$, $g(x) = \sqrt{x}$, and $h(x) = \frac{1}{x}$. Find an equation defining each function and state the domain of the function.*

**51.** $f \circ g$ **52.** $g \circ f$ **53.** $f \circ h$

**54.** $h \circ f$ **55.** $h \circ g$ **56.** $g \circ h$

**57.** $f \circ f$ **58.** $g \circ g$ **59.** $h \circ g \circ f$

**60.** $f \circ g \circ h$ **61.** $h \circ f \circ g$ **62.** $g \circ h \circ f$

*Let $f(x) = |x|$, $g(x) = x - 7$, and $h(x) = x^2$. Write each of the following functions as a composition of functions chosen from $f$, $g$, and $h$.*

**63.** $F(x) = x^2 - 7$ **64.** $G(x) = |x| - 7$

**65.** $H(x) = (x - 7)^2$ **66.** $M(x) = |x - 7|$

**67.** $N(x) = (|x| - 7)^2$ **68.** $R(x) = |x^2 - 7|$

**69.** $P(x) = |x - 7| - 7$ **70.** $Q(x) = (x^2 - 7)^2$

**71.** $S(x) = x - 14$ **72.** $T(x) = x^4$

*For each given function $f(x)$, find two functions $g(x)$ and $h(x)$ such that $f = h \circ g$. Answers may vary.*

**73.** $f(x) = x^3 - 2$ **74.** $f(x) = (x - 2)^3$

**75.** $f(x) = \sqrt{x + 5}$ **76.** $f(x) = \sqrt{x} + 5$

**77.** $f(x) = \sqrt{3x - 1}$ **78.** $f(x) = 3\sqrt{x} - 1$

**79.** $f(x) = 4|x| + 5$ **80.** $f(x) = |4x + 5|$

*Use the two given functions to write $y$ as a function of $x$.*

**81.** $y = 2a - 3$, $a = 3x + 1$

**82.** $y = -4d - 1$, $d = -3x - 2$

**83.** $y = w^2 - 2$, $w = x + 3$

**84.** $y = 3t^2 - 3$, $t = x - 1$

**85.** $y = 3m - 1$, $m = \frac{x+1}{3}$ **86.** $y = 2z + 5$, $z = \frac{1}{2}x - \frac{5}{2}$

*Find each function from the given verbal description of the function.*

**87.** If $m$ is $n$ minus 4, and $y$ is the square of $m$, then write $y$ as a function of $n$.

**88.** If $u$ is the sum of $t$ and 9, and $v$ is $u$ divided by 3, then write $v$ as a function of $t$.

**89.** If $w$ is equal to the sum of $x$ and 16, $z$ is the square root of $w$, and $y$ is $z$ divided by 8, then write $y$ as a function of $x$.

**90.** If $a$ is the cube of $b$, $c$ is the sum of $a$ and 25, and $d$ is the square root of $c$, then write $d$ as a function of $b$.

*Graph each function on a graphing calculator without simplifying the given expression. Examine the graph and write a function for the graph. Then simplify the original function and see whether it is the same as your function. Is the domain of the original function equal to the domain of the function after it is simplified?*

**91.** $y = ((x - 1)/(x + 1) + 1)/((x - 1)/(x + 1) - 1)$

**92.** $y = ((3x + 1)/(x - 1) + 1)/((3x + 1)/(x - 1) - 3)$

*Define $y_1 = \sqrt{x + 1}$ and $y_2 = 3x - 4$ on your graphing calculator. For each function $y_3$, defined in terms of $y_1$ and $y_2$, determine the domain and range of $y_3$ from its graph on your calculator and explain what each graph illustrates.*

**93.** $y_3 = y_1 + y_2$ **94.** $y_3 = 3y_1 - 4$

**95.** $y_3 = \sqrt{y_2 + 1}$ **96.** $y_3 = \sqrt{y_1 + 1}$

*Define $y_1 = \sqrt[3]{x}$, $y_2 = \sqrt{x}$, and $y_3 = x + 4$. For each function $y_4$, determine the domain and range of $y_4$ from its graph on your calculator and explain what each graph illustrates.*

**97.** $y_4 = y_1 + y_2 + y_3$ **98.** $y_4 = \sqrt{y_1 + 4}$

*Solve each problem.*

**99.** *Profitable Business* Charles buys factory-reconditioned hedge trimmers for \$40 each and sells them on eBay for \$68 each. He has a fixed cost of \$200 per month. If $x$ is the number of hedge trimmers he sells per month, then his revenue and cost (in dollars) are given by $R(x) = 68x$ and $C(x) = 40x + 200$. Find a formula for the function $P(x) = R(x) - C(x)$. For what values of $x$ is his profit positive?

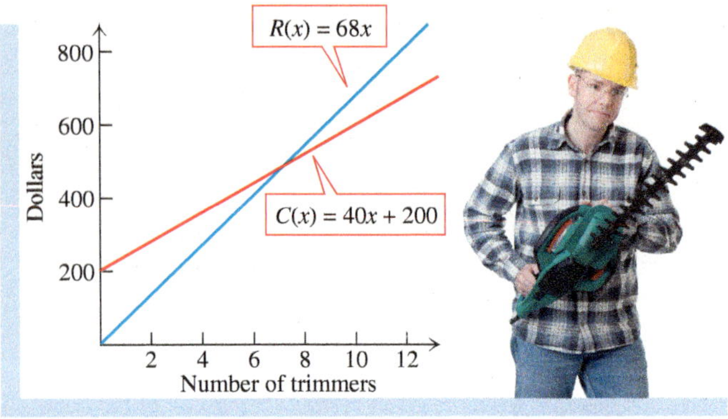


**Figure for Exercise 99**

**100.** *Profit* The revenue in dollars that a company receives for installing $x$ alarm systems per month is given by $R(x) = 3000x - 20x^2$, while the cost in dollars is given by $C(x) = 600x + 4000$. The function $P(x) = R(x) - C(x)$ gives the profit for installing $x$ alarm systems per month. Find $P(x)$ and simplify it.

**101.** Write the area $A$ of a square with a side of length $s$ as a function of its diagonal $d$.
HINT Write down the formulas involving the area, perimeter, and diagonal of square.

**102.** Write the perimeter of a square $P$ as a function of the area $A$.

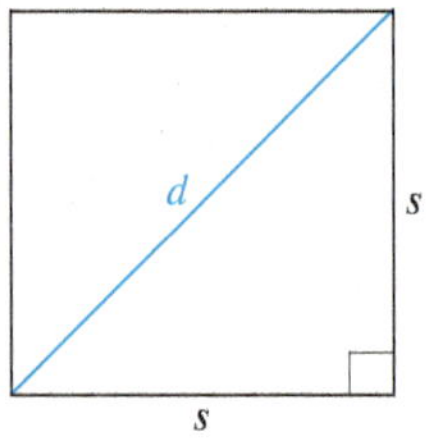


**Figure for Exercises 101 and 102**

**103.** *Deforestation in Nigeria* In Nigeria deforestation occurs at the rate of about 5.2% per year. If $x$ is the total forest area of Nigeria at the start of 2000, then $f(x) = 0.948x$ is the amount of forest land in Nigeria at the start of 2001. Find and interpret $(f \circ f)(x)$. Find and interpret $(f \circ f \circ f)(x)$.

**104.** *Doubling Your Money* On the average, money invested in bonds doubles every 12 years. So $f(x) = 2x$ gives the value of $x$ dollars invested in bonds after 12 years. Find and interpret $(f \circ f)(x)$. Find and interpret $(f \circ f \circ f)(x)$.

**105.** *Hamburgers* If hamburgers are \$1.20 each, then $C(x) = 1.20x$ gives the pre-tax cost in dollars for $x$ hamburgers. If sales tax is 5%, then $T(x) = 1.05x$ gives the total cost when the pre-tax cost is $x$ dollars. Write the total cost as a function of the number of hamburgers.

**106.** *Laying Sod* Southern Sod will deliver and install 20 pallets of St. Augustine sod for \$2200 or 30 pallets for \$3200 not including tax.

**a.** Write the cost (not including tax) as a linear function of $x$, where $x$ is the number of pallets.

**b.** Write a function that gives the total cost (including tax at 9%) as a function of $x$, where $x$ is the cost (not including tax).

**c.** Find the function that gives the total cost as a function of $x$, where $x$ is the number of pallets.

**107.** *Displacement-Length Ratio* The displacement-length ratio $D$ indicates whether a sailboat is relatively heavy or relatively light:

$$D = (d \div 2240) \div x$$

where $d$ is the displacement in pounds and

$$x = (L \div 100)^3$$

where $L$ is the length at the waterline in feet (*Sailing*, www.sailing.com). Assuming that the displacement is 26,000 pounds, write $D$ as a function of $L$ and simplify it.

**108.** *Sail Area-Displacement Ratio* The sail area-displacement ratio $S$ measures the sail power available to drive a sailboat:

$$S = A \div y$$

where $A$ is the sail area in square feet and

$$y = (d \div 64)^{2/3}$$

where $d$ is the displacement in pounds. Assuming that the sail area is 6500 square feet, write $S$ as a function of $d$ and simplify it.

**109.** *Area of a Window* A window is in the shape of a square with a side of length $s$, with a semicircle of diameter $s$ adjoining the top of the square. Write the total area of the window $W$ as a function of $s$.
HINT Start with the formulas for the area of a square and the area of a circle.

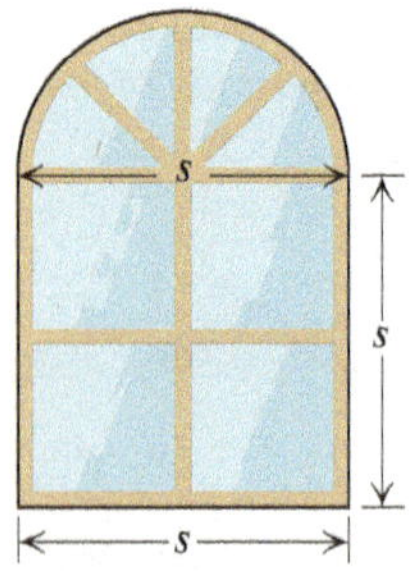


**Figure for Exercises 109 and 110**

**110.** *Area of a Window* Using the window of Exercise 109, write the area of the square $A$ as a function of the area of the semicircle $S$.

**111.** *Packing a Square Piece of Glass* A glass prism with a square cross section is to be shipped in a cylindrical cardboard tube that has an inside diameter of $d$ inches. Given that $s$ is the length of a side of the square and the glass fits snugly into the tube, write $s$ as a function of $d$.

HINT Use the Pythagorean theorem and the formula for the area of a circle.

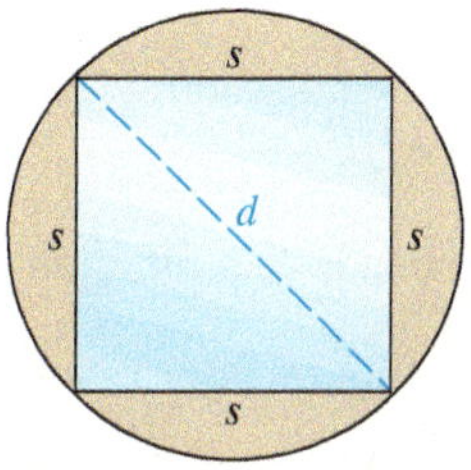


**Figure for Exercise 111**

**112.** *Packing a Triangular Piece of Glass* A glass prism is to be shipped in a cylindrical cardboard tube that has an inside diameter of $d$ inches. Given that the cross section of the prism is an equilateral triangle with side of length $p$ and the prism fits snugly into the tube, write $p$ as a function of $d$.

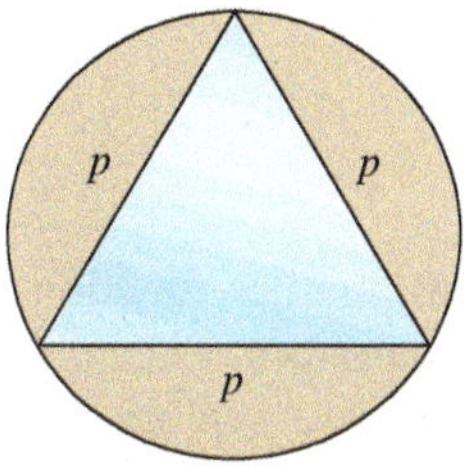


**Figure for Exercise 112**

## FOR WRITING/DISCUSSION

**113.** Sears often has Super Saturday sales during which everything is 10% off, even items already on sale. If a coat is on sale at 25%, then is it 35% off on Super Saturday? What operation with functions is at work here? Explain your answers.

**114.** We can combine two functions to obtain a new function by addition, subtraction, multiplication, division, and composition. Are any of these operations commutative? Associative? Explain your answers and give examples.

## ▶ RETHINKING

**115.** Find and simplify the difference quotient for the function $f(x) = 1 + \frac{3}{x}$.

**116.** Let $f(x) = \begin{cases} x & \text{for} \quad x \geq 1 \\ -x & \text{for} \quad x < 1 \end{cases}$. Find the domain, the range, and the interval on which the function is increasing.

**117.** Find the domain and range for $f(x) = -5\sqrt{x - 3} + 2$.

**118.** Solve $4(x - 3)^2 - 6 = 0$.

**119.** Solve $5 - 3x < 6 + 2x$.

**120.** Find the slope of the line $3x - 9y = 7$.

## THINKING OUTSIDE THE BOX XVI & XVII

*Thirty-Percent Reduction* If each of the numbers in the expression $ab^2c^3d^4$ is decreased by 30%, then the value of the expression is decreased by what percent? Give the answer to the nearest tenth of a percent.

*Whole-Number Expression* What is the largest whole number $N$ that cannot be expressed as $N = 3x + 11y$ where $x$ and $y$ are whole numbers?

## POP QUIZ 2.4

**1.** Write the area of a circle as a function of its diameter.

*Let $f(x) = x^2$ and $g(x) = x - 2$. Find and simplify.*

**2.** $(f + g)(3)$ **3.** $(f \cdot g)(4)$ **4.** $(f \circ g)(5)$

*Let $m = \{(1, 3), (4, 8)\}$ and $n = \{(3, 5), (4, 9)\}$. Find each function.*

**5.** $m + n$ **6.** $n \circ m$

*Let $h(x) = x^2$ and $j(x) = \sqrt{x + 2}$. Find the domain of each function.*

**7.** $h + j$ **8.** $h \circ j$ **9.** $j \circ h$

## LINKING concepts... For Individual or Group Explorations

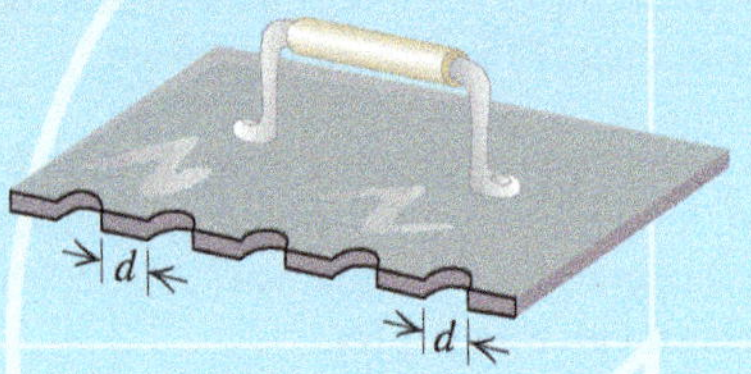


### Modeling Glue Coverage

*Frank is designing a notched trowel for spreading glue. The notches on the trowel are in the shape of semicircles. The diameter of each notch and the space between consecutive notches are d inches as shown in the figure. Suppose that the trowel is used to make parallel beads of glue on one square foot of floor.*

**a)** Write the number of parallel beads made as a function of $d$.

**b)** The cross section of each parallel bead is a semicircle with diameter $d$. Write the area of a cross section as a function of $d$.

**c)** Write the volume of one of the parallel beads as a function of $d$.

**d)** Write the volume of glue on one square foot of floor as a function of $d$.

**e)** Write the number of square feet that one gallon of glue will cover as a function of $d$. (Use $1 \text{ ft}^3 = 7.5$ gal.)

**f)** Suppose that the trowel has square notches where the sides of the squares are $d$ inches in length and the distance between consecutive notches is $d$ inches. Write the number of square feet that one gallon of glue will cover as a function of $d$.

# 2.5 Inverse Functions

It is possible for one function to undo what another function does. For example, squaring undoes the operation of taking a square root. The composition of two such functions is the identity function. In this section we explore this idea in detail.

## One-to-One Functions

Consider a medium pizza that costs \$5 plus \$2 per topping. Table 2.1 shows the ordered pairs of the function that determines the cost. Note that for every number of toppings there is a unique cost and for every cost there is a unique number of toppings. There is a **one-to-one correspondence** between the domain and range of this function and the function is a *one-to-one function.* For a function that is not one-to-one, consider a Wendy's menu. Every item corresponds to a unique price, but the price \$0.99 corresponds to many different items.

**Table 2.1**

| Toppings $x$ | Cost $y$ |
|---|---|
| 0 | \$ 5 |
| 1 | 7 |
| 2 | 9 |
| 3 | 11 |
| 4 | 13 |

**Definition: One-to-one Function**

If a function has no two ordered pairs with different first coordinates and the same second coordinate, then the function is called **one-to-one.**

The definition of function requires that every element of the domain corresponds to exactly one element of the range. For a one-to-one function we have the additional requirement that every element of the range corresponds to exactly one element of the domain.

### EXAMPLE 1 One-to-one with ordered pairs

Determine whether each function is one-to-one.

**a.** $\{(1, 3), (2, 5), (3, 4), (4, 9), (5, 0)\}$
**b.** $\{(4, 16), (-4, 16), (2, 4), (-2, 4), (5, 25)\}$
**c.** $\{(3, 0.99), (5, 1.99), (7, 2.99), (8, 3.99), (9, 0.99)\}$

#### Solution

**a.** This function is one-to-one because no two ordered pairs have different first coordinates and the same second coordinate.
**b.** This function is not one-to-one because the ordered pairs $(4, 16)$ and $(-4, 16)$ have different first coordinates and the same second coordinate.
**c.** This function is not one-to-one because the ordered pairs $(3, 0.99)$ and $(9, 0.99)$ have different first coordinates and the same second coordinate.

▶**TRY THIS.** Determine whether the function $\{(-1, \pi), (0, 3.14), (1, 22/7)\}$ is one-to-one. ■

If a function is given as a short list of ordered pairs, then it is easy to determine whether the function is one-to-one. A graph of a function can also be used to determine whether a function is one-to-one using the **horizontal line test.**

**Horizontal Line Test**

If each horizontal line crosses the graph of a function at no more than one point, then the function is one-to-one.

The graph of a one-to-one function never has the same $y$-coordinate for two different $x$-coordinates on the graph. So if it is possible to draw a horizontal line that crosses the graph of a function two or more times, then the function is not one-to-one.

### EXAMPLE 2 The horizontal line test

Use the horizontal line test to determine whether the functions shown in Fig. 2.57 are one-to-one.

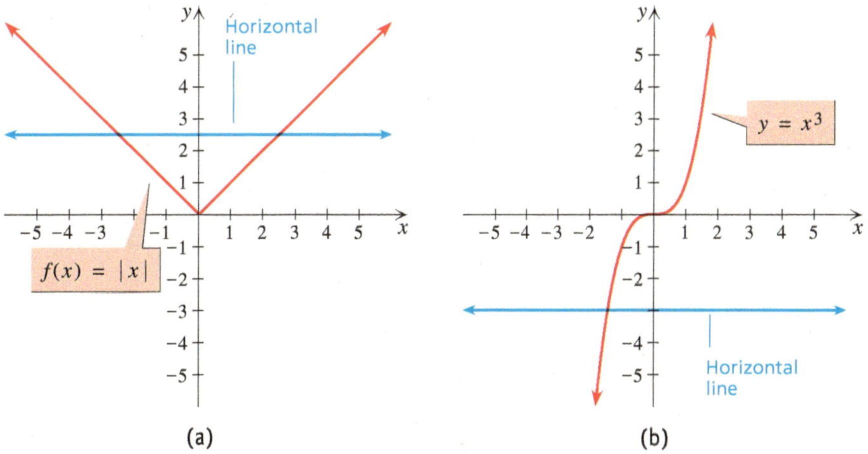


Figure 2.57

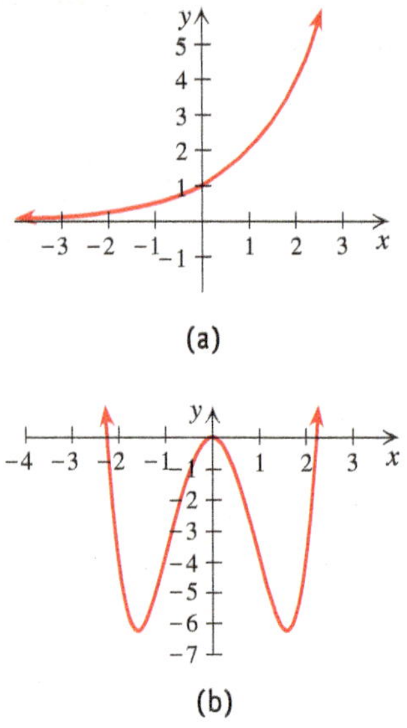


Figure 2.58

### Solution

The function $f(x) = |x|$ is not one-to-one because it is possible to draw a horizontal line that crosses the graph twice as shown in Fig. 2.57(a). The function $y = x^3$ in Fig. 2.57(b) is one-to-one because it appears to be impossible to draw a horizontal line that crosses the graph more than once.

▸**TRY THIS.** Use the horizontal line test to determine whether the functions shown in Fig. 2.58 are one-to-one. ■

The horizontal line test explains the visual difference between the graph of a one-to-one function and the graph of a function that is not one-to-one. Because no graph of a function is perfectly accurate, conclusions made from a graph alone may not be correct. For example, from the graph of $y = x^3 - 0.01x$ shown in Fig. 2.59(a) we would conclude that the function is one-to-one. However, another view of the same function in Fig. 2.59(b) shows that the function is not one-to-one.

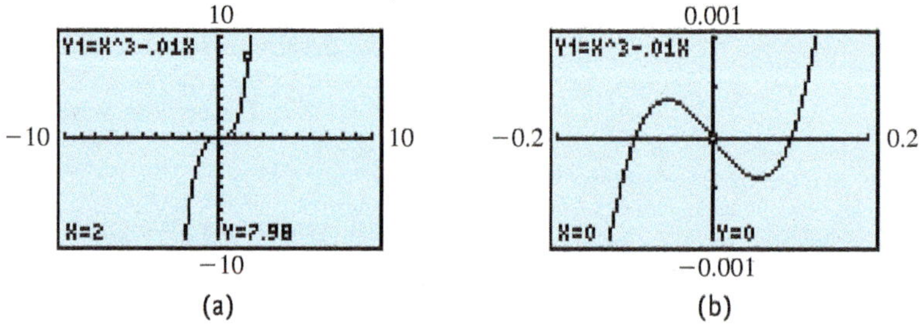


Figure 2.59

Using the equation for a function we can often determine conclusively whether the function is one-to-one. A function $f$ is not one-to-one if it is possible to find two different numbers $x_1$ and $x_2$ such that $f(x_1) = f(x_2)$. For example, for the function $f(x) = x^2$, it is possible to find two different numbers, 2 and $-2$, such that $f(2) = f(-2)$. So $f(x) = x^2$ is not one-to-one. To prove that a function $f$ is one-to-one, we must show that $f(x_1) = f(x_2)$ implies that $x_1 = x_2$.

## EXAMPLE 3 Using the definition of one-to-one

Determine whether each function is one-to-one.

**a.** $f(x) = \dfrac{2x + 1}{x - 3}$ **b.** $g(x) = |x|$

### Solution

**a.** If $f(x_1) = f(x_2)$, then we have the following equation:

$$\frac{2x_1 + 1}{x_1 - 3} = \frac{2x_2 + 1}{x_2 - 3}$$

$$(2x_1 + 1)(x_2 - 3) = (2x_2 + 1)(x_1 - 3) \quad \text{Multiply by the LCD.}$$

$$2x_1x_2 - 6x_1 + x_2 - 3 = 2x_2x_1 - 6x_2 + x_1 - 3$$

$$7x_2 = 7x_1$$

$$x_2 = x_1$$

Since $f(x_1) = f(x_2)$ implies that $x_1 = x_2$, $f(x)$ is a one-to-one function.

**b.** If $g(x_1) = g(x_2)$, then $|x_1| = |x_2|$. But this does not imply that $x_1 = x_2$, because two different numbers can have the same absolute value. For example $|3| = |-3|$ but $3 \neq -3$. So $g$ is not one-to-one.

▶**TRY THIS.** Determine whether $h(x) = 5x^2$ is one-to-one. ■

## Inverse Functions

Consider again the function given in Table 2.1, which determines the cost of a pizza from the number of toppings. Because that function is one-to-one we can make a table in which the number of toppings is determined from the cost, as shown in Table 2.2. Of course we could just read Table 2.1 backward, but we make a new table to emphasize that there is a new function under discussion. Table 2.2 is the *inverse function* for the function in Table 2.1.

**Table 2.2**

| Cost $x$ | Toppings $y$ |
|---|---|
| \$5 | 0 |
| 7 | 1 |
| 9 | 2 |
| 11 | 3 |
| 13 | 4 |

A function is a set of ordered pairs in which no two ordered pairs have the same first coordinates and different second coordinates. If we interchange the $x$- and $y$-coordinates in each ordered pair of a function, as in Tables 2.1 and 2.2, the resulting set of ordered pairs might or might not be a function. If the original function is one-to-one, then the set obtained by interchanging the coordinates in each ordered pair is a function, the inverse function. If a function is one-to-one, then it has an inverse function or it is **invertible.**

**Definition: Inverse Function**

The **inverse** of a one-to-one function $f$ is the function $f^{-1}$ (read "$f$ inverse"), where the ordered pairs of $f^{-1}$ are obtained by interchanging the coordinates in each ordered pair of $f$.

In this notation, the number $-1$ in $f^{-1}$ does not represent a negative exponent. It is merely a symbol for denoting the inverse function.

### EXAMPLE 4 Finding an inverse function

For each function, determine whether it is invertible. If it is invertible, then find the inverse.

**a.** $f = \{(-2, 3), (4, 5), (2, 3)\}$ **b.** $g = \{(3, 1), (5, 2), (7, 4), (9, 8)\}$

#### Solution

**a.** This function $f$ is *not* one-to-one because of the ordered pairs $(-2, 3)$ and $(2, 3)$. So $f$ is not invertible.

**b.** The function $g$ is one-to-one, and so $g$ is invertible. The inverse of $g$ is the function $g^{-1} = \{(1, 3), (2, 5), (4, 7), (8, 9)\}$.

▶**TRY THIS.** Determine whether $h = \{(2, 1), (3, 4), (4, 0)\}$ is invertible. If it is invertible, then find the inverse function. ■

### EXAMPLE 5 Using inverse function notation

Let $f = \{(1, 3), (2, 4), (5, 7)\}$. Find $f^{-1}$, $f^{-1}(3)$, and $(f^{-1} \circ f)(1)$.

**Solution**

Interchange the $x$- and $y$-coordinates of each ordered pair of $f$ to find $f^{-1}$:

$$f^{-1} = \{(3, 1), (4, 2), (7, 5)\}$$

To find the value of $f^{-1}(3)$, notice that $f^{-1}(3)$ is the second coordinate when the first coordinate is 3 in the function $f^{-1}$. So $f^{-1}(3) = 1$. To find the composition, use the definition of composition of functions:

$$(f^{-1} \circ f)(1) = f^{-1}(f(1)) = f^{-1}(3) = 1$$

►**TRY THIS.** Let $h = \{(2, 1), (3, 4), (4, 0)\}$. Find $h(3)$, $h^{-1}(4)$, and $(h \circ h^{-1})(1)$. ■

Since the coordinates in the ordered pairs are interchanged, the domain of $f^{-1}$ is the range of $f$, and the range of $f^{-1}$ is the domain of $f$. If $f^{-1}$ is the inverse function of $f$, then certainly $f$ is the inverse of $f^{-1}$. The functions $f$ and $f^{-1}$ are inverses of each other.

## Inverse Functions Using Function Notation

Suppose that the cost of a pizza is \$5 plus \$2 per topping. The cost $C$ can be determined from the number of toppings $T$ by the function $C = 2T + 5$. We can easily solve this formula for $T$ to get $T = \frac{C - 5}{2}$. This formula can be used to determine the number of toppings from the cost. The functions $C = 2T + 5$ and $T = \frac{C - 5}{2}$ are inverse functions.

When we use the variables $x$ and $y$, we always like to keep $x$ as the independent variable and $y$ as the dependent variable. So in the pizza situation $y = 2x + 5$ is a function that determines the cost $y$ from the number of toppings $x$. The inverse function $y = \frac{x - 5}{2}$ is used to determine the number of toppings from the cost, but now $x$ is the cost and $y$ is the number of toppings. The roles of $x$ and $y$ are switched. Using function notation we write $f(x) = 2x + 5$ and $f^{-1}(x) = \frac{x - 5}{2}$. The one-to-one function $f$ pairs members of the domain of $f$ with members of the range of $f$, and the inverse function $f^{-1}$ exactly reverses those pairings, as shown in Fig. 2.60.

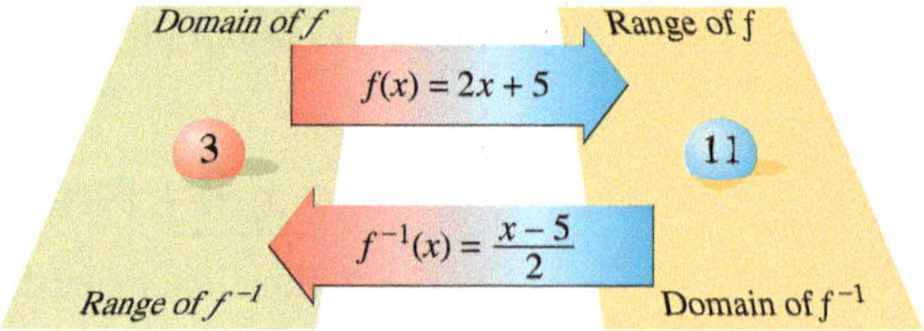


**Figure 2.60**

Note that the inverse of $f(x) = 2x + 5$ can be obtained by thinking of $f$ as a composition of multiplying by 2 and adding 5 to the result. To reverse this composition you must subtract 5 and then divide the result by 2. Think of putting on your socks and then your shoes. To get back to bare feet, you must remove your shoes and then remove your socks. So $f^{-1}(x) = \frac{x - 5}{2}$. Note that $f^{-1}(x) \neq \frac{x}{2} - 5$. The *quantity* $x - 5$ must be divided by 2. The inverses of some of the common functions that we use are shown in Table 2.3.

**Table 2.3**

| Function | Inverse |
|---|---|
| $f(x) = 2x$ | $f^{-1}(x) = x/2$ |
| $f(x) = x - 5$ | $f^{-1}(x) = x + 5$ |
| $f(x) = x^3$ | $f^{-1}(x) = \sqrt[3]{x}$ |
| $f(x) = 1/x$ | $f^{-1}(x) = 1/x$ |
| $f(x) = -x$ | $f^{-1}(x) = -x$ |

## EXAMPLE 6 Finding an inverse by reversing a composition

Find the inverse of each function.

**a.** $f(x) = 2x + 1$ **b.** $g(x) = \dfrac{x^3 + 5}{2}$

**Solution**

**a.** The function $f(x) = 2x + 1$ is a composition of multiplying $x$ by 2 and then adding 1. So the inverse function is a composition of subtracting 1 and then dividing by 2: $f^{-1}(x) = (x - 1)/2$. See Fig. 2.61.

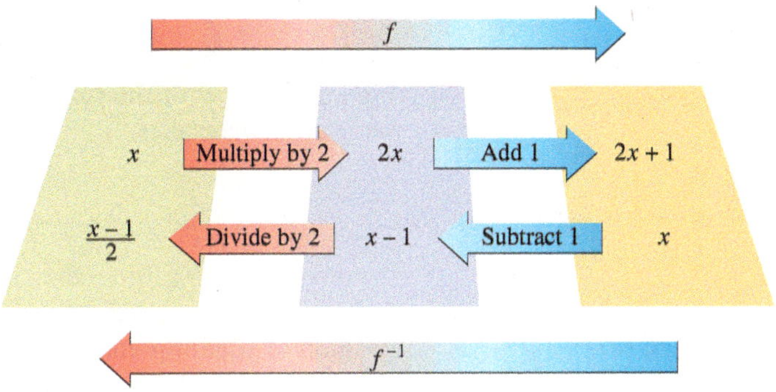


Figure 2.61

**b.** The function $g(x) = (x^3 + 5)/2$ is a composition of cubing $x$, adding 5, and dividing by 2. The inverse is a composition of multiplying by 2, subtracting 5, and then taking the cube root: $g^{-1}(x) = \sqrt[3]{2x - 5}$. See Fig. 2.62.

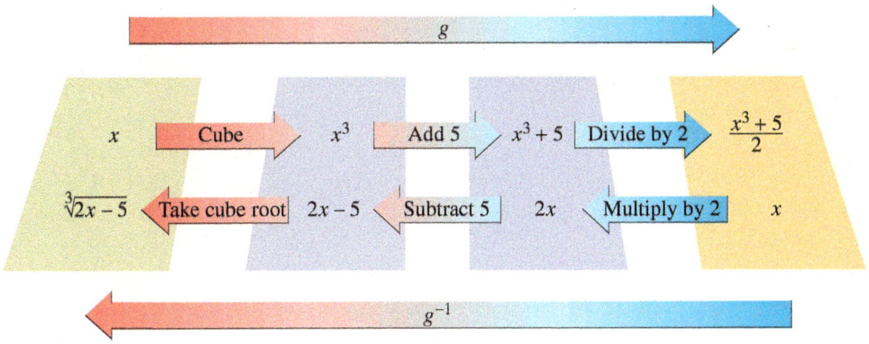


Figure 2.62

▶**TRY THIS.** Find the inverse of $f(x) = \frac{2}{3}x + 6$. ■

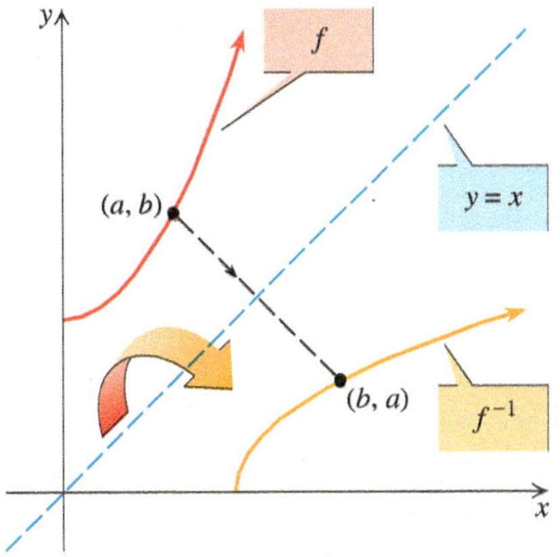


Figure 2.63

## Graphs of $f$ and $f^{-1}$

The inverse of a one-to-one function contains exactly the same ordered pairs as the function except that the coordinates are reversed. So for every point $(a, b)$ on the graph of an invertible function $f$, there is a corresponding point $(b, a)$ on the graph of $f^{-1}$. See Fig. 2.63. This fact causes the graph of $f^{-1}$ to be a reflection of the graph of $f$ with respect to the line $y = x$.

**Reflection Property of Inverse Functions**

> If $f$ is a one-to-one function, then the graph of $f^{-1}$ is a reflection of the graph of $f$ with respect to the line $y = x$.

Only one-to-one functions are invertible. However, sometimes it is possible to restrict the domain of a function that is not one-to-one so that it is one-to-one on the restricted domain. For example, $f(x) = x^2$ for $x$ in $(-\infty, \infty)$ is not one-to-one and not invertible. But $f(x) = x^2$ for $x \geq 0$ is one-to-one and invertible. The inverse of $f(x) = x^2$ for $x \geq 0$ is the square root function $f^{-1}(x) = \sqrt{x}$. Graphing will help you to see when the domain of an inverse function must be restricted.

## EXAMPLE 7 Graphing a function and its inverse

Find the inverse of the function $f(x) = \sqrt{x - 1}$ and graph both $f$ and $f^{-1}$ on the same coordinate axes.

### Solution

The graph of $f(x) = \sqrt{x - 1}$ is shown in Fig. 2.64. The domain of $f$ is $[1, \infty)$ and the range is $[0, \infty)$. From the graph, we see that $f$ is one-to-one and invertible. Since $f$ is a composition of subtracting 1 and then taking a square root, $f^{-1}$ must be a composition of squaring and then adding 1. Since the domain of $f^{-1}$ must be the range of $f$, we have $f^{-1}(x) = x^2 + 1$ for $x \geq 0$. The graph of $f^{-1}$ is half of a parabola, as shown in Fig. 2.64, and it is a reflection of the graph of $f$ with respect to the line $y = x$. Note that without the restriction on the domain of $f^{-1}$ the reflection property would not hold true.

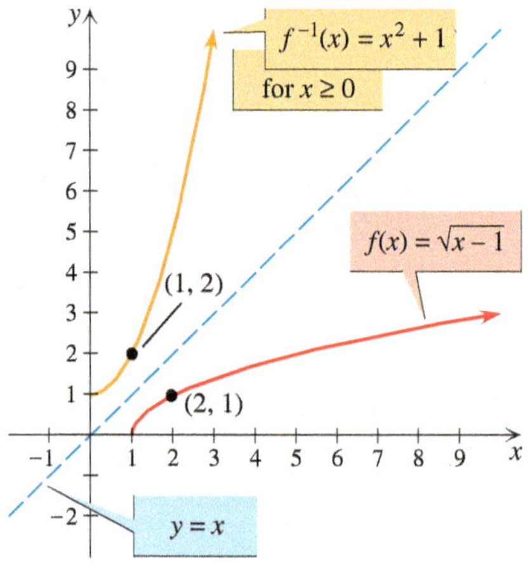


Figure 2.64

▸**TRY THIS.** Find the inverse of $f(x) = \sqrt{x + 2}$ and graph $f$ and $f^{-1}$. ■

## The Switch-and-Solve Method

In the inverse function, the roles of $x$ and $y$ are switched. So we can actually switch $x$ and $y$ in the formula, do a bit of rewriting, and we have the inverse function. As you will see, this method is similar to finding an inverse by reversing a composition. Its advantage is that it works on some functions that are not readily viewed as a composition as in Example 8(b).

PROCEDURE

**Finding $f^{-1}(x)$ by the Switch-and-Solve Method**

To find the inverse of a one-to-one function given in function notation:

1. Replace $f(x)$ by $y$.
2. Interchange $x$ and $y$.
3. Solve the equation for $y$.
4. Replace $y$ by $f^{-1}(x)$.
5. Check that the domain of $f$ is the range of $f^{-1}$ and the range of $f$ is the domain of $f^{-1}$.

## EXAMPLE 8 The switch-and-solve method

Find the inverse of each function.

**a.** $f(x) = 4x - 1$ **b.** $f(x) = \dfrac{2x + 1}{x - 3}$ **c.** $f(x) = \sqrt{x + 2} - 3$

### Solution

**a.** The graph of $f(x) = 4x - 1$ is a line with slope 4. By the horizontal line test the function is one-to-one and invertible. Replace $f(x)$ with $y$ to get $y = 4x - 1$. Next, interchange $x$ and $y$ to get $x = 4y - 1$. Now solve for $y$:

$$x = 4y - 1$$

$$x + 1 = 4y$$

$$\frac{x + 1}{4} = y$$

Replace $y$ by $f^{-1}(x)$ to get $f^{-1}(x) = \frac{x + 1}{4}$. The domain of $f$ is $(-\infty, \infty)$ and that is the range of $f^{-1}$. The range of $f$ is $(-\infty, \infty)$ and that is the domain of $f^{-1}$.

**b.** In Example 3(a) we showed that $f$ is a one-to-one function. So we can find $f^{-1}$ by interchanging $x$ and $y$ and solving for $y$:

$$y = \frac{2x + 1}{x - 3} \quad \text{Replace } f(x) \text{ by } y.$$

$$x = \frac{2y + 1}{y - 3} \quad \text{Interchange } x \text{ and } y.$$

$$x(y - 3) = 2y + 1 \quad \text{Solve for } y.$$

$$xy - 3x = 2y + 1$$

$$xy - 2y = 3x + 1$$

$$y(x - 2) = 3x + 1$$

$$y = \frac{3x + 1}{x - 2}$$

Replace $y$ by $f^{-1}(x)$ to get $f^{-1}(x) = \frac{3x + 1}{x - 2}$. The domain of $f$ is all real numbers except 3. The range of $f^{-1}$ is the set of all real numbers except 3. We exclude 3 because $\frac{3x + 1}{x - 2} = 3$ has no solution. Check that the range of $f$ is equal to the domain of $f^{-1}$.

**c.** Replace $f(x)$ with $y$ to get $y = \sqrt{x + 2} - 3$. Interchange $x$ and $y$ to get $x = \sqrt{y + 2} - 3$. Now solve for $y$:

$$x = \sqrt{y + 2} - 3$$
$$x + 3 = \sqrt{y + 2}$$
$$(x + 3)^2 = y + 2 \qquad \text{Square both sides.}$$
$$(x + 3)^2 - 2 = y$$
$$x^2 + 6x + 7 = y$$

The equation obtained by squaring both sides is not equivalent to the one above it. So we must make sure that the domain of the inverse function is the same as the range of the original function. The range of $f(x) = \sqrt{x + 2} - 3$ is the interval $[-3, \infty)$ and so that must be the domain of $f^{-1}$. So replace $y$ by $f^{-1}(x)$ to get $f^{-1}(x) = x^2 + 6x + 7$ for $x \geq -3$.

▶**TRY THIS.** Use switch-and-solve to find the inverse of $h(x) = x^3 - 5$. ■

If a function $f$ is defined by a short list of ordered pairs, then we simply write all of the pairs in reverse to find $f^{-1}$. If two functions are defined by formulas, it may not be obvious whether the ordered pairs of one are the reverse of the ordered pairs of the other. However, we can use the compositions of the functions to make the determination as stated in the following theorem.

**Theorem: Verifying Whether $f$ and $g$ Are Inverses**

The functions $f$ and $g$ are inverses of each other if and only if

1. $g(f(x)) = x$ for every $x$ in the domain of $f$ and
2. $f(g(x)) = x$ for every $x$ in the domain of $g$.

## EXAMPLE 9 Using composition to verify inverse functions

Determine whether the functions $f(x) = x^3 - 1$ and $g(x) = \sqrt[3]{x + 1}$ are inverse functions.

### Solution

Find $f(g(x))$ and $g(f(x))$:

$$f(g(x)) = f(\sqrt[3]{x + 1}) = (\sqrt[3]{x + 1})^3 - 1 = x + 1 - 1 = x$$
$$g(f(x)) = g(x^3 - 1) = \sqrt[3]{x^3 - 1 + 1} = \sqrt[3]{x^3} = x$$

Since $f(g(x)) = x$ is true for any real number (the domain of $g$) and $g(f(x)) = x$ is true for any real number (the domain of $f$), the functions $f$ and $g$ are inverses of each other.

▶**TRY THIS.** Determine whether $f(x) = 2x + 1$ and $g(x) = \frac{x - 1}{2}$ are inverse functions.

# FUNCTION gallery... Some Inverse Functions

## Linear

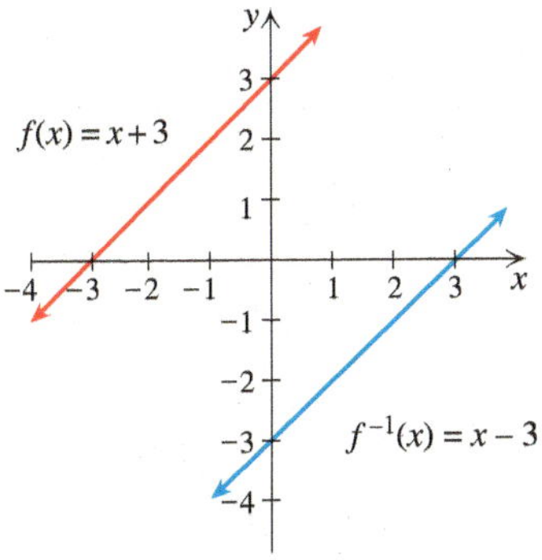


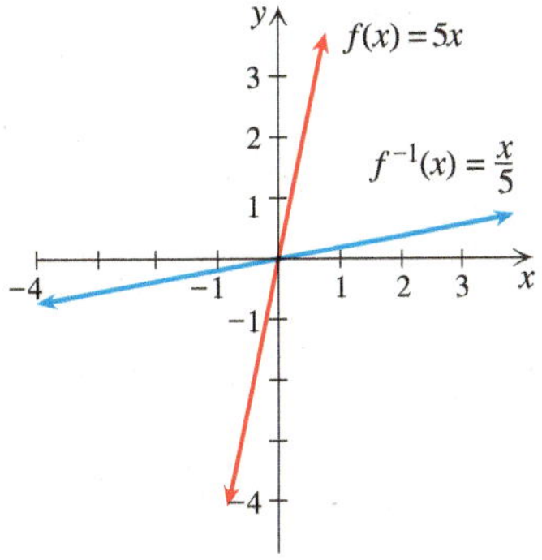


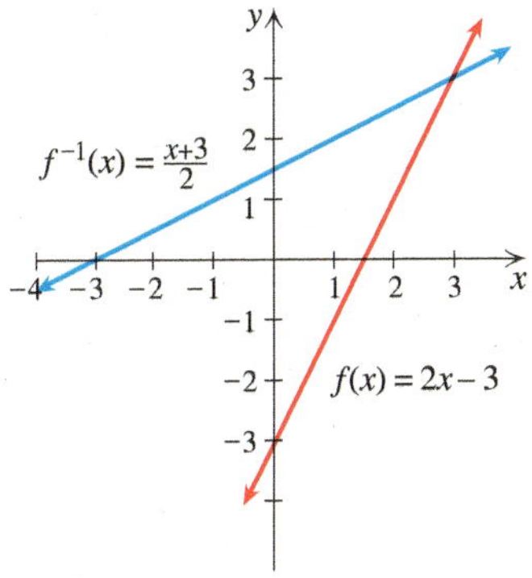


## Powers and Roots

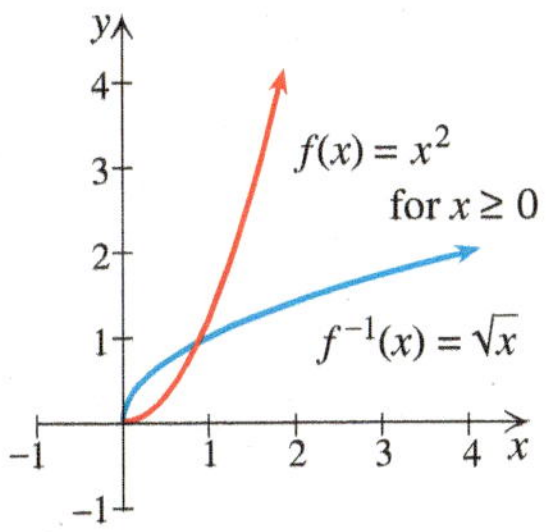


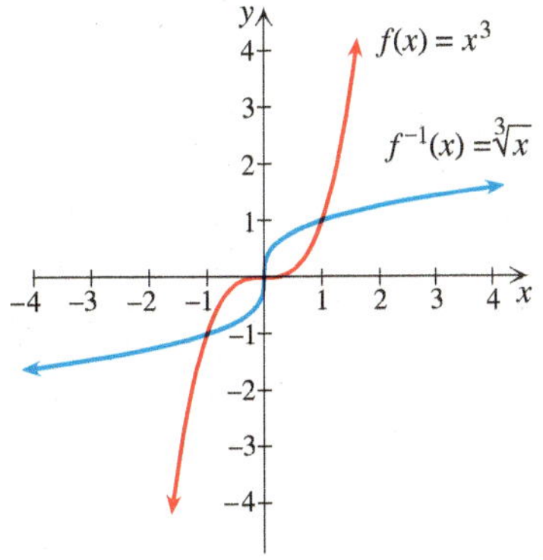


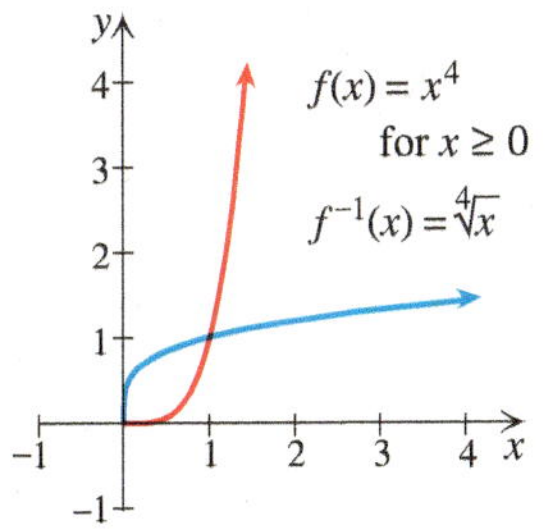


# FOR thought... True or False? Explain.

1. The inverse of the function $\{(2, 3), (5, 5)\}$ is $\{(5, 2), (5, 3)\}$.
2. The function $f(x) = -2$ is an invertible function.
3. If $g(x) = x^2$, then $g^{-1}(x) = \sqrt{x}$.
4. The only functions that are invertible are the one-to-one functions.
5. Every function has an inverse function.
6. The function $f(x) = x^4$ is invertible.
7. If $f(x) = 3\sqrt{x - 2}$, then $f^{-1}(x) = \frac{(x + 2)^2}{3}$ for $x \geq 0$.
8. If $f(x) = |x - 3|$, then $f^{-1}(x) = |x| + 3$.
9. According to the horizontal line test, $y = |x|$ is one-to-one.
10. The function $g(x) = -x$ is the inverse of the function $f(x) = -x$.

## EXERCISES 2.5

*Fill in the blank.*

**1.** If a function has no two ordered pairs with different first coordinates and the same second coordinate then the function is ________.

**2.** If a function is one-to-one, then it is ________.

**3.** If $f$ is one-to-one, then the function obtained by interchanging the coordinates in each ordered pair of $f$ is the ________ of $f$.

**4.** The graphs of $f$ and $f^{-1}$ are ________ with respect to the line $y = x$.

*Determine whether each function is one-to-one.*

**5.** $\{(3, 3), (5, 5), (6, 6), (9, 9)\}$

**6.** $\{(3, 4), (5, 6), (7, 8), (9, 10), (11, 15)\}$

**7.** $\{(-1, 1), (1, 1), (-2, 4), (2, 4)\}$

**8.** $\{(3, 2), (5, 2), (7, 2)\}$

**9.** $\{(1, 99), (2, 98), (3, 97), (4, 96), (5, 99)\}$

**10.** $\{(-1, 9), (-2, 8), (-3, 7), (1, 9), (2, 8), (3, 7)\}$

*Use the horizontal line test to determine whether each function is one-to-one.*

**11.** $f(x) = x^2 - 3x$

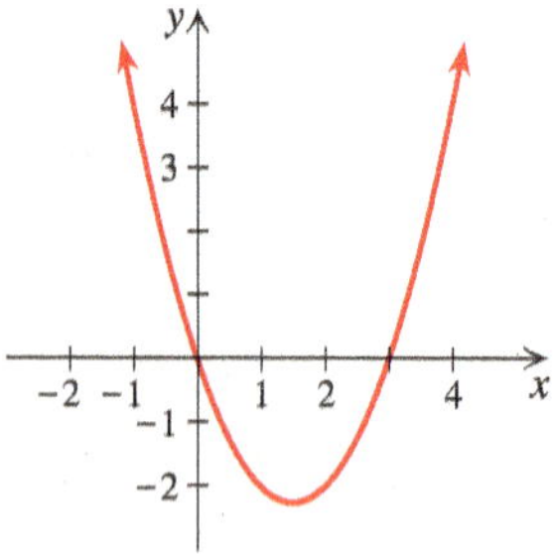

**12.** $g(x) = |x - 2| + 1$

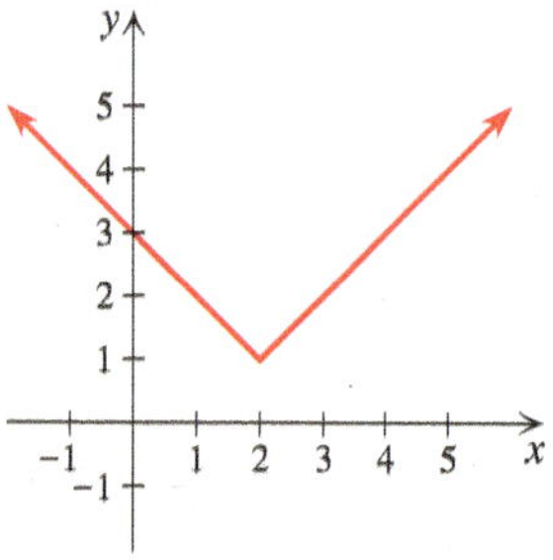

**13.** $y = \sqrt[3]{x} + 2$

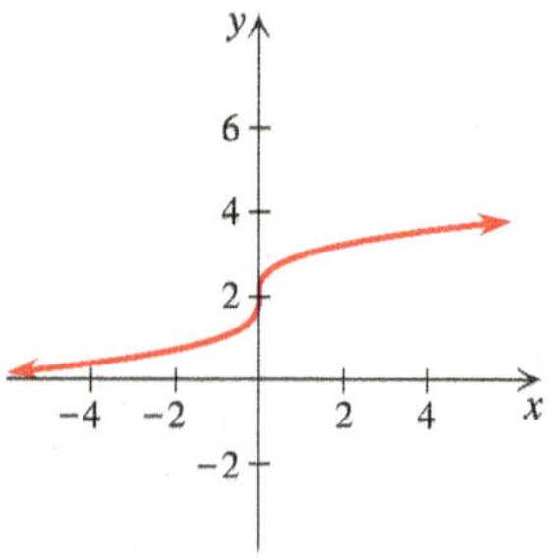

**14.** $y = (x - 2)^3$

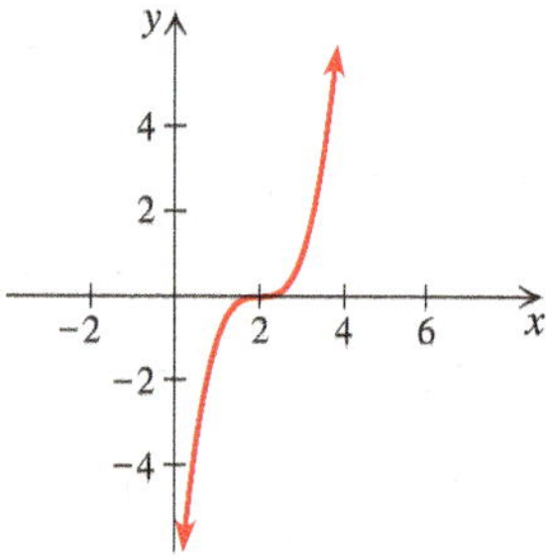

**15.** $y = x^3 - x$

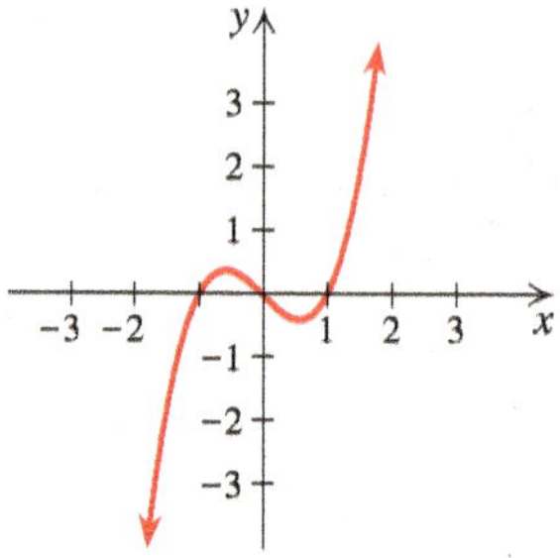

**16.** $y = \dfrac{1}{x}$

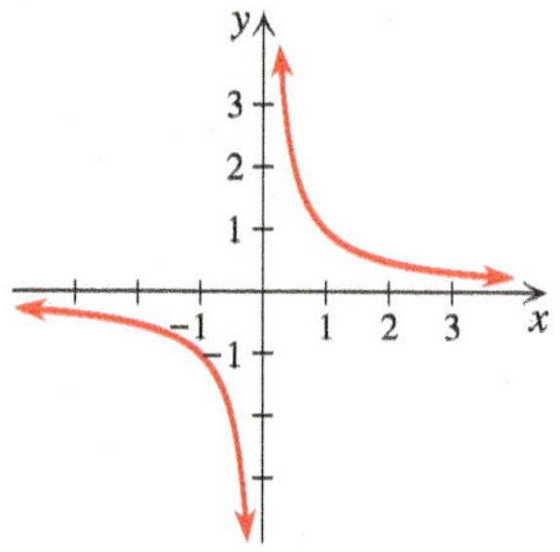

*Determine whether each function is one-to-one.*

**17.** $f(x) = 2x - 3$

**18.** $h(x) = 4x - 9$

**19.** $q(x) = \dfrac{1 - x}{x - 5}$

**20.** $g(x) = \dfrac{x + 2}{x - 3}$

**21.** $p(x) = |x + 1|$

**22.** $r(x) = 2|x - 1|$

**23.** $w(x) = x^2 + 3$

**24.** $v(x) = 2x^2 - 1$

**25.** $k(x) = \sqrt[3]{x + 9}$

**26.** $t(x) = \sqrt{x + 3}$

*Determine whether each function is invertible. If it is invertible, find the inverse.*

**27.** $\{(9, 3), (2, 2)\}$

**28.** $\{(4, 5), (5, 6)\}$

**29.** $\{(-1, 0), (1, 0), (5, 0)\}$

**30.** $\{(1, 2), (5, 2), (6, 7)\}$

**31.** $\{(3, 3), (2, 2), (4, 4), (7, 7)\}$

**32.** $\{(1, 1), (2, 4), (4, 16), (7, 49)\}$

**33.** $\{(1, 1), (2, 2), (4.5, 2), (5, 5)\}$

**34.** $\{(0, 2), (2, 0), (1, 0), (4, 6)\}$

*Determine whether each function is invertible and explain your answer.*

**35.** The function that pairs the universal product code of an item at Sears with a price.

**36.** The function that pairs the number of days since your birth with your age in years.

**37.** The function that pairs the length of a VCR tape in feet with the playing time in minutes.

**38.** The function that pairs the speed of your car in miles per hour with the speed in kilometers per hour.

**39.** The function that pairs the number of days of a hotel stay with the total cost for the stay.

**40.** The function that pairs the number of days that a deposit of $100 earns interest at 6% compounded daily with the amount of interest.

*For each function f, find $f^{-1}$, $f^{-1}(5)$, and $(f^{-1} \circ f)(2)$.*

**41.** $f = \{(2, 1), (3, 5)\}$

**42.** $f = \{(-1, 5), (0, 0), (2, 6)\}$

**43.** $f = \{(-3, -3), (0, 5), (2, -7)\}$

**44.** $f = \{(3.2, 5), (2, 1.99)\}$

*Determine whether each function is invertible by inspecting its graph on a graphing calculator.*

**45.** $f(x) = (x + 0.01)(x + 0.02)(x + 0.03)$

**46.** $f(x) = x^3 - 0.6x^2 + 0.11x - 0.006$

**47.** $f(x) = |x - 2| - |5 - x|$

**48.** $f(x) = \sqrt[3]{0.1x + 3} + \sqrt[3]{-0.1x}$

*Find the inverse of each function by reversing a composition.*

**49.** **a.** $f(x) = 5x + 1$ **b.** $f(x) = 3x - 88$

**c.** $f(x) = 3x - 7$ **d.** $f(x) = 4 - 3x$

**e.** $f(x) = \frac{1}{2}x - 9$ **f.** $f(x) = -x$

**g.** $f(x) = \sqrt[3]{x} - 9$ **h.** $f(x) = 3x^3 - 7$

**i.** $f(x) = \sqrt[3]{x - 1} + 5$ **j.** $f(x) = 2\sqrt[3]{x - 7}$

**50.** **a.** $f(x) = \frac{x}{2}$ **b.** $f(x) = x + 99$

**c.** $f(x) = 5x + 1$ **d.** $f(x) = 5 - 2x$

**e.** $f(x) = \frac{x}{3} + 6$ **f.** $f(x) = \frac{1}{x}$

**g.** $f(x) = \sqrt[3]{x - 9}$ **h.** $f(x) = -x^3 + 4$

**i.** $f(x) = 3\sqrt[3]{x + 4}$ **j.** $f(x) = \sqrt[3]{x + 3} - 9$

*Determine whether each pair of functions f and g are inverses of each other.*

**51.**

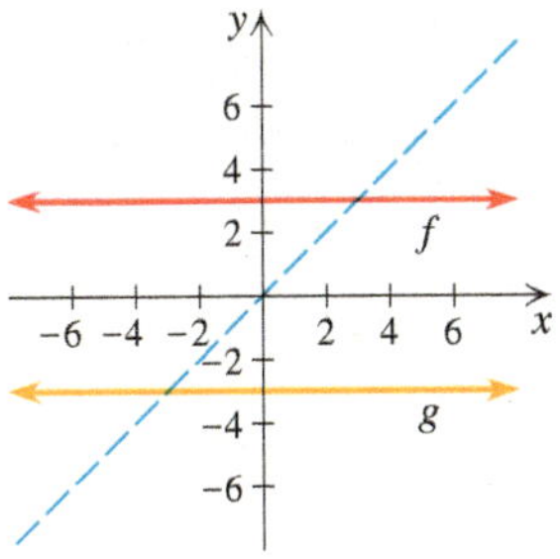


**52.**

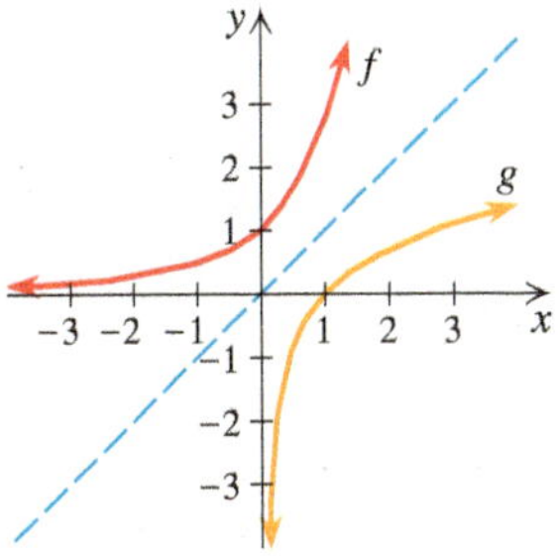


**53.**

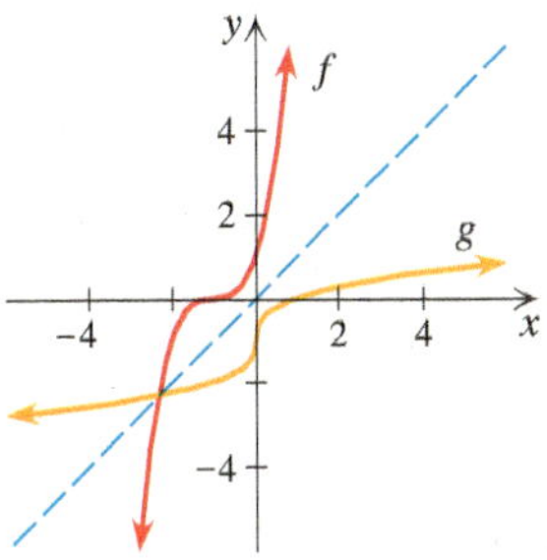


**54.**

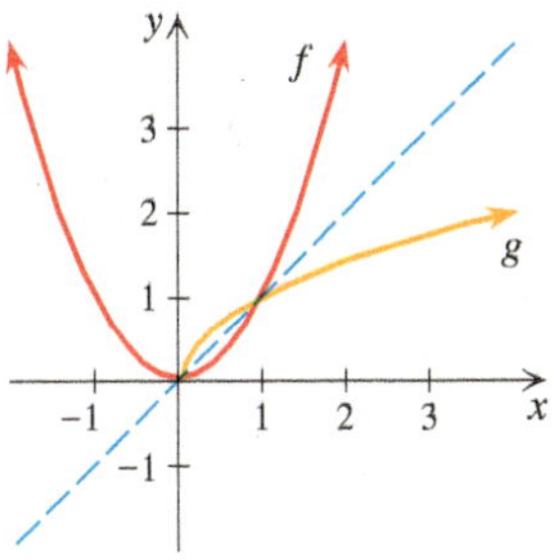


*For each function f, sketch the graph of $f^{-1}$.*

**55.**

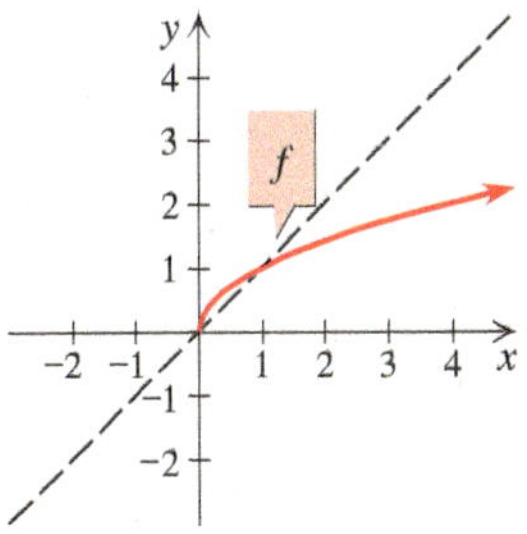


**56.**

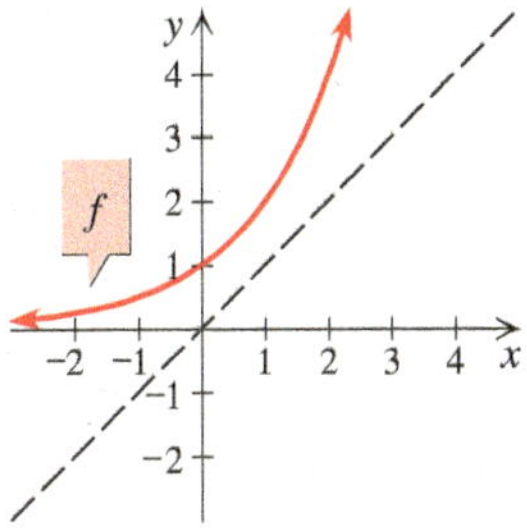


**57.**

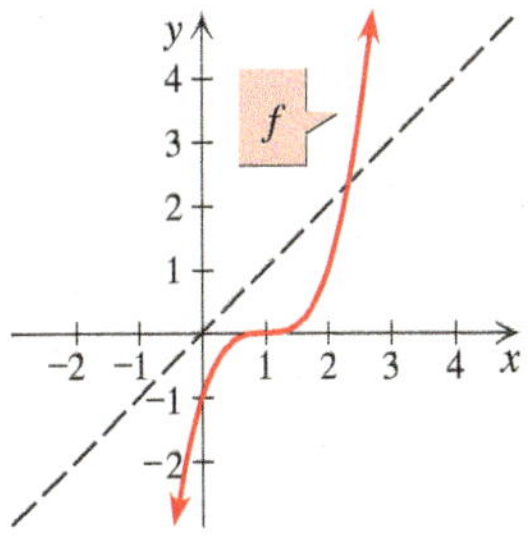


**58.**

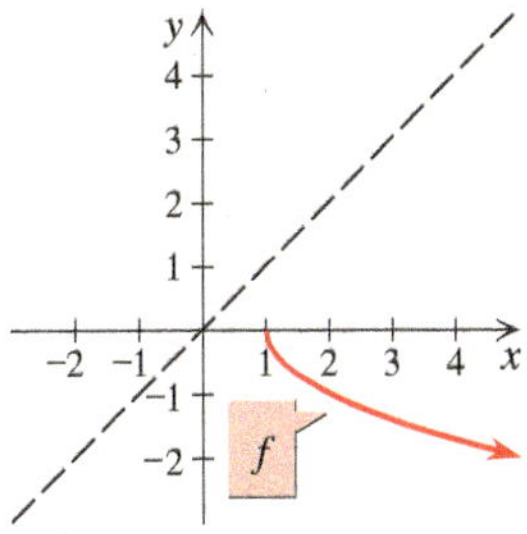


*Find the inverse of each function and graph both f and $f^{-1}$ on the same coordinate plane.*

**59.** $f(x) = 3x + 2$

**60.** $f(x) = -x - 8$

**61.** $f(x) = x^2 - 4$ for $x \geq 0$

**62.** $f(x) = 1 - x^2$ for $x \geq 0$

**63.** $f(x) = x^3$

**64.** $f(x) = -x^3$

**65.** $f(x) = \sqrt{x} - 3$

**66.** $f(x) = \sqrt{x - 3}$

*Find the inverse of each function using the procedure for the switch-and-solve method on page 163.*

**67.** $f(x) = 3x - 7$

**68.** $f(x) = -2x + 5$

**69.** $f(x) = 2 + \sqrt{x - 3}$

**70.** $f(x) = \sqrt{3x - 1}$

**71.** $f(x) = -x - 9$

**72.** $f(x) = -x + 3$

**73.** $f(x) = \frac{x + 3}{x - 5}$

**74.** $f(x) = \frac{2x - 1}{x - 6}$

**75.** $f(x) = -\frac{1}{x}$

**76.** $f(x) = x$

**77.** $f(x) = \sqrt[3]{x - 9} + 5$ **78.** $f(x) = \sqrt[3]{\dfrac{x}{2}} + 5$

**79.** $f(x) = (x - 2)^2$ for $x \geq 2$ **80.** $f(x) = x^2$ for $x \leq 0$

*In each case find $f(g(x))$ and $g(f(x))$. Then determine whether $g$ and $f$ are inverse functions.*

**81.** $f(x) = 4x + 4, g(x) = 0.25x - 1$

**82.** $f(x) = 20 - 5x, g(x) = -0.2x + 4$

**83.** $f(x) = x^2 + 1, g(x) = \sqrt{x - 1}$

**84.** $f(x) = \sqrt[4]{x}, g(x) = x^4$

**85.** $f(x) = \dfrac{1}{x} + 3, g(x) = \dfrac{1}{x - 3}$

**86.** $f(x) = 4 - \dfrac{1}{x}, g(x) = \dfrac{1}{4 - x}$

**87.** $f(x) = \sqrt[3]{\dfrac{x-2}{5}}, g(x) = 5x^3 + 2$

**88.** $f(x) = x^3 - 27, g(x) = \sqrt[3]{x} + 3$

*For each exercise, graph the three functions on the same screen of a graphing calculator. Enter these functions as shown without simplifying any expressions. Explain what these exercises illustrate.*

**89.** $y_1 = \sqrt[3]{x} - 1,\ y_2 = (x + 1)^3,\ y_3 = (\sqrt[3]{x} - 1 + 1)^3$

**90.** $y_1 = (2x - 1)^{1/3},\ y_2 = (x^3 + 1)/2,$
$y_3 = (2((x^3 + 1)/2) - 1)^{1/3}$

*Solve each problem.*

**91.** *Price of a Car* The tax on a new car is 8% of the purchase price $P$. Express the total cost $C$ as a function of the purchase price. Express the purchase price $P$ as a function of the total cost $C$.

**92.** *Volume of a Cube* Express the volume of a cube $V(x)$ as a function of the length of a side $x$. Express the length of a side of a cube $S(x)$ as a function of the volume $x$.

**93.** *Rowers and Speed* The world record times in the 2000-m race are a function of the number of rowers as shown in the figure (www.cbs.sportsline.com). If $r$ is the number of rowers and $t$ is the time in minutes, then the formula $t = -0.39r + 7.89$ models this relationship. Is this function invertible? Find the inverse. If the time for a 2000-m race was 5.55 min, then how many rowers were probably in the boat?

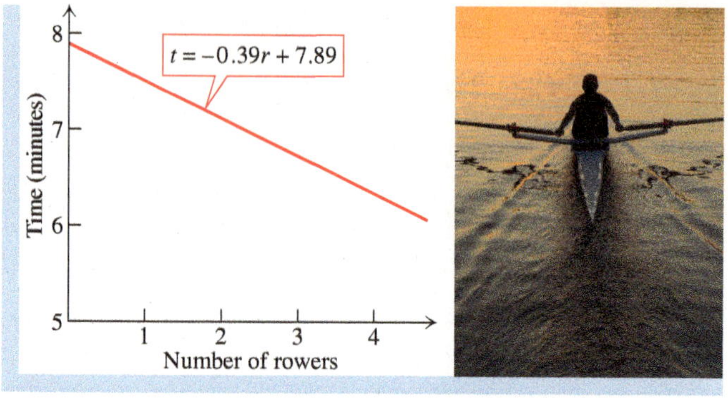


**Figure for Exercise 93**

**94.** *Temperature* The function $C = \frac{5}{9}(F - 32)$ expresses the Celsius temperature as a function of the Fahrenheit temperature. Find the inverse function. What is it used for?

**95.** *Landing Speed* The function $V = \sqrt{1.496w}$ expresses the landing speed $V$ (in feet per second) as a function of the gross weight (in pounds) of the Piper Cheyenne aircraft. Find the inverse function. Use the inverse function to find the gross weight for a Piper Cheyenne for which the proper landing speed is 115 ft/sec.

**96.** *Poiseuille's Law* Under certain conditions, the velocity $V$ of blood in a vessel at distance $r$ from the center of the vessel is given by $V = 500(5.625 \times 10^{-5} - r^2)$ where $0 \leq r \leq 7.5 \times 10^{-3}$. Write $r$ as a function of $V$.

**97.** *Depreciation Rate* The depreciation rate $r$ for a \$50,000 new car is given by the function $r = 1 - \left(\frac{V}{50{,}000}\right)^{1/5}$, where $V$ is the value of the car when it is five years old.

**a.** What is the depreciation rate for a \$50,000 BMW that is worth \$28,000 after 5 years?

**b.** Write $V$ as a function of $r$.

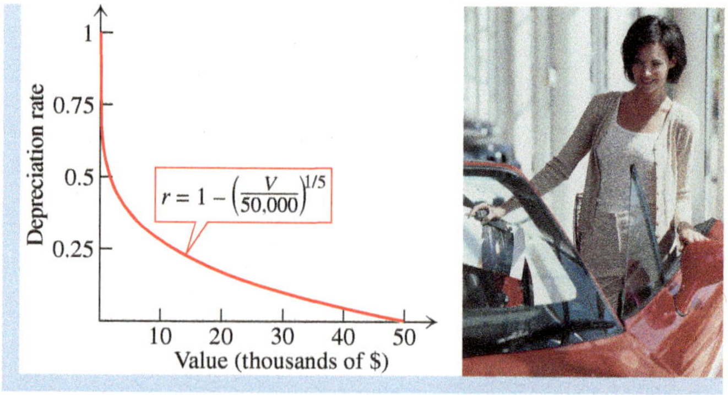


**Figure for Exercise 97**

**98.** *Annual Growth Rate* One measurement of the quality of a mutual fund is its average annual growth rate over the last 10 years. The function $r = \left(\frac{P}{10{,}000}\right)^{1/10} - 1$ expresses the average annual growth rate $r$ as a function of the present value $P$ of an investment of \$10,000 made 10 years ago. An investment of \$10,000 in Fidelity's Contrafund in 1998 was worth \$22,402 in 2008 (Fidelity Investments, www.fidelity.com). What was the annual growth rate for that period? Write $P$ as a function of $r$.

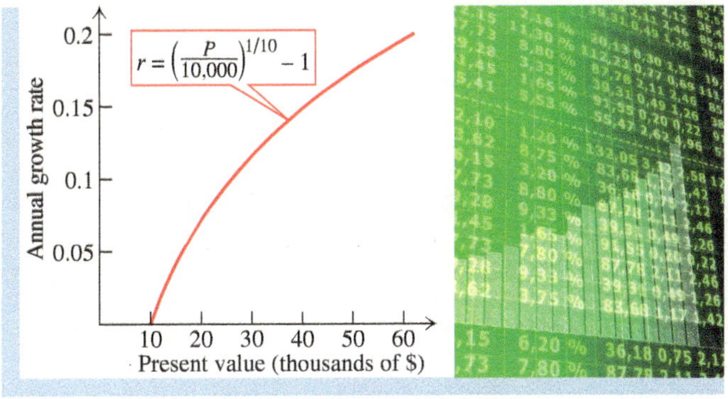


Figure for Exercise 98

## FOR WRITING/DISCUSSION

**99.** If $f(x) = 2x + 1$ and $g(x) = 3x - 5$, verify the equation $(f \circ g)^{-1} = g^{-1} \circ f^{-1}$.
HINT Find formulas for $f \circ g$, $(f \circ g)^{-1}$, $f^{-1}$, $g^{-1}$, and $g^{-1} \circ f^{-1}$.

**100.** Explain why $(f \circ g)^{-1} = g^{-1} \circ f^{-1}$ for any invertible functions $f$ and $g$. Discuss any restrictions on the domains and ranges of $f$ and $g$ for this equation to be correct.

**101.** Use a geometric argument to prove that if $a \neq b$, then $(a, b)$ and $(b, a)$ lie on a line perpendicular to the line $y = x$ and are equidistant from $y = x$.

**102.** Why is it difficult to find the inverse of a function such as $f(x) = (x - 3)/(x + 2)$ mentally?

**103.** Find an expression equivalent to $(x - 3)/(x + 2)$ in which $x$ appears only once. Explain.

**104.** Use the result of Exercise 103 to find the inverse of $f(x) = (x - 3)/(x + 2)$ mentally. Explain.

## ▶ RETHINKING

**105.** Let $f(x) = \dfrac{2x + 3}{5}$ and $g(x) = 5x - 9$. Find $(f \circ g)(2)$ and $(f \cdot g)(2)$.

**106.** The graph of $y = \sqrt{x}$ is stretched by a factor of 2, reflected in the $x$-axis, then translated 5 units to the right. Write the equation of the graph in its final position.

**107.** Let $f(x) = -\sqrt{9 - x^2}$. Find the domain, range, and the interval on which the function is increasing.

**108.** Is the set of ordered pairs $\{(1, 3), (3, 5), (9, 0), (1, 4)\}$ a function?

**109.** Solve $0.125x + 0.75 = 0.225x - 0.8$.

**110.** Find the equation of the line through $(2, 4)$ that is perpendicular to the line $2x - y = 19$. Write the answer in slope-intercept form.

## THINKING OUTSIDE THE BOX XVIII

*Costly Computers* A school district purchased $x$ computers at $y$ dollars each for a total of \$640,000. Both $x$ and $y$ are whole numbers and neither is a multiple of 10. Find the absolute value of the difference between $x$ and $y$.

## POP QUIZ 2.5

1. Is the function $\{(1, 3), (4, 5), (2, 3)\}$ invertible?
2. If $f = \{(5, 4), (3, 6), (2, 5)\}$, then what is $f^{-1}(5)$?
3. If $f(x) = 2x$, then what is $f^{-1}(8)$?
4. Is $f(x) = x^4$ a one-to-one function?
5. If $f(x) = 2x - 1$, then what is $f^{-1}(x)$?
6. If $g(x) = \sqrt[3]{x + 1} - 4$, then what is $g^{-1}(x)$?
7. Find $(h \circ j)(x)$ if $h(x) = x^3 - 5$ and $j(x) = \sqrt[3]{x + 5}$.

## LINKING concepts... For Individual or Group Explorations

### Pay Day Loans

*Star Financial Corporation offers short-term loans as shown in the accompanying advertisement. If you borrow P dollars for n days at the annual percentage rate r, then you pay back A dollars where*

$$A = P\left(1 + \frac{r}{365}\right)^n.$$

**a)** Solve the formula for $r$ in terms of $A$, $P$, and $n$.

**b)** Find the annual percentage rate (APR) for each of the loans described in the advertisement.

**c)** If you borrowed \$200, then how much would you have to pay back in one year using the APR that you determined in part (b)?

**d)** Find the current annual percentage rates for a car loan, house mortgage, and a credit card.

**e)** Why do you think Star Financial Corporation charges such high rates?

# 2.6 Constructing Functions with Variation

The area of a circle is a function of the radius, $A = \pi r^2$. As the values of $r$ change or vary, the values of $A$ vary also. Certain relationships are traditionally described by indicating how one variable changes according to the values of another variable. In this section we will construct functions for those relationships.

## Direct Variation

If we save 3 cubic yards of landfill space for every ton of paper that we recycle, then the volume $V$ of space saved is a linear function of the number of tons recycled $n$, $V = 3n$. If $n$ is 20 tons, then $V$ is 60 cubic yards. If $n$ is doubled to 40 tons, then $V$ is also doubled to 120 cubic yards. If $n$ is tripled, $V$ is also tripled. This relationship is called **direct variation.**

**Definition: Direct Variation**

The statement ***y* varies directly as *x*** or ***y* is directly proportional to *x*** means that

$$y = kx$$

for some fixed nonzero real number $k$. The constant $k$ is called the **variation constant** or **proportionality constant.**

Compare the landfill savings to a pizza that costs \$5 plus \$2 per topping. The cost is a linear function of the number of toppings $n$, $C = 2n + 5$. If $n = 2$,

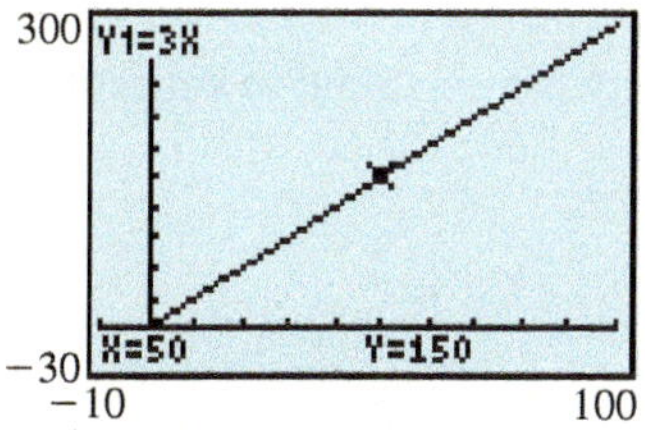


Figure 2.65

then $C = \$9$ and if $n = 4$, then $C = \$13$. The cost increases as $n$ increases, but doubling $n$ does not double $C$. The savings in landfill space is directly proportional to the number of tons of paper that is recycled, but the cost of the pizza is not directly proportional to the number of toppings. The graph of $V = 3n$ in Fig. 2.65 is a straight line through the origin. In general, if $y$ varies directly as $x$, then $y$ is a linear function of $x$ whose graph is a straight line through the origin. We are merely introducing some new terms to describe an old idea.

### EXAMPLE 1 Direct variation

The cost $C$ of a house in Wedgewood Estates is directly proportional to the size of the house $s$. If a 2850-square-foot house costs \$182,400, then what is the cost of a 3640-square-foot house?

**Solution**

Because $C$ is directly proportional to $s$, there is a constant $k$ such that

$$C = ks.$$

We can find the constant by using $C = \$182{,}400$ when $s = 2850$:

$$182{,}400 = k(2850)$$
$$64 = k$$

The cost for a house is \$64 per square foot. To find the cost of a 3640-square-foot house, use the formula $C = 64s$:

$$C = 64(3640)$$
$$= 232{,}960$$

The 3640-square-foot house costs \$232,960.

▶**TRY THIS.** The cost of a smoothie is directly proportional to its size. If a 12-ounce smoothie is \$3.60, then what is the cost of a 16-ounce smoothie? ■

Since $y = kx$ in a direct variation, we have $\frac{y}{x} = k$. So the ratio of $y$ to $x$ is constant. If $(x_1, y_1)$ and $(x_2, y_2)$ are two ordered pairs in a direct variation, then

$$\frac{y_1}{x_1} = \frac{y_2}{x_2}.$$

We could have used this proportion in Example 1. If a 2850-square-foot house costs \$182,400 and a 3640-square-foot house costs an unknown amount $C$, we have

$$\frac{182{,}400}{2850} = \frac{C}{3640}$$

or

$$C = \frac{182{,}400}{2850} \cdot 3640 = \$232{,}960.$$

Cruising speed: 675 mph

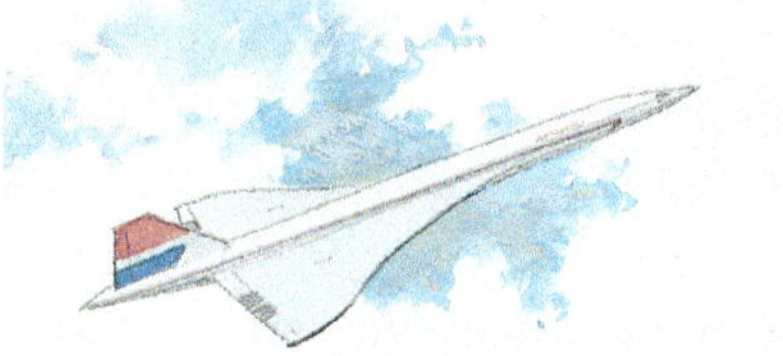

Cruising speed: 1350 mph

## Inverse Variation

The time that it takes to make the 3470-mile trip by air from New York to London is a function of the rate of the airplane, $T = 3470/R$. A Boeing 747 averaging 675 mph can make the trip in about 5 hours. The Concorde, traveling twice as fast at 1350 mph, made it in half the time, about 2.5 hours. If you travel three times as fast as the 747 you could make the trip in one-third the time. This relationship is called **inverse variation.**

### Definition: Inverse Variation

The statement ***y* varies inversely as *x*** or ***y* is inversely proportional to *x*** means that

$$y = \frac{k}{x}$$

for a fixed nonzero real number $k$.

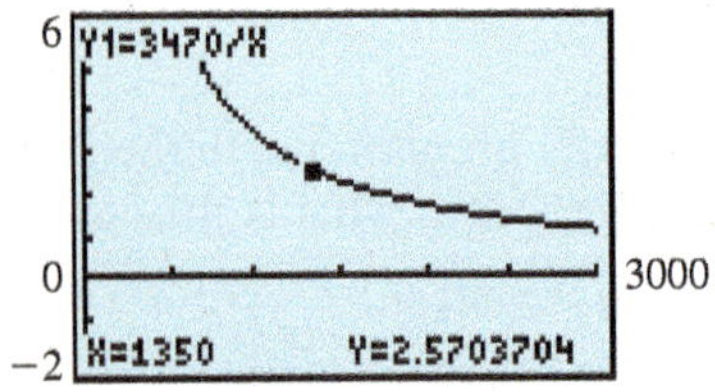


Figure 2.66

The graph of $T = 3470/R$ is shown in Fig. 2.66. As the speed increases the time decreases, but it does not reach zero for any speed.

### EXAMPLE 2 Inverse variation

The time it takes to remove all of the campaign signs in Bakersfield after an election varies inversely with the number of volunteers working on the job. If 12 volunteers could complete the job in 7 days, then how many days would it take for 15 volunteers?

**Solution**

Since the time $T$ varies inversely as the number of volunteers $n$, we have

$$T = \frac{k}{n}.$$

Since $T = 7$ when $n = 12$, we can find $k$:

$$7 = \frac{k}{12}$$

$$84 = k$$

So the formula is $T = 84/n$. Now if $n = 15$, we get

$$T = \frac{84}{15} = 5.6.$$

It would take 15 volunteers 5.6 days to remove all of the political signs.

▸**TRY THIS.** The time required to rake the grounds at Rockwood Manor varies inversely with the number of rakers. If 4 rakers can complete the job in 12 hours, then how long would it take for 6 rakers to complete the job? ■

Since $y = k/x$ in an inverse variation, we have $xy = k$. So the product of $x$ and $y$ is constant. If $(x_1, y_1)$ and $(x_2, y_2)$ are two ordered pairs in an inverse variation, then $x_1y_1 = x_2y_2$ or

$$\frac{y_1}{y_2} = \frac{x_2}{x_1}.$$

We could have used this inverse proportion in Example 2. If 12 volunteers take 7 days and 15 volunteers take $T$ days, then we have

$$\frac{7}{T} = \frac{15}{12}$$

or

$$T = \frac{7 \cdot 12}{15} = 5.6 \text{ days}.$$

## Joint Variation

We have been studying functions of one variable, but situations often arise in which one variable depends on the values of two or more variables. For example, the area of a rectangular room depends on the length *and* the width, $A = LW$. If carpeting costs \$25 per square yard, then the total cost is a function of the length and width, $C = 25LW$. This example illustrates **joint variation.**

**Definition: Joint Variation**

> The statement **$y$ varies jointly as $x$ and $z$** or **$y$ is jointly proportional to $x$ and $z$** means that
>
> $$y = kxz$$
>
> for a fixed nonzero real number $k$.

### EXAMPLE 3 Joint variation

The cost of constructing a 9-foot by 12-foot patio is \$734.40. If the cost varies jointly as the length and width, then what does an 8-foot by 14-foot patio cost?

**Solution**

Since the cost varies jointly as $L$ and $W$, we have $C = kLW$ for some constant $k$. Use $W = 9$, $L = 12$, and $C = \$734.40$ to find $k$:

$$734.40 = k(12)(9)$$
$$6.80 = k$$

So the formula is $C = 6.80LW$. Use $W = 8$ and $L = 14$ to find $C$:

$$C = 6.80(14)(8)$$
$$= 761.60$$

The cost of constructing an 8-foot by 14-foot patio is \$761.60.

▶**TRY THIS.** The cost of a fence that is 200 feet long and 5 feet high is \$3000. If the cost varies jointly with the length and height, then what is the cost of a fence that is 250 feet long and 6 feet high? ■

**HISTORICAL NOTE**

Johannes Kepler (1571–1630) was a German mathematician, astronomer, astrologer, and author of science fiction stories. He is best known for his laws of planetary motion. (See *Linking Concepts* at the end of this section for Kepler's harmonic law.) He also did fundamental work in the field of optics and helped to legitimize the telescopic discoveries of his contemporary Galileo Galilei.

## Combined Variation

Some relationships are combinations of direct and inverse variation. For example, the formula $V = kT/P$ expresses the volume of a gas in terms of its temperature $T$ and its pressure $P$. The volume varies directly as the temperature and inversely as the pressure. The next example gives more illustrations of combining direct, inverse, and joint variation.

### EXAMPLE 4 Constructing functions for combined variation

Write each sentence as a function involving a constant of variation $k$.

**a.** $y$ varies directly as $x$ and inversely as $z$.
**b.** $y$ varies jointly as $x$ and the square root of $z$.
**c.** $y$ varies jointly as $x$ and the square of $z$ and inversely as the cube of $w$.

**Solution**

**a.** $y = \frac{kx}{z}$ **b.** $y = kx\sqrt{z}$ **c.** $y = \frac{kxz^2}{w^3}$

▶**TRY THIS.** Suppose that $M$ varies directly as $w$ and the cube of $z$. Write a formula that describes this variation with $k$ as the constant. ■

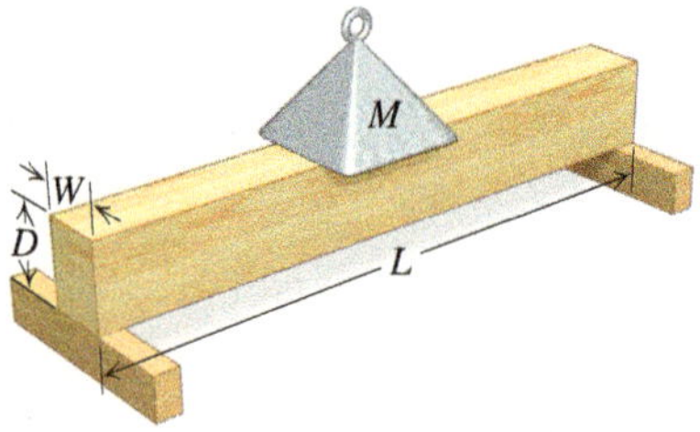


Figure 2.67

The language of variation evolved as a means of describing functions that involve only multiplication and/or division. The terms "varies directly," "directly proportional," and "jointly proportional" indicate multiplication. The term "inverse" indicates division. The term "variation" is *never* used in this text to refer to a formula involving addition or subtraction. Note that the formula for the cost of a pizza $C = 2n + 5$ is *not* the kind of formula that is described in terms of variation.

### EXAMPLE 5 Solving a combined variation problem

The recommended maximum load for the ceiling joist with rectangular cross section shown in Fig. 2.67 varies directly as the product of the width and the square of the depth of the cross section, and inversely as the length of the joist. If the recommended maximum load is 800 pounds for a 2-inch by 6-inch joist that is 12 feet long, then what is the recommended maximum load for a 2-inch by 12-inch joist that is 16 feet long?

#### Solution

The maximum load $M$ varies directly as the product of the width $W$ and the square of the depth $D$, and inversely as the length $L$:

$$M = \frac{kWD^2}{L}$$

Because the depth is squared, the maximum load is always obtained by positioning the joist so that the larger dimension is the depth. Use $W = 2$, $D = 6$, $L = 12$, and $M = 800$ to find $k$:

$$800 = \frac{k \cdot 2 \cdot 6^2}{12}$$

$$800 = 6k$$

$$\frac{400}{3} = k$$

Now use $W = 2$, $D = 12$, $L = 16$, and $k = 400/3$ in $M = \frac{kWD^2}{L}$ to find $M$:

$$M = \frac{400 \cdot 2 \cdot 12^2}{3 \cdot 16} = 2400$$

The maximum load on the joist is 2400 pounds. Notice that feet and inches were used in the formula without converting one to the other. If we solve the problem using only feet or only inches, then the value of $k$ is different but we get the same value for $M$.

▸**TRY THIS.** $W$ varies directly with $m$ and inversely as the square of $t$. If $W = 480$ when $m = 5$ and $t = 4$, then what is $W$ when $m = 3$ and $t = 6$? ■

## FOR thought... True or False? Explain.

1. If the cost of an 800 number is \$29.95 per month plus 28 cents per minute, then the monthly cost varies directly as the number of minutes.
2. If bananas are \$0.39 per pound, then your cost for a bunch of bananas varies directly with the number of bananas purchased.
3. The area of a circle varies directly as the square of the radius.
4. The area of a triangle varies jointly as the length of the base and the height.
5. The number $\pi$ is a constant of proportionality.

6. If $y$ is inversely proportional to $x$, then for $x = 0$, we get $y = 0$.

7. If the cost of potatoes varies directly with the number of pounds purchased, then the proportionality constant is the price per pound.

8. The height of a student in centimeters is directly proportional to the height of the same student in inches.

9. The surface area of a cube varies directly as the square of the length of an edge.

10. The surface area of a rectangular box varies jointly as its length, width, and height.

## EXERCISES 2.6

*Fill in the blank.*

1. If $y = kx$, then $y$ ________ as $x$.

2. If $y = kx$, then $k$ is the ________ constant.

3. If $y = k/x$, then $y$ ________ as $x$.

4. If $y = kxz$, then $y$ ________ as $x$ and $z$.

*Construct a function that expresses the relationship in each statement. Use k as the constant of variation.*

5. Your grade on the next test $G$ varies directly with the number of hours $n$ that you study for it.

6. The amount of sales tax on a new car is directly proportional to the purchase price of the car.

7. The volume of a gas in a cylinder $V$ is inversely proportional to the pressure on the gas $P$.

8. For two lines that are perpendicular and have slopes, the slope of one is inversely proportional to the slope of the other.

9. The cost of constructing a silo varies jointly as the height and the radius.

10. The volume of grain that a silo can hold varies jointly as the height and the square of the radius.

11. $Y$ varies directly as $x$ and inversely as the square root of $z$.

12. $W$ varies directly with $r$ and $t$ and inversely with $v$.

*For each formula that illustrates a type of variation, write the formula in words using the vocabulary of variation. The letters k and π represent constants, and all other letters represent variables.*

13. $A = \pi r^2$

14. $C = \pi D$

15. $y = \dfrac{1}{x}$

16. $m_1 = \dfrac{-1}{m_2}$

17. $T = 3W + 5$

18. $y = k\sqrt{x - 3}$

19. $a = kzw$

20. $V = LWH$

21. $H = \dfrac{k\sqrt{t}}{s}$

22. $B = \dfrac{y^2}{2\sqrt{x}}$

23. $D = \dfrac{2LJ}{W}$

24. $E = mc^2$

*Find the constant of variation and construct the function that is expressed in each statement.*

25. $y$ varies directly as $x$, and $y = 5$ when $x = 9$.

26. $h$ is directly proportional to $z$, and $h = 210$ when $z = 200$.

27. $T$ is inversely proportional to $y$, and $T = -30$ when $y = 5$.

28. $H$ is inversely proportional to $n$, and $H = 9$ when $n = -6$.

29. $m$ varies directly as the square of $t$, and $m = 54$ when $t = 3\sqrt{2}$.

30. $p$ varies directly as the cube root of $w$, and $p = \sqrt[3]{2}/2$ when $w = 4$.

31. $y$ varies directly as $x$ and inversely as the square root of $z$, and $y = 2.192$ when $x = 2.4$ and $z = 2.25$.

32. $n$ is jointly proportional to $x$ and the square root of $b$, and $n = -18.954$ when $x = -1.35$ and $b = 15.21$.

*Solve each variation problem.*

33. If $y$ varies directly as $x$, and $y = 9$ when $x = 2$, what is $y$ when $x = -3$?

34. If $y$ is directly proportional to $z$, and $y = 6$ when $z = \sqrt{12}$, what is $y$ when $z = \sqrt{75}$?

35. If $P$ is inversely proportional to $w$, and $P = 2/3$ when $w = 1/4$, what is $P$ when $w = 1/6$?

36. If $H$ varies inversely as $q$, and $H = 0.03$ when $q = 0.01$, what is $H$ when $q = 0.05$?

37. If $A$ varies jointly as $L$ and $W$, and $A = 30$ when $L = 3$ and $W = 5\sqrt{2}$, what is $A$ when $L = 2\sqrt{3}$ and $W = 1/2$?

38. If $J$ is jointly proportional to $G$ and $V$, and $J = \sqrt{3}$ when $G = \sqrt{2}$ and $V = \sqrt{8}$, what is $J$ when $G = \sqrt{6}$ and $V = 8$?

**39.** If $y$ is directly proportional to $u$ and inversely proportional to the square of $v$, and $y = 7$ when $u = 9$ and $v = 6$, what is $y$ when $u = 4$ and $v = 8$?

**40.** If $q$ is directly proportional to the square root of $h$ and inversely proportional to the cube of $j$, and $q = 18$ when $h = 9$ and $j = 2$, what is $q$ when $h = 16$ and $j = 1/2$?

*Determine whether the first variable varies directly or inversely with the other variable. Construct a function for the variation using the appropriate variation constant.*

**41.** The length of a car in inches, the length of the same car in feet

**42.** The time in seconds to pop a bag of microwave popcorn, the time in minutes to pop the same bag of popcorn

**43.** The price for each person sharing a \$20 pizza, the number of hungry people

**44.** The number of identical steel rods with a total weight of 40,000 lb, the weight of one of those rods

**45.** The speed of a car in miles per hour, the speed of the same car at the same time in kilometers per hour

**46.** The weight of a sumo wrestler in pounds, the weight of the same sumo wrestler in kilograms. (Chad Rowan of the United States, at 455 lb, or 206 kg, was the first-ever foreign grand champion in Tokyo's annual tournament.)

**Figure for Exercise 46**

**47.** The temperature in degrees Celsius, the temperature at the same time and place in degrees Fahrenheit

**48.** The perimeter of a rectangle with a length of 4 ft, the width of the same rectangle

**49.** The area of a rectangle with a length of 30 in., the width of the same rectangle

**50.** The area of a triangle with a height of 10 cm, the base of the same triangle

**51.** The number of gallons of gasoline that you can buy for \$50, the price per gallon of that same gasoline

**52.** The length of Yolanda's 40 ft$^2$ closet, the width of that closet

*Solve each problem.*

**53.** *First-Class Diver* The pressure exerted by water at a point below the surface varies directly with the depth. The pressure is 4.34 lb/in.$^2$ at a depth of 10 ft. What pressure does the sperm whale (the deepest diver among the air-breathing mammals) experience when it dives 6000 ft. below the surface?

**54.** *Second Place* The elephant seal (second only to the sperm whale) experiences a pressure of 2170 lb/in.$^2$ when it makes its deepest dives. At what depth does the elephant seal experience this pressure? See the previous exercise.

**55.** *Processing Oysters* The time required to process a shipment of oysters varies directly with the number of pounds in the shipment and inversely with the number of workers assigned. If 3000 lb can be processed by 6 workers in 8 hr, then how long would it take 5 workers to process 4000 lb?

**56.** *View from an Airplane* The view $V$ from the air is directly proportional to the square root of the altitude $A$. If the view from horizon to horizon at an altitude of 16,000 ft is approximately 154 mi, then what is the view from 36,000 ft?

**57.** *Simple Interest* The amount of simple interest on a deposit varies jointly with the principal and the time in days. If \$20.80 is earned on a deposit of \$4000 for 16 days, how much interest would be earned on a deposit of \$6500 for 24 days?

**58.** *Free Fall at Six Flags* Visitors at Six Flags in Atlanta get a brief thrill in the park's "free fall" ride. Passengers seated in a small cab drop from a 10-story tower down a track that eventually curves level with the ground. The distance the cab falls varies directly with the square of the time it is falling. If the cab falls 16 ft in the first second, then how far has it fallen after 2 sec?

**59.** *Cost of Plastic Sewer Pipe* The cost of a plastic sewer pipe varies jointly as its diameter and its length. If a 20-ft pipe with a diameter of 6 in. costs \$18.60, then what is the cost of a 16-ft pipe with a diameter of 8 in.?

**60.** *Cost of Copper Tubing* A plumber purchasing materials for a renovation observed that the cost of copper tubing varies jointly as the length and the diameter of the tubing. If 20 ft of $\frac{1}{2}$-in.-diameter tubing costs \$36.60, then what is the cost of 100 ft of $\frac{3}{4}$-in.-diameter tubing?

**61.** *Weight of a Can* The weight of a can of baked beans varies jointly with the height and the square of the diameter. If a 4-in.-high can with a 3-in. radius weighs 14.5 oz, then what is the weight of a 5-in.-high can with a diameter of 6 in.?

**62.** *Nitrogen Gas Shock Absorber* The volume of gas in a nitrogen gas shock absorber varies directly with the temperature of the gas and inversely with the amount of weight on the piston. If the volume is 10 in.$^3$ at 80°F with a weight of 600 lb, then what is the volume at 90°F with a weight of 800 lb?

**63.** *Velocity of Underground Water* Darcy's law states that the velocity $V$ of underground water through sandstone varies directly as the head $h$ and inversely as the length $l$ of the flow. The head is the vertical distance between the point of intake into the rock and the point of discharge such as a spring, and the length is the length of the flow from intake to discharge. In a certain sandstone a velocity of 10 ft per year has been recorded with a head of 50 ft and length of 200 ft. What would we expect the velocity to be if the head is 60 ft and the length is 300 ft?

**64.** *Cross-Sectional Area of a Well* The rate of discharge of a well, $V$, varies jointly as the hydraulic gradient, $i$, and the cross-sectional area of the well wall, $A$. Suppose that a well with a cross-sectional area of 10 $ft^2$ discharges 3 gal of water per minute in an area where the hydraulic gradient is 0.3. If we dig another well nearby where the hydraulic gradient is 0.4, and we want a discharge of 5 gal/min, then what should be the cross-sectional area for the well?

**65.** *Dollars Per Death* The accompanying table shows the annual amount of money that the National Institutes of Health spends on disease research and the annual number of deaths for three diseases (www.nih.gov). Is the annual spending directly proportional to the annual number of deaths? Should it be?

**Table for Exercise 65**

| Disease | Research Spending | Annual Deaths |
|---|---|---|
| AIDS | \$1.34 billion | 42,506 |
| Diabetes | \$295 million | 59,085 |
| Heart disease | \$958 million | 738,781 |

**66.** *Bicycle Gear Ratio* The gear ratio $r$ for a bicycle varies jointly with the number of teeth on the chainring $n$, by the pedals, and the diameter of the wheel $w$ (in inches), and inversely with the number of teeth on the cog $c$, by the wheel (www.harriscyclery.com). Find the missing entries in the accompanying table.

**Table for Exercise 66**

| $n$ | $w$ | $c$ | $r$ |
|---|---|---|---|
| 50 | 27 | 25 | 54 |
| 40 | 26 | 13 | |
| 45 | 27 | | 67.5 |

**67.** *Grade on an Algebra Test* Calvin believes that his grade on a college algebra test varies directly with the number of hours spent studying during the week prior to the test and inversely with the number of hours spent at the Beach Club playing volleyball during the week prior to the test. If he scored 76 on a test when he studied 12 hr and played 10 hr during the week prior to the test, then what score should he expect if he studies 9 hr and plays 15 hr?

**68.** *Carpeting a Room* The cost of carpeting a room varies jointly as the length and width. Julie advertises a price of \$263.40 to carpet a 9-ft by 12-ft room with Dupont Stainmaster. Julie gave a customer a price of \$482.90 for carpeting a room with a width of 4 yd with that same carpeting, but she has forgotten the length of the room. What is the length of the room?

**69.** *Pole-Vault Principle* The height a pole-vaulter attains is directly proportional to the square of his velocity on the runway. Given that a speed of 32 ft/sec will loft a vaulter to 16 ft, find the speed necessary for world record holder Sergie Bubka to reach a record height of 20 ft 2.5 in.

## FOR WRITING/DISCUSSION

**70.** *Cooperative Learning* Working in groups, drop a ball from various heights and measure the distance that the ball rebounds after hitting the floor. Enter your pairs of data into a graphing calculator and calculate the regression line. What is the value of the correlation coefficient, $r$? Do you think that the rebound distance is directly proportional to the drop distance? Compare your results with other groups.

**71.** *Moving Melons* To move more watermelons, a retailer decided to make the price of each watermelon purchased by a single customer inversely proportional to the number of watermelons purchased by the customer. Is this a good idea? Explain.

**72.** *Common Units* In general we do not perform computations involving inches and feet until they are converted to a common unit of measure. Explain how it is possible to use inches and feet in Example 5 without converting to a common unit of measure.

## ▶RETHINKING

**73.** Find the inverse of the function $f(x) = \sqrt[3]{x - 9} + 1$.

**74.** Write a formula that expresses the diagonal $d$ of a square as a function of the area $A$.

**75.** Galdino drove his truck from 8 A.M. to 11 A.M. in the rain. From 11 A.M. to 4 P.M. the skies were clear and he averaged 5 mph more than he did in the rain. If the total distance traveled was 425 miles, then what was his average speed in the rain?

**76.** Determine the symmetry of the graph of the function $f(x) = x^3 - 8x$.

**77.** Find the equation of the line through $(-4, 2)$ that is parallel to the line $5x - 10y = 31$. Write the answer in standard form.

**78.** Solve $|3x - 9| < 6$.

## THINKING OUTSIDE THE BOX XIX

*Good Timing* Sharon leaves at the same time every day and walks to her 8 o'clock class. On Monday she averaged 4 miles per hour and was one minute late. On Wednesday she averaged 5 miles per hour and was one minute early. Find the exact speed that she should walk on Friday to get to class on time.

## POP QUIZ 2.6

1. Find the constant of variation if $y$ varies directly as $x$ and $y = 4$ when $x = 20$.

2. If $a$ varies inversely as $b$ and the constant is 10, then what is $a$ when $b = 2$?

3. The cost of a round rug varies directly as the square of its radius. If a rug with a 3-foot radius costs \$108, then what is the cost of a rug with a 4-foot radius?

4. The cost of a rectangular tapestry varies jointly as its length and width. If a $6 \times 2$ foot tapestry costs \$180, then what is the cost of a $5 \times 4$ foot tapestry?

## LINKING concepts... For Individual or Group Explorations

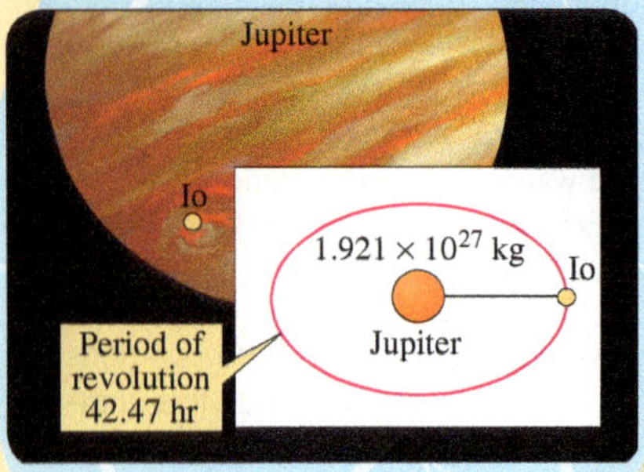


### Determining the Mass of a Planet

*A result of Kepler's harmonic law is that the mass of a planet with a satellite is directly proportional to the cube of the mean distance from the satellite to the planet, and inversely proportional to the square of the period of revolution. Early astronomers estimated the mass of the earth to be* $5.976 \times 10^{24}$ kg *and observed that the moon orbited the earth with a period of* 27.322 *days at a mean distance of* $384.4 \times 10^3$ km.

**a)** Write a formula for the mass of a planet according to Kepler.

**b)** Find the proportionality constant using the observations and estimates for the earth and the moon.

**c)** Extend your knowledge across the solar system to find the mass of Mars based on observations that Phobos orbits Mars in 7.65 hr at a mean distance of 9330 km.

**d)** What is the approximate ratio of the mass of the earth to the mass of Mars?

**e)** If the mass of Jupiter is $1.921 \times 10^{27}$ kg and the period of revolution of Io around Jupiter is 42.47 hr, then what is the mean distance between Io and Jupiter?

# Highlights

## 2.1 Functions

| | | |
|---|---|---|
| **Relation** | Any set of ordered pairs | $\{(1, 5), (1, 3), (3, 5)\}$ |
| **Function** | A relation in which no two ordered pairs have the same first coordinate and different second coordinates | $\{(1, 5), (2, 3), (3, 5)\}$ |
| **Vertical Line Test** | If no vertical line crosses a graph more than once, then the graph is a function. | Function 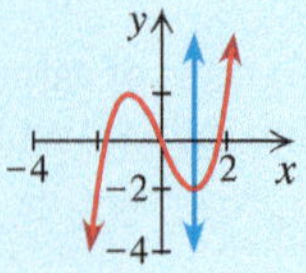 Not a function 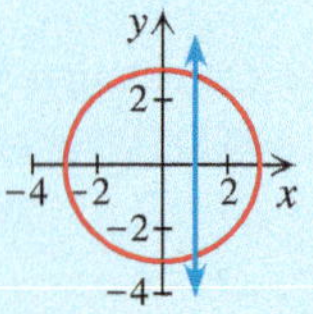 |

| | | |
|---|---|---|
| **Average Rate of Change** | The slope of the line through $(a, f(a))$ and $(b, f(b))$: $\frac{f(b) - f(a)}{b - a}$ | Average rate of change of $f(x) = x^2$ on $[2, 9]$ is $\frac{9^2 - 2^2}{9 - 2}$. |
| **Difference Quotient** | Average rate of change of $f$ on $[x, x + h]$: $\frac{f(x + h) - f(x)}{h}$ | $f(x) = x^2$, difference quotient $= \frac{(x + h)^2 - x^2}{h} = 2x + h$ |

## 2.2 Graphs of Relations and Functions

| | | |
|---|---|---|
| **Graph of a Relation** | An illustration of all ordered pairs of a relation | Graph of $y = x + 1$ shows all ordered pairs in this function. |
| **Circle** | A circle is not the graph of a function. | Since $(0, \pm 2)$ satisfies $x^2 + y^2 = 4$, it is not a function. |
| **Increasing, Decreasing, or Constant** | When going from left to right, a function is increasing if its graph is rising, decreasing if its graph is falling, constant if its graph is staying the same. | $y = \lvert x \rvert$ is increasing on $[0, \infty)$ and decreasing on $(-\infty, 0]$. $y = 5$ is constant on $(-\infty, \infty)$. |

## 2.3 Families of Functions, Transformations, and Symmetry

| | | |
|---|---|---|
| **Transformations of $y = f(x)$** | Horizontal: $y = f(x - h)$<br>Vertical: $y = f(x) + k$<br>Stretching: $y = af(x)$ for $a > 1$<br>Shrinking: $y = af(x)$ for $0 < a < 1$<br>Reflection: $y = -f(x)$ | $y = (x - 4)^2$<br>$y = x^2 + 9$<br>$y = 3x^2$<br>$y = 0.5x^2$<br>$y = -x^2$ |
| **Family of Functions** | All functions of the form $f(x) = af(x - h) + k$ $(a \neq 0)$ for a given function $y = f(x)$ | The square root family: $y = a\sqrt{x - h} + k$ |
| **Even Function** | Graph is symmetric with respect to $y$-axis. $f(-x) = f(x)$ | $f(x) = x^2$, $g(x) = \lvert x \rvert$ |
| **Odd Function** | Graph is symmetric about the origin. $f(-x) = -f(x)$ | $f(x) = x$, $g(x) = x^3$ |
| **Inequalities** | $f(x) > 0$ is satisfied on all intervals where the graph of $y = f(x)$ is above the $x$-axis.<br>$f(x) < 0$ is satisfied on all intervals where the graph of $y = f(x)$ is below the $x$-axis. | $4 - x^2 > 0$ on $(-2, 2)$<br>$4 - x^2 < 0$ on $(-\infty, -2) \cup (2, \infty)$ |

## 2.4 Operations with Functions

| | | |
|---|---|---|
| **Sum** | $(f + g)(x) = f(x) + g(x)$ | $f(x) = x^2 - 4$, $g(x) = x + 2$<br>$(f + g)(x) = x^2 + x - 2$ |
| **Difference** | $(f - g)(x) = f(x) - g(x)$ | $(f - g)(x) = x^2 - x - 6$ |
| **Product** | $(f \cdot g)(x) = f(x) \cdot g(x)$ | $(f \cdot g)(x) = x^3 + 2x^2 - 4x - 8$ |
| **Quotient** | $(f/g)(x) = f(x)/g(x)$ | $(f/g)(x) = x - 2$ |
| **Composition** | $(f \circ g)(x) = f(g(x))$ | $(f \circ g)(x) = (x + 2)^2 - 4$<br>$(g \circ f)(x) = x^2 - 2$ |

## 2.5 Inverse Functions

| | | |
|---|---|---|
| **One-to-One Function** | A function that has no two ordered pairs with different first coordinates and the same second coordinate | $\{(1, 2), (3, 5), (6, 9)\}$ <br> $g(x) = x + 3$ is one-to-one. <br> $f(x) = x^2$ is not one-to-one. |
| **Inverse Function** | A one-to-one function has an inverse. The inverse function has the same ordered pairs, with the coordinates reversed. | $f = \{(1, 2), (3, 5), (6, 9)\}$ <br> $f^{-1} = \{(2, 1), (5, 3), (9, 6)\}$ <br> $g(x) = x + 3, g^{-1}(x) = x - 3$ |
| **Horizontal Line Test** | If there is a horizontal line that crosses the graph of $f$ more than once, then $f$ is not invertible. | $y = 4$ crosses $f(x) = x^2$ twice, so $f$ is not invertible. |
| **Graph of $f^{-1}$** | Reflect the graph of $f$ about the line $y = x$ to get the graph of $f^{-1}$. | 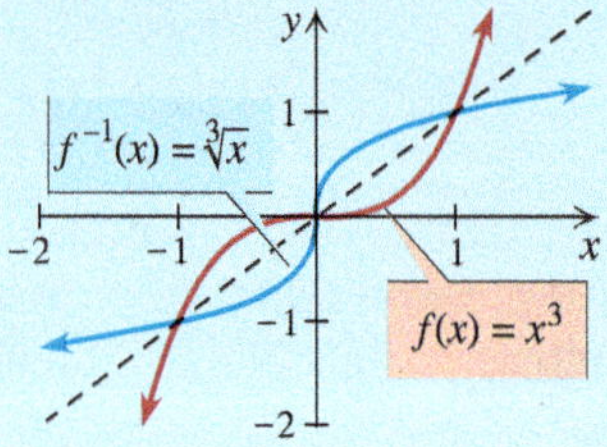 |



## 2.6 Constructing Functions with Variation

| | | |
|---|---|---|
| **Direct Variation** | $y = kx$ for a constant $k \neq 0$. | $D = 40T$ |
| **Inverse Variation** | $y = k/x$ for a constant $k \neq 0$. | $T = 200/R$ |
| **Joint Variation** | $y = kxz$ for a constant $k \neq 0$. | $C = 25LW$ |

# Chapter 2 Review Exercises

*Graph each set of ordered pairs. State the domain and range of each relation. Determine whether each set is a function.*

1. $\{(0, 0), (1, 1), (-2, -2)\}$
2. $\{(0, -3), (1, -1), (2, 1), (2, 3)\}$
3. $\{(x, y) \mid y = 3 - x\}$
4. $\{(x, y) \mid 2x + y = 5\}$
5. $\{(x, y) \mid x = 2\}$
6. $\{(x, y) \mid y = 3\}$
7. $\{(x, y) \mid x^2 + y^2 = 0.01\}$
8. $\{(x, y) \mid x^2 + y^2 = 2x - 4y\}$
9. $\{(x, y) \mid x = y^2 + 1\}$
10. $\{(x, y) \mid y = |x - 2|\}$
11. $\{(x, y) \mid y = \sqrt{x} - 3\}$
12. $\{(x, y) \mid x = \sqrt{y}\}$

*Let $f(x) = x^2 + 3$ and $g(x) = 2x - 7$. Find and simplify each expression.*

13. $f(-3)$
14. $g(3)$
15. $g(12)$
16. $f(-1)$
17. $x$, if $f(x) = 19$
18. $x$, if $g(x) = 9$
19. $(g \circ f)(-3)$
20. $(f \circ g)(3)$
21. $(f + g)(2)$
22. $(f - g)(-2)$
23. $(f \cdot g)(-1)$
24. $(f/g)(4)$
25. $f(g(2))$
26. $g(f(-2))$
27. $(f \circ g)(x)$
28. $(g \circ f)(x)$
29. $(f \circ f)(x)$
30. $(g \circ g)(x)$
31. $f(a + 1)$
32. $g(a + 2)$
33. $\dfrac{f(3 + h) - f(3)}{h}$
34. $\dfrac{g(5 + h) - g(5)}{h}$
35. $\dfrac{f(x + h) - f(x)}{h}$
36. $\dfrac{g(x + h) - g(x)}{h}$
37. $g\left(\dfrac{x + 7}{2}\right)$
38. $f(\sqrt{x - 3})$
39. $g^{-1}(x)$
40. $g^{-1}(-3)$

*Use transformations to graph each pair of functions on the same coordinate plane.*

41. $f(x) = \sqrt{x}, g(x) = 2\sqrt{x + 3}$
42. $f(x) = \sqrt{x}, g(x) = -2\sqrt{x} + 3$

**43.** $f(x) = |x|, g(x) = -2|x + 2| + 4$

**44.** $f(x) = |x|, g(x) = \frac{1}{2}|x - 1| - 3$

**45.** $f(x) = x^2, g(x) = \frac{1}{2}(x - 2)^2 + 1$

**46.** $f(x) = x^2, g(x) = -2x^2 + 4$

*Let $f(x) = x - 4$, $g(x) = \sqrt{x}$ and $h(x) = x^2$. Find each composition and state its domain.*

**47.** $f \circ g$

**48.** $g \circ f$

**49.** $f \circ h$

**50.** $h \circ f$

**51.** $g \circ f \circ h$

**52.** $h \circ f \circ g$

*For each exercise, graph the function by transforming the given graph of $y = f(x)$.*

**53.** $y = 2f(x - 2) + 1$

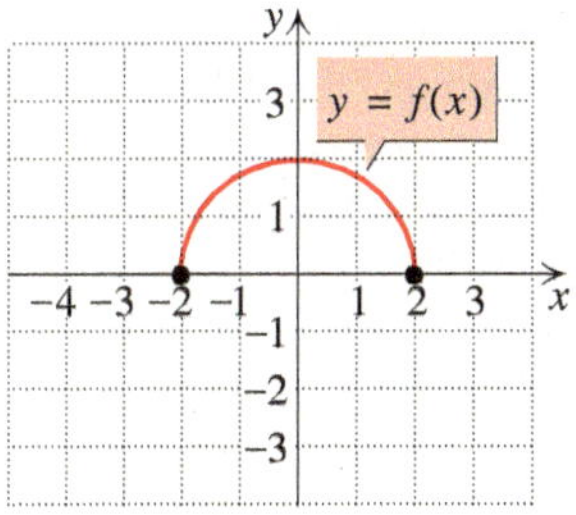


**54.** $y = 2f(x + 3) - 1$

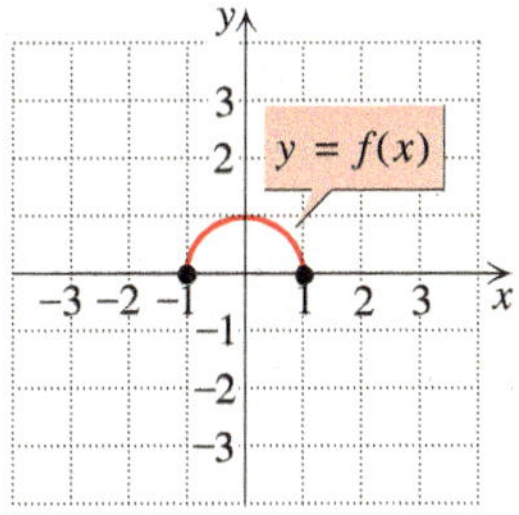


**55.** $y = -f(x + 1) - 3$

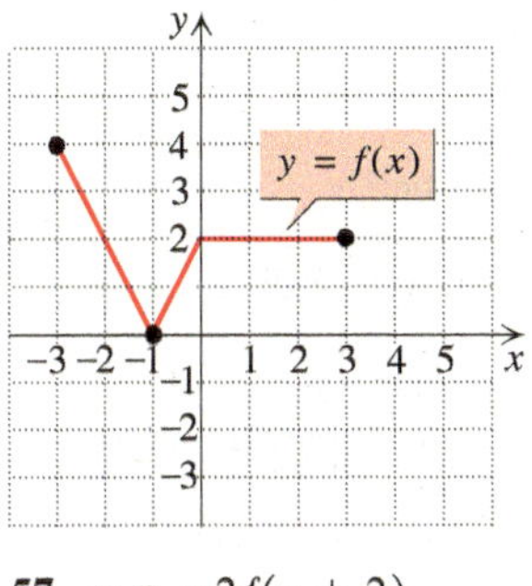


**56.** $y = -f(x - 1) + 2$

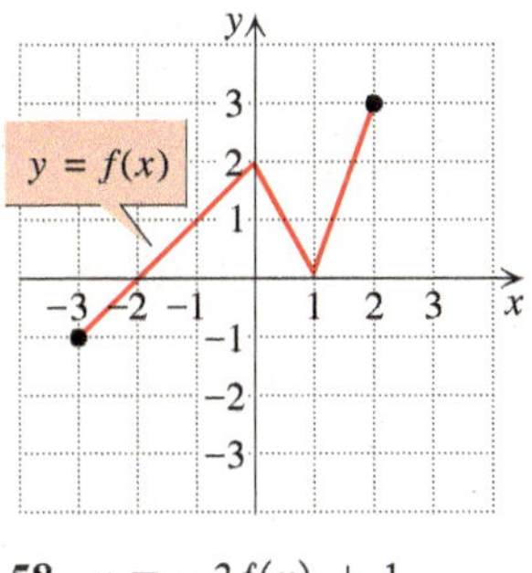


**57.** $y = -2f(x + 2)$

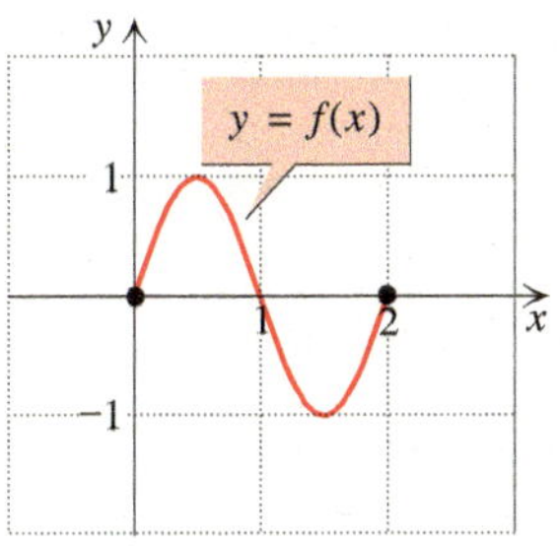


**58.** $y = -3f(x) + 1$

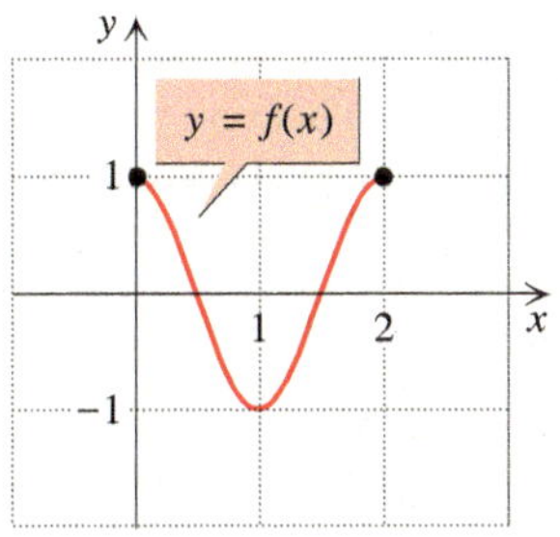


**59.** $y = -2f(x) + 3$

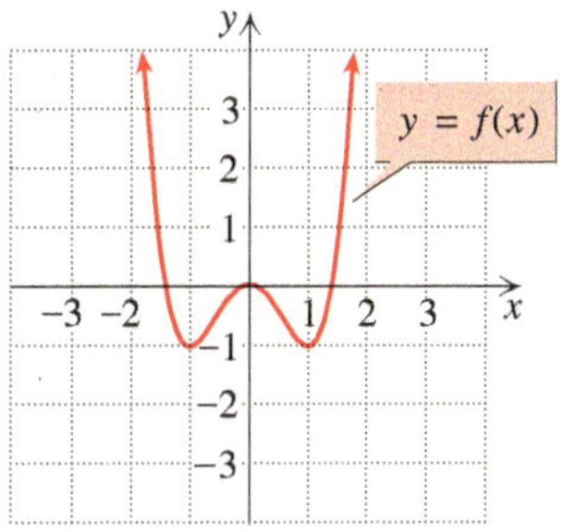


**60.** $y = 4f(x - 1) + 3$

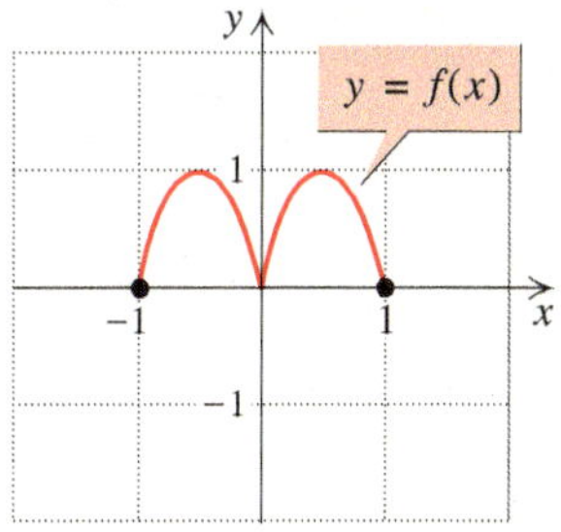


*Let $f(x) = \sqrt[3]{x}$, $g(x) = x - 4$, $h(x) = x/3$, and $j(x) = x^2$. Write each function as a composition of the appropriate functions chosen from f, g, h, and j.*

**61.** $F(x) = \sqrt[3]{x - 4}$

**62.** $G(x) = -4 + \sqrt[3]{x}$

**63.** $H(x) = \sqrt[3]{\frac{x^2 - 4}{3}}$

**64.** $M(x) = \left(\frac{x - 4}{3}\right)^2$

**65.** $N(x) = \frac{1}{3}x^{2/3}$

**66.** $P(x) = x^2 - 16$

**67.** $R(x) = \frac{x^2}{3} - 4$

**68.** $Q(x) = x^2 - 8x + 16$

*Find the difference quotient for each function and simplify it.*

**69.** $f(x) = -5x + 9$

**70.** $f(x) = \sqrt{x - 7}$

**71.** $f(x) = \frac{1}{2x}$

**72.** $f(x) = -5x^2 + x$

*Sketch the graph of each function and state its domain and range. Determine the intervals on which the function is increasing, decreasing, or constant.*

**73.** $f(x) = \sqrt{100 - x^2}$

**74.** $f(x) = -\sqrt{7 - x^2}$

**75.** $f(x) = \begin{cases} -x^2 & \text{for } x \le 0 \\ x^2 & \text{for } x > 0 \end{cases}$

**76.** $f(x) = \begin{cases} x^2 & \text{for } x \le 0 \\ x & \text{for } 0 < x \le 4 \end{cases}$

**77.** $f(x) = \begin{cases} -x - 4 & \text{for } x \le -2 \\ -|x| & \text{for } -2 < x < 2 \\ x - 4 & \text{for } x \ge 2 \end{cases}$

**78.** $f(x) = \begin{cases} -x - 4 & \text{for } x \le -2 \\ -|x| & \text{for } -2 < x < 2 \\ x - 4 & \text{for } x \ge 2 \end{cases}$

*Each of the following graphs is from the absolute-value family. Construct a function for each graph and state the domain and range of the function.*

**79.**

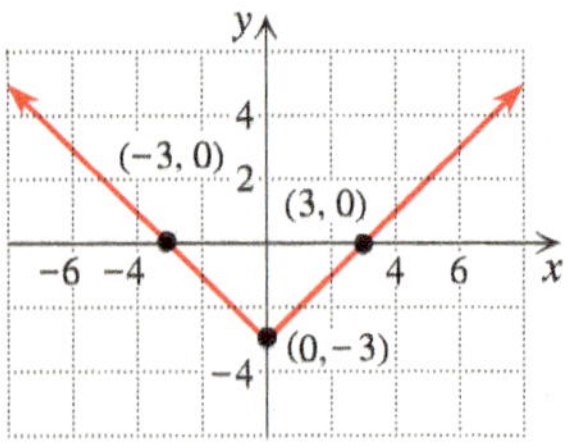


**80.**

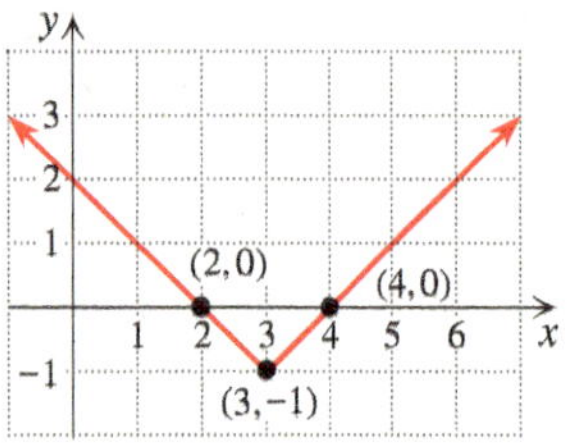


**81.**

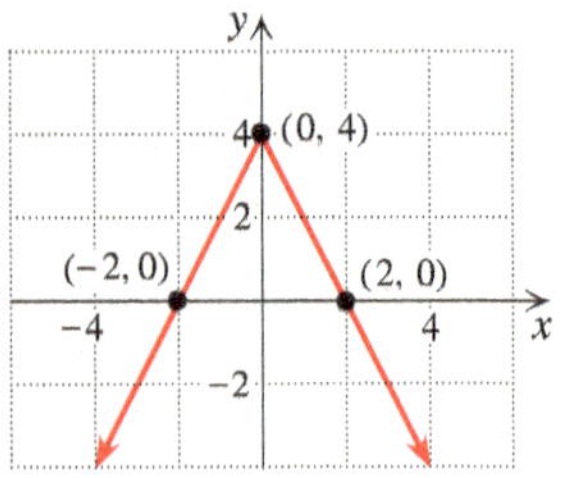


**82.**

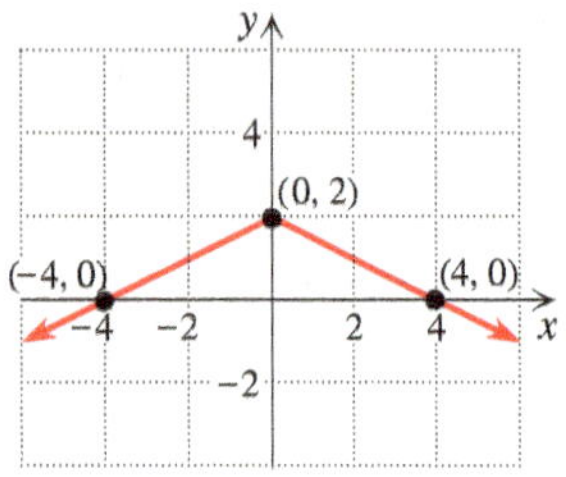


**83.**

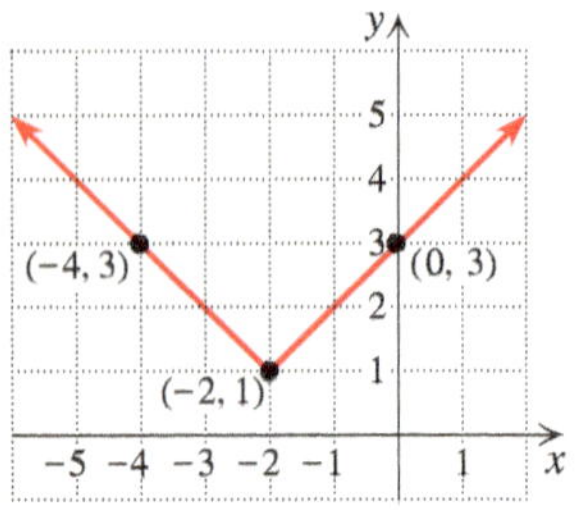


**84.**

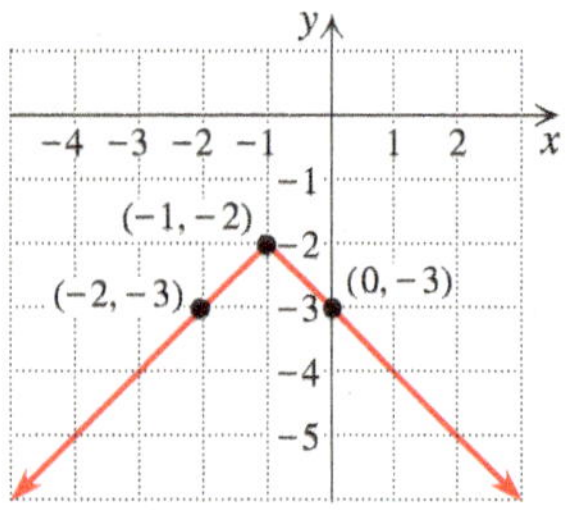


*Determine whether the graph of each function is symmetric with respect to the y-axis or the origin.*

**85.** $f(x) = x^4 - 9x^2$

**86.** $f(x) = |x| - 99$

**87.** $f(x) = -x^3 - 5x$

**88.** $f(x) = \dfrac{15}{x}$

**89.** $f(x) = -x + 1$

**90.** $f(x) = |x - 1|$

**91.** $f(x) = \sqrt{x^2}$

**92.** $f(x) = \sqrt{16 - x^2}$

*Use the graph of an appropriate function to find the solution set to each inequality.*

**93.** $|x - 3| \geq 1$

**94.** $|x + 2| + 1 < 0$

**95.** $-2x^2 + 4 > 0$

**96.** $-\dfrac{1}{2}x^2 + 2 \leq 0$

**97.** $-\sqrt{x + 1} - 2 > 0$

**98.** $\sqrt{x} - 3 < 0$

*Graph each pair of functions on the same coordinate plane. What is the relationship between the functions in each pair?*

**99.** $f(x) = \sqrt{x + 3}$, $g(x) = x^2 - 3$ for $x \geq 0$

**100.** $f(x) = (x - 2)^3$, $g(x) = \sqrt[3]{x} + 2$

**101.** $f(x) = 2x - 4$, $g(x) = \dfrac{1}{2}x + 2$

**102.** $f(x) = -\dfrac{1}{2}x + 4$, $g(x) = -2x + 8$

*Determine whether each function is invertible. If the function is invertible, then find the inverse function, and state the domain and range of the inverse function.*

**103.** $\{(\pi, 0), (\pi/2, 1), (2\pi/3, 0)\}$

**104.** $\{(-2, 1/3), (-3, 1/4), (-4, 1/5)\}$

**105.** $f(x) = 3x - 21$

**106.** $f(x) = 3|x|$

**107.** $y = \sqrt{9 - x^2}$

**108.** $y = 7 - x$

**109.** $f(x) = \sqrt{x - 9}$

**110.** $f(x) = \sqrt{x} - 9$

**111.** $f(x) = \dfrac{x - 7}{x + 5}$

**112.** $f(x) = \dfrac{2x - 3}{5 - x}$

**113.** $f(x) = x^2 + 1$ for $x \leq 0$

**114.** $f(x) = (x + 3)^4$ for $x \geq 0$

*Solve each problem.*

**115.** *Turning a Profit* Mary Beth buys roses for \$1.20 each and sells them for \$2 each at an outdoor market where she rents space for \$40 per day. She buys and sells $x$ roses per day. Construct her daily cost, revenue, and profit functions. How many roses must she sell to make a profit?

**116.** *Tin Can* The surface area $S$ and volume $V$ for a can with a top and bottom are given by

$$S = 2\pi r^2 + 2\pi rh \quad \text{and} \quad V = \pi r^2 h,$$

where $r$ is the radius and $h$ is the height. Suppose that a can has a volume of 1 $\text{ft}^3$.

**a.** Write its height as a function of its radius.

**b.** Write its radius as a function of its height.

**c.** Write its surface area as a function of its radius.

**117.** *Dropping the Ball* A ball is dropped from a height of 64 feet. Its height $h$ (in feet) is a function of time $t$ (in seconds), where $h = -16t^2 + 64$ for $t$ in the interval $[0, 2]$. Find the inverse of this function and state the domain of the inverse function.

**118.** *Sales Tax Function* If $S$ is the subtotal in dollars of your groceries before a 5% sales tax, then the function $T = 1.05S$ gives the total cost in dollars including tax. Write $S$ as a function of $T$.

**119.** Write the diameter of a circle $d$ as a function of its area $A$.

**120.** *Circle Inscribed in a Square* A cylindrical pipe with an outer radius $r$ must fit snugly through a square hole in a wall. Write the area of the square as a function of the radius of the pipe.

**121.** *Load on a Spring* When a load of 5 lb is placed on a spring, its length is 6 in., and when a load of 9 lb is placed on the spring, its length is 8 in. What is the average rate of change of the length of the spring as the load varies from 5 lb to 9 lb?

**122.** *Changing Speed of a Dragster* Suppose that 2 sec after starting, a dragster is traveling 40 mph, and 5 sec after starting, the dragster is traveling 130 mph. What is the average rate of change of the speed of the dragster over the time interval from 2 sec to 5 sec? What are the units for this measurement?

**123.** Given that $D$ is directly proportional to $W$ and $D = 9$ when $W = 25$, find $D$ when $W = 100$.

**124.** Given that $t$ varies directly as $u$ and inversely as $v$, and $t = 6$ when $u = 8$ and $v = 2$, find $t$ when $u = 19$ and $v = 3$.

**125.** *Dinosaur Speed* R. McNeill Alexander, a British paleontologist, uses observations of living animals to estimate the speed of dinosaurs. Alexander believes that two animals of different sizes but geometrically similar shapes will move in a similar fashion. For animals of similar shapes, their velocity is directly proportional to the square root of their hip height. Alexander has compared a white rhinoceros, with a hip height of 1.5 m, and a member of the genus *Triceratops*, with a hip height of 2.8 m. If a white rhinoceros can move at 45 km/hr, then what is the estimated velocity of the dinosaur?

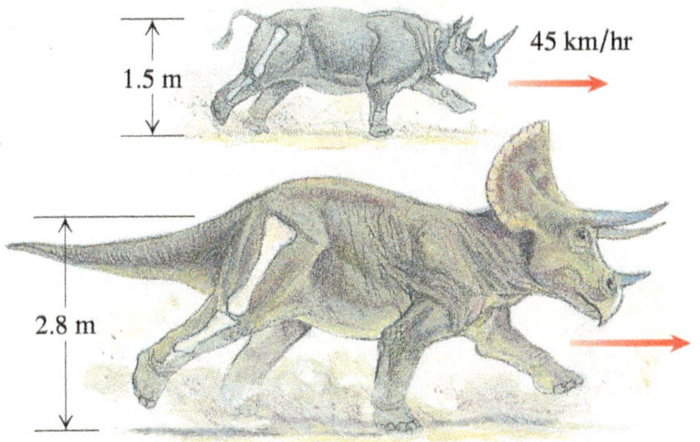


**Figure for Exercise 125**

**126.** *Arranging Plants in a Row* Timothy has a shipment of strawberry plants to plant in his field, which is prepared with rows of equal length. He knows that the number of rows that he will need varies inversely with the number of plants per row. If he plants 240 plants per row, then he needs 21 rows. How many rows does he need if he plants 224 plants per row?

**127.** *Handling Charge for a Globe* The shipping and handling charge for a globe from Mercator's Map Company varies directly as the square of the diameter. If the charge is \$4.32 for a 6-in.-diameter globe, then what is the charge for a 16-in.-diameter globe?

**128.** *Newton's Law of Gravity* Newton's law of gravity states that the force of gravity acting between two objects varies directly as the product of the masses of the two objects and inversely as the square of the distance between their centers. Write an equation for Newton's law of gravity.

### FOR WRITING/DISCUSSION

**129.** Explain what the vertical and horizontal line tests are and how to apply them.

**130.** Is a vertical line the graph of a linear function? Explain.

### THINKING OUTSIDE THE BOX XX & XXI

*Area of a Polygon* Find the exact area of the polygon bounded by the $x$-axis, the $y$-axis, the line $2x + 3y = 7$, and the line $3x - 4y = 18$.

*Whole Lotta Shakin'* At the start of an economic conference between the eastern delegation and the western delegation, each delegate shook hands with every other member of his own delegation for a total of 466 handshakes. Next, each delegate shook hands with every person in the other delegation for 480 more handshakes. What was the total number of delegates at the conference?

# Chapter 2 Test

*Determine whether each equation defines y as a function of x.*

**1.** $x^2 + y^2 = 25$

**2.** $3x - 5y = 20$

**3.** $x = |y + 2|$

**4.** $y = x^3 - 3x^2 + 2x - 1$

*State the domain and range of each relation.*

**5.** $\{(2, -3), (2, -4), (5, 7)\}$

**6.** $y = \sqrt{x - 9}$

**7.** $x = |y - 1|$

*Sketch the graph of each function.*

**8.** $3x - 4y = 12$

**9.** $y = 2x - 3$

**10.** $y = \sqrt{25 - x^2}$

**11.** $y = -(x - 2)^2 + 5$

**12.** $y = 2|x| - 4$

**13.** $y = \sqrt{x + 3} - 5$

**14.** $f(x) = \begin{cases} x & \text{for } x < 0 \\ 2 & \text{for } x \geq 0 \end{cases}$

*Let $f(x) = \sqrt{x + 2}$ and $g(x) = 3x - 1$. Find and simplify each of the following expressions.*

**15.** $f(7)$

**16.** $(f \circ g)(2)$

**17.** $(f \circ g)(x)$

**18.** $g^{-1}(x)$

**19.** $(f + g)(14)$

**20.** $\dfrac{g(x + h) - g(x)}{h}$

*Solve each problem.*

**21.** State the intervals on which $f(x) = (x - 3)^2$ is increasing and those on which $f(x)$ is decreasing.

**22.** Discuss the symmetry of the graph of the function $f(x) = x^4 - 3x^2 + 9$.

**23.** State the solution set to the inequality $|x - 1| - 3 < 0$ using interval notation.

**24.** Find the inverse of the function $g(x) = 3 + \sqrt[3]{x - 2}$.

**25.** Find the inverse of the function $f(x) = \sqrt{x - 5}$.

**26.** Jang's Postal Service charges \$35 for addressing 200 envelopes, and \$60 for addressing 400 envelopes. What is the average rate of change of the cost as the number of envelopes goes from 200 to 400?

**27.** The intensity of illumination $I$ from a light source varies inversely as the square of the distance $d$ from the source. If a flash on a camera has an intensity of 300 candlepower at a distance of 2 m, then what is the intensity of the flash at a distance of 10 m?

**28.** Construct a function expressing the volume $V$ of a cube as a function of the length of the diagonal $d$ of a side of the cube.

# TYING IT ALL together...

## Chapters 1–2

*Find all real solutions to each equation.*

**1.** $2x - 3 = 0$

**2.** $-2x + 6 = 0$

**3.** $|x| - 100 = 0$

**4.** $1 - 2|x + 90| = 0$

**5.** $3 - 2\sqrt{x + 30} = 0$

**6.** $\sqrt{x - 3} + 15 = 0$

**7.** $-2(x - 2)^2 + 1 = 0$

**8.** $4(x + 2)^2 - 1 = 0$

**9.** $\sqrt{9 - x^2} - 2 = 0$

**10.** $\sqrt{49 - x^2} + 3 = 0$

*Graph each function. Identify the domain, range, and x-intercepts.*

**11.** $y = 2x - 3$

**12.** $y = -2x + 6$

**13.** $y = |x| - 100$

**14.** $y = 1 - 2|x + 90|$

**15.** $y = 3 - 2\sqrt{x + 30}$

**16.** $y = \sqrt{x - 3} + 15$

**17.** $y = -2(x - 2)^2 + 1$

**18.** $y = 4(x + 2)^2 - 1$

**19.** $y = \sqrt{9 - x^2} - 2$

**20.** $y = \sqrt{49 - x^2} + 3$

*Solve each inequality. State the solution set using interval notation.*

**21.** $2x - 3 > 0$

**22.** $-2x + 6 \le 0$

**23.** $|x| - 100 \ge 0$

**24.** $1 - 2|x + 90| > 0$

**25.** $3 - 2\sqrt{x + 30} \le 0$

**26.** $\sqrt{x - 3} + 15 > 0$

**27.** $-2(x - 2)^2 + 1 < 0$

**28.** $4(x + 2)^2 - 1 \ge 0$

**29.** $\sqrt{9 - x^2} - 2 \ge 0$

**30.** $\sqrt{49 - x^2} + 3 \le 0$

*Let $f(x) = 3(x + 1)^3 - 24$ for the following exercises.*

**31.** Find $f(2)$ without using a calculator.

**32.** Graph the function $f$.

**33.** Express $f$ as a composition using the functions $F(x) = 3x$, $G(x) = x + 1$, $H(x) = x^3$, and $K(x) = x - 24$.

**34.** Solve the equation $3(x + 1)^3 - 24 = 0$.

**35.** Solve $3(x + 1)^3 - 24 \ge 0$.

**36.** Solve $y = 3(x + 1)^3 - 24$ for $x$.

**37.** Find the inverse of the function $f$.

**38.** Graph the function $f^{-1}$.

**39.** Solve the inequality $f^{-1}(x) > 0$.

**40.** Express $f^{-1}$ as a composition using the functions $F^{-1}(x) = x/3$, $G^{-1}(x) = x - 1$, $H^{-1}(x) = \sqrt[3]{x}$, and $K^{-1}(x) = x + 24$.

# CONCEPTS OF calculus...

## Instantaneous rate of change

*Suppose that you start at Amarillo and drive west on* I-40 *for* 500 *miles. If this trip takes you* 10 *hours, then the average rate at which your location is changing is* 50 *miles per hour. At every instant of your trip your speedometer shows the instantaneous rate at which your location is changing. Your instantaneous rate of change might range from* 0 *miles per hour to* 70 *miles per hour or more.*

*We defined the average rate of change of a function $f(x)$ on the interval $[x, x + h]$ as $\frac{f(x+h) - f(x)}{h}$. We cannot use a speedometer to find the instantaneous rate of change of a function. We use the idea of limits, which was discussed in the Concepts of Calculus at the end of Chapter* 1. *For the instantaneous rate of change we shrink the interval to nothing. So we define the instantaneous rate of change of $f(x)$ as $\lim_{h \to 0} \frac{f(x+h) - f(x)}{h}$.*

## Exercises

**1.** Let $f(x) = x^2$.

**a.** Find $\frac{f(2+h) - f(2)}{h}$.

**b.** Find $\lim_{h \to 0} \frac{f(2+h) - f(2)}{h}$.

**2.** Let $f(x) = x^2 - 2x$.

**a.** Find $\frac{f(x+h) - f(x)}{h}$.

**b.** Find $\lim_{h \to 0} \frac{f(x+h) - f(x)}{h}$.

**c.** Find the instantaneous rate of change of $f(x) = x^2 - 2x$ when $x = 5$.

**3.** Let $f(x) = \sqrt{x}$.

**a.** Find $\frac{f(x+h) - f(x)}{h}$ and write your answer with a rational numerator.

**b.** Find $\lim_{h \to 0} \frac{f(x+h) - f(x)}{h}$.

**c.** Find the instantaneous rate of change of $f(x) = \sqrt{x}$ when $x = 9$.

**4.** A ball is tossed into the air from ground level at 128 feet per second. The function $f(t) = -16t^2 + 128t$ gives the height above ground (in feet) as a function of time (in seconds).

**a.** Find $f(0)$ and $f(3)$.

**b.** How far did the ball travel in the first three seconds of its flight?

**c.** Find the average rate of change of the height for the time interval $[0, 3]$.

**d.** Find $\frac{f(t+h) - f(t)}{h}$.

**e.** Find $\lim_{h \to 0} \frac{f(t+h) - f(t)}{h}$.

**f.** Find the instantaneous rate of change of the height (or the *instantaneous velocity*) of the ball at times $t = 0, 2, 4, 6,$ and 8 seconds.

*The average rate of change of the function $f$ on the interval $[c, x]$ is $\frac{f(x) - f(c)}{x - c}$.*

**5.** Let $f(x) = \frac{1}{x}$.

**a.** Find and simplify $\frac{f(x) - f(2)}{x - 2}$.

**b.** Find $\lim_{x \to 2} \frac{f(x) - f(2)}{x - 2}$.

**c.** Find the instantaneous rate of change of $f(x)$ when $x = 2$.

**6.** Let $f(x) = x^3$.

**a.** Find and simplify $\frac{f(x) - f(c)}{x - c}$.

**b.** Find $\lim_{x \to c} \frac{f(x) - f(c)}{x - c}$.

**c.** Find the instantaneous rate of change of $f(x)$ when $x = c$.

# 3 Polynomial and Rational Functions

Others had died trying, but in May of 1927, Charles Lindbergh claimed the $25,000 prize for flying across the Atlantic Ocean. He was the first aviator to make the 3610-mile nonstop solo flight from New York to Paris.

Lindbergh did not just get in a plane and start flying. He spent years planning, designing, and practicing for his attempt to cross the Atlantic. Lindbergh collected and analyzed a vast amount of data to determine relationships between wind velocity, air speed, fuel consumption, and time elapsed. As he burned fuel, the plane got lighter and he flew slower.

**WHAT YOU WILL learn...** In this chapter we explore polynomial and rational functions and their graphs. We'll learn how to use these basic functions of algebra to model many real-life situations, and we'll see how Lindbergh used them in planning his historic flight.

# 3.1 Quadratic Functions and Inequalities

A **polynomial function** is defined by a polynomial. A **quadratic function** is defined by a quadratic or second-degree polynomial. So a quadratic function has the form $f(x) = ax^2 + bx + c$ where $a \neq 0$. We studied functions of the form $f(x) = a(x - h)^2 + k$ in Section 2.3 (though we did not call them quadratic functions there). Once we see that these forms are equivalent, we can use what we learned in Section 2.3 in our study of quadratic functions here.

## Two Forms for Quadratic Functions

It is easy to convert $f(x) = a(x - h)^2 + k$ to the form $f(x) = ax^2 + bx + c$ by squaring the binomial. To convert in the other direction we use completing the square from Section 1.3. Recall that the perfect square trinomial whose first two terms are $x^2 + bx$ is $x^2 + bx + \left(\frac{b}{2}\right)^2$. So to get the last term you *take one-half of the coefficient of the middle term and square it.*

When completing the square in Section 1.3, we added the last term of the perfect square trinomial to both sides of the equation. Here we want to keep $f(x)$ on the left side. To complete the square and change only the right side, we add and subtract on the right side as shown in Example 1(a). Another complication here is that the coefficient of $x^2$ is not always 1. To overcome this problem, you must factor before completing the square as shown in Example 1(b).

### EXAMPLE 1 Completing the square for a quadratic function

Rewrite each function in the form $f(x) = a(x - h)^2 + k$.

**a.** $f(x) = x^2 + 6x$ **b.** $f(x) = 2x^2 - 20x + 3$

**Solution**

**a.** One-half of 6 is 3 and $3^2$ is 9. Adding and subtracting 9 on the right side of $f(x) = x^2 + 6x$ does not change the function.

$$\begin{aligned} f(x) &= x^2 + 6x \\ &= x^2 + 6x + 9 - 9 && 9 = \left(\tfrac{1}{2}\cdot 6\right)^2 \\ &= (x + 3)^2 - 9 && \text{Factor.} \end{aligned}$$

**b.** First factor 2 out of the first two terms, because the leading coefficient in $(x - h)^2$ is 1.

$$\begin{aligned} f(x) &= 2x^2 - 20x + 3 \\ &= 2(x^2 - 10x) + 3 \\ &= 2(x^2 - 10x + 25 - 25) + 3 && 25 = \left(\tfrac{1}{2}\cdot 10\right)^2 \\ &= 2(x^2 - 10x + 25) - 50 + 3 && \text{Remove } -25 \text{ from the parentheses.} \\ &= 2(x - 5)^2 - 47 \end{aligned}$$

Because of the 2 preceding the parentheses, the second 25 was doubled when it was removed from the parentheses.

▸**TRY THIS.** Rewrite $f(x) = 2x^2 - 8x + 9$ in the form $f(x) = a(x - h)^2 + k$. ■

In the next example we complete the square and graph a quadratic function.

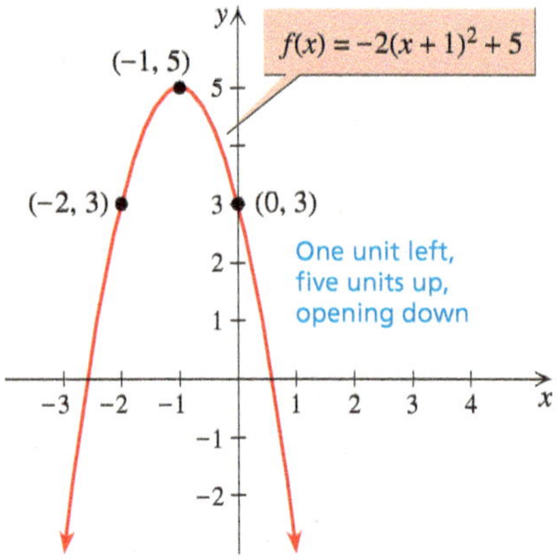


**Figure 3.1**

## EXAMPLE 2 Graphing a quadratic function

Rewrite $f(x) = -2x^2 - 4x + 3$ in the form $f(x) = a(x - h)^2 + k$ and sketch its graph.

### Solution

Start by completing the square:

$$\begin{aligned} f(x) &= -2(x^2 + 2x) + 3 && \text{Factor out } -2 \text{ from the first two terms.} \\ &= -2(x^2 + 2x + 1 - 1) + 3 && \text{Complete the square for } x^2 + 2x. \\ &= -2(x^2 + 2x + 1) + 2 + 3 && \text{Remove } -1 \text{ from the parentheses.} \\ &= -2(x + 1)^2 + 5 \end{aligned}$$

The function is now in the form $f(x) = a(x - h)^2 + k$. The number 1 indicates that the graph of $f(x) = x^2$ is translated one unit to the left. Since $a = -2$, the graph is stretched by a factor of 2 and reflected below the $x$-axis. Finally, the graph is translated five units upward. The graph shown in Fig. 3.1 includes the points $(0, 3)$, $(-1, 5)$, and $(-2, 3)$.

▸**TRY THIS.** Rewrite $f(x) = -2x^2 - 4x + 1$ in the form $f(x) = a(x - h)^2 + k$ and sketch its graph. ■

We can use completing the square as in Example 2 to prove the following theorem.

**Theorem: Quadratic Functions**

> The graph of any quadratic function is a transformation of the graph of $f(x) = x^2$.

**PROOF** Since $f(x) = a(x - h)^2 + k$ is a transformation of $f(x) = x^2$, we will show that $f(x) = ax^2 + bx + c$ can be written in that form:

$$\begin{aligned} f(x) &= ax^2 + bx + c \\ &= a\left(x^2 + \frac{b}{a}x\right) + c && \text{Factor } a \text{ out of the first two terms.} \end{aligned}$$

To complete the square for $x^2 + \frac{b}{a}x$, add and subtract $\frac{b^2}{4a^2}$ inside the parentheses:

$$\begin{aligned} f(x) &= a\left(x^2 + \frac{b}{a}x + \frac{b^2}{4a^2} - \frac{b^2}{4a^2}\right) + c \\ &= a\left(x^2 + \frac{b}{a}x + \frac{b^2}{4a^2}\right) - \frac{b^2}{4a} + c && \text{Remove } -\tfrac{b^2}{4a^2} \text{ from the parentheses.} \\ &= a\left(x + \frac{b}{2a}\right)^2 + \frac{4ac - b^2}{4a} && \text{Factor and get a common denominator.} \\ &= a(x - h)^2 + k && \text{Let } h = -\tfrac{b}{2a} \text{ and } k = \tfrac{4ac - b^2}{4a}. \end{aligned}$$

So the graph of any quadratic function is a transformation of $f(x) = x^2$. ■

In Section 2.3 we stated that any transformation of $f(x) = x^2$ is called a parabola. Therefore, the graph of any quadratic function is a parabola.

## Opening, Vertex, and Axis of Symmetry

If $a > 0$, the graph of $f(x) = a(x - h)^2 + k$ **opens upward;** if $a < 0$, the graph **opens downward** as shown in Fig. 3.2. Notice that $h$ determines the amount of horizontal translation and $k$ determines the amount of vertical translation of the graph of $f(x) = x^2$. Because of the translations, the point $(0, 0)$ on the graph of $f(x) = x^2$ moves to the point $(h, k)$ on the graph of $f(x) = a(x - h)^2 + k$. Since $(0, 0)$ is the lowest point on the graph of $f(x) = x^2$, the lowest point on any parabola that opens upward is $(h, k)$. Since $(0, 0)$ is the highest point on the graph of $f(x) = -x^2$, the highest point on any parabola that opens downward is $(h, k)$. The point $(h, k)$ is called the **vertex** of the parabola. Since $(h, k)$ is the vertex, $f(x) = a(x - h)^2 + k$ is called the **vertex form** of the equation of a parabola. The form $f(x) = ax^2 + bx + c$ is the **general form** of the equation of a parabola.

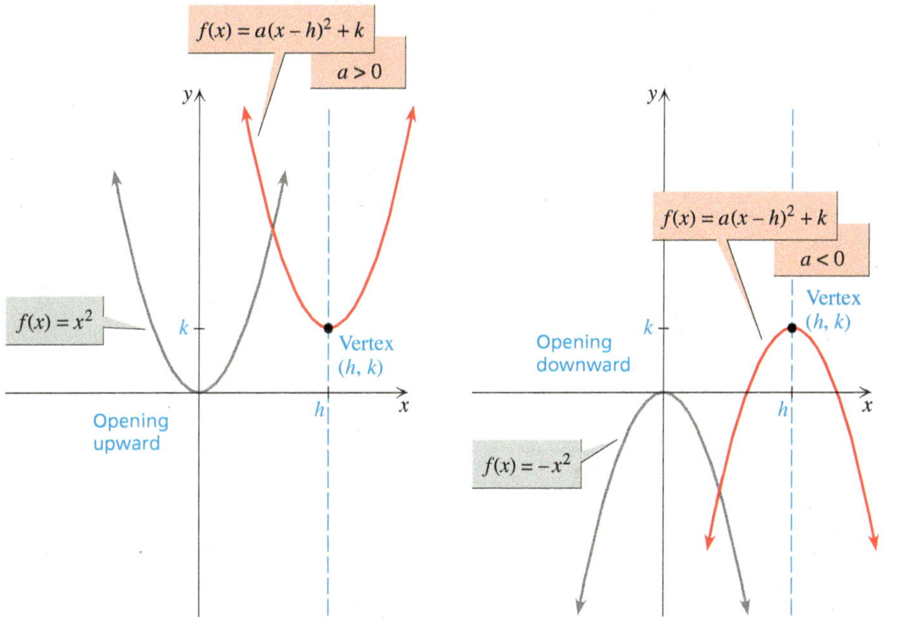


Figure 3.2

The graph of $f(x) = x^2$ is symmetric about the $y$-axis, which runs vertically through the vertex of the parabola. Since this symmetry is preserved in transformations, the graph of any quadratic function is symmetric about the vertical line through its vertex, the **axis of symmetry.** See Fig. 3.2. The equation of the axis of symmetry is $x = h$ or $x = -b/(2a)$, because we used $h = -b/(2a)$ when we converted general form to vertex form.

You can use a graphing calculator to experiment with different values for $a$, $h$, and $k$. Two possibilities are shown in Fig. 3.3(a) and (b). The graphing calculator allows you to see results quickly and to use values that you would not use otherwise. □

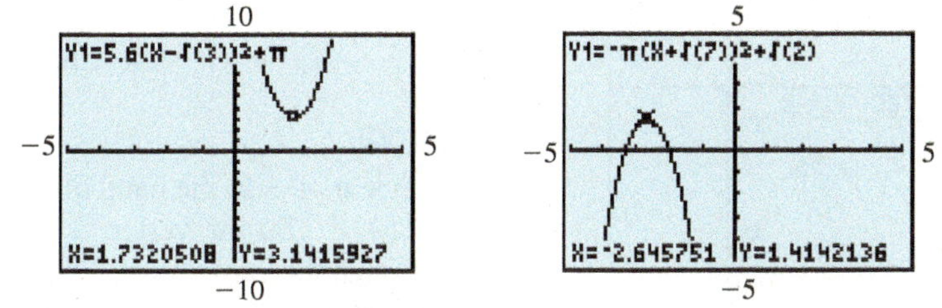


(a) (b)

Figure 3.3

It is easy to find the vertex $(h, k)$ if the equation of the parabola is given as $f(x) = a(x - h)^2 + k$. For the parabola $f(x) = ax^2 + bx + c$ we can find the vertex using $x = -b/(2a)$, because the axis of symmetry goes through the vertex.

**SUMMARY**

### Vertex of a Parabola

1. For a quadratic function in the form $f(x) = a(x - h)^2 + k$ the vertex of the parabola is $(h, k)$.
2. For the form $f(x) = ax^2 + bx + c$, the $x$-coordinate of the vertex is $\frac{-b}{2a}$. The $y$-coordinate is $f\left(\frac{-b}{2a}\right)$.

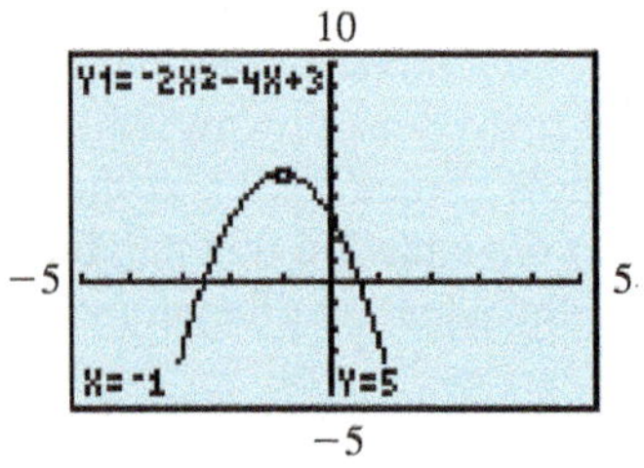


Figure 3.4

## EXAMPLE 3 Finding the vertex

Find the vertex of the graph of $f(x) = -2x^2 - 4x + 3$.

**Solution**

Use $x = -b/(2a)$ to find the $x$-coordinate of the vertex:

$$x = \frac{-b}{2a} = \frac{-(-4)}{2(-2)} = -1$$

Now find $f(-1)$:

$$f(-1) = -2(-1)^2 - 4(-1) + 3 = 5$$

The vertex is $(-1, 5)$. In Example 2, $f(x) = -2x^2 - 4x + 3$ was rewritten as $f(x) = -2(x + 1)^2 + 5$. In this form we see immediately that the vertex is $(-1, 5)$. The graph in Fig. 3.4 supports these conclusions.

▶**TRY THIS.** Find the vertex of the graph of $f(x) = 3x^2 - 12x + 5$. ■

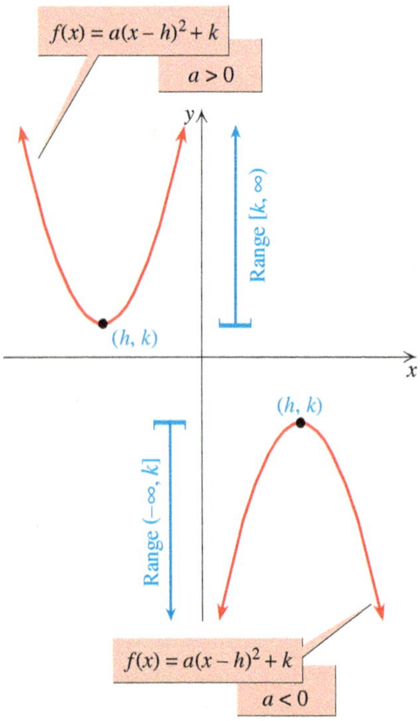


Figure 3.5

The domain of every quadratic function $f(x) = a(x - h)^2 + k$ is the set of real numbers, $(-\infty, \infty)$. The range of a quadratic function is determined from the second coordinate of the vertex. If $a > 0$, the range is $[k, \infty)$ and $k$ is called the **minimum value of the function.** The function is decreasing on $(-\infty, h]$ and increasing on $[h, \infty)$. See Fig. 3.5. If $a < 0$, the range is $(-\infty, k]$ and $k$ is called the **maximum value of the function.** The function is increasing on $(-\infty, h]$ and decreasing on $[h, \infty)$.

## EXAMPLE 4 Identifying the characteristics of a parabola

For each parabola, determine whether the parabola opens upward or downward, and find the vertex, axis of symmetry, and range of the function. Find the maximum or minimum value of the function and the intervals on which the function is increasing or decreasing.

**a.** $y = -2(x + 4)^2 - 8$ **b.** $y = 2x^2 - 4x - 9$

**Solution**

**a.** Since $a = -2$, the parabola opens downward. In $y = a(x - h)^2 + k$, the vertex is $(h, k)$. So the vertex is $(-4, -8)$. The axis of symmetry is the vertical line through the vertex, $x = -4$. Since the parabola opens downward from $(-4, -8)$, the range of the function is $(-\infty, -8]$. The maximum value of the function is $-8$, and the function is increasing on $(-\infty, -4]$ and decreasing on $[-4, \infty)$. The graph in Fig. 3.6 supports these results. □

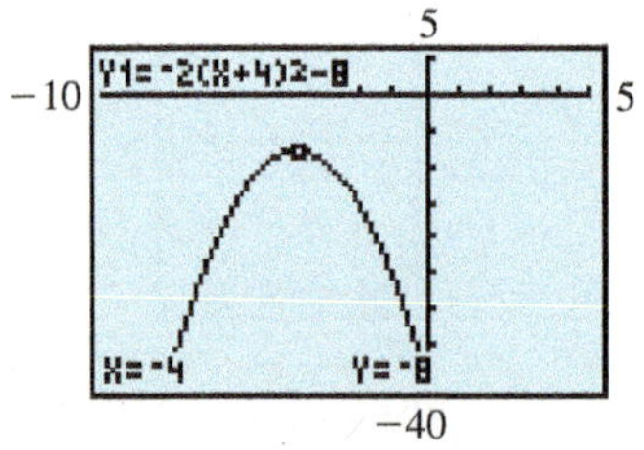


Figure 3.6

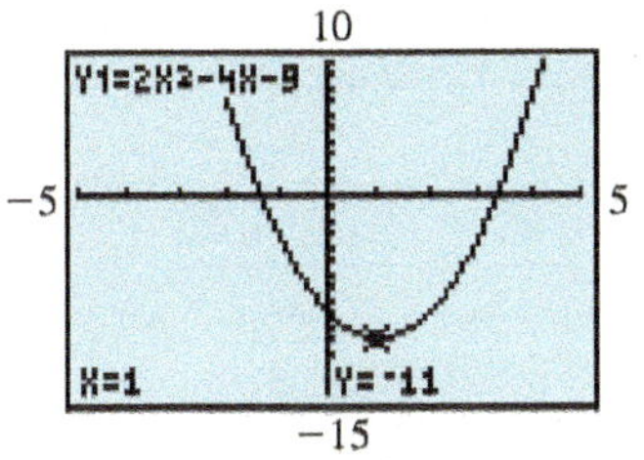


Figure 3.7

**b.** Since $a = 2$, the parabola opens upward. In the form $y = ax^2 + bx + c$, the $x$-coordinate of the vertex is $x = -b/(2a)$. In this case,

$$x = \frac{-b}{2a} = \frac{-(-4)}{2(2)} = 1.$$

Use $x = 1$ to find $y = 2(1)^2 - 4(1) - 9 = -11$. So the vertex is $(1, -11)$, and the axis of symmetry is the vertical line $x = 1$. Since the parabola opens upward, the vertex is the lowest point and the range is $[-11, \infty)$. The function is decreasing on $(-\infty, 1]$ and increasing on $[1, \infty)$. The minimum value of the function is $-11$. The graph in Fig. 3.7 supports these results.

▶**TRY THIS.** Identify all of the characteristics of the parabola $y = 3(x - 5)^2 - 4$. ■

## Intercepts

The graph of $y = ax^2 + bx + c$ always has a $y$-intercept. The $y$-intercept $(0, c)$ is found by replacing $x$ with 0 in $y = ax^2 + bx + c$. Whether the graph has any $x$-intercepts depends on whether the quadratic equation $ax^2 + bx + c = 0$ has any solutions. In Section 1.7 we learned that any quadratic equation could be solved using the quadratic formula:

$$x = \frac{-b \pm \sqrt{b^2 - 4ac}}{2a} \quad \text{Quadratic formula}$$

Since the quadratic formula could produce 0, 1, or 2 real solutions to the equation, there might be 0, 1, or 2 $x$-intercepts. Note how the axis of symmetry, $x = -b/(2a)$, appears in this formula. If the graph has two $x$-intercepts, then they are on opposite sides of the graph and are equidistant from the axis of symmetry as shown in Fig. 3.8.

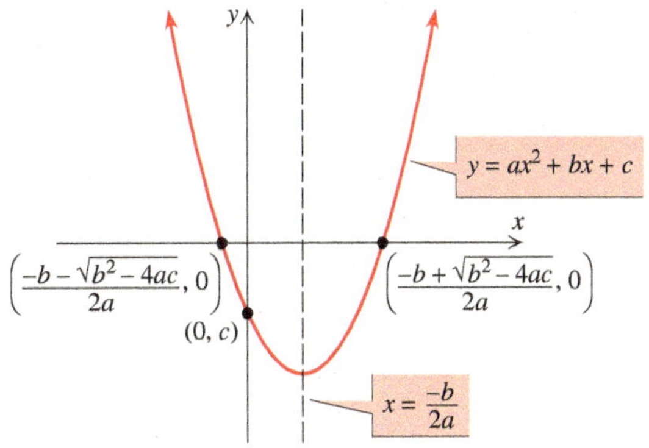


Figure 3.8

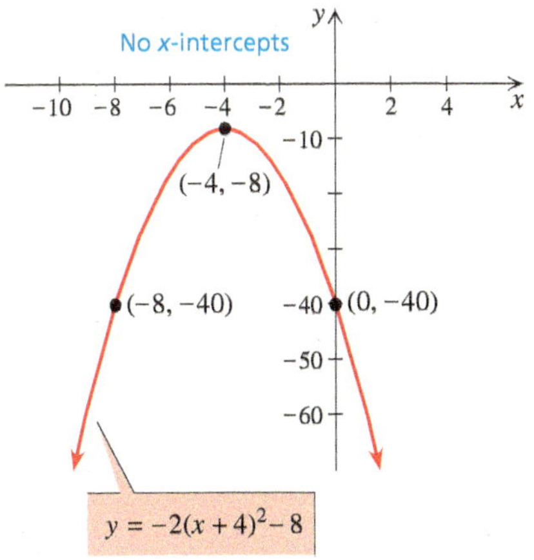


Figure 3.9

## EXAMPLE 5 Finding the intercepts

Find the $y$-intercept and the $x$-intercepts for each parabola and sketch the graph of each parabola.

**a.** $y = -2(x + 4)^2 - 8$ **b.** $y = 2x^2 - 4x - 9$

### Solution

**a.** If $x = 0$, $y = -2(0 + 4)^2 - 8 = -40$. The $y$-intercept is $(0, -40)$. Because the graph is symmetric about the line $x = -4$, the point $(-8, -40)$ is also on the graph. Since the parabola opens downward from $(-4, -8)$, which is below the $x$-axis, there are no $x$-intercepts. If we try to solve $-2(x + 4)^2 - 8 = 0$ to find the $x$-intercepts, we get $(x + 4)^2 = -4$, which has no real solution. The graph is shown in Fig. 3.9.

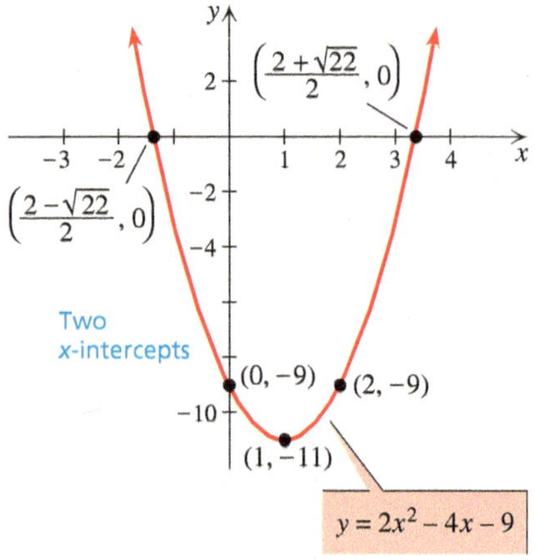


Figure 3.10

**b.** If $x = 0$, $y = 2(0)^2 - 4(0) - 9 = -9$. The $y$-intercept is $(0, -9)$. From Example 4(b), the vertex is $(1, -11)$. Because the graph is symmetric about the line $x = 1$, the point $(2, -9)$ is also on the graph. The $x$-intercepts are found by solving $2x^2 - 4x - 9 = 0$:

$$x = \frac{4 \pm \sqrt{(-4)^2 - 4(2)(-9)}}{2(2)} = \frac{4 \pm \sqrt{88}}{4} = \frac{2 \pm \sqrt{22}}{2}$$

The $x$-intercepts are $\left(\frac{2 + \sqrt{22}}{2}, 0\right)$ and $\left(\frac{2 - \sqrt{22}}{2}, 0\right)$. See Fig. 3.10.

▶**TRY THIS.** Find all intercepts and sketch the graph of $y = 3(x - 5)^2 - 4$. ■

## Quadratic Inequalities

A **quadratic inequality** is an inequality that involves a quadratic polynomial. To solve a quadratic inequality such as $ax^2 + bx + c > 0$, we can use the graph of the quadratic function $y = ax^2 + bx + c$. The values of $x$ that satisfy the inequality are the same as the values of $x$ for which $y > 0$ on the graph of the function $y = ax^2 + bx + c$. You can use the following strategy for this **graphical method.**

**STRATEGY**

### Solving a Quadratic Inequality by the Graphical Method

1. Get 0 on one side of the inequality and a quadratic polynomial on the other side.
2. Find all roots to the quadratic polynomial.
3. Graph the corresponding quadratic function. The roots from (2) determine the $x$-intercepts.
4. Read the solution set to the inequality from the graph of the parabola.

### EXAMPLE 6 Graphical method for solving quadratic inequalities

Solve each inequality. Write the solution set in interval notation and graph it.

**a.** $x^2 - x > 6$ **b.** $x^2 - x \le 6$

**Solution**

**a.** Rewrite the inequality as $x^2 - x - 6 > 0$. Then find all roots to the quadratic polynomial:

$$x^2 - x - 6 = 0$$
$$(x - 3)(x + 2) = 0$$
$$x - 3 = 0 \quad \text{or} \quad x + 2 = 0$$
$$x = 3 \quad \text{or} \quad x = -2$$

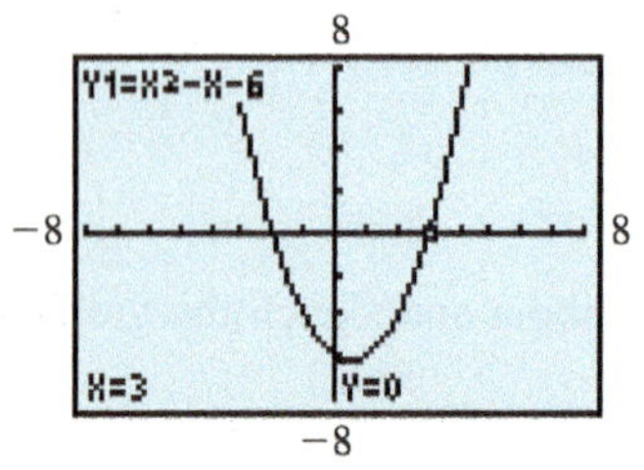


Figure 3.11

The graph of $y = x^2 - x - 6$ is a parabola that opens upward with $x$-intercepts at $(-2, 0)$ and $(3, 0)$. You can picture this graph in your mind, sketch it on paper, or use a graphing calculator as in Fig. 3.11. In any case, the $y$-coordinates on the parabola are negative between the intercepts and positive outside the intercepts. So this inequality is satisfied outside the intercepts. The solution set is $(-\infty, -2) \cup (3, \infty)$. The graph of the solution set is shown in Fig. 3.12.

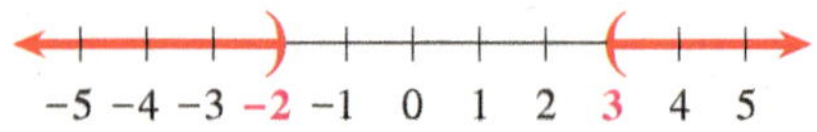


Figure 3.12

**b.** The inequality $x^2 - x \le 6$ is equivalent to $x^2 - x - 6 \le 0$. We already found the roots to $x^2 - x - 6 = 0$ and the graph of $y = x^2 - x - 6$ in part (a). The

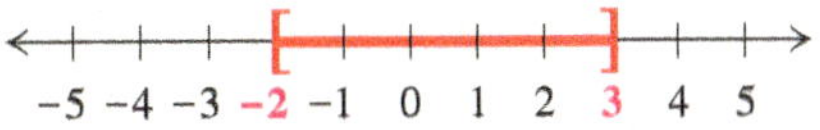


Figure 3.13

$y$-coordinates are less than zero between the intercepts and equal to zero at the intercepts. So the endpoints of the interval are included in the solution set. The solution set is $[-2, 3]$ and its graph is shown in Fig. 3.13.

▶**TRY THIS.** Solve $x^2 + 6x + 8 \geq 0$ by the graphical method. ■

Note that it is usually not a good idea to make conclusions from a graph. The graph may not show all of the important features of the function. However, we did use the graph in Example 6, because we know that the graph of every quadratic function is a parabola. Once we have found the $x$-intercepts, we know the precise location of the parabola.

The **test-point method** for solving a quadratic inequality does not involve the graph of the quadratic function. In place of reading the graph, we test points in the intervals determined by the roots to the quadratic polynomial. This method will be used again for polynomial and rational inequalities later in this chapter.

**STRATEGY**

**Solving a Quadratic Inequality Using Test Points**

1. Get 0 on one side of the inequality and a quadratic polynomial on the other side.
2. Find all roots to the quadratic polynomial and plot them on a number line.
3. Select a test point in each interval determined by the roots.
4. Evaluate the quadratic function at each test point.
5. Determine the solution set from the signs of the function at the test points.

### EXAMPLE 7 Solving a quadratic inequality using the test-point method

Solve $2x^2 - 4x - 9 < 0$. Write the solution set in interval notation.

**Solution**

The roots to $2x^2 - 4x - 9 = 0$ were found to be $\frac{2 \pm \sqrt{22}}{2}$ in Example 5(b). Since $\frac{2 - \sqrt{22}}{2} \approx -1.3$ and $\frac{2 + \sqrt{22}}{2} \approx 3.3$, we make the number line as in Fig. 3.14. Select a convenient test point in each of the three intervals determined by the roots. Our selections $-2$, 1, and 5 are shown in red on the number line. Now evaluate $2x^2 - 4x - 9$ for each test point:

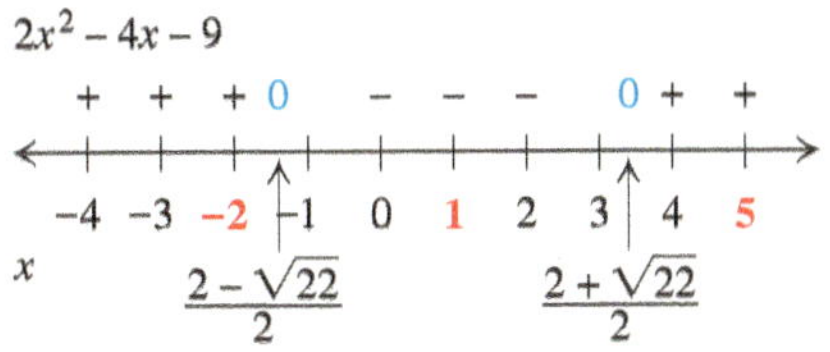


Figure 3.14

$$2(-2)^2 - 4(-2) - 9 = 7 \quad \text{Positive}$$
$$2(1)^2 - 4(1) - 9 = -11 \quad \text{Negative}$$
$$2(5)^2 - 4(5) - 9 = 21 \quad \text{Positive}$$

The signs of these results are shown on the number line in Fig. 3.14. Since $2x^2 - 4x - 9 < 0$ between the roots, the solution set is

$$\left(\frac{2 - \sqrt{22}}{2}, \frac{2 + \sqrt{22}}{2}\right).$$

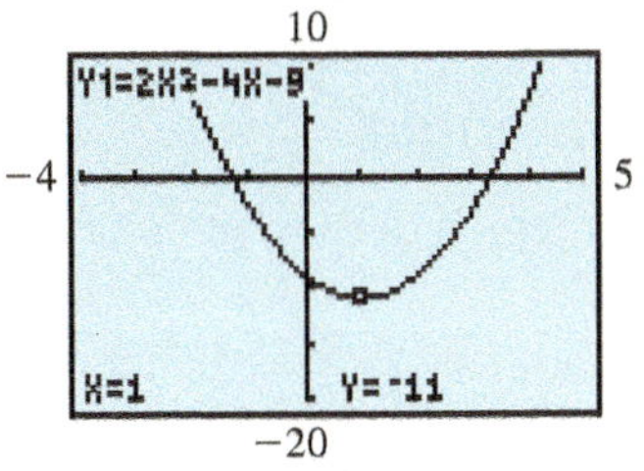


Figure 3.15

The graph of $y = 2x^2 - 4x - 9$ in Fig. 3.15 is below the $x$-axis between the $x$-intercepts and supports our conclusion. Note that every point that appears on the graph can be viewed as a test point.

▶**TRY THIS.** Solve $x^2 - 2x - 4 < 0$ using the test-point method. ■

If there are no real solutions to the quadratic equation that corresponds to a quadratic inequality, then the solution set to the inequality is either the empty set or all real numbers. For example, $x^2 + 4 = 0$ has no real solutions. Use any number as a test point to find that the solution set to $x^2 + 4 > 0$ is $(-\infty, \infty)$ and the solution set to $x^2 + 4 < 0$ is the empty set $\varnothing$.

## Applications of Maximum and Minimum

If one variable is a quadratic function of another, then the maximum or minimum value of the dependent variable occurs at the vertex of the graph of the quadratic function. A nice application of this idea occurs in modeling projectile motion. The formula $h(t) = -16t^2 + v_0 t + h_0$ is used to find the height in feet at time $t$ in seconds for a projectile that is launched in a vertical direction with initial velocity of $v_0$ feet per second from an initial height of $h_0$ feet. Since the height is a quadratic function of time, the maximum height occurs at the vertex of the parabola.

### EXAMPLE 8 Finding maximum height of a projectile

A ball is tossed straight upward with an initial velocity of 80 feet per second from a rooftop that is 12 feet above ground level. The height of the ball in feet at time $t$ in seconds is given by $h(t) = -16t^2 + 80t + 12$. Find the maximum height above ground level for the ball.

**Solution**

Since the height is a quadratic function of $t$ with a negative leading coefficient, the height has a maximum value at the vertex of the parabola. Use $-b/(2a)$ to find the $t$-coordinate of the vertex. Since $a = -16$ and $b = 80$

$$\frac{-b}{2a} = \frac{-80}{2(-16)} = 2.5.$$

So the ball reaches its maximum height at time $t = 2.5$ seconds. Now

$$h(2.5) = -16(2.5)^2 + 80(2.5) + 12 = 112.$$

So the maximum height of the ball is 112 feet above the ground.

▶**TRY THIS.** The height (in feet) above the ground for a ball that is tossed straight upward is given by $h(t) = -16t^2 + 96t + 40$, where $t$ is the time in seconds. Find the maximum height above the ground for this ball. ■

In Example 8, the quadratic function was given. In the next example we need to write the quadratic function and then find its maximum value.

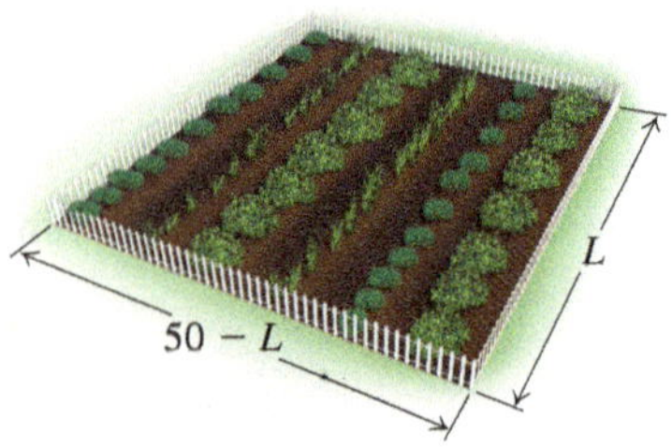


Figure 3.16

### EXAMPLE 9 Maximizing area of a rectangle

If 100 m of fencing will be used to fence a rectangular region, then what dimensions for the rectangle will maximize the area of the region?

**Solution**

Since the 100 m of fencing forms the perimeter of a rectangle as shown in Fig. 3.16, we have $2L + 2W = 100$, where $W$ is its width and $L$ is its length. Dividing by 2 we get $L + W = 50$ or $W = 50 - L$. Since $A = LW$ for a rectangle, by substituting we get

$$A = LW = L(50 - L) = -L^2 + 50L.$$

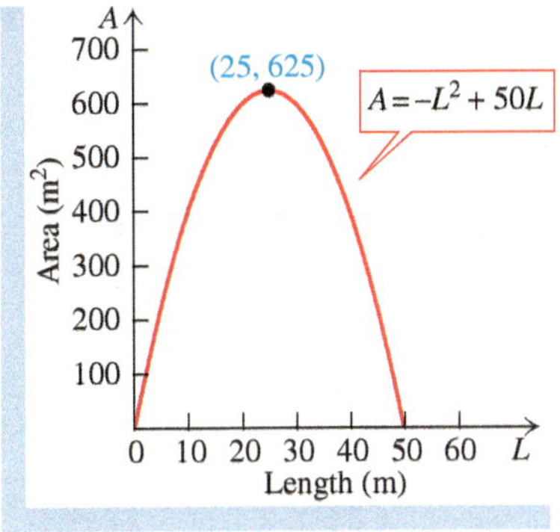


Figure 3.17

So the area is a quadratic function of the length. The graph of this function is the parabola in Fig. 3.17. Since the parabola opens downward, the maximum value of $A$ occurs when

$$L = \frac{-b}{2a} = \frac{-50}{2(-1)} = 25.$$

If $L = 25$ then $W = 25$ also, since $W = 50 - L$. So the length should be 25 meters and the width 25 meters to get the maximum area. The rectangle that gives the maximum area is actually a square with an area of 625 $m^2$.

▶**TRY THIS.** If a rectangle has a perimeter of 50 meters, then what dimensions will maximize its area? ■

## FOR thought... True or False? Explain.

1. The domain and range of a quadratic function are $(-\infty, \infty)$.
2. The vertex of the graph of $y = 2(x - 3)^2 - 1$ is $(3, 1)$.
3. The graph of $y = -3(x + 2)^2 - 9$ has no $x$-intercepts.
4. The maximum value of $y$ in the function $y = -4(x - 1)^2 + 9$ is 9.
5. For $y = 3x^2 - 6x + 7$, the value of $y$ is at its minimum when $x = 1$.
6. The graph of $f(x) = 9x^2 + 12x + 4$ has one $x$-intercept and one $y$-intercept.
7. The graph of every quadratic function has exactly one $y$-intercept.
8. The inequality $\pi(x - \sqrt{3})^2 + \pi/2 \leq 0$ has no solution.
9. The maximum area of a rectangle with fixed perimeter $p$ is $p^2/16$.
10. The function $f(x) = (x - 3)^2$ is increasing on the interval $[-3, \infty)$.

## EXERCISES 3.1

*Fill in the blank.*

1. If $a > 0$, the graph of $f(x) = a(x - h)^2 + k$ opens ________.
2. If $a < 0$, the graph of $f(x) = a(x - h)^2 + k$ opens ________.
3. The point $(h, k)$ is the ________ of the parabola $y = a(x - h)^2 + k$.
4. For $f(x) = ax^2 + bx + c$ $(a \neq 0)$, the $x$-coordinate of the ________ is $-b/(2a)$.
5. If $a > 0$ and $f(x) = a(x - h)^2 + k$, then $k$ is the ________ value of the function.
6. If $a < 0$ and $f(x) = a(x - h)^2 + k$, then $k$ is the ________ value of the function.
7. The line $x = -b/(2a)$ is the ________ for the function $f(x) = ax^2 + bx + c$ $(a \neq 0)$.
8. The point $(0, c)$ is the ________ for the function $f(x) = ax^2 + bx + c$ $(a \neq 0)$.

*Write each quadratic function in the form $y = a(x - h)^2 + k$ and sketch its graph.*

9. $y = x^2 + 4x$
10. $y = x^2 - 6x$
11. $y = x^2 - 3x$
12. $y = x^2 + 5x$
13. $y = 2x^2 - 12x + 22$
14. $y = 3x^2 - 12x + 1$
15. $y = -3x^2 + 6x - 3$
16. $y = -2x^2 - 4x + 8$
17. $y = x^2 + 3x + \frac{5}{2}$
18. $y = x^2 - x + 1$
19. $y = -2x^2 + 3x - 1$
20. $y = 3x^2 + 4x + 2$

*Find the vertex of the graph of each quadratic function. See the summary on finding the vertex on page 190.*

**21.** $f(x) = 3x^2 - 12x + 1$

**22.** $f(x) = -2x^2 - 8x + 9$

**23.** $f(x) = -3(x - 4)^2 + 1$

**24.** $f(x) = \frac{1}{2}(x + 6)^2 - \frac{1}{4}$

**25.** $y = -\frac{1}{2}x^2 - \frac{1}{3}x$

**26.** $y = \frac{1}{4}x^2 + \frac{1}{2}x - 1$

*From the graph of each parabola, determine whether the parabola opens upward or downward, and find the vertex, axis of symmetry, and range of the function. Find the maximum or minimum value of the function and the intervals on which the function is increasing or decreasing.*

**27.**

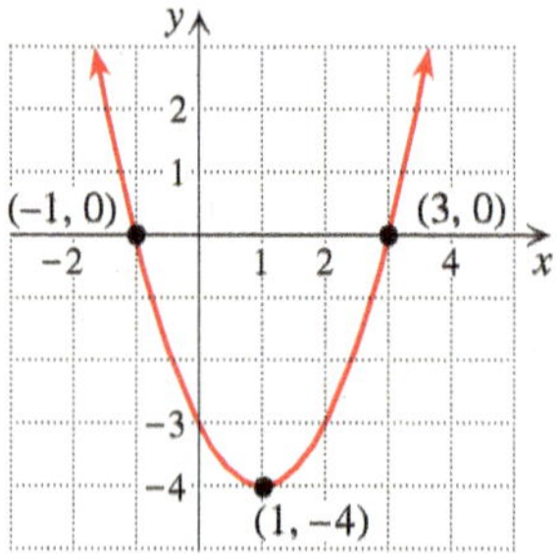


**28.**

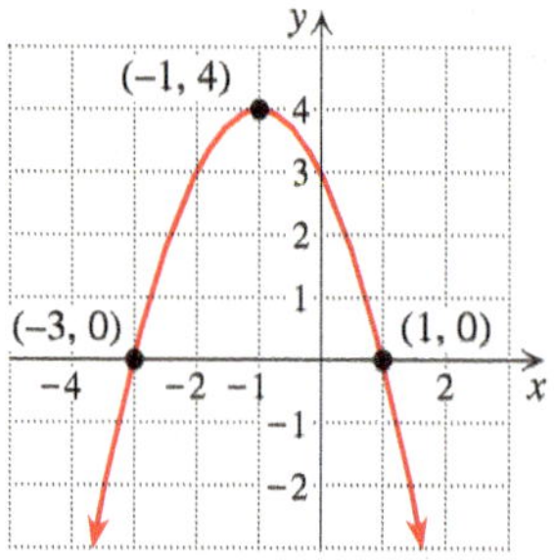


*Find the range of each quadratic function and the maximum or minimum value of the function. Identify the intervals on which each function is increasing or decreasing.*

**29.** $f(x) = 3 - x^2$

**30.** $f(x) = 5 - x^2$

**31.** $y = (x - 1)^2 - 1$

**32.** $y = (x + 3)^2 + 4$

**33.** $y = x^2 + 8x - 2$

**34.** $y = x^2 - 2x - 3$

**35.** $y = \frac{1}{2}(x - 3)^2 + 4$

**36.** $y = -\frac{1}{3}(x + 6)^2 + 37$

**37.** $f(x) = -2x^2 + 6x + 9$

**38.** $f(x) = -3x^2 - 9x + 4$

**39.** $y = -\frac{3}{4}\left(x - \frac{1}{2}\right)^2 + 9$

**40.** $y = \frac{3}{2}\left(x - \frac{1}{3}\right)^2 - 6$

*Identify the vertex, axis of symmetry, y-intercept, x-intercepts, and opening of each parabola, then sketch the graph.*

**41.** $y = x^2 - 3$

**42.** $y = 8 - x^2$

**43.** $y = x^2 - x$

**44.** $y = 2x - x^2$

**45.** $f(x) = x^2 + 6x + 9$

**46.** $f(x) = x^2 - 6x$

**47.** $f(x) = (x - 3)^2 - 4$

**48.** $f(x) = (x + 1)^2 - 9$

**49.** $y = -3(x - 2)^2 + 12$

**50.** $y = -2(x + 3)^2 + 8$

**51.** $y = -2x^2 + 4x + 1$

**52.** $y = -x^2 + 2x - 6$

*Solve each inequality by using the graphical method. State the solution set in interval notation.*

**53.** $x^2 - 2x - 3 > 0$

**54.** $x^2 + x - 2 \ge 0$

**55.** $x^2 - 4x + 1 < 0$

**56.** $x^2 - 2x - 1 \le 0$

**57.** $x + 1 < 6x^2$

**58.** $x + 6 > 5x^2$

*Identify the solution set to each quadratic inequality by inspecting the graphs of $y = x^2 - 2x - 3$ and $y = -x^2 - 2x + 3$ as shown.*

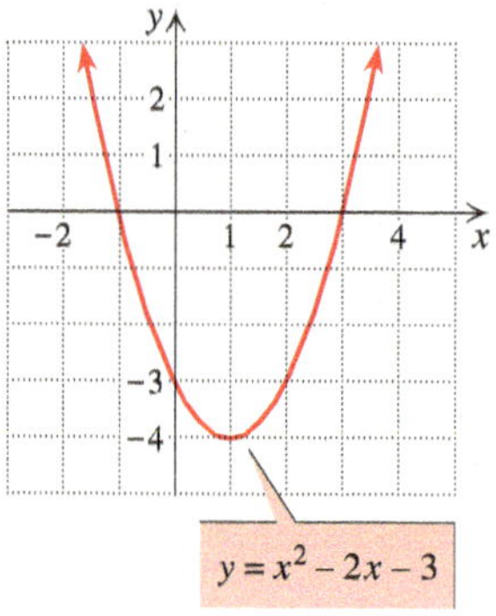


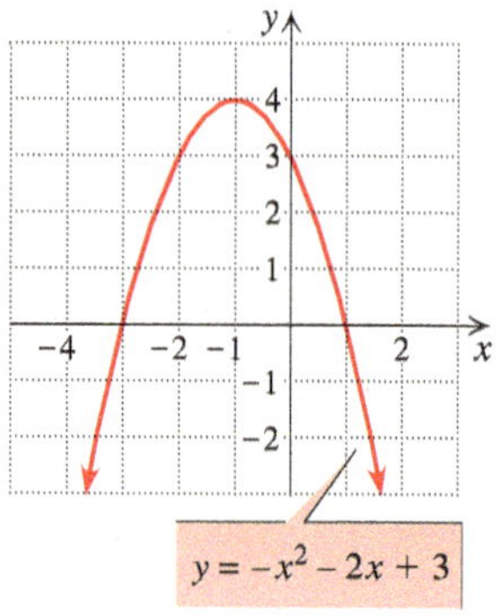


**59.** $x^2 - 2x - 3 \ge 0$

**60.** $x^2 - 2x - 3 < 0$

**61.** $-x^2 - 2x + 3 > 0$

**62.** $-x^2 - 2x + 3 \le 0$

**63.** $x^2 + 2x \le 3$

**64.** $x^2 \le 2x + 3$

*Solve each inequality by using the test-point method. State the solution set in interval notation and graph it.*

**65.** $x^2 - 4x + 2 < 0$

**66.** $x^2 - 4x + 1 \le 0$

**67.** $x^2 - 9 > 1$

**68.** $6 < x^2 - 1$

**69.** $y^2 + 18 > 10y$

**70.** $y^2 + 3 \ge 6y$

**71.** $p^2 + 9 > 0$

**72.** $-5 - s^2 < 0$

**73.** $a^2 + 20 \le 8a$

**74.** $6t \le t^2 + 25$

**75.** $-2w^2 + 5w < 6$

**76.** $-3z^2 - 5 > 2z$

*Solve each inequality by using the method of your choice. State the solution set in interval notation and graph it.*

**77.** $2x^2 - x - 3 < 0$

**78.** $3x^2 - 4x - 4 \le 0$

**79.** $2x + 15 < x^2$

**80.** $5x - x^2 < 4$

**81.** $w^2 - 4w - 12 \ge 0$

**82.** $y^2 + 8y + 15 \le 0$

**83.** $t^2 \le 16$

**84.** $36 \le h^2$

**85.** $a^2 + 6a + 9 \le 0$

**86.** $c^2 + 4 \le 4c$

**87.** $4z^2 - 12z + 9 > 0$

**88.** $9s^2 + 6s + 1 \ge 0$

*The next two exercises incorporate many concepts of quadratics.*

**89.** Let $f(x) = x^2 - 3x - 10$.
  **a.** Solve $f(x) = 0$.
  **b.** Solve $f(x) = -10$.
  **c.** Solve $f(x) > 0$.
  **d.** Solve $f(x) \leq 0$.
  **e.** Write $f$ in the form $f(x) = a(x - h)^2 + k$ and describe the graph of $f$ as a transformation of the graph of $y = x^2$.
  **f.** Graph $f$ and state the domain, range, and the maximum or minimum $y$-coordinate on the graph.
  **g.** What is the relationship between the graph of $f$ and the answers to parts (c) and (d)?
  **h.** Find the intercepts, axis of symmetry, vertex, opening, and intervals on which $f$ is increasing or decreasing.

**90.** Repeat parts (a) through (h) from the previous exercise for $f(x) = -x^2 + 2x + 1$.

*Solve each problem*

**91.** *Maximum Height of a Football* If a football is kicked straight up with an initial velocity of 128 ft/sec from a height of 5 ft, then its height above the earth is a function of time given by $h(t) = -16t^2 + 128t + 5$. What is the maximum height reached by this ball?
HINT Find the vertex of the graph of the quadratic function.

**92.** *Maximum Height of a Ball* If a juggler can toss a ball into the air at a velocity of 64 ft/sec from a height of 6 ft, then what is the maximum height reached by the ball?

**93.** *Shooting an Arrow* If an archer shoots an arrow straight upward with an initial velocity of 160 ft/sec from a height of 8 ft, then its height above the ground in feet at time $t$ in seconds is given by the function

$$h(t) = -16t^2 + 160t + 8.$$

  **a.** What is the maximum height reached by the arrow?
  **b.** How long does it take for the arrow to reach the ground?

**94.** *Rocket-Propelled Grenade* If a soldier in basic training fires a rocket-propelled grenade (RPG) straight up from ground level with an initial velocity of 256 ft/sec, then its height above the ground in feet at time $t$ in seconds is given by the function

$$h(t) = -16t^2 + 256t.$$

  **a.** What is the maximum height reached by the RPG?
  **b.** How long does it take for the RPG to reach the ground?

**95.** *Lindbergh's Air Speed* Flying too fast or slow wastes fuel. For the *Spirit of St. Louis,* miles per pound of fuel $M$ was a function of air speed $A$ in miles per hour, modeled by the formula

$$M = -0.000653A^2 + 0.127A - 5.01.$$

  **a.** Use the accompanying graph to estimate the most economical air speed.
  **b.** Use the formula to find the value of $A$ that maximizes $M$.
  **c.** How many gallons of fuel did he burn per hour if he flew at 97 mph and got 1.2 mi/lb of fuel, which weighed 6.12 lb/gal?

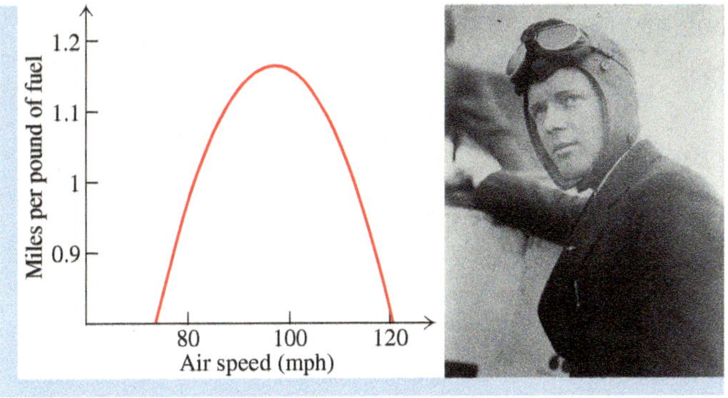


**Figure for Exercise 95**

**96.** *Average Farm Size* The average size in acres of a farm in the United States in the year 1994 + $x$ can be modeled by the quadratic function

$$A(x) = 0.37x^2 - 3.06x + 440.3.$$

  **a.** Find the year in which the average size of a farm reached its maximum.
  **b.** Find the year in which the average size was 470 acres.
  **c.** For what years shown in the accompanying figure is the average size decreasing? Increasing?
  **d.** When was the average size less than 435 acres?

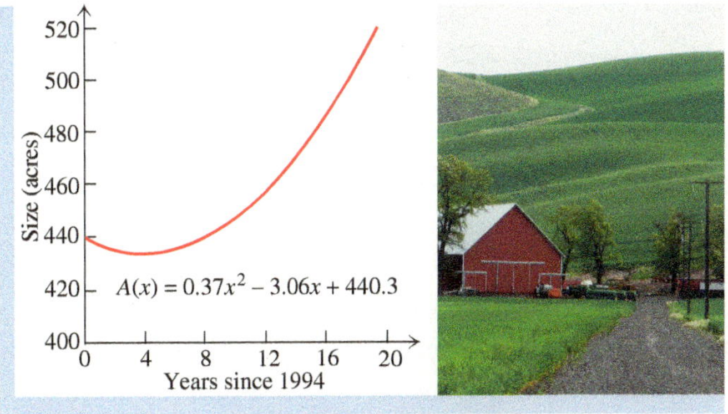


**Figure for Exercise 96**

**97.** *Maximum Area* Shondra wants to enclose a rectangular garden with 200 yards of fencing. What dimensions for the garden will maximize its area?
HINT Write a quadratic function for the area and find the vertex.

**98.** *Mirror Mirror* Chantel wants to make a rectangular frame for a mirror using 10 feet of frame molding. What dimensions will maximize the area of the mirror assuming that there is no waste?

**99.** *Twin Kennels* Martin plans to construct a rectangular kennel for two dogs using 120 feet of chain-link fencing. He plans to fence all four sides and down the middle to keep the dogs separate. What overall dimensions will maximize the total area fenced?
HINT Write a quadratic function for the area and find the vertex.

**100.** *Cross Fenced* Kim wants to construct rectangular pens for four animals with 400 feet of fencing. To get four separate pens she will fence a large rectangle and then fence through the middle of the rectangle parallel to the length and parallel to the width. What overall dimensions will maximize the total area of the pens?

**101.** *Big Barn* Mike wants to enclose a rectangular area for his rabbits alongside his large barn using 30 feet of fencing. What dimensions will maximize the area fenced if the barn is used for one side of the rectangle?

**102.** *Maximum Area* Kevin wants to enclose a rectangular garden using 14 eight-ft railroad ties, which he cannot cut. What are the dimensions of the rectangle that maximize the area enclosed?

**103.** *Cross Section of a Gutter* Seth has a piece of aluminum that is 10 in. wide and 12 ft long. He plans to form a gutter with a rectangular cross section and an open top by folding up the sides as shown in the figure. What dimensions of the gutter would maximize the amount of water that it can hold.
HINT Write a quadratic function for the cross-sectional area and find the vertex.

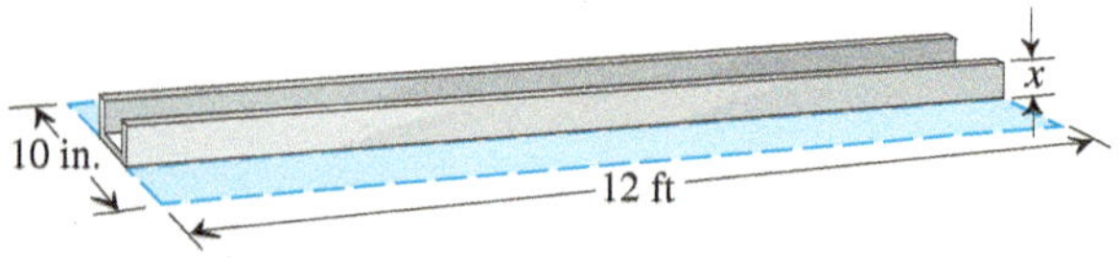


Figure for Exercise 103

## Foreshadowing Calculus

In algebra we can maximize or minimize a quadratic function because we can easily find the vertex of a parabola. However, max/min problems involving other functions are usually solved using techniques of calculus.

**104.** *Maximum Volume of a Cage* Sharon has a 12-ft board that is 12 in. wide. She wants to cut it into five pieces to make a cage for two pigeons, as shown in the figure. The front and back will be covered with chicken wire. What should be the dimensions of the cage to maximize the volume and use all of the 12-ft board?

Figure for Exercise 104

**105.** *Maximizing Revenue* Mona Kalini gives a walking tour of Honolulu to one person for \$49. To increase her business, she advertised at the National Orthodontist Convention that she would lower the price by \$1 per person for each additional person, up to 49 people.

**a.** Write the price per person $p$ as a function of the number of people $n$.
HINT See Exercise 101 of Section 1.4.

**b.** Write her revenue as a function of the number of people on the tour.
HINT Her revenue is the product of $n$ and $p$.

**c.** What is the maximum revenue for her tour?

**106.** *Concert Tickets* At \$10 per ticket, Willie Williams and the Wranglers will fill all 8000 seats in the Assembly Center. The manager knows that for every \$1 increase in the price, 500 tickets will go unsold.

**a.** Write the number of tickets sold $n$ as a function of ticket price $p$.
HINT See Exercise 102 of Section 1.4.

**b.** Write the total revenue as a function of the ticket price.
HINT Revenue is the product of $n$ and $p$.

**c.** What ticket price will maximize the revenue?

**107.** *Variance of the Number of Smokers* If $p$ is the probability that a randomly selected person in Chicago is a smoker, then $1 - p$ is the probability that the person is not a smoker. The variance of the number of smokers in a random sample of 50 Chicagoans is $50p(1 - p)$. What value of $p$ maximizes the variance?

**108.** *Rate of Flu Infection* In a town of 5000 people the daily rate of infection with a flu virus varies directly with the product of the number of people who have been infected and the number of people not infected. When 1000 people have been infected, the flu is spreading at a rate of 40 new cases per day. For what number of people infected is the daily rate of infection at its maximum?

**109.** *Altitude and Pressure* At an altitude of $h$ feet the atmospheric pressure $a$ (atm) is given by

$$a = 3.89 \times 10^{-10}h^2 - 3.48 \times 10^{-5}h + 1$$

(Sportscience, www.sportsci.org).

**a.** Graph this quadratic function.

**b.** In 1953, Edmund Hillary and Tenzing Norgay were the first persons to climb Mount Everest (see accompanying figure). Was the atmospheric pressure increasing or decreasing as they went to the 29,029-ft summit?

**c.** Where is the function decreasing? Increasing?

**d.** Does your answer to part (c) make sense?

**e.** For what altitudes do you think the function is valid?

Figure for Exercise 109

**110.** *Increasing Revenue* A company's weekly revenue in dollars is given by $R(x) = 2000x - 2x^2$, where $x$ is the number of items produced during a week.

**a.** For what $x$ is $R(x) > 0$?

**b.** On what interval is $R(x)$ increasing? Decreasing?

**c.** For what number of items is the revenue maximized?

**d.** What is the maximum revenue?

**e.** Find the marginal revenue function $MR(x)$, where $MR(x)$ is defined by

$$MR(x) = R(x + 1) - R(x).$$

**f.** On what interval is $MR(x)$ positive? Negative?

*Use a calculator or a computer for the following regression problems.*

**111.** *Quadratic Versus Linear* The average retail price for a used basic 2-door hatchback Corvette depends on the age of the car, as shown in the accompanying table for the summer of 2008 (Edmund's, www.edmunds.com).

Table for Exercise 111

| Age (yr) | Price ($) |
|---|---|
| 1 | 35,243 |
| 2 | 32,102 |
| 3 | 29,616 |
| 4 | 25,215 |
| 5 | 22,648 |
| 6 | 21,051 |
| 7 | 19,942 |
| 8 | 17,585 |
| 9 | 15,924 |

**a.** Use both linear regression and quadratic regression on a graphing calculator to express the price as a function of the age of the car.

**b.** Plot the data, the linear function, and the quadratic function on your calculator. Judging from what you see, which function appears to fit the data better?

**c.** Predict the price of an eleven-year-old car using both the linear function and the quadratic function.

**112.** *Homicides in the United States* The accompanying table gives the number of homicides per 100,000 population in the United States (Federal Bureau of Investigation, www.fbi.gov).

**a.** Use quadratic regression on a graphing calculator to express the number of homicides as a function of $x$, where $x$ is the number of years after 1994.

**b.** Plot the data and the quadratic function on your calculator. Judging from what you see, does the function appear to be a good model for the data?

**c.** Use the quadratic function to find the year in which the number of homicides was at a minimum.

**d.** Use the quadratic function to find the year in which there will be 10 homicides per 100,000 population.

Table for Exercise 112

| Year | Homicides per 100,000 | Year | Homicides per 100,000 |
|---|---|---|---|
| 1994 | 9.0 | 2001 | 5.6 |
| 1995 | 8.2 | 2002 | 5.6 |
| 1996 | 7.4 | 2003 | 5.7 |
| 1997 | 6.8 | 2004 | 5.5 |
| 1998 | 6.3 | 2005 | 5.6 |
| 1999 | 5.7 | 2006 | 5.7 |
| 2000 | 5.5 | 2007 | 5.5 |

## ▶ RETHINKING

**113.** If $y$ varies inversely as $x$ and $y = 3$ when $x = 12$, find $y$ when $x = 40$.

**114.** Is the function $f(x) = x^2 - 4$ one-to-one?

**115.** If $f(x) = 2x - 9$ and $g(x) = \sqrt{x}$, find $(f + g)(9)$.

**116.** The graph of $y = x^2$ is reflected in the $x$-axis, translated 4 units to the left and 5 units upward. What is the equation of the curve in its final position?

**117.** Find the domain and range for the function $\{(1, 2), (3, 5), (4, 2), (-3, 5)\}$.

**118.** Find the domain and range for $f(x) = -\sqrt{x - 1} + 2$.

### THINKING OUTSIDE THE BOX XXII

*Overlapping Region* A right triangle with sides of length 3, 4, and 5 is drawn so that the endpoints of the side of length 5 are $(0, 0)$ and $(5, 0)$. A square with sides of length 1 is drawn so that its center is the vertex of the right angle and its sides are parallel to the $x$- and $y$-axes. What is the area of the region where the square and the triangle overlap?

## POP QUIZ 3.1

**1.** Write $y = 2x^2 + 16x - 1$ in the form $y = a(x - h)^2 + k$.

**2.** Find the vertex of the graph of $y = 3(x + 4)^2 + 8$.

**3.** Find the range of $f(x) = -x^2 - 4x + 9$.

**4.** Find the minimum $y$-value for $y = x^2 - 3x$.

**5.** Find the $x$-intercepts and axis of symmetry for $y = x^2 - 2x - 8$.

**6.** Solve the inequality $x^2 + 4x < 0$.

## LINKING concepts... For Individual or Group Explorations

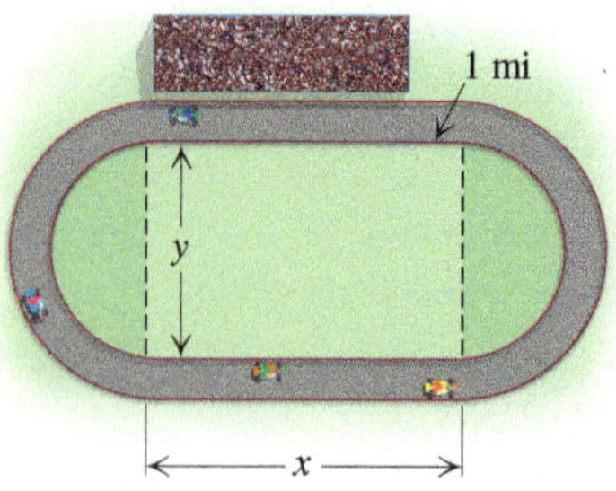


### Designing a Race Track

*An architect is designing a race track that will be 60 feet wide and have semicircular ends, as shown in the accompanying figure. The length of the track is to be one mile, measured on the inside edge of the track.*

**a)** Find the exact dimensions for $x$ and $y$ in the figure that will maximize the area of the rectangular center section.

**b)** Round the answers to part (a) to the nearest tenth of a foot and make an accurate drawing of the track.

**c)** Find the exact dimensions for $x$ and $y$ that will maximize the total area enclosed by the track.

**d)** Repeat parts (a), (b), and (c) assuming that the one-mile length is measured on the outside edge of the track.

# 3.2 Zeros of Polynomial Functions

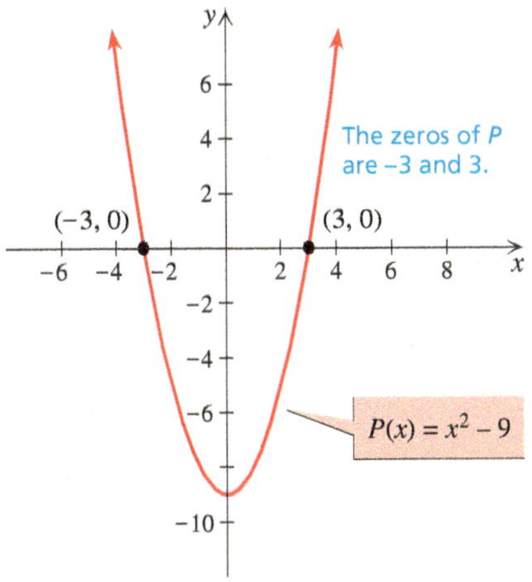


Figure 3.18

We have studied linear and quadratic functions extensively. In this section we will study general polynomial functions.

## The Remainder Theorem

If $y = P(x)$ is a polynomial function, then a value of $x$ that satisfies $P(x) = 0$ is called a **zero** of the polynomial function or a zero of the polynomial. For example, 3 and $-3$ are zeros of the function $P(x) = x^2 - 9$, because $P(3) = 0$ and $P(-3) = 0$. Note that the zeros of $P(x) = x^2 - 9$ are the same as the solutions to the equation $x^2 - 9 = 0$. The real zeros of a polynomial function appear on the graph of the function as the $x$-coordinates of the $x$-intercepts. The $x$-intercepts of the graph of $P(x) = x^2 - 9$ shown in Fig. 3.18 are $(-3, 0)$ and $(3, 0)$.

For polynomial functions of degree 2 or less, the zeros can be found by solving quadratic or linear equations. Our goal in this section is to find all of the zeros of a polynomial function when possible. For polynomials of degree higher than 2, the difficulty of this task ranges from easy to impossible, but we have some theorems to assist us. The remainder theorem relates evaluating a polynomial to division of polynomials.

**The Remainder Theorem**

If $R$ is the remainder when a polynomial $P(x)$ is divided by $x - c$, then $R = P(c)$.

**PROOF** Let $Q(x)$ be the quotient and $R$ be the remainder when $P(x)$ is divided by $x - c$. Since the dividend is equal to the divisor times the quotient plus the remainder, we have

$$P(x) = (x - c)Q(x) + R.$$

This statement is true for any value of $x$, and so it is also true for $x = c$:

$$\begin{aligned} P(c) &= (c - c)Q(c) + R \\ &= 0 \cdot Q(c) + R \\ &= R \end{aligned}$$

So $P(c)$ is equal to the remainder when $P(x)$ is divided by $x - c$. ■

To illustrate the remainder theorem, we will now use long division to evaluate a polynomial. Long division is discussed in Section A.3 of Appendix A.

### EXAMPLE 1 Using the remainder theorem to evaluate a polynomial

Use the remainder theorem to find $P(3)$ if $P(x) = 2x^3 - 5x^2 + 4x - 6$.

**Solution**

By the remainder theorem $P(3)$ is the remainder when $P(x)$ is divided by $x - 3$:

$$\begin{array}{r} 2x^2 + x + 7 \\ x - 3 \overline{\smash{)}\, 2x^3 - 5x^2 + 4x - 6} \\ \underline{2x^3 - 6x^2} \qquad\qquad \\ x^2 + 4x \qquad \\ \underline{x^2 - 3x} \qquad \\ 7x - 6 \\ \underline{7x - 21} \\ 15 \end{array}$$

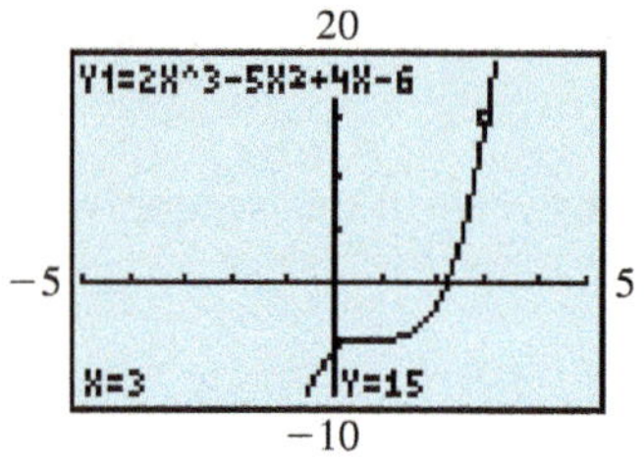


Figure 3.19

The remainder is 15 and therefore $P(3) = 15$. We can check by finding $P(3)$ in the usual manner:

$$P(3) = 2 \cdot 3^3 - 5 \cdot 3^2 + 4 \cdot 3 - 6 = 54 - 45 + 12 - 6 = 15$$

You can check this with a graphing calculator as shown in Fig. 3.19.

▸**TRY THIS.** Use the remainder theorem to find $P(1)$ if $P(x) = x^2 - 7x + 9$. ■

## Synthetic Division

In Example 1 we found $P(3) = 15$ in two different ways. Certainly, evaluating $P(x)$ for $x = 3$ in the usual manner is faster than dividing $P(x)$ by $x - 3$ using the ordinary method of dividing polynomials. However, there is a faster method, called **synthetic division,** for dividing by $x - 3$. Compare the two methods side by side, both showing $2x^3 - 5x^2 + 4x - 6$ divided by $x - 3$:

***Ordinary Division of Polynomials***

$$\begin{array}{r}
2x^2 + x + 7 \leftarrow \text{Quotient} \\
x - 3\overline{)2x^3 - 5x^2 + 4x - 6} \\
\underline{2x^3 - 6x^2} \qquad\qquad \\
x^2 + 4x \qquad \\
\underline{x^2 - 3x} \qquad \\
7x - 6 \\
\underline{7x - 21} \\
15 \leftarrow \text{Remainder}
\end{array}$$

***Synthetic Division***

$$\begin{array}{r|rrrr}
3 & 2 & -5 & 4 & -6 \\
 & & 6 & 3 & 21 \\
\hline
 & 2 & 1 & 7 & 15 \leftarrow \text{Remainder}
\end{array}$$

(2 1 7: Quotient)

Synthetic division certainly looks easier than ordinary division, and in general it is faster than evaluating the polynomial by substitution. Synthetic division is used as a quick means of dividing a polynomial by a binomial of the form $x - c$.

In synthetic division we write just the necessary parts of the ordinary division. Instead of writing $2x^3 - 5x^2 + 4x - 6$, write the coefficients 2, −5, 4, and −6. For $x - 3$, write only the 3. The bottom row in synthetic division gives the coefficients of the quotient and the remainder.

To actually perform the synthetic division, start with the following arrangement of coefficients:

$$\begin{array}{r|rrrr}
3 & 2 & -5 & 4 & -6 \\
 & & & & \\
\hline
\end{array}$$

Bring down the first coefficient, 2. Multiply 2 by 3 and write the answer beneath −5. Then add:

$$\begin{array}{r|rrrr}
3 & 2 & -5 & 4 & -6 \\
 & \downarrow & 6 & & \\
\hline
\text{Multiply} \rightarrow & 2 & 1 & & \\
 & & \uparrow & & \\
 & & \text{Add} & &
\end{array}$$

Using 3 rather than −3 when dividing by $x - 3$ allows us to multiply and add rather than multiply and subtract as in ordinary division. Now multiply 1 by 3 and write the answer beneath 4. Then add. Repeat the multiply-and-add step for the remaining column:

$$\begin{array}{r|rrrr}
3 & 2 & -5 & 4 & -6 \\
 & & 6 & 3 & 21 \\
\hline
 & 2 & 1 & 7 & 15
\end{array}$$

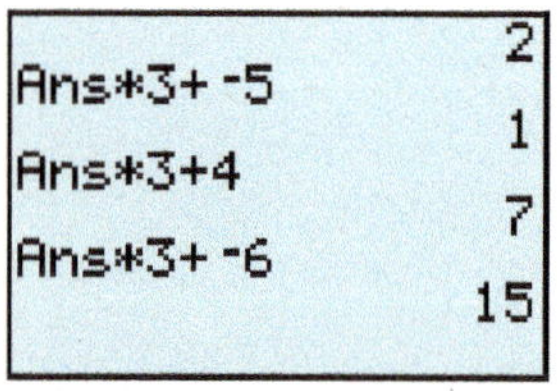


Figure 3.20

To perform this arithmetic on a graphing calculator, start with the leading coefficient 2 as the answer. Then repeatedly multiply the answer by 3 and add the next coefficient as shown in Fig. 3.20. □

The quotient is $2x^2 + x + 7$, and the remainder is 15. Since the divisor in synthetic division is of the form $x - c$, the degree of the quotient is always one less than the degree of the dividend.

## EXAMPLE 2 Synthetic division

Use synthetic division to find the quotient and remainder when $x^4 - 14x^2 + 5x - 9$ is divided by $x + 4$.

### Solution

Since $x + 4 = x - (-4)$, we use $-4$ in the synthetic division. Use $1, 0, -14, 5$, and $-9$ as the coefficients of the polynomial. We use 0 for the coefficient of the missing $x^3$-term, as we would in ordinary division of polynomials.

| | | | | | |
|---|---|---|---|---|---|
| −4 | 1 | 0 | −14 | 5 | −9 |
| | | −4 | 16 | −8 | 12 |
| Multiply → | 1 | −4 | 2 | −3 | 3 |
| | | ↑ Add | | | |

The quotient is $x^3 - 4x^2 + 2x - 3$ and the remainder is 3.

▶**TRY THIS.** Use synthetic division to find the quotient and remainder when $x^3 - 7x + 5$ is divided by $x + 2$. ■

To find the value of a polynomial $P(x)$ for $x = c$, we can divide $P(x)$ by $x - c$ or we can substitute $c$ for $x$ and compute. Using long division to divide by $x - c$ is not an efficient way to evaluate a polynomial. However, if we use synthetic division to divide by $x - c$ we can actually evaluate some polynomials using fewer arithmetic operations than we use in the substitution method. Note that in part (a) of the next example, synthetic division takes more steps than substitution, but in part (b) synthetic division takes fewer steps. To gain a better understanding of why synthetic division gives the value of a polynomial, see Linking Concepts at the end of this section.

## EXAMPLE 3 Using synthetic division to evaluate a polynomial

Let $f(x) = x^3$ and $g(x) = x^3 - 3x^2 + 5x - 12$. Use synthetic division to find the following function values.

**a.** $f(-2)$ **b.** $g(4)$

### Solution

**a.** To find $f(-2)$, divide the polynomial $x^3$ by $x - (-2)$ or $x + 2$ using synthetic division. Write $x^3$ as $x^3 + 0x^2 + 0x + 0$, and use 1, 0, 0, and 0 as the coefficients. We use a zero for each power of $x$ below $x^3$.

| | | | | |
|---|---|---|---|---|
| −2 | 1 | 0 | 0 | 0 |
| | | −2 | 4 | −8 |
| | 1 | −2 | 4 | −8 |

The remainder is $-8$, so $f(-2) = -8$. To check, find $f(-2) = (-2)^3 = -8$.

**b.** To find $g(4)$, use synthetic division to divide $x^3 - 3x^2 + 5x - 12$ by $x - 4$:

| | | | | |
|---|---|---|---|---|
| 4 | 1 | −3 | 5 | −12 |
| | | 4 | 4 | 36 |
| | 1 | 1 | 9 | 24 |

The remainder is 24, so $g(4) = 24$. Check this answer by finding $g(4) = 4^3 - 3(4^2) + 5(4) - 12 = 24$.

▶**TRY THIS.** Use synthetic division to find $P(3)$ if $P(x) = x^3 + x^2 - 9$. ■

## The Factor Theorem

Consider the polynomial function $P(x) = x^2 - x - 6$. We can find the zeros of the function by solving $x^2 - x - 6 = 0$ by factoring:

$$(x - 3)(x + 2) = 0$$

$$x - 3 = 0 \quad \text{or} \quad x + 2 = 0$$

$$x = 3 \quad \text{or} \quad x = -2$$

Both 3 and $-2$ are zeros of the function $P(x) = x^2 - x - 6$. Note how each factor of the polynomial corresponds to a zero of the function. This example suggests the following theorem.

**The Factor Theorem**

The number $c$ is a zero of the polynomial function $y = P(x)$ if and only if $x - c$ is a factor of the polynomial $P(x)$.

**PROOF** If $c$ is a zero of the polynomial function $y = P(x)$, then $P(c) = 0$. If $P(x)$ is divided by $x - c$, we get a quotient $Q(x)$ and a remainder $R$ such that

$$P(x) = (x - c)Q(x) + R.$$

By the remainder theorem, $R = P(c)$. Since $P(c) = 0$, we have $R = 0$ and $P(x) = (x - c)Q(x)$, which proves that $x - c$ is a factor of $P(x)$.

In Exercise 92 you will be asked to prove that if $x - c$ is a factor of $P(x)$, then $c$ is a zero of the polynomial function. These two arguments together establish the truth of the factor theorem. ■

Synthetic division can be used in conjunction with the factor theorem. If the remainder of dividing $P(x)$ by $x - c$ is 0, then $P(c) = 0$ and $c$ is a zero of the polynomial function. By the factor theorem, $x - c$ is a factor of $P(x)$.

### EXAMPLE 4 Using the factor theorem to factor a polynomial

Determine whether $x + 4$ is a factor of the polynomial $P(x) = x^3 - 13x + 12$. If it is a factor, then factor $P(x)$ completely.

**Solution**

By the factor theorem, $x + 4$ is a factor of $P(x)$ if and only if $P(-4) = 0$. We can find $P(-4)$ using synthetic division:

$$\begin{array}{r|rrrr} -4 & 1 & 0 & -13 & 12 \\ & & -4 & 16 & -12 \\ \hline & 1 & -4 & 3 & 0 \end{array}$$

Since $P(-4)$ is equal to the remainder, $P(-4) = 0$ and $-4$ is a zero of $P(x)$. By the factor theorem, $x + 4$ is a factor of $P(x)$. Since the other factor is the quotient from the synthetic division, $P(x) = (x + 4)(x^2 - 4x + 3)$. Factor the quadratic polynomial to get $P(x) = (x + 4)(x - 1)(x - 3)$.

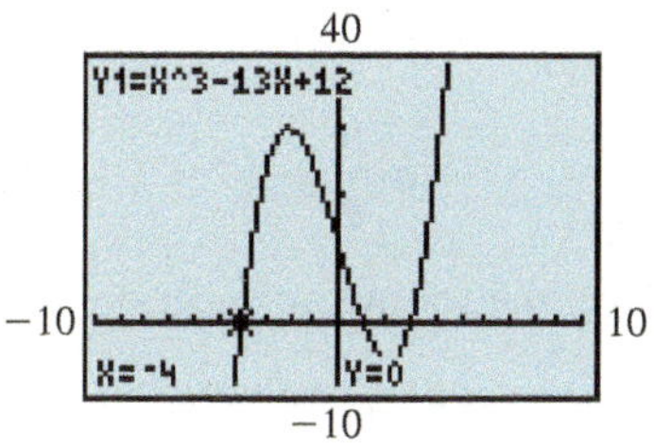


Figure 3.21

You can check this result by examining the calculator graph of $y = x^3 - 13x + 12$ shown in Fig. 3.21. The graph appears to cross the $x$-axis at $-4$, 1, and 3, supporting the conclusion that $P(x) = (x + 4)(x - 1)(x - 3)$.

▶**TRY THIS.** Is $x - 1$ a factor of $P(x) = x^3 - 3x + 2$? If it is a factor, then factor $P(x)$ completely. ■

## The Fundamental Theorem of Algebra

Whether a number is a zero of a polynomial function can be determined by synthetic division. But does every polynomial function have a zero? This question was answered in the affirmative by Carl F. Gauss when he proved the fundamental theorem of algebra in his doctoral thesis in 1799 at the age of 22.

**The Fundamental Theorem of Algebra**

If $y = P(x)$ is a polynomial function of positive degree, then $y = P(x)$ has at least one zero in the set of complex numbers.

Gauss also proved the $n$-root theorem of Section 3.3 that says that the number of zeros of a polynomial (or polynomial function) of degree $n$ is at most $n$. For example, a fifth-degree polynomial function has at least one zero and at most five.

Note that the zeros guaranteed by Gauss are in the set of complex numbers. So the zero might be real or imaginary. The theorem applies only to polynomial functions of degree 1 or more, because a polynomial function of zero degree such as $P(x) = 7$ has no zeros. A polynomial function of degree 1, $f(x) = ax + b$, has exactly one zero, which is found by solving $ax + b = 0$. A polynomial function of degree 2, $f(x) = ax^2 + bx + c$, has one or two zeros that can be found by the quadratic formula. For higher-degree polynomials the situation is not as simple. The fundamental theorem tells us that a polynomial function has at least one zero but not how to find it. For this purpose we have some other theorems.

**HISTORICAL NOTE**

Carl Friedrich Gauss (1777–1855) was a German mathematician and scientist of profound genius who contributed significantly to many fields, including number theory, analysis, differential geometry, geodesy, magnetism, astronomy, and optics. Gauss had a remarkable influence in many fields of mathematics and science and is ranked as one of history's most influential mathematicians.

## The Rational Zero Theorem

Zeros or roots that are rational numbers, the **rational zeros,** are generally the easiest to find. The polynomial function $f(x) = 6x^2 - x - 35$ has two rational zeros that can be found as follows:

$$6x^2 - x - 35 = 0$$

$$(2x - 5)(3x + 7) = 0 \quad \text{Factor.}$$

$$x = \frac{5}{2} \quad \text{or} \quad x = -\frac{7}{3}$$

Note that in $5/2$, 5 is a factor of $-35$ (the constant term) and 2 is a factor of 6 (the leading coefficient). For the zero $-7/3$, $-7$ is a factor of $-35$ and 3 is a factor of 6. Of course, these observations are not surprising, because we used these facts to factor the quadratic polynomial in the first place. Note that there are a lot of other factors of $-35$ and 6 for which the ratio is *not* a zero of this function. This example illustrates the rational zero theorem, which is also called the rational root theorem.

**The Rational Zero Theorem**

If $f(x) = a_n x^n + a_{n-1}x^{n-1} + a_{n-2}x^{n-2} + \cdots + a_1 x + a_0$ is a polynomial function with integral coefficients ($a_n \neq 0$ and $a_0 \neq 0$) and $p/q$ (in lowest terms) is a rational zero of $f(x)$, then $p$ is a factor of the constant term $a_0$ and $q$ is a factor of the leading coefficient $a_n$.

**PROOF** If $p/q$ is a zero of $f(x)$, then $f(p/q) = 0$:

$$a_n\left(\frac{p}{q}\right)^n + a_{n-1}\left(\frac{p}{q}\right)^{n-1} + a_{n-2}\left(\frac{p}{q}\right)^{n-2} + \cdots + a_1\frac{p}{q} + a_0 = 0$$

Subtract $a_0$ from each side and multiply by $q^n$ to get the following equation:

$$a_np^n + a_{n-1}p^{n-1}q + a_{n-2}p^{n-2}q^2 + \cdots + a_1pq^{n-1} = -a_0q^n$$

Because the coefficients are integers and $p$ and $q$ are integers, both sides of this equation are integers. Since $p$ is a factor of the left side, $p$ must be a factor of the right side, which is the same integer. Since $p/q$ is in lowest terms, $p$ is not a factor of $q$. So $p$ must be a factor of $a_0$.

To prove that $q$ is a factor of $a_n$, rearrange the preceding equation as follows:

$$a_{n-1}p^{n-1}q + a_{n-2}\,p^{n-2}q^2 + \cdots + a_1pq^{n-1} + a_0q^n = -a_np^n$$

Now $q$ is a factor of $a_n$ by the same argument used previously. ■

The rational zero theorem does not identify exactly which rational numbers are zeros of a function; it only gives *possibilities* for the rational zeros.

## EXAMPLE 5 Using the rational zero theorem

Find all possible rational zeros for each polynomial function.

**a.** $f(x) = 2x^3 - 3x^2 - 11x + 6$ **b.** $g(x) = 3x^3 - 8x^2 - 8x + 8$

### Solution

**a.** If the rational number $p/q$ is a zero of $f(x)$, then $p$ is a factor of 6 and $q$ is a factor of 2. The positive factors of 6 are 1, 2, 3, and 6. The positive factors of 2 are 1 and 2. Take each factor of 6 and divide by 1 to get 1/1, 2/1, 3/1, and 6/1. Take each factor of 6 and divide by 2 to get 1/2, 2/2, 3/2, and 6/2. Simplify the ratios, eliminate duplications, and put in the negative factors to get

$$\pm 1, \quad \pm 2, \quad \pm 3, \quad \pm 6, \quad \pm\frac{1}{2}, \quad \text{and} \quad \pm\frac{3}{2}$$

as the possible rational zeros to the function $f(x)$.

**b.** If the rational number $p/q$ is a zero of $g(x)$, then $p$ is a factor of 8 and $q$ is a factor of 3. The factors of 8 are 1, 2, 4, and 8. The factors of 3 are 1 and 3. If we take all possible ratios of a factor of 8 over a factor of 3, we get

$$\pm 1, \quad \pm 2, \quad \pm 4, \quad \pm 8, \quad \pm\frac{1}{3}, \quad \pm\frac{2}{3}, \quad \pm\frac{4}{3}, \quad \text{and} \quad \pm\frac{8}{3}$$

as the possible rational zeros of the function $g(x)$.

▸**TRY THIS.** Find all possible rational zeros for $h(x) = 2x^3 + x^2 - 4x - 3$. ■

Our goal is to find all of the zeros to a polynomial function. The zeros to a polynomial function might be rational, irrational, or imaginary. We can determine the rational zeros by simply evaluating the polynomial function for every number in the list of possible rational zeros. If the list is long, looking at a graph of the function can speed up the process. We will use synthetic division to evaluate the polynomial, because synthetic division gives the quotient polynomial as well as the value of the polynomial.

## EXAMPLE 6 Finding all zeros of a polynomial function

Find all of the real and imaginary zeros for each polynomial function of Example 5.

**a.** $f(x) = 2x^3 - 3x^2 - 11x + 6$ **b.** $g(x) = 3x^3 - 8x^2 - 8x + 8$

### Solution

**a.** The possible rational zeros of $f(x)$ are listed in Example 5(a). Use synthetic division to check each possible zero to see whether it is actually a zero. Try 1 first.

$$\begin{array}{r|rrrr} 1 & 2 & -3 & -11 & 6 \\ & & 2 & -1 & -12 \\ \hline & 2 & -1 & -12 & -6 \end{array}$$

Since the remainder is $-6$, 1 is not a zero of the function. Keep on trying numbers from the list of possible rational zeros. To save space, we will not show any more failures. So try 1/2 next.

$$\begin{array}{r|rrrr} \frac{1}{2} & 2 & -3 & -11 & 6 \\ & & 1 & -1 & -6 \\ \hline & 2 & -2 & -12 & 0 \end{array}$$

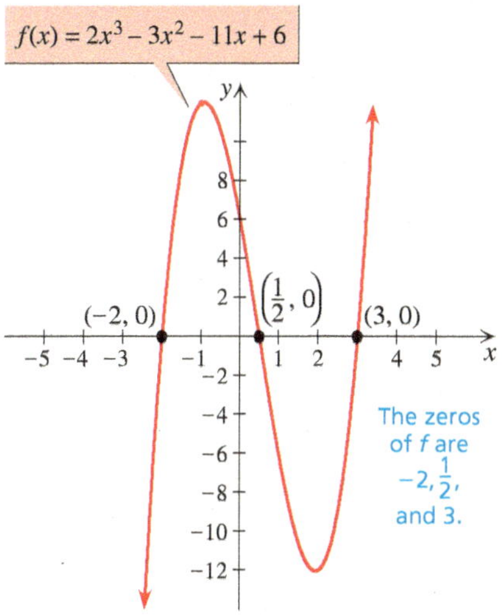


Figure 3.22

Since the remainder in the synthetic division is 0, 1/2 is a zero of $f(x)$. By the factor theorem, $x - 1/2$ is a factor of the polynomial. The quotient is the other factor.

$$2x^3 - 3x^2 - 11x + 6 = 0$$

$$\left(x - \frac{1}{2}\right)(2x^2 - 2x - 12) = 0 \quad \text{Factor.}$$

$$(2x - 1)(x^2 - x - 6) = 0 \quad \text{Factor 2 out of the second "factor" and distribute it into the first factor.}$$

$$(2x - 1)(x - 3)(x + 2) = 0 \quad \text{Factor completely.}$$

$$2x - 1 = 0 \quad \text{or} \quad x - 3 = 0 \quad \text{or} \quad x + 2 = 0$$

$$x = \frac{1}{2} \quad \text{or} \quad x = 3 \quad \text{or} \quad x = -2$$

The zeros of the function $f$ are 1/2, 3, and $-2$. Note that each zero of $f$ corresponds to an $x$-intercept on the graph of $f$ shown in Fig. 3.22. Because this polynomial had three rational zeros, we could have found them all by using synthetic division or by examining the calculator graph. However, it is good to factor the polynomial to see the correspondence between the three zeros, the three factors, and the three $x$-intercepts.

You could speed up this process of finding a zero with the calculator graph shown in Fig. 3.23. It is not too hard to discover that 1/2 is a zero by looking at the graph and the possible rational zeros that we listed in Example 5. If you use a graph to find that 1/2 is a zero, you still need to do synthetic division to factor the polynomial. □

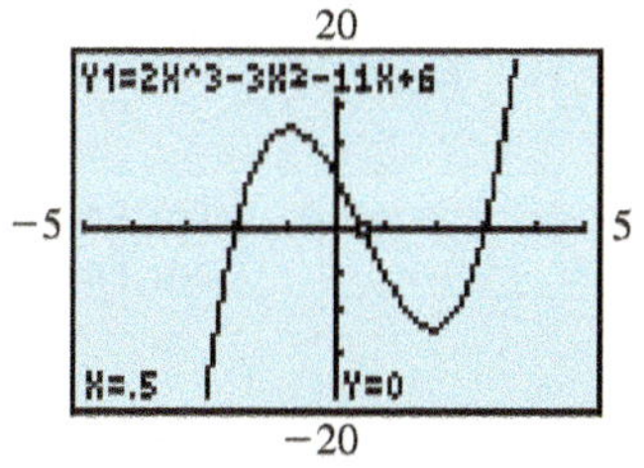


Figure 3.23

**b.** The possible rational zeros of $g(x)$ are listed in Example 5(b). First check 2/3 to see whether it produces a remainder of 0.

$$\begin{array}{r|rrrr} \frac{2}{3} & 3 & -8 & -8 & 8 \\ & & 2 & -4 & -8 \\ \hline & 3 & -6 & -12 & 0 \end{array}$$

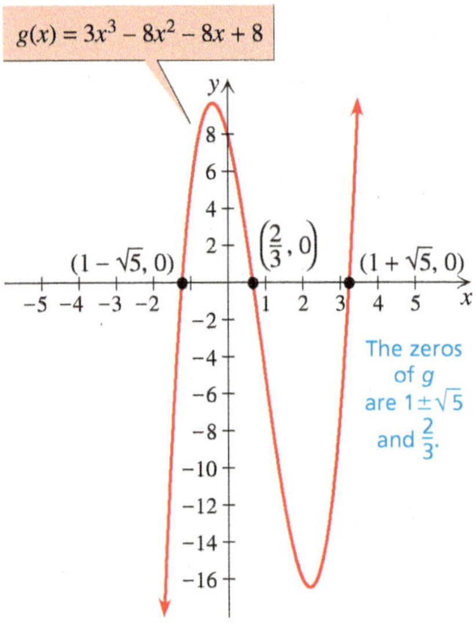


**Figure 3.24**

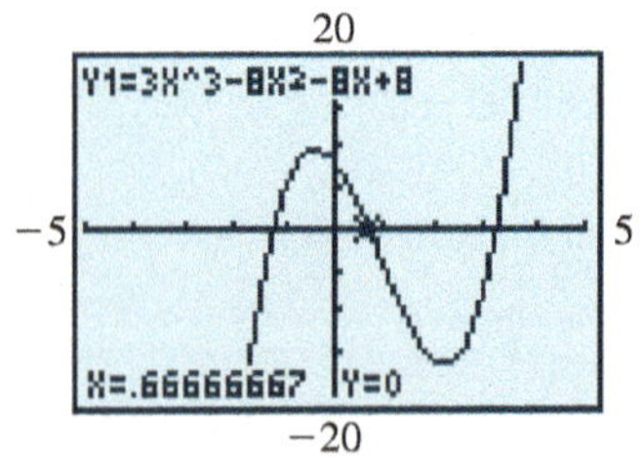


**Figure 3.25**

Since the remainder in the synthetic division is 0, $2/3$ is a zero of $g(x)$. By the factor theorem, $x - 2/3$ is a factor of the polynomial. The quotient is the other factor.

$$3x^3 - 8x^2 - 8x + 8 = 0$$

$$\left(x - \frac{2}{3}\right)(3x^2 - 6x - 12) = 0$$

$$(3x - 2)(x^2 - 2x - 4) = 0$$

$$3x - 2 = 0 \quad \text{or} \quad x^2 - 2x - 4 = 0$$

$$x = \frac{2}{3} \quad \text{or} \quad x = \frac{2 \pm \sqrt{20}}{2}$$

$$x = \frac{2}{3} \quad \text{or} \quad x = 1 \pm \sqrt{5}$$

There are one rational and two irrational roots to the equation. So the zeros of the function $g$ are $2/3$, $1 + \sqrt{5}$, and $1 - \sqrt{5}$. Each zero corresponds to an $x$-intercept on the graph of $g$ shown in Fig. 3.24.

You could graph the function with a calculator as shown in Fig. 3.25. Keeping in mind the list of possible rational zeros, it is not hard to discover that $2/3$ is a zero.

▶**TRY THIS.** Find all zeros for $h(x) = 2x^3 + x^2 - 4x - 3$. ■

Note that in Example 6(a) all of the zeros were rational. All three could have been found by continuing to check the possible rational zeros using synthetic division. In Example 6(b) we would be wasting time if we continued to check the possible rational zeros, because there is only one. When we get a quadratic polynomial, it is best to either factor the quadratic polynomial or use the quadratic formula to find the remaining zeros. Of course, the remaining zeros could be imaginary, as is demonstrated in the next example.

## EXAMPLE 7 Finding all zeros of a polynomial function

Find all real and imaginary zeros for each function.

**a.** $f(x) = 2x^3 - 11x^2 + 22x - 15$ **b.** $g(x) = x^4 + 3x^3 + x^2 + 15x - 20$

### Solution

**a.** The possible rational zeros consist of all possible factors of 15 over factors of 2:

$$\pm 1, \pm 3, \pm 5, \pm 15, \pm\frac{1}{2}, \pm\frac{3}{2}, \pm\frac{5}{2}, \pm\frac{15}{2}$$

Now use synthetic division to check if any of these is actually a zero of the function. To save space, we will try $3/2$ first:

$$\begin{array}{c|cccc} \frac{3}{2} & 2 & -11 & 22 & -15 \\ & & 3 & -12 & 15 \\ \hline & 2 & -8 & 10 & 0 \end{array}$$

Since 0 is the remainder, $3/2$ is a zero of $f(x)$. By the factor theorem $x - 3/2$ is a factor of the polynomial. The quotient is the other factor.

$$2x^3 - 11x^2 + 22x - 15 = 0$$

$$\left(x - \frac{3}{2}\right)(2x^2 - 8x + 10) = 0$$

$$(2x - 3)(x^2 - 4x + 5) = 0$$

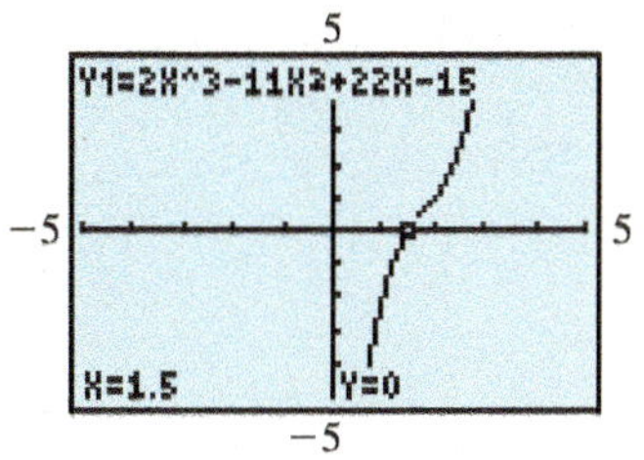


Figure 3.26

Apply the quadratic formula to $x^2 - 4x + 5 = 0$ to get

$$x = \frac{4 \pm \sqrt{(-4)^2 - 4(1)(5)}}{2(1)} = \frac{4 \pm \sqrt{-4}}{2} = \frac{4 \pm 2i}{2} = 2 \pm i.$$

So the zeros of the function are 3/2, $2 - i$, and $2 + i$.

Note that the graph of $f(x) = 2x^3 - 11x^2 + 22x - 15$ in Fig. 3.26 has only one $x$-intercept because there is only one real zero to the function. □

**b.** The possible rational zeros consist of all possible factors of 20 over factors of 1:

$$\pm 1, \pm 2, \pm 4, \pm 5, \pm 10, \pm 20$$

Use synthetic division to check if any of these numbers is actually a zero of the function. To save space, we will try 1 and −4 first:

| | | | | | |
|---|---|---|---|---|---|
| 1 | 1 | 3 | 1 | 15 | −20 |
| | | 1 | 4 | 5 | 20 |
| −4 | 1 | 4 | 5 | 20 | 0 |
| | | −4 | 0 | −20 | |
| | 1 | 0 | 5 | 0 | |

Now we have $g(x) = (x - 1)(x + 4)(x^2 + 5)$. If $x^2 + 5 = 0$, then $x = \pm i\sqrt{5}$. So the zeros of $g(x)$ are 1, −4, and $\pm i\sqrt{5}$.

▶**TRY THIS.** Find all real and imaginary zeros to $f(x) = 3x^3 - 2x^2 + 12x - 8$. ■

## FOR thought... True or False? Explain.

**1.** The function $f(x) = 1/x$ has at least one zero.

**2.** If $P(x) = x^4 - 6x^2 - 8$ is divided by $x^2 - 2$, then the remainder is $P(\sqrt{2})$.

**3.** If $1 - 2i$ and $1 + 2i$ are zeros of $P(x) = x^3 - 5x^2 + 11x - 15$, then $x - 1 - 2i$ and $x - 1 + 2i$ are factors of $P(x)$.

**4.** If we divide $x^5 - 1$ by $x - 2$, then the remainder is 31.

**5.** Every polynomial function has at least one zero.

**6.** If $P(x) = x^3 - x^2 + 4x - 5$ and $b$ is the remainder from division of $P(x)$ by $x - c$, then $b^3 - b^2 + 4b - 5 = c$.

**7.** If $P(x) = x^3 - 5x^2 + 4x - 15$, then $P(4) = 0$.

**8.** The equation $\pi^2 x^4 - \frac{1}{\sqrt{2}}x^3 + \frac{1}{\sqrt{7} + \pi} = 0$ has at least one complex solution.

**9.** The binomial $x - 1$ is a factor of $x^5 + x^4 - x^3 - x^2 - x + 1$.

**10.** The binomial $x + 3$ is a factor of $3x^4 - 5x^3 + 7x^2 - 9x - 2$.

## EXERCISES 3.2

*Fill in the blank.*

**1.** If $y = P(x)$ is a polynomial function and $P(w) = 0$, then $w$ is a(n) ________ of the function.

**2.** If $R$ is the ________ when a polynomial $P(x)$ is divided by $x - c$, then $R = P(c)$.

**3.** The number $c$ is a(n) ________ of the polynomial function $y = P(x)$ if and only if $x - c$ is a(n) ________ of the polynomial $P(x)$.

**4.** If $y = P(x)$ is a polynomial function of positive degree, then $y = P(x)$ has at least one ________ in the set of ________.

*Use ordinary division of polynomials to find the quotient and remainder when the first polynomial is divided by the second.*

**5.** $x^2 - 5x + 7, x - 2$

**6.** $x^2 - 3x + 9, x - 4$

**7.** $-2x^3 + 4x - 9, x + 3$

**8.** $-4w^3 + 5w^2 - 7, w - 3$

**9.** $s^4 - 3s^2 + 6, s^2 - 5$

**10.** $h^4 + 3h^3 + h - 5, h^2 - 3$

*Use synthetic division to find the quotient and remainder when the first polynomial is divided by the second.*

**11.** $x^2 + 4x + 1, x - 2$

**12.** $2x^2 - 3x + 6, x - 5$

**13.** $-x^3 + x^2 - 4x + 9, x + 3$

**14.** $-3x^3 + 5x^2 - 6x + 1, x + 1$

**15.** $4x^3 - 5x + 2, x - \frac{1}{2}$

**16.** $-6x^3 + 25x^2 - 9, x - \frac{3}{2}$

**17.** $2a^3 - 3a^2 + 4a + 3, a + \frac{1}{2}$

**18.** $-3b^3 - b^2 - 3b - 1, b + \frac{1}{3}$

**19.** $x^4 - 3, x - 1$

**20.** $x^4 - 16, x - 2$

**21.** $x^5 - 6x^3 + 4x - 5, x - 2$

**22.** $2x^5 - 5x^4 - 5x + 7, x - 3$

*Let $f(x) = x^5 - 1$, $g(x) = x^3 - 4x^2 + 8$, and $h(x) = 2x^4 + x^3 - x^2 + 3x + 3$. Find the following function values by using synthetic division. Check by using substitution.*

**23.** $f(1)$ **24.** $f(-1)$ **25.** $f(-2)$ **26.** $f(3)$

**27.** $g(1)$ **28.** $g(-1)$ **29.** $g\left(-\frac{1}{2}\right)$ **30.** $g\left(\frac{1}{2}\right)$

**31.** $h(-1)$ **32.** $h(2)$ **33.** $h(1)$ **34.** $h(-3)$

*Determine whether the given binomial is a factor of the polynomial following it. If it is a factor, then factor the polynomial completely.*

**35.** $x + 3, x^3 + 4x^2 + x - 6$

**36.** $x + 5, x^3 + 8x^2 + 11x - 20$

**37.** $x - 4, x^3 + 4x^2 - 17x - 60$

**38.** $x - 2, x^3 - 12x^2 + 44x - 48$

*Determine whether each given number is a zero of the polynomial function following the number.*

**39.** $3, f(x) = 2x^3 - 5x^2 - 4x + 3$

**40.** $-2, g(x) = 3x^3 - 6x^2 - 3x - 19$

**41.** $-2, g(d) = d^3 + 2d^2 + 3d + 1$

**42.** $-1, w(x) = 3x^3 + 2x^2 - 2x - 1$

**43.** $-1, P(x) = x^4 + 2x^3 + 4x^2 + 6x + 3$

**44.** $3, G(r) = r^4 + 4r^3 + 5r^2 + 3r + 17$

**45.** $\frac{1}{2}, H(x) = x^3 + 3x^2 - 5x + 7$

**46.** $-\frac{1}{2}, T(x) = 2x^3 + 3x^2 - 3x - 2$

*Use the rational zero theorem to find all possible rational zeros for each polynomial function.*

**47.** $f(x) = x^3 - 9x^2 + 26x - 24$

**48.** $g(x) = x^3 - 2x^2 - 5x + 6$

**49.** $h(x) = x^3 - x^2 - 7x + 15$

**50.** $m(x) = x^3 + 4x^2 + 4x + 3$

**51.** $P(x) = 8x^3 - 36x^2 + 46x - 15$

**52.** $T(x) = 18x^3 - 9x^2 - 5x + 2$

**53.** $M(x) = 18x^3 - 21x^2 + 10x - 2$

**54.** $N(x) = 4x^3 - 10x^2 + 4x + 5$

*Find all of the real and imaginary zeros for each polynomial function.*

**55.** $f(x) = x^3 - 9x^2 + 26x - 24$

**56.** $g(x) = x^3 - 2x^2 - 5x + 6$

**57.** $h(x) = x^3 - x^2 - 7x + 15$

**58.** $m(x) = x^3 + 4x^2 + 4x + 3$

**59.** $P(a) = 8a^3 - 36a^2 + 46a - 15$

**60.** $T(b) = 18b^3 - 9b^2 - 5b + 2$

**61.** $M(t) = 18t^3 - 21t^2 + 10t - 2$

**62.** $N(t) = 4t^3 - 10t^2 + 4t + 5$

**63.** $y = x^3 - 26x + 60$

**64.** $y = x^3 - x^2 + 2$

**65.** $S(w) = w^4 + w^3 - w^2 + w - 2$

**66.** $W(v) = 2v^4 + 5v^3 + 3v^2 + 15v - 9$

**67.** $V(x) = x^4 + 2x^3 - x^2 - 4x - 2$

**68.** $U(x) = x^4 - 4x^3 + x^2 + 12x - 12$

**69.** $f(x) = 24x^3 - 26x^2 + 9x - 1$

**70.** $f(x) = 30x^3 - 47x^2 - x + 6$

**71.** $y = 16x^3 - 33x^2 + 82x - 5$

**72.** $y = 15x^3 - 37x^2 + 44x - 14$

**73.** $f(x) = 21x^4 - 31x^3 - 21x^2 - 31x - 42$

**74.** $f(x) = 119x^4 - 5x^3 + 214x^2 - 10x - 48$

**75.** $f(x) = (x^2 + 9)(x^3 + 6x^2 + 3x - 10)$

**76.** $f(x) = (x^2 - 5)(x^3 - 5x^2 - 12x + 36)$

**77.** $f(x) = (x^2 - 4x + 1)(x^3 - 9x^2 + 23x - 15)$

**78.** $f(x) = (x^2 - 4x + 13)(x^3 - 4x^2 - 17x + 60)$

*Use division to write each rational expression in the form quotient + remainder/divisor. Use synthetic division when possible.*

**79.** $\dfrac{2x + 1}{x - 2}$ **80.** $\dfrac{x - 1}{x + 3}$ **81.** $\dfrac{a^2 - 3a + 5}{a - 3}$

**82.** $\dfrac{2b^2 - 3b + 1}{b + 2}$ **83.** $\dfrac{c^2 - 3c - 4}{c^2 - 4}$ **84.** $\dfrac{2h^2 + h - 2}{h^2 - 1}$

**85.** $\dfrac{4t - 5}{2t + 1}$ **86.** $\dfrac{6y - 1}{3y - 1}$

*Solve each problem.*

**87.** *Drug Testing* The concentration of a drug (in parts per million) in a patient's bloodstream $t$ hours after administration of the drug is given by the function

$$P(t) = -t^4 + 12t^3 - 58t^2 + 132t.$$

**a.** Use the formula to determine when the drug will be totally eliminated from the bloodstream.

**b.** Use the graph to estimate the maximum concentration of the drug.

**c.** Use the graph to estimate the time at which the maximum concentration occurs.

**d.** Use the graph to estimate the amount of time for which the concentration is above 80 ppm.

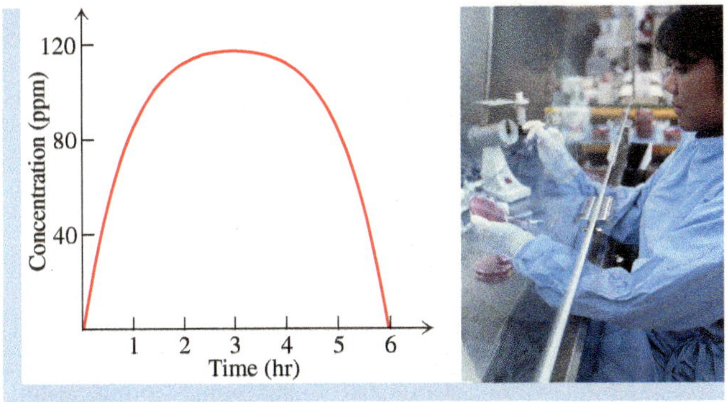


Figure for Exercise 87

**88.** *Open-Top Box* Joan intends to make an 18-in.$^3$ open-top box out of a 6 in. by 7 in. piece of copper by cutting equal squares ($x$ in. by $x$ in.) from the corners and folding up the sides. Write the difference between the intended volume and the actual volume as a function of $x$. For what value of $x$ is there no difference between the intended volume and the actual volume?

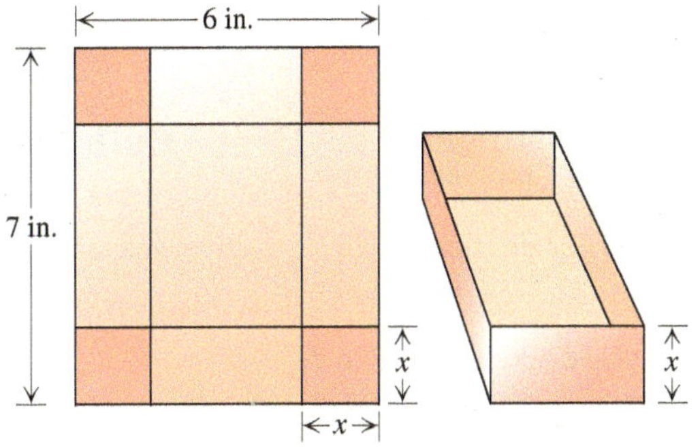


Figure for Exercise 88

**89.** *Cartridge Box* The height of a box containing an HP Laser Jet III printer cartridge is 4 in. more than the width and the length is 9 in. more than the width. If the volume of the box is 630 in.$^3$, then what are the dimensions of the box?

**90.** *Computer Case* The width of the case for a 733-megahertz Pentium computer is 4 in. more than twice the height and the depth is 1 in. more than the width. If the volume of the case is 1632 in.$^3$, then what are the dimensions of the case?

## FOR WRITING/DISCUSSION

**91.** *Synthetic Division* Explain how synthetic division can be used to find the quotient and remainder when $3x^3 + 4x^2 + 2x - 4$ is divided by $3x - 2$.

**92.** Prove that if $x - c$ is a factor of $P(x)$, then $c$ is a zero of the polynomial function.

## ▶ RETHINKING

**93.** Write the function $f(x) = 2x^2 - 3x + 1$ in the form $f(x) = a(x - h)^2 + k$.

**94.** Solve $2x^2 - 3x + 1 = 0$.

**95.** Solve $|3x + 7| \geq 0$. Write the solution set in interval notation.

**96.** Solve $2x - 9 = 5 - 4(3x + 2)$.

**97.** Factor completely.
**a.** $24a^3 + 18a^2 - 60a$

**b.** $x^5 - 16x$

**98.** Solve $8 - 6(x - 4) \leq 2(x - 5) - 4(2x - 1)$.

## THINKING OUTSIDE THE BOX XXIII

*Moving a Refrigerator* A box containing a refrigerator is 3 ft wide, 3 ft deep, and 6 ft high. To move it, Wally lays it on its side, then on its top, then on its other side, and finally stands it upright as shown in the figure. Exactly how far has point $A$ traveled in going from its initial location to its final location?

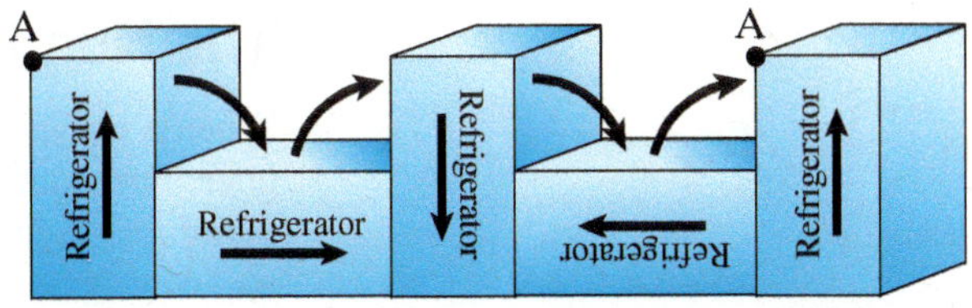


Figure for Thinking Outside the Box XXIII

## POP QUIZ 3.2

1. Use ordinary division to find the quotient and remainder when $x^3 - 5x + 7$ is divided by $x + 4$.

2. Use synthetic division to find the quotient and remainder when $x^2 - 3x + 9$ is divided by $x - 5$.

3. Use synthetic division to find $f(3)$ if $f(x) = x^3 - 2x^2 + 4x - 1$.

4. List the possible rational zeros for $f(x) = 2x^3 - 3x + 8$.

5. Find all real and imaginary zeros for $f(x) = 2x^3 - x^2 + 18x - 9$.

## LINKING concepts... For Individual or Group Explorations

### Horner's Method

*A fourth-degree polynomial in x such as* $3x^4 + 5x^3 + 4x^2 + 3x + 1$ *contains all of the powers of x from the first through the fourth. However, any polynomial can be written without powers of x. Evaluating a polynomial without powers of x (Horner's method) is somewhat easier than evaluating a polynomial with powers.*

**a)** Show that $\{[(3x + 5)x + 4]x + 3\}x + 1 = 3x^4 + 5x^3 + 4x^2 + 3x + 1$ is an identity.

**b)** Rewrite the polynomial $P(x) = 6x^5 - 3x^4 + 9x^3 + 6x^2 - 8x + 12$ without powers of $x$ as in part (a).

**c)** Find $P(2)$ without a calculator using both forms of the polynomial.

**d)** For which form did you perform fewer arithmetic operations?

**e)** Explain in detail how to rewrite any polynomial without powers of $x$.

**f)** Explain how this new form relates to synthetic division and the remainder theorem.

# 3.3 The Theory of Equations

One of the main goals in algebra is to keep expanding our knowledge of solving equations. The solutions (roots) of a polynomial equation $P(x) = 0$ are precisely the zeros of a polynomial function $y = P(x)$. Therefore the theorems of Section 3.2 concerning zeros of polynomial functions apply also to the roots of polynomial equations. In this section we study several additional theorems that are useful in solving polynomial equations.

## The Number of Roots of a Polynomial Equation

When a polynomial equation is solved by factoring, a factor may occur more than once. For example, $x^2 - 10x + 25 = 0$ is equivalent to $(x - 5)^2 = 0$. Since the factor $x - 5$ occurs twice, we say that 5 is a root of the equation with *multiplicity* 2.

**Definition: Multiplicity**

If the factor $x - c$ occurs $k$ times in the complete factorization of the polynomial $P(x)$, then $c$ is called a root of $P(x) = 0$ with **multiplicity** $k$.

If a quadratic equation has a single root, as in $x^2 - 10x + 25 = 0$, then that root has multiplicity 2. If a root with multiplicity 2 is counted as two roots, then every quadratic equation has two roots in the set of complex numbers. This situation is generalized in the following theorem, where the phrase "when multiplicity is considered" means that a root with multiplicity $k$ is counted as $k$ individual roots.

***n*-Root Theorem**

If $P(x) = 0$ is a polynomial equation with real or complex coefficients and positive degree $n$, then, when multiplicity is considered, $P(x) = 0$ has $n$ roots.

**PROOF** By the fundamental theorem of algebra, the polynomial equation $P(x) = 0$ with degree $n$ has at least one complex root $c_1$. By the factor theorem, $P(x) = 0$ is equivalent to

$$(x - c_1)Q_1(x) = 0,$$

where $Q_1(x)$ is a polynomial with degree $n - 1$ (the quotient when $P(x)$ is divided by $x - c_1$). Again, by the fundamental theorem of algebra, there is at least one complex root $c_2$ of $Q_1(x) = 0$. By the factor theorem, $P(x) = 0$ can be written as

$$(x - c_1)(x - c_2)Q_2(x) = 0,$$

where $Q_2(x)$ is a polynomial with degree $n - 2$. Reasoning in this manner $n$ times, we get a quotient polynomial that has 0 degree, $n$ factors for $P(x)$, and $n$ complex roots, not necessarily all different. ■

## EXAMPLE 1 Finding all roots of a polynomial equation

State the degree of each polynomial equation. Find all real and imaginary roots of each equation, stating multiplicity when it is greater than one.

**a.** $6x^5 + 24x^3 = 0$ **b.** $(x - 3)^2(x + 14)^5 = 0$

### Solution

**a.** This fifth-degree equation can be solved by factoring:

$$6x^3(x^2 + 4) = 0$$

$$6x^3 = 0 \quad \text{or} \quad x^2 + 4 = 0$$

$$x^3 = 0 \quad \text{or} \quad x^2 = -4$$

$$x = 0 \quad \text{or} \quad x = \pm 2i$$

The roots are $\pm 2i$ and 0. Since there are two imaginary roots and 0 is a root with multiplicity 3, there are five roots when multiplicity is considered.

Because 0 is the only real root, the graph of $y = 6x^5 + 24x^3$ has only one $x$-intercept at (0, 0) as shown in Fig. 3.27. □

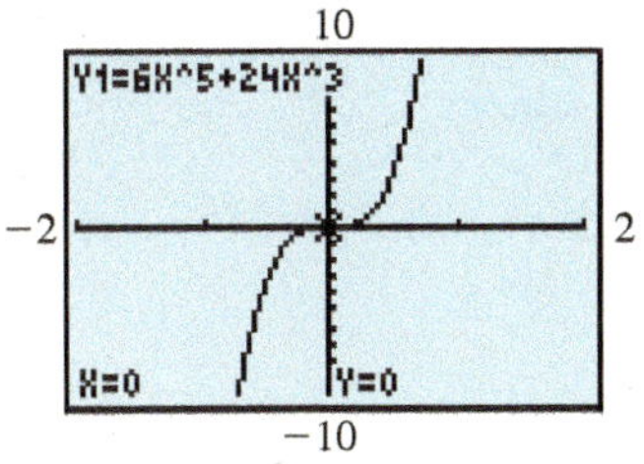

Figure 3.27

**b.** The highest power of $x$ in $(x - 3)^2$ is 2, and in $(x + 14)^5$ is 5. By the product rule for exponents, the highest power of $x$ in this equation is 7. The only roots of this seventh-degree equation are 3 and $-14$. The root 3 has multiplicity 2, and $-14$ has multiplicity 5. So there are seven roots when multiplicity is considered.

Because the equation has two real solutions, the graph of $y = (x - 3)^2(x + 14)^5$ has two $x$-intercepts at (3, 0) and $(-14, 0)$ as shown in Fig. 3.28.

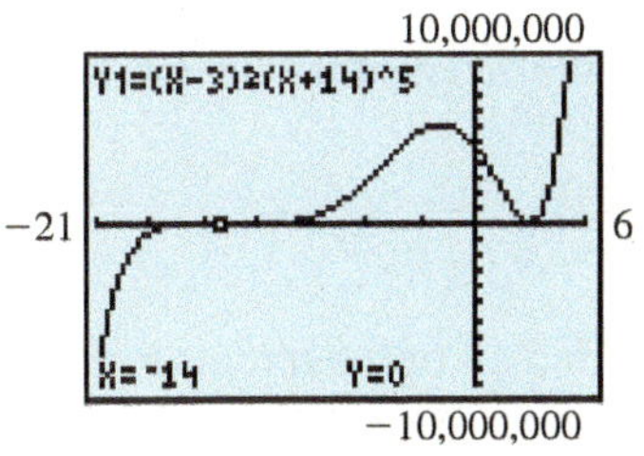

Figure 3.28

▶**TRY THIS.** Find all roots to $x^3(x + 2)^2(2x - 5) = 0$, including multiplicities. ■

Note that graphing polynomial functions and solving polynomial equations go hand in hand. The solutions to the equation can help us find an appropriate viewing window for the graph as they did in Example 1, and the graph can help us find solutions to the equation.

## The Conjugate Pairs Theorem

For second-degree polynomial equations, the imaginary roots occur in pairs. For example, the roots of $x^2 - 2x + 5 = 0$ are

$$x = \frac{2 \pm \sqrt{(-2)^2 - 4(1)(5)}}{2} = 1 \pm 2i.$$

The roots $1 - 2i$ and $1 + 2i$ are complex conjugates. The $\pm$ symbol in the quadratic formula causes the complex solutions of a quadratic equation with real coefficients to occur in conjugate pairs. The conjugate pairs theorem indicates that this situation occurs also for polynomial equations of higher degree.

**Conjugate Pairs Theorem**

If $P(x) = 0$ is a polynomial equation with real coefficients and the complex number $a + bi$ $(b \neq 0)$ is a root, then $a - bi$ is also a root.

The proof for this theorem is left for the exercises.

### EXAMPLE 2 Using the conjugate pairs theorem

Find a polynomial equation with real coefficients that has 2 and $1 - i$ as roots.

**Solution**

If the polynomial has real coefficients, then its imaginary roots occur in conjugate pairs. So a polynomial with these two roots must actually have at least three roots: 2, $1 - i$, and $1 + i$. Since each root of the equation corresponds to a factor of the polynomial, we can write the following equation.

$$(x - 2)[x - (1 - i)][x - (1 + i)] = 0$$
$$(x - 2)[(x - 1) + i][(x - 1) - i] = 0 \quad \text{Regroup.}$$
$$(x - 2)[(x - 1)^2 - i^2] = 0 \quad (a + b)(a - b) = a^2 - b^2$$
$$(x - 2)[x^2 - 2x + 1 + 1] = 0 \quad i^2 = -1$$
$$(x - 2)(x^2 - 2x + 2) = 0$$
$$x^3 - 4x^2 + 6x - 4 = 0$$

This equation has the required roots and the smallest degree. Any multiple of this equation would also have the required roots but would not be as simple.

**▸TRY THIS.** Find a polynomial equation with real coefficients that has 3 and $-i$ as roots. ■

**HISTORICAL NOTE**

Évariste Galois (1811–1832) was a French mathematician. While still in his teens, he was able to determine a necessary and sufficient condition for a polynomial equation to be solvable by radicals, thereby solving a long-standing problem. His work laid the fundamental foundations for Galois theory, a major branch of abstract algebra.

Galois died in a duel at the age of twenty.

## Descartes's Rule of Signs

None of the theorems in this chapter tells us how to find all of the $n$ roots to a polynomial equation of degree $n$. However, the theorems and rules presented here add to our knowledge of polynomial equations and help us to predict the type and number of solutions to expect for a particular equation. Descartes's rule of signs is a method for determining the number of positive, negative, and imaginary solutions. For this rule, a solution with multiplicity $k$ is counted as $k$ solutions.

When a polynomial is written in descending order, a **variation of sign** occurs when the signs of consecutive terms change. For example, if

$$P(x) = 3x^5 - 7x^4 - 8x^3 - x^2 + 3x - 9,$$

there are sign changes in going from the first to the second term, from the fourth to the fifth term, and from the fifth to the sixth term. So there are three variations of sign for $P(x)$. This information determines the number of positive real solutions to $P(x) = 0$. Descartes's rule requires that we look at $P(-x)$ and also count the variations of sign after it is simplified:

$$\begin{aligned} P(-x) &= 3(-x)^5 - 7(-x)^4 - 8(-x)^3 - (-x)^2 + 3(-x) - 9 \\ &= -3x^5 - 7x^4 + 8x^3 - x^2 - 3x - 9 \end{aligned}$$

In $P(-x)$ the signs of the terms change from the second to the third term and again from the third to the fourth term. So there are two variations of sign for $P(-x)$. This information determines the number of negative real solutions to $P(x) = 0$.

### Descartes's Rule of Signs

Suppose $P(x) = 0$ is a polynomial equation with real coefficients and with terms written in descending order.

- The number of positive real roots of the equation is either equal to the number of variations of sign of $P(x)$ or less than that by an even number.
- The number of negative real roots of the equation is either equal to the number of variations of sign of $P(-x)$ or less than that by an even number.

The proof of Descartes's rule of signs is beyond the scope of this text, but we can apply the rule to polynomial equations. Descartes's rule of signs is especially helpful when the number of variations of sign is 0 or 1.

### EXAMPLE 3 Using Descartes's rule of signs

Discuss the possibilities for the roots to $2x^3 - 5x^2 - 6x + 4 = 0$.

#### Solution

The number of variations of sign in

$$P(x) = 2x^3 - 5x^2 - 6x + 4$$

is 2. By Descartes's rule, the number of positive real roots is either 2 or 0. Since

$$\begin{aligned} P(-x) &= 2(-x)^3 - 5(-x)^2 - 6(-x) + 4 \\ &= -2x^3 - 5x^2 + 6x + 4, \end{aligned}$$

there is one variation of sign in $P(-x)$. So there is exactly one negative real root.

The equation must have three roots, because it is a third-degree polynomial equation. Since there must be three roots and one is negative, the other two roots must be either both imaginary numbers or both positive real numbers. Table 3.1 summarizes these two possibilities.

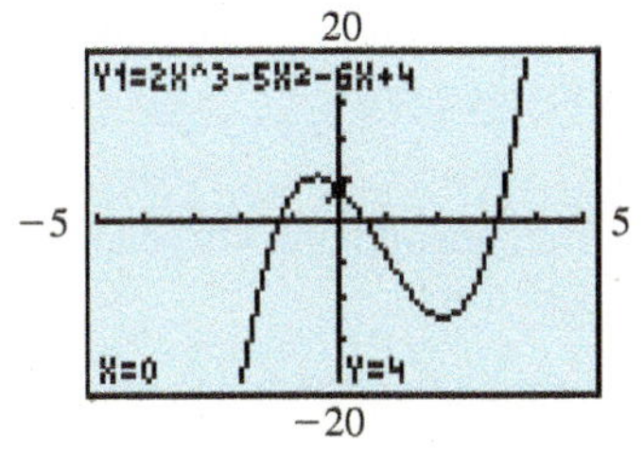


Figure 3.29

**Table 3.1** Number of roots

| Positive | Negative | Imaginary |
|---|---|---|
| 2 | 1 | 0 |
| 0 | 1 | 2 |

The graph of $y = 2x^3 - 5x^2 - 6x + 4$ shown in Fig. 3.29 crosses the positive $x$-axis twice and the negative $x$-axis once. So the first case in Table 3.1 is actually correct.

▸**TRY THIS.** Discuss the possibilities for the roots to $x^3 - 5x^2 + 4x + 3 = 0$. ■

## EXAMPLE 4 Using Descartes's rule of signs

Discuss the possibilities for the roots to each equation.

**a.** $3x^4 - 5x^3 - x^2 - 8x + 4 = 0$ **b.** $3x^3 + 4x^2 + 5 = 0$

### Solution

**a.** There are two variations of sign in the polynomial

$$P(x) = 3x^4 - 5x^3 - x^2 - 8x + 4.$$

According to Descartes's rule, there are either two or zero positive real roots to the equation. Since

$$P(-x) = 3(-x)^4 - 5(-x)^3 - (-x)^2 - 8(-x) + 4$$
$$= 3x^4 + 5x^3 - x^2 + 8x + 4,$$

there are two variations of sign in $P(-x)$. So the number of negative real roots is either two or zero. Since the degree of the polynomial is 4, there must be four roots. Each line of Table 3.2 gives a possible distribution of the type of those four roots. Note that the number of imaginary roots is even in each case, as we would expect from the conjugate pairs theorem.

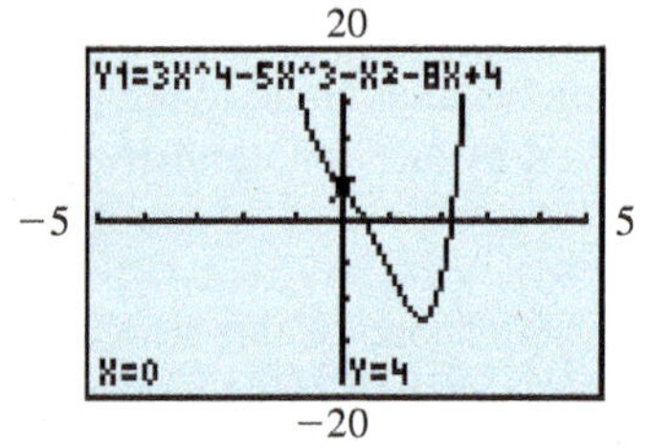


Figure 3.30

**Table 3.2** Number of roots

| Positive | Negative | Imaginary |
|---|---|---|
| 2 | 2 | 0 |
| 2 | 0 | 2 |
| 0 | 2 | 2 |
| 0 | 0 | 4 |

The calculator graph of $y = 3x^4 - 5x^3 - x^2 - 8x + 4$ in Fig. 3.30 shows two positive intercepts and no negative intercepts. However, we might not have the appropriate viewing window. The negative intercepts might be less than $-5$. In this case the graph did not allow us to conclude which line in Table 3.2 is correct. □

**b.** There are no variations of sign in $P(x) = 3x^3 + 4x^2 + 5$. So there are no positive real roots to the equation. Since

$$P(-x) = -3x^3 + 4x^2 + 5,$$

there is one variation of sign in $P(-x)$. So there is exactly one negative real root. Note that zero is not a root. Since the degree of the polynomial is 3, there must be 3 complex roots. Thus there is one negative real root and two imaginary roots.

▶**TRY THIS.** Discuss the possibilities for the roots to $x^4 - 6x^2 + 10 = 0$. ■

## Bounds on the Roots

If a polynomial equation has no roots greater than $c$, then $c$ is called an **upper bound** for the roots. If there are no roots less than $c$, then $c$ is called a **lower bound** for the roots. The next theorem is used to determine upper and lower bounds for the roots of a polynomial equation. We will not prove this theorem.

**Theorem on Bounds**

Suppose that $P(x)$ is a polynomial with real coefficients and a positive leading coefficient, and synthetic division with $x - c$ is performed.

- If $c > 0$ and all terms in the bottom row are nonnegative, then $c$ is an upper bound for the roots of $P(x) = 0$.
- If $c < 0$ and the terms in the bottom row alternate in sign, then $c$ is a lower bound for the roots of $P(x) = 0$.

If 0 appears in the bottom row of the synthetic division, then it may be assigned either a positive or negative sign in determining whether the signs alternate. For example, the numbers 3, 0, 5, and −6 would be alternating in sign if we assign a negative sign to 0. The numbers −7, 5, −8, 0, and −2 would be alternating in sign if we assign a positive sign to 0.

## EXAMPLE 5 Finding bounds for the roots

Use the theorem on bounds to establish the best integral bounds for the roots of $2x^3 - 5x^2 - 6x + 4 = 0$.

### Solution

Try synthetic division with the integers 1, 2, 3, and so on. The first integer for which all terms on the bottom row are nonnegative is the best upper bound for the roots according to the theorem on bounds. Note that we are not trying fractions because we are looking for integral bounds for the roots and not the actual roots.

| | | | | |
|---|---|---|---|---|
| 1 | 2 | −5 | −6 | 4 |
| | | 2 | −3 | −9 |
| | 2 | −3 | −9 | −5 |

| | | | | |
|---|---|---|---|---|
| 2 | 2 | −5 | −6 | 4 |
| | | 4 | −2 | −16 |
| | 2 | −1 | −8 | −12 |

| | | | | |
|---|---|---|---|---|
| 3 | 2 | −5 | −6 | 4 |
| | | 6 | 3 | −9 |
| | 2 | 1 | −3 | −5 |

| | | | | |
|---|---|---|---|---|
| 4 | 2 | −5 | −6 | 4 |
| | | 8 | 12 | 24 |
| | 2 | 3 | 6 | 28 |

By the theorem on bounds, no number greater than 4 can be a root to the equation. Now try synthetic division with the integers −1, −2, −3, and so on. The first negative integer for which the terms on the bottom row alternate in sign is the best lower bound for the roots.

| | | | | |
|---|---|---|---|---|
| −1 | 2 | −5 | −6 | 4 |
| | | −2 | 7 | −1 |
| | 2 | −7 | 1 | 3 |

| | | | | |
|---|---|---|---|---|
| −2 | 2 | −5 | −6 | 4 |
| | | −4 | 18 | −24 |
| | 2 | −9 | 12 | −20 |

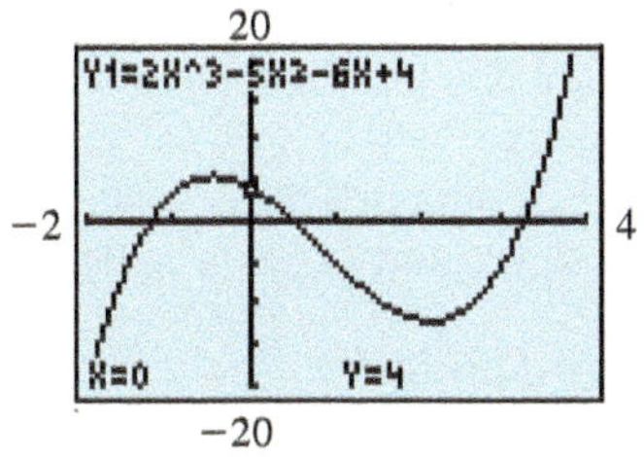

Figure 3.31

By the theorem on bounds, no number less than −2 can be a root to the equation. So all of the real roots to this equation are between −2 and 4.

The graph of $y = 2x^3 - 5x^2 - 6x + 4$ in Fig. 3.31 shows three $x$-intercepts between −2 and 4, which supports our conclusion about the bounds for the roots. Note that the bounds for the roots help you determine a good window.

**▸TRY THIS.** Use the theorem on bounds to establish the best integral bounds for the roots to $x^3 + 3x^2 - 5x - 15 = 0$. ■

In the next example we use all of the available information about roots.

## EXAMPLE 6 Using all of the theorems about roots

Find all of the solutions to $2x^3 - 5x^2 - 6x + 4 = 0$.

### Solution

In Example 3 we used Descartes's rule of signs on this equation to determine that it has either two positive roots and one negative root or one negative root and two imaginary roots. In Example 5 we used the theorem on bounds to determine that all of the real roots to this equation are between −2 and 4. From the rational zero theorem, the possible rational roots are ±1, ±2, ±4, and ±1/2. Since there must be one negative root and it must be greater than −2, the only possible numbers from the list are $-1$ and $-1/2$. So start by checking $-1/2$ and $-1$ with synthetic division.

$$\begin{array}{r|rrrr} -\frac{1}{2} & 2 & -5 & -6 & 4 \\ & & -1 & 3 & \frac{3}{2} \\ \hline & 2 & -6 & -3 & \frac{11}{2} \end{array} \qquad \begin{array}{r|rrrr} -1 & 2 & -5 & -6 & 4 \\ & & -2 & 7 & -1 \\ \hline & 2 & -7 & 1 & 3 \end{array}$$

Since neither $-1$ nor $-1/2$ is a root, the negative root must be irrational. The only rational possibilities for the two positive roots smaller than 4 are $1/2$, 1, and 2.

$$\begin{array}{r|rrrr} \frac{1}{2} & 2 & -5 & -6 & 4 \\ & & 1 & -2 & -4 \\ \hline & 2 & -4 & -8 & 0 \end{array}$$

Since $1/2$ is a root of the equation, $x - 1/2$ is a factor of the polynomial. The last line in the synthetic division indicates that the other factor is $2x^2 - 4x - 8$.

$$\left(x - \frac{1}{2}\right)(2x^2 - 4x - 8) = 0$$

$$(2x - 1)(x^2 - 2x - 4) = 0$$

$$2x - 1 = 0 \quad \text{or} \quad x^2 - 2x - 4 = 0$$

$$x = \frac{1}{2} \quad \text{or} \quad x = \frac{2 \pm \sqrt{4 - 4(1)(-4)}}{2} = 1 \pm \sqrt{5}$$

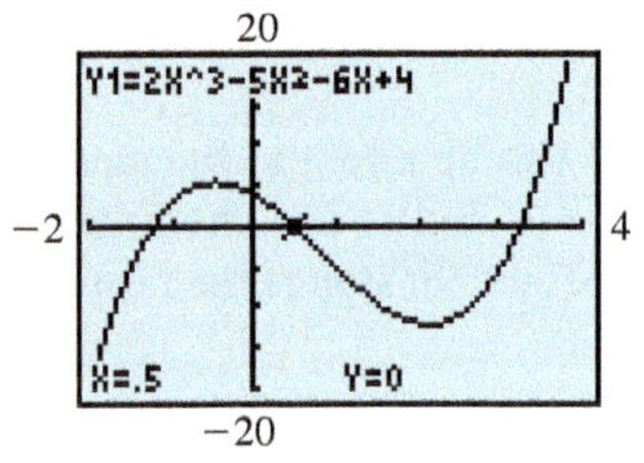

**Figure 3.32**

There are two positive roots, $1/2$ and $1 + \sqrt{5}$. The negative root is $1 - \sqrt{5}$. Note that the roots guaranteed by Descartes's rule of signs are real numbers but not necessarily rational numbers.

The graph of $y = 2x^3 - 5x^2 - 6x + 4$ in Fig. 3.32 supports these conclusions, because its $x$-intercepts appear to be $(1 - \sqrt{5}, 0)$, $(1/2, 0)$, and $(1 + \sqrt{5}, 0)$.

▶**TRY THIS.** Find all solutions to $2x^3 - 5x^2 - 8x + 5 = 0$. ■

## FOR thought... True or False? Explain.

**1.** The number 1 is a root of $x^3 - 1 = 0$ with multiplicity 3.

**2.** The equation $x^3 = 125$ has three complex number solutions.

**3.** For $(x + 1)^3(x^2 - 2x + 1) = 0$, $-1$ is a root with multiplicity 3.

**4.** For $(x - 5)^3(x^2 - 3x - 10) = 0$, 5 is a root with multiplicity 3.

**5.** If $4 - 5i$ is a solution to a polynomial equation with real coefficients, then $5i - 4$ is also a solution to the equation.

**6.** If $P(x) = 0$ is a polynomial equation with real coefficients and $i$, $2 - 3i$, and $5 + 7i$ are roots, then the degree of $P(x)$ is at least 6.

**7.** Both $-3 - i\sqrt{5}$ and $3 - i\sqrt{5}$ are solutions to $5x^3 - 9x^2 + 17x - 23 = 0$.

**8.** Both $3/2$ and 2 are solutions to $2x^5 - 4x^3 - 6x^2 - 3x - 6 = 0$.

**9.** The equation $x^3 - 5x^2 + 6x - 1 = 0$ has no negative roots.

**10.** The equation $5x^3 - 171 = 0$ has two imaginary solutions.

# EXERCISES 3.3

*Fill in the blank.*

1. If the factor $x - c$ occurs $k$ times in the complete factorization of the polynomial $P(x)$, then $c$ is a root of $P(x) = 0$ with ________ $k$.

2. If $P(x) = 0$ is a polynomial equation with real or complex coefficients and positive degree $n$, then, counting multiplicity, $P(x) = 0$ has ________ roots.

3. If $P(x) = 0$ is a polynomial equation with real coefficients and the complex number $a + bi$ $(b \neq 0)$ is a root, then ________ is also a root.

4. If a polynomial equation has no root greater than $c$, then $c$ is a(n) ________ for the roots.

*State the degree of each polynomial equation. Find all of the real and imaginary roots of each equation, stating multiplicity when it is greater than one.*

5. $x^2 - 10x + 25 = 0$
6. $x^2 - 18x + 81 = 0$
7. $x^5 - 9x^3 = 0$
8. $x^6 + x^4 = 0$
9. $x^4 - 2x^3 + x^2 = 0$
10. $x^5 - 6x^4 + 9x^3 = 0$
11. $(2x - 3)^2(3x + 4)^2 = 0$
12. $(2x^2 + x)^2(3x - 1)^4 = 0$
13. $x^3 - 4x^2 - 6x = 0$
14. $-x^3 + 8x^2 - 14x = 0$

*Find each product.*

15. $(x - 3i)(x + 3i)$
16. $(x + 6i)(x - 6i)$
17. $[x - (1 + \sqrt{2})][x - (1 - \sqrt{2})]$
18. $[x - (3 - \sqrt{5})][x - (3 + \sqrt{5})]$
19. $[x - (3 + 2i)][x - (3 - 2i)]$
20. $[x - (3 - i)][x - (3 + i)]$
21. $(x - 2)[x - (3 + 4i)][x - (3 - 4i)]$
22. $(x + 1)[x - (1 - i)][x - (1 + i)]$

*Find a polynomial equation with real coefficients that has the given roots.*

23. $-3, 5$
24. $6, -1$
25. $-4i, 4i$
26. $-9i, 9i$
27. $3 - i$
28. $4 + i$
29. $-2, i$
30. $4, -i$
31. $0, i\sqrt{3}$
32. $-2, i\sqrt{2}$
33. $3, 1 - i$
34. $5, 4 - 3i$
35. $1, 2, 3$
36. $-1, 2, -3$
37. $1, 2 - 3i$
38. $-1, 4 - 2i$
39. $\frac{1}{2}, \frac{1}{3}, \frac{1}{4}$
40. $-\frac{1}{2}, -\frac{1}{3}, 1$
41. $i, 1 + i$
42. $3i, 3 - i$

*Use Descartes's rule of signs to discuss the possibilities for the roots of each equation. Do not solve the equation.*

43. $x^3 + 5x^2 + 7x + 1 = 0$
44. $2x^3 - 3x^2 + 5x - 6 = 0$
45. $-x^3 - x^2 + 7x + 6 = 0$
46. $-x^4 - 5x^2 - x + 7 = 0$
47. $y^4 + 5y^2 + 7 = 0$
48. $-3y^4 - 6y^2 + 7 = 0$
49. $t^4 - 3t^3 + 2t^2 - 5t + 7 = 0$
50. $-5r^4 + 4r^3 + 7r - 16 = 0$
51. $x^5 + x^3 + 5x = 0$
52. $x^4 - x^2 + 1 = 0$

*Use the theorem on bounds to establish the best integral bounds for the roots of each equation.*

53. $2x^3 - 5x^2 + 6 = 0$
54. $2x^3 - x^2 - 5x + 3 = 0$
55. $4x^3 + 8x^2 - 11x - 15 = 0$
56. $6x^3 + 5x^2 - 36x - 35 = 0$
57. $w^4 - 5w^3 + 3w^2 + 2w - 1 = 0$
58. $3z^4 - 7z^2 + 5 = 0$
59. $-2x^3 + 5x^2 - 3x + 9 = 0$
60. $-x^3 + 8x - 12 = 0$

*Use the rational zero theorem, Descartes's rule of signs, and the theorem on bounds as aids in finding all real and imaginary roots to each equation.*

61. $x^3 - 4x^2 - 7x + 10 = 0$
62. $x^3 + 9x^2 + 26x + 24 = 0$
63. $x^3 - 10x - 3 = 0$
64. $2x^3 - 7x^2 - 16 = 0$
65. $x^4 + 2x^3 - 7x^2 + 2x - 8 = 0$
66. $x^4 - 4x^3 + 7x^2 - 16x + 12 = 0$
67. $6x^3 + 25x^2 - 24x + 5 = 0$
68. $6x^3 - 11x^2 - 46x - 24 = 0$
69. $x^4 + 2x^3 - 3x^2 - 4x + 4 = 0$
70. $x^5 + 3x^3 + 2x = 0$
71. $x^4 - 6x^3 + 12x^2 - 8x = 0$
72. $x^4 + 9x^3 + 27x^2 + 27x = 0$
73. $x^6 - x^5 - x^4 + x^3 - 12x^2 + 12x = 0$
74. $2x^7 - 2x^6 + 7x^5 - 7x^4 - 4x^3 + 4x^2 = 0$
75. $8x^5 + 2x^4 - 33x^3 + 4x^2 + 25x - 6 = 0$
76. $6x^5 + x^4 - 28x^3 - 3x^2 + 16x - 4 = 0$

*For each of the following functions use synthetic division and the theorem on bounds to find integers a and b, such that the interval $(a, b)$ contains all real zeros of the function. This method does not necessarily give the shortest interval containing all real zeros. By inspecting the graph of each function, find the shortest interval $(c, d)$ that contains all real zeros of the function with c and d integers. The second interval should be a subinterval of the first.*

**77.** $y = 2x^3 - 3x^2 - 50x + 18$ **78.** $f(x) = x^3 - 33x - 58$

**79.** $f(x) = x^4 - 26x^2 + 153$

**80.** $y = x^4 + x^3 - 16x^2 - 10x + 60$

**81.** $y = 4x^3 - 90x^2 - 2x + 45$

**82.** $f(x) = x^4 - 12x^3 + 27x^2 - 6x - 8$

*Solve each problem.*

**83.** *Growth Rate for Bacteria* The instantaneous growth rate of a population is the rate at which it is growing at every instant in time. The instantaneous growth rate $r$ of a colony of bacteria $t$ hours after the start of an experiment is given by the function

$$r = 0.01t^3 - 0.08t^2 + 0.11t + 0.20$$

for $0 \le t \le 7$. Find the times for which the instantaneous growth rate is zero.

**84.** *Retail Store Profit* The manager of a retail store has figured that her monthly profit $P$ (in thousands of dollars) is determined by her monthly advertising expense $x$ (in tens of thousands of dollars) according to the formula

$$P = x^3 - 20x^2 + 100x \quad \text{for} \quad 0 \le x \le 4.$$

For what value of $x$ does she get \$147,000 in profit?

**85.** *Designing Fireworks* Marshall is designing a rocket for the Red Rocket Fireworks Company. The rocket will consist of a cardboard circular cylinder with a height that is four times as large as the radius. On top of the cylinder will be a cone with a height of 2 in. and a radius equal to the radius of the base as shown in the figure. If he wants to fill the cone and the cylinder with a total of $114\pi$ in.$^3$ of powder, then what should be the radius of the cylinder?

**HINT** Use the formulas for the volume of a cone and a cylinder, which can be found inside the back cover of this text.

Figure for Exercise 85

**86.** *Heating and Air* An observatory is built in the shape of a right circular cylinder with a hemispherical roof as shown in the figure. The heating and air contractor has figured the volume of the structure as $3168\pi$ ft$^3$. If the height of the cylindrical walls is 2 ft more than the radius of the building, then what is the radius of the building?

**HINT** Use the formulas for the volume of a sphere and a cylinder.

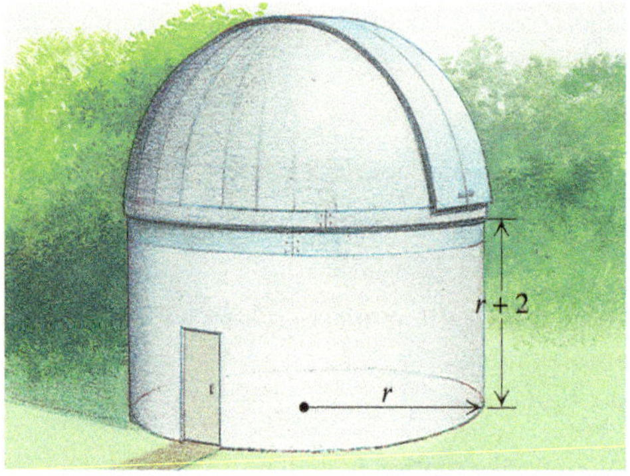


Figure for Exercise 86

## FOR WRITING/DISCUSSION

**87.** *Conjugate of a Sum* Show that the conjugate of the sum of two complex numbers is equal to the sum of their conjugates.

**88.** *Conjugate of a Product* Show that the conjugate of the product of two complex numbers is equal to the product of their conjugates.

**89.** *Conjugate of a Real Number* Show that $\overline{a} = a$ for any real number $a$, where $\overline{a}$ is the conjugate of $a$.

**90.** *Conjugate Pairs* Assume that $a + bi$ is a root of $a_nx^n + a_{n-1}x^{n-1} + \cdots + a_1x + a_0 = 0$ and substitute $a + bi$ for $x$. Take the conjugate of each side of the resulting equation and use the results of Exercises 87–89 to simplify it. Your final equation should show that $a - bi$ is a root of $a_nx^n + a_{n-1}x^{n-1} + \cdots + a_1x + a_0 = 0$. This proves the conjugate pairs theorem.

**91.** *Finding Polynomials* Find a third-degree polynomial function such that $f(0) = 3$ and whose zeros are 1, 2, and 3. Explain how you found it.

**92.** *Finding Polynomials* Is there a third-degree polynomial function such that $f(0) = 6$ and $f(-1) = 12$ and whose zeros are 1, 2, and 3? Explain.

## RETHINKING

**93.** Use the rational zeros theorem to list the possible rational zeros to the function $f(x) = 2x^3 - 5x^2 + 7x - 6$.

**94.** Solve the inequality $x^2 - 4x + 3 \ge 0$. Write the solution set using interval notation.

**95.** Suppose that $b$ is the number of prime numbers in the interval $(0, a)$ where $a$ is a positive integer. Determine whether $a$ is a function of $b$, $b$ is a function of $a$, or neither.

**96.** A gambler drove from Newark to Atlantic City. On the return trip he was broke and decreased his average speed by one-third to save gas. By what percent did his time increase for the return trip?

**97.** Let $f(x) = 2x^2 - 9$ and $g(x) = 2x - 4$. Find $(f \circ g)(x)$.

**98.** Find the equation of the line through $(2, 4)$ that is parallel to the line $x = 6$.

### THINKING OUTSIDE THE BOX XXIV

*Packing Billiard Balls* There are several ways to tightly pack nine billiard balls each with radius 1 into a rectangular box. Find the volume of the box in each of the following cases and determine which box has the least volume.

**a.** Four balls are placed so that they just fit into the bottom of the box, then another layer of four, then one ball in the middle tangent to all four in the second layer, as shown in this side view.

**b.** Four balls are placed so that they just fit into the bottom of the box as in (a), then one is placed in the middle on top of the first four. Finally, four more are placed so that they just fit at the top of the box.

**c.** The box is packed with layers of four, one, and four as in (b), but the box is required to be cubic. In this case, the four balls in the bottom layer will not touch each other and the four balls in the top layer will not touch each other. The ball in the middle will be tangent to all of the other eight balls.

**Figure for Thinking Outside the Box XXIV**

## POP QUIZ 3.3

**1.** Find all real and imaginary roots to $x^5 - x^3 = 0$, including multiplicities.

**2.** Find a polynomial equation with real coefficients that has the roots $-4i$ and 5.

**3.** By Descartes's rule of signs, how many positive roots can $x^4 + x^3 - 3x^2 + 5x + 9 = 0$ have?

**4.** By Descartes's rule of signs, how many negative roots can $3x^3 + 5x^2 - x + 9 = 0$ have?

**5.** Find all real and imaginary roots to $x^3 - 3x^2 - 6x + 8 = 0$.

## LINKING concepts... For Individual or Group Explorations

### Designing a Crystal Ball

*The wizard Gandalf is creating a massive crystal ball. To achieve maximum power, the orb must be a perfect solid sphere of clear crystal mounted on a square solid silver base, as shown in the figure.*

**a)** Write a formula for the total volume of material used in terms of the radius of the sphere, the thickness of the base, and the length of the side of the base.

**b)** The wizard has determined that the diameter of the sphere must equal the length of the side of the square base, and the thickness of the base must be $\pi$ in. Find the exact radius of the sphere if the total amount of material used in the sphere and base must be $1296\pi$ in.$^3$.

**c)** If it turns out that the sphere in part (b) is not powerful enough, Gandalf plans to make a sphere and base of solid dilithium crystal. For this project the diameter of the sphere must be 2 in. less than the length of the side of the base and the thickness of the base must be 2 in. Find the approximate radius of the sphere if the total volume of material used in the sphere and base must be 5000 in.$^3$.

# 3.4 Miscellaneous Equations

In Section 3.3 we learned that an $n$th-degree polynomial equation has $n$ roots. However, it is not always obvious how to find them. In this section we will solve polynomial equations using some new techniques and we will solve several other types of equations. Unfortunately, we cannot generally predict the number of roots to nonpolynomial equations.

## Factoring Higher-Degree Equations

In Section 1.7 we solved quadratic equations by factoring and setting the factors equal to zero. We used the zero factor property, which we restate here.

**Zero Factor Property**

> If $A$ and $B$ are algebraic expressions, then the equation $AB = 0$ is equivalent to the compound statement $A = 0$ or $B = 0$.

This property holds true also for any number of factors. So if we can factor a higher-degree polynomial equation, we can set each factor equal to zero and solve the resulting equations. This is often the easiest method for solving a higher-degree polynomial equation.

### EXAMPLE 1 Solving an equation by factoring

Solving each equation.

**a.** $x^3 + 3x^2 + x + 3 = 0$ **b.** $30x^3 + 16x^2 - 24x = 0$

**Solution**

**a.** Factor the polynomial on the left-hand side by grouping.

$$x^2(x + 3) + 1(x + 3) = 0 \quad \text{Factor by grouping.}$$
$$(x^2 + 1)(x + 3) = 0 \quad \text{Factor out } x + 3.$$
$$x^2 + 1 = 0 \quad \text{or} \quad x + 3 = 0 \quad \text{Zero factor property}$$
$$x^2 = -1 \quad \text{or} \quad x = -3$$
$$x = \pm i \quad \text{or} \quad x = -3$$

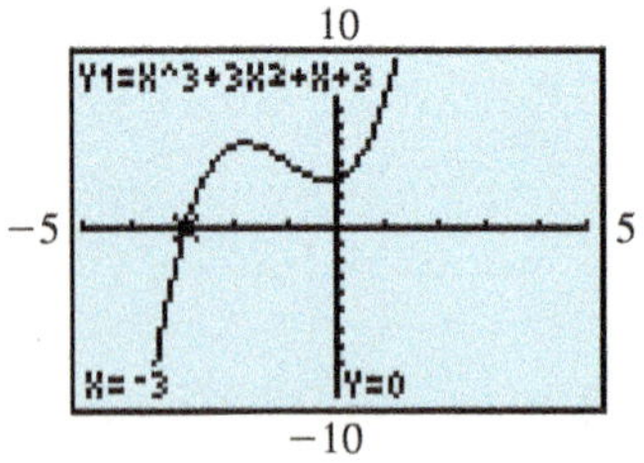

**Figure 3.33**

The solution set is $\{-3, -i, i\}$.

The graph in Fig. 3.33 supports these solutions. Because there is only one real solution, the graph crosses the $x$-axis only once. □

**b.** First factor out the common factor $2x$:

$$30x^3 + 16x^2 - 24x = 0$$
$$2x(15x^2 + 8x - 12) = 0 \quad \text{Factor out the GCF } 2x.$$
$$2x(3x - 2)(5x + 6) = 0 \quad \text{Factor the trinomial.}$$
$$2x = 0 \quad \text{or} \quad 3x - 2 = 0 \quad \text{or} \quad 5x + 6 = 0 \quad \text{Zero factor property}$$
$$x = 0 \quad \text{or} \quad x = \frac{2}{3} \quad \text{or} \quad x = -\frac{6}{5}$$

The solution set is $\left\{-\frac{6}{5}, 0, \frac{2}{3}\right\}$. You should check by graphing the corresponding polynomial function. It should have three $x$-intercepts.

▶**TRY THIS.** Solve $x^3 - 2x^2 + 5x - 10 = 0$ by factoring. ■

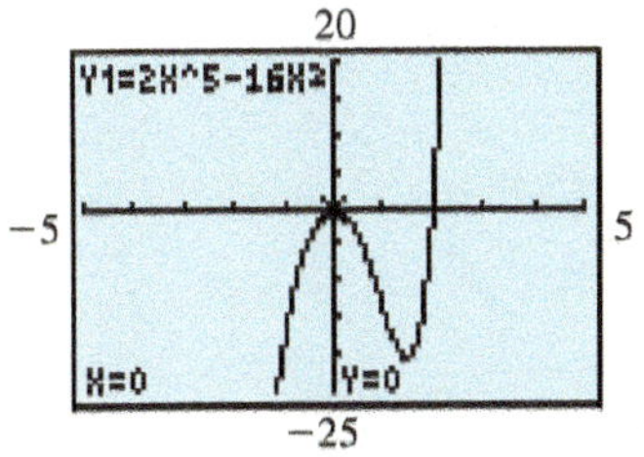


Figure 3.34

**EXAMPLE 2** Solving an equation by factoring

Solve $2x^5 = 16x^2$.

**Solution**

Write the equation with 0 on the right-hand side, then factor completely.

$$2x^5 - 16x^2 = 0$$

$$2x^2(x^3 - 8) = 0 \quad \text{Factor out the greatest common factor.}$$

$$2x^2(x - 2)(x^2 + 2x + 4) = 0 \quad \text{Factor the difference of two cubes.}$$

$$2x^2 = 0 \quad \text{or} \quad x - 2 = 0 \quad \text{or} \quad x^2 + 2x + 4 = 0$$

$$x = 0 \quad \text{or} \quad x = 2 \quad \text{or} \quad x = \frac{-2 \pm \sqrt{-12}}{2} = -1 \pm i\sqrt{3}$$

The solution set is $\{0, 2, -1 \pm i\sqrt{3}\}$. Since 0 is a root with multiplicity 2, there are five roots, counting multiplicity, to this fifth-degree equation.

The graph in Fig. 3.34 supports this solution.

▶**TRY THIS.** Solve $x^5 = 27x^2$. ■

Note that in Example 2, if we had divided each side by $x^2$ as our first step, we would have lost the solution $x = 0$. *We do not usually divide each side of an equation by a variable expression.* Instead, bring all expressions to the same side and factor out the common factors.

## Equations Involving Square Roots

Recall that $\sqrt{x}$ represents the nonnegative square root of $x$. To solve $\sqrt{x} = 3$ we can use the definition of square root. Since the nonnegative square root of 9 is 3, the solution to $\sqrt{x} = 3$ is 9. To solve $\sqrt{x} = -3$ we again use the definition of square root. Because $\sqrt{x}$ is nonnegative while $-3$ is negative, this equation has no solution.

More complicated equations involving square roots are usually solved by squaring both sides. However, squaring both sides does not always lead to an equivalent equation. If we square both sides of $\sqrt{x} = 3$, we get $x = 9$, which is equivalent to $\sqrt{x} = 3$. But if we square both sides of $\sqrt{x} = -3$, we also get $x = 9$, which is not equivalent to $\sqrt{x} = -3$. Because 9 appeared in the attempt to solve $\sqrt{x} = -3$, but does not satisfy the equation, it is called an *extraneous root.* This same situation can occur with an equation involving a fourth root or any other even root. So if you raise each side of an equation to an even power, you must check for extraneous roots.

**EXAMPLE 3** Squaring each side to solve an equation

Solve $\sqrt{x} + 2 = x$.

**Solution**

Isolate the radical before squaring each side.

$$\sqrt{x} = x - 2$$

$$(\sqrt{x})^2 = (x - 2)^2 \quad \text{Square each side.}$$

$$x = x^2 - 4x + 4 \quad \text{Use the special product } (a - b)^2 = a^2 - 2ab + b^2.$$

$$0 = x^2 - 5x + 4 \quad \text{Write in the form } ax^2 + bx + c = 0.$$

$$0 = (x - 4)(x - 1) \quad \text{Factor the quadratic polynomial.}$$

$$x - 4 = 0 \quad \text{or} \quad x - 1 = 0 \quad \text{Zero factor property}$$

$$x = 4 \quad \text{or} \quad x = 1$$

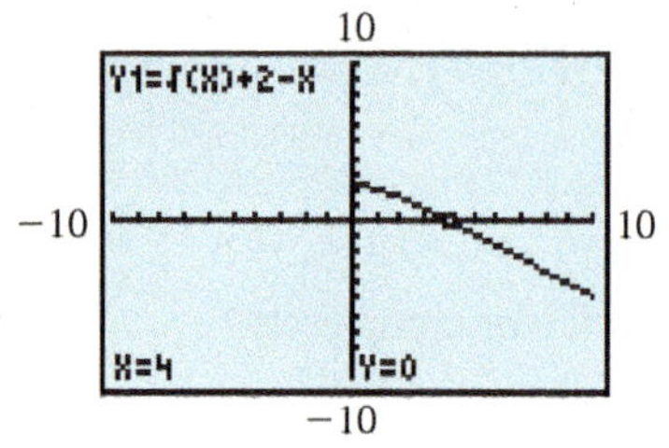


Figure 3.35

Checking $x = 4$, we get $\sqrt{4} + 2 = 4$, which is correct. Checking $x = 1$, we get $\sqrt{1} + 2 = 1$, which is incorrect. So 1 is an extraneous root and the solution set is $\{4\}$.

The graph in Fig. 3.35 supports this solution.

**▶TRY THIS.** Solve $\sqrt{x} + 12 = x$. ■

The next example involves two radicals. In this example we will isolate the more complicated radical before squaring each side. But not all radicals are eliminated upon squaring each side. So we isolate the remaining radical and square each side again.

## EXAMPLE 4 Squaring each side twice

Solve $\sqrt{2x + 1} - \sqrt{x} = 1$.

### Solution

First we write the equation so that the more complicated radical is isolated. Then we square each side. On the left side, when we square $\sqrt{2x + 1}$, we get $2x + 1$. On the right side, when we square $1 + \sqrt{x}$, we use the special product rule $(a + b)^2 = a^2 + 2ab + b^2$.

$$\sqrt{2x + 1} = 1 + \sqrt{x}$$

$$(\sqrt{2x + 1})^2 = (1 + \sqrt{x})^2 \quad \text{Square each side.}$$

$$2x + 1 = 1 + 2\sqrt{x} + x$$

$$x = 2\sqrt{x} \quad \text{All radicals are not eliminated by the first squaring.}$$

$$x^2 = (2\sqrt{x})^2 \quad \text{Square each side a second time.}$$

$$x^2 = 4x$$

$$x^2 - 4x = 0$$

$$x(x - 4) = 0$$

$$x = 0 \quad \text{or} \quad x - 4 = 0$$

$$x = 0 \quad \text{or} \quad x = 4$$

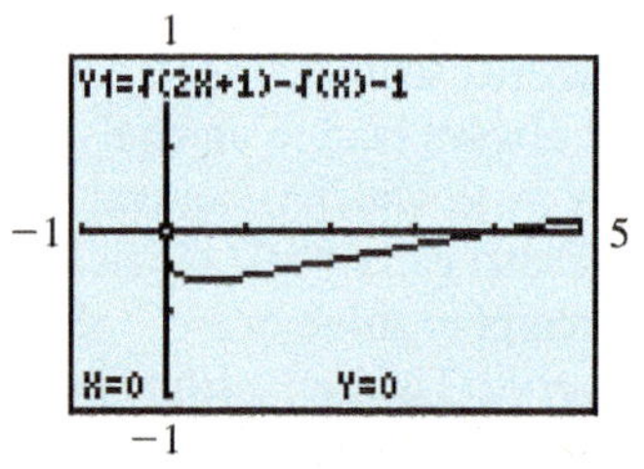


Figure 3.36

Both 0 and 4 satisfy the original equation. So the solution set is $\{0, 4\}$.

The graph in Fig. 3.36 supports this solution.

**▶TRY THIS.** Solve $\sqrt{3x - 2} - \sqrt{x} = 2$. ■

Use the following strategy when solving equations involving square roots.

**STRATEGY**

**Solving Equations Involving Square Roots**

1. Isolate the radical if there is only one. Separate the radicals on opposite sides of the equation if there is more than one.
2. Square both sides and simplify.
3. Isolate or separate any remaining radicals and square again.
4. Check all solutions because squaring can produce extraneous solutions.

## Equations with Rational Exponents

To solve equations of the form $x^{m/n} = k$ in which $m$ and $n$ are positive integers and $m/n$ is in lowest terms, we adapt the methods of Examples 3 and 4 of raising each side to a power. Cubing each side of $x^{2/3} = 4$, yields $(x^{2/3})^3 = 4^3$ or $x^2 = 64$. By the square root property, $x = \pm 8$. We can shorten this solution by raising each side of the equation to the power 3/2 (the reciprocal of 2/3) and inserting the $\pm$ symbol to obtain the two square roots.

$$x^{2/3} = 4$$

$$(x^{2/3})^{3/2} = \pm 4^{3/2} \quad \text{Raise each side to the power 3/2 and insert } \pm.$$

$$x = \pm 8$$

The equation $x^{2/3} = 4$ has two solutions because the numerator of the exponent 2/3 is an even number. An equation such as $x^{-3/2} = 1/8$ has only one solution because the numerator of the exponent $-3/2$ is odd. To solve $x^{-3/2} = 1/8$, raise each side to the power $-2/3$ (the reciprocal of $-3/2$).

$$x^{-3/2} = \frac{1}{8}$$

$$(x^{-3/2})^{-2/3} = \left(\frac{1}{8}\right)^{-2/3} \quad \text{Raise each side to the power } -2/3.$$

$$x = 4$$

To solve equations of the form $x^{m/n} = k$ we can use the following strategy.

**STRATEGY**

**Solving $x^{m/n} = k$ ($k \neq 0$, $m/n$ in lowest terms)**

1. Raise each side of the equation to the reciprocal power $n/m$. Recall that $n/m$ means the $n$th power of the $m$th root.
2. Remember that there are two real even roots of any positive real number and there is exactly one real odd root of any real number.
3. If $m$ is even and $k > 0$, then $x = \pm k^{n/m}$. If $m$ is even and $k < 0$ there is no real solution.
4. If $m$ is odd, then there is only one real solution, $x = k^{n/m}$.

### EXAMPLE 5 Equations with rational exponents

Solve each equation.

**a.** $x^{4/3} = 625$ **b.** $(y - 2)^{-5/2} = 32$

**Solution**

**a.** Raise each side of the equation to the power 3/4. Use the $\pm$ symbol because the numerator of 4/3 is even.

$$x^{4/3} = 625$$

$$(x^{4/3})^{3/4} = \pm 625^{3/4}$$

$$x = \pm 125$$

Check in the original equation. The solution set is $\{-125, 125\}$. The graph in Fig. 3.37 supports this solution. □

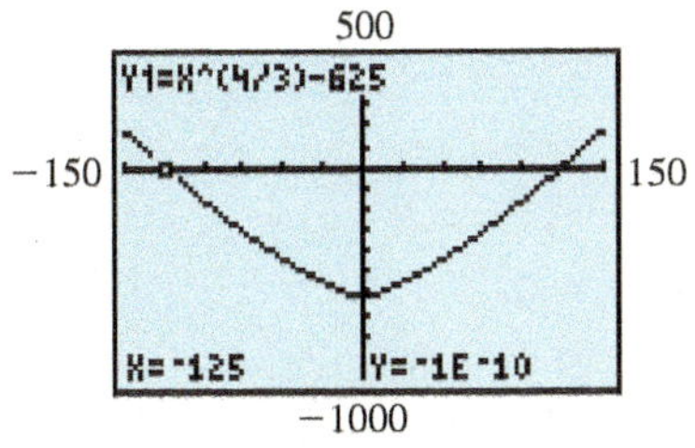


Figure 3.37

**b.** Raise each side to the power $-2/5$. Because the numerator in $-5/2$ is an odd number, there is only one solution.

$$(y-2)^{-5/2} = 32$$

$$((y-2)^{-5/2})^{-2/5} = 32^{-2/5} \quad \text{Raise each side to the power } -2/5.$$

$$y - 2 = \frac{1}{4}$$

$$y = 2 + \frac{1}{4} = \frac{9}{4}$$

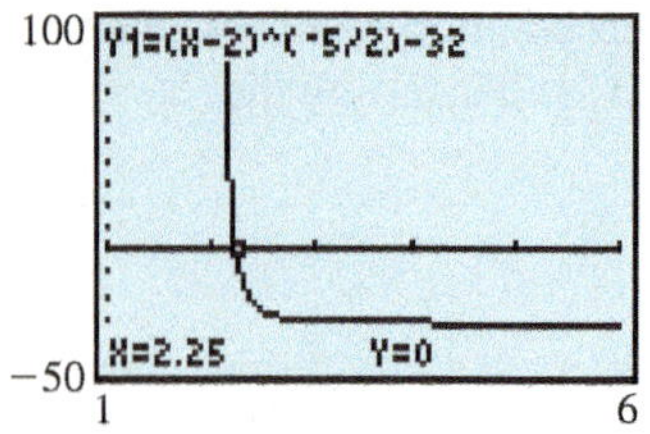


Figure 3.38

Check 9/4 in the original equation. The solution set is $\left\{\frac{9}{4}\right\}$.
The graph in Fig. 3.38 supports this solution.

▶**TRY THIS.** Solve $x^{-4/5} = 16$. ■

## Equations of Quadratic Type

In some cases, an equation can be converted to a quadratic equation by substituting a single variable for a more complicated expression. Such equations are called **equations of quadratic type.** An equation of quadratic type has the form $au^2 + bu + c = 0$, where $a \neq 0$ and $u$ is an algebraic expression.

In the next example, the expression $x^2$ in a fourth-degree equation is replaced by $u$, yielding a quadratic equation. After the quadratic equation is solved, $u$ is replaced by $x^2$ so that we find values for $x$ that satisfy the original fourth-degree equation.

### EXAMPLE 6 Solving a fourth-degree polynomial equation

Find all real and imaginary solutions to each equation.

**a.** $x^4 - 14x^2 + 45 = 0$ **b.** $x^4 + 14x^2 - 32 = 0$

**Solution**

**a.** We let $u = x^2$ so that $u^2 = (x^2)^2 = x^4$.

$$(x^2)^2 - 14x^2 + 45 = 0$$

$$u^2 - 14u + 45 = 0 \quad \text{Replace } x^2 \text{ by } u.$$

$$(u-9)(u-5) = 0$$

$$u - 9 = 0 \quad \text{or} \quad u - 5 = 0$$

$$u = 9 \quad \text{or} \quad u = 5$$

$$x^2 = 9 \quad \text{or} \quad x^2 = 5 \quad \text{Replace } u \text{ by}$$

$$x = \pm 3 \quad \text{or} \quad x = \pm\sqrt{5}$$

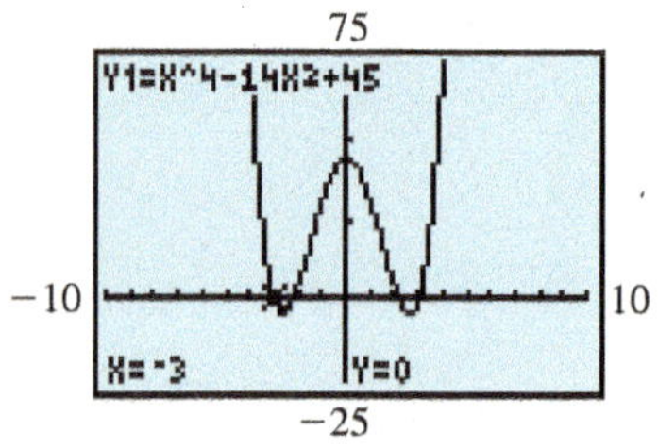


Figure 3.39

Check in the original equation. The solution set is $\{-3, -\sqrt{5}, \sqrt{5}, 3\}$.
The graph in Fig. 3.39 supports this solution. □

**b.** We could let $u = x^2$ as in part (a) or we can just factor the polynomial as follows.

$$x^4 + 14x^2 - 32 = 0$$

$$(x^2 + 16)(x^2 - 2) = 0$$

$$x^2 + 16 = 0 \quad \text{or} \quad x^2 - 2 = 0$$

$$x^2 = -16 \quad \text{or} \quad x^2 = 2$$

$$x = \pm 4i \quad \text{or} \quad x = \pm\sqrt{2}$$

The solution set is $\{-\sqrt{2}, \sqrt{2}, -4i, 4i\}$. Check in the original equation.

▶**TRY THIS.** Solve $x^4 - 9x^2 + 20 = 0$. ■

Note that the equation of Example 6 could be solved by factoring without doing substitution, because $x^4 - 14x^2 + 45 = (x^2 - 9)(x^2 - 5)$. Since the next example involves a more complicated algebraic expression, we use substitution to simplify it, although it too could be solved by factoring, without substitution.

## EXAMPLE 7 Another equation of quadratic type

Solve $(x^2 - x)^2 - 18(x^2 - x) + 72 = 0$.

### Solution

If we let $u = x^2 - x$, then the equation becomes a quadratic equation.

$$(x^2 - x)^2 - 18(x^2 - x) + 72 = 0$$

$$u^2 - 18u + 72 = 0 \quad \text{Replace } x^2 - x \text{ by } u.$$

$$(u - 6)(u - 12) = 0$$

$$u - 6 = 0 \quad \text{or} \quad u - 12 = 0$$

$$u = 6 \quad \text{or} \quad u = 12$$

$$x^2 - x = 6 \quad \text{or} \quad x^2 - x = 12 \quad \text{Replace } u \text{ by } x^2 - x.$$

$$x^2 - x - 6 = 0 \quad \text{or} \quad x^2 - x - 12 = 0$$

$$(x - 3)(x + 2) = 0 \quad \text{or} \quad (x - 4)(x + 3) = 0$$

$$x = 3 \quad \text{or} \quad x = -2 \quad \text{or} \quad x = 4 \quad \text{or} \quad x = -3$$

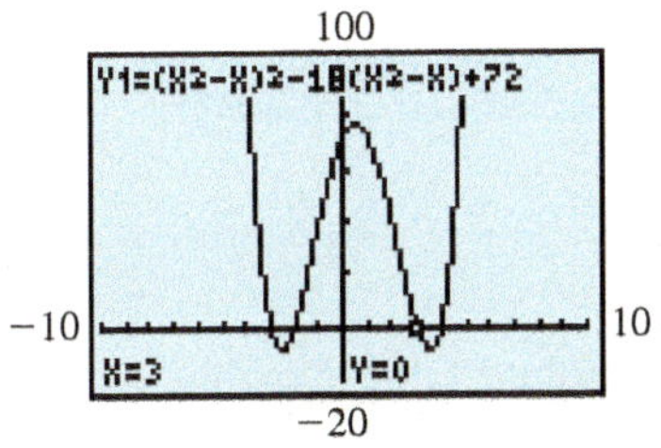


Figure 3.40

Check in the original equation. The solution set is $\{-3, -2, 3, 4\}$.

The graph in Fig. 3.40 supports this solution.

**▶TRY THIS.** Solve $(x^2 + x)^2 - 8(x^2 + x) + 12 = 0$. ■

The next equations of quadratic type have rational exponents.

## EXAMPLE 8 Quadratic type and rational exponents

Find all real solutions to each equation.

**a.** $x^{2/3} - 9x^{1/3} + 8 = 0$ **b.** $(11x^2 - 18)^{1/4} = x$

### Solution

**a.** If we let $u = x^{1/3}$, then $u^2 = (x^{1/3})^2 = x^{2/3}$.

$$u^2 - 9u + 8 = 0 \quad \text{Replace } x^{2/3} \text{ by } u^2 \text{and } x^{1/3} \text{ by } u.$$

$$(u - 8)(u - 1) = 0$$

$$u = 8 \quad \text{or} \quad u = 1$$

$$x^{1/3} = 8 \quad \text{or} \quad x^{1/3} = 1 \quad \text{Replace } u \text{ by } x^{1/3}.$$

$$(x^{1/3})^3 = 8^3 \quad \text{or} \quad (x^{1/3})^3 = 1^3$$

$$x = 512 \quad \text{or} \quad x = 1$$

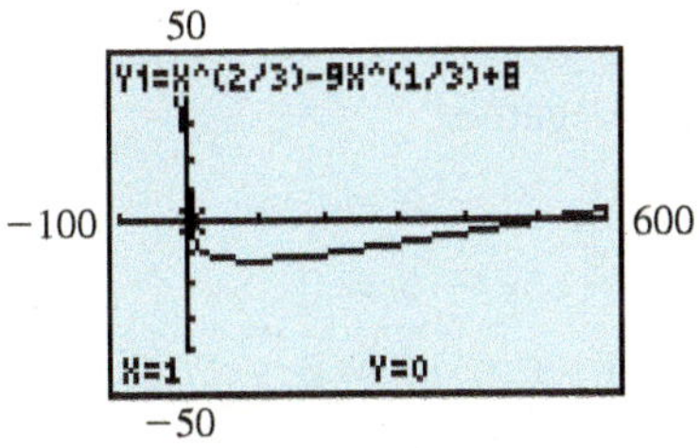


Figure 3.41

Check in the original equation. The solution set is $\{1, 512\}$.

The graph in Fig. 3.41 supports this solution. □

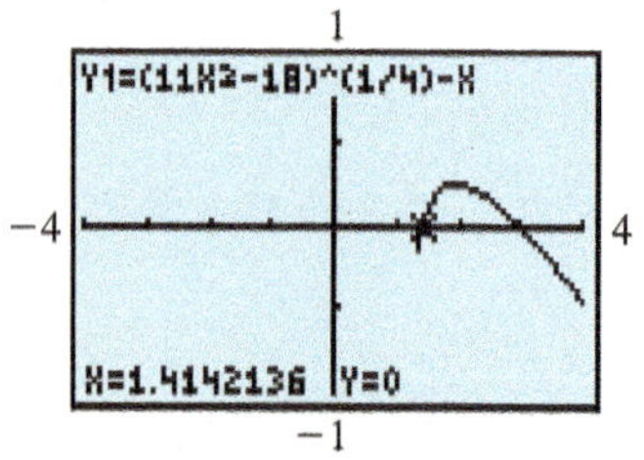


**Figure 3.42**

**b.**

$$(11x^2 - 18)^{1/4} = x$$

$$((11x^2 - 18)^{1/4})^4 = x^4 \quad \text{Raise each side to the power 4.}$$

$$11x^2 - 18 = x^4$$

$$x^4 - 11x^2 + 18 = 0$$

$$(x^2 - 9)(x^2 - 2) = 0$$

$$x^2 = 9 \quad \text{or} \quad x^2 = 2$$

$$x = \pm 3 \quad \text{or} \quad x = \pm\sqrt{2}$$

Since the exponent 1/4 means principal fourth root, the right-hand side of the equation cannot be negative. So $-3$ and $-\sqrt{2}$ are extraneous roots. Since 3 and $\sqrt{2}$ satisfy the original equation, the solution set is $\{\sqrt{2}, 3\}$.

The graph in Fig. 3.42 supports this solution.

▶**TRY THIS.** Solve $x^{2/3} - x^{1/3} - 6 = 0$. ■

## Equations Involving Absolute Value

We solved basic absolute value equations in Section 1.1. In the next two examples we solve some more complicated absolute value equations.

### EXAMPLE 9 An equation involving absolute value

Solve $|x^2 - 2x - 16| = 8$.

#### Solution

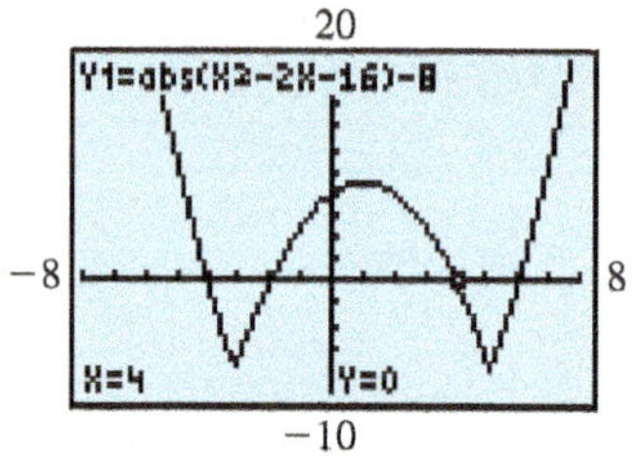


**Figure 3.43**

First write an equivalent statement without using absolute value symbols.

$$x^2 - 2x - 16 = 8 \quad \text{or} \quad x^2 - 2x - 16 = -8$$

$$x^2 - 2x - 24 = 0 \quad \text{or} \quad x^2 - 2x - 8 = 0$$

$$(x - 6)(x + 4) = 0 \quad \text{or} \quad (x - 4)(x + 2) = 0$$

$$x = 6 \quad \text{or} \quad x = -4 \quad \text{or} \quad x = 4 \quad \text{or} \quad x = -2$$

The solution set is $\{-4, -2, 4, 6\}$.

The graph in Fig. 3.43 supports this solution.

▶**TRY THIS.** Solve $|x^2 - x - 4| = 2$. ■

In the next example we have an equation in which an absolute value expression is equal to an expression that could be positive or negative and an equation with two absolute value expressions.

### EXAMPLE 10 More equations involving absolute value

Solve each equation.

**a.** $|x^2 - 6| = 5x$ **b.** $|a - 1| = |2a - 3|$

#### Solution

**a.** Since $|x^2 - 6|$ is nonnegative for any value of $x$, $5x$ must be nonnegative. Write the equivalent statement assuming that $5x$ is nonnegative:

$$x^2 - 6 = 5x \quad \text{or} \quad x^2 - 6 = -5x$$

$$x^2 - 5x - 6 = 0 \quad \text{or} \quad x^2 + 5x - 6 = 0$$

$$(x - 6)(x + 1) = 0 \quad \text{or} \quad (x + 6)(x - 1) = 0$$

$$x = 6 \quad \text{or} \quad x = -1 \quad \text{or} \quad x = -6 \quad \text{or} \quad x = 1$$

The expression $|x^2 - 6|$ is nonnegative for any real number $x$. But $5x$ is negative if $x = -1$ or if $x = -6$. So $-1$ and $-6$ are extraneous roots. They do not satisfy the original equation. The solution set is $\{1, 6\}$.

The graph in Fig. 3.44 supports this solution. □

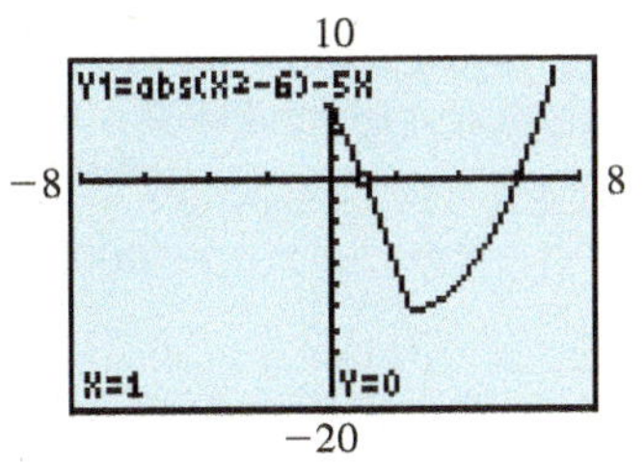


**Figure 3.44**

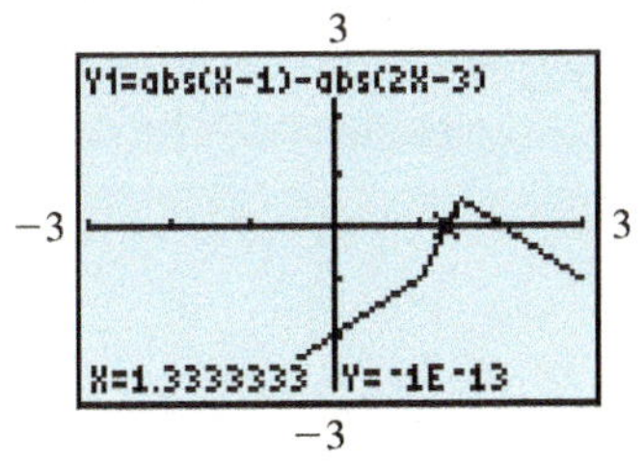


**Figure 3.45**

**b.** The equation $|a - 1| = |2a - 3|$ indicates that $a - 1$ and $2a - 3$ have the same absolute value. If two quantities have the same absolute value, they are either equal or opposites. Use this fact to write an equivalent statement without absolute value signs.

$$a - 1 = 2a - 3 \quad \text{or} \quad a - 1 = -(2a - 3)$$

$$a + 2 = 2a \quad \text{or} \quad a - 1 = -2a + 3$$

$$2 = a \quad \text{or} \quad a = \frac{4}{3}$$

Check that both 2 and $\frac{4}{3}$ satisfy the original absolute value equation. The solution set is $\left\{\frac{4}{3}, 2\right\}$.

The graph in Fig. 3.45 supports this solution.

▶**TRY THIS.** Solve $|x| = |x - 1|$. ■

## Applications

The **break-even point** for a business is the point at which the cost of doing business is equal to the revenue generated by the business. The business is profitable when the revenue is greater than the cost.

### EXAMPLE 11 Break-even point for a bus tour

A tour operator uses the equation $C = 3x + \sqrt{50x + 9000}$ to find his cost in dollars for taking $x$ people on a tour of San Francisco.

**a.** For what value of $x$ is the cost \$160?
**b.** If he charges \$10 per person for the tour, then what is his break-even point?

#### Solution

**a.** Replace $C$ by \$160 and solve the equation.

$$160 = 3x + \sqrt{50x + 9000}$$

$$160 - 3x = \sqrt{50x + 9000} \quad \text{Isolate the radical.}$$

$$25{,}600 - 960x + 9x^2 = 50x + 9000 \quad \text{Square each side.}$$

$$9x^2 - 1010x + 16{,}600 = 0$$

$$x = \frac{-(-1010) \pm \sqrt{1010^2 - 4(9)(16{,}600)}}{2(9)}$$

$$x = 20 \quad \text{or} \quad x \approx 92.2$$

Check that 20 satisfies the original equation but 92.2 does not. The cost is \$160 when 20 people take the tour.

**b.** At \$10 per person, the revenue in dollars is given by $R = 10x$. When the revenue is equal to the cost, as shown in Fig. 3.46, the operator breaks even.

$$10x = 3x + \sqrt{50x + 9000}$$
$$7x = \sqrt{50x + 9000}$$
$$49x^2 = 50x + 9000 \quad \text{Square each side.}$$
$$49x^2 - 50x - 9000 = 0$$
$$x = \frac{50 \pm \sqrt{50^2 - 4(49)(-9000)}}{2(49)}$$
$$x \approx -13.052 \quad \text{or} \quad x \approx 14.072$$

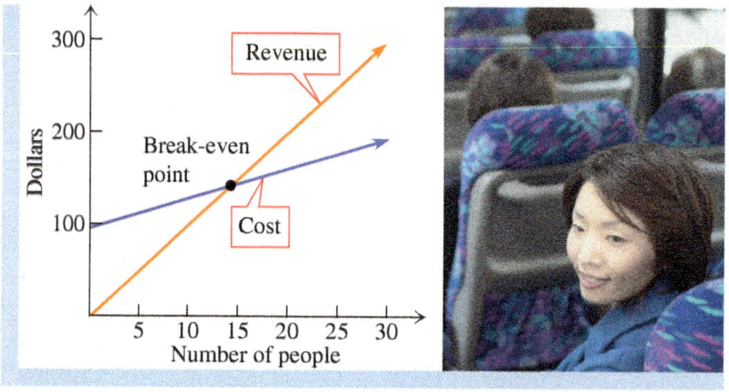


**Figure 3.46**

If 14.072 people took the tour, the operator would break even. Since the break-even point is not a whole number, the operator actually needs 15 people to make a profit.

▸**TRY THIS.** The function $D = \sqrt{P} - \sqrt{P - 400}$ gives the monthly demand $D$ for custom-made suits in terms of the price $P$ in dollars. For what price is the demand 10 suits? ■

## FOR thought... True or False? Explain.

1. Squaring each side of $\sqrt{x - 1} + \sqrt{x} = 6$ yields $x - 1 + x = 36$.
2. The equations $(2x - 1)^2 = 9$ and $2x - 1 = 3$ are equivalent.
3. The equations $x^{2/3} = 9$ and $x = 27$ have the same solution set.
4. To solve $2x^{1/4} - x^{1/2} + 3 = 0$, we let $u = x^{1/2}$ and $u^2 = x^{1/4}$.
5. If $(x - 1)^{-2/3} = 4$, then $x = 1 \pm 4^{-3/2}$.
6. No negative number satisfies $x^{-2/5} = 4$.
7. The solution set to $|2x + 10| = 3x$ is $\{-2, 10\}$.
8. No negative number satisfies $|x^2 - 3x + 2| = 7x$.
9. The equation $|2x + 1| = |x|$ is equivalent to $2x + 1 = x$ or $2x + 1 = -x$.
10. The equation $x^9 - 5x^3 + 6 = 0$ is an equation of quadratic type.

## EXERCISES 3.4

*Find all real and imaginary solutions to each equation. Check your answers.*

**1.** $x^3 + 3x^2 - 4x - 12 = 0$

**2.** $x^3 - x^2 - 5x + 5 = 0$

**3.** $2x^3 + 1000x^2 - x - 500 = 0$

**4.** $3x^3 - 1200x^2 - 2x + 800 = 0$

**5.** $a^3 + 5a = 15a^2$

**6.** $b^3 + 20b = 9b^2$

**7.** $3y^4 - 12y^2 = 0$

**8.** $5m^4 - 10m^3 + 5m^2 = 0$

**9.** $a^4 - 16 = 0$

**10.** $w^4 + 8w = 0$

*Find all real solutions to each equation. Check your answers.*

**11.** $\sqrt{x + 1} = x - 5$

**12.** $\sqrt{x - 1} = x - 7$

**13.** $\sqrt{x} - 2 = x - 22$

**14.** $3 + \sqrt{x} = 1 + x$

**15.** $w = \dfrac{\sqrt{1 - 3w}}{2}$

**16.** $t = \dfrac{\sqrt{2 - 3t}}{3}$

**17.** $\dfrac{1}{z} = \dfrac{3}{\sqrt{4z + 1}}$

**18.** $\dfrac{1}{p} - \dfrac{2}{\sqrt{9p + 1}} = 0$

**19.** $\sqrt{x^2 - 2x - 15} = 3$

**20.** $\sqrt{3x^2 + 5x - 3} = x$

**21.** $\sqrt{x + 40} - \sqrt{x} = 4$

**22.** $\sqrt{x} + \sqrt{x - 36} = 2$

**23.** $\sqrt{n + 4} + \sqrt{n - 1} = 5$

**24.** $\sqrt{y + 10} - \sqrt{y - 2} = 2$

**25.** $\sqrt{2x + 5} + \sqrt{x + 6} = 9$

**26.** $\sqrt{3x - 2} - \sqrt{x - 2} = 2$

*Find all real solutions to each equation. Check your answers.*

**27.** $x^{2/3} = 2$

**28.** $x^{2/3} = \dfrac{1}{2}$

**29.** $w^{-4/3} = 16$

**30.** $w^{-3/2} = 27$

**31.** $t^{-1/2} = 7$

**32.** $t^{-1/2} = \dfrac{1}{2}$

**33.** $(s - 1)^{-1/2} = 2$

**34.** $(s - 2)^{-1/2} = \dfrac{1}{3}$

*Find all real and imaginary solutions to each equation. Check your answers.*

**35.** $x^4 - 12x^2 + 27 = 0$

**36.** $x^4 + 10 = 7x^2$

**37.** $x^4 + 6x^2 - 7 = 0$

**38.** $x^4 - x^2 - 12 = 0$

**39.** $x^4 - 81 = 0$

**40.** $x^4 - 625 = 0$

**41.** $\left(\dfrac{2c - 3}{5}\right)^2 + 2\left(\dfrac{2c - 3}{5}\right) = 8$

**42.** $\left(\dfrac{b - 5}{6}\right)^2 - \left(\dfrac{b - 5}{6}\right) - 6 = 0$

**43.** $\dfrac{1}{(5x - 1)^2} + \dfrac{1}{5x - 1} - 12 = 0$

**44.** $\dfrac{1}{(x - 3)^2} + \dfrac{2}{x - 3} - 24 = 0$

**45.** $(v^2 - 4v)^2 - 17(v^2 - 4v) + 60 = 0$

**46.** $(u^2 + 2u)^2 - 2(u^2 + 2u) - 3 = 0$

**47.** $x - 4\sqrt{x} + 3 = 0$

**48.** $2x + 3\sqrt{x} - 20 = 0$

**49.** $q - 7q^{1/2} + 12 = 0$

**50.** $h + 1 = 2h^{1/2}$

**51.** $x^{2/3} + 10 = 7x^{1/3}$

**52.** $x^{1/2} - 3x^{1/4} + 2 = 0$

*Solve each absolute value equation.*

**53.** $|w^2 - 4| = 3$

**54.** $|a^2 - 1| = 1$

**55.** $|v^2 - 3v| = 5v$

**56.** $|z^2 - 12| = z$

**57.** $|x^2 - x - 6| = 6$

**58.** $|2x^2 - x - 2| = 1$

**59.** $|x + 5| = |2x + 1|$

**60.** $|3x - 4| = |x|$

**61.** $|x - 2| + 1 = 5x$

**62.** $|x - 4| - 1 = -4x$

**63.** $|x - 4| = |x - 2|$

**64.** $|2x - 3| = |2x + 7|$

*Solve each equation. Find imaginary solutions when possible.*

**65.** $\sqrt{16x + 1} - \sqrt{6x + 13} = -1$

**66.** $\sqrt{16x + 1} - \sqrt{6x + 13} = 1$

**67.** $v^6 - 64 = 0$

**68.** $t^4 - 1 = 0$

**69.** $(7x^2 - 12)^{1/4} = x$

**70.** $(10x^2 - 1)^{1/4} = 2x$

**71.** $\sqrt[3]{2 + x - 2x^2} = x$

**72.** $\sqrt{48 + \sqrt{x}} - 4 = \sqrt[4]{x}$

**73.** $\left(\dfrac{x - 2}{3}\right)^2 - 2\left(\dfrac{x - 2}{3}\right) + 10 = 0$

**74.** $\dfrac{1}{(x + 1)^2} - \dfrac{2}{x + 1} + 2 = 0$

**75.** $(3u - 1)^{2/5} = 2$

**76.** $(2u + 1)^{2/3} = 3$

**77.** $x^2 - 11\sqrt{x^2 + 1} + 31 = 0$

**78.** $2x^2 - 3\sqrt{2x^2 - 3} - 1 = 0$

**79.** $|x^2 - 2x| = |3x - 6|$

**80.** $|x^2 + 5x| = |3 - x^2|$

**81.** $(3m + 1)^{-3/5} = -\frac{1}{8}$

**82.** $(1 - 2m)^{-5/3} = -\frac{1}{32}$

**83.** $|x^2 - 4| = x - 2$

**84.** $|x^2 + 7x| = x^2 - 4$

*Solve each problem.*

**85.** *Maximum Sail Area* According to the International America's Cup Rules, the maximum sail area $S$ for a boat with length $L$ (in meters) and displacement $D$ (in cubic meters) is determined by the equation

$$L + 1.25S^{1/2} - 9.8D^{1/3} = 16.296$$

(America's Cup, www.americascup.org). Find $S$ for a boat with length 21.24 m and displacement 18.34 m$^3$.

**Figure for Exercises 85 and 86**

**86.** *Minimum Displacement for a Yacht* The minimum displacement $D$ for a boat with length 21.52 m and a sail area of 310.64 m$^2$ is determined by the equation $L + 1.25S^{1/2} - 9.8D^{1/3} = 16.296$. Find this boat's minimum displacement.

**87.** *Cost of Baking Bread* The daily cost for baking $x$ loaves of bread at Juanita's Bakery is given in dollars by $C = 0.5x + \sqrt{8x + 5000}$. Find the number of loaves for which the cost is \$83.50.

**88.** *Break-Even Analysis* If the bread in Exercise 87 sells for \$4.49 per loaf, then what is the minimum number of loaves that Juanita must bake and sell to make a profit?

**89.** *Square Roots* Find two numbers that differ by 6 and whose square roots differ by 1.

**90.** *Right Triangle* One leg of a right triangle is 1 cm longer than the other leg. What is the length of the short leg if the total length of the hypotenuse and the short leg is 10 cm?

**91.** *Perimeter of a Right Triangle* A sign in the shape of a right triangle has one leg that is 7 in. longer than the other leg. What is the length of the shorter leg if the perimeter is 30 in.?

**92.** *Right Triangle Inscribed in a Semicircle* One leg of a right triangle is 1 ft longer than the other leg. If the triangle is inscribed in a circle and the hypotenuse of the triangle is the diameter of the circle, then what is the length of the radius for which the length of the radius is equal to the length of the shortest side?

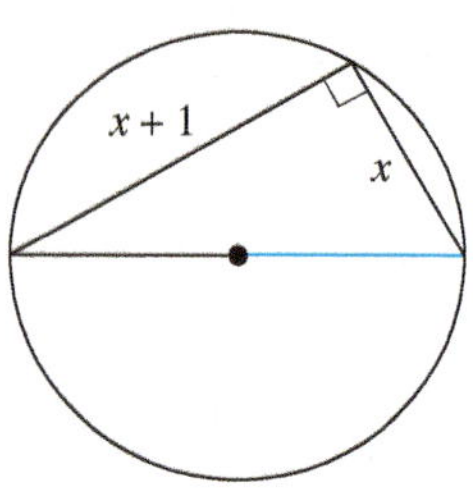


**Figure for Exercise 92**

**93.** *Area of a Foundation* The original plans for Jennifer's house called for a square foundation. After increasing one side by 30 ft and decreasing the other by 10 ft, the area of the rectangular foundation was 2100 ft$^2$. What was the area of the original square foundation?

**94.** *Shipping Carton* Heloise designed a cubic box for shipping paper. The height of the box was acceptable, but the paper would not fit into the box. After increasing the length by 0.5 in. and the width by 6 in. the area of the bottom was 119 in.$^2$ and the paper would fit. What was the original volume of the cubic box?

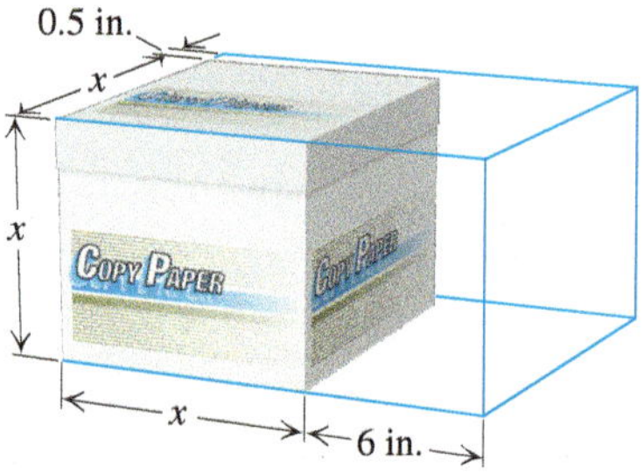


**Figure for Exercise 94**

**95.** *Sail Area-Displacement Ratio* The sail area-displacement ratio $S$ is defined as

$$S = A\left(\frac{d}{64}\right)^{-2/3},$$

where $A$ is the sail area in square feet and $d$ is the displacement in pounds. The Oceanis 381 is a 39-ft sailboat with a sail area-displacement ratio of 14.26 and a sail area of 598.9 ft$^2$. Find the displacement for the Oceanis 381.

**96.** *Capsize Screening Value* The capsize screening value $C$ is defined as

$$C = b\left(\frac{d}{64}\right)^{-1/3},$$

where $b$ is the beam (or width) in feet and $d$ is the displacement in pounds. The Bahia 46 is a 46-ft catamaran with a capsize screening value of 3.91 and a beam of 26.1 ft. Find the displacement for the Bahia 46.

**97.** *Insulated Carton* Nina is designing a box for shipping frozen shrimp. The box is to have a square base and a height that is 2 in. greater than the width of the base. The box will be surrounded with a 1-in.-thick layer of styrofoam. If the volume of the inside of the box must be equal to the volume of the styrofoam used, then what volume of shrimp can be shipped in the box?

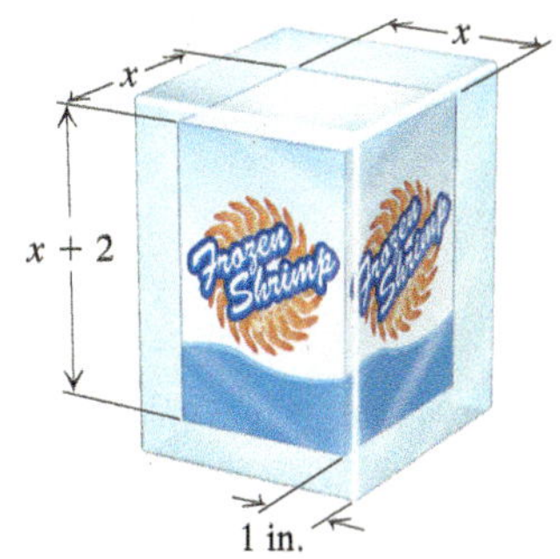


**Figure for Exercise 97**

**98.** *Volume of a Cubic Box* If the width of the base of a cubic container is increased by 3 m and the length of the base decreased by 1 m, then the volume of the new container is 6 m$^3$. What is the height of the cubic container?

**99.** *Hiking Time* William, Nancy, and Edgar met at the lodge at 8 A.M., and William began hiking west at 4 mph. At 10 A.M., Nancy began hiking north at 5 mph and Edgar went east on a three-wheeler at 12 mph. At what time was the distance between Nancy and Edgar 14 mi greater than the distance between Nancy and William?

**100.** *Accuracy of Transducers* Setra Systems Inc., of Acton, MA, calculates the accuracy $A$ of its pressure transducers using the formula

$$A = \sqrt{(NL)^2 + (HY)^2 + (NR)^2}$$

where $NL$ represents nonlinearity, $HY$ represents hysteresis, and $NR$ represents nonrepeatability. If the nonlinearity is 0.1%, the hysteresis is 0.05%, and the nonrepeatability is 0.02%, then what is the accuracy? Solve the formula for hysteresis.

**101.** *Geometric Mean* The geometric mean of the numbers $x_1, x_2, \ldots, x_n$ is defined as

$$GM = \sqrt[n]{x_1 \cdot x_2 \cdot \cdots \cdot x_n}.$$

The accompanying graph shows the quarterly net income for Apple Computer for the first three quarters of 2006.

**a.** Find the geometric mean of the net income for these three quarters.

**b.** What would the net income have to be for the next quarter so that the geometric mean for the four quarters would be $500 million?

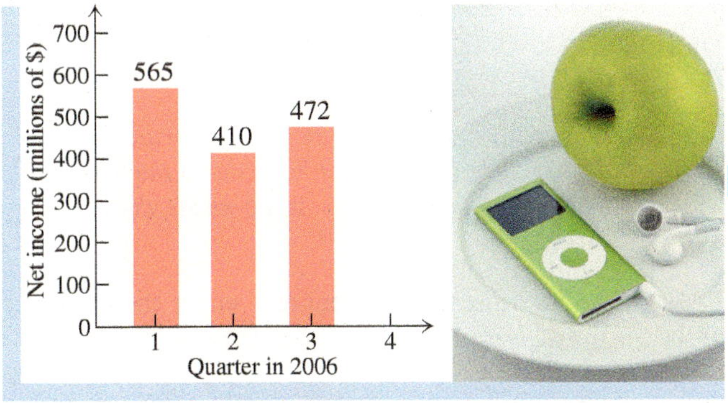


**Figure for Exercise 101**

## Foreshadowing calculus

The situation in the next exercise is studied also in calculus. However, in calculus we find the route that minimizes the total time for the trip.

**102.** *Time Swimming and Running* Lauren is competing in her town's cross-country competition. Early in the event, she must race from point $A$ on the Greenbriar River to point $B$, which is 5 mi downstream and on the opposite bank. The Greenbriar is 1 mi wide. In planning her strategy, Lauren knows she can use any combination of running and swimming. She can run 10 mph and swim 8 mph. How long would it take if she ran 5 mi downstream and then swam across? Find the time it would take if she swam diagonally from $A$ to $B$. Find $x$ so that she could run $x$ miles along the bank, swim diagonally to $B$, and complete the race in 36 min. (Ignore the current in the river.)

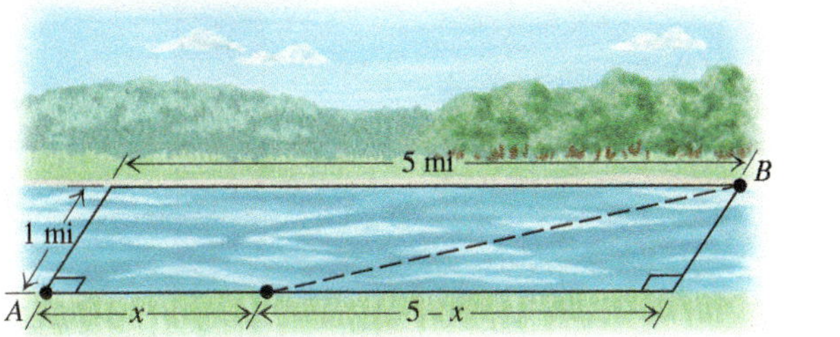


**Figure for Exercise 102**

**103.** *Boston Molasses Disaster* In the city of Boston, during the afternoon of January 15, 1919, a cylindrical metal tank containing 25,850,000 kg of molasses ruptured. The sticky liquid poured into the streets in a 9-m-deep stream that knocked down buildings and killed pedestrians and horses (www.discovery.com). If the diameter of the tank was equal to its height and the weight of molasses is 1600 kg/m$^3$, then what was the height of the tank in meters?

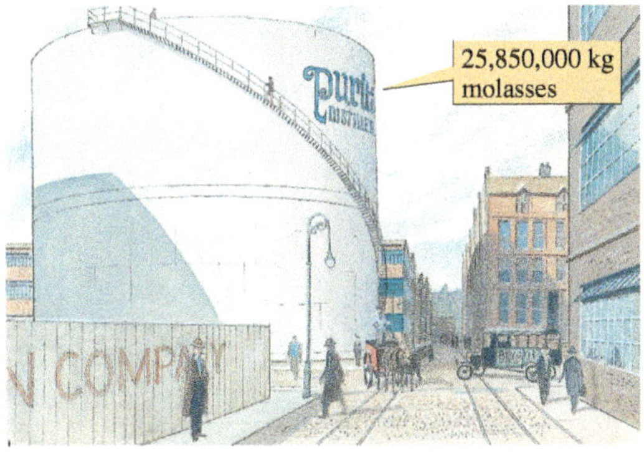


**Figure for Exercise 103**

**104.** *Storing Supplies* An army sergeant wants to use a 20-ft by 40-ft piece of canvas to make a two-sided tent for holding supplies as shown in the figure on the next page.

**a.** Write the volume of the tent as a function of $b$.

**b.** For what value of $b$ is the volume 1600 ft$^3$?

**c.** Use a graphing calculator to find the values for $b$ and $h$ that will maximize the volume of the tent.

**HINT** The volume is the area of the triangular end times the length of the tent.

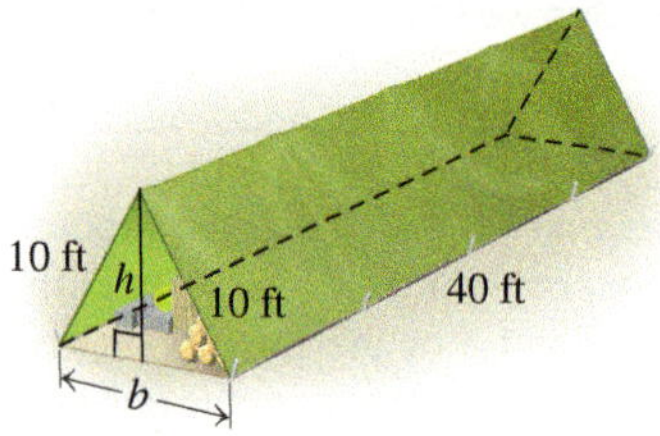


**Figure for Exercise 104**

**105.** *Radius of a Pipe* A large pipe is placed next to a wall and a 1-foot-high block is placed 5 feet from the wall to keep the pipe in place as shown in the accompanying figure. What is the radius of the pipe?

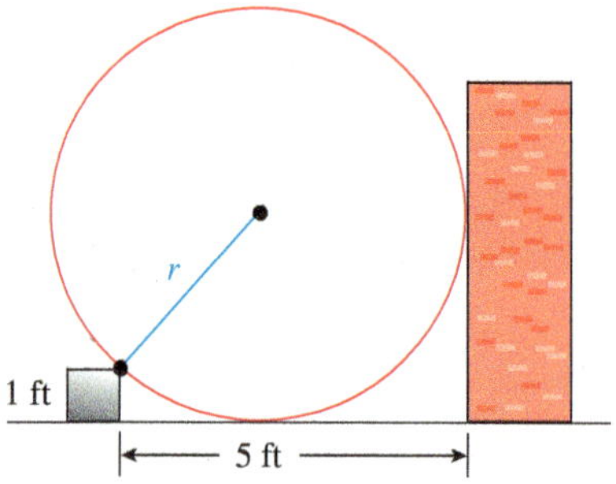


**Figure for Exercise 105**

**106.** *Radius of a Pipe* A large pipe is held in place on level ground by using a 1-foot-high block on one side and a 2-foot-high block on the other side. If the distance between the blocks is 6 feet, then what is the radius of the pipe?

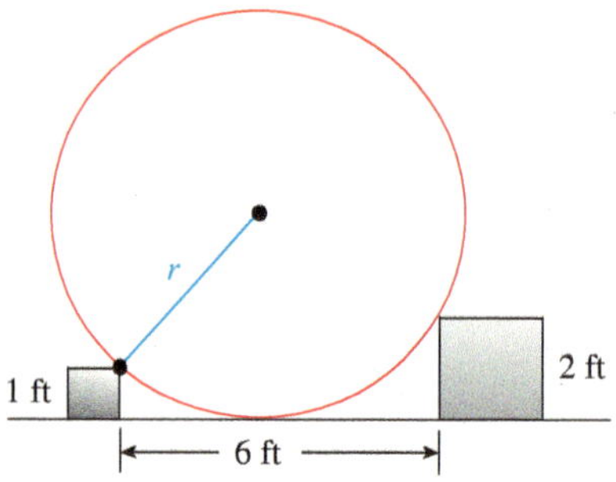


**Figure for Exercise 106**

## ▶ RETHINKING

**107.** Find a polynomial equation with real coefficients, and leading coefficient 1, that has $-5$ and $2i$ as roots.

**108.** Find all zeros to the function $f(x) = x^3 + 2x^2 - 13x + 10$.

**109.** Solve $-x^2 + 2x + 8 > 0$. Write the solution set in interval notation.

**110.** Let $f(x) = [\![x - 3]\!]$. Find $f(-5.2)$, $f(6.2)$, and $f(6.9)$.

**111.** Find the equation (in slope-intercept form) for the line through $(9, 4)$ that is perpendicular to the line $y = \frac{3}{2}x + 7$.

**112.** Find the additive inverse and multiplicative inverse of $-8i$.

## THINKING OUTSIDE THE BOX XXV

*Painting Problem* A painter has seven 3-ft by 5-ft rectangular drop cloths. If he lays each drop cloth on the carpet as a 3-ft by 5-ft rectangle, without folding, cutting, or tearing them, then what is the maximum area that he can cover with these drop cloths in an 8-ft by 13-ft room?

## POP QUIZ 3.4

*Solve each equation. Find imaginary solutions when possible.*

**1.** $x^3 + x^2 + x + 1 = 0$

**2.** $\sqrt{x + 4} = x - 2$

**3.** $x^{-2/3} = 4$

**4.** $x^4 - 3x^2 = 4$

**5.** $|x + 3| = |2x - 5|$

## LINKING concepts... For Individual or Group Explorations

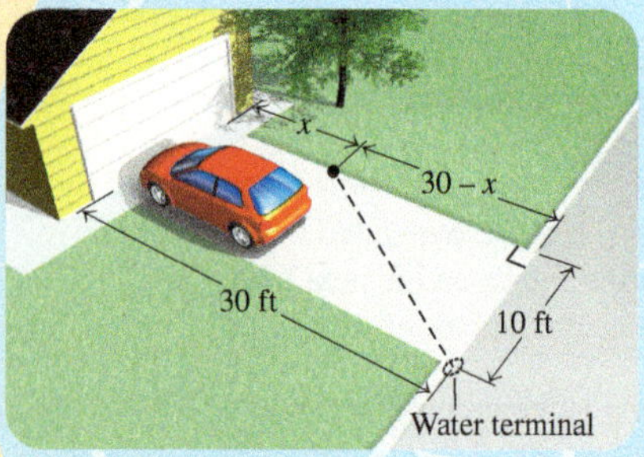


### Minimizing Construction Cost

*A homeowner needs to run a new water pipe from his house to a water terminal as shown in the accompanying diagram. The terminal is* 30 ft *down the* 10-ft-*wide driveway and on the other side. A contractor charges* \$3/ft *alongside the driveway and* \$4/ft *for underneath the driveway.*

**a)** What will it cost if the contractor runs the pipe entirely under the driveway along the diagonal of the 30-ft by 10-ft rectangle?

**b)** What will it cost if the contractor runs the pipe 30 ft alongside the driveway and then 10 ft straight across?

c) The contractor claims that he can do the job for \$120 by going alongside the driveway for some distance and then going under the drive diagonally to the terminal. Find $x$, the distance alongside the driveway.

d) Write the cost as a function of $x$ and sketch the graph of the function.

e) Use the minimum feature of a graphing calculator to find the approximate value for $x$ that will minimize the cost.

f) What is the minimum cost (to the nearest cent) for which the job can be done?

# 3.5 Graphs of Polynomial Functions

In Chapter 1 we learned that the graph of a polynomial function of degree 0 or 1 is a straight line. In Section 3.1 we learned that the graph of a second-degree polynomial function is a parabola. In this section we will concentrate on graphs of polynomial functions of degree greater than 2.

## Drawing Good Graphs

A graph of an equation is a picture of all of the ordered pairs that satisfy the equation. However, it is impossible to draw a perfect picture of any set of ordered pairs. We usually find a few important features of the graph and make sure that our picture brings out those features. For example, you can make a good graph of a linear function by drawing a line through the intercepts using a ruler and a sharp pencil. A good parabola should look smooth and symmetric and pass through the vertex and intercepts.

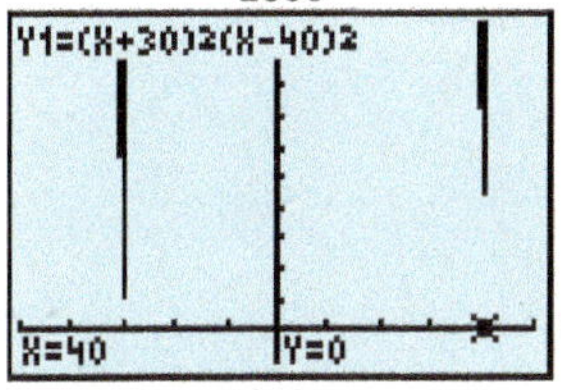


Figure 3.47

A graphing calculator is a tremendous aid in graphing because it can quickly plot many points. However, the calculator does not know if it has drawn a graph that shows the important features. For example, the graph of $y = (x + 30)^2\,(x - 40)^2$ has $x$-intercepts at $(-30, 0)$ and $(40, 0)$, but they do not appear on the graph in Fig. 3.47. □

An important theorem for understanding the graphs of polynomial functions is the intermediate value theorem (IVT). The IVT says that *a polynomial function takes on every value between any two of its values.* For example, consider $f(x) = x^3$, for which $f(1) = 1$ and $f(2) = 8$. Select any number between 1 and 8, say 5. By the IVT there is a number $c$ in the interval $(1, 2)$ such that $f(c) = 5$. In this case, $c$ is easy to find: $c = \sqrt[3]{5}$. The IVT allows us to "connect the dots" when drawing a graph. The curve cannot go from $(1, 1)$ to $(2, 8)$ without hitting every $y$-coordinate between 1 and 8. If one of the values is positive and the other negative, the IVT guarantees that the curve crosses the $x$-axis on the interval. For example, $f(-2) = -8$ and $f(2) = 8$. By the IVT there is a $c$ in the interval $(-2, 2)$ for which $f(c) = 0$. Of course, in this case $c = 0$ and the $x$-intercept is $(0, 0)$.

Note that the greatest integer function, which is not a polynomial function, does not obey the IVT. The greatest integer function jumps from one integer to the next without taking on any values between the integers.

We will not prove the IVT here, as it is proved in calculus. The theorem is stated symbolically as follows.

**The Intermediate Value Theorem**

Suppose that $f$ is a polynomial function and $[a, b]$ is an interval for which $f(a) \neq f(b)$. If $k$ is a number between $f(a)$ and $f(b)$, then there is a number $c$ in the interval $(a, b)$ such that $f(c) = k$.

## Symmetry

Symmetry is a very special property of graphs of some functions but not others. Recognizing that the graph of a function has some symmetry usually cuts in half the work required to obtain the graph and also helps cut down on errors in graphing. So far we have discussed the following types of symmetry.

**SUMMARY**

### Types of Symmetry

1. The graph of a function $f(x)$ is *symmetric about the y-axis* and $f$ is an even function if $f(-x) = f(x)$ for any value of $x$ in the domain of the function. (Section 2.3)
2. The graph of a function $f(x)$ is *symmetric about the origin* and $f$ is an odd function if $f(-x) = -f(x)$ for any value of $x$ in the domain of the function. (Section 2.3)
3. The graph of a quadratic function $f(x) = ax^2 + bx + c$ is *symmetric about its axis of symmetry,* $x = -b/(2a)$. (Section 3.1)

The graphs of $f(x) = x^2$ and $f(x) = x^3$ shown in Figs. 3.48 and 3.49 are nice examples of symmetry about the $y$-axis and symmetry about the origin, respectively. The axis of symmetry of $f(x) = x^2$ is the $y$-axis. The symmetry of the other quadratic functions comes from the fact that the graph of every quadratic function is a transformation of the graph of $f(x) = x^2$.

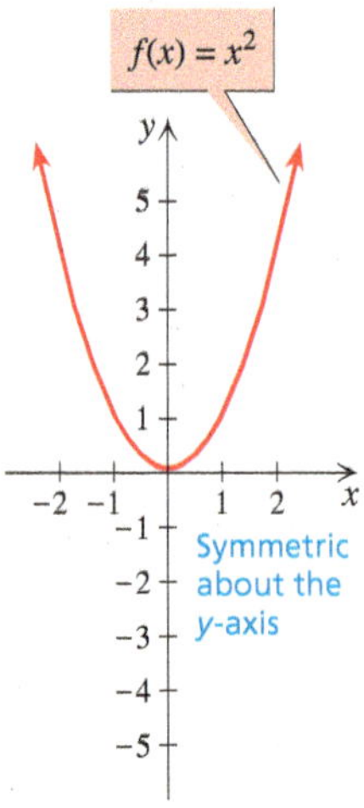


Figure 3.48

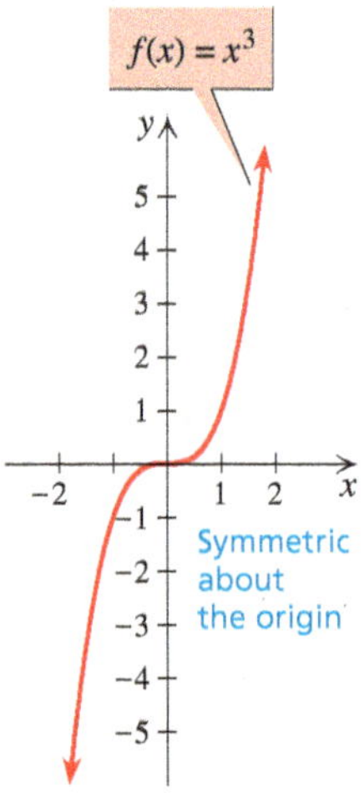


Figure 3.49

**EXAMPLE 1** Determining the symmetry of a graph

Discuss the symmetry of the graph of each polynomial function.

**a.** $f(x) = 5x^3 - x$ **b.** $g(x) = 2x^4 - 3x^2$ **c.** $h(x) = x^2 - 3x + 6$
**d.** $j(x) = x^4 - x^3$

### Solution

**a.** Replace $x$ by $-x$ in $f(x) = 5x^3 - x$ and simplify:

$$f(-x) = 5(-x)^3 - (-x)$$
$$= -5x^3 + x$$

Since $f(-x)$ is the opposite of $f(x)$, the graph is symmetric about the origin. Figure 3.50 supports this conclusion. □

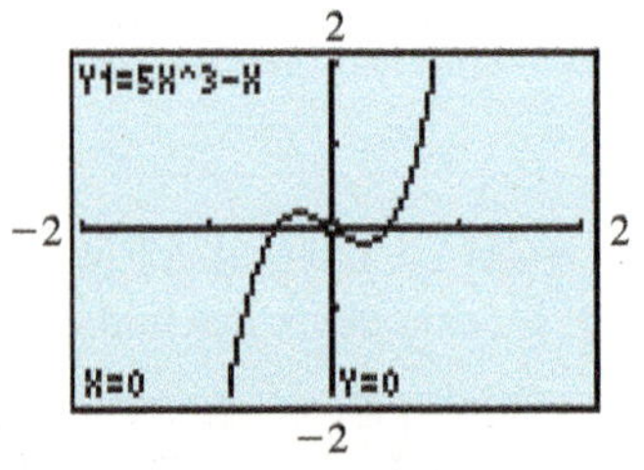


Figure 3.50

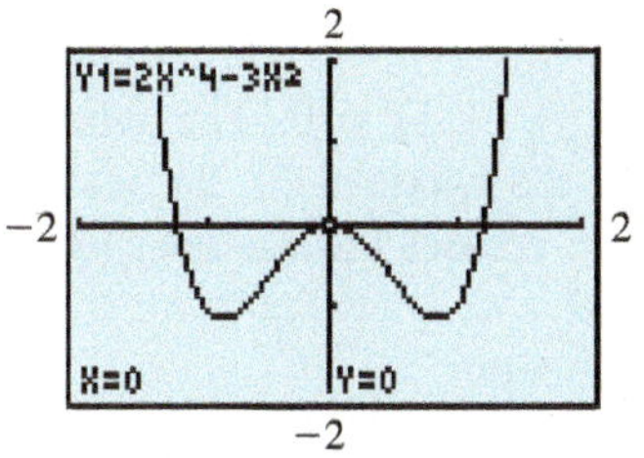


Figure 3.51

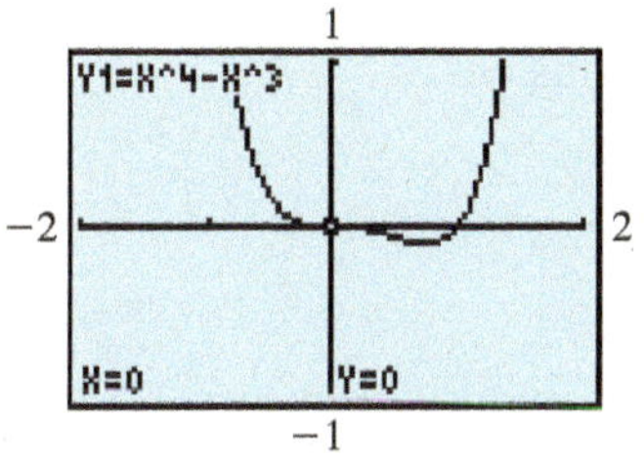


Figure 3.52

**b.** Replace $x$ by $-x$ in $g(x) = 2x^4 - 3x^2$ and simplify:

$$g(-x) = 2(-x)^4 - 3(-x)^2$$
$$= 2x^4 - 3x^2$$

Since $g(-x) = g(x)$, the graph is symmetric about the $y$-axis. Figure 3.51 supports this conclusion. □

**c.** Because $h$ is a quadratic function, its graph is symmetric about the line $x = -b/(2a)$, which in this case is the line $x = 3/2$.

**d.** In this case

$$j(-x) = (-x)^4 - (-x)^3$$
$$= x^4 + x^3.$$

So $j(-x) \neq j(x)$ and $j(-x) \neq -j(x)$. The graph of $j$ is not symmetric about the $y$-axis and is not symmetric about the origin. Figure 3.52 supports this conclusion.

►**TRY THIS.** Discuss the symmetry of $f(x) = -x^3 - 4x$. ■

## Behavior at the x-Intercepts

The $x$-intercepts are key points for the graph of a polynomial function, as they are for any function. Consider the graph of $y = (x - 2)^2(x + 1)$ in Fig. 3.53. Near 2 the values of $y$ are positive, as shown in Fig. 3.54, and the graph does not cross the $x$-axis. Near $-1$ the values of $y$ are negative for $x < -1$ and positive for $x > -1$, as shown in Fig. 3.55, and the graph crosses the $x$-axis. The reason for this behavior is the exponents in $(x - 2)^2$ and $(x + 1)^1$. Because $(x - 2)^2$ has an even exponent, $(x - 2)^2$ cannot be negative. Because $(x + 1)^1$ has an odd exponent, $(x + 1)^1$ changes sign at $-1$ but does not change sign at 2. So the product $(x - 2)^2(x + 1)$ does not change sign at 2, but does change sign at $-1$.

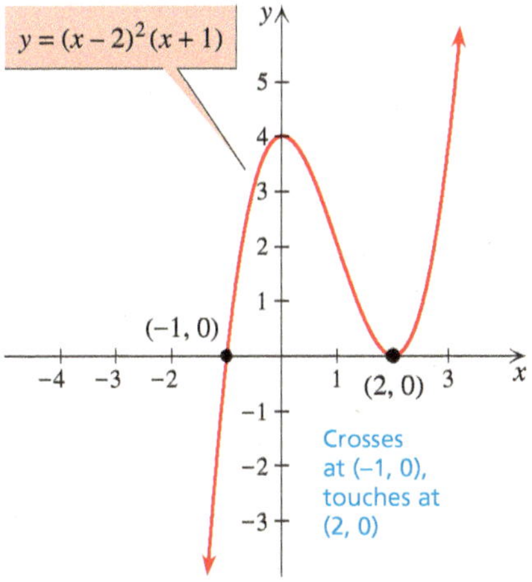


Figure 3.53

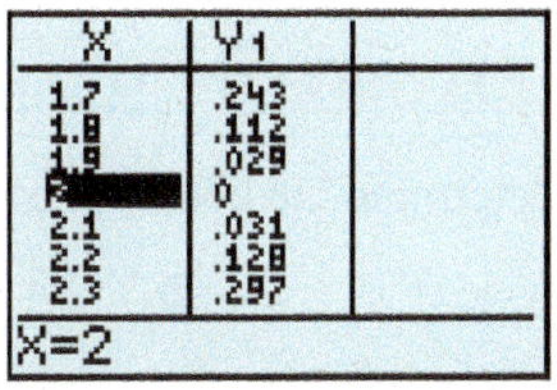


Figure 3.54

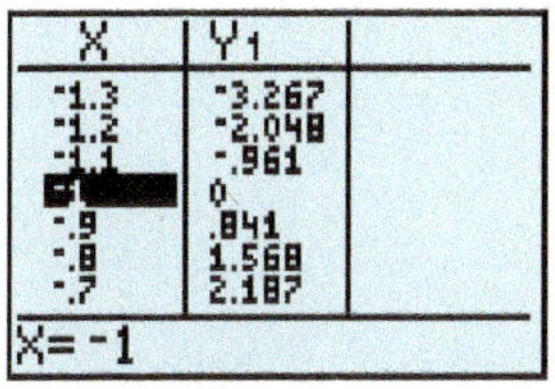


Figure 3.55

Every $x$-intercept corresponds to a factor of the polynomial and to a root of the polynomial function. Whether the multiplicity of the root is odd or even determines the behavior of the graph at the $x$-intercept.

**Theorem: Behavior at the x-Intercepts**

Let $a$ be a root with multiplicity $k$ for a polynomial function.
If $k$ is odd, then the graph crosses the $x$-axis at $(a, 0)$.
If $k$ is even, then the graph touches but does not cross the $x$-axis at $(a, 0)$.

As another example, consider the graphs of $f(x) = x^2$ and $f(x) = x^3$ shown in Figs. 3.48 and 3.49. Zero is a root with odd multiplicity for $f(x) = x^3$ and even multiplicity for $f(x) = x^2$. The graph of $f(x) = x^3$ crosses the $x$-axis at $(0, 0)$ but $f(x) = x^2$ does not cross the $x$-axis at $(0, 0)$.

### EXAMPLE 2 Crossing at the x-intercepts

Find the $x$-intercepts and determine whether the graph of the function crosses the $x$-axis at each $x$-intercept.

**a.** $f(x) = (x - 1)^2(x - 3)$ **b.** $f(x) = x^3 + 2x^2 - 3x$

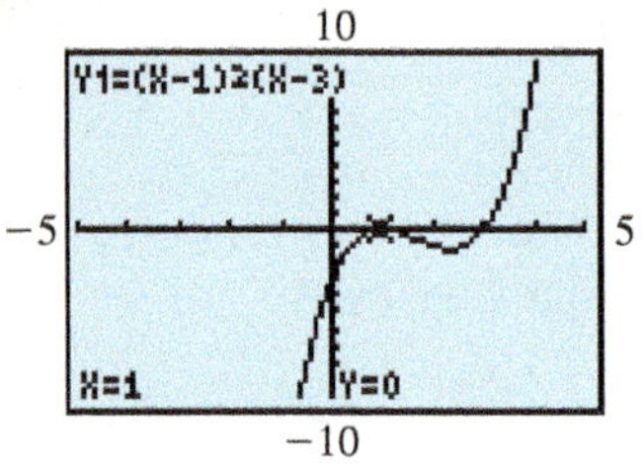


Figure 3.56

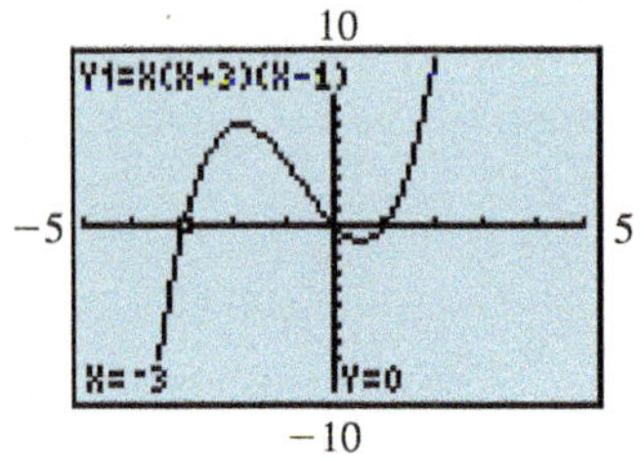


Figure 3.57

### Solution

**a.** The $x$-intercepts are found by solving $(x - 1)^2(x - 3) = 0$. The $x$-intercepts are $(1, 0)$ and $(3, 0)$. The graph does not cross the $x$-axis at $(1, 0)$ because the factor $x - 1$ occurs to an even power. The graph crosses the $x$-axis at $(3, 0)$ because $x - 3$ occurs to an odd power.

The graph in Fig. 3.56 supports these conclusions. □

**b.** The $x$-intercepts are found by solving $x^3 + 2x^2 - 3x = 0$. By factoring, we get $x(x + 3)(x - 1) = 0$. The $x$-intercepts are $(0, 0)$, $(-3, 0)$, and $(1, 0)$. Since each factor occurs an odd number of times (once), the graph crosses the $x$-axis at each of the $x$-intercepts.

The graph in Fig. 3.57 supports these conclusions.

▶**TRY THIS.** Determine whether the graph of $f(x) = (x - 1)^3(x + 5)^2$ crosses the $x$-axis at each of its $x$-intercepts. ■

## The Leading Coefficient Test

We now consider the behavior of a polynomial function as the $x$-coordinate goes to or approaches infinity or negative infinity. In symbols, $x \to \infty$ or $x \to -\infty$. Since we seek only an intuitive understanding of the ideas presented here, we will not give precise definitions of these terms. Precise definitions are given in a calculus course.

To say that $x \to \infty$ means that $x$ gets larger and larger without bound. For our purposes we can think of $x$ assuming the values 1, 2, 3, and so on, without end. Similarly, $x \to -\infty$ means that $x$ gets smaller and smaller without bound. Think of $x$ assuming the values $-1, -2, -3$, and so on, without end.

As $x$ approaches $\infty$ the $y$-coordinates of any polynomial function approach $\infty$ or $-\infty$. Likewise, when $x \to -\infty$ the $y$-coordinates approach $\infty$ or $-\infty$. The direction that $y$ goes is determined by the degree of the polynomial and the sign of the leading coefficient. The four possible types of behavior are illustrated in the next example.

### EXAMPLE 3 Behavior as $x \to \infty$ or $x \to -\infty$

Determine the behavior of the graph of each function as $x \to \infty$ or $x \to -\infty$.

**a.** $y = x^3 - x$ **b.** $y = -x^3 + 1$ **c.** $y = x^4 - 4x^2$ **d.** $y = -x^4 + 4x^2 + x$

### Solution

**a.** Consider the following table. If you have a graphing calculator, make a table like this and scroll through it.

| $x$ | $-30$ | $-20$ | $-10$ | 0 | 10 | 20 | 30 |
|---|---|---|---|---|---|---|---|
| $y = x^3 - x$ | $-26{,}970$ | $-7980$ | $-990$ | 0 | 990 | 7980 | 26,970 |

⟸ $y \to -\infty$ $y \to \infty$ ⟹

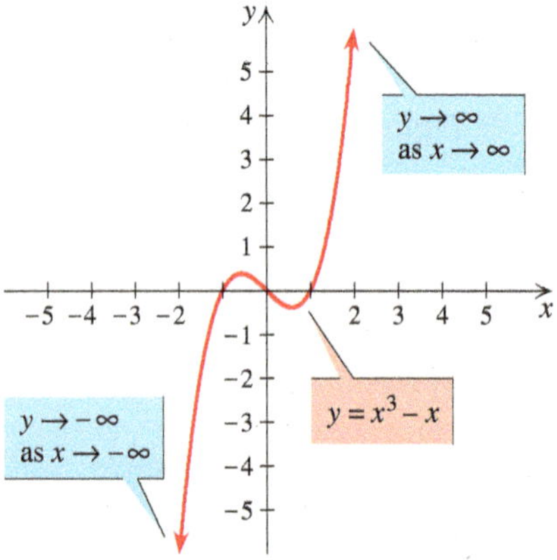


Figure 3.58

The graph of $y = x^3 - x$ is shown in Fig. 3.58. As $x$ gets larger and larger $(x \to \infty)$, $y$ increases without bound $(y \to \infty)$. As $x$ gets smaller and smaller $(x \to -\infty)$, $y$ decreases without bound $(y \to -\infty)$. Notice that the degree of the polynomial is odd and the sign of the leading coefficient is positive. The behavior of this function is stated with limit notation as

$$\lim_{x \to \infty} x^3 - x = \infty \quad \text{and} \quad \lim_{x \to -\infty} x^3 - x = -\infty.$$

The notation $\lim_{x \to \infty} x^3 - x = \infty$ is read as "the limit as $x$ approaches $\infty$ of $x^3 - x$ is $\infty$." We could also write $\lim_{x \to \infty} y = \infty$ if it is clear that $y = x^3 - x$.

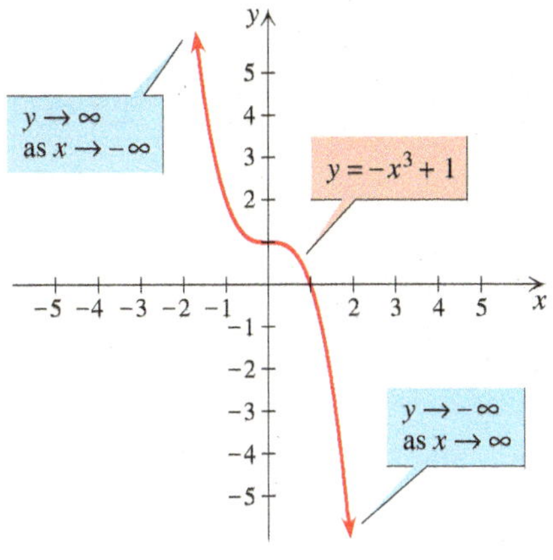


**Figure 3.59**

**b.** Consider the following table. If you have a graphing calculator, make a table like this and scroll through it.

| $x$ | $-30$ | $-20$ | $-10$ | 0 | 10 | 20 | 30 |
|---|---|---|---|---|---|---|---|
| $y = -x^3 + 1$ | 27,001 | 8001 | 1001 | 1 | $-999$ | $-7999$ | $-26{,}999$ |

$\leftarrow y \to \infty$ $\qquad$ $y \to -\infty \rightarrow$

The graph of $y = -x^3 + 1$ is shown in Fig. 3.59. As $x$ gets larger and larger ($x \to \infty$), $y$ decreases without bound ($y \to -\infty$). As $x$ gets smaller and smaller ($x \to -\infty$), $y$ increases without bound ($y \to \infty$). Notice that the degree of this polynomial is odd and the sign of the leading coefficient is negative. The behavior of this function is stated with limit notation as

$$\lim_{x\to\infty} -x^3 + 1 = -\infty \quad \text{and} \quad \lim_{x\to-\infty} -x^3 + 1 = \infty.$$

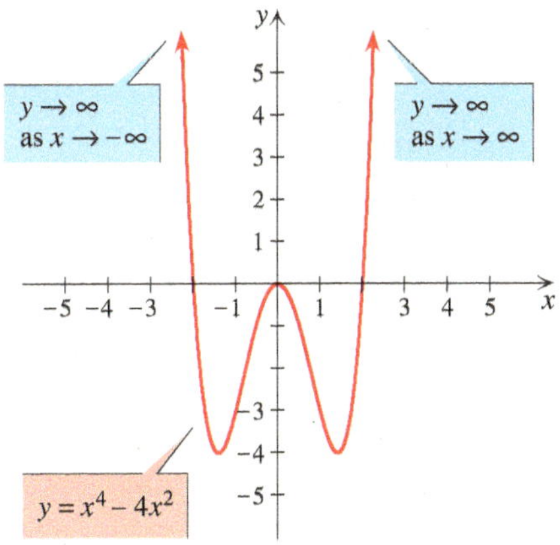


**Figure 3.60**

**c.** Consider the following table. If you have a graphing calculator, make a table like this and scroll through it.

| $x$ | 0 | $\pm 10$ | $\pm 20$ | $\pm 30$ | $\pm 40$ |
|---|---|---|---|---|---|
| $y = x^4 - 4x^2$ | 0 | 9600 | 158,400 | 806,400 | 2,553,600 |

$y \to \infty \rightarrow$

The graph of $y = x^4 - 4x^2$ is shown in Fig. 3.60. As $x \to \infty$ or $x \to -\infty$, $y$ increases without bound ($y \to \infty$). Notice that the degree of this polynomial is even and the sign of the leading coefficient is positive. The behavior of this function is stated with limit notation as

$$\lim_{x\to\infty} x^4 - 4x^2 = \infty \quad \text{and} \quad \lim_{x\to-\infty} x^4 - 4x^2 = \infty.$$

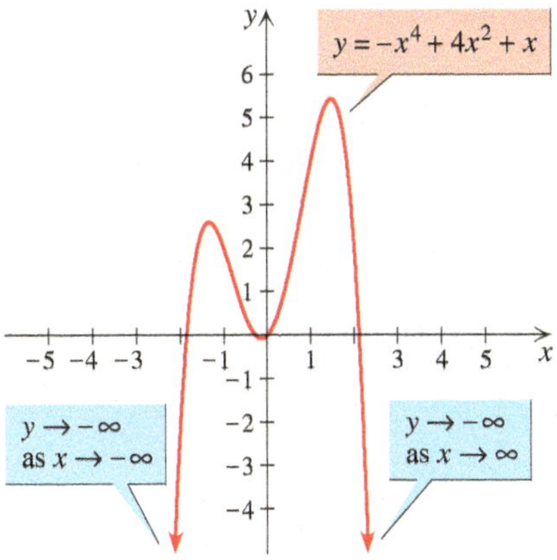


**Figure 3.61**

**d.** Consider the following table. If you have a graphing calculator, make a table like this and scroll through it.

| $x$ | $-20$ | $-10$ | 0 | 10 | 20 |
|---|---|---|---|---|---|
| $y = -x^4 + 4x^2 + x$ | $-158{,}420$ | $-9610$ | 0 | $-9590$ | $-158{,}380$ |

$\leftarrow y \to -\infty$ $\qquad$ $y \to -\infty \rightarrow$

The graph of $y = -x^4 + 4x^2 + x$ is shown in Fig. 3.61. As $x \to \infty$ or $x \to -\infty$, $y$ decreases without bound ($y \to -\infty$). Notice that the degree of this polynomial is even and the sign of the leading coefficient is negative. The behavior of this function is stated with limit notation as

$$\lim_{x\to\infty} -x^4 + 4x^2 + x = -\infty \quad \text{and} \quad \lim_{x\to-\infty} -x^4 + 4x^2 + x = -\infty.$$

►**TRY THIS.** Discuss the behavior of the graph of $f(x) = -x^3 - 4x$. ■

As $x \to \infty$ or $x \to -\infty$ all first-degree polynomial functions have "end behavior" like the lines $y = x$ or $y = -x$, all second-degree polynomial functions behave like the parabolas $y = x^2$ or $y = -x^2$, and all third-degree polynomial functions behave like $y = x^3$ or $y = -x^3$, and so on. The end behavior is determined by the degree and the sign of the first term. The smaller-degree terms in the polynomial determine the number of "hills" and "valleys" between the ends of the curve. With degree $n$ there are at most $n - 1$ hills and valleys. The end behavior of polynomial functions is summarized in the **leading coefficient test.**

**Leading Coefficient Test**

If $f(x) = a_nx^n + a_{n-1}x^{n-1} + \cdots + a_1x + a_0$, the behavior of the graph of $f$ to the left and right is determined as follows:

For $n$ odd and $a_n > 0$, $\lim_{x\to\infty} f(x) = \infty$ and $\lim_{x\to-\infty} f(x) = -\infty$.
For $n$ odd and $a_n < 0$, $\lim_{x\to\infty} f(x) = -\infty$ and $\lim_{x\to-\infty} f(x) = \infty$.
For $n$ even and $a_n > 0$, $\lim_{x\to\infty} f(x) = \infty$ and $\lim_{x\to-\infty} f(x) = \infty$.
For $n$ even and $a_n < 0$, $\lim_{x\to\infty} f(x) = -\infty$ and $\lim_{x\to-\infty} f(x) = -\infty$.

The leading coefficient test is presented visually in Fig. 3.62, which shows only the "ends" of the graphs of the polynomial functions.

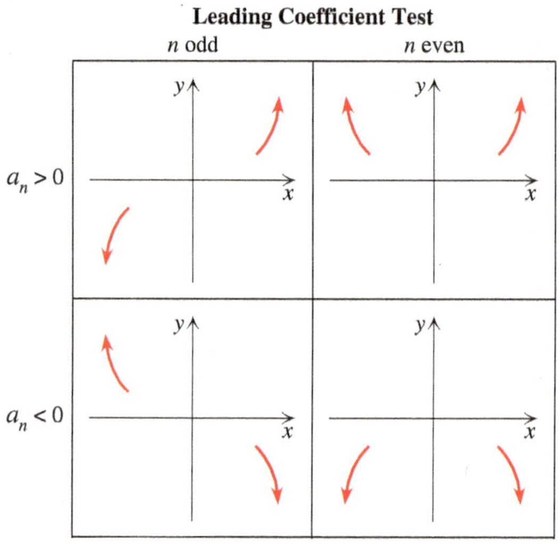


Figure 3.62

## Sketching Graphs of Polynomial Functions

A good graph of a polynomial function should include the features that we have been discussing. The following strategy will help you graph polynomial functions.

**STRATEGY**

**Graphing a Polynomial Function**

1. Check for symmetry.
2. Find all real zeros of the polynomial function.
3. Determine the behavior at the corresponding $x$-intercepts.
4. Determine the behavior as $x \to \infty$ and as $x \to -\infty$.
5. Calculate several ordered pairs including the $y$-intercept to verify your suspicions about the shape of the graph.
6. Draw a smooth curve through the points to make the graph.

**EXAMPLE 4** Graphing polynomial functions

Sketch the graph of each polynomial function.

**a.** $f(x) = x^3 - 5x^2 + 7x - 3$ **b.** $f(x) = x^4 - 200x^2 + 10{,}000$

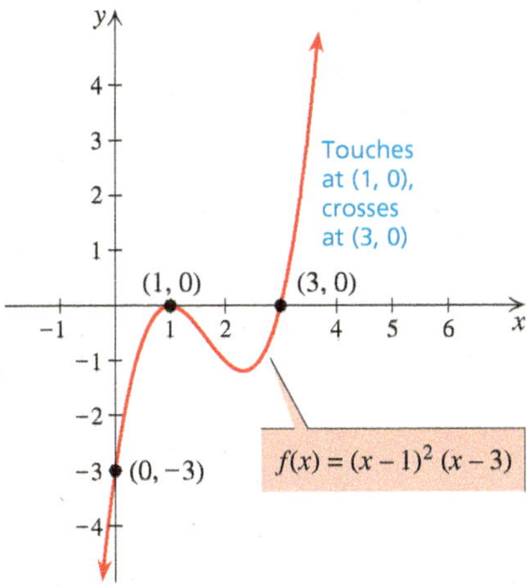


Figure 3.63

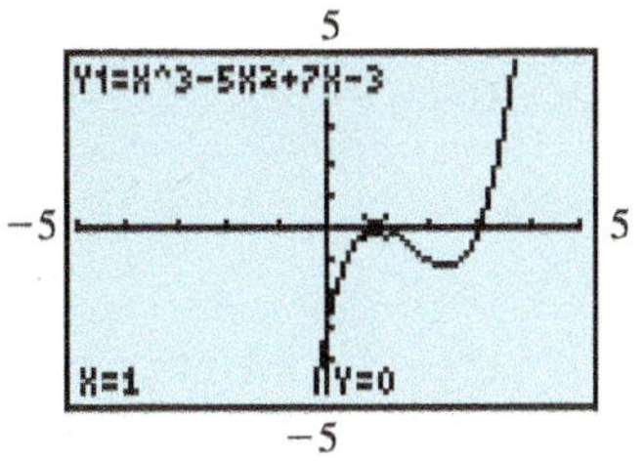


Figure 3.64

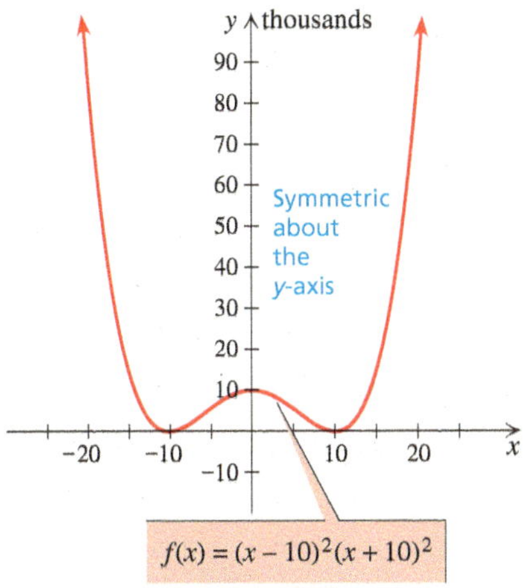


Figure 3.65

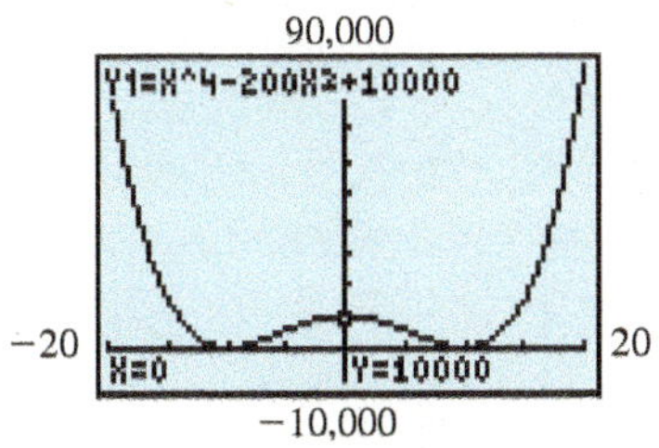


Figure 3.66

### Solution

**a.** First find $f(-x)$ to determine symmetry and the number of negative roots.

$$f(-x) = (-x)^3 - 5(-x)^2 + 7(-x) - 3$$
$$= -x^3 - 5x^2 - 7x - 3$$

From $f(-x)$, we see that the graph has neither type of symmetry. Because $f(-x)$ has no sign changes, $x^3 - 5x^2 + 7x - 3 = 0$ has no negative roots by Descartes's rule of signs. The only possible rational roots are 1 and 3.

$$\begin{array}{r|rrrr} 1 & 1 & -5 & 7 & -3 \\ & & 1 & -4 & 3 \\ \hline & 1 & -4 & 3 & 0 \end{array}$$

From the synthetic division we know that 1 is a root and we can factor $f(x)$:

$$f(x) = (x-1)(x^2 - 4x + 3)$$
$$= (x-1)^2(x-3) \qquad \text{Factor completely.}$$

The $x$-intercepts are $(1, 0)$ and $(3, 0)$. The graph of $f$ does not cross the $x$-axis at $(1, 0)$ because $x - 1$ occurs to an even power, while the graph crosses at $(3, 0)$ because $x - 3$ occurs to an odd power. The $y$-intercept is $(0, -3)$. Since the leading coefficient is positive and the degree is odd, $y \to \infty$ as $x \to \infty$ and $y \to -\infty$ as $x \to -\infty$. Calculate two more ordered pairs for accuracy, say $(2, -1)$ and $(4, 9)$. Draw a smooth curve as in Fig. 3.63.

The calculator graph shown in Fig. 3.64 supports these conclusions. □

**b.** First find $f(-x)$:

$$f(-x) = (-x)^4 - 200(-x)^2 + 10{,}000$$
$$= x^4 - 200x^2 + 10{,}000$$

Since $f(x) = f(-x)$, the graph is symmetric about the $y$-axis. We can factor the polynomial as follows.

$$f(x) = x^4 - 200x^2 + 10{,}000$$
$$= (x^2 - 100)(x^2 - 100)$$
$$= (x-10)(x+10)(x-10)(x+10)$$
$$= (x-10)^2(x+10)^2$$

The $x$-intercepts are $(10, 0)$ and $(-10, 0)$. Because each factor for these intercepts has an even power, the graph does not cross the $x$-axis at the intercepts. The $y$-intercept is $(0, 10{,}000)$. Since the leading coefficient is positive and the degree is even, $y \to \infty$ as $x \to \infty$ or as $x \to -\infty$. The graph also goes through $(-20, 90{,}000)$ and $(20, 90{,}000)$. Draw a smooth curve through these points and the intercepts as shown in Fig. 3.65.

The calculator graph shown in Fig. 3.66 supports these conclusions.

**▸TRY THIS.** Sketch the graph of $f(x) = x^4 - 4x^2$. ■

## Polynomial Inequalities

We can solve polynomial inequalities using the graphical method or test-point method that we used on quadratic inequalities in Section 3.1. You should recall that for either of these methods we find all real roots to the corresponding polynomial function. The real roots divide the number line into intervals. With the graphical method we determine the sign of the polynomial in each interval from a graph, whereas with the test-point method we determine the sign of the polynomial by testing an arbitrary point in each interval. We will not use the graphical method here because it is difficult to draw a good graph of some polynomial functions, especially if the $x$-intercepts are really close together or very far apart. The test-point method is more reliable.

## EXAMPLE 5 Solving a polynomial inequality with test points

Solve $x^4 + x^3 - 15x^2 - 3x + 36 < 0$ using the test-point method.

### Solution

Use synthetic division to see that 3 and $-4$ are zeros of the function $f(x) = x^4 + x^3 - 15x^2 - 3x + 36$:

| | | | | | |
|---|---|---|---|---|---|
| 3 | 1 | 1 | −15 | −3 | 36 |
| | | 3 | 12 | −9 | −36 |
| −4 | 1 | 4 | −3 | −12 | 0 |
| | | −4 | 0 | 12 | |
| | 1 | 0 | −3 | 0 | |

Since $x^4 + x^3 - 15x^2 - 3x + 36 = (x - 3)(x + 4)(x^2 - 3)$, the other two zeros are $\pm\sqrt{3}$. The four zeros determine five intervals on the number line in Fig. 3.67. Select an arbitrary test point in each of these intervals. The selected points $-5$, $-3$, 0, 2, and 4 are shown in red in the figure:

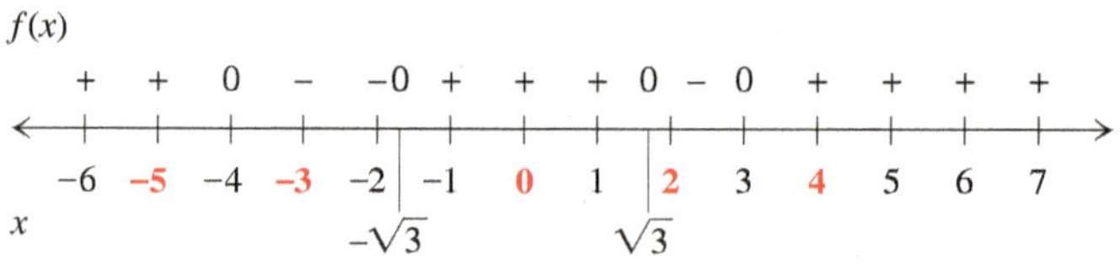


**Figure 3.67**

Now evaluate $f(x) = x^4 + x^3 - 15x^2 - 3x + 36$ at each test point:

$$f(-5) = 176, \quad f(-3) = -36, \quad f(0) = 36, \quad f(2) = -6, \quad f(4) = 104$$

These values indicate that the signs of the function are $+$, $-$, $+$, $-$, and $+$ on the intervals shown in Fig. 3.67. The values of $x$ that satisfy the original inequality are the values of $x$ for which $f(x)$ is negative. So the solution set is $(-4, -\sqrt{3}) \cup (\sqrt{3}, 3)$.

The calculator graph of $y = x^4 + x^3 - 15x^2 - 3x + 36$ in Fig. 3.68 confirms that $y$ is negative for $x$ in $(-4, -\sqrt{3}) \cup (\sqrt{3}, 3)$. Since the multiplicity of each zero of the function is one, the graph crosses the $x$-axis at each intercept and the $y$-coordinates change sign at each intercept.

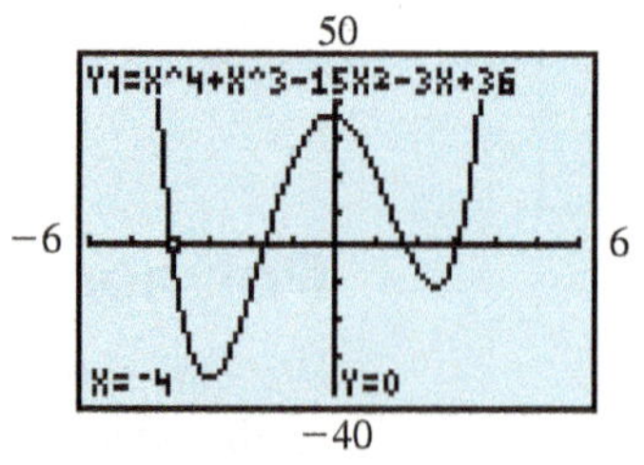


**Figure 3.68**

▸**TRY THIS.** Solve $x^4 - 4x^2 < 0$ using test points. ■

# FUNCTION gallery... Polynomial Functions

**Linear:** $f(x) = mx + b$

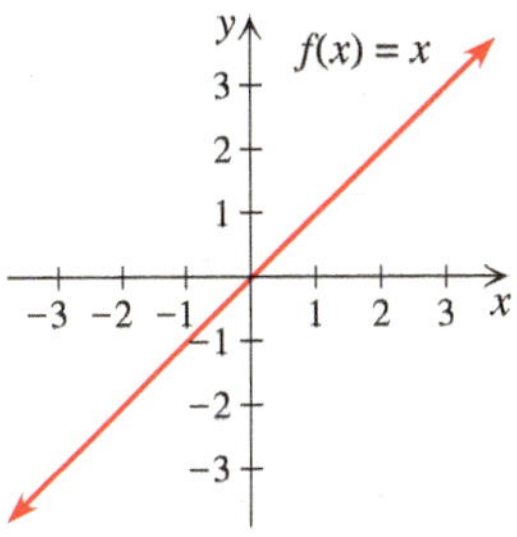


Slope 1, $y$-intercept $(0, 0)$

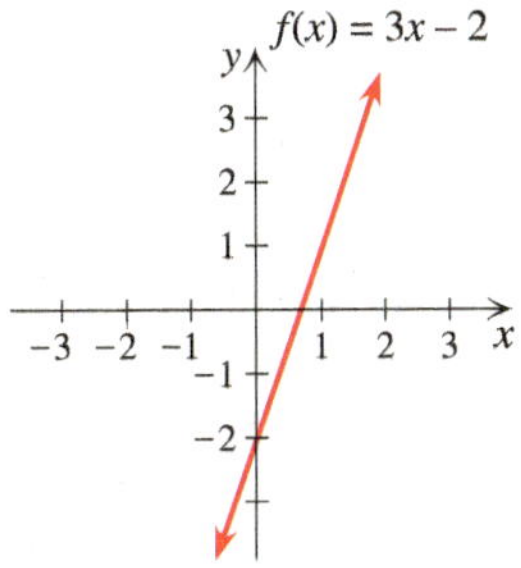


Slope 3, $y$-intercept $(0, -2)$

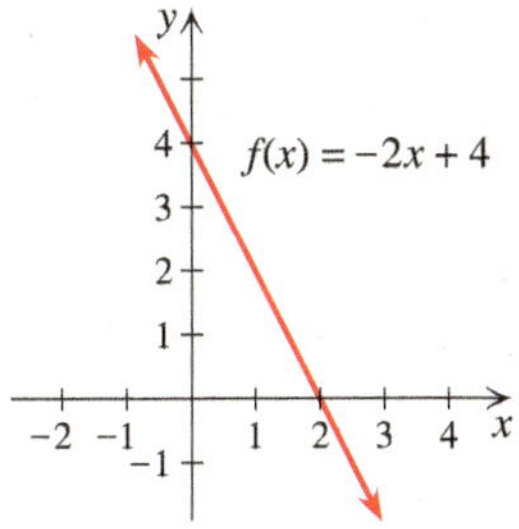


Slope $-2$, $y$-intercept $(0, 4)$

**Quadratic:** $f(x) = ax^2 + bx + c$ or $f(x) = a(x - h)^2 + k$

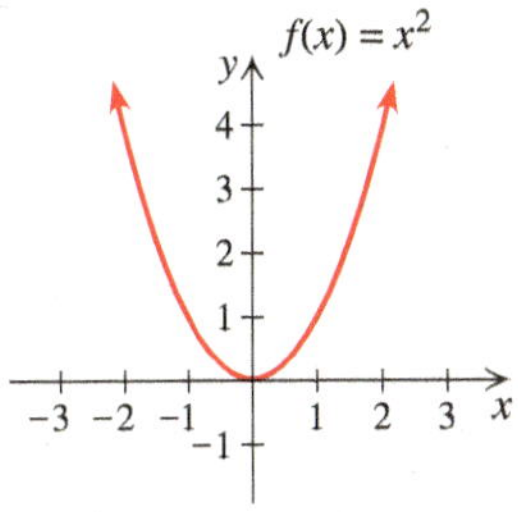


Vertex $(0, 0)$
Range $[0, \infty)$

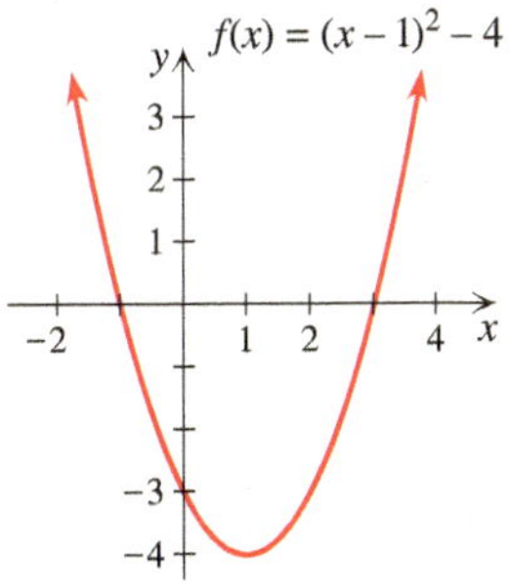


Vertex $(1, -4)$
Range $[-4, \infty)$

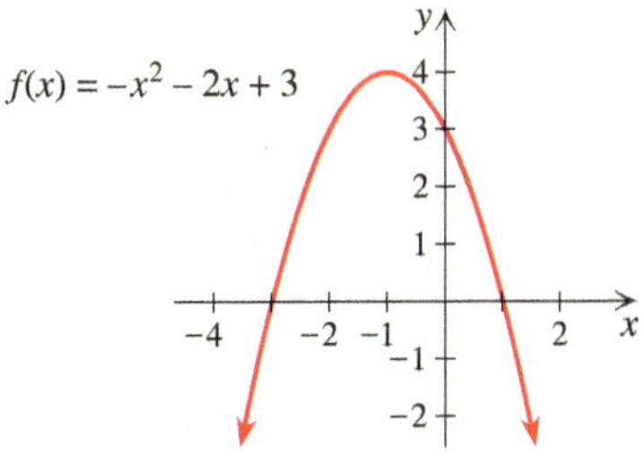


Vertex $(-1, 4)$
Range $(-\infty, 4]$

**Cubic:** $f(x) = ax^3 + bx^2 + cx + d$

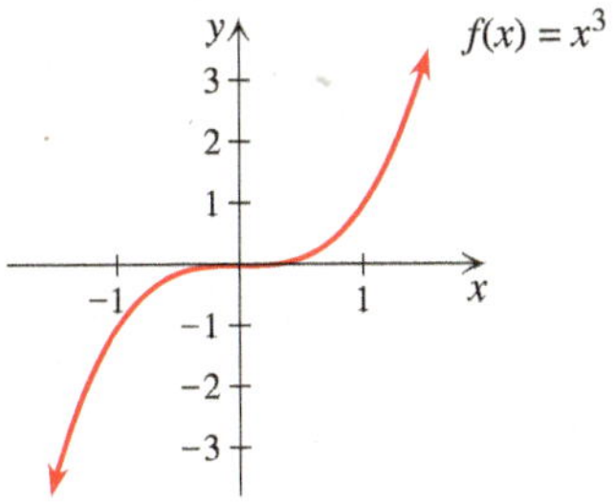


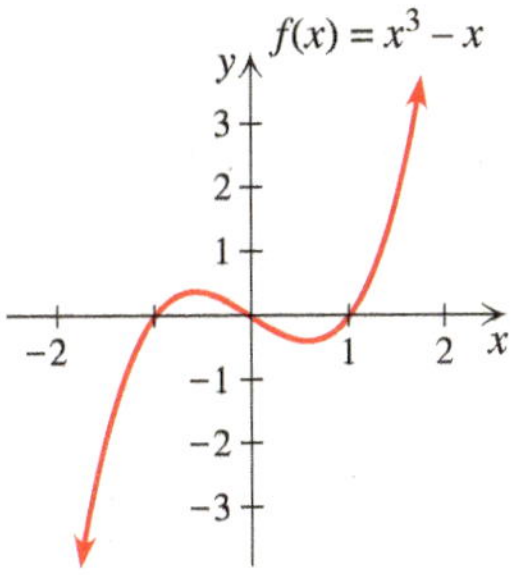


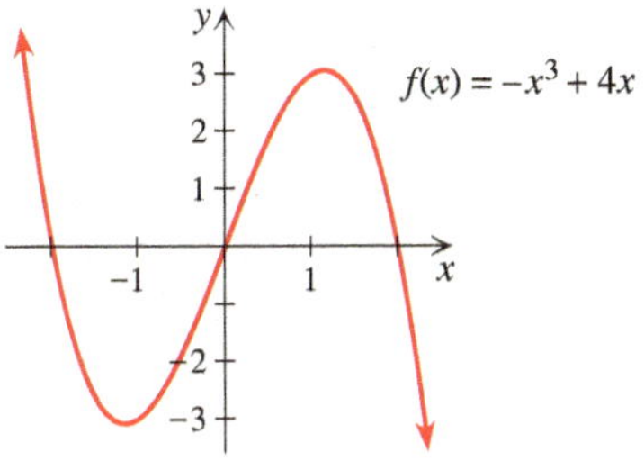


**Quartic or Fourth-Degree:** $f(x) = ax^4 + bx^3 + cx^2 + dx + e$

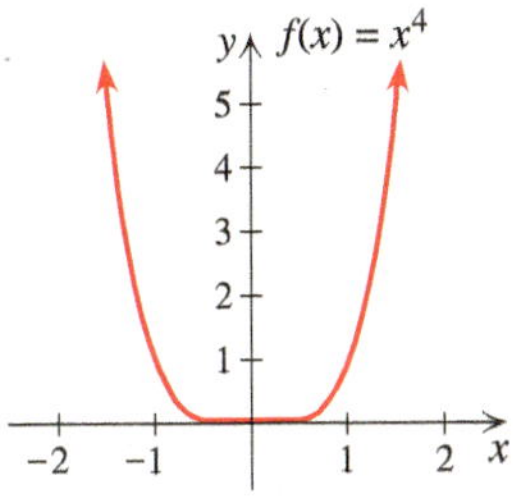


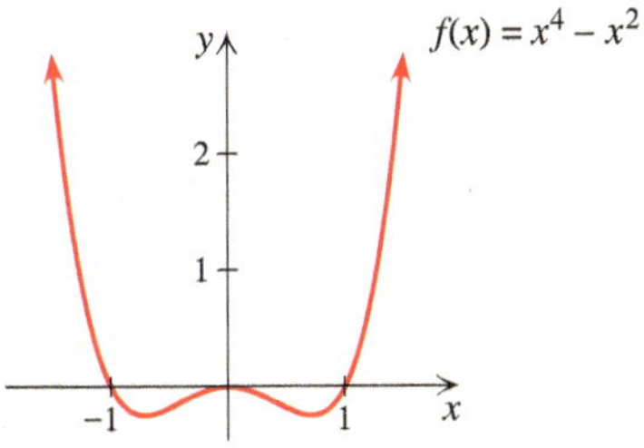


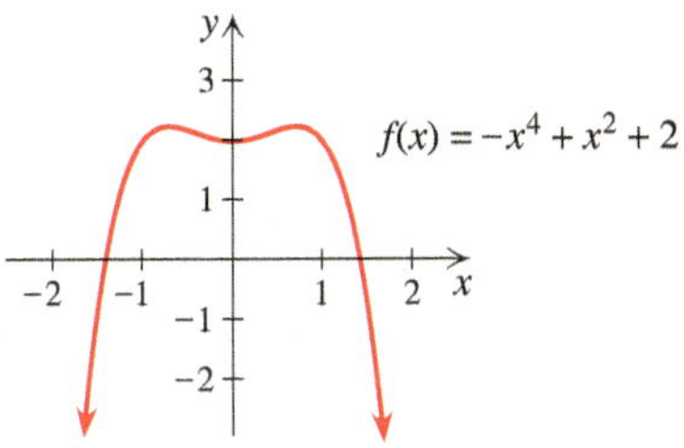

## FOR thought... True or False? Explain.

1. If $P$ is a function for which $P(2) = 8$ and $P(-2) = -8$, then the graph of $P$ is symmetric about the origin.

2. If $y = -3x^3 + 4x^2 - 6x + 9$, then $y \to -\infty$ as $x \to \infty$.

3. If the graph of $y = P(x)$ is symmetric about the origin and $P(8) = 4$, then $-P(-8) = 4$.

4. If $f(x) = x^3 - 3x$, then $f(x) = f(-x)$ for any value of $x$.

5. If $f(x) = x^4 - x^3 + x^2 - 6x + 7$, then $f(-x) = x^4 + x^3 + x^2 + 6x + 7$.

6. The graph of $f(x) = x^2 - 6x + 9$ has only one $x$-intercept.

7. The $x$-intercepts for $P(x) = (x - 1)^2(x + 1)$ are $(0, 1)$ and $(0, -1)$.

8. The $y$-intercept for $P(x) = 4(x - 3)^2 + 2$ is $(0, 2)$.

9. The graph of $f(x) = x^2(x + 8)^2$ has no points in quadrants III and IV.

10. The graph of $f(x) = x^3 - 1$ has three $x$-intercepts.

## EXERCISES 3.5

*Fill in the blank.*

1. If $f(-x) = f(x)$ for every value of $x$ in the domain of the function, then the graph of $f(x)$ is symmetric about the ________.

2. If $f(-x) = -f(x)$ for every value of $x$ in the domain of the function, then the graph of $f(x)$ is symmetric about the ________.

3. The graph of a quadratic function $f(x) = ax^2 + bx + c$ is symmetric about the line $x =$ ________.

4. If $a$ is a root with multiplicity 3 for a polynomial function, then the graph of the function ________ the $x$-axis at $(a, 0)$.

*Discuss the symmetry of the graph of each polynomial function. See the summary of the types of symmetry on page 236.*

5. $f(x) = x^6$
6. $f(x) = x^5 - x$
7. $f(x) = x^2 - 3x + 5$
8. $f(x) = 5x^2 + 10x + 1$
9. $f(x) = 3x^6 - 5x^2 + 3x$
10. $f(x) = x^6 - x^4 + x^2 - 8$
11. $f(x) = 4x^3 - x$
12. $f(x) = 7x^3 + x^2$
13. $f(x) = (x - 5)^2$
14. $f(x) = (x^2 - 1)^2$
15. $f(x) = -x$
16. $f(x) = 3x$

*Find the x-intercepts and discuss the behavior of the graph of each polynomial function at its x-intercepts.*

17. $f(x) = (x - 4)^2$
18. $f(x) = (x - 1)^2(x + 3)^2$
19. $f(x) = (2x - 1)^3$
20. $f(x) = x^6$
21. $f(x) = 4x - 1$
22. $f(x) = x^2 - 5x - 6$
23. $f(x) = x^2 - 3x + 10$
24. $f(x) = x^4 - 16$
25. $f(x) = x^3 - 3x^2$
26. $f(x) = x^3 - x^2 - x + 1$
27. $f(x) = 2x^3 - 5x^2 + 4x - 1$
28. $f(x) = x^3 - 3x^2 + 4$
29. $f(x) = -2x^3 - 8x^2 + 6x + 36$
30. $f(x) = -x^3 + 7x - 6$

*For each function use the leading coefficient test to determine whether $y \to \infty$ or $y \to -\infty$ as $x \to \infty$.*

31. $y = 2x^3 - x^2 + 9$
32. $y = -3x + 7$
33. $y = -3x^4 + 5$
34. $y = 6x^4 - 5x^2 - 1$
35. $y = x - 3x^3$
36. $y = 5x - 7x^4$

*For each function use the leading coefficient test to determine whether $y \to \infty$ or $y \to -\infty$ as $x \to -\infty$.*

**37.** $y = -2x^5 - 3x^2$

**38.** $y = x^3 + 8x + \sqrt{2}$

**39.** $y = 3x^6 - 999x^3$

**40.** $y = -12x^4 - 5x$

*For each graph discuss its symmetry, indicate whether the graph crosses the x-axis at each x-intercept, and determine whether $y \to \infty$ or $y \to -\infty$ as $x \to \infty$ and $x \to -\infty$.*

**41.**

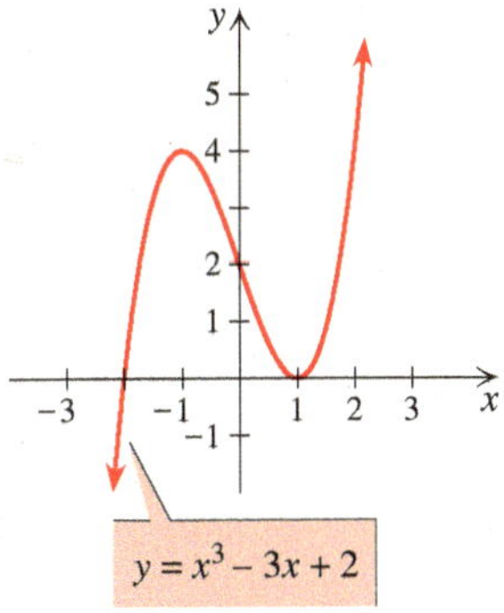


**42.**

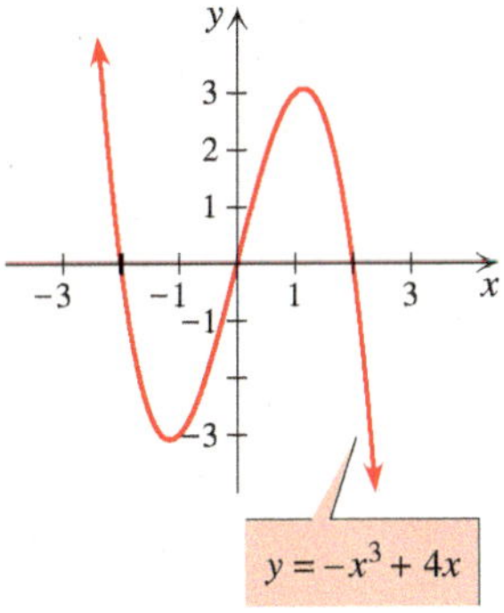


**43.**

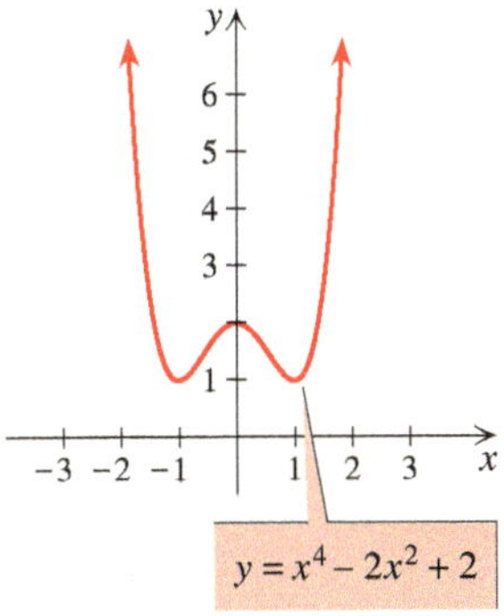


**44.**

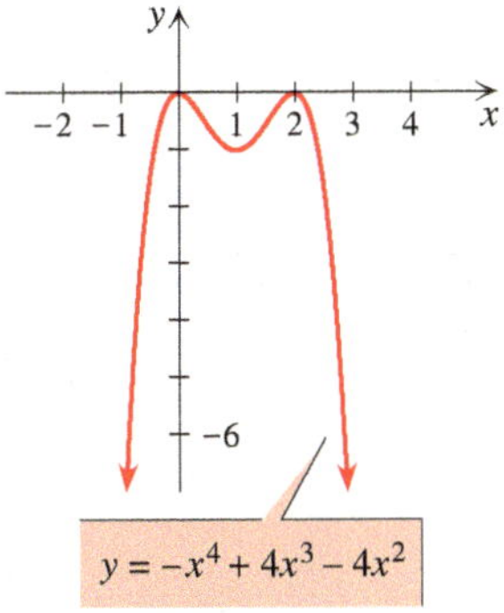


*Determine whether each limit is equal to $\infty$ or $-\infty$.*

**45.** $\lim_{x \to \infty} x^2 - 4$

**46.** $\lim_{x \to \infty} x^3 + x$

**47.** $\lim_{x \to \infty} -x^5 - x^2$

**48.** $\lim_{x \to \infty} -3x^4 + 9x^2$

**49.** $\lim_{x \to -\infty} -3x$

**50.** $\lim_{x \to -\infty} x^3 - 5$

**51.** $\lim_{x \to -\infty} -2x^2 + 1$

**52.** $\lim_{x \to -\infty} 6x^4 - x$

*For each given function make a rough sketch of the graph that shows the behavior at the x-intercepts and the behavior as x approaches $\infty$ and $-\infty$.*

**53.** $f(x) = (x - 1)^2(x + 3)$

**54.** $f(x) = (x + 2)^2(x - 5)^2$

**55.** $f(x) = -2(2x - 1)^2(x + 1)^3$

**56.** $f(x) = -3(3x - 4)^2(2x + 1)^4$

*Match each polynomial function with its graph (a)–(h).*

**57.** $f(x) = -2x + 1$

**58.** $f(x) = -2x^2 + 1$

**59.** $f(x) = -2x^3 + 1$

**60.** $f(x) = -2x^2 + 4x - 1$

**61.** $f(x) = -2x^4 + 6$

**62.** $f(x) = -2x^4 + 6x^2$

**63.** $f(x) = x^3 + 4x^2 - x - 4$

**64.** $f(x) = (x - 2)^4$

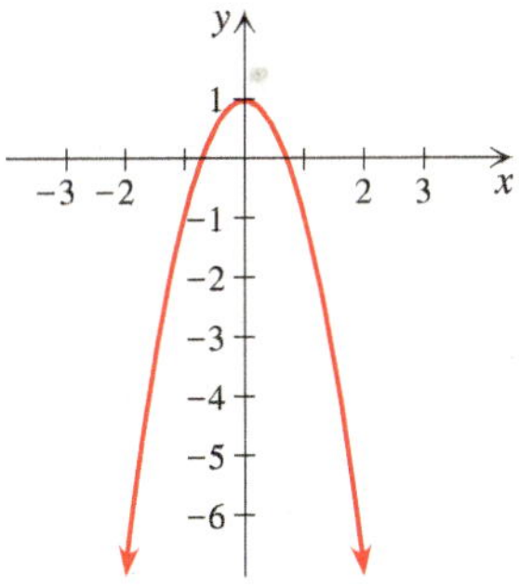

(a)

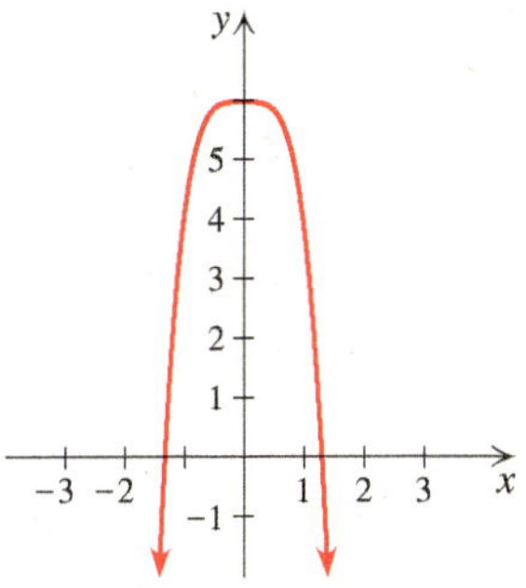

(b)

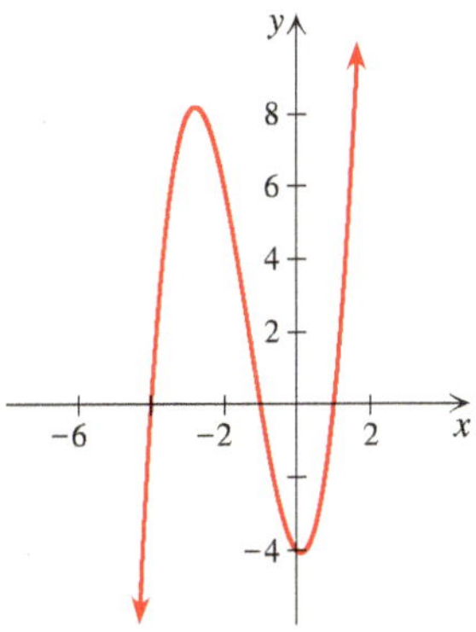

(c)

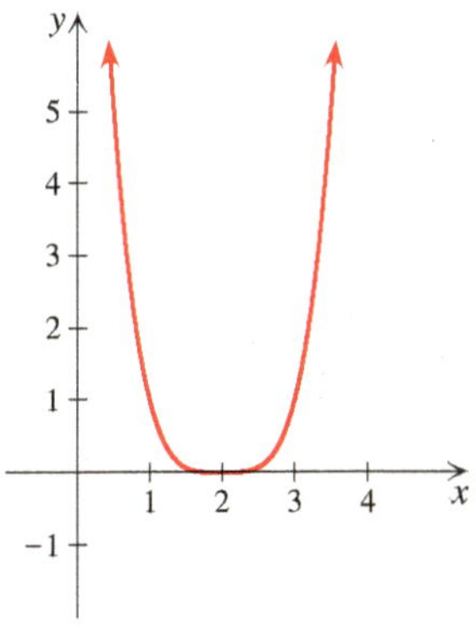

(d)

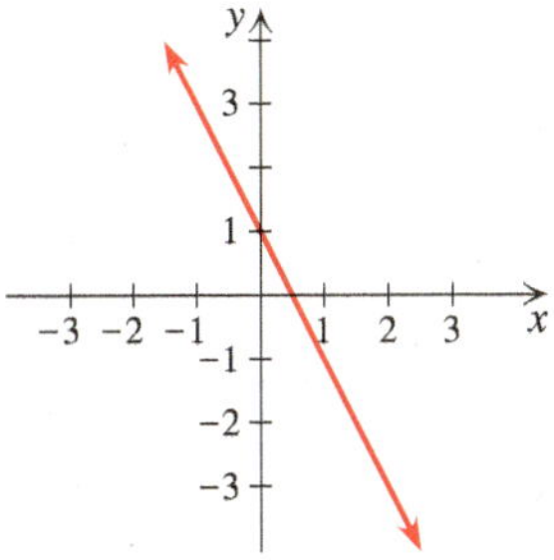

(e)

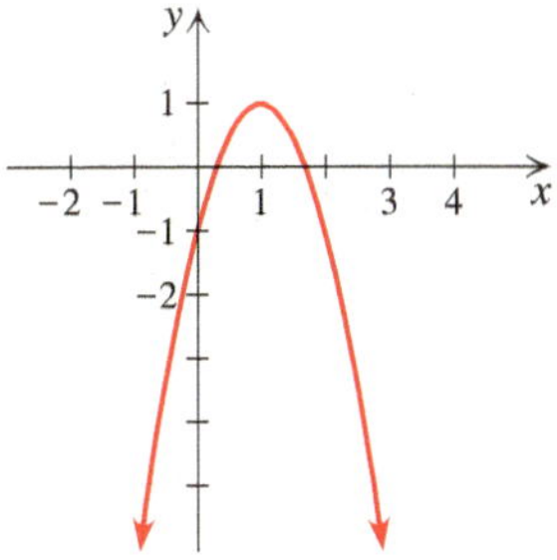

(f)

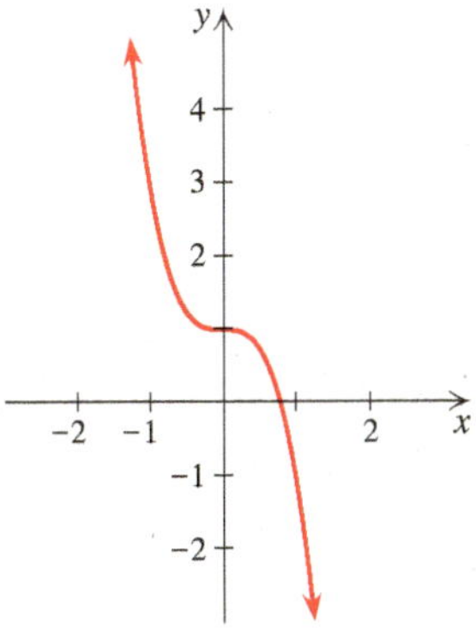

(g)

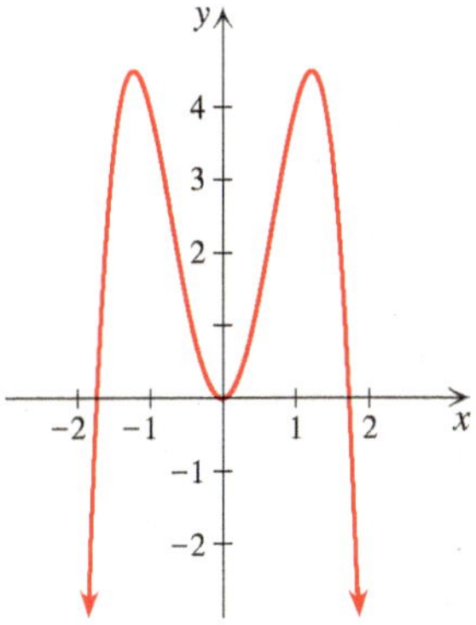

(h)

*Sketch the graph of each function. See the strategy for graphing polynomial functions on page 240.*

**65.** $f(x) = x - 30$

**66.** $f(x) = 40 - x$

**67.** $f(x) = (x - 30)^2$

**68.** $f(x) = (40 - x)^2$

**69.** $f(x) = x^3 - 40x^2$

**70.** $f(x) = x^3 - 900x$

**71.** $f(x) = (x - 20)^2(x + 20)^2$

**72.** $f(x) = (x - 20)^2(x + 12)$

**73.** $f(x) = -x^3 - x^2 + 5x - 3$

**74.** $f(x) = -x^4 + 6x^3 - 9x^2$

**75.** $f(x) = x^3 - 10x^2 - 600x$

**76.** $f(x) = -x^4 + 24x^3 - 144x^2$

**77.** $f(x) = x^3 + 18x^2 - 37x + 60$

**78.** $f(x) = x^3 - 7x^2 - 25x - 50$

**79.** $f(x) = -x^4 + 196x^2$

**80.** $f(x) = -x^4 + x^2 + 12$

**81.** $f(x) = x^3 + 3x^2 + 3x + 1$

**82.** $f(x) = -x^3 + 3x + 2$

**83.** $f(x) = (x - 3)^2(x + 5)^2(x + 7)$

**84.** $f(x) = x(x + 6)^2(x^2 - x - 12)$

*Solve each polynomial inequality using the test-point method.*

**85.** $x^3 - 3x > 0$

**86.** $-x^3 + 3x + 2 < 0$

**87.** $2x^2 - x^4 \le 0$

**88.** $-x^4 + x^2 + 12 \ge 0$

**89.** $x^3 + 4x^2 - x - 4 > 0$

**90.** $x^3 + 2x^2 - 2x - 4 < 0$

**91.** $x^3 - 4x^2 - 20x + 48 \ge 0$

**92.** $x^3 + 7x^2 - 36 \le 0$

**93.** $x^3 - x^2 + x - 1 < 0$

**94.** $x^3 + x^2 + 2x - 4 > 0$

**95.** $x^4 - 19x^2 + 90 \le 0$

**96.** $x^4 - 5x^3 + 3x^2 + 15x - 18 \ge 0$

*State the solution sets to the inequalities in Exercises 97–104 by reading the following graphs.*

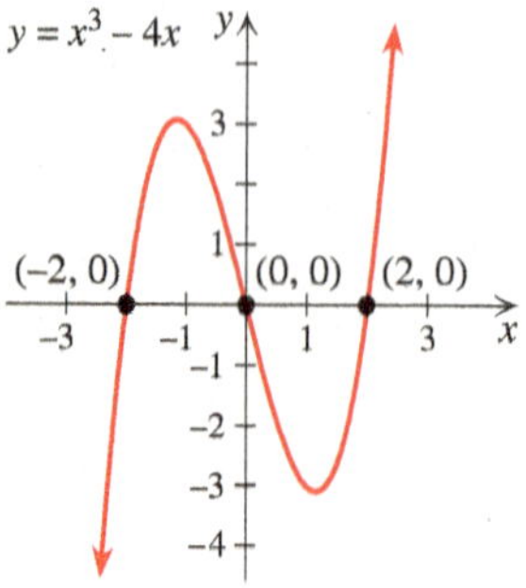


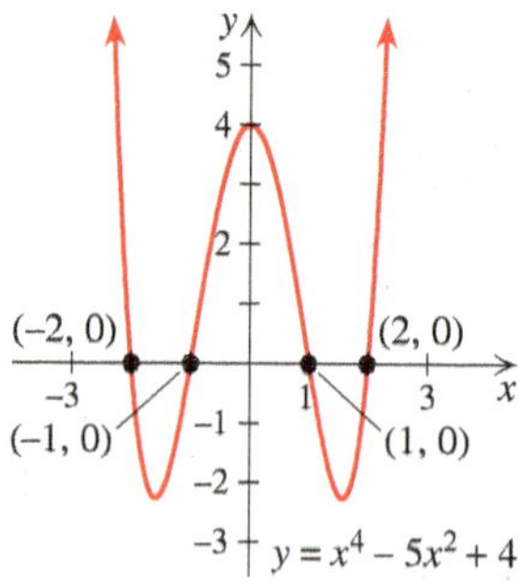


**97.** $x^3 - 4x > 0$

**98.** $x^3 - 4x \le 0$

**99.** $x^3 - 4x \ge 0$

**100.** $x^3 - 4x < 0$

**101.** $x^4 - 5x^2 + 4 < 0$

**102.** $x^4 - 5x^2 + 4 > 0$

**103.** $x^4 + 4 \ge 5x^2$

**104.** $x^4 + 4 \le 5x^2$

*Determine which of the given functions is shown in the accompanying graph.*

**105.** **a.** $f(x) = (x - 3)(x + 2)$ **b.** $f(x) = (x + 3)(x - 2)$

**c.** $f(x) = \frac{1}{3}(x - 3)(x + 2)$ **d.** $f(x) = \frac{1}{3}(x + 3)(x - 2)$

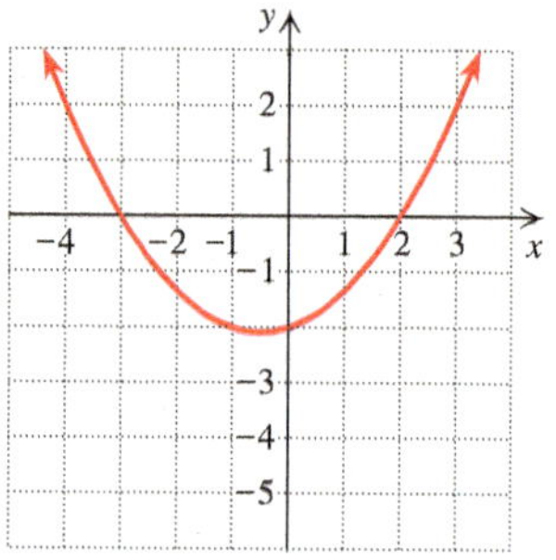


**Figure for Exercise 105**

**106.** **a.** $f(x) = (x + 3)(x + 1)(x - 1)$

**b.** $f(x) = (x - 3)(x^2 - 1)$

**c.** $f(x) = \frac{2}{3}(x + 3)(x + 1)(x - 1)$

**d.** $f(x) = -\frac{2}{3}(x + 3)(x^2 - 1)$

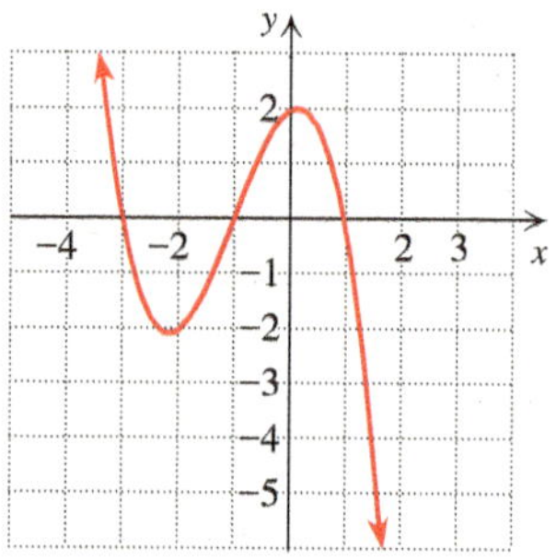


**Figure for Exercise 106**

*Find the equation and sketch the graph for each function.*

**107.** A quadratic function with $x$-intercepts $(-5, 0)$ and $(4, 0)$ and $y$-intercept $(0, 3)$

**108.** A quadratic function with $x$-intercepts $(3, 0)$ and $(-6, 0)$ and $y$-intercept $(0, -5)$

**109.** A quadratic function that passes through $(1, 2)$ and has $x$-intercepts $(-1, 0)$ and $(3, 0)$

**110.** A quadratic function that passes through $(-1, -3)$ and has $x$-intercepts $(2, 0)$ and $(4, 0)$

**111.** A cubic function (a third-degree polynomial function) with $x$-intercepts $(2, 0)$, $(-3, 0)$, and $(4, 0)$ and $y$-intercept $(0, 6)$

**112.** A cubic function with $x$-intercepts $(1, 0)$, $(-1/2, 0)$, and $(1/4, 0)$ and $y$-intercept $(0, 2)$

**113.** A quartic function (a fourth-degree polynomial function) that passes through the point $(1, 3)$ and has $x$-intercepts $(\pm 2, 0)$ and $(\pm 4, 0)$

**114.** A quartic function that passes through the point $(1, 5)$ and has $x$-intercepts $(1/2, 0)$, $(1/3, 0)$, $(1/4, 0)$, and $(-2/3, 0)$

*Solve each problem.*

**115.** *Maximum Profit* A company's weekly profit (in thousands of dollars) is given by the function $P(x) = x^3 - 3x^2 + 2x + 3$, where $x$ is the amount (in thousands of dollars) spent per week on advertising. Use a graphing calculator to estimate the local maximum and the local minimum value for the profit. What amount must be spent on advertising to get the profit higher than the local maximum profit?

**116.** *Maximum Volume* An open-top box is to be made from a 6 in. by 7 in. piece of copper by cutting equal squares ($x$ in. by $x$ in.) from each corner and folding up the sides. Write the volume of the box as a function of $x$. Use a graphing calculator to find the maximum possible volume to the nearest hundredth of a cubic inch.

**117.** *Economic Forecast* The total annual profit (in thousands of dollars) for a department store chain is determined by $P = 90x - 6x^2 + 0.1x^3$, where $x$ is the number of stores in the chain. The company now has 10 stores in operation and plans to pursue an aggressive expansion program. What happens to the total annual profit as the company opens more and more stores? For what values of $x$ is profit increasing?

**118.** *Contaminated Chicken* The number of bacteria of a certain type found on a chicken $t$ minutes after processing in a contaminated plant is given by the function

$$N = 30t + 25t^2 + 44t^3 - 0.01t^4.$$

Find $N$ for $t = 30$, 300, and 3000. What happens to $N$ as $t$ keeps increasing?

**119.** *Packing Cheese* Workers at the Green Bay Cheese Factory are trying to cover a block of cheese with an 8 in. by 12 in. piece of foil paper as shown in the figure. The ratio of the length and width of the block must be 4 to 3 to accommodate the label. Find a polynomial function that gives the volume of the block of cheese covered in this manner as a function of the thickness $x$. Use a graphing calculator to find the dimensions of the block that will maximize the volume.

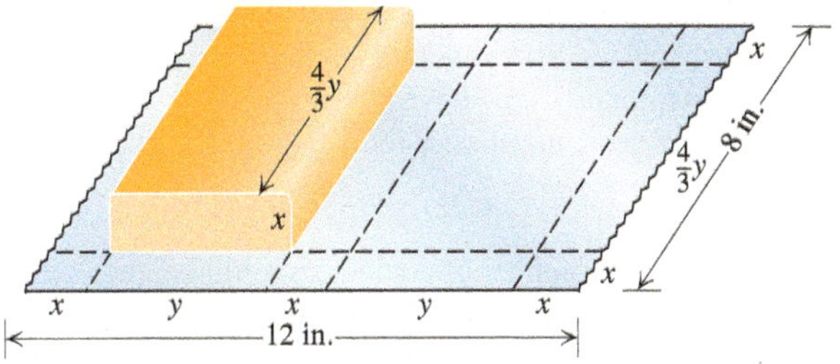


Figure for Exercise 119

**120.** *Giant Teepee* A casino designer is planning a giant teepee that is 80 ft in diameter and 120 ft high as shown in the figure. Inside the teepee is to be a cylindrical room for slot machines. Write the volume of the cylindrical room as a function of its radius. Use a graphing calculator to find the radius that maximizes the volume of the cylindrical room.

HINT If $h$ is the height of the room and $r$ the radius, then the ratio of $h$ to $40 - r$ is 3 to 1.

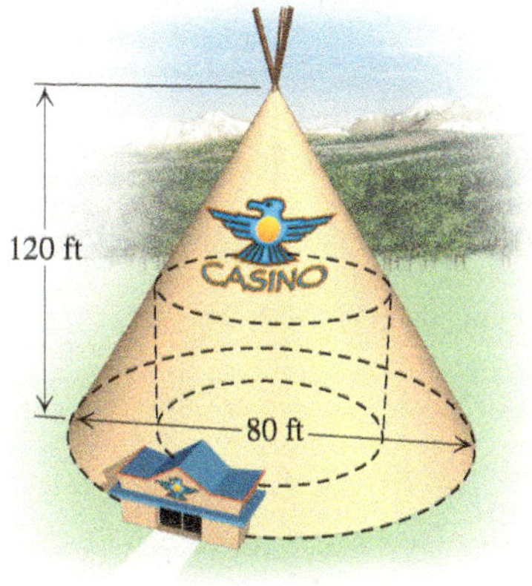


Figure for Exercise 120

**121.** *Paint Coverage* A one-pint can of spray paint will cover $50\text{ ft}^2$. Of course a small amount of paint coats the inside of the cylindrical can and it is not used. What would happen if the inside of the can had a surface area of $50\text{ ft}^2$? Find the dimensions for the two one-pint cans that have a surface area of $50\text{ ft}^2$.

HINT For a cylinder, $V = \pi r^2 h$ and $S = 2\pi r^2 + 2\pi rh$. Use $1\text{ ft}^3 \approx 7.5$ gallons.

Figure for Exercise 121

**122.** *Booming Business* When Computer Recyclers opened its doors, business started booming. After a few months, there was a temporary slowdown in sales, after which sales took off again. We can model sales for this business with the function

$$N = 8t^3 - 133t^2 + 653t,$$

where $N$ is the number of computers sold in month $t$ ($t = 0$ corresponds to the opening of the business).

**a.** Use a graphing calculator to estimate the month in which the temporary slowdown was the worst.

**b.** What percentage drop in sales occurred at the bottom of the slowdown compared to the previous high point in sales?

## RETHINKING

**123.** Solve $\sqrt{x + 12} - \sqrt{x - 9} = 3$.

**124.** Find a polynomial equation with real coefficients that has the roots $-5, 2 + i$.

**125.** Use synthetic division to find the quotient and remainder when $x^4 + 3x^3 - 9x^2 + 2x - 5$ is divided by $x + 2$.

**126.** For the function $y = -4x^2 + 8x + 9$ find the domain, range, and the maximum value for $y$.

**127.** Find the domain and range of the relation $x = |y - 5|$.

**128.** Is the function $\{(0, 3), (-9, 0), (-3, 5), (9, 7)\}$ one-to-one?

## THINKING OUTSIDE THE BOX XXVI

*Leaning Ladder* A 7-ft ladder is leaning against a vertical wall. There is a point near the bottom of the ladder that is 1 ft from the ground and 1 ft from the wall. Find the exact or approximate distance from the top of the ladder to the ground.

**Figure for Thinking Outside the Box XXVI**

## POP QUIZ 3.5

1. Discuss the symmetry of the graph of $y = x^4 - 3x^2$.
2. Discuss the symmetry of the graph of $y = x^3 - 3x$.
3. Does $f(x) = (x - 4)^3$ cross the $x$-axis at $(4, 0)$?
4. If $y = x^4 - 3x^3$, does $y$ go to $\infty$ or $-\infty$ as $x \to \infty$?
5. If $y = -2x^4 + 5x^2$, does $y$ go to $\infty$ or $-\infty$ as $x \to -\infty$?
6. Solve $(x - 3)^2(x + 1)^3 > 0$.

## LINKING concepts... For Individual or Group Explorations

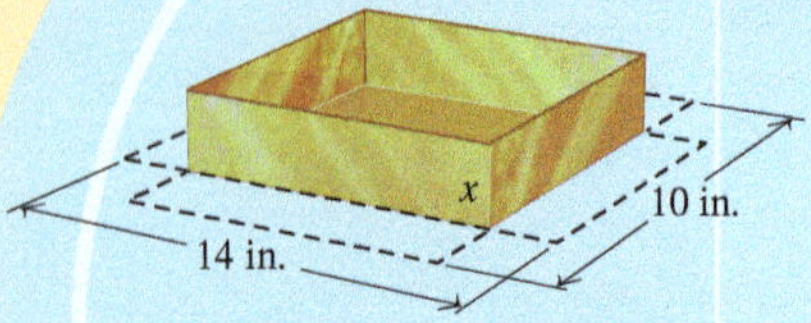


### Maximizing Volume

*A sheet metal worker is planning to make an open-top box by cutting equal squares (x-in. by x-in.) from the corners of a* 10-in. *by* 14-in. *piece of copper. A second box is to be made in the same manner from an* 8-in. *by* 10-in. *piece of aluminum, but its height is to be one-half that of the first box.*

**a)** Find polynomial functions for the volume of each box.

**b)** Find the values of $x$ for which the copper box is 72 in.$^3$ larger than the aluminum box.

**c)** Write the difference between the two volumes $d$ as a function of $x$ and graph it.

**d)** Find $d$ for $x = 1.5$ in. and $x = 8$ in.

**e)** For what value of $x$ is the difference between the two volumes the largest?

**f)** For what value of $x$ is the total volume the largest?

# 3.6 Rational Functions and Inequalities

In this section we will use our knowledge of polynomial functions to study functions that are ratios of polynomial functions.

## Rational Functions and Their Domains

Functions such as

$$y = \frac{1}{x}, \qquad f(x) = \frac{x - 3}{x - 1}, \qquad \text{and} \qquad g(x) = \frac{2x - 3}{x^2 - 4}$$

are rational functions.

**Definition: Rational Function**

> If $P(x)$ and $Q(x)$ are polynomials, then a function of the form
> $$f(x) = \frac{P(x)}{Q(x)}$$
> is called a **rational function,** provided that $Q(x)$ is not the zero polynomial.

To simplify discussions of rational functions we will assume that $f(x)$ is in lowest terms ($P(x)$ and $Q(x)$ have no common factors) unless it is stated otherwise.

The domain of a polynomial function is the set of all real numbers, while the domain of a rational function is restricted to real numbers that do not cause the denominator to have a value of 0. The domain of $y = 1/x$ is the set of all real numbers except 0.

### EXAMPLE 1 The domain of a rational function

Find the domain of each rational function.

**a.** $f(x) = \dfrac{x - 3}{x - 1}$ **b.** $g(x) = \dfrac{2x - 3}{x^2 - 4}$

**Solution**

**a.** Since $x - 1 = 0$ only for $x = 1$, the domain of $f$ is the set of all real numbers except 1. The domain is written in interval notation as $(-\infty, 1) \cup (1, \infty)$.

**b.** Since $x^2 - 4 = 0$ for $x = \pm 2$, any real number except 2 and $-2$ can be used for $x$. So the domain of $g$ is $(-\infty, -2) \cup (-2, 2) \cup (2, \infty)$.

▶**TRY THIS.** Find the domain of $f(x) = \frac{x - 1}{x^2 - 9}$. ■

## Horizontal and Vertical Asymptotes

The graph of a rational function such as $f(x) = 1/x$ does not look like the graph of a polynomial function. The domain of $f(x) = 1/x$ is the set of all real numbers except 0. However, 0 is an important number for the graph of this function because of the behavior of the graph when $x$ is close to 0. The following table shows ordered pairs in which $x$ is close to 0.

$x \to 0$ from left ⇒ ⇐ $x \to 0$ from right

| $x$ | $-0.1$ | $-0.01$ | $-0.001$ | 0 | 0.001 | 0.01 | 0.1 |
|---|---|---|---|---|---|---|---|
| $y = 1/x$ | $-10$ | $-100$ | $-1000$ | | 1000 | 100 | 10 |

$y \to -\infty$ ⇒ ⇐ $y \to \infty$

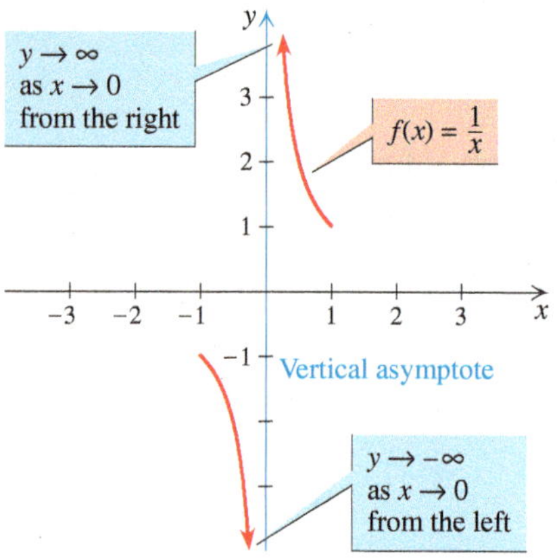


**Figure 3.69**

Notice that the closer $x$ is to 0, the farther $y$ is from 0. In symbols, $y \to \infty$ as $x \to 0$ from the right. Using limit notation we write $\lim_{x \to 0^+} \frac{1}{x} = \infty$, where the plus symbol indicates that $x$ is approaching 0 from above or from the right. If $x \to 0$ from the left, $y \to -\infty$. Using limit notation we write $\lim_{x \to 0^-} \frac{1}{x} = -\infty$, where the negative symbol indicates that $x$ is approaching 0 from below or from the left. Plotting the ordered pairs from the table suggests a curve that gets closer and closer to the vertical line $x = 0$ (the $y$-axis) but never touches it, as shown in Fig. 3.69. The $y$-axis is called a *vertical asymptote* for this curve.

The following table shows ordered pairs in which $x$ is far from 0.

| | ← $x \to -\infty$ | | | | | $x \to \infty$ → | |
|---|---|---|---|---|---|---|---|
| $x$ | −1000 | −100 | −10 | 0 | 10 | 100 | 1000 |
| $y = 1/x$ | −0.001 | −0.01 | −0.1 | | 0.1 | 0.01 | 0.001 |
| | ← $y \to 0$ from below | | | | | $y \to 0$ from above → | |

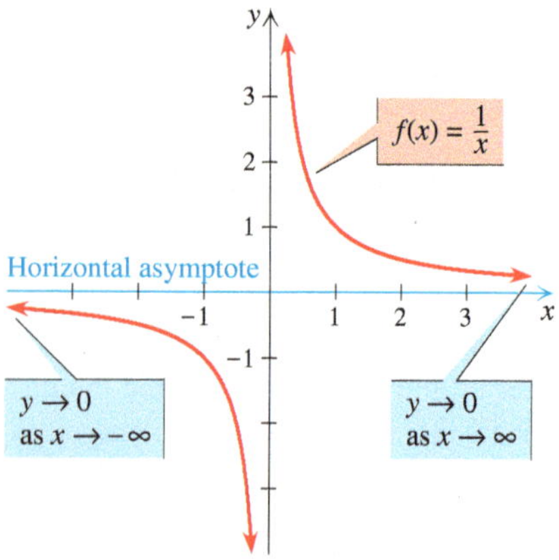


**Figure 3.70**

Notice that the farther $x$ is from 0, the closer $y$ is to 0. Using limit notation we write $\lim_{x \to \infty} \frac{1}{x} = 0$ and $\lim_{x \to -\infty} \frac{1}{x} = 0$. Plotting the ordered pairs from the table suggests a curve that lies just above the positive $x$-axis and just below the negative $x$-axis. The $x$-axis is called a *horizontal asymptote* for the graph of $f$. The complete graph of $f(x) = 1/x$ is shown in Fig. 3.70. Since $f(-x) = 1/(-x) = -f(x)$, the graph is symmetric about the origin.

For any rational function expressed in lowest terms, a horizontal asymptote is determined by the value approached by the rational expression as $|x| \to \infty$ ($x \to \infty$ or $x \to -\infty$). A vertical asymptote occurs for every number that causes the denominator of the function to have a value of 0, provided the rational function is in lowest terms. As we will learn shortly, not every rational function has a vertical and a horizontal asymptote. We can give a formal definition of asymptotes as described below.

## Definition: Vertical and Horizontal Asymptotes

Let $f(x) = P(x)/Q(x)$ be a rational function written in lowest terms.

If $|f(x)| \to \infty$ as $x \to a$, then the vertical line $x = a$ is a **vertical asymptote.** Using limit notation, $x = a$ is a vertical asymptote if $\lim_{x \to a} |f(x)| = \infty$.

The line $y = a$ is a **horizontal asymptote** if $f(x) \to a$ as $x \to \infty$ or $x \to -\infty$. Using limit notation, $y = a$ is a horizontal asymptote if $\lim_{x \to \infty} f(x) = a$ or $\lim_{x \to -\infty} f(x) = a$.

To find a horizontal asymptote we need to approximate the value of a rational expression when $x$ is arbitrarily large. If $x$ is large, then expressions such as

$$\frac{500}{x}, \quad -\frac{14}{x}, \quad \frac{3}{x^2}, \quad \frac{6}{x^3}, \quad \text{and} \quad \frac{4}{x-5},$$

which consist of a fixed number over a polynomial, are approximately zero. To approximate a ratio of two polynomials that both involve $x$, such as $\frac{x-2}{2x+3}$, we rewrite the expression by dividing by the highest power of $x$:

$$\frac{x-2}{2x+3} = \frac{\frac{x}{x} - \frac{2}{x}}{\frac{2x}{x} + \frac{3}{x}} = \frac{1 - \frac{2}{x}}{2 + \frac{3}{x}}$$

Since $2/x$ and $3/x$ are approximately zero when $x$ is large, the approximate value of this expression is $1/2$. We use this idea in the next example.

## EXAMPLE 2 Identifying horizontal and vertical asymptotes

Find the horizontal and vertical asymptotes for each rational function.

**a.** $f(x) = \dfrac{3}{x^2 - 1}$ **b.** $g(x) = \dfrac{x}{x^2 - 4}$ **c.** $h(x) = \dfrac{2x + 1}{x + 3}$

### Solution

Figure 3.71

**a.** The denominator $x^2 - 1$ has a value of 0 if $x = \pm 1$. So the lines $x = 1$ and $x = -1$ are vertical asymptotes. As $x \to \infty$ or $x \to -\infty$, $x^2 - 1$ gets larger, making $3/(x^2 - 1)$ the ratio of 3 and a large number. Thus $3/(x^2 - 1) \to 0$ and the $x$-axis is a horizontal asymptote.

The calculator table shown in Fig. 3.71 supports the conclusion that the $x$-axis is a horizontal asymptote. □

**b.** The denominator $x^2 - 4$ has a value of 0 if $x = \pm 2$. So the lines $x = 2$ and $x = -2$ are vertical asymptotes. As $x \to \infty$, $x/(x^2 - 4)$ is a ratio of two large numbers. The approximate value of this ratio is not clear. However, if we divide the numerator and denominator by $x^2$, the highest power of $x$, we can get a clearer picture of the value of this ratio:

$$g(x) = \frac{x}{x^2 - 4} = \frac{\frac{x}{x^2}}{\frac{x^2}{x^2} - \frac{4}{x^2}} = \frac{\frac{1}{x}}{1 - \frac{4}{x^2}}$$

Figure 3.72

As $|x| \to \infty$, the values of $1/x$ and $4/x^2$ go to 0. So

$$g(x) \to \frac{0}{1 - 0} = 0.$$

Thus the $x$-axis is a horizontal asymptote.

The calculator table shown in Fig. 3.72 supports the conclusion that the $x$-axis is a horizontal asymptote. □

**c.** The denominator $x + 3$ has a value of 0 if $x = -3$. So the line $x = -3$ is a vertical asymptote. To find any horizontal asymptotes, rewrite the rational expression by dividing the numerator and denominator by the highest power of $x$:

$$h(x) = \frac{2x + 1}{x + 3} = \frac{\frac{2x}{x} + \frac{1}{x}}{\frac{x}{x} + \frac{3}{x}} = \frac{2 + \frac{1}{x}}{1 + \frac{3}{x}}$$

Figure 3.73

As $x \to \infty$ or $x \to -\infty$ the values of $1/x$ and $3/x$ approach 0. So

$$h(x) \to \frac{2 + 0}{1 + 0} = 2.$$

Thus the line $y = 2$ is a horizontal asymptote.

The calculator table shown in Fig. 3.73 supports the conclusion that $y = 2$ is a horizontal asymptote.

▶**TRY THIS.** Find all asymptotes for $f(x) = \frac{x - 1}{x^2 - 9}$. ■

If the degree of the numerator of a rational function is less than the degree of the denominator, as in Examples 2(a) and (b), then the $x$-axis is a horizontal asymptote

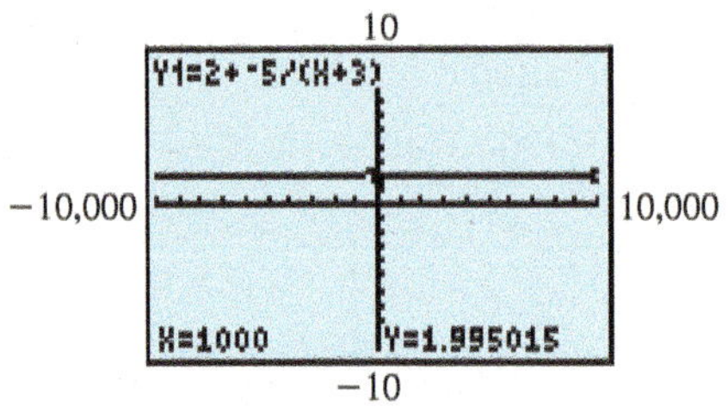


Figure 3.74

for the graph of the function. If the degree of the numerator is equal to the degree of the denominator, as in Example 2(c), then the $x$-axis is not a horizontal asymptote. We can see this clearly if we use division to rewrite the expression as quotient + remainder/divisor. For the function of Example 2(c), we get

$$h(x) = \frac{2x + 1}{x + 3} = 2 + \frac{-5}{x + 3}.$$

The graph of $h$ is a translation two units upward of the graph of $y = -5/(x + 3)$, which has the $x$-axis as a horizontal asymptote. So $y = 2$ is a horizontal asymptote for $h$. Note that 2 is simply the ratio of the leading coefficients.

Since the graph of $h$ is very close to $y = 2$, the view of $h$ in Fig. 3.74 looks like $y = 2$. Note how this choice of viewing window causes some of the important features of the graph to disappear. □

## Oblique Asymptotes

Each rational function of Example 2 had one horizontal asymptote and had a vertical asymptote for each zero of the polynomial in the denominator. The horizontal asymptote $y = 0$ occurs because the $y$-coordinate gets closer and closer to 0 as $x \to \infty$ or $x \to -\infty$. Some rational functions have a nonhorizontal line for an asymptote. An asymptote that is neither horizontal nor vertical is called an **oblique asymptote** or **slant asymptote.** Oblique asymptotes are determined by using long division or synthetic division of polynomials.

### EXAMPLE 3 A rational function with an oblique asymptote

Determine all of the asymptotes for $g(x) = \frac{2x^2 + 3x - 5}{x + 2}$.

**Solution**

If $x + 2 = 0$, then $x = -2$. So the line $x = -2$ is a vertical asymptote. Because the degree of the numerator is larger than the degree of the denominator, we use long division (or in this case, synthetic division) to rewrite the function as quotient + remainder/divisor (dividing the numerator and denominator by $x^2$ will not work in this case):

$$g(x) = \frac{2x^2 + 3x - 5}{x + 2} = 2x - 1 + \frac{-3}{x + 2}$$

If $|x| \to \infty$, then $-3/(x + 2) \to 0$. So the value of $g(x)$ approaches $2x - 1$ as $|x| \to \infty$. The line $y = 2x - 1$ is an oblique asymptote for the graph of $g$. You may look ahead to Fig. 3.83 to see the graph of this function with its oblique asymptote.

Note that the wide view of $g$ in Fig. 3.75 looks like the line $y = 2x - 1$.

▶**TRY THIS.** Find all asymptotes for $f(x) = \frac{3x^2 - 4}{x - 1}$.

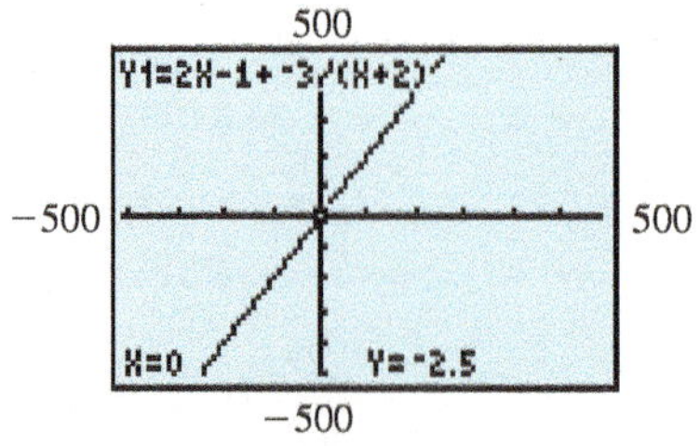


Figure 3.75 ■

If the degree of $P(x)$ is 1 greater than the degree of $Q(x)$ and the degree of $Q(x)$ is at least 1, then the rational function has an oblique asymptote. In this case use division to rewrite the function as quotient + remainder/divisor. The graph of the equation formed by setting $y$ equal to the quotient is an oblique asymptote. We conclude this discussion of asymptotes with a summary.

**SUMMARY**

### Finding Asymptotes for a Rational Function

Let $f(x) = P(x)/Q(x)$ be a rational function in lowest terms with the degree of $Q(x)$ at least 1.

1. The graph of $f$ has a vertical asymptote corresponding to each root of $Q(x) = 0$.
2. If the degree of $P(x)$ is less than the degree of $Q(x)$, then the $x$-axis is a horizontal asymptote.
3. If the degree of $P(x)$ equals the degree of $Q(x)$, then the horizontal asymptote is determined by the ratio of the leading coefficients.
4. If the degree of $P(x)$ is greater than the degree of $Q(x)$, then use division to rewrite the function as quotient + remainder/divisor. The graph of the equation formed by setting $y$ equal to the quotient is an asymptote. This asymptote is an oblique or slant asymptote if the degree of $P(x)$ is 1 larger than the degree of $Q(x)$.

## Sketching Graphs of Rational Functions

We now use asymptotes and symmetry to help us sketch the graphs of the rational functions discussed in Examples 2 and 3. Use the following steps to graph a rational function.

**PROCEDURE**

### Graphing a Rational Function

To graph a rational function in lowest terms:

1. Determine the asymptotes and draw them as dashed lines.
2. Check for symmetry.
3. Find any intercepts.
4. Plot several selected points to determine how the graph approaches the asymptotes.
5. Draw curves through the selected points, approaching the asymptotes.

### EXAMPLE 4 Functions with horizontal and vertical asymptotes

Sketch the graph of each rational function.

**a.** $f(x) = \dfrac{3}{x^2 - 1}$ **b.** $g(x) = \dfrac{x}{x^2 - 4}$ **c.** $h(x) = \dfrac{2x + 1}{x + 3}$

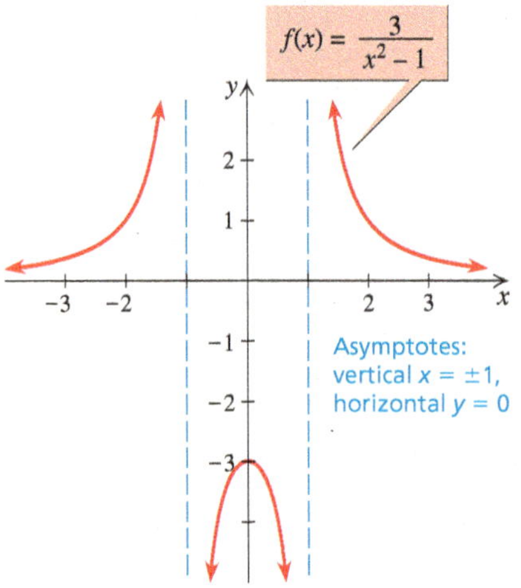


Figure 3.76

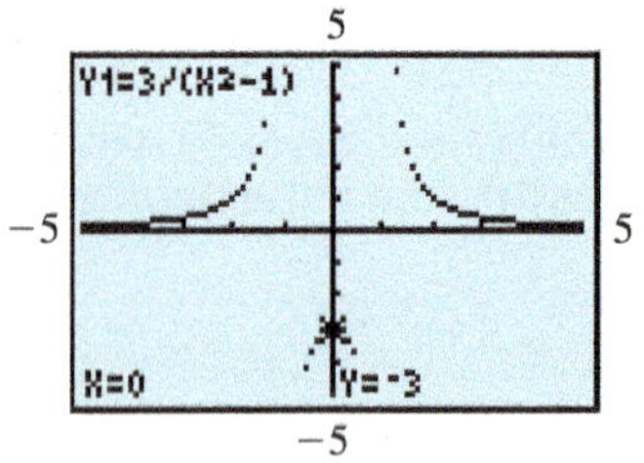


Figure 3.77

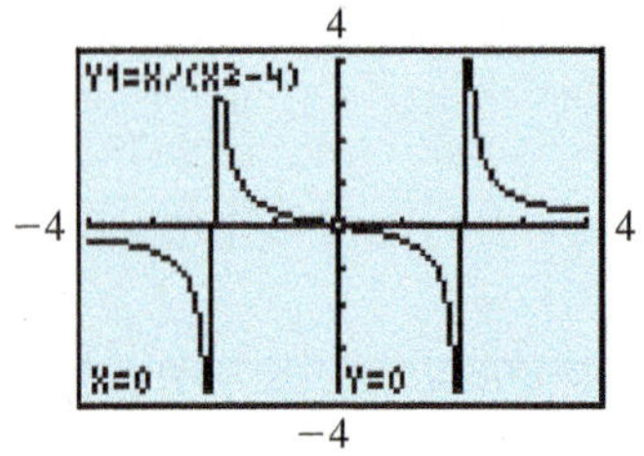


Figure 3.79

### Solution

**a.** From Example 2(a), $x = 1$ and $x = -1$ are vertical asymptotes, and the $x$-axis is a horizontal asymptote. Draw the vertical asymptotes using dashed lines. Since all of the powers of $x$ are even, $f(-x) = f(x)$ and the graph is symmetric about the $y$-axis. The $y$-intercept is $(0, -3)$. There are no $x$-intercepts because $f(x) = 0$ has no solution. Evaluate the function at $x = 0.9$ and $x = 1.1$ to see how the curve approaches the asymptote at $x = 1$. Evaluate at $x = 2$ and $x = 3$ to see whether the curve approaches the horizontal asymptote from above or below. We get $(0.9, -15.789)$, $(1.1, 14.286)$, $(2, 1)$, and $(3, 3/8)$. From these points you can see that the curve is going downward toward $x = 1$ from the left and upward toward $x = 1$ from the right. It is approaching its horizontal asymptote from above. Now use the symmetry with respect to the $y$-axis to draw the curve approaching its asymptotes as shown in Fig. 3.76.

The calculator graph in dot mode in Fig. 3.77 supports these conclusions. □

**b.** Draw the vertical asymptotes $x = 2$ and $x = -2$ from Example 2(b) as dashed lines. The $x$-axis is a horizontal asymptote. Because $f(-x) = -f(x)$, the graph is symmetric about the origin. The $x$-intercept is $(0, 0)$. Evaluate the function near the vertical asymptote $x = 2$, and for larger values of $x$ to see how the curve approaches the horizontal asymptote. We get $(1.9, -4.872)$, $(2.1, 5.122)$, $(3, 3/5)$, and $(4, 1/3)$. From these points you can see that the curve is going downward toward $x = 2$ from the left and upward toward $x = 2$ from the right. It is approaching its horizontal asymptote from above. Now use the symmetry with respect to the origin to draw the curve approaching its asymptotes as shown in Fig. 3.78.

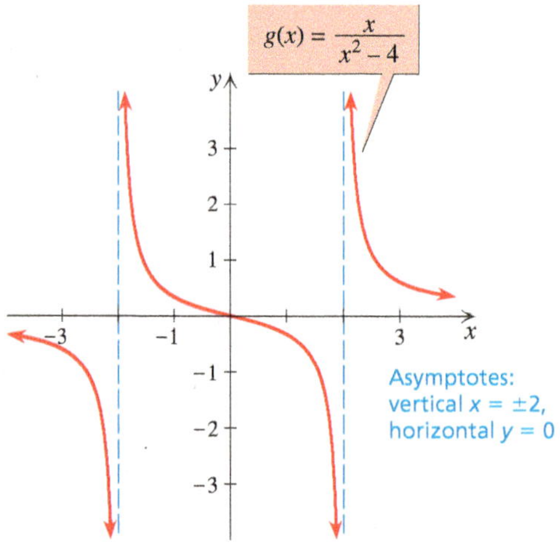


Figure 3.78

The calculator graph in connected mode in Fig. 3.79 confirms these conclusions. Note how the calculator appears to draw the vertical asymptotes as it connects the points that are close to but on opposite sides of the asymptotes. □

**c.** Draw the vertical asymptote $x = -3$ and the horizontal asymptote $y = 2$ from Example 2(c) as dashed lines. The $x$-intercept is $(-1/2, 0)$ and the $y$-intercept is $(0, 1/3)$. The points $(-2, -3)$, $(7, 1.5)$, $(-4, 7)$, and $(-13, 2.5)$ are also on the graph. From these points we can conclude that the curve goes upward toward $x = -3$ from the left and downward toward $x = -3$ from the right. It approaches

$y = 2$ from above as $x$ goes to $-\infty$ and from below as $x$ goes to $\infty$. Draw the graph approaching its asymptotes as shown in Fig. 3.80.

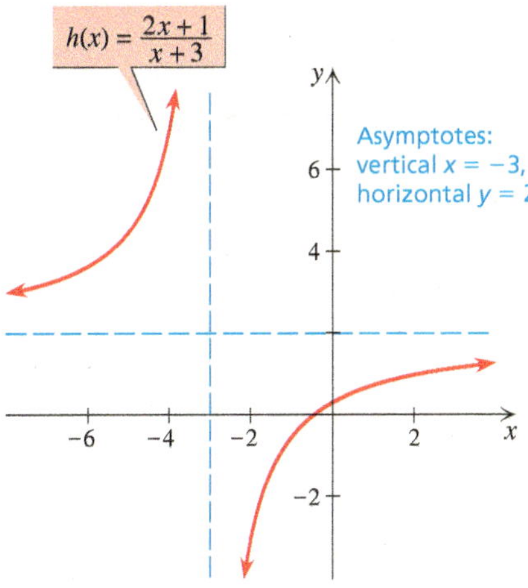


Figure 3.80

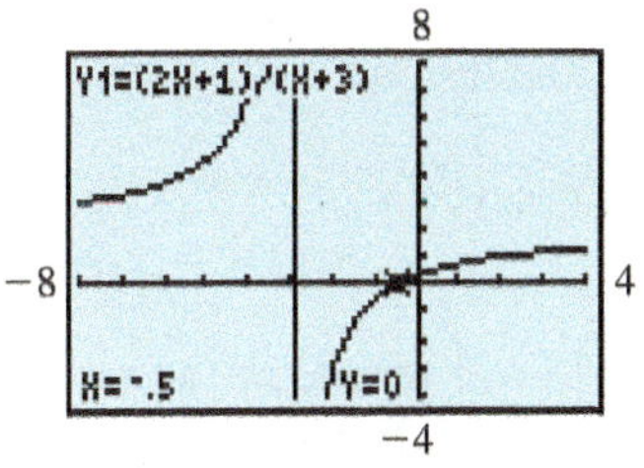


Figure 3.81

The calculator graph in Fig. 3.81 supports these conclusions.

▶**TRY THIS.** Sketch the graph of $f(x) = \frac{x-1}{x^2-9}$. ■

The graph of a rational function cannot cross a vertical asymptote, but it can cross a nonvertical asymptote. The graph of a rational function gets closer and closer to a nonvertical asymptote as $|x| \to \infty$, but it can also cross a nonvertical asymptote as illustrated in the next example.

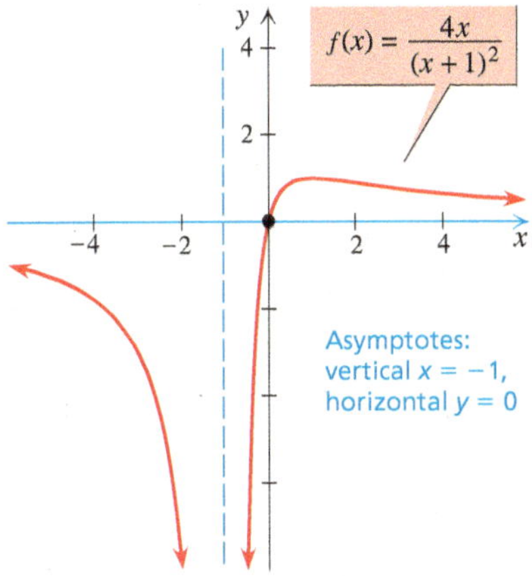


Figure 3.82

## EXAMPLE 5 A rational function that crosses an asymptote

Sketch the graph of $f(x) = \frac{4x}{(x+1)^2}$.

### Solution

The graph of $f$ has a vertical asymptote at $x = -1$, and since the degree of the numerator is less than the degree of the denominator, the $x$-axis is a horizontal asymptote. If $x > 0$ then $f(x) > 0$, and if $x < 0$ then $f(x) < 0$. So the graph approaches the horizontal axis from above for $x > 0$, and from below for $x < 0$. However, the $x$-intercept is $(0, 0)$. Therefore, the graph crosses its horizontal asymptote at $(0, 0)$. The graph shown in Fig. 3.82 goes through $(1, 1)$, $(2, 8/9)$, $(-2, -8)$, and $(-3, -3)$.

▶**TRY THIS.** Sketch the graph of $f(x) = \frac{x}{(x-2)^2}$. ■

The graph of any function should illustrate the function's most important features. For a rational function, the most important feature is its asymptotic behavior. The graph of a rational function does not cross its vertical asymptotes, and the graph approaches but does not touch its nonvertical asymptotes as $x$ approaches $\infty$ or $-\infty$. So be careful with calculator graphs. On some calculators the graph will cross its vertical asymptote in connected mode and, unless the viewing window is selected carefully, the graph will appear to touch its nonvertical asymptotes.

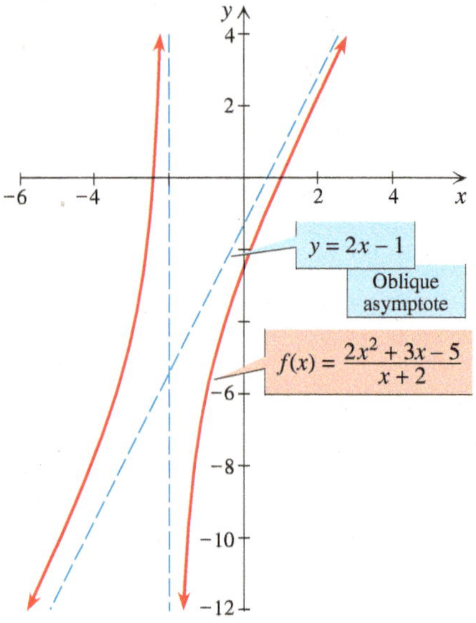


Figure 3.83

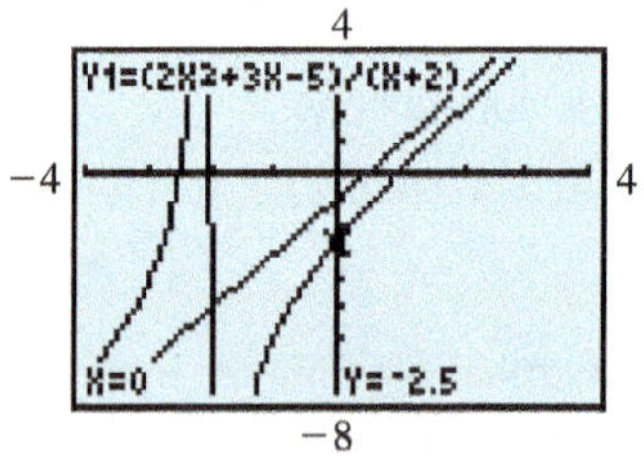


Figure 3.84

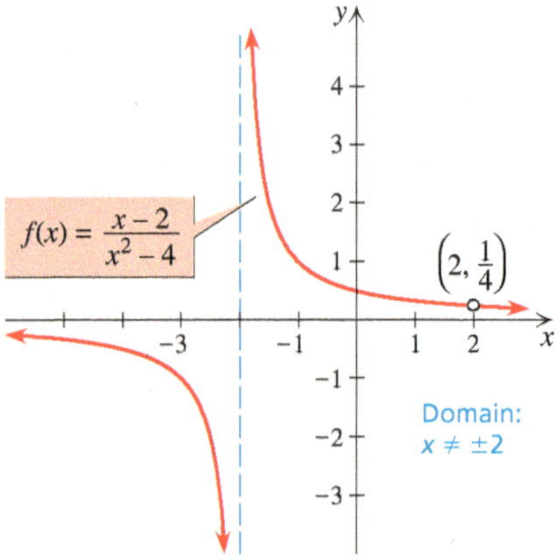


Figure 3.85

## EXAMPLE 6 Graphing a function with an oblique asymptote

Sketch the graph of $f(x) = \frac{2x^2 + 3x - 5}{x + 2}$.

### Solution

Draw the vertical asymptote $x = -2$ and the oblique asymptote $y = 2x - 1$ determined in Example 3 as dashed lines. The $x$-intercepts, $(1, 0)$ and $(-2.5, 0)$, are found by solving $2x^2 + 3x - 5 = 0$. The $y$-intercept is $(0, -2.5)$. The points $(-1, -6)$, $(4, 6.5)$, and $(-3, -4)$ are also on the graph. From these points we can conclude that the curve goes upward toward $x = -2$ from the left and downward toward $x = -2$ from the right. It approaches $y = 2x - 1$ from above as $x$ goes to $-\infty$ and from below as $x$ goes to $\infty$. Draw the graph approaching its asymptotes as shown in Fig. 3.83.

The calculator graph in Fig. 3.84 supports these conclusions.

▶**TRY THIS.** Sketch the graph of $f(x) = \frac{3x^2 - 4}{x - 1}$. ■

All rational functions so far have been given in lowest terms. In the next example the numerator and denominator have a common factor. In this case the graph does not have as many vertical asymptotes as you might expect. The graph is almost identical to the graph of the rational function obtained by reducing the expression to lowest terms.

## EXAMPLE 7 A graph with a hole in it

Sketch the graph of $f(x) = \frac{x - 2}{x^2 - 4}$.

### Solution

Since $x^2 - 4 = (x - 2)(x + 2)$, the domain of $f$ is the set of all real numbers except 2 and $-2$. Because the rational expression can be reduced, the function $f$ could also be defined as

$$f(x) = \frac{1}{x + 2} \quad \text{for } x \neq 2 \text{ and } x \neq -2.$$

Note that the domain is determined before the function is simplified. The graph of $y = 1/(x + 2)$ has only one vertical asymptote $x = -2$. In fact, the graph of $y = 1/(x + 2)$ is a translation two units to the left of $y = 1/x$. Since $f(2)$ is undefined, the point $(2, 1/4)$ that would normally be on the graph of $y = 1/(x + 2)$ is omitted. The missing point is indicated on the graph in Fig. 3.85 as a small open circle. Note that the graph made by a graphing calculator will usually not show the missing point.

▶**TRY THIS.** Sketch the graph of $f(x) = \frac{x - 1}{x^2 - 1}$. ■

## Rational Inequalities

An inequality that involves a rational expression, such as $\frac{x + 3}{x - 2} \leq 2$, is called a **rational inequality.** Our first thought for solving this inequality might be to clear the denominator by multiplying each side by $x - 2$. But, when we multiply each side of an inequality by a real number, we must know whether the real number is positive or negative. Whether $x - 2$ is positive or negative depends on $x$. So multiplying by $x - 2$ is not a good idea. *We usually do not multiply a rational inequality* *by an expression that involves a variable.* However, we can solve rational inequalities using the test-point method as we did for quadratic inequalities in Section 3.1 and polynomial inequalities in Section 3.5.

We saw in Section 3.5 that a polynomial function can change sign only at a zero of the function. A rational function can change sign only at a zero of the function or at a vertical asymptote. So to solve a rational inequality we use the following strategy.

**STRATEGY**

### Solving a Rational Inequality with Test Points

1. Rewrite the inequality (if necessary) with zero on one side and a ratio of two polynomials in lowest terms on the other side.
2. Find all zeros for the polynomial functions in the numerator and denominator.
3. The rational function is undefined at the zeros of the denominator. They give the locations of the vertical asymptotes.
4. Locate the zeros from (2) on a number line and test a point in each interval to see whether the rational function is positive or negative on the interval.

## EXAMPLE 8 Solving a rational inequality with test points

Solve $\frac{x+3}{x-2} \leq 2$. State the solution set using interval notation.

### Solution

As with quadratic inequalities, we must have 0 on one side. Note that we do not multiply each side by $x - 2$.

$$\frac{x+3}{x-2} \leq 2$$

$$\frac{x+3}{x-2} - 2 \leq 0$$

$$\frac{x+3}{x-2} - \frac{2(x-2)}{x-2} \leq 0 \quad \text{Get a common denominator.}$$

$$\frac{x+3-2(x-2)}{x-2} \leq 0 \quad \text{Subtract.}$$

$$\frac{7-x}{x-2} \leq 0 \quad \text{Combine like terms.}$$

Now $7 - x = 0$ if $x = 7$ and $x - 2 = 0$ if $x = 2$. Put a 0 above 7 and a "U" (for undefined) above 2 as shown in Fig. 3.86.

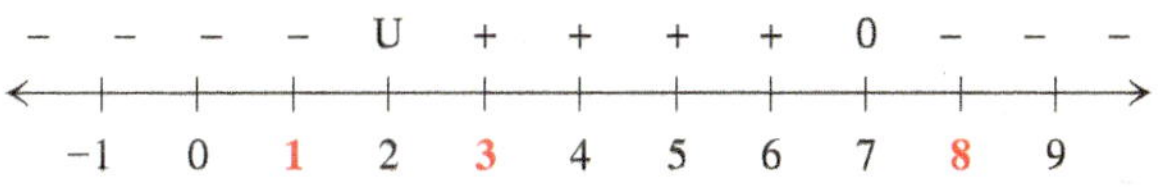


Figure 3.86

The numbers 2 and 7 divide the number line into three intervals. Select a test point in each interval. We will choose 1, 3, and 8, which are shown in red in Fig. 3.86. Let $R(x) = \frac{7-x}{x-2}$ and evaluate $R(x)$ at each test point:

$$R(1) = \frac{7-1}{1-2} = -6, \qquad R(3) = \frac{7-3}{3-2} = 4, \qquad R(8) = \frac{7-8}{8-2} = -\frac{1}{6}$$

So $-$, $+$, and $-$ are the signs of the rational function in the three intervals in Fig. 3.86. The inequality $\frac{7-x}{x-2} \le 0$ is satisfied whenever $R(x) \le 0$. Note that $R(7) = 0$ but $R(2)$ is undefined. So 7 is included in the solution set, but 2 is not. According to the test points, the solution set is $(-\infty, 2) \cup [7, \infty)$.

The graph of $y = (7-x)/(x-2)$ in Fig. 3.87 supports our conclusion that the inequality is satisfied if $x < 2$ or if $x \ge 7$.

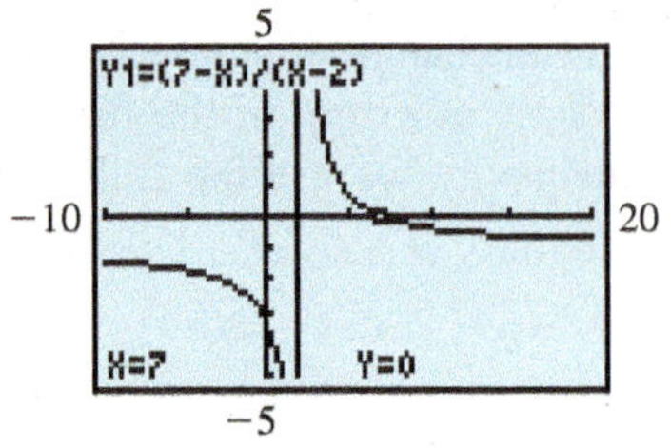


**Figure 3.87**

▶**TRY THIS.** Solve $\frac{1}{x-1} \ge 1$ using test points. ■

The test-point method works on any rational inequality for which we can find all zeros of the numerator and denominator. The test-point method depends on the fact that the graph of a rational function can go from one side of the $x$-axis to the other only at a vertical asymptote (where the function is undefined) or at an $x$-intercept (where the value of the function is 0).

## EXAMPLE 9 Solving a rational inequality with the test-point method

Solve $\frac{x+3}{x^2-1} \ge 0$. State the solution set using interval notation.

### Solution

Since the denominator can be factored, the inequality is equivalent to

$$\frac{x+3}{(x-1)(x+1)} \ge 0.$$

The rational expression is undefined if $x = \pm 1$ and has value 0 if $x = -3$. Above each of these numbers on a number line put a 0 or a "U" (for undefined) as shown in Fig. 3.88. Now we select $-4$, $-2$, 0, and 3 as test points (shown in red in Fig. 3.88). Let $R(x) = \frac{x+3}{x^2-1}$ and evaluate $R(x)$ at each test point to determine the sign of $R(x)$ in the interval of the test point:

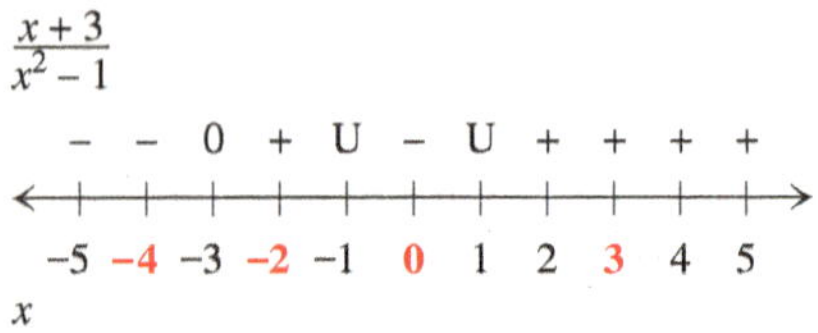


**Figure 3.88**

$$R(-4) = -\frac{1}{15}, \qquad R(-2) = \frac{1}{3}, \qquad R(0) = -3, \qquad R(3) = \frac{3}{4}$$

So $-$, $+$, $-$, and $+$ are the signs of the rational expression in the four intervals in Fig. 3.88. The inequality is satisfied whenever the rational expression has a positive or 0 value. Since $R(-1)$ and $R(1)$ are undefined, $-1$ and 1 are not in the solution set. The solution set is $[-3, -1) \cup (1, \infty)$.

The graph of $y = (x+3)/(x^2-1)$ in Fig. 3.89 supports the conclusion that the inequality is satisfied if $-3 \le x < -1$ or $x > 1$.

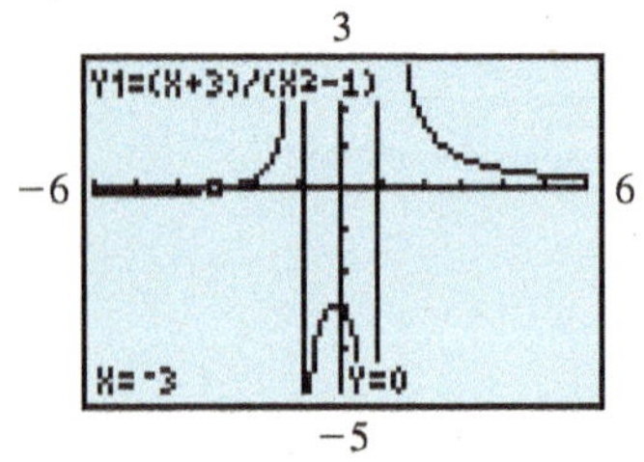


**Figure 3.89**

▶**TRY THIS.** Solve $\frac{x+3}{x-1} \ge 0$ with test points. ■

## Applications

Rational functions can occur in many applied situations. A horizontal asymptote might indicate that the average cost of producing a product approaches a fixed value in the long run. A vertical asymptote might show that the cost of a project goes up astronomically as a certain barrier is approached.

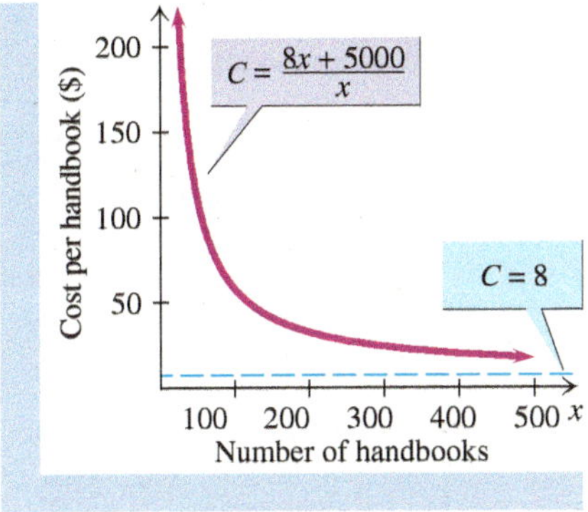


Figure 3.90

## EXAMPLE 10 Average cost of a handbook

Eco Publishing spent \$5000 to produce an environmental handbook and \$8 each for printing. Write a function that gives the average cost to the company per printed handbook. Graph the function for $0 < x \leq 500$. What happens to the average cost if the book becomes very, very popular?

### Solution

The total cost of producing and printing $x$ handbooks is $8x + 5000$. To find the average cost per book, divide the total cost by the number of books:

$$C = \frac{8x + 5000}{x}$$

This rational function has a vertical asymptote at $x = 0$ and a horizontal asymptote $C = 8$. The graph is shown in Fig. 3.90. As $x$ gets larger and larger, the \$5000 production cost is spread out over more and more books, and the average cost per book approaches \$8, as shown by the horizontal asymptote. Using limit notation we write

$$\lim_{x \to \infty} C = 8.$$

**▸TRY THIS.** A truck rents for \$100 per day plus \$0.80 per mile that it is driven. Write a function that gives the average cost per mile for a one-day rental. What happens to the average cost per mile as the number of miles gets very large? ■

## FUNCTION gallery... Some Basic Rational Functions

### Horizontal Asymptote *x*-axis and Vertical Asymptote *y*-axis

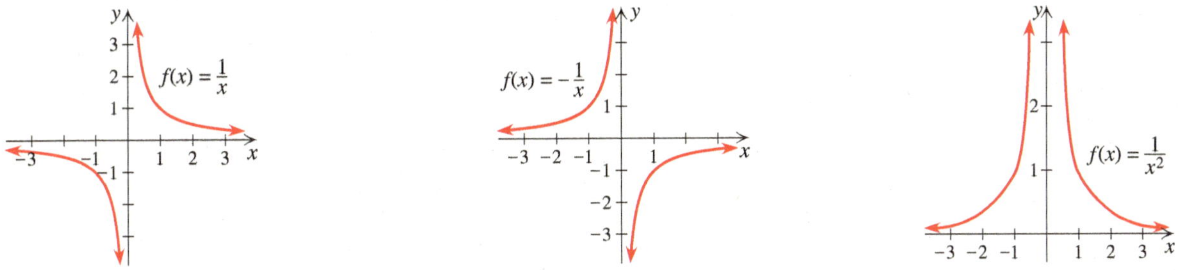


### Various Asymptotes

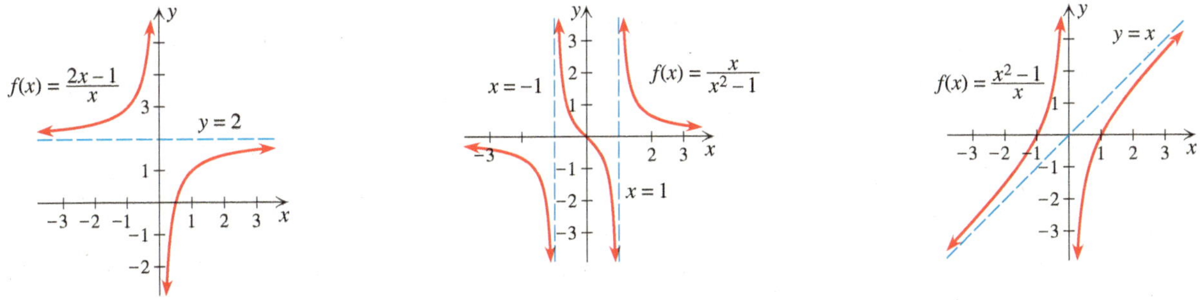

## FOR thought... True or False? Explain.

1. The function $f(x) = \frac{1}{\sqrt{x} - 3}$ is a rational function.

2. The domain of $f(x) = \frac{x+2}{x-2}$ is $(-\infty, -2) \cup (-2, 2) \cup (2, \infty)$.

3. The number of vertical asymptotes for a rational function equals the degree of the denominator.

4. The graph of $f(x) = \frac{1}{x^4 - 4x^2}$ has four vertical asymptotes.

5. The $x$-axis is a horizontal asymptote for $f(x) = \frac{x^2 - 2x + 7}{x^3 - 4x}$.

6. The $x$-axis is a horizontal asymptote for the graph of $f(x) = \frac{5x - 1}{x + 2}$.

7. The line $y = x - 3$ is an asymptote for the graph of $f(x) = x - 3 + \frac{1}{x - 1}$.

8. The graph of a rational function cannot intersect its asymptote.

9. The graph of $f(x) = \frac{4}{x^2 - 16}$ is symmetric about the $y$-axis.

10. The graph of $f(x) = \frac{2x + 6}{x^2 - 9}$ has only one vertical asymptote.

## EXERCISES 3.6

*Fill in the blank.*

1. If $P(x)$ and $Q(x)$ are polynomials, then $f(x) = P(x)/Q(x)$ is a(n) ________ function provided $Q(x)$ is not the zero polynomial.

2. If $\lim_{x\to\infty} f(x) = c$, then the line $y = c$ is a(n) ________.

3. If $\lim_{x\to a} |f(x)| = \infty$, then the line $x = a$ is a(n) ________.

4. An asymptote that is neither horizontal nor vertical is a(n) ________ or ________ asymptote.

*Find the domain of each rational function.*

5. $f(x) = \dfrac{4}{x + 2}$

6. $f(x) = \dfrac{-1}{x - 2}$

7. $f(x) = \dfrac{-x}{x^2 - 4}$

8. $f(x) = \dfrac{2}{x^2 - x - 2}$

9. $f(x) = \dfrac{2x + 3}{x - 3}$

10. $f(x) = \dfrac{4 - x}{x + 2}$

11. $f(x) = \dfrac{x^2 - 2x + 4}{x}$

12. $f(x) = \dfrac{x^3 + 2}{x^2}$

13. $f(x) = \dfrac{3x^2 - 1}{x^3 - x}$

14. $f(x) = \dfrac{x^2 + 1}{8x^3 - 2x}$

15. $f(x) = \dfrac{-x^2 + x}{x^2 + 5x + 6}$

16. $f(x) = \dfrac{-x^2 + 2x - 3}{x^2 + x - 12}$

*Determine the domain and the equations of the asymptotes for the graph of each rational function.*

17.

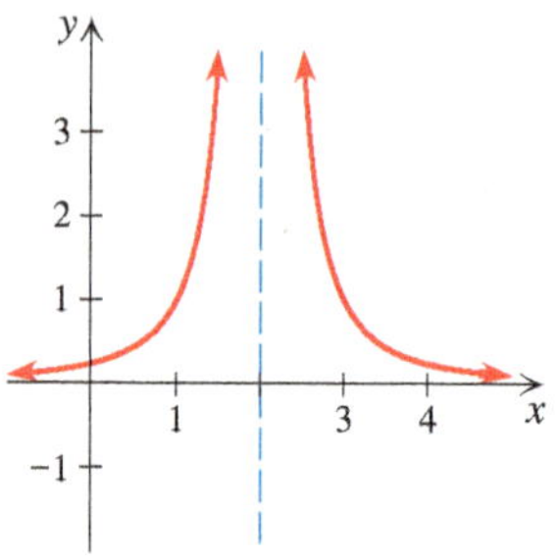


18.

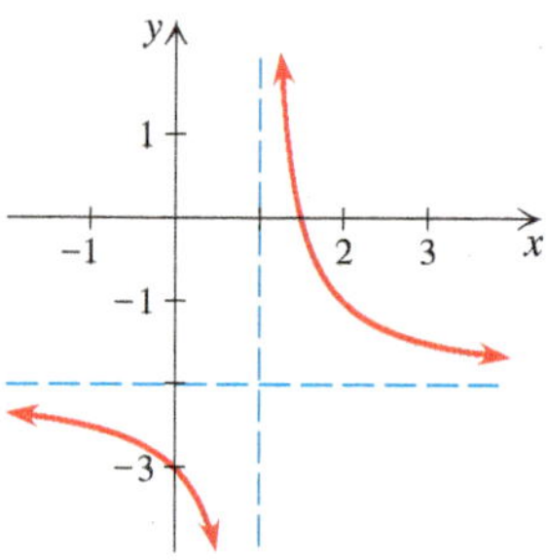


19.

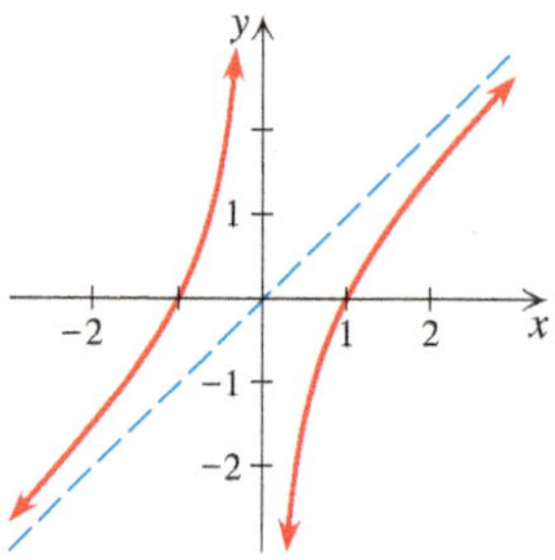


20.

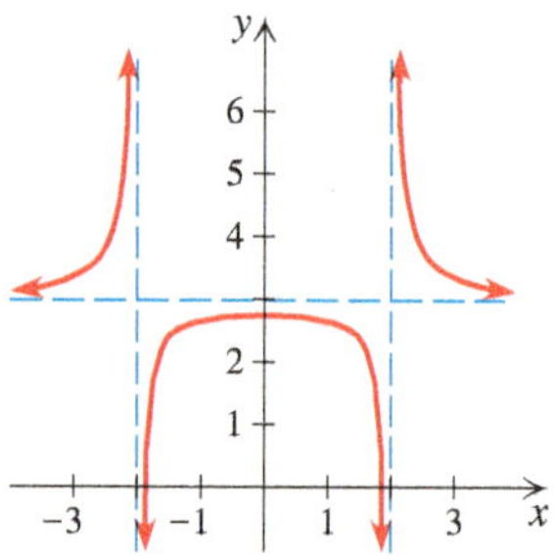


*Determine the equations of all asymptotes for the graph of each function. See the summary for finding asymptotes for a rational function on page 253.*

21. $f(x) = \dfrac{5}{x - 2}$

22. $f(x) = \dfrac{-1}{x + 12}$

23. $f(x) = \dfrac{-x}{x^2 - 9}$

24. $f(x) = \dfrac{-2}{x^2 - 5x + 6}$

25. $f(x) = \dfrac{2x + 4}{x - 1}$

26. $f(x) = \dfrac{5 - x}{x + 5}$

27. $f(x) = \dfrac{x^2 - 2x + 1}{x}$

28. $f(x) = \dfrac{x^3 - 8}{x^2}$

29. $f(x) = \dfrac{3x^2 + 4}{x + 1}$

30. $f(x) = \dfrac{x^2}{x - 9}$

31. $f(x) = \dfrac{-x^2 + 4x}{x + 2}$

32. $f(x) = \dfrac{-x^2 + 3x - 7}{x - 3}$

*Find all asymptotes, x-intercepts, and y-intercepts for the graph of each rational function and sketch the graph of the function. See the procedure for graphing a rational function on page 253.*

33. $f(x) = \dfrac{-1}{x}$

34. $f(x) = \dfrac{1}{x^2}$

35. $f(x) = \dfrac{1}{x - 2}$

36. $f(x) = \dfrac{-1}{x + 1}$

37. $f(x) = \dfrac{1}{x^2 - 4}$

38. $f(x) = \dfrac{1}{x^2 - 2x + 1}$

39. $f(x) = \dfrac{-1}{(x + 1)^2}$

40. $f(x) = \dfrac{-2}{x^2 - 9}$

41. $f(x) = \dfrac{2x + 1}{x - 1}$

42. $f(x) = \dfrac{3x - 1}{x + 1}$

43. $f(x) = \dfrac{x - 3}{x + 2}$

44. $f(x) = \dfrac{2 - x}{x + 2}$

45. $f(x) = \dfrac{x}{x^2 - 1}$

46. $f(x) = \dfrac{-x}{x^2 - 9}$

47. $f(x) = \dfrac{4x}{x^2 - 2x + 1}$

48. $f(x) = \dfrac{-2x}{x^2 + 6x + 9}$

49. $f(x) = \dfrac{8 - x^2}{x^2 - 9}$

50. $f(x) = \dfrac{2x^2 + x - 8}{x^2 - 4}$

51. $f(x) = \dfrac{2x^2 + 8x + 2}{x^2 + 2x + 1}$

52. $f(x) = \dfrac{-x^2 + 7x - 9}{x^2 - 6x + 9}$

*Use a graph or a table to find each limit.*

53. $\lim\limits_{x \to \infty} \dfrac{1}{x^2}$

54. $\lim\limits_{x \to -\infty} \dfrac{1}{x^2}$

55. $\lim\limits_{x \to \infty} \dfrac{2x - 3}{x - 1}$

56. $\lim\limits_{x \to \infty} \dfrac{3x^2 - 1}{x^2 - x}$

57. $\lim\limits_{x \to 0^+} \dfrac{1}{x^2}$

58. $\lim\limits_{x \to 0^-} \dfrac{1}{x^2}$

59. $\lim\limits_{x \to 1^+} \dfrac{2}{x - 1}$

60. $\lim\limits_{x \to 1^-} \dfrac{2}{x - 1}$

*Find the oblique asymptote and sketch the graph of each rational function.*

61. $f(x) = \dfrac{x^2 + 1}{x}$

62. $f(x) = \dfrac{x^2 - 1}{x}$

63. $f(x) = \dfrac{x^3 - 1}{x^2}$

64. $f(x) = \dfrac{x^3 + 1}{x^2}$

65. $f(x) = \dfrac{x^2}{x + 1}$

66. $f(x) = \dfrac{x^2}{x - 1}$

67. $f(x) = \dfrac{2x^2 - x}{x - 1}$

68. $f(x) = \dfrac{-x^2 + x + 1}{x + 1}$

69. $f(x) = \dfrac{x^3 - x^2 - 4x + 5}{x^2 - 4}$

70. $f(x) = \dfrac{x^3 + 2x^2 + x - 2}{x^2 - 1}$

71. $f(x) = \dfrac{-x^3 + x^2 + 5x - 4}{x^2 + x - 2}$

72. $f(x) = \dfrac{x^3 + x^2 - 16x - 24}{x^2 - 2x - 8}$

*Match each rational function with its graph (a)–(h), without using a graphing calculator.*

73. $f(x) = -\dfrac{3}{x}$

74. $f(x) = \dfrac{1}{3 - x}$

75. $f(x) = \dfrac{x}{x - 3}$

76. $f(x) = \dfrac{x - 3}{x}$

77. $f(x) = \dfrac{1}{x^2 - 3x}$

78. $f(x) = \dfrac{x^2}{x^2 - 9}$

79. $f(x) = \dfrac{x^2 - 3}{x}$

80. $f(x) = \dfrac{-x^3 + 1}{x^2}$

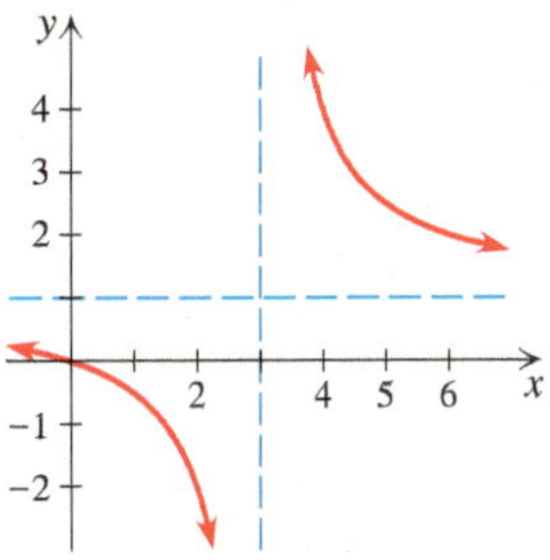

(a)

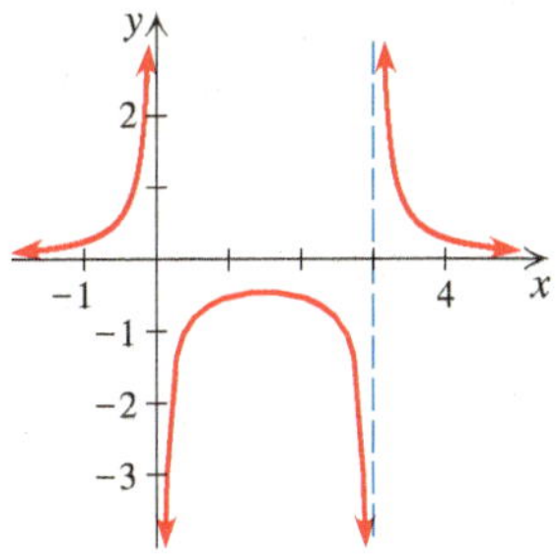

(b)

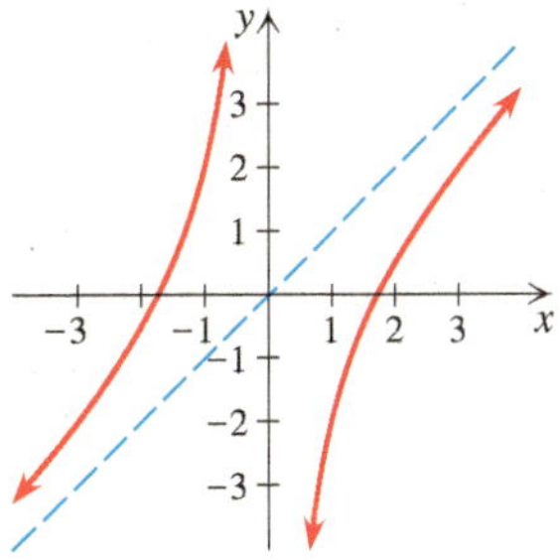

(c)

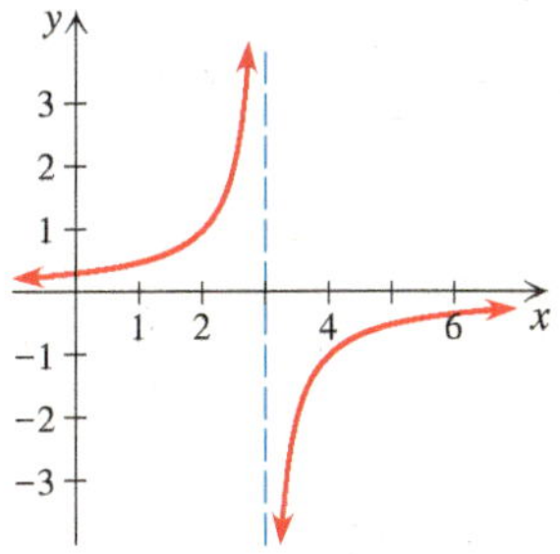

(d)

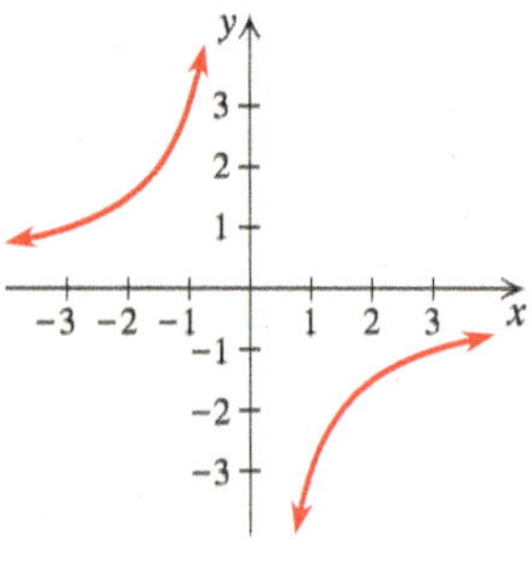

(e)

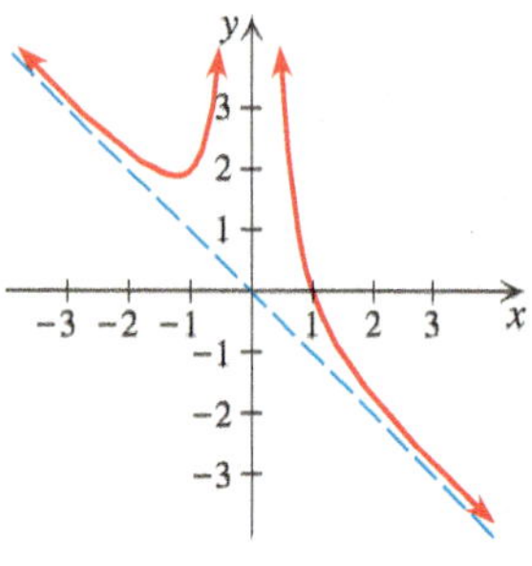

(f)

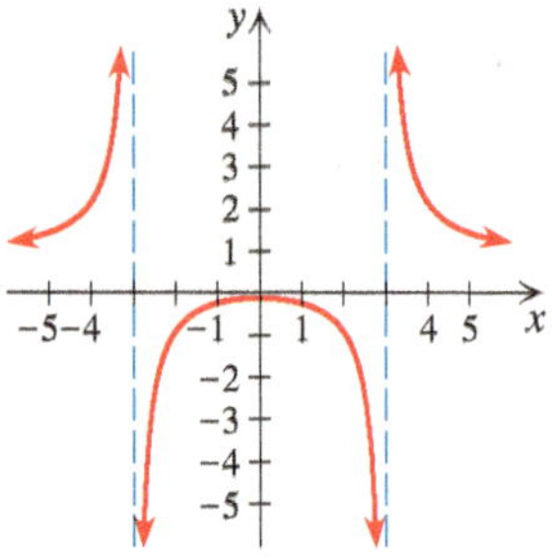

(g)

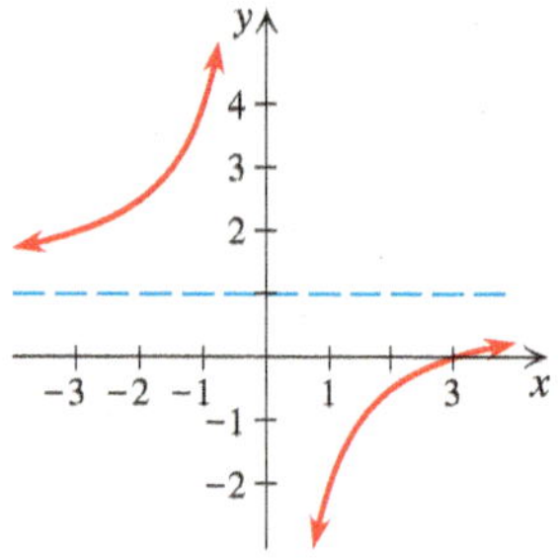

(h)

*Sketch the graph of each rational function. Note that the functions are not in lowest terms. Find the domain first.*

**81.** $f(x) = \dfrac{x + 1}{x^2 - 1}$

**82.** $f(x) = \dfrac{x}{x^2 + 2x}$

**83.** $f(x) = \dfrac{x^2 - 1}{x - 1}$

**84.** $f(x) = \dfrac{x^2 - 5x + 6}{x - 2}$

**85.** $f(x) = \dfrac{x^4 - x^2}{x^2 - 1}$

**86.** $f(x) = \dfrac{-x^5 + x^3}{x^3 - x}$

**87.** $f(x) = \dfrac{x^2 - 4}{x^4 - 4x^2}$

**88.** $f(x) = \dfrac{-x^2 + 1}{x^4 - x^2}$

*Sketch the graph of each rational function.*

**89.** $f(x) = \dfrac{2}{x^2 + 1}$

**90.** $f(x) = \dfrac{x}{x^2 + 1}$

**91.** $f(x) = \dfrac{x - 1}{x^3 - 9x}$

**92.** $f(x) = \dfrac{x^2 + 1}{x^3 - 4x}$

**93.** $f(x) = \dfrac{x + 1}{x^2}$

**94.** $f(x) = \dfrac{x - 1}{x^2}$

*Solve with the test-point method. State the solution set using interval notation. See the strategy for solving a rational inequality on page 257.*

**95.** $\dfrac{x - 4}{x + 2} \le 0$

**96.** $\dfrac{x + 3}{x + 5} \ge 0$

**97.** $\dfrac{q - 2}{q + 3} < 2$

**98.** $\dfrac{p + 1}{2p - 1} \ge 1$

**99.** $\dfrac{w^2 - w - 6}{w - 6} \ge 0$

**100.** $\dfrac{z - 5}{z^2 + 2z - 8} \le 0$

**101.** $\dfrac{1}{x + 2} > \dfrac{1}{x - 3}$

**102.** $\dfrac{1}{x + 1} > \dfrac{2}{x - 1}$

**103.** $x < \dfrac{3x - 8}{5 - x}$

**104.** $\dfrac{2}{x + 3} \ge \dfrac{1}{x - 1}$

**105.** $\dfrac{(x - 3)(x + 1)}{x - 5} \ge 0$

**106.** $\dfrac{(x + 2)(x - 1)}{(x + 4)^2} \le 0$

**107.** $\dfrac{x^2 - 7}{2 - x^2} \le 0$

**108.** $\dfrac{x^2 + 1}{5 - x^2} \ge 0$

**109.** $\dfrac{x^2 + 2x + 1}{x^2 - 2x - 15} \ge 0$

**110.** $\dfrac{x^2 - 2x - 8}{x^2 + 10x + 25} \le 0$

**111.** $\dfrac{1}{w} > \dfrac{1}{w^2}$

**112.** $\dfrac{1}{w} > w^2$

**113.** $w > \dfrac{w - 5}{w - 3}$

**114.** $w < \dfrac{w - 2}{w + 1}$

*State the solution sets to the inequalities in Exercises 115–122 by reading the following graphs.*

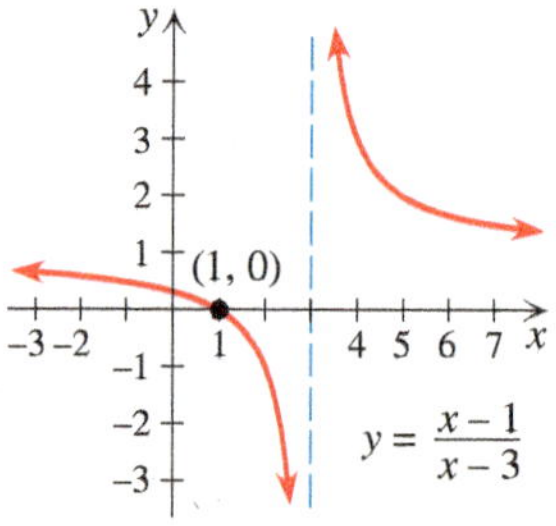


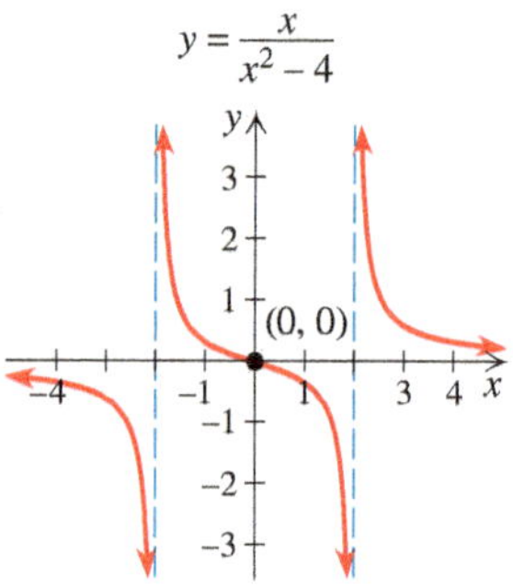


**115.** $\dfrac{x - 1}{x - 3} > 0$

**116.** $\dfrac{x - 1}{x - 3} \le 0$

**117.** $\dfrac{x - 1}{x - 3} \ge 0$

**118.** $\dfrac{x - 1}{x - 3} < 0$

**119.** $\dfrac{x}{x^2 - 4} < 0$

**120.** $\dfrac{x}{x^2 - 4} > 0$

**121.** $\dfrac{x}{x^2 - 4} \ge 0$

**122.** $\dfrac{x}{x^2 - 4} \le 0$

*Find the equation and sketch the graph of each function.*

**123.** A rational function that passes through $(3, 1)$, has the $x$-axis as a horizontal asymptote, and has the line $x = 1$ as its only vertical asymptote

**124.** A rational function that passes through $(0, 4)$, has the $x$-axis as a horizontal asymptote, and has the line $x = 3$ as its only vertical asymptote

**125.** A rational function that passes through $(0, 5)$, has the line $y = 1$ as a horizontal asymptote, and has the line $x = 2$ as its only vertical asymptote

**126.** A rational function that passes through $(-1, 2)$, has the line $y = -2$ as a horizontal asymptote, and has the line $x = -3$ as its only vertical asymptote

**127.** A rational function that passes through $(0, 0)$ and $(4, 8/7)$, has the $x$-axis as a horizontal asymptote, and has two vertical asymptotes $x = 3$ and $x = -3$

**128.** A rational function that passes through $(0, 1)$ and $(2, 1)$, has the $x$-axis as a horizontal asymptote, and has two vertical asymptotes $x = 3$ and $x = -3$

**129.** A rational function that passes through $(0, 3)$, has $y = 2x + 1$ as an oblique asymptote, and has $x = 1$ as its only vertical asymptote

**130.** A rational function that passes through $(3, 3)$, has $y = x - 2$ as an oblique asymptote, and has $x = -5$ as its only vertical asymptote

*Solve each problem.*

**131.** *Admission to the Zoo* Winona paid \$100 for a lifetime membership to Friends of the Zoo, so that she could gain admittance to the zoo for only \$1 per visit. Write Winona's average cost per visit $C$ as a function of the number of visits when she has visited $x$ times. What is her average cost per visit when she has visited the zoo 100 times? Graph the function for $x > 0$. What happens to her average cost per visit if she starts when she is young and visits the zoo every day?

**132.** *Renting a Car* The cost of renting a car for one day is \$39 plus 30 cents per mile. Write the average cost per mile $C$ as a function of the number of miles driven in one day $x$. Graph the function for $x > 0$. What happens to $C$ as the number of miles gets very large?

**133.** *Average Speed of an Auto Trip* A 200-mi trip by an electric car must be completed in 4 hr. Let $x$ be the number of hours it takes to travel the first half of the distance and write the average speed for the second half of the trip as a function of $x$. Graph this function for $0 < x < 4$. What is the significance of the vertical asymptote?

HINT Average speed is distance divided by time.

**134.** *Billboard Advertising* An Atlanta marketing agency figures that the monthly cost of a billboard advertising campaign depends on the fraction of the market $p$ that the client wishes to reach. For $0 \le p < 1$ the cost in dollars is determined by the formula $C = (4p - 1200)/(p - 1)$. What is the monthly cost for a campaign intended to reach 95% of the market? Graph this function for $0 \le p < 1$. What happens to the cost for a client who wants to reach 100% of the market?

**135.** *Carbon Monoxide Poisoning* The accompanying graph shows the relationship between carbon monoxide levels and time for which permanent brain damage will occur *(American Sensors, Carbon Monoxide Detector Owner's Manual)*.

**a.** At 1800 PPM, how long will it take for permanent brain damage to occur?

**b.** What level of carbon monoxide takes 240 minutes to produce permanent brain damage?

**c.** What are the horizontal and vertical asymptotes for the curve? Explain what these asymptotes mean in simple terms.

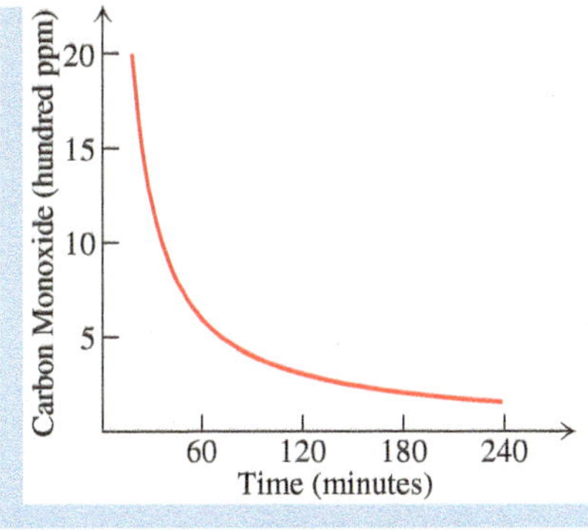


**Figure for Exercise 135**

**136.** *Balancing the Costs* A furniture maker buys foam rubber $x$ times per year. The delivery charge is \$400 per purchase regardless of the amount purchased. The annual cost of storage is figured as $\$10{,}000/x$, because the more frequent the purchase, the less it costs for storage. So the annual cost of delivery and storage is given by

$$C = 400x + \frac{10{,}000}{x}.$$

**a.** Graph the function with a graphing calculator.

**b.** Find the number of purchases per year that minimizes the annual cost of delivery and storage.

**137.** *Making a Gas Tank* An engineer is designing a cylindrical metal tank that is to hold 500 $\text{ft}^3$ of gasoline.

**a.** Write the height $h$ as a function of the radius $r$.

HINT The volume is given by $V = \pi r^2 h$.

**b.** Use the result of part (a) to write the surface area $S$ as a function of $r$ and graph it.

HINT The surface area is given by $S = 2\pi r^2 + 2\pi rh$.

**c.** Use the minimum feature of a graphing calculator to find the radius to the nearest tenth of a foot that minimizes the surface area. Ignore the thickness of the metal.

**d.** If the tank costs \$8 per square foot to construct, then what is the minimum cost for making the tank?

**138.** *Making a Glass Tank* An architect for the Aquarium of the Americas is designing a cylindrical fish tank to hold 1000 $\text{ft}^3$ of water. The bottom and side of the tank will be 3-in.-thick glass and the tank will have no top.

**a.** Write the inside surface area of the tank as a function of the inside radius and graph the function.

**b.** Use the minimum feature of a graphing calculator to find the inside radius that minimizes that inside surface area.

**c.** If Owens Corning will build the tank for \$250 per cubic foot of glass used, then what will be the cost for the minimal tank?

## FOR WRITING/DISCUSSION

**139.** *Approximate Value* Given an arbitrary rational function $R(x)$, explain how you can find the approximate value of $R(x)$ when the absolute value of $x$ is very large.

**140.** *Cooperative Learning* Each student in your small group should write the equations of the horizontal, vertical, and/or oblique asymptotes of an unknown rational function. Then work as a group to find a rational function that has the specified asymptotes. Is there a rational function with any specified asymptotes?

## ▶ RETHINKING

**141.** Solve the inequality $(x - 1)^2(x + 5) > 0$.

**142.** Find all real solutions to $(x - 1)^{2/3} = 8$.

**143.** Find all real and imaginary solutions to $x^5 + 13x^3 + 36x = 0$.

**144.** Use Descartes's rule of signs to discuss the possibilities for the roots of the equation $-x^4 - 6x^2 - 3x + 9 = 0$.

**145.** Find the vertex of the graph of $f(x) = -x^2 + 3x - 10$ and determine the interval on which the function is increasing.

**146.** Find the inverse of the function $f(x) = -8\sqrt[3]{x - 5} + 4$.

## THINKING OUTSIDE THE BOX XXVII

*Filling a Triangle* Fiber-optic cables just fit inside a triangular pipe as shown in the figure. The cables have circular cross sections and the cross section of the pipe is an equilateral triangle with sides of length 1. Suppose that there are $n$ cables of the same size in the bottom row, $n - 1$ of that size in the next row, and so on.

**a.** Write the total cross-sectional area of the cables as a function of $n$.

**b.** As $n$ approaches infinity, will the triangular pipe get totally filled with cables? Explain.

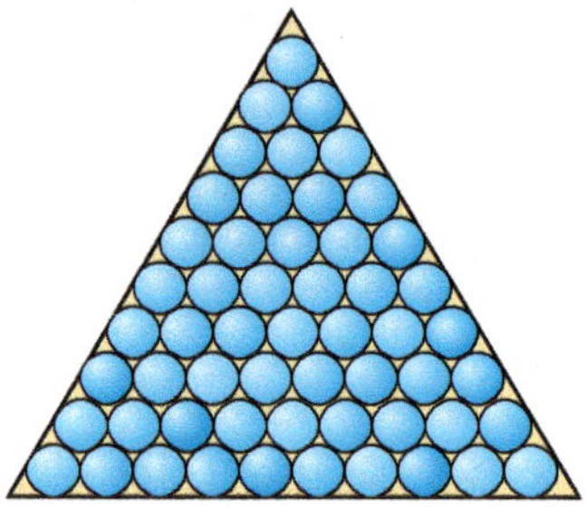

**Figure for Thinking Outside the Box XXVII**

## POP QUIZ 3.6

**1.** What is the domain of $f(x) = \frac{x - 1}{x + 4}$?

**2.** Find the equations of all asymptotes for the graph of $y = \frac{x - 5}{x + 2}$.

**3.** Find the $x$-intercept and $y$-intercept for $f(x) = \frac{x^2 - 9}{x^2 - 1}$.

**4.** What is the horizontal asymptote for $y = \frac{x - 8}{x + 3}$?

**5.** What is the oblique asymptote for $f(x) = \frac{x^2 - 2x}{x - 3}$?

**6.** Solve $\frac{x - 1}{x + 4} \geq 0$.

## LINKING concepts... For Individual or Group Explorations

### Modeling Capital and Operating Cost

*To decide when to replace company cars, an accountant looks at two cost components: capital cost and operating cost. The capital cost is the difference between the original cost and the salvage value, and the operating cost is the cost of operating the vehicle.*

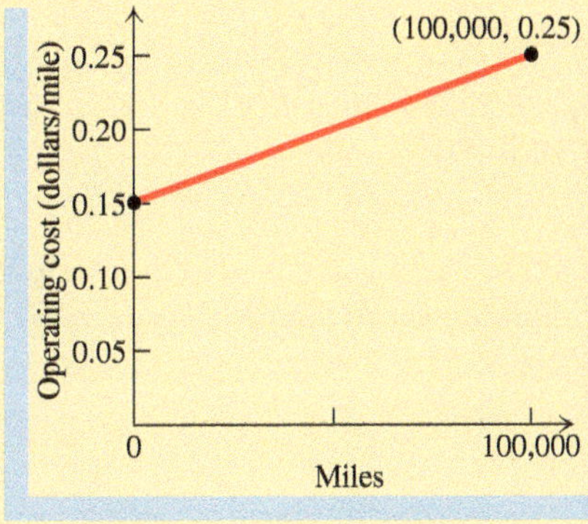

**a)** For a new Ford Fusion, the capital cost is \$3000 plus \$0.12 for each mile that the car is driven. Write the capital cost $C$ as a function of the number of miles the car is driven $x$.

**b)** If a Ford Fusion is purchased for \$19,846 and driven 80,000 miles, then what is its salvage value?

**c)** The operating cost $P$ for the Fusion starts at \$0.15 per mile when the car is new and increases linearly to \$0.25 per mile when the car reaches 100,000 miles. Write $P$ as a function of $x$.

**d)** What is the operating cost at 70,000 miles?

**e)** The total cost per mile is given by $T = \frac{C}{x} + P$. Identify all asymptotes and graph $T$ as a function of $x$.

**f)** Explain each asymptote in terms of this application.

**g)** Find $T$ for $x = 20{,}000$, 30,000, and 90,000 miles.

**h)** The senior vice president has told the accountant that the goal is to keep the total operating cost less than \$0.38 per mile. Is this goal practical?

# Highlights

## 3.1 Quadratic Functions and Inequalities

| | | |
|---|---|---|
| **Quadratic Function** | $y = ax^2 + bx + c$ where $a \neq 0$<br>Graph is a parabola opening upward for $a > 0$ and downward for $a < 0$. | $y = 2x^2 + 8x - 1$<br>Opens upward |
| **Vertex** | $x = -b/(2a)$ is the $x$-coordinate of the vertex.<br>Use $y = ax^2 + bx + c$ to find the $y$-coordinate. | $x = -8/(2 \cdot 2) = -2$<br>Vertex: $(-2, -9)$ |
| **Maximum / Minimum** | The $y$-coordinate of the vertex is the max value of the function if $a < 0$ and the min if $a > 0$. | $y = 2x^2 + 8x - 1$<br>Min $y$-value is $-9$. |
| **Two Forms** | General form: $y = ax^2 + bx + c$<br>Vertex form: $y = a(x - h)^2 + k$, vertex $(h, k)$<br>By completing the square the general form can be rewritten into vertex form. | $y = 2x^2 + 8x - 1$<br>$y = 2(x + 2)^2 - 9$ |
| **Inequalities** | To solve $ax^2 + bx + c > 0$ find all roots to $ax^2 + bx + c = 0$, then test a point in each interval determined by the roots. | $(x - 2)(x + 3) > 0$<br>Roots are $-3$ and 2.<br>Solution set: $(-\infty, -3) \cup (2, \infty)$ |

## 3.2 Zeros of Polynomial Functions

| | | |
|---|---|---|
| **Remainder Theorem** | The remainder when $P(x)$ is divided by $x - c$ is $P(c)$. | $P(x) = x^2 + 3x - 4$<br>$P(x)$ divided by $x + 1$ has remainder $P(-1)$ or $-6$. |
| **Synthetic Division** | An abbreviated version of long division, used only for dividing a polynomial by $x - c$. | $\begin{array}{r\|rrr} -1 & 1 & 3 & -4 \\ & & -1 & -2 \\ \hline & 1 & 2 & -6 \end{array}$<br>Dividend $P(x)$, divisor $x + 1$ quotient $x + 2$, remainder $-6$ |
| **Factor Theorem** | $c$ is a zero of $y = P(x)$ if and only if $x - c$ is a factor of $P(x)$. | Since $x - 2$ is a factor of $P(x) = x^2 + x - 6$, $P(2) = 0$. |
| **Fundamental Theorem of Algebra** | If $y = P(x)$ is a polynomial function of positive degree, then $y = P(x)$ has at least one zero in the set of complex numbers. | $f(x) = x^7 - x^5 + 3x^2 - 9$ has at least one complex zero and in fact has seven of them. |
| **Rational Zero Theorem** | If $p/q$ is a rational zero in lowest terms for $y = P(x)$ with integral coefficients, then $p$ is a factor of the constant term and $q$ is a factor of the leading coefficient. | $P(x) = 6x^2 + x - 15$<br>$P(3/2) = 0$<br>3 is a factor of $-15$;<br>2 is a factor of 6. |

## 3.3 The Theory of Equations

| | | |
|---|---|---|
| ***n*-Root Theorem** | If $P(x)$ has positive degree $n$ and complex coefficients, then $P(x) = 0$ has $n$ roots counting multiplicity. | $(x - 2)^3(x^4 - 9) = 0$ has seven complex solutions counting multiplicity. |
| **Conjugate Pairs Theorem** | If $P(x)$ has real coefficients, then the imaginary roots of $P(x) = 0$ occur in conjugate pairs. | $x^2 - 4x + 5 = 0$<br>$x = 2 \pm i$ |
| **Descartes's Rule of Signs** | The changes in sign of $P(x)$ determine the number of positive roots and the changes in sign of $P(-x)$ determine the number of negative roots to $P(x) = 0$. | $P(x) = x^5 - x^2$<br>has 1 positive root, no negative roots |
| **Theorem on Bounds** | Use synthetic division to obtain an upper bound and lower bound for the roots of $P(x) = 0$. | See Example 5 in Section 3.3. |

## 3.4 Miscellaneous Equations

| | | |
|---|---|---|
| **Higher Degree** | Solve by factoring. | $x^4 - x^2 - 6 = 0$<br>$(x^2 - 3)(x^2 + 2) = 0$ |
| **Squaring Each Side** | Possibly extraneous roots | $\sqrt{x - 3} = -2$, $x - 3 = 4$<br>$x = 7$ extraneous root |
| **Rational Exponents** | Raise each side to a whole-number power and then apply the even or odd root property. | $x^{2/3} = 4$, $x^2 = 64$, $x = \pm 8$ |
| **Quadratic Type** | Make a substitution to get a quadratic equation. | $x^{1/3} + x^{1/6} - 12 = 0$<br>$a^2 + a - 12 = 0$ if $a = x^{1/6}$ |
| **Absolute Value** | Write equivalent equations without absolute value. | $\|x^2 - 4\| = 2$<br>$x^2 - 4 = 2$ or $x^2 - 4 = -2$ |

## 3.5 Graphs of Polynomial Functions

| | | |
|---|---|---|
| **Axis of Symmetry** | The parabola $y = ax^2 + bx + c$ is symmetric about the line $x = -b/(2a)$. | Axis of symmetry for $y = x^2 - 4x$ is $x = 2$. |
| **Behavior at the $x$-Intercepts** | A polynomial function crosses the $x$-axis at $(c, 0)$ if $x - c$ occurs with an odd power or touches the $x$-axis if $x - c$ occurs with an even power. | $f(x) = (x - 3)^2(x + 1)^3$ crosses at $(-1, 0)$, touches but does not cross at $(3, 0)$. |
| **End Behavior** | The leading coefficient and the degree of the polynomial determine the behavior as $x \to -\infty$ or $x \to \infty$. | $y = x^3$<br>$y \to \infty$ as $x \to \infty$<br>$y \to -\infty$ as $x \to -\infty$ |
| **Polynomial Inequality** | Locate all zeros and then test a point in each interval determined by the zeros. | $x^3 - 4x > 0$ when $x$ is in $(-2, 0) \cup (2, \infty)$. |

## 3.6 Rational Functions and Inequalities

| | | |
|---|---|---|
| **Vertical Asymptote** | A rational function in lowest terms has a vertical asymptote wherever the denominator is zero. | $y = x/(x^2 - 4)$ has vertical asymptotes $x = -2$ and $x = 2$. |
| **Horizontal Asymptote** | If degree of denominator exceeds degree of numerator, then the $x$-axis is the horizontal asymptote.<br>If degree of numerator equals degree of denominator, then the ratio of leading coefficients is the horizontal asymptote. | $y = x/(x^2 - 4)$<br>Horizontal asymptote: $y = 0$<br>$y = (2x - 1)/(3x - 2)$<br>Horizontal asymptote: $y = 2/3$ |
| **Oblique or Slant Asymptote** | If degree of numerator exceeds degree of denominator by 1, then use division to determine the slant asymptote. | $y = \frac{x^2}{x - 1} = x + 1 + \frac{1}{x - 1}$<br>Slant asymptote: $y = x + 1$ |
| **Rational Inequality** | Test a point in each interval determined by the zeros and the vertical asymptotes. | $\frac{x - 2}{x + 3} \le 0$ when $x$ is in the interval $(-3, 2]$. |

# Chapter 3 Review Exercises

*Solve each problem.*

1. Write the function $f(x) = 3x^2 - 2x + 1$ in the form $f(x) = a(x - h)^2 + k$.

2. Write the function $f(x) = -4\left(x - \frac{1}{3}\right)^2 - \frac{1}{2}$ in the form $f(x) = ax^2 + bx + c$.

3. Find the vertex, axis of symmetry, $x$-intercepts, and $y$-intercept for the parabola $y = 2x^2 - 4x - 1$.

4. Find the maximum value of the function $y = -x^2 + 3x - 5$.

5. Write the equation of a parabola that has $x$-intercepts $(-1, 0)$ and $(3, 0)$ and $y$-intercept $(0, 6)$.

6. Write the equation of the parabola that has vertex $(1, 2)$ and $y$-intercept $(0, 5)$.

*Find all the real and imaginary zeros for each polynomial function.*

7. $f(x) = 3x - 1$
8. $g(x) = 7$
9. $h(x) = x^2 - 8$
10. $m(x) = x^3 - 8$
11. $n(x) = 8x^3 - 1$
12. $C(x) = 3x^2 - 2$
13. $P(t) = t^4 - 100$
14. $S(t) = 25t^4 - 1$
15. $R(s) = 8s^3 - 4s^2 - 2s + 1$
16. $W(s) = s^3 + s^2 + s + 1$
17. $f(x) = x^3 + 2x^2 - 6x$
18. $f(x) = 2x^3 - 4x^2 + 3x$

*For each polynomial, find the indicated value in two different ways.*

**19.** $P(x) = 4x^3 - 3x^2 + x - 1, P(3)$

**20.** $P(x) = 2x^3 + 5x^2 - 3x - 2, P(-2)$

**21.** $P(x) = -8x^5 + 2x^3 - 6x + 2, P\left(-\frac{1}{2}\right)$

**22.** $P(x) = -4x^4 + 3x^2 + 1, P\left(\frac{1}{2}\right)$

*Use the rational zero theorem to list all possible rational zeros for each polynomial function.*

**23.** $f(x) = -3x^3 + 6x^2 + 5x - 2$

**24.** $f(x) = 2x^4 + 9x^2 - 8x - 3$

**25.** $f(x) = 6x^4 - x^2 - 9x + 3$

**26.** $f(x) = 4x^3 - 5x^2 - 13x - 8$

*Find a polynomial equation with integral coefficients (and lowest degree) that has the given roots.*

**27.** $-\frac{1}{2}, 3$

**28.** $\frac{1}{2}, -5$

**29.** $3 - 2i$

**30.** $4 + 2i$

**31.** $2, 1 - 2i$

**32.** $-3, 3 - 4i$

**33.** $2 - \sqrt{3}$

**34.** $1 + \sqrt{2}$

*Use Descartes's rule of signs to discuss the possibilities for the roots to each equation. Do not solve the equation.*

**35.** $x^8 + x^6 + 2x^2 = 0$

**36.** $-x^3 - x - 3 = 0$

**37.** $4x^3 - 3x^2 + 2x - 9 = 0$

**38.** $5x^5 + x^3 + 5x = 0$

**39.** $x^3 + 2x^2 + 2x + 1 = 0$

**40.** $-x^4 - x^3 + 3x^2 + 5x - 8 = 0$

*Establish the best integral bounds for the roots of each equation according to the theorem on bounds.*

**41.** $6x^2 + 5x - 50 = 0$

**42.** $4x^2 - 12x - 27 = 0$

**43.** $2x^3 - 15x^2 + 31x - 12 = 0$

**44.** $x^3 + 6x^2 + 11x + 7 = 0$

**45.** $12x^3 - 4x^2 - 3x + 1 = 0$

**46.** $4x^3 - 4x^2 - 9x + 10 = 0$

*Find all real and imaginary solutions to each equation, stating multiplicity when it is greater than one.*

**47.** $x^3 - 6x^2 + 11x - 6 = 0$

**48.** $x^3 + 7x^2 + 16x + 12 = 0$

**49.** $6x^4 - 5x^3 + 7x^2 - 5x + 1 = 0$

**50.** $6x^4 + 5x^3 + 25x^2 + 20x + 4 = 0$

**51.** $x^3 - 9x^2 + 28x - 30 = 0$

**52.** $x^3 - 4x^2 + 6x - 4 = 0$

**53.** $x^3 - 4x^2 + 7x - 6 = 0$

**54.** $2x^3 - 5x^2 + 10x - 4 = 0$

**55.** $2x^4 - 5x^3 - 2x^2 + 2x = 0$

**56.** $2x^5 - 15x^4 + 26x^3 - 12x^2 = 0$

*Find all real solutions to each equation.*

**57.** $|2v - 1| = 3v$

**58.** $|2h - 3| = |h|$

**59.** $x^4 + 7x^2 = 18$

**60.** $2x^{-2} + 5x^{-1} = 12$

**61.** $\sqrt{x + 6} - \sqrt{x - 5} = 1$

**62.** $\sqrt{2x - 1} = \sqrt{x - 1} + 1$

**63.** $\sqrt{y} + \sqrt[4]{y} = 6$

**64.** $\sqrt[3]{x^2} + \sqrt[3]{x} = 2$

**65.** $x^4 - 3x^2 - 4 = 0$

**66.** $(y - 1)^2 - (y - 1) = 2$

**67.** $(x - 1)^{2/3} = 4$

**68.** $(2x - 3)^{-1/2} = \frac{1}{2}$

**69.** $(x + 3)^{-3/4} = -8$

**70.** $\left(\frac{1}{x - 3}\right)^{-1/4} = \frac{1}{2}$

**71.** $\sqrt[3]{3x - 7} = \sqrt[3]{4 - x}$

**72.** $\sqrt[3]{x + 1} = \sqrt[6]{4x + 9}$

*Discuss the symmetry of the graph of each function.*

**73.** $f(x) = 2x^2 - 3x + 9$

**74.** $f(x) = -3x^2 + 12x - 1$

**75.** $f(x) = -3x^4 - 2$

**76.** $f(x) = \frac{-x^3}{x^2 - 1}$

**77.** $f(x) = \frac{x}{x^2 + 1}$

**78.** $f(x) = 2x^4 + 3x^2 + 1$

*Find the domain of each rational function.*

**79.** $f(x) = \frac{x^2 - 4}{2x + 5}$

**80.** $f(x) = \frac{4x + 1}{x^2 - x - 6}$

**81.** $f(x) = \frac{1}{x^2 + 1}$

**82.** $f(x) = \frac{x - 9}{x^2 - 1}$

*Find the x-intercepts, y-intercept, and asymptotes for the graph of each function and sketch the graph.*

**83.** $f(x) = x^2 - x - 2$

**84.** $f(x) = -2(x - 1)^2 + 6$

**85.** $f(x) = x^3 - 3x - 2$

**86.** $f(x) = x^3 - 3x^2 + 4$

**87.** $f(x) = \frac{1}{2}x^3 - \frac{1}{2}x^2 - 2x + 2$

**88.** $f(x) = \frac{1}{2}x^3 - 3x^2 + 4x$ **89.** $f(x) = \frac{1}{4}x^4 - 2x^2 + 4$

**90.** $f(x) = \frac{1}{2}x^4 + 2x^3 + 2x^2$

**91.** $f(x) = \frac{2}{x + 3}$ **92.** $f(x) = \frac{1}{2 - x}$

**93.** $f(x) = \frac{2x}{x^2 - 4}$ **94.** $f(x) = \frac{2x^2}{x^2 - 4}$

**95.** $f(x) = \frac{x^2 - 2x + 1}{x - 2}$ **96.** $f(x) = \frac{-x^2 + x + 2}{x - 1}$

**97.** $f(x) = \frac{2x - 1}{2 - x}$ **98.** $f(x) = \frac{1 - x}{x + 1}$

**99.** $f(x) = \frac{x^2 - 4}{x - 2}$ **100.** $f(x) = \frac{x^3 + x}{x}$

*Solve each inequality. State the solution set using interval notation.*

**101.** $8x^2 + 1 < 6x$ **102.** $x^2 + 2x < 63$

**103.** $(3 - x)(x + 5) \geq 0$ **104.** $-x^2 - 2x + 15 < 0$

**105.** $4x^3 - 400x^2 - x + 100 \geq 0$

**106.** $x^3 - 49x^2 - 52x + 100 < 0$

**107.** $\frac{x + 10}{x + 2} < 5$ **108.** $\frac{x - 6}{2x + 1} \geq 1$

**109.** $\frac{12 - 7x}{x^2} > -1$ **110.** $x - \frac{2}{x} \leq -1$

**111.** $\frac{x^2 - 3x + 2}{x^2 - 7x + 12} \geq 0$ **112.** $\frac{x^2 + 4x + 3}{x^2 - 2x - 15} \leq 0$

*Solve each problem.*

**113.** Find the quotient and remainder when $x^3 - 6x^2 + 9x - 15$ is divided by $x - 3$.

**114.** Find the quotient and remainder when $3x^3 + 4x^2 + 2x - 4$ is divided by $3x - 2$.

**115.** *Altitude of a Rocket* If the altitude in feet of a model rocket is given by the equation $S = -16t^2 + 156t$, where $t$ is the time in seconds after ignition, then what is the maximum height attained by the rocket?

**116.** *Antique Saw* Willard is making a reproduction of an antique saw. The handle consists of two pieces of wood joined at a right angle, with the blade being the hypotenuse of the right triangle as shown in the figure. If the total length of the handle is to be 36 in., then what length for each piece would minimize the square of the length of the blade?

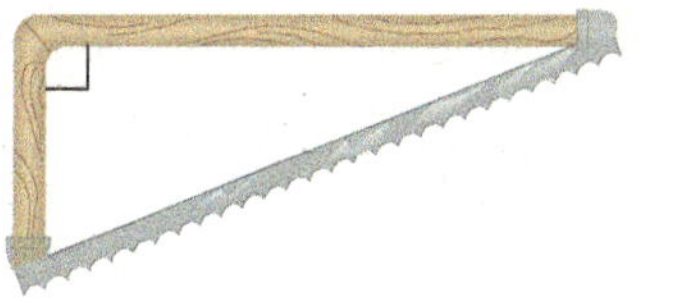

**Figure for Exercise 116**

**117.** *Bonus Room* A homeowner wants to put a room in the attic of her house. The house is 48 ft wide and the roof has a 7–12 pitch. (The roof rises 7 ft in a horizontal distance of 12 ft.) Find the dimensions of the room that will maximize the area of the cross section shown in the figure.

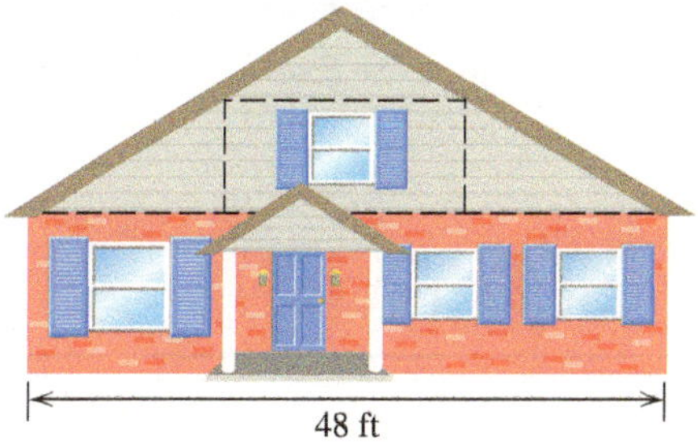


**Figure for Exercise 117**

**118.** *Maximizing Area* An isosceles triangle has one vertex at the origin and a horizontal base below the $x$-axis with its endpoints on the curve $y = x^2 - 16$. See the figure. Let $(a, b)$ be the vertex in the fourth quadrant and write the area of the triangle as a function of $a$. Use a graphing calculator to find the point $(a, b)$ for which the triangle has the maximum possible area.

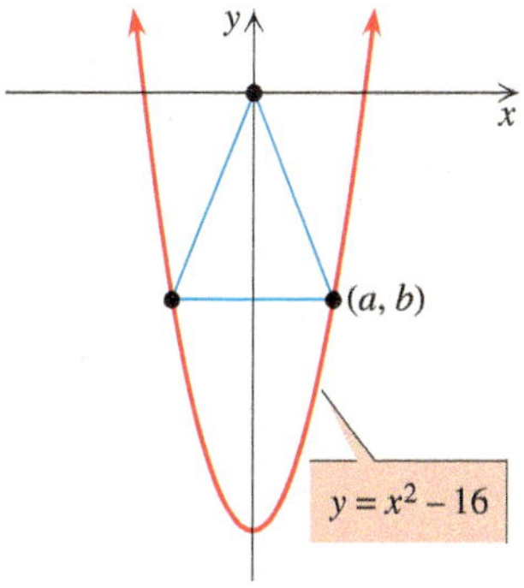


**Figure for Exercise 118**

**119.** *Limiting Velocity* As a skydiver falls, his velocity keeps increasing. However, because of air resistance, the rate at which the velocity increases keeps decreasing and there is a limit velocity that the skydiver cannot exceed. To see this behavior, graph

$$V = \frac{1000t}{5t + 8}$$

where $V$ is the velocity in feet per second and $t$ is the time in seconds.

**a.** What is the velocity at time $t = 10$ sec?

**b.** What is the horizontal asymptote for this graph?

**c.** What is the limiting velocity that cannot be exceeded?

**Figure for Exercise 119**

### THINKING OUTSIDE THE BOX XXVIII & XXIX

*Polynomial Equation* Find a polynomial equation with integral coefficients for which $\sqrt{3} + \sqrt{5}$ is a root.

*Bad Arithmetic* Each letter in the following addition problem represents a different digit. Find the digits that will make the addition problem correct. There is more than one possibility.

```
  SEVEN
 +EIGHT
 TWELVE
```

# Chapter 3 Test

*Solve each problem.*

**1.** Write $y = 3x^2 - 12x + 1$ in the form $y = a(x - h)^2 + k$.

**2.** Identify the vertex, axis of symmetry, $y$-intercept, $x$-intercepts, and range for $y = 3x^2 - 12x + 1$.

**3.** What is the minimum value of $y$ in the function $y = 3x^2 - 12x + 1$?

**4.** Use synthetic division to find the quotient and remainder when $2x^3 - 4x + 5$ is divided by $x + 3$.

**5.** What is the remainder when

$$x^{98} - 19x^{73} + 17x^{44} - 12x^{23} + 2x^9 - 3$$

is divided by $x - 1$?

**6.** List the possible rational zeros for the function $f(x) = 3x^3 - 4x^2 + 5x - 6$ according to the rational zero theorem.

**7.** Find a polynomial equation with real coefficients that has the roots $-3$ and $4i$.

**8.** Use Descartes's rule of signs to discuss the possibilities for the roots of $x^3 - 3x^2 + 5x + 7 = 0$.

**9.** The altitude in feet of a toy rocket $t$ seconds after launch is given by the function $S(t) = -16t^2 + 128t$. Find the maximum altitude reached by the rocket.

*Find all real and imaginary zeros for each polynomial function.*

**10.** $f(x) = x^2 - 9$

**11.** $f(x) = x^4 - 16$

**12.** $f(x) = x^3 - 4x^2 - x + 10$

*Find all real and imaginary roots of each equation. State the multiplicity of a root when it is greater than one.*

**13.** $x^4 + 2x^2 + 1 = 0$

**14.** $(x^3 - 2x^2)(2x + 3)^3 = 0$

**15.** $2x^3 - 9x^2 + 14x - 5 = 0$

*Sketch the graph of each function.*

**16.** $y = 2(x - 3)^2 + 1$

**17.** $y = (x - 2)^2(x + 1)$

**18.** $y = x^3 - 4x$

*Find all asymptotes and sketch the graph of each function.*

**19.** $y = \dfrac{1}{x - 2}$

**20.** $y = \dfrac{2x - 3}{x - 2}$

**21.** $f(x) = \dfrac{x^2 + 1}{x}$

**22.** $y = \dfrac{4}{x^2 - 4}$

**23.** $y = \dfrac{x^2 - 2x + 1}{x - 1}$

*Solve each inequality.*

**24.** $x^2 - 2x < 8$

**25.** $\dfrac{x + 2}{x - 3} > -1$

**26.** $\dfrac{x + 3}{(4 - x)(x + 1)} \geq 0$

**27.** $x^3 - 7x > 0$

*Find all real solutions to each equation.*

**28.** $(x - 3)^{-2/3} = \dfrac{1}{3}$

**29.** $\sqrt{x} - \sqrt{x - 7} = 1$

# TYING IT ALL together...

## Chapters 1–3

*Find all real and imaginary solutions to each equation.*

**1.** $2x - 3 = 0$

**2.** $2x^2 - 3x = 0$

**3.** $x^3 + 3x^2 + 3x + 1 = 0$

**4.** $\dfrac{x - 1}{x - 2} = 0$

**5.** $x^{2/3} - 9 = 0$

**6.** $x^2 + 1 = 0$

*For Exercises 7–10, use the given functions and their graphs to fill in the table below. If a function does not have a feature, put a minus sign in the table.*

**7.** $f(x) = 2x - 4$

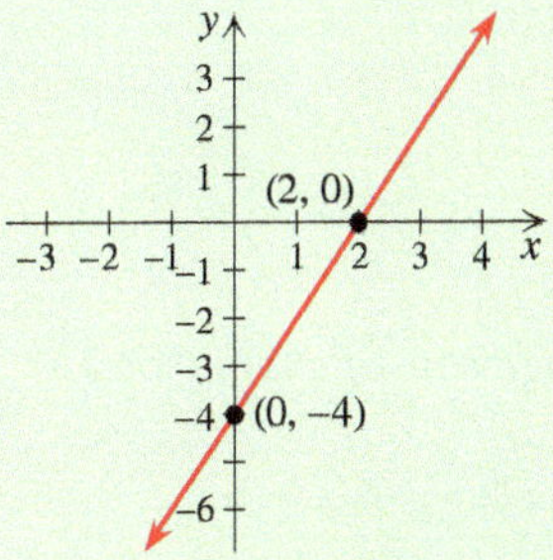


**8.** $f(x) = -x^2 + 2x + 3$

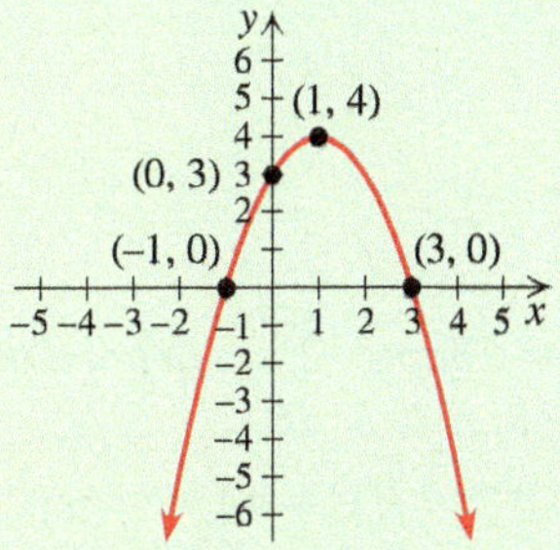


**9.** $f(x) = x^3 - 3x^2 - 9x + 27$

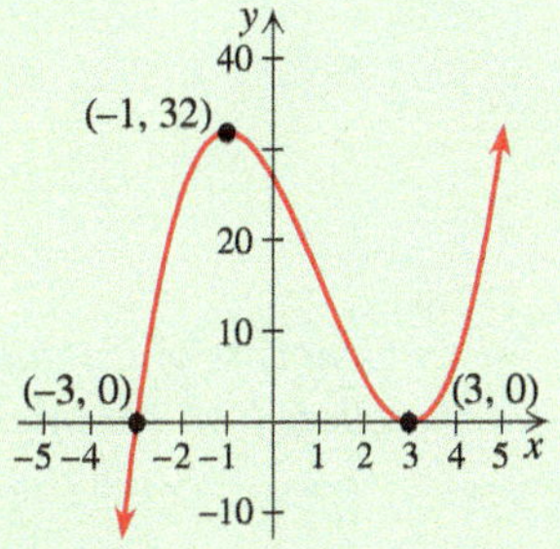


**10.** $f(x) = \dfrac{x^2 - 1}{x^2 - 4}$

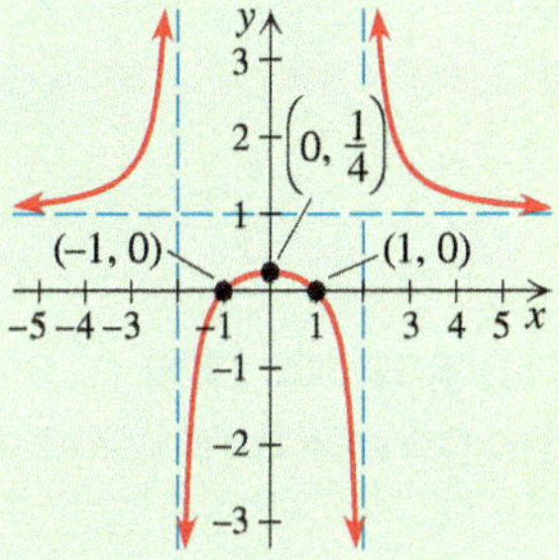


| | **7.** | **8.** | **9.** | **10.** |
|---|---|---|---|---|
| $x$-intercepts | | | | |
| $y$-intercept | | | | |
| Where is $f(x)$ increasing? | | | | |
| Where is $f(x)$ decreasing? | | | | |
| Domain of $f(x)$ | | | | |
| Range of $f(x)$ | | | | |
| Solution set to $f(x) = 0$ | | | | |
| Solution set to $f(x) > 0$ | | | | |
| Solution set to $f(x) \le 0$ | | | | |
| Vertex | | | | |
| Vertical asymptotes | | | | |
| Horizontal asymptote | | | | |

*Solve each inequality.*

**11.** $2x - 3 < 0$

**12.** $2x^2 - 3x < 0$

**13.** $x^3 + 3x^2 + 3x + 1 > 0$

**14.** $\dfrac{x - 1}{x - 2} > 0$

**15.** $x^{2/3} - 9 < 0$

**16.** $x^2 + 1 < 0$

*Find the inverse for each function that has an inverse.*

**17.** $f(x) = 2x - 3$

**18.** $f(x) = 2x^2 - 3x$

**19.** $f(x) = x^3 + 3x^2 + 3x + 1$

**20.** $f(x) = \dfrac{x - 1}{x - 2}$

**21.** $f(x) = x^{2/3} - 9$

**22.** $f(x) = x^2 + 1$

*Fill in the blanks.*

**23.** A polynomial with three terms is called a(n) ________.

**24.** A polynomial with two terms is called a(n) ________.

**25.** The expressions $1 - \sqrt{2}$ and $1 + \sqrt{2}$ are called ________.

**26.** Let $P(x) = Q(x)D(x) + R(x)$ where $P$, $Q$, $D$, and $R$ are polynomial functions where the degree of $R(x)$ is less than the degree of $D(x)$ and $D(x) \neq 0$. $P(x)$ is called the ________, $Q(x)$ is called the ________, $D(x)$ is called the ________, and $R(x)$ is called the ________.

**27.** A monomial that includes every number and variable that is a factor of all terms of a polynomial is called the ________ of the polynomial.

**28.** A polynomial that cannot be factored using integral coefficients is called ________.

**29.** In the complex number $a + bi$, $a$ is called the ________ and $b$ is called the ________.

**30.** A(n) ________ is an equation involving two or more variables.

**31.** If $x$ and $y$ are real numbers, then $(x, y)$ could be a(n) ________ or a(n) ________.

**32.** The $xy$-coordinate system is also called the ________.

# CONCEPTS OF calculus...

## Instantaneous rate of change of the power functions

*In Concepts of Calculus for Chapter 2, we defined the average rate of change of a function f on the interval* $[x, x + h]$ *as*

$$\frac{f(x + h) - f(x)}{h}.$$

*We then defined the instantaneous rate of change of the function as the limit as h approaches zero of the average rate of change. The notation $f'(x)$, read "f prime of x," is used for the instantaneous rate of change. So*

$$f'(x) = \lim_{h \to 0} \frac{f(x + h) - f(x)}{h}.$$

*In the following exercises we will find the instantaneous rate of change of the power functions and discover the power rule.*

## Exercises

**1.** Let $f(x) = x$.

**a.** Find $\frac{f(x + h) - f(x)}{h}$ and simplify it.

**b.** Find $\lim_{h \to 0} \frac{f(x + h) - f(x)}{h}$.

**c.** Find $f'(x)$ and $f'(2)$.

**2.** Let $f(x) = x^2$.

**a.** Find $\frac{f(x + h) - f(x)}{h}$ and simplify it.

**b.** Find $\lim_{h \to 0} \frac{f(x + h) - f(x)}{h}$.

**c.** Find $f'(x)$ and $f'(2)$.

**3.** Let $f(x) = x^3$.

**a.** Find $\frac{f(x + h) - f(x)}{h}$ and simplify it. (*Hint:* To find $(x + h)^3$, multiply $x^2 + 2xh + h^2$ by $x + h$.)

**b.** Find $\lim_{h \to 0} \frac{f(x + h) - f(x)}{h}$.

**c.** Find $f'(x)$ and $f'(2)$.

**4.** Let $f(x) = x^4$.

**a.** Find $\frac{f(x + h) - f(x)}{h}$ and simplify it.

**b.** Find $\lim_{h \to 0} \frac{f(x + h) - f(x)}{h}$.

**c.** Find $f'(x)$ and $f'(2)$.

**5.** Look for a pattern in the results for $f'(x)$ in Exercises 1–4. Use the pattern to determine $f'(x)$ for $f(x) = x^5$.

**6.** Let $f'(x) = x^n$ where $n$ is a positive integer. Use the pattern established in Exercises 1–4 to determine $f'(x)$. This result is known as the *power rule*.

# 4 Exponential and Logarithmic Functions

In the summer of 2008, scientists discovered a new batch of well-preserved dinosaur bones, petrified trees, and even freshwater clams in southeastern Utah. An excavation revealed at least four sauropods, which are long-necked, long-tailed plant-eating dinosaurs. Although the excavation did not unearth any new species, scientists say that it could provide new clues about life in the region some 150 million years ago.

Paleontologists use data gathered from fossils to determine when and how dinosaurs lived. Sauropods were the largest land animals ever to have lived. Sauropods could be as long as 130 feet and weigh as much as 100 tons. Their remains have been found on every continent except Antarctica.

**WHAT YOU WILL learn...** In this chapter we learn how scientists use exponential functions to date archaeological finds that are millions of years old. Through applications ranging from population growth to growth of financial investments, we'll see how exponential functions can help us predict the future and discover the past.

# 4.1 Exponential Functions and Their Applications

The functions that involve some combination of basic arithmetic operations, powers, or roots are called **algebraic functions.** Most of the functions studied so far are algebraic functions. In this chapter we turn to the exponential and logarithmic functions. These functions are used to describe phenomena ranging from growth of investments to the decay of radioactive materials, which cannot be described with algebraic functions. Since the exponential and logarithmic functions transcend what can be described with algebraic functions, they are called **transcendental functions.**

## The Definition

In algebraic functions such as

$$j(x) = x^2, \quad p(x) = x^5, \quad \text{and} \quad m(x) = x^{1/3}$$

the base is a variable and the exponent is constant. For the exponential functions the base is constant and the exponent is a variable. The functions

$$f(x) = 2^x, \quad g(x) = 5^x, \quad \text{and} \quad h(x) = \left(\frac{1}{3}\right)^x$$

are exponential functions.

**Definition: Exponential Function**

An **exponential function** with **base *a*** is a function of the form

$$f(x) = a^x$$

where $a$ and $x$ are real numbers such that $a > 0$ and $a \neq 1$.

We rule out the base $a = 1$ in the definition because $f(x) = 1^x$ is the constant function $f(x) = 1$. Negative numbers and 0 are not used as bases because powers such as $(-4)^{1/2}$ and $0^{-2}$ are not real numbers or not defined. Other functions with variable exponents, such as $f(x) = 3^{2x+1}$ or $f(x) = 10 \cdot 7^{-x^2}$, may also be called exponential functions.

To evaluate an exponential function we use our knowledge of exponents. A review of exponents can be found in Section A.2 of Appendix A.

### EXAMPLE 1 Evaluating exponential functions

Let $f(x) = 4^x$, $g(x) = 5^{2-x}$, and $h(x) = \left(\frac{1}{3}\right)^x$. Find the following values. Check your answers with a calculator.

**a.** $f(3/2)$ **b.** $g(3)$ **c.** $h(-2)$

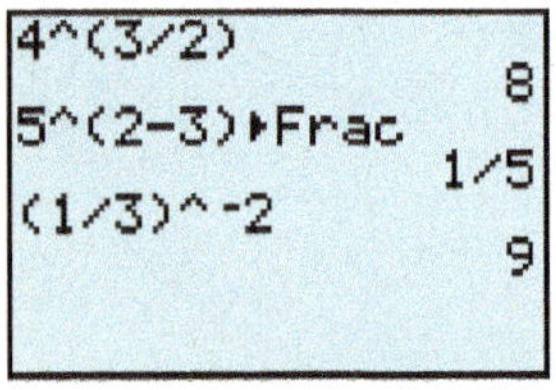


Figure 4.1

**Solution**

**a.** $f(3/2) = 4^{3/2} = (\sqrt{4})^3 = 2^3 = 8$ Take the square root, then cube.

**b.** $g(3) = 5^{2-3} = 5^{-1} = \dfrac{1}{5}$ Exponent $-1$ means reciprocal.

**c.** $h(-2) = \left(\dfrac{1}{3}\right)^{-2} = 3^2 = 9$ Find the reciprocal, then square.

Evaluating these functions with a calculator, as in Fig. 4.1, yields the same results.

▶**TRY THIS.** Find $f(-2)$ and $f(1/2)$ if $f(x) = 9^x$. ■

## Domain of Exponential Functions

The domain of the exponential function $f(x) = 2^x$ is the set of all real numbers. To understand how this function can be evaluated for any *real* number, first note that the function is evaluated for any *rational* number using powers and roots. For example,

$$f(1.7) = f(17/10) = 2^{17/10} = \sqrt[10]{2^{17}} \approx 3.249009585.$$

Until now, we have not considered *irrational* exponents.

Consider the expression $2^{\sqrt{3}}$ and recall that the irrational number $\sqrt{3}$ is an infinite nonterminating nonrepeating decimal number:

$$\sqrt{3} = 1.7320508075\ldots$$

If we use rational approximations to $\sqrt{3}$ as exponents, we see a pattern:

$$2^{1.7} = 3.249009585\ldots$$
$$2^{1.73} = 3.317278183\ldots$$
$$2^{1.732} = 3.321880096\ldots$$
$$2^{1.73205} = 3.321995226\ldots$$
$$2^{1.7320508} = 3.321997068\ldots$$

As the exponents get closer and closer to $\sqrt{3}$ we get results that are approaching some number. We define $2^{\sqrt{3}}$ to be that number. Of course, it is impossible to write the exact value of $\sqrt{3}$ or $2^{\sqrt{3}}$ as a decimal, but you can use a calculator to get $2^{\sqrt{3}} \approx 3.321997085$. Since any irrational number can be approximated by rational numbers in this same manner, $2^x$ is defined similarly for any irrational number. This idea extends to any exponential function.

**Domain of an Exponential Function**

The domain of $f(x) = a^x$ for $a > 0$ and $a \neq 1$ is the set of all real numbers.

## Graphing Exponential Functions

Even though the domain of an exponential function is the set of real numbers, for ease of computation, we generally choose only rational numbers for $x$ to find ordered pairs on the graph of the function.

### EXAMPLE 2 Graphing an exponential function ($a > 1$)

Sketch the graph of each exponential function and state the domain and range.

**a.** $f(x) = 2^x$ **b.** $g(x) = 10^x$

**Solution**

**a.** Find some ordered pairs satisfying $f(x) = 2^x$ as follows:

| $x$ | $-2$ | $-1$ | 0 | 1 | 2 | 3 |
|---|---|---|---|---|---|---|
| $y = 2^x$ | 1/4 | 1/2 | 1 | 2 | 4 | 8 |

*y*-values increasing →

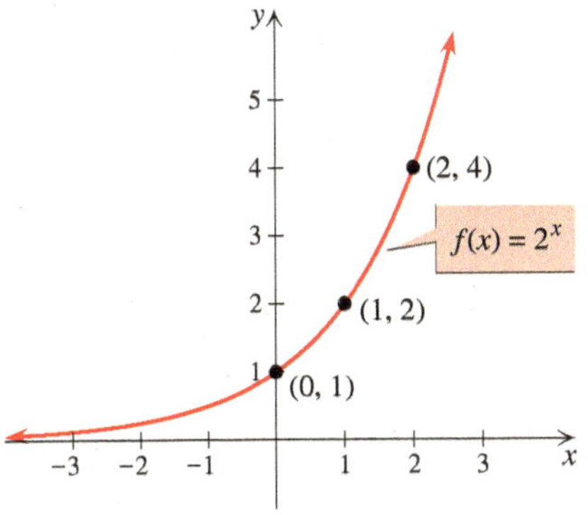


Figure 4.2

As $x \to \infty$, so does $2^x$. Using limit notation, $\lim_{x\to\infty} 2^x = \infty$. As $x \to -\infty$, $2^x$ approaches but never reaches 0. Using limit notation, $\lim_{x\to-\infty} 2^x = 0$. So the $x$-axis is a horizontal asymptote for the curve. It can be shown that the graph of $f(x) = 2^x$ increases in a continuous manner, with no "jumps" or "breaks". So the graph is a smooth curve through the points shown in Fig. 4.2. The domain of $f(x) = 2^x$ is $(-\infty, \infty)$ and the range is $(0, \infty)$. The function is increasing on $(-\infty, \infty)$.

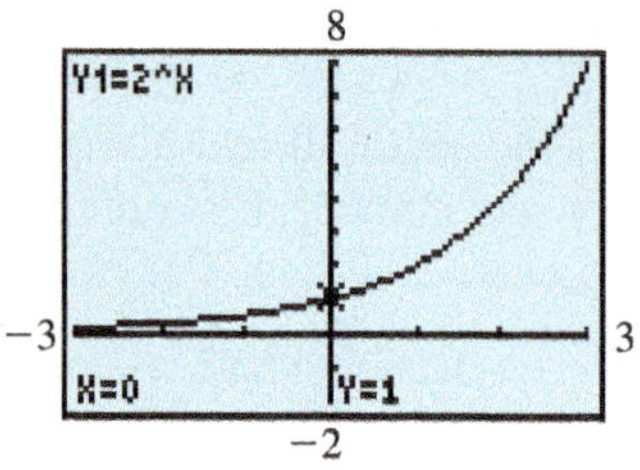


Figure 4.3

The calculator graph shown in Fig. 4.3 supports these conclusions. Because of the limited resolution of the calculator screen, the calculator graph appears to touch its horizontal asymptote in this window. You can have a mental picture of a curve that gets closer and closer but never touches its asymptote, but it is impossible to accurately draw such a picture. □

**b.** Find some ordered pairs satisfying $g(x) = 10^x$ as follows:

| $x$ | $-2$ | $-1$ | $0$ | $1$ | $2$ | $3$ |
|---|---|---|---|---|---|---|
| $y = 10^x$ | 0.01 | 0.1 | 1 | 10 | 100 | 1000 |

*y*-values increasing →

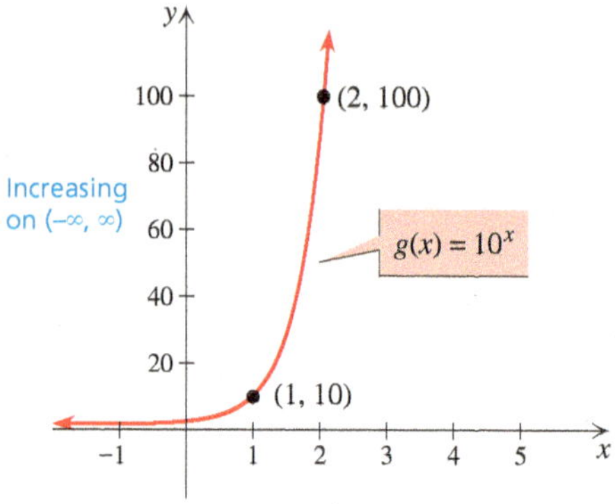


Figure 4.4

These points indicate that the graph of $g(x) = 10^x$ behaves in the same manner as the graph of $f(x) = 2^x$. We have $\lim_{x\to\infty} 10^x = \infty$ and $\lim_{x\to-\infty} 10^x = 0$. So the $x$-axis is a horizontal asymptote. The domain of $g(x) = 10^x$ is $(-\infty, \infty)$ and the range is $(0, \infty)$. The function is increasing on $(-\infty, \infty)$. Plotting the points in the table yields the curve shown in Fig. 4.4. In theory the curve never touches the $x$-axis. So we usually try to keep our hand-drawn curve from touching the $x$-axis. The calculator graph in Fig. 4.5 supports these conclusions.

►**TRY THIS.** Graph $f(x) = 9^x$ and state the domain and range. ■

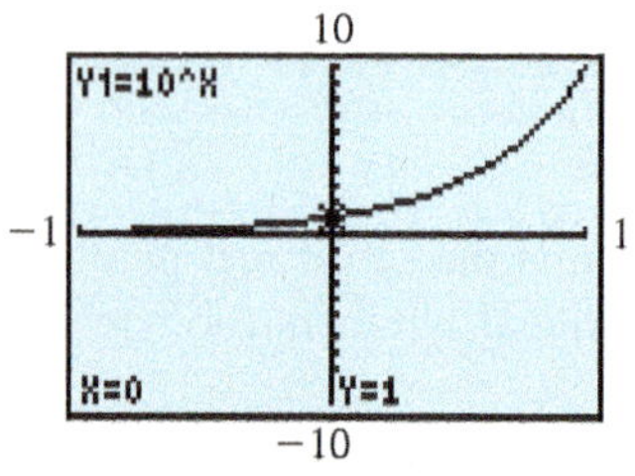


Figure 4.5

Note the similarities between the graphs shown in Figs. 4.2 and 4.4. Both pass through the point $(0, 1)$, both functions are increasing, and both have the $x$-axis as a horizontal asymptote. Since a horizontal line can cross these graphs only once, the functions $f(x) = 2^x$ and $g(x) = 10^x$ are one-to-one by the horizontal line test. All functions of the form $f(x) = a^x$ for $a > 1$ have graphs similar to those in Figs. 4.2 and 4.4.

The phrase "growing exponentially" is often used to describe a population or other quantity whose growth can be modeled with an increasing exponential function. For example, the earth's population is said to be growing exponentially because of the shape of the graph shown in Fig. 4.6

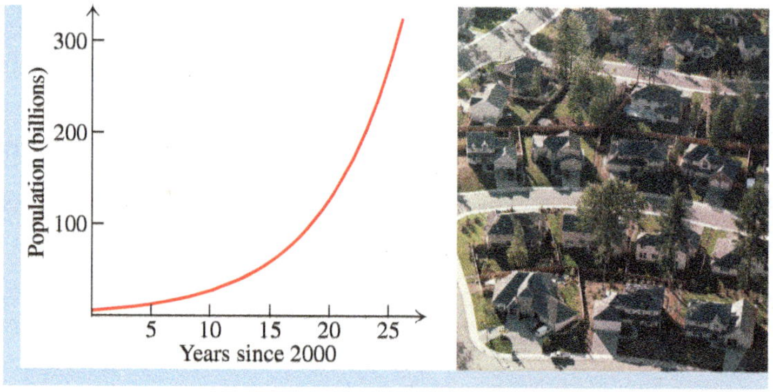


Figure 4.6

If $a > 1$, then the graph of $f(x) = a^x$ is increasing. In the next example we graph exponential functions in which $0 < a < 1$ and we will see that these functions are decreasing.

## EXAMPLE 3 Graphing an exponential function ($0 < a < 1$)

Sketch the graph of each function and state the domain and range of the function.

**a.** $f(x) = (1/2)^x$ **b.** $g(x) = 3^{-x}$

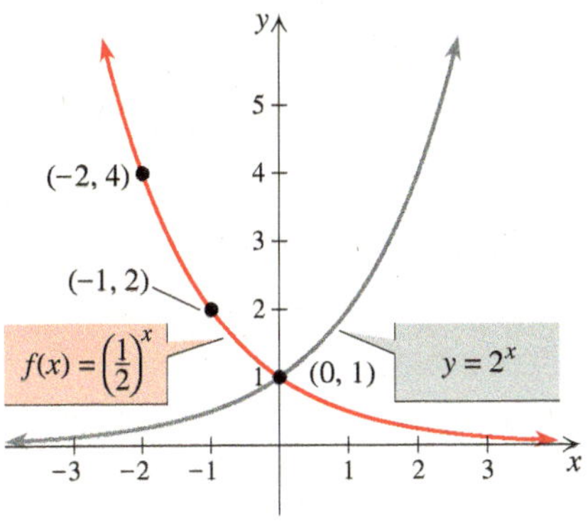


Figure 4.7

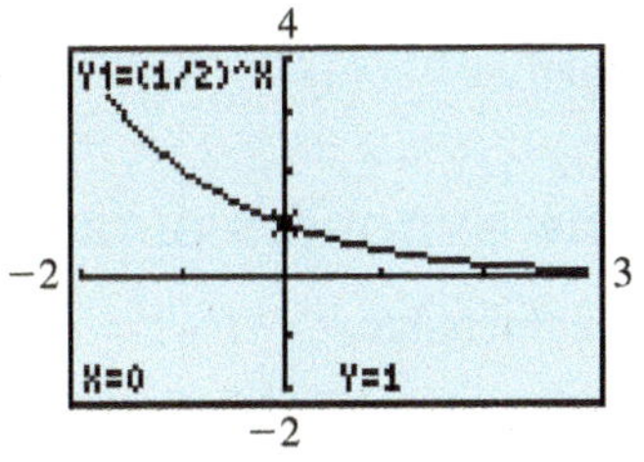


Figure 4.8

### Solution

**a.** Find some ordered pairs satisfying $f(x) = (1/2)^x$ as follows:

| $x$ | −2 | −1 | 0 | 1 | 2 | 3 |
|---|---|---|---|---|---|---|
| $y = \left(\frac{1}{2}\right)^x$ | 4 | 2 | 1 | 1/2 | 1/4 | 1/8 |

*y*-values decreasing

Plot these points and draw a smooth curve through them as in Fig. 4.7 Note that $\lim_{x\to\infty}\left(\frac{1}{2}\right)^x = 0$ and $\lim_{x\to-\infty}\left(\frac{1}{2}\right)^x = \infty$. The $x$-axis is a horizontal asymptote. The domain is $(-\infty, \infty)$ and the range is $(0, \infty)$. The function is decreasing on $(-\infty, \infty)$. Because $(1/2)^x = 2^{-x}$ the graph of $f(x) = (1/2)^x$ is a reflection in the $y$-axis of the graph of $y = 2^x$.

The calculator graph in Fig. 4.8 supports these conclusions. □

**b.** Since $3^{-x} = 1/3^x = (1/3)^x$, this function is of the form $y = a^x$ for $0 < a < 1$. Find some ordered pairs satisfying $g(x) = 3^{-x}$ or $g(x) = (1/3)^x$ as follows:

| $x$ | −2 | −1 | 0 | 1 | 2 | 3 |
|---|---|---|---|---|---|---|
| $y = 3^{-x}$ | 9 | 3 | 1 | 1/3 | 1/9 | 1/27 |

*y*-values decreasing

Plot these points and draw a smooth curve through them as shown in Fig. 4.9. Note that $\lim_{x\to\infty} 3^{-x} = 0$ and $\lim_{x\to-\infty} 3^{-x} = \infty$. The domain is $(-\infty, \infty)$ and the range is $(0, \infty)$. The function is decreasing on its entire domain, $(-\infty, \infty)$. The $x$-axis is again a horizontal asymptote.

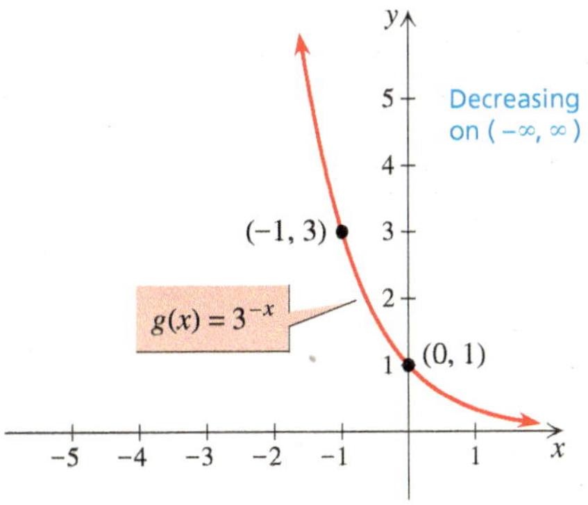


Figure 4.9

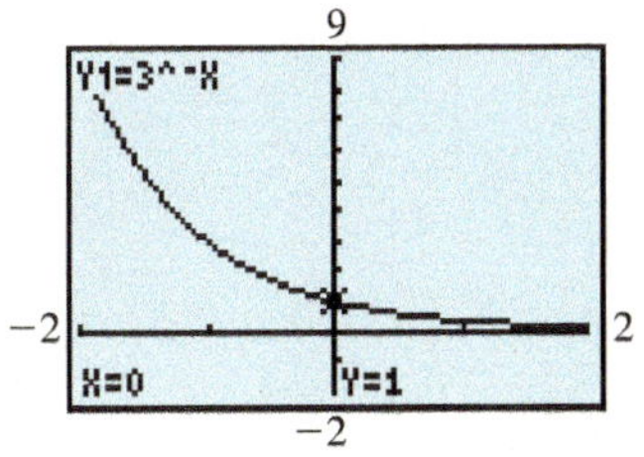


Figure 4.10

The calculator graph in Fig. 4.10 supports these conclusions.

▶**TRY THIS.** Graph $f(x) = 9^{-x}$ and state the domain and range. ■

Again note the similarities between the graphs in Figs. 4.7 and 4.9. Both pass through $(0, 1)$, both functions are decreasing, and both have the $x$-axis as a horizontal asymptote. By the horizontal line test, these functions are also one-to-one. The base determines whether the exponential function is increasing or decreasing. In general, we have the following properties of exponential functions.

### Properties of Exponential Functions

The exponential function $f(x) = a^x$ has the following properties:

1. The function $f$ is increasing for $a > 1$ and decreasing for $0 < a < 1$.
2. The $y$-intercept of the graph of $f$ is $(0, 1)$.
3. The graph has the $x$-axis as a horizontal asymptote.
4. The domain of $f$ is $(-\infty, \infty)$, and the range of $f$ is $(0, \infty)$.
5. The function $f$ is one-to-one.

## The Exponential Family of Functions

A function of the form $f(x) = a^x$ is an exponential function. Any function of the form $g(x) = b \cdot a^{x-h} + k$ is a member of the **exponential family** of functions. The graph of $g$ is a transformation of the graph of $f$, as discussed in Section 2.3:

The graph of $f$ moves to the left if $h < 0$ or to the right if $h > 0$.

The graph of $f$ moves upward if $k > 0$ or downward if $k < 0$.

The graph of $f$ is stretched if $b > 1$ and shrunk if $0 < b < 1$.

The graph of $f$ is reflected in the $x$-axis if $b$ is negative.

### EXAMPLE 4 Graphing members of the exponential family

Sketch the graph of each function and state its domain and range.

**a.** $y = 2^{x-3}$ **b.** $f(x) = -4 + 3^{x+2}$

#### Solution

**a.** The graph of $y = 2^{x-3}$ is a translation three units to the right of the graph of $y = 2^x$. For accuracy, find a few ordered pairs that satisfy $y = 2^{x-3}$:

| $x$ | 3 | 4 | 5 |
|---|---|---|---|
| $y = 2^{x-3}$ | 1 | 2 | 4 |

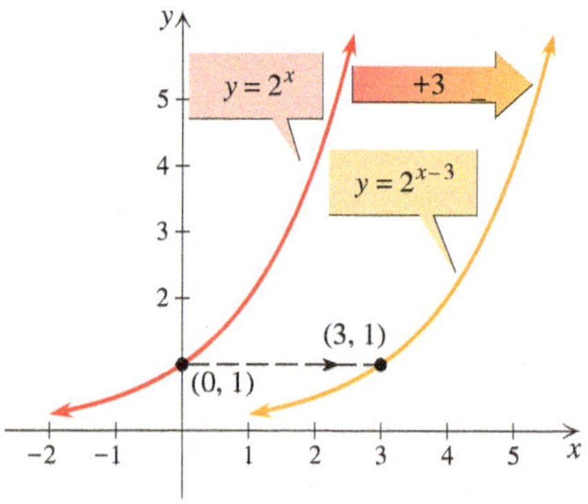


Figure 4.11

Draw the graph of $y = 2^{x-3}$ through these points as shown in Fig. 4.11 The domain of $y = 2^{x-3}$ is $(-\infty, \infty)$, and the range is $(0, \infty)$.

The calculator graph in Fig. 4.12 supports these conclusions. □

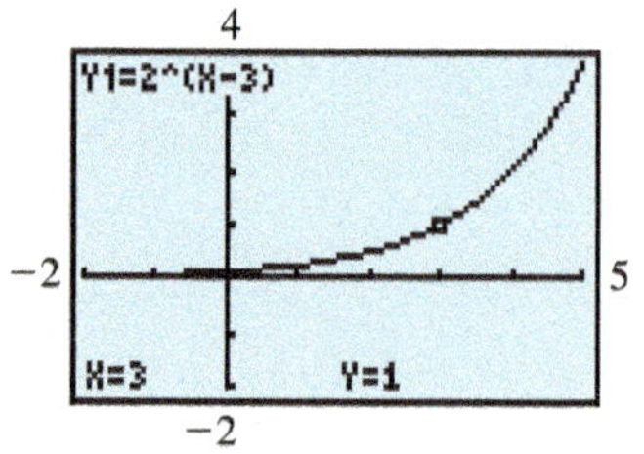


Figure 4.12

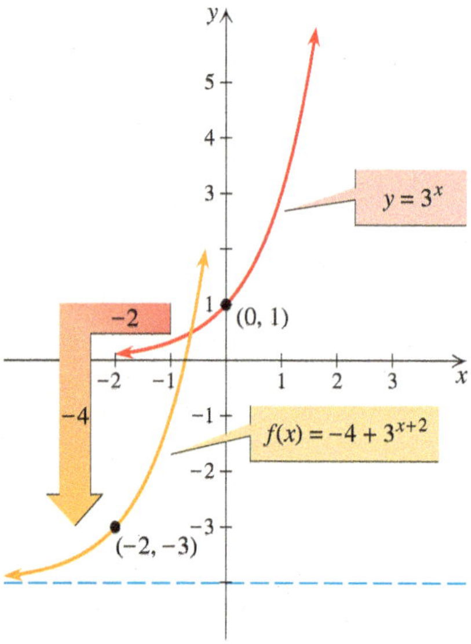


Figure 4.13

**b.** The graph of $f(x) = -4 + 3^{x+2}$ is obtained by translating $y = 3^x$ to the left two units and downward four units. For accuracy, find a few points on the graph of $f(x) = -4 + 3^{x+2}$:

| $x$ | −3 | −2 | −1 | 0 |
|---|---|---|---|---|
| $y = -4 + 3^{x+2}$ | −11/3 | −3 | −1 | 5 |

Sketch a smooth curve through these points as shown in Fig. 4.13 The horizontal asymptote is the line $y = -4$. The domain of $f$ is $(-\infty, \infty)$, and the range is $(-4, \infty)$.

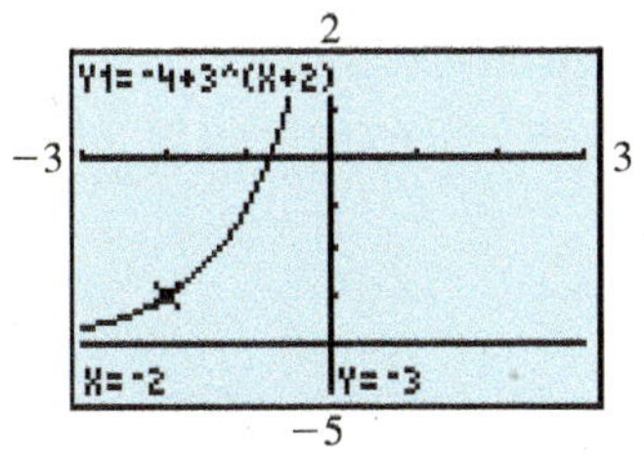


Figure 4.14

The calculator graph of $y = -4 + 3^{x+2}$ and the asymptote $y = -4$ in Fig. 4.14 support these conclusions.

▶**TRY THIS.** Graph $f(x) = 2^{x-2} + 1$ and state the domain and range. ■

If you recognize that a function is a transformation of a simpler function, then you know what its graph looks like. This information along with a few ordered pairs will help you to make accurate graphs.

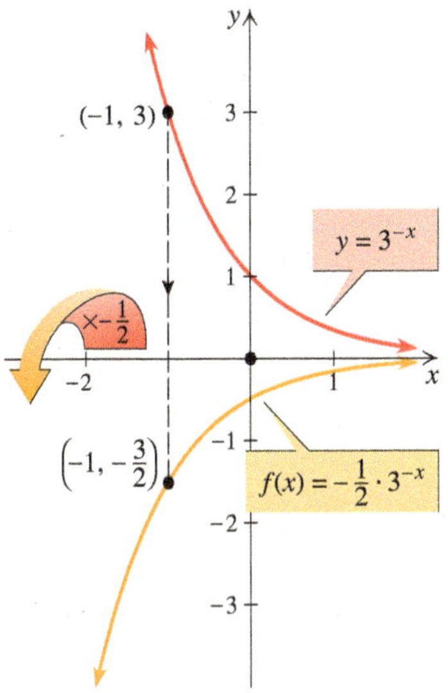


Figure 4.15

## EXAMPLE 5 Graphing an exponential function with a reflection

Sketch the graph of $f(x) = -\frac{1}{2} \cdot 3^{-x}$ and state the domain and range.

### Solution

The graph of $y = 3^{-x}$ is shown in Fig. 4.9. Multiplying by $-1/2$ shrinks the graph and reflects it below the $x$-axis. The $x$-axis is the horizontal asymptote for the graph of $f$. Find a few ordered pairs as follows:

| $x$ | $-2$ | $-1$ | $0$ | $1$ | $2$ |
|---|---|---|---|---|---|
| $y = -\frac{1}{2} \cdot 3^{-x}$ | $-9/2$ | $-3/2$ | $-1/2$ | $-1/6$ | $-1/18$ |

$y$-values increasing →

Sketch the curve through these points as in Fig. 4.15. The domain is $(-\infty, \infty)$ and the range is $(-\infty, 0)$.

▶**TRY THIS.** Graph $f(x) = -4^{-x}$ and state the domain and range. ■

The form $f(x) = b \cdot a^x$ is often used in applications. In this case $b$ is the initial value of the function, because $f(0) = b \cdot a^0 = b$. For example, if you discover 3 ants in your cookie jar (time 0) and the number of ants doubles after each minute, then the number of ants $x$ minutes after your initial discovery is $3 \cdot 2^x$. If it makes sense to start at $x = 1$, then use $x - 1$ for the exponent. For example, if you get \$2 in tips on your first day at work and your tips increase by 50% on each succeeding day, then the amount on day $x$ is $2 \cdot 1.5^{x-1}$ dollars.

## Exponential Equations

From the graphs of exponential functions, we observed that they are one-to-one. For an exponential function, one-to-one means that if two exponential expressions with the same base are equal, then the exponents are equal. For example, if $2^m = 2^n$, then $m = n$.

**One-to-One Property of Exponential Functions**

For $a > 0$ and $a \neq 1$,

$$\text{if} \quad a^{x_1} = a^{x_2}, \quad \text{then} \quad x_1 = x_2.$$

The one-to-one property is used in solving simple exponential equations. For example, to solve $2^x = 8$, we recall that $8 = 2^3$. So the equation becomes $2^x = 2^3$. By the one-to-one property, $x = 3$ is the only solution. The one-to-one property applies only to equations in which each side is a power of the same base.

### EXAMPLE 6 Solving exponential equations

Solve each exponential equation.

**a.** $4^x = \frac{1}{4}$ **b.** $\left(\frac{1}{10}\right)^x = 100$

#### Solution

**a.** Since $1/4 = 4^{-1}$, we can write the right-hand side as a power of 4 and use the one-to-one property:

$$4^x = \frac{1}{4} = 4^{-1} \quad \text{Both sides are powers of 4.}$$

$$x = -1 \quad \text{One-to-one property}$$

**b.** Since $(1/10)^x = 10^{-x}$ and $100 = 10^2$, we can write each side as a power of 10:

$$\left(\frac{1}{10}\right)^x = 100 \quad \text{Original equation}$$

$$10^{-x} = 10^2 \quad \text{Both sides are powers of 10.}$$

$$-x = 2 \quad \text{One-to-one property}$$

$$x = -2$$

▶**TRY THIS.** Solve $2^{-x} = \frac{1}{8}$. ■

The type of equation that we solved in Example 6 arises naturally when we try to find the first coordinate of an ordered pair of an exponential function when given the second coordinate, as shown in the next example.

### EXAMPLE 7 Finding the first coordinate given the second

Let $f(x) = 5^{2-x}$. Find $x$ such that $f(x) = 125$.

#### Solution

To find $x$ such that $f(x) = 125$, we must solve $5^{2-x} = 125$:

$$5^{2-x} = 5^3 \quad \text{Since } 125 = 5^3$$

$$2 - x = 3 \quad \text{One-to-one property}$$

$$x = -1$$

▶**TRY THIS.** Let $f(x) = 3^{4-x}$. Find $x$ such that $f(x) = \frac{1}{9}$. ■

## The Compound Interest Model

Exponential functions are used to model phenomena such as population growth, radioactive decay, and compound interest. Here we will show how these functions are used to determine the amount of an investment earning compound interest.

If $P$ dollars are deposited in an account with a simple annual interest rate $r$ for $t$ years, then the amount $A$ in the account at the end of $t$ years is found by using the formula $A = P + Prt$ or $A = P(1 + rt)$. If \$1000 is deposited at 6% annual rate for one quarter of a year, then at the end of three months the amount is

$$A = 1000\left(1 + 0.06 \cdot \frac{1}{4}\right) = 1000(1.015) = \$1015.$$

If the account begins the next quarter with \$1015, then at the end of the second quarter we again multiply by 1.015 to get the amount

$$A = 1000(1.015)^2 \approx \$1030.23.$$

This process is referred to as compound interest because interest is put back into the account and the interest also earns interest. At the end of 20 years, or 80 quarters, the amount is

$$A = 1000(1.015)^{80} \approx \$3290.66.$$

Compound interest can be thought of as simple interest computed over and over. The general **compound interest formula** follows.

**Compound Interest Formula**

> If a principal $P$ is invested for $t$ years at an annual rate $r$ compounded $n$ times per year, then the amount $A$, or ending balance, is given by
> $$A = P\left(1 + \frac{r}{n}\right)^{nt}.$$

Figure 4.16

The principal $P$ is also called **present value** and the amount $A$ is called **future value.**

### EXAMPLE 8 Using the compound interest formula

Find the amount or future value when a principal of \$20,000 is invested at 6% compounded daily for three years.

**Solution**

Use $P = \$20,000$, $r = 0.06$, $n = 365$, and $t = 3$ in the compound interest formula:

$$A = \$20,000\left(1 + \frac{0.06}{365}\right)^{365 \cdot 3} \approx \$23,943.99$$

Fig. 4.16 shows this expression on a graphing calculator.

**▶TRY THIS.** Find the amount when \$9000 is invested at 5.4% compounded monthly for 6 years. ■

**HISTORICAL NOTE**

Leonhard Euler (1707–1783) was a Swiss mathematician and physicist. Euler introduced and popularized several notational conventions through his numerous textbooks. He introduced the concept of a function and was the first to write $f(x)$. He also introduced the modern notation for the trigonometric functions: the letter $e$ (Euler's number) for the base of the natural logarithm, the Greek letter sigma for summations, and the letter $i$ to denote the imaginary unit.

When interest is compounded daily, financial institutions usually use the exact number of days. To keep our discussion of interest simple, we assume that all years have 365 days and ignore leap years. When discussing months we assume that all months have 30 days. There may be other rules involved in computing interest that are specific to the institution doing the computing. One popular method is to compound quarterly and only give interest on money that is on deposit for full quarters. With this rule you could have money on deposit for nearly six months (say January 5 to June 28) and receive no interest.

## Continuous Compounding and the Number *e*

The more often that interest is figured during the year, the more interest an investment will earn. The first five lines of Table 4.1 show the future value of \$10,000 invested at 12% for one year for more and more frequent compounding. The last line shows the limiting amount \$11,274.97, which we cannot exceed no matter how often we compound the interest on this investment for one year.

**Table 4.1**

| Compounding | Future Value in One Year |
|---|---|
| Annually | $\$10{,}000\left(1 + \frac{0.12}{1}\right)^{1} = \$11{,}200$ |
| Quarterly | $\$10{,}000\left(1 + \frac{0.12}{4}\right)^{4} \approx \$11{,}255.09$ |
| Monthly | $\$10{,}000\left(1 + \frac{0.12}{12}\right)^{12} \approx \$11{,}268.25$ |
| Daily | $\$10{,}000\left(1 + \frac{0.12}{365}\right)^{365} \approx \$11{,}274.75$ |
| Hourly | $\$10{,}000\left(1 + \frac{0.12}{8760}\right)^{8760} \approx \$11{,}274.96$ |
| Continuously | $\$10{,}000e^{0.12(1)} \approx \$11{,}274.97$ |

### Foreshadowing Calculus

There are several ways to define the number e and the function $f(x) = e^x$ in calculus. Calculus can deepen your understanding of many of the concepts that you study in algebra.

To better understand Table 4.1, we need to use a fact from calculus. The expression $[1 + r/n]^{nt}$ used in calculating the first five values in the table can be shown to approach $e^{rt}$ as $n$ goes to $\infty$. Using limit notation,

$$\lim_{n\to\infty}\left[1 + \frac{r}{n}\right]^{nt} = e^{rt}.$$

The number $e$ (like $\pi$) is an irrational number that occurs in many areas of mathematics. Using $r = 1$ and $t = 1$ in this limit gives us $\lim_{n\to\infty}[1 + 1/n]^n = e$, which can be used as the definition of the number $e$. You can find approximate values for powers of $e$ using a calculator with an $e^x$-key. To find an approximation for $e$ itself, find $e^1$ on a calculator:

$$e \approx 2.718281828459$$

Figure 4.17

Figure 4.17 shows the graphing calculator computation of the future value of \$10,000 at 12% compounded continuously for one year and the value of $e^1$. □

So there is a limit to the results obtained by increasingly frequent compounding. The limit is the product of the principal and $e^{rt}$. Using $A = P \cdot e^{rt}$ to find the amount is called **continuous compounding.**

### Continuous Compounding Formula

If a principal $P$ is invested for $t$ years at an annual rate $r$ compounded continuously, then the amount $A$, or ending balance, is given by

$$A = P \cdot e^{rt}.$$

## EXAMPLE 9 Interest compounded continuously

Find the amount when a principal of \$5600 is invested at $6\frac{1}{4}$% annual rate compounded continuously for 5 years and 9 months.

### Solution

Convert 5 years and 9 months to 5.75 years. Use $r = 0.0625$, $t = 5.75$, and $P = \$5600$ in the continuous compounding formula:

$$A = 5600 \cdot e^{(0.0625)(5.75)} \approx \$8021.63$$

▶**TRY THIS.** Find the amount when \$8000 is invested at 6.3% compounded continuously for 7 years and 3 months. ■

The function $f(x) = e^x$ is called the **base-*e* exponential function.** Variations of this function are used to model many types of growth and decay. The graph of

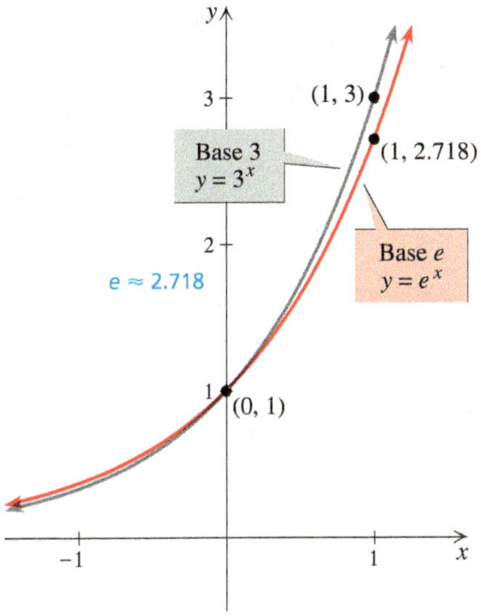


Figure 4.18

$y = e^x$ looks like the graph of $y = 3^x$, because the value of $e$ is close to 3. Use your calculator to check that the graph of $y = e^x$ shown in Fig. 4.18 goes approximately through the points $(-1, 0.368)$, $(0, 1)$, $(1, 2.718)$, and $(2, 7.389)$.

## The Radioactive Decay Model

The mathematical model of radioactive decay is based on the formula

$$A = A_0 e^{rt},$$

which gives the amount $A$ of a radioactive substance remaining after $t$ years, where $A_0$ is the initial amount present and $r$ is the annual rate of decay. The only difference between this formula and the continuous compounding formula is that when decay is involved, the rate $r$ is negative.

### EXAMPLE 10 Radioactive decay of carbon-14

Finely chipped spear points shaped by nomadic hunters in thc Ice Age were found at the scene of a large kill in Lubbock, Texas. From the amount of decay of carbon-14 in the bone fragments of the giant bison, scientists determined that the kill took place about 9833 years ago (*National Geographic*, December 1955). Carbon-14 is a radioactive substance that is absorbed by an organism while it is alive but begins to decay upon the death of the organism. The number of grams of carbon-14 remaining in a fragment of charred bone from the giant bison after $t$ years is given by the formula $A = 3.6e^{rt}$, where $r = -1.21 \times 10^{-4}$. Find the amount present initially and after 9833 years.

#### Solution

To find the initial amount, let $t = 0$ and $r = -1.21 \times 10^{-4}$ in the formula $A = A_0 e^{rt}$:

$$A = 3.6e^{(-1.21\times10^{-4})(0)} = 3.6 \text{ grams}$$

To find the amount present after 9833 years, let $t = 9833$ and $r = -1.21 \times 10^{-4}$:

$$A = 3.6e^{(-1.21\times10^{-4})(9833)}$$

$$\approx 1.1 \text{ grams}$$

The initial amount of carbon-14 was 3.6 grams, and after 9833 years, approximately 1.1 grams of carbon-14 remained. See Fig. 4.19.

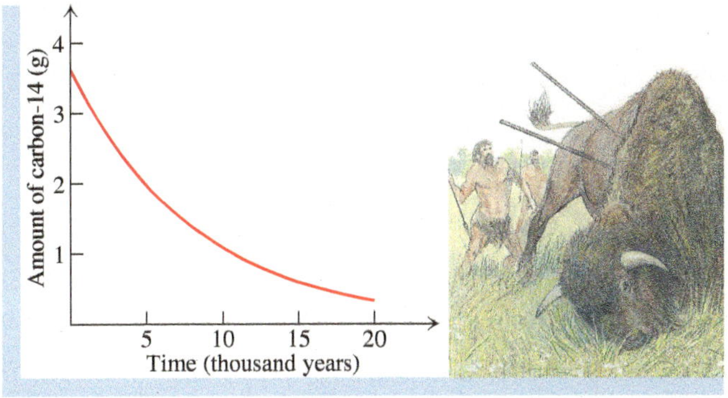


Figure 4.19

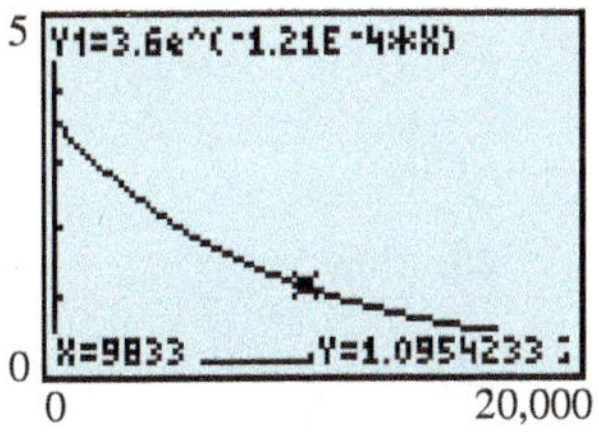


Figure 4.20

The calculator graph in Fig. 4.20 can be used to find the amount of radioactive substance present after 9833 years.

▸**TRY THIS.** If the amount (in grams) of carbon-14 present after $t$ years is given by $A = 2.9e^{rt}$, where $r = -1.21 \times 10^{-4}$, then what is the initial amount and what amount is present after 6500 years? ■

The formula for radioactive decay is used to determine the age of ancient objects such as bones, campfire charcoal, and meteorites. To date objects, we must solve the radioactive decay formula for $t$, which will be done in Section 4.4 after we have studied logarithms.

## FOR thought... True or False? Explain.

1. The function $f(x) = (-2)^x$ is an exponential function.
2. The function $f(x) = 2^x$ is invertible.
3. If $2^x = \frac{1}{8}$, then $x = -3$.
4. If $f(x) = 3^x$, then $f(0.5) = \sqrt{3}$.
5. If $f(x) = e^x$, then $f(0) = 1$.
6. If $f(x) = e^x$ and $f(t) = e^2$, then $t = 2$.
7. The $x$-axis is a horizontal asymptote for the graph of $y = e^x$.
8. The function $f(x) = (0.5)^x$ is increasing.
9. The functions $f(x) = 4^{x-1}$ and $g(x) = (0.25)^{1-x}$ have the same graph.
10. $2^{1.73} = \sqrt[100]{2^{173}}$

## EXERCISES 4.1

*Fill in the blank.*

1. Functions that involve some combination of basic arithmetic operations, powers, or roots are ________ functions.
2. Exponential and logarithmic functions are ________ functions.
3. A function of the form $f(x) = a^x$ where $a$ and $x$ are real numbers with $a > 0$ and $a \neq 1$ is a(n) ________ function.
4. The domain of $f(x) = a^x$ for $a > 0$ is the set of ________.
5. The function $f(x) = a^x$ is ________ if $a > 1$ and ________ if $0 < a < 1$.
6. The graph of $f(x) = a^x$ has the $x$-axis as a(n) ________.
7. The ________ of $f(x) = a^x$ is $(0, \infty)$.
8. The ________ of functions consists of all functions of the form $f(x) = b \cdot a^{x-h} + k$.

*Evaluate each exponential expression without using a calculator.*

9. $3^3$
10. $2^5$
11. $-2^0$
12. $-4^0$
13. $2^{-3}$
14. $3^{-2}$
15. $\left(\frac{1}{2}\right)^{-4}$
16. $\left(\frac{1}{3}\right)^{-2}$
17. $8^{2/3}$
18. $9^{3/2}$
19. $-9^{-3/2}$
20. $-4^{-3/2}$

*Let $f(x) = 3^x$, $g(x) = 2^{1-x}$, and $h(x) = (1/4)^x$. Find the following values.*

21. $f(2)$
22. $f(4)$
23. $f(-2)$
24. $f(-3)$
25. $g(2)$
26. $g(1)$
27. $g(-2)$
28. $g(-3)$
29. $h(-1)$
30. $h(-2)$
31. $h(-1/2)$
32. $h(3/2)$

*Sketch the graph of each function by finding at least three ordered pairs on the graph. State the domain, the range, and whether the function is increasing or decreasing.*

33. $f(x) = 5^x$
34. $f(x) = 4^x$
35. $y = 10^{-x}$
36. $y = e^{-x}$
37. $f(x) = (1/4)^x$
38. $f(x) = (0.2)^x$

*Use a graph or a table to find each limit.*

39. $\lim_{x\to\infty} 3^x$
40. $\lim_{x\to-\infty} 3^x$
41. $\lim_{x\to\infty} 5^{-x}$
42. $\lim_{x\to-\infty} 5^{-x}$
43. $\lim_{x\to\infty} \left(\frac{1}{3}\right)^x$
44. $\lim_{x\to-\infty} \left(\frac{1}{3}\right)^x$
45. $\lim_{x\to-\infty} e^{-x}$
46. $\lim_{x\to\infty} e^{-x}$

*Use transformations to help you graph each function. Identify the domain, range, and horizontal asymptote. Determine whether the function is increasing or decreasing.*

**47.** $f(x) = 2^x - 3$

**48.** $f(x) = 3^{-x} + 1$

**49.** $f(x) = 2^{x+3} - 5$

**50.** $f(x) = 3^{1-x} - 4$

**51.** $y = -2^{-x}$

**52.** $y = -10^{-x}$

**53.** $y = 1 - 2^x$

**54.** $y = -1 - 2^{-x}$

**55.** $f(x) = 0.5 \cdot 3^{x-2}$

**56.** $f(x) = -0.1 \cdot 5^{x+4}$

**57.** $y = 500(0.5)^x$

**58.** $y = 100 \cdot 2^x$

*Write the equation of each graph in its final position.*

**59.** The graph of $y = 2^x$ is translated five units to the right and then two units downward.

**60.** The graph of $y = e^x$ is translated three units to the left and then one unit upward.

**61.** The graph of $y = (1/4)^x$ is translated one unit to the right, reflected in the $x$-axis, and then translated two units downward.

**62.** The graph of $y = 10^x$ is translated three units upward, two units to the left, and then reflected in the $x$-axis.

*Solve each equation.*

**63.** $2^x = 64$

**64.** $5^x = 1$

**65.** $10^x = 0.1$

**66.** $10^{2x} = 1000$

**67.** $-3^x = -27$

**68.** $-2^x = -\dfrac{1}{2}$

**69.** $3^{-x} = 9$

**70.** $2^x = \dfrac{1}{8}$

**71.** $8^x = 2$

**72.** $9^x = 3$

**73.** $e^x = \dfrac{1}{e^2}$

**74.** $e^{-x} = \dfrac{1}{e}$

**75.** $\left(\dfrac{1}{2}\right)^x = 8$

**76.** $\left(\dfrac{2}{3}\right)^x = \dfrac{9}{4}$

**77.** $10^{x-1} = 0.01$

**78.** $10^{|x|} = 1000$

**79.** $2 \cdot 2^{2x} = 4^x + 64$

**80.** $\left(\dfrac{4}{9}\right)^x \cdot \left(\dfrac{8}{27}\right)^{1-x} = \dfrac{2}{3}$

*Let $f(x) = 2^x$, $g(x) = (1/3)^x$, $h(x) = 10^x$, and $m(x) = e^x$. Find the value of $x$ in each equation.*

**81.** $f(x) = 4$

**82.** $f(x) = 32$

**83.** $f(x) = \dfrac{1}{2}$

**84.** $f(x) = 1$

**85.** $g(x) = 1$

**86.** $g(x) = 9$

**87.** $g(x) = 27$

**88.** $g(x) = \dfrac{1}{9}$

**89.** $h(x) = 1000$

**90.** $h(x) = 10^5$

**91.** $h(x) = 0.1$

**92.** $h(x) = 0.0001$

**93.** $m(x) = e$

**94.** $m(x) = e^3$

**95.** $m(x) = \dfrac{1}{e}$

**96.** $m(x) = 1$

*Fill in the missing coordinate in each ordered pair so that the pair is a solution to the given equation.*

**97.** $y = 3^x$ $(2,\ ),(\ ,3),(-1,\ ),(\ ,1/9)$

**98.** $y = 10^x$ $(3,\ ),(\ ,1),(-1,\ ),(\ ,0.01)$

**99.** $f(x) = 5^{-x}$ $(0,\ ),(\ ,25),(-1,\ ),(\ ,1/5)$

**100.** $f(x) = e^{-x}$ $(1,\ ),(\ ,e),(0,\ ),(\ ,e^2)$

**101.** $f(x) = -2^x$ $(4,\ ),(\ ,-1/4),(-1,\ ),(\ ,-32)$

**102.** $f(x) = -(1/4)^{x-1}$ $(3,\ ),(\ ,-4),(-1,\ ),(\ ,-1/16)$

*Solve each problem. When needed, use 365 days per year and 30 days per month.*

**103.** *Periodic Compounding* A deposit of \$5000 earns 8% annual interest. Find the amount in the account at the end of 6 years and the amount of interest earned during the 6 years if the interest is compounded

**a.** annually

**b.** quarterly

**c.** monthly

**d.** daily.

**104.** *Periodic Compounding* Melinda invests her \$80,000 winnings from Publishers Clearing House at a 9% annual percentage rate. Find the amount of the investment at the end of 20 years and the amount of interest earned during the 20 years if the interest is compounded

**a.** annually

**b.** quarterly

**c.** monthly

**d.** daily.

**105.** *Compounded Continuously* The Lakewood Savings Bank pays 8% annual interest compounded continuously. How much will a deposit of \$5000 amount to for each time period?

HINT Convert months to days.

**a.** 6 years

**b.** 8 years 3 months

**c.** 5 years 4 months 22 days

**d.** 20 years 321 days

**106.** *Compounding Continuously* The Commercial Federal Credit Union pays $6\frac{3}{4}$% annual interest compounded continuously. How much will a deposit of \$9000 amount to for each time period?

HINT Convert months to days.

**a.** 13 years

**b.** 12 years 8 months

**c.** 10 years 6 months 14 days

**d.** 40 years 66 days

**107.** *Present Value Compounding Daily* A credit union pays 6.5% annual interest compounded daily. What deposit today (present value) would amount to \$3000 in 5 years and 4 months?
HINT Solve the compound interest formula for $P$.

**108.** *Higher Yields* Federal Savings and Loan offers $6\frac{3}{4}\%$ annual interest compounded daily on certificates of deposit. What dollar amount deposited now would amount to \$40,000 in 3 years and 2 months?

**109.** *Present Value Compounding Continuously* Peoples Bank offers 5.42% compounded continuously on CDs. What amount invested now would grow to \$20,000 in 30 years.
HINT Solve the continuous compounding formula for $P$.

**110.** *Saving for Retirement* An investor wants to have a retirement nest egg of \$100,000 and estimates that her investment now will grow at 3% compounded continuously for 40 years. What amount should she invest now to achieve this goal?

**111.** *Working by the Hour* One million dollars is deposited in an account paying 6% compounded continuously.

**a.** What amount of interest will it earn in its first hour on deposit?
HINT One hour is 1/8760 of a year.

**b.** What amount of interest will it earn during its 500th hour on deposit?
HINT Subtract the amount for $t = 499$ from the amount for $t = 500$.

**112.** *National Debt* The national debt was about \$10 trillion in 2008.

**a.** If the United States paid 5.5% interest compounded continuously on the debt, then what amount of interest does the government pay in one day?

**b.** How much is saved in one day if the interest were 5.5% compounded daily?

**113.** *Radioactive Decay* The number of grams of a certain radioactive substance present at time $t$ is given by the formula $A = 200e^{-0.001t}$, where $t$ is the number of years. How many grams are present at time $t = 0$? How many grams are present at time $t = 500$?

**114.** *Population Growth* The function $P = 2.4e^{0.03t}$ models the size of the population of a small country, where $P$ is in millions of people in the year $2000 + t$.

**a.** What was the population in 2000?

**b.** Use the formula to estimate the population to the nearest tenth of a million in 2020.

**115.** *Facebook Users* The function $U = 0.1e^{1.45t}$ models the number of Facebook users, where $U$ is in millions of people and $t$ is the number of years since 2005.

**a.** How many Facebook users (to the nearest hundred thousand) were there in 2006?

**b.** How many Facebook users (to the nearest million) were there in 2011?

**116.** *Facebook Revenue* The function $R = 52e^{0.91t}$ models the revenue of Facebook, where $R$ is in millions of dollars and $t$ is the number of years since 2006.

**a.** What was Facebook's revenue in 2006?

**b.** What was Facebook's revenue (to the nearest million) in 2010?

**117.** *Cell Phone Subscribers* The number of cell phone subscribers (in millions) in the United States is given in the accompanying table (www.infoplease.com).

**Table for Exercise 117**

| Year | Cell Phone Subscribers (millions) |
|---|---|
| 1990 | 5.3 |
| 1995 | 33.8 |
| 2000 | 109.4 |
| 2005 | 207.9 |
| 2006 | 233.0 |
| 2007 | 255.4 |
| 2008 | 271.6 |

**a.** Use the exponential regression feature of a graphing calculator to find an equation of the form $y = ab^x$ that fits the data, where $x$ is the years since 1990.

**b.** Judging from the graph of the data and the curve, does the exponential model look like a good model?

**c.** Use the equation to estimate the number of cell phone subscribers in 2010.

**118.** *Newspaper Circulation* As more and more people turn to Internet and 24-hour cable network news, newspaper circulation has declined. The accompanying table gives the total number of daily papers sold in the United States (www.naa.org).

**Table for Exercise 118**

| Year | Papers Sold (millions) |
|---|---|
| 1990 | 62.3 |
| 1995 | 58.1 |
| 2000 | 55.8 |
| 2005 | 53.3 |
| 2006 | 52.3 |
| 2007 | 51.2 |
| 2008 | 49.6 |

**a.** Use the exponential regression feature of a graphing calculator to find an equation of the form $y = ab^x$ that fits the data, where $x$ is the years since 1990.

**b.** Judging from the graph of the data and the curve, does the exponential model look like a good model for these data?

**c.** Use the equation to estimate the daily newspaper circulation in 2012.

**119.** *Position of a Football* The football is on the 10-yard line. Several penalties in a row are given, and each penalty moves the ball half the distance to the closer goal line. Write a formula that gives the ball's position $P$ after the $n$th such penalty.
**HINT** First write some ordered pairs starting with $(1, 5)$.

**120.** *Cost of a Parking Ticket* The cost of a parking ticket on campus is \$15 for the first offense. Given that the cost doubles for each additional offense, write a formula for the cost $C$ as a function of the number of tickets $n$.
**HINT** First write some ordered pairs starting with $(1, 15)$.

**121.** *Challenger Disaster* Using data on O-ring damage from 24 previous space shuttle launches, Professor Edward R. Tufte of Yale University concluded that the number of O-rings damaged per launch is an exponential function of the temperature at the time of the launch. If NASA had used a model such as $n = 644e^{-0.15t}$, where $t$ is the Fahrenheit temperature at the time of launch and $n$ is the number of O-rings damaged, then the tragic end to the flight of the space shuttle *Challenger* might have been avoided. Using this model, find the number of O-rings that would be expected to fail at 31°F, the temperature at the time of the *Challenger* launch on January 28, 1986.

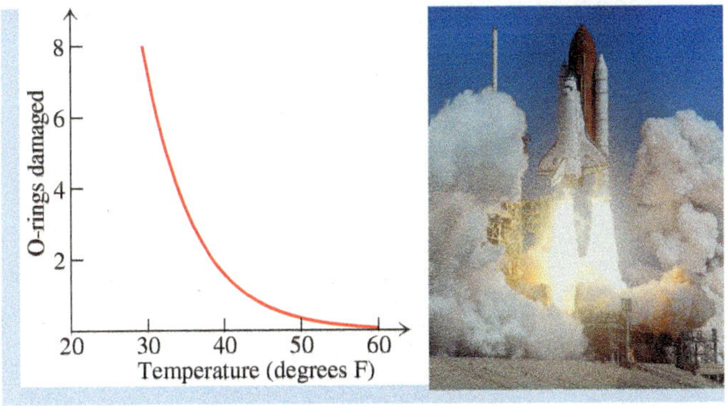


**Figure for Exercise 121**

**122.** *Manufacturing Cost* The cost in dollars for manufacturing $x$ units of a certain drug is given by the function $C(x) = xe^{0.001x}$. Find the cost for manufacturing 500 units. Find the function $AC(x)$ that gives the average cost per unit for manufacturing $x$ units. What happens to the average cost per unit as $x$ gets larger and larger?

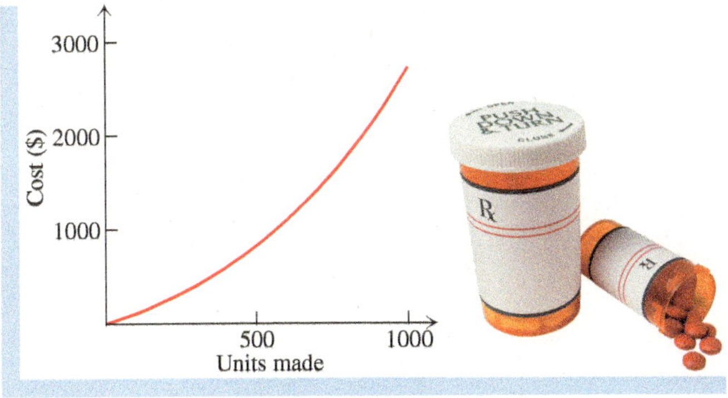


**Figure for Exercise 122**

## FOR WRITING/DISCUSSION

**123.** *Cooperative Learning* Work in a small group to write a summary (including drawings) of the types of graphs that can be obtained for exponential functions of the form $y = a^x$ for $a > 0$ and $a \neq 1$.

**124.** *Group Toss* Have all students in your class stand and toss a coin. Those that get heads sit down. Those that are left standing toss again and all who obtain heads must sit down. Repeat until no one is left standing. For each toss record the number of coins that are tossed—for example, $(1, 30)$, $(2, 14)$, and so on. Do not include a pair with zero coins tossed. Enter the data into a graphing calculator and use exponential regression to find an equation of the form $y = a \cdot b^x$ that fits the data. What percent of those standing did you expect would sit down after each toss? Is the value of $b$ close to this number?

## ▶ RETHINKING

**125.** Find the equations of the horizontal and vertical asymptotes for the graph of the function $f(x) = \dfrac{2x - 3}{x + 7}$.

**126.** Solve $|2x - 5| = 7x$.

**127.** Solve $|2x - 5| > 7$.

**128.** Solve $2x - 5 = 7x$.

**129.** Solve $2x - 5 = y$ for $x$.

**130.** Find the equation of the line through $(4, 3)$ that is parallel to $y = 2x - 99$

## THINKING OUTSIDE THE BOX XXX & XXXI

*Swim Meet* Two swimmers start out from opposite sides of a pool. Each swims the length of the pool and back at a constant rate. They pass each other for the first time 40 feet from one side of the pool and for the second time 45 feet from the other side of the pool. What is the length of the pool?

*One-seventh to One* Six matches are arranged to form the fraction one-seventh as shown here:

$$\frac{\text{I}}{\text{VII}}$$

Move one match, but not the fraction bar, to make a fraction that is equal to one.

## POP QUIZ 4.1

1. What is $f(4)$ if $f(x) = -2^x$?
2. Is $f(x) = 3^{-x}$ increasing or decreasing?
3. Find the domain and range for $y = e^{x-1} + 2$.
4. What is the horizontal asymptote for $y = 2^x - 1$?
5. Solve $(1/4)^x = 64$.
6. What is $a$, if $f(a) = 8$ and $f(x) = 2^{-x}$?
7. If \$1000 earns 4% annual interest compounded quarterly, then what is the amount after 20 years?
8. Find the amount in the last problem if the interest is compounded continuously.

## LINKING concepts... For Individual or Group Explorations

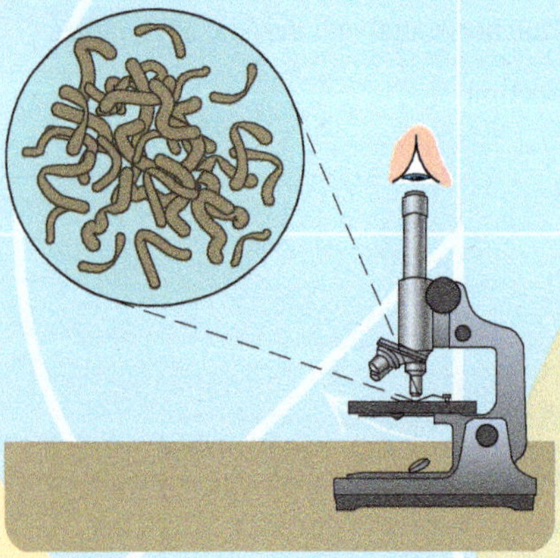

### Comparing Exponential and Linear Models

*The function $f(t) = 300e^{0.5t}$ gives the number of bacteria present in a culture t hours after the start of an experiment in which the bacteria are growing continuously at a rate of 50% per hour. Let $A[a, b]$ represent the average rate of change of f on the time interval $[a, b]$.*

**a)** Fill in the following table.

| Interval $[a, b]$ | $f(a)$ | $A[a, b]$ | $A[a, b]/f(a)$ |
|---|---|---|---|
| $[3.00, 3.05]$ | | | |
| $[7.50, 7.51]$ | | | |
| $[8.623, 8.624]$ | | | |

**b)** What are the units for the quantity $A[a, b]$?

**c)** What can you conjecture about the ratio $A[a, b]/f(a)$?

**d)** Test your conjecture on a few more intervals of various lengths. Explain how the length of the interval affects the ratio.

**e)** State your conjecture in terms of variation (Section 2.6). What is the constant of proportionality?

**f)** Suppose the bacteria were growing in a linear manner, say $f(t) = 800t + 300$. Make a table like the given table and make a conjecture about the ratio $A[a, b]/f(a)$.

**g)** Explain how your conclusions about average rate of change can be used to justify an exponential model as better than a linear model for modeling growth of a bacteria (or human) population.

# 4.2 Logarithmic Functions and Their Applications

Since exponential functions are one-to-one functions (Section 4.1), they are invertible. In this section we will study the inverses of the exponential functions.

## The Definition

The inverses of the exponential functions are called **logarithmic functions.** Since $f^{-1}$ is a general name for an inverse function, we adopt a more descriptive notation for these inverses. If $f(x) = a^x$, then instead of $f^{-1}(x)$, we write $\log_a(x)$ for the inverse of the base-$a$ exponential function. We read $\log_a(x)$ as "log of $x$ with base $a$," and we call the expression $\log_a(x)$ a **logarithm.**

The meaning of $\log_a(x)$ will be clearer if we consider the exponential function

$$f(x) = 2^x$$

as an example. Since $f(3) = 2^3 = 8$, the base-2 exponential function pairs the exponent 3 with the value of the exponential expression 8. Since the function $\log_2(x)$ reverses that pairing, we have $\log_2(8) = 3$. So $\log_2(8)$ is the exponent that is used on the base 2 to obtain 8. *In general, $\log_a(x)$ is the exponent that is used on the base $a$ to obtain the value $x$.*

### Foreshadowing Calculus

In calculus the logarithm functions can be defined geometrically (as area under a certain curve). Then exponential functions are defined as the inverses of the logarithm functions.

**Definition: Logarithmic Function**

For $a > 0$ and $a \neq 1$, the **logarithmic function with base $a$** is denoted $f(x) = \log_a(x)$, where

$$y = \log_a(x) \quad \text{if and only if} \quad a^y = x.$$

### EXAMPLE 1 Evaluating logarithmic functions

Find the indicated values of the logarithmic functions.

**a.** $\log_3(9)$ **b.** $\log_2\left(\frac{1}{4}\right)$ **c.** $\log_{1/2}(8)$

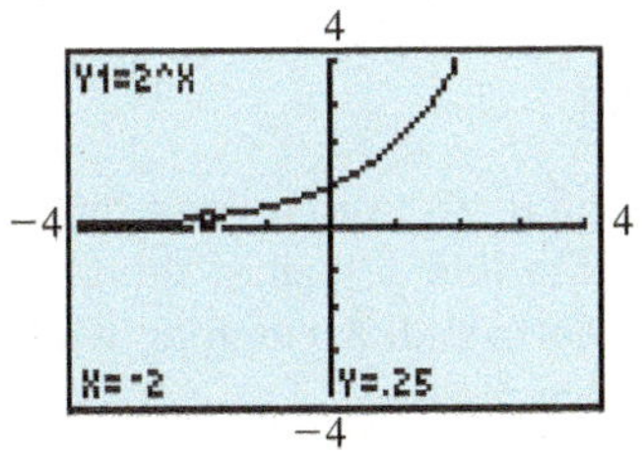


Figure 4.21

**Solution**

**a.** By the definition of logarithm, $\log_3(9)$ is the exponent that is used on the base 3 to obtain 9. Since $3^2 = 9$, we have $\log_3(9) = 2$.

**b.** Since $\log_2(1/4)$ is the exponent that is used on the base 2 to obtain $1/4$, we try various powers of 2 until we find $2^{-2} = 1/4$. So $\log_2(1/4) = -2$.

You can use the graph of $y = 2^x$ as in Fig. 4.21 to check that $2^{-2} = 1/4$. □

**c.** Since $\log_{1/2}(8)$ is the exponent that is used on a base of $1/2$ to obtain 8, we try various powers of $1/2$ until we find $(1/2)^{-3} = 8$. So $\log_{1/2}(8) = -3$.

**▸TRY THIS.** Find $\log_4(1/16)$. ■

Since the exponential function $f(x) = a^x$ has domain $(-\infty, \infty)$ and range $(0, \infty)$, the logarithmic function $f(x) = \log_a(x)$ has domain $(0, \infty)$ and range $(-\infty, \infty)$. So there are no logarithms of negative numbers or zero. Expressions such as $\log_2(-4)$ and $\log_3(0)$ are undefined. Note that $\log_a(1) = 0$ for any base $a$, because $a^0 = 1$ for any base $a$.

There are two bases that are used more frequently than the others; they are 10 and $e$. The notation $\log_{10}(x)$ is abbreviated $\log(x)$, and $\log_e(x)$ is abbreviated $\ln(x)$.

### Definition: Common and Natural Logarithms

The **common logarithmic function** is denoted as $f(x) = \log(x)$, where

$$y = \log(x) \quad \text{if and only if} \quad 10^y = x.$$

The **natural logarithmic function** is denoted as $f(x) = \ln(x)$, where

$$y = \ln(x) \quad \text{if and only if} \quad e^y = x.$$

Figure 4.22

Most scientific calculators have function keys for the exponential functions $10^x$ and $e^x$ and their inverses $\log(x)$ and $\ln(x)$. Natural logarithms are also called **Napierian logarithms** after John Napier (1550–1617).

Note that $\log(76)$ is approximately 1.8808 and $10^{1.8808}$ is approximately 76 as shown in Fig. 4.22. If you use more digits for $\log(76)$ as the power of 10, then the calculator gets closer to 76. □

You can use a calculator to find common or natural logarithms, but you should know how to find the values of logarithms such as those in Examples 1 and 2 without using a calculator.

## EXAMPLE 2 Evaluating logarithmic functions

Find the indicated values of the logarithmic functions without a calculator. Use a calculator to check.

**a.** $\log(1000)$ **b.** $\ln(1)$ **c.** $\ln(-6)$

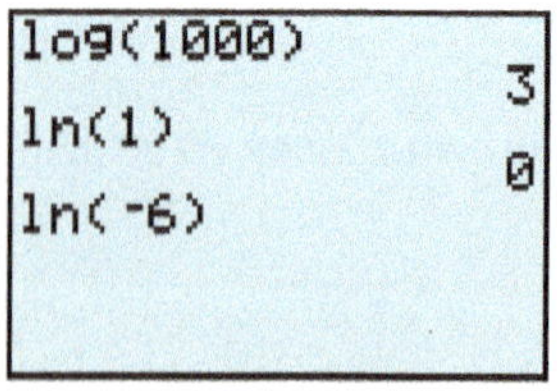


Figure 4.23

### Solution

**a.** To find $\log(1000)$, we must find the exponent that is used on the base 10 to obtain 1000. Since $10^3 = 1000$, we have $\log(1000) = 3$.

**b.** Since $e^0 = 1$, $\ln(1) = 0$.

**c.** The expression $\ln(-6)$ is undefined because $-6$ is not in the domain of the natural logarithm function. There is no power of $e$ that results in $-6$.

The calculator results for parts (a) and (b) are shown in Fig. 4.23. If you ask a calculator for $\ln(-6)$, it will give you an error message.

▸**TRY THIS.** Find $\log(100)$. ■

**HISTORICAL NOTE**

John Napier (1550–1617) was a Scottish mathematician, physicist, and astronomer/astrologer. He is best remembered as the inventor of logarithms and Napier's bones (a calculating device) and for popularizing the use of the decimal point. Napier described a method of multiplication and division using metal plates, which was the direct antecedent of the slide rule.

## Graphs of Logarithmic Functions

The functions $y = a^x$ and $y = \log_a(x)$ for $a > 0$ and $a \neq 1$ are inverse functions. So the graph of $y = \log_a(x)$ is a reflection about the line $y = x$ of the graph of $y = a^x$. The graph of $y = a^x$ has the $x$-axis as its horizontal asymptote, while the graph of $y = \log_a(x)$ has the $y$-axis as its vertical asymptote.

## EXAMPLE 3 Graph of a base-$a$ logarithmic function with $a > 1$

Sketch the graphs of $y = 2^x$ and $y = \log_2(x)$ on the same coordinate system. State the domain and range of each function.

### Solution

Since these two functions are inverses of each other, the graph $y = \log_2(x)$ is a reflection of the graph of $y = 2^x$ about the line $y = x$. Make a table of ordered pairs for each function:

| $x$ | $-1$ | 0 | 1 | 2 |
|---|---|---|---|---|
| $y = 2^x$ | 1/2 | 1 | 2 | 4 |

| $x$ | 1/2 | 1 | 2 | 4 |
|---|---|---|---|---|
| $y = \log_2(x)$ | $-1$ | 0 | 1 | 2 |

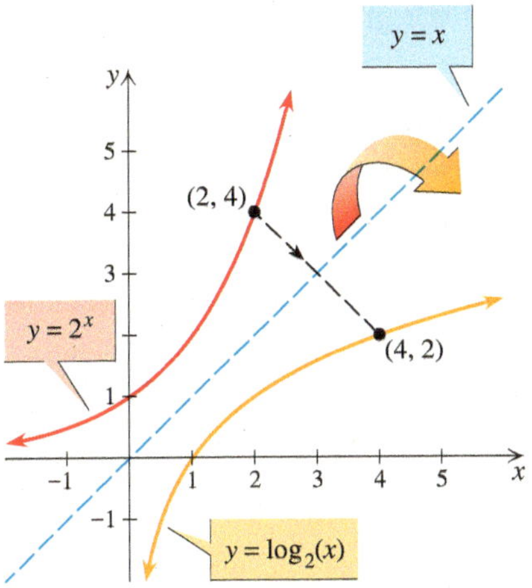


Figure 4.24

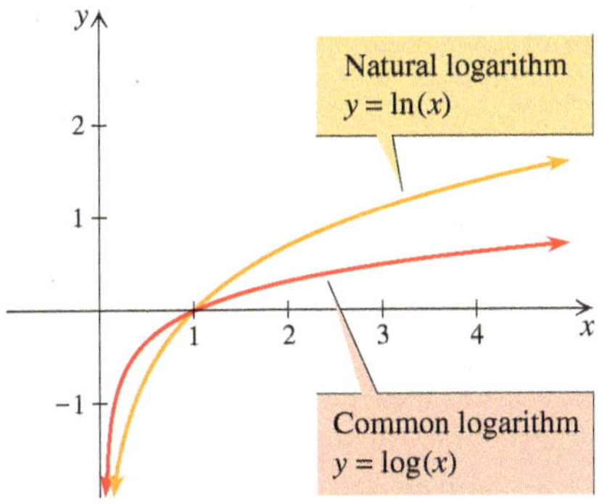


Figure 4.25

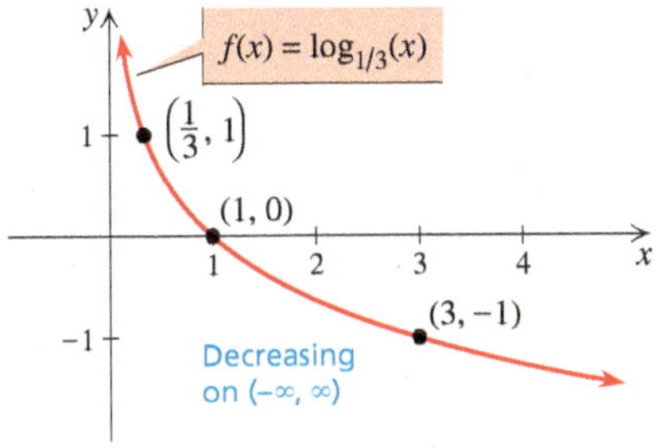


Figure 4.26

Sketch $y = 2^x$ and $y = \log_2(x)$ through the points given in the table as shown in Fig. 4.24. Keep in mind that $y = \log_2(x)$ is a reflection of $y = 2^x$ about the line $y = x$. The important features of the two functions are given in the following table:

| | Domain | Range | Asymptotes | Limits |
|---|---|---|---|---|
| $y = 2^x$ | $(-\infty, \infty)$ | $(0, \infty)$ | $x$-axis | $\lim_{x\to-\infty} 2^x = 0$, $\lim_{x\to\infty} 2^x = \infty$ |
| $y = \log_2(x)$ | $(0, \infty)$ | $(-\infty, \infty)$ | $y$-axis | $\lim_{x\to0^+} \log_2(x) = -\infty$, $\lim_{x\to\infty} \log_2(x) = \infty$ |

▸**TRY THIS.** Graph $f(x) = \log_6(x)$ and state the domain and range. ■

Note that the function $y = \log_2(x)$ graphed in Fig. 4.24 is one-to-one by the horizontal line test and it is an increasing function. The graphs of the common logarithm function $y = \log(x)$ and the natural logarithm function $y = \ln(x)$ are shown in Fig. 4.25 and they are also increasing functions. The function $y = \log_a(x)$ is increasing if $a > 1$. By contrast, if $0 < a < 1$, the function $y = \log_a(x)$ is decreasing, as illustrated in the next example.

### EXAMPLE 4 Graph of a base-*a* logarithmic function with $0 < a < 1$

Sketch the graph of $f(x) = \log_{1/3}(x)$ and state its domain and range.

#### Solution

Make a table of ordered pairs for the function:

| $x$ | 1/9 | 1/3 | 1 | 3 | 9 |
|---|---|---|---|---|---|
| $y = \log_{1/3}(x)$ | 2 | 1 | 0 | −1 | −2 |

Sketch the curve through these points as shown in Fig. 4.26 The $y$-axis is a vertical asymptote for the curve. In terms of limits, $\lim_{x\to0^+} \log_{1/3}(x) = \infty$ and $\lim_{x\to\infty} \log_{1/3}(x) = -\infty$. The domain is $(0, \infty)$, and its range is $(-\infty, \infty)$.

▸**TRY THIS.** Graph $f(x) = \log_{1/6}(x)$ and state the domain and range. ■

The logarithmic functions have properties corresponding to the properties of the exponential functions stated in Section 4.1.

## Properties of Logarithmic Functions

The logarithmic function $f(x) = \log_a(x)$ has the following properties:

1. The function $f$ is increasing for $a > 1$ and decreasing for $0 < a < 1$.
2. The $x$-intercept of the graph of $f$ is $(1, 0)$.
3. The graph has the $y$-axis as a vertical asymptote.
4. The domain of $f$ is $(0, \infty)$, and the range of $f$ is $(-\infty, \infty)$.
5. The function $f$ is one-to-one.
6. The functions $f(x) = \log_a(x)$ and $f(x) = a^x$ are inverse functions.

## The Logarithmic Family of Functions

A function of the form $f(x) = \log_a(x)$ is a logarithmic function. Any function of the form $g(x) = b \cdot \log_a(x - h) + k$ is a member of the **logarithmic family** of functions. The graph of $g$ is a transformation of the graph of $f$, as discussed in Section 2.3:

The graph of $f$ moves to the left if $h < 0$ or to the right if $h > 0$.
The graph of $f$ moves upward if $k > 0$ or downward if $k < 0$.
The graph of $f$ is stretched if $b > 1$ and shrunk if $0 < b < 1$.
The graph of $f$ is reflected in the $x$-axis if $b$ is negative.

## EXAMPLE 5 Graphing members of the logarithmic family

Sketch the graph of each function and state its domain and range.

**a.** $y = \log_2(x - 1)$ **b.** $f(x) = -\frac{1}{2}\log_2(x + 3)$

### Solution

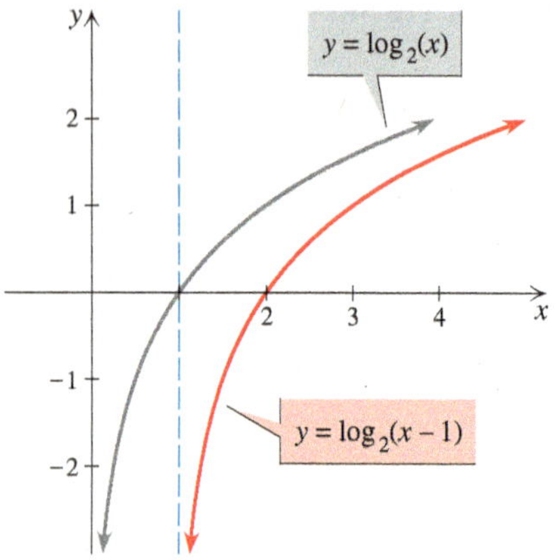


Figure 4.27

**a.** The graph of $y = \log_2(x - 1)$ is obtained by translating the graph of $y = \log_2(x)$ to the right one unit. Since the domain of $y = \log_2(x)$ is $(0, \infty)$, the domain of $y = \log_2(x - 1)$ is $(1, \infty)$. The line $x = 1$ is the vertical asymptote. Calculate a few ordered pairs to get an accurate graph.

$x > 1$

| $x$ | 1.5 | 2 | 3 | 5 |
|---|---|---|---|---|
| $y = \log_2(x - 1)$ | $-1$ | 0 | 1 | 2 |

$y$ is any real number

Sketch the curve as shown in Fig. 4.27 The range is $(-\infty, \infty)$.

**b.** The graph of $f$ is obtained by translating the graph of $y = \log_2(x)$ to the left three units. The vertical asymptote for $f$ is the line $x = -3$. Multiplication by $-1/2$ shrinks the graph and reflects it in the $x$-axis. Calculate a few ordered pairs for accuracy.

$x > -3$

| $x$ | $-2$ | $-1$ | 1 | 5 |
|---|---|---|---|---|
| $y = -\frac{1}{2}\log_2(x + 3)$ | 0 | $-1/2$ | $-1$ | $-3/2$ |

$y$ is any real number

Sketch the curve through these points as shown in Fig. 4.28 The domain of $f$ is $(-3, \infty)$, and the range is $(-\infty, \infty)$.

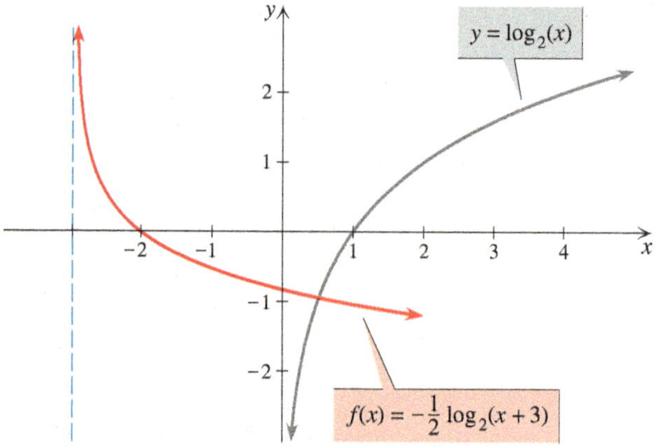


Figure 4.28

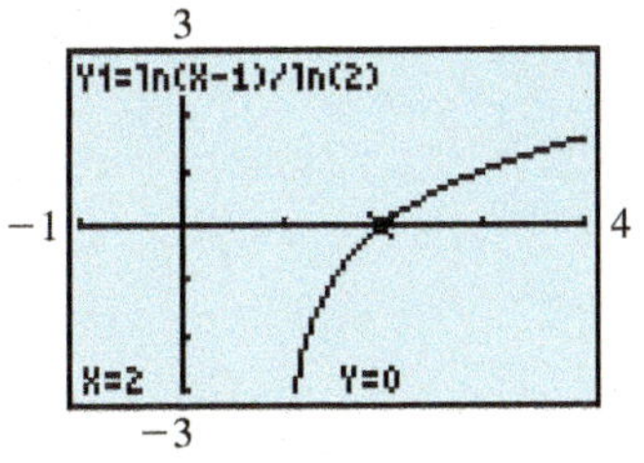


Figure 4.29

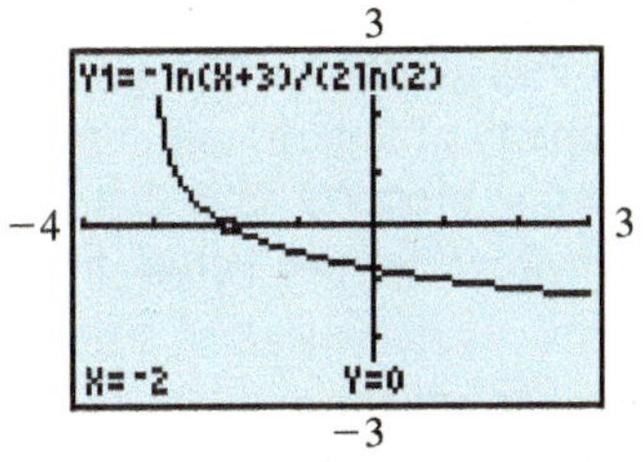


Figure 4.30

▶**TRY THIS.** Graph $f(x) = -\log_2(x + 1)$ and state the domain and range. ■

A graphing calculator typically has keys for log and ln only. To graph logarithmic functions with other bases, use the formula $\log_a(x) = \frac{\ln(x)}{\ln(a)}$, which will be proven in Section 4.3. So to check Example 5(a), graph $y = \frac{\ln(x - 1)}{\ln(2)}$ as shown in Fig. 4.29. For Example 5(b) graph $y = -\frac{1}{2} \cdot \frac{\ln(x + 3)}{\ln(2)}$ or $y = \frac{-\ln(x + 3)}{2 \cdot \ln(2)}$ as in Fig. 4.30.

## Logarithmic and Exponential Equations

Some equations involving logarithms can be solved by writing an equivalent exponential equation, and some equations involving exponents can be solved by writing an equivalent logarithmic equation. Rewriting logarithmic and exponential equations is possible because of the definition of logarithms:

$$y = \log_a(x) \quad \text{if and only if} \quad a^y = x$$

### EXAMPLE 6 Rewriting logarithmic and exponential equations

Write each equation involving logarithms as an equivalent exponential equation, and write each equation involving exponents as an equivalent logarithmic equation.

**a.** $\log_5(625) = 4$ **b.** $\log_3(n) = 5$ **c.** $3^x = 50$ **d.** $e^{x-1} = 9$

#### Solution

**a.** $\log_5(625) = 4$ is equivalent to $5^4 = 625$.
**b.** $\log_3(n) = 5$ is equivalent to $3^5 = n$.
**c.** $3^x = 50$ is equivalent to $x = \log_3(50)$.
**d.** $e^{x-1} = 9$ is equivalent to $x - 1 = \ln(9)$.

▶**TRY THIS.** Write $\log_4(w) = 7$ as an exponential equation. ■

In the next example we find the inverses of an exponential function and a logarithmic function.

### EXAMPLE 7 Finding inverse functions

**a.** Find $f^{-1}$ for $f(x) = \frac{1}{2} \cdot 6^{x-3}$.
**b.** Find $f^{-1}$ for $f(x) = \log_2(x - 1) + 4$.

#### Solution

**a.** Using the switch-and-solve method, we switch $x$ and $y$, then solve for $y$:

$$x = \frac{1}{2} \cdot 6^{y-3} \quad \text{Switch } x \text{ and } y \text{ in } y = \tfrac{1}{2} \cdot 6^{x-3}.$$

$$2x = 6^{y-3} \quad \text{Multiply each side by 2.}$$

$$y - 3 = \log_6(2x) \quad \text{Definition of logarithm}$$

$$y = \log_6(2x) + 3 \quad \text{Add 3 to each side.}$$

So $f^{-1}(x) = \log_6(2x) + 3$.

**b.** Switch $x$ and $y$, then solve for $y$:

$$x = \log_2(y - 1) + 4 \quad \text{Switch } x \text{ and } y \text{ in } y = \log_2(x - 1) + 4.$$

$$x - 4 = \log_2(y - 1) \quad \text{Subtract 4 from each side.}$$

$$y - 1 = 2^{x-4} \quad \text{Definition of logarithm}$$

$$y = 2^{x-4} + 1 \quad \text{Add 1 to each side.}$$

So $f^{-1}(x) = 2^{x-4} + 1.$

▶**TRY THIS.** Find $f^{-1}$ for $f(x) = 3 \cdot e^{5x-1}$. ■

The one-to-one property of exponential functions was used to solve exponential equations in Section 4.1. Likewise, the one-to-one property of logarithmic functions is used in solving logarithmic equations. The one-to-one property says that *if two quantities have the same base-a logarithm, then the quantities are equal.* For example, if $\log_2(m) = \log_2(n)$, then $m = n$.

**One-to-One Property of Logarithms**

For $a > 0$ and $a \neq 1$,

$$\text{if} \quad \log_a(x_1) = \log_a(x_2), \quad \text{then} \quad x_1 = x_2.$$

The one-to-one properties and the definition of logarithm are at present the only new tools that we have for solving equations. Later in this chapter we will develop more properties of logarithms and solve more complicated equations.

## EXAMPLE 8 Solving equations involving logarithms

Solve each equation.

**a.** $\log_3(x) = -2$ **b.** $\log_x(5) = 2$ **c.** $5^x = 9$ **d.** $\ln(x^2) = \ln(3x)$

### Solution

**a.** Use the definition of logarithm to write the equivalent exponential equation.

$$\log_3(x) = -2 \quad \text{Original equation}$$

$$x = 3^{-2} \quad \text{Definition of logarithm}$$

$$= \frac{1}{9}$$

Since $\log_3(1/9) = -2$ is correct, the solution to the equation is $1/9$.

**b.** $\log_x(5) = 2$ Original equation

$$x^2 = 5 \quad \text{Definition of logarithm}$$

$$x = \pm\sqrt{5}$$

Since the base of a logarithm is always nonnegative, the only solution is $\sqrt{5}$.

**c.** $5^x = 9$ Original equation

$$x = \log_5(9) \quad \text{Definition of logarithm}$$

The exact solution to $5^x = 9$ is the irrational number $\log_5(9)$. In the next section we will learn how to find a rational approximation for $\log_5(9)$.

**d.** $\ln(x^2) = \ln(3x)$ Original equation

$$x^2 = 3x \quad \text{One-to-one property of logarithms}$$

$$x^2 - 3x = 0$$

$$x(x - 3) = 0$$

$$x = 0 \quad \text{or} \quad x = 3$$

Checking $x = 0$ in the original equation, we get the undefined expression $\ln(0)$. So the only solution to the equation is 3.

▸**TRY THIS.** Solve $\log_5(2x) = -3$. ■

## EXAMPLE 9 Equations involving common and natural logarithms

Use a calculator to find the value of $x$ rounded to four decimal places.

**a.** $10^x = 50$ **b.** $2e^{0.5x} = 6$

### Solution

Figure 4.31

**a.** $10^x = 50$ Original equation

$x = \log(50)$ Definition of logarithm

$x \approx 1.6990$

Figure 4.31 shows how to find the approximate value of $\log(50)$ and how to check the approximate answer and exact answer. □

**b.** $2e^{0.5x} = 6$ Original equation

$e^{0.5x} = 3$ Divide by 2 to get the form $a^x = y$.

$0.5x = \ln(3)$ Definition of logarithm

$$x = \frac{\ln(3)}{0.5}$$

$x \approx 2.1972$

Figure 4.32

Fig. 4.32 shows how to find the approximate value of $x$ and how to check it in the original equation.

▸**TRY THIS.** Solve $10^{3x} = 70$. ■

## FUNCTION gallery... Exponential and Logarithmic Functions

**Exponential:** $f(x) = a^x$, domain $(-\infty, \infty)$, range $(0, \infty)$

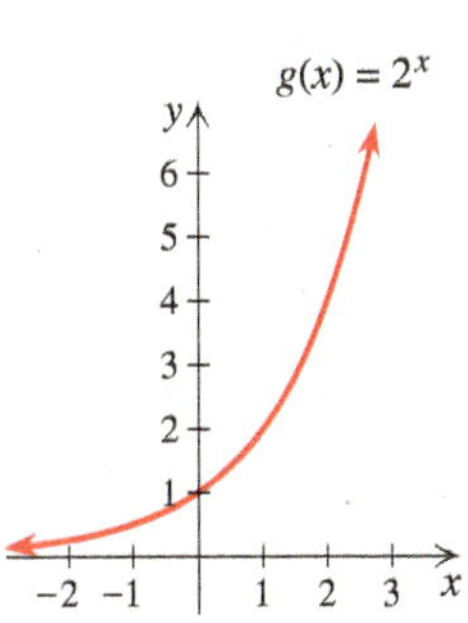


Increasing on $(-\infty, \infty)$
$y$-intercept $(0, 1)$

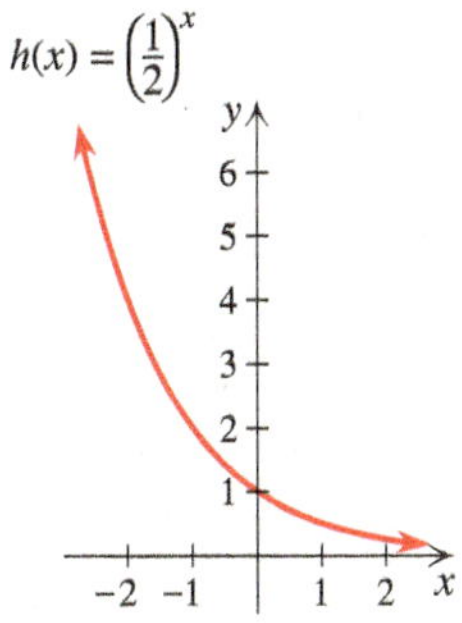


Decreasing on $(-\infty, \infty)$
$y$-intercept $(0, 1)$

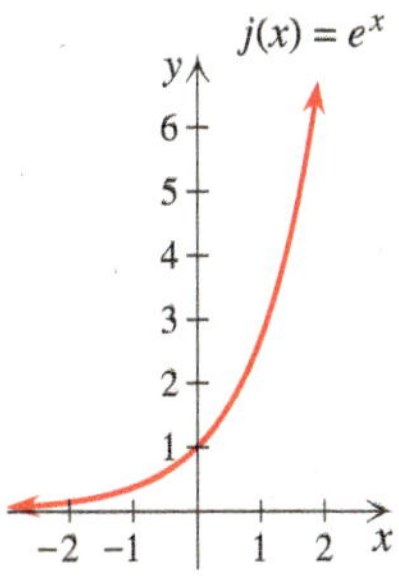


Increasing on $(-\infty, \infty)$
$y$-intercept $(0, 1)$

**Logarithmic:** $f^{-1}(x) = \log_a(x)$, domain $(0, \infty)$, range $(-\infty, \infty)$

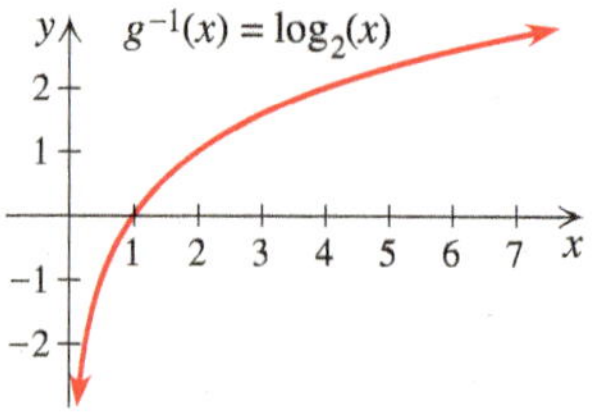


Increasing on $(0, \infty)$
$x$-intercept $(1, 0)$

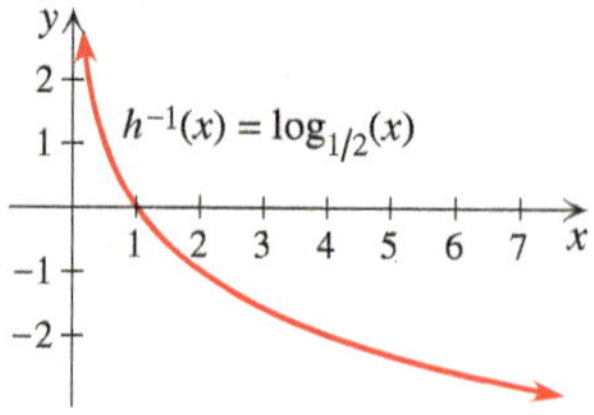


Decreasing on $(0, \infty)$
$x$-intercept $(1, 0)$

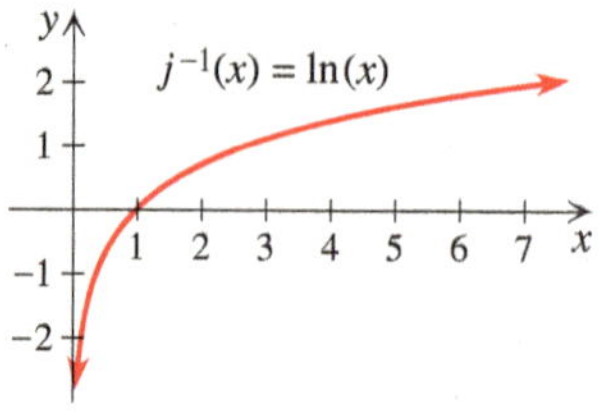


Increasing on $(0, \infty)$
$x$-intercept $(1, 0)$

## Applications

We saw in Section 4.1 that if a principal of $P$ dollars earns interest at an annual rate $r$ compounded continuously, then the amount after $t$ years is given by

$$A = Pe^{rt}.$$

If either the rate or the time is the only unknown in this formula, then the definition of logarithms can be used to solve for the rate or the time.

### EXAMPLE 10 Finding the time in a continuous compounding problem

If \$8000 is invested at 9% compounded continuously, then how long will it take for the investment to grow to \$20,000?

**Solution**

Use the formula $A = Pe^{rt}$ with $A = \$20{,}000$, $P = \$8000$, and $r = 0.09$:

$$20{,}000 = 8000e^{0.09t}$$

$$2.5 = e^{0.09t} \qquad \text{Divide by 8000 to get the form } y = a^x.$$

$$0.09t = \ln(2.5) \qquad \text{Definition of logarithm: } y = a^x \text{ if and only if } x = \log_a(y)$$

$$t = \frac{\ln(2.5)}{0.09}$$

$$\approx 10.181 \text{ years}$$

We can multiply 365 by 0.181 to get approximately 66 days. So the investment grows to \$20,000 in approximately 10 years and 66 days.

You can use a graphing calculator to check as shown in Fig. 4.33.

**Figure 4.33**

▶**TRY THIS.** How long does it take for \$4 to grow to \$12 at 5% compounded continuously? ■

The formula $A = Pe^{rt}$ was introduced to model the continuous growth of money. However, this type of formula is used in a wide variety of applications. In the following exercises you will find problems involving population growth, declining forests, and global warming.

## FOR thought... True or False? Explain.

1. The first coordinate of an ordered pair in an exponential function is a logarithm.
2. $\log_{100}(10) = 2$
3. If $f(x) = \log_3(x)$, then $f^{-1}(x) = 3^x$.
4. $10^{\log(1000)} = 1000$
5. The domain of $f(x) = \ln(x)$ is $(-\infty, \infty)$.
6. $\ln(e^{2.451}) = 2.451$
7. For any positive real number $x$, $e^{\ln(x)} = x$.
8. For any base $a$, where $a > 0$ and $a \neq 1$, $\log_a(0) = 1$.
9. $\log(10^3) + \log(10^5) = \log(10^8)$
10. $\log_2(32) - \log_2(8) = \log_2(4)$

## EXERCISES 4.2

*Fill in the blank.*

1. The inverse of an exponential function is a(n) ________ function.
2. A base 10 logarithm is a(n) ________ logarithm.
3. Base $e$ logarithm is a(n) ________ logarithm.
4. The function $f(x) = \log_a(x)$ is ________ if $a > 1$ and ________ if $0 < a < 1$.
5. The $y$-axis is a(n) ________ for the graph of $f(x) = \log_a(x)$.
6. The ________ of the function $f(x) = \log_a(x)$ is $(0, \infty)$.
7. The ________ of functions consists of all functions of the form $f(x) = b \cdot \log_a(x - h) + k$.
8. The ________ property of logarithms indicates that if $\log_a(m) = \log_a(n)$, then $m = n$.

*Determine the number that can be used in place of the question mark to make the equation true.*

9. $2^? = 64$
10. $2^? = 16$
11. $3^? = \frac{1}{81}$
12. $3^? = 1$
13. $16^? = 2$
14. $16^? = 16$
15. $\left(\frac{1}{5}\right)^? = 125$
16. $\left(\frac{1}{5}\right)^? = \frac{1}{125}$

*Find the indicated value of the logarithmic functions.*

17. $\log_2(64)$
18. $\log_2(16)$
19. $\log_3(1/81)$
20. $\log_3(1)$
21. $\log_{16}(2)$
22. $\log_{16}(16)$
23. $\log_{1/5}(125)$
24. $\log_{1/5}(1/125)$
25. $\log(0.1)$
26. $\log(10^6)$
27. $\log(1)$
28. $\log(10)$
29. $\ln(e)$
30. $\ln(0)$
31. $\ln(e^{-5})$
32. $\ln(e^9)$

*Sketch the graph of each function, and state the domain and range of each function.*

33. $y = \log_3(x)$
34. $y = \log_4(x)$
35. $f(x) = \log_5(x)$
36. $g(x) = \log_8(x)$
37. $y = \log_{1/2}(x)$
38. $y = \log_{1/4}(x)$
39. $h(x) = \log_{1/5}(x)$
40. $s(x) = \log_{1/10}(x)$
41. $f(x) = \ln(x - 1)$
42. $f(x) = \log_3(x + 2)$
43. $f(x) = -3 + \log(x + 2)$
44. $f(x) = 4 - \log(x + 6)$
45. $f(x) = -\frac{1}{2}\log(x - 1)$
46. $f(x) = -2 \cdot \log_2(x + 2)$

*Use a graph or a table to find each limit.*

47. $\lim_{x \to \infty} \log_3(x)$
48. $\lim_{x \to 0^+} \log_3(x)$
49. $\lim_{x \to 0^+} \log_{1/2}(x)$
50. $\lim_{x \to \infty} \log_{1/2}(x)$
51. $\lim_{x \to 0^+} \ln(x)$
52. $\lim_{x \to \infty} \ln(x)$
53. $\lim_{x \to \infty} \log(x)$
54. $\lim_{x \to 0^+} \log(x)$

*Write the equation of each graph in its final position.*

55. The graph of $y = \ln(x)$ is translated three units to the right and then four units downward.
56. The graph of $y = \log(x)$ is translated five units to the left and then seven units upward.
57. The graph of $y = \log_2(x)$ is translated five units to the right, reflected in the $x$-axis, and then translated one unit downward.
58. The graph of $y = \log_3(x)$ is translated four units upward, six units to the left, and then reflected in the $x$-axis.

*Write each equation as an equivalent exponential equation.*

**59.** $\log_2(32) = 5$

**60.** $\log_3(81) = 4$

**61.** $\log_5(x) = y$

**62.** $\log_4(a) = b$

**63.** $\log(1000) = z$

**64.** $\ln(y) = 3$

**65.** $\ln(5) = x$

**66.** $\log(y) = 2$

**67.** $\log_a(x) = m$

**68.** $\log_b(q) = t$

*Write each equation as an equivalent logarithmic equation.*

**69.** $5^3 = 125$

**70.** $2^7 = 128$

**71.** $e^3 = y$

**72.** $10^5 = w$

**73.** $y = 10^m$

**74.** $p = e^x$

**75.** $y = a^z$

**76.** $w = b^k$

**77.** $a^{x-1} = n$

**78.** $w^{x+2} = r$

*For each function, find $f^{-1}$.*

**79.** $f(x) = 2^x$

**80.** $f(x) = 5^x$

**81.** $f(x) = \log_7(x)$

**82.** $f(x) = \log(x)$

**83.** $f(x) = \ln(x - 1)$

**84.** $f(x) = \log(x + 4)$

**85.** $f(x) = 3^{x+2}$

**86.** $f(x) = 6^{x-1}$

**87.** $f(x) = \frac{1}{2} \cdot 10^{x-1} + 5$

**88.** $f(x) = 2^{3x+1} - 6$

*Solve each equation. Find the exact solutions.*

**89.** $\log_2(x) = 8$

**90.** $\log_5(x) = 3$

**91.** $\log_3(x) = \frac{1}{2}$

**92.** $\log_4(x) = \frac{1}{3}$

**93.** $\log_x(16) = 2$

**94.** $\log_x(16) = 4$

**95.** $3^x = 77$

**96.** $\frac{1}{2^x} = 5$

**97.** $\ln(x - 3) = \ln(2x - 9)$

**98.** $\log_2(4x) = \log_2(x + 6)$

**99.** $\log_x(18) = 2$

**100.** $\log_x(9) = \frac{1}{2}$

**101.** $3^{x+1} = 7$

**102.** $5^{3-x} = 12$

**103.** $\log(x) = \log(6 - x^2)$

**104.** $\log_3(2x) = \log_3(24 - x^2)$

**105.** $\log_x\left(\frac{1}{9}\right) = -\frac{2}{3}$

**106.** $\log_x\left(\frac{1}{16}\right) = \frac{4}{3}$

**107.** $4^{2x-1} = \frac{1}{2}$

**108.** $e^{3x-4} = 1$

**109.** $\log_{32}(64) = x$

**110.** $\ln\left(\frac{1}{\sqrt{e}}\right) = x$

**111.** $\log_2(\log_3(\log_4(x))) = 0$

**112.** $\log_2(\log_2(\log_2(2^{(4^x)}))) = 3$

*Find the approximate solution to each equation. Round to four decimal places.*

**113.** $10^x = 25$

**114.** $e^x = 2$

**115.** $e^{2x} = 3$

**116.** $10^{3x} = 5$

**117.** $5e^x = 4$

**118.** $10^x - 3 = 5$

**119.** $\frac{1}{10^x} = 2$

**120.** $\frac{1}{e^{x-1}} = 5$

*Solve each problem.*

**121.** *Finding Time* Find the amount of time to the nearest tenth of a year that it would take for \$10 to grow to \$20 at each of the following annual rates compounded continuously.
**a.** 2% **b.** 4%
**c.** 8% **d.** 16%

**122.** *Finding Time* Find the amount of time to the nearest tenth of a year that it would take for \$10 to grow to \$40 at each of the following annual rates compounded continuously.
**a.** 1% **b.** 2%
**c.** 8.327% **d.** $7\frac{2}{3}\%$

**123.** *Finding Rate* Find the annual percentage rate compounded continuously to the nearest tenth of a percent for which \$10 would grow to \$30 for each of the following time periods.
**a.** 5 years **b.** 10 years
**c.** 20 years **d.** 40 years

**124.** *Finding Rate* Find the annual percentage rate to the nearest tenth of a percent for which \$10 would grow to \$50 for each of the following time periods.
**a.** 4 years **b.** 8 years
**c.** 16 years **d.** 32 years

**125.** *Becoming a Millionaire* Find the amount of time to the nearest day that it would take for a deposit of \$1000 to grow to \$1 million at 14% compounded continuously.

**126.** *Doubling Your Money* How long does it take for a deposit of \$1000 to double at 8% compounded continuously?

**127.** *Finding the Rate* Solve the equation $A = Pe^{rt}$ for $r$, then find the rate at which a deposit of \$1000 would double in 3 years compounded continuously.

**128.** *Finding the Rate* At what interest rate would a deposit of \$30,000 grow to \$2,540,689 in 40 years with continuous compounding?

**129.** *Rule of 70*
**a.** Find the time that it takes for an investment to double at 10% compounded continuously.
**b.** The time that it takes for an investment to double at $r$% is approximately 70 divided by $r$ (the rule of 70). So at 10%, an investment will double in about 7 years. Explain why this rule works.

**130.** *Using the Rule of 70* Find approximate answers to these questions without using a calculator. See Exercise 129.

**a.** Connie deposits \$1000 in a bank at an annual interest rate of 2%. How long does it take for her money to double?

**b.** Celeste invests \$1000 in the stock market and her money grows at an annual rate of 10% (the historical average rate of return for the stock market). How long does it take for her money to double?

**c.** What is the ratio of the value of Celeste's investment after 35 years to the value of Connie's investment after 35 years?

**131.** *Miracle in Boston* To illustrate the "miracle" of compound interest, Ben Franklin bequeathed \$4000 to the city of Boston in 1790. The fund grew to \$4.5 million in 200 years. Find the annual rate compounded continuously that would cause this "miracle" to happen.

**132.** *Miracle in Philadelphia* Ben Franklin's gift of \$4000 to the city of Philadelphia in 1790 was not managed as well as his gift to Boston. The Philadelphia fund grew to only \$2 million in 200 years. Find the annual rate compounded continuously that would yield this total value.

**133.** *Deforestation in Nigeria* In Nigeria, deforestation occurs at the rate of about 5.2% per year. Assume that the amount of forest remaining is determined by the function

$$F = F_0e^{-0.052t},$$

where $F_0$ is the present acreage of forest land and $t$ is the time in years from the present. In how many years will there be only 60% of the present acreage remaining?

**HINT** Find the amount of time it takes for $F_0$ to become $0.60F_0$.

**134.** *Deforestation in El Salvador* It is estimated that at the present rate of deforestation in El Salvador, in 20 years only 53% of the present forest will be remaining. Use the exponential model $F = F_0e^{rt}$ to determine the annual rate of deforestation in El Salvador.

**135.** *World Population* The population of the world doubled from 1950 to 1987, going from 2.5 billion to 5 billion people. Using the exponential model,

$$P = P_0e^{rt},$$

find the annual growth rate $r$ for that period. Although the annual growth rate has declined slightly to 1.63% annually, the population of the world is still growing at a tremendous rate. Using the initial population of 5 billion in 1987 and an annual rate of 1.63%, estimate the world population in the year 2010.

**136.** *Black Death* Because of the Black Death, or plague, the only substantial period in recorded history when the earth's population was not increasing was from 1348 to 1400. During that period the world population decreased by about 100 million people. Use the exponential model $P = P_0e^{rt}$ and the data from the accompanying table to find the annual growth rate for the period 1400 to 2000. If the 100 million people had not been lost, then how many people would they have grown to in 600 years using the growth rate that you just found?

**Table for Exercise 136**

| Year | World Population |
|---|---|
| 1348 | $0.47 \times 10^9$ |
| 1400 | $0.37 \times 10^9$ |
| 1900 | $1.60 \times 10^9$ |
| 2000 | $6.07 \times 10^9$ |

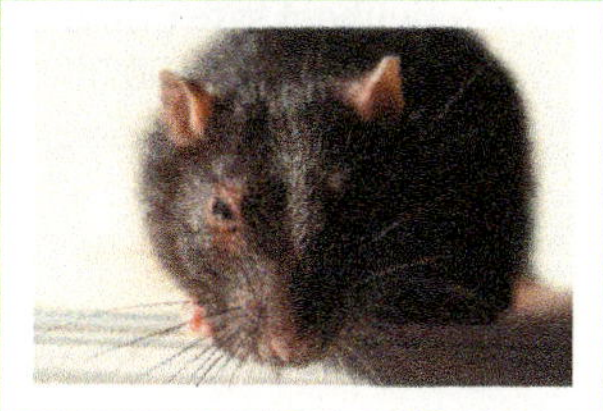

**137.** *Castle Rock Fire* The 2007 Castle Rock fire in Idaho was said to be growing exponentially after it expanded from 30 acres to 12,058 acres in 5 days (*Idaho Mountain Express*, August 22, 2007). See the accompanying figure. This exponential growth can be modeled by the function

$$a = 30e^{1.2t}$$

where $t$ is the number of days since the start of the fire and $a$ is the size of the fire in acres.

**a.** According to the exponential model, what will be the size of the fire after 7 days?

**b.** If the fire continues to grow exponentially, then in how many days will the size of the fire be 53,480,960 acres (the size of the state of Idaho)?

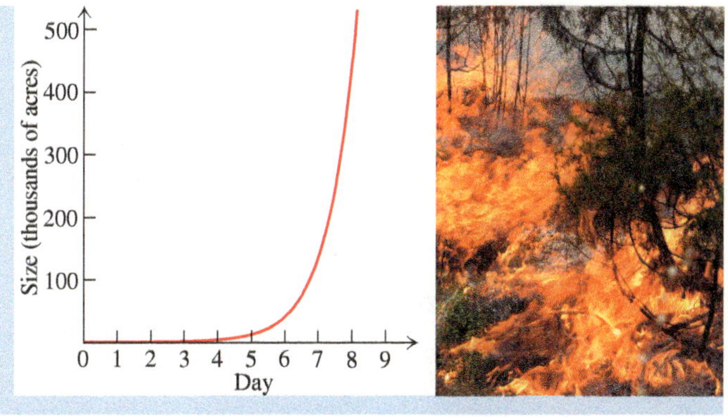


**Figure for Exercise 137**

**138.** *Global Warming* The increasing global temperature can be modeled by the function

$$I = 0.1e^{0.02t},$$

where $I$ is the increase in global temperature in degrees Celsius since 1900, and $t$ is the number of years since 1900 (NASA, www.science.nasa.gov).

**a.** How much warmer was it in 2010 than it was in 1950?

**b.** In what year will the global temperature be 4° greater than the global temperature in 2000?

**139.** *Safe Water* The graph on the next page shows the percentage of population without safe water $p$ as a function of per capita income $I$ (World Resources Institute, www.wri.org). Because the horizontal axis has a *log scale* (each mark is 10 times the previous mark), the relationship looks linear but it is not.

**a.** Find a formula for this function.

**HINT** First find the equation of the line through (2, 100) and (4, 10) in the form $p = mx + b$, then replace $x$ with $\log(I)$.

**b.** What percent of the population would be expected to be without safe drinking water in a city with a per capita income of \$100,000?

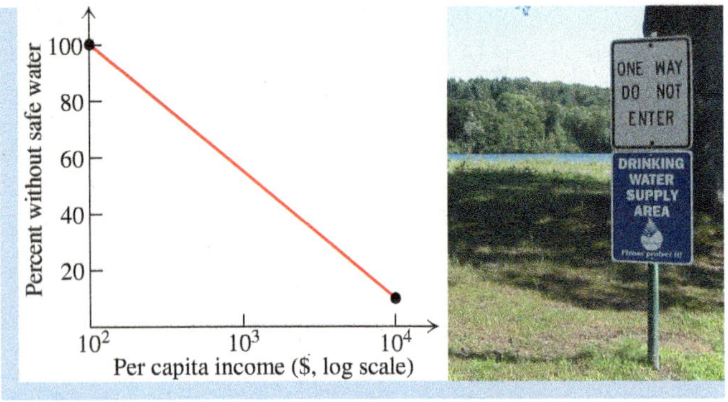


**Figure for Exercise 139**

**140.** *Municipal Waste* The accompanying graph shows the amount of municipal waste per capita $w$ as a function of per capita income $I$ (World Resources Institute, www.wri.org).

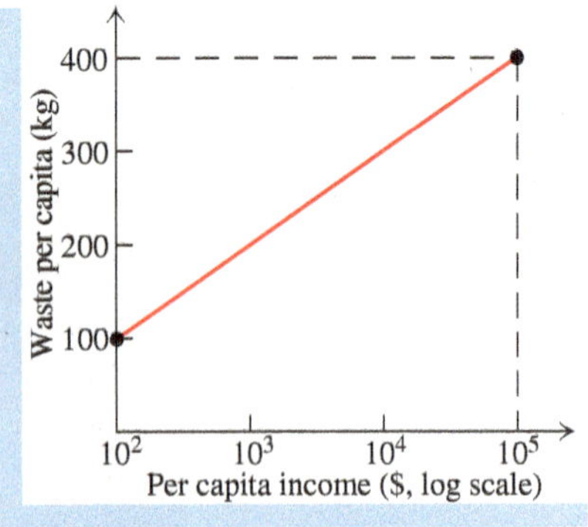


**Figure for Exercise 140**

**a.** Use the technique of the previous exercise to find a formula for $w$ as a function of $I$.

**b.** How much municipal waste per capita would you expect in a city where the per capita income is \$200,000?

*Use the following definition for Exercises 141–144.*

In chemistry, the pH of a solution is defined to be

$$\text{pH} = -\log[H^+],$$

where $H^+$ is the hydrogen ion concentration of the solution in moles per liter. Distilled water has a pH of approximately 7. A substance with a pH under 7 is called an acid, and one with a pH over 7 is called a base.

**141.** *Acidity of Tomato Juice* Tomato juice has a hydrogen ion concentration of $10^{-4.1}$ moles per liter. Find the pH of tomato juice.

**142.** *Acidity in Your Stomach* The gastric juices in your stomach have a hydrogen ion concentration of $10^{-1}$ moles per liter. Find the pH of your gastric juices.

**143.** *Acidity of Orange Juice* The hydrogen ion concentration of orange juice is $10^{-3.7}$ moles per liter. Find the pH of orange juice.

**144.** *Acidosis* A healthy body maintains the hydrogen ion concentration of human blood at $10^{-7.4}$ moles per liter. What is the pH of normal healthy blood? The condition of low blood pH is called acidosis. Its symptoms are sickly sweet breath, headache, and nausea.

## FOR WRITING/DISCUSSION

**145.** *Different Models* Calculators that perform exponential regression often use $y = a \cdot b^x$ as the exponential growth model instead of $y = a \cdot e^{cx}$. For what value of $c$ is $a \cdot b^x = a \cdot e^{cx}$? If a calculator gives $y = 500(1.036)^x$ for a growth model, then what is the continuous growth rate to the nearest hundredth of a percent?

**146.** *Making Conjectures* Consider the function $y = \log(10^n \cdot x)$ where $n$ is an integer. Use a graphing calculator to graph this function for several choices of $n$. Make a conjecture about the relationship between the graph of $y = \log(10^n \cdot x)$ and the graph of $y = \log(x)$. Save your conjecture and attempt to prove it after you have studied the properties of logarithms, which are coming in Section 4.3. Repeat this exercise with $y = \log(x^n)$ where $n$ is an integer.

**147.** *Increasing or Decreasing* Which exponential and logarithmic functions are increasing? Decreasing? Is the inverse of an increasing function increasing or decreasing? Is the inverse of a decreasing function increasing or decreasing? Explain.

**148.** *Cooperative Learning* Work in a small group to write a summary (including drawings) of the types of graphs that can be obtained for logarithmic functions of the form $y = \log_a(x)$ for $a > 0$ and $a \neq 1$.

## ▶ RETHINKING

**149.** Find the domain and range of the function
$$f(x) = -\frac{1}{2}3^{x-5} + 7.$$

**150.** Solve $2^{x-3} = 4^{5x-1}$.

**151.** Evaluate $(2 \times 10^{-9})^3(5 \times 10^3)^2$ without a calculator. Write the answer in scientific notation.

**152.** A pond contains 2000 fish of which 10% are bass. How many bass must be added so that 20% of the fish in the pond are bass?

**153.** Find all real and imaginary solutions to $x^3 + 13x = 4x^2$.

**154.** The cost of installing an oak floor varies jointly with the length and width of the room. If the cost is \$875.60 for a room that is 8 feet by 11 feet, what is the cost for a room that is 10 feet by 14 feet?

## THINKING OUTSIDE THE BOX XXXII

*Seven-Eleven* A convenience store sells a gallon of milk for \$7 and a loaf of bread for \$11. You are allowed to buy any combination of milk and bread, including only milk or only bread. Your total bill is always a whole number of dollars, but there are many whole numbers that cannot be the total. For example, the total cannot be \$15. What is the largest whole number of dollars that cannot be the total?

## POP QUIZ 4.2

1. What is $x$ if $2^x = 32$?
2. Find $\log_2(32)$.
3. Is $y = \log_3(x)$ increasing or decreasing?
4. Find all asymptotes for $f(x) = \ln(x - 1)$.
5. Write $3^a = b$ as a logarithmic equation.
6. Find $f^{-1}$ if $f(x) = \log(x + 3)$.
7. Solve $\log_5(x) = 3$.
8. Solve $\log_x(36) = 2$.
9. How long (to the nearest day) does it take for \$2 to grow to \$4 at 5% compounded continuously?

## LINKING concepts... For Individual or Group Explorations

### Modeling the U.S. Population

*To effectively plan for the future one must attempt to predict the future. Government agencies use data about the past to construct a model and predict the future. The following table gives the population of the United States in millions every* 10 *years since* 1900 *(Census Bureau, www.census.gov).*

| Year | Pop. | Year | Pop. |
|---|---|---|---|
| 1900 | 76 | 1960 | 180 |
| 1910 | 92 | 1970 | 204 |
| 1920 | 106 | 1980 | 227 |
| 1930 | 123 | 1990 | 249 |
| 1940 | 132 | 2000 | 279 |
| 1950 | 152 | 2010 | 309 |

**a)** Draw a bar graph of the data in the accompanying table. Use a computer graphics program if one is available.

**b)** Enter the data into your calculator and use exponential regression to find an exponential model of the form $y = a \cdot b^x$, where $x = 0$ corresponds to 1900.

**c)** Make a table (like the given table) that shows the predicted population according to the exponential model rather than the actual population.

**d)** Plot the points from part (c) on your bar graph and sketch an exponential curve through the points.

**e)** Use the given population data with linear regression to find a linear model of the form $y = ax + b$ and graph the line on your bar graph.

**f)** Predict the population in the year 2020 using the exponential model and the linear model.

**g)** Judging from your exponential curve and the line on your bar graph, in which prediction do you have the most confidence?

**h)** Use only the fact that the population grew from 76 million at time $t = 0$ (the year 1900) to 309 million at time $t = 110$ (the year 2010) to find a formula of the form $P(t) = P_0 \cdot e^{rt}$ for the population at any time $t$. Do not use regression.

**i)** Use the formula from part (h) to predict the population in the year 2020. Compare this prediction to the prediction that you found using exponential regression.

# 4.3 Rules of Logarithms

The rules of logarithms are closely related to the rules of exponents, because logarithms are exponents. In this section we use the rules of exponents to develop some rules of logarithms. With these rules of logarithms we will be able to solve more equations involving exponents and logarithms. The rules of exponents can be found in Section A.2 of Appendix A.

## The Inverse Rules

The definition of logarithms leads to two rules that are useful in solving equations. If $f(x) = a^x$ and $g(x) = \log_a(x)$, then

$$g(f(x)) = g(a^x) = \log_a(a^x) = x$$

for any real number $x$. The result of this composition is $x$ because the functions are inverses. If we compose in the opposite order we get

$$f(g(x)) = f(\log_a(x)) = a^{\log_a(x)} = x$$

for any positive real number $x$. The results are called the **inverse rules.**

**Inverse Rules**

If $a > 0$ and $a \neq 1$, then

1. $\log_a(a^x) = x$ for any real number $x$
2. $a^{\log_a(x)} = x$ for $x > 0$.

Figure 4.34

The inverse rules are easy to use if you remember that $\log_a(x)$ is the power of $a$ that produces $x$. For example, $\log_2(67)$ is the power of 2 that produces 67. So $2^{\log_2(67)} = 67$. Similarly, $\ln(e^{99})$ is the power of $e$ that produces $e^{99}$. So $\ln(e^{99}) = 99$. You can illustrate the inverse rules with a calculator, as shown in Fig. 4.34.

### EXAMPLE 1 Using the inverse rules

Simplify each expression.

**a.** $e^{\ln(x^2)}$ **b.** $\log_7(7^{2x-1})$

**Solution**

By the inverse rules, $e^{\ln(x^2)} = x^2$ and $\log_7(7^{2x-1}) = 2x - 1$.

▶**TRY THIS.** Simplify $10^{\log(5w)}$ and $\log(10^p)$. ■

## The Logarithm of a Product

By the product rule for exponents, we add exponents when multiplying exponential expressions having the same base. To find a corresponding rule for logarithms, let's examine the equation $2^3 \cdot 2^2 = 2^5$. Notice that the exponents 3, 2, and 5 are the base-2 logarithms of 8, 4, and 32, respectively.

$$\begin{array}{ccc} \log_2(8) & \log_2(4) & \log_2(32) \\ \searrow & \downarrow & \swarrow \end{array}$$
$$2^3 \cdot 2^2 = 2^5$$

When we add the exponents 3 and 2 to get 5, we are adding logarithms and getting another logarithm as the result. So the base-2 logarithm of 32 (the product of 8 and 4) is the sum of the base-2 logarithms of 8 and 4:

$$\log_2(8 \cdot 4) = \log_2(8) + \log_2(4)$$

This example suggests the **product rule for logarithms.**

**Product Rule for Logarithms**

For $M > 0$ and $N > 0$,

$$\log_a(MN) = \log_a(M) + \log_a(N).$$

**PROOF**

$$a^{\log_a M + \log_a N} = a^{\log_a M} \cdot a^{\log_a N} \quad \text{Product rule for exponents}$$

$$= M \cdot N \quad \text{Inverse rule}$$

Now by the definition of logarithm ($a^x = y \Leftrightarrow \log_a y = x$) we have

$$\log_a(MN) = \log_a(M) + \log_a(N).$$ ■

Figure 4.35

The product rule for logarithms says that *the logarithm of a product of two numbers is equal to the sum of their logarithms*, provided that all of the logarithms are defined and all have the same base. There is no rule about the logarithm of a sum, and the logarithm of a sum is generally *not* equal to the sum of the logarithms. For example, $\log_2(8 + 8) \neq \log_2(8) + \log_2(8)$ because $\log_2(8 + 8) = 4$ while $\log_2(8) + \log_2(8) = 6$.

You can use a calculator to illustrate the product rule, as in Fig. 4.35 □

## EXAMPLE 2 Using the product rule for logarithms

Write each expression as a single logarithm. All variables represent positive real numbers.

**a.** $\log_3(x) + \log_3(6)$ **b.** $\ln(3) + \ln(x^2) + \ln(y)$

### Solution

**a.** By the product rule for logarithms, $\log_3(x) + \log_3(6) = \log_3(6x)$.

**b.** By the product rule for logarithms, $\ln(3) + \ln(x^2) + \ln(y) = \ln(3x^2y)$.

▶**TRY THIS.** Write $\log(x) + \log(y)$ as a single logarithm. ■

## The Logarithm of a Quotient

By the quotient rule for exponents, we subtract the exponents when dividing exponential expressions having the same base. To find a corresponding rule for logarithms, examine the equation $2^5/2^2 = 2^3$. Notice that the exponents 5, 2, and 3 are the base-2 logarithms of 32, 4, and 8, respectively:

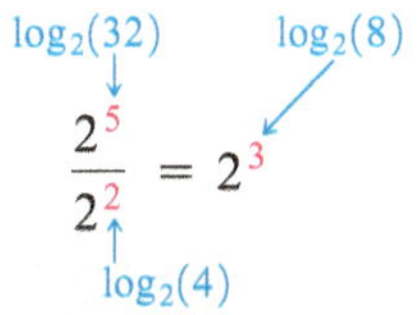


When we subtract the exponents 5 and 2 to get 3, we are subtracting logarithms and getting another logarithm as the result. So the base-2 logarithm of 8 (the quotient of 32 and 4) is the difference of the base-2 logarithms of 32 and 4:

$$\log_2\left(\frac{32}{4}\right) = \log_2(32) - \log_2(4)$$

This example suggests the **quotient rule for logarithms.**

**Quotient Rule for Logarithms**

For $M > 0$ and $N > 0$,

$$\log_a\left(\frac{M}{N}\right) = \log_a(M) - \log_a(N).$$

The quotient rule for logarithms says that *the logarithm of a quotient of two numbers is equal to the difference of their logarithms*, provided that all logarithms are defined and all have the same base. (The proof of the quotient rule is similar to that of the product rule and so it is left as an exercise.) Note that the quotient rule does not apply to division of logarithms. For example,

$$\frac{\log_2(32)}{\log_2(4)} \neq \log_2(32) - \log_2(4),$$

because $\log_2(32)/\log_2(4) = 5/2$, while $\log_2(32) - \log_2(4) = 3$.

You can use a calculator to illustrate the quotient rule, as shown in Fig. 4.36. □

**Figure 4.36**

## EXAMPLE 3 Using the quotient rule for logarithms

Write each expression as a single logarithm. All variables represent positive real numbers.

**a.** $\log_3(24) - \log_3(4)$ **b.** $\ln(x^6) - \ln(x^2)$

### Solution

**a.** By the quotient rule, $\log_3(24) - \log_3(4) = \log_3(24/4) = \log_3(6)$.

**b.** $\ln(x^6) - \ln(x^2) = \ln\left(\frac{x^6}{x^2}\right)$ By the quotient rule for logarithms

$= \ln(x^4)$ By the quotient rule for exponents

▶**TRY THIS.** Write $\log(5x) - \log(5)$ as a single logarithm. ■

## The Logarithm of a Power

By the power rule for exponents, we multiply the exponents when finding a power of an exponential expression. For example, $(2^3)^2 = 2^6$. Notice that the exponents 3 and 6 are the base-2 logarithms of 8 and 64, respectively.

$$\underset{\log_2(8)}{} \qquad \underset{\log_2(64)}{}$$
$$(2^{3})^2 = 2^{6}$$

(arrows: $\log_2(8) \downarrow$ points to the exponent 3, $\log_2(64) \downarrow$ points to the exponent 6)

So the base-2 logarithm of 64 (the second power of 8) is twice the base-2 logarithm of 8:

$$\log_2(8^2) = 2 \cdot \log_2(8)$$

This example suggests the **power rule for logarithms.**

**Power Rule for Logarithms**

For $M > 0$ and any real number $N$,

$$\log_a(M^N) = N \cdot \log_a(M).$$

**Figure 4.37**

The power rule for logarithms says that *the logarithm of a power of a number is equal to the power times the logarithm of the number*, provided that all logarithms are defined and have the same base. The proof is left as an exercise.

You can illustrate the power rule on a calculator, as shown in Fig. 4.37. □

### EXAMPLE 4 Using the power rule for logarithms

Rewrite each expression in terms of $\log(3)$.

**a.** $\log(3^8)$ **b.** $\log(\sqrt{3})$ **c.** $\log\left(\frac{1}{3}\right)$

**Solution**

**a.** $\log(3^8) = 8 \cdot \log(3)$ By the power rule for logarithms

**b.** $\log(\sqrt{3}) = \log(3^{1/2}) = \frac{1}{2}\log(3)$ By the power rule for logarithms

**c.** $\log\left(\frac{1}{3}\right) = \log(3^{-1}) = -\log(3)$ By the power rule for logarithms

▶**TRY THIS.** Write $\ln(27)$ in terms of $\ln(3)$. ■

## Using the Rules

When simplifying or rewriting expressions, we often apply several rules. In the following box we list all of the available rules of logarithms.

**Rules of Logarithms with Base $a$**

If $M$, $N$, and $a$ are positive real numbers with $a \neq 1$, and $x$ is any real number, then

| | |
|---|---|
| **1.** $\log_a(a) = 1$ | **2.** $\log_a(1) = 0$ |
| **3.** $\log_a(a^x) = x$ | **4.** $a^{\log_a(N)} = N$ |
| **5.** $\log_a(MN) = \log_a(M) + \log_a(N)$ | **6.** $\log_a(M/N) = \log_a(M) - \log_a(N)$ |
| **7.** $\log_a(M^x) = x \cdot \log_a(M)$ | **8.** $\log_a(1/N) = -\log_a(N)$ |

Note that rule 1 is a special case of rule 3 with $x = 1$, rule 2 follows from the fact that $a^0 = 1$ for any nonzero base $a$, and rule 8 is a special case of rule 6 with $M = 1$.

The rules for logarithms with base $a$ in the preceding box apply to all logarithms, including common logarithms (base 10) and natural logarithms (base $e$). Since natural logarithms are very popular, we list the rules of logarithms again for base $e$ in the following box for easy reference.

**Rules of Natural Logarithms**

If $M$ and $N$ are positive real numbers and $x$ is any real number, then

| | |
|---|---|
| **1.** $\ln(e) = 1$ | **2.** $\ln(1) = 0$ |
| **3.** $\ln(e^x) = x$ | **4.** $e^{\ln(N)} = N$ |
| **5.** $\ln(MN) = \ln(M) + \ln(N)$ | **6.** $\ln(M/N) = \ln(M) - \ln(N)$ |
| **7.** $\ln(M^x) = x \cdot \ln(M)$ | **8.** $\ln(1/N) = -\ln(N)$ |

### EXAMPLE 5 Using the rules of logarithms

Rewrite each expression in terms of $\log(2)$ and $\log(3)$.

**a.** $\log(6)$ **b.** $\log\left(\frac{16}{3}\right)$ **c.** $\log\left(\frac{1}{3}\right)$

**Solution**

**a.** $\log(6) = \log(2 \cdot 3)$ Factor.

$= \log(2) + \log(3)$ Rule 5

Figure 4.38

**b.** $\log\left(\frac{16}{3}\right) = \log(16) - \log(3)$ Rule 6

$= \log(2^4) - \log(3)$ Write 16 as a power of 2.

$= 4 \cdot \log(2) - \log(3)$ Rule 7

You can check this answer with a calculator, as shown in Fig. 4.38. □

**c.** $\log\left(\frac{1}{3}\right) = -\log(3)$ Rule 8

▶**TRY THIS.** Write $\ln(45)$ in terms of $\ln(3)$ and $\ln(5)$. ■

Be careful to use the rules of logarithms exactly as stated. For example, the logarithm of a product is the sum of the logarithms. The logarithm of a product is generally *not* equal to the product of the logarithms. That is,

$$\log(2 \cdot 3) = \log(2) + \log(3) \quad \text{but} \quad \log(2 \cdot 3) \neq \log(2) \cdot \log(3).$$

## EXAMPLE 6 Rewriting a logarithmic expression

Rewrite each expression using a sum or difference of multiples of logarithms.

**a.** $\ln\left(\frac{3x^2}{yz}\right)$ **b.** $\log_3\left(\frac{(x-1)^2}{z^{3/2}}\right)$

### Solution

**a.** $\ln\left(\frac{3x^2}{yz}\right) = \ln(3x^2) - \ln(yz)$ Quotient rule for logarithms

$= \ln(3) + \ln(x^2) - [\ln(y) + \ln(z)]$ Product rule for logarithms

$= \ln(3) + 2 \cdot \ln(x) - \ln(y) - \ln(z)$ Power rule for logarithms

Note that $\ln(y) + \ln(z)$ must be in brackets (or parentheses) because of the subtraction symbol preceding it.

**b.** $\log_3\left(\frac{(x-1)^2}{z^{3/2}}\right) = \log_3((x-1)^2) - \log_3(z^{3/2})$ Quotient rule for logarithms

$= 2 \cdot \log_3(x-1) - \frac{3}{2}\log_3(z)$ Power rule for logarithms

▶**TRY THIS.** Write $\log\left(\frac{x^2}{5a}\right)$ using sums and/or differences of multiples of logarithms. ■

In Example 7 we use the rules "in reverse" of the way we did in Example 6.

## EXAMPLE 7 Rewriting as a single logarithm

Rewrite each expression as a single logarithm.

**a.** $\ln(x-1) + \ln(3) - 3 \cdot \ln(x)$ **b.** $\frac{1}{2}\log(y) - \frac{1}{3}\log(z)$

### Solution

**a.** Use the product rule, the power rule, and then the quotient rule for logarithms:

$$\begin{aligned}\ln(x-1) + \ln(3) - 3 \cdot \ln(x) &= \ln[3(x-1)] - \ln(x^3)\\ &= \ln[3x-3] - \ln(x^3)\\ &= \ln\left(\frac{3x-3}{x^3}\right)\end{aligned}$$

**b.** $\frac{1}{2}\log(y) - \frac{1}{3}\log(z) = \log(y^{1/2}) - \log(z^{1/3})$

$$= \log(\sqrt{y}) - \log(\sqrt[3]{z})$$

$$= \log\left(\frac{\sqrt{y}}{\sqrt[3]{z}}\right)$$

▶**TRY THIS.** Write $\ln(x) - \ln(y) - 2 \cdot \ln(z)$ as a single logarithm. ■

The rules for logarithms correspond to the rules for exponents because *logarithms are exponents*. The following table shows the rules side by side.

| | Exponent Rule | Logarithm Rule |
|---|---|---|
| **Product Rule** | $a^x a^y = a^{x+y}$ | $\log_a(MN) = \log_a(M) + \log_a(N)$ |
| **Quotient Rule** | $\frac{a^x}{a^y} = a^{x-y}$ | $\log_a\left(\frac{M}{N}\right) = \log_a(M) - \log_a(N)$ |
| **Power Rule** | $(a^x)^y = a^{xy}$ | $\log_a(M^N) = N \cdot \log_a(M)$ |
| **Zero Exponent** | $a^0 = 1$ | $\log_a(1) = 0$ |

## The Base-Change Formula

The exact solution to an exponential equation is often expressed in terms of logarithms. For example, the exact solution to $(1.03)^x = 5$ is $x = \log_{1.03}(5)$. But how do we calculate $\log_{1.03}(5)$? The next example shows how to find a rational approximation for $\log_{1.03}(5)$ using rules of logarithms. In this example we also introduce a new idea in solving equations, *taking the logarithm of each side.* The base-*a* logarithms of two equal quantities are equal because logarithms are functions.

### EXAMPLE 8 A rational approximation for a logarithm

Find an approximate rational solution to $(1.03)^x = 5$. Round to four decimal places.

#### Solution

Take the logarithm of each side using one of the bases available on a calculator.

$$(1.03)^x = 5$$

$$\log((1.03)^x) = \log(5) \quad \text{Take the logarithm of each side.}$$

$$x \cdot \log(1.03) = \log(5) \quad \text{Power rule for logarithms}$$

$$x = \frac{\log(5)}{\log(1.03)} \quad \text{Divide each side by } \log(1.03).$$

$$\approx 54.4487$$

Divide the logarithms and check with a calculator, as shown in Fig. 4.39.

**Figure 4.39**

▶**TRY THIS.** Solve $5.44^x = 2.3$. Round to four decimal places. ■

The solution to the equation of Example 8 is a base-1.03 logarithm, but we obtained a rational approximation for it using base-10 logarithms. We can use the procedure of Example 8 to write a base-$a$ logarithm in terms of a base-$b$ logarithm for any bases $a$ and $b$:

$$a^x = M \qquad \text{Equivalent equation: } x = \log_a(M)$$

$$\log_b(a^x) = \log_b(M) \qquad \text{Take the base-}b\text{ logarithm of each side.}$$

$$x \cdot \log_b(a) = \log_b(M) \qquad \text{Power rule for logarithms}$$

$$x = \frac{\log_b(M)}{\log_b(a)} \qquad \text{Divide each side by } \log_b(a).$$

Since $x = \log_a(M)$, we have the following **base-change formula.**

### Base-Change Formula

If $a > 0$, $b > 0$, $a \neq 1$, $b \neq 1$, and $M > 0$, then

$$\log_a(M) = \frac{\log_b(M)}{\log_b(a)}.$$

The base-change formula says that the logarithm of a number in one base is equal to the logarithm of the number in the new base divided by the logarithm of the old base. Using this formula, a logarithm such as $\log_3(7)$ can be easily found with a calculator. Let $a = 3$, $b = 10$, and $M = 7$ in the formula to get

$$\log_3(7) = \frac{\log(7)}{\log(3)} \approx 1.7712.$$

Figure 4.40

Note that you get the same value for $\log_3(7)$ using natural logarithms, as shown in Fig. 4.40. □

## EXAMPLE 9 Using the base-change formula with compound interest

If \$2500 is invested at 6% compounded daily, then how long (to the nearest day) would it take for the investment to double in value?

### Solution

We want the number of years for \$2500 to grow to \$5000 at 6% compounded daily. Use $P = \$2500$, $A = \$5000$, $n = 365$, and $r = 0.06$ in the formula for compound interest:

$$A = P\left(1 + \frac{r}{n}\right)^{nt}$$

$$5000 = 2500\left(1 + \frac{0.06}{365}\right)^{365t}$$

$$2 = \left(1 + \frac{0.06}{365}\right)^{365t} \qquad \text{Divide by 2500 to get the form } y = a^x.$$

$$2 \approx (1.000164384)^{365t} \qquad \text{Approximate the base with a decimal number.}$$

$$365t \approx \log_{1.000164384}(2) \qquad \text{Definition of logarithm}$$

$$t \approx \frac{1}{365} \cdot \frac{\ln(2)}{\ln(1.000164384)} \qquad \text{Base-change formula (Use either ln or log.)}$$

$$\approx 11.553 \text{ years}$$

The investment of \$2500 will double in about 11 years, 202 days.

►**TRY THIS.** How long (to the nearest day) does it take for \$100 to grow to \$400 at 5% compounded daily? ■

Note that Example 9 can also be solved by taking the log (or ln) of each side and applying the power rule for logarithms.

$$\log(2) = \log((1 + 0.06/365)^{365t})$$

$$\log(2) = 365t \cdot \log(1 + 0.06/365)$$

$$\frac{\log(2)}{365 \cdot \log(1 + 0.06/365)} = t$$

$$11.553 \approx t$$

## Finding the Rate

If a variable is a function of time, we are often interested in finding the rate of growth or decay. In the continuous model, $y = ae^{rt}$, the rate $r$ appears as an exponent. If we know the values of the other variables, we find the rate by solving for $r$ using natural logarithms. If $y = ab^t$ is used instead of $y = ae^{rt}$, the rate does not appear in the formula. However, we can write $y = ab^t$ as $y = ae^{\ln(b) \cdot t}$, because $b = e^{\ln(b)}$. Comparing $e^{\ln(b) \cdot t}$ to $e^{rt}$, we see that the rate is $\ln(b)$.

In the case of interest compounded periodically, the growth is not continuous and we use the formula $A = P(1 + r/n)^{nt}$. Because the rate appears in the base, it can be found without using logarithms, as shown in the next example.

### EXAMPLE 10 Finding the rate with interest compounded periodically

For what annual percentage rate would \$1000 grow to \$3500 in 20 years compounded monthly?

**Solution**

Use $A = 3500$, $P = 1000$, $n = 12$, and $t = 20$ in the compound interest formula.

$$A = P\left(1 + \frac{r}{n}\right)^{nt}$$

$$3500 = 1000\left(1 + \frac{r}{12}\right)^{240}$$

$$3.5 = \left(1 + \frac{r}{12}\right)^{240} \quad \text{Divide each side by 1000.}$$

Since 240 is even, there is a positive and a negative 240th root of 3.5. Ignore the negative root in this application because it gives a negative interest rate.

$$1 + \frac{r}{12} = (3.5)^{1/240} \quad \text{Logarithms are not needed because the exponent is a constant.}$$

$$\frac{r}{12} = (3.5)^{1/240} - 1$$

$$r = 12((3.5)^{1/240} - 1)$$

$$\approx 0.063$$

If \$1000 earns approximately 6.3% compounded monthly, then it will grow to \$3500 in 20 years.

▶**TRY THIS.** For what annual rate compounded daily would \$100 grow to \$300 in 20 years? ■

## FOR thought... True or False? Explain.

1. $\frac{\log(8)}{\log(3)} = \log(8) - \log(3)$
2. $\ln(\sqrt{3}) = \frac{\ln(3)}{2}$
3. $\frac{\log_{19}(8)}{\log_{19}(2)} = \log_3(27)$
4. $\frac{\log_2(7)}{\log_2(5)} = \frac{\log_3(7)}{\log_3(5)}$
5. $e^{\ln(x)} = x$ for any real number $x$.
6. The equations $\log(x - 2) = 4$ and $\log(x) - \log(2) = 4$ are equivalent.
7. The equations $x + 1 = 2x + 3$ and $\log(x + 1) = \log(2x + 3)$ are equivalent.
8. $\ln(e^x) = x$ for any real number $x$.
9. If $30 = x^{50}$, then $x$ is between 1 and 2.
10. If $20 = a^4$, then $\ln(a) = \frac{1}{4}\ln(20)$.

## EXERCISES 4.3

*Fill in the blank.*

1. The logarithm of a product of two numbers is equal to the __________ of their logarithms.
2. The logarithm of a(n) __________ of two numbers is equal to the difference of their logarithms.
3. The logarithm of a power of a number is equal to the __________ times the logarithm of the number.
4. According to the __________ formula, the base-$a$ logarithm of a number is equal to the base-$b$ logarithm of the number divided by $\log_b(a)$.

*Simplify each expression.*

5. $e^{\ln(\sqrt{y})}$
6. $10^{\log(3x+1)}$
7. $\log(10^{y+1})$
8. $\ln(e^{2k})$
9. $7^{\log_7(999)}$
10. $\log_4(2^{300})$

*Rewrite each expression as a single logarithm.*

11. $\log(5) + \log(3)$
12. $\ln(6) + \ln(2)$
13. $\log_2(x - 1) + \log_2(x)$
14. $\log_3(x + 2) + \log_3(x - 1)$
15. $\log_4(12) - \log_4(2)$
16. $\log_2(25) - \log_2(5)$
17. $\ln(x^8) - \ln(x^3)$
18. $\log(x^2 - 4) - \log(x - 2)$

*Rewrite each expression as a sum or difference of logarithms.*

19. $\log_2(3x)$
20. $\log_3(xy)$
21. $\log\left(\frac{x}{2}\right)$
22. $\log\left(\frac{a}{b}\right)$
23. $\log(x^2 - 1)$
24. $\log(a^2 - 9)$
25. $\ln\left(\frac{x - 1}{x}\right)$
26. $\ln\left(\frac{a + b}{b}\right)$

*Rewrite each expression in terms of* $\log_a(5)$.

27. $\log_a(5^3)$
28. $\log_a(25)$
29. $\log_a(\sqrt{5})$
30. $\log_a(\sqrt[3]{5})$
31. $\log_a\left(\frac{1}{5}\right)$
32. $\log_a\left(\frac{1}{125}\right)$

*Rewrite each expression in terms of* $\log_a(2)$ *and* $\log_a(5)$.

33. $\log_a(10)$
34. $\log_a(0.4)$
35. $\log_a(2.5)$
36. $\log_a(250)$
37. $\log_a(\sqrt{20})$
38. $\log_a(0.0005)$
39. $\log_a\left(\frac{4}{25}\right)$
40. $\log_a(0.1)$

*Rewrite each expression as a sum or difference of multiples of logarithms.*

41. $\log_3(5x)$
42. $\log_2(xyz)$
43. $\log_2\left(\frac{5}{2y}\right)$
44. $\log\left(\frac{4a}{3b}\right)$
45. $\log(3\sqrt{x})$
46. $\ln(\sqrt{x/4})$
47. $\log(3 \cdot 2^{x-1})$
48. $\ln\left(\frac{5^{-x}}{2}\right)$
49. $\ln\left(\frac{\sqrt[3]{xy}}{t^{4/3}}\right)$
50. $\log\left(\frac{3x^2}{(ab)^{2/3}}\right)$

**51.** $\ln\left(\frac{6\sqrt{x-1}}{5x^3}\right)$

**52.** $\log_4\left(\frac{3x\sqrt{y}}{\sqrt[3]{x-1}}\right)$

*Rewrite each expression as a single logarithm.*

**53.** $\log_2(5) + 3 \cdot \log_2(x)$

**54.** $\log(x) + 5 \cdot \log(x)$

**55.** $\log_7(x^5) - 4 \cdot \log_7(x^2)$

**56.** $\frac{1}{3}\ln(6) - \frac{1}{3}\ln(2)$

**57.** $\log(2) + \log(x) + \log(y) - \log(z)$

**58.** $\ln(2) + \ln(3) + \ln(5) - \ln(7)$

**59.** $\frac{1}{2}\log(x) - \log(y) + \log(z) - \frac{1}{3}\log(w)$

**60.** $\frac{5}{6}\log_2(x) + \frac{2}{3}\log_2(y) - \frac{1}{2}\log_2(x) - \log_2(y)$

**61.** $3 \cdot \log_4(x^2) - 4 \cdot \log_4(x^{-3}) + 2 \cdot \log_4(x)$

**62.** $\frac{1}{2}[\log(x) + \log(y)] - \log(z)$

*Find an approximate rational solution to each equation. Round answers to four decimal places.*

**63.** $2^x = 9$

**64.** $3^x = 12$

**65.** $(0.56)^x = 8$

**66.** $(0.23)^x = 18.4$

**67.** $(1.06)^x = 2$

**68.** $(1.09)^x = 3$

**69.** $(0.73)^x = 0.5$

**70.** $(0.62)^x = 0.25$

*Use a calculator and the base-change formula to find each logarithm to four decimal places.*

**71.** $\log_4(9)$

**72.** $\log_3(4.78)$

**73.** $\log_{9.1}(2.3)$

**74.** $\log_{1.2}(13.7)$

**75.** $\log_{1/2}(12)$

**76.** $\log_{1.05}(3.66)$

*Solve each equation. Round answers to four decimal places.*

**77.** $(1.02)^{4t} = 3$

**78.** $(1.025)^{12t} = 3$

**79.** $(1.0001)^{365t} = 3.5$

**80.** $(1.00012)^{365t} = 2.4$

**81.** $(1 + r)^3 = 2.3$

**82.** $\left(1 + \frac{r}{4}\right)^{20} = 3$

**83.** $2\left(1 + \frac{r}{12}\right)^{360} = 8.4$

**84.** $5\left(1 + \frac{r}{360}\right)^{720} = 12$

*Solve each equation. Round answers to four decimal places.*

**85.** $\log_x(33.4) = 5$

**86.** $\log_x(12.33) = 2.3$

**87.** $\log_x(0.546) = -1.3$

**88.** $\log_x(0.915) = -3.2$

*For Exercises 89–94 find the time or rate required for each investment given in the table to grow to the specified amount. The letter W represents an unknown principal.*

| | Principal | Ending Balance | Rate | Compounded | Time |
|---|---|---|---|---|---|
| **89.** | \$800 | \$2000 | 8% | Daily | ? |
| **90.** | \$10,000 | \$1,000,000 | 7.75% | Annually | ? |
| **91.** | \$$W$ | \$3$W$ | 10% | Quarterly | ? |
| **92.** | \$$W$ | \$2$W$ | 12% | Monthly | ? |
| **93.** | \$500 | \$2000 | ? | Annually | 25 yr |
| **94.** | \$1000 | \$2500 | ? | Monthly | 8 yr |

**95.** *Ben's Gift to Boston* Ben Franklin's gift of \$4000 to Boston grew to \$4.5 million in 200 years. At what interest rate compounded annually would this growth occur?

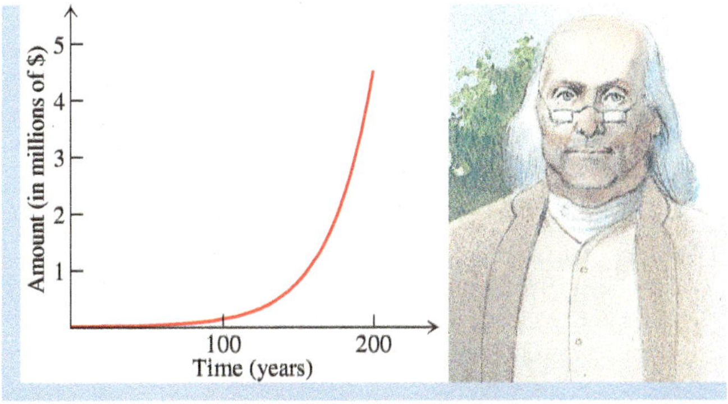


**Figure for Exercise 95**

**96.** *Ben's Gift to Philadelphia* Ben Franklin's gift of \$4000 to Philadelphia grew to \$2 million in 200 years. At what interest rate compounded monthly would this growth occur?

**97.** *Cheap Cars* One of the cheapest new cars you could buy in 1970 was the Ford Maverick at \$1995. In 2009 the cheapest new car was probably the Kia Rio at \$10,890. Use the formula for compound interest to find the annual growth rate (to the nearest tenth of a percent) for the price of the cheapest new car.

**98.** *Price of Milk* In 1960 a gallon of milk was around \$0.49. In 2009 a gallon of milk was around \$4.59. Use the formula for compound interest to find the annual growth rate (to the nearest tenth of a percent) for the price of a gallon to milk.

**99.** *Richter Scale* The common logarithm is used to measure the intensity of an earthquake on the Richter scale. The Richter scale rating of an earthquake of intensity $I$ is given by $\log(I) - \log(I_0)$, where $I_0$ is the intensity of a small "benchmark" earthquake. Write the Richter scale rating as a single logarithm. What is the Richter scale rating of an earthquake for which $I = 1000 \cdot I_0$?

**100.** *Colombian Earthquake of 1906* At 8.6 on the Richter scale, the Colombian earthquake of January 31, 1906, was one of the strongest earthquakes on record. Use the formula from Exercise 99 to write $I$ as a multiple of $I_0$ for this earthquake.

**101.** *Time for Growth* The time in years for a population of size $P_0$ to grow to size $P$ at the annual growth rate $r$ is given by $t = \ln((P/P_0)^{1/r})$. Use the rules of logarithms to express $t$ in terms of $\ln(P)$ and $\ln(P_0)$.

**102.** *Formula for pH* The pH of a solution is given by $\text{pH} = \log(1/H^+)$, where $H^+$ is the hydrogen ion concentration of the solution. Use the rules of logarithms to express the pH in terms of $\log(H^+)$.

**103.** *Rollover Time* The probability that a \$1 ticket does not win the Louisiana Lottery is $\frac{7{,}059{,}051}{7{,}059{,}052}$. The probability $p$ that $n$ independently sold tickets are all losers and the lottery rolls over is given by

$$p = \left(\frac{7{,}059{,}051}{7{,}059{,}052}\right)^n.$$

**a.** As $n$ increases is $p$ increasing or decreasing?

**b.** For what number of tickets sold is the probability of a rollover greater than 50%?

**104.** *Economic Impact* An economist estimates that 75% of the money spent in Chattanooga is respent in four days on the average in Chattanooga. So if $P$ dollars are spent, then the economic impact $I$ in dollars after $n$ respendings is given by

$$I = P(0.75)^n.$$

When $I < 0.02P$, then $P$ no longer has an impact on the economy. If the Telephone Psychics Convention brings \$1.3 million to the city, then in how many days will that money no longer have an impact on the city?

**105.** *Marginal Revenue* The revenue in dollars from the sale of $x$ items is given by the function $R(x) = 500 \cdot \log(x + 1)$. The marginal revenue function $MR(x)$ is the difference quotient for $R(x)$ when $h = 1$. Find $MR(x)$ and write it as a single logarithm. What happens to the marginal revenue as $x$ gets larger and larger?

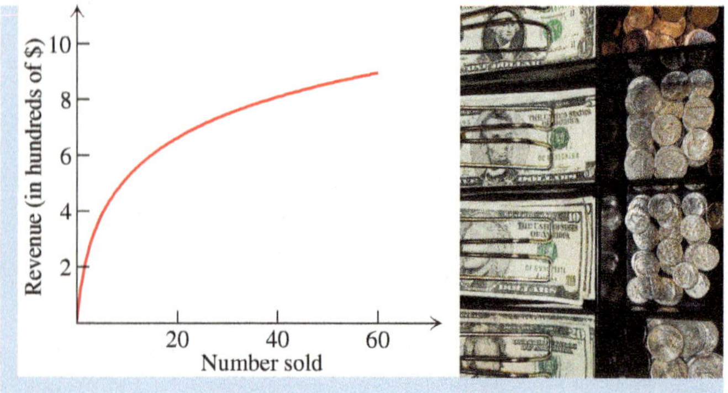


**Figure for Exercise 105**

**106.** *Human Memory Model* A class of college algebra students was given a test on college algebra concepts every month for one year after completing a college algebra course. The mean score for the class $t$ months after completing the course can be modeled by the function $m = \ln[e^{80}/(t + 1)^7]$ for $0 \le t \le 12$. Find the mean score of the class for $t = 0, 5$, and 12. Use the rules of logarithms to rewrite the function.

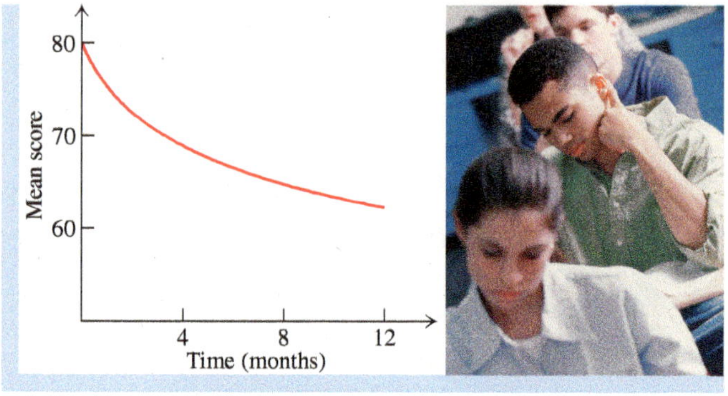


**Figure for Exercise 106**

**107.** *Computers per Capita* The number of personal computers per 1000 people in the United States from 1990 through 2010 is given in the accompanying table (*Consumer Industry Almanac,* www.c-i-a.com).

**a.** Use exponential regression on a graphing calculator to find the best-fitting curve of the form $y = a \cdot b^x$, where $x = 0$ corresponds to 1990.

**b.** Write your equation in the form $y = ae^{cx}$.

**c.** Assuming that the number of computers per 1000 people is growing continuously, what is the annual percentage rate?

**d.** In what year will the number of computers per 1000 people reach 1500?

**e.** Judging from the graph of the data and the curve, does the exponential model look like a good model?

**Table for Exercise 107**

| Year | Computers per 1000 |
|---|---|
| 1990 | 192 |
| 1995 | 321 |
| 2000 | 628 |
| 2005 | 778 |
| 2010 | 932 |

**108.** *Crude Oil Prices* The average cost in dollars of a barrel of domestic crude oil for each year from 2000 to 2008 is shown in the accompanying table (www.inflationdata.com).

Table for Exercise 108

| Year | Cost per Barrel |
|---|---|
| 2000 | 27 |
| 2001 | 23 |
| 2002 | 23 |
| 2003 | 28 |
| 2004 | 38 |
| 2005 | 50 |
| 2006 | 58 |
| 2007 | 64 |
| 2008 | 130 |

**a.** Use exponential regression on a graphing calculator to find the best-fitting curve of the form $y = a \cdot b^x$, where $x = 0$ corresponds to 2000.

**b.** Use the exponential model from part (a) to predict the average price of a barrel of domestic crude in 2015.

## FOR WRITING/DISCUSSION

**109.** *Power Rule* Applying the power rule to $y = \log(x^2)$ yields $y = 2 \cdot \log(x)$, but are these functions the same? What is the domain of each function? Graph the functions. Find another example of a function whose domain changes after application of a rule for logarithms.

**110.** *Quotient Rule* Write a proof for the quotient rule for logarithms that is similar to the proof given in the text for the product rule for logarithms.

**111.** *Cooperative Learning* Work in a small group to write a proof for the power rule for logarithms.

**112.** *Finding Relationships* Graph each of the following pairs of functions on the same screen of a graphing calculator. (Use the base-change formula to graph with bases other than 10 or *e*.) Explain how the functions in each pair are related.

**a.** $y_1 = \log_3(\sqrt{3x}), y_2 = 0.5 + \log_3(x)$

**b.** $y_1 = \log_2(1/x), y_2 = -\log_2(x)$

**c.** $y_1 = 3^{x-1}, y_2 = \log_3(x) + 1$

**d.** $y_1 = 3 + 2^{x-4}, y_2 = \log_2(x - 3) + 4$

## RETHINKING

**113.** Determine the domain and range of the function $y = \log_2(x - 1)$.

**114.** Solve $\log_2(x - 3) = 8$.

**115.** Find the amount when \$10,000 is invested for 5 years and 3 months at 4.3% compounded continuously.

**116.** Find the amount when \$10,000 is invested for 5 years and 3 months at 4.3% compounded daily.

**117.** Solve $\frac{3}{8}(x - 5) = \frac{3}{2}x + \frac{5}{6}$.

**118.** Evaluate $-4^{3/2}$ and $4^{-3/2}$.

## THINKING OUTSIDE THE BOX XXXIII

*Unit Fractions* The fractions $\frac{1}{2}, \frac{1}{3}, \frac{1}{4}, \frac{1}{5}, \frac{1}{6}, \frac{1}{7}$, etc., are called *unit fractions*. Some rational numbers can be expressed as sums of unit fractions. For example,

$$\frac{4}{7} = \frac{1}{2} + \frac{1}{14} \quad \text{and} \quad \frac{11}{18} = \frac{1}{3} + \frac{1}{6} + \frac{1}{9}.$$

Write each of the following fractions as a sum of unit fractions using the fewest number of unit fractions, and keep the total of all of the denominators in the unit fractions as small as possible.

**a.** $\frac{6}{23}$ **b.** $\frac{14}{15}$ **c.** $\frac{7}{11}$

## POP QUIZ 4.3

**1.** Write $\log(9) + \log(3)$ as a single logarithm.

**2.** Write $\log(9) - \log(3)$ as a single logarithm.

**3.** Write $3\ln(x) + \ln(y)$ as a single logarithm.

**4.** Write $\ln(5x)$ as a sum of logarithms.

**5.** Write $\ln(\sqrt{18})$ in terms of $\ln(2)$ and $\ln(3)$.

*Solve each equation. Round answers to four decimal places.*

**6.** $3^x = 11$

**2.** $\log_x(22.5) = 3$

## LINKING concepts... For Individual or Group Explorations

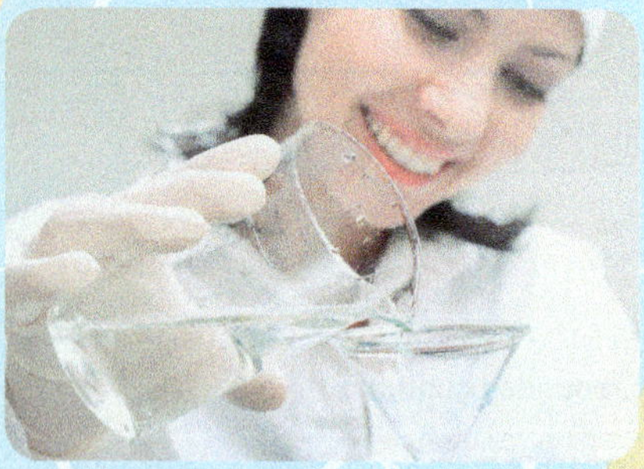

### A Logarithmic Model for Water Quality

*The United States Geological Survey measures the quality of a water sample by using the diversity index d, given by*

$$d = -[p_1 \cdot \log_2(p_1) + p_2 \cdot \log_2(p_2) + \cdots + p_n \cdot \log_2(p_n)],$$

*where n is the number of different taxons (biological classifications) represented in the sample and $p_1$ through $p_n$ are the percentages of organisms in each of the n taxons. For example, if* 10% *of the organisms in a water sample are E. coli and* 90% *are fecal coliform, then*

$$d = -[0.1 \cdot \log_2(0.1) + 0.9 \cdot \log_2(0.9)] \approx 0.5.$$

**a)** Find the value of $d$ when the only organism found in a sample is *E. coli* bacteria.

**b)** Let $n = 3$ in the formula and write the diversity index as a single logarithm.

**c)** If a water sample is found to contain 20% of its organisms of one type, 30% of another type, and 50% of a third type, then what is the diversity index for the water sample?

**d)** If the organisms in a water sample are equally distributed among 100 different taxons, then what is the diversity index?

**e)** If 99% of the organisms in a water sample are from one taxon and the other 1% are equally distributed among 99 other taxons, then what is the diversity index?

**f)** The diversity index can be found for populations other than organisms in a water sample. Find the diversity index for the dogs in the movie *101 Dalmatians* (cartoon version).

**g)** Identify a population of your choice and different classifications within the population. (For example, the trees on campus can be classified as pine, maple, spruce, or elm.) Gather real data and calculate the diversity index for your population.

# 4.4 More Equations and Applications

The rules of Section 4.3 combined with the techniques that we have already used in Sections 4.1 and 4.2 allow us to solve several new types of equations involving exponents and logarithms.

## Logarithmic Equations

An equation involving a single logarithm can usually be solved by using the definition of logarithm as we did in Section 4.2.

### EXAMPLE 1 An equation involving a single logarithm

Solve the equation $\log(x - 3) = 4$.

### Solution

Write the equivalent equation using the definition of logarithm:

$$\log(x - 3) = 4 \qquad \text{Original equation } \log_a(y) = x \Leftrightarrow y = a^x$$
$$x - 3 = 10^4$$
$$x = 10{,}003$$

Check this number in the original equation. The solution is 10,003.

▶**TRY THIS.** Solve $\log(2x + 1) = 3$. ■

When more than one logarithm is present, we can use the one-to-one property as in Section 4.2 or use the other rules of logarithms to combine logarithms.

## EXAMPLE 2 Equations involving more than one logarithm

Solve each equation.

**a.** $\log_2(x) + \log_2(x + 2) = \log_2(6x + 1)$ **b.** $\log(x) - \log(x - 1) = 2$
**c.** $2 \cdot \ln(x) = \ln(x + 3) + \ln(x - 1)$

### Solution

**a.** Since the sum of two logarithms is equal to the logarithm of a product, we can rewrite the left-hand side of the equation.

$$\log_2(x) + \log_2(x + 2) = \log_2(6x + 1)$$
$$\log_2(x(x + 2)) = \log_2(6x + 1) \qquad \text{Product rule for logarithms}$$
$$x^2 + 2x = 6x + 1 \qquad \text{One-to-one property of logarithms}$$
$$x^2 - 4x - 1 = 0 \qquad \text{Solve quadratic equation.}$$
$$x = \frac{4 \pm \sqrt{16 - 4(-1)}}{2} = 2 \pm \sqrt{5}$$

Figure 4.41

Since $2 - \sqrt{5}$ is a negative number, $\log_2(2 - \sqrt{5})$ is undefined and $2 - \sqrt{5}$ is not a solution. The only solution to the equation is $2 + \sqrt{5}$. Check this solution by using a calculator and the base-change formula.

The check is shown with a graphing calculator in Fig. 4.41. □

**b.** Since the difference of two logarithms is equal to the logarithm of a quotient, we can rewrite the left-hand side of the equation:

$$\log(x) - \log(x - 1) = 2$$
$$\log\left(\frac{x}{x - 1}\right) = 2 \qquad \text{Quotient rule for logarithms}$$
$$\frac{x}{x - 1} = 10^2 \qquad \text{Definition of logarithm}$$
$$x = 100x - 100 \qquad \text{Solve for } x.$$
$$-99x = -100$$
$$x = \frac{100}{99}$$

Figure 4.42

Check 100/99 in the original equation, as shown in Fig. 4.42. □

**c.** $2 \cdot \ln(x) = \ln(x + 3) + \ln(x - 1)$

$$\ln(x^2) = \ln(x^2 + 2x - 3) \qquad \text{Power rule, product rule}$$
$$x^2 = x^2 + 2x - 3 \qquad \text{One-to-one property}$$
$$0 = 2x - 3$$
$$\frac{3}{2} = x$$

Use a calculator to check 33/22 in the original equation. The only solution to the equation is 3/2.

▸**TRY THIS.** Solve $\log(x - 1) + \log(3) = \log(x) - \log(4)$. ■

All solutions to logarithmic and exponential equations should be checked in the original equation, because extraneous roots can occur, as in Example 2(a). Use your calculator to check every solution and you will increase your proficiency with your calculator.

## Exponential Equations

An exponential equation with a single exponential expression can usually be solved by using the definition of logarithm, as in Section 4.2.

### EXAMPLE 3 Equations involving a single exponential expression

Solve the equation $(1.02)^{4t-1} = 5$.

**Solution**

Write an equivalent equation, using the definition of logarithm.

$$4t - 1 = \log_{1.02}(5)$$
$$t = \frac{1 + \log_{1.02}(5)}{4} \qquad \text{The exact solution}$$
$$= \frac{1 + \dfrac{\ln(5)}{\ln(1.02)}}{4} \qquad \text{Base-change formula}$$
$$\approx 20.5685$$

Figure 4.43

The approximate solution is 20.5685.
Check with a graphing calculator, as shown in Fig. 4.43.

▸**TRY THIS.** Solve $(1.05)^{3t} = 8$. ■

Note that Example 3 can also be solved by taking the ln (or log) of each side and applying the power rule for logarithms:

$$\ln(1.02^{4t-1}) = \ln(5)$$
$$(4t - 1)\ln(1.02) = \ln(5)$$
$$4t \cdot \ln(1.02) - \ln(1.02) = \ln(5)$$
$$t = \frac{\ln(5) + \ln(1.02)}{4 \cdot \ln(1.02)} \approx 20.5685$$

If an equation has an exponential expression on each side, as in Example 4, then it is best to take the log or ln of each side to solve it.

## EXAMPLE 4 Equations involving two exponential expressions

Find the exact and approximate solutions to each equation.

**a.** $6^x = 7^{x-1}$ **b.** $3^{2x-1} = 5^x$

### Solution

**a.**

$$\begin{aligned}
\log(6^x) &= \log(7^{x-1}) && \text{Take the log of each side.}\\
x \cdot \log(6) &= (x-1)\log(7) && \text{Power rule for logarithms}\\
x \cdot \log(6) &= x \cdot \log(7) - \log(7) && \text{Distributive property}\\
x \cdot \log(6) - x \cdot \log(7) &= -\log(7)\\
x(\log(6) - \log(7)) &= -\log(7) && \text{Factor.}\\
x &= \frac{-\log(7)}{\log(6/7)} && \log(6) - \log(7) = \log(6/7)\\
x &\approx 12.6234
\end{aligned}$$

**b.**

$$\begin{aligned}
\ln(3^{2x-1}) &= \ln(5^x) && \text{Take the natural logarithm of each side.}\\
(2x-1)\ln(3) &= x \cdot \ln(5) && \text{Power rule for logarithms}\\
2x \cdot \ln(3) - \ln(3) &= x \cdot \ln(5) && \text{Distributive property}\\
2x \cdot \ln(3) - x \cdot \ln(5) &= \ln(3)\\
x[2 \cdot \ln(3) - \ln(5)] &= \ln(3)\\
x &= \frac{\ln(3)}{2 \cdot \ln(3) - \ln(5)} && \text{Exact solution}\\
&\approx 1.8691
\end{aligned}$$

Figure 4.44

Check the solution in the original equation, as shown in Fig. 4.44.

▶**TRY THIS.** Solve $3^{x-1} = 2^x$. ■

The technique of Example 4 can be used on any equation of the form $a^M = b^N$. Take the natural logarithm of each side and apply the power rule, to get an equation of the form $M \cdot \ln(a) = N \cdot \ln(b)$. This last equation usually has no exponents and is relatively easy to solve.

## Strategy for Solving Equations

We solved equations involving exponential and logarithmic functions in Sections 4.1 through 4.4. There is no formula that will solve every exponential or logarithmic equation, but the following strategy will help you solve exponential and logarithmic equations.

**STRATEGY**

**Solving Exponential and Logarithmic Equations**

1. If the equation involves a single logarithm or a single exponential expression, then use the definition of logarithm: $y = \log_a(x)$ if and only if $a^y = x$.
2. Use the one-to-one properties when applicable:
   a) if $a^M = a^N$, then $M = N$.
   b) if $\log_a(M) = \log_a(N)$, then $M = N$.

3. If an equation has several logarithms with the same base, then combine them using the product and quotient rules:
   a) $\log_a(M) + \log_a(N) = \log_a(MN)$
   b) $\log_a(M) - \log_a(N) = \log_a(M/N)$
4. If an equation has only exponential expressions with different bases on each side, then take the natural or common logarithm of each side and use the power rule: $a^M = b^N$ is equivalent to $\ln(a^M) = \ln(b^N)$ or $M \cdot \ln(a) = N \cdot \ln(b)$.

## Radioactive Dating

In Section 4.1, we stated that the amount $A$ of a radioactive substance remaining after $t$ years is given by

$$A = A_0 e^{rt},$$

where $A_0$ is the initial amount present and $r$ is the annual rate of decay for that particular substance. A standard measurement of the speed at which a radioactive substance decays is its **half-life.** The half-life of a radioactive substance is the amount of time that it takes for one-half of the substance to decay. Of course, when one-half has decayed, one-half remains.

Now that we have studied logarithms, we can use the formula for radioactive decay to determine the age of an ancient object that contains a radioactive substance. One such substance is potassium-40, which is found in rocks. Once the rock is formed, the potassium-40 begins to decay. The amount of time that has passed since the formation of the rock can be determined by measuring the amount of potassium-40 that has decayed into argon-40. Dating rocks using potassium-40 is known as **potassium-argon dating.**

### EXAMPLE 5 Sauropods in Utah

Our chapter opener topic was a recent dinosaur find in Utah. Since dinosaur bones are too old to contain organic material for radiocarbon dating, paleontologists estimate the age of bones like those of the sauropods found in Utah by dating volcanic debris in surrounding rock using potassium-argon dating. The half-life of potassium-40 is 1.31 billion years. If 92.4% of the original amount of potassium-40 is still present in the rock, then how old is the rock?

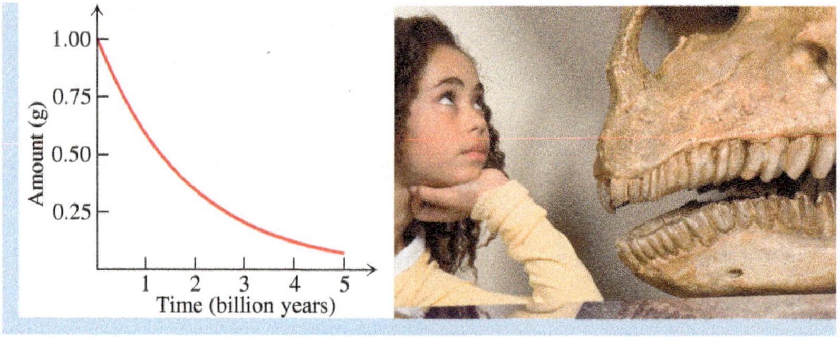


**Figure 4.45**

### Solution

The half-life is the amount of time that it takes for 1 gram to decay to 0.5 gram. Use $A_0 = 1$, $A = 0.5$, and $t = 1.31 \times 10^9$ in the formula $A = A_0 e^{rt}$ to find $r$.

$$0.5 = 1 \cdot e^{(1.31\times10^9)(r)}$$

$$(1.31 \times 10^9)(r) = \ln(0.5) \qquad \text{Definition of logarithm}$$

$$r = \frac{\ln(0.5)}{1.31 \times 10^9}$$

$$\approx -5.29 \times 10^{-10}$$

Now we can find the amount of time that it takes for 1 gram to decay to 0.924 gram. Use $r \approx -5.29 \times 10^{-10}$, $A_0 = 1$, and $A = 0.924$ in the formula.

$$0.924 = 1 \cdot e^{(-5.29\times 10^{-10})(t)}$$

$$(-5.29 \times 10^{-10})(t) = \ln(0.924) \quad \text{Definition of logarithm}$$

$$t = \frac{\ln(0.924)}{-5.29 \times 10^{-10}}$$

$$\approx 150 \text{ million years}$$

So the rock and the sauropods are about 150 million years old.

▸**TRY THIS.** If the half-life of a substance is 1 million years and 40% of the original amount is still present in a rock, then how old is the rock? ■

The radioactive decay model is also used to model the manner in which the human body eliminates a drug. If $A_0$ is the initial dose of a drug, then the amount remaining in the body at time $t$ is given by $A = A_0e^{rt}$ where $r$ is the elimination rate.

### EXAMPLE 6 Drug elimination

The half-life of the antidepressant Prozac in an average man is 2 days. How long does it take for 95% of the original dose to be eliminated from a man's body?

#### Solution

First find the elimination rate by using the fact that the initial amount $A_0$ is reduced to $\frac{1}{2}A_0$ in 2 days:

$$\frac{1}{2}A_0 = A_0e^{2r} \quad \text{Use } A = A_0e^{rt}.$$

$$\frac{1}{2} = e^{2r} \quad \text{Divide each side by } A_0.$$

$$2r = \ln(1/2) \quad \text{Definition of logarithm}$$

$$r = \frac{\ln(1/2)}{2} \approx -0.34657$$

Now find the time. When 95% of the drug is eliminated, only 5% remains. So we find the time that it takes for $A_0$ to be reduced to $0.05A_0$ using the elimination rate $-0.34657$:

$$0.05A_0 = A_0e^{-0.34657t} \quad \text{Use } A = A_0e^{rt}.$$

$$0.05 = e^{-0.34657t} \quad \text{Divide each side by } A_0.$$

$$-0.34657t = \ln(0.05) \quad \text{Definition of logarithm}$$

$$t = \frac{\ln(0.05)}{-0.34657} \approx 8.6$$

Thus, 95% of the original dose is eliminated in about 8.6 days.

▸**TRY THIS.** In a certain woman the half-life of Prozac is 4 days. How long does it take for 99% of the original dose to be eliminated from the woman's body? ■

## Newton's Model for Cooling

Newton's law of cooling states that when a warm object is placed in colder surroundings or a cold object is placed in warmer surroundings, then the difference between the two temperatures decreases in an exponential manner. If $D_0$ is the initial difference in temperature, then the difference $D$ at time $t$ is modeled by the formula

$$D = D_0 e^{kt},$$

where $k$ is a constant that depends on the object and the surroundings. In the next example we use Newton's law of cooling to answer a question that you might have asked yourself as the appetizers were running low.

### EXAMPLE 7 Using Newton's law of cooling

A turkey with a temperature of 40°F is moved to a 350° oven. After 4 hours the internal temperature of the turkey is 170°F. If the turkey is done when its temperature reaches 185°, then how much longer must it cook?

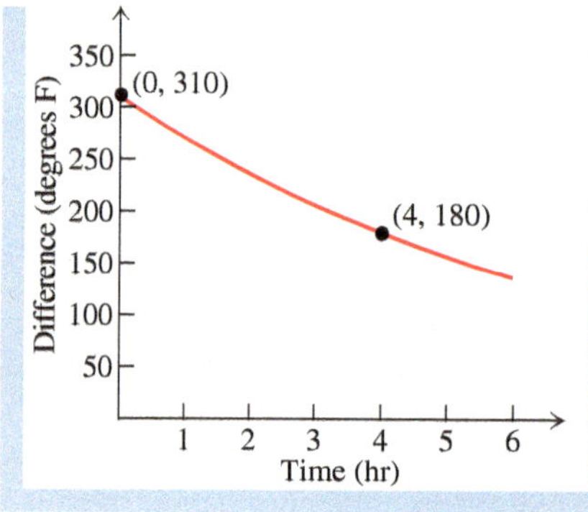

**Figure 4.46**

#### Solution

The initial difference of 310° (350° − 40°) has dropped to a difference of 180° (350° − 170°) after 4 hours. See Fig. 4.45. With this information we can find $k$:

$$180 = 310e^{4k}$$

$$e^{4k} = \frac{180}{310} \qquad \text{Isolate } e^{4k} \text{ by dividing by 310.}$$

$$4k = \ln(18/31) \qquad \text{Definition of logarithm: } e^x = y \Leftrightarrow x = \ln(y)$$

$$k = \frac{\ln(18/31)}{4} \approx -0.1359$$

The turkey is done when the difference in temperature between the turkey and the oven is 165° (the oven temperature 350° minus the turkey temperature 185°). Now find the time for which the difference will be 165°:

$$165 = 310e^{-0.1359t}$$

$$e^{-0.1359t} = \frac{165}{310}$$

$$-0.1359t = \ln(165/310) \qquad \text{Definition of logarithm: } e^x = y \Leftrightarrow x = \ln(y)$$

$$t = \frac{\ln(165/310)}{-0.1359} \approx 4.6404$$

The difference in temperature between the turkey and the oven will be 165° when the turkey has cooked approximately 4.6404 hours. So the turkey must cook about 0.6404 hour (38.4 minutes) longer.

▸**TRY THIS.** A cup of boiling water (212°F) is placed outside, where the temperature is 12°F. In 4 minutes the water temperature is 102°F. How much longer will it take for the water temperature to reach 82°F? ■

## Paying off a Loan

If $n$ is the number of periods per year, $r$ is the annual percentage rate (APR), $t$ is the number of years, and $i$ is the interest rate per period ($i = r/n$), then the periodic payment $R$ that will pay off a loan of $P$ dollars is given by

$$R = P\frac{i}{1 - (1 + i)^{-nt}}.$$

The process of paying off a loan using this formula for $R$ is called **amortization.** Homeowners are often interested in how long it will take to pay off an existing loan if they increase the monthly payment. We can answer that question by using logarithms to solve the formula for $t$.

### EXAMPLE 8 Finding the time

A couple still owes \$90,000 on a house that is financed at 8% annual percentage rate compounded monthly. If they start paying \$1200 per month, when will the loan be paid off?

**Solution**

Use $R = 1200$, $P = 90{,}000$, $n = 12$, and $i = 0.08/12$ in the formula for the monthly payment:

$$\begin{aligned}
1200 &= 90{,}000\frac{0.08/12}{1 - (1 + 0.08/12)^{-12t}}\\
1200(1 - (1 + 0.08/12)^{-12t}) &= 600\\
1 - (1 + 0.08/12)^{-12t} &= 0.5\\
0.5 &= (1 + 0.08/12)^{-12t}\\
\ln(0.5) &= -12t \cdot \ln(1 + 0.08/12)\\
\frac{\ln(0.5)}{-12 \cdot \ln(1 + 0.08/12)} &= t\\
8.6932 &\approx t
\end{aligned}$$

So in approximately 8.6932 years, or about 8 years and 8 months, the loan will be paid off.

▸**TRY THIS.** How long (to the nearest month) does it take to pay off \$120,000 financed at 7.5% APR compounded monthly with payments of \$2000 per month? ■

## FOR thought... True or False? Explain.

**1.** The equation $3(1.02)^x = 21$ is equivalent to $x = \log_{1.02}(7)$.

**2.** If $x - x \cdot \ln(3) = 8$, then $x = \frac{8}{1 - \ln(3)}$.

**3.** The solution to $\ln(x) - \ln(x - 1) = 6$ is $1 - \sqrt{6}$.

**4.** If $2^{x-3} = 3^{2x+1}$, then $x - 3 = \log_2(3^{2x+1})$.

**5.** The exact solution to $3^x = 17$ is 2.5789.

**6.** The equation $\log(x) + \log(x - 3) = 1$ is equivalent to $\log(x^2 - 3x) = 1$.

**7.** The equation $4^x = 2^{x-1}$ is equivalent to $2x = x - 1$.

**8.** The equation $(1.09)^x = 2.3$ is equivalent to $x \cdot \ln(1.09) = \ln(2.3)$.

**9.** $\ln(2) \cdot \log(7) = \log(2) \cdot \ln(7)$

**10.** $\log(e) \cdot \ln(10) = 1$

## EXERCISES 4.4

*Each of these equations involves a single logarithm. Solve each equation. See the strategy for solving exponential and logarithmic equations on pages 317 and 318.*

**1.** $\log_2(x) = 3$

**2.** $\log_3(x) = 0$

**3.** $\log(x + 20) = 2$

**4.** $\log(x - 6) = 1$

**5.** $\log(x^2 - 15) = 1$

**6.** $\log(x^2 - 5x + 16) = 1$

**7.** $\log_x(9) = 2$

**8.** $\log_x(16) = 4$

**9.** $-2 = \log_x(4)$

**10.** $-\frac{1}{2} = \log_x(9)$

**11.** $\log_x(10) = 3$

**12.** $\log_x(5) = 2$

**13.** $\log_8(x) = -\frac{2}{3}$

**14.** $\log_4(x) = -\frac{5}{2}$

*Each of these equations involves more than one logarithm. Solve each equation. Give exact solutions.*

**15.** $\log_2(x + 2) + \log_2(x - 2) = 5$

**16.** $\log_6(w - 1) + \log_6(w - 2) = 1$

**17.** $\log\left(\frac{x - 3}{2}\right) + \log\left(\frac{x + 2}{7}\right) = 0$

**18.** $\log_2\left(\frac{a - 2}{5}\right) + \log_2\left(\frac{a + 3}{10}\right) = 0$

**19.** $\log(x + 1) - \log(x) = 3$

**20.** $\log_5(x) - \log_5(x - 2) = 3$

**21.** $\log_4(x) - \log_4(x + 2) = 2$

**22.** $\log_3(x - 6) - \log_3(2x) = 4$

**23.** $\log(5) = 2 - \log(x)$

**24.** $\log(4) = 1 + \log(x - 1)$

**25.** $\ln(x) + \ln(x + 2) = \ln(8)$

**26.** $\log_3(x) = \log_3(2) - \log_3(x - 2)$

**27.** $\log(4) + \log(x) = \log(5) - \log(x)$

**28.** $\ln(x) - \ln(x + 1) = \ln(x + 3) - \ln(x + 5)$

**29.** $\log_2(x) - \log_2(3x - 1) = 0$

**30.** $\log_3(x) + \log_3(1/x) = 0$

**31.** $x \cdot \ln(3) = 2 - x \cdot \ln(2)$

**32.** $x \cdot \log(5) + x \cdot \log(7) = \log(9)$

*Each of these equations involves a single exponential expression. Solve each equation. Round approximate solutions to four decimal places.*

**33.** $2^{x-1} = 7$

**34.** $5^{3x} = 29$

**35.** $(1.09)^{4x} = 3.4$

**36.** $(1.04)^{2x} = 2.5$

**37.** $3^{-x} = 30$

**38.** $10^{-x+3} = 102$

**39.** $9 = e^{-3x^2}$

**40.** $25 = 10^{-2x}$

*Each of these equations involves more than one exponential expression. Solve each equation. Round approximate solutions to four decimal places.*

**41.** $6^x = 3^{x+1}$

**42.** $2^x = 7^{x-1}$

**43.** $e^{x+1} = 10^x$

**44.** $e^x = 2^{x+1}$

**45.** $2^{x-1} = 4^{3x}$

**46.** $3^{3x-4} = 9^x$

**47.** $6^{x+1} = 12^x$

**48.** $2^x \cdot 2^{x+1} = 4^{x^2+x}$

*Solve each equation. Round approximate solutions to four decimal places.*

**49.** $e^{-\ln(w)} = 3$

**50.** $10^{2 \cdot \log(y)} = 4$

**51.** $(\log(z))^2 = \log(z^2)$

**52.** $\ln(e^x) - \ln(e^6) = \ln(e^2)$

**53.** $4(1.02)^x = 3(1.03)^x$

**54.** $500(1.06)^x = 400(1.02)^{4x}$

**55.** $e^{3\cdot\ln(x^2)-2\cdot\ln(x)} = \ln(e^{16})$

**56.** $\sqrt{\log(x) - 3} = \log(x) - 3$

**57.** $\left(\frac{1}{2}\right)^{2x-1} = \left(\frac{1}{4}\right)^{3x+2}$

**58.** $\left(\frac{2}{3}\right)^{x+1} = \left(\frac{9}{4}\right)^{x+2}$

*Find the approximate solution to each equation by graphing an appropriate function on a graphing calculator and locating the x-intercept. Note that these equations cannot be solved by the techniques that we have learned in this chapter.*

**59.** $2^x = 3^{x-1} + 5^{-x}$

**60.** $2^x = \log(x + 4)$

**61.** $\ln(x + 51) = \log(-48 - x)$

**62.** $2^x = 5 - 3^{x+1}$

**63.** $x^2 = 2^x$

**64.** $x^3 = e^x$

*Solve each problem.*

**65.** *Finding the Rate* If the half-life of a radioactive substance is 10,000 years, then at what rate is it decaying?
HINT The amount goes from $A_0$ to $\frac{1}{2}A_0$ in 10,000 years.

**66.** *Finding the Rate* If the half-life of a drug is 12 hours, then at what rate is it being eliminated from the body?

**67.** *Dating a Bone* A piece of bone from an organism is found to contain 10% of the carbon-14 that it contained when the organism was living. If the half-life of carbon-14 is 5730 years, then how long ago was the organism alive?
HINT First find the rate of decay for carbon-14.

**68.** *Old Clothes* If only 15% of the carbon-14 in a remnant of cloth has decayed, then how old is the cloth?
HINT Use the decay rate for carbon-14 from the previous problem.

**69.** *Dating a Tree* How long does it take for 12 g of carbon-14 in a tree trunk to be reduced to 10 g of carbon-14 by radioactive decay?

**70.** *Carbon-14 Dating* How long does it take for 2.4 g of carbon-14 to be reduced to 1.3 g of carbon-14 by radioactive decay?

**71.** *Radioactive Waste* If 25 g of radioactive waste reduces to 20 g of radioactive waste after 8000 years, then what is the half-life for this radioactive element?
HINT First find the rate of decay, then find the time for half of it to decay.

**72.** *Finding the Half-Life* If 80% of a radioactive element remains radioactive after 250 million years, then what percent remains radioactive after 600 million years? What is the half-life of this element?

**73.** *Lorazepam* The drug lorazepam, used to relieve anxiety and nervousness, has a half-life of 14 hours. Its chemical structure is shown in the accompanying figure. If a doctor prescribes one 2.5-milligram tablet every 24 hours, then what percentage of the last dosage remains in the patient's body when the next dosage is taken?

Lorazepam

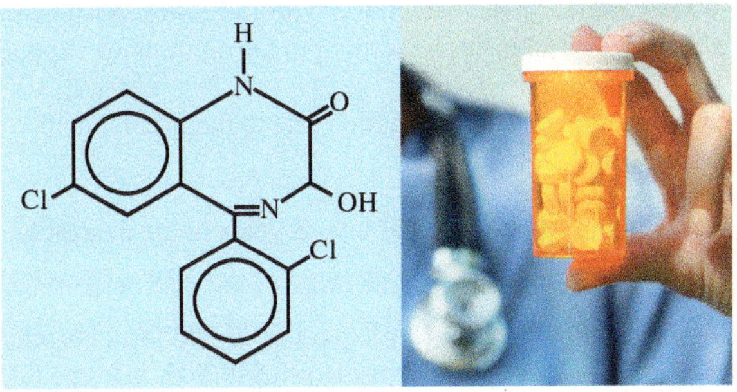


**Figure for Exercise 73**

**74.** *Drug Build-Up* The level of a prescription drug in the human body over time can be found using the formula

$$L = \frac{D}{1 - (0.5)^{n/h}}$$

where $D$ is the amount taken every $n$ hours and $h$ is the drug's half-life in hours.

**a.** If 2.5 milligrams of lorazepam with a half-life of 14 hours is taken every 24 hours, then to what level does the drug build up over time?

**b.** If a doctor wants the level of lorazepam to build up to a level of 5.58 milligrams in a patient taking 2.5 milligram doses, then how often should the doses be taken?

**c.** What is the difference between taking 2.5 milligrams of lorazepam every 12 hours and taking 5 milligrams every 24 hours?

**75.** *Dead Sea Scrolls* Willard Libby, a nuclear chemist from the University of Chicago, developed radiocarbon dating in the 1940s. This dating method, effective on specimens up to about 40,000 years old, works best on objects like shells, charred bones, and plants that contain organic matter (carbon). Libby's first great success came in 1951 when he dated the Dead Sea Scrolls. Carbon-14 has a half-life of 5730 years. If Libby found 79.3% of the original carbon-14 still present, then in about what year were the scrolls made?

**76.** *Leakeys Date Zinjanthropus* In 1959, archaeologists Louis and Mary Leakey were exploring Olduvai Gorge, Tanzania, when they uncovered the remains of *Zinjanthropus*, an early hominid with traits of both ape and man. Dating of the volcanic rock revealed that 91.2% of the original potassium had not decayed into argon. What age was assigned to the rock and the bones of *Zinjanthropus*? The radioactive decay of potassium-40 to argon-40 occurs with a half-life of 1.31 billion years.

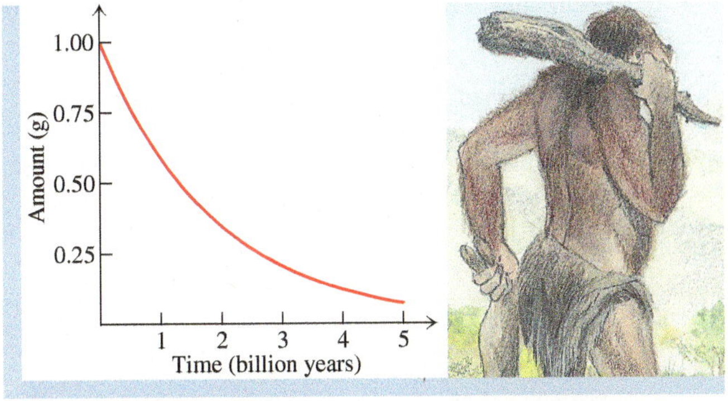


**Figure for Exercise 76**

**77.** *Depreciating Camry* Antoinette paid \$21,075 for a Toyota Camry. After three years it was worth \$11,981. Assume that the price is decreasing according to the continuous exponential decay model $P = P_0e^{rt}$.

**a.** Find the annual depreciation rate to the nearest tenth of a percent.

**b.** Find the value of the car after 5 years to the nearest hundred dollars.

**78.** *Depreciating Porsche* Byron paid \$107,500 for a Porsche 911 Turbo. After four years it was worth \$49,900. Assume that the price is decreasing according to the continuous exponential decay model $P = P_0e^{rt}$.

**a.** Find the annual depreciation rate to the nearest tenth of a percent.

**b.** Find the value of the car after 7 years to the nearest hundred dollars.

**79.** *Growth of Blogging* In 2009 there were about 0.1 million blog sites on the Internet. In 2011 there were 4.8 million. Assuming that the number of blog sites is experiencing continuous exponential growth, predict the number of blog sites (to the nearest million) in 2014.

**80.** *Growth of the Internet* Only 1.7% of the world's population used the Internet in 1997, whereas 28.8% of the world's population used it in 2010. Assuming continuous exponential growth, find the year in which the percentage will reach 100%.

**81.** *Cooking a Roast* James knows that to get well-done beef, it should be brought to a temperature of 170°F. He placed a sirloin tip roast with a temperature of 35°F in an oven with a temperature of 325°, and after 3 hr the temperature of the roast was 140°. How much longer must the roast be in the oven to get it well done? If the oven temperature is set at 170°, how long will it take to get the roast well done?

**HINT** The difference between the roast temperature and the oven temperature decreases exponentially.

**82.** *Room Temperature* Marlene brought a can of polyurethane varnish that was stored at 40°F into her shop, where the temperature was 74°. After 2 hr the temperature of the varnish was 58°. If the varnish must be 68° for best results, then how much longer must Marlene wait until she uses the varnish?

**83.** *Time of Death* A detective discovered a body in a vacant lot at 7 A.M. and found that the body temperature was 80°F. The county coroner examined the body at 8 A.M. and found that the body temperature was 72°. Assuming that the body temperature was 98° when the person died and that the air temperature was a constant 40° all night, what was the approximate time of death?

**84.** *Cooling Hot Steel* A blacksmith immersed a piece of steel at 600°F into a large bucket of water with a temperature of 65°. After 1 min the temperature of the steel was 200°. How much longer must the steel remain immersed to bring its temperature down to 100°?

**85.** *Paying off a Loan* Find the time (to the nearest month) that it takes to pay off a loan of \$100,000 at 9% APR compounded monthly with payments of \$1250 per month.

**86.** *Solving for Time* Solve the formula

$$R = P\frac{i}{1 - (1 + i)^{-nt}}$$

for $t$. Then use the result to find the time (to the nearest month) that it takes to pay off a loan of \$48,265 at $8\frac{3}{4}$% APR compounded monthly with payments of \$700 per month.

**87.** *Equality of Investments* Fiona invested \$1000 at 6% compounded continuously. At the same time, Maria invested \$1100 at 6% compounded daily. How long will it take (to the nearest day) for their investments to be equal in value?

**88.** *Depreciation and Inflation* Boris won a \$35,000 luxury car on *Wheel of Fortune*. He plans to keep it until he can trade it evenly for a new compact car that currently costs \$10,000. If the value of the luxury car decreases by 8% each year and the cost of the compact car increases by 5% each year, then in how many years will he be able to make the trade?

**89.** *Population of Rabbits* The population of rabbits in a national forest is modeled by the formula $P = 12{,}300 + 1000 \cdot \ln(t + 1)$, where $t$ is the number of years from the present.

**a.** How many rabbits are now in the forest?

**b.** Use the accompanying graph to estimate the number of years that it will take for the rabbit population to reach 15,000.

**c.** Use the formula to determine the number of years that it will take for the rabbit population to reach 15,000.

**Figure for Exercise 89**

**90.** *Population of Foxes* The population of foxes in the forest of Exercise 89 appears to be growing according to the formula $P = 400 + 50 \cdot \ln(90t + 1)$. When the population of foxes is equal to 5% of the population of rabbits, the system is considered to be out of ecological balance. In how many years will the system be out of balance?

**91.** *Habitat Destruction* Biologists use the species-area curve $n = k \log(A)$ to estimate the number of species $n$ that live in a region of area $A$, where $k$ is a constant.

**a.** If 2500 species live in a rainforest of 400 square kilometers, then how many species will be left when half of this rainforest is destroyed by logging?

**b.** A rainforest of 1200 square kilometers supported 3500 species in 1950. Due to intensive logging, what remains of this forest supported only 1000 species in 2000. What percent of this rainforest has been destroyed?

**Figure for Exercise 91**

**92.** *Extinction of Species* In 1980 an Amazon rainforest of area $A$ contained $n$ species. In the year 2000, after intensive logging, this rainforest was reduced to an area that contained only half as many species. Use the species-area formula from the previous exercise to find the area in 2000.

**93.** *Visual Magnitude of a Star* If all stars were at the same distance, it would be a simple matter to compare their brightness. However, the brightness that we see, the apparent visual magnitude $m$, depends on a star's intrinsic brightness, or absolute visual magnitude $M_V$, and the distance $d$ from the observer in parsecs (1 parsec = 3.262 light years), according to the formula $m = M_V - 5 + 5 \cdot \log(d)$. The values of $M_V$ range from $-8$ for the intrinsically brightest stars to $+15$ for the intrinsically faintest stars. The nearest star to the sun, Alpha Centauri, has an apparent visual magnitude of 0 and an absolute visual magnitude of 4.39. Find the distance $d$ in parsecs to Alpha Centauri.

**94.** *Visual Magnitude of Deneb* The star Deneb is 490 parsecs away and has an apparent visual magnitude of 1.26, which means that it is harder to see than Alpha Centauri. Use the formula from Exercise 93 to find the absolute visual magnitude of Deneb. Is Deneb intrinsically very bright or very faint? If Deneb and Alpha Centauri were both 490 parsecs away, then which would appear brighter?

**95.** *Price per Gigabyte* According to Moore's law, the performance of electronic parts grows exponentially while the cost of those parts declines exponentially. The accompanying figure shows the decline in price $P$ for a gigabyte of hard drive storage for the years 1982 through 2005. The relationship looks linear because the price axis has a log scale.

**a.** Find a formula for $P$ in terms of $x$, where $x$ is the number of years since 1982. See Exercise 139 of Section 4.2.

**b.** Find $P$ in 1991.

**c.** Find the year in which the price will be \$0.25.

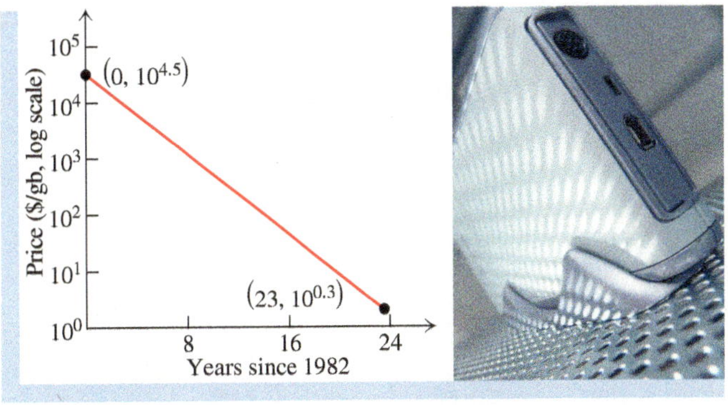


**Figure for Exercise 95**

**96.** *Hard Drive Capacity* The accompanying figure shows how hard drive capacity $C$ has increased for the years 1982 through 2005. The relationship looks linear because the capacity axis has a log scale.

**a.** Find a formula for $C$ in terms of $x$, where $x$ is the number of years since 1982. See the previous exercise.

**b.** Find $C$ in 1994.

**c.** Find the year in which the capacity will be 1000 gigabytes.

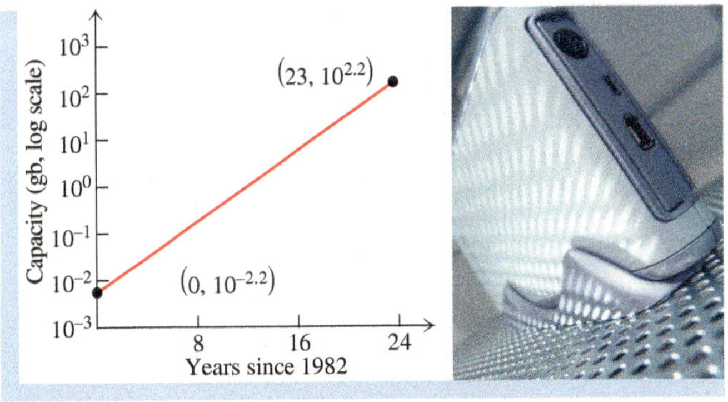


**Figure for Exercise 96**

**97.** *Noise Pollution* The level of a sound in decibels (db) is determined by the formula

$$\text{sound level} = 10 \cdot \log(I \times 10^{12})\,\text{db},$$

where $I$ is the intensity of the sound in watts per square meter. To combat noise pollution, a city has an ordinance prohibiting sounds above 90 db on a city street. What value of $I$ gives a sound of 90 db?

**98.** *Doubling the Sound Level* A small stereo amplifier produces a sound of 50 db at a distance of 20 ft from the speakers. Use the formula from Exercise 97 to find the intensity of the sound at this point in the room. If the intensity just found is doubled, what happens to the sound level? What must the intensity be to double the sound level to 100 db?

**99.** *Present Value of a CD* What amount (present value) must be deposited today in a certificate of deposit so that the investment will grow to \$20,000 in 18 years at 6% compounded continuously.

**100.** *Present Value of a Bond* A \$50 U.S. Savings Bond paying 6.22% compounded monthly matures in 11 years 2 months. What is the present value of the bond?

**101.** *Two-Parent Families* The percentage of households with children that consist of two-parent families is shown in the following table (U.S. Census Bureau, www.census.gov).

**Table for Exercise 101**

| Year | Two Parents |
|---|---|
| 1970 | 85% |
| 1980 | 77 |
| 1990 | 73 |
| 1992 | 71 |
| 1994 | 69 |
| 1996 | 68 |
| 1998 | 68 |
| 2000 | 68 |

**a.** Use logarithmic regression on a graphing calculator to find the best-fitting curve of the form $y = a + b \cdot \ln(x)$, where $x = 0$ corresponds to 1960.

**b.** Use your equation to predict the percentage of two-parent families in 2010.

**c.** In what year will the percentage of two-parent families reach 50%?

**d.** Graph your equation and the data on your graphing calculator. Does this logarithmic model look like another model that we have used?

**102.** *Thermistor Resistance* A thermistor is a resistor whose resistance varies with the temperature, as shown in the figure. The relationship between temperature $T$ in °C and resistance $R$ in ohms for a BetaTHERM thermistor model 100K6A1 is given by the Steinhart-Hart equation

$$T = \frac{1}{A + B \cdot \ln(R) + C \cdot [\ln(R)]^3} - 273.15,$$

where

$$A = 8.27153 \times 10^{-4}$$
$$B = 2.08796 \times 10^{-4}$$
$$C = 8.060985 \times 10^{-8}.$$

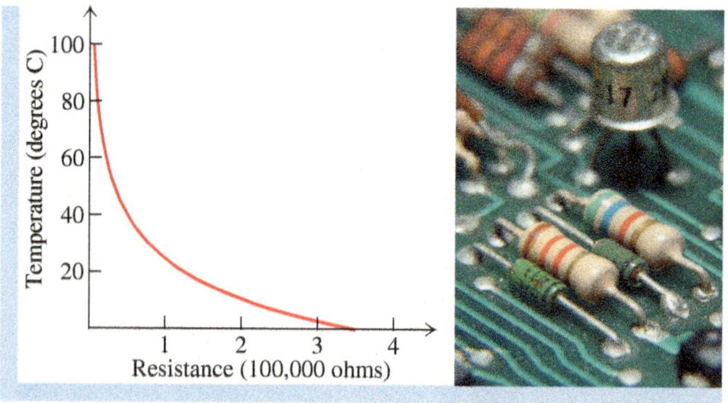


**Figure for Exercise 102**

At what temperature is the resistance $1 \times 10^5$ ohms? Use a graphing calculator to find the resistance when $T = 35°C$.

## Foreshadowing Calculus

To evaluate an exponential or logarithmic function we simply press a button on a calculator. But what does the calculator do to find the answer? The next two exercises show formulas from calculus that are used to evaluate $e^x$ and $\ln(1 + x)$.

**103.** *Infinite Series for $e^x$* The following formula from calculus is used to compute values of $e^x$:

$$e^x = 1 + x + \frac{x^2}{2!} + \frac{x^3}{3!} + \frac{x^4}{4!} + \cdots + \frac{x^n}{n!} + \cdots,$$

where $n! = 1 \cdot 2 \cdot 3 \cdot \cdots \cdot n$ for any positive integer $n$. The notation $n!$ is read "$n$ factorial." For example, $3! = 1 \cdot 2 \cdot 3 = 6$. In calculating $e^x$, the more terms that we use from the formula, the closer we get to the true value of $e^x$. Use the first five terms of the formula to estimate the value of $e^{0.1}$ and compare your result to the value of $e^{0.1}$ obtained using the $e^x$-key on your calculator.

**104.** *Infinite Series for Logarithms* The following formula from calculus can be used to compute values of natural logarithms:

$$\ln(1 + x) = x - \frac{x^2}{2} + \frac{x^3}{3} - \frac{x^4}{4} + \cdots,$$

where $-1 < x < 1$. The more terms that we use from the formula, the closer we get to the true value of $\ln(1 + x)$. Find $\ln(1.4)$ by using the first five terms of the series and compare your result to the calculator value for $\ln(1.4)$.

## RETHINKING

**105.** Simplify $\log_6(4) + \log_6(9)$.

**106.** Find the approximate rational solution to the equation $1.56^{x-1} = 9.8$. Round the answer to four decimal places.

**107.** Solve $\log_x(9.8) = 2.4$. Round to four decimal places.

**108.** The graph of $y = 2^x$ is reflected in the $x$-axis, translated 5 units to the right, and then 9 units upward. Find the equation of the curve in its final position.

**109.** Let $f(x) = 3^{x-5}$ and $g(x) = \log_3(x) + 5$. Find $(g \circ f)(x)$.

**110.** Factor $3x - 9 + wx - 3w$.

## THINKING OUTSIDE THE BOX XXXIV

*Pile of Pipes* Six pipes are placed in a pile as shown in the diagram. The three pipes on the bottom each have a radius of 2 feet. The two pipes on top of those each have a radius of 1 foot. At the top of the pile is a pipe that is tangent to the two smaller pipes and one of the larger pipes. What is its exact radius?

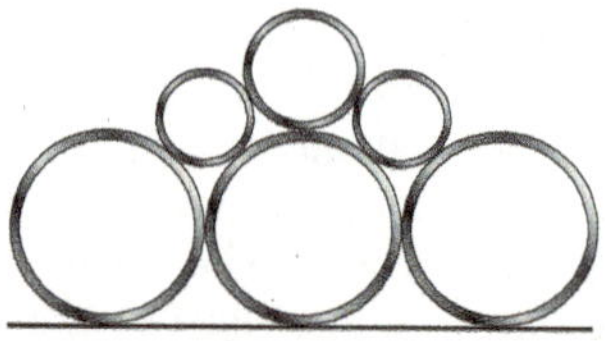

**Figure for Thinking Outside the Box XXXIV**

## POP QUIZ 4.4

*Solve each equation. Round approximate solutions to four decimal places.*

1. $\log_5(x) = 4$
2. $\log(x + 1) = 3$
3. $\ln(x) + \ln(x - 1) = \ln(12)$
4. $3^{x-5} = 10$
5. $8^x = 3^{x+5}$

## LINKING concepts... For Individual or Group Explorations

### The Logistic Growth Model

*When a virus infects a finite population of size P in which no one is immune, the virus spreads slowly at first, then more rapidly as more people are infected, and finally slows down when nearly everyone has been infected. This situation is modeled by a* ***logistic curve*** *of the form*

$$n = \frac{P}{1 + (P - 1) \cdot e^{-ct}}$$

*where n is the number of people who have caught the virus on or before day t, and c is a positive constant.*

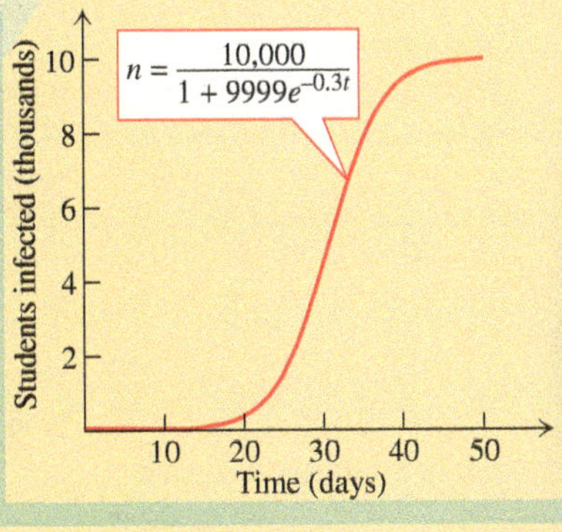

**a)** According to the model, how many people have caught the virus at time $t = 0$?

**b)** Now consider what happens when one student carrying a flu virus returns from spring break to a university of 10,000 students. For $c = 0.1, 0.2$, and so on through $c = 0.9$, graph the logistic curve

$$y_1 = 10000/(1 + 9999e\wedge(-cx))$$

and the daily number of new cases

$$y_2 = y_1(x) - y_1(x - 1).$$

For each value of $c$ use the graph of $y_2$ to find the day on which the flu is spreading most rapidly.

**c)** The Health Center estimated that the greatest number of new cases of the flu occurred on the 19th day after the students returned from spring break. What value of $c$ should be used to model this situation? How many new cases occurred on the 19th day?

**d)** Algebraically find the day on which the number of infected students reached 9000.

**e)** The Health Center has a team of doctors from Atlanta arriving on the 30th day to help with this three-day flu epidemic. What do you think of this plan?

# Highlights

## 4.1 Exponential Functions and Their Applications

| | | |
|---|---|---|
| **Exponential Function** | $f(x) = a^x$ for $a > 0$ and $a \neq 1$<br>Domain: $(-\infty, \infty)$, Range: $(0, \infty)$<br>Horizontal asymptote: $y = 0$ | $f(x) = 2^x, g(x) = e^x$ |
| **Increasing and Decreasing** | $f(x) = a^x$ is increasing on $(-\infty, \infty)$ if $a > 1$, decreasing on $(-\infty, \infty)$ if $0 < a < 1$. | $f(x) = 3^x$ is increasing.<br>$g(x) = (0.4)^x$ is decreasing. |
| **One-to-One** | If $a^{x_1} = a^{x_2}$, then $x_1 = x_2$. | $2^{x-1} = 2^3 \Rightarrow x - 1 = 3$ |
| **Amount Formula** | $P$ dollars, annual interest rate $r$, for $t$ years<br>Compounded periodically: $n$ periods per year,<br>$A = P\left(1 + \frac{r}{n}\right)^{nt}$<br>Compounded continuously:<br>$A = Pe^{rt} (e \approx 2.718)$ | $P = \$1000, r = 5\%, t = 10$ yr<br>Compounded monthly:<br>$A = 1000\left(1 + \frac{0.05}{12}\right)^{120}$<br>Compounded continuously:<br>$A = 1000e^{0.05(10)}$ |

## 4.2 Logarithmic Functions and Their Applications

| | | |
|---|---|---|
| **Logarithmic Function** | $f(x) = \log_a(x)$ for $a > 0$ and $a \neq 1$<br>$y = \log_a(x) \Leftrightarrow a^y = x$<br>Domain: $(0, \infty)$, Range: $(-\infty, \infty)$<br>Vertical asymptote: $x = 0$<br>Common log: base 10, $f(x) = \log(x)$<br>Natural log: base $e$, $f(x) = \ln(x)$ | $f(x) = \log_2(x)$<br>$f(32) = \log_2(32) = 5$<br>$f(1) = \log_2(1) = 0$<br>$f(1/4) = \log_2(1/4) = -2$ |
| **Increasing and Decreasing** | $f(x) = \log_a(x)$ is increasing on $(0, \infty)$ if $a > 1$, decreasing on $(0, \infty)$ if $0 < a < 1$. | $f(x) = \ln(x)$ is increasing.<br>$g(x) = \log_{1/2}(x)$ is decreasing. |
| **One-to-One** | If $\log_a(x_1) = \log_a(x_2)$, then $x_1 = x_2$. | $\log_2(3x) = \log_2(4) \Rightarrow 3x = 4$ |
| **Inverse Functions** | The inverse of a logarithmic function is an exponential function (and vice versa). Find the inverse using switch-and-solve. | $f(x) = \log_2(x) + 3$<br>$y = \log_2(x) + 3$<br>$x = \log_2(y) + 3$<br>$x - 3 = \log_2(y)$<br>$y = 2^{x-3}$<br>$f^{-1}(x) = 2^{x-3}$ |

## 4.3 Rules of Logarithms

| | | |
|---|---|---|
| **Inverse Rules** | $\log_a(a^x) = x$ and $a^{\log_a(x)} = x$ | $\ln(e^5) = 5, 10^{\log(7)} = 7$ |
| **Product Rule** | $\log_a(MN) = \log_a(M) + \log_a(N)$ | $\log_2(8x) = 3 + \log_2(x)$ |
| **Quotient Rule** | $\log_a(M/N) = \log_a(M) - \log_a(N)$ | $\log(1/2) = \log(1) - \log(2)$ |
| **Power Rule** | $\log_a(M^N) = N \cdot \log_a(M)$ | $\ln(e^3) = 3 \cdot \ln(e)$ |
| **Base-Change Formula** | $\log_a(M) = \frac{\log_b(M)}{\log_b(a)}$ | $\log_3(7) = \frac{\ln(7)}{\ln(3)} = \frac{\log(7)}{\log(3)}$ |

## 4.4 More Equations and Applications

| | | |
|---|---|---|
| **Equation-Solving Strategy** | **1.** Use $y = \log_a(x) \Leftrightarrow a^y = x$ on equations with a single logarithm or single exponential. | $\log_2(x) = 3 \Leftrightarrow x = 2^3$ |
| | **2.** Use the one-to-one properties to eliminate logarithms or exponential expressions. | $\log(x^2) = \log(5) \Leftrightarrow x^2 = 5$<br>$e^x = e^{2x-1} \Leftrightarrow x = 2x - 1$ |
| | **3.** Combine logarithms using the product and quotient rules. | $\ln(x) + \ln(2) = 3$<br>$\ln(2x) = 3$ |
| | **4.** Take a logarithm of each side and use the power rule. | $3^x = 4^{2x}$<br>$x \cdot \ln(3) = 2x \cdot \ln(4)$ |

# Chapter 4 Review Exercises

*Simplify each expression.*

**1.** $2^6$ **2.** $\ln(e^2)$ **3.** $\log_2(64)$

**4.** $3 + 2 \cdot \log(10)$ **5.** $\log_9(1)$ **6.** $5^{\log_5(99)}$

**7.** $\log_2(2^{17})$ **8.** $\log_2(\log_2(16))$

**9.** $\sqrt{4^{2/\log_3(4)+2\cdot\log_9(3)}}$ **10.** $\sqrt[3]{5^{1/\log_2(5)} + 3^{1/\log_6(3)}}$

*Let* $f(x) = 2^x$, $g(x) = 10^x$, *and* $h(x) = \log_2(x)$. *Simplify each expression.*

**11.** $f(5)$ **12.** $g(-1)$ **13.** $\log(g(3))$

**14.** $g(\log(5))$ **15.** $(h \circ f)(9)$ **16.** $(f \circ h)(7)$

**17.** $g^{-1}(1000)$ **18.** $g^{-1}(1)$ **19.** $h(1/8)$

**20.** $f(1/2)$ **22.** $f^{-1}(8)$ **22.** $(f^{-1} \circ f)(13)$

*Rewrite each expression as a single logarithm.*

**23.** $\log(x - 3) + \log(x)$ **24.** $(1/2)\ln(x) - 2 \cdot \ln(y)$

**25.** $2 \cdot \ln(x) + \ln(y) + \ln(3)$

**26.** $3 \cdot \log_2(x) - 2 \cdot \log_2(y) + \log_2(z)$

*Rewrite each expression as a sum or difference of multiples of logarithms.*

**27.** $\log(3x^4)$ **28.** $\ln\left(\frac{x^5}{y^3}\right)$

**29.** $\log_3\left(\frac{5\sqrt{x}}{y^4}\right)$ **30.** $\log_2\left(\sqrt{xy^3}\right)$

*Rewrite each expression in terms of* $\ln(2)$ *and* $\ln(5)$.

**31.** $\ln(10)$ **32.** $\ln(0.4)$ **33.** $\ln(50)$ **34.** $\ln(\sqrt{20})$

*Find the exact solution to each equation.*

**35.** $\log(x) = 10$ **36.** $\log_3(x + 1) = -1$

**37.** $\log_x(81) = 4$ **38.** $\log_x(1) = 0$

**39.** $\log_{1/3}(27) = x + 2$ **40.** $\log_{1/2}(4) = x - 1$

**41.** $3^{x+2} = \frac{1}{9}$ **42.** $2^{x-1} = \frac{1}{4}$

**43.** $e^{x-2} = 9$ **44.** $\frac{1}{2^{1-x}} = 3$

**45.** $4^{x+3} = \frac{1}{2^x}$ **46.** $3^{2x-1} \cdot 9^x = 1$

**47.** $\log(x) + \log(2x) = 5$ **48.** $\log(x + 90) - \log(x) = 1$

**49.** $\log_2(x) + \log_2(x - 4) = \log_2(x + 24)$

**50.** $\log_5(x + 18) + \log_5(x - 6) = 2 \cdot \log_5(x)$

**51.** $2 \cdot \ln(x + 2) = 3 \cdot \ln(4)$

**52.** $x \cdot \log_2(12) = x \cdot \log_2(3) + 1$

**53.** $x \cdot \log(4) = 6 - x \cdot \log(25)$

**54.** $\log(\log(x)) = 1$

*Find the missing coordinate so that each ordered pair satisfies the given equation.*

**55.** $y = \left(\frac{1}{3}\right)^x$: $(-1, \ )$, $( \ , 27)$, $(-1/2, \ )$, $( \ , 1)$

**56.** $y = \log_9(x - 1)$: $(2, \ )$, $(4, \ )$, $( \ , 3/2)$, $( \ , -1)$

*Match each equation to one of the graphs (a)–(h).*

**57.** $y = 2^x$ **58.** $y = 2^{-x}$

**59.** $y = \log_2(x)$ **60.** $y = \log_{1/2}(x)$

**61.** $y = 2^{x+2}$ **62.** $y = \log_2(x + 2)$

**63.** $y = 2^x + 2$ **64.** $y = 2 + \log_2(x)$

**a.**

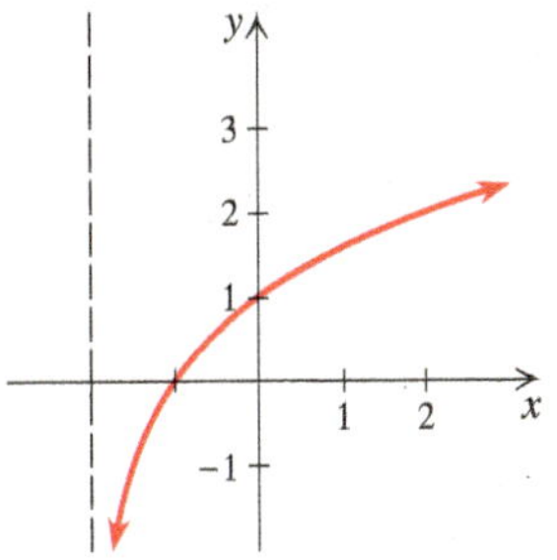


**b.**

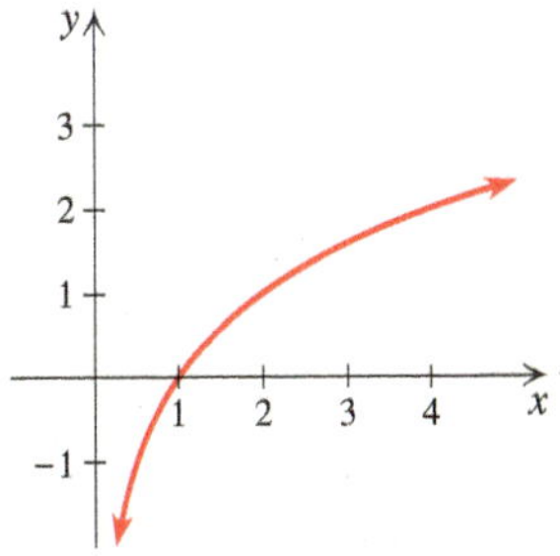


**c.**

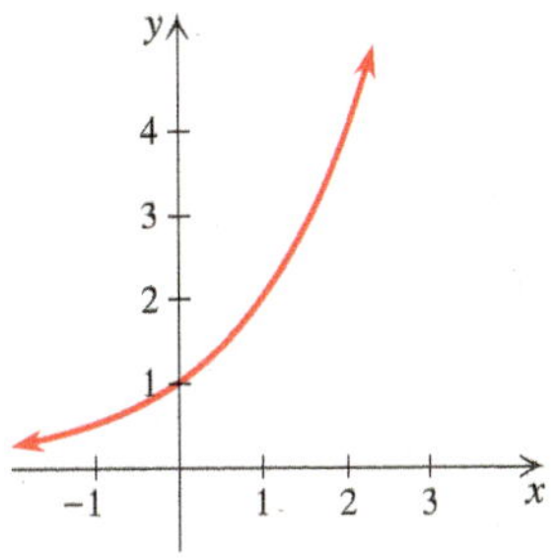


**d.**

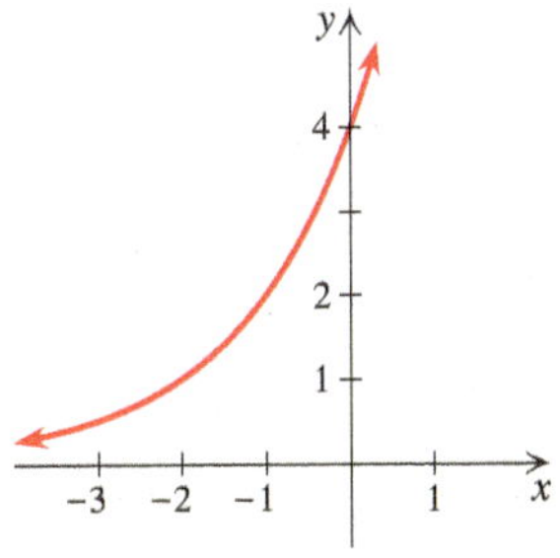

**e.**

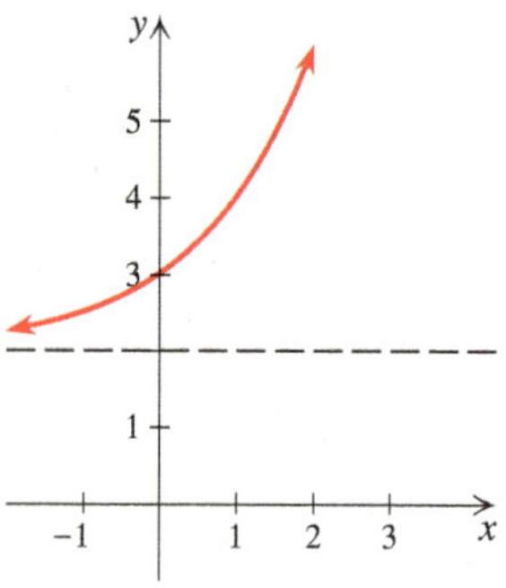


**f.**

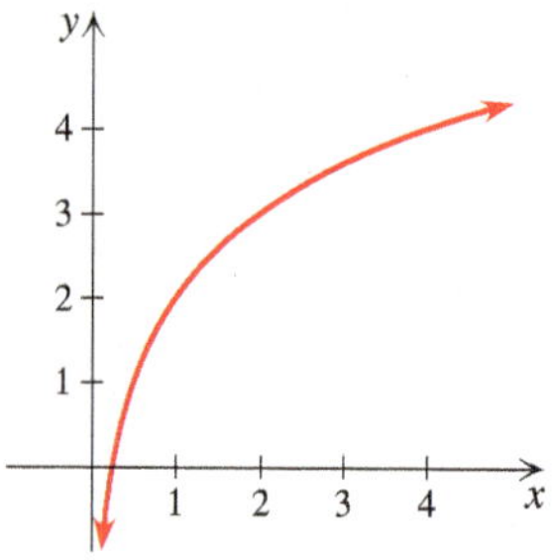


**g.**

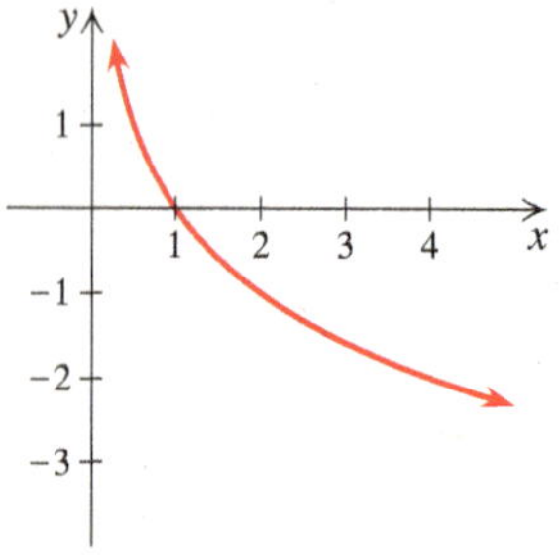


**h.**

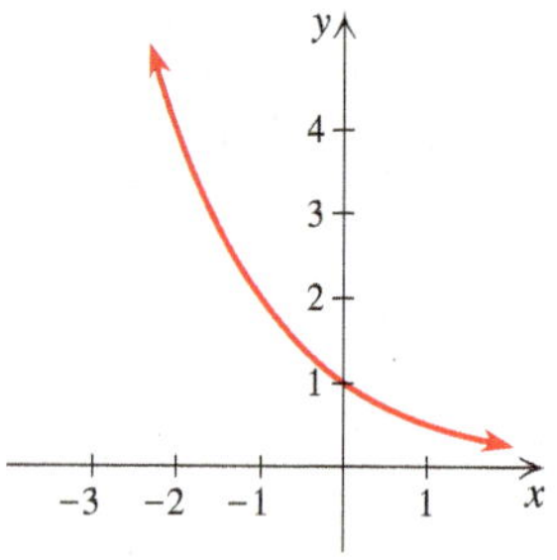


*Sketch the graph of each function. State the domain, the range, and whether the function is increasing or decreasing. Identify any asymptotes.*

**65.** $f(x) = 5^x$

**66.** $f(x) = e^x$

**67.** $f(x) = 10^{-x}$

**68.** $f(x) = (1/2)^x$

**69.** $y = \log_3(x)$

**70.** $y = \log_5(x)$

**71.** $y = 1 + \ln(x + 3)$

**72.** $y = 3 - \log_2(x)$

**73.** $f(x) = 1 + 2^{x-1}$

**74.** $f(x) = 3 - 2^{x+1}$

**75.** $y = \log_3(-x + 2)$

**76.** $y = 1 + \log_3(x + 2)$

*For each function $f$, find $f^{-1}$.*

**77.** $f(x) = 7^x$

**78.** $f(x) = 3^x$

**79.** $f(x) = \log_5(x)$

**80.** $f(x) = \log_8(x)$

**81.** $f(x) = 3 \cdot \log(x - 1)$

**82.** $f(x) = \log_2(x + 3) - 5$

**83.** $f(x) = e^{x+2} - 3$

**84.** $f(x) = 2 \cdot 3^x + 1$

*Use a calculator to find an approximate solution to each equation. Round answers to four decimal places.*

**85.** $3^x = 10$

**86.** $4^{2x} = 12$

**87.** $\log_3(x) = 1.876$

**88.** $\log_5(x + 2) = 2.7$

**89.** $5^x = 8^{x+1}$

**90.** $3^x = e^{x+1}$

*Use the rules of logarithms to determine whether each equation is true or false. Explain your answer.*

**91.** $\log_3(81) = \log_3(9) \cdot \log_3(9)$

**92.** $\log(81) = \log(9) \cdot \log(9)$

**93.** $\ln(3^2) = (\ln(3))^2$

**94.** $\ln\left(\frac{5}{8}\right) = \frac{\ln(5)}{\ln(8)}$

**95.** $\log_2(8^4) = 12$

**96.** $\log(8.2 \times 10^{-9}) = -9 + \log(8.2)$

**97.** $\log(1006) = 3 + \log(6)$

**98.** $\frac{\log_2(16)}{\log_2(4)} = \log_2(16) - \log_2(4)$

**99.** $\frac{\log_2(8)}{\log_2(16)} = \log_2(8) - \log_2(16)$

**100.** $\ln(e^{(e^x)}) = e^x$

**101.** $\log_2(25) = 2 \cdot \log(5)$

**102.** $\log_3\left(\frac{5}{7}\right) = \log(5) - \log(7)$

**103.** $\log_2(7) = \frac{\log(7)}{\log(2)}$

**104.** $\frac{\ln(17)}{\ln(3)} = \frac{\log(17)}{\log(3)}$

*Solve each problem.*

**105.** *Comparing pH* If the hydrogen ion concentration of liquid A is 10 times the hydrogen ion concentration of liquid B, then how does the pH of A compare with the pH of B? See Exercise 141 in Section 4.2 and Exercise 102 in Section 4.3.

**106.** *Solving a Formula* Solve the formula $A = P + Ce^{-kt}$ for $t$.

**107.** *Future Value* If \$50,000 is deposited in a bank account paying 5% compounded quarterly, then what will be the value of the account at the end of 18 years?

**108.** *Future Value* If \$30,000 is deposited in First American Savings and Loan in an account paying 6.18% compounded continuously, then what will be the value of the account after 12 years and 3 months?

**109.** *Doubling Time with Quarterly Compounding* How long (to the nearest quarter) will it take for the investment of Exercise 107 to double?

**110.** *Doubling Time with Continuous Compounding* How long (to the nearest day) will it take for the investment of Exercise 108 to double?

**111.** *Finding the Half-Life* The number of grams $A$ of a certain radioactive substance present at time $t$ is given by the formula $A = 25e^{-0.00032t}$, where $t$ is the number of years from the present. How many grams are present initially? How many grams are present after 1000 years? What is the half-life of this substance?

**112.** *Comparing Investments* Shinichi invested \$800 at 6% compounded continuously. At the same time, Toshio invested \$1000 at 5% compounded monthly. How long (to the nearest month) will it take for their investments to be equal in value?

**113.** *Learning Curve* According to an educational psychologist, the number of words learned by a student of a foreign language after $t$ hours in the language laboratory is given by $f(t) = 40{,}000(1 - e^{-0.0001t})$. How many hours would it take to learn 10,000 words?

**114.** *Pediatric Tuberculosis* Health researchers divided the state of Maryland into regions according to per capita income and found a relationship between per capita income $i$ (in thousands) and the percentage of pediatric TB cases $P$. The equation $P = 31.5(0.935)^i$ can be used to model the data.

**a.** What percentage of TB cases would you expect to find in the region with a per capita income of \$8000?

**b.** What per capita income would you expect in the region where only 2% of the TB cases occurred?

**115.** *Sea Worthiness* Three measures of a boat's sea worthiness are given in the accompanying table, where $A$ is the sail area in square feet, $d$ is the displacement in pounds, $L$ is the length at the water line in feet, and $b$ is the beam in feet. The Freedom 40 is a \$291,560 sailboat with a sail area of 1026 $ft^2$, a displacement of 25,005 lb, a beam of 13 ft 6 in., and a length at the water line of 35 ft 1 in. Its sail area-displacement ratio is 19.20, its displacement-length ratio is 258.51, and its capsize screening value is 1.85. Find $x$, $y$, and $z$.

**Table for Exercise 115**

| Measure | Expression |
|---|---|
| Sail area-displacement ratio | $A\left(\dfrac{d}{64}\right)^x$ |
| Displacement-length ratio | $\dfrac{d}{2240}\left(\dfrac{L}{100}\right)^y$ |
| Capsize-screening value | $b\left(\dfrac{d}{64}\right)^z$ |

**116.** *Doubling the Bet* A strategy used by some gamblers when they lose is to play again and double the bet, assuming that the amount won is equal to the amount bet. So if a gambler loses \$2, then the gambler plays again and bets \$4, then \$8, and so on.

**a.** Why would the gambler use this strategy?

**b.** On the $n$th bet, the gambler risks $2^n$ dollars. Estimate the first value of $n$ for which the amount bet is more than \$1 million using the accompanying graph of $y = 2^n$.

**c.** Use logarithms to find the answer to part (b).

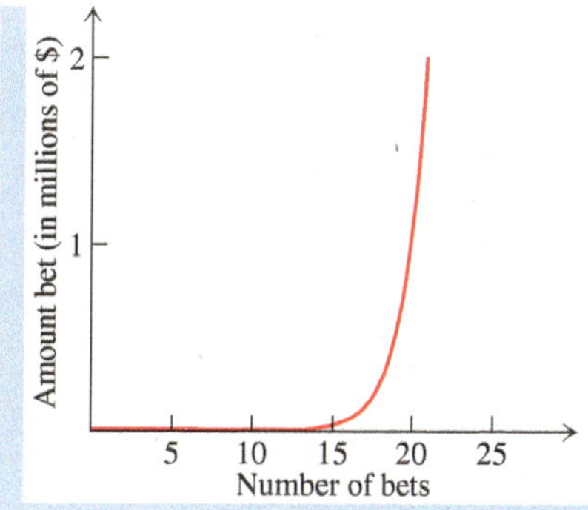

**Figure for Exercise 116**

## THINKING OUTSIDE THE BOX XXXV

*Crescent City* Three semicircles are drawn so that their diameters are the three sides of a right triangle as shown in the diagram. What is the ratio of the total area of the two crescents $A$ and $B$ to the area of the triangle $T$?

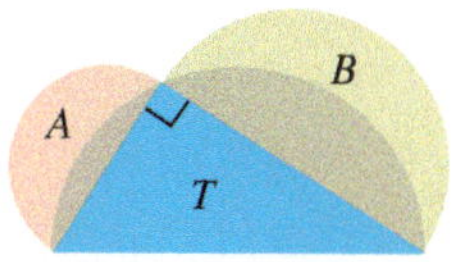


**Figure for Thinking Outside the Box XXXV**

# Chapter 4 Test

*Simplify each expression.*

**1.** $\log(1000)$

**2.** $\log_7(1/49)$

**3.** $3^{\log_3(6.47)}$

**4.** $\ln(e^{\sqrt{2}})$

*Find the inverse of each function.*

**5.** $f(x) = \ln(x)$

**6.** $f(x) = 8^{x+1} - 3$

*Write each expression as a single logarithm.*

**7.** $\log(x) + 3 \cdot \log(y)$

**8.** $\frac{1}{2} \cdot \ln(x - 1) - \ln(33)$

*Write each expression in terms of* $\log_a(2)$ *and* $\log_a(7)$.

**9.** $\log_a(28)$

**10.** $\log_a(3.5)$

*Find the exact solution to each equation.*

**11.** $\log_2(x) + \log_2(x - 2) = 3$

**12.** $\log(10x) - \log(x + 2) = 2 \cdot \log(3)$

*Find the exact solution and an approximate solution (to four decimal places) for each equation.*

**13.** $3^x = 5^{x-1}$

**14.** $\log_3(x - 1) = 5.46$

*Graph each equation in the xy-plane. State the domain and range of the function and whether the function is increasing or decreasing. Identify any asymptotes.*

**15.** $f(x) = 2^x + 1$

**16.** $y = \log_{1/2}(x - 1)$

*Solve each problem.*

**17.** What is the only ordered pair that satisfies both $y = \log_2(x)$ and $y = \ln(x)$?

**18.** A student invested \$2000 at 8% annual percentage rate for 20 years. What is the amount of the investment if the interest is compounded quarterly? Compounded continuously?

**19.** A satellite has a radioisotope power supply. The power supply in watts is given by the formula $P = 50e^{-t/250}$, where $t$ is the time in days. How much power is available at the end of 200 days? What is the half-life of the power supply? If the equipment aboard the satellite requires 9 watts of power to operate properly, then what is the operational life of the satellite?

**20.** If \$4000 is invested at 6% compounded quarterly, then how long will it take for the investment to grow to \$10,000?

**21.** An educational psychologist uses the model $t = -50 \cdot \ln(1 - p)$ for $0 \le p < 1$ to predict the number of hours $t$ that it will take for a child to reach level $p$ in the new video game Mario Goes to Mars (MGM). Level $p = 0$ means that the child knows nothing about MGM. What is the predicted level of a child for 100 hr of playing MGM? If $p = 1$ corresponds to mastery of MGM, then is it possible to master MGM?

## TYING IT ALL together...

### Chapters 1–4

*Solve each equation.*

**1.** $(x - 3)^2 = 4$

**2.** $2 \cdot \log(x - 3) = \log(4)$

**3.** $\log_2(x - 3) = 4$

**4.** $2^{x-3} = 4$

**5.** $\sqrt{x - 3} = 4$

**6.** $|x - 3| = 4$

**7.** $x^2 - 4x = -2$

**8.** $2^{x-3} = 4^x$

**9.** $\sqrt[3]{x - 5} = 5$

**10.** $2^x = 3$

**11.** $\log(x - 3) + \log(4) = \log(x)$

**12.** $x^3 - 4x^2 + x + 6 = 0$

*Sketch the graph of each function.*

**13.** $y = x^2$

**14.** $y = (x - 2)^2$

**15.** $y = 2^x$

**16.** $y = x^{-2}$

**17.** $y = \log_2(x - 2)$

**18.** $y = x - 2$

**19.** $y = 2x$

**20.** $y = \log(2^x)$

**21.** $y = e^2$

**22.** $y = 2 - x^2$

**23.** $y = \dfrac{2}{x}$

**24.** $y = \dfrac{1}{x - 2}$

*Find the inverse of each function.*

**25.** $f(x) = \dfrac{1}{3}x$

**26.** $f(x) = \dfrac{1}{3^x}$

**27.** $f(x) = \sqrt{x - 2}$

**28.** $f(x) = 2 + (x - 5)^3$

**29.** $f(x) = \log(\sqrt{x} - 3)$

**30.** $f = \{(3, 1), (5, 4)\}$

**31.** $f(x) = 3 + \dfrac{1}{x - 5}$

**32.** $f(x) = 3 - e^{\sqrt{x}}$

*Find a formula for each composition function given that $p(x) = e^x$, $m(x) = x + 5$, $q(x) = \sqrt{x}$, and $r(x) = \ln(x)$. State the domain and range of the composition function.*

**33.** $p \circ m$

**34.** $p \circ q$

**35.** $q \circ p \circ m$

**36.** $m \circ r \circ q$

**37.** $p \circ r \circ m$

**38.** $r \circ q \circ p$

*Express each of the following functions as a composition of the functions f, g, and h, where* $f(x) = \log_2(x)$, $g(x) = x - 4$, *and* $h(x) = x^3$.

**39.** $F(x) = \log_2(x^3 - 4)$ **40.** $H(x) = (\log_2(x))^3 - 4$ **41.** $G(x) = (\log_2(x) - 4)^3$ **42.** $M(x) = (\log_2(x - 4))^3$

*Fill in the blanks.*

**43.** The point at the intersection of the $x$-axis and $y$-axis is called the ________.

**44.** The set of all points in the plane that lie at a fixed distance from a fixed point is called a(n) ________.

**45.** The slope of a line is the ratio of its ________ to its ________.

**46.** The equation $y - y_1 = m(x - x_1)$ is called the ________ of the equation of a line.

**47.** The equation $y = mx + b$ is called the ________ of the equation of a line.

**48.** Two lines with slopes $m_1$ and $m_2$ are ________ if and only if $m_1 m_2 = -1$.

**49.** Two lines with slopes $m_1$ and $m_2$ are ________ if and only if $m_1 = m_2$.

**50.** An equation of the form $ax^2 + bx + c = 0$ with $a \neq 0$ is called a(n) ________ equation.

**51.** The formula $x = \dfrac{-b \pm \sqrt{b^2 - 4ac}}{2a}$ is called the ________ formula.

**52.** Two equations with the same solution set are called ________ equations.

## FUNCTION gallery... Some Basic Functions of Algebra with Transformations

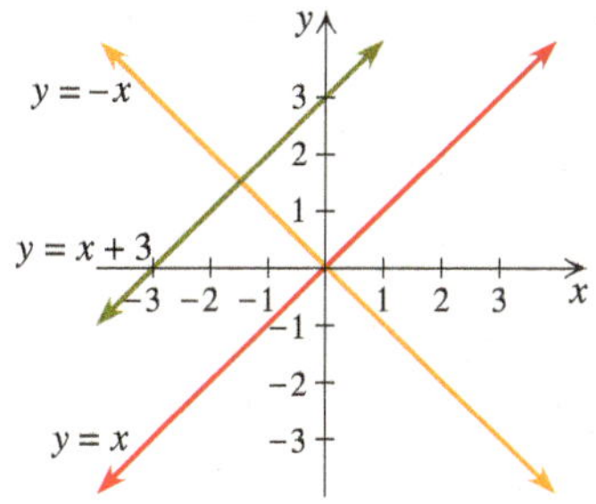

**Linear**

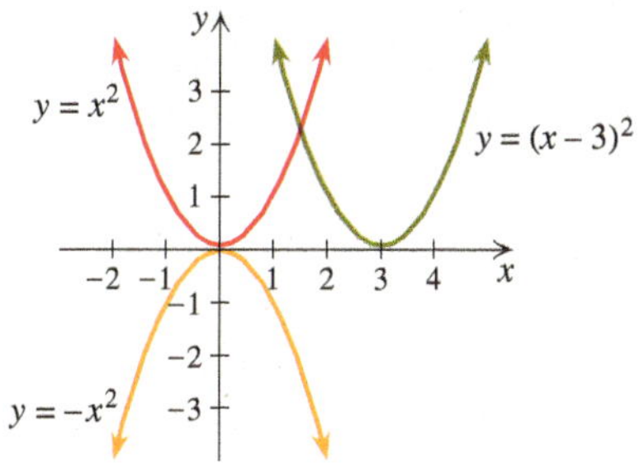

**Quadratic**

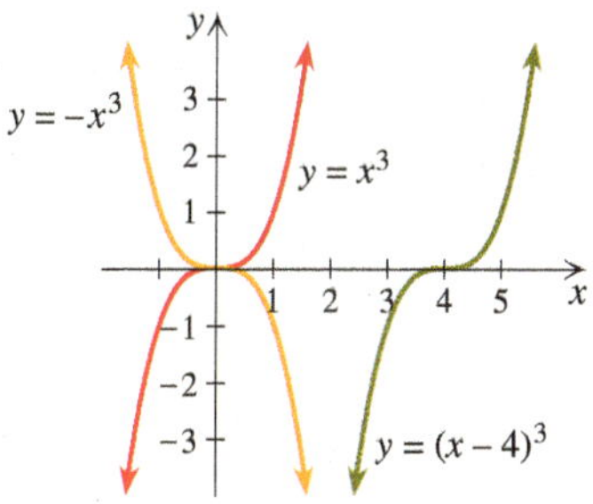

**Cubic**

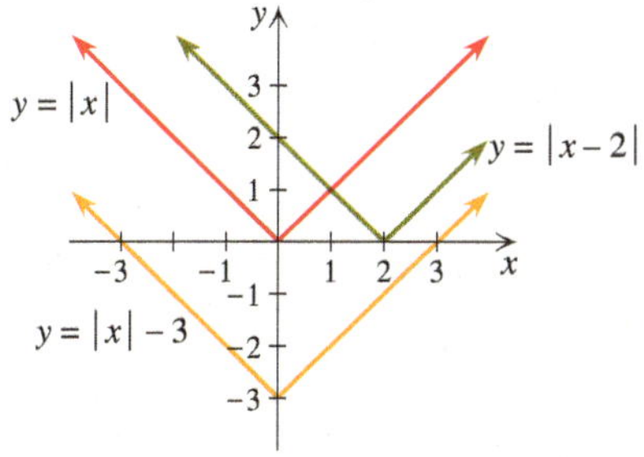

**Absolute value**

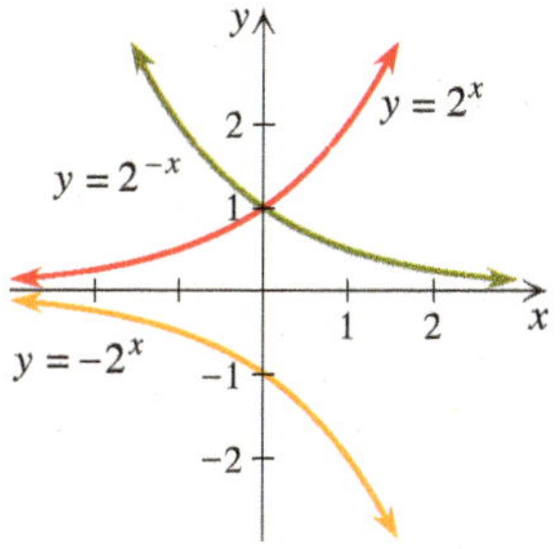

**Exponential**

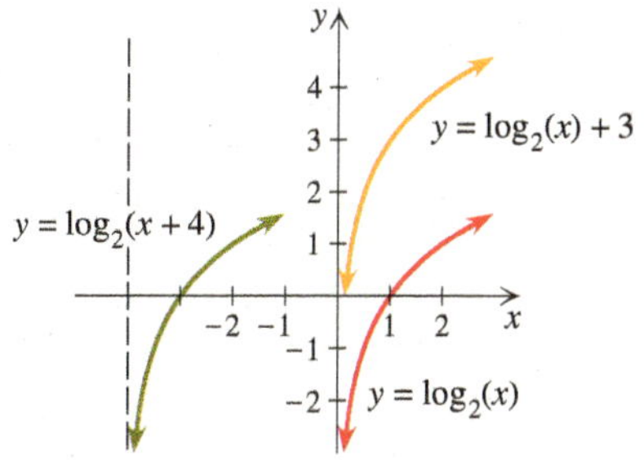

**Logarithmic**

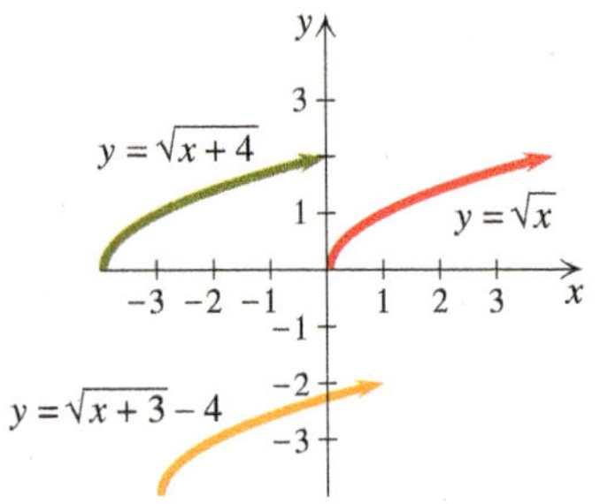


**Square root**

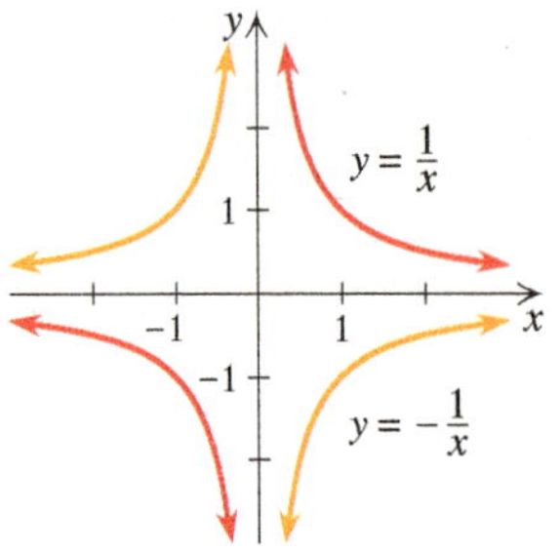


**Reciprocal**

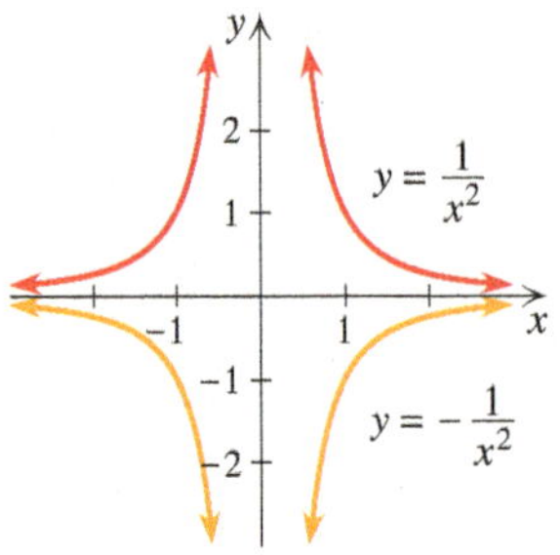


**Rational**

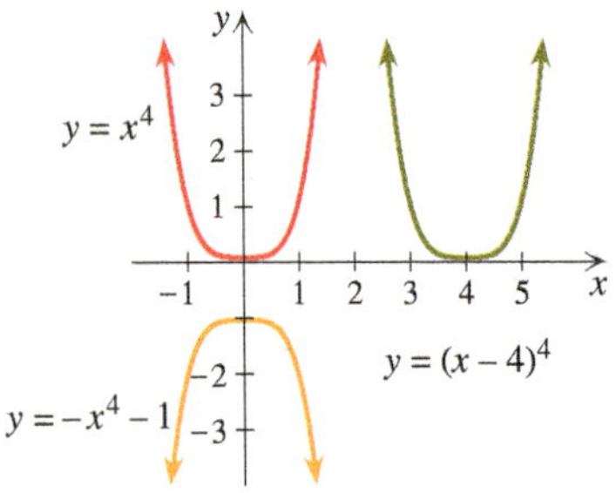


**Fourth degree**

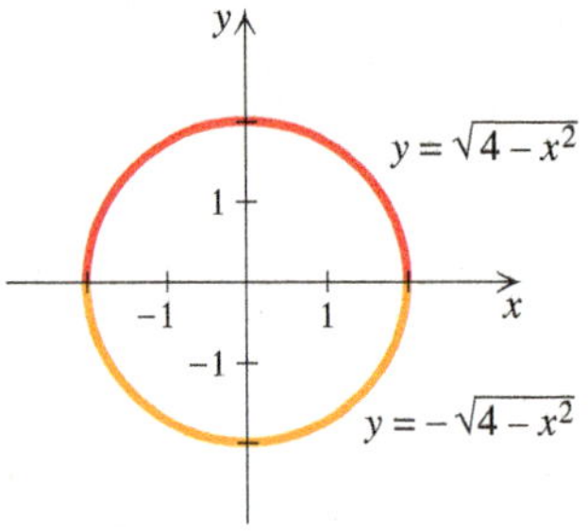


**Semicircle**

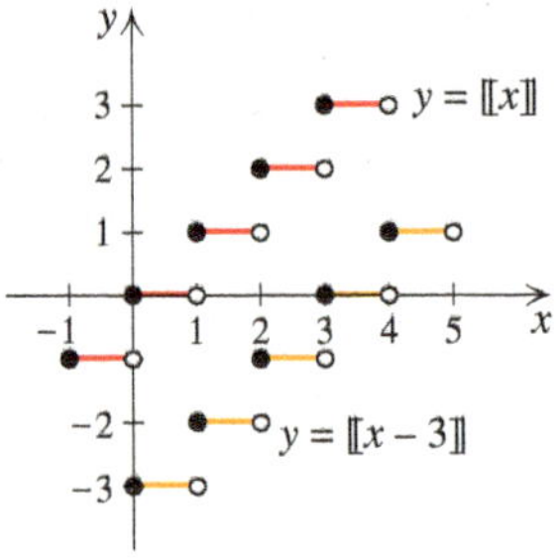


**Greatest integer**

## CONCEPTS OF calculus...

### The instantaneous rate of change of $f(x) = e^x$

*In the Concepts of Calculus for Chapter 2, we defined the average rate of change of a function f on the interval $[x, x + h]$ as*

$$\frac{f(x + h) - f(x)}{h}.$$

*We then defined the instantaneous rate of change of the function as the limit as h approaches zero of the average rate of change. The notation $f'(x)$, read "f prime of x," is used for the instantaneous rate of change. So*

$$f'(x) = \lim_{h \to 0} \frac{f(x + h) - f(x)}{h}.$$

*In the following exercises we will find the instantaneous rate of change of the base e exponential function. It will be necessary here to find a limit using a table of values as we did in the Concepts of Calculus for Chapter 1.*

### Exercises

**1. a.** Fill in the following table.

| $h$ | 0.1 | 0.01 | 0.001 | 0.0001 |
|---|---|---|---|---|
| $\frac{e^h - 1}{h}$ | | | | |

**b.** Can you evaluate $\frac{e^h - 1}{h}$ if $h = 0$?

**c.** Find $\lim_{h \to 0} \frac{e^h - 1}{h}$.

**2.** Let $f(x) = e^x$.

**a.** Find $\frac{f(x + h) - f(x)}{h}$ and simplify it by factoring out $e^x$.

**b.** Use the result of Exercise 1 to find $\lim_{h \to 0} \frac{f(x + h) - f(x)}{h}$.

**c.** Find $f'(x)$.

**3.** Let $f(x) = e^{2x}$. Use the ideas presented in Exercises 1 and 2 to find $f'(x)$.

**4.** Let $f(x) = e^{3x}$. Use the ideas presented in Exercises 1 and 2 to find $f'(x)$.

**5.** Let $f(x) = e^{nx}$, where $n$ is a positive integer. What can you conjecture about $f'(x)$?

# 5 The Trigonometric Functions

Since the beginning of human flight, humans have been using the air to spy on their neighbors. In 1960 the U-2 spy plane flew at an altitude of 13 miles to photograph the Soviet Union and Cuba. Today, spy planes have been replaced with unmanned aerial vehicles (UAVs) and satellites. At least 10 types of UAVs have been used in Iraq. Pictures taken from UAVs and satellites from an altitude of 700 miles are used to determine sizes of buildings and troop movements.

Determining sizes of objects without measuring them physically is one of the principal applications of trigonometry. Long before we dreamed of traveling to the moon, trigonometry was used to calculate the distance to the moon.

**WHAT YOU WILL learn...** In this chapter we explore many applications of trigonometric functions, ranging from finding the velocity of a lawnmower blade to estimating the size of a building in an aerial photograph.

# 5.1 Angles and Their Measurements

Trigonometry was first studied by the Greeks, Egyptians, and Babylonians and used in surveying, navigation, and astronomy. Using trigonometry, they had a powerful tool for finding areas of triangular plots of land, as well as lengths of sides and measures of angles, without physically measuring them. We begin our study of trigonometry by studying angles and their measurements.

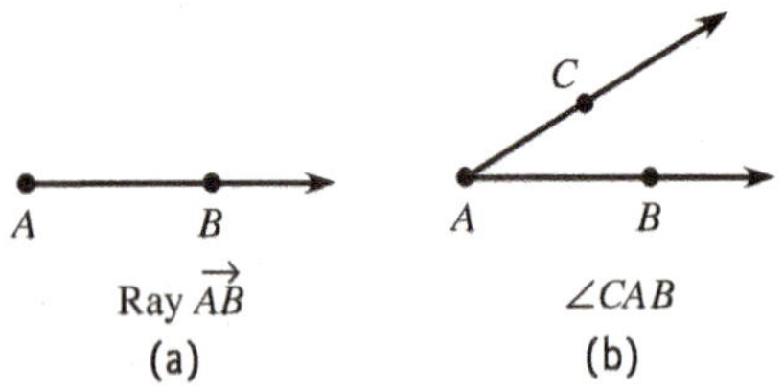


Figure 5.1

## Degree Measure of Angles

In geometry a **ray** is defined as a point on a line together with all points of the line on one side of that point. Figure 5.1(a) shows ray $\overrightarrow{AB}$. An **angle** is defined as the union of two rays with a common endpoint, the **vertex.** The angle shown in Fig. 5.1(b) is named $\angle A$, $\angle BAC$, or $\angle CAB$. (Read the symbol $\angle$ as "angle.") Angles are also named using Greek letters such as $\alpha$ (alpha), $\beta$ (beta), $\gamma$ (gamma), or $\theta$ (theta).

An angle is often thought of as being formed by rotating one ray away from a fixed ray as indicated by angle $\alpha$ and the arrow in Fig. 5.2(a). The fixed ray is the **initial side** and the rotated ray is the **terminal side.** An angle whose vertex is the center of a circle, as shown in Fig. 5.2(b), is a **central angle,** and the arc of the circle through which the terminal side moves is the **intercepted arc.** An angle in **standard position** is located in a rectangular coordinate system with the vertex at the origin and the initial side on the positive $x$-axis as shown in Fig. 5.2(c).

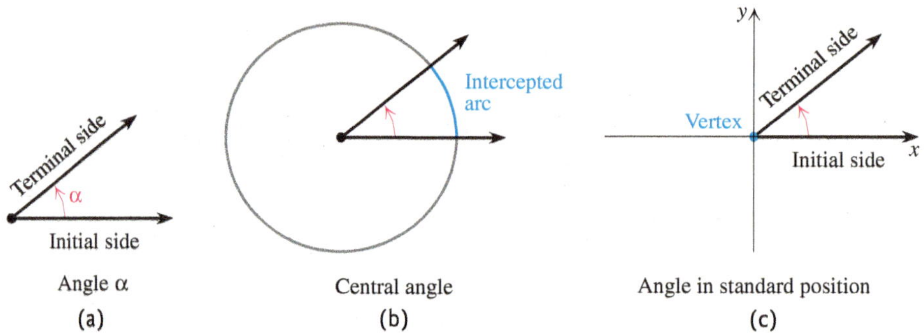


Figure 5.2

The measure $m(\alpha)$ of an angle $\alpha$ indicates the amount of rotation of the terminal side from the initial position. It is found using any circle centered at the vertex. The circle is divided into 360 equal arcs and each arc is one **degree** $(1°)$.

**Definition: Degree Measure**

> The **degree measure of an angle** is the number of degrees in the intercepted arc of a circle centered at the vertex. The degree measure is positive if the rotation is counterclockwise and negative if the rotation is clockwise.

Figure 5.3 shows the positions of the terminal sides of some angles in standard position with common positive measures between 0° and 360°. An angle with measure between 0° and 90° is an **acute angle.** An angle with measure between 90° and 180° is an **obtuse angle.** An angle of exactly 180° is a **straight angle.**

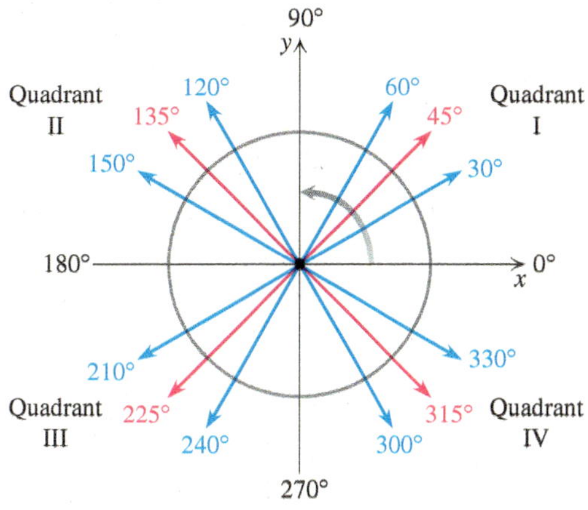


Figure 5.3

See Fig. 5.4. A 90° angle is a **right angle.** An angle in standard position is said to lie in the quadrant where its terminal side lies. If the terminal side is on an axis, the angle is a **quadrantal angle.** We often think of the degree measure of an angle as the angle itself. For example, we write $m(\alpha) = 60°$ or $\alpha = 60°$, and we say that 60° is an acute angle.

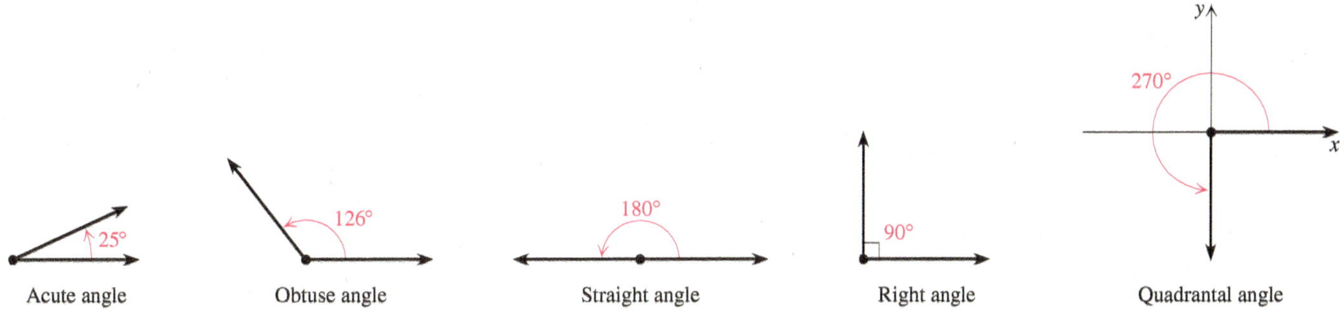


Figure 5.4

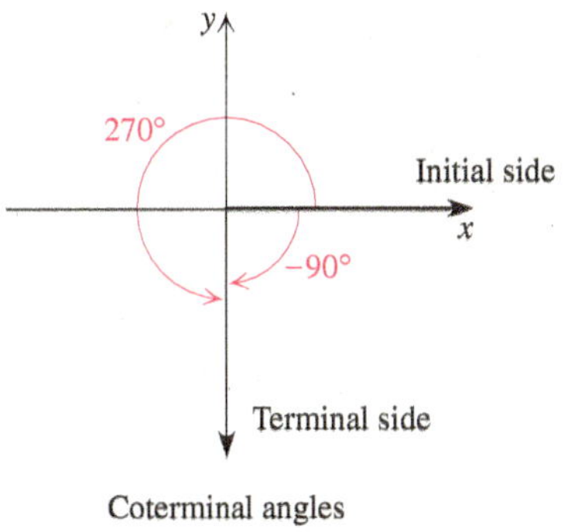


Coterminal angles

Figure 5.5

The initial side of an angle may be rotated in a positive or negative direction to get to the position of the terminal side. For example, if the initial side shown in Fig. 5.5 rotates clockwise for one-quarter of a revolution to get to the terminal position, then the measure of the angle is −90°. If the initial side had rotated counterclockwise to get to the position of the terminal side, then the measure of the angle would be 270°. If the initial side had rotated clockwise for one and a quarter revolutions to get to the terminal position, then the angle would be −450°. **Coterminal angles** are angles in standard position that have the same initial side and the same terminal side. The angles −90°, −450°, and 270° are examples of coterminal angles. Any two coterminal angles have degree measures that differ by a multiple of 360° (one complete revolution).

## Coterminal Angles

> Angles $\alpha$ and $\beta$ in standard position are coterminal if and only if there is an integer $k$ such that $m(\beta) = m(\alpha) + k360°$.

Note that any angle formed by two rays can be thought of as one angle with infinitely many different measures, or infinitely many different coterminal angles.

## EXAMPLE 1 Finding coterminal angles

Find two positive angles and two negative angles that are coterminal with $-50°$.

### Solution

Since any angle of the form $-50° + k360°$ is coterminal with $-50°$, there are infinitely many possible answers. For simplicity, we choose the positive integers 1 and 2 and the negative integers $-1$ and $-2$ for $k$ to get the following angles:

$$-50° + 1 \cdot 360° = 310°$$

$$-50° + 2 \cdot 360° = 670°$$

$$-50° + (-1)360° = -410°$$

$$-50° + (-2)360° = -770°$$

The angles $310°$, $670°$, $-410°$, and $-770°$ are coterminal with $-50°$.

►**TRY THIS.** Find two positive and two negative angles coterminal with $10°$. ■

To determine whether two angles are coterminal we must determine whether they differ by a multiple of $360°$.

## EXAMPLE 2 Determining whether angles are coterminal

Determine whether angles in standard position with the given measures are coterminal.

**a.** $m(\alpha) = 190°, m(\beta) = -170°$ **b.** $m(\alpha) = 150°, m(\beta) = 880°$

### Solution

**a.** If there is an integer $k$ such that $190 + 360k = -170$, then $\alpha$ and $\beta$ are coterminal.

$$190 + 360k = -170$$

$$360k = -360$$

$$k = -1$$

Since the equation has an integral solution, $\alpha$ and $\beta$ are coterminal.

**b.** If there is an integer $k$ such that $150 + 360k = 880$, then $\alpha$ and $\beta$ are coterminal.

$$150 + 360k = 880$$

$$360k = 730$$

$$k = \frac{73}{36}$$

Since there is no integral solution to the equation, $\alpha$ and $\beta$ are not coterminal.

►**TRY THIS.** Determine whether $-690°$ and $390°$ are coterminal. ■

The quadrantal angles, such as $90°$, $180°$, $270°$, and $360°$, have terminal sides on an axis and do not lie in any quadrant. Any angle that is not coterminal with a quadrantal angle lies in one of the four quadrants. To determine the quadrant in which an angle lies, add or subtract multiples of $360°$ (one revolution) to obtain a coterminal angle with a measure between $0°$ and $360°$. A nonquadrantal angle is in the quadrant in which its terminal side lies.

## EXAMPLE 3 Determining in which quadrant an angle lies

Name the quadrant in which each angle lies.

**a.** $230°$ **b.** $-580°$ **c.** $1380°$

### Solution

**a.** Since $180° < 230° < 270°$, a $230°$ angle lies in quadrant III.

**b.** We must add $2(360°)$ to $-580°$ to get an angle between $0°$ and $360°$:

$$-580° + 2(360°) = 140°$$

So $140°$ and $-580°$ are coterminal. Since $90° < 140° < 180°$, $140°$ lies in quadrant II and so does a $-580°$ angle.

**c.** From $1380°$ we must subtract $3(360°)$ to obtain an angle between $0°$ and $360°$:

$$1380° - 3(360°) = 300°$$

So $1380°$ and $300°$ are coterminal. Since $270° < 300° < 360°$, $300°$ lies in quadrant IV and so does $1380°$.

▶**TRY THIS.** Name the quadrant in which $-890°$ lies. ■

Each degree is divided into 60 equal parts called **minutes,** and each minute is divided into 60 equal parts called **seconds.** A minute (min) is 1/60 of a degree (deg), and a second (sec) is 1/60 of a minute or 1/3600 of a degree. An angle with measure $44°12'30''$ is an angle with a measure of 44 degrees, 12 minutes, and 30 seconds. Historically, angles were measured by using degrees-minutes-seconds, but with calculators it is convenient to have the fractional parts of a degree written as a decimal number such as $7.218°$. Some calculators can handle angles in degrees-minutes-seconds and even convert them to decimal degrees.

## EXAMPLE 4 Converting degrees-minutes-seconds to decimal degrees

Convert the measure $44°12'30''$ to decimal degrees.

### Solution

Since 1 degree = 60 minutes and 1 degree = 3600 seconds, we get

$$12 \text{ min} = 12 \text{ min} \cdot \frac{1 \text{ deg}}{60 \text{ min}} = \frac{12}{60} \text{ deg} \quad \text{and}$$

$$30 \text{ sec} = 30 \text{ sec} \cdot \frac{1 \text{ deg}}{3600 \text{ sec}} = \frac{30}{3600} \text{ deg.}$$

So

$$44°12'30'' = \left(44 + \frac{12}{60} + \frac{30}{3600}\right)^{\circ} \approx 44.2083°.$$

Figure 5.6

A graphing calculator can convert degrees-minutes-seconds to decimal degrees as shown in Fig. 5.6.

▶**TRY THIS.** Convert $35°15'12''$ to decimal degrees. ■

Note that the conversion of Example 4 was done by "cancellation of units." Minutes in the numerator canceled with minutes in the denominator to give the result in degrees, and seconds in the numerator canceled with seconds in the denominator to give the result in degrees.

### EXAMPLE 5 Converting decimal degrees to degrees-minutes-seconds

Convert the measure 44.235° to degrees-minutes-seconds.

**Solution**

First convert 0.235° to minutes. Since 1 degree = 60 minutes,

$$0.235 \text{ deg} = 0.235 \text{ deg} \cdot \frac{60 \text{ min}}{1 \text{ deg}} = 14.1 \text{ min}.$$

So $44.235° = 44°14.1'$. Now convert $0.1'$ to seconds. Since 1 minute = 60 seconds,

$$0.1 \text{ min} = 0.1 \text{ min} \cdot \frac{60 \text{ sec}}{1 \text{ min}} = 6 \text{ sec}.$$

So $44.235° = 44°14'6''$.

A graphing calculator can convert to degrees-minutes-seconds as shown in Fig. 5.7.

Figure 5.7

**▸TRY THIS.** Convert 56.321° to degrees-minutes-seconds. ■

Note again how the original units canceled in Example 5, giving the result in the desired units of measurement.

## Radian Measure of Angles

Degree measure of angles is used mostly in applied areas such as surveying, navigation, and engineering. Radian measure of angles is used more in scientific fields and results in simpler formulas in trigonometry and calculus.

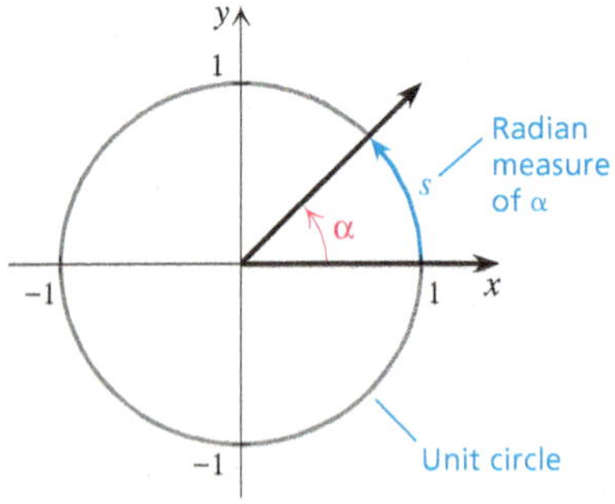


Figure 5.8

For radian measures of angles we use a **unit circle** (a circle with radius 1) centered at the origin. The radian measure of an angle in standard position is simply the length of the intercepted arc on the unit circle. See Fig. 5.8. Since the radius of the unit circle is the real number 1 without any dimension (such as feet or inches), the length of an intercepted arc is a real number without any dimension and so the radian measure of an angle is also a real number without any dimension. One **radian** (abbreviated 1 rad) is the real number 1.

**Definition: Radian Measure**

To find the **radian measure** of the angle $\alpha$ in standard position, find the length of the intercepted arc on the unit circle. If the rotation is counterclockwise, the radian measure is the length of the arc. If the rotation is clockwise, the radian measure is the opposite of the length of the arc.

Radian measure is called a **directed length** because it is positive or negative depending on the direction of rotation of the initial side. If $s$ is the length of the intercepted arc on the unit circle for an angle $\alpha$, as shown in Fig. 5.8, we write $m(\alpha) = s$. To emphasize that $s$ is the length of an arc on a unit circle, we may write $m(\alpha) = s$ radians.

Because the circumference of a circle with radius $r$ is $2\pi r$, the circumference of the unit circle is $2\pi$. If the initial side rotates 360° (one complete revolution), then the length of the intercepted arc is $2\pi$. So an angle of 360° has a radian measure of $2\pi$ radians. We express this relationship as $360° = 2\pi$ rad or simply $360° = 2\pi$. Dividing each side by 2 yields $180° = \pi$, which is the basic relationship to remember for conversion of one unit of measurement to the other.

Use the MODE key on a graphing calculator to set the calculator to radian or degree mode as shown in Fig. 5.9. □

Figure 5.9

**Degree-Radian Conversion**

> Conversion from degrees to radians or radians to degrees is based on
>
> $$180 \text{ degrees} = \pi \text{ radians}.$$

To convert degrees to radians or radians to degrees, we use 180 deg $= \pi$ rad and cancellation of units. For example,

$$1 \text{ deg} = 1 \text{ deg} \cdot \frac{\pi \text{ rad}}{180 \text{ deg}} = \frac{\pi}{180} \text{ rad} \approx 0.01745 \text{ rad}$$

and

$$1 \text{ rad} = 1 \text{ rad} \cdot \frac{180 \text{ deg}}{\pi \text{ rad}} = \frac{180}{\pi} \text{ deg} \approx 57.3 \text{ deg}.$$

If your calculator is in radian mode, as in Fig. 5.10(a), pressing ENTER converts degrees to radians. When the calculator is in degree mode, as in Fig. 5.10(b), pressing ENTER converts radians to degrees. □

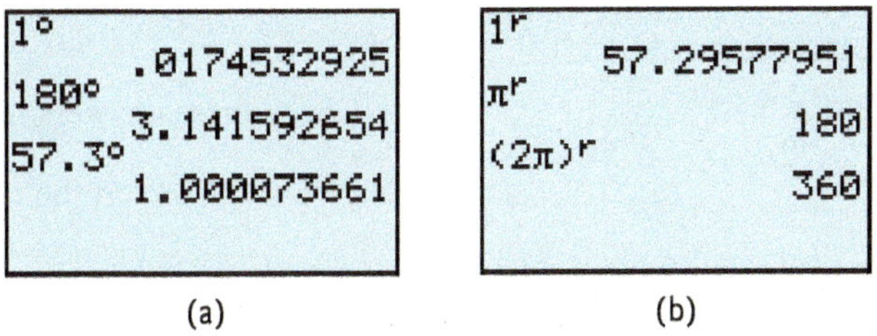


**Figure 5.10**

Figure 5.11(a) shows an angle of 1°, and Fig. 5.11(b) shows an angle of 1 radian. An angle of 1 radian intercepts an arc on the unit circle equal in length to the radius of the unit circle.

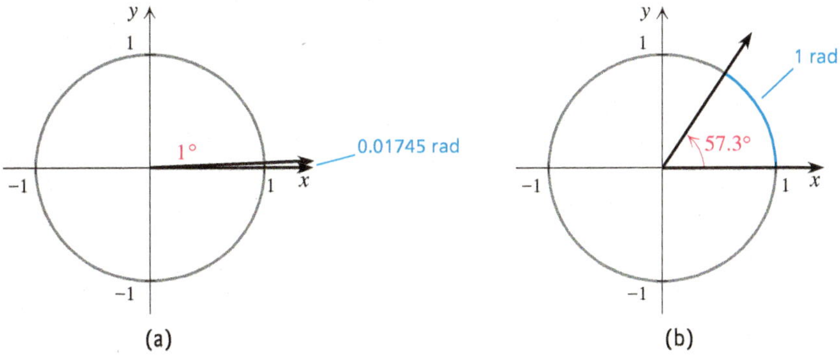


**Figure 5.11**

## EXAMPLE 6 Converting from degrees to radians

Convert the degree measures to radians.

**a.** 270° **b.** −23.6°

### Solution

**a.** To convert degrees to radians, multiply the degree measure by $\pi$ rad/180 deg:

$$270° = 270 \text{ deg} \cdot \frac{\pi \text{ rad}}{180 \text{ deg}} = \frac{3\pi}{2} \text{ rad}$$

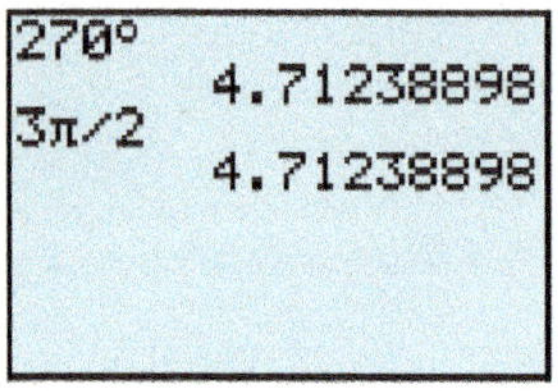


Figure 5.12

The exact value, $3\pi/2$ rad, is approximately 4.71 rad; but when a measure in radians is a simple multiple of $\pi$, we usually write the exact value.

Check this result with a calculator in radian mode as shown in Fig. 5.12. □

**b.** $-23.6° = -23.6 \text{ deg} \cdot \dfrac{\pi \text{ rad}}{180 \text{ deg}} \approx -0.412 \text{ rad}.$

▶**TRY THIS.** Convert 210° to radians. ■

## EXAMPLE 7 Converting from radians to degrees

Convert the radian measures to degrees.

**a.** $\dfrac{7\pi}{6}$ **b.** 12.3

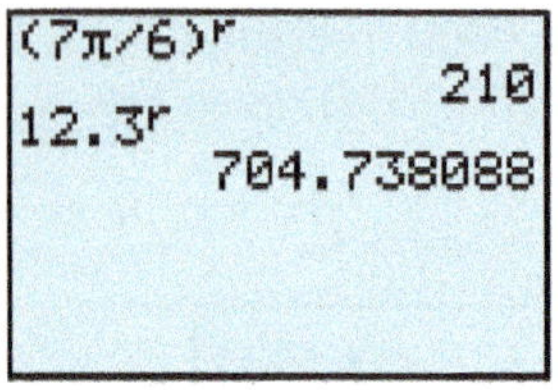


Figure 5.13

### Solution

Multiply the radian measure by $180 \text{ deg}/\pi$ rad:

**a.** $\dfrac{7\pi}{6} = \dfrac{7\pi}{6} \text{ rad} \cdot \dfrac{180 \text{ deg}}{\pi \text{ rad}} = 210°$ **b.** $12.3 = 12.3 \text{ rad} \cdot \dfrac{180 \text{ deg}}{\pi \text{ rad}} \approx 704.7°$

Check these answers in degree mode as shown in Fig. 5.13.

▶**TRY THIS.** Convert the radian measure $\frac{5\pi}{3}$ to degrees. ■

Figure 5.14 shows angles with common measures in standard position. Coterminal angles in standard position have radian measures that differ by an integral multiple of $2\pi$ (their degree measures differ by an integral multiple of 360°).

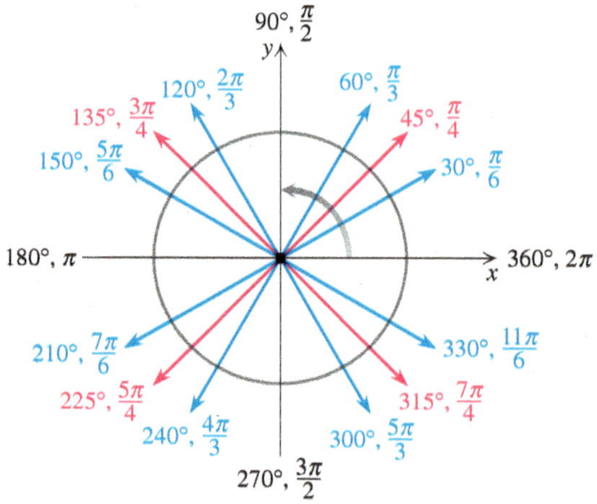


Figure 5.14

## EXAMPLE 8 Finding coterminal angles using radian measure

Find two positive and two negative angles that are coterminal with $\pi/6$.

### Solution

All angles coterminal with $\pi/6$ have a radian measure of the form $\pi/6 + k(2\pi)$, where $k$ is an integer.

$$\frac{\pi}{6} + 1(2\pi) = \frac{\pi}{6} + \frac{12\pi}{6} = \frac{13\pi}{6}$$

$$\frac{\pi}{6} + 2(2\pi) = \frac{\pi}{6} + \frac{24\pi}{6} = \frac{25\pi}{6}$$

$$\frac{\pi}{6} + (-1)(2\pi) = \frac{\pi}{6} - \frac{12\pi}{6} = -\frac{11\pi}{6}$$

$$\frac{\pi}{6} + (-2)(2\pi) = \frac{\pi}{6} - \frac{24\pi}{6} = -\frac{23\pi}{6}$$

The angles $13\pi/6$, $25\pi/6$, $-11\pi/6$, and $-23\pi/6$ are coterminal with $\pi/6$.

▸**TRY THIS.** Find two positive and two negative angles coterminal with $-\pi/3$. ■

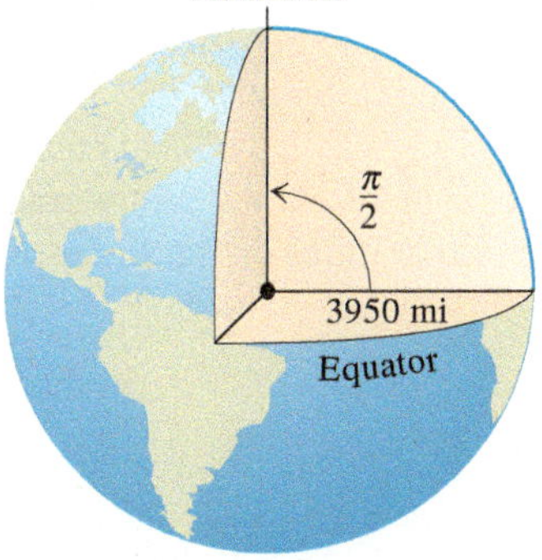


Figure 5.15

## Arc Length

Radian measure of a central angle of a circle can be used to easily find the length of the intercepted arc of the circle. For example, an angle of $\pi/2$ radians positioned at the center of the earth, as shown in Fig. 5.15, intercepts an arc on the surface of the earth that runs from the equator to the North Pole. Using 3950 miles as the approximate radius of the earth yields a circumference of $2\pi(3950) = 7900\pi$ miles. Since $\pi/2$ radians is 1/4 of a complete circle, the length of the intercepted arc is 1/4 of the circumference. So the distance from the equator to the North Pole is $7900\pi/4$ miles, or about 6205 miles.

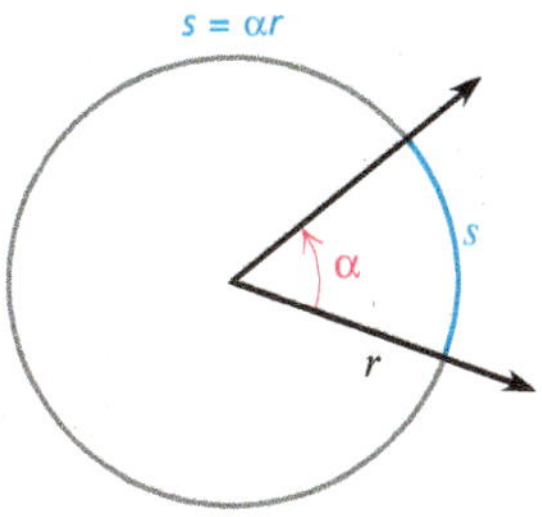


Figure 5.16

In general, a central angle $\alpha$ in a circle of radius $r$ intercepts an arc whose length $s$ is a fraction of the circumference of the circle, as shown in Fig. 5.16. Since a complete revolution is $2\pi$ radians, that fraction is $\alpha/(2\pi)$. Since the circumference is $2\pi r$, we get

$$s = \frac{\alpha}{2\pi} \cdot 2\pi r = \alpha r.$$

**Theorem: Length of an Arc**

The length $s$ of an arc intercepted by a central angle of $\alpha$ *radians* on a circle of radius $r$ is given by

$$s = \alpha r.$$

If $\alpha$ is negative, the formula $s = \alpha r$ gives a negative number for the length of the arc. So $s$ is a directed length. Note that the formula $s = \alpha r$ applies only if $\alpha$ is in radians.

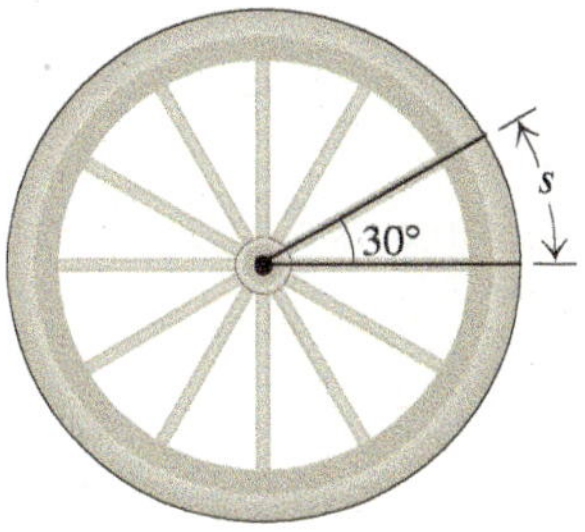


Figure 5.17

### EXAMPLE 9 Finding the length of an arc

The wagon wheel shown in Fig. 5.17 has a diameter of 28 inches and an angle of 30° between the spokes. What is the length of the arc $s$ between two adjacent spokes?

**Solution**

First convert 30° to radians:

$$30° = 30 \text{ deg} \cdot \frac{\pi \text{ rad}}{180 \text{ deg}} = \frac{\pi}{6} \text{ rad}$$

Now use $r = 14$ inches and $\alpha = \pi/6$ in the formula for arc length $s = \alpha r$:

$$s = \frac{\pi}{6} \cdot 14 \text{ in.} = \frac{7\pi}{3} \text{ in.} \approx 7.33 \text{ in.}$$

Since the radian measure of an angle is a dimensionless real number, the product of radians and inches is given in inches.

**►TRY THIS.** Find the length of an arc intercepted by a central angle of 45° on a circle of radius 8 inches. ■

### EXAMPLE 10 Finding the central angle

For four years, U-2 spy planes flew reconnaissance over the Soviet Union, collecting information for the United States. The U-2, designed to fly at 70,000 feet—well out of range of Soviet guns—carried 12,000 feet of film in a camera that could photograph a path 2000 miles long and about 600 miles wide. Find the central angle, to the nearest tenth of a degree, that intercepts an arc of 600 miles on the surface of the earth (radius 3950 miles), as shown in Fig. 5.18.

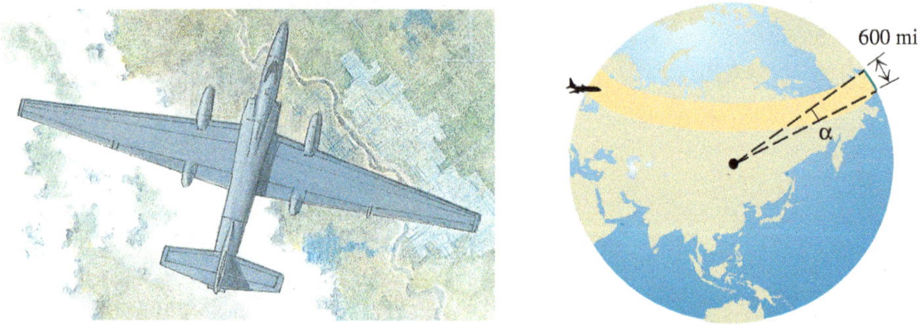


Figure 5.18

#### Solution

From the formula $s = \alpha r$ we get $\alpha = s/r$:

$$\alpha = \frac{600}{3950} \approx 0.1519 \text{ rad}$$

Multiply by 180 deg/$\pi$ rad to find the angle in degrees:

$$\alpha = 0.1519 \text{ rad} \cdot \frac{180 \text{ deg}}{\pi \text{ rad}} \approx 8.7°$$

**►TRY THIS.** Find the central angle in degrees that intercepts an arc of length 3 cm on a circle of radius 50 cm. ■

**HISTORICAL NOTE**

Eratosthenes (276 B.C.–194 B.C.) was a mathematician, geographer, and astronomer. His contemporaries nick-named him "Beta" because he was believed to be second in the Mediterranean world in many fields. He is noted for devising a system of latitude and longitude, and for being the first person to calculate the circumference of the earth. Exercise 123 in this section shows how he found the circumference of the earth using the formula for the length of an arc.

## Linear and Angular Velocity

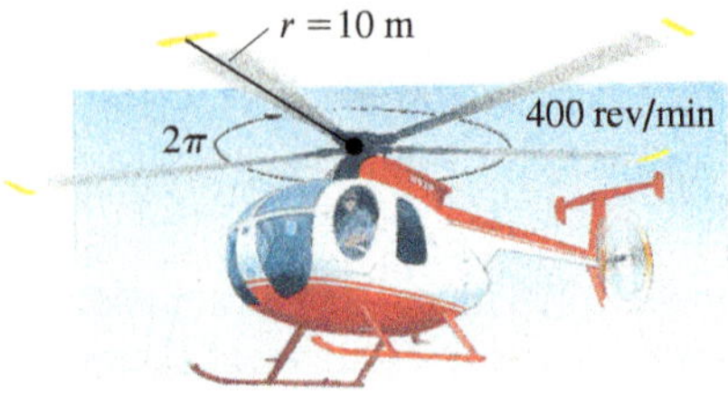


Figure 5.19

Suppose that a point is in motion on a circle. The distance traveled by the point in some unit of time is the length of an arc on the circle. The *linear velocity* of the point is the rate at which that distance is changing. For example, suppose that the helicopter blade of length 10 meters shown in Fig. 5.19 is rotating at 400 revolutions per minute about its center. A point on the tip of the blade (10 meters from the center) travels $20\pi$ meters for each revolution of the blade, because $C = 2\pi r$. Since the blade is rotating at 400 revolutions per minute, the point has a linear velocity of $8000\pi$ meters per minute.

If we draw a ray from the center of the circle through the point on the tip of the helicopter blade, then an angle is formed by the initial and terminal positions of the ray over some unit of time. The *angular velocity* of the point is the rate at which that angle is changing. For example, the ray shown in Fig. 5.19 rotates through an angle of $2\pi$ rad for each revolution of the blade. Since the blade is rotating at 400 revolutions per minute, the angular velocity of the point is $800\pi$ radians per minute. The angular velocity does not depend on the length of the blade (or the radius of the circular path), but only on the number of revolutions per unit of time.

We use the letter $v$ to represent linear velocity for a point in circular motion and the Greek letter $\omega$ (omega) to represent angular velocity, and define them as follows.

**Definition: Linear Velocity and Angular Velocity**

If a point is in motion on a circle of radius $r$ through an angle of $\alpha$ radians in time $t$, then its **linear velocity** is

$$v = \frac{s}{t},$$

where $s$ is the arc length determined by $s = \alpha r$, and its **angular velocity** is

$$\omega = \frac{\alpha}{t}.$$

Since $v = s/t = \alpha r/t$ and $\omega = \alpha/t$, we get $v = r\omega$. We have proved the following theorem.

**Theorem: Linear Velocity in Terms of Angular Velocity**

If $v$ is the linear velocity of a point on a circle of radius $r$, and $\omega$ is its angular velocity, then

$$v = r\omega.$$

It is important to recognize that both linear and angular velocity can be expressed with many different units. For example, an angular velocity of 240 radians per hour can be expressed in radians per minute by the cancellation of units method:

$$240 \text{ rad/hr} = \frac{240 \text{ rad}}{\text{hr}} \cdot \frac{1 \text{ hr}}{60 \text{ min}} = 4 \text{ rad/min}$$

A linear velocity of 45 ft/sec can be expressed in yards per minute:

$$45 \text{ ft/sec} = \frac{45 \text{ ft}}{\text{sec}} \cdot \frac{60 \text{ sec}}{1 \text{ min}} \cdot \frac{1 \text{ yd}}{3 \text{ ft}} = 900 \text{ yd/min}$$

### EXAMPLE 11 Finding angular and linear velocity

What are the angular velocity in radians per second and the linear velocity in miles per hour of the tip of a 22-inch lawnmower blade that is rotating at 2500 revolutions per minute?

#### Solution

Use the fact that $2\pi$ radians = 1 revolution and 60 seconds = 1 minute to find the angular velocity:

$$\omega = \frac{2500 \text{ rev}}{\text{min}} = \frac{2500 \text{ rev}}{\text{min}} \cdot \frac{2\pi \text{ rad}}{1 \text{ rev}} \cdot \frac{1 \text{ min}}{60 \text{ sec}} \approx 261.799 \text{ rad/sec}$$

To find the linear velocity, use $v = r\omega$ with $r = 11$ inches:

$$v = 11 \text{ in.} \cdot \frac{261.799 \text{ rad}}{\text{sec}} \approx 2879.789 \text{ in./sec}$$

Convert to miles per hour:

$$v = \frac{2879.789 \text{ in.}}{\text{sec}} \cdot \frac{1 \text{ ft}}{12 \text{ in.}} \cdot \frac{1 \text{ mi}}{5280 \text{ ft}} \cdot \frac{3600 \text{ sec}}{1 \text{ hr}} \approx 163.624 \text{ mi/hr}$$

▶**TRY THIS.** Find the angular velocity in radians per second and the linear velocity in miles per hour for a particle that is moving in a circular path at 5 revolutions per second on a circle of radius 10 feet. ■

Any point on the surface of the earth (except at the poles) makes one revolution about the axis of the earth in 24 hours. So the angular velocity of a point on the earth is $\pi/12$ radians per hour. The linear velocity of a point on the surface of the earth depends on its distance from the axis of the earth.

### EXAMPLE 12 Linear velocity on the surface of the earth

What is the linear velocity in miles per hour of a point on the equator?

#### Solution

A point on the equator has an angular velocity of $\pi/12$ radians per hour on a circle of radius 3950 miles. Using the formula $v = \omega r$, we get

$$v = \frac{\pi}{12} \text{ rad/hr} \cdot 3950 \text{ mi} = \frac{3950\pi}{12} \text{ mi/hr} \approx 1034 \text{ mi/hr}.$$

Note that we write angular velocity as radians per hour, but radians are omitted from the answer in miles per hour because radians are simply real numbers.

▶**TRY THIS.** Find the linear velocity in feet per second for a point on the equator? ■

## FOR thought... True or False? Explain.

1. An angle is a union of two rays with a common endpoint.
2. The lengths of the rays of $\angle A$ determine the degree measure of $\angle A$.
3. Angles of 5° and −365° are coterminal.
4. The radian measure of an angle cannot be negative.
5. An angle of $38\pi/4$ radians is a quadrantal angle.
6. If $m(\angle A) = 210°$, then $m(\angle A) = 5\pi/6$ radians.
7. $25°20'40'' = 25.34°$
8. The angular velocity of Seattle is $\pi/12$ radians per hour.
9. Seattle and Los Angeles have the same linear velocity.
10. A central angle of 1 rad in a circle of radius $r$ intercepts an arc of length $r$.

## EXERCISES 5.1

*Fill in the blank.*

1. An ________ is a union of two rays with a common endpoint.
2. An angle whose vertex is the center of a circle is a ________ angle.
3. An angle in ________ is located in a rectangular coordinate system with the vertex at the origin and the initial side on the positive $x$-axis.
4. An angle with a measure between 0° and 90° is an ________ angle.

**5.** An angle with a measure between 90° and 180° is an ________ angle.

**6.** An angle with a measure of 90° is a ________ angle.

**7.** Angles in standard position with the same initial side and the same terminal side are ________ angles.

**8.** An angle in standard position whose terminal side is on an axis is a ________ angle.

**9.** A ________ is one sixtieth of a degree.

**10.** A ________ is one sixtieth of a minute.

**11.** A circle with radius 1 is a ________ circle.

**12.** If $v$ is the ________ velocity of a point on a circle of radius $r$ and $\omega$ is its ________ velocity, then $v = r\omega$.

*Find two positive angles and two negative angles that are coterminal with each given angle.*

**13.** 60° **14.** 45° **15.** −16° **16.** −90°

*Determine whether the angles in each given pair are coterminal.*

**17.** 123.4°, −236.6° **18.** 744°, −336°

**19.** 1055°, 155° **20.** 0°, 359.9°

*Name the quadrant in which each angle lies.*

**21.** 85° **22.** 110° **23.** −125° **24.** −200°

**25.** 300° **26.** 205° **27.** 750° **28.** −980°

*Match each of the following angles with one of the degree measures:* 30°, 45°, 60°, 120°, 135°, 150°.

**29.**

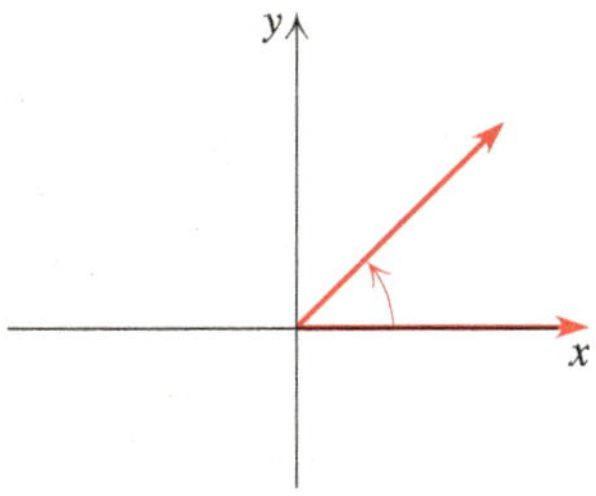


**30.**

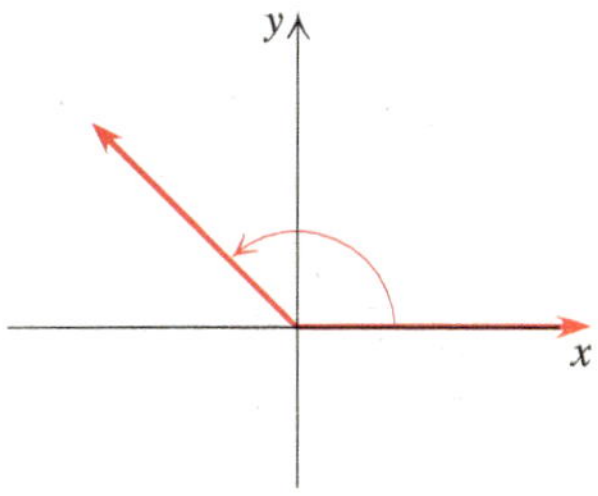


**31.**

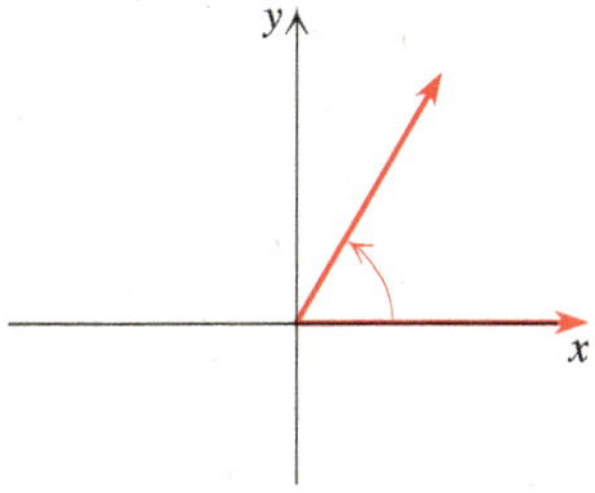


**32.**

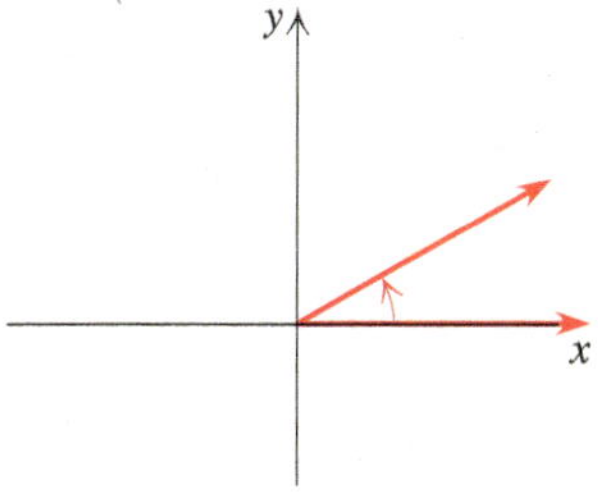


**33.**

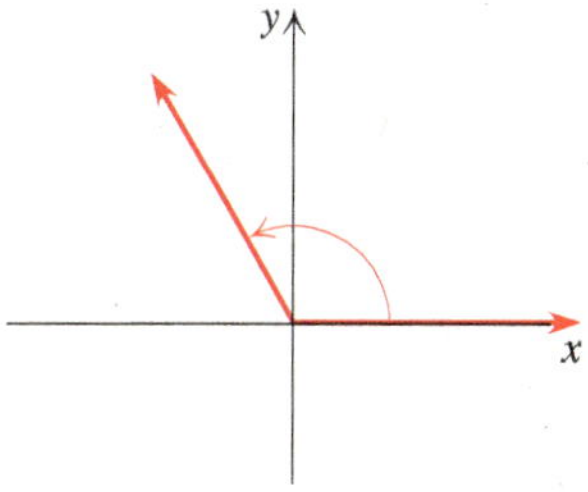


**34.**

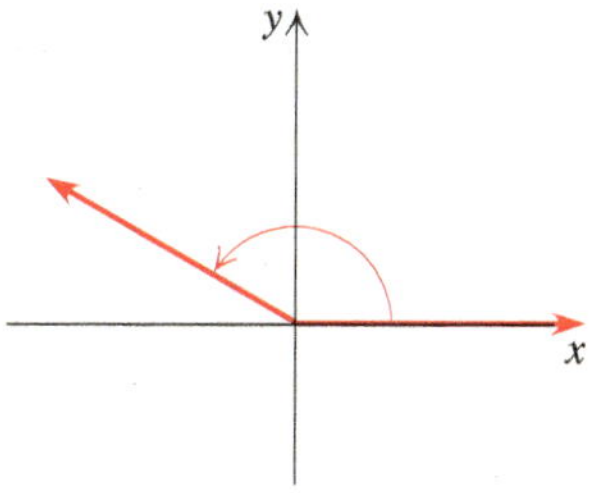


*Find the measure in degrees of the least positive angle that is coterminal with each given angle.*

**35.** 400° **36.** 540° **37.** −340°

**38.** −180° **39.** −1100° **40.** −840°

*Convert each angle to decimal degrees. When necessary, round to four decimal places.*

**41.** 13°12′ **42.** 45°6′ **43.** −8°30′18″

**44.** −5°45′30″ **45.** 28°5′9″ **46.** 44°19′32″

*Convert each angle to degrees-minutes-seconds. Round to the nearest whole number of seconds.*

**47.** 75.5° **48.** 39.4° **49.** −17.33°

**50.** −9.12° **51.** 18.123° **52.** 122.786°

*Convert each degree measure to radian measure. Give exact answers.*

**53.** 30° **54.** 45° **55.** 18°

**56.** 48° **57.** −67.5° **58.** −105°

**59.** 630° **60.** 495°

*Convert each degree measure to radian measure. Use the value of $\pi$ found on a calculator and round answers to three decimal places.*

**61.** 37.4° **62.** 125.3° **63.** −13°47′

**64.** −99°15′ **65.** −53°37′6″ **66.** 187°49′36″

*Convert each radian measure to degree measure. Use the value of $\pi$ found on a calculator and round approximate answers to three decimal places.*

**67.** $\frac{5\pi}{12}$ **68.** $\frac{17\pi}{12}$ **69.** $\frac{7\pi}{4}$ **70.** $\frac{13\pi}{6}$

**71.** $-6\pi$ **72.** $-9\pi$ **73.** 2.39 **74.** 0.452

*Using radian measure, find two positive angles and two negative angles that are coterminal with each given angle.*

**75.** $\frac{\pi}{3}$ **76.** $\frac{\pi}{4}$ **77.** $-\frac{\pi}{6}$ **78.** $-\frac{2\pi}{3}$

*Find the measure in radians of the least positive angle that is coterminal with each given angle.*

**79.** $3\pi$ **80.** $6\pi$ **81.** $\frac{9\pi}{2}$ **82.** $\frac{19\pi}{2}$

**83.** $-\frac{5\pi}{3}$ **84.** $-\frac{7\pi}{6}$ **85.** $-\frac{13\pi}{3}$ **86.** $-\frac{19\pi}{4}$

**87.** 8.32 **88.** $-23.55$

*Determine whether the angles in each given pair are coterminal.*

**89.** $\frac{3\pi}{4}, \frac{29\pi}{4}$ **90.** $-\frac{\pi}{3}, \frac{5\pi}{3}$

**91.** $\frac{7\pi}{6}, -\frac{5\pi}{6}$ **92.** $\frac{3\pi}{2}, -\frac{9\pi}{2}$

*Name the quadrant in which each angle lies.*

**93.** $\frac{5\pi}{12}$ **94.** $\frac{13\pi}{12}$ **95.** $-\frac{6\pi}{7}$ **96.** $-\frac{39\pi}{20}$

**97.** $\frac{13\pi}{8}$ **98.** $-\frac{11\pi}{8}$ **99.** $-7.3$ **100.** 23.1

*Find the missing degree or radian measure for each position of the terminal side shown. For Exercise 101, use degrees between 0° and 360° and radians between 0 and $2\pi$. For Exercise 102, use degrees between −360° and 0° and radians between $-2\pi$ and 0. Practice these two exercises until you have memorized the degree and radian measures corresponding to these common angles.*

**101.**

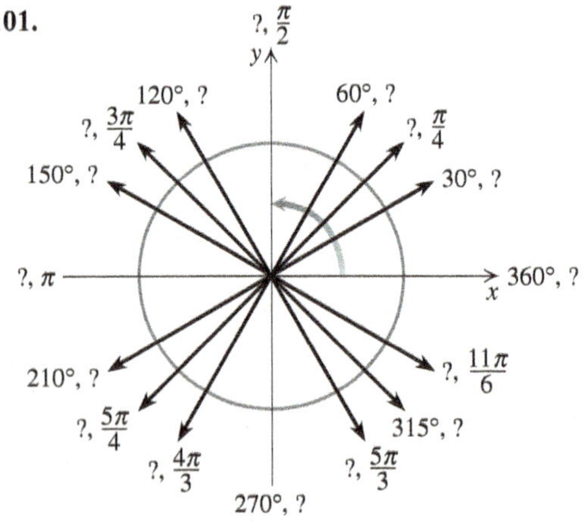


**102.**

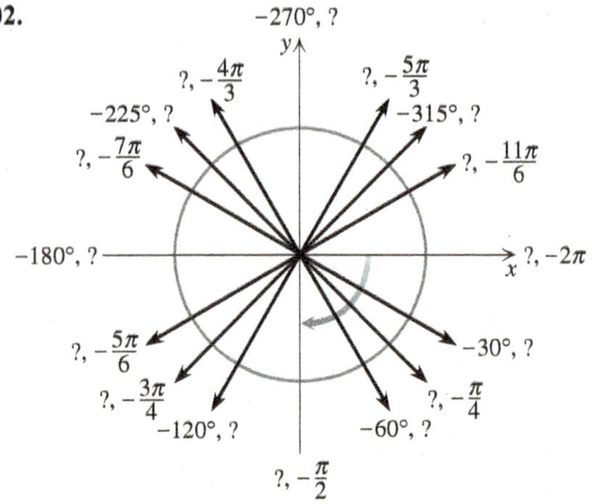


*Find the length of the arc intercepted by the given central angle α in a circle of radius r.*

**103.** $\alpha = \frac{\pi}{4}, r = 12$ ft **104.** $\alpha = 1, r = 4$ cm

**105.** $\alpha = 3°, r = 4000$ mi **106.** $\alpha = 60°, r = 2$ m

*Find the radius of the circle in which the given central angle α intercepts an arc of the given length s.*

**107.** $\alpha = 1, s = 1$ mi **108.** $\alpha = 0.004, s = 99$ km

**109.** $\alpha = 180°, s = 10$ km **110.** $\alpha = 360°, s = 8$ m

*Solve each problem.*

**111.** *Distance to the North Pole* Peshtigo, Wisconsin, is on the 45th parallel. This means that an arc from Peshtigo to the North Pole subtends a central angle of 45° as shown in the figure. If the radius of the earth is 3950 mi, then how far is it from Peshtigo to the North Pole?

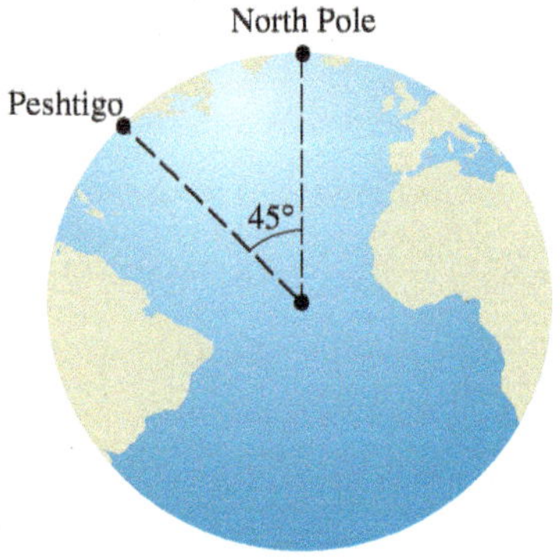


**Figure for Exercise 111**

**112.** *Distance to the Helper* A surveyor sights her 6-ft 2-in. helper on a nearby hill as shown in the figure. If the angle of sight between the helper's feet and head is 0°37′, then approximately how far away is the helper?

**HINT** Assume that the helper is a 6-ft 2-in. arc on a circle of radius $r$.

**Figure for Exercise 112**

**113.** *Photographing the Earth* From an altitude of 161 mi, the Orbiting Geophysical Observatory (OGO-1) can photograph a path on the surface of the earth that is approximately 2000 mi wide. Find the central angle, to the nearest tenth of a degree, that intercepts an arc of 2000 mi on the surface of the earth (radius 3950 mi).

**114.** *Margin of Error in a Field Goal* Assume that the distance between the goal posts, 18.5 ft, is the length of an arc on a circle of radius 50 yd as shown in the figure. The kicker aims to kick the ball midway between the uprights. To score a field goal, what is the maximum number of degrees that the actual trajectory can deviate from the intended trajectory? What is the maximum number of degrees for a 20-yd field goal?

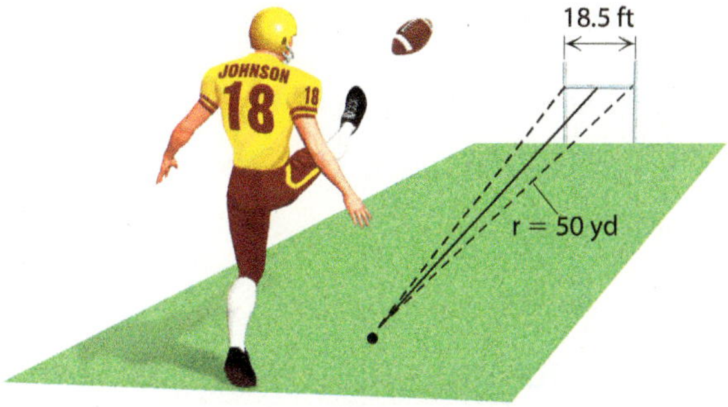


**Figure for Exercise 114**

**115.** *Linear Velocity* What is the linear velocity in cm/min for any edge point of a 12-cm-diameter CD (compact disc) spinning at 10,350 rev/min?

**116.** *Angular Velocity* What is the angular velocity in radians per minute for any point on a CD that is rotating at 10,350 revolutions per minute.

**117.** *Lawnmower Blade* What is the linear velocity in miles per hour of the tip of a lawnmower blade spinning at 2800 rev/min for a lawnmower that cuts a 20-in.-wide path?

**118.** *Router Bit* A router bit makes 45,000 rev/min. What is the linear velocity in miles per hour of the outside edge of a bit that cuts a 1-in.-wide path?

**119.** *Table Saw* The blade on a table saw rotates at 3450 revolutions per minute. What is the difference in the linear velocity in miles per hour for a point on the edge of a 12-in.-diameter blade and a 10-in.-diameter blade?

**120.** *Automobile Tire* If a car runs over a nail at 55 mph and the nail is lodged in the tire tread 13 in. from the center of the wheel, then what is the angular velocity of the nail in radians per hour?

**121.** *Linear Velocity Near the North Pole* Find the linear velocity for a point on the surface of the earth that is one mile from the North Pole.

HINT First determine how far the point travels in 24 hours.

**122.** *Linear Velocity of Peshtigo* What are the linear and angular velocities for Peshtigo, Wisconsin (on the 45th parallel), with respect to its rotation around the axis of the earth? (See Exercise 111.)

**123.** *Eratosthenes Measures the Earth* Over 2200 years ago Eratosthenes read in the Alexandria library that at noon on June 21 a vertical stick in Syene cast no shadow. So on June 21 at noon Eratosthenes set out a vertical stick in Alexandria and found an angle of 7° in the position shown in the accompanying drawing. Eratosthenes reasoned that since the sun is so far away, sunlight must be arriving at the earth in parallel rays. With this assumption he concluded that the earth is round and the central angle in the drawing must also be 7°. He then paid a man to pace off the distance between Syene and Alexandria and found it to be 800 km. From these facts, calculate the circumference of the earth as Eratosthenes did and compare his answer with the circumference calculated by using the currently accepted radius of 6378 km.

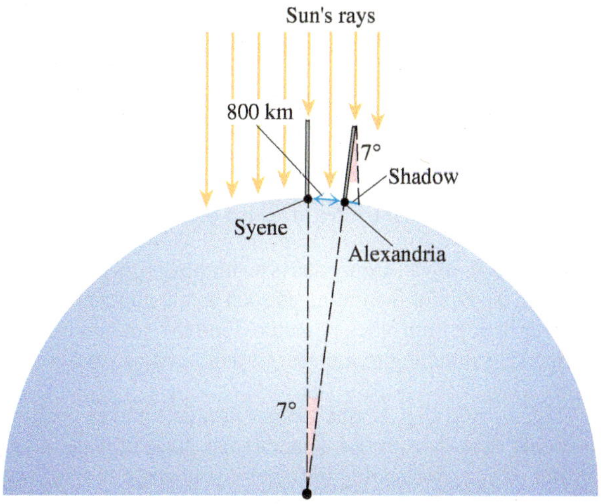


**Figure for Exercise 123**

**124.** *Achieving Synchronous Orbit* The space shuttle orbits the earth in about 90 min at an altitude of about 125 mi, but a communication satellite must always remain above a fixed location on the earth, in synchronous orbit with the earth. Since the earth rotates 15° per hour, the angular velocity of a communication satellite must be 15° per hour. To achieve a synchronous orbit with the earth, the radius of the orbit of a satellite must be 6.5 times the radius of the earth (3950 mi). What is the linear velocity in miles per hour of such a satellite?

**125.** *Area of a Slice of Pizza* If a 16-in.-diameter pizza is cut into six slices of the same size, then what is the area of each piece?

**126.** *Area of a Sector of a Circle* If a slice with central angle $\alpha$ radians is cut from a pizza of radius $r$, then what is the area of the slice?

**127.** *Belt and Pulleys* A belt connects two pulleys with radii 3 in. and 5 in. as shown in the diagram on the next page. The velocity of a point $A$ on the belt is 10 ft/sec. What is the linear velocity (ft/sec) and angular velocity (rad/sec) for point $B$? What is the linear velocity (ft/sec) and angular velocity (rad/sec) for point $C$?

**128.** *Belt and Pulleys* A belt connects two pulleys with radii 3 in. and 5 in. as shown in the accompanying diagram. Point $B$ is rotating at 1000 revolutions per minute. What is the linear velocity (ft/sec) for points $A$, $B$, and $C$? How many revolutions per minute is point $C$ making?

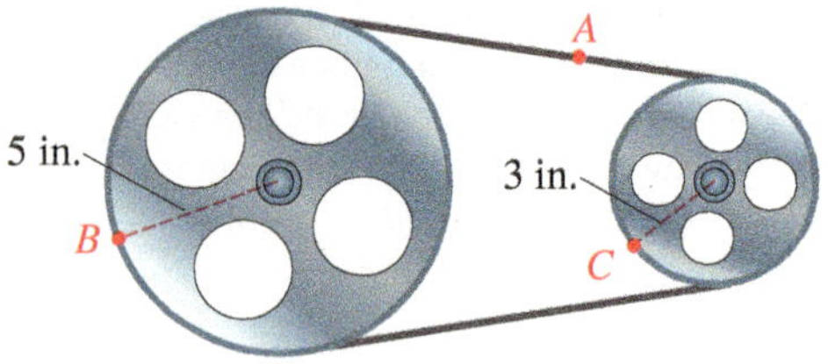


**Figure for Exercises 127 and 128**

**129.** *Time After Time* What are the first two times (to the nearest second) after 12 noon for which the minute hand and the hour hand of a clock are perpendicular to each other?

**HINT** First find the rates at which the hands are moving.

**130.** *Perfect Timing* Tammy starts swimming laps at 4 P.M. exactly. She quits swimming as soon as the minute hand and hour hand of the clock coincide. Find the amount of time she spends swimming to the nearest tenth of a second.

**131.** *Volume of a Cup* Sunbelt Paper Products makes conical paper cups by cutting a 4-in. circular piece of paper on the radius and overlapping the paper by an angle $\alpha$ as shown in the accompanying figure.

**a.** Find the volume of the cup if $\alpha = 30°$.

**b.** Write the volume of the cup as a function of $\alpha$.

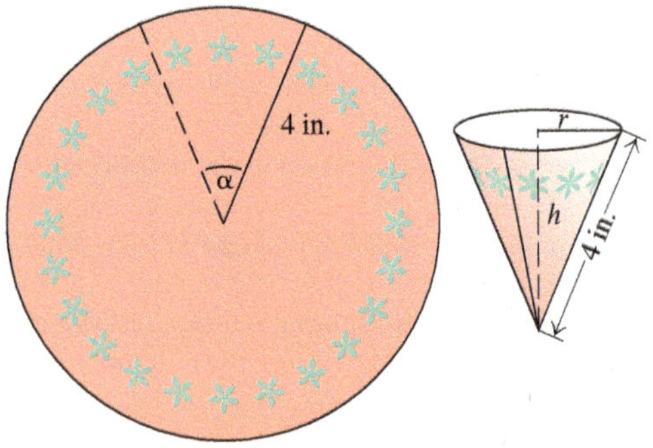


**Figure for Exercise 131**

**132.** *Maximizing the Volume of a Cup* Use the maximum feature of a graphing calculator to find the angle $\alpha$ that gives the maximum volume for the cup in the previous exercise. What is the maximum volume for the cup?

## RETHINKING

**133.** Solve $x(x + 3) = 10$.

**134.** Solve $\log(x) + \log(x + 3) = 1$.

**135.** Solve $\log_2(x) - \log_2(x + 3) = -3$.

**136.** Find the exact solution to $5^{3x-1} = \log_4(64)$.

**137.** Solve $\frac{2x - 1}{x + 3} \geq 1$.

**138.** Solve $(x + 2)(x + 1) > 6$.

## THINKING OUTSIDE THE BOX XXXVI

*Triangles and Circles* Each of the five circles in the accompanying diagram has radius $r$. The four right triangles are congruent and each hypotenuse has length 1. Each line segment that appears to intersect a circle intersects it at exactly one point (a point of tangency). Find $r$.

**Figure for Thinking Outside the Box XXXVI**

## POP QUIZ 5.1

1. Find the degree measure of the least positive angle that is coterminal with 1267°.
2. In which quadrant does −120° lie?
3. Convert 70°30′36″ to decimal degrees.
4. Convert 270° to radian measure.
5. Convert $7\pi/4$ to degree measure.
6. Are $-3\pi/4$ and $5\pi/4$ coterminal?
7. Find the exact length of the arc intercepted by a central angle of 60° in a circle with radius 30 feet.

## LINKING concepts... For Individual or Group Explorations

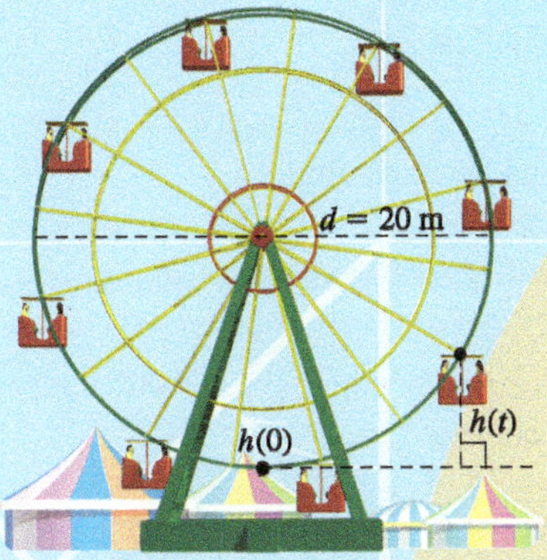


### Constructing the Sine Function

*Imagine that you are riding on a 20-meter-diameter Ferris wheel that is making three revolutions per minute.*

**a)** What is your linear velocity?

**b)** What is your angular velocity?

**c)** If there are eight equally spaced seats on the Ferris wheel, then what is the length of the arc between two adjacent seats?

**d)** Use a compass to draw a circle with diameter 20 centimeters to represent the Ferris wheel. Let $h(t)$ be your height in meters at time $t$ seconds, where $h(0) = 0$. Locate your position on the Ferris wheel for each time in the following table and find $h(t)$ by measuring your drawing.

| $t$ | 0 | 2.5 | 5 | 7.5 | 10 | 12.5 | 15 | 17.5 | 20 |
|---|---|---|---|---|---|---|---|---|---|
| $h(t)$ | | | | | | | | | |

**e)** Sketch the graph of $h(t)$ on graph paper for $t$ ranging from 0 to 60 seconds.

**f)** How many solutions are there to $h(t) = 18$ in the interval $[0, 60]$?

**g)** Find the exact values of $h(2.5)$ and $h(7.5)$.

# 5.2 The Sine and Cosine Functions

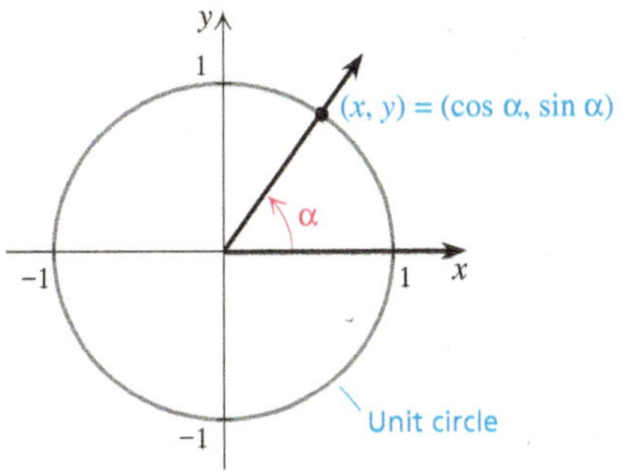


Figure 5.20

In Section 5.1 we learned that angles can be measured in degrees or radians. Now we define two trigonometric functions whose domain is the set of all angles (measured in degrees or radians). These two functions are unlike any functions defined in algebra, but they form the foundation of trigonometry.

## Definition

If $\alpha$ is an angle in standard position whose terminal side intersects the unit circle at point $(x, y)$, as shown in Fig. 5.20, then **sine of $\alpha$**—abbreviated $\sin(\alpha)$ or $\sin\alpha$—is the $y$-coordinate of that point, and **cosine of $\alpha$**—abbreviated $\cos(\alpha)$ or $\cos\alpha$—is the $x$-coordinate. Sine and cosine are called **trigonometric functions.**

**Definition: Sine and Cosine**

If $\alpha$ is an angle in standard position and $(x, y)$ is the point of intersection of the terminal side and the unit circle, then

$$\sin\alpha = y \quad \text{and} \quad \cos\alpha = x.$$

The domain of the sine function and the cosine function is the set of angles in standard position, but since each angle has a measure in degrees or radians, we generally

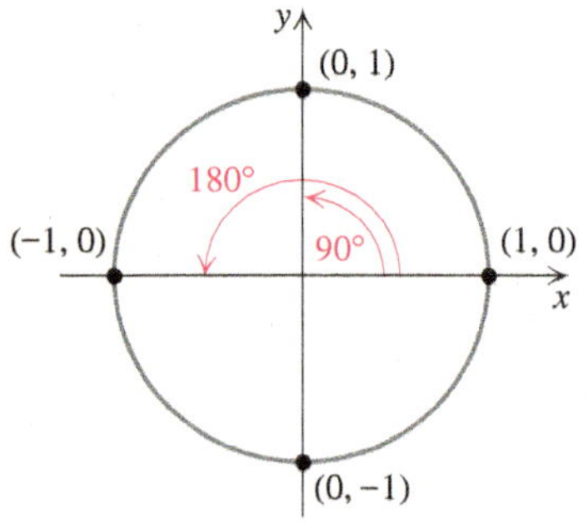


Figure 5.21

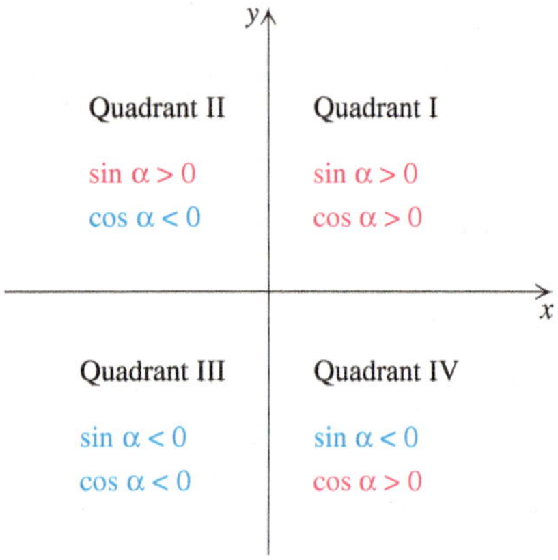


Figure 5.22

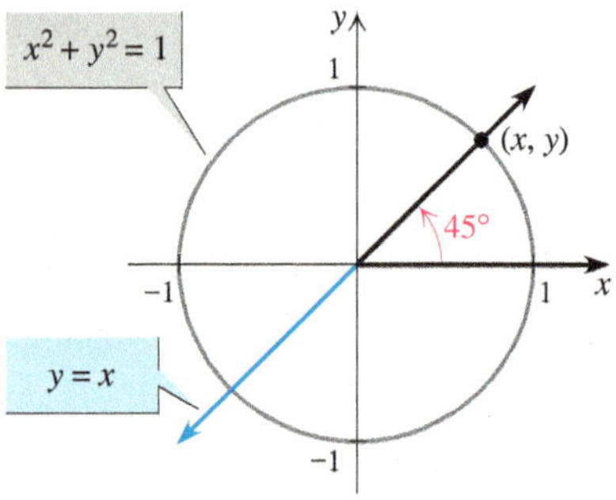


Figure 5.23

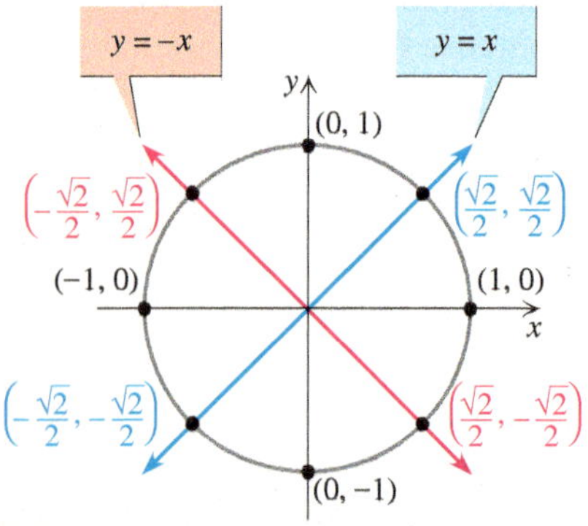


Figure 5.24

use the set of degree measures or the set of radian measures as the domain. If $(x, y)$ is on the unit circle, then $-1 \le x \le 1$ and $-1 \le y \le 1$, so the range of each of these functions is the interval $[-1, 1]$.

### EXAMPLE 1 Evaluating sine and cosine at a multiple of 90°

Find the exact values of the sine and cosine functions for each angle.

**a.** 90° **b.** $\pi/2$ **c.** 180° **d.** $-5\pi/2$ **e.** $-720°$

#### Solution

**a.** Consider the unit circle shown in Fig. 5.21. Since the terminal side of 90° intersects the unit circle at $(0, 1)$, $\sin(90°) = 1$ and $\cos(90°) = 0$.
**b.** Since $\pi/2$ is coterminal with 90°, we have $\sin(\pi/2) = 1$ and $\cos(\pi/2) = 0$.
**c.** The terminal side for 180° intersects the unit circle at $(-1, 0)$. So $\sin(180°) = 0$ and $\cos(180°) = -1$.
**d.** Since $-5\pi/2$ is coterminal with $3\pi/2$, the terminal side of $-5\pi/2$ lies on the negative $y$-axis and intersects the unit circle at $(0, -1)$. So $\sin(-5\pi/2) = -1$ and $\cos(-5\pi/2) = 0$.
**e.** Since $-720°$ is coterminal with 0°, the terminal side of $-720°$ lies on the positive $x$-axis and intersects the unit circle at $(1, 0)$. So $\sin(-720°) = 0$ and $\cos(-720°) = 1$.

▸**TRY THIS.** Find the exact values of $\sin(-\pi/2)$ and $\cos(-\pi/2)$. ■

The *signs* of the sine and cosine functions depend on the quadrant in which the angle lies. For any point $(x, y)$ on the unit circle in quadrant I, the $x$- and $y$-coordinates are positive. So if $\alpha$ is an angle in quadrant I, $\sin\alpha > 0$ and $\cos\alpha > 0$. Since the $x$-coordinate of any point in quadrant II is negative, the cosine of any angle in quadrant II is negative. Figure 5.22 gives the signs of sine and cosine for each of the four quadrants.

## Sine and Cosine of a Multiple of 45°

In Example 1, exact values of the sine and cosine functions were found for some angles that were multiples of 90° (quadrantal angles). We can also find the exact values of $\sin(45°)$ and $\cos(45°)$ and then use that information to find the exact values of these functions for any multiple of 45°.

Since the terminal side for 45° lies on the line $y = x$, as shown in Fig. 5.23, the $x$- and $y$-coordinates at the point of intersection with the unit circle are equal. From Section 1.3, the equation of the unit circle is $x^2 + y^2 = 1$. Because $y = x$, we can write $x^2 + x^2 = 1$ and solve for $x$:

$$2x^2 = 1$$

$$x^2 = \frac{1}{2}$$

$$x = \pm\sqrt{\frac{1}{2}} = \pm\frac{\sqrt{2}}{2}$$

For a 45° angle in quadrant I, $x$ and $y$ are both positive numbers. So $\sin(45°) = \cos(45°) = \sqrt{2}/2$.

There are four points where the lines $y = x$ and $y = -x$ intersect the unit circle. The coordinates of these points, shown in Fig. 5.24, can all be found as above or by the symmetry of the unit circle. Figure 5.24 shows the coordinates of the key points for determining the sine and cosine of any angle that is an integral multiple of 45°.

## EXAMPLE 2 Evaluating sine and cosine at a multiple of 45°

Find the exact value of each expression.

**a.** $\sin(135°)$ **b.** $\sin\left(\frac{5\pi}{4}\right)$ **c.** $2\sin\left(-\frac{9\pi}{4}\right)\cos\left(-\frac{9\pi}{4}\right)$

### Solution

**a.** The terminal side for 135° lies in quadrant II, halfway between 90° and 180°. As shown in Fig. 5.24, the point on the unit circle halfway between 90° and 180° is $(-\sqrt{2}/2, \sqrt{2}/2)$. So $\sin(135°) = \sqrt{2}/2$.

**b.** The terminal side for $5\pi/4$ is in quadrant III, halfway between $4\pi/4$ and $6\pi/4$. From Fig. 5.24, $\sin(5\pi/4)$ is the $y$-coordinate of $(-\sqrt{2}/2, -\sqrt{2}/2)$. So $\sin(5\pi/4) = -\sqrt{2}/2$.

**c.** Since $-8\pi/4$ is one clockwise revolution, $-9\pi/4$ is coterminal with $-\pi/4$. From Fig. 5.24, we have $\cos(-9\pi/4) = \cos(-\pi/4) = \sqrt{2}/2$ and $\sin(-9\pi/4) = -\sqrt{2}/2$. So

$$2\sin\left(-\frac{9\pi}{4}\right)\cos\left(-\frac{9\pi}{4}\right) = 2\cdot\left(-\frac{\sqrt{2}}{2}\right)\cdot\frac{\sqrt{2}}{2} = -1.$$

▶**TRY THIS.** Find the exact values of $\sin(-5\pi/4)$ and $\cos(-5\pi/4)$. ■

## Sine and Cosine of a Multiple of 30°

Exact values for the sine and cosine of 60° and the sine and cosine of 30° are found with a little help from geometry. The terminal side of a 60° angle intersects the unit circle at a point $(x, y)$, as shown in Fig. 5.25(a), where $x$ and $y$ are the lengths of the legs of a 30-60-90 triangle whose hypotenuse is length 1.

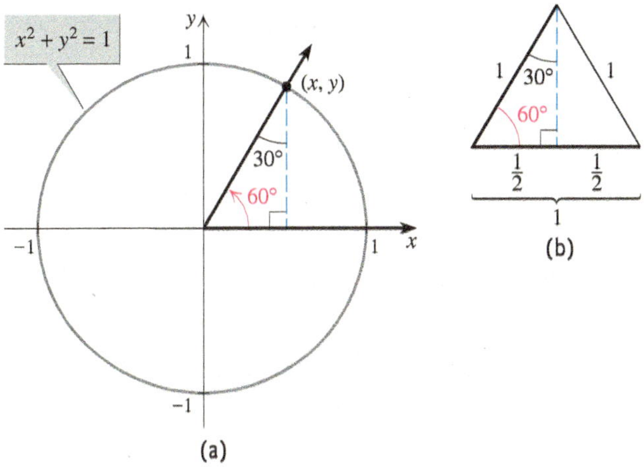


**Figure 5.25**

Since two congruent 30-60-90 triangles are formed by the altitude of an equilateral triangle, as shown in Fig. 5.25(b), the side opposite the 30° angle is half the length of the hypotenuse. Since the length of the hypotenuse is 1, the side opposite 30° is 1/2, that is, $x = 1/2$. Use the fact that $x^2 + y^2 = 1$ for the unit circle to find $y$:

$$\left(\frac{1}{2}\right)^2 + y^2 = 1 \qquad \text{Replace } x \text{ with } \tfrac{1}{2} \text{ in } x^2 + y^2 = 1.$$

$$y^2 = \frac{3}{4}$$

$$y = \pm\sqrt{\frac{3}{4}} = \pm\frac{\sqrt{3}}{2}$$

Since $y > 0$ (in quadrant I), $\cos(60°) = 1/2$ and $\sin(60°) = \sqrt{3}/2$.

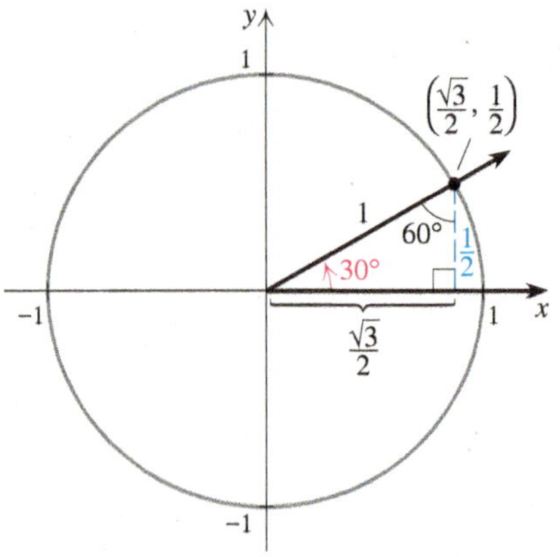


**Figure 5.26**

The point of intersection of the terminal side for a 30° angle with the unit circle determines a 30-60-90 triangle exactly the same size as the one described above. In this case the longer leg is on the $x$-axis, as shown in Fig. 5.26, and the point on the unit circle is $(\sqrt{3}/2, 1/2)$. So $\cos(30°) = \sqrt{3}/2$, and $\sin(30°) = 1/2$.

Using the symmetry of the unit circle and the values just found for the sine and cosine of 30° and 60°, we can label points on the unit circle, as shown in Fig. 5.27. This figure shows the coordinates of every point on the unit circle where the terminal side of a multiple of 30° intersects the unit circle. For example, the terminal side of a 120° angle in standard position intersects the unit circle at $(-1/2, \sqrt{3}/2)$, because $120° = 4 \cdot 30°$.

In the next example, we use Fig. 5.27 to evaluate expressions involving multiples of 30° ($\pi/6$). Note that the square of $\sin\alpha$ could be written as $(\sin\alpha)^2$, but for simplicity it is written as $\sin^2\alpha$. Likewise, $\cos^2\alpha$ is used for $(\cos\alpha)^2$.

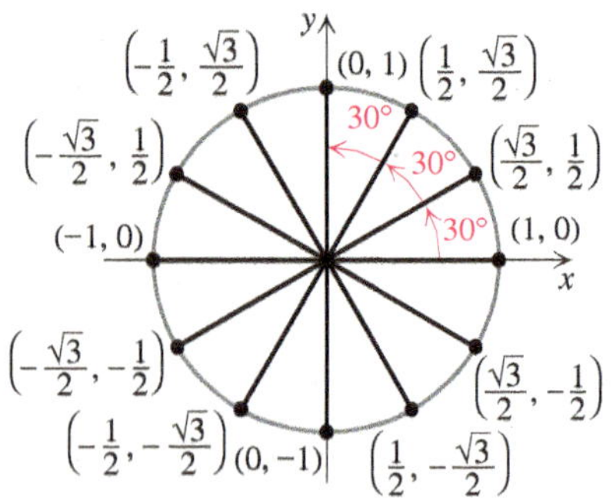


**Figure 5.27**

### EXAMPLE 3 Evaluating sine and cosine at a multiple of 30°

Find the exact value of each expression.

**a.** $\sin\left(\frac{7\pi}{6}\right)$ **b.** $\cos^2(-240°) - \sin^2(-240°)$

**Solution**

**a.** Since $\pi/6 = 30°$, we have $7\pi/6 = 210°$. To determine the location of the terminal side of $7\pi/6$, notice that $7\pi/6$ is 30° larger than the straight angle 180° and so it lies in quadrant III. From Fig. 5.27, $7\pi/6$ intersects the unit circle at $(-\sqrt{3}/2, -1/2)$ and $\sin(7\pi/6) = -1/2$.

**b.** Since $-240°$ is coterminal with 120°, the terminal side for $-240°$ lies in quadrant II and intersects the unit circle at $(-1/2, \sqrt{3}/2)$, as indicated in Fig. 5.27. So $\cos(-240°) = -1/2$ and $\sin(-240°) = \sqrt{3}/2$. Now use these values in the original expression:

$$\cos^2(-240°) - \sin^2(-240°) = \left(-\frac{1}{2}\right)^2 - \left(\frac{\sqrt{3}}{2}\right)^2 = \frac{1}{4} - \frac{3}{4} = -\frac{1}{2}$$

►**TRY THIS.** Find the exact values of $\sin(-\pi/6)$ and $\cos(-\pi/6)$. ■

A good way to remember the sines and cosines for the common angles is to note the following patterns:

$$\sin(30°) = \frac{\sqrt{1}}{2}, \quad \sin(45°) = \frac{\sqrt{2}}{2}, \quad \sin(60°) = \frac{\sqrt{3}}{2}$$

$$\cos(30°) = \frac{\sqrt{3}}{2}, \quad \cos(45°) = \frac{\sqrt{2}}{2}, \quad \cos(60°) = \frac{\sqrt{1}}{2}$$

## Reference Angles

One way to find $\sin\alpha$ and $\cos\alpha$ is to determine the coordinates of the intersection of the terminal side of $\alpha$ and the unit circle, as we did in Examples 2 and 3. Another way to find $\sin\alpha$ and $\cos\alpha$ is to find the sine and cosine of the corresponding *reference angle,* which is in the interval $[0, \pi/2]$, and attach the appropriate sign. Reference angles are essential for evaluating trigonometric functions if you are using tables that give values only for angles in the interval $[0, \pi/2]$.

**Definition: Reference Angle**

If $\theta$ is a nonquadrantal angle in standard position, then the **reference angle** for $\theta$ is the positive acute angle $\theta'$ (read "theta prime") formed by the terminal side of $\theta$ and the positive or negative $x$-axis.

Figure 5.28 shows the reference angle $\theta'$ for an angle $\theta$ with terminal side in each of the four quadrants. Notice that in each case $\theta'$ is the acute angle formed by the terminal side of $\theta$ and the $x$-axis.

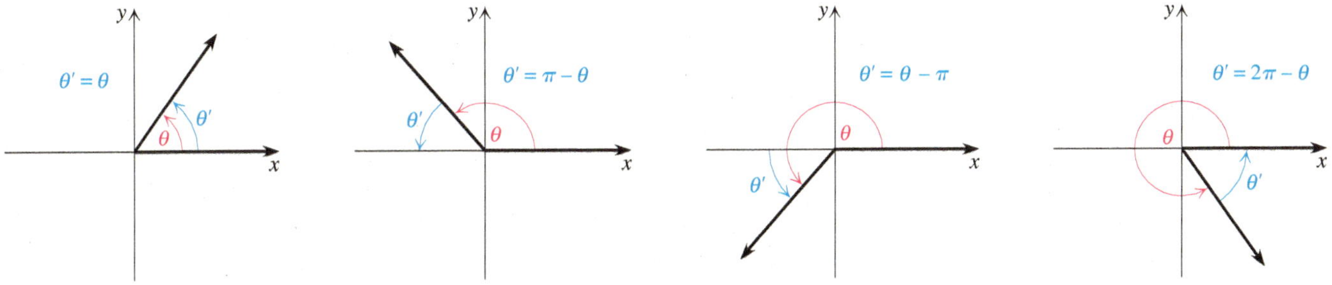


Figure 5.28

## EXAMPLE 4 Finding reference angles

For each given angle $\theta$, sketch the reference angle $\theta'$ and give the measure of $\theta'$ in both radians and degrees.

**a.** $\theta = 120°$ **b.** $\theta = 7\pi/6$ **c.** $\theta = 690°$ **d.** $\theta = -7\pi/4$

### Solution

**a.** The terminal side of 120° is in quadrant II, as shown in Fig. 5.29(a). The reference angle $\theta'$ is $180° - 120° = 60°$ or $\pi/3$.

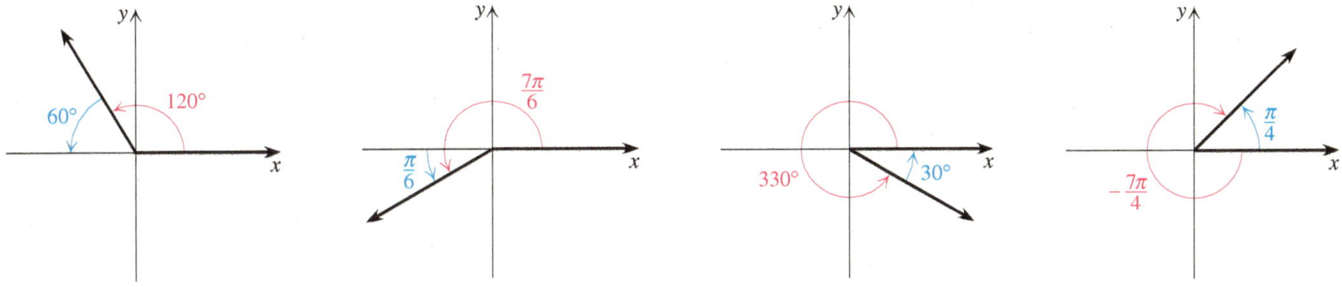


Figure 5.29

**b.** The terminal side of $7\pi/6$ is in quadrant III, as shown in Fig. 5.29(b). The reference angle $\theta'$ is $7\pi/6 - \pi = \pi/6$ or 30°.

**c.** Since 690° is coterminal with 330°, the terminal side is in quadrant IV, as shown in Fig. 5.29(c). The reference angle $\theta'$ is $360° - 330° = 30°$ or $\pi/6$.

**d.** An angle of $-7\pi/4$ radians has a terminal side in quadrant I, as shown in Fig. 5.29(d), and the reference angle is $\pi/4$ or 45°.

▸**TRY THIS.** Find the reference angle for 135° in degrees and radians. ■

Because the measure of any reference angle is between 0 and $\pi/2$, any reference angle can be positioned in the first quadrant. Figure 5.30(a) shows a typical reference angle $\theta'$ in standard position with its terminal side intersecting the unit circle at $(x, y)$. So $\cos\theta' = x$ and $\sin\theta' = y$.

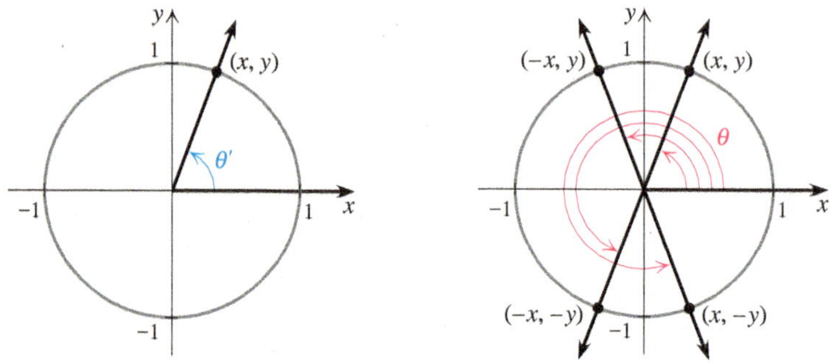


Reference angle $\theta'$
(a)

Possible $\theta$ corresponding to $\theta'$
(b)

**Figure 5.30**

For what angle $\theta$ could $\theta'$ be the reference angle? The angle $\theta$ (in standard position) that corresponds to $\theta'$ could have terminated in quadrant I, II, III, or IV, where it would intersect the unit circle at $(x, y)$, $(-x, y)$, $(-x, -y)$, or $(x, -y)$, as shown in Fig. 5.30(b). So $\cos\theta = \pm x$ and $\sin\theta = \pm y$. Therefore, we can find $\cos\theta'$ and $\sin\theta'$ and prefix the appropriate sign to get $\cos\theta$ and $\sin\theta$. The appropriate sign is determined by the quadrant in which the terminal side of $\theta$ lies. The signs of $\sin\theta$ and $\cos\theta$ for the four quadrants were summarized in Fig. 5.22.

**Theorem: Evaluating Trigonometric Functions Using Reference Angles**

For an angle $\theta$ in standard position that is not a quadrantal angle, the value of a trigonometric function of $\theta$ can be found by finding the value of the function for its reference angle $\theta'$ and prefixing the appropriate sign.

## EXAMPLE 5 Trigonometric functions using reference angles

Find the sine and cosine for each angle using reference angles.

**a.** $\theta = 120°$ **b.** $\theta = 7\pi/6$ **c.** $\theta = 690°$ **d.** $\theta = -7\pi/4$

### Solution

**a.** Figure 5.29(a) shows the terminal side of 120° in quadrant II and its reference angle 60°. Since $\sin\theta > 0$ and $\cos\theta < 0$ for any angle $\theta$ in quadrant II, $\sin(120°) > 0$ and $\cos(120°) < 0$. So to get $\cos(120°)$ from $\cos(60°)$, we must prefix a negative sign. We have

$$\sin(120°) = \sin(60°) = \frac{\sqrt{3}}{2} \quad \text{and} \quad \cos(120°) = -\cos(60°) = -\frac{1}{2}.$$

**b.** Figure 5.29(b) shows the terminal side of $7\pi/6$ in quadrant III and its reference angle $\pi/6$. Since $\sin\theta < 0$ and $\cos\theta < 0$ for any angle $\theta$ in quadrant III, we must prefix a negative sign to both $\sin(\pi/6)$ and $\cos(\pi/6)$. So

$$\sin\left(\frac{7\pi}{6}\right) = -\sin\left(\frac{\pi}{6}\right) = -\frac{1}{2} \quad \text{and} \quad \cos\left(\frac{7\pi}{6}\right) = -\cos\left(\frac{\pi}{6}\right) = -\frac{\sqrt{3}}{2}.$$

**c.** The angle 690° terminates in quadrant IV (where $\sin\theta < 0$ and $\cos\theta > 0$) and its reference angle is 30°, as shown in Fig. 5.29(c). So

$$\sin(690°) = -\sin(30°) = -\frac{1}{2} \quad \text{and} \quad \cos(690°) = \cos(30°) = \frac{\sqrt{3}}{2}.$$

**d.** The angle $-7\pi/4$ terminates in quadrant I (where $\sin\theta > 0$ and $\cos\theta > 0$) and its reference angle is $\pi/4$, as shown in Fig. 5.29(d). So

$$\sin\left(-\frac{7\pi}{4}\right) = \sin\left(\frac{\pi}{4}\right) = \frac{\sqrt{2}}{2} \quad \text{and} \quad \cos\left(-\frac{7\pi}{4}\right) = \cos\left(\frac{\pi}{4}\right) = \frac{\sqrt{2}}{2}.$$

▸**TRY THIS.** Find sin(135°) and cos(135°) using reference angles. ■

## The Sine and Cosine of a Real Number

### ■ Foreshadowing Calculus

In calculus we study functions such as $f(x) = x^2 + \sin x$. For this function to make sense, $x$ must be a real number and not degrees. So in calculus we generally use real numbers or radians as the input of the trigonometric functions.

The domain of sine and cosine can be thought of as the set of all angles, degree measures of angles, or radian measures of angles. However, since radian measures are just real numbers, the domain of sine and cosine is the set of real numbers. The set of real numbers is often visualized as the domain of sine and cosine by placing a number line (the set of real numbers) next to the unit circle, as shown in Fig. 5.31(a). To find $\sin s$ or $\cos s$ for $s > 0$, wrap the top half of the number line around the unit circle, as shown in Fig. 5.31(b). The real number $s$ then corresponds to an arc on the unit circle of length $s$, ending at $(x, y)$. Of course, $\sin s = y$ and $\cos s = x$. Arcs corresponding to negative real numbers are found by wrapping the negative half of the number line around the unit circle in a clockwise direction, as shown in Fig. 5.31(c). Because of their relationship to the unit circle, sine and cosine are often called **circular functions.**

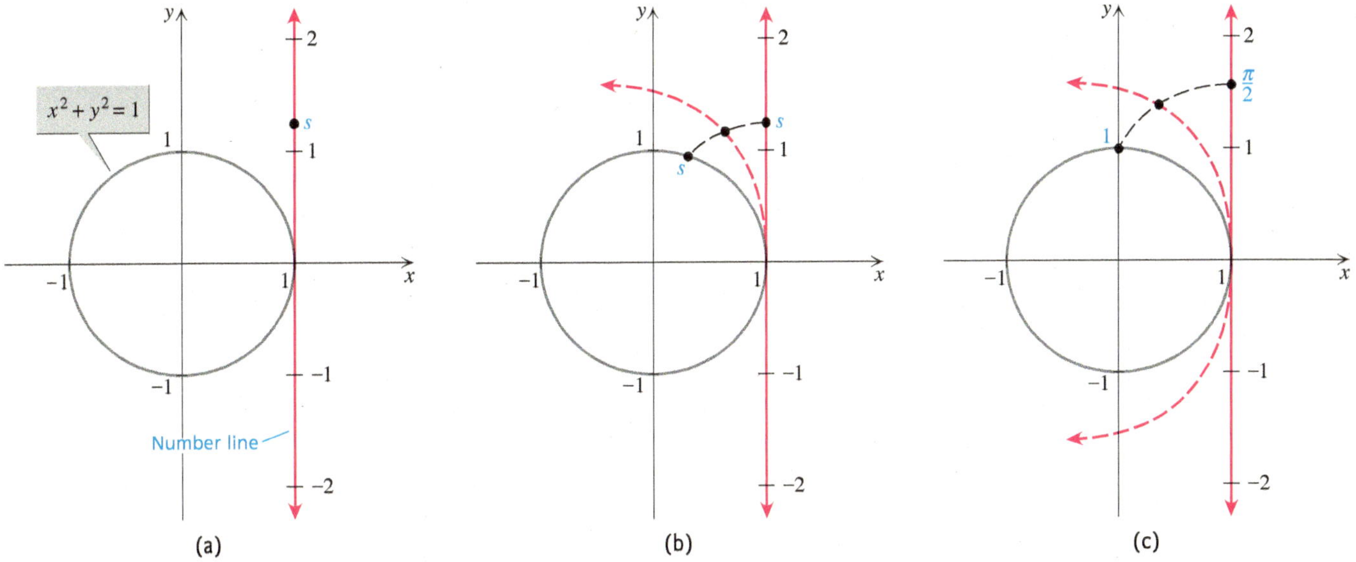


Figure 5.31

## Approximate Values for Sine and Cosine

The sine and cosine for any angle that is a multiple of 30° or 45° can be found exactly. These angles are so common that it is important to know these exact values. However, for most other angles or real numbers we use approximate values for the sine and cosine, found with the help of a scientific calculator or a graphing calculator. Calculators can evaluate $\sin\alpha$ or $\cos\alpha$ if $\alpha$ is a real number (radian) or $\alpha$ is the degree measure of an angle. Generally, there is a mode key that sets the calculator to degree mode or radian mode. Consult your calculator manual.

### EXAMPLE 6 Evaluating sine and cosine with a calculator

Find each function value rounded to four decimal places.

**a.** sin(4.27) **b.** cos(−39.46°)

Figure 5.32

### Solution

**a.** With the calculator in radian mode, we get $\sin(4.27) \approx -0.9037$.
**b.** With the calculator in degree mode, we get $\cos(-39.46°) \approx 0.7721$.

If you use the symbol for radians or degrees, as shown in Fig. 5.32, then it is not necessary to change the mode with a graphing calculator.

▸**TRY THIS.** Find $\sin(55.6)$ and $\sin(34.2°)$ with a calculator. ■

## The Fundamental Identity

An identity is an equation that is satisfied for all values of the variable for which both sides are defined. The most fundamental identity in trigonometry involves the squares of the sine and cosine functions. For convenience we write $(\sin\alpha)^2$ as $\sin^2\alpha$ and $(\cos\alpha)^2$ as $\cos^2\alpha$. If $\alpha$ is an angle in standard position then $\sin\alpha = y$ and $\cos\alpha = x$, where $(x, y)$ is the point of intersection of the terminal side of $\alpha$ and the unit circle. Since every point on the unit circle satisfies the equation $x^2 + y^2 = 1$, we get the following identity.

**The Fundamental Identity of Trigonometry**

If $\alpha$ is any angle or real number,

$$\sin^2\alpha + \cos^2\alpha = 1.$$

Figure 5.33

The fundamental identity can be illustrated with a graphing calculator as shown in Fig. 5.33 □

If we know the value of the sine or cosine of an angle, then we can use the fundamental identity to find the value of the other function of the angle.

### EXAMPLE 7 Using the fundamental identity

Find $\cos\alpha$, given that $\sin\alpha = 3/5$ and $\alpha$ is an angle in quadrant II.

### Solution

Use the fundamental identity $\sin^2\alpha + \cos^2\alpha = 1$ to find $\cos\alpha$:

$$\left(\frac{3}{5}\right)^2 + \cos^2\alpha = 1 \qquad \text{Replace } \sin\alpha \text{ with } 3/5.$$

$$\cos^2\alpha = \frac{16}{25} \qquad 1 - \frac{9}{25} = \frac{16}{25}$$

$$\cos\alpha = \pm\sqrt{\frac{16}{25}} = \pm\frac{4}{5}$$

Since $\cos\alpha < 0$ for any $\alpha$ in quadrant II, we choose the negative sign and get $\cos\alpha = -4/5$.

▸**TRY THIS.** Find $\sin\alpha$ given that $\cos\alpha = 1/4$ and $\alpha$ is in quadrant IV. ■

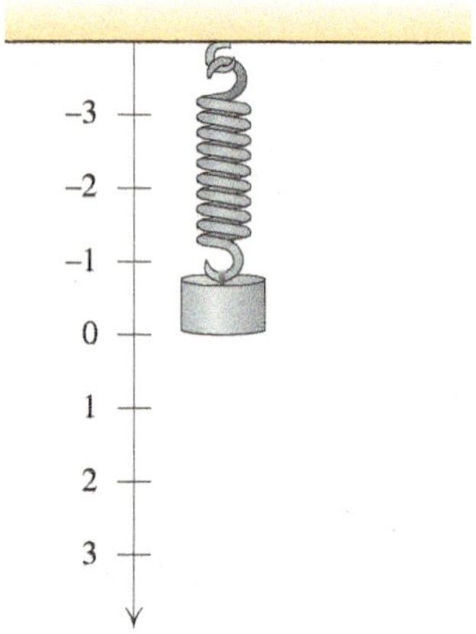


Figure 5.34

## Modeling the Motion of a Spring

The sine and cosine functions are used in modeling the motion of a spring. If a weight is at rest while hanging from a spring, as shown in Fig. 5.34, then it is at the **equilibrium** position, or 0 on a vertical number line. If the weight is set in motion with an initial velocity $v_0$ from location $x_0$, then the location at time $t$ is given by

$$x = \frac{v_0}{\omega}\sin(\omega t) + x_0\cos(\omega t).$$

The letter $\omega$ (omega) is a constant that depends on the stiffness of the spring and the amount of weight on the spring. For positive values of $x$ the weight is below equilibrium, and for negative values it is above equilibrium. The initial velocity is considered to be positive if it is in the downward direction and negative if it is upward. Note that the domain of the sine and cosine functions in this formula is the nonnegative real numbers, and this application has nothing to do with angles.

### EXAMPLE 8 Motion of a spring

A weight on a certain spring is set in motion with an upward velocity of 3 centimeters per second from a position 2 centimeters below equilibrium. Assume that for this spring and weight combination the constant $\omega$ has a value of 1. Write a formula that gives the location of the weight in centimeters as a function of the time $t$ in seconds, and find the location of the weight 2 seconds after the weight is set in motion.

#### Solution

Since upward velocity is negative and locations below equilibrium are positive, use $v_0 = -3$, $x_0 = 2$, and $\omega = 1$ in the formula for the motion of a spring:

$$x = -3 \sin t + 2 \cos t$$

If $t = 2$ seconds, then $x = -3 \sin(2) + 2 \cos(2) \approx -3.6$ centimeters, which is 3.6 centimeters above the equilibrium position.

Figure 5.35

Figure 5.35 shows the calculation for $x$ with a graphing calculator in radian mode.

**▶TRY THIS.** If $x = -2 \sin(2t) + 3 \cos(2t)$ gives the location in centimeters of a weight on a spring $t$ seconds after it is set in motion, then what is the location at time $t = 3$? ■

## FOR thought... True or False? Explain.

**1.** $\cos(90°) = 1$

**2.** $\cos(90) = 0$

**3.** $\sin(45°) = 1/\sqrt{2}$

**4.** $\sin(-\pi/3) = \sin(\pi/3)$

**5.** $\cos(-\pi/3) = -\cos(\pi/3)$

**6.** $\sin(390°) = \sin(30°)$

**7.** If $\sin(\alpha) < 0$ and $\cos(\alpha) > 0$, then $\alpha$ is an angle in quadrant III.

**8.** If $\sin^2(\alpha) = 1/4$ and $\alpha$ is in quadrant III, then $\sin(\alpha) = 1/2$.

**9.** If $\cos^2(\alpha) = 3/4$ and $\alpha$ is an angle in quadrant I, then $\alpha = \pi/6$.

**10.** For any angle $\alpha$, $\cos^2(\alpha) = (1 - \sin(\alpha))(1 + \sin(\alpha))$.

## EXERCISES 5.2

*Fill in the blank.*

**1.** If $\alpha$ is an angle in standard position whose terminal side intersects the unit circle at $(x, y)$, then $y =$ ________ and $x =$ ________.

**2.** The identity $\sin^2(\alpha) + \cos^2(\alpha) = 1$ is the ________ identity.

*Redraw each diagram and label the indicated points with the proper coordinates. Exercise 3 shows the points where the terminal side of every multiple of 45° intersects the unit circle. Exercise 4 shows the points where the terminal side of every multiple of 30° intersects the unit circle. Repeat Exercises 3 and 4 until you can do them from memory.*

**3.**

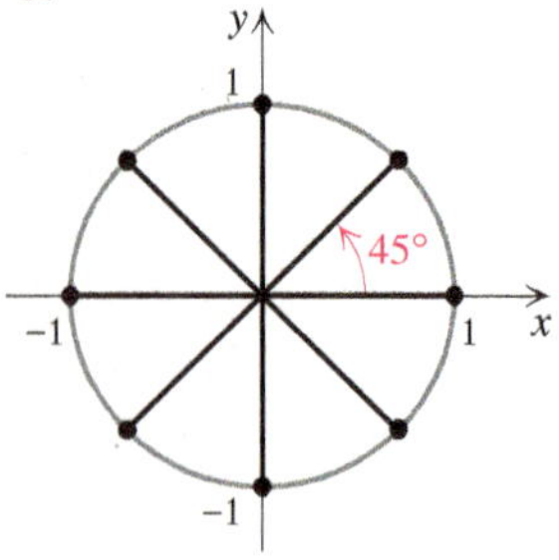


**4.**

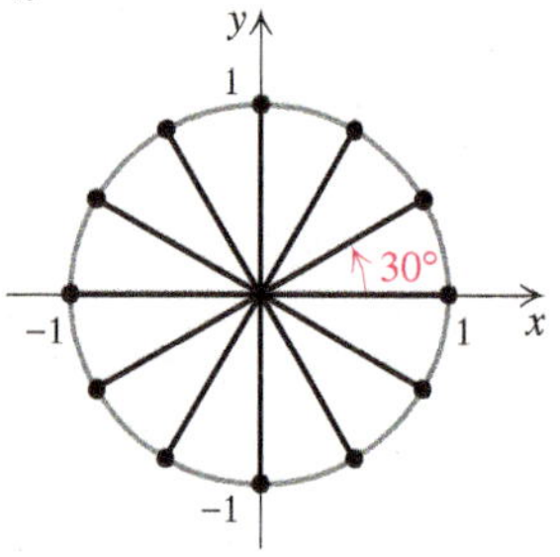


*Use the diagrams drawn in Exercises 3 and 4 to help you find the exact value of each function. Do not use a calculator.*

**5.** $\sin(0)$ **6.** $\cos(\pi)$ **7.** $\cos(-90°)$

**8.** $\sin(0°)$ **9.** $\sin(2\pi)$ **10.** $\cos(-\pi)$

**11.** $\cos(3\pi/2)$ **12.** $\sin(-5\pi/2)$ **13.** $\sin(135°)$

**14.** $\cos(-135°)$ **15.** $\sin(-\pi/4)$ **16.** $\sin(3\pi/4)$

**17.** $\sin(30°)$ **18.** $\sin(120°)$ **19.** $\cos(-60°)$

**20.** $\cos(-120°)$ **21.** $\cos(7\pi/6)$ **22.** $\cos(-2\pi/3)$

**23.** $\sin(-4\pi/3)$ **24.** $\sin(5\pi/6)$ **25.** $\sin(390°)$

**26.** $\sin(765°)$ **27.** $\cos(-420°)$ **28.** $\cos(-450°)$

**29.** $\cos(13\pi/6)$ **30.** $\cos(-7\pi/3)$

*Sketch the given angle in standard position and find its reference angle in degrees and radians.*

**31.** 210° **32.** 330° **33.** $5\pi/3$ **34.** $7\pi/6$

**35.** −300° **36.** −210° **37.** $5\pi/6$ **38.** $-13\pi/6$

**39.** 405° **40.** 390° **41.** $-3\pi/4$ **42.** $-2\pi/3$

*Determine whether each of the following expressions is positive (+) or negative (−) without using a calculator.*

**43.** $\sin(121°)$ **44.** $\cos(157°)$ **45.** $\cos(359°)$

**46.** $\sin(213°)$ **47.** $\sin(7\pi/6)$ **48.** $\cos(7\pi/3)$

**49.** $\cos(-3\pi/4)$ **50.** $\sin(-7\pi/4)$

*Use reference angles to find the exact value of each expression.*

**51.** $\sin(135°)$ **52.** $\sin(420°)$ **53.** $\cos(5\pi/3)$

**54.** $\cos(11\pi/6)$ **55.** $\sin(7\pi/4)$ **56.** $\sin(-13\pi/4)$

**57.** $\cos(-17\pi/6)$ **58.** $\cos(-5\pi/3)$ **59.** $\sin(-45°)$

**60.** $\cos(-120°)$ **61.** $\cos(-240°)$ **62.** $\sin(-225°)$

*Find the exact value of each expression without using a calculator. Check your answer with a calculator.*

**63.** $\dfrac{\cos(\pi/3)}{\sin(\pi/3)}$ **64.** $\dfrac{\sin(-5\pi/6)}{\cos(-5\pi/6)}$

**65.** $\dfrac{\sin(7\pi/4)}{\cos(7\pi/4)}$ **66.** $\dfrac{\sin(-3\pi/4)}{\cos(-3\pi/4)}$

**67.** $\sin\left(\dfrac{\pi}{3} + \dfrac{\pi}{6}\right)$ **68.** $\cos\left(\dfrac{\pi}{3} - \dfrac{\pi}{6}\right)$

**69.** $\dfrac{1 - \cos(5\pi/6)}{\sin(5\pi/6)}$ **70.** $\dfrac{\sin(5\pi/6)}{1 + \cos(5\pi/6)}$

**71.** $\sin(\pi/4) + \cos(\pi/4)$ **72.** $\sin^2(\pi/6) + \cos^2(\pi/6)$

*Use a calculator to find the value of each function. Round answers to four decimal places.*

**73.** $\cos(-359.4°)$ **74.** $\sin(344.1°)$

**75.** $\sin(23°48')$ **76.** $\cos(49°13')$

**77.** $\sin(-48°3'12'')$ **78.** $\cos(-9°4'7'')$

**79.** $\sin(1.57)$ **80.** $\cos(3.14)$

**81.** $\cos(7\pi/12)$ **82.** $\sin(-13\pi/8)$

*Find the exact value of each expression for the given value of θ. Do not use a calculator.*

**83.** $\sin(2\theta)$ if $\theta = \pi/4$ **84.** $\sin(2\theta)$ if $\theta = \pi/6$

**85.** $\cos(2\theta)$ if $\theta = \pi/6$ **86.** $\cos(2\theta)$ if $\theta = \pi/3$

**87.** $\sin(\theta/2)$ if $\theta = 3\pi/2$ **88.** $\sin(\theta/2)$ if $\theta = 2\pi/3$

**89.** $\cos(\theta/2)$ if $\theta = \pi/3$ **90.** $\cos(\theta/2)$ if $\theta = \pi/2$

*Solve each problem.*

**91.** Find $\cos(\alpha)$, given that $\sin(\alpha) = 5/13$ and $\alpha$ is in quadrant II.

**92.** Find $\sin(\alpha)$, given that $\cos(\alpha) = -4/5$ and $\alpha$ is in quadrant III.

**93.** Find $\sin(\alpha)$, given that $\cos(\alpha) = 3/5$ and $\alpha$ is in quadrant IV.

**94.** Find $\cos(\alpha)$, given that $\sin(\alpha) = -12/13$ and $\alpha$ is in quadrant IV.

**95.** Find $\cos(\alpha)$, given that $\sin(\alpha) = 1/3$ and $\cos(\alpha) > 0$.

**96.** Find $\sin(\alpha)$, given that $\cos(\alpha) = 2/5$ and $\sin(\alpha) < 0$.

*Solve each problem.*

**97.** *Motion of a Spring* A weight on a vertical spring is given an initial downward velocity of 4 cm/sec from a point 3 cm above equilibrium. Assuming that the constant $\omega$ has a value of 1, write the formula for the location of the weight at time $t$, and find its location 3 sec after it is set in motion.

**98.** *Motion of a Spring* A weight on a vertical spring is given an initial upward velocity of 3 in./sec from a point 1 in. below equilibrium. Assuming that the constant $\omega$ has a value of $\sqrt{3}$, write the formula for the location of the weight at time $t$, and find its location 2 sec after it is set in motion.

**99.** *Spacing Between Teeth* The length of an arc intercepted by a central angle of $\theta$ radians in a circle of radius $r$ is $r\theta$. The length of the chord, $c$, joining the endpoints of that arc is given by $c = r\sqrt{2 - 2\cos\theta}$. Find the actual distance between the tips of two adjacent teeth on a 12-in.-diameter carbide-tipped circular saw blade with 22 equally spaced teeth. Compare your answer with the length of a circular arc joining two adjacent teeth on a circle 12 in. in diameter. Round to three decimal places.

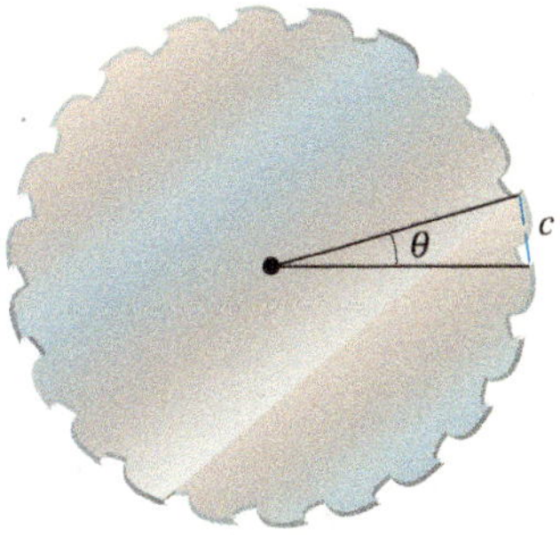


**Figure for Exercise 99**

**100.** *Shipping Stop Signs* A manufacturer of stop signs ships 100 signs in a circular drum. Use the formula from Exercise 99 to find the radius of the inside of the circular drum, given that the length of an edge of the stop sign is 10 in. and the signs just fit as shown in the figure.

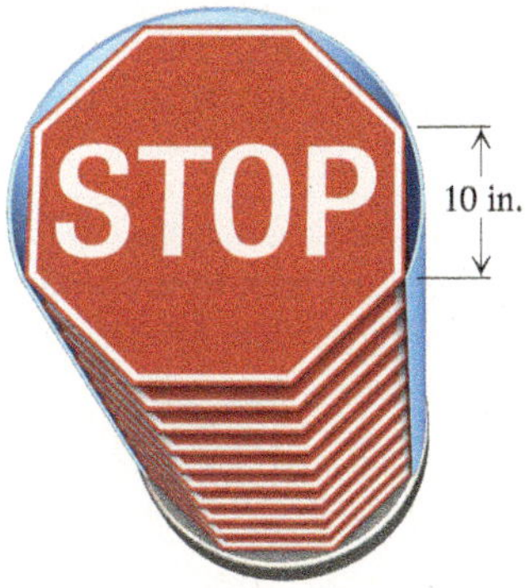


**Figure for Exercise 100**

**101.** *Cosine in Terms of Sine* Use the fundamental identity to write two formulas for $\cos\alpha$ in terms of $\sin\alpha$. Indicate which formula to use for a given value of $\alpha$.

**102.** *Throwing a Javelin* The formula

$$d = \frac{1}{32}v_0^2\sin(2\theta)$$

gives the distance $d$ in feet that a projectile will travel when its launch angle is $\theta$ radians and its initial velocity is $v_0$ feet per second. Approximately what initial velocity in miles per hour does it take to throw a javelin 367 feet with launch angle 43°, which is a typical launch angle (www.canthrow.com)?

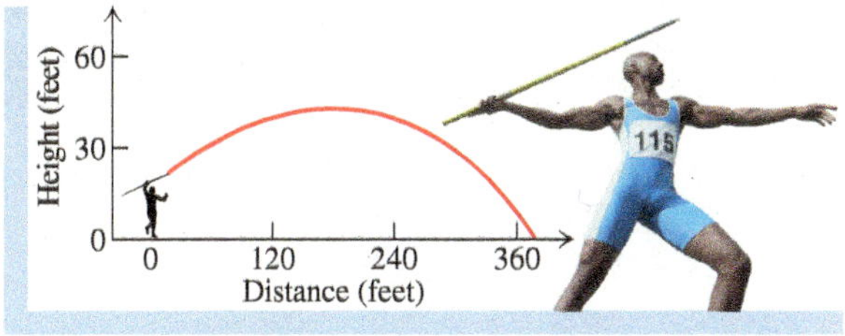


**Figure for Exercise 102**

## ▶ RETHINKING

**103.** Convert 48.23° to degrees-minutes-seconds format.

**104.** Convert 135° to radians.

**105.** List all possible rational solutions to $4x^3 - 7x^2 + 8x - 6 = 0$ according to the rational roots theorem.

**106.** Solve $x^{2/3} - 4x^{1/3} - 5 = 0$.

**107.** Find the domain and range of $y = 1 - \sqrt{3x - 9}$ and determine whether the relation is a function.

**108.** Find the inverse of the function $f(x) = x^2$ for $x \geq 0$.

## THINKING OUTSIDE THE BOX XXXVII

*Telling Time* At 12 noon the hour hand, minute hand, and second hand of a clock are all pointing straight upward.

**a.** Find the first time after 12 noon to the nearest tenth of a second at which the angle between the hour hand and minute hand is 120°.

**b.** Is there a time between 12 noon and 1 P.M. at which the three hands divide the face of the clock into thirds (that is, the angles between the hour hand, minute hand, and second hand are equal and each is 120°)?

**c.** Does the alignment of the three hands described in part (b) ever occur?

## POP QUIZ 5.2

**1.** What is the reference angle for 120°?

*Find the exact value.*

**2.** $\sin(0°)$

**3.** $\sin(-60°)$

**4.** $\sin(3\pi/4)$

**5.** $\cos(90°)$

**6.** $\cos(-2\pi/3)$

**7.** $\cos(11\pi/6)$

**8.** $\sin(\alpha)$, if $\cos(\alpha) = -3/5$ and $\alpha$ is in quadrant III

**LINKING concepts...** For Individual or Group Explorations

### Domain and Range

*The domain for both the sine and cosine functions is* $(-\infty, \infty)$ *and the range for both functions is* $[-1, 1]$. *When we combine these trigonometric functions with other functions that we have studied, determining domain and range can get interesting. For example, the range for* $f(x) = \ln(\sin(x))$ *is* $(-\infty, 0]$. *For some functions the domain or range can only be approximated. Find the domain and range for each of the following functions. Use a calculator table or graph if necessary.*

**a)** $f(x) = \sin^2(x)$

**b)** $f(x) = x^2 + \sin(x)$

**c)** $f(x) = \sin^2(x) + \cos^2(x)$

**d)** $f(x) = x \cdot \cos(x)$

**e)** $f(x) = x + \sin(x)$

**f)** $f(x) = \sin(\cos(x))$

**g)** $f(x) = \ln(\cos(x))$

**h)** $f(x) = e^{\sin(x)}$

# 5.3 The Graphs of the Sine and Cosine Functions

| X | Y1 |
|---|---|
| 0 | 0 |
| .7854 | .70711 |
| 1.5708 | 1 |
| 2.3562 | .70711 |
| 3.1416 | 0 |
| 3.927 | -.7071 |
| 4.7124 | -1 |

X=0

**Figure 5.36**

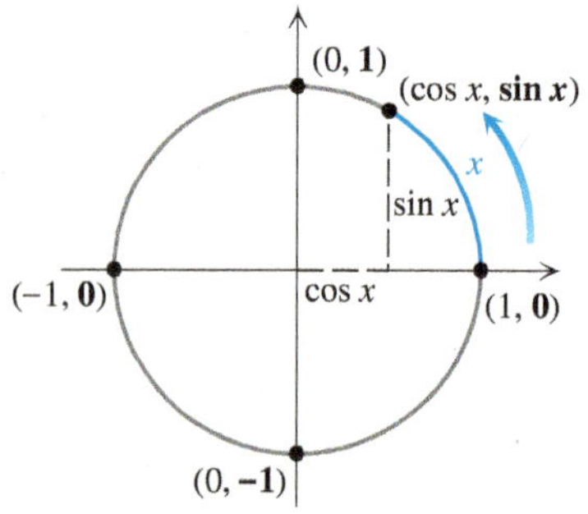


**Figure 5.37**

In Section 5.2 we studied the sine and cosine functions. In this section we will study their graphs. The graphs of trigonometric functions are important for understanding their use in modeling physical phenomena such as radio, sound, and light waves, and the motion of a spring or a pendulum.

## The Graph of $y = \sin(x)$

Until now, $s$ or $\alpha$ has been used as the independent variable in writing $\sin(s)$ or $\sin(\alpha)$. When graphing functions in an $xy$-coordinate system, it is customary to use $x$ as the independent variable and $y$ as the dependent variable, as in $y = \sin(x)$. We assume that $x$ is a real number or radian measure unless it is stated that $x$ is the degree measure of an angle. We often omit the parentheses and write $\sin x$ for $\sin(x)$.

Consider some ordered pairs that satisfy $y = \sin x$ for $x$ in $[0, 2\pi]$, as shown in Fig. 5.36. These ordered pairs are best understood by recalling that $\sin x$ is the second coordinate of the terminal point on the unit circle for an arc of length $x$, as shown in Fig. 5.37. As the arc length $x$ increases from length 0 to $\pi/2$, the second coordinate of the endpoint increases to 1. As the length of $x$ increases from $\pi/2$ to $\pi$, the second coordinate of the endpoint decreases to 0. As the arc length $x$ increases from $\pi$ to $2\pi$, the second coordinate decreases to $-1$ and then increases to 0. The calculator graph in Fig. 5.38 shows 95 accurately plotted points on the graph of $y = \sin(x)$ between 0 and $2\pi$. The actual graph of $y = \sin(x)$ is a smooth curve through those 95 points, as shown in Fig. 5.39.

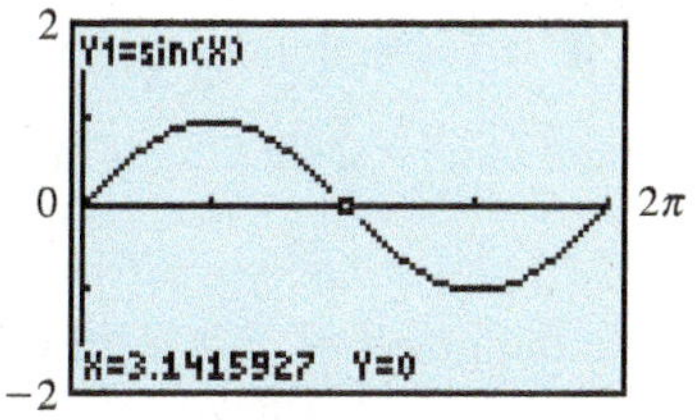


Figure 5.38

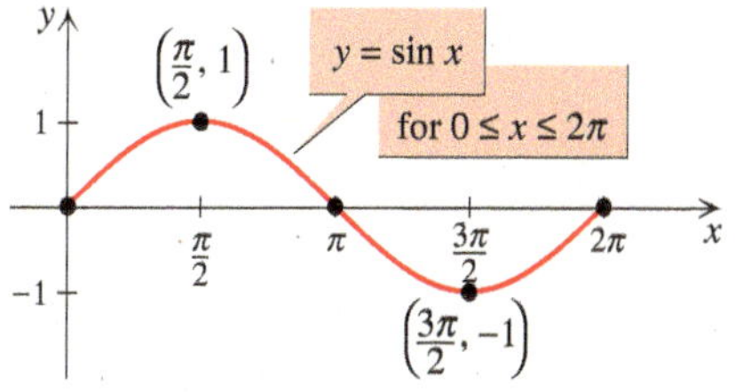


Figure 5.39

Since the $x$-intercepts and the maximum and minimum values of the function occur at multiples of $\pi$, we usually label the $x$-axis with multiples of $\pi$, as in Fig. 5.39. There are five key points on the graph of $y = \sin x$ between 0 and $2\pi$. Their exact coordinates are given in the following table.

| $x$ | 0 | $\pi/2$ | $\pi$ | $3\pi/2$ | $2\pi$ |
|---|---|---|---|---|---|
| $y = \sin x$ | 0 | 1 | 0 | $-1$ | 0 |

Note that these five points divide the interval $[0, 2\pi]$ into four equal parts.

Since the domain of $y = \sin x$ is the set of all real numbers (or radian measures of angles), we must consider values of $x$ outside the interval $[0, 2\pi]$. As $x$ increases from $2\pi$ to $4\pi$, the values of $\sin x$ again increase from 0 to 1, decrease to $-1$, then increase to 0. Because $\sin(x + 2\pi) = \sin x$, the exact shape that we saw in Fig. 5.39 is repeated for $x$ in intervals such as $[2\pi, 4\pi]$, $[-2\pi, 0]$, $[-4\pi, -2\pi]$, and so on. So the curve shown in Fig. 5.40 continues indefinitely to the left and right. The range of $y = \sin x$ is $[-1, 1]$. The graph $y = \sin x$ or any transformation of $y = \sin x$ is called a **sine wave,** a **sinusoidal wave,** or a **sinusoid.**

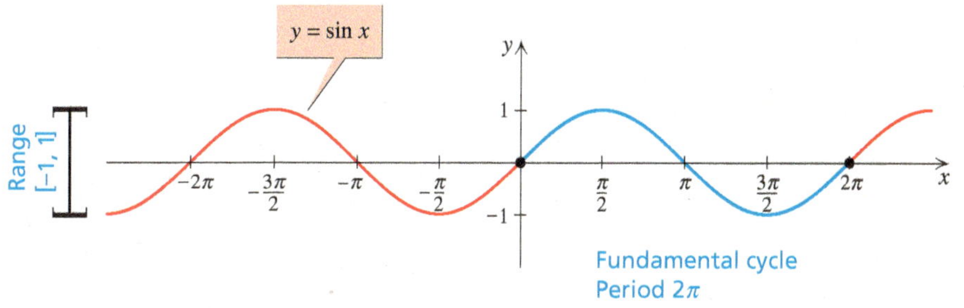


Figure 5.40

Because the shape of $y = \sin x$ for $x$ in $[0, 2\pi]$ is repeated infinitely often, $y = \sin x$ is called a *periodic function.*

**Definition: Periodic Function**

If $y = f(x)$ is a function and $a$ is a nonzero constant such that $f(x) = f(x + a)$ for every $x$ in the domain of $f$, then $f$ is called a **periodic function.** The smallest such positive constant $a$ is the **period** of the function.

For the sine function, the smallest value of $a$ such that $\sin x = \sin(x + a)$ is $a = 2\pi$, and so the period of $y = \sin x$ is $2\pi$. To prove that $2\pi$ is actually the *smallest* value of $a$ for which $\sin x = \sin(x + a)$ is a bit complicated, and we will omit the proof. The graph of $y = \sin x$ over any interval of length $2\pi$ is called a **cycle** of the sine wave. The graph of $y = \sin x$ over $[0, 2\pi]$ is called the **fundamental cycle** of $y = \sin x$.

## EXAMPLE 1 Graphing a periodic function

Sketch the graph of $y = 2\sin x$ for $x$ in the interval $[-2\pi, 2\pi]$.

### Solution

We can obtain the graph of $y = 2\sin x$ from the graph of $y = \sin x$ by stretching the graph of $y = \sin x$ by a factor of 2. In other words, double the $y$-coordinates of the five key points on the graph of $y = \sin x$ to obtain five key points on the graph of $y = 2\sin x$ for the interval $[0, 2\pi]$, as shown in the following table.

| $x$ | 0 | $\pi/2$ | $\pi$ | $3\pi/2$ | $2\pi$ |
|---|---|---|---|---|---|
| $y = 2\sin x$ | 0 | 2 | 0 | $-2$ | 0 |

The $x$-intercepts for $y = 2\sin x$ on $[0, 2\pi]$ are $(0, 0)$, $(\pi, 0)$, and $(2\pi, 0)$. Midway between the intercepts are found the highest point $(\pi/2, 2)$ and the lowest point $(3\pi/2, -2)$. Draw one cycle of $y = 2\sin x$ through these five points, as shown in Fig. 5.41. Draw another cycle of the sine wave for $x$ in $[-2\pi, 0]$ to complete the graph of $y = 2\sin x$ for $x$ in $[-2\pi, 2\pi]$.

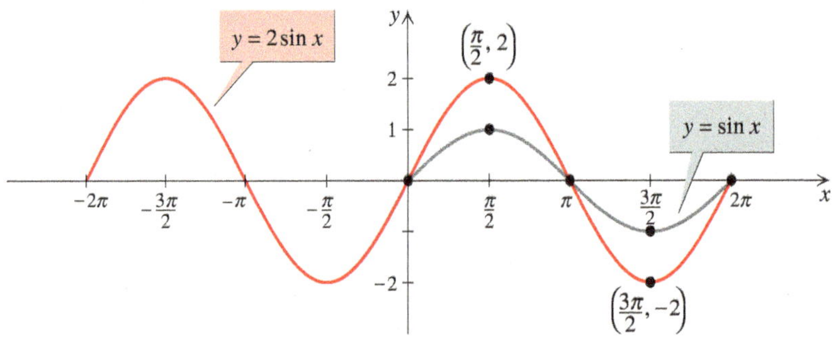


Figure 5.41

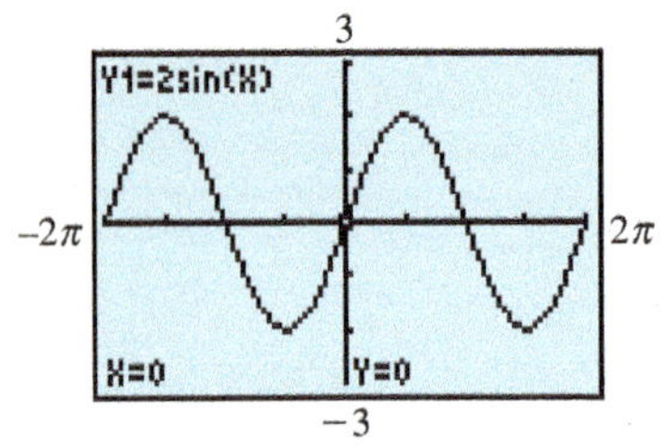


Figure 5.42

Use $\pi/2$ as the $x$-scale on your calculator, as we did in Fig. 5.41. Set the viewing window to show the same two cycles as Fig. 5.41. Note that you need some idea of what the graph looks like to get a good calculator graph. The calculator graph in Fig. 5.42 supports our conclusions about this function.

▶**TRY THIS.** Graph $y = 4\sin x$ for $x$ in the interval $[0, 2\pi]$. ■

The amplitude of a sine wave is a measure of the "height" of the wave. When an oscilloscope is used to get a picture of the sine wave corresponding to a sound, the amplitude of the sine wave corresponds to the intensity or loudness of the sound.

**Definition: Amplitude**

The **amplitude** of a sine wave, or the amplitude of the function, is the absolute value of half the difference between the maximum and minimum $y$-coordinates on the wave.

## EXAMPLE 2 Finding amplitude

Find the amplitude of the functions $y = \sin x$ and $y = 2\sin x$.

### Solution

For $y = \sin x$, the maximum $y$-coordinate is 1 and the minimum is $-1$. So the amplitude is

$$\left|\frac{1}{2}[1 - (-1)]\right| = 1.$$

For $y = 2 \sin x$, the maximum $y$-coordinate is 2 and the minimum is $-2$. So the amplitude is

$$\left|\frac{1}{2}[2 - (-2)]\right| = 2.$$

▸**TRY THIS.** Find the amplitude for $y = 4 \sin(x)$. ■

Figure 5.43

## The Graph of $y = \cos(x)$

The graph of $y = \cos x$ is best understood by recalling that $\cos x$ is the first coordinate of the terminal point on the unit circle for an arc of length $x$, as shown in Fig. 5.37. As the arc length $x$ increases from length 0 to $\pi/2$, the first coordinate of the endpoint decreases from 1 to 0. Some ordered pairs that satisfy $y = \cos x$ are shown in Fig. 5.43. As the length of $x$ increases from $\pi/2$ to $\pi$, the first coordinate decreases from 0 to $-1$. As the length increases from $\pi$ to $2\pi$, the first coordinate increases from $-1$ to 1. The graph of $y = \cos x$ has exactly the same shape as the graph of $y = \sin x$. If the graph of $y = \sin x$ is shifted a distance of $\pi/2$ to the left, the graphs would coincide. For this reason the graph of $y = \cos x$ is also called a sine wave with amplitude 1 and period $2\pi$. The graph of $y = \cos x$ over $[0, 2\pi]$ is called the **fundamental cycle** of $y = \cos x$. Since $\cos x = \cos(x + 2\pi)$, the fundamental cycle of $y = \cos x$ is repeated on $[2\pi, 4\pi]$, $[-2\pi, 0]$, and so on. The graph of $y = \cos x$ is shown in Fig. 5.44.

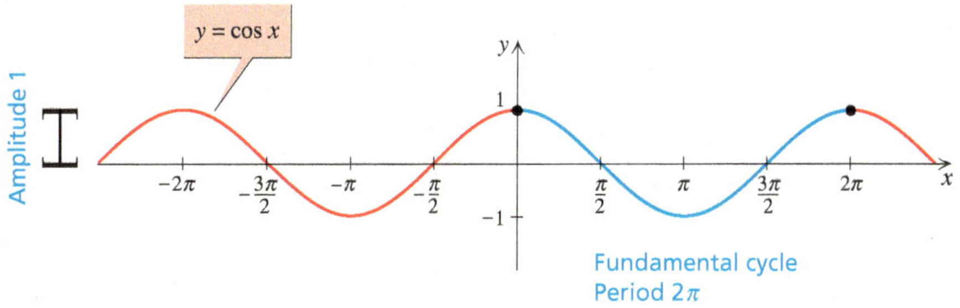


Figure 5.44

Note that there are five key points on the graph of $y = \cos x$ between 0 and $2\pi$. These points give us the highest and lowest points in the cycle as well as the $x$-intercepts. The exact coordinates are given in the following table.

| $x$ | 0 | $\pi/2$ | $\pi$ | $3\pi/2$ | $2\pi$ |
|---|---|---|---|---|---|
| $y = \cos x$ | 1 | 0 | $-1$ | 0 | 1 |

These five points divide the fundamental cycle into four equal parts.

### EXAMPLE 3 Graphing another periodic function

Sketch the graph of $y = -3 \cos x$ for $x$ in the interval $[-2\pi, 2\pi]$ and find its amplitude.

#### Solution

Make a table of ordered pairs for $x$ in $[0, 2\pi]$ to get one cycle of the graph. Note that the five $x$-coordinates in the table divide the interval $[0, 2\pi]$ into four equal parts. Multiply the $y$-coordinates of $y = \cos x$ by $-3$ to obtain the $y$-coordinates for $y = -3 \cos x$.

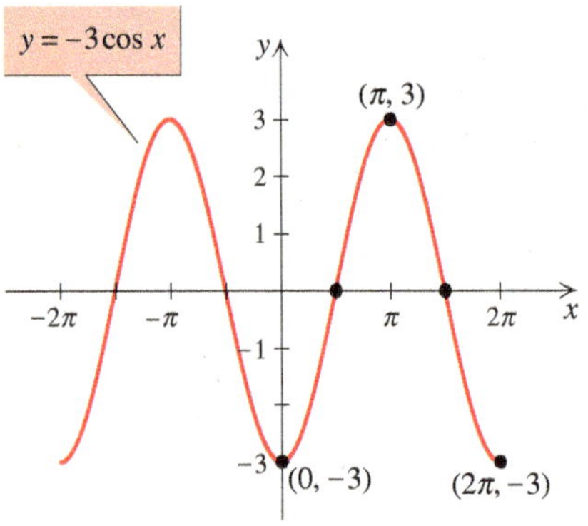


Figure 5.45

| $x$ | 0 | $\pi/2$ | $\pi$ | $3\pi/2$ | $2\pi$ |
|---|---|---|---|---|---|
| $y = -3\cos x$ | $-3$ | 0 | 3 | 0 | $-3$ |

Draw one cycle of $y = -3\cos x$ through these five points, as shown in Fig. 5.45. Repeat the same shape for $x$ in the interval $[-2\pi, 0]$ to get the graph of $y = -3\cos x$ for $x$ in $[-2\pi, 2\pi]$. The amplitude is $|0.5(3 - (-3))|$ or 3.

▸**TRY THIS.** Graph $y = 4\cos x$ for $x$ in $[0, 2\pi]$ and find the amplitude. ■

## Transformations of Sine and Cosine

In Section 2.3 we discussed how various changes in a formula affect the graph of the function. We know the changes that cause horizontal or vertical translations, reflections, and stretching or shrinking. The graph of $y = 2\sin x$ from Example 1 can be obtained by stretching the graph of $y = \sin x$. The graph of $y = -3\cos x$ in Example 3 can be obtained by stretching and reflecting the graph of $y = \cos x$. The amplitude of $y = 2\sin x$ is 2 and the amplitude of $y = -3\sin x$ is 3. In general, the amplitude is determined by the coefficient of the sine or cosine function.

**Theorem: Amplitude**

The amplitude of $y = A\sin x$ or $y = A\cos x$ is $|A|$.

In Section 2.3 we saw that the graph of $y = f(x - C)$ is a horizontal translation of the graph of $y = f(x)$, to the right if $C > 0$ and to the left if $C < 0$. So the graphs of $y = \sin(x - C)$ and $y = \cos(x - C)$ are horizontal translations of $y = \sin x$ and $y = \cos x$, respectively, to the right if $C > 0$ and to the left if $C < 0$.

**Definition: Phase Shift**

The **phase shift** of the graph of $y = \sin(x - C)$ or $y = \cos(x - C)$ is $C$.

### EXAMPLE 4 Horizontal translation

Graph two cycles of $y = \sin(x + \pi/6)$, and determine the phase shift of the graph.

#### Solution

Since $x + \pi/6 = x - (-\pi/6)$ and $C < 0$, the graph of $y = \sin(x + \pi/6)$ is obtained by moving $y = \sin x$ a distance of $\pi/6$ to the left. Since the phase shift is $-\pi/6$, label the $x$-axis with multiples of $\pi/6$, as shown in Fig. 5.46. Concentrate on moving the fundamental cycle of $y = \sin x$. The three $x$-intercepts $(0, 0)$, $(\pi, 0)$, and $(2\pi, 0)$ move to $(-\pi/6, 0)$, $(5\pi/6, 0)$, and $(11\pi/6, 0)$. The high and low

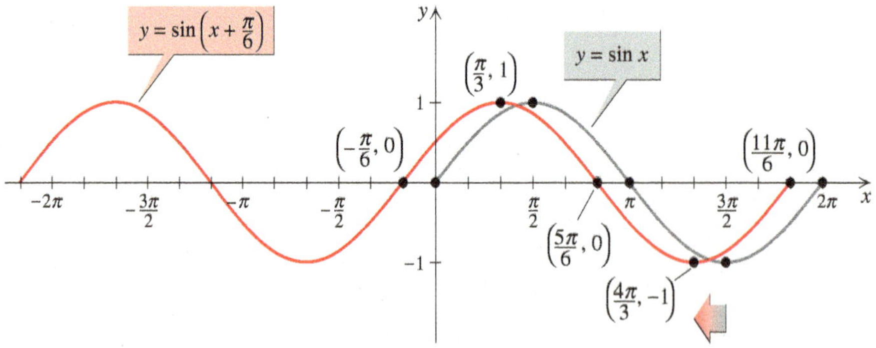


Figure 5.46

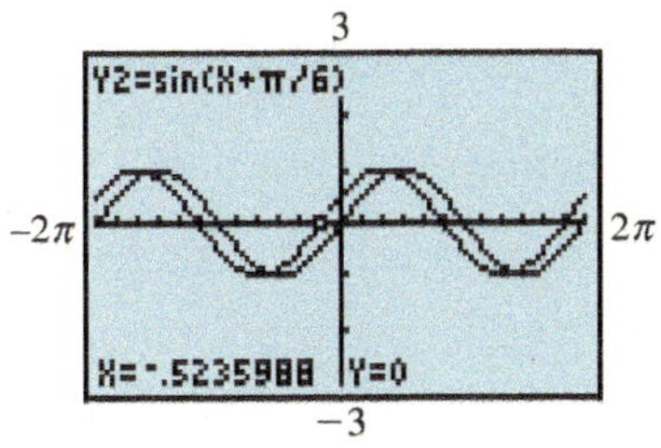


Figure 5.47

points, $(\pi/2, 1)$ and $(3\pi/2, -1)$, move to $(\pi/3, 1)$ and $(4\pi/3, -1)$. Draw one cycle through these five points and continue the pattern for another cycle, as shown in Fig. 5.46. The second cycle could be drawn to the right or left of the first cycle.

Use $\pi/6$ as the $x$-scale on your calculator, as we did in Fig. 5.46. Set the viewing window to show approximately the same two cycles as Fig. 5.46. The graphs of $y_1 = \sin(x)$ and $y_2 = \sin(x + \pi/6)$ in Fig. 5.47 support the conclusion that the shift is $\pi/6$ to the left.

▸**TRY THIS.** Graph one cycle of $y = \cos(x - \pi/6)$ and find the phase shift. ■

The graphs of $y = \sin(x) + D$ and $y = \cos(x) + D$ are vertical translations of $y = \sin x$ and $y = \cos x$, respectively. The translation is upward for $D > 0$ and downward for $D < 0$. The next example combines a phase shift and a vertical translation. Note how we follow the five basic points of the fundamental cycle to see where they go in the transformation.

### EXAMPLE 5 Horizontal and vertical translation

Graph two cycles of $y = \cos(x - \pi/4) + 2$, and determine the phase shift of the graph.

#### Solution

The graph of $y = \cos(x - \pi/4) + 2$ is obtained by moving $y = \cos x$ a distance of $\pi/4$ to the right and two units upward. Since the phase shift is $\pi/4$, label the $x$-axis with multiples of $\pi/4$, as shown in Fig. 5.48. Concentrate on moving the fundamental cycle of $y = \cos x$. The points $(0, 1)$, $(\pi, -1)$, and $(2\pi, 1)$ move to $(\pi/4, 3)$, $(5\pi/4, 1)$, and $(9\pi/4, 3)$. The $x$-intercepts $(\pi/2, 0)$ and $(3\pi/2, 0)$ move to $(3\pi/4, 2)$ and $(7\pi/4, 2)$. Draw one cycle through these five points and continue the pattern for another cycle, as shown in Fig. 5.48.

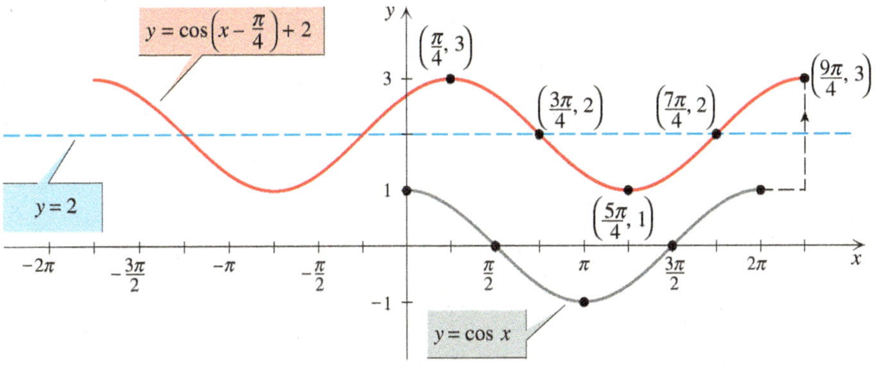


Figure 5.48

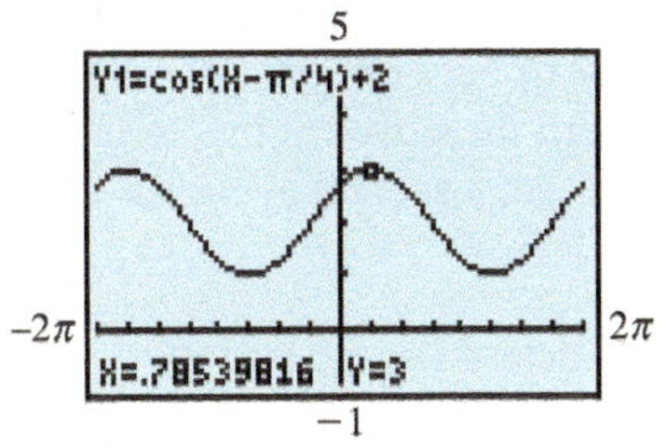


Figure 5.49

Use $\pi/4$ as the $x$-scale on your calculator, as we did in Fig. 5.48. Set the viewing window about the same as in Fig. 5.48. The calculator graph in Fig. 5.49 supports the conclusion that the shift is $\pi/4$ to the right and two units upward.

▸**TRY THIS.** Graph one cycle of $y = \sin(x + \pi/4) + 1$ and find the phase shift. ■

## Changing the Period

We can alter the sine and cosine functions in a way that we did not alter the functions of algebra in Chapter 2. The period of a periodic function can be changed by replacing $x$ by a multiple of $x$.

## EXAMPLE 6 Changing the period

Graph two cycles of $y = \sin(2x)$ and determine the period of the function.

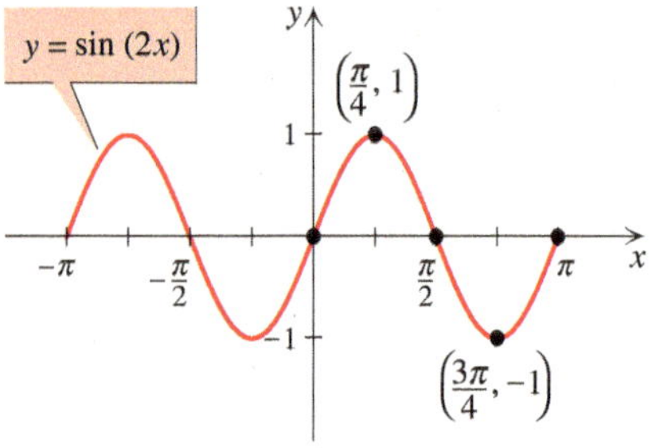


Figure 5.50

### Solution

The graph of $y = \sin x$ completes its fundamental cycle for $0 \le x \le 2\pi$. So $y = \sin(2x)$ completes one cycle for $0 \le 2x \le 2\pi$, or $0 \le x \le \pi$. Thus the period is $\pi$. For $0 \le x \le \pi$, the $x$-intercepts are $(0, 0)$, $(\pi/2, 0)$, and $(\pi, 0)$. Midway between 0 and $\pi/2$, at $x = \pi/4$, the function reaches a maximum value of 1, and the function attains its minimum value of $-1$ at $x = 3\pi/4$. Draw one cycle of the graph through the $x$-intercepts and through $(\pi/4, 1)$ and $(3\pi/4, -1)$. Then graph another cycle of $y = \sin(2x)$ as shown in Fig. 5.50.

▶**TRY THIS.** Graph two cycles of $y = \cos(2x)$ and determine the period. ■

Note that in Example 6 the period of $y = \sin(2x)$ is the period of $y = \sin x$ divided by 2. In general, one complete cycle of $y = \sin(Bx)$ or $y = \cos(Bx)$ for $B > 0$ occurs for $0 \le Bx \le 2\pi$, or $0 \le x \le 2\pi/B$.

**Theorem: Period of $y = \sin(Bx)$ and $y = \cos(Bx)$**

The period $P$ of $y = \sin(Bx)$ and $y = \cos(Bx)$ for $B > 0$ is given by

$$P = \frac{2\pi}{B}.$$

Note that the period is a natural number (that is, not a multiple of $\pi$) when $B$ is a multiple of $\pi$.

## EXAMPLE 7 A period that is not a multiple of $\pi$

Determine the period of $y = \cos\left(\frac{\pi}{2}x\right)$ and graph two cycles of the function.

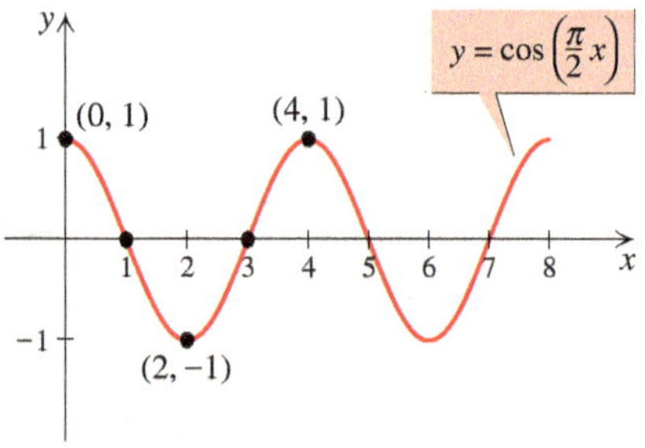


Figure 5.51

### Solution

For this function, $B = \pi/2$. To find the period, use $P = 2\pi/B$:

$$P = \frac{2\pi}{\pi/2} = 4$$

So one cycle of $y = \cos\left(\frac{\pi}{2}x\right)$ is completed for $0 \le x \le 4$. The cycle starts at $(0, 1)$ and ends at $(4, 1)$. A minimum point occurs halfway in between, at $(2, -1)$. The $x$-intercepts are $(1, 0)$ and $(3, 0)$. Draw a curve through these five points to get one cycle of the graph. Continue this pattern from 4 to 8 to get a second cycle, as shown in Fig. 5.51.

▶**TRY THIS.** Graph two cycles of $y = \cos(\pi x)$ and determine the period. ■

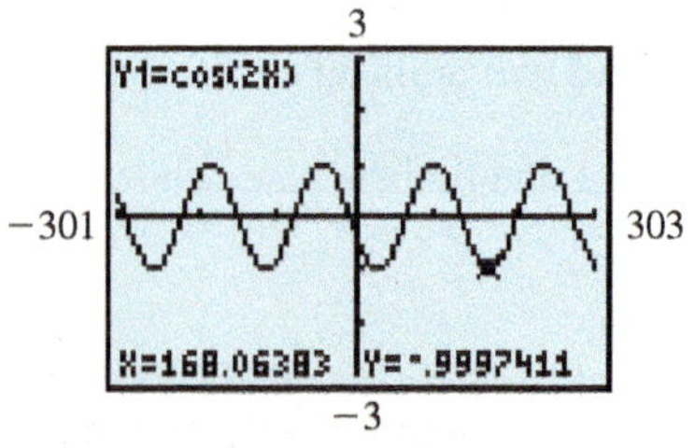


Figure 5.52

A calculator graph for a periodic function can be very misleading. For example, consider $y = \cos(2x)$, shown in Fig. 5.52. From the figure it appears that $y = \cos(2x)$ has a period of about 150 and is a left shift of $y = \cos(x)$. However, we know that the period is $\pi$ and that there is no shift. What we see in Fig. 5.52 is the pattern formed by choosing 95 equally spaced points on the graph of $y = \cos(2x)$. Equally spaced points on the graph of a periodic function will usually have some kind of pattern, but the pattern may not be a good graph of the function. Because $x$ ranges from $-301$ to 303 in Fig. 5.52, the spaces between these points are approximately 6 units each, which is enough for about two cycles of $y = \cos(2x)$. So the viewing window is much too large to show the relatively small features of $y = \cos(2x)$. □

## The General Sine Wave

We can use any combination of translating, reflecting, phase shifting, stretching, shrinking, or period changing in a single trigonometric function.

**The General Sine Wave**

The graph of

$$y = A\sin[B(x - C)] + D \quad \text{or} \quad y = A\cos[B(x - C)] + D$$

is a sine wave with an amplitude $|A|$, period $2\pi/B$ $(B > 0)$, phase shift $C$, and vertical translation $D$.

We assume that $B > 0$, because any general sine or cosine function can be rewritten with $B > 0$ using identities from Chapter 6. Notice that $A$ and $B$ affect the shape of the curve, while $C$ and $D$ determine its location.

**PROCEDURE**

### Graphing a Sine Wave

To graph $y = A\sin[B(x - C)] + D$ or $y = A\cos[B(x - C)] + D$:

1. Sketch one cycle of $y = \sin Bx$ or $y = \cos Bx$ on $[0, 2\pi/B]$.
2. Change the amplitude of the cycle according to the value of $A$.
3. If $A < 0$, reflect the curve in the $x$-axis.
4. Translate the cycle $|C|$ units to the right if $C > 0$ or to the left if $C < 0$.
5. Translate the cycle $|D|$ units upward if $D > 0$ or downward if $D < 0$.

### EXAMPLE 8 A transformation of $y = \sin(x)$

Determine amplitude, period, and phase shift, and sketch two cycles of $y = 2\sin(3x + \pi) + 1$.

**Solution**

First we rewrite the function in the form $y = A\sin[B(x - C)] + D$ by factoring 3 out of $3x + \pi$:

$$y = 2\sin\left[3\left(x + \frac{\pi}{3}\right)\right] + 1$$

From this equation we get $A = 2$, $B = 3$, and $C = -\pi/3$. So the amplitude is 2, the period is $2\pi/3$, and the phase shift is $-\pi/3$. The period change causes the fundamental cycle of $y = \sin x$ on $[0, 2\pi]$ to shrink to the interval $[0, 2\pi/3]$. Now draw one cycle of $y = \sin 3x$ on $[0, 2\pi/3]$, as shown in Fig. 5.53. Stretch the

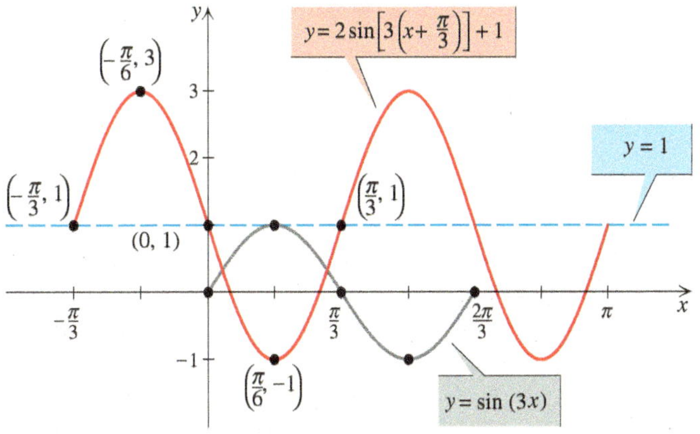


Figure 5.53

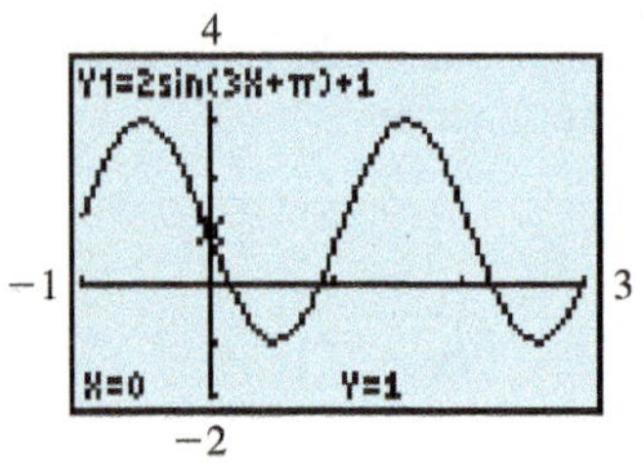


Figure 5.54

cycle vertically so that it has an amplitude of 2. The numbers $\pi/3$ and 1 shift the cycle a distance of $\pi/3$ to the left and up one unit. So one cycle of the function occurs on $[-\pi/3, \pi/3]$. Check by evaluating the function at the endpoints and midpoint of the interval $[-\pi/3, \pi/3]$ to get $(-\pi/3, 1)$, $(0, 1)$, and $(\pi/3, 1)$. Since the graph is shifted one unit upward, these points are the points where the curve intersects the line $y = 1$. Evaluate the function midway between these points to get $(-\pi/6, 3)$ and $(\pi/6, -1)$, the highest and lowest points of this cycle. One cycle is drawn through these five points and continued for another cycle, as shown in Fig. 5.53.

The calculator graph in Fig. 5.54 supports our conclusions about amplitude, period, and phase shift. Note that it is easier to obtain the amplitude, period, and phase shift from the equation than from the calculator graph.

▶**TRY THIS.** Determine the amplitude, period, and phase shift, and graph one cycle of $y = 3\sin(2x - \pi) - 1$. ■

## EXAMPLE 9 A transformation of $y = \cos(x)$

Determine amplitude, period, and phase shift, and sketch one cycle of $y = -3\cos(2x - \pi) - 1$.

### Solution

Rewrite the function in the general form as

$$y = -3\cos[2(x - \pi/2)] - 1.$$

The amplitude is 3. Since the period is $2\pi/B$, the period is $2\pi/2$ or $\pi$. The fundamental cycle is shrunk to the interval $[0, \pi]$. Sketch one cycle of $y = \cos 2x$ on $[0, \pi]$, as shown in Fig. 5.55. This cycle is reflected in the $x$-axis and stretched by a factor of 3. A shift of $\pi/2$ to the right means that one cycle of the original function occurs for $\pi/2 \le x \le 3\pi/2$. Evaluate the function at the endpoints and midpoint of the interval $[\pi/2, 3\pi/2]$ to get $(\pi/2, -4)$, $(\pi, 2)$, and $(3\pi/2, -4)$ for the endpoints and midpoint of this cycle. Since the graph is translated downward one unit, midway between these maximum and minimum points we get points where the graph intersects the line $y = -1$. These points are $(3\pi/4, -1)$ and $(5\pi/4, -1)$. Draw one cycle of the graph through these five points, as shown in Fig. 5.55.

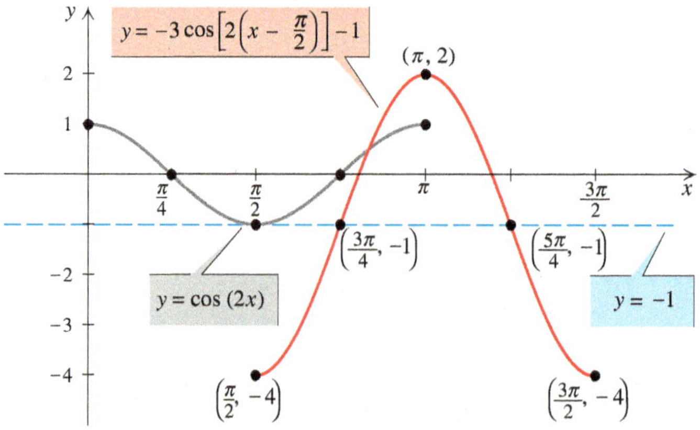


Figure 5.55

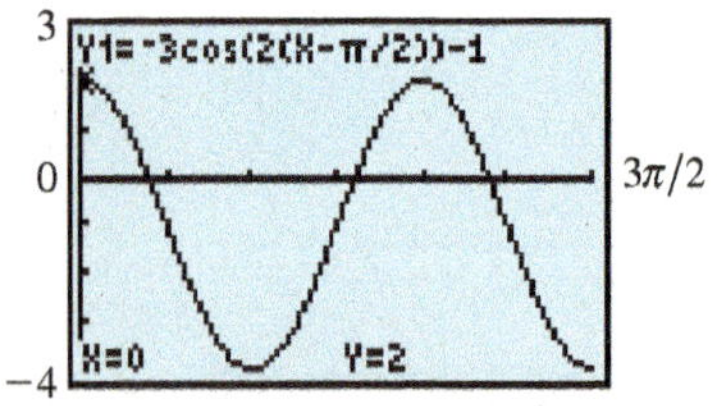


Figure 5.56

The calculator graph in Fig. 5.56 supports our conclusions.

▶**TRY THIS.** Determine the amplitude, period, and phase shift, and graph one cycle of $y = -2\cos(2x + \pi) + 1$. ■

## Frequency

Sine waves are used to model physical phenomena such as radio, sound, or light waves. A high-frequency radio wave is a wave that has a large number of cycles per second. If we think of the $x$-axis as a time axis, then the period of a sine wave is the amount of time required for the wave to complete one cycle, and the reciprocal of the period is the number of cycles per unit of time. For example, the sound wave for middle C on a piano completes 262 cycles per second. The period of the wave is $1/262$ second, which means that one cycle is completed in $1/262$ second.

**Definition: Frequency**

> The **frequency** $F$ of a sine wave with period $P$ is defined by $F = 1/P$.

### EXAMPLE 10 Frequency of a sine wave

Find the frequency of the sine wave given by $y = \sin(524\pi x)$.

**Solution**

First find the period:

$$P = \frac{2\pi}{524\pi} = \frac{1}{262} \approx 0.004$$

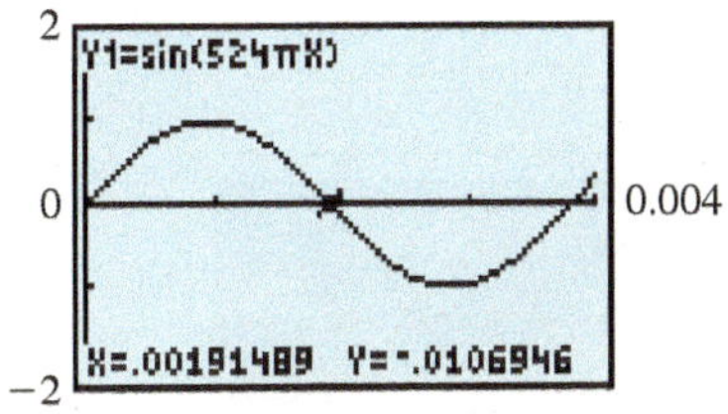


Figure 5.57

Since $F = 1/P$, the frequency is 262. A sine wave with this frequency completes 262 cycles for $x$ in $[0, 1]$ or approximately one cycle in $[0, 0.004]$. We cannot draw a graph showing 262 cycles in an interval of length one, but we can see the cycle that occurs in the interval $[0, 0.004]$ by looking at Fig. 5.57.

▸**TRY THIS.** Find the frequency of the sine wave given by $y = \cos(100\pi x)$. ■

Note that if $B$ is a large positive number in $y = \sin(Bx)$ or $y = \cos(Bx)$, then the period is short and the frequency is high.

## Sinusoidal Curve Fitting

We now consider the problem of finding an equation for a sinusoid that passes through some given points. If we know the five key points on one cycle of a sinusoid, then we can write an equation for the curve in the form $y = A\sin[B(x - C)] + D$.

### EXAMPLE 11 Modeling room temperature

Ten minutes after a furnace is turned on, the temperature in a room reaches 74° and the furnace turns off. It takes two minutes for the room to cool to 70° and two minutes for the furnace to bring it back to 74° as shown in the following table.

| Time (min) | 10 | 11 | 12 | 13 | 14 | 15 | 16 | 17 | 18 |
|---|---|---|---|---|---|---|---|---|---|
| Temperature (°F) | 74 | 72 | 70 | 72 | 74 | 72 | 70 | 72 | 74 |

Assuming that the temperature (after time 10) is a sine function of the time, find the function and graph it.

**Solution**

Because the temperature ranges from 70° to 74°, the amplitude of the sine curve is 2. Because the temperature goes from its maximum of 74° back to 74° in 4 minutes, the period is 4. Since the period is $2\pi/B$, we get $B = \pi/2$. Now concentrate on one

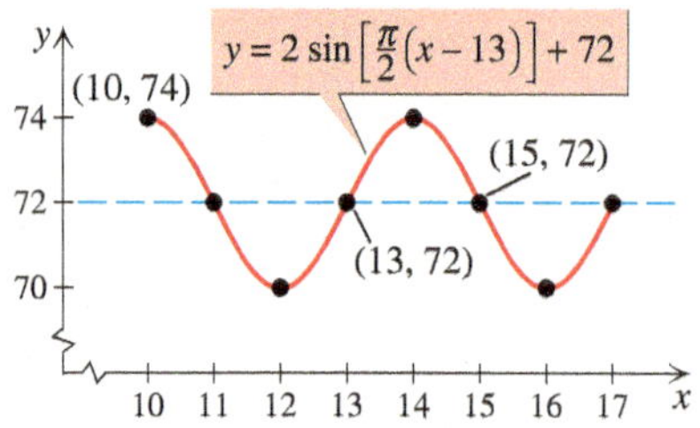


**Figure 5.58**

cycle of the function. Starting at $(13, 72)$ the temperature increases to its maximum of 74°, decreases to its minimum of 70°, and then ends in the middle at $(17, 72)$. We choose this cycle because it duplicates the behavior of $y = \sin(x)$ on its fundamental cycle $[0, 2\pi]$. So shifting $y = \sin(x)$ to the right 13 and up 72 gives us $y = 2\sin\left[\frac{\pi}{2}(x - 13)\right] + 72$. Its graph is shown in Fig. 5.58. Changing this equation with any right or left shift by a multiple of 4 gives an equivalent equation. For example, $y = 2\sin\left[\frac{\pi}{2}(x - 1)\right] + 72$ is an equivalent equation.

▸**TRY THIS.** The points $(1, 2)$, $(2, 5)$, $(3, 2)$, $(4, -1)$, and $(5, 2)$ are on one cycle of a sine wave. Find an equation for the curve. ■

If we have real data, the points will usually not fit exactly on a sine curve as they did in the last example. In this case we can use the sinusoidal regression feature of a graphing calculator to find a sine curve that approximates the data.

## EXAMPLE 12 Modeling the time of sunrise

The times of sunrise in Miami, Florida, on the first of every month for one year are shown in the following table (U.S. Naval Observatory, http://aa.usno.navy.mil). The time is the number of minutes after 5 A.M.

| Month | Time | Month | Time |
|---|---|---|---|
| 1 | 127 | 7 | 33 |
| 2 | 125 | 8 | 47 |
| 3 | 104 | 9 | 61 |
| 4 | 72 | 10 | 73 |
| 5 | 44 | 11 | 90 |
| 6 | 29 | 12 | 111 |

## FUNCTION gallery... The Sine and Cosine Functions

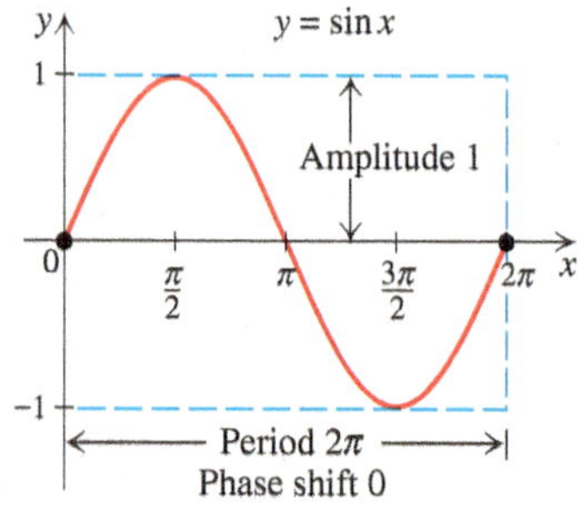


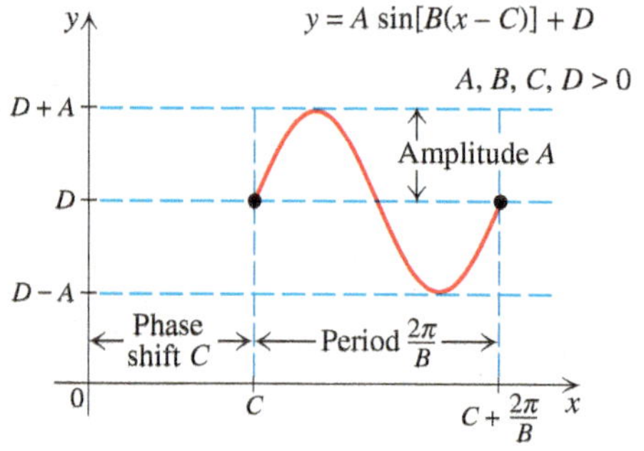


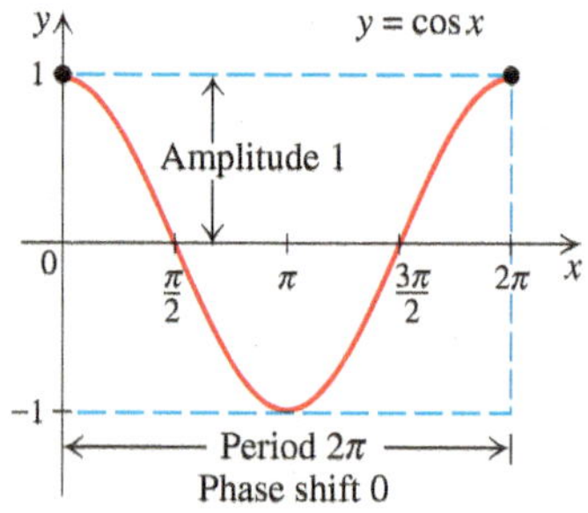


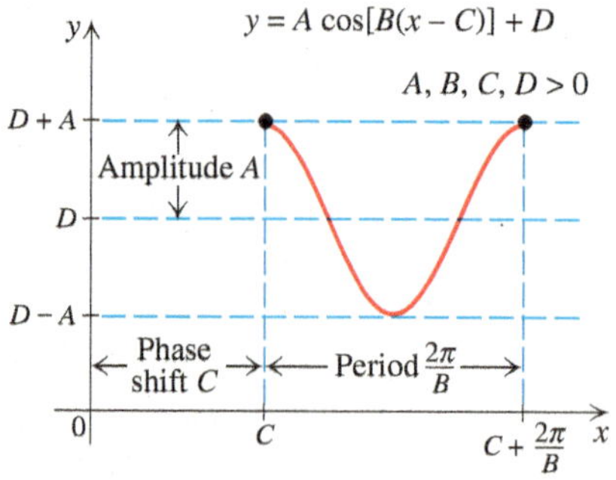

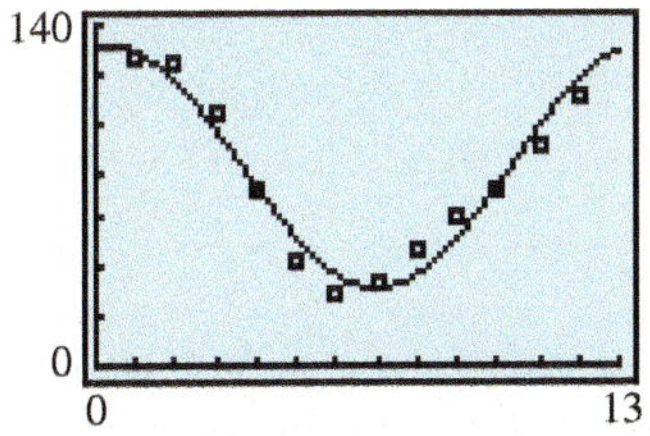


**Figure 5.59**

Use the sinusoidal regression feature of a graphing calculator to find an equation that fits the data. Graph the data and the curve on your graphing calculator. Find the period from the equation.

**Solution**

Enter the data and use the sinusoidal regression feature (SinReg) to get $y = 49.92\sin(0.47x + 1.42) + 81.54$, where $x$ is the month and $y$ is the number of minutes after 5 A.M. Figure 5.59 shows the data and the sine curve. The period is $2\pi/0.47$, or approximately 13.4 months. Since the period should be 12 months, the sine curve does not fit the data very well. See Exercises 99 and 100 for some data that really look like a sine curve.

▸**TRY THIS.** Use sinusoidal regression to find an equation that fits the points $(1, 2.2)$, $(2, 4.9)$, $(3, 1.9)$, $(4, -1.1)$, and $(5, 2.1)$. ■

## FOR thought... True or False? Explain.

1. The period of $y = \cos(2\pi x)$ is $\pi$.
2. The range of $y = 4\sin(x) + 3$ is $[-4, 4]$.
3. The graph of $y = \sin(2x + \pi/6)$ has a period of $\pi$ and a phase shift $\pi/6$.
4. The points $(5\pi/6, 0)$ and $(11\pi/6, 0)$ are on the graph of $y = \cos(x - \pi/3)$.
5. The frequency of the sine wave $y = \sin x$ is $1/(2\pi)$.
6. The period for $y = \sin(0.1\pi x)$ is 20.
7. The graphs of $y = \sin x$ and $y = \cos(x + \pi/2)$ are identical.
8. The period of $y = \cos(4x)$ is $\pi/2$.
9. The maximum value of the function $y = -2\cos(3x) + 4$ is 6.
10. The range of the function $y = 3\sin(5x - \pi) + 2$ is $[-1, 5]$.

## EXERCISES 5.3

*Fill in the blank.*

1. The graph of $y = \sin x$ is a ________.
2. The graph of $y = \sin x$ over $[0, 2\pi]$ is the ________ of $y = \sin x$.
3. The ________ of a sine wave is the absolute value of half the difference between the maximum and minimum $y$-coordinates on the wave.
4. The ________ of the graph of $y = \sin(x - C)$ is $C$.
5. The ________ of $y = \cos(Bx)$ is $2\pi/B$.
6. The ________ of a sine wave with period $P$ is $1/P$.

*Match each graph with one of the functions $y = 3\sin(x)$, $y = 3\cos(x)$, $y = -2\sin(x)$, and $y = -2\cos(x)$. Determine the amplitude of each function.*

**7.**

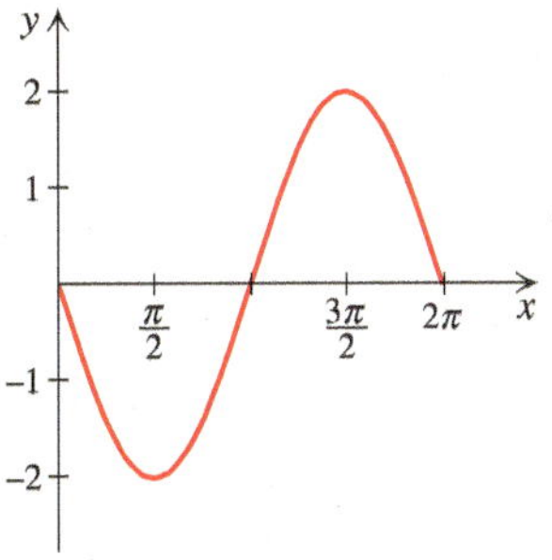


**8.**

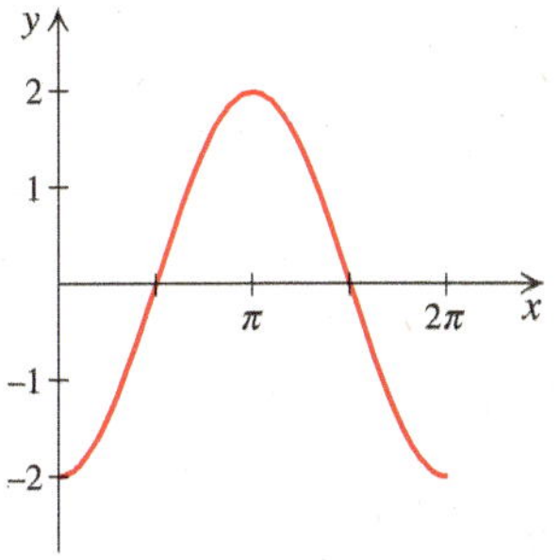

**9.**

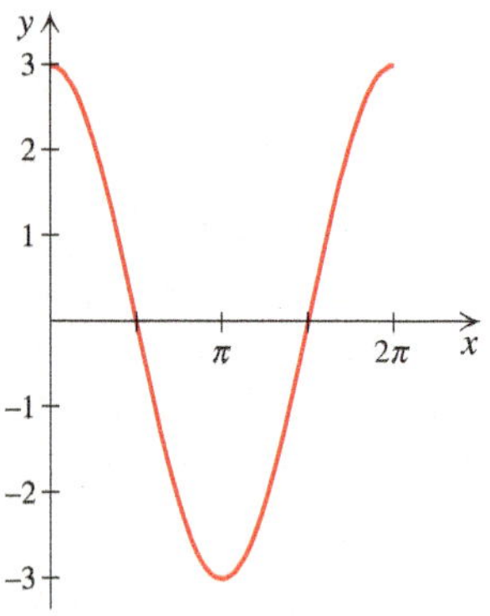


**10.**

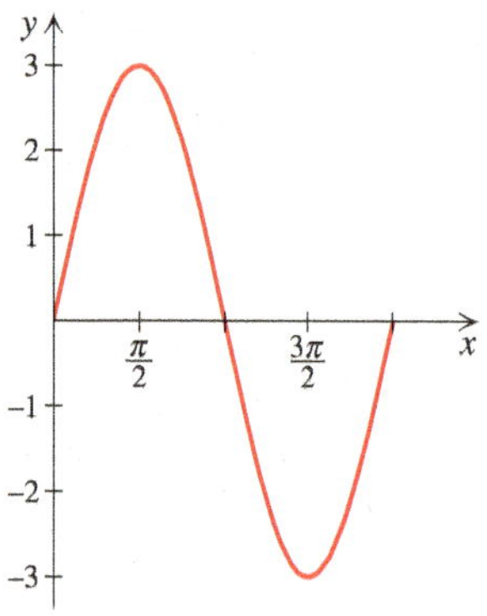


*Determine the amplitude, period, and phase shift for each function.*

**11.** $y = -2\cos x$

**12.** $y = -4\cos x$

**13.** $f(x) = \cos(x - \pi/2)$

**14.** $f(x) = \sin(x + \pi/2)$

**15.** $y = -2\sin(x + \pi/3)$

**16.** $y = -3\sin(x - \pi/6)$

*Determine the amplitude and phase shift for each function, and sketch at least one cycle of the graph. Label five points as done in the examples.*

**17.** $y = -\sin x$

**18.** $y = -\cos x$

**19.** $y = -3\sin x$

**20.** $y = 4\sin x$

**21.** $y = \dfrac{1}{2}\cos x$

**22.** $y = \dfrac{1}{3}\cos x$

**23.** $y = \sin(x + \pi)$

**24.** $y = \cos(x - \pi)$

**25.** $y = \cos(x - \pi/3)$

**26.** $y = \cos(x + \pi/4)$

**27.** $f(x) = \cos(x) + 2$

**28.** $f(x) = \cos(x) - 3$

**29.** $y = -\sin(x) - 1$

**30.** $y = -\sin(x) + 2$

**31.** $y = \sin(x + \pi/4) + 2$

**32.** $y = \sin(x - \pi/2) - 2$

**33.** $y = 2\cos(x + \pi/6) + 1$

**34.** $y = 3\cos(x + 2\pi/3) - 2$

**35.** $f(x) = -2\sin(x - \pi/3) + 1$

**36.** $f(x) = -3\cos(x + \pi/3) - 1$

*Determine the amplitude, period, and phase shift for each function.*

**37.** $y = 3\sin(4x)$

**38.** $y = -2\cos(3x)$

**39.** $y = -\cos\left(\dfrac{x}{2}\right) + 3$

**40.** $y = \sin\left(\dfrac{x}{3}\right) - 5$

**41.** $y = 2\sin(x - \pi) + 3$

**42.** $y = -5\cos(x + 4) + \pi$

**43.** $y = -2\cos\left(2x + \dfrac{\pi}{2}\right) - 1$

**44.** $y = 4\cos\left(3x - \dfrac{\pi}{4}\right)$

**45.** $y = -2\cos\left(\dfrac{\pi}{2}x + \pi\right)$

**46.** $y = 8\sin\left(\dfrac{\pi}{3}x - \dfrac{\pi}{2}\right)$

*Find a function of the form $y = A\sin[B(x - C)] + D$ with the given period, phase shift, and range. Answers may vary.*

**47.** $\pi, -\pi/2, [3, 7]$

**48.** $\pi/2, \pi, [-2, 4]$

**49.** $2, 2, [-1, 9]$

**50.** $4, 7, [5, 25]$

**51.** $\dfrac{1}{2}, -\pi, [-9, 3]$

**52.** $\dfrac{1}{3}, 2\pi, [-6, 2]$

*Find the equation for each curve in its final position.*

**53.** The graph of $y = \sin(x)$ is shifted a distance of $\pi/4$ to the right, reflected in the $x$-axis, then translated one unit upward.

**54.** The graph of $y = \cos(x)$ is shifted a distance of $\pi/6$ to the left, reflected in the $x$-axis, then translated two units downward.

**55.** The graph of $y = \cos(x)$ is stretched by a factor of 3, shifted a distance of $\pi$ to the right, translated two units downward, then reflected in the $x$-axis.

**56.** The graph of $y = \sin(x)$ is shifted a distance of $\pi/2$ to the left, translated one unit upward, stretched by a factor of 4, then reflected in the $x$-axis.

*Let $f(x) = \sin(x)$, $g(x) = x - \pi/4$, and $h(x) = 3x$. Find a formula for $F$ in each case.*

**57.** $F = f \circ g \circ h$

**58.** $F = h \circ f \circ g$

**59.** $F = f \circ h \circ g$

**60.** $F = g \circ f \circ h$

*Sketch at least one cycle of the graph of each function. Determine the period, phase shift, and range of the function. Label five points on the graph as done in the examples. See the procedure for graphing a sine wave on page 369.*

**61.** $y = \sin(3x)$

**62.** $y = \cos(x/3)$

**63.** $y = -\sin(2x)$

**64.** $y = -\cos(3x)$

**65.** $y = \cos(4x) + 2$

**66.** $y = \sin(3x) - 1$

**67.** $y = 2 - \sin(x/4)$

**68.** $y = 3 - \cos(x/5)$

**69.** $y = \sin\left(\dfrac{\pi}{3}x\right)$

**70.** $y = \sin\left(\dfrac{\pi}{4}x\right)$

**71.** $f(x) = \sin\left[2\left(x - \dfrac{\pi}{2}\right)\right]$

**72.** $f(x) = \sin\left[3\left(x + \dfrac{\pi}{3}\right)\right]$

**73.** $f(x) = \sin\left(\dfrac{\pi}{2}x + \dfrac{3\pi}{2}\right)$

**74.** $f(x) = \cos\left(\frac{\pi}{3}x - \frac{\pi}{3}\right)$

**75.** $y = 2\cos\left[2\left(x + \frac{\pi}{6}\right)\right] + 1$

**76.** $y = 3\cos\left[4\left(x - \frac{\pi}{2}\right)\right] - 1$

**77.** $y = -\frac{1}{2}\sin\left[3\left(x - \frac{\pi}{6}\right)\right] - 1$

**78.** $y = -\frac{1}{2}\sin\left[4\left(x + \frac{\pi}{4}\right)\right] + 1$

*Write an equation of the form* $y = A\sin[B(x - C)] + D$ *whose graph is the given sine wave.*

**79.**

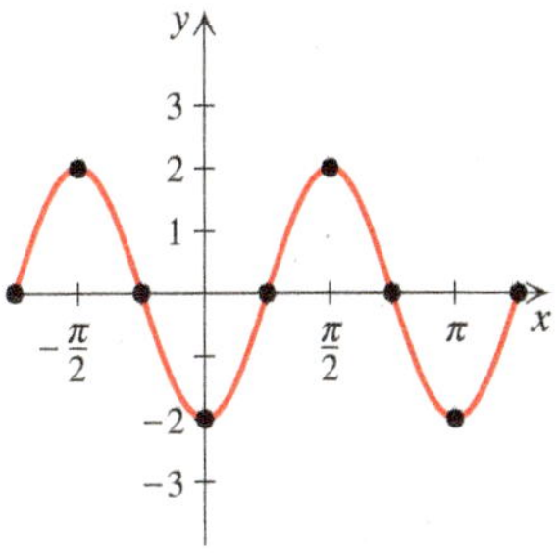


**80.**

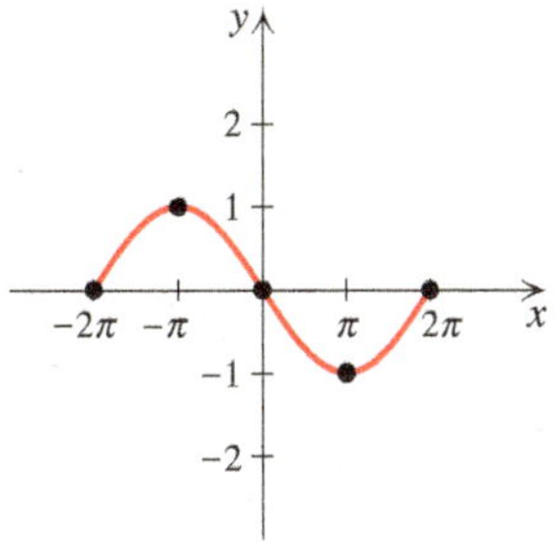


**81.**

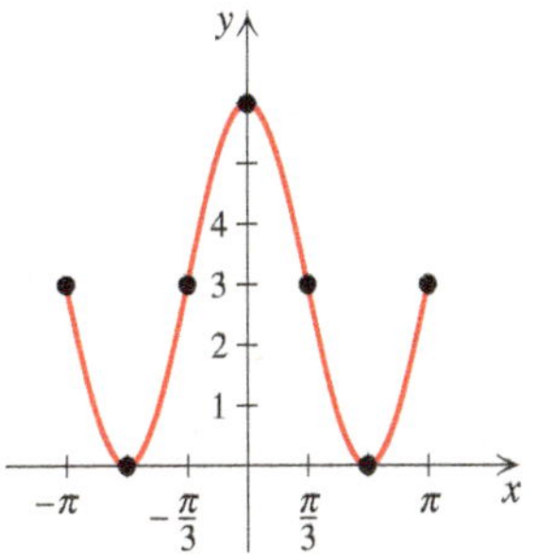


**82.**

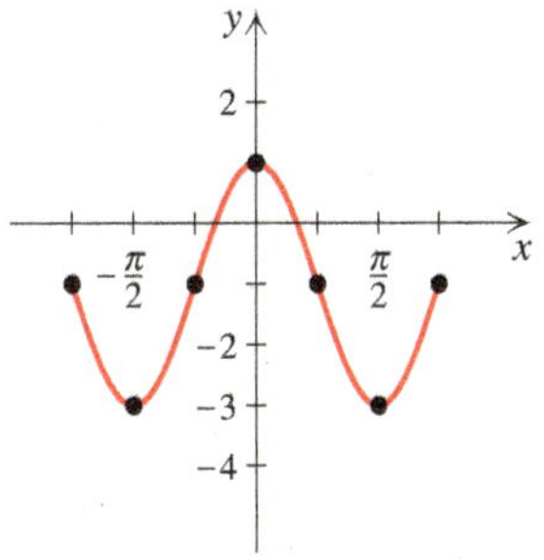


*Solve each problem.*

**83.** What is the frequency of the sine wave determined by $y = \sin(200\pi x)$, where $x$ is time in seconds?

**84.** What is the frequency of the sine wave determined by $y = \cos(0.001\pi x)$, where $x$ is time in seconds?

**85.** If the period of a sine wave is 0.025 hr, then what is the frequency?

**86.** If the frequency of a sine wave is 40,000 cycles per second, then what is the period?

**87.** *Motion of a Spring* A weight hanging on a vertical spring is set in motion with a downward velocity of 6 cm/sec from its equilibrium position. Assume that the constant $\omega$ for this particular spring and weight combination is 2. Write the formula that gives the location of the weight in centimeters as a function of the time $t$ in seconds. Find the amplitude and period of the function and sketch its graph for $t$ in the interval $[0, 2\pi]$. (See Section 5.2 for the general formula that describes the motion of a spring.)

**88.** *Motion of a Spring* A weight hanging on a vertical spring is set in motion with an upward velocity of 4 cm/sec from its equilibrium position. Assume that the constant $\omega$ for this particular spring and weight combination is $\pi$. Write the formula that gives the location of the weight in centimeters as a function of the time $t$ in seconds. Find the period of the function and sketch its graph for $t$ in the interval $[0, 4]$.

**89.** *Sunspots* Astronomers have been recording sunspot activity for over 130 years. The number of sunspots per year varies like a periodic function over time, as shown in the graph. What is the approximate period of this function?

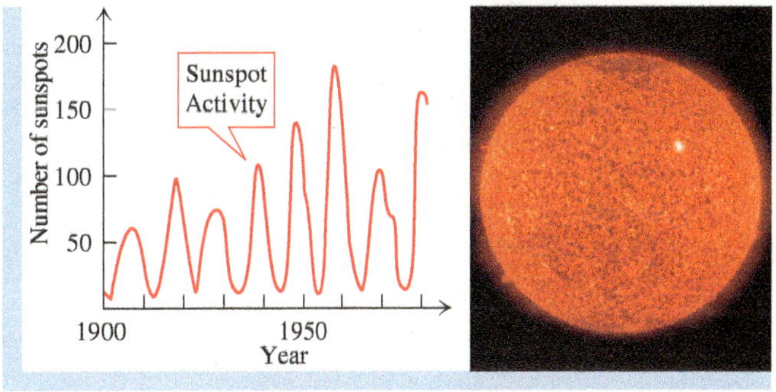


**Figure for Exercise 89**

**90.** *First Pulsar* In 1967, Jocelyn Bell, a graduate student at Cambridge University, England, found the peculiar pattern shown in the graph on a paper chart from a radio telescope. She had made the first discovery of a pulsar, a very small neutron star that emits beams of radiation as it rotates as fast as 1000 times per second. From the graph shown here, estimate the period of the first discovered pulsar, now known as CP 1919.

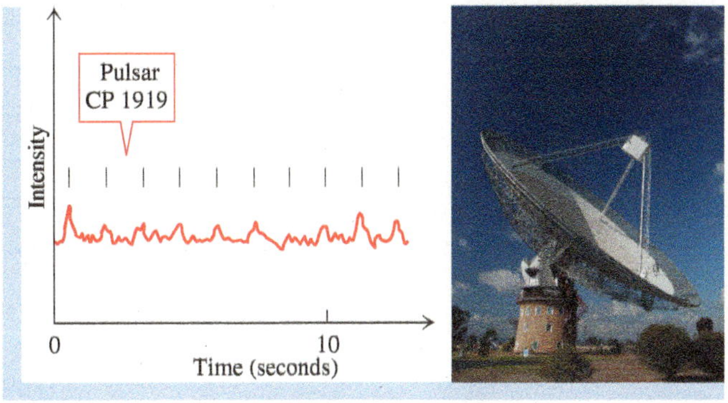


**Figure for Exercise 90**

**91.** *Lung Capacity* The volume of air $v$ in cubic centimeters in the lungs of a certain distance runner is modeled by the equation $v = 400\sin(60\pi t) + 900$, where $t$ is time in minutes.

**a.** What are the maximum and minimum volumes of air in the runner's lungs at any time?

**b.** How many breaths does the runner take per minute?

**92.** *Blood Velocity* The velocity $v$ of blood at a valve in the heart of a certain rodent is modeled by the equation $v = -4\cos(6\pi t) + 4$, where $v$ is in centimeters per second and $t$ is time in seconds.

**a.** What are the maximum and minimum velocities of the blood at this valve?

**b.** What is the rodent's heart rate in beats per minute?

**93.** *Periodic Revenue* For the past three years, the manager of The Toggery Shop has observed that revenue reaches a high of about \$40,000 in December and a low of about \$10,000 in June, and that a graph of the revenue looks like a sinusoid. If the months are numbered 1 through 36 with 1 corresponding to January, then what are the period, amplitude, and phase shift for this sinusoid? What is the vertical translation? Write a formula for the curve and find the approximate revenue for April.

**94.** *Periodic Cost* For the past three years, the manager of The Toggery Shop has observed that the utility bill reaches a high of about \$500 in January and a low of about \$200 in July, and the graph of the utility bill looks like a sinusoid. If the months are numbered 1 through 36 with 1 corresponding to January, then what are the period, amplitude, and phase shift for this sinusoid? What is the vertical translation? Write a formula for the curve and find the approximate utility bill for November.

**95.** *Discovering a Planet* On April 26, 1997, astronomers announced the discovery of a new planet orbiting the star Rho Coronae Borealis (*Sky and Telescope,* July 1997). The astronomers deduced the existence of the planet by measuring the change in line-of-sight velocity of the star over a period of ten months. The measurements appear to fall along a sine wave, as shown in the accompanying figure.

**a.** What are the period, amplitude, and equation for the sine wave?

**b.** How many earth days does it take for the planet to orbit Rho Coronae Borealis?

**c.** Use the equation to estimate the change in line-of-sight velocity on day 36.

**d.** On day 36 do you think that the planet is between the earth and Rho or that Rho is between the earth and the planet?

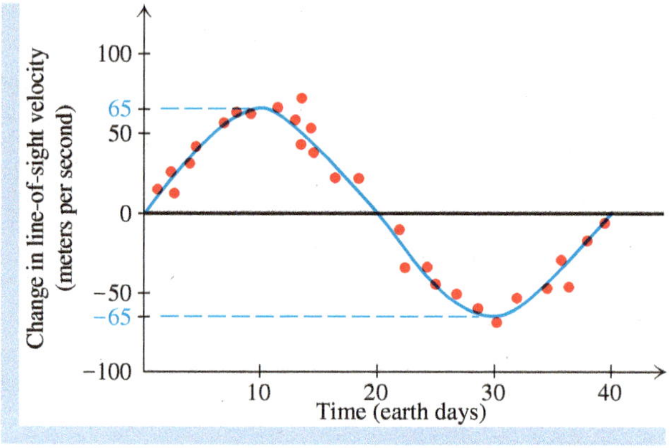


**Figure for Exercise 95**

**96.** *Jupiter's Satellites* The accompanying graph shows the positions in June of Ganymede, Callisto, Io, and Europa, the four bright satellites of Jupiter (*Sky and Telescope*). Jupiter itself is the center horizontal bar. The paths of the satellites are nearly sine waves because the orbits of the satellites are nearly circular. The distances from Jupiter to Io, Europa, Ganymede, and Callisto are 262,000, 417,000, 666,000, and 1,170,000 miles, respectively.

**a.** From the graph estimate the period of revolution to the nearest hour for each satellite. For which satellite can you obtain the period with the most accuracy?

**b.** What is the amplitude of each sine wave?

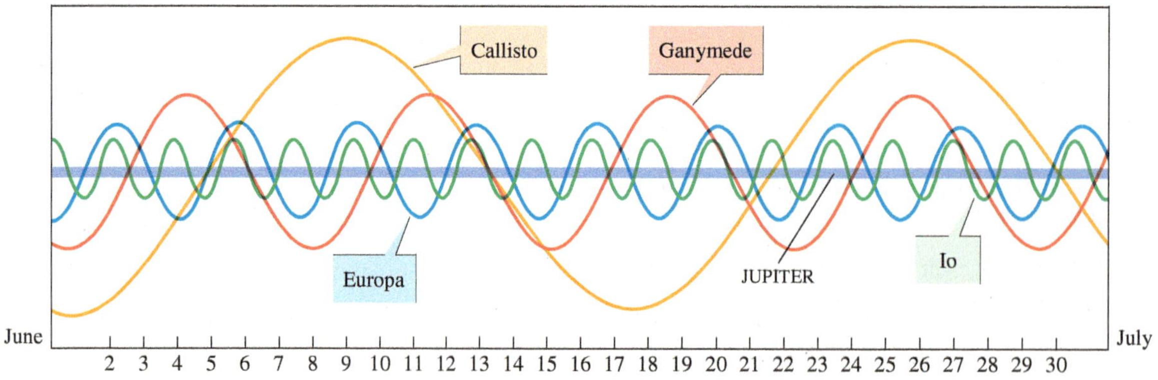


**Figure for Exercise 96**

**97.** *Ocean Waves* Scientists use the same types of terms to describe ocean waves that we use to describe sine waves. The *wave period* is the time between crests and the *wavelength* is the distance between crests. The wave *height* is the vertical distance from the trough to the crest. The accompanying figure shows a *swell* in a coordinate system. Write an equation for the swell, assuming that its shape is that of a sinusoid.

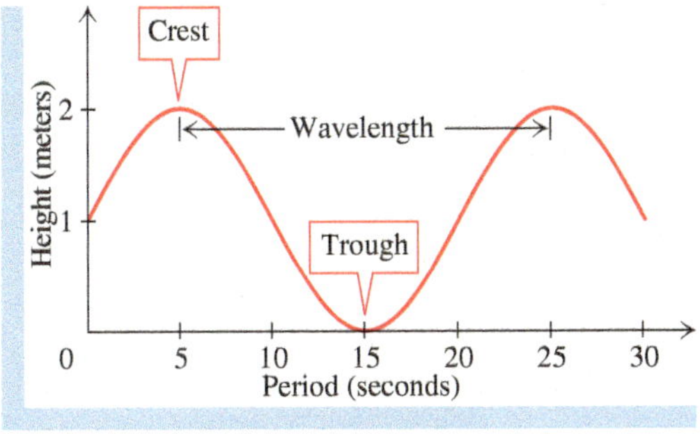


**Figure for Exercise 97**

**98.** *Large Ocean Waves* A *tsunami* is a series of large waves caused by an earthquake. The wavelength for a tsunami can be as long as several hundred kilometers. The accompanying figure shows a tsunami in a coordinate system. Write an equation for the tsunami, assuming that its shape is that of a sinusoid.

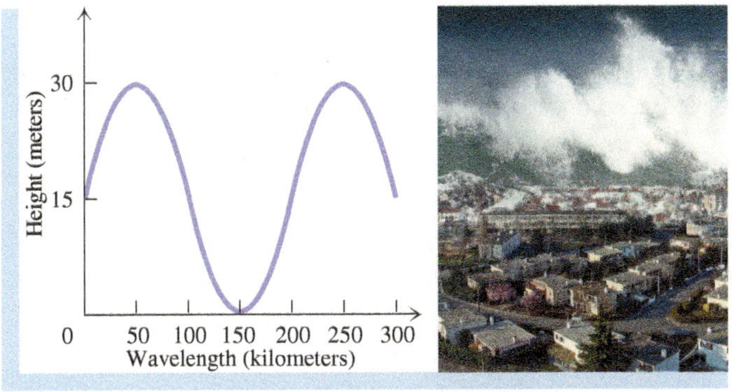


**Figure for Exercise 98**

**99.** *Moon Illumination* The accompanying table shows the percentage of the moon that will be illuminated at midnight for the 31 days of January 2020 (U.S. Naval Observatory, http://aa.usno.navy.mil).

| Day | % | Day | % | Day | % |
|---|---|---|---|---|---|
| 1 | 32 | 12 | 97 | 23 | 3 |
| 2 | 41 | 13 | 92 | 24 | 1 |
| 3 | 50 | 14 | 85 | 25 | 0 |
| 4 | 60 | 15 | 76 | 26 | 2 |
| 5 | 69 | 16 | 65 | 27 | 5 |
| 6 | 78 | 17 | 54 | 28 | 10 |
| 7 | 85 | 18 | 43 | 29 | 17 |
| 8 | 92 | 19 | 32 | 30 | 24 |
| 9 | 97 | 20 | 22 | 31 | 33 |
| 10 | 100 | 21 | 14 | | |
| 11 | 100 | 22 | 7 | | |

Use the sinusoidal regression feature of a graphing calculator to find the equation for a sine curve that fits the data. Graph the data and the curve on your graphing calculator. Find the period from the equation. Use the equation to predict the percentage of the moon that will be illuminated on February 8, 2020.

**100.** *Amount of Daylight* The accompanying table gives the number of minutes between sunrise and sunset for the first day of each month in the year 2020 in Miami, Florida.

| Month | Time | Month | Time |
|---|---|---|---|
| 1 | 634 | 7 | 823 |
| 2 | 659 | 8 | 799 |
| 3 | 699 | 9 | 759 |
| 4 | 746 | 10 | 714 |
| 5 | 788 | 11 | 669 |
| 6 | 819 | 12 | 638 |

Use the sinusoidal regression feature of a graphing calculator to find the equation for a sine curve that fits the data. Graph the data and the curve on your graphing calculator. Find the period from the equation. Use the equation to predict the number of minutes between sunrise and sunset on February 1, 2021.

## FOR WRITING/DISCUSSION

**101.** *Periodic Temperature* Air temperature $T$ generally varies in a periodic manner, with highs during the day and lows during the night. Assume that no drastic changes in the weather are expected and let $t$ be time in hours with $t = 0$ at midnight tonight. Graph $T$ for $t$ in the interval $[0, 48]$ and write a function of the form $T = A\sin[B(t - C)] + D$ for your graph. Explain your choices for $A$, $B$, $C$, and $D$.

**102.** *Even or Odd* Determine whether $y = \sin x$ and $y = \cos x$ are even or odd functions and explain your answers.

## ► RETHINKING

**103.** Convert the degree measure 225° to radians.

**104.** Convert the radian measure $7\pi/6$ to degrees.

**105.** A 30-inch lawnmower blade is rotating at 2000 revolutions per minute. Find the linear velocity of the tip of the blade in miles per hour (to the nearest tenth).

**106.** A wheel is rotating at 200 rev/sec. Find the angular velocity in radians per minute (to the nearest tenth).

**107.** Find the exact value of each function.

**a.** $\sin(45°)$

**b.** $\cos(-\pi/4)$

**108.** Find the exact value of each function.

**a.** $\sin(60°)$

**b.** $\cos(-5\pi/6)$

## THINKING OUTSIDE THE BOX XXXVIII

*The Survivor* There are 13 contestants on a reality television show. They are instructed to each take a seat at a circular table containing 13 chairs that are numbered consecutively with the numbers 1 through 13. The producer then starts at number 1 and tells that contestant that he is a survivor. That contestant then leaves the table. The producer skips a contestant and tells the contestant in chair number 3 that he is a survivor. That contestant then leaves the table. The producer continues around the table skipping a contestant and telling the next contestant that he is a survivor. Each survivor leaves the table. The last person left at the table is *not* a survivor and must leave the show.

**a.** For $n = 13$ find the unlucky number $k$, for which the person sitting in chair $k$ must leave the show.

**b.** Find $k$ for $n = 8$, 16, and 41.

**c.** Find a formula for $k$.

## POP QUIZ 5.3

1. Determine the amplitude, period, and phase shift for $y = -5\sin(2x + 2\pi/3)$.
2. List the coordinates for the five key points for one cycle of $y = 3\sin(2x)$.
3. If $y = \cos(x)$ is shifted $\pi/2$ to the right, reflected in the $x$-axis, and shifted 3 units upward, then what is the equation of the curve in its final position?
4. What is the range of $f(x) = -4\sin(x - 3) + 2$?
5. What is the frequency of the sine wave determined by $y = \sin(500\pi x)$, where $x$ is time in minutes?

## LINKING concepts... For Individual or Group Explorations

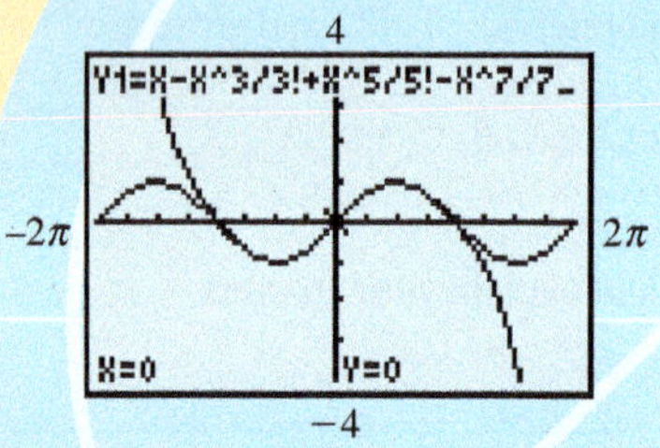


### Taylor Polynomials

*In calculus it is proved that many functions can be approximated to any degree of accuracy by polynomial functions called Taylor polynomials. The Taylor polynomials for* $\sin(x)$ *are*

$$y = x,$$

$$y = x - \frac{x^3}{3!},$$

$$y = x - \frac{x^3}{3!} + \frac{x^5}{5!},$$

$$y = x - \frac{x^3}{3!} + \frac{x^5}{5!} - \frac{x^7}{7!}$$

*and so on. As shown in the accompanying figure from a graphing calculator, the seventh-degree Taylor polynomial approximates the sine curve very well between* $-\pi$ *and* $\pi$.

**a)** Graph $y_1 = x$ and $y_2 = \sin(x)$ on a graphing calculator and determine approximately the values for $x$ for which $y_1$ differs from $y_2$ by less than 0.1.

**b)** Repeat part (a) using the fifth-, ninth-, and 19th-degree Taylor polynomials for $y_1$.

c) The Taylor polynomials for $\cos(x)$ are $y = 1$, $y = 1 - x^2/2!$, $y = 1 - x^2/2! + x^4/4!$, $y = 1 - x^2/2! + x^4/4! - x^6/6!$, and so on. Graph the zero-, fourth-, eighth-, and 18th-degree Taylor polynomials and determine approximately the intervals on which each of them differs from $\cos(x)$ by less than 0.1. Compare your answers with the answers obtained for parts (a) and (b).

d) Suppose that each operation $(+, -, \times, \div)$ takes one second. Explain how to find $\sin(x)$ to the nearest tenth for any real number $x$ in the shortest amount of time by using Taylor polynomials.

# 5.4 The Other Trigonometric Functions and Their Graphs

So far we have studied two of the six trigonometric functions. In this section we will define the remaining four functions.

## Definitions

The tangent, cotangent, secant, and cosecant functions are all defined in terms of the unit circle. We use the abbreviation tan for tangent, cot for cotangent, sec for secant, and csc for cosecant. As usual, we may think of $\alpha$ as an angle, the measure of an angle in degrees, the measure of an angle in radians, or a real number.

**Definition: Tangent, Cotangent, Secant, and Cosecant Functions**

If $\alpha$ is an angle in standard position and $(x, y)$ is the point of intersection of the terminal side and the unit circle, we define the **tangent, cotangent, secant,** and **cosecant** functions as

$$\tan\alpha = \frac{y}{x}, \quad \cot\alpha = \frac{x}{y}, \quad \sec\alpha = \frac{1}{x}, \quad \text{and} \quad \csc\alpha = \frac{1}{y}.$$

We exclude from the domain of each function any value of $\alpha$ for which the denominator is 0.

Since $\sin\alpha = y$ and $\cos\alpha = x$, we can rewrite the definitions of tangent, cotangent, secant, and cosecant to get the following identities for these four functions.

**Identities from the Definitions**

If $\alpha$ is any angle or real number

$$\tan\alpha = \frac{\sin\alpha}{\cos\alpha}, \quad \cot\alpha = \frac{\cos\alpha}{\sin\alpha}, \quad \sec\alpha = \frac{1}{\cos\alpha}, \quad \text{and} \quad \csc\alpha = \frac{1}{\sin\alpha},$$

provided no denominator is zero.

**Foreshadowing Calculus**

In trigonometry we see that all of the trigonometric functions are related to each other. These relationships are important in calculus when we study rates of change of the trigonometric functions.

The domain of the tangent and secant functions is the set of angles except those for which $\cos\alpha = 0$. The only points on the unit circle where the first coordinate is 0 are $(0, 1)$ and $(0, -1)$. Angles such as $\pi/2$, $3\pi/2$, $5\pi/2$, and so on, have terminal sides through either $(0, 1)$ or $(0, -1)$. These angles are of the form $\pi/2 + k\pi$, where $k$ is any integer. So the domain of tangent and secant is

$$\left\{\alpha \,\middle|\, \alpha \neq \frac{\pi}{2} + k\pi, \text{ where } k \text{ is an integer}\right\}.$$

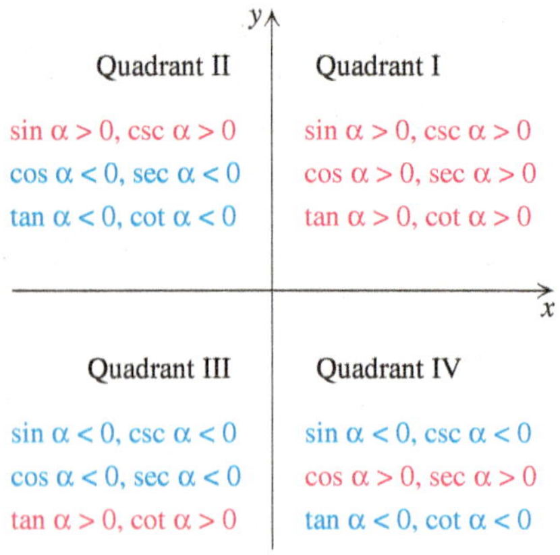


**Figure 5.60**

The only points on the unit circle where the second coordinate is 0 are $(1, 0)$ and $(-1, 0)$. Angles that are multiples of $\pi$, such as $0$, $\pi$, $2\pi$, and so on, have terminal sides that go through one of these points. Any angle of the form $k\pi$, where $k$ is any integer, has a terminal side that goes through either $(1, 0)$ or $(-1, 0)$. So $\sin(k\pi) = 0$ for any integer $k$. By definition, the zeros of the sine function are excluded from the domain of cotangent and cosecant; thus the domain of cotangent and cosecant is

$$\{\alpha \mid \alpha \neq k\pi, \text{ where } k \text{ is an integer}\}.$$

To find the values of the six trigonometric functions for an angle $\alpha$, first find $\sin\alpha$ and $\cos\alpha$. Then use the identities from the definitions to find values of the other four functions. Of course, $\cot\alpha$ can be found also by using $1/\tan\alpha$ but not if $\tan\alpha$ is undefined as it is for $\alpha = \pm\pi/2$. If $\tan\alpha$ is undefined, then $\cot\alpha = 0$, and if $\tan\alpha = 0$, then $\cot\alpha$ is undefined.

The signs of the six trigonometric functions for angles in each quadrant are shown in Fig. 5.60. It is not necessary to memorize these signs, because they can be easily obtained by knowing the signs of the sine and cosine functions in each quadrant.

## EXAMPLE 1 Evaluating the trigonometric functions

Find the values of all six trigonometric functions for each angle.

**a.** $\pi/4$ **b.** $150°$

### Solution

**a.** Use $\sin(\pi/4) = \sqrt{2}/2$ and $\cos(\pi/4) = \sqrt{2}/2$ to find the other values:

$$\tan(\pi/4) = \frac{\sin(\pi/4)}{\cos(\pi/4)} = \frac{\sqrt{2}/2}{\sqrt{2}/2} = 1 \qquad \cot(\pi/4) = \frac{\cos(\pi/4)}{\sin(\pi/4)} = 1$$

$$\sec(\pi/4) = \frac{1}{\cos(\pi/4)} = \frac{2}{\sqrt{2}} = \sqrt{2} \qquad \csc(\pi/4) = \frac{1}{\sin(\pi/4)} = \frac{2}{\sqrt{2}} = \sqrt{2}$$

**b.** The reference angle for $150°$ is $30°$. So $\sin(150°) = \sin(30°) = 1/2$ and $\cos(150°) = -\cos(30°) = -\sqrt{3}/2$. Use these values to find the other four values:

$$\tan(150°) = \frac{1/2}{-\sqrt{3}/2} = -\frac{1}{\sqrt{3}} = -\frac{\sqrt{3}}{3} \qquad \cot(150°) = -\frac{3}{\sqrt{3}} = -\sqrt{3}$$

$$\sec(150°) = \frac{1}{\cos(150°)} = -\frac{2}{\sqrt{3}} = -\frac{2\sqrt{3}}{3} \qquad \csc(150°) = \frac{1}{\sin(150°)} = 2$$

▸**TRY THIS.** Find the values of all six trigonometric functions for $-\pi/4$. ■

Most scientific calculators have keys for the sine, cosine, and tangent functions only. To find values for the other three functions, we use the definitions. Keys labeled $\sin^{-1}$, $\cos^{-1}$, and $\tan^{-1}$ on a calculator are for the inverse trigonometric functions, which we will study in the next section. These keys do *not* give the values of $1/\sin x$, $1/\cos x$, or $1/\tan x$.

## EXAMPLE 2 Evaluating with a calculator

Use a calculator to find approximate values rounded to four decimal places.

**a.** $\sec(\pi/12)$ **b.** $\csc(123°)$ **c.** $\cot(-12.4)$

Figure 5.61

### Solution

**a.** $\sec(\pi/12) = \dfrac{1}{\cos(\pi/12)} \approx 1.0353$

**b.** $\csc(123°) = \dfrac{1}{\sin(123°)} \approx 1.1924$

**c.** $\cot(-12.4) = \dfrac{1}{\tan(-12.4)} \approx 5.9551$

These expressions are evaluated with a graphing calculator in radian mode as shown in Fig. 5.61. Note that in radian mode the degree symbol is used for part (b).

▶**TRY THIS.** Find $\sec(\pi/22)$ and $\csc(-14.5)$ to four decimal places. ■

## Graph of $y = \tan(x)$

Consider some ordered pairs that satisfy $y = \tan x$ for $x$ in $(-\pi/2, \pi/2)$. Note that $-\pi/2$ and $\pi/2$ are not in the domain of $y = \tan x$, but $x$ can be chosen close to $\pm\pi/2$.

$x \to -\pi/2$ from the right ⟵ ⟶ $x \to \pi/2$ from the left

| $x$ | $-1.56$ | $-1.5$ | $-1.2$ | $-\pi/4$ | 0 | $\pi/4$ | 1.2 | 1.5 | 1.56 |
|---|---|---|---|---|---|---|---|---|---|
| $y = \tan x$ | $-92.6$ | $-14.1$ | $-2.6$ | $-1$ | 0 | 1 | 2.6 | 14.1 | 92.6 |

$y \to -\infty$ ⟵ ⟶ $y \to \infty$

The graph of $y = \tan x$ includes the points $(-\pi/4, -1)$, $(0, 0)$, and $(\pi/4, 1)$. Since $\tan x = \sin x/\cos x$, the graph of $y = \tan x$ has a vertical asymptote for every zero of the cosine function. So the vertical lines $x = \pi/2 + k\pi$ for any integer $k$ are the **vertical asymptotes.** The behavior of $y = \tan x$ near the asymptotes $x = \pm\pi/2$ can be seen from the table of ordered pairs. As $x$ approaches $\pi/2$ (approximately 1.57) from the left, $\tan x \to \infty$. Using limit notation, $\lim_{x \to \pi/2^-} \tan(x) = \infty$. As $x$ approaches $-\pi/2$ from the right, $\tan x \to -\infty$. Using limit notation, $\lim_{x \to -\pi/2^+} \tan(x) = -\infty$. In the interval $(-\pi/2, \pi/2)$, the function is increasing. The graph of $y = \tan x$ is shown in Fig. 5.62. The shape of the tangent curve between $-\pi/2$ and $\pi/2$ is repeated between each pair of consecutive asymptotes, as shown in the figure. The period of $y = \tan x$ is $\pi$, and the fundamental cycle is the portion of the graph between $-\pi/2$ and $\pi/2$. Since the range of $y = \tan x$ is $(-\infty, \infty)$, the concept of amplitude is not defined for this function.

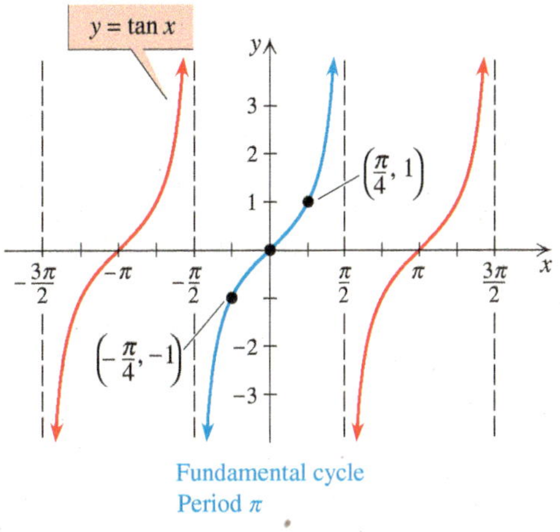


Figure 5.62

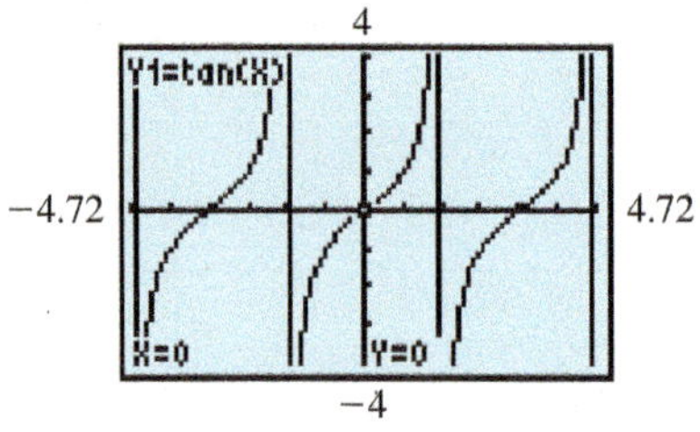


Figure 5.63

The calculator graph of $y = \tan(x)$ in connected mode is shown in Fig. 5.63. In connected mode the calculator connects two points on opposite sides of each asymptote and appears to draw the vertical asymptotes. □

To help us understand the sine and cosine curve, we identified five key points on the fundamental cycle. For the tangent curve, we have three key points and the asymptotes (where the function is undefined). The exact coordinates are given in the following table.

| $x$ | $-\pi/2$ | $-\pi/4$ | 0 | $\pi/4$ | $\pi/2$ |
|---|---|---|---|---|---|
| $y = \tan x$ | undefined | $-1$ | 0 | 1 | undefined |

We can transform the graph of the tangent function by using the same techniques that we used for the sine and cosine functions.

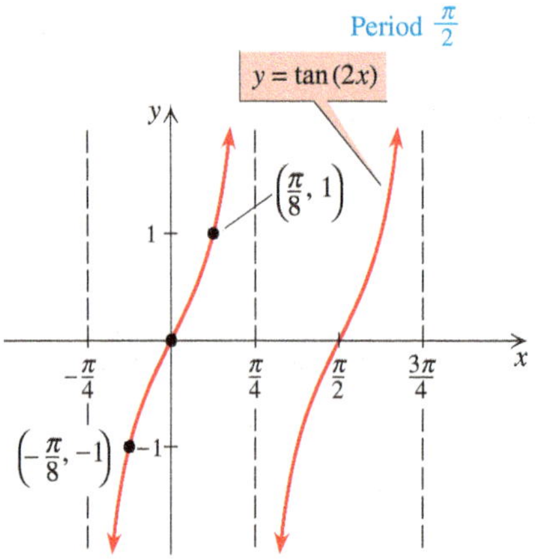


Figure 5.64

## EXAMPLE 3 A tangent function with a transformation

Sketch two cycles of the function $y = \tan(2x)$ and determine the period.

### Solution

The graph of $y = \tan x$ completes one cycle for $-\pi/2 < x < \pi/2$. So the graph of $y = \tan(2x)$ completes one cycle for $-\pi/2 < 2x < \pi/2$, or $-\pi/4 < x < \pi/4$. The graph includes the points $(-\pi/8, -1)$, $(0, 0)$, and $(\pi/8, 1)$. The graph is similar to the graph of $y = \tan x$, but the asymptotes are $x = \pi/4 + k\pi/2$ for any integer $k$, and the period of $y = \tan(2x)$ is $\pi/2$. Two cycles of the graph are shown in Fig. 5.64.

▶**TRY THIS.** Graph two cycles of $y = \tan(2\pi x)$ and determine the period. ■

By understanding what happens to the fundamental cycles of the sine, cosine, and tangent functions when the period changes, we can easily determine the location of the new function and sketch its graph quickly. We start with the fundamental cycles of $y = \sin x$, $y = \cos x$, and $y = \tan x$, which occur over the intervals $[0, 2\pi]$, $[0, 2\pi]$, and $[-\pi/2, \pi/2]$, respectively. For $y = \sin Bx$, $y = \cos Bx$, and $y = \tan Bx$ (for $B > 0$) these fundamental cycles move to the intervals $[0, 2\pi/B]$, $[0, 2\pi/B]$, and $[-\pi/(2B), \pi/(2B)]$, respectively. In every case, divide the old period by $B$ to get the new period. Note that only the nonzero endpoints of the intervals are changed. The following Function Gallery summarizes the period change for the sine, cosine, and tangent functions with $B > 1$.

## FUNCTION gallery... Periods of Sine, Cosine, and Tangent ($B > 1$)

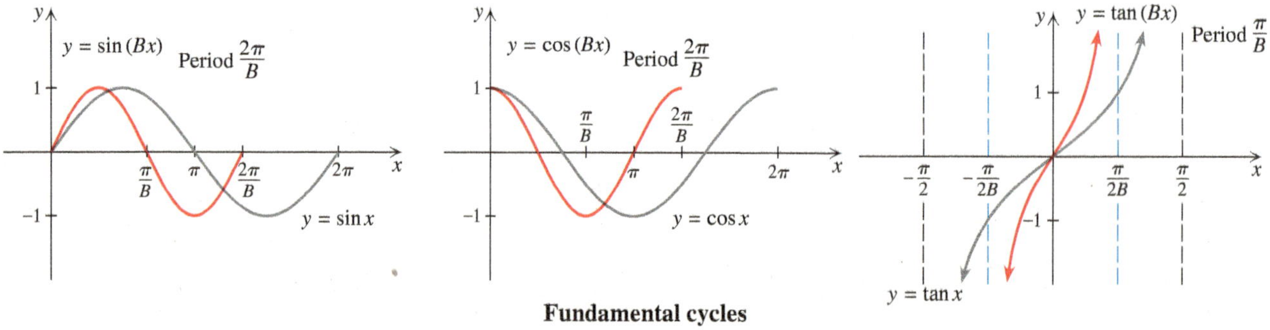


Fundamental cycles

## Graph of $y = \cot(x)$

We know that $\cot x = 1/\tan x$, provided $\tan x$ is defined and not 0. So we can use $y = \tan x$ to graph $y = \cot x$. Since $\tan x = 0$ for $x = k\pi$, where $k$ is any integer, the vertical lines $x = k\pi$ for an integer $k$ are the vertical asymptotes of $y = \cot x$. Consider some ordered pairs that satisfy $y = \cot x$ between the asymptotes $x = 0$ and $x = \pi$:

$x \to 0$ from the right | $x \to \pi$ from the left

| $x$ | 0.01 | 0.02 | $\pi/4$ | $\pi/2$ | $3\pi/4$ | 3.1 | 3.14 |
|---|---|---|---|---|---|---|---|
| $y = \cot x$ | 100.00 | 49.99 | 1 | 0 | $-1$ | $-24.0$ | $-627.9$ |

$y \to \infty$ | $y \to -\infty$

As $x \to 0$ from the right, $\cot x \to \infty$. Using limit notation, $\lim_{x\to 0^+} \cot(x) = \infty$. As $x \to \pi$ from the left, $\cot x \to -\infty$. Using limit notation, $\lim_{x\to \pi^-} \cot(x) = -\infty$. In the interval $(0, \pi)$, the function is decreasing. The graph of $y = \cot x$ is shown in Fig. 5.65. The shape of the curve between 0 and $\pi$ is repeated between each pair of consecutive asymptotes, as shown in the figure. The period of $y = \cot x$ is $\pi$, and the fundamental cycle is the portion of the graph between 0 and $\pi$. The range of $y = \cot x$ is $(-\infty, \infty)$.

Because $\cot x = 1/\tan x$, $\cot x$ is large when $\tan x$ is small, and vice versa. The graph of $y = \cot x$ has an $x$-intercept wherever $y = \tan x$ has a vertical asymptote, and a vertical asymptote wherever $y = \tan x$ has an $x$-intercept. So for every integer $k$, $(\pi/2 + k\pi, 0)$ is an $x$-intercept of $y = \cot x$, and the vertical line $x = k\pi$ is an asymptote, as shown in Fig. 5.65.

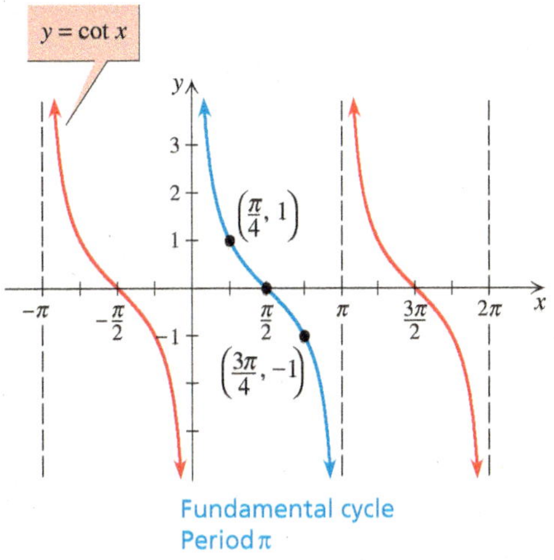


**Figure 5.65**

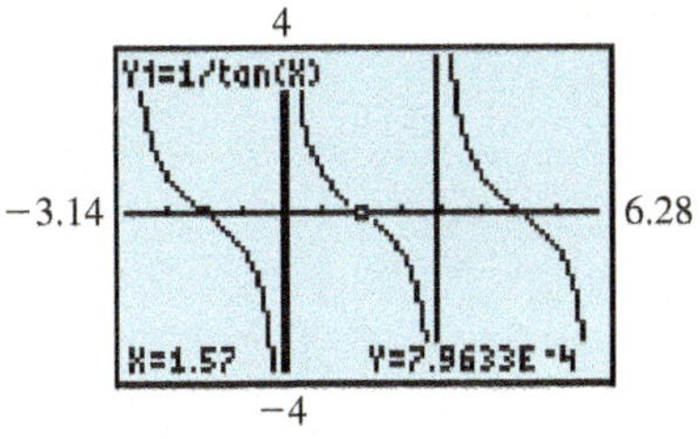


**Figure 5.66**

To see the graph of $y = \cot(x)$ on a graphing calculator, you can graph $y = 1/\tan(x)$, as shown in Fig. 5.66. □

For the cotangent curve, note the three key points and the asymptotes (where the function is undefined). The exact coordinates are given in the following table.

| $x$ | 0 | $\pi/4$ | $\pi/2$ | $3\pi/4$ | $\pi$ |
|---|---|---|---|---|---|
| $y = \cot x$ | undefined | 1 | 0 | $-1$ | undefined |

When we graph a transformation of the cotangent function, as in the next example, we must determine what happens to these five features.

## EXAMPLE 4 A cotangent function with a transformation

Sketch two cycles of the function $y = 0.3\cot(2x + \pi/2)$, and determine the period.

### Solution

Factor out 2 to write the function as

$$y = 0.3\cot\left[2\left(x + \frac{\pi}{4}\right)\right].$$

Since $y = \cot x$ completes one cycle for $0 < x < \pi$, the graph of $y = 0.3\cot[2(x + \pi/4)]$ completes one cycle for

$$0 < 2\left(x + \frac{\pi}{4}\right) < \pi, \quad \text{or} \quad 0 < x + \frac{\pi}{4} < \frac{\pi}{2}, \quad \text{or} \quad -\frac{\pi}{4} < x < \frac{\pi}{4}.$$

The interval $(-\pi/4, \pi/4)$ for the fundamental cycle can also be obtained by dividing the period $\pi$ of $y = \cot x$ by 2 to get $\pi/2$ as the period, and then shifting the interval $(0, \pi/2)$ a distance of $\pi/4$ to the left. The factor 0.3 shrinks the $y$-coordinates. The graph goes through $(-\pi/8, 0.3)$, $(0, 0)$, and $(\pi/8, -0.3)$ as it approaches its vertical asymptotes $x = \pm\pi/4$. The graph for two cycles is shown in Fig. 5.67.

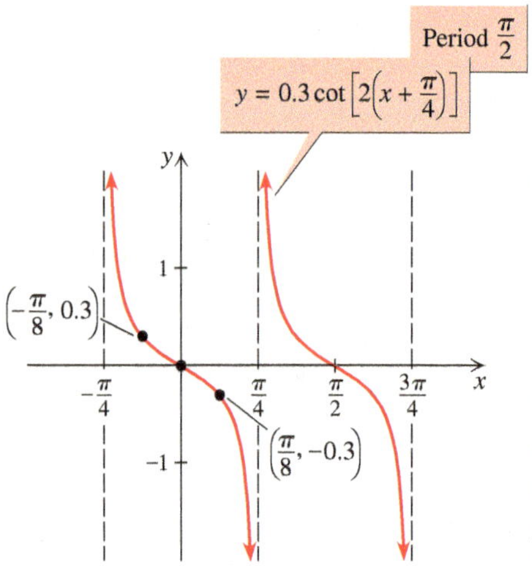


**Figure 5.67**

The calculator graph in Fig. 5.68 supports these conclusions.

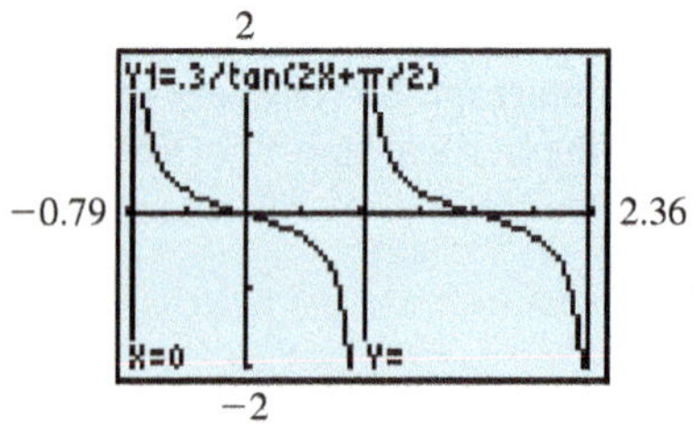


**Figure 5.68**

**▸TRY THIS.** Graph two cycles of $y = 3\cot(2x + \pi)$ and determine the period. ■

Since the period for $y = \cot x$ is $\pi$, the period for $y = \cot(Bx)$ with $B > 0$ is $\pi/B$ and $y = \cot(Bx)$ completes one cycle on the interval $(0, \pi/B)$. The left-hand asymptote for the fundamental cycle of $y = \cot x$ remains fixed at $x = 0$, while the right-hand asymptote changes to $x = \pi/B$.

## Graph of $y = \sec(x)$

Since $\sec x = 1/\cos x$, the values of $\sec x$ are large when the values of $\cos x$ are small. For any $x$ such that $\cos x$ is 0, $\sec x$ is undefined and the graph of $y = \sec x$ has a vertical asymptote. Because of the reciprocal relationship between $\sec x$ and $\cos x$, we first draw the graph of $y = \cos x$ for reference when graphing $y = \sec x$. At every $x$-intercept of $y = \cos x$, we draw a vertical asymptote, as shown in Fig. 5.69. If $\cos x = \pm 1$, then $\sec x = \pm 1$. So every maximum or minimum point on the

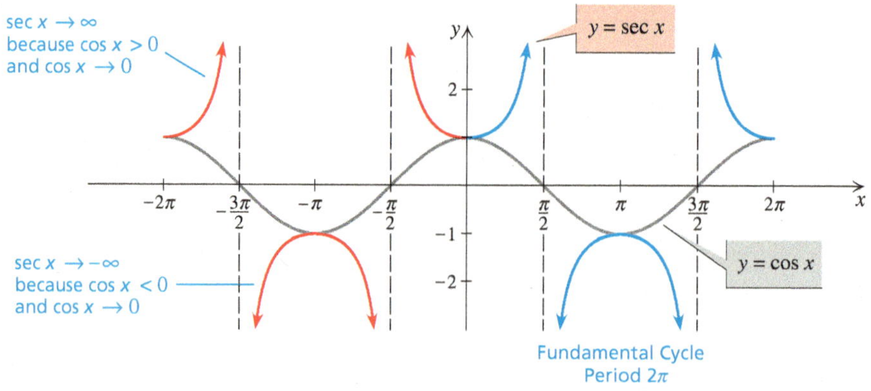


**Figure 5.69**

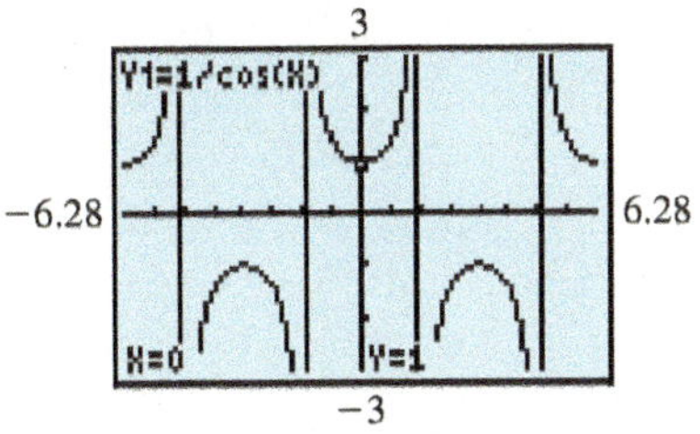


Figure 5.70

graph of $y = \cos x$ is also on the graph of $y = \sec x$. If $\cos x > 0$ and $\cos x \to 0$, then $\sec x \to \infty$. If $\cos x < 0$ and $\cos x \to 0$, then $\sec x \to -\infty$. These two facts cause the graph of $y = \sec x$ to approach its asymptotes in the manner shown in Fig. 5.69. The period of $y = \sec x$ is $2\pi$, the same as the period for $y = \cos x$. The range of $y = \sec x$ is $(-\infty, -1] \cup [1, \infty)$.

The calculator graph in Fig. 5.70 supports these conclusions. □

## EXAMPLE 5 A secant function with a transformation

Sketch two cycles of the function $y = 2\sec(x - \pi/2)$, and determine the period and the range of the function.

### Solution

Since

$$y = 2\sec\left(x - \frac{\pi}{2}\right) = \frac{2}{\cos(x - \pi/2)}$$

we first graph $y = \cos(x - \pi/2)$, as shown in Fig. 5.71. The function $y = \sec(x - \pi/2)$ goes through the maximum and minimum points on the graph of $y = \cos(x - \pi/2)$, but $y = 2\sec(x - \pi/2)$ stretches $y = \sec(x - \pi/2)$ by a factor of 2. So the portions of the curve that open up do not go lower than 2, and the portions that open down do not go higher than $-2$, as shown in the figure. The period is $2\pi$, and the range is $(-\infty, -2] \cup [2, \infty)$.

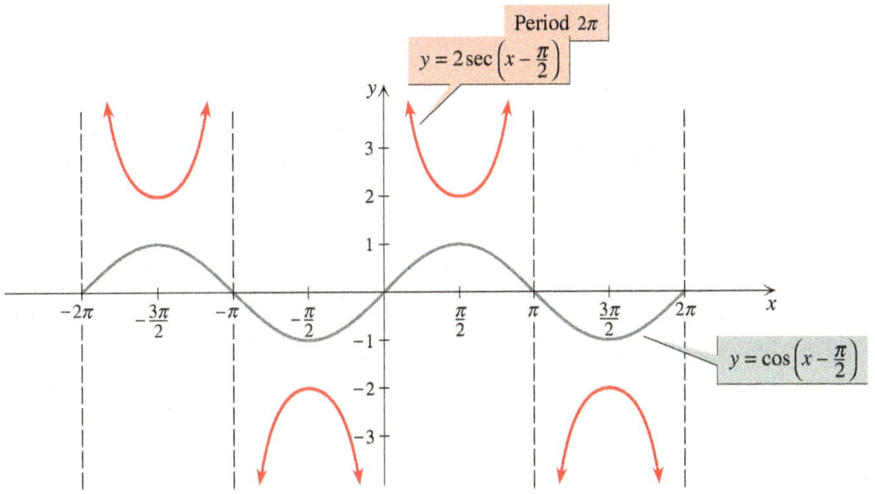


Figure 5.71

▶**TRY THIS.** Graph one cycle of $y = 3\sec(2x)$ and determine the period and range. ■

## Graph of $y = \csc(x)$

Since $\csc x = 1/\sin x$, the graph of $y = \csc x$ is related to the graph of $y = \sin x$ in the same way that the graphs of $y = \sec x$ and $y = \cos x$ are related. To graph $y = \csc x$, first draw the graph of $y = \sin x$ and a vertical asymptote at each $x$-intercept. Since the graph of $y = \sin x$ can be obtained by shifting $y = \cos x$ a distance of $\pi/2$ to the right, the graph of $y = \csc x$ is obtained from $y = \sec x$ by shifting a distance of $\pi/2$ to the right, as shown in Fig. 5.72. The period of $y = \csc x$ is $2\pi$, the same as the period for $y = \sin x$.

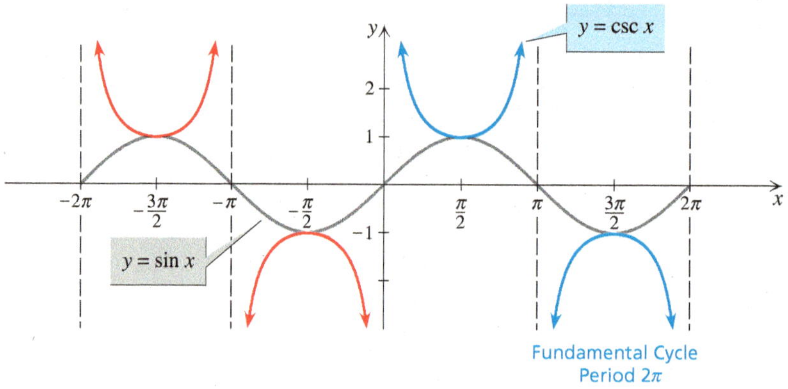


**Figure 5.72**

## EXAMPLE 6 A cosecant function with a transformation

Sketch two cycles of the graph of $y = \csc(2x - 2\pi/3)$ and determine the period and the range of the function.

### Solution

Since

$$y = \csc(2x - 2\pi/3)$$
$$= \frac{1}{\sin[2(x - \pi/3)]}$$

we first graph $y = \sin[2(x - \pi/3)]$. The period for $y = \sin[2(x - \pi/3)]$ is $\pi$ with phase shift of $\pi/3$. So the fundamental cycle of $y = \sin x$ is transformed to occur on the interval $[\pi/3, 4\pi/3]$. Draw at least two cycles of $y = \sin[2(x - \pi/3)]$ with a vertical asymptote (for the cosecant function) at every $x$-intercept, as shown in Fig. 5.73. Each portion of $y = \csc[2(x - \pi/3)]$ that opens up has a minimum value of 1, and each portion that opens down has a maximum value of $-1$. The period of the function is $\pi$, and the range is $(-\infty, -1] \cup [1, \infty)$.

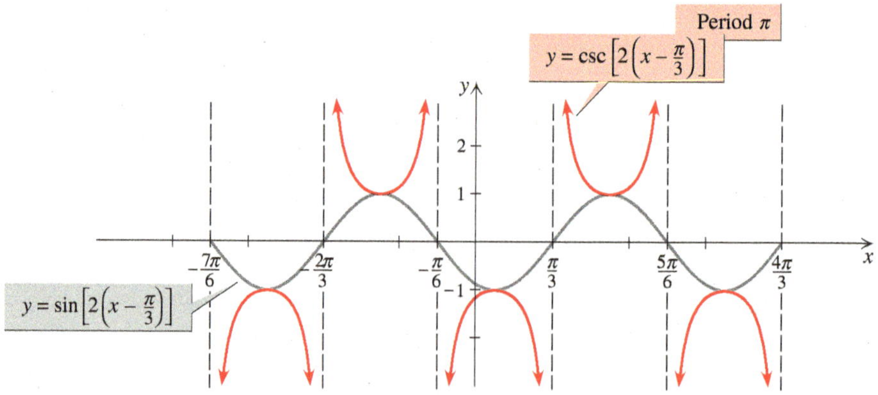


**Figure 5.73**

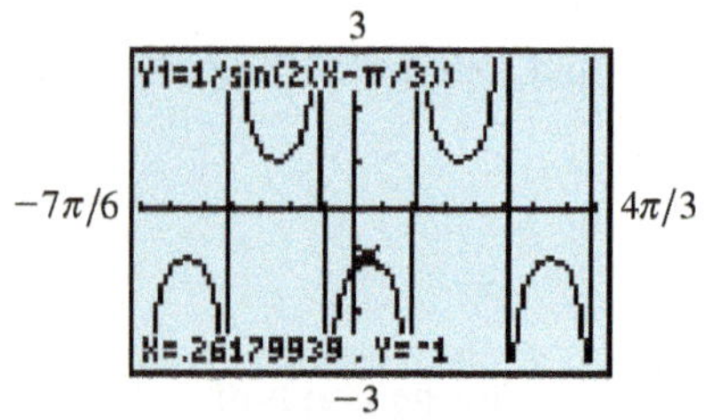


**Figure 5.74**

The calculator graph in Fig. 5.74 supports these conclusions.

▶**TRY THIS.** Graph one cycle of $y = \csc(2x - \pi)$ and determine the period and range. ■

The Function Gallery summarizes some of the facts that we have learned about the six trigonometric functions. Also, one cycle of the graph of each trigonometric function is shown.

## FUNCTION gallery... Trigonometric Functions

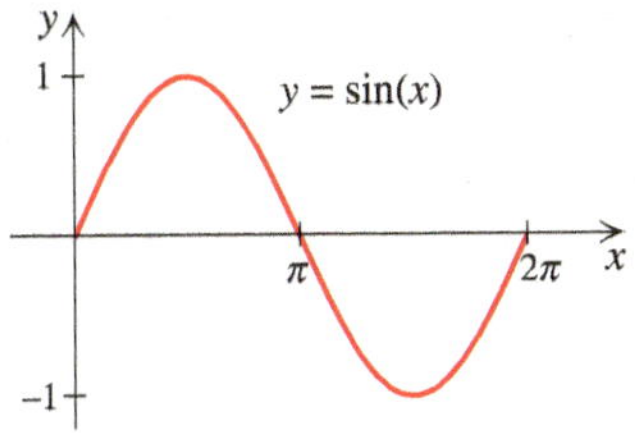


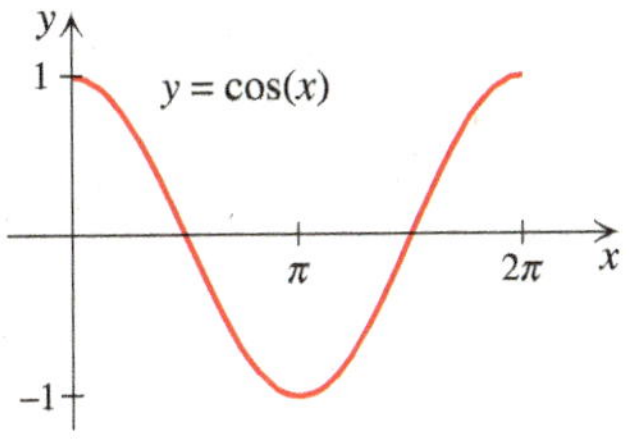


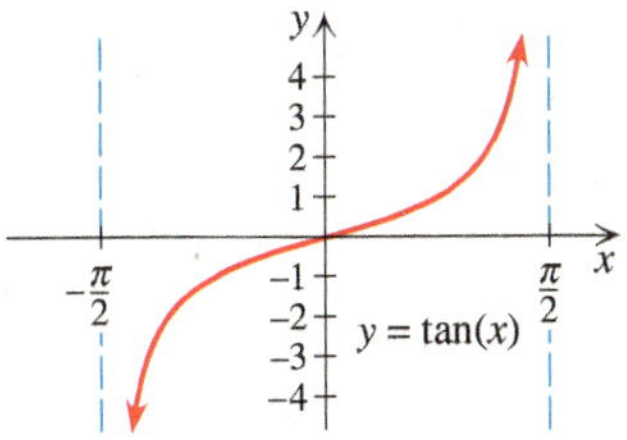


| | $y = \sin(x)$ | $y = \cos(x)$ | $y = \tan(x)$ |
|---|---|---|---|
| Domain (*k* any integer) | $(-\infty, \infty)$ | $(-\infty, \infty)$ | $x \neq \frac{\pi}{2} + k\pi$ |
| Range | $[-1, 1]$ | $[-1, 1]$ | $(-\infty, \infty)$ |
| Period | $2\pi$ | $2\pi$ | $\pi$ |
| Fundamental cycle | $[0, 2\pi]$ | $[0, 2\pi]$ | $\left[-\frac{\pi}{2}, \frac{\pi}{2}\right]$ |

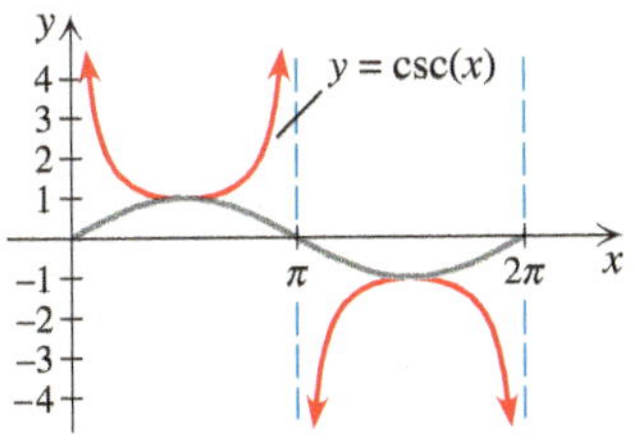


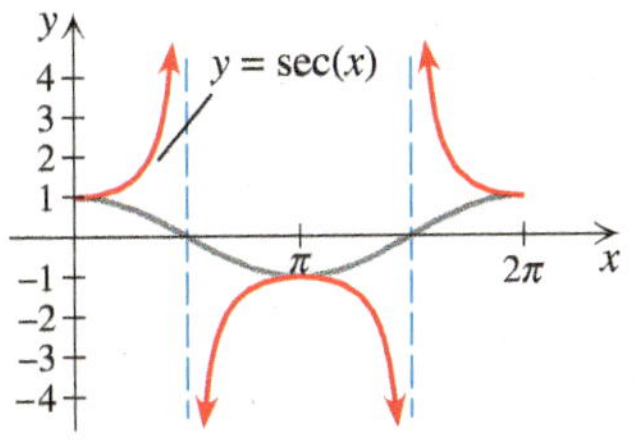


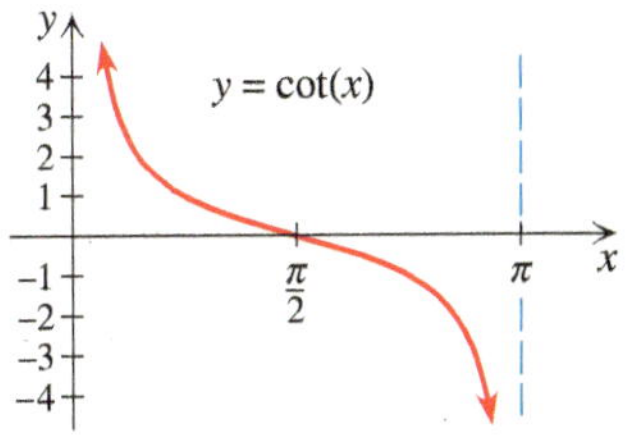


| | $y = \csc(x)$ | $y = \sec(x)$ | $y = \cot(x)$ |
|---|---|---|---|
| Domain (*k* any integer) | $x \neq k\pi$ | $x \neq \frac{\pi}{2} + k\pi$ | $x \neq k\pi$ |
| Range | $(-\infty, -1] \cup [1, \infty)$ | $(-\infty, -1] \cup [1, \infty)$ | $(-\infty, \infty)$ |
| Period | $2\pi$ | $2\pi$ | $\pi$ |
| Fundamental cycle | $[0, 2\pi]$ | $[0, 2\pi]$ | $[0, \pi]$ |

## FOR thought... True or False? Explain.

1. $\sec(\pi/4) = 1/\sin(\pi/4)$

2. $\cot(\pi/2) = 1/\tan(\pi/2)$

3. $\csc(60^\circ) = 2\sqrt{3}/3$

4. $\tan(5\pi/2) = 0$

5. $\sec(95^\circ) = \sqrt{5}$

6. $\csc(120^\circ) = 2/\sqrt{3}$

7. The graphs of $y = 2 \csc x$ and $y = 1/(2 \sin x)$ are identical.

8. The range of $y = 0.5 \csc(13x - 5\pi)$ is $(-\infty, -0.5] \cup [0.5, \infty)$.

9. The graph of $y = \tan(3x)$ has vertical asymptotes at $x = \pm\pi/6$.

10. The graph of $y = \cot(4x)$ has vertical asymptotes at $x = \pm\pi/4$.

## EXERCISES 5.4

*Fill in the blank.*

**1.** If $\alpha$ is an angle in standard position whose terminal side intersects the unit circle at $(x, y)$ then $y/x =$ ________ and $1/x =$ ________ provided $x \neq 0$.

**2.** The line $x = \pi/2$ is a ________ for the graph of $y = \tan x$.

*Redraw each unit circle and label each indicated point with the proper value of the tangent function. The points in Exercise 3 are the terminal points for arcs with lengths that are multiples of $\pi/4$. The points in Exercise 4 are the terminal points for arcs with lengths that are multiples of $\pi/6$. Repeat Exercises 3 and 4 until you can do them from memory.*

**3.**

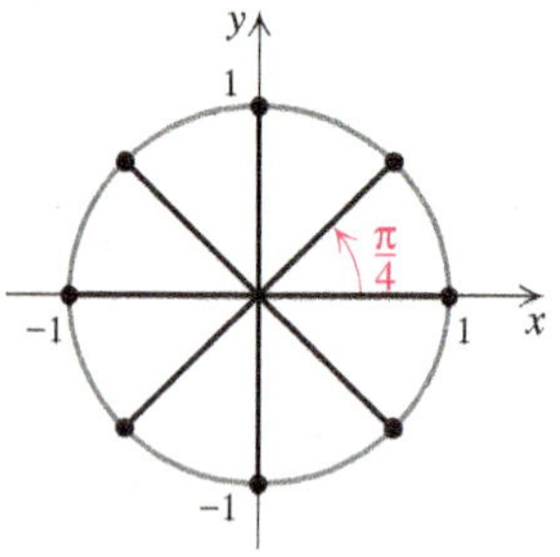


**4.**

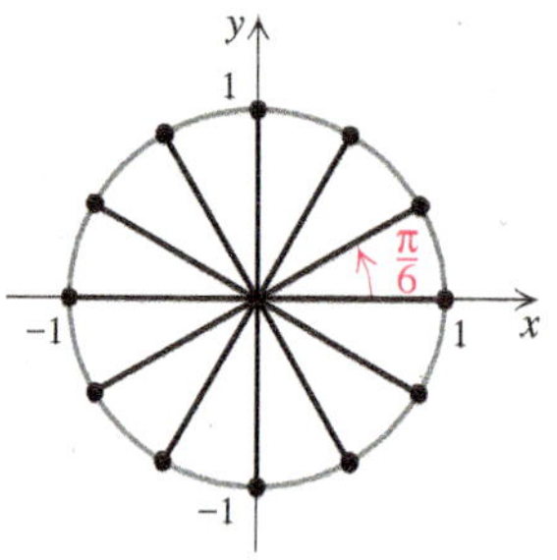


*Find the exact value of each of the following expressions without using a calculator.*

**5.** $\tan(\pi/3)$ **6.** $\tan(\pi/4)$

**7.** $\tan(-\pi/4)$ **8.** $\tan(\pi/6)$

**9.** $\cot(\pi/2)$ **10.** $\cot(2\pi/3)$

**11.** $\cot(-\pi/3)$ **12.** $\cot(0)$

**13.** $\sec(\pi/6)$ **14.** $\sec(\pi/3)$

**15.** $\sec(\pi/2)$ **16.** $\sec(\pi)$

**17.** $\csc(-\pi)$ **18.** $\csc(\pi/6)$

**19.** $\csc(3\pi/4)$ **20.** $\csc(-\pi/3)$

**21.** $\tan(135°)$ **22.** $\tan(270°)$

**23.** $\cot(210°)$ **24.** $\cot(120°)$

**25.** $\sec(-120°)$ **26.** $\sec(-90°)$

**27.** $\csc(315°)$ **28.** $\csc(240°)$

**29.** $\cot(-90°)$ **30.** $\cot(270°)$

*Find the approximate value of each expression. Round to four decimal places.*

**31.** $\tan(1.55)$ **32.** $\tan(1.6)$

**33.** $\cot(-3.48)$ **34.** $\cot(22.4)$

**35.** $\csc(0.002)$ **36.** $\csc(1.54)$

**37.** $\sec(\pi/12)$ **38.** $\sec(-\pi/8)$

**39.** $\cot(0.09°)$ **40.** $\cot(179.4°)$

**41.** $\csc(-44.3°)$ **42.** $\csc(-124.5°)$

**43.** $\sec(89.2°)$ **44.** $\sec(-0.024°)$

**45.** $\tan(-44.6°)$ **46.** $\tan(138°)$

*Find the exact value of each expression for the given value of $\theta$. Do not use a calculator.*

**47.** $\sec^2(2\theta)$ if $\theta = \pi/6$ **48.** $\csc^2(2\theta)$ if $\theta = \pi/8$

**49.** $\tan(\theta/2)$ if $\theta = \pi/3$ **50.** $\csc(\theta/2)$ if $\theta = \pi/2$

**51.** $\sec(\theta/2)$ if $\theta = 3\pi/2$ **52.** $\cot(\theta/2)$ if $\theta = 2\pi/3$

*Determine the period and sketch at least one cycle of the graph of each function.*

**53.** $y = \tan(3x)$ **54.** $y = \tan(4x)$

**55.** $y = \cot(x + \pi/4)$ **56.** $y = \cot(x - \pi/6)$

**57.** $y = \cot(x/2)$ **58.** $y = \cot(x/3)$

**59.** $y = \tan(\pi x)$ **60.** $y = \tan(\pi x/2)$

**61.** $y = -2\tan x$ **62.** $y = -\tan(x - \pi/2)$

**63.** $y = -\cot(x + \pi/2)$ **64.** $y = 2 + \cot x$

**65.** $y = \cot(2x - \pi/2)$ **66.** $y = \cot(3x + \pi)$

**67.** $y = \tan\left(\frac{\pi}{2}x - \frac{\pi}{2}\right)$ **68.** $y = \tan\left(\frac{\pi}{4}x + \frac{3\pi}{4}\right)$

*Determine the period and sketch at least one cycle of the graph of each function. State the range of each function.*

**69.** $y = \sec(2x)$ **70.** $y = \sec(3x)$

**71.** $y = \csc(x - \pi/2)$ **72.** $y = \csc(x + \pi/4)$

**73.** $y = \csc(x/2)$ **74.** $y = \csc(x/4)$

**75.** $y = \sec(\pi x/2)$ **76.** $y = \sec(\pi x)$

**77.** $y = 2\sec x$ **78.** $y = \frac{1}{2}\sec x$

**79.** $y = \csc(2x - \pi/2)$ **80.** $y = \csc(3x + \pi)$

**81.** $y = -\csc\left(\frac{\pi}{2}x + \frac{\pi}{2}\right)$ **82.** $y = -2\csc(\pi x - \pi)$

**83.** $y = 2 + 2\sec(2x)$ **84.** $y = 2 - 2\sec\left(\frac{x}{2}\right)$

*Determine the period and range of each function.*

**85.** $y = \tan(2x - \pi) + 3$

**86.** $y = 2\cot(3x + \pi) - 8$

**87.** $y = 2\sec(x/2 - 1) - 1$

**88.** $y = -2\sec(x/3 - 6) + 3$

**89.** $y = -3\csc(2x - \pi) - 4$

**90.** $y = 4\csc(3x - \pi) + 5$

*Write the equation of each curve in its final position.*

**91.** The graph of $y = \tan(x)$ is shifted $\pi/4$ units to the right, stretched by a factor of 3, then translated 2 units upward.

**92.** The graph of $y = \cot(x)$ is shifted $\pi/2$ units to the left, reflected in the $x$-axis, then translated 1 unit upward.

**93.** The graph of $y = \sec(x)$ is shifted $\pi$ units to the left, reflected in the $x$-axis, then shifted 2 units upward.

**94.** The graph of $y = \csc(x)$ is shifted 2 units to the right, translated 3 units downward, then reflected in the $x$-axis.

## FOR WRITING/DISCUSSION

**95.** Graph $y = x + \sin x$ for $-100 \le x \le 100$ and $-100 \le y \le 100$. Explain your results.

**96.** Graph $y = x + \tan x$ for $-6 \le x \le 6$ and $-10 \le y \le 10$. Explain your results.

**97.** *Average Rate of Change* The average rate of change of a function on a short interval $[x, x + h]$ for a fixed value of $h$ is a function itself. Sometimes it is a function that we can recognize by its graph.

**a.** Graph $y_1 = \sin(x)$ and its average rate of change

$$y_2 = (y_1(x + 0.1) - y_1(x))/0.1$$

for $-2\pi \le x \le 2\pi$. What familiar function does $y_2$ look like?

**b.** Repeat part (a) for $y_1 = \cos(x)$, $y_1 = e^x$, $y_1 = \ln(x)$, and $y_1 = x^2$.

**98.** *Discovering Planets* On October 23, 1996, astronomers announced that they had discovered a planet orbiting the star 16 Cygni B (*Sky and Telescope*, January 1997). They did not see the planet itself. Rather, for 8 years they detected the periodic wobbling motion toward and away from earth that the planet induced in the star as it orbited the star. See the accompanying figure.

**a.** Estimate the period of the function shown in the figure.

**b.** Which trigonometric function has a graph similar to the graph in the figure?

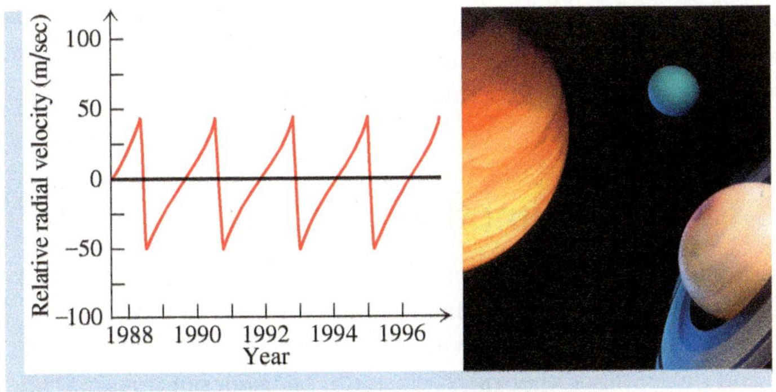


**Figure for Exercise 98**

## RETHINKING

**99.** Find the negative angle between 0° and −360° that is coterminal with 510°.

**100.** Name the quadrant in which the terminal side of $17\pi/12$ lies.

**101.** A point on the earth's equator makes one revolution in 24 hours. Find the linear velocity in feet per second for such a point using 3950 miles as the radius of the earth.

**102.** Determine the amplitude, period, phase shift, and range for the function $f(x) = \frac{1}{2}\sin(x - \pi/2) + 3$.

**103.** The graph of $y = \cos(x)$ is shifted $\pi$ units to the left, reflected in the $x$-axis, and then shifted 2 units upward. What is the equation of the curve in its final position?

**104.** Find $\sin(\alpha)$ given that $\cos(\alpha) = 1/3$ and $\alpha$ lies in quadrant IV.

## THINKING OUTSIDE THE BOX XXXIX

*Counting Votes* Fifteen experts are voting to determine the best convertible of the year. The choices are a Porsche Carrera, a Chrysler Crossfire, and a Nissan Roadster. The experts will rank the three cars 1st, 2nd, and 3rd. There are three common ways to determine the winner.

1. *Plurality:* The car with the most first place votes (preferences) is the winner.
2. *Instant runoff:* The car with the least number of preferences is eliminated. Then the ballots where the eliminated car is first are revised so that the second place car is moved to first. Finally, the car with the most preferences is the winner.
3. *The point system:* Two points are given for each time a car occurs in first place on a ballot, one point for each time the car appears in second place on a ballot, and no points for third place.

When the ballots were cast, the Porsche won when plurality was used, the Chrysler won when instant runoff was used, and the Nissan won when the point system was used. Determine 15 actual votes for which this result would occur.

## POP QUIZ 5.4

1. What is the period for $y = \tan(3x)$?
2. Find the equations of all asymptotes for $y = \cot(2x)$.
3. Find the equations of all asymptotes for $y = \sec(2x)$.
4. What is the range of $y = 3\csc(2x)$?

*Find the exact value.*

5. $\tan(\pi/4)$
6. $\tan(120°)$
7. $\cot(-\pi/3)$
8. $\sec(60°)$
9. $\csc(-3\pi/4)$

## LINKING concepts... For Individual or Group Explorations

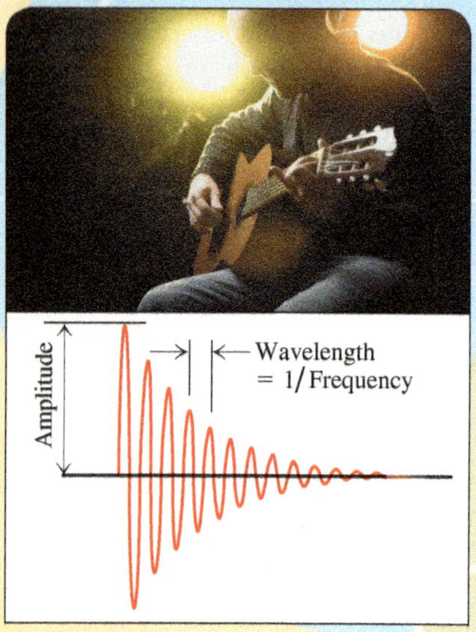


### Modeling a Guitar Note

*The waveform of a guitar note is characterized by an initial sharp peak that falls off rapidly, as shown in the figure. Guitar effects pedals electronically modify that waveform. We can mathematically modify the oscillations of the sine wave to get the waveform of a guitar note as well as many others. The basic sine wave oscillates between the two horizontal lines $y = 1$ and $y = -1$. If we multiply $\sin(x)$ by any other function $g(x)$ we will get a curve that oscillates between the graphs of $y = g(x)$ and $y = -g(x)$.*

**a.** Graph $y_1 = x\sin(x)$, $y_2 = x$, and $y_3 = -x$. For what exact values of $x$ is $x\sin(x) = x$? For what exact values of $x$ is $x\sin(x) = -x$?

**b.** For what exact values of $x$ is $x^2\sin(x) = x^2$? For what exact values of $x$ is $x^2\sin(x) = -x^2$? Support your conclusions with a graph.

**c.** Graph $y_1 = \frac{1}{x}\sin(x)$ for $0 \le x \le 10$ and $-2 \le y \le 2$. Is it true that for $x > 0$, $-\frac{1}{x} < \frac{1}{x}\sin(x) < \frac{1}{x}$? Prove your answer.

**d.** Graph $f(x) = \frac{1}{x}\sin(x)$ for $-2 \le x \le 2$ and $-2 \le y \le 2$. What is $f(0)$? Is it true that for all $x$ in the interval $[-0.1, 0.1]$ for which $f(x)$ is defined, $f(x)$ satisfies $0.99 < f(x) < 1$? Explain.

**e.** Experiment with functions until you get one that looks like the one in the figure.

# 5.5 The Inverse Trigonometric Functions

We have learned how to find the values of the trigonometric functions for angles or real numbers, but to make the trigonometric functions really useful we must be able to reverse this process. In this section we define the inverses of the trigonometric functions.

## The Inverse of the Sine Function

In Chapter 2 we learned that only one-to-one functions are invertible. Since $y = \sin x$ with domain $(-\infty, \infty)$ is a periodic function, it is certainly not one-to-one. However, if we restrict the domain to the interval $[-\pi/2, \pi/2]$, then the restricted function is one-to-one and invertible. Other intervals could be used, but this interval is chosen to keep the inverse function as simple as possible.

The graph of the sine function with domain $[-\pi/2, \pi/2]$ is shown in Fig. 5.75(a). Its range is $[-1, 1]$. We now define the inverse sine function and denote it as $f^{-1}(x) = \sin^{-1} x$ (read "inverse sine of $x$") or $f^{-1}(x) = \arcsin x$ (read "arc sine of $x$").

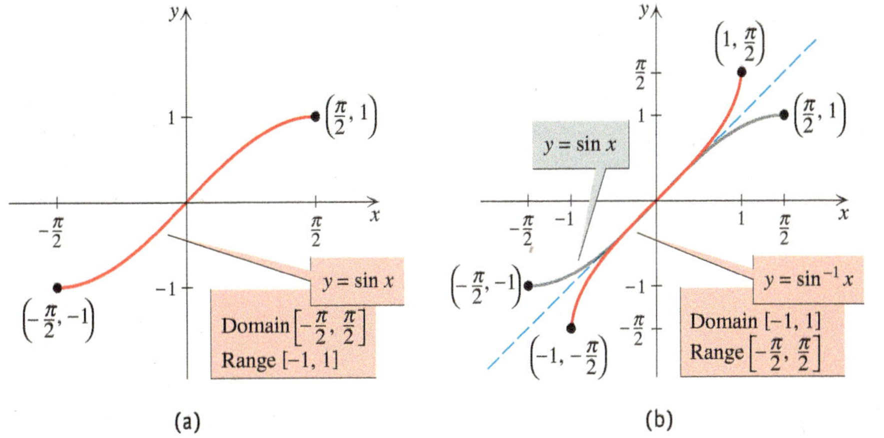


Figure 5.75

**Definition: The Inverse Sine Function**

> The function $y = \sin^{-1} x$ or $y = \arcsin x$ is the inverse of the function $y = \sin x$ restricted to $[-\pi/2, \pi/2]$. The domain of $y = \sin^{-1} x$ is $[-1, 1]$ and its range is $[-\pi/2, \pi/2]$.

The graph of $y = \sin^{-1} x$ is a reflection about the line $y = x$ of the graph of $y = \sin x$ on $[-\pi/2, \pi/2]$, as shown in Fig. 5.75(b).

If $y = \sin^{-1} x$, then $y$ is the real number such that $-\pi/2 \le y \le \pi/2$ and $\sin y = x$. Depending on the context, $\sin^{-1} x$ might also be an angle, a measure of an angle in degrees or radians, or the length of an arc of the unit circle. The expression $\sin^{-1} x$ can be read as "the angle whose sine is $x$" or "the arc length whose sine is $x$." The notation $y = \arcsin x$ reminds us that $y$ is the arc length whose sine is $x$. For example, $\arcsin(1)$ is the arc length in $[-\pi/2, \pi/2]$ whose sine is 1. Since we know that $\sin(\pi/2) = 1$, we have $\arcsin(1) = \pi/2$. We will assume that $\sin^{-1} x$ is a real number unless indicated otherwise.

Note that the $n$th power of the sine function is usually written as $\sin^n(x)$ as a shorthand notation for $(\sin x)^n$, provided $n \neq -1$. The $-1$ used in $\sin^{-1} x$ indicates the inverse function and does *not* mean reciprocal. To write $1/\sin x$ using exponents, we must write $(\sin x)^{-1}$.

**Foreshadowing Calculus**

In calculus we discover some interesting relationships between functions. The rates of change of the trigonometric functions are other trigonometric functions, but the rates of change of the inverse trigonometric functions are algebraic functions.

## EXAMPLE 1 Evaluating the inverse sine function

Find the exact value of each expression without using a table or a calculator.

**a.** $\sin^{-1}(1/2)$ **b.** $\arcsin(-\sqrt{3}/2)$

### Solution

**a.** The value of $\sin^{-1}(1/2)$ is the number $\alpha$ in the interval $[-\pi/2, \pi/2]$ such that $\sin(\alpha) = 1/2$. We recall that $\sin(\pi/6) = 1/2$, and so $\sin^{-1}(1/2) = \pi/6$. Note that $\pi/6$ is the only value of $\alpha$ in $[-\pi/2, \pi/2]$ for which $\sin(\alpha) = 1/2$.

**b.** The value of $\arcsin(-\sqrt{3}/2)$ is the number $\alpha$ in $[-\pi/2, \pi/2]$ such that $\sin(\alpha) = -\sqrt{3}/2$. Since $\sin(-\pi/3) = -\sqrt{3}/2$, we have $\arcsin(-\sqrt{3}/2) = -\pi/3$. Note that $-\pi/3$ is the only value of $\alpha$ in $[-\pi/2, \pi/2]$ for which $\sin(\alpha) = -\sqrt{3}/2$.

▸**TRY THIS.** Find the exact value of $\sin^{-1}(-\sqrt{2}/2)$. ■

### EXAMPLE 2 Evaluating the inverse sine function

Find the exact value of each expression in degrees without using a table or a calculator.

**a.** $\sin^{-1}(\sqrt{2}/2)$ **b.** $\arcsin(0)$

#### Solution

**a.** The value of $\sin^{-1}(\sqrt{2}/2)$ in degrees is the angle $\alpha$ in the interval $[-90°, 90°]$ such that $\sin(\alpha) = \sqrt{2}/2$. We recall that $\sin(45°) = \sqrt{2}/2$, and so $\sin^{-1}(\sqrt{2}/2) = 45°$.

**b.** The value of $\arcsin(0)$ in degrees is the angle $\alpha$ in the interval $[-90°, 90°]$ for which $\sin(\alpha) = 0$. Since $\sin(0°) = 0$, we have $\arcsin(0) = 0°$.

▸**TRY THIS.** Find the exact value in degrees of $\sin^{-1}(-\sqrt{3}/2)$. ■

In the next example, we use a calculator to find the degree measure of an angle whose sine is given. To obtain degree measure, make sure the calculator is in degree mode. Scientific calculators usually have a key labeled $\sin^{-1}$ that gives values for the inverse sine function.

### EXAMPLE 3 Finding an angle given its sine

Let $\alpha$ be an angle such that $-90° < \alpha < 90°$. In each case, find $\alpha$ to the nearest tenth of a degree.

**a.** $\sin\alpha = 0.88$ **b.** $\sin\alpha = -0.27$

Figure 5.76

#### Solution

**a.** The value of $\sin^{-1}(0.88)$ is the only angle in $[-90°, 90°]$ with a sine of 0.88. Use a calculator in degree mode to get $\alpha = \sin^{-1}(0.88) \approx 61.6°$.

**b.** Use a calculator to get $\alpha = \sin^{-1}(-0.27) \approx -15.7°$.

Figure 5.76 shows how to find the angle in part (a) on a graphing calculator and how to check. Make sure that the mode is degrees.

▸**TRY THIS.** Find $\alpha$ to the nearest tenth of a degree such that $\sin\alpha = 0.3$ and $-90° < \alpha < 90°$. ■

## The Inverse Cosine Function

Since the cosine function is not one-to-one on $(-\infty, \infty)$, we restrict the domain to $[0, \pi]$, where the cosine function is one-to-one and invertible. The graph of the cosine function with this restricted domain is shown in Fig. 5.77(a). Note that the range of

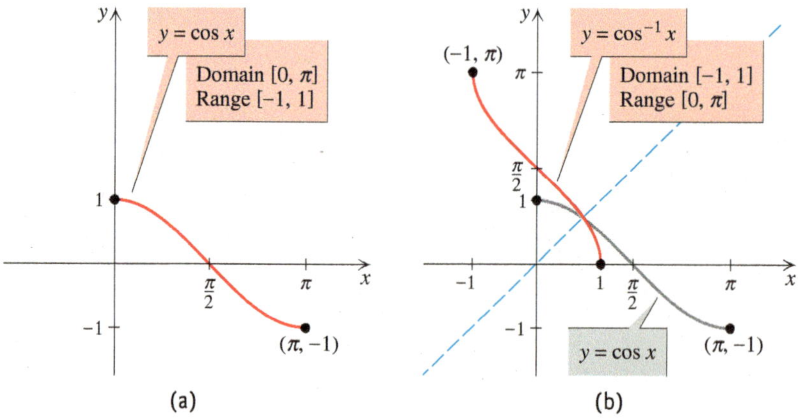


Figure 5.77

the restricted function is $[-1, 1]$. We now define the inverse of $f(x) = \cos x$ for $x$ in $[0, \pi]$ and denote it as $f^{-1}(x) = \cos^{-1} x$ or $f^{-1}(x) = \arccos x$.

**Definition: The Inverse Cosine Function**

> The function $\mathbf{y = cos^{-1} x}$ or $\mathbf{y = arccos\, x}$ is the inverse of the function $y = \cos x$ restricted to $[0, \pi]$. The domain of $y = \cos^{-1} x$ is $[-1, 1]$ and its range is $[0, \pi]$.

If $y = \cos^{-1} x$, then $y$ is the real number in $[0, \pi]$ such that $\cos y = x$. The expression $\cos^{-1} x$ can be read as "the angle whose cosine is $x$" or "the arc length whose cosine is $x$." The graph of $y = \cos^{-1} x$, shown in Fig. 5.77(b), is obtained by reflecting the graph of $y = \cos x$ (restricted to $[0, \pi]$) about the line $y = x$. We will assume that $\cos^{-1} x$ is a real number unless indicated otherwise.

### EXAMPLE 4 Evaluating the inverse cosine function

Find the exact value of each expression without using a table or a calculator.

**a.** $\cos^{-1}(-1)$ **b.** $\arccos(-1/2)$ **c.** $\cos^{-1}(\sqrt{2}/2)$

**Solution**

**a.** The value of $\cos^{-1}(-1)$ is the number $\alpha$ in $[0, \pi]$ such that $\cos(\alpha) = -1$. We recall that $\cos(\pi) = -1$, and so $\cos^{-1}(-1) = \pi$.

**b.** The value of $\arccos(-1/2)$ is the number $\alpha$ in $[0, \pi]$ such that $\cos(\alpha) = -1/2$. We recall that $\cos(2\pi/3) = -1/2$, and so $\arccos(-1/2) = 2\pi/3$.

**c.** Since $\cos(\pi/4) = \sqrt{2}/2$, we have $\cos^{-1}(\sqrt{2}/2) = \pi/4$.

▸**TRY THIS.** Find the exact value of $\cos^{-1}(-\sqrt{3}/2)$. ■

In the next example, we use a calculator to find the degree measure of an angle whose cosine is given. Most scientific calculators have a key labeled $\cos^{-1}$ that gives values for the inverse cosine function. To get the degree measure, make sure the calculator is in degree mode.

### EXAMPLE 5 Finding an angle given its cosine

In each case find the angle $\alpha$ to the nearest tenth of a degree, given that $0° < \alpha < 180°$.

**a.** $\cos \alpha = 0.23$ **b.** $\cos \alpha = -0.82$

**Solution**

**a.** Since $\cos^{-1}(0.23)$ is the unique angle in $[0°, 180°]$ with a cosine of 0.23, $\alpha = \cos^{-1}(0.23) \approx 76.7°$.

**b.** Use a calculator to get $\alpha = \cos^{-1}(-0.82) \approx 145.1°$.

Figure 5.78

Figure 5.78 shows how to find $\cos^{-1}(0.23)$ on a graphing calculator and how to check. Make sure that the mode is degrees.

▸**TRY THIS.** Find $\alpha$ to the nearest tenth of a degree such that $\cos \alpha = -0.4$ and $0° < \alpha < 180°$. ■

## Inverses of Tangent, Cotangent, Secant, and Cosecant

Since all of the trigonometric functions are periodic, they must all be restricted to a domain where they are one-to-one before the inverse functions can be defined. There is more than one way to choose a domain to get a one-to-one function, but we will use the most common restrictions. The restricted domain for $y = \tan x$ is $(-\pi/2, \pi/2)$, for $y = \csc x$ it is $[-\pi/2, 0) \cup (0, \pi/2]$, for $y = \sec x$ it is $[0, \pi/2) \cup (\pi/2, \pi]$, and for $y = \cot x$ it is $(0, \pi)$. The functions $\tan^{-1}$, $\cot^{-1}$, $\sec^{-1}$, and $\csc^{-1}$ are defined to be the inverses of these restricted functions. The notations arctan, arccot, arcsec, and arccsc are also used for these inverse functions. The graphs of all six inverse functions, with their domains and ranges, are shown in the accompanying Function Gallery.

When studying inverse trigonometric functions, you should first learn to evaluate $\sin^{-1}$, $\cos^{-1}$, and $\tan^{-1}$. Then use the identities

$$\csc \alpha = \frac{1}{\sin \alpha}, \qquad \sec \alpha = \frac{1}{\cos \alpha}, \qquad \text{and} \qquad \cot \alpha = \frac{1}{\tan \alpha}$$

to evaluate $\csc^{-1}$, $\sec^{-1}$, and $\cot^{-1}$. For example, $\sin(\pi/6) = 1/2$ and $\csc(\pi/6) = 2$. So the angle whose cosecant is 2 is the same as the angle whose sine is $1/2$. In symbols,

$$\csc^{-1}(2) = \sin^{-1}\left(\frac{1}{2}\right) = \frac{\pi}{6}.$$

In general, $\csc^{-1}x = \sin^{-1}(1/x)$. Likewise, $\sec^{-1}x = \cos^{-1}(1/x)$. For the inverse cotangent, $\cot^{-1}x = \tan^{-1}(1/x)$ only for $x > 0$, because of the choice of $(0, \pi)$ as the range of the inverse cotangent. We have $\cot^{-1}(0) = \pi/2$ and, if $x < 0$, $\cot^{-1}(x) = \tan^{-1}(1/x) + \pi$. These identities are summarized following the Function Gallery.

### FUNCTION gallery... Inverse Trigonometric Functions

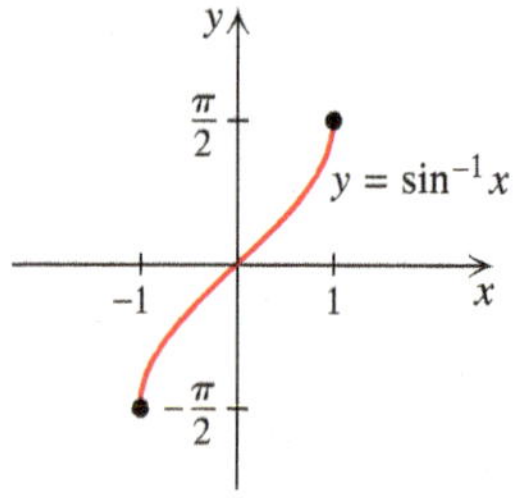


Domain $[-1, 1]$
Range $\left[-\frac{\pi}{2}, \frac{\pi}{2}\right]$

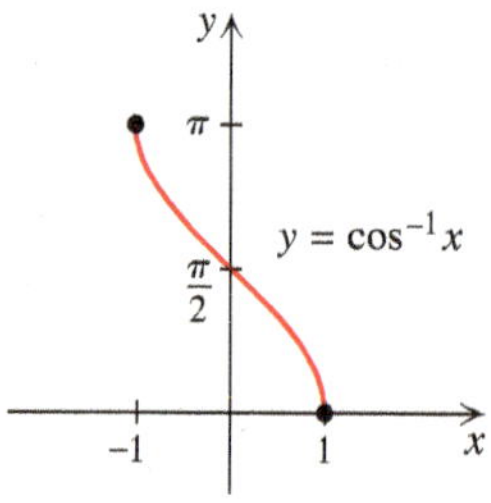


Domain $[-1, 1]$
Range $[0, \pi]$

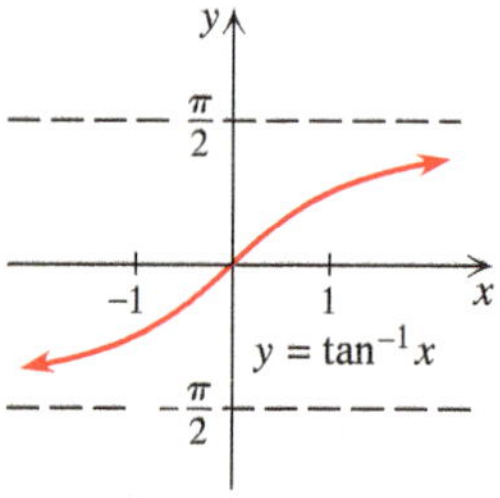


Domain $(-\infty, \infty)$
Range $\left(-\frac{\pi}{2}, \frac{\pi}{2}\right)$

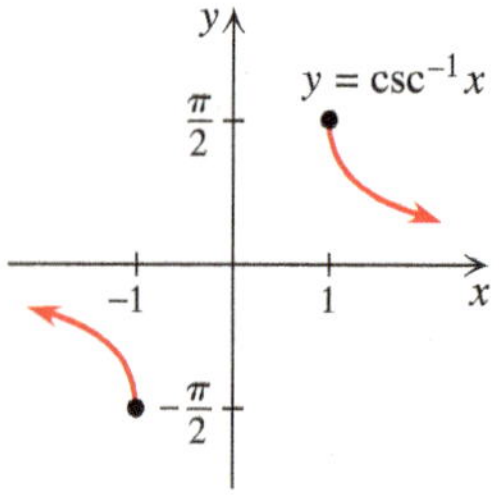


Domain $(-\infty, -1] \cup [1, \infty)$
Range $\left[-\frac{\pi}{2}, 0\right) \cup \left(0, \frac{\pi}{2}\right]$

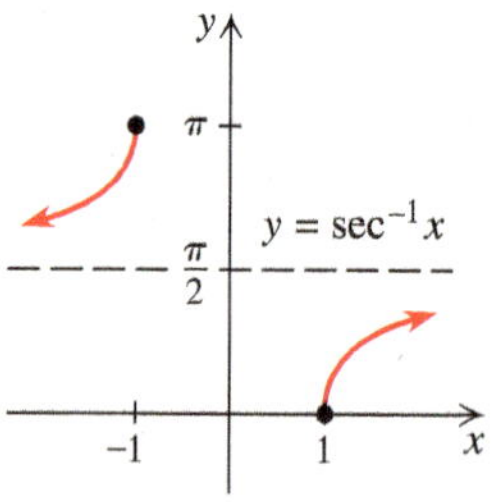


Domain $(-\infty, -1] \cup [1, \infty)$
Range $\left[0, \frac{\pi}{2}\right) \cup \left(\frac{\pi}{2}, \pi\right]$

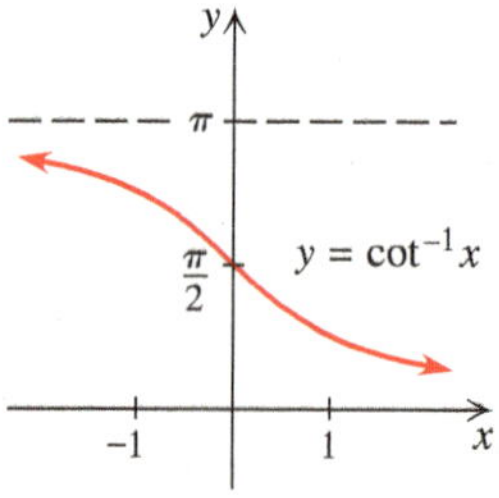


Domain $(-\infty, \infty)$
Range $(0, \pi)$

We can see another relationship between $\cot^{-1}$ and $\tan^{-1}$ from their graphs in the Function Gallery. Notice that the graph of $y = \cot^{-1}(x)$ can be obtained by reflecting the graph of $y = \tan^{-1}(x)$ about the $x$-axis and then translating $\pi/2$ units upward. So $\cot^{-1}(x) = -\tan^{-1}(x) + \pi/2$ or $\cot^{-1}(x) = \pi/2 - \tan^{-1}(x)$.

**Identities for the Inverse Functions**

$$\csc^{-1}(x) = \sin^{-1}(1/x) \text{ for } |x| \geq 1$$

$$\sec^{-1}(x) = \cos^{-1}(1/x) \text{ for } |x| \geq 1$$

$$\cot^{-1}(x) = \begin{cases} \tan^{-1}(1/x) & \text{for } x > 0 \\ \tan^{-1}(1/x) + \pi & \text{for } x < 0 \\ \pi/2 & \text{for } x = 0 \end{cases}$$

$$\cot^{-1}(x) = \pi/2 - \tan^{-1}(x)$$

## EXAMPLE 6 Evaluating the inverse functions

Find the exact value of each expression without using a table or a calculator.

**a.** $\tan^{-1}(1)$ **b.** $\text{arcsec}(-2)$ **c.** $\csc^{-1}(\sqrt{2})$ **d.** $\text{arccot}(-1/\sqrt{3})$

### Solution

**a.** Since $\tan(\pi/4) = 1$ and since $\pi/4$ is in the range of $\tan^{-1}$, we have

$$\tan^{-1}(1) = \frac{\pi}{4}.$$

**b.** To evaluate the inverse secant, we use the identity $\sec^{-1}(x) = \cos^{-1}(1/x)$. In this case, the arc whose secant is $-2$ is the same as the arc whose cosine is $-1/2$. So we must find $\cos^{-1}(-1/2)$. Since $\cos(2\pi/3) = -1/2$ and since $2\pi/3$ is in the range of arccos, we have

$$\text{arcsec}(-2) = \arccos\left(-\frac{1}{2}\right) = \frac{2\pi}{3}.$$

**c.** To evaluate $\csc^{-1}(\sqrt{2})$, we use the identity $\csc^{-1}(x) = \sin^{-1}(1/x)$ with $x = \sqrt{2}$. So we must find $\sin^{-1}(1/\sqrt{2})$. Since $\sin(\pi/4) = 1/\sqrt{2}$ and since $\pi/4$ is in the range of $\csc^{-1}$,

$$\csc^{-1}(\sqrt{2}) = \sin^{-1}\left(\frac{1}{\sqrt{2}}\right) = \frac{\pi}{4}.$$

**d.** If $x$ is negative, we use the identity $\cot^{-1}(x) = \tan^{-1}(1/x) + \pi$. Since $x = -1/\sqrt{3}$, we must find $\tan^{-1}(-\sqrt{3})$. Since $\tan^{-1}(-\sqrt{3}) = -\pi/3$, we get

$$\text{arccot}\left(\frac{-1}{\sqrt{3}}\right) = \tan^{-1}(-\sqrt{3}) + \pi = -\frac{\pi}{3} + \pi = \frac{2\pi}{3}.$$

Note that $\cot(-\pi/3) = \cot(2\pi/3) = -1/\sqrt{3}$, but $\text{arccot}(-1/\sqrt{3}) = 2\pi/3$ because $2\pi/3$ is in the range of the function arccot.

▶**TRY THIS.** Find the exact value of $\arctan(\sqrt{3})$. ■

The functions $\sin^{-1}$, $\cos^{-1}$, and $\tan^{-1}$ are available on scientific and graphing calculators. The calculator values of the inverse functions are given in degrees or radians, depending on the mode setting. We will assume that the values of the inverse functions are to be in radians unless indicated otherwise. Calculators use the domains and ranges of these functions defined here. The functions $\sec^{-1}$, $\csc^{-1}$, and $\cot^{-1}$ are generally not available on a calculator, and expressions involving these functions must be written in terms of $\sin^{-1}$, $\cos^{-1}$, and $\tan^{-1}$ by using the identities.

### EXAMPLE 7 Evaluating the inverse functions with a calculator

Find the approximate value of each expression rounded to four decimal places.

**a.** $\sin^{-1}(0.88)$ **b.** $\text{arccot}(2.4)$ **c.** $\csc^{-1}(4)$ **d.** $\cot^{-1}(0)$

**Solution**

**a.** Use the inverse sine function in radian mode to get $\sin^{-1}(0.88) \approx 1.0759$.
**b.** Use the identity $\cot^{-1}(x) = \tan^{-1}(1/x)$ because $2.4 > 0$:

$$\text{arccot}(2.4) = \tan^{-1}\left(\frac{1}{2.4}\right) \approx 0.3948$$

Figure 5.79

**c.** $\csc^{-1}(4) = \sin^{-1}(0.25) \approx 0.2527$
**d.** The value of $\cot^{-1}(0)$ cannot be found on a calculator. But we know that $\cot(\pi/2) = 0$ and $\pi/2$ is in the range $(0, \pi)$ for $\cot^{-1}$, so we have $\cot^{-1}(0) = \pi/2$.

The expressions in parts (a), (b), and (c) are shown on a graphing calculator in Fig. 5.79.

▶**TRY THIS.** Find $\cot^{-1}(1.3)$ to four decimal places. ■

## Compositions of Functions

One trigonometric function can be followed by another to form a composition of functions. We can evaluate an expression such as $\sin(\sin(\alpha))$ because the sine of the real number $\sin(\alpha)$ is defined. However, it is more common to have a composition of a trigonometric function and an inverse trigonometric function. For example, $\tan^{-1}(\alpha)$ is the angle whose tangent is $\alpha$, and so $\sin(\tan^{-1}(\alpha))$ is the sine of the angle whose tangent is $\alpha$.

### EXAMPLE 8 Evaluating compositions of functions

Find the exact value of each composition without using a table or a calculator.

**a.** $\sin(\tan^{-1}(0))$ **b.** $\arcsin(\cos(\pi/6))$ **c.** $\tan(\sec^{-1}(\sqrt{2}))$
**d.** $\arcsin(\sin(4\pi/3))$

**Solution**

**a.** Since $\tan(0) = 0$, $\tan^{-1}(0) = 0$. Therefore,

$$\sin(\tan^{-1}(0)) = \sin(0) = 0.$$

**b.** Since $\cos(\pi/6) = \sqrt{3}/2$, we have

$$\arcsin\left(\cos\left(\frac{\pi}{6}\right)\right) = \arcsin\left(\frac{\sqrt{3}}{2}\right) = \frac{\pi}{3}.$$

**Figure 5.80**

**c.** To find $\sec^{-1}(\sqrt{2})$, we use the identity $\sec^{-1}(x) = \cos^{-1}(1/x)$. Since $\cos(\pi/4) = 1/\sqrt{2}$, we have $\cos^{-1}(1/\sqrt{2}) = \pi/4$ and $\sec^{-1}(\sqrt{2}) = \pi/4$. Therefore,

$$\tan(\sec^{-1}(\sqrt{2})) = \tan\left(\frac{\pi}{4}\right) = 1.$$

You can check parts (a), (b), and (c) with a graphing calculator, as shown in Fig. 5.80. Note that $\pi/3 \approx 1.047$. □

**d.** Since $\sin(4\pi/3) = -\sin(\pi/3)$ we have $\sin(4\pi/3) = -\sqrt{3}/2$. Now the angle in the interval $[-\pi/2, \pi/2]$ whose sine is $-\sqrt{3}/2$ is $-\pi/3$. So

$$\arcsin(\sin(4\pi/3)) = \arcsin(-\sqrt{3}/2) = -\pi/3.$$

Note that $\arcsin(\sin(4\pi/3)) \neq 4\pi/3$.

▸**TRY THIS.** Find the exact value of $\sin^{-1}(\sin(5\pi/6))$. ■

The composition of $y = \sin x$ restricted to $[-\pi/2, \pi/2]$ and $y = \sin^{-1} x$ is the identity function, because they are inverse functions of each other. So

$$\sin^{-1}(\sin x) = x \quad \text{for } x \text{ in } [-\pi/2, \pi/2]$$

and

$$\sin(\sin^{-1} x) = x \quad \text{for } x \text{ in } [-1, 1].$$

Note that if $x$ is not in $[-\pi/2, \pi/2]$, then the sine function followed by the inverse sine function is not the identity function. For example, $\sin^{-1}(\sin(2\pi/3)) = \pi/3$.

The general sine function $f(x) = A\sin[B(x - C)] + D$ is a composition of a trigonometric function and several algebraic functions. Since the sine function has an inverse and the algebraic functions have inverses, we can find the inverse for a general sine function provided it is restricted to a suitable domain.

## EXAMPLE 9 The inverse of a general sine function

Find the inverse of $f(x) = 3\sin(2x) + 5$, where $-\pi/4 \le x \le \pi/4$, and determine the domain of $f^{-1}$.

### Solution

Interchange $x$ and $y$ in $y = 3\sin(2x) + 5$, and then solve for $y$:

$$x = 3\sin(2y) + 5 \qquad \text{Switch } x \text{ and } y.$$

$$\frac{x-5}{3} = \sin(2y)$$

$$2y = \sin^{-1}\left(\frac{x-5}{3}\right) \qquad \text{Definition of } \sin^{-1}$$

$$y = \frac{1}{2}\sin^{-1}\left(\frac{x-5}{3}\right)$$

$$f^{-1}(x) = \frac{1}{2}\sin^{-1}\left(\frac{x-5}{3}\right)$$

Since $\sin(2x)$ is between $-1$ and 1, the range of $f$ is $[2, 8]$ and the domain of $f^{-1}$ is $[2, 8]$.

▸**TRY THIS.** Find the inverse of $f(x) = 2\cos(3x) - 1$ where $0 \le x \le \pi/3$ and find the domain of $f^{-1}$. ■

## FOR thought... True or False? Explain.

1. $\sin^{-1}(0) = \sin(0)$
2. $\sin(3\pi/4) = 1/\sqrt{2}$
3. $\cos^{-1}(0) = 1$
4. $\sin^{-1}(\sqrt{2}/2) = 135°$
5. $\cot^{-1}(5) = \dfrac{1}{\tan^{-1}(5)}$
6. $\sec^{-1}(5) = \cos^{-1}(0.2)$
7. $\sin(\cos^{-1}(\sqrt{2}/2)) = 1/\sqrt{2}$
8. $\sec(\sec^{-1}(2)) = 2$
9. The functions $f(x) = \sin^{-1}x$ and $f^{-1}(x) = \sin x$ are inverse functions.
10. The secant and cosecant functions are inverses of each other.

## EXERCISE 5.5

*Fill in the blank.*

1. The ________ of $y = \arcsin(x)$ is $[-1, 1]$.
2. The ________ of $y = \arccos(x)$ is $[0, \pi]$.
3. The ________ of $y = \arctan(x)$ is $(-\infty, \infty)$.
4. The ________ of $y = \tan^{-1}(x)$ is $(-\pi/2, \pi/2)$.

*Find the exact value of each expression without using a calculator or table.*

5. $\sin^{-1}(-1/2)$
6. $\sin^{-1}(0)$
7. $\arcsin(1/2)$
8. $\arcsin(\sqrt{3}/2)$
9. $\arcsin(\sqrt{2}/2)$
10. $\arcsin(1)$

*Find the exact value of each expression in degrees without using a calculator or table.*

11. $\sin^{-1}(-1/\sqrt{2})$
12. $\sin^{-1}(\sqrt{3}/2)$
13. $\arcsin(1/2)$
14. $\arcsin(-1)$
15. $\sin^{-1}(0)$
16. $\sin^{-1}(\sqrt{2}/2)$

*In each case find $\alpha$ to the nearest tenth of a degree, where $-90° \le \alpha \le 90°$.*

17. $\sin\alpha = -1/3$
18. $\sin\alpha = 0.4138$
19. $\sin\alpha = 0.5682$
20. $\sin\alpha = -0.34$

*Find the exact value of each expression without using a calculator or table.*

21. $\cos^{-1}(-\sqrt{2}/2)$
22. $\cos^{-1}(1)$
23. $\arccos(1/2)$
24. $\arccos(-\sqrt{3}/2)$
25. $\arccos(-1)$
26. $\arccos(0)$

*Find the exact value of each expression in degrees without using a calculator or table.*

27. $\cos^{-1}(-\sqrt{2}/2)$
28. $\cos^{-1}(\sqrt{3}/2)$
29. $\arccos(-1)$
30. $\arccos(0)$
31. $\cos^{-1}(-1/2)$
32. $\cos^{-1}(1)$

*In each case find $\alpha$ to the nearest tenth of a degree, where $0° \le \alpha \le 180°$.*

33. $\cos\alpha = -0.993$
34. $\cos\alpha = 0.7392$
35. $\cos\alpha = 0.001$
36. $\cos\alpha = -0.499$

*Find the exact value of each expression without using a calculator or table.*

37. $\tan^{-1}(-1)$
38. $\cot^{-1}(1/\sqrt{3})$
39. $\sec^{-1}(2)$
40. $\csc^{-1}(2/\sqrt{3})$
41. $\text{arcsec}(\sqrt{2})$
42. $\arctan(-1/\sqrt{3})$
43. $\text{arccsc}(-2)$
44. $\text{arccot}(-\sqrt{3})$
45. $\tan^{-1}(0)$
46. $\sec^{-1}(1)$
47. $\csc^{-1}(1)$
48. $\csc^{-1}(-1)$
49. $\cot^{-1}(-1)$
50. $\cot^{-1}(0)$
51. $\cot^{-1}(-\sqrt{3}/3)$
52. $\cot^{-1}(1)$

*Find the approximate value of each expression with a calculator. Round answers to two decimal places.*

53. $\arcsin(0.5682)$
54. $\sin^{-1}(-0.4138)$
55. $\cos^{-1}(-0.993)$
56. $\cos^{-1}(0.7392)$

57. $\tan^{-1}(-0.1396)$
58. $\cot^{-1}(4.32)$
59. $\sec^{-1}(-3.44)$
60. $\csc^{-1}(6.8212)$
61. $\text{arcsec}(\sqrt{6})$
62. $\arctan(-2\sqrt{7})$
63. $\text{arccsc}(-2\sqrt{2})$
64. $\text{arccot}(-\sqrt{5})$
65. $\text{arccot}(-12)$
66. $\text{arccot}(0.001)$
67. $\cot^{-1}(15.6)$
68. $\cot^{-1}(-1.01)$

*Find the exact value of each composition without using a calculator or table.*

69. $\tan(\arccos(1/2))$
70. $\sec(\arcsin(1/\sqrt{2}))$
71. $\sin^{-1}(\cos(2\pi/3))$
72. $\tan^{-1}(\sin(\pi/2))$
73. $\cot^{-1}(\cot(\pi/6))$
74. $\sec^{-1}(\sec(\pi/3))$
75. $\arcsin(\sin(3\pi/4))$
76. $\arccos(\cos(-\pi/3))$
77. $\tan(\arctan(1))$
78. $\cot(\text{arccot}(0))$
79. $\cos^{-1}(\cos(3\pi/2))$
80. $\sin(\csc^{-1}(-2))$
81. $\cos(2\sin^{-1}(\sqrt{2}/2))$
82. $\tan(2\cos^{-1}(1/2))$
83. $\sin^{-1}(2\sin(\pi/6))$
84. $\cos^{-1}(0.5\tan(\pi/4))$

*Use a calculator to find the approximate value of each composition. Round answers to four decimal places. Some of these expressions are undefined.*

85. $\sin(\cos^{-1}(0.45))$
86. $\cos(\tan^{-1}(44.33))$
87. $\sec^{-1}(\cos(\pi/9))$
88. $\csc^{-1}(\sec(\pi/8))$
89. $\tan(\sin^{-1}(-0.7))$
90. $\csc^{-1}(\sin(3\pi/7))$
91. $\cot(\cos^{-1}(-1/\sqrt{7}))$
92. $\sin^{-1}(\csc(1.08))$
93. $\cot(\csc^{-1}(3.6))$
94. $\csc(\sec^{-1}(-2.4))$
95. $\sec(\cot^{-1}(5.2))$
96. $\csc(\cot^{-1}(-3.1))$

*Find the inverse of each function and state its domain.*

97. $f(x) = \sin(2x)$ for $-\frac{\pi}{4} \le x \le \frac{\pi}{4}$

98. $f(x) = \cos(3x)$ for $0 \le x \le \frac{\pi}{3}$

99. $f(x) = 3 + \tan(\pi x)$ for $-\frac{1}{2} < x < \frac{1}{2}$

100. $f(x) = 2 - \sin(\pi x - \pi)$ for $\frac{1}{2} \le x \le \frac{3}{2}$

101. $f(x) = \sin^{-1}(x/2) + 3$ for $-2 \le x \le 2$

102. $f(x) = 2\cos^{-1}(5x) + 3$ for $-\frac{1}{5} \le x \le \frac{1}{5}$

*In a circle with radius r, a central angle θ intercepts a chord of length c, where $\theta = \cos^{-1}\left(1 - \frac{c^2}{2r^2}\right)$. Use this formula for Exercises 103 and 104.*

103. An airplane at 2000 feet flies directly over a gun that has a range of 2400 feet, as shown in the figure. What is the measure in degrees of the angle for which the airplane is within range of the gun?

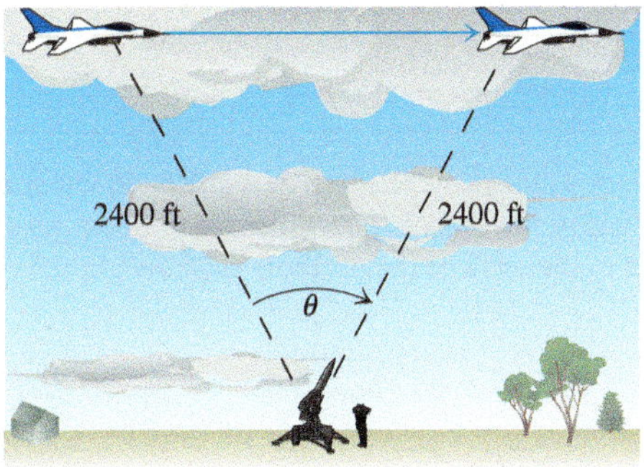


**Figure for Exercise 103**

104. A triangle has two sides that are both 5.2 meters long and one side with a length of 1.3 meters. Find the measure in degrees for the smallest angle of the triangle.

## FOR WRITING/DISCUSSION

105. Graph the function $y = \sin(\sin^{-1} x)$ for $-2\pi \le x \le 2\pi$ and explain your result.

106. Graph the function $y = \sin^{-1}(\sin x)$ for $-2\pi \le x \le 2\pi$ and explain your result.

107. Graph $y = \sin^{-1}(1/x)$ and explain why the graph looks like the graph of $y = \csc^{-1} x$ shown in the Function Gallery on page 394.

108. Graph $y = \tan^{-1}(1/x)$ and explain why the graph does *not* look like the graph of $y = \cot^{-1} x$ shown in the Function Gallery on page 394.

## ▶ RETHINKING

109. Evaluate without a calculator. Some of these expressions are undefined.
   a. $\cos(\pi)$
   b. $\sin(3\pi/4)$
   c. $\tan(\pi/3)$
   d. $\tan(\pi/2)$
   e. $\sec(2\pi/3)$
   f. $\csc(\pi)$
   g. $\cot(5\pi/6)$
   h. $\sin(-\pi/4)$

**110.** The earth travels around the sun on a path that is roughly circular with a radius of 93 million miles. How fast is the earth moving through space? Round to the nearest thousand miles per hour.

**111.** Find the smallest positive angle in radians that is coterminal with $-23\pi/6$.

**112.** Convert the radian measure $7\pi/4$ to degrees.

**113.** Determine the period, asymptotes, and range for the graph of $y = 3\sec(2x - \pi)$.

**114.** If $\beta$ is an angle in standard position such that $\sin(\beta) = 1/4$ and $\beta$ terminates in quadrant II, then what is the exact value of $\cos(\beta)$?

## THINKING OUTSIDE THE BOX XL

*Clear Sailing* A sailor plans to install a windshield wiper on a porthole that has radius 1 foot. The wiper blade of length $x$ feet is to be attached to the edge of the porthole as shown in the figure. The area cleaned by the blade is a sector of a circle centered at the point of attachment.

**a.** Write the area cleaned by the blade as a function of $x$.

**b.** If the blade cleans half of the window, then what is the exact length of the blade?

**c.** Use a graphing calculator to find the length for the blade that would maximize the area cleaned?

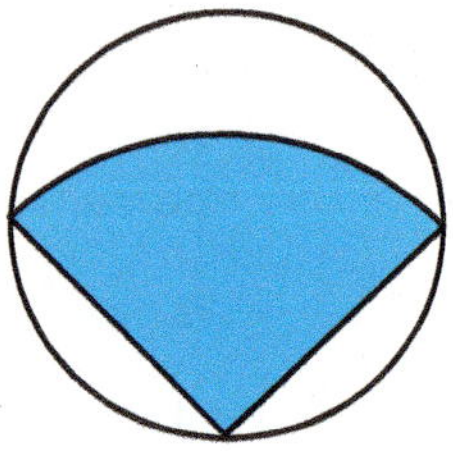

**Figure for Thinking Outside the Box XL**

## POP QUIZ 5.5

**1.** Find $f^{-1}(x)$ if $f(x) = \cos(2x)$ for $0 \le x \le \pi/2$.

*Find the exact value.*

**2.** $\sin^{-1}(-1)$

**3.** $\sin^{-1}(1/2)$

**4.** $\arccos(-1)$

**5.** $\arctan(-1)$

**6.** $\tan(\arcsin(1/2))$

**7.** $\sin^{-1}(\sin(3\pi/4))$

## LINKING concepts... For Individual or Group Explorations

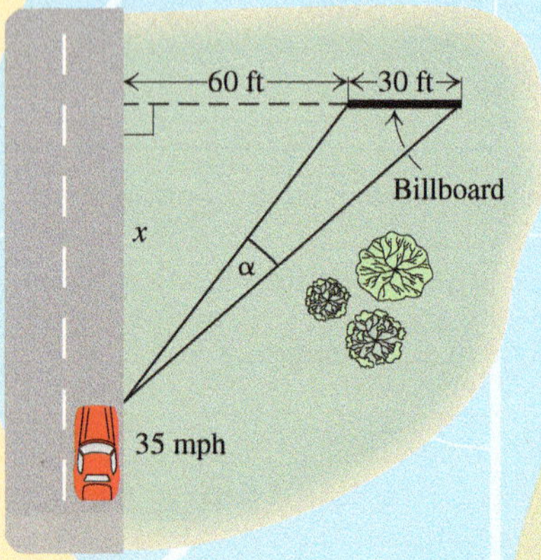


### Maximizing the Viewing Angle

*A billboard that is* 30 ft *wide is placed* 60 ft *from a highway, as shown in the figure. The billboard is easiest to read from the highway when the viewing angle* $\alpha$ *is larger than* 7°.

**a)** Using results from the next section, we can show that

$$\alpha = \tan^{-1}(90/x) - \tan^{-1}(60/x).$$

Graph this function.

**b)** For what approximate values of $x$ is $\alpha$ greater than 7°?

**c)** For approximately how long does a motorist traveling at 35 mph have a viewing angle larger than 7°?

# 5.6 Right Triangle Trigonometry

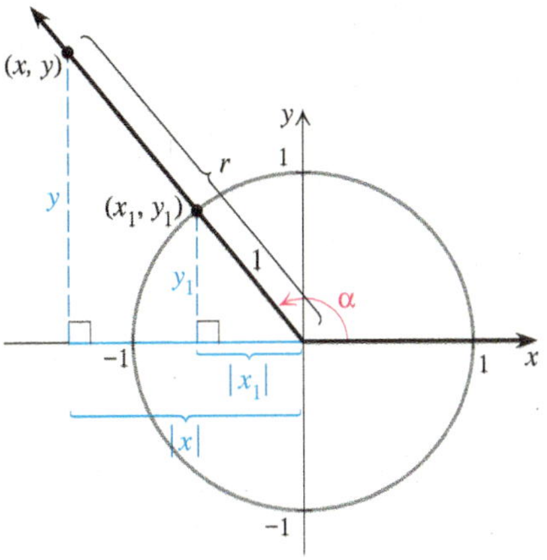


Figure 5.81

One reason trigonometry was invented was to determine the measures of sides and angles of geometric figures without actually measuring them. In this section we study right triangles (triangles that have a 90° angle) and see what information is needed to determine the measures of all unknown sides and angles of a right triangle.

## Trigonometric Ratios

We defined the sine and cosine functions for an angle in standard position in terms of the point at which the terminal side intersects the unit circle. However, it is often the case that we do not know that point, but we do know a different point on the terminal side. When this situation occurs, we can find the values for the sine and cosine of the angle using trigonometric ratios.

Figure 5.81 shows an angle $\alpha$ in quadrant II with the terminal side passing through the point $(x, y)$ and intersecting the unit circle at $(x_1, y_1)$. If we draw vertical line segments down to the $x$-axis, we form two similar right triangles, as shown in Fig. 5.81. If $r$ is the distance from $(x, y)$ to the origin, then $r = \sqrt{x^2 + y^2}$. The lengths of the legs in the larger triangle are $y$ and $|x|$, and its hypotenuse is $r$. The lengths of the legs in the smaller triangle are $y_1$ and $|x_1|$, and its hypotenuse is 1. Since ratios of the lengths of corresponding sides of similar triangles are equal, we have

$$\frac{y_1}{y} = \frac{1}{r} \quad \text{and} \quad \frac{|x_1|}{|x|} = \frac{1}{r}.$$

The first equation can be written as $y_1 = y/r$ and the second as $x_1 = x/r$. (Since $x_1$ and $x$ are both negative in this case, the absolute value symbols can be omitted.) Since $y_1 = \sin\alpha$ and $x_1 = \cos\alpha$, we get

$$\sin\alpha = \frac{y}{r} \quad \text{and} \quad \cos\alpha = \frac{x}{r}.$$

This argument can be repeated in each quadrant with $(x, y)$ chosen inside, outside, or on the unit circle. These ratios also give the correct sine or cosine if $\alpha$ is a quadrantal angle. Since all other trigonometric functions are related to the sine and cosine, their values can also be obtained from $x$, $y$, and $r$.

**Theorem: Trigonometric Ratios**

If $(x, y)$ is any point other than the origin on the terminal side of an angle $\alpha$ in standard position and $r = \sqrt{x^2 + y^2}$, then

$$\sin\alpha = \frac{y}{r}, \quad \cos\alpha = \frac{x}{r}, \quad \text{and} \quad \tan\alpha = \frac{y}{x}\ (x \neq 0).$$

### EXAMPLE 1 Trigonometric ratios

Find the values of the six trigonometric functions of the angle $\alpha$ in standard position whose terminal side passes through $(4, -2)$.

**Solution**

Use $x = 4$, $y = -2$, and $r = \sqrt{4^2 + (-2)^2} = \sqrt{20} = 2\sqrt{5}$ to get

$$\sin\alpha = \frac{-2}{2\sqrt{5}} = -\frac{\sqrt{5}}{5}, \quad \cos\alpha = \frac{4}{2\sqrt{5}} = \frac{2\sqrt{5}}{5}, \quad \text{and}$$

$$\tan\alpha = \frac{-2}{4} = -\frac{1}{2}.$$

Since cosecant, secant, and cotangent are the reciprocals of sine, cosine, and tangent,

$$\csc\alpha = \frac{1}{\sin\alpha} = -\frac{5}{\sqrt{5}} = -\sqrt{5}, \qquad \sec\alpha = \frac{1}{\cos\alpha} = \frac{5}{2\sqrt{5}} = \frac{\sqrt{5}}{2}, \qquad \text{and}$$

$$\cot\alpha = \frac{1}{\tan\alpha} = -2.$$

▸**TRY THIS.** Find $\sin\alpha$, $\cos\alpha$, and $\tan\alpha$ if $\alpha$ is an angle in standard position whose terminal side passes through $(2, 3)$. ■

## Right Triangles

So far the trigonometric functions have been tied to a coordinate system and an angle in standard position. Trigonometric ratios can also be used to evaluate the trigonometric functions for an acute angle of a right triangle without having the angle or the triangle located in a coordinate system.

Consider a right triangle with acute angle $\alpha$, legs of length $x$ and $y$, and hypotenuse $r$, as shown in Fig. 5.82(a). If this triangle is positioned in a coordinate system, as in Fig. 5.82(b), then $(x, y)$ is a point on the terminal side of $\alpha$ and

$$\sin(\alpha) = \frac{y}{r}, \qquad \cos(\alpha) = \frac{x}{r}, \qquad \text{and} \qquad \tan(\alpha) = \frac{y}{x}.$$

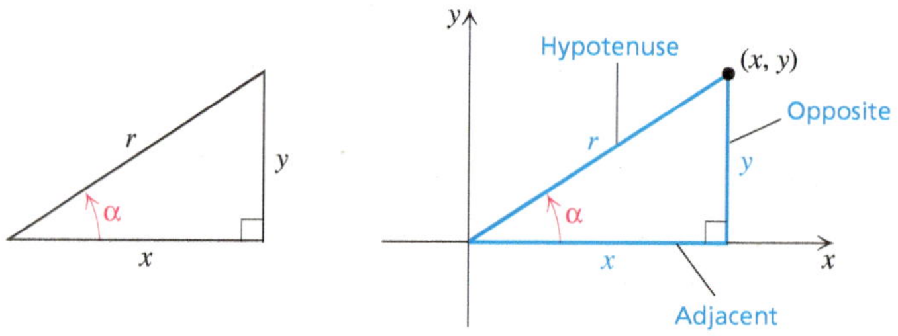


Figure 5.82

However, the values of the trigonometric functions are simply ratios of the lengths of the sides of the right triangle and it is not necessary to move the triangle to a coordinate system to find them. Notice that $y$ is the length of the side **opposite** the angle $\alpha$, $x$ is the length of the side **adjacent** to $\alpha$, and $r$ is the length of the **hypotenuse.** We use the abbreviations opp, adj, and hyp to represent the lengths of these sides in the following theorem.

**Theorem: Trigonometric Functions of an Acute Angle of a Right Triangle**

If $\alpha$ is an acute angle of a right triangle, then

$$\sin\alpha = \frac{\text{opp}}{\text{hyp}}, \qquad \cos\alpha = \frac{\text{adj}}{\text{hyp}}, \qquad \text{and} \qquad \tan\alpha = \frac{\text{opp}}{\text{adj}}.$$

Students often use SOH-CAH-TOA to help them remember this theorem.

Since the cosecant, secant, and cotangent are the reciprocals of the sine, cosine, and tangent, respectively, the values of all six trigonometric functions can be found for an acute angle of a right triangle.

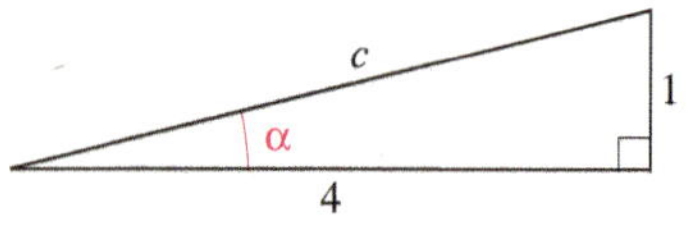


Figure 5.83

### EXAMPLE 2 Trigonometric functions in a right triangle

Find the values of all six trigonometric functions for the angle $\alpha$ of the right triangle with legs of length 1 and 4, as shown in Fig. 5.83.

### Solution

The length of the hypotenuse is $c = \sqrt{4^2 + 1^2} = \sqrt{17}$. Since the length of the side opposite $\alpha$ is 1 and the length of the adjacent side is 4, we have

$$\sin\alpha = \frac{\text{opp}}{\text{hyp}} = \frac{1}{\sqrt{17}} = \frac{\sqrt{17}}{17}, \qquad \cos\alpha = \frac{\text{adj}}{\text{hyp}} = \frac{4}{\sqrt{17}} = \frac{4\sqrt{17}}{17},$$

$$\tan\alpha = \frac{\text{opp}}{\text{adj}} = \frac{1}{4}, \qquad \csc\alpha = \frac{1}{\sin\alpha} = \sqrt{17},$$

$$\sec\alpha = \frac{1}{\cos\alpha} = \frac{\sqrt{17}}{4}, \qquad \cot\alpha = \frac{1}{\tan\alpha} = 4.$$

▸**TRY THIS.** A right triangle has legs with lengths 2 and 4. Find $\sin\alpha$, $\cos\alpha$, and $\tan\alpha$ if $\alpha$ is the angle opposite the side of length 4. ■

## Solving a Right Triangle

The values of the trigonometric functions for an acute angle of a right triangle are determined by ratios of lengths of sides of the triangle. We can use those ratios along with the inverse trigonometric functions to find missing parts of a right triangle in which some of the measures of angles or lengths of sides are known. Finding all of the unknown lengths of sides or measures of angles is called **solving the triangle.** A triangle can be solved only if enough information is given to determine a unique triangle. For example, the lengths of the sides in a 30-60-90 triangle cannot be found because there are infinitely many such triangles of different sizes. However, the lengths of the missing sides in a 30-60-90 triangle with a hypotenuse of 6 will be found in Example 3.

In solving right triangles, we usually name the acute angles $\alpha$ and $\beta$ and the lengths of the sides opposite those angles $a$ and $b$. The 90° angle is $\gamma$, and the length of the side opposite $\gamma$ is $c$.

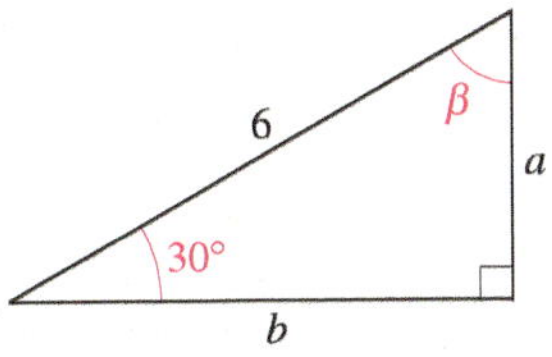

Figure 5.84

### EXAMPLE 3 Solving a right triangle

Solve the right triangle in which $\alpha = 30°$ and $c = 6$.

### Solution

The triangle is shown in Fig. 5.84. Since $\alpha = 30°$, $\gamma = 90°$, and the sum of the measures of the angles of any triangle is 180°, we have $\beta = 60°$. Since $\sin\alpha = \text{opp}/\text{hyp}$, we get $\sin 30° = a/6$ and

$$a = 6 \cdot \sin 30° = 6 \cdot \frac{1}{2} = 3.$$

Since $\cos\alpha = \text{adj}/\text{hyp}$, we get $\cos 30° = b/6$ and

$$b = 6 \cdot \cos 30° = 6 \cdot \frac{\sqrt{3}}{2} = 3\sqrt{3}.$$

The angles of the right triangle are 30°, 60°, and 90°, and the sides opposite those angles are 3, $3\sqrt{3}$, and 6, respectively.

▸**TRY THIS.** Solve the right triangle in which $\alpha = 60°$ and $c = 2$. ■

### EXAMPLE 4 Solving a right triangle

Solve the right triangle in which $a = 4$ and $b = 6$. Find the acute angles to the nearest tenth of a degree.

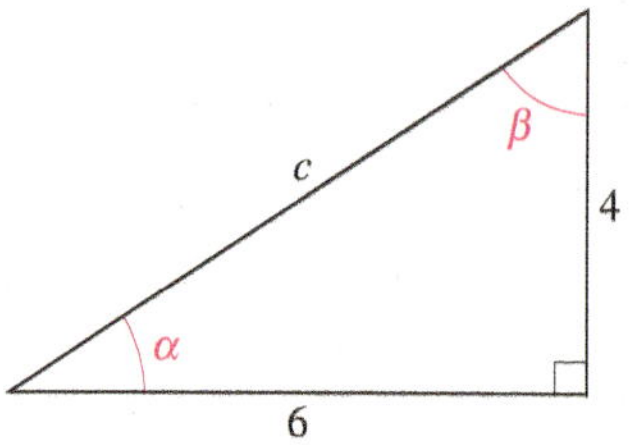


**Figure 5.85**

**Solution**

The triangle is shown in Fig. 5.85. By the Pythagorean theorem, $c^2 = 4^2 + 6^2$, or $c = \sqrt{52} = 2\sqrt{13}$. To find $\alpha$, first find $\sin \alpha$:

$$\sin \alpha = \frac{\text{opp}}{\text{hyp}} = \frac{4}{2\sqrt{13}} = \frac{2}{\sqrt{13}}$$

Now, $\alpha$ is the angle whose sine is $2/\sqrt{13}$:

$$\alpha = \sin^{-1}\left(\frac{2}{\sqrt{13}}\right) \approx 33.7°$$

Since $\alpha + \beta = 90°$, $\beta = 90° - 33.7° = 56.3°$. The angles of the triangle are 33.7°, 56.3°, and 90°, and the sides opposite those angles are 4, 6, and $2\sqrt{13}$, respectively.

▸**TRY THIS.** Solve the right triangle in which $a = 2$ and $b = 5$. ■

In Example 4 we could have found $\alpha$ by using $\alpha = \tan^{-1}(4/6) \approx 33.7°$. We could then have found $c$ by using $\cos(33.7°) = 6/c$, or $c = 6/\cos(33.7°) \approx 7.2$. There are many ways to solve a right triangle, but the basic strategy is always the same.

**STRATEGY**

### Solving a Right Triangle

1. Use the Pythagorean theorem to find the length of a third side when the lengths of two sides are known.
2. Use the trigonometric ratios to find missing sides or angles.
3. Use the fact that the sum of the measures of the angles of a triangle is 180° to determine a third angle when two are known.

## Significant Digits

In solving right triangles and in most other problems in algebra and trigonometry, we usually assume that the given numbers are exact. Starting with exact numbers, exact answers or approximations to any degree of accuracy desired can be obtained. However, in real applications, lengths of sides in triangles or measures of angles are determined by actual measurement. In such cases, we must be concerned with the accuracy of our measurements and the values obtained from those measurements using trigonometry. For example, an angle of 21.75° at the center of a circle with a one-mile radius intercepts an arc of length 2004.3 feet, but if the same angle is given to the nearest tenth of a degree, 21.8°, then the length of the intercepted arc is 2008.9 feet, a difference of 4.6 feet.

To say that an angle is measured at 21.8° means that the angle is actually between 21.75° and 21.85°. Since all digits in 21.8° are meaningful, we say that 21.8° has three **significant digits.** If a number has no nonzero digits to the left of the decimal point, then zeros that follow the decimal point but precede nonzero digits are not significant. A measurement of 0.00034 centimeters has two significant digits. The number 44.006 has five significant digits. In the measurement 44.0 pounds, the zero means that the measurement was made to the nearest tenth, and so it is significant. Zeros that follow the decimal point but are not followed by nonzero digits are significant. For whole-number measurements without decimal points, we assume that all digits are significant. So measurements of 4782 feet and 8000 miles both have four significant digits. In scientific notation, only significant digits are written. For example, $8 \times 10^3$ has only one significant digit, while $8.000 \times 10^3$ has four.

When arithmetic operations or trigonometric functions are applied to measurements, the answers are only as accurate as the *least accurate* measurement involved. For example, if it is 23 miles from your house to the airport and $8 \times 10^3$ miles (to the nearest thousand) to Hong Kong, then to say that it is 8023 miles to Hong Kong from your house implies an accuracy that is not justified. Since $8 \times 10^3$ miles is a less accurate measurement than 23 miles and $8 \times 10^3$ miles has one significant digit, any answer obtained from these two measurements has only one significant digit. So only the first digit of 8023 is significant, and it is $8 \times 10^3$ miles (to the nearest thousand) from your house to Hong Kong. A "rule of thumb" that works well in most cases is that an answer should be given with the same number of significant digits as is in the least accurate measurement in the computation.

### EXAMPLE 5 Significant digits

Assume that the given numbers are measurements. Perform each computation, and give the answer with the appropriate number of significant digits.

**a.** $\sin(123.4°)$ **b.** $\dfrac{\cos(3°)}{\sin(1.9°)}$ **c.** $544 \cdot \sin(25°)$

#### Solution

**a.** Since 123.4° has four significant digits, we round the calculator value to four decimal places. So $\sin(123.4°) \approx 0.8348$.

**b.** Since the least number of significant digits in 3° and 1.9° is one, round the value obtained on a calculator to one significant digit:

$$\frac{\cos(3°)}{\sin(1.9°)} \approx 3 \times 10^1 \text{ (to the nearest ten)}$$

**c.** The least number of significant digits in 544 and 25° is two. So

$$544 \cdot \sin(25°) \approx 230 \text{ (to the nearest ten).}$$

In scientific notation, $544 \cdot \sin(25°) \approx 2.3 \times 10^2$.

▶**TRY THIS.** Find $24 \sin(5.461°)$ using the appropriate number of significant digits in the answer. ■

## Applications

Using trigonometry, we can find the size of an object without actually measuring the object but by measuring an angle. Two common terms used in this regard are **angle of elevation** and **angle of depression.** The angle of elevation $\alpha$ for a point above a horizontal line is the angle formed by the horizontal line and the observer's line of sight through the point, as shown in Fig. 5.86. The angle of depression $\beta$ for a point

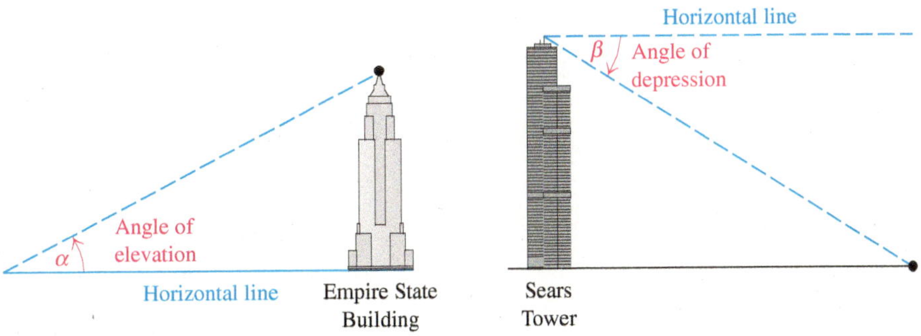


**Figure 5.86**

below a horizontal line is the angle formed by the horizontal line and the observer's line of sight through the point, as shown in Fig. 5.86. We use these angles and our skills in solving triangles to find the sizes of objects that would be inconvenient to measure.

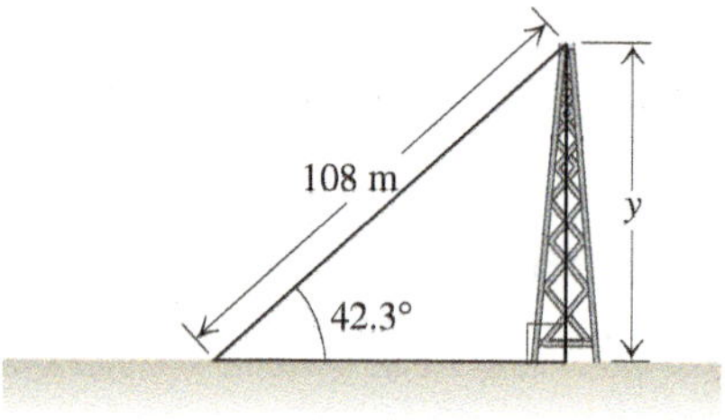


Figure 5.87

### EXAMPLE 6 Finding the height of an object

A guy wire of length 108 meters runs from the top of an antenna to the ground. If the angle of elevation of the top of the antenna, sighting along the guy wire, is 42.3°, then what is the height of the antenna?

**Solution**

Let $y$ represent the height of the antenna, as shown in Fig. 5.87. Since $\sin(42.3°) = y/108$,

$$y = 108 \cdot \sin(42.3°) \approx 72.7 \text{ meters.}$$

Three significant digits are used in the answer because both of the given measurements contain three significant digits.

▶**TRY THIS.** The angle of elevation of the top of a cell phone tower is 38.2° at a distance of 344 feet from the tower. What is the height of the tower? ■

In Example 6 we knew the distance to the top of the antenna, and we found the height of the antenna. If we knew the distance on the ground to the base of the antenna and the angle of elevation of the guy wire, we could still have found the height of the antenna. Both cases involve knowing the distance to the antenna either on the ground or through the air. However, one of the biggest triumphs of trigonometry is being able to find the size of an object or the distance to an object (such as the moon) without going to the object. The next example shows one way to find the height of an object without actually going to it. In Section 7.1, Example 7, we will show another (slightly simpler) solution to the same problem using the law of sines.

### EXAMPLE 7 Finding the height of an object from a distance

The angle of elevation of the top of a water tower from point $A$ on the ground is 19.9°. From point $B$, 50.0 feet closer to the tower, the angle of elevation is 21.8°. What is the height of the tower?

**Solution**

Let $y$ represent the height of the tower and $x$ represent the distance from point $B$ to the base of the tower, as shown in Fig. 5.88.

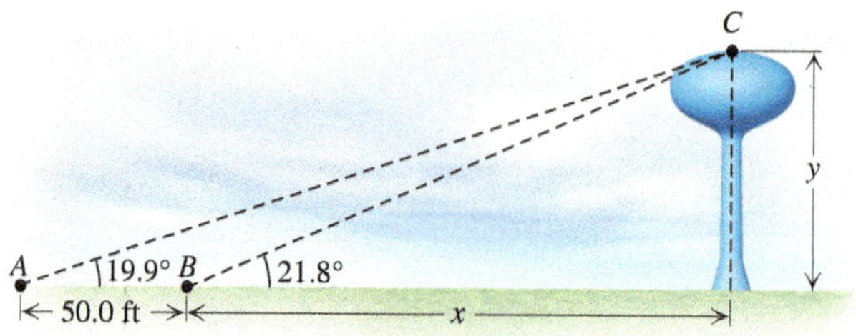


Figure 5.88

At point $B$, $\tan 21.8° = y/x$ or

$$x = \frac{y}{\tan 21.8°}.$$

Since the distance to the base of the tower from point $A$ is $x + 50$,

$$\tan 19.9° = \frac{y}{x + 50}$$

or

$$y = (x + 50)\tan 19.9°.$$

To find the value of $y$ we must write an equation that involves only $y$. Since $x = y/\tan 21.8°$, we can substitute $y/\tan 21.8°$ for $x$ in the last equation:

$$y = \left(\frac{y}{\tan 21.8°} + 50\right)\tan 19.9°$$

$$y = \frac{y \cdot \tan 19.9°}{\tan 21.8°} + 50 \tan 19.9° \quad \text{Distributive property}$$

$$y - \frac{y \cdot \tan 19.9°}{\tan 21.8°} = 50 \tan 19.9°$$

$$y\left(1 - \frac{\tan 19.9°}{\tan 21.8°}\right) = 50 \tan 19.9° \quad \text{Factor out } y.$$

$$y = \frac{50 \tan 19.9°}{1 - \dfrac{\tan 19.9°}{\tan 21.8°}} \approx 191 \text{ feet}$$

Figure 5.89

This computation is shown on a graphing calculator in Fig. 5.89.

**TRY THIS.** At one location, the angle of elevation of the top of an antenna is 44.2°. At a point that is 100 feet closer to the antenna, the angle of elevation is 63.1°. What is the height of the antenna? ■

In the next example we combine the solution of a right triangle with the arc length of a circle from Section 5.1 to solve a problem of aerial photography.

## EXAMPLE 8 Photography from a spy plane

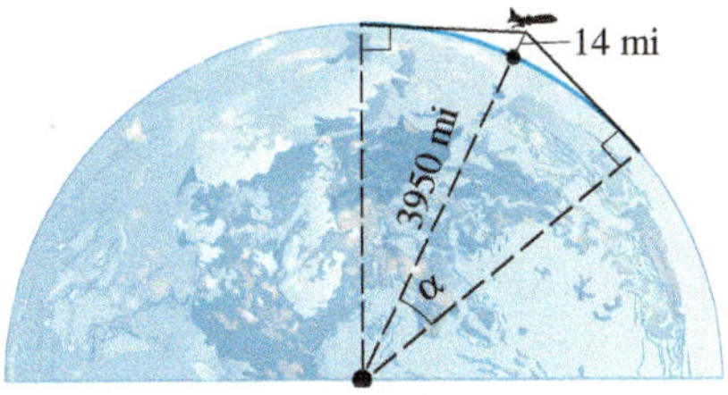


Figure 5.90

In the late 1950s, the Soviets labored to develop a missile that could stop the U-2 spy plane. On May 1, 1960, Nikita S. Khrushchev announced to the world that the Soviets had shot down Francis Gary Powers while Powers was photographing the Soviet Union from a U-2 at an altitude of 14 miles. How wide a path on the earth's surface could Powers see from that altitude? (Use 3950 miles as the earth's radius.)

### Solution

Figure 5.90 shows the line of sight to the horizon on the left-hand side and right-hand side of the airplane while flying at the altitude of 14 miles.

Since a line tangent to a circle (the line of sight) is perpendicular to the radius at the point of tangency, the angle $\alpha$ at the center of the earth in Fig. 5.90 is an acute angle of a right triangle with hypotenuse 3950 + 14 or 3964. So we have

$$\cos \alpha = \frac{3950}{3964}$$

$$\alpha = \cos^{-1}\left(\frac{3950}{3964}\right) \approx 4.8°.$$

The width of the path seen by Powers is the length of the arc intercepted by the central angle $2\alpha$ or 9.6°. Using the formula $s = \alpha r$ from Section 5.1, where $\alpha$ is in radians, we get

$$s = 9.6 \text{ deg} \cdot \frac{\pi \text{ rad}}{180 \text{ deg}} \cdot 3950 \text{ miles} \approx 661.8 \text{ miles}.$$

From an altitude of 14 miles, Powers could see a path that was 661.8 miles wide. Actually, he photographed a path that was somewhat narrower, because parts of the photographs near the horizon were not usable.

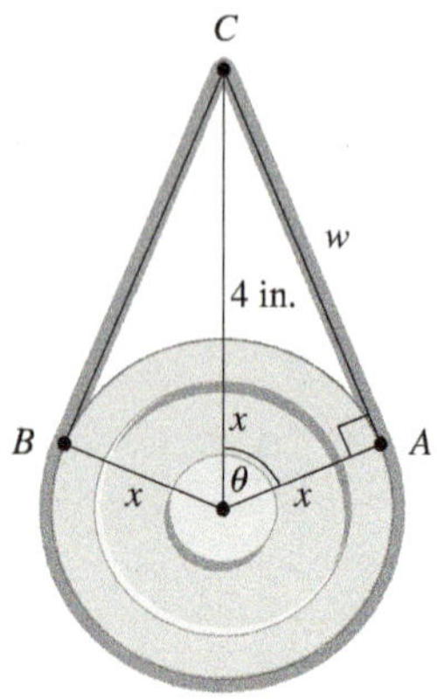


Figure 5.91

**▶TRY THIS.** How wide a path (to the nearest mile) on the earth's surface can be seen from a commercial airliner flying at an altitude of 7 miles? ■

Example 9 involves a circle and tangent lines, and appears to be similar to Example 8. However, in Example 9 we will write an equation that can only be solved using a graphing calculator or computer. Notice that in Example 9 we will use right triangle trigonometry, the arc length formula, the circumference formula, the Pythagorean theorem, and the inverse cosine function.

### EXAMPLE 9 Graphing calculator required

A 20-inch belt connects a pulley with a very small shaft on a motor, as shown in Fig. 5.91. The distance between the shaft and the large pulley is 4 inches. Assuming the shaft is a single point, find the radius of the large pulley to the nearest tenth of an inch.

#### Solution

Let $x$ be the radius of the circle and $w$ be the distance from the shaft to point $A$, as shown in the figure. Since the radius is perpendicular to the belt at $A$, we have $\theta = \cos^{-1}\left(\frac{x}{x+4}\right)$. Assuming $\theta$ is in radians, the length of the arc of the circle intercepted by $\theta$ is $x \cdot \cos^{-1}\left(\frac{x}{x+4}\right)$. Since the circumference of the circle is $2\pi x$, the length of the long arc from $A$ to $B$ is

$$2\pi x - 2x \cdot \cos^{-1}\left(\frac{x}{x+4}\right).$$

By the Pythagorean theorem, $x^2 + w^2 = (x+4)^2$ or $w = \sqrt{8x+16}$. Since the total length of the belt is 20 inches, we have

$$2\pi x - 2x \cdot \cos^{-1}\left(\frac{x}{x+4}\right) + 2\sqrt{8x+16} = 20.$$

One way to solve this equation with a graphing calculator is to graph

$$y = 2\pi x - 2x \cdot \cos^{-1}\left(\frac{x}{x+4}\right) + 2\sqrt{8x+16} - 20$$

and find the $x$-intercept, as shown in Fig. 5.92. The radius of the large pulley is approximately 2.2 inches.

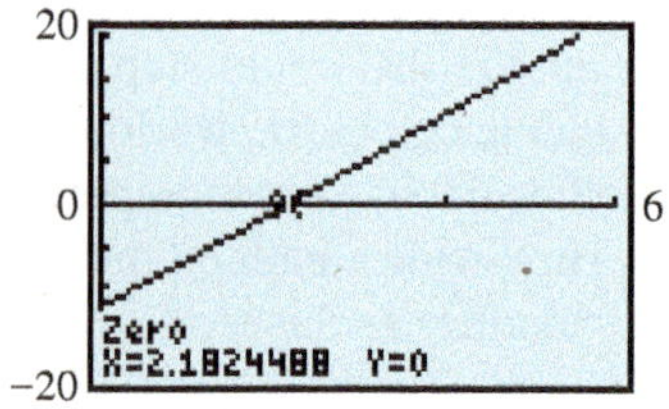


Figure 5.92

**▶TRY THIS.** It is about 200 miles from Houston to San Antonio on interstate I-10. A pilot at an altitude of 5 miles over Houston spots San Antonio on the horizon. From this information calculate the radius of the earth. ■

## FOR thought... 

**True or False? Explain. For Exercises 1–4, $\alpha$ is an angle in standard position.**

**1.** If the terminal side of $\alpha$ goes through $(5, -10)$, then $\sin \alpha = 10/\sqrt{125}$.

**2.** If the terminal side of $\alpha$ goes through $(-1, 2)$, then $\sec \alpha = -\sqrt{5}$.

**3.** If the terminal side of $\alpha$ goes through $(-2, 3)$, then $\alpha = \sin^{-1}(3/\sqrt{13})$.

**4.** If the terminal side of $\alpha$ goes through $(3, 1)$, then $\alpha = \cos^{-1}(3/\sqrt{10})$.

**5.** In a right triangle, $\sin \alpha = \cos \beta$, $\sec \alpha = \csc \beta$, and $\tan \alpha = \cot \beta$.

**6.** If $a = 4$ and $b = 2$ in a right triangle, then $c = \sqrt{6}$.

**7.** If $a = 6$ and $b = 2$ in a right triangle, then $\beta = \tan^{-1}(3)$.

**8.** If $a = 8$ and $\alpha = 55°$ in a right triangle, then $b = 8/\tan(55°)$.

**9.** In a right triangle with sides of length 3, 4, and 5, the smallest angle is $\cos^{-1}(0.8)$.

**10.** In a right triangle, $\sin(90°) = \text{hyp}/\text{adj}$.

## EXERCISES 5.6

*Fill in the blank.*

**1.** If $\alpha$ is an acute angle of a right triangle, then $\sin(\alpha)$ is the _________ divided by the _________.

**2.** If $\alpha$ is an acute angle of a right triangle, then $\cos(\alpha)$ is the _________ divided by the _________.

**3.** If $\alpha$ is an acute angle of a right triangle, then $\tan(\alpha)$ is the _________ divided by the _________.

**4.** The _________ of the measures of the three angles of any triangle is _________.

*Assume that $\alpha$ is an angle in standard position whose terminal side contains the given point. Find the exact values of* $\sin \alpha$, $\cos \alpha$, $\tan \alpha$, $\csc \alpha$, $\sec \alpha$, *and* $\cot \alpha$.

**5.** $(3, 4)$  **6.** $(4, 4)$

**7.** $(-2, 6)$  **8.** $(-3, 6)$

**9.** $(-2, -\sqrt{2})$  **10.** $(-1, -\sqrt{3})$

**11.** $(\sqrt{6}, -\sqrt{2})$  **12.** $(2\sqrt{3}, -2)$

*For Exercises 13–18 find exact values of* $\sin \alpha$, $\cos \alpha$, $\tan \alpha$, $\sin \beta$, $\cos \beta$, *and* $\tan \beta$ *for the given right triangle.*

**13.**

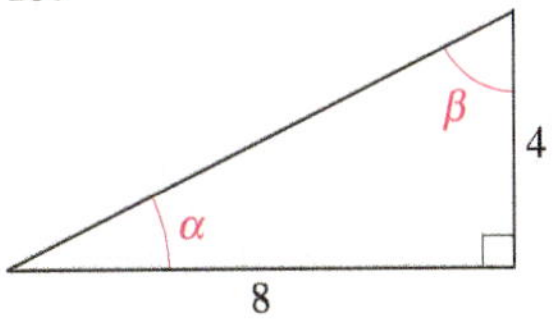


**14.**

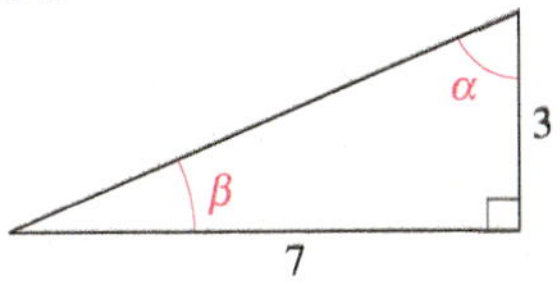


**15.**

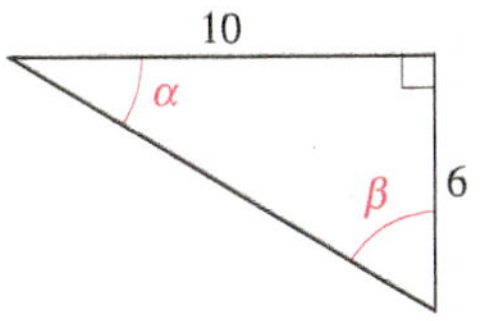


**16.**

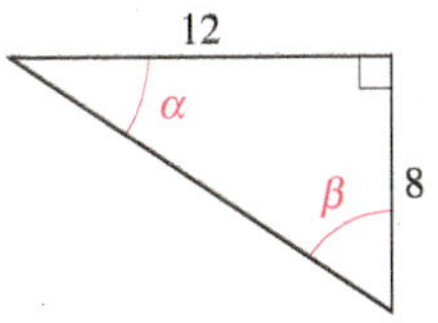


**17.**

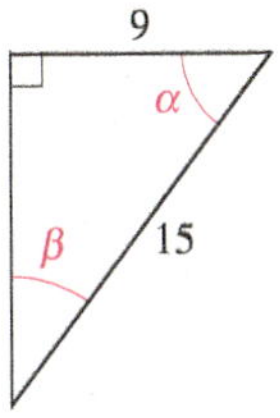


**18.**

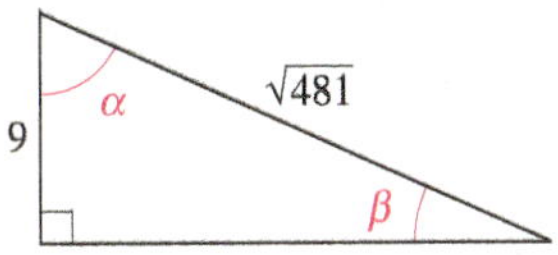


*Assume that $\alpha$ is an angle in standard position whose terminal side contains the given point and that $0° < \alpha < 90°$. Find the degree measure of $\alpha$ to the nearest tenth of a degree.*

**19.** $(1.5, 9)$  **20.** $(4, 5)$

**21.** $(\sqrt{2}, \sqrt{6})$  **22.** $(4.3, 6.9)$

*Assume that $\alpha$ is an angle in standard position whose terminal side contains the given point and that $0 < \alpha < \pi/2$. Find the radian measure of $\alpha$ to the nearest tenth of a radian.*

**23.** $(4, 6.3)$  **24.** $(1/3, 1/2)$

**25.** $(\sqrt{5}, 1)$  **26.** $(\sqrt{7}, \sqrt{3})$

*Solve each right triangle with the given sides and angles. In each case, make a sketch. Note that $\alpha$ is the acute angle opposite leg a and $\beta$ is the acute angle opposite leg b. The hypotenuse is c. See the strategy for solving a right triangle on page 404.*

**27.** $\alpha = 60°, c = 20$

**28.** $\beta = 45°, c = 10$

**29.** $a = 6, b = 8$

**30.** $a = 10, c = 12$

**31.** $b = 6, c = 8.3$

**32.** $\alpha = 32.4°, b = 10$

**33.** $\alpha = 16°, c = 20$

**34.** $\beta = 47°, a = 3$

**35.** $\alpha = 39°9', a = 9$

**36.** $\beta = 19°12', b = 60$

*Perform each computation with the given measurements, and give your answers with the appropriate number of significant digits.*

**37.** $456 \cdot \tan 3.2°$

**38.** $123.32 \cdot \tan 36.3°$

**39.** $\sin^2 65.7°$

**40.** $4 + 6.33 \sin 79.7°$

**41.** $7.50/\sin 23.559°$

**42.** $12.3/\cot 0.03°$

**43.** $500/\sec 125.3°$

**44.** $16.3 - 8.31 \cdot \csc 8.76°$

*Solve each problem.*

**45.** *Aerial Photography* An aerial photograph from a U-2 spy plane is taken of a building suspected of housing nuclear warheads. The photograph is made when the angle of elevation of the sun is 32°. By comparing the shadow cast by the building to objects of known size in the illustration, analysts determine that the shadow is 80 ft long. How tall is the building?

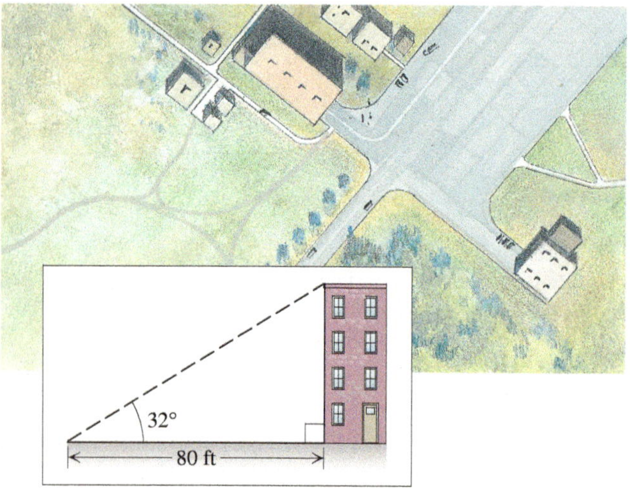


**Figure for Exercise 45**

**46.** *Giant Redwood* A hiker stands 80 feet from a giant redwood tree and sights the top with an angle of elevation of 75°. How tall is the tree to the nearest foot?

**47.** *Avoiding a Swamp* Muriel was hiking directly toward a long, straight road when she encountered a swamp. She turned 65° to the right and hiked 4 mi in that direction to reach the road. How far was she from the road when she encountered the swamp?

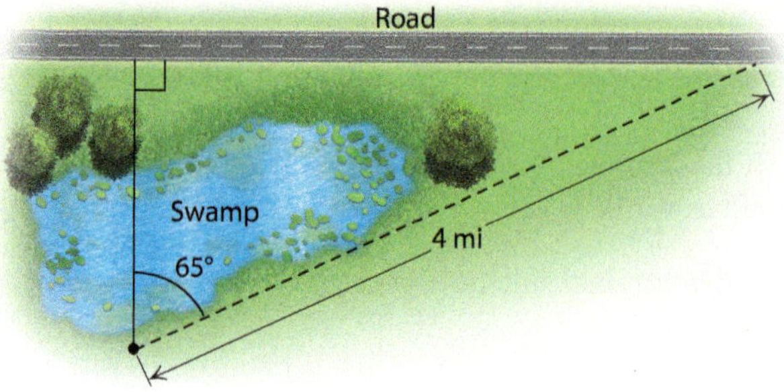


**Figure for Exercise 47**

**48.** *Tall Antenna* A 100-foot guy wire is attached to the top of an antenna. The angle between the guy wire and the ground is 62°. How tall is the antenna to the nearest foot?

**49.** *Angle of Depression* From a highway overpass, 14.3 m above the road, the angle of depression of an oncoming car is measured at 18.3°. How far is the car from a point on the highway directly below the observer?

**Figure for Exercise 49**

**50.** *Length of a Tunnel* A tunnel under a river is 196.8 ft below the surface at its lowest point, as shown in the drawing. If the angle of depression of the tunnel is 4.962°, then how far apart on the surface are the entrances to the tunnel? How long is the tunnel?

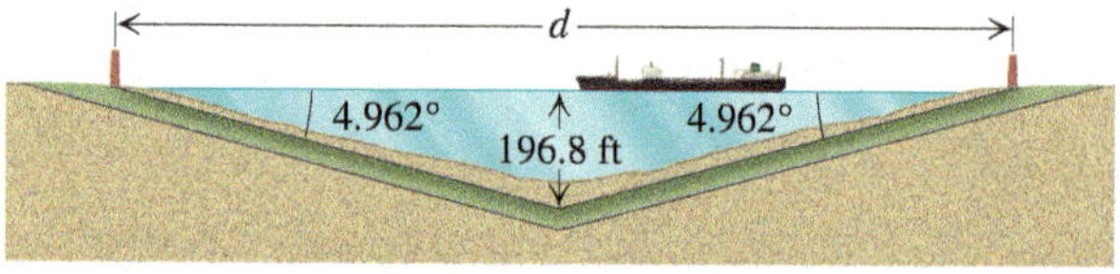


**Figure for Exercise 50**

**51.** *Height of a Crosswalk* The angle of elevation of a pedestrian crosswalk over a busy highway is 8.34°, as shown in the drawing. If the distance between the ends of the crosswalk measured on the ground is 342 ft, then what is the height $h$ of the crosswalk at the center?

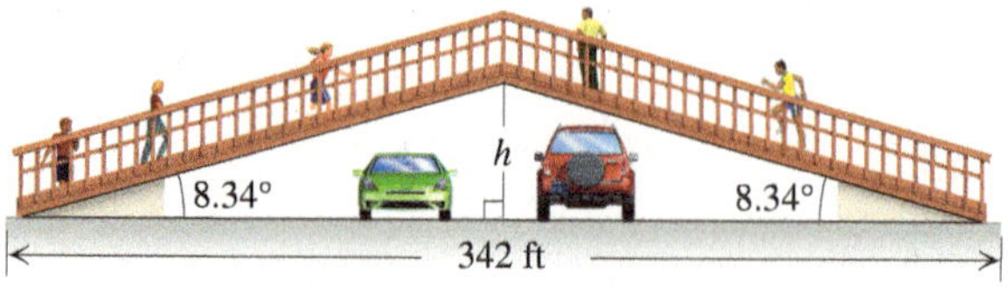


**Figure for Exercise 51**

**52.** *Shortcut to Snyder* To get from Muleshoe to Snyder, Harry drives 50 mph for 178 mi south on route 214 to Seminole, then goes east on route 180 to Snyder. Harriet leaves Muleshoe one hour later at 55 mph, but takes US 84, which goes straight from Muleshoe to Snyder through Lubbock. If US 84 intersects route 180 at a 50° angle, then how many more miles does Harry drive?

**53.** *Installing a Guy Wire* A 41-m guy wire is attached to the top of a 34.6-m antenna and to a point on the ground. How far is the point on the ground from the base of the antenna, and what angle does the guy wire make with the ground?

**54.** *Robin and Marian* Robin Hood plans to use a 30-ft ladder to reach the castle window of Maid Marian. Little John, who made the ladder, advised Robin that the angle of elevation of the ladder must be between 55° and 70° for safety. What are the minimum and maximum heights that can safely be reached by the top of the ladder when it is placed against the 50-ft castle wall?

**55.** *Detecting a Speeder* A policewoman has positioned herself 500 ft from the intersection of two roads. She has carefully measured the angles of the lines of sight to points $A$ and $B$ as shown in the drawing. If a car passes from $A$ to $B$ in 1.75 sec and the speed limit is 55 mph, is the car speeding?
**HINT** Find the distance from $B$ to $A$ and use $R = D/T$.

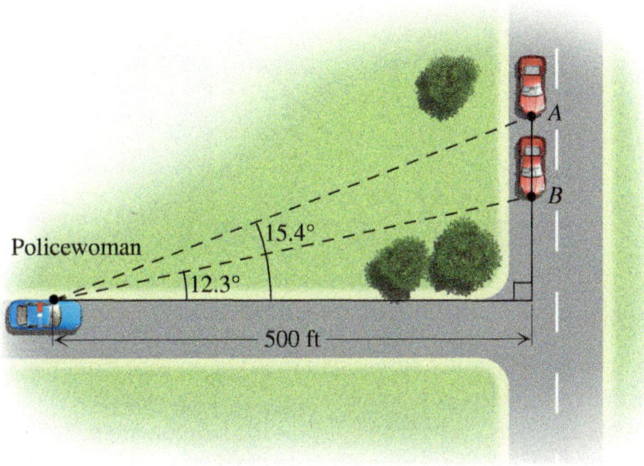


**Figure for Exercise 55**

**56.** *Progress of a Forest Fire* A forest ranger atop a 3248-ft mesa is watching the progress of a forest fire spreading in her direction. In 5 minutes the angle of depression of the leading edge of the fire changed from 11.34° to 13.51°. At what speed in miles per hour is the fire spreading in the direction of the ranger?

**57.** *Height of a Rock* Inscription Rock rises almost straight upward from the valley floor. From one point the angle of elevation of the top of the rock is 16.7°. From a point 168 m closer to the rock, the angle of elevation of the top of the rock is 24.1°. How high is Inscription Rock?

**58.** *Height of a Balloon* A hot air balloon is between two spotters who are 1.2 mi apart. One spotter reports that the angle of elevation of the balloon is 76°, and the other reports that it is 68°. What is the altitude of the balloon in miles?

**59.** *Passing in the Night* A boat sailing north sights a lighthouse to the east at an angle of 32° from the north, as shown in the drawing. After the boat travels one more kilometer, the angle of the lighthouse from the north is 36°. If the boat continues to sail north, then how close will the boat come to the lighthouse?

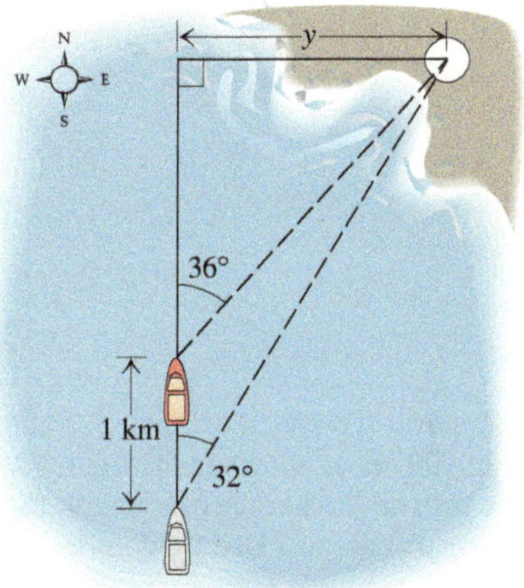


**Figure for Exercise 59**

**60.** *Height of a Skyscraper* For years the Woolworth skyscraper in New York held the record for the world's tallest office building. If the length of the shadow of the Woolworth building increases by 17.4 m as the angle of elevation of the sun changes from 44° to 42°, then how tall is the building?

**61.** *Parsecs* In astronomy the light year (abbreviated ly) and the parsec (abbreviated pc) are the two units used to measure distances. The distance in space at which a line from the earth to the sun subtends an angle of 1 second is 1 parsec. Find the number of miles in 1 parsec by using a right triangle positioned in space with its right angle at the sun as shown in the figure. How many years does it take light to travel 1 parsec? (One light year [ly] is the distance that light travels in one year, one astronomical unit [AU] is the distance from the earth to the sun, and 1 ly = 63,240 AU.)

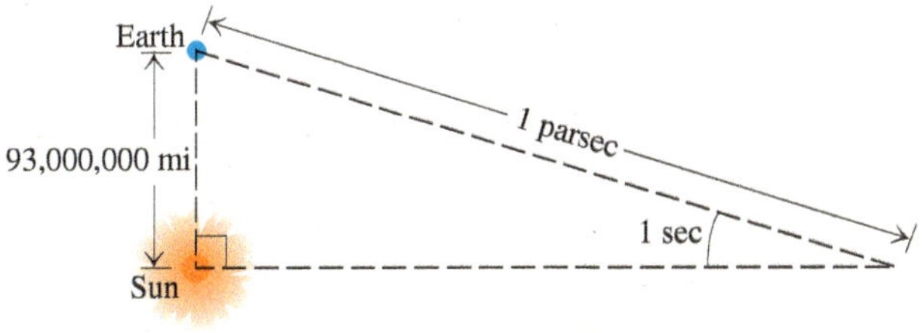


**Figure for Exercise 61**

**62.** *Catching a Green Flash* Ed and Diane are on the beach to observe a green flash. As the very last sliver of the sun sinks into the ocean, refraction bends the final ray of sunlight into a rainbow of colors. The red and orange set first, and the green light or *green flash* is the last bit of sunlight seen before sunset.

**a.** If Diane's eyes are 2 ft above the ocean, then what is the angle $\alpha$ shown in the figure?

**b.** If Ed's eyes are 6 ft above the ocean, then what is the angle $\alpha$ for Ed?

**c.** Diane sees the green flash first. The difference between the two alphas is how far the earth must rotate before Ed sees it. How many seconds after Diane sees the green flash does Ed see it?

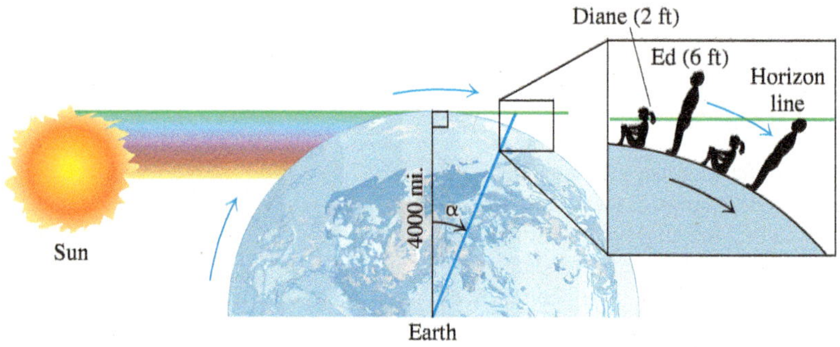


**Figure for Exercise 62**

**63.** *View from Landsat* The satellite Landsat orbits the earth at an altitude of 700 mi, as shown in the figure. What is the width of the path on the surface of the earth that can be seen by the cameras of Landsat? Use 3950 mi for the radius of the earth.

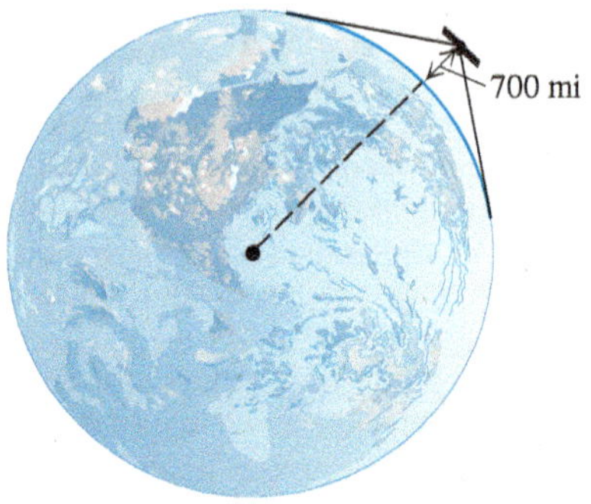


**Figure for Exercise 63**

**64.** *Communicating Via Satellite* A communication satellite is usually put into a synchronous orbit with the earth, which means that it stays above a fixed point on the surface of the earth at all times. The radius of the orbit of such a satellite is 6.5 times the radius of the earth (3950 mi). The satellite is used to relay a signal from one point on the earth to another point on the earth. The sender and receiver of a signal must be in a line of sight with the satellite, as shown in the figure. What is the maximum distance on the surface of the earth between the sender and receiver for this type of satellite?

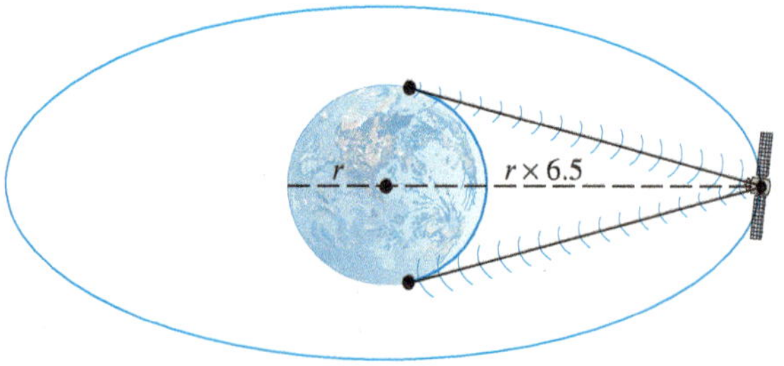


**Figure for Exercise 64**

**65.** *Angle of Elevation* From point $A$ the angle of elevation to the top of the building is 30°, as shown in the accompanying figure. From point $B$, 20 meters closer to the building, the angle of elevation is 45°. Find the angle of elevation of the building from point $C$, which is another 20 meters closer to the building.

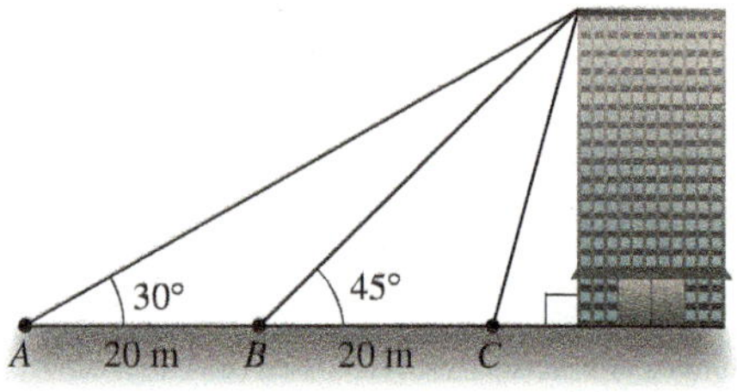


**Figure for Exercise 65**

**66.** *Rate of Ascent* A hot air balloon is rising upward from the earth at a constant rate, as shown in the accompanying figure. An observer 250 meters away spots the balloon at an angle of elevation of 24°. Two minutes later the angle of elevation of the balloon is 58°. At what rate is the balloon ascending? Answer to the nearest tenth of a meter per second.

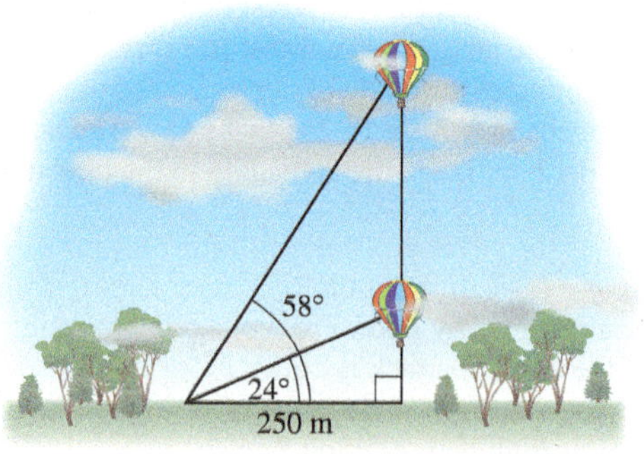


**Figure for Exercise 66**

**67.** *Watering the Lawn* Four lawn sprinklers are placed to water a square lawn, as shown in the accompanying figure. The region watered by each sprinkler is circular with a radius of 6 m.

**a.** How many square meters of lawn are not reached by any of the sprinklers? Find the exact answer.

**b.** What percent of the lawn is watered by at least one sprinkler? Round to the nearest tenth of a percent.

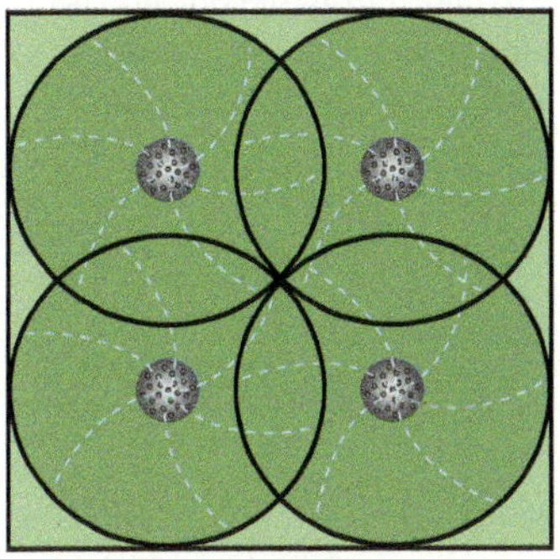

**Figure for Exercise 67**

**68.** *Shipping Pipes* Six identical pipes, each with a radius of 1 foot, are tied tightly together with a metal band, as shown in the accompanying figure. Find the exact length of the metal band.

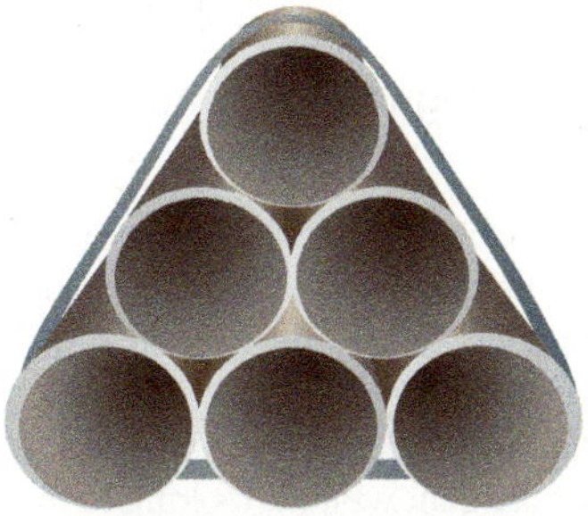

**Figure for Exercise 68**

**69.** *Maximizing Area* A builder wants to build a rectangular house on a triangular lot, as shown in the figure. Find the dimensions of the house that will maximize the area of the house.

**HINT** Write the length in terms of the width. Then write a quadratic function for the area.

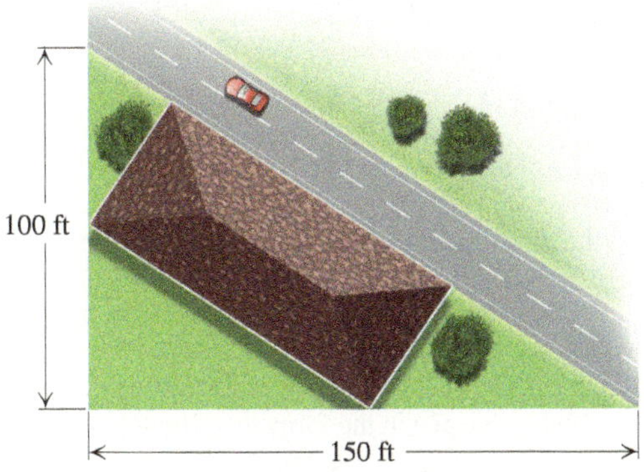


**Figure for Exercises 69 and 70**

**70.** *Figuring the Setback* The builder in the previous exercise just remembered that the house must be set back 10 ft from each side of the property and 40 ft from the street. Now find the dimensions of the house that will maximize the area of the house.

**71.** *Solving a Lot* The diagonals of the property shown in the figure are 90 ft and 120 ft. The diagonals cross 30 ft from the street. Find the width, $w$, of the property.

**HINT** Write an equation involving $w$ and solve it with a graphing calculator.

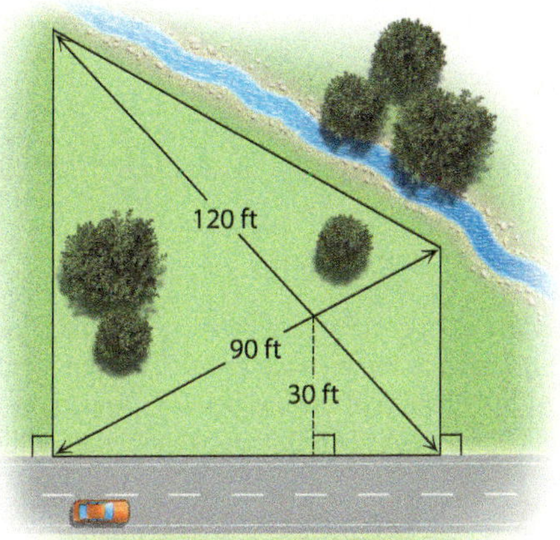


**Figure for Exercise 71**

**72.** *Finding the Radius of the Earth* Assume that it takes 4 sec for the green flash in Exercise 62 to travel from Diane's eyes at 2 ft to Ed's eyes at 6 ft. Find the radius of the earth to the nearest mile.

**73.** *Hanging a Pipe* A contractor wants to pick up a 10-ft-diameter pipe with a chain of length 40 ft. The chain encircles the pipe and is attached to a hook on a crane. What is the distance between the hook and the pipe? (This is an actual problem that an engineer was asked to solve on the job.)

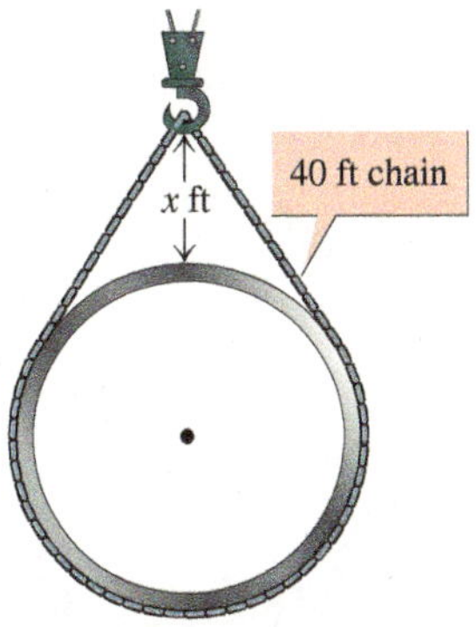


**Figure for Exercise 73**

**74.** *Circling the Earth* A rope that is 1 meter longer than the circumference of the earth is placed around the earth at the equator. If the rope is pulled up tightly to point $A$, as shown in the accompanying figure, then what is the distance from point $A$ to the surface of the earth? Use 6400 kilometers as the radius of the earth.

**Figure for Exercise 74**

**75.** *Blocking a Pipe* A large pipe is held in place by using a 1-ft-high block on one side and a 2-ft-high block on the other side. If the length of the arc between the points where the pipe touches the blocks is 6 ft, then what is the radius of the pipe?

**HINT** Ignore the thickness of the pipe.

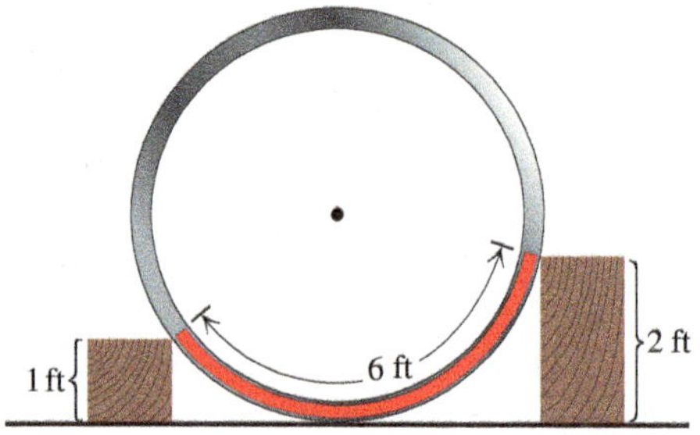


**Figure for Exercise 75**

**76.** *Pipe on an Inclined Plane* A large pipe on an inclined plane is held in place by sticks of length 2 ft and 6 ft, as shown in the figure. The sticks are in line with the center of the pipe. The extended lines of the sticks intersect at an 18° angle at the center of the circle. What is the radius of the pipe? Find the radius of the pipe if the angle is 19°.

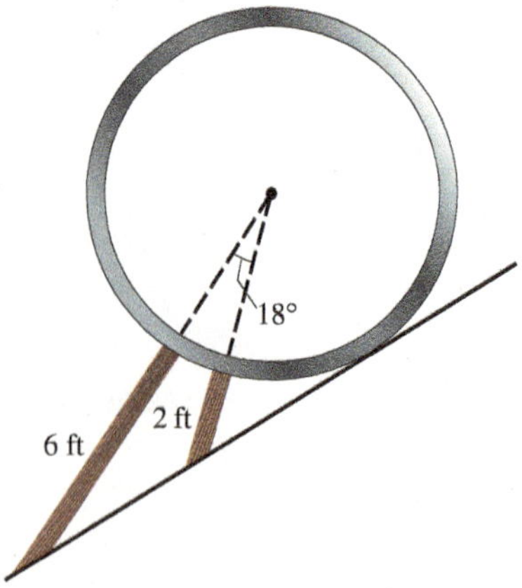


**Figure for Exercise 76**

## FOR WRITING/DISCUSSION

**77.** *Cooperative Learning* Construct a device for measuring angle of elevation or angle of depression by attaching a weighted string to a protractor, as shown in the figure. Work with a group to select an object such as a light pole or tree. Measure the angle of elevation to the top of the object and the angle of depression to the bottom of the object while standing at some appropriate distance from the object. Use these angles, the distance to the object, and the distance of your eye from the ground to find the height of the object.

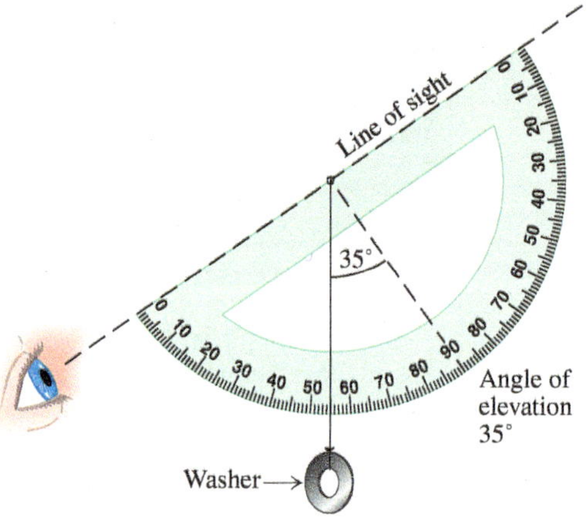


**Figure for Exercise 77**

**78.** *Cooperative Learning* Work in a group to select a local building and find its height without going up to the building. Use the difference in the lengths of its shadow at two different times of day and the technique of Example 7. Use the protractor and string from the last exercise to measure the angle of elevation of the sun, without looking at the sun.

## ▶ RETHINKING

**79.** Evaluate each function. Give the result in degrees.

**a.** $\sin^{-1}(-1/2)$ **b.** $\cos^{-1}(-1/2)$

**c.** $\tan^{-1}(-1)$

**80.** Evaluate each trigonometric function if possible.

**a.** $\sin(7\pi/6)$ **b.** $\cos(3\pi)$

**c.** $\tan(-3\pi/4)$ **d.** $\sec(\pi/2)$

**e.** $\csc(\pi/3)$ **f.** $\cot(-\pi/4)$

**81.** Find the amplitude, period, phase shift, and range for the function $y = -3\sin(\pi x/2 - \pi/2) + 7$.

**82.** The five key points on one cycle of a sine wave are $(\pi/4, 2)$, $(\pi/2, 5)$, $(3\pi/4, 2)$, $(\pi, -1)$, and $(5\pi/4, 2)$. Write the equation of the sine wave in the form $y = A\cos(B(x - C)) + D$.

**83.** Find the period and range for the function $y = 5\sec(\pi x)$.

**84.** Find the domain of the function $y = \csc(2x)$.

### THINKING OUTSIDE THE BOX XLI

*Kicking a Field Goal* In professional football the ball must be placed between the left and right hash mark when a field goal is to be kicked. The hash marks are 9.25 ft from the center line of the field. The goal post is 30 ft past the goal line. Suppose that $\theta$ is the angle between the lines of sight from the ball to the left and right uprights as shown in the figure. The larger the value of $\theta$, the easier it is to kick the ball between the uprights. What is the difference (to the nearest thousandth of a degree) between the values of $\theta$ at the right hash mark and the value of $\theta$ at the center of the field when the ball is 60 ft from the goal line? The two vertical bars on the goal are 18.5 ft apart and the horizontal bar is 10 ft above the ground.

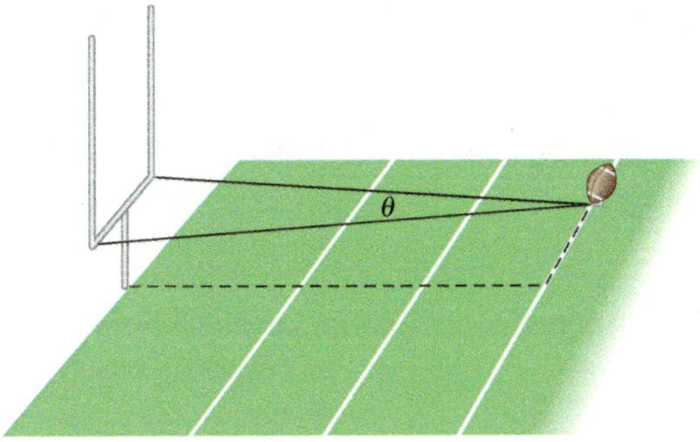


**Figure for Thinking Outside the Box XLI**

### POP QUIZ 5.6

1. Find $\sin\alpha$, $\cos\alpha$, and $\tan\alpha$ if the terminal side of $\alpha$ in standard position goes through $(-3, 4)$.

2. A right triangle has legs with lengths 3 and 6, and $\alpha$ is the acute angle opposite the smallest leg. Find exact values for $\sin\alpha$, $\cos\alpha$, and $\tan\alpha$.

3. At a distance of 1000 feet from a building the angle of elevation to the top of the building is 36°. Find the height of the building to the nearest foot.

# Highlights

## 5.1 Angles and Their Measurements

| | | |
|---|---|---|
| **Degree Measure** | Divide the circumference of a circle into 360 equal arcs. The degree measure is the number of degrees through which the initial side of an angle is rotated to get to the terminal side (positive for counterclockwise, negative for clockwise). | 90° is 1/4 of a circle.<br>180° is 1/2 of a circle. |
| **Radian Measure** | The length of the arc on the unit circle through which the initial side rotates to get to the terminal side | $\pi/2$ is 1/4 of the unit circle.<br>$\pi$ is 1/2 of the unit circle. |
| **Converting** | Use $\pi$ radians = 180° and cancellation of units. | $90° \cdot \dfrac{\pi \text{ rad}}{180°} = \dfrac{\pi}{2}$ rad |
| **Arc Length** | $s = \alpha r$ where $s$ is the arc intercepted by a central angle of $\alpha$ radians on a circle of radius $r$. | $\alpha = 90°, r = 10$ ft<br>$s = \alpha r = \dfrac{\pi}{2} \cdot 10 \text{ ft} = 5\pi$ ft |
| **Angular and Linear Velocity** | A point in motion on a circle of radius $r$ through an angle of $\alpha$ radians or arc length $s$ in time $t$ has angular velocity $\omega$, where $\omega = \alpha/t$, and linear velocity $v$, where $v = s/t$. | $r = 10$ ft, $\alpha = \pi/2$<br>$s = 5\pi$ ft, $t = 2$ sec<br>$\omega = \pi/4$ rad/sec<br>$v = 2.5\pi$ ft/sec |

## 5.2 The Sine and Cosine Functions

| | | |
|---|---|---|
| **Sine and Cosine Functions** | If $\alpha$ is an angle in standard position and $(x, y)$ is the point of intersection of the terminal side and the unit circle, then $\sin\alpha = y$ and $\cos\alpha = x$. | $\sin(90°) = 1$<br>$\cos(\pi/2) = 0$<br>$\sin(0) = 0$ |

| | | |
|---|---|---|
| **Domain** | The domain for sine or cosine can be the set of angles in standard position, the measures of those angles, or the set of real numbers. | $\sin(30°) = 1/2$<br>$\sin(\pi/4) = \sqrt{2}/2$<br>$\sin(72.6) \approx -0.3367$ |
| **Fundamental Identity** | For any angle $\alpha$ (or real number $\alpha$), $\sin^2(\alpha) + \cos^2(\alpha) = 1$. | $\sin^2(30°) + \cos^2(30°) = 1$<br>$\sin^2(\sqrt[3]{2}) + \cos^2(\sqrt[3]{2}) = 1$ |
| **Reference Angles** | To find the sine (or cosine) of an angle, first find the sine or cosine of its reference angle and prefix the appropriate sign. | $\sin(2\pi/3) = \sin(\pi/3)$<br>$\cos(2\pi/3) = -\cos(\pi/3)$ |

## 5.3 The Graphs of the Sine and Cosine Functions

| | | |
|---|---|---|
| **Graphs** | The graphs of $y = \sin x$ and $y = \cos x$ are sine waves, each with period $2\pi$. | |
| **General Sine and Cosine Functions** | The graph of $y = A \sin[B(x - C)] + D$ or $y = A \cos[B(x - C)] + D$ is a sine wave with amplitude $\lvert A \rvert$, period $2\pi/\lvert B \rvert$, phase shift $C$, and vertical translation $D$. | $y = -3\sin(2(x - \pi)) + 1$<br>amplitude 3, period $\pi$, phase shift $\pi$, vertical translation 1 |
| **Graphing a General Sine Function** | Start with the five key points on $y = \sin x$:<br>$(0, 0), \left(\frac{\pi}{2}, 1\right), (\pi, 0), \left(\frac{3\pi}{2}, -1\right), (2\pi, 0)$<br>Divide each $x$-coordinate by $B$ and add $C$. Multiply each $y$-coordinate by $A$ and add $D$. Sketch one cycle of the general function through the five new points. Graph a general cosine function in the same manner. | $y = 3\sin\left(2\left(x - \frac{\pi}{2}\right)\right) + 1$<br>Five new points: $\left(\frac{\pi}{2}, 1\right), \left(\frac{3\pi}{4}, 4\right),$ $(\pi, 1), \left(\frac{5\pi}{4}, -2\right), \left(\frac{3\pi}{2}, 1\right)$<br>Draw one cycle through these points. |
| **Frequency** | $F = 1/P$; $P$ is the period and $F$ is the frequency. | $y = \sin(24\pi x)$, $F = 12$ |

## 5.4 The Other Trigonometric Functions and Their Graphs

| | | |
|---|---|---|
| **Tangent, Cotangent, Secant, and Cosecant** | If $(x, y)$ is the point of intersection of the unit circle and the terminal side of $\alpha$, then $\tan\alpha = y/x$, $\cot\alpha = x/y$, $\sec\alpha = 1/x$, and $\csc\alpha = 1/y$.<br>If 0 occurs in a denominator the function is undefined. | $\tan(\pi/4) = 1$<br>$\cot(\pi/2) = 0$<br>$\sec(30°) = 2/\sqrt{3}$<br>$\csc(0)$ is undefined. |
| **Identities** | $\tan\alpha = \dfrac{\sin\alpha}{\cos\alpha}$, $\cot\alpha = \dfrac{\cos\alpha}{\sin\alpha}$, $\sec\alpha = \dfrac{1}{\cos\alpha}$, $\csc\alpha = \dfrac{1}{\sin\alpha}$ | |
| **Tangent** | Period $\pi$, fundamental cycle on $(-\pi/2, \pi/2)$, vertical asymptotes $x = \pi/2 + k\pi$ | |
| **Cotangent** | Period $\pi$, fundamental cycle on $(0, \pi)$, vertical asymptotes $x = k\pi$ | |
| **Secant** | Period $2\pi$, vertical asymptotes $x = \pi/2 + k\pi$ | |
| **Cosecant** | Period $2\pi$, vertical asymptotes $x = k\pi$ | |

## 5.5 The Inverse Trigonometric Functions

| | | |
|---|---|---|
| **Inverse Sine** | If $y = \sin^{-1}x$ for $x$ in $[-1, 1]$, then $y$ is the real number in $[-\pi/2, \pi/2]$ such that $\sin y = x$. | $\sin^{-1}(1) = \pi/2$<br>$\sin^{-1}(-1/2) = -\pi/6$ |
| **Inverse Cosine** | If $y = \cos^{-1}x$ for $x$ in $[-1, 1]$, then $y$ is the real number in $[0, \pi]$ such that $\cos y = x$. | $\cos^{-1}(-1) = \pi$<br>$\cos^{-1}(-1/2) = 2\pi/3$ |
| **Inverse Tangent** | If $y = \tan^{-1}x$ for $x$ in $(-\infty, \infty)$, then $y$ is the real number in $(-\pi/2, \pi/2)$ such that $\tan y = x$. | $\tan^{-1}(1) = \pi/4$<br>$\tan^{-1}(-1) = -\pi/4$ |

## 5.6 Right Triangle Trigonometry

| | | |
|---|---|---|
| **Trigonometric Ratios** | If $(x, y)$ is any point other than $(0, 0)$ on the terminal side of $\alpha$ in standard position and $r = \sqrt{x^2 + y^2}$, then $\sin \alpha = y/r$, $\cos \alpha = x/r$, and $\tan \alpha = y/x$ $(x \neq 0)$. | Terminal side of $\alpha$ through $(3, 4)$ gives $r = 5$ and $\sin \alpha = 4/5$, $\cos \alpha = 3/5$, and $\tan \alpha = 4/3$. |
| **Trigonometric Functions in a Right Triangle** | If $\alpha$ is an acute angle of a right triangle, then $\sin \alpha = \text{opp}/\text{hyp}$, $\cos \alpha = \text{adj}/\text{hyp}$, and $\tan \alpha = \text{opp}/\text{adj}$. | $\sin \alpha = 5/13$, $\cos \alpha = 12/13$, $\tan \alpha = 5/12$ 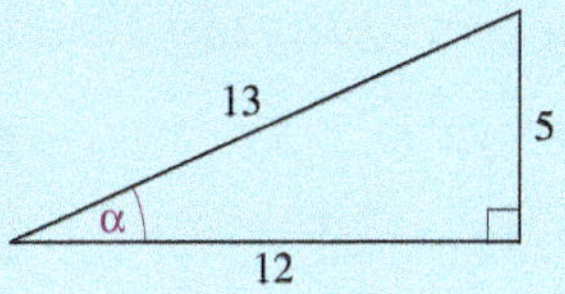 |



# Chapter 5 Review Exercises

*Find the measure in degrees of the least positive angle that is coterminal with each given angle.*

**1.** $388°$ **2.** $-840°$ **3.** $-153°14'27''$

**4.** $455°39'24''$ **5.** $-\pi$ **6.** $-35\pi/6$

**7.** $13\pi/5$ **8.** $29\pi/12$

*Convert each radian measure to degree measure. Do not use a calculator.*

**9.** $5\pi/3$ **10.** $-3\pi/4$ **11.** $3\pi/2$ **12.** $5\pi/6$

*Convert each degree measure to radian measure. Do not use a calculator.*

**13.** $330°$ **14.** $405°$ **15.** $-300°$ **16.** $-210°$

*Fill in the tables. Do not use a calculator.*

**17.**

| $\theta$ deg | 0 | 30 | 45 | 60 | 90 | 120 | 135 | 150 | 180 |
|---|---|---|---|---|---|---|---|---|---|
| $\theta$ rad | | | | | | | | | |
| $\sin \theta$ | | | | | | | | | |
| $\cos \theta$ | | | | | | | | | |
| $\tan \theta$ | | | | | | | | | |

**18.**

| $\theta$ rad | 0 | $\frac{\pi}{6}$ | $\frac{\pi}{4}$ | $\frac{\pi}{3}$ | $\frac{\pi}{2}$ | $\frac{2\pi}{3}$ | $\frac{3\pi}{4}$ | $\frac{5\pi}{6}$ | $\pi$ |
|---|---|---|---|---|---|---|---|---|---|
| $\theta$ deg | | | | | | | | | |
| $\sin \theta$ | | | | | | | | | |
| $\cos \theta$ | | | | | | | | | |
| $\tan \theta$ | | | | | | | | | |

*Give the exact values of each of the following expressions. Do not use a calculator.*

**19.** $\sin(-\pi/4)$ **20.** $\cos(-2\pi/3)$ **21.** $\tan(\pi/3)$

**22.** $\sec(\pi/6)$ **23.** $\csc(-120°)$ **24.** $\cot(135°)$

**25.** $\sin(180°)$ **26.** $\tan(0°)$ **27.** $\cos(3\pi/2)$

**28.** $\csc(5\pi/6)$ **29.** $\sec(-\pi)$ **30.** $\cot(-4\pi/3)$

**31.** $\cot(420°)$ **32.** $\sin(390°)$ **33.** $\cos(-135°)$

**34.** $\tan(225°)$ **35.** $\sec(2\pi/3)$ **36.** $\csc(-3\pi/4)$

**37.** $\tan(5\pi/6)$ **38.** $\sin(7\pi/6)$

*For each triangle shown below, find the exact values of* $\sin \alpha$, $\cos \alpha$, $\tan \alpha$, $\csc \alpha$, $\sec \alpha$, *and* $\cot \alpha$.

**39.**

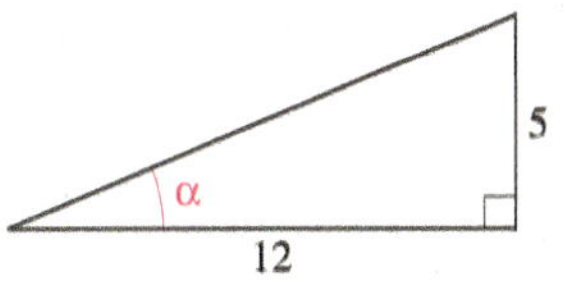



**40.**

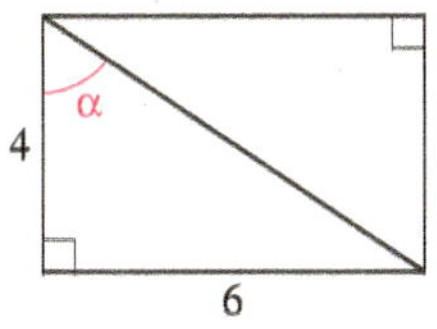



*Find an approximate value for each expression. Round to four decimal places.*

**41.** $\sin(44°)$ **42.** $\cos(-205°)$

**43.** $\cos(4.62)$ **44.** $\sin(3.14)$

**45.** $\tan(\pi/17)$ **46.** $\sec(2.33)$

**47.** $\csc(105°4')$ **48.** $\cot(55°3'12'')$

*Find the exact value of each expression.*

**49.** $\sin^{-1}(-0.5)$

**50.** $\cos^{-1}(-0.5)$

**51.** $\arctan(-1)$

**52.** $\text{arccot}(1/\sqrt{3})$

**53.** $\sec^{-1}(\sqrt{2})$

**54.** $\csc^{-1}(\sec(\pi/3))$

**55.** $\sin^{-1}(\sin(5\pi/6))$

**56.** $\cos(\cos^{-1}(-\sqrt{3}/2))$

*Find the exact value of each expression in degrees.*

**57.** $\sin^{-1}(1)$

**58.** $\tan^{-1}(1)$

**59.** $\arccos(-1/\sqrt{2})$

**60.** $\text{arcsec}(2)$

**61.** $\cot^{-1}(\sqrt{3})$

**62.** $\cot^{-1}(-\sqrt{3})$

**63.** $\text{arccot}(0)$

**64.** $\text{arccot}(-\sqrt{3}/3)$

*Solve each right triangle with the given parts.*

**65.** $a = 2, b = 3$

**66.** $a = 3, c = 7$

**67.** $a = 3.2, \alpha = 21.3°$

**68.** $\alpha = 34.6°, c = 9.4$

*Sketch at least one cycle of the graph of each function and determine the period and range.*

**69.** $f(x) = 2\sin(3x)$

**70.** $f(x) = 1 + \cos(x + \pi/4)$

**71.** $y = \tan(2x + \pi)$

**72.** $y = \cot(x - \pi/4)$

**73.** $y = \sec\left(\frac{1}{2}x\right)$

**74.** $y = \csc\left(\frac{\pi}{2}x\right)$

**75.** $y = \frac{1}{2}\cos(2x)$

**76.** $y = 1 - \sin(x - \pi/3)$

**77.** $y = \cot(2x + \pi/3)$

**78.** $y = 2\tan(x + \pi/4)$

**79.** $y = \frac{1}{3}\csc(2x + \pi)$

**80.** $y = 1 + 2\sec(x - \pi/4)$

*For each of the following sine curves find an equation of the form $y = A\sin[B(x - C)] + D$.*

**81.**

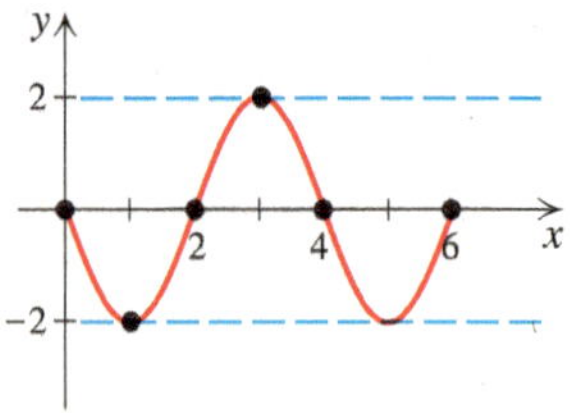


**82.**

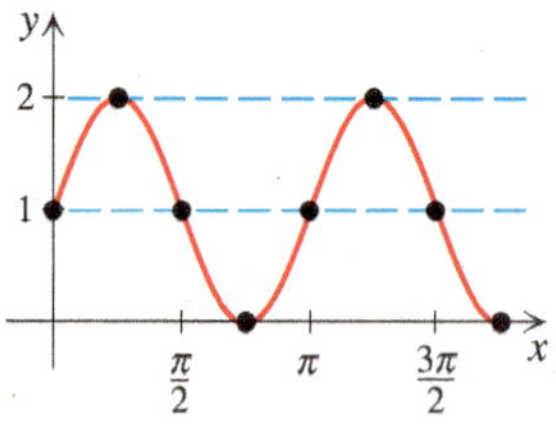


**83.**

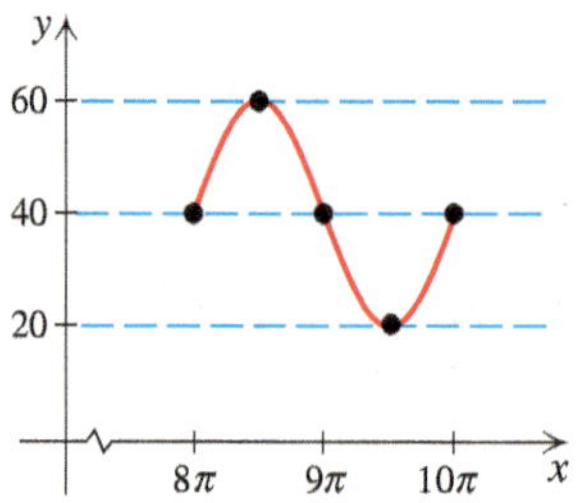


**84.**

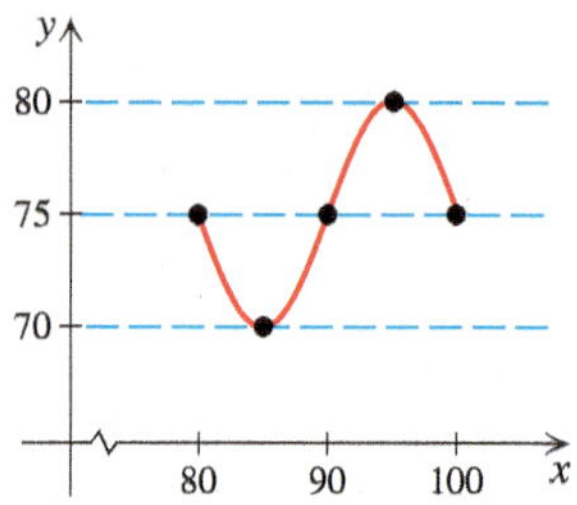


*Solve each problem.*

**85.** Find $\alpha$ if $\alpha$ is the angle between 90° and 180° whose sine is 1/2.

**86.** Find $\beta$ if $\beta$ is the angle between 180° and 270° whose cosine is $-\sqrt{3}/2$.

**87.** Find $\sin\alpha$, given that $\cos\alpha = 1/5$ and $\alpha$ is in quadrant IV.

**88.** Find $\tan\alpha$, given that $\sin\alpha = 1/3$ and $\alpha$ is in quadrant II.

**89.** Find $x$ if $x$ is the length of the shortest side of a right triangle that has a 16° angle and a hypotenuse of 24 ft.

**90.** Find $x$ if $x$ is the length of the hypotenuse of a right triangle that has a 22° angle and a shortest side of length 12 m.

**91.** Find $\alpha$ if $\alpha$ is the largest acute angle of a right triangle that has legs with lengths 6 cm and 8 cm.

**92.** Find $\beta$ if $\beta$ is the largest acute angle of a right triangle that has a hypotenuse with length 19 yd and a leg with length 8 yd.

**93.** *Broadcasting the Oldies* If radio station Q92 is broadcasting its oldies at 92.3 FM, then it is broadcasting at a frequency of 92.3 megahertz, or $92.3 \times 10^6$ cycles per second. What is the period of a wave with this frequency? (FM stands for frequency modulation.)

**94.** *AM Radio* If WLS in Chicago is broadcasting at 890 AM (amplitude modulation), then its signal has a frequency of 890 kilohertz, or $890 \times 10^3$ cycles per second. What is the period of a wave with this frequency?

**95.** *Irrigation* A center-pivot irrigation system waters a circular field by rotating about the center of the field. If the system makes one revolution in 8 hr, what is the linear velocity in feet per hour of a nozzle that is 120 ft from the center?

**96.** *Angular Velocity* If a bicycle with a 26-in.-diameter wheel is traveling 16 mph, then what is the angular velocity of the valve stem in radians per hour?

**97.** *Crooked Man* A man is standing 1000 ft from a surveyor. The surveyor measures an angle of 0.4° sighting from the man's feet to his head. Assume that the man can be represented by the arc intercepted by a central angle of 0.4° in a circle of radius 1000 ft. Find the height of the man.

**98.** *Straight Man* Assume that the man of Exercise 97 can be represented by the side $a$ of a right triangle with angle $\alpha = 0.4°$ and $b = 1000$ ft. Find the height of the man, and compare your answer with the answer of the previous exercise.

**99.** *Oscillating Depth* The depth of water in a tank oscillates between 12 ft and 16 ft. It takes 10 min for the depth to go from 12 to 16 ft and 10 min for the depth to go from 16 to 12 ft. Express the depth as a function of time in the form $y = A\sin[B(x - C)] + D$, where the depth is 16 ft at time 0. Graph one cycle of the function.

**100.** *Oscillating Temperature* The temperature of the water in a tank oscillates between 100°F and 120°F. It takes 30 min for the temperature to go from 100° to 120° and 30 min for the temperature to go from 120° to 100°. Express the temperature as a function of time in the form $y = A\sin[B(x - C)] + D$, where the temperature is 100° at time 20 min. Graph one cycle of the function.

**101.** *Shooting a Target* Judy is standing 200 ft from a circular target with a radius of 3 in. To hit the center of the circle, she must hold the gun perfectly level, as shown in the figure. Will she hit the target if her aim is off by one-tenth of a degree in any direction?

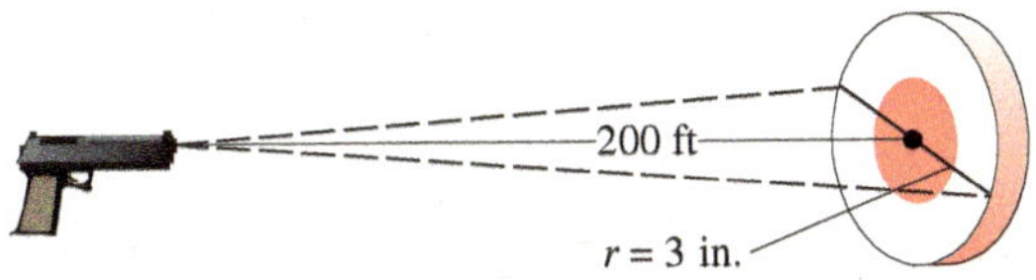


**Figure for Exercise 101**

**102.** *Height of Buildings* Two buildings are 300 ft apart. From the top of the shorter building, the angle of elevation of the top of the taller building is 23°, and the angle of depression of the base of the taller building is 36°, as shown in the figure. How tall is each building?

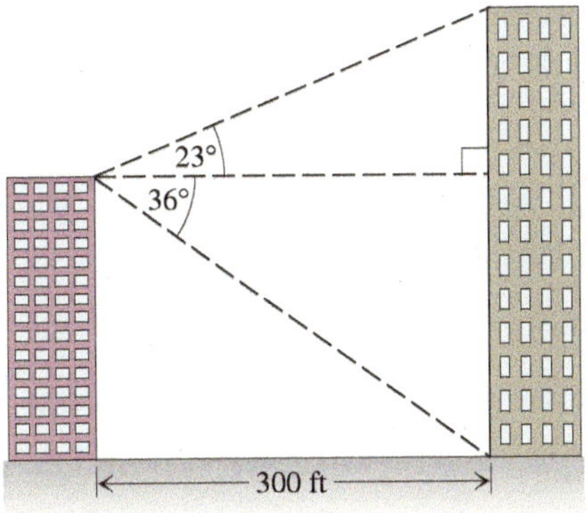


**Figure for Exercise 102**

**103.** *Rotating Space Station* One design for a space station is a giant rotating "doughnut" where humans live and work on the outside wall, as shown in the figure. The inhabitants would have an artificial gravity equal to the normal gravity on the earth when the time $T$ in seconds for one revolution is related to the radius $r$ in meters by the equation $T^2g = 4\pi^2r$. If $g$ is acceleration of gravity (9.8 m/sec$^2$) and the station is built with $r = 90$ m, then what angular velocity in radians per second would be necessary to produce artificial gravity?

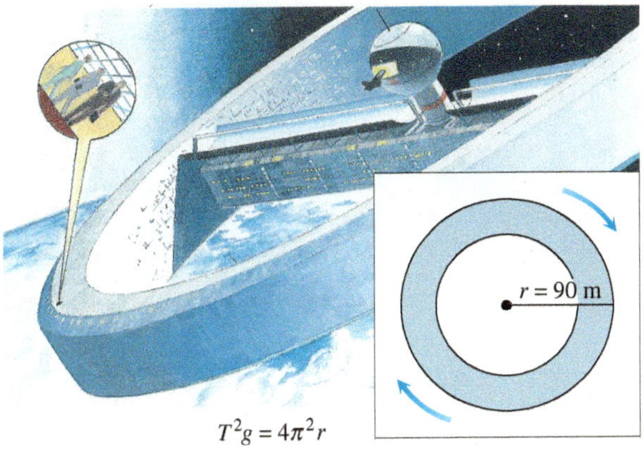


**Figure for Exercise 103**

**104.** *Cloud Height* Visual Flight Rules require that the height of the clouds be more than 1000 ft for a pilot to fly without instrumentation. At night, cloud height can be determined from the ground by using a searchlight and an observer, as shown in the figure. If the beam of light is aimed straight upward and the observer 500 ft away sights the cloud with an angle of elevation of 55°, then what is the height of the cloud cover?

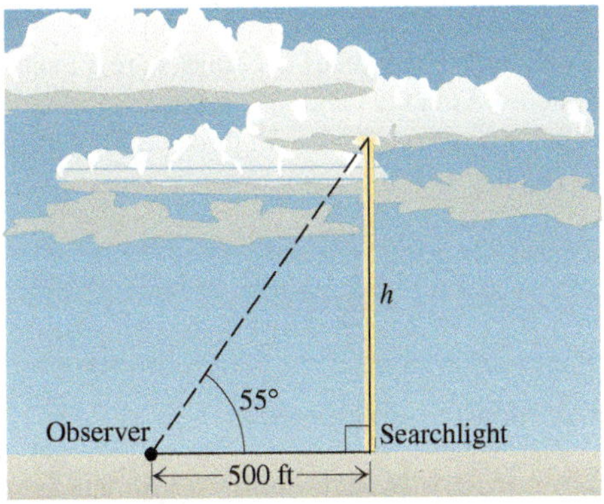


**Figure for Exercise 104**

**105.** *Radar Range* The radar antenna on a cargo ship is located 90 feet above the water, as shown in the accompanying figure. Find the distance to the horizon from that height. Use 3950 miles as the radius of the earth.

**106.** *Increasing Visibility* To increase its radar visibility, a sailboat has a radar reflector mounted on the mast, 25 feet above the water. Find the distance to the horizon from that height. Assuming that radar waves travel in a straight line, at what distance can the cargo ship of the previous exercise get a radar image of the sailboat?

**Figure for Exercises 105 and 106**

**107.** *Two Pulleys and a Belt* The distance between the centers of two circular pulleys is 12 inches. The radii of the pulleys are 3 inches and 6 inches. Find the length (to the nearest tenth of an inch) of the belt that wraps around the two pulleys, as shown in the accompanying figure.

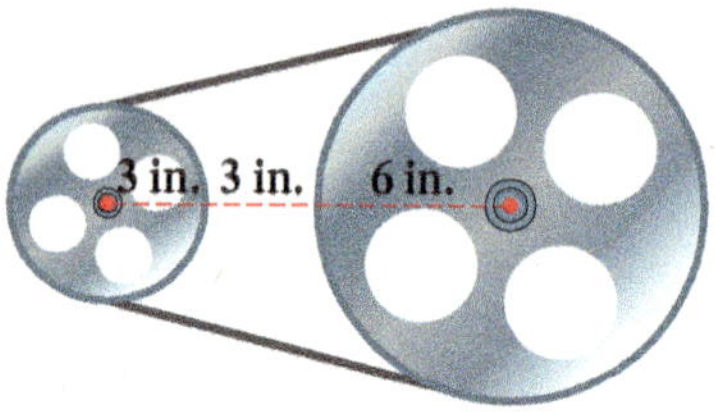


**Figure for Exercise 107**

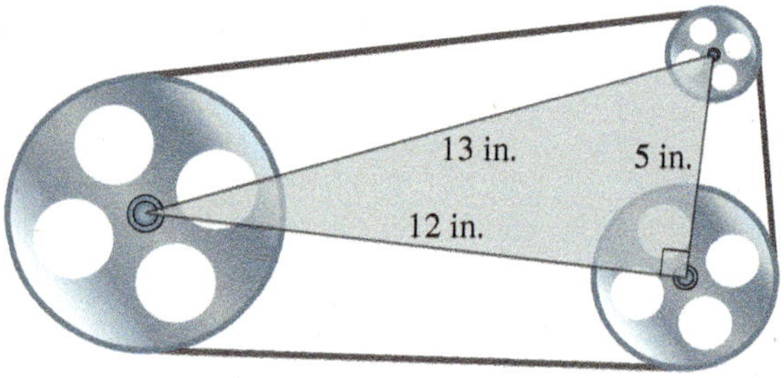


**Figure for Exercise 108**

**108.** *Three Pulleys and a Belt* The centers of three pulleys are located at the vertices of a right triangle whose sides are 5 inches, 12 inches, and 13 inches, as shown in the accompanying diagram. The radii of the pulleys are 1 inch, 2 inches, and 3 inches. Find the length (to the nearest tenth of an inch) of the belt that wraps around all three pulleys, as shown in the diagram.

### THINKING OUTSIDE THE BOX XLII

*Buckling Bridge* A 100-ft bridge expands 1 in. during the heat of the day. Since the ends of the bridge are embedded in rock, the bridge buckles upward and forms an arc of a circle for which the original bridge is a chord. What is the approximate distance moved by the center of the bridge?

# Chapter 5 Test

*Find the exact value of each expression.*

**1.** $\cos 420°$ **2.** $\sin(-390°)$ **3.** $\tan(3\pi/4)$

**4.** $\sec(-\pi/3)$ **5.** $\csc(7\pi/6)$ **6.** $\cot(-2\pi/3)$

**7.** $\sin^{-1}(-1/2)$ **8.** $\cos^{-1}(-1/2)$ **9.** $\arctan(-1)$

**10.** $\sec^{-1}(\sqrt{3}/2)$ **11.** $\csc^{-1}(-\sqrt{2}/2)$

**12.** $\sin(\cos^{-1}(-1/3))$

*Sketch at least one cycle of the graph of each function. Determine the period, range, and amplitude for each function.*

**13.** $y = \sin(3x) - 2$ **14.** $y = \cos(x + \pi/2)$

**15.** $y = \tan(\pi x/2)$ **16.** $y = 2\sin(2x + \pi)$

**17.** $y = 2\sec(x - \pi)$ **18.** $y = \csc(x - \pi/2)$

**19.** $y = \cot(2x)$ **20.** $y = -\cos(x - \pi/2)$

*Solve each problem.*

**21.** Find the arc length intercepted by a central angle of $46°24'6''$ in a circle with a radius of 35.62 m.

**22.** Find the degree measure of an angle of 2.34 radians.

**23.** Find the exact value of $\cos \alpha$, given that $\sin \alpha = 1/4$ and $\alpha$ is in quadrant II.

**24.** Find the exact values of all six trigonometric functions for an angle $\alpha$ in standard position whose terminal side contains the point $(5, -2)$.

**25.** If a bicycle wheel with a 26-in. diameter is making 103 revolutions per minute, then what is the angular velocity in radians per minute for a point on the tire?

**26.** At what speed in miles per hour will a bicycle travel if the rider can cause the 26-in.-diameter wheel to rotate 103 revolutions per minute?

**27.** To estimate the blood pressure of *Brachiosaurus,* Professor Ostrom wanted to estimate the height of the head of the famed *Brachiosaurus* skeleton at Humboldt University in Berlin. From a distance of 11 m from a point directly below the head, the angle of elevation of the head was approximately 48°. Use this information to find the height of the head of *Brachiosaurus*.

**28.** From a point on the street the angle of elevation of the top of the John Hancock Building is 65.7°. From a point on the street that is 100 ft closer to the building the angle of elevation is 70.1°. Find the height of the building.

**29.** The pH of a water supply oscillates between 7.2 and 7.8. It takes 2 days for the pH to go from 7.2 to 7.8 and 2 days for the pH to go from 7.8 to 7.2. Express the pH as a function of time in the form $y = A \sin[B(x - C)] + D$ where the pH is 7.2 on day 13. Graph one cycle of the function.

## TYING IT ALL together...

### Chapters 1–5

*Sketch the graph of each function. State the domain and range of each function.*

**1.** $y = 2 + e^x$ **2.** $y = 2 + x^2$ **3.** $y = 2 + \sin x$ **4.** $y = 2 + \ln(x)$

**5.** $y = \ln(x - \pi/4)$ **6.** $y = \sin(x - \pi/4)$ **7.** $y = \log_2(2x)$ **8.** $y = \cos(2x)$

*Determine whether each function is even, odd, or neither.*

**9.** $f(x) = e^x$ **10.** $f(x) = \sin x$ **11.** $f(x) = \cos x$ **12.** $f(x) = x^4 - x^2 + 1$

**13.** $f(x) = \tan x$ **14.** $f(x) = \sec x$ **15.** $f(x) = \ln(x)$ **16.** $f(x) = x^3 - 3x$

*Determine whether each function is increasing or decreasing on the given interval.*

**17.** $f(x) = \sin x$, $(0, \pi/2)$ **18.** $f(x) = 2^x$, $(-\infty, \infty)$ **19.** $f(x) = \tan x$, $(-\pi/2, \pi/2)$

**20.** $f(x) = \cos x$, $(0, \pi)$ **21.** $f(x) = x^2$, $(-\infty, 0)$ **22.** $f(x) = \sec x$, $(0, \pi/2)$

*Fill in the blanks.*

**23.** A trinomial of the form $a^2 + 2ab + b^2$ is called a(n) ________ trinomial.

**24.** A function of the form $f(x) = a^x$ where $a$ and $x$ are real numbers such that $a > 0$ and $a \neq 1$ is called a(n) ________ function.

**25.** The inverse of an exponential function is called a(n) ________ function.

**26.** A base-10 logarithm is called a(n) ________ logarithm.

**27.** A base-$e$ logarithm is called a(n) ________ logarithm.

**28.** A(n) ________ is a union of two rays with a common endpoint.

**29.** An angle whose vertex is the center of a circle is called a(n) ________ angle.

**30.** A(n) ________ angle has a degree measure between 0° and 90°.

**31.** An angle in standard position whose terminal side lies on an axis is a(n) ________ angle.

**32.** Angles in standard position that have the same initial side and same terminal side are called ________ angles.

## CONCEPTS OF calculus...

### Evaluating transcendental functions

*If we want the value of* $\sin(5)$, $e^3$, *or* $\ln(6)$, *we simply press a few buttons on our calculators and instantly get the answer. But there are no algebraic formulas that will produce values of transcendental functions like the trigonometric, exponential, and logarithmic functions. So how does a calculator do it? A calculator uses programs and formulas, but not the usual kind of formulas. A calculator uses infinite algebraic formulas called infinite series. The following infinite series are studied in calculus:*

$$e^x = 1 + x + \frac{x^2}{2!} + \frac{x^3}{3!} + \frac{x^4}{4!} + \cdots \text{ where } -\infty < x < \infty$$

$$\ln(1 + x) = x - \frac{x^2}{2} + \frac{x^3}{3} - \frac{x^4}{4} + \cdots \text{ where } -1 < x \le 1$$

$$\sin(x) = x - \frac{x^3}{3!} + \frac{x^5}{5!} - \frac{x^7}{7!} + \cdots \text{ where } -\infty < x < \infty$$

$$\cos(x) = 1 - \frac{x^2}{2!} + \frac{x^4}{4!} - \frac{x^6}{6!} + \cdots \text{ where } -\infty < x < \infty$$

*If the value of one of these functions is an irrational number, we cannot calculate it exactly. The more terms that we use from the infinite formula, the closer we get to the true value of the function.*

## Exercises

**1.** Use the first five terms of the formula for $e^x$ to calculate $e^{0.1}$ and compare your result to the value of $e^{0.1}$ obtained using the $e^x$-key on your calculator.

**2.** Find $\ln(2)$ by using the first 65 terms of the series and compare your result to the calculator value for $\ln(2)$.

**3.** Why can't we use the series for $\ln(1 + x)$ with $x = 4$ to calculate $\ln(5)$? Assuming that we know $\ln(2)$ correct to nine decimal places, find $\ln(5)$ by using the infinite series and the equation $\ln(5) = \ln(2^2 \cdot 1.25)$.

**4.** Find $\ln(9)$ using the method of the previous exercise.

**5.** Let $x = \pi/6$ and evaluate

$x$,

$x - x^3/3!$,

$x - x^3/3! + x^5/5!$,

$x - x^3/3! + x^5/5! - x^7/7!$,

and so on. Which is the first expression to give 0.5 on your calculator. What is $\sin(\pi/6)$?

**6.** Repeat the previous exercise with $x = 13\pi/6$.

**7.** Let $x = \pi/4$ and evaluate

$1 - x^2/2!$

$1 - x^2/2! + x^4/4!$,

$1 - x^2/2! + x^4/4! - x^6/6!$,

and so on. Which is the first expression to agree with your calculator's value for $\cos(\pi/4)$?

**8.** Repeat the previous exercise with $x = 9\pi/4$.

**9.** Make a conjecture about the size of $x$ and the number of terms needed to give an accurate value for $\sin(x)$ or $\cos(x)$.

# 6 Trigonometric Identities and Conditional Equations

People have been hurling missiles of one kind or another since the ice age. Whether a ball, a rock, or a bullet is thrown, we've learned through trial and error that a projectile's path is determined by its initial velocity and the angle at which it's launched.

Athletes achieve a balance between accuracy and distance through years of practice. Operators of modern artillery achieve accuracy by calculating velocity and angle of trajectory with mathematical precision.

**WHAT YOU WILL learn...** In this chapter, we will use trigonometric equations to analyze the flight of a projectile. Using trigonometry we can find the velocity and angle of trajectory required to launch a projectile a given distance. Then, using a trigonometric identity, we will find the proper angle of trajectory to achieve the maximum distance for a projectile.

# 6.1 Basic Identities

All of the trigonometric functions are related to each other because they are all defined in terms of the coordinates on a unit circle. For this reason, any expression involving them can be written in many different forms. Identities are used to simplify expressions and determine whether expressions are equivalent. Recall that an identity is an equation that is satisfied by *every* number for which both sides are defined. There are infinitely many trigonometric identities, but only the most common identities should be memorized. Here we review the most basic ones.

## Identities from Definitions

We saw in Section 5.4 that the following identities follow directly from the unit circle definitions of tangent, cotangent, secant, and cosecant.

**Identities from the Definitions**

If $\alpha$ is any angle or real number

$$\tan\alpha = \frac{\sin\alpha}{\cos\alpha},\quad \cot\alpha = \frac{\cos\alpha}{\sin\alpha},\quad \sec\alpha = \frac{1}{\cos\alpha},\quad \text{and}\quad \csc\alpha = \frac{1}{\sin\alpha},$$

provided no denominator is zero.

From these identities it follows that tangent and cotangent are reciprocals of each other, secant and cosine are reciprocals of each other, and cosecant and sine are reciprocals of each other.

**Reciprocal Identities**

$$\sin\alpha = \frac{1}{\csc\alpha}\qquad \cos\alpha = \frac{1}{\sec\alpha}\qquad \tan\alpha = \frac{1}{\cot\alpha}$$

$$\csc\alpha = \frac{1}{\sin\alpha}\qquad \sec\alpha = \frac{1}{\cos\alpha}\qquad \cot\alpha = \frac{1}{\tan\alpha}$$

The fundamental identity $\sin^2 x + \cos^2 x = 1$ from Section 5.2 is based on the definitions of $\sin x$ and $\cos x$ as coordinates of a point on the unit circle. If we divide each side of this identity by $\sin^2 x$, we get a new identity:

$$\frac{\sin^2 x}{\sin^2 x} + \frac{\cos^2 x}{\sin^2 x} = \frac{1}{\sin^2 x}$$

$$1 + \left(\frac{\cos x}{\sin x}\right)^2 = \left(\frac{1}{\sin x}\right)^2$$

$$1 + \cot^2 x = \csc^2 x$$

If we divide each side of the fundamental identity by $\cos^2 x$, we get another new identity:

$$\frac{\sin^2 x}{\cos^2 x} + \frac{\cos^2 x}{\cos^2 x} = \frac{1}{\cos^2 x}$$

$$\tan^2 x + 1 = \sec^2 x$$

Since these three identities are derived from the Pythagorean theorem, they are called the **Pythagorean identities.**

**Pythagorean Identities**

$$\sin^2 x + \cos^2 x = 1\qquad 1 + \cot^2 x = \csc^2 x\qquad \tan^2 x + 1 = \sec^2 x$$

We can support our conclusion that an equation is an identity by examining a graph. For example, consider the graph of $y = \sin^2 x + \cos^2 x$ in Fig. 6.1. The fact that $\sin^2 x + \cos^2 x = 1$ for 95 values of $x$ supports the conclusion that $\sin^2 x + \cos^2 x = 1$ for all real numbers. Because a calculator evaluates a function at a finite number of values only, a calculator graph does not prove that an equation is an identity. Note that $\sin x = 1$ is satisfied for 95 values of $x$ in Fig. 6.2 but it is certainly not an identity.

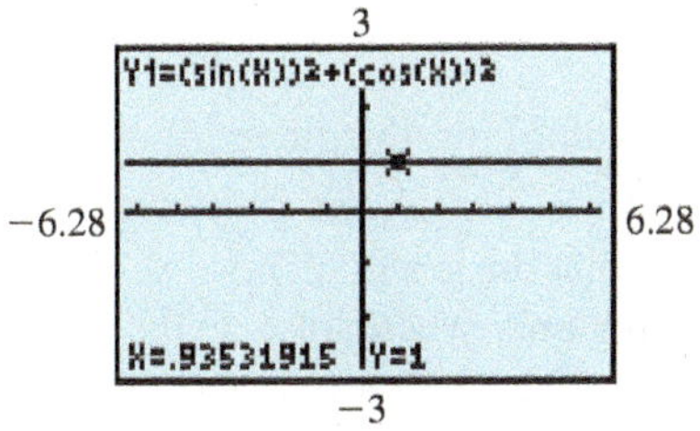


Figure 6.1

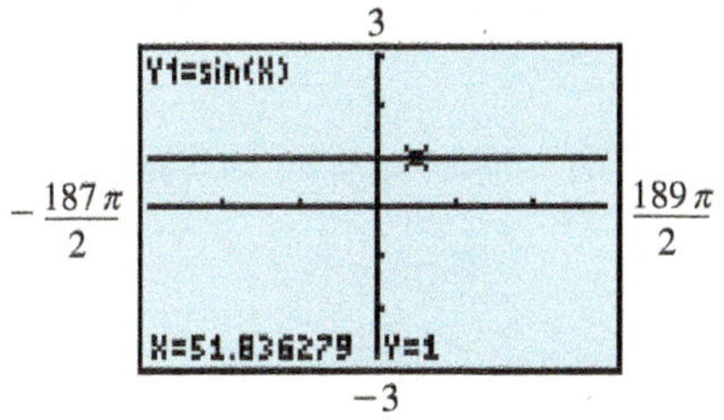


Figure 6.2

■ **Foreshadowing Calculus**

Trigonometric identities are used extensively in calculus. We often use identities to change the form or simplify a function when solving calculus problems involving trigonometric functions.

## Using Identities

Identities are used in many ways. One way is to simplify expressions involving trigonometric functions. In the next example we follow the strategy of writing each expression in terms of sines and cosines only and then simplifying the expression.

### EXAMPLE 1 Using identities to simplify

Write each expression in terms of sines and/or cosines, and then simplify.

**a.** $\dfrac{\tan x}{\sec x}$ **b.** $\sin x + \cot x \cos x$

**Solution**

**a.** Rewrite each trigonometric function in terms of sine and cosine:

$$\frac{\tan x}{\sec x} = \frac{\frac{\sin x}{\cos x}}{\frac{1}{\cos x}} = \frac{\sin x}{\cos x} \cdot \frac{\cos x}{1} \qquad \text{Invert and multiply.}$$

$$= \sin x$$

**b.** $$\sin x + \cot x \cos x = \sin x + \frac{\cos x}{\sin x} \cdot \cos x \qquad \text{Rewrite using sines and cosines.}$$

$$= \sin x + \frac{\cos^2 x}{\sin x}$$

$$= \frac{\sin^2 x}{\sin x} + \frac{\cos^2 x}{\sin x} \qquad \text{Multiply sin x by } \sin x/\sin x.$$

$$= \frac{\sin^2 x + \cos^2 x}{\sin x}$$

$$= \frac{1}{\sin x} \qquad \sin^2 x + \cos^2 x = 1$$

$$= \csc x \qquad \text{Use a basic identity.}$$

▶**TRY THIS.** Simplify $\dfrac{\tan x \csc x}{\sec x}$. ■

Using identities we can write any one of the six trigonometric functions in terms of any other. For example, the identity

$$\sin x = \frac{1}{\csc x}$$

expresses the sine function in terms of the cosecant. To write sine in terms of cosine, we rewrite the identity $\sin^2 x + \cos^2 x = 1$:

$$\sin^2 x = 1 - \cos^2 x$$
$$\sin x = \pm\sqrt{1 - \cos^2 x}$$

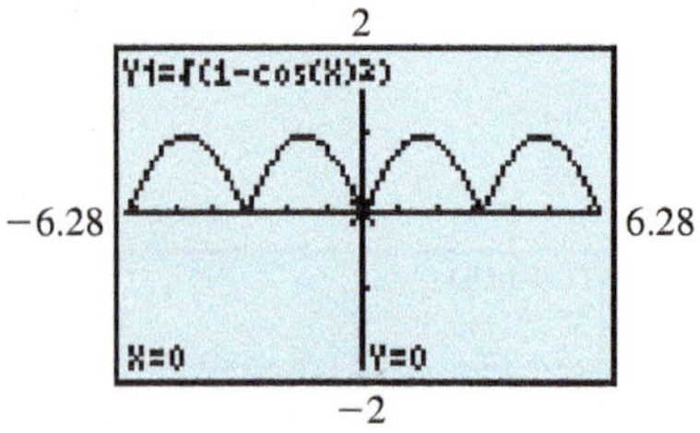


**Figure 6.3**

The $\pm$ symbol in this identity means that if $\sin x > 0$, then the identity is $\sin x = \sqrt{1 - \cos^2 x}$, and if $\sin x < 0$, then the identity is $\sin x = -\sqrt{1 - \cos^2 x}$. Note that the graph of $y = \sqrt{1 - (\cos x)^2}$ in Fig. 6.3 is the same as the graph of $y = \sin x$ only when $\sin x > 0$. □

## EXAMPLE 2 Writing one function in terms of another

Write an identity that expresses the tangent function in terms of the sine function.

### Solution

The Pythagorean identity $\sin^2 x + \cos^2 x = 1$ yields $\cos x = \pm\sqrt{1 - \sin^2 x}$. Since $\tan x = \sin x/\cos x$, we can write

$$\tan x = \pm\frac{\sin x}{\sqrt{1 - \sin^2 x}}.$$

We use the positive or negative sign, depending on the value of $x$.

▸**TRY THIS.** Write an identity that expresses the cotangent function in terms of the sine function. ■

Because all of the trigonometric functions are related by identities, we can find their values for an angle if we know the value of any one of them and the quadrant of the angle.

## EXAMPLE 3 Using identities to find function values

Given that $\tan \alpha = -2/3$ and $\alpha$ is in quadrant IV, find the values of the remaining five trigonometric functions at $\alpha$ by using identities.

### Solution

Use $\tan \alpha = -2/3$ in the identity $\sec^2 \alpha = 1 + \tan^2 \alpha$:

$$\sec^2 \alpha = 1 + \left(-\frac{2}{3}\right)^2 = \frac{13}{9}$$
$$\sec \alpha = \pm\frac{\sqrt{13}}{3}$$

Because $\alpha$ is in quadrant IV, $\cos \alpha > 0$ and $\sec \alpha > 0$. So $\sec \alpha = \sqrt{13}/3$. Since cosine is the reciprocal of the secant and cotangent is the reciprocal of tangent,

$$\cos \alpha = \frac{3}{\sqrt{13}} = \frac{3\sqrt{13}}{13} \quad \text{and} \quad \cot \alpha = -\frac{3}{2}.$$

Since $\sin \alpha$ is negative in quadrant IV, $\sin \alpha = -\sqrt{1 - \cos^2 \alpha}$:

$$\sin \alpha = -\sqrt{1 - \frac{9}{13}} = -\sqrt{\frac{4}{13}} = -\frac{2\sqrt{13}}{13}$$

Since cosecant is the reciprocal of sine, $\csc \alpha = -\sqrt{13}/2$.

►**TRY THIS.** Suppose that $\cot \alpha = -1/3$ and $\alpha$ is in quadrant II. Find $\sin \alpha$ and $\cos \alpha$. ■

In the next example we use the Pythagorean identities to rewrite a composition of trigonometric functions as an algebraic function.

### EXAMPLE 4 Converting compositions to algebraic functions

Find an equivalent algebraic expression for $\sin(\arctan(x))$.

**Solution**

Let $\theta = \arctan(x)$. Since $\tan(\theta) = x$ and $\tan^2(\theta) + 1 = \sec^2(\theta)$, we have $\sec^2(\theta) = x^2 + 1$ and

$$\cos^2(\theta) = \frac{1}{x^2 + 1}.$$

Since $\sin^2(\theta) = 1 - \cos^2(\theta)$, we have

$$\sin^2(\theta) = 1 - \frac{1}{x^2+1} = \frac{x^2+1}{x^2+1} - \frac{1}{x^2+1} = \frac{x^2}{x^2+1}$$

and

$$\sin(\theta) = \pm\sqrt{\frac{x^2}{x^2+1}}.$$

To simplify the last equation we use the fact that $\theta = \arctan(x)$. If $x > 0$, then $0 < \theta < \frac{\pi}{2}$ and $\sin(\theta) > 0$. If $x < 0$, then $-\frac{\pi}{2} < \theta < 0$ and $\sin(\theta) < 0$. In either case $x$ and $\sin(\theta)$ have the same sign. So we have

$$\sin(\theta) = \frac{x}{\sqrt{x^2+1}}$$

and

$$\sin(\arctan(x)) = \frac{x}{\sqrt{x^2+1}}.$$

►**TRY THIS.** Find an equivalent algebraic expression for $\sec(\arccos(x))$. ■

**■ Foreshadowing Calculus**

Example 4 shows how to write $\sin(\theta)$ in terms of $x$ when it is known that $\tan(\theta) = x$. This relationship between algebraic and trigonometric functions is important in calculus.

The identities discussed so far give relationships between the different trigonometric functions. The next identities show how the value of each trigonometric function for $-x$ is related to its value for $x$.

## Odd and Even Identities

In Section 2.3 we defined an **odd function** as one for which $f(-x) = -f(x)$ and an **even function** as one for which $f(-x) = f(x)$. In algebra most of the odd functions have odd exponents and the even functions have even exponents. We can also classify each of the trigonometric functions as either odd or even, but we cannot make the determination based on exponents. Instead, we examine the definitions of the functions in terms of the unit circle.

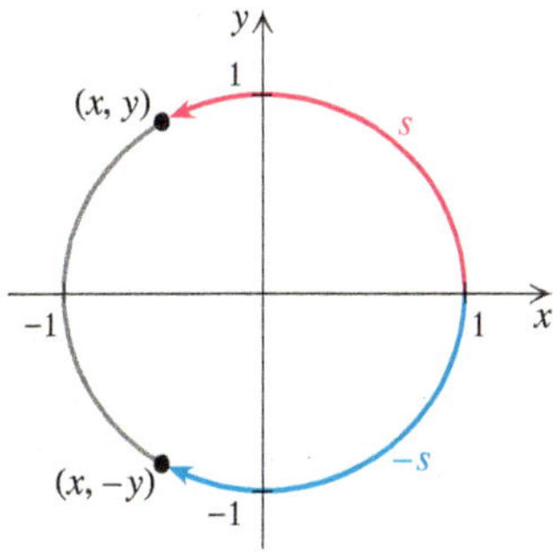


Figure 6.4

Figure 6.4 shows the real numbers $s$ and $-s$ as arcs on the unit circle. If the terminal point of $s$ is $(x, y)$, then the terminal point of $-s$ is $(x, -y)$. The $y$-coordinate for the terminal point of $-s$ is the opposite of the $y$-coordinate for the terminal point of $s$. So $\sin(-s) = -\sin(s)$, and sine is an odd function. Since the $x$-coordinates for $s$ and $-s$ are equal, $\cos(-s) = \cos(s)$, and cosine is an even function. Because sine is an odd function, cosecant is also an odd function:

$$\csc(-s) = \frac{1}{\sin(-s)} = \frac{1}{-\sin(s)} = -\csc(s)$$

Likewise we can establish that secant is an even function and tangent and cotangent are both odd functions. These arguments establish the following identities.

## Odd and Even Identities

| | | |
|---|---|---|
| **Odd:** | $\sin(-x) = -\sin(x)$ | $\csc(-x) = -\csc(x)$ |
| | $\tan(-x) = -\tan(x)$ | $\cot(-x) = -\cot(x)$ |
| **Even:** | $\cos(-x) = \cos(x)$ | $\sec(-x) = \sec(x)$ |

To help you remember which of the trigonometric functions are odd and which are even, remember that the graph of an odd function is symmetric about the origin and the graph of an even function is symmetric about the $y$-axis. The graphs of $y = \sin x$, $y = \csc x$, $y = \tan x$, and $y = \cot x$ are symmetric about the origin, while $y = \cos x$ and $y = \sec x$ are symmetric about the $y$-axis.

### EXAMPLE 5 Using odd and even identities

Simplify each expression.

**a.** $\sin(-x)\cot(-x)$ **b.** $\dfrac{1}{1 + \cos(-x)} + \dfrac{1}{1 - \cos x}$

**Solution**

**a.** $\sin(-x)\cot(-x) = (-\sin x)(-\cot x)$

$$= \sin x \cdot \cot x = \sin x \cdot \frac{\cos x}{\sin x} = \cos x$$

**b.** To add these expressions, we need a common denominator.

$$\frac{1}{1 + \cos(-x)} + \frac{1}{1 - \cos x} = \frac{1}{1 + \cos x} + \frac{1}{1 - \cos x} \quad \text{Because } \cos(-x) = \cos x$$

$$= \frac{1(1 - \cos x)}{(1 + \cos x)(1 - \cos x)} + \frac{1(1 + \cos x)}{(1 - \cos x)(1 + \cos x)}$$

$$= \frac{1 - \cos x}{1 - \cos^2 x} + \frac{1 + \cos x}{1 - \cos^2 x}$$

$$= \frac{1 - \cos x}{\sin^2 x} + \frac{1 + \cos x}{\sin^2 x} \quad \text{Pythagorean identity}$$

$$= \frac{1 - \cos x + 1 + \cos x}{\sin^2 x} = \frac{2}{\sin^2 x}$$

$$= 2 \cdot \frac{1}{\sin^2 x} = 2\csc^2 x$$

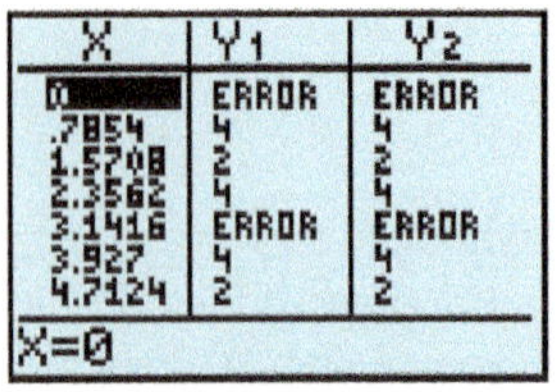


Figure 6.5

We can check this result by making a table for $y_1 = 1/(1 + \cos(-x)) + 1/(1 - \cos(x))$ and $y_2 = 2/(\sin(x))^2$, as shown in Fig. 6.5. The identical values for $y_1$ and $y_2$ support the result.

▸**TRY THIS.** Simplify $\csc(-x)\tan(-x)$. ■

### EXAMPLE 6 Odd or even functions

Determine whether each function is odd, even, or neither.

**a.** $f(x) = \dfrac{\cos(2x)}{x}$ **b.** $g(t) = \sin t + \cos t$

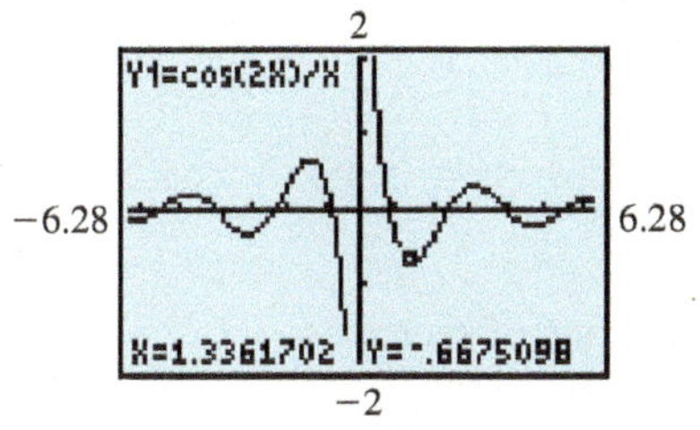


Figure 6.6

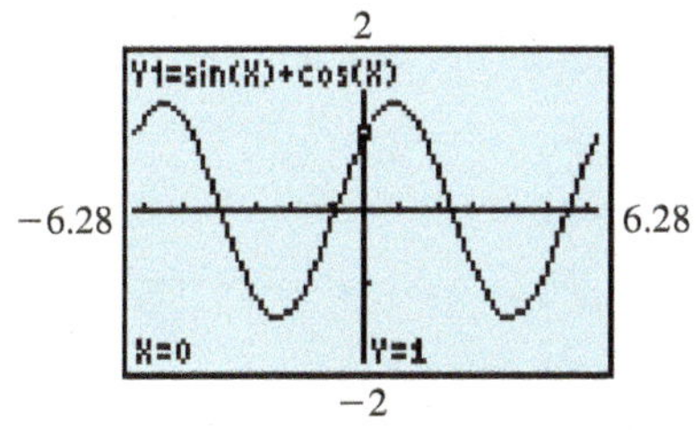


Figure 6.7

### Solution

**a.** We replace $x$ by $-x$ and see whether $f(-x)$ is equal to $f(x)$ or $-f(x)$:

$$\begin{aligned} f(-x) &= \frac{\cos(2(-x))}{-x} && \text{Replace } x \text{ by } -x. \\ &= \frac{\cos(-2x)}{-x} \\ &= \frac{\cos(2x)}{-x} && \text{Since cosine is an even function} \\ &= -\frac{\cos(2x)}{x} \end{aligned}$$

Since $f(-x) = -f(x)$, the function $f$ is an odd function.

The graph of $y = \cos(2x)/x$ in Fig. 6.6 appears to be symmetric about the origin, which supports the conclusion that the function is odd.

**b.** 

$$\begin{aligned} g(-t) &= \sin(-t) + \cos(-t) && \text{Replace } t \text{ by } -t. \\ &= -\sin t + \cos t && \text{Sine is odd and cosine is even.} \end{aligned}$$

Since $g(-t) \neq g(t)$ and $g(-t) \neq -g(t)$, the function is neither odd nor even.

Since the graph of $y = \sin(x) + \cos(x)$ in Fig. 6.7 is neither symmetric about the origin nor symmetric about the $y$-axis, the graph supports the conclusion that the function is neither odd nor even.

▸**TRY THIS.** Determine whether $f(x) = \csc(x) + \tan(x)$ is an odd or even function.

## Identity or Not?

In this section we have seen the basic identities, the reciprocal identities, the Pythagorean identities, and the odd-even identities and we are just getting started. With so many identities in trigonometry it might seem difficult to determine whether a given equation is an identity. Remember that an identity is satisfied by every value of the variable for which both sides are defined. If you can find one number for which the left-hand side of the equation has a value different from the right-hand side, then the equation is not an identity.

### EXAMPLE 7 Proving that an equation is not an identity

Show that $\sin(2t) = 2\sin(t)$ is not an identity.

### Solution

If $t = \pi/4$, then

$$\sin(2t) = \sin\left(2 \cdot \frac{\pi}{4}\right) = \sin\left(\frac{\pi}{2}\right) = 1$$

and

$$2\sin(t) = 2\sin\left(\frac{\pi}{4}\right) = 2 \cdot \frac{\sqrt{2}}{2} = \sqrt{2}.$$

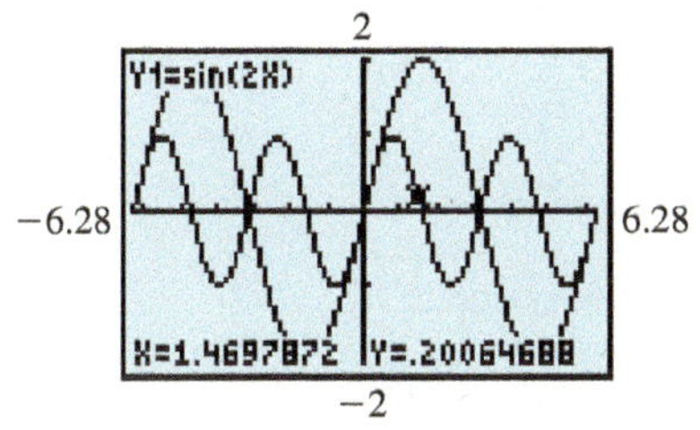


Figure 6.8

Since the values of $\sin(2t)$ and $2\sin(t)$ are unequal for $t = \pi/4$, the equation is not an identity.

The graphs of $y_1 = \sin(2x)$ and $y_2 = 2\sin(x)$ in Fig. 6.8 support the conclusion that $\sin(2t) = 2\sin(t)$ is not an identity.

▸**TRY THIS.** Show that $\cos(3t) = 3\cos(t)$ is not an identity.

An equation that is not an identity may have many solutions. So you might have to try more than one value for the variable to find one that fails to satisfy the equation. Section 6.2 will be devoted to proving that given equations are identities.

## FOR thought... True or False? Explain.

1. The equation $\sin x = \cos x$ is an identity.
2. If we simplify the expression $(\tan x)(\cot x)$, we get 1.
3. The function $f(x) = \sin^2 x$ is an even function.
4. The function $f(x) = \cos^3 x$ is an odd function.
5. If $\sin(2x) = 2\sin(x)\cos(x)$ is an identity, then so is $\dfrac{\sin(2x)}{\sin(x)} = 2\cos(x)$.
6. The equation $(\sin x + \cos x)^2 = \sin^2 x + \cos^2 x$ is an identity.
7. Tangent is written in terms of secant as $\tan x = \pm\sqrt{1 - \sec^2 x}$.
8. $\sin(-3)\cos(-3)\tan(-3)\sec(3)\csc(-3)\cot(3) = -1$
9. $\sin^2(-\pi/9) + \cos^2(-\pi/9) = -1$
10. $(1 - \sin(\pi/7))(1 + \sin(\pi/7)) = \cos^2(-\pi/7)$

## EXERCISES 6.1

*Fill in the blank.*

1. If $f(-x) = f(x)$, then $f$ is a(n) ________ function.
2. If $f(-x) = -f(x)$, then $f$ is a(n) ________ function.
3. The identify $1 + \cot^2 x = \csc^2 x$ is one of the ________ identities.
4. An equation that is true for all values of the variable for which both sides are defined is a(n) ________.
5. Sine, cosecant, tangent, and cotangent are ________ functions.
6. Cosine and secant are ________ functions.

*Write each expression in terms of sines and/or cosines, and then simplify.*

7. $\tan x \cos x$
8. $\sin x \cot x$
9. $\sec x \cos x$
10. $\sin x \csc x$
11. $\dfrac{\sec x}{\tan x}$
12. $\dfrac{\cot x}{\csc x}$
13. $\dfrac{\sin x}{\csc x} + \cos^2 x$
14. $\dfrac{\cos x}{\sec x} + \sin^2 x$
15. $\dfrac{\sin x}{\csc x} + \dfrac{\cos x}{\sec x}$
16. $\dfrac{1}{\sin^2 x} - \dfrac{1}{\tan^2 x}$
17. $(1 - \sin\alpha)(1 + \sin\alpha)$
18. $(\sec\alpha - 1)(\sec\alpha + 1)$
19. $(\cos\beta\tan\beta + 1)(\sin\beta - 1)$
20. $(1 + \cos\beta)(1 - \cot\beta\sin\beta)$
21. $\dfrac{1 + \cos\alpha\tan\alpha\csc\alpha}{\csc\alpha}$
22. $\dfrac{(\cos\alpha\tan\alpha + 1)(\sin\alpha - 1)}{\cos^2\alpha}$

*Write an identity that expresses the first function in terms of the second.*

23. $\cot(x)$, in terms of $\csc(x)$
24. $\sec(x)$, in terms of $\tan(x)$
25. $\sin(x)$, in terms of $\cot(x)$
26. $\cos(x)$, in terms of $\tan(x)$
27. $\tan(x)$, in terms of $\csc(x)$
28. $\cot(x)$, in terms of $\sec(x)$

*In each exercise, use identities to find the exact values at $\alpha$ for the remaining five trigonometric functions.*

29. $\tan\alpha = 1/2$ and $0 < \alpha < \pi/2$
30. $\sin\alpha = 3/4$ and $\pi/2 < \alpha < \pi$
31. $\cos\alpha = -\sqrt{3}/5$ and $\alpha$ is in quadrant III
32. $\sec\alpha = -4\sqrt{5}/5$ and $\alpha$ is in quadrant II
33. $\cot\alpha = -1/3$ and $-\pi/2 < \alpha < 0$
34. $\csc\alpha = \sqrt{3}$ and $0 < \alpha < \pi/2$

*Find an equivalent algebraic expression for each composition.*

**35.** $\sin(\arccos(x))$

**36.** $\cos(\arcsin(x))$

**37.** $\cos(\arctan(x))$

**38.** $\tan(\arccos(x))$

**39.** $\tan(\arcsin(x))$

**40.** $\sec(\arcsin(x))$

**41.** $\sec(\arctan(x))$

**42.** $\csc(\arcsin(x))$

*Simplify each expression.*

**43.** $\sin(-x)\cot(-x)$

**44.** $\sec(-x) - \sec(x)$

**45.** $\sin(y) + \sin(-y)$

**46.** $\cos(y) + \cos(-y)$

**47.** $\dfrac{\sin(x)}{\cos(-x)} + \dfrac{\sin(-x)}{\cos(x)}$

**48.** $\dfrac{\cos(-x)}{\sin(-x)} - \dfrac{\cos(-x)}{\sin(x)}$

**49.** $(1 + \sin(\alpha))(1 + \sin(-\alpha))$

**50.** $(1 - \cos(-\alpha))(1 + \cos(\alpha))$

**51.** $\sin(-\beta)\cos(-\beta)\csc(\beta)$

**52.** $\tan(-\beta)\csc(-\beta)\cos(\beta)$

*Determine whether each function is odd, even, or neither.*

**53.** $f(x) = \sin(2x)$

**54.** $f(x) = \cos(2x)$

**55.** $f(x) = \cos x + \sin x$

**56.** $f(x) = 2\sin x\cos x$

**57.** $f(t) = \sec^2(t) - 1$

**58.** $f(t) = 2 + \tan(t)$

**59.** $f(\alpha) = 1 + \sec\alpha$

**60.** $f(\beta) = 1 + \csc\beta$

**61.** $f(x) = \dfrac{\sin x}{x}$

**62.** $f(x) = x\cos x$

**63.** $f(x) = x + \sin x$

**64.** $f(x) = \csc(x^2)$

*Match each given expression with an equivalent expression (a)–(p).*

**65.** $\sin x$ **66.** $\cos x$ **67.** $\tan x$

**68.** $\cot x$ **69.** $\sec x$ **70.** $\csc x$

**71.** $\sin^2 x$ **72.** $\cos^2 x$ **73.** $\tan^2 x$

**74.** $\sin(-x)$ **75.** $\cos(-x)$ **76.** $\tan(-x)$

**77.** $\cot(-x)$ **78.** $\sec(-x)$ **79.** $\csc(-x)$

**80.** $\cot^2 x$ **81.** $\sec^2 x$ **82.** $\csc^2 x$

**a.** $-\sin x$ **b.** $-\cot x$ **c.** $-\tan x$

**d.** $1 + \tan^2 x$ **e.** $1 + \cot^2 x$ **f.** $-\csc x$

**g.** $\dfrac{1}{\sec x}$ **h.** $\dfrac{1}{\csc x}$ **i.** $\dfrac{1}{\sin x}$

**j.** $1 - \sin^2 x$ **k.** $1 - \cos^2 x$ **l.** $\sec^2(x) - 1$

**m.** $\dfrac{1}{\cos x}$ **n.** $\dfrac{\sin x}{\cos x}$ **o.** $\dfrac{\cos x}{\sin x}$

**p.** $\csc^2(x) - 1$

*Show that each equation is not an identity. Write your explanation in paragraph form.*

**83.** $(\sin\gamma + \cos\gamma)^2 = \sin^2\gamma + \cos^2\gamma$

**84.** $\tan^2 x - 1 = \sec^2 x$

**85.** $(1 + \sin\beta)^2 = 1 + \sin^2\beta$

**86.** $\sin(2\alpha) = \sin\alpha\cos\alpha$

**87.** $\sin\alpha = \sqrt{1 - \cos^2\alpha}$

**88.** $\tan\alpha = \sqrt{\sec^2\alpha - 1}$

**89.** $\sin(y) = \sin(-y)$

**90.** $\cos(-y) = -\cos y$

**91.** $\cos^2 y - \sin^2 y = \sin(2y)$

**92.** $\cos(2x) = 2\cos x\sin x$

*Use identities to simplify each expression.*

**93.** $1 - \dfrac{1}{\cos^2 x}$

**94.** $\dfrac{\sin^4 x - \sin^2 x}{\sec x}$

**95.** $\dfrac{-\tan^2 t - 1}{\sec^2 t}$

**96.** $\dfrac{\cos w\sin^2 w + \cos^3 w}{\sec w}$

**97.** $\dfrac{\sin^2\alpha - \cos^2\alpha}{1 - 2\cos^2\alpha}$

**98.** $\dfrac{\sin^3(-\theta)}{\sin^3\theta - \sin\theta}$

**99.** $\dfrac{\tan^3 x - \sec^2 x\tan x}{\cot(-x)}$

**100.** $\sin x + \dfrac{\cos^2 x}{\sin x}$

**101.** $\dfrac{1}{\sin^3 x} - \dfrac{\cot^2 x}{\sin x}$

**102.** $1 - \dfrac{\sec^2 x}{\tan^2 x}$

**103.** $\sin^4 x - \cos^4 x$

**104.** $\csc^4 x - \cot^4 x$

*Solve each problem.*

**105.** Suppose $\sin\theta = 1/3$. What are the possible values for $\cos\theta$?

**106.** Suppose $\cos\theta = 4/5$. What are the possible values for $\tan\theta$?

**107.** Suppose $\sin\theta = u$. Write $\cos\theta$ in terms of $u$.

**108.** Suppose $\cos\theta = u$. Write $\tan\theta$ in terms of $u$.

**109.** For what values of $x$ is the identity $\tan x = \dfrac{\sin x}{\cos x}$ *not* valid?

**110.** For what values of $x$ is the identity $\cot x = \dfrac{\cos x}{\sin x}$ *not* valid?

## FOR WRITING/DISCUSSION

**111.** *Algebraic Identities* List as many algebraic identities as you can and explain why each one is an identity.

**112.** *Trigonometric Identities* List as many trigonometric identities as you can and explain why each one is an identity.

## RETHINKING

**113.** At a distance of 2000 feet from a building, the angle of elevation to the top of the building is 30°. Find the height of the building to the nearest foot.

**114.** Find the central angle (to the nearest tenth of a degree) that intercepts an arc of length 5 feet on a circle of radius 60 feet.

**115.** Find the amplitude, period, phase shift, and range for the function $f(x) = 5\cos(2x - \pi) + 3$.

**116.** If the period of a sine wave is 0.125 second, then what is the frequency?

**117.** If $\beta$ is an angle in standard position such that $\sin(\beta) = 1$, then what is $\cos(\beta)$?

**118.** Evaluate each function. Give the result in degrees.

**a.** $\sin^{-1}(-1/2)$

**b.** $\cos^{-1}(-1/2)$

**c.** $\tan^{-1}(-1)$

## THINKING OUTSIDE THE BOX XLIII

*Tangent Circles* The three large circles in the accompanying diagram are tangent to each other and each has radius 1. The small circle in the middle is tangent to each of the three large circles. Find its radius.

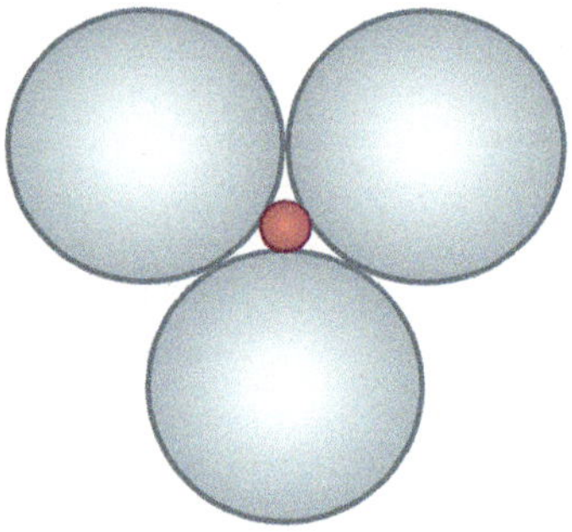

**Figure for Thinking Outside the Box XLIII**

## POP QUIZ 6.1

**1.** Simplify the expression $\cot x \sec x$.

**2.** Find exact values for $\cos \alpha$ and $\cot \alpha$ if $\sin \alpha = 1/3$ and $0 < \alpha < \pi/2$.

**3.** Is $f(x) = \cos(3x)$ even or odd?

**4.** Write an equivalent algebraic expression for $\cos(\arcsin(w))$.

**5.** Simplify $\dfrac{2}{\sec^2 \alpha} + \dfrac{2}{\csc^2 \alpha}$.

## LINKING concepts... For Individual or Group Explorations

### Modeling the Motion of a Projectile

*A projectile is fired with initial velocity of $v_0$ feet per second. The projectile can be pictured as being fired from the origin into the first quadrant, making an angle $\theta$ with the positive x-axis, as shown in the figure. If there is no air resistance, then at t seconds the coordinates of the projectile (in feet) are $x = v_0 t \cos\theta$ and $y = -16t^2 + v_0 t \sin\theta$. Suppose a projectile leaves the gun at 100 ft/sec and $\theta = 60°$.*

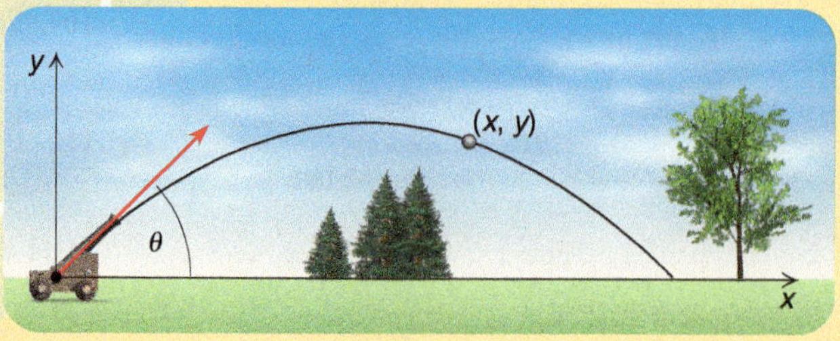


**a)** What are the coordinates of the projectile at time $t = 4$ sec?

**b)** For how many seconds is the projectile in the air?

**c)** How far from the gun does the projectile land?

**d)** What is the maximum height attained by the projectile?

**e)** Find expressions in terms of $v_0$ and $\theta$ for the time in the air, the distance from the gun, and the maximum height.

**f)** Show that $y = -\frac{16 \sec^2 \theta}{v_0^2} x^2 + x \tan \theta$.

# 6.2 Verifying Identities

In Section 6.1 we saw that any trigonometric function can be expressed in terms of any other trigonometric function. So an expression involving one or more trigonometric functions could be written in many different equivalent forms. In trigonometry and calculus, where trigonometric functions are used, we must often decide whether two expressions are equivalent. The fact that two expressions are equivalent is expressed as an identity. In this section we concentrate on techniques for verifying that a given equation is an identity.

## Developing a Strategy

In simplifying expressions in Section 6.1, we used known identities and properties of algebra to rewrite a complicated expression as a simpler equivalent expression. It may not always be clear that one expression is simpler than another, but an expression that involves fewer symbols is generally considered simpler. To prove or verify that an equation is an identity, we start with the expression on one side of the equation and use known identities and properties of algebra to convert it into the expression on the other side of the equation. It is the same process as simplifying expressions, except that in this case we know exactly what we want as the final expression. If one side appears to be more complicated than the other, we usually start with the more complicated side and simplify it.

### EXAMPLE 1 Simplifying the complicated side

Verify that the following equation is an identity:

$$1 + \sec x \sin x \tan x = \sec^2 x$$

**Solution**

We start with the left-hand side, the more complicated side, and write it in terms of sine and cosine. Our goal is to get $\sec^2 x$ as the final simplified expression.

$$\begin{aligned} 1 + \sec x \sin x \tan x &= 1 + \frac{1}{\cos x} \sin x \frac{\sin x}{\cos x} \\ &= 1 + \frac{\sin^2 x}{\cos^2 x} \\ &= 1 + \tan^2 x \\ &= \sec^2 x && \text{Pythagorean identity} \end{aligned}$$

▶**TRY THIS.** Verify that $1 - \sec(x)\csc(x)\tan(x) = -\tan^2(x)$ is an identity. ■

In verifying identities it is often necessary to multiply binomials or factor trinomials involving trigonometric functions.

## EXAMPLE 2 Multiplying binomials

Find each product.

**a.** $(1 + \tan x)(1 - \tan x)$ **b.** $(2 \sin \alpha + 1)^2$

### Solution

**a.** The product of a sum and a difference is equal to the difference of two squares:

$$(1 + \tan x)(1 - \tan x) = 1 - \tan^2 x$$

**b.** Use $(a + b)^2 = a^2 + 2ab + b^2$ with $a = 2 \sin \alpha$ and $b = 1$:

$$(2 \sin \alpha + 1)^2 = 4 \sin^2 \alpha + 4 \sin \alpha + 1$$

▸**TRY THIS.** Find the product $(1 - 2 \sin x)(1 + 2 \sin x)$. ■

## EXAMPLE 3 Factoring with trigonometric functions

Factor each expression.

**a.** $\sec^2 x - \tan^2 x$ **b.** $\sin^2 \beta + \sin \beta - 2$

### Solution

**a.** A difference of two squares is equal to a product of a sum and a difference:

$$\sec^2 x - \tan^2 x = (\sec x + \tan x)(\sec x - \tan x)$$

**b.** Factor this expression as you would factor $x^2 + x - 2 = (x + 2)(x - 1)$:

$$\sin^2 \beta + \sin \beta - 2 = (\sin \beta + 2)(\sin \beta - 1)$$

▸**TRY THIS.** Factor $\sin^2 x + 4 \sin x + 4$. ■

The equations that appear in Examples 2 and 3 are identities. They are not the kind of identities that are memorized, because they can be obtained at any time by the principles of multiplying or factoring. The next example uses the identity $(1 - \sin \alpha)(1 + \sin \alpha) = 1 - \sin^2 \alpha$.

## EXAMPLE 4 An identity with equal fractions

Prove that the following equation is an identity:

$$\frac{\cos \alpha}{1 - \sin \alpha} = \frac{1 + \sin \alpha}{\cos \alpha}$$

### Solution

Before attempting to prove an identity, it is a good idea to check the calculator graphs of the opposite sides. The graphs of $y_1 = \cos(x)/(1 - \sin(x))$ and $y_2 = (1 + \sin(x))/\cos(x)$ in Fig. 6.9 appear to coincide. So the equation is probably an identity. □

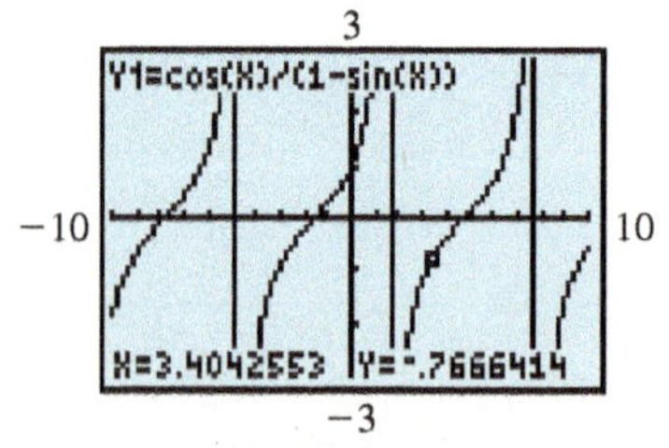


Figure 6.9

To prove that the equation is an identity, multiply the numerator and denominator of the left-hand side by $1 + \sin\alpha$, because $1 + \sin\alpha$ appears in the numerator of the right-hand side.

$$\frac{\cos\alpha}{1-\sin\alpha} = \frac{\cos\alpha(1+\sin\alpha)}{(1-\sin\alpha)(1+\sin\alpha)}$$

$$= \frac{\cos\alpha(1+\sin\alpha)}{1-\sin^2\alpha} \quad \text{Leave numerator in factored form.}$$

$$= \frac{\cos\alpha(1+\sin\alpha)}{\cos^2\alpha} \quad \text{Pythagorean identity}$$

$$= \frac{1+\sin\alpha}{\cos\alpha} \quad \text{Reduce.}$$

▸**TRY THIS.** Prove that $\frac{\csc x - 1}{\cot x} = \frac{\cot x}{\cot x + 1}$ is an identity. ■

In Example 4 the numerator and denominator were multiplied by $1 + \sin\alpha$ because the expression on the right-hand side has $1 + \sin\alpha$ in its numerator. We could have multiplied the numerator and denominator of the left-hand side by $\cos\alpha$, because the expression on the right-hand side has $\cos\alpha$ in its denominator. You should prove that the equation of Example 4 is an identity by multiplying the numerator and denominator by $\cos\alpha$.

Another strategy to use when a fraction has a sum or difference in its numerator is to write the fraction as a sum or difference of two fractions, as in the next example.

## EXAMPLE 5 Writing one fraction as two fractions

Prove that the following equation is an identity:

$$\frac{\csc x - \sin x}{\sin x} = \cot^2 x$$

### Solution

First rewrite the left-hand side as a difference of two rational expressions.

$$\frac{\csc x - \sin x}{\sin x} = \frac{\csc x}{\sin x} - \frac{\sin x}{\sin x}$$

$$= \csc x \cdot \frac{1}{\sin x} - 1$$

$$= \csc^2 x - 1 \quad \text{Since } 1/\sin x = \csc x$$

$$= \cot^2 x \quad \text{Pythagorean identity}$$

▸**TRY THIS.** Prove that $\frac{\sec x - \cos x}{\cos x} = \tan^2 x$ is an identity. ■

In verifying an identity, you can start with the expression on either side of the equation and use known identities and properties of algebra to get the expression on the other side. Note that we do *not* start with the given equation and work on it to get simpler and simpler equations as we do when solving an equation, because we do not yet know that the given equation is actually an identity. So when an expression involves the sum or difference of fractions, as in the next example, get a common denominator and combine the fractions. Do *not* multiply each side by the LCD as is done when solving an equation involving fractions.

## EXAMPLE 6 Starting with the right-hand side

Prove that the following equation is an identity:

$$2\tan^2 x = \frac{1}{\csc x - 1} - \frac{1}{\csc x + 1}$$

### Solution

In this equation, the right-hand side is the more complicated one. We will simplify the right-hand side by getting a common denominator and combining the fractions, keeping $2\tan^2 x$ in mind as our goal.

$$\frac{1}{\csc x - 1} - \frac{1}{\csc x + 1} = \frac{1(\csc x + 1)}{(\csc x - 1)(\csc x + 1)} - \frac{1(\csc x - 1)}{(\csc x + 1)(\csc x - 1)}$$

$$= \frac{\csc x + 1}{\csc^2 x - 1} - \frac{\csc x - 1}{\csc^2 x - 1}$$

$$= \frac{2}{\csc^2 x - 1} \quad \text{Subtract the numerators.}$$

$$= \frac{2}{\cot^2 x} \quad \text{Pythagorean identity}$$

$$= 2 \cdot \frac{1}{\cot^2 x} = 2\tan^2 x$$

▸**TRY THIS.** Prove that $-2\cot^2 x = \frac{1}{1 - \sec x} + \frac{1}{1 + \sec x}$ is an identity. ■

In every example so far, we started with one side and converted it into the other side. Although all identities can be verified by that method, sometimes it is simpler to convert both sides into the same expression. If both sides are shown to be equivalent to a common expression, then the identity is certainly verified. This technique works best when both sides are rather complicated and we simply set out to simplify them.

## EXAMPLE 7 Working on both sides

Prove that the following equation is an identity:

$$\frac{1 - \sin^2 t}{1 - \csc(-t)} = \frac{1 + \sin(-t)}{\csc t}$$

### Solution

Use $\csc(-t) = -\csc t$ to simplify the left-hand side:

$$\frac{1 - \sin^2 t}{1 - \csc(-t)} = \frac{1 - \sin^2 t}{1 + \csc t} = \frac{\cos^2 t}{1 + \csc t}$$

Now look for a way to get the right-hand side of the original equation equivalent to $\cos^2 t/(1 + \csc t)$. Since $\sin(-t) = -\sin t$, we can write $1 + \sin(-t)$ as $1 - \sin t$ and then get $\cos^2 t$ in the numerator by multiplying by $1 + \sin t$.

$$\frac{1 + \sin(-t)}{\csc t} = \frac{1 - \sin t}{\csc t} = \frac{(1 - \sin t)(1 + \sin t)}{\csc t(1 + \sin t)}$$

$$= \frac{1 - \sin^2 t}{\csc t + \csc t \sin t} = \frac{\cos^2 t}{\csc t + \frac{1}{\sin t}\sin t}$$

$$= \frac{\cos^2 t}{1 + \csc t}$$

Since both sides of the equation are equivalent to the same expression, the equation is an identity.

▶**TRY THIS.** Prove that $\frac{1 - \cos^2(-t)}{\sin(-t)} = \tan(-t)\cos(-t)$ is an identity. ■

## The Strategy

Verifying identities takes practice. There are many different ways to verify a particular identity, so just go ahead and get started. If you do not seem to be getting anywhere, try another approach. The strategy that we used in the examples is stated below. Other techniques are sometimes needed to prove identities, but you can prove most of the identities in this text using the following strategy.

**STRATEGY**

### Verifying Identities

1. Work on one side of the equation (usually the more complicated side), keeping in mind the expression on the other side as your goal.
2. Some expressions can be simplified quickly if they are rewritten in terms of sines and cosines only. (See Example 1.)
3. To convert one rational expression into another, multiply the numerator and denominator of the first by either the numerator or the denominator of the desired expression. (See Example 4.)
4. If the numerator of a rational expression is a sum or difference, convert the rational expression into a sum or difference of two rational expressions. (See Example 5.)
5. If a sum or difference of two rational expressions occurs on one side of the equation, then find a common denominator and combine them into one rational expression. (See Example 6.)

## FOR thought... True or False?

*Answer true if the equation is an identity and false if it is not. If false, find a value of x for which the two sides have different values.*

**1.** $\frac{\sin x}{\csc x} = \sin^2 x$

**2.** $\frac{\cot x}{\tan x} = \tan^2 x$

**3.** $\frac{\sec x}{\csc x} = \tan x$

**4.** $\sin x \sec x = \tan x$

**5.** $\frac{\cos x + \sin x}{\cos x} = 1 + \tan x$

**6.** $\sec x + \frac{\sin x}{\cos x} = \frac{1 + \sin x \cos x}{\cos x}$

**7.** $\frac{1}{1 - \sin x} = \frac{1 + \sin x}{\cos^2 x}$

**8.** $\tan x \cot x = 1$

**9.** $(1 - \cos x)(1 - \cos x) = \sin^2 x$

**10.** $(1 - \csc x)(1 + \csc x) = \cot^2 x$

## EXERCISES 6.2

*Match each expression on the left with one on the right that completes each equation as an identity.*

1. $\cos x \tan x = ?$ — **A.** $1$
2. $\sec x \cot x = ?$ — **B.** $-\tan^2 x$
3. $(\csc x - \cot x)(\csc x + \cot x) = ?$ — **C.** $\cot^2 x$
4. $\dfrac{\sin x + \cos x}{\sin x} = ?$ — **D.** $\sin x$
5. $(1 - \sec x)(1 + \sec x) = ?$ — **E.** $-\sin^2 x$
6. $(\csc x - 1)(\csc x + 1) = ?$ — **F.** $\dfrac{1}{\sin x \cos x}$
7. $\dfrac{\csc x - \sin x}{\csc x} = ?$ — **G.** $1 + \cot^2 x$
8. $\dfrac{\cos x - \sec x}{\sec x} = ?$ — **H.** $\cos^2 x$
9. $\dfrac{\csc x}{\sin x} = ?$ — **I.** $\csc x$
10. $\dfrac{\sin x}{\cos x} + \dfrac{\cos x}{\sin x} = ?$ — **J.** $1 + \cot x$

*Find the products.*

11. $(2\cos\beta + 1)(\cos\beta - 1)$
12. $(2\csc\beta - 1)(\csc\beta - 3)$
13. $(\csc x + \sin x)^2$
14. $(2\cos x - \sec x)^2$
15. $(2\sin\theta - 1)(2\sin\theta + 1)$
16. $(3\sec\theta - 2)(3\sec\theta + 2)$
17. $(3\sin\theta + 2)^2$
18. $(3\cos\theta - 2)^2$
19. $(2\sin^2 y - \csc^2 y)^2$
20. $(\tan^2 y + \cot^2 y)^2$

*Find the products and simplify your answers.*

21. $(1 - \sin\alpha)(1 + \sin\alpha)$
22. $(1 - \cos\alpha)(1 + \cos\alpha)$
23. $(\csc\alpha - 1)(\csc\alpha + 1)$
24. $(\sec\alpha - 1)(\sec\alpha + 1)$
25. $(\tan\alpha - \sec\alpha)(\tan\alpha + \sec\alpha)$
26. $(\cot\alpha - \csc\alpha)(\cot\alpha + \csc\alpha)$

*Factor each trigonometric expression.*

27. $2\sin^2\gamma - 5\sin\gamma - 3$
28. $\cos^2\gamma - \cos\gamma - 6$
29. $\tan^2\alpha - 6\tan\alpha + 8$
30. $2\cot^2\alpha + \cot\alpha - 3$
31. $4\sec^2\beta + 4\sec\beta + 1$
32. $9\csc^2\theta - 12\csc\theta + 4$
33. $\tan^2\alpha - \sec^2\beta$
34. $\sin^4 y - \cos^4 x$
35. $\sin^2\beta\cos\beta + \sin\beta\cos\beta - 2\cos\beta$
36. $\cos^2\theta\tan\theta - 2\cos\theta\tan\theta - 3\tan\theta$
37. $4\sec^4 x - 4\sec^2 x + 1$
38. $\cos^4 x - 2\cos^2 x + 1$
39. $\sin\alpha\cos\alpha + \cos\alpha + \sin\alpha + 1$
40. $2\sin^2\theta + \sin\theta - 2\sin\theta\cos\theta - \cos\theta$

*Simplify each expression.*

41. $\dfrac{1}{a} - \dfrac{\cos^2 x}{a}$
42. $\dfrac{1}{\cos x} - \dfrac{\sin^2 x}{\cos x}$
43. $\dfrac{\sin(2x)}{2} + \sin(2x)$
44. $\cos(2x) - \dfrac{\cos(2x)}{2}$
45. $\dfrac{\tan x}{3} + \dfrac{\tan x}{2}$
46. $\dfrac{\sin x}{b} + \dfrac{\sin x}{3b}$
47. $\dfrac{\sin x - \sin^2 x}{\sin x}$
48. $\dfrac{\cos^3 x - \cos x}{-\cos x}$
49. $\dfrac{\sin^2 x - \cos^2 x}{\sin x - \cos x}$
50. $\dfrac{1 - \cos^2 x}{1 - \cos x}$
51. $\dfrac{\sin^2 x - \sin x - 2}{\sin^2 x - 4}$
52. $\dfrac{\tan^2 x - 2\tan x + 1}{1 + \tan(-x)}$
53. $\dfrac{\sin^2(-x) - \sin(-x)}{1 - \sin(-x)}$
54. $\dfrac{\cos^2(-x) - \cos(-x)}{1 - \cos(-x)}$

*Prove that each of the following equations is an identity. See the strategy for verifying identities on page 437.*

55. $\sin(x)\cot(x) = \cos(x)$
56. $\cos^2(x)\tan^2(x) = \sin^2(x)$
57. $1 - \sec(x)\cos^3(x) = \sin^2(x)$
58. $1 - \csc(x)\sin^3(x) = \cos^2(x)$
59. $1 + \sec^2(x)\sin^2(x) = \sec^2(x)$
60. $1 + \csc^2(x)\cos^2(x) = \csc^2(x)$
61. $\dfrac{\sin^3(x) + \sin(x)\cos^2(x)}{\cos(x)} = \tan(x)$
62. $\dfrac{\cos(x)\sin^2(x) + \cos^3(x)}{\sin(x)} = \cot(x)$
63. $\dfrac{\sin(x)}{\csc(x)} + \dfrac{\cos(x)}{\sec(x)} = 1$
64. $\sin^3(x)\csc(x) + \cos^3(x)\sec(x) = 1$
65. $\dfrac{1}{\csc\theta - \cot\theta} = \dfrac{1 + \cos\theta}{\sin\theta}$
66. $\dfrac{-1}{\tan\theta - \sec\theta} = \dfrac{1 + \sin\theta}{\cos\theta}$

**67.** $\dfrac{\sec x - \cos x}{\sec x} = \sin^2 x$

**68.** $\dfrac{\sec x - \cos x}{\cos x} = \tan^2 x$

**69.** $1 - \sin x = \dfrac{1 - \sin^2(-x)}{1 - \sin(-x)}$

**70.** $\tan^2 x = \dfrac{1 - \sin^2 x \csc^2 x + \sin^2 x}{\cos^2 x}$

**71.** $\sin^4 w = \dfrac{1 - \cot^2 w + \cos^2 w \cot^2 w}{\csc^2 w}$

**72.** $\tan^4 z = \dfrac{\sec^2 z - \csc^2 z + \csc^2 z \cos^2 z}{\cot^2 z}$

**73.** $1 + \csc x \sec x = \dfrac{\cos(-x) - \csc(-x)}{\cos(x)}$

**74.** $\tan^2(-x) - \dfrac{\sin(-x)}{\sin x} = \sec^2 x$

**75.** $\tan x \cos x + \csc x \sin^2 x = 2 \sin x$

**76.** $\cot x \sin x - \cos^2 x \sec x = 0$

**77.** $(1 + \sin \alpha)^2 + \cos^2 \alpha = 2 + 2 \sin \alpha$

**78.** $(1 + \cot \alpha)^2 - 2 \cot \alpha = \dfrac{1}{(1 - \cos \alpha)(1 + \cos \alpha)}$

**79.** $\dfrac{\sin^2 \beta + \sin \beta - 2}{2 \sin \beta - 2} = \dfrac{\sin \beta + 2}{2}$

**80.** $\dfrac{4 \sec^2 \beta + 4 \sec \beta + 1}{2 \sec \beta + 1} = \dfrac{2}{\cos \beta} + 1$

**81.** $2 - \csc \beta \sin \beta = \sin^2 \beta + \cos^2 \beta$

**82.** $(1 - \sin^2 \beta)(1 + \sin^2 \beta) = 2 \cos^2 \beta - \cos^4 \beta$

**83.** $\tan x + \cot x = \sec x \csc x$

**84.** $\dfrac{\csc x}{\cot x} - \dfrac{\cot x}{\csc x} = \dfrac{\tan x}{\csc x}$

**85.** $\dfrac{\sec x}{\tan x} - \dfrac{\tan x}{\sec x} = \cos x \cot x$

**86.** $\dfrac{1 - \sin^2 x}{1 - \sin x} = \dfrac{\csc x + 1}{\csc x}$

**87.** $\sec^2 x = \dfrac{\csc x}{\csc x - \sin x}$

**88.** $\dfrac{\sin x}{\sin x + 1} = \dfrac{\csc x - 1}{\cot^2 x}$

**89.** $\dfrac{\csc y + 1}{\csc y - 1} = \dfrac{1 + \sin y}{1 - \sin y}$

**90.** $\dfrac{1 - 2 \cos^2 y}{1 - 2 \cos y \sin y} = \dfrac{\sin y + \cos y}{\sin y - \cos y}$

**91.** $\ln(\sec \theta) = -\ln(\cos \theta)$

**92.** $\ln(\tan \theta) = \ln(\sin \theta) + \ln(\sec \theta)$

**93.** $\ln|\sec \alpha + \tan \alpha| = -\ln|\sec \alpha - \tan \alpha|$

**94.** $\ln|\csc \alpha + \cot \alpha| = -\ln|\csc \alpha - \cot \alpha|$

*The equation $f_1(x) = f_2(x)$ is an identity if and only if the graphs of $y = f_1(x)$ and $y = f_2(x)$ coincide at all values of $x$ for which both sides are defined. Graph $y = f_1(x)$ and $y = f_2(x)$ on the same screen of your calculator for each of the following equations. From the graphs, make a conjecture as to whether each equation is an identity, then prove your conjecture.*

**95.** $\dfrac{\sin \theta + \cos \theta}{\sin \theta} = 1 + \cot \theta$

**96.** $\dfrac{\sin \theta + \cos \theta}{\cos \theta} = \cot \theta + 1$

**97.** $(\sin x + \csc x)^2 = \sin^2 x + \csc^2 x$

**98.** $\tan x + \sec x = \dfrac{\sin^2 x + 1}{\cos x}$

**99.** $\cot x + \sin x = \dfrac{1 + \cos x - \cos^2 x}{\sin x}$

**100.** $1 - 2 \cos^2 x + \cos^4 x = \sin^4 x$

**101.** $\dfrac{\sin x}{\cos x} - \dfrac{\cos x}{\sin x} = \dfrac{2 \cos^2 x - 1}{\sin x \cos x}$

**102.** $\dfrac{1}{1 - \sin x} + \dfrac{1}{1 + \sin x} = \dfrac{2}{\cos^2 x}$

**103.** $\dfrac{\cos(-x)}{1 - \sin x} = \dfrac{1 - \sin(-x)}{\cos x}$

**104.** $\dfrac{\sin^2 x}{1 - \cos x} = 0.99 + \cos x$

## FOR WRITING/DISCUSSION

**105.** Find functions $f_1(x)$ and $f_2(x)$ such that $f_1(x) = f_2(x)$ for infinitely many values of $x$, but yet $f_1(x) = f_2(x)$ is not an identity. Explain your example.

**106.** Find functions $f_1(x)$ and $f_2(x)$ such that $f_1(x) \neq f_2(x)$ for infinitely many values of $x$, but yet $f_1(x) = f_2(x)$ is an identity. Explain your example.

## ▶ RETHINKING

**107.** State the three Pythagorean identities.

**108.** Use identities to simplify $\dfrac{1}{\cos^2 x} - \tan^2 x$.

**109.** Use identities to simplify the expression $\dfrac{\csc x}{\sec x}$.

**110.** Find the point that lies midway between $(\pi/3, 1)$ and $(\pi/2, 1)$.

**111.** Determine the amplitude, period, and phase shift for the function $y = -4\sin(2\pi x/3 - \pi/3)$.

**112.** Determine the period, asymptotes, and range for the function $y = 2\sec(x/4)$.

**113.** The formula $d = \frac{1}{32}v_0^2\sin(2\theta)$ gives the distance $d$ in feet that a projectile will travel when its launch angle is $\theta$ and its initial velocity is $v_0$ feet per second. What initial velocity in miles per hour does it take to throw a baseball 200 feet with $\theta = 33°$. Round to the nearest tenth.

## THINKING OUTSIDE THE BOX XLIV

*Cutting Cardboard* A cardboard tube has height 1 and diameter 1. Suppose that it is cut at a 45° angle as shown in the first figure. The tube is then laid flat in the first quadrant as shown in the second figure. What is the equation of the curve? Assume that the curve is a sine wave.

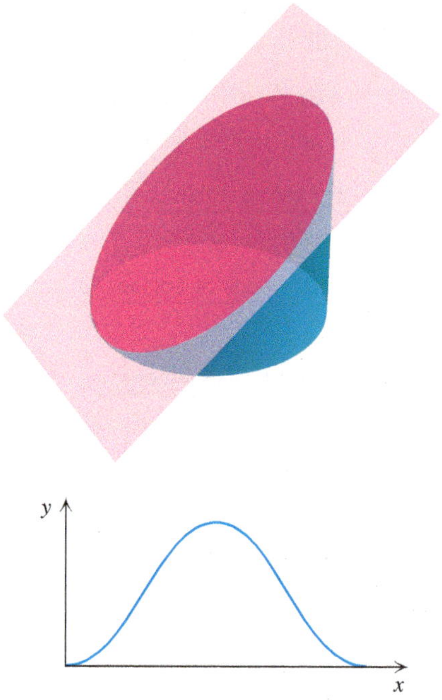


**Figure for Thinking Outside the Box XLIV**

## POP QUIZ 6.2

**1.** Find the product $(2\sin x + 1)(\sin x - 1)$.

**2.** Factor $2\cos^2 x + \cos x - 1$.

**3.** Simplify $\dfrac{1}{\cos(-x)} - \dfrac{\sin^2(-x)}{\cos(x)}$.

**4.** Prove that $\dfrac{\cos(-x) - \sec(-x)}{\sec(x)} = -\sin^2(x)$ is an identity.

## LINKING concepts... For Individual or Group Explorations

### Constructing an Eyebrow Window

*A company that manufactures custom aluminum windows makes an eyebrow window that is placed on top of a rectangular window as shown in the diagram. When a customer orders an eyebrow window, the customer gives the width w and height h of the eyebrow. To make the window, the shop needs to know the radius of the circular arc and the length of the circular arc. (This is an actual problem given to the author by a former student who worked for the company.)*

**a)** If the width and height of the eyebrow are 36 in. and 10 in., respectively, then what is the radius of the circular arc?

**b)** Find the length of the circular arc for a width of 36 in. and a height of 10 in.

**c)** Find a formula that expresses the radius of the circular arc $r$ in terms of the width of the eyebrow $w$ and the height of the eyebrow $h$.

**d)** Find a formula that expresses the length of the circular arc $L$ in terms of $w$ and $h$.

# 6.3 Sum and Difference Identities

In every identity discussed in Sections 6.1 and 6.2, the trigonometric functions were functions of only a single variable, such as $\theta$ or $x$. In this section we establish several new identities involving sine, cosine, and tangent of a sum or difference of two variables. These identities are used in solving equations and in simplifying expressions. They do not follow from the known identities but rather from the geometry of the unit circle.

## The Cosine of a Sum

With so many identities showing relationships between the trigonometric functions, we might be fooled into thinking that almost any equation is an identity. For example, consider the equation

$$\cos(\alpha + \beta) = \cos\alpha + \cos\beta.$$

This equation looks nice, but is it an identity? It is easy to check with a calculator that if $\alpha = 30°$ and $\beta = 45°$, we get

$$\cos(30° + 45°) = \cos(75°) \approx 0.2588 \quad \text{and} \quad \cos(30°) + \cos(45°) \approx 1.573.$$

So the equation $\cos(\alpha + \beta) = \cos\alpha + \cos\beta$ is *not* an identity. However, there is an identity for the cosine of a sum of two angles.

To derive an identity for $\cos(\alpha + \beta)$, consider angles $\alpha$, $\alpha + \beta$, and $-\beta$ in standard position, as shown in Fig. 6.10. Assume that $\alpha$ and $\beta$ are acute angles, although any angles $\alpha$ and $\beta$ may be used. The terminal side of $\alpha$ intersects the unit circle at the point $A(\cos\alpha, \sin\alpha)$. The terminal side of $\alpha + \beta$ intersects the unit circle at $B(\cos(\alpha + \beta), \sin(\alpha + \beta))$. The terminal side of $-\beta$ intersects the unit circle at $C(\cos(-\beta), \sin(-\beta))$ or $C(\cos\beta, -\sin\beta)$. Let $D$ be the point $(1, 0)$. Since $\angle BOD = \alpha + \beta$ and $\angle AOC = \alpha + \beta$, the chords $\overline{BD}$ and $\overline{AC}$ are equal in length. Using the distance formula to find the lengths, we get the following equation:

$$\sqrt{(\cos(\alpha + \beta) - 1)^2 + (\sin(\alpha + \beta) - 0)^2} = \sqrt{(\cos\alpha - \cos\beta)^2 + (\sin\alpha - (-\sin\beta))^2}$$

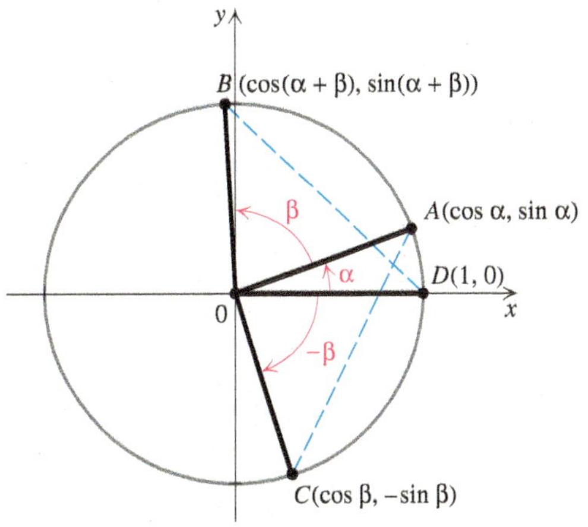


**Figure 6.10**

Square each side and simplify:

$$(\cos(\alpha + \beta) - 1)^2 + \sin^2(\alpha + \beta) = (\cos\alpha - \cos\beta)^2 + (\sin\alpha + \sin\beta)^2$$

Square the binomials:

$$\cos^2(\alpha + \beta) - 2\cos(\alpha + \beta) + 1 + \sin^2(\alpha + \beta)$$
$$= \cos^2\alpha - 2\cos\alpha\cos\beta + \cos^2\beta + \sin^2\alpha + 2\sin\alpha\sin\beta + \sin^2\beta$$

Using the identity $\cos^2 x + \sin^2 x = 1$ once on the left-hand side and twice on the right-hand side, we get the following equation:

$$2 - 2\cos(\alpha + \beta) = 2 - 2\cos\alpha\cos\beta + 2\sin\alpha\sin\beta$$

Subtract 2 from each side and divide each side by $-2$ to get the identity for the cosine of a sum:

**Identity: Cosine of a Sum**

$$\cos(\alpha + \beta) = \cos\alpha\cos\beta - \sin\alpha\sin\beta$$

If an angle is the sum of two angles for which we know the exact values of the trigonometric functions, we can use this new identity to find the exact value of the cosine of the sum. For example, we can find the exact values of expressions such as $\cos(75°)$, $\cos(7\pi/12)$, and $\cos(195°)$ because

$$75° = 45° + 30°, \qquad \frac{7\pi}{12} = \frac{\pi}{3} + \frac{\pi}{4}, \qquad \text{and} \qquad 195° = 150° + 45°.$$

Note that we do not need identities to find approximate values for $\cos(75°)$, $\cos(7\pi/12)$, and $\cos(195°)$. All we need is a calculator.

### EXAMPLE 1 The exact value of a cosine

Find the exact value of $\cos(75°)$.

**Solution**

Use $75° = 30° + 45°$ and the identity for the cosine of a sum.

$$\begin{aligned}\cos 75° &= \cos(30° + 45°)\\ &= \cos(30°)\cos(45°) - \sin(30°)\sin(45°)\\ &= \frac{\sqrt{3}}{2}\cdot\frac{\sqrt{2}}{2} - \frac{1}{2}\cdot\frac{\sqrt{2}}{2}\\ &= \frac{\sqrt{6} - \sqrt{2}}{4}\end{aligned}$$

To check, evaluate $\cos 75°$ and $(\sqrt{6} - \sqrt{2})/4$ using a calculator.

▸**TRY THIS.** Find the exact value of $\cos(105°)$ using the cosine of a sum identity. ■

Note that we found the exact value of $\cos(75°)$ only to illustrate an identity. If we need $\cos(75°)$ in an application, we can use the calculator value for it.

## The Cosine of a Difference

To derive an identity for the cosine of a difference, we write $\alpha - \beta = \alpha + (-\beta)$ and use the identity for the cosine of a sum:

$$\begin{aligned}\cos(\alpha - \beta) &= \cos(\alpha + (-\beta))\\ &= \cos\alpha\cos(-\beta) - \sin\alpha\sin(-\beta)\end{aligned}$$

Now use the identities $\cos(-\beta) = \cos\beta$ and $\sin(-\beta) = -\sin\beta$ to get the identity for the cosine of a difference.

**Identity: Cosine of a Difference**

$$\cos(\alpha - \beta) = \cos\alpha\cos\beta + \sin\alpha\sin\beta$$

If an angle is the difference of two angles for which we know the exact values of sine and cosine, then we can find the exact value for the cosine of the angle.

### EXAMPLE 2 The exact value of a cosine

Find the exact value of $\cos(\pi/12)$.

**Solution**

Since $\frac{\pi}{3} = \frac{4\pi}{12}$ and $\frac{\pi}{4} = \frac{3\pi}{12}$, we can use $\frac{\pi}{12} = \frac{\pi}{3} - \frac{\pi}{4}$ and the identity for the cosine of a difference:

$$\begin{aligned}\cos\left(\frac{\pi}{12}\right) &= \cos\left(\frac{\pi}{3} - \frac{\pi}{4}\right)\\ &= \cos\frac{\pi}{3}\cos\frac{\pi}{4} + \sin\frac{\pi}{3}\sin\frac{\pi}{4}\\ &= \frac{1}{2}\cdot\frac{\sqrt{2}}{2} + \frac{\sqrt{3}}{2}\cdot\frac{\sqrt{2}}{2}\\ &= \frac{\sqrt{2} + \sqrt{6}}{4}\end{aligned}$$

To check, evaluate $\cos(\pi/12)$ and $(\sqrt{2} + \sqrt{6})/4$ using a calculator.

▶**TRY THIS.** Find the exact value of $\cos(75°)$ using the cosine of a difference identity. ■

The identities for the cosine of a sum or a difference can be used to find identities for the sine of a sum or difference. To develop those identities, we first find identities that relate each trigonometric function and its **cofunction.** Sine and cosine are cofunctions, tangent and cotangent are cofunctions, and secant and cosecant are cofunctions.

## Cofunction Identities

If $\alpha = \pi/2$ is substituted into the identity for $\cos(\alpha - \beta)$, we get another identity:

$$\begin{aligned}\cos\left(\frac{\pi}{2} - \beta\right) &= \cos\frac{\pi}{2}\cos\beta + \sin\frac{\pi}{2}\sin\beta\\ &= 0\cdot\cos\beta + 1\cdot\sin\beta\\ &= \sin\beta\end{aligned}$$

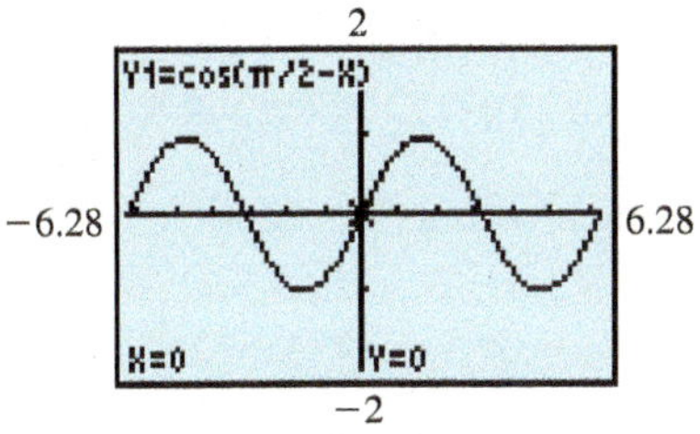

Figure 6.11

The identity $\cos(\pi/2 - \beta) = \sin\beta$ is a **cofunction identity.**

The graphing calculator graph in Fig. 6.11 supports this conclusion because $y_1 = \cos(\pi/2 - x)$ appears to coincide with $y_2 = \sin(x)$. □

If we let $u = \pi/2 - \beta$ or $\beta = \pi/2 - u$ in the identity $\sin\beta = \cos(\pi/2 - \beta)$, we get the identity

$$\sin\left(\frac{\pi}{2} - u\right) = \cos u.$$

Using these identities, we can establish the cofunction identities for secant, cosecant, tangent, and cotangent. For the cofunction identities, *the value of any trigonometric function at u is equal to the value of its cofunction at* $(\pi/2 - u)$.

**Cofunction Identities**

$$\sin\left(\frac{\pi}{2} - u\right) = \cos u \qquad \cos\left(\frac{\pi}{2} - u\right) = \sin u$$

$$\tan\left(\frac{\pi}{2} - u\right) = \cot u \qquad \cot\left(\frac{\pi}{2} - u\right) = \tan u$$

$$\sec\left(\frac{\pi}{2} - u\right) = \csc u \qquad \csc\left(\frac{\pi}{2} - u\right) = \sec u$$

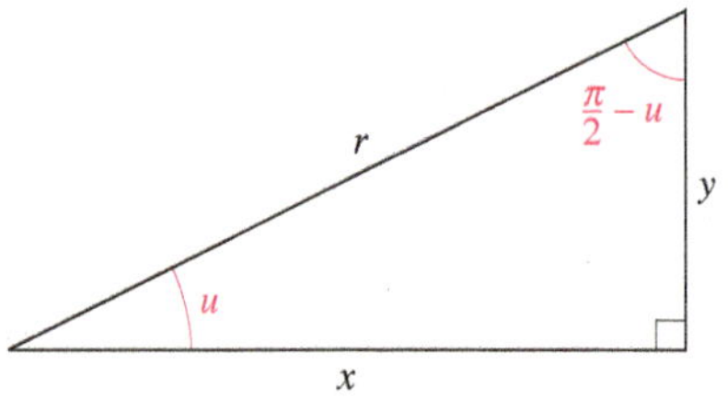


**Figure 6.12**

If $0 < u < \pi/2$, then $u$ and $(\pi/2 - u)$ are the measures of the acute angles of a right triangle, as shown in Fig. 6.12. The term *cofunction* comes from the fact that these angles are complementary. The cofunction identities indicate that the value of a trigonometric function of one acute angle of a right triangle is equal to the value of the cofunction of the other. Since the cofunction identities hold also for complementary angles measured in degrees, we can write equations such as

$$\sin(20°) = \cos(70°), \qquad \cot(89°) = \tan(1°), \qquad \text{and} \qquad \sec(50°) = \csc(40°).$$

The cofunction identities are consistent with the values obtained for the trigonometric functions for an acute angle using the ratios of the sides of a right triangle. For example, the cofunction identity $\sin(u) = \cos(\pi/2 - u)$ is correct for angles $u$ and $(\pi/2 - u)$ shown in Fig. 6.12, because

$$\sin(u) = \frac{\text{opp}}{\text{hyp}} = \frac{y}{r}$$

and

$$\cos\left(\frac{\pi}{2} - u\right) = \frac{\text{adj}}{\text{hyp}} = \frac{y}{r}.$$

## EXAMPLE 3 Using the cofunction identities

Use a cofunction identity to find the exact value of $\sin(5\pi/12)$.

### Solution

Since $\sin(u) = \cos(\pi/2 - u)$,

$$\sin\left(\frac{5\pi}{12}\right) = \cos\left(\frac{\pi}{2} - \frac{5\pi}{12}\right) = \cos\left(\frac{\pi}{12}\right).$$

From Example 2, $\cos(\pi/12) = (\sqrt{2} + \sqrt{6})/4$. So $\sin(5\pi/12) = (\sqrt{2} + \sqrt{6})/4$.

▶**TRY THIS.** Find the exact value of $\sin(15°)$ using a cofunction identity. ■

## Sine of a Sum or a Difference

We now use a cofunction identity to find an identity for $\sin(\alpha + \beta)$:

$$\begin{aligned}
\sin(\alpha + \beta) &= \cos\left(\frac{\pi}{2} - (\alpha + \beta)\right) \\
&= \cos\left(\left(\frac{\pi}{2} - \alpha\right) - \beta\right) \\
&= \cos\left(\frac{\pi}{2} - \alpha\right)\cos\beta + \sin\left(\frac{\pi}{2} - \alpha\right)\sin\beta \\
&= \sin\alpha\cos\beta + \cos\alpha\sin\beta
\end{aligned}$$

To find an identity for $\sin(\alpha - \beta)$, use the identity for $\sin(\alpha + \beta)$:

$$\begin{aligned}\sin(\alpha - \beta) &= \sin(\alpha + (-\beta)) \\ &= \sin\alpha\cos(-\beta) + \cos\alpha\sin(-\beta) \\ &= \sin\alpha\cos\beta - \cos\alpha\sin\beta\end{aligned}$$

The identities for the sine of a sum or difference are stated as follows.

**Identities: Sine of a Sum or Difference**

$$\sin(\alpha + \beta) = \sin\alpha\cos\beta + \cos\alpha\sin\beta$$
$$\sin(\alpha - \beta) = \sin\alpha\cos\beta - \cos\alpha\sin\beta$$

### EXAMPLE 4 The exact value of a sine

Find the exact value of $\sin(195°)$.

**Solution**

Use $195° = 150° + 45°$ and the identity for the sine of a sum:

$$\begin{aligned}\sin(195°) &= \sin(150° + 45°) \\ &= \sin(150°)\cos(45°) + \cos(150°)\sin(45°) \\ &= \frac{1}{2}\cdot\frac{\sqrt{2}}{2} + \left(-\frac{\sqrt{3}}{2}\right)\cdot\frac{\sqrt{2}}{2} \\ &= \frac{\sqrt{2} - \sqrt{6}}{4}\end{aligned}$$

To check, evaluate $\sin(195°)$ and $(\sqrt{2} - \sqrt{6})/4$ using a calculator.

▶**TRY THIS.** Find the exact value of $\sin(75°)$ using the sine of a sum identity. ■

## Tangent of a Sum or Difference

We can use the identities for $\sin(\alpha + \beta)$ and $\cos(\alpha + \beta)$ to find an identity for $\tan(\alpha + \beta)$:

$$\begin{aligned}\tan(\alpha + \beta) &= \frac{\sin(\alpha + \beta)}{\cos(\alpha + \beta)} \\ &= \frac{\sin\alpha\cos\beta + \cos\alpha\sin\beta}{\cos\alpha\cos\beta - \sin\alpha\sin\beta}\end{aligned}$$

To express the right-hand side in terms of tangent, multiply the numerator and denominator by $1/(\cos\alpha\cos\beta)$ and use the identity $\tan x = \sin x/\cos x$:

$$\begin{aligned}\tan(\alpha + \beta) &= \frac{\dfrac{\sin\alpha\cos\beta}{\cos\alpha\cos\beta} + \dfrac{\cos\alpha\sin\beta}{\cos\alpha\cos\beta}}{\dfrac{\cos\alpha\cos\beta}{\cos\alpha\cos\beta} - \dfrac{\sin\alpha\sin\beta}{\cos\alpha\cos\beta}} \\ &= \frac{\tan\alpha + \tan\beta}{1 - \tan\alpha\tan\beta}\end{aligned}$$

Use the identity $\tan(-\beta) = -\tan\beta$ to get a similar identity for $\tan(\alpha - \beta)$.

**Identities: Tangent of a Sum or Difference**

$$\tan(\alpha + \beta) = \frac{\tan\alpha + \tan\beta}{1 - \tan\alpha\tan\beta} \qquad \tan(\alpha - \beta) = \frac{\tan\alpha - \tan\beta}{1 + \tan\alpha\tan\beta}$$

## EXAMPLE 5 The exact value of a tangent

Find the exact value of $\tan(\pi/12)$.

### Solution

Use $\pi/12 = \pi/3 - \pi/4$ and the identity for the tangent of a difference:

$$\tan\left(\frac{\pi}{12}\right) = \tan\left(\frac{\pi}{3} - \frac{\pi}{4}\right)$$

$$= \frac{\tan\frac{\pi}{3} - \tan\frac{\pi}{4}}{1 + \tan\frac{\pi}{3}\tan\frac{\pi}{4}} = \frac{\sqrt{3} - 1}{1 + \sqrt{3}\cdot 1}$$

$$= \frac{(\sqrt{3} - 1)(\sqrt{3} - 1)}{(\sqrt{3} + 1)(\sqrt{3} - 1)}$$

$$= \frac{3 - 2\sqrt{3} + 1}{2} = \frac{4 - 2\sqrt{3}}{2} = 2 - \sqrt{3}$$

▸**TRY THIS.** Find the exact value of $\tan(75°)$ using the tangent of a sum identity. ■

In the next example we use sum and difference identities in reverse to simplify an expression.

## EXAMPLE 6 Simplifying with sum and difference identities

Use an appropriate identity to simplify each expression.

**a.** $\cos(47°)\cos(2°) + \sin(47°)\sin(2°)$ **b.** $\sin t \cos 2t - \cos t \sin 2t$

### Solution

**a.** This expression is the right-hand side of the identity for the cosine of a difference, $\cos(\alpha - \beta) = \cos\alpha\cos\beta + \sin\alpha\sin\beta$, with $\alpha = 47°$ and $\beta = 2°$:

$$\cos(47°)\cos(2°) + \sin(47°)\sin(2°) = \cos(47° - 2°) = \cos(45°) = \frac{\sqrt{2}}{2}$$

Figure 6.13

Check this result with a calculator, as in Fig. 6.13. □

**b.** We recognize this expression as the right-hand side of the identity for the sine of a difference, $\sin(\alpha - \beta) = \sin\alpha\cos\beta - \cos\alpha\sin\beta$. So

$$\sin t \cos 2t - \cos t \sin 2t = \sin(t - 2t) = \sin(-t) = -\sin t.$$

▸**TRY THIS.** Simplify $\sin\alpha\cos 3\alpha + \cos\alpha\sin 3\alpha$. ■

As shown next, we use identities to find $\sin(\alpha + \beta)$ without knowing $\alpha$ or $\beta$.

## EXAMPLE 7 Evaluating a sine of a sum

Find the exact value of $\sin(\alpha + \beta)$, given that $\sin\alpha = -3/5$ and $\cos\beta = -1/3$, with $\alpha$ in quadrant IV and $\beta$ in quadrant III.

### Solution

To use the identity for $\sin(\alpha + \beta)$, we need $\cos\alpha$ and $\sin\beta$ in addition to the given values. Use $\sin\alpha = -3/5$ in the identity $\sin^2 x + \cos^2 x = 1$ to find $\cos\alpha$:

$$\left(-\frac{3}{5}\right)^2 + \cos^2\alpha = 1, \qquad \cos^2\alpha = \frac{16}{25}, \qquad \cos\alpha = \pm\frac{4}{5}$$

(a)

(b)

**Figure 6.14**

Since cosine is positive in quadrant IV, $\cos\alpha = 4/5$. Use $\cos\beta = -1/3$ in $\sin^2 x + \cos^2 x = 1$ to find $\sin\beta$:

$$\sin^2\beta + \left(-\frac{1}{3}\right)^2 = 1, \qquad \sin^2\beta = \frac{8}{9}, \qquad \sin\beta = \pm\frac{2\sqrt{2}}{3}$$

Since sine is negative in quadrant III, $\sin\beta = -2\sqrt{2}/3$. Now use the appropriate values in the identity $\sin(\alpha + \beta) = \sin\alpha\cos\beta + \cos\alpha\sin\beta$:

$$\sin(\alpha+\beta) = -\frac{3}{5}\left(-\frac{1}{3}\right) + \frac{4}{5}\left(-\frac{2\sqrt{2}}{3}\right) = \frac{3}{15} - \frac{8\sqrt{2}}{15} = \frac{3 - 8\sqrt{2}}{15}$$

To check, we use the inverse functions to find $\alpha$ and $\beta$ and then $\sin(\alpha + \beta)$ on a graphing calculator. Since $\cos^{-1}(-1/3)$ is in quadrant II, $\beta$ in degrees is $360 - \cos^{-1}(-1/3)$. The angle $\alpha$ is $\sin^{-1}(-3/5)$, as shown in Fig. 6.14(a). Now use these values for $\alpha$ and $\beta$ to check that $\sin(\alpha + \beta) = (3 - 8\sqrt{2})/15$, as shown in Fig. 6.14(b).

►**TRY THIS.** Find the exact value of $\cos(\alpha - \beta)$ given that $\sin\alpha = 1/3$, $\cos\beta = -1/3$, and both angles terminate in quadrant II. ■

## FOR thought... True or False? Explain.

1. $\cos(3°)\cos(2°) - \sin(3°)\sin(2°) = \cos(1°)$
2. $\cos(4)\cos(5) + \sin(4)\sin(5) = \cos(-1)$
3. For any real number $t$, $\cos(t - \pi/2) = \sin t$.
4. For any radian measure $\alpha$, $\sin(\alpha - \pi/2) = \cos\alpha$.
5. $\sec(\pi/3) = \csc(\pi/6)$
6. $\sin(5\pi/6) = \sin(2\pi/3) + \sin(\pi/6)$
7. $\sin(5\pi/12) = \sin(\pi/3)\cos(\pi/4) + \cos(\pi/3)\sin(\pi/4)$
8. $\tan(21°30'5'') = \cot(68°29'55'')$
9. For any real number $x$, $\sec(x) - \csc(\pi/2 - x) = \tan(x) - \cot(\pi/2 - x)$.
10. $\dfrac{\tan(13°) + \tan(-20°)}{1 - \tan(13°)\tan(-20°)} = \dfrac{\tan(45°) + \tan(-52°)}{1 - \tan(45°)\tan(-52°)}$

## EXERCISES 6.3

*Fill in the blank.*

1. Sine and ________ are cofunctions.
2. The value of any trigonometric function at $x$ is equal to the value of its ________ at $(\pi/2 - x)$.

*Use an identity to find the exact value of each expression. Use a calculator to check.*

3. $\cos\left(\frac{\pi}{3} + \frac{\pi}{4}\right)$
4. $\cos\left(\frac{2\pi}{3} + \frac{\pi}{4}\right)$
5. $\cos(60° - 45°)$
6. $\cos(120° - 45°)$

*Use a cofunction identity to fill in the blank.*

7. $\sin(20°) = \cos(\quad)$
2. $\cos(15°) = \sin(\quad)$
9. $\tan\left(\frac{\pi}{6}\right) = \cot(\quad)$
10. $\cot\left(\frac{\pi}{3}\right) = \tan(\quad)$
11. $\sec(90° - 6°) = \csc(\quad)$
6. $\csc(90° - 17°) = \sec(\quad)$

*Use an identity to find the exact value of each expression. Use a calculator to check.*

13. $\sin\left(\frac{\pi}{4} + \frac{\pi}{3}\right)$
14. $\sin\left(\frac{\pi}{4} + \frac{2\pi}{3}\right)$
15. $\sin(60° - 45°)$
16. $\sin(45° - 120°)$
17. $\tan\left(\frac{3\pi}{4} + \frac{\pi}{3}\right)$
18. $\tan\left(\frac{\pi}{4} + \frac{\pi}{3}\right)$
19. $\tan(210° - 45°)$
20. $\tan(45° - 150°)$

*Find the following sums or differences in terms of $\pi$.*

**21.** $\frac{\pi}{4} + \frac{\pi}{3}$

**22.** $\frac{2\pi}{3} - \frac{\pi}{4}$

**23.** $\frac{3\pi}{4} + \frac{\pi}{3}$

**24.** $\frac{\pi}{6} + \frac{\pi}{4}$

*Simplify each expression by using sum or difference identities.*

**25.** $\sin(23°)\cos(67°) + \cos(23°)\sin(67°)$

**26.** $\sin(55°)\cos(10°) - \cos(55°)\sin(10°)$

**27.** $\cos\left(\frac{\pi}{6}\right)\cos\left(\frac{\pi}{3}\right) - \sin\left(\frac{\pi}{6}\right)\sin\left(\frac{\pi}{3}\right)$

**28.** $\cos\left(\frac{7\pi}{12}\right)\cos\left(\frac{\pi}{3}\right) + \sin\left(\frac{7\pi}{12}\right)\sin\left(\frac{\pi}{3}\right)$

**29.** $\dfrac{\tan(\pi/12) + \tan(\pi/6)}{1 - \tan(\pi/12)\tan(\pi/6)}$

**30.** $\dfrac{\tan(5\pi/12) - \tan(\pi/12)}{1 + \tan(5\pi/12)\tan(\pi/12)}$

**31.** $\sin(2k)\cos(k) + \cos(2k)\sin(k)$

**32.** $\cos(3y)\cos(y) - \sin(3y)\sin(y)$

*Express each given angle as $\alpha + \beta$ or $\alpha - \beta$, where $\sin\alpha$ and $\sin\beta$ are known exactly.*

**33.** $75°$ **34.** $15°$ **35.** $165°$ **36.** $195°$

*Use appropriate identities to find the exact value of each expression. Do not use a calculator.*

**37.** $\cos(5\pi/12)$

**38.** $\cos(7\pi/12)$

**39.** $\sin(7\pi/12)$

**40.** $\sin(5\pi/12)$

**41.** $\tan(75°)$

**42.** $\tan(-15°)$

**43.** $\sin(-15°)$

**44.** $\sin(165°)$

**45.** $\cos(195°)$

**46.** $\cos(-75°)$

**47.** $\tan(-13\pi/12)$

**48.** $\tan(7\pi/12)$

*Simplify each expression by using appropriate identities. Do not use a calculator.*

**49.** $\sin(3°)\cos(-87°) + \cos(3°)\sin(87°)$

**50.** $\sin(34°)\cos(13°) + \cos(-34°)\sin(-13°)$

**51.** $\cos(-\pi/2)\cos(\pi/5) + \sin(\pi/2)\sin(\pi/5)$

**52.** $\cos(12°)\cos(-3°) - \sin(12°)\sin(-3°)$

**53.** $\dfrac{\tan(\pi/7) + \cot(\pi/2 - \pi/6)}{1 + \tan(-\pi/7)\tan(\pi/6)}$

**54.** $\dfrac{\cot(\pi/2 - \pi/3) + \tan(-\pi/6)}{1 + \tan(\pi/3)\cot(\pi/2 - \pi/6)}$

**55.** $\sin(14°)\cos(35°) + \cos(14°)\cos(55°)$

**56.** $\cos(10°)\cos(20°) + \cos(-80°)\sin(-20°)$

*Match each expression on the left with an expression on the right that has exactly the same value. Do not use a calculator.*

| | |
|---|---|
| **57.** $\cos(44°)$ | **A.** $\cos(0)$ |
| **58.** $\sin(-46°)$ | **B.** $-\cos(44°)$ |
| **59.** $\cos(46°)$ | **C.** $\cos(36°)$ |
| **60.** $\sin(136°)$ | **D.** $\cot\left(\frac{5\pi}{14}\right)$ |
| **61.** $\sec(1)$ | **E.** $-\cos(46°)$ |
| **62.** $\tan\left(\frac{\pi}{7}\right)$ | **F.** $\csc\left(\frac{\pi - 2}{2}\right)$ |
| **63.** $\csc\left(\frac{\pi}{2}\right)$ | **G.** $\sin(46°)$ |
| **64.** $\sin(-44°)$ | **H.** $\sin(44°)$ |

*Solve each problem.*

**65.** Find the exact value of $\sin(\alpha + \beta)$, given that $\sin\alpha = 3/5$ and $\sin\beta = 5/13$, with $\alpha$ in quadrant II and $\beta$ in quadrant I.

**66.** Find the exact value of $\sin(\alpha - \beta)$, given that $\sin\alpha = -4/5$ and $\cos\beta = 12/13$, with $\alpha$ in quadrant III and $\beta$ in quadrant IV.

**67.** Find the exact value of $\cos(\alpha + \beta)$, given that $\sin\alpha = 2/3$ and $\sin\beta = -1/2$, with $\alpha$ in quadrant I and $\beta$ in quadrant III.

**68.** Find the exact value of $\cos(\alpha - \beta)$, given that $\cos\alpha = \sqrt{3}/4$ and $\cos\beta = -\sqrt{2}/3$, with $\alpha$ in quadrant I and $\beta$ in quadrant II.

**69.** Find the exact value of $\sin(\alpha - \beta)$ if $\sin\alpha = -24/25$ and $\cos\beta = -8/17$ with $\alpha$ in quadrant III and $\beta$ in quadrant II.

**70.** Find the exact value of $\sin(\alpha + \beta)$ if $\sin\alpha = 7/25$ and $\sin\beta = -8/17$ with $\alpha$ in quadrant II and $\beta$ in quadrant III.

**71.** Find the exact value of $\cos(\alpha - \beta)$ if $\sin\alpha = 24/25$ and $\cos\beta = 8/17$ with $\alpha$ in quadrant II and $\beta$ in quadrant IV.

**72.** Find the exact value of $\cos(\alpha + \beta)$ if $\sin\alpha = -7/25$ and $\sin\beta = 8/17$ with $\alpha$ in quadrant IV and $\beta$ in quadrant II.

*Write each expression as a function of $\alpha$ alone.*

**73.** $\cos(\pi/2 + \alpha)$

**74.** $\sin(\alpha - \pi)$

**75.** $\cos(180° - \alpha)$

**76.** $\sin(180° - \alpha)$

**77.** $\sin(360° - \alpha)$

**78.** $\cos(\alpha - \pi)$

**79.** $\sin(90° + \alpha)$

**80.** $\cos(360° - \alpha)$

*Verify that each equation is an identity.*

**81.** $\sin(180° - \alpha) = \dfrac{1 - \cos^2 \alpha}{\sin \alpha}$

**82.** $\cos(x - \pi/2) = \cos x \tan x$

**83.** $\dfrac{\cos(x + y)}{\cos x \cos y} = 1 - \tan x \tan y$

**84.** $\dfrac{\sin(x + y)}{\sin x \cos y} = 1 + \cot x \tan y$

**85.** $\sin(\alpha + \beta)\sin(\alpha - \beta) = \sin^2 \alpha - \sin^2 \beta$

**86.** $\cos(\alpha + \beta)\cos(\alpha - \beta) = \cos^2 \beta - \sin^2 \alpha$

**87.** $\cos(2x) = \cos^2 x - \sin^2 x$

**88.** $\sin(2x) = 2 \sin x \cos x$

**89.** $\sin(x - y) - \sin(y - x) = 2 \sin x \cos y - 2 \cos x \sin y$

**90.** $\cos(x - y) + \cos(y - x) = 2 \cos x \cos y + 2 \sin x \sin y$

**91.** $\tan(s + t)\tan(s - t) = \dfrac{\tan^2 s - \tan^2 t}{1 - \tan^2 s \tan^2 t}$

**92.** $\tan(\pi/4 + x) = \cot(\pi/4 - x)$

**93.** $\dfrac{\cos(\alpha + \beta)}{\sin(\alpha - \beta)} = \dfrac{1 - \tan \alpha \tan \beta}{\tan \alpha - \tan \beta}$

**94.** $\dfrac{\cos(\alpha - \beta)}{\sin(\alpha + \beta)} = \dfrac{1 + \tan \alpha \tan \beta}{\tan \alpha + \tan \beta}$

**95.** $\sec(v + t) = \dfrac{\cos v \cos t + \sin v \sin t}{\cos^2 v - \sin^2 t}$

**96.** $\csc(v - t) = \dfrac{\sin v \cos t + \cos v \sin t}{\sin^2 v - \sin^2 t}$

**97.** $\dfrac{\cos(x + y)}{\cos(x - y)} = \dfrac{\cot y - \tan x}{\cot y + \tan x}$

**98.** $\dfrac{\sin(x + y)}{\sin(x - y)} = \dfrac{\cot y + \cot x}{\cot y - \cot x}$

**99.** $\dfrac{\sin(\alpha + \beta)}{\sin \alpha + \sin \beta} = \dfrac{\sin \alpha - \sin \beta}{\sin(\alpha - \beta)}$

**100.** $\dfrac{\cos(\alpha + \beta)}{\cos \alpha + \sin \beta} = \dfrac{\cos \alpha - \sin \beta}{\cos(\beta - \alpha)}$

## FOR WRITING/DISCUSSION

**101.** Verify the four cofunction identities that were not proved in the text.

**102.** Verify the identity for the tangent of a difference.

**103.** Show that $\sin(\alpha + \beta) = \sin \alpha + \sin \beta$ is not an identity.

**104.** Without using a calculator, find the exact value of

$$\frac{\sin^2(1°) + \sin^2(2°) + \sin^2(3°) + \cdots + \sin^2(90°)}{\cos^2(1°) + \cos^2(2°) + \cos^2(3°) + \cdots + \cos^2(90°)}.$$

## ▶ RETHINKING

**105.** Simplify $(1 - \sin \alpha)(1 + \sin \alpha)$.

**106.** Simplify $\dfrac{\sin x}{\csc x - \sin x}$.

**107.** Find the period and equations of the asymptotes for the function $y = \cot(2x) + 1$.

**108.** Let $f(x) = \sin(x)$, $g(x) = x + 2$, and $h(x) = 3x$. Find $g(f(h(x)))$ and $h(g(f(x)))$.

**109.** Suppose that $\sin \alpha = 1/4$ and $\alpha$ is in quadrant II. Use identities to find the exact values of the other five trigonometric functions.

**110.** Complete the following odd and even identities.

**a.** $\sin(-x) =$ ______

**b.** $\cos(-x) =$ ______

**c.** $\tan(-x) =$ ______

**d.** $\csc(-x) =$ ______

**e.** $\sec(-x) =$ ______

**f.** $\cot(-x) =$ ______

## THINKING OUTSIDE THE BOX XLV

*Spiral of Triangles* The hypotenuse in the first right triangle shown in the figure has length $\sqrt{2}$. In the second the hypotenuse is $\sqrt{3}$. In the third it is $\sqrt{4}$, and so on. In this sequence of right triangles, which right triangle (after the first one) intersects the first one at more than one point?

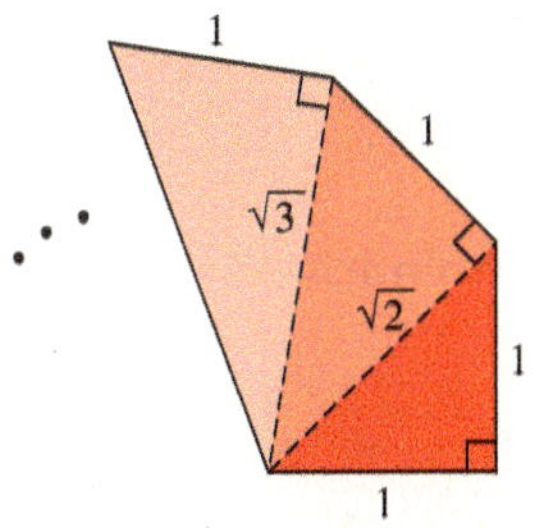


**Figure for Thinking Outside the Box XLV**

## POP QUIZ 6.3

1. Use an identity to find the exact value of $\cos(135° - 120°)$.
2. Use a cofunction identity to find an angle $\alpha$ for which $\sin(10°) = \cos(\alpha)$.
3. Simplify $\sin(2x)\cos(x) + \cos(2x)\sin(x)$.
4. Find the exact value of $\sin(\alpha - \beta)$, if $\sin \alpha = 4/5$ and $\cos \beta = 1/2$, with $\alpha$ and $\beta$ in quadrant I.

# 6.4 Double-Angle and Half-Angle Identities

In Section 6.3 we studied identities for functions of sums and differences of angles. The double-angle and half-angle identities, which we develop next, are special cases of those identities. These special cases occur so often that they are remembered as separate identities.

## Double-Angle Identities

To get an identity for $\sin 2x$, replace both $\alpha$ and $\beta$ by $x$ in the identity for $\sin(\alpha + \beta)$:

$$\sin 2x = \sin(x + x) = \sin x \cos x + \cos x \sin x = 2 \sin x \cos x$$

To find an identity for $\cos 2x$, replace both $\alpha$ and $\beta$ by $x$ in the identity for $\cos(\alpha + \beta)$:

$$\cos 2x = \cos(x + x) = \cos x \cos x - \sin x \sin x = \cos^2 x - \sin^2 x$$

To get a second form of the identity for $\cos 2x$, replace $\sin^2 x$ with $1 - \cos^2 x$:

$$\cos(2x) = \cos^2 x - \sin^2 x = \cos^2 x - (1 - \cos^2 x) = 2\cos^2 x - 1$$

Replacing $\cos^2 x$ with $1 - \sin^2 x$ produces a third form of the identity for $\cos 2x$:

$$\cos(2x) = 1 - 2\sin^2 x$$

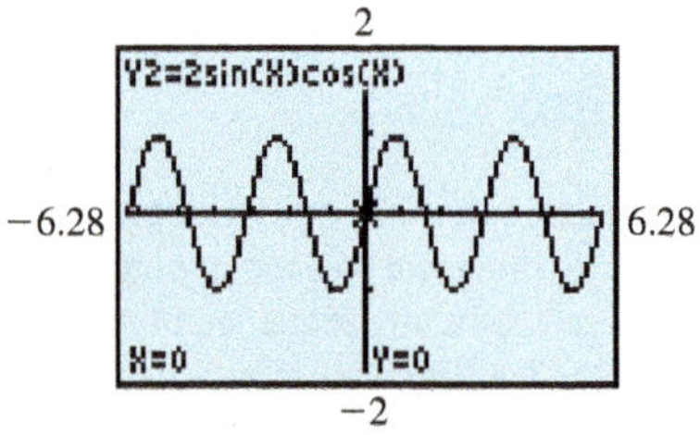

Figure 6.15

To find an identity for $\tan 2x$, we can replace both $\alpha$ and $\beta$ by $x$ in the identity for $\tan(\alpha + \beta)$:

$$\tan 2x = \tan(x + x) = \frac{\tan x + \tan x}{1 - \tan x \tan x} = \frac{2\tan x}{1 - \tan^2 x}$$

You can visually check identities by graphing both sides. For example, $y_1 = \sin(2x)$ and $y_2 = 2\sin(x)\cos(x)$ appear to coincide in Fig. 6.15. □

These identities are known as the **double-angle identities.**

**Double-Angle Identities**

$$\sin 2x = 2 \sin x \cos x \qquad \tan 2x = \frac{2\tan x}{1 - \tan^2 x}$$
$$\cos 2x = \cos^2 x - \sin^2 x$$
$$\cos 2x = 2\cos^2 x - 1$$
$$\cos 2x = 1 - 2\sin^2 x$$

Be careful to learn the double-angle identities exactly as they are written. A "nice looking" equation such as $\cos 2x = 2\cos x$ could be mistaken for an identity if you are not careful. [Since $\cos(\pi/2) \neq 2\cos(\pi/4)$, the "nice looking" equation is not an identity.] Remember that an equation is not an identity if at least one permissible value of the variable fails to satisfy the equation.

## EXAMPLE 1 Evaluating with double-angle identities

Find sin(120°), cos(120°), and tan(120°) by using double-angle identities.

### Solution

Note that $120° = 2 \cdot 60°$ and use the values $\sin(60°) = \sqrt{3}/2$, $\cos(60°) = 1/2$, and $\tan(60°) = \sqrt{3}$ in the appropriate identities.

$$\sin(120°) = 2\sin(60°)\cos(60°) = 2 \cdot \frac{\sqrt{3}}{2} \cdot \frac{1}{2} = \frac{\sqrt{3}}{2}$$

$$\cos(120°) = \cos^2(60°) - \sin^2(60°) = \left(\frac{1}{2}\right)^2 - \left(\frac{\sqrt{3}}{2}\right)^2 = -\frac{1}{2}$$

$$\tan(120°) = \frac{2\tan(60°)}{1 - \tan^2(60°)} = \frac{2 \cdot \sqrt{3}}{1 - (\sqrt{3})^2} = \frac{2\sqrt{3}}{-2} = -\sqrt{3}$$

These results are the well-known values of sin(120°), cos(120°), and tan(120°).

►**TRY THIS.** Find sin(60°) using a double-angle identity. ■

## Half-Angle Identities

The double-angle identities can be used to get identities for $\sin(x/2)$, $\cos(x/2)$, and $\tan(x/2)$. These identities are known as **half-angle identities.**

To get an identity for $\cos(x/2)$, start by solving the double-angle identity $\cos 2x = 2\cos^2 x - 1$ for $\cos x$:

$$2\cos^2 x - 1 = \cos 2x$$

$$\cos^2 x = \frac{1 + \cos 2x}{2}$$

$$\cos x = \pm\sqrt{\frac{1 + \cos 2x}{2}}$$

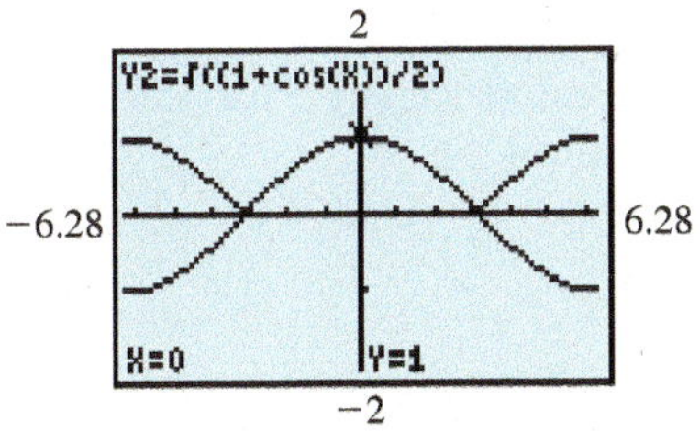


Figure 6.16

Because the last equation is correct for any value of $x$, it is also correct if $x$ is replaced by $x/2$. Replacing $x$ by $x/2$ yields the half-angle identity

$$\cos\frac{x}{2} = \pm\sqrt{\frac{1 + \cos x}{2}}.$$

So for any value of $x$, $\cos(x/2)$ is equal to either the positive or the negative square root of $(1 + \cos x)/2$.

The graphs of $y_1 = \cos(x/2)$ and $y_2 = \sqrt{(1 + \cos x)/2}$ in Fig. 6.16 illustrate this identity. Graph these curves on your calculator to see that the graph of $y_2$ does not go below the $x$-axis. Then use TRACE to see that $y_1 = y_2$ for $y_1 \geq 0$ and $y_1 = -y_2$ for $y_1 < 0$. □

## EXAMPLE 2 Evaluating with half-angle identities

Use the half-angle identity to find the exact value of $\cos(\pi/8)$.

### Solution

Use $x = \pi/4$ in the half-angle identity for cosine:

$$\cos\frac{\pi}{8} = \cos\frac{\pi/4}{2} = \sqrt{\frac{1 + \cos\pi/4}{2}} = \sqrt{\frac{1 + \frac{\sqrt{2}}{2}}{2}}$$

$$= \sqrt{\frac{2 + \sqrt{2}}{4}} = \frac{\sqrt{2 + \sqrt{2}}}{2}$$

The positive square root is used because the angle $\pi/8$ is in the first quadrant, where cosine is positive.

▶**TRY THIS.** Find the exact value of $\sin(-22.5°)$ using a half-angle identity. ■

To get an identity for $\sin(x/2)$, solve $1 - 2\sin^2 x = \cos 2x$ for $\sin x$:

$$1 - 2\sin^2 x = \cos 2x$$
$$-2\sin^2 x = \cos 2x - 1$$
$$\sin^2 x = \frac{1 - \cos 2x}{2}$$
$$\sin x = \pm\sqrt{\frac{1 - \cos 2x}{2}}$$

Replacing $x$ by $x/2$ yields the half-angle identity for sine:

$$\sin\frac{x}{2} = \pm\sqrt{\frac{1 - \cos x}{2}}$$

The half-angle identity for sine can be used to find the exact value for any angle that is one-half of an angle for which we know the exact value of cosine.

## EXAMPLE 3 Evaluating with half-angle identities

Use the half-angle identity to find the exact value of $\sin(75°)$.

### Solution

Use $x = 150°$ in the half-angle identity for sine:

$$\sin(75°) = \sin\left(\frac{150°}{2}\right) = \pm\sqrt{\frac{1 - \cos(150°)}{2}}$$

Since 75° is in quadrant I, $\sin(75°)$ is positive. The reference angle for 150° is 30° and $\cos(150°)$ is negative, because 150° is in quadrant II. So

$$\cos(150°) = -\cos(30°) = -\frac{\sqrt{3}}{2}$$

and

$$\sin(75°) = \sqrt{\frac{1 + \frac{\sqrt{3}}{2}}{2}} = \sqrt{\frac{2 + \sqrt{3}}{4}} = \frac{\sqrt{2 + \sqrt{3}}}{2}.$$

**Figure 6.17**

Check with a calculator, as shown in Fig. 6.17.

▶**TRY THIS.** Find the exact value of $\cos 75°$ using the half-angle identity. ■

To get a half-angle identity for tangent, use the half-angle identities for sine and cosine:

$$\tan\frac{x}{2} = \frac{\sin\frac{x}{2}}{\cos\frac{x}{2}} = \frac{\pm\sqrt{\frac{1 - \cos x}{2}}}{\pm\sqrt{\frac{1 + \cos x}{2}}} = \pm\sqrt{\frac{1 - \cos x}{1 + \cos x}}$$

To get another form of this identity, multiply the numerator and denominator inside the radical by $1 + \cos x$. Use the fact that $\sqrt{\sin^2 x} = |\sin x|$ but $\sqrt{(1 + \cos x)^2} = 1 + \cos x$ because $1 + \cos x$ is nonnegative:

$$\tan\frac{x}{2} = \pm\sqrt{\frac{(1 - \cos x)(1 + \cos x)}{(1 + \cos x)^2}} = \pm\sqrt{\frac{\sin^2 x}{(1 + \cos x)^2}} = \pm\frac{|\sin x|}{1 + \cos x}$$

We use the positive or negative sign depending on the sign of $\tan(x/2)$. It can be shown that $\sin x$ has the same sign as $\tan(x/2)$. (See Exercise 83.) With this fact we can omit the absolute value and the $\pm$ symbol and write the identity as

$$\tan\frac{x}{2} = \frac{\sin x}{1 + \cos x}.$$

To get a third form of the identity for $\tan(x/2)$, multiply the numerator and denominator of the right-hand side by $1 - \cos x$:

$$\tan\frac{x}{2} = \frac{\sin x(1 - \cos x)}{(1 + \cos x)(1 - \cos x)} = \frac{\sin x(1 - \cos x)}{\sin^2 x} = \frac{1 - \cos x}{\sin x}$$

The half-angle identities are summarized as follows.

### Half-Angle Identities

$$\sin\frac{x}{2} = \pm\sqrt{\frac{1 - \cos x}{2}} \qquad \cos\frac{x}{2} = \pm\sqrt{\frac{1 + \cos x}{2}}$$

$$\tan\frac{x}{2} = \pm\sqrt{\frac{1 - \cos x}{1 + \cos x}} \qquad \tan\frac{x}{2} = \frac{\sin x}{1 + \cos x} \qquad \tan\frac{x}{2} = \frac{1 - \cos x}{\sin x}$$

## EXAMPLE 4 Evaluating with half-angle identities

Use the half-angle identity to find the exact value of $\tan(-15°)$.

### Solution

Use $x = -30°$ in the half-angle identity $\tan\frac{x}{2} = \frac{\sin x}{1 + \cos x}$.

$$\begin{aligned}\tan(-15°) &= \tan\left(\frac{-30°}{2}\right)\\ &= \frac{\sin(-30°)}{1 + \cos(-30°)}\\ &= \frac{-\frac{1}{2}}{1 + \frac{\sqrt{3}}{2}} = \frac{-1}{2 + \sqrt{3}} = \frac{-1(2 - \sqrt{3})}{(2 + \sqrt{3})(2 - \sqrt{3})}\\ &= -2 + \sqrt{3}\end{aligned}$$

▶**TRY THIS.** Find the exact value of $\tan(-22.5°)$ using the half-angle identity. ■

## EXAMPLE 5 Using the identities

Find $\sin(\alpha/2)$, given that $\sin \alpha = 3/5$ and $\pi/2 < \alpha < \pi$.

### Solution

Use the identity $\sin^2\alpha + \cos^2\alpha = 1$ to find $\cos\alpha$:

$$\left(\frac{3}{5}\right)^2 + \cos^2\alpha = 1$$

$$\cos^2\alpha = \frac{16}{25}$$

$$\cos\alpha = \pm\frac{4}{5}$$

For $\pi/2 < \alpha < \pi$, we have $\cos\alpha < 0$. So $\cos\alpha = -4/5$. Now use the half-angle identity for sine:

$$\sin\frac{\alpha}{2} = \pm\sqrt{\frac{1 - \cos\alpha}{2}} = \pm\sqrt{\frac{1 + \frac{4}{5}}{2}} = \pm\sqrt{\frac{9}{10}} = \pm\frac{\sqrt{90}}{10} = \pm\frac{3\sqrt{10}}{10}$$

If $\pi/2 < \alpha < \pi$, then $\pi/4 < \alpha/2 < \pi/2$. So $\sin(\alpha/2) > 0$ and $\sin(\alpha/2) = 3\sqrt{10}/10$.

▸**TRY THIS.** Find $\cos(\alpha/2)$ given that $\sin\alpha = 1/3$ and $\pi/2 < \alpha < \pi$. ■

## Verifying Identities

### EXAMPLE 6 A triple-angle identity

Prove that the following equation is an identity:

$$\sin(3x) = \sin x(3\cos^2 x - \sin^2 x)$$

### Solution

Write $3x$ as $2x + x$ and use the identity for the sine of a sum:

$$\begin{aligned}
\sin(3x) &= \sin(2x + x) \\
&= \sin 2x \cos x + \cos 2x \sin x && \text{Sine of a sum identity} \\
&= 2\sin x\cos x \cos x + (\cos^2 x - \sin^2 x)\sin x && \text{Double-angle identities} \\
&= 2\sin x\cos^2 x + \sin x\cos^2 x - \sin^3 x && \text{Distributive property} \\
&= 3\sin x\cos^2 x - \sin^3 x \\
&= \sin x(3\cos^2 x - \sin^2 x) && \text{Factor.}
\end{aligned}$$

▸**TRY THIS.** Prove $\cos(3x) = \cos^3 x - 3\cos x\sin^2 x$ is an identity. ■

Note that in Example 6 we used identities to expand the simpler side of the equation. In this case, simplifying the more complicated side is more difficult.

### EXAMPLE 7 Verifying identities involving half-angles

Prove that the following equation is an identity:

$$\tan\frac{x}{2} + \cot\frac{x}{2} = 2\csc x$$

### Solution

Write the left-hand side in terms of $\tan(x/2)$ and then use two different half-angle identities for $\tan(x/2)$:

$$\begin{aligned}\tan\frac{x}{2} + \cot\frac{x}{2} &= \tan\frac{x}{2} + \frac{1}{\tan\frac{x}{2}} \\ &= \frac{\sin x}{1 + \cos x} + \frac{\sin x}{1 - \cos x} \\ &= \frac{\sin x(1 - \cos x)}{(1 + \cos x)(1 - \cos x)} + \frac{\sin x(1 + \cos x)}{(1 - \cos x)(1 + \cos x)} \\ &= \frac{2\sin x}{1 - \cos^2 x} \\ &= \frac{2\sin x}{\sin^2 x} = 2\frac{1}{\sin x} = 2\csc x\end{aligned}$$

▸**TRY THIS.** Prove $\sin^2\left(\frac{x}{2}\right)\cos^2\left(\frac{x}{2}\right) = \frac{\sin^2 x}{4}$ is an identity. ■

## FOR thought... True or False? Explain.

1. $\frac{\sin 42°}{2} = \sin 21° \cos 21°$
2. $\cos(\sqrt{8}) = 2\cos^2(\sqrt{2}) - 1$
3. $\sin 150° = \sqrt{\frac{1 - \cos 75°}{2}}$
4. $\sin 200° = -\sqrt{\frac{1 - \cos 40°}{2}}$
5. $\tan\frac{7\pi}{8} = \sqrt{\frac{1 - \cos(7\pi/4)}{1 + \cos(7\pi/4)}}$
6. $\tan(-\pi/8) = \frac{1 - \cos(\pi/4)}{\sin(-\pi/4)}$
7. For any real number $x$, $\frac{\sin 2x}{2} = \sin x$.
8. The equation $\cos x = \sqrt{\frac{1 + \cos 2x}{2}}$ is an identity.
9. The equation $\sqrt{(1 - \cos x)^2} = 1 - \cos x$ is an identity.
10. If $180° < \alpha < 360°$, then $\sin \alpha < 0$ and $\tan(\alpha/2) < 0$.

## EXERCISES 6.4

*Find the exact value of each expression using double-angle identities.*

1. $\sin(90°)$
2. $\cos(60°)$
3. $\tan(60°)$
4. $\cos(180°)$
5. $\sin(3\pi/2)$
6. $\cos(4\pi/3)$
7. $\tan(4\pi/3)$
8. $\sin(2\pi/3)$

*Find the exact value of each expression using the half-angle identities.*

9. $\cos(15°)$
10. $\cos(\pi/8)$
11. $\sin(15°)$
12. $\sin(-\pi/6)$
13. $\tan(15°)$
14. $\tan(3\pi/8)$
15. $\sin(22.5°)$
16. $\tan(75°)$

*For each equation determine whether the positive or negative sign makes the equation correct. Do not use a calculator.*

17. $\sin 118.5° = \pm\sqrt{\frac{1 - \cos 237°}{2}}$
18. $\sin 222.5° = \pm\sqrt{\frac{1 - \cos 445°}{2}}$
19. $\cos 100° = \pm\sqrt{\frac{1 + \cos 200°}{2}}$
20. $\cos\frac{9\pi}{7} = \pm\sqrt{\frac{1 + \cos(18\pi/7)}{2}}$

21. $\tan\dfrac{-5\pi}{12} = \pm\sqrt{\dfrac{1 - \cos(-5\pi/6)}{1 + \cos(-5\pi/6)}}$

22. $\tan\dfrac{17\pi}{12} = \pm\sqrt{\dfrac{1 - \cos(17\pi/6)}{1 + \cos(17\pi/6)}}$

*Use identities to simplify each expression. Do not use a calculator.*

23. $2 \sin 13° \cos 13°$

24. $\sin^2\left(\dfrac{\pi}{5}\right) - \cos^2\left(\dfrac{\pi}{5}\right)$

25. $2\cos^2(22.5°) - 1$

26. $1 - 2\sin^2\left(-\dfrac{\pi}{8}\right)$

27. $\dfrac{\tan 15°}{1 - \tan^2(15°)}$

28. $\dfrac{\tan 30°}{1 - \tan^2(30°)}$

29. $\dfrac{\sin 12°}{1 + \cos 12°}$

30. $\csc 8°(1 - \cos 8°)$

31. $2\sin\left(\dfrac{\pi}{9} - \dfrac{\pi}{2}\right)\cos\left(\dfrac{\pi}{2} - \dfrac{\pi}{9}\right)$

32. $2\cos^2\left(\dfrac{\pi}{5} - \dfrac{\pi}{2}\right) - 1$

33. $\cos^2\left(\dfrac{\pi}{9}\right) - \sin^2\left(\dfrac{\pi}{9}\right)$

34. $\dfrac{2}{\cot 5(1 - \tan^2 5)}$

*Match each given expression with an equivalent expression (a)–(j).*

35. $1 - \cos^2 2$

36. $\dfrac{2\tan 2}{1 - \tan^2 2}$

37. $\dfrac{1 + \cos 2}{2}$

38. $\dfrac{1 + \cos 2}{\sin 2}$

39. $2 \sin 2 \cos 2$

40. $\dfrac{1}{\cos^2 2} - 1$

41. $\dfrac{1 - \cos 2}{1 + \cos 2}$

42. $\cos^2 2 - \sin^2 2$

43. $\dfrac{1 - \cos 2}{2}$

44. $1 - \sin^2 2$

a. $\sin 4$  b. $\cos 4$  c. $\sin^2 2$

d. $\cos^2 2$  e. $\tan 4$  f. $\sin^2 1$

g. $\cos^2 1$  h. $\tan^2 1$  i. $\cot 1$

j. $\tan^2 2$

*In each case, find* $\sin\alpha$, $\cos\alpha$, $\tan\alpha$, $\csc\alpha$, $\sec\alpha$, *and* $\cot\alpha$.

45. $\cos(2\alpha) = 3/5$ and $0° < 2\alpha < 90°$

46. $\cos(2\alpha) = 1/3$ and $360° < 2\alpha < 450°$

47. $\sin(2\alpha) = 5/13$ and $0° < \alpha < 45°$

48. $\sin(2\alpha) = -8/17$ and $180° < 2\alpha < 270°$

49. $\cos(\alpha/2) = -1/4$ and $\pi/2 < \alpha/2 < 3\pi/4$

50. $\sin(\alpha/2) = -1/3$ and $7\pi/4 < \alpha/2 < 2\pi$

51. $\sin(\alpha/2) = 4/5$ and $\alpha/2$ is in quadrant II

52. $\sin(\alpha/2) = 1/5$ and $\alpha/2$ is in quadrant II

*Verify that each equation is an identity.*

53. $\cos^4 s - \sin^4 s = \cos 2s$

54. $\sin 2s = -2\sin s \sin(s - \pi/2)$

55. $\cos 3t = \cos^3 t - 3\sin^2 t \cos t$

56. $\dfrac{\sin 4t}{4} = \cos^3 t \sin t - \sin^3 t \cos t$

57. $\dfrac{\cos 2x + \cos 2y}{\sin x + \cos y} = 2\cos y - 2\sin x$

58. $(\sin\alpha - \cos\alpha)^2 = 1 - \sin 2\alpha$

59. $\dfrac{\cos 2x}{\sin^2 x} = \csc^2 x - 2$

60. $\dfrac{\cos 2s}{\cos^2 s} = \sec^2 s - 2\tan^2 s$

61. $2\sin^2\left(\dfrac{u}{2}\right) = \dfrac{\sin^2 u}{1 + \cos u}$

62. $\cos 2y = \dfrac{1 - \tan^2 y}{1 + \tan^2 y}$

63. $\tan^2\left(\dfrac{x}{2}\right) = \dfrac{\sec x + \cos x - 2}{\sec x - \cos x}$

64. $\sec^2\left(\dfrac{x}{2}\right) = \dfrac{2\sec x + 2}{\sec x + 2 + \cos x}$

65. $\dfrac{1 - \sin^2\left(\dfrac{x}{2}\right)}{1 + \sin^2\left(\dfrac{x}{2}\right)} = \dfrac{1 + \cos x}{3 - \cos x}$

66. $\dfrac{1 - \cos^2\left(\dfrac{x}{2}\right)}{1 - \sin^2\left(\dfrac{x}{2}\right)} = \dfrac{1 - \cos x}{1 + \cos x}$

*For each equation, either prove that it is an identity or prove that it is not an identity.*

67. $\sin(2x) = 2\sin x$

68. $\dfrac{\cos 2x}{2} = \cos x$

69. $\tan\left(\dfrac{x}{2}\right) = \dfrac{1}{2}\tan x$

70. $\tan\left(\dfrac{x}{2}\right) = \sqrt{\dfrac{1 - \cos x}{1 + \cos x}}$

71. $\sin(2x)\cdot\sin\left(\dfrac{x}{2}\right) = \sin^2 x$

72. $\tan x + \tan x = \tan 2x$

73. $\cot\dfrac{x}{2} - \tan\dfrac{x}{2} = \dfrac{\sin 2x}{\sin^2 x}$

74. $\csc^2\left(\dfrac{x}{2}\right) + \sec^2\left(\dfrac{x}{2}\right) = 4\csc^2 x$

*Solve each problem.*

**75.** Find the exact value of $\sin(2\alpha)$ given that $\sin(\alpha) = 3/5$ and $\alpha$ is in quadrant II.

**76.** Find the exact value of $\sin(2\alpha)$ given that $\tan(\alpha) = -8/15$ and $\alpha$ is in quadrant IV.

**77.** Find the exact value of $\cos(2\alpha)$ given that $\sin(\alpha) = 8/17$ and $\alpha$ is in quadrant II.

**78.** Find the exact value of $\tan(2\alpha)$ given that $\sin(\alpha) = -4/5$ and $\alpha$ is in quadrant III.

**79.** In the figure, $\angle B$ is a right angle and the line segment $\overline{AD}$ bisects $\angle A$. If $AB = 5$ and $BC = 3$, what is the exact value of $BD$?

HINT Use the half-angle identity for tangent.

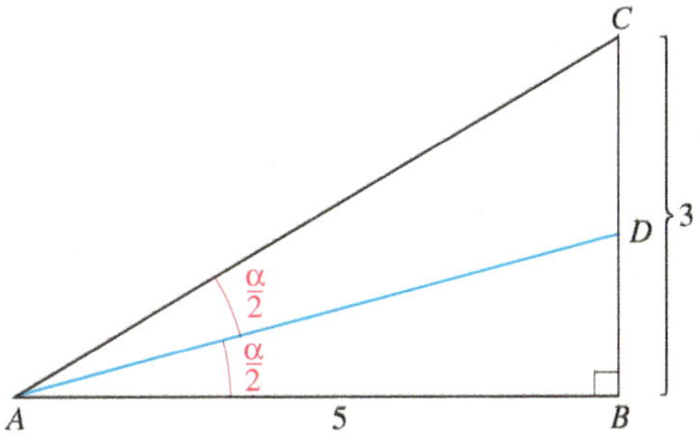


**Figure for Exercise 79**

**80.** In the figure, $\angle B$ is a right angle and the line segment $\overline{AD}$ bisects $\angle A$. If $AB = 10$ and $BD = 2$, what is the exact value of $CD$?

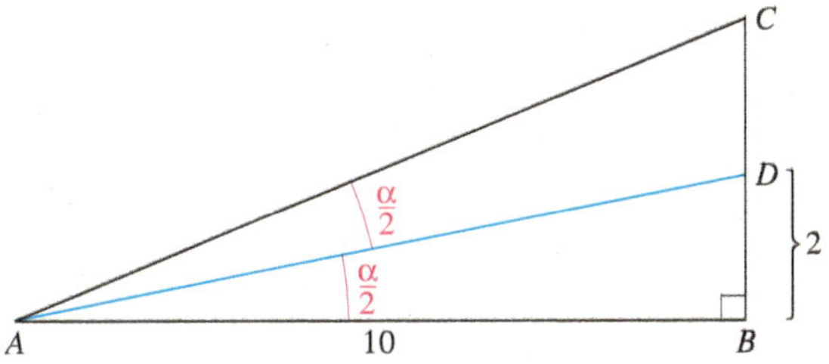


**Figure for Exercise 80**

**81.** *Viewing Area* Find a formula for the viewing area of a television screen in terms of its diagonal and the angle $\alpha$ shown in the figure. Rewrite the formula using a single trigonometric function.

**82.** Use the formula from the previous exercise to find the viewing area for a 32-in. diagonal measure television for which $\alpha = 37.2°$.

**Figure for Exercises 81 and 82**

## FOR WRITING/DISCUSSION

**83.** Show that $\tan(x/2)$ has the same sign as $\sin x$ for any real number $x$.

**84.** Explain why $1 + \cos x \geq 0$ for any real number $x$.

## ▶ RETHINKING

**85.** Simplify $\dfrac{1}{1 + \sin(-x)} + \dfrac{1}{1 + \sin(x)}$.

**86.** Simplify $\dfrac{\cos^3(x) + \cos(x)\sin^2(x)}{\sin(x)}$.

**87.** Complete the sum and difference identities.

**a.** $\cos(x + y) =$ ________

**b.** $\cos(x - y) =$ ________

**88.** One of the acute angles of a right triangle is 26° and its hypotenuse is 38.6 inches. Find the lengths of its legs to the nearest tenth of an inch.

**89.** Evaluate each trigonometric function if possible.

**a.** $\sin(5\pi/6)$ **b.** $\cos(\pi)$

**c.** $\tan(3\pi/2)$ **d.** $\sec(-\pi/3)$

**e.** $\csc(-\pi/2)$ **f.** $\cot(5\pi/4)$

**90.** A race car is averaging 180 mph on a circular track with radius 1/4 mile. Find its angular velocity in radians per minute.

## THINKING OUTSIDE THE BOX XLVI

*Completely Saturated* Four lawn sprinklers are positioned at the vertices of a square that is 2 meters on each side. If each sprinkler waters a circular area with radius 2 meters, then what is the exact area of the region that gets watered by all four sprinklers?

## POP QUIZ 6.4

1. Find $\sin(2\alpha)$ exactly if $\sin\alpha = 1/4$ and $\pi/2 < \alpha < \pi$.

2. Find $\sin(\alpha/2)$ exactly if $\sin\alpha = -4/5$ and $\pi < \alpha < 3\pi/2$.

3. Prove that $\sin^4 x - \cos^4 x = -\cos(2x)$

## LINKING concepts... For Individual or Group Explorations

### Modeling the Motion of a Football

*A good field-goal kicker must learn through experience how to get the maximum distance in a kick. However, without touching a football, we can find the angle at which the football should be kicked to achieve the maximum distance. Assuming no air resistance, the distance in feet that a projectile (such as a football) travels when launched from the ground with an initial velocity of $v_0$ ft/sec is given by*

$$x = (v_0^2 \sin\theta \cos\theta)/16,$$

*where $\theta$ is the angle between the trajectory and the ground, as shown in the figure. (See 6.1 Linking Concepts.)*

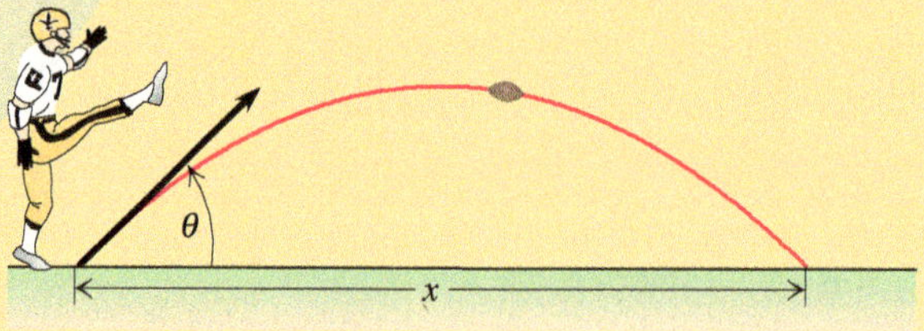


**a)** Use an identity to write the distance $x$ as a function of $2\theta$.

**b)** Graph the function that you found in part (a) using an initial velocity of 50 ft/sec.

**c)** From your graph, determine the value of $\theta$ that maximizes $x$. Does this value of $\theta$ maximize $x$ for any velocity? Explain.

**d)** When a kicker kicks a football 55 yd using the angle determined in part (c), what is the initial velocity of the football in miles per hour?

**e)** Do you think that the actual initial velocity for a 55-yd field goal is larger or smaller than that found in part (d)?

# 6.5 Product and Sum Identities

In Section 5.2 we saw that the location $x$ at time $t$ for an object on a spring, given an initial velocity $v_0$ from an initial location $x_0$, is given by

$$x = \frac{v_0}{\omega}\sin(\omega t) + x_0 \cos(\omega t),$$

where $\omega$ is a constant. Even though $x$ is determined by a combination of sine and cosine, an object in motion on a spring oscillates in a "periodic" fashion. In this section we will develop identities that involve sums of trigonometric functions. These identities are used in solving equations, graphing functions, and explaining how the combination of the sine and cosine functions in the spring equation can work together to produce a periodic motion.

## Product-to-Sum Identities

In Section 6.3 we learned identities for sine and cosine of a sum or difference:

$$\sin(A + B) = \sin A \cos B + \cos A \sin B$$
$$\sin(A - B) = \sin A \cos B - \cos A \sin B$$
$$\cos(A + B) = \cos A \cos B - \sin A \sin B$$
$$\cos(A - B) = \cos A \cos B + \sin A \sin B$$

Now we can add the first two of these identities to get a new identity:

$$\begin{aligned} \sin(A + B) &= \sin A \cos B + \cos A \sin B \\ \sin(A - B) &= \sin A \cos B - \cos A \sin B \\ \hline \sin(A + B) + \sin(A - B) &= 2 \sin A \cos B \end{aligned}$$

When we divide each side by 2, we get an identity expressing a product in terms of a sum:

$$\sin A \cos B = \frac{1}{2}[\sin(A + B) + \sin(A - B)]$$

We can produce three other, similar identities from the sum and difference identities. The four identities, known as the **product-to-sum identities,** are listed below. These identities are not used as often as the other identities that we study in this chapter. It is not necessary to memorize these identities. Just remember them by name and look them up as necessary.

**Product-to-Sum Identities**

$$\sin A \cos B = \frac{1}{2}[\sin(A + B) + \sin(A - B)]$$

$$\sin A \sin B = \frac{1}{2}[\cos(A - B) - \cos(A + B)]$$

$$\cos A \sin B = \frac{1}{2}[\sin(A + B) - \sin(A - B)]$$

$$\cos A \cos B = \frac{1}{2}[\cos(A - B) + \cos(A + B)]$$

**Foreshadowing Calculus**

The product-to-sum identities are important identities for some electrical engineers. They are used along with calculus in the wave theory to design computer modems, which combine and take apart sine waves.

### EXAMPLE 1 Expressing a product as a sum

Use the product-to-sum identities to rewrite each expression.

**a.** $\sin 12° \cos 9°$ **b.** $\sin(\pi/12)\sin(\pi/8)$

**Solution**

**a.** Use the product-to-sum identity for $\sin A \cos B$:

$$\begin{aligned} \sin 12° \cos 9° &= \frac{1}{2}[\sin(12° + 9°) + \sin(12° - 9°)] \\ &= \frac{1}{2}[\sin 21° + \sin 3°] \end{aligned}$$

**Figure 6.18**

**b.** Use the product-to-sum identity for $\sin A \sin B$:

$$\sin\left(\frac{\pi}{12}\right)\sin\left(\frac{\pi}{8}\right) = \frac{1}{2}\left[\cos\left(\frac{\pi}{12} - \frac{\pi}{8}\right) - \cos\left(\frac{\pi}{12} + \frac{\pi}{8}\right)\right]$$

$$= \frac{1}{2}\left[\cos\left(-\frac{\pi}{24}\right) - \cos\left(\frac{5\pi}{24}\right)\right]$$

$$= \frac{1}{2}\left[\cos\frac{\pi}{24} - \cos\frac{5\pi}{24}\right]$$

Use a calculator to check, as shown in Fig. 6.18.

**▸TRY THIS.** Use a product-to-sum identity to rewrite $\sin(4x)\cos(3x)$. ■

## EXAMPLE 2 Evaluating a product

Use a product-to-sum identity to find the exact value of $\cos(67.5°)\sin(112.5°)$.

### Solution

Use the identity $\cos A \sin B = \frac{1}{2}[\sin(A + B) - \sin(A - B)]$:

$$\cos(67.5°)\sin(112.5°) = \frac{1}{2}[\sin(67.5° + 112.5°) - \sin(67.5° - 112.5°)]$$

$$= \frac{1}{2}[\sin(180°) - \sin(-45°)]$$

$$= \frac{1}{2}[0 + \sin(45°)]$$

$$= \frac{1}{2} \cdot \frac{\sqrt{2}}{2} = \frac{\sqrt{2}}{4}$$

**▸TRY THIS.** Find the exact value of $\sin(52.5°)\cos(7.5°)$. ■

## Sum-to-Product Identities

It is sometimes useful to write a sum of two trigonometric functions as a product. Consider the product-to-sum formula

$$\sin A \cos B = \frac{1}{2}[\sin(A + B) + \sin(A - B)].$$

To make the right-hand side look simpler, we let $A + B = x$ and $A - B = y$. From these equations we get $x + y = 2A$ and $x - y = 2B$, or

$$A = \frac{x + y}{2} \qquad \text{and} \qquad B = \frac{x - y}{2}.$$

Substitute these values into the identity for $\sin A \cos B$:

$$\sin\left(\frac{x + y}{2}\right)\cos\left(\frac{x - y}{2}\right) = \frac{1}{2}[\sin x + \sin y]$$

Multiply each side by 2 to get

$$\sin x + \sin y = 2\sin\left(\frac{x + y}{2}\right)\cos\left(\frac{x - y}{2}\right).$$

The other three product-to-sum formulas can also be rewritten as sum-to-product identities by using similar procedures.

### Sum-to-Product Identities

$$\sin x + \sin y = 2\sin\left(\frac{x+y}{2}\right)\cos\left(\frac{x-y}{2}\right)$$

$$\sin x - \sin y = 2\cos\left(\frac{x+y}{2}\right)\sin\left(\frac{x-y}{2}\right)$$

$$\cos x + \cos y = 2\cos\left(\frac{x+y}{2}\right)\cos\left(\frac{x-y}{2}\right)$$

$$\cos x - \cos y = -2\sin\left(\frac{x+y}{2}\right)\sin\left(\frac{x-y}{2}\right)$$

## EXAMPLE 3 Expressing a sum or difference as a product

Use the sum-to-product identities to rewrite each expression.

**a.** $\sin 8° + \sin 6°$ **b.** $\cos(\pi/5) - \cos(\pi/8)$ **c.** $\sin(6t) - \sin(4t)$

### Solution

**a.** Use the sum-to-product identity for $\sin x + \sin y$:

$$\sin 8° + \sin 6° = 2\sin\left(\frac{8° + 6°}{2}\right)\cos\left(\frac{8° - 6°}{2}\right) = 2\sin 7° \cos 1°$$

**b.** Use the sum-to-product identity for $\cos x - \cos y$:

$$\cos\left(\frac{\pi}{5}\right) - \cos\left(\frac{\pi}{8}\right) = -2\sin\left(\frac{\pi/5 + \pi/8}{2}\right)\sin\left(\frac{\pi/5 - \pi/8}{2}\right)$$

$$= -2\sin\left(\frac{13\pi}{80}\right)\sin\left(\frac{3\pi}{80}\right)$$

**c.** Use the sum-to-product identity for $\sin x - \sin y$:

$$\sin 6t - \sin 4t = 2\cos\left(\frac{6t + 4t}{2}\right)\sin\left(\frac{6t - 4t}{2}\right)$$

$$= 2\cos 5t \sin t$$

▸**TRY THIS.** Use a sum-to-product identity to rewrite $\cos(4x) + \cos(2x)$. ■

## EXAMPLE 4 Evaluating a sum

Use a sum-to-product identity to find the exact value of $\cos(112.5°) + \cos(67.5°)$.

### Solution

Use the sum-to-product identity $\cos x + \cos y = 2\cos\left(\frac{x+y}{2}\right)\cos\left(\frac{x-y}{2}\right)$:

$$\cos(112.5°) + \cos(67.5°) = 2\cos\left(\frac{112.5° + 67.5°}{2}\right)\cos\left(\frac{112.5° - 67.5°}{2}\right)$$

$$= 2\cos(90°)\cos(22.5°)$$

$$= 0 \quad \text{Because } \cos(90°) = 0.$$

▸**TRY THIS.** Use a sum-to-product identity to evaluate $\sin(105°) + \sin(15°)$. ■

## The Function $y = a \sin x + b \cos x$

The function $y = a \sin x + b \cos x$ involves an expression similar to those in the sum-to-product identities, but it is not covered by the sum-to-product identities. Functions of this type occur in applications such as the position of a weight in motion due to the force of a spring, the position of a swinging pendulum, and the current in an electrical circuit. In these applications it is important to express this function in terms of a single trigonometric function.

To get an identity for $a \sin x + b \cos x$ for nonzero values of $a$ and $b$, we let $\alpha$ be an angle in standard position whose terminal side contains the point $(a, b)$. Using trigonometric ratios from Section 5.6, we get

$$\sin \alpha = \frac{b}{r} = \frac{b}{\sqrt{a^2 + b^2}} \quad \text{and} \quad \cos \alpha = \frac{a}{r} = \frac{a}{\sqrt{a^2 + b^2}}.$$

Now write $a \sin x + b \cos x$ with the trigonometric ratios, and then replace the ratios by $\sin \alpha$ and $\cos \alpha$:

$$\begin{aligned} a \sin x + b \cos x &= \sqrt{a^2 + b^2}\left((\sin x)\frac{a}{\sqrt{a^2 + b^2}} + (\cos x)\frac{b}{\sqrt{a^2 + b^2}}\right) \\ &= \sqrt{a^2 + b^2}\,(\sin x \cos \alpha + \cos x \sin \alpha) \\ &= \sqrt{a^2 + b^2}\,\sin(x + \alpha) \quad \text{By the identity for the sine of a sum} \end{aligned}$$

This identity is called the **reduction formula** because it reduces two trigonometric functions to one.

### Theorem: Reduction Formula

If $\alpha$ is an angle in standard position whose terminal side contains $(a, b)$, then

$$a \sin x + b \cos x = \sqrt{a^2 + b^2}\,\sin(x + \alpha) \text{ for any real number } x.$$

To rewrite an expression of the form $a \sin x + b \cos x$ by using the reduction formula, we need to find $\alpha$ so that the terminal side of $\alpha$ goes through $(a, b)$. By using trigonometric ratios, we have

$$\sin \alpha = \frac{b}{\sqrt{a^2 + b^2}}, \quad \cos \alpha = \frac{a}{\sqrt{a^2 + b^2}}, \quad \text{and} \quad \tan \alpha = \frac{b}{a}.$$

Since we know $a$ and $b$, we can find $\sin \alpha$, $\cos \alpha$, or $\tan \alpha$, and then use an inverse trigonometric function to find $\alpha$. However, because of the ranges of the inverse functions, the angle obtained from an inverse function might not have its terminal side through $(a, b)$ as required.

### EXAMPLE 5 Using the reduction formula

Use the reduction formula to rewrite $-3 \sin x - 3 \cos x$ in the form $A \sin(x + C)$.

#### Solution

Because $a = -3$ and $b = -3$, we have

$$\sqrt{a^2 + b^2} = \sqrt{18} = 3\sqrt{2}.$$

Since the terminal side of $\alpha$ must go through $(-3, -3)$, we have

$$\cos \alpha = \frac{-3}{3\sqrt{2}} = -\frac{\sqrt{2}}{2}.$$

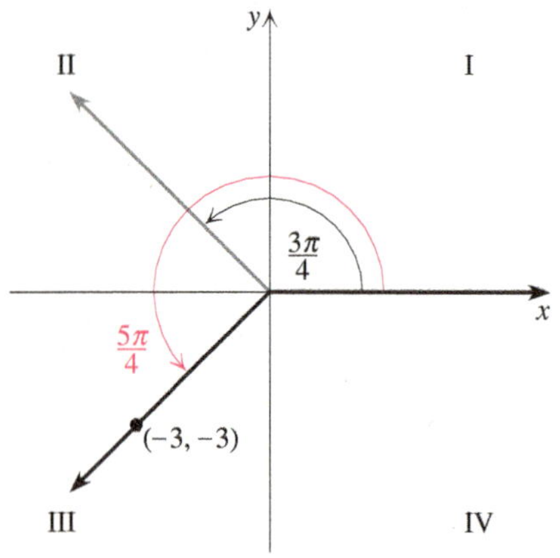

Figure 6.19

Now $\cos^{-1}(-\sqrt{2}/2) = 3\pi/4$, but the terminal side of $3\pi/4$ is in quadrant II, as shown in Fig. 6.19. However, we also have $\cos(5\pi/4) = -\sqrt{2}/2$, and the terminal

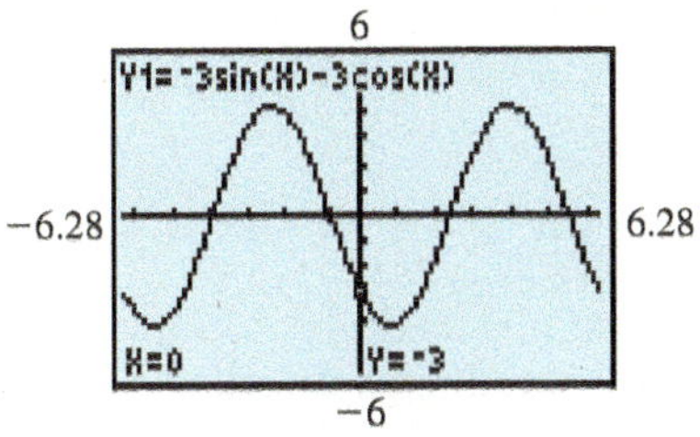


**Figure 6.20**

side for $5\pi/4$ does pass through $(-3, -3)$ in quadrant III. So $\alpha = 5\pi/4$. By the reduction formula,

$$-3 \sin x - 3 \cos x = 3\sqrt{2} \sin\left(x + \frac{5\pi}{4}\right).$$

The reduction formula explains why the calculator graph of $y = -3 \sin x - 3 \cos x$ in Fig. 6.20 appears to be a sine wave with amplitude $3\sqrt{2}$.

▶**TRY THIS.** Rewrite $-2 \sin x + 2 \cos x$ in the form $A \sin(x + C)$. ■

In the next example, we see that the reduction formula makes it easier to understand functions of the form $y = a \sin x + b \cos x$.

## EXAMPLE 6 Using the reduction formula in graphing

Graph one cycle of the function

$$y = \sqrt{3} \sin x + \cos x$$

and state the amplitude, period, and phase shift.

### Solution

Because $a = \sqrt{3}$ and $b = 1$, we have $\sqrt{a^2 + b^2} = \sqrt{4} = 2$. The terminal side of $\alpha$ must go through $(\sqrt{3}, 1)$. So $\sin \alpha = 1/2$, and $\sin^{-1}(1/2) = \pi/6$. Since the terminal side of $\pi/6$ is in quadrant I and goes through $(\sqrt{3}, 1)$, we can rewrite the function as

$$y = \sqrt{3} \sin x + \cos x = 2 \sin\left(x + \frac{\pi}{6}\right).$$

From the new form, we see that the graph is a sine wave with amplitude 2, period $2\pi$, and phase shift $-\pi/6$, as shown in Fig. 6.21.

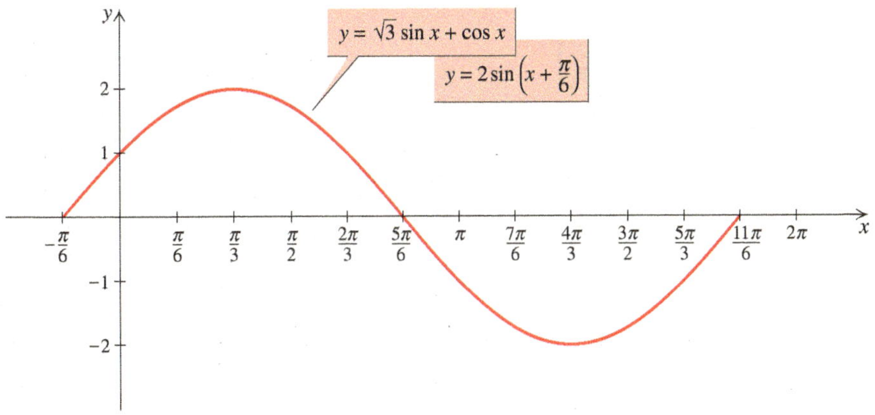


**Figure 6.21**

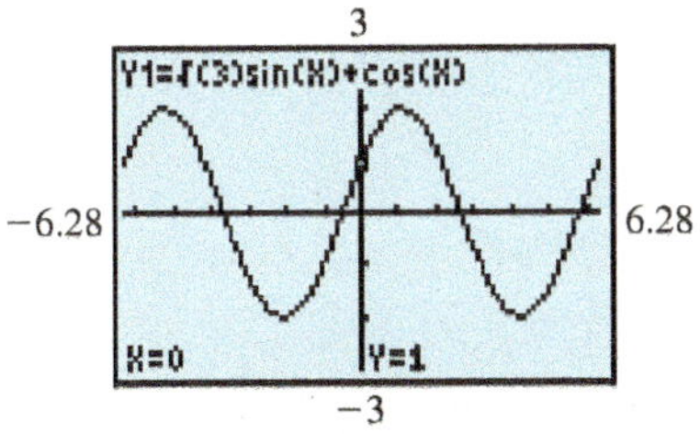


**Figure 6.22**

The calculator graph shown in Fig. 6.22 supports these conclusions.

▶**TRY THIS.** Determine the amplitude and phase shift for $y = \sin x - \cos x$. ■

Since $y = a \sin x + b \cos x$ and $y = \sqrt{a^2 + b^2} \sin(x + \alpha)$ are the same function, the graph of $y = a \sin x + b \cos x$ is a transformation of $y = \sin x$. So, for nonzero $a$ and $b$, the graph of $y = a \sin x + b \cos x$ is a sine wave with amplitude $\sqrt{a^2 + b^2}$, period $2\pi$, and phase shift $-\alpha$.

## Modeling the Motion of a Spring

In Section 5.2, the function

$$x = \frac{v_0}{\omega}\sin(\omega t) + x_0 \cos(\omega t)$$

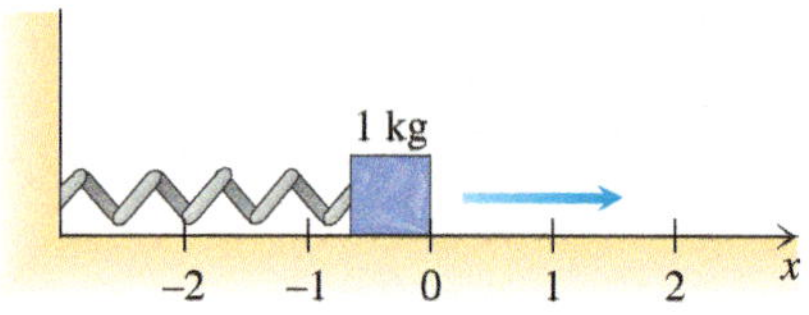

Figure 6.23

was used to give the position $x$ at time $t$ for a weight in motion on a vertical spring. In this formula, $v_0$ is the initial velocity, $x_0$ is the initial position, and $\omega$ is a constant. The same equation is used for a weight attached to a horizontal spring and set in motion on a frictionless surface, as shown in Fig. 6.23.

We use the reduction formula on the spring equation to determine how far a spring actually stretches and compresses when the block is set in motion with a particular velocity.

### EXAMPLE 7 Using the reduction formula with springs

At time $t = 0$ the 1-kilogram block shown in Fig. 6.23 is moved (and the spring compressed) to a position 1 meter to the left of the resting position. From this position the block is given a velocity of 2 meters per second to the right. Use the reduction formula to find the maximum distance reached by the block from rest. Assume that $\omega = 1$.

#### Solution

Since $v_0 = 2$, $x_0 = -1$, and $\omega = 1$, the position of the block at time $t$ is given by

$$x = 2 \sin t - \cos t.$$

Use the reduction formula to rewrite this equation. If $a = 2$ and $b = -1$, then

$$\sqrt{a^2 + b^2} = \sqrt{2^2 + (-1)^2} = \sqrt{5}.$$

If the terminal side of $\alpha$ goes through $(2, -1)$, then $\tan \alpha = -1/2$ and

$$\tan^{-1}(-1/2) \approx -0.46.$$

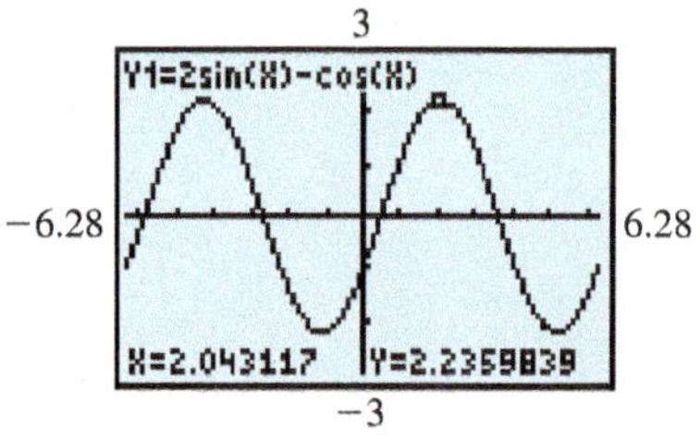

Figure 6.24

An angle of $-0.46$ rad lies in quadrant IV and goes through $(2, -1)$. Use $\alpha = -0.46$ in the reduction formula to get

$$x = \sqrt{5}\sin(t - 0.46).$$

Since the amplitude of this function is $\sqrt{5}$, the block oscillates between $x = \sqrt{5}$ meters and $x = -\sqrt{5}$ meters. The maximum distance from $x = 0$ is $\sqrt{5}$ meters. The maximum $y$ value on the calculator graph in Fig. 6.24 supports the conclusion that the amplitude is $\sqrt{5}$.

▶**TRY THIS.** The location in centimeters of a block on a spring is given by $x = \sin t - 3 \cos t$. Find the maximum distance reached by the block from rest. ■

## FOR thought... True or False? Explain.

1. $\sin 45° \cos 15° = 0.5[\sin 60° + \sin 30°]$
2. $\cos(\pi/8)\sin(\pi/4) = 0.5[\sin(3\pi/8) - \sin(\pi/8)]$
3. $2 \cos 6° \cos 8° = \cos 2° + \cos 14°$
4. $\sin 5° - \sin 9° = 2 \cos 7° \sin 2°$
5. $\cos 4 + \cos 12 = 2 \cos 8 \cos 4$
6. $\cos(\pi/3) - \cos(\pi/2) = -2 \sin(5\pi/12)\sin(\pi/12)$
7. $\sin(\pi/6) + \cos(\pi/6) = \sqrt{2}\sin(\pi/6 + \pi/4)$
8. $\frac{1}{2}\sin(\pi/6) + \frac{\sqrt{3}}{2}\cos(\pi/6) = \sin(\pi/2)$
9. The graph of $y = \frac{1}{2}\sin x + \frac{\sqrt{3}}{2}\cos x$ is a sine wave with amplitude 1.
10. The equation $\frac{1}{\sqrt{2}}\sin x + \frac{1}{\sqrt{2}}\cos x = \sin\left(x + \frac{\pi}{4}\right)$ is an identity.

# EXERCISES 6.5

*Use the product-to-sum identities to rewrite each expression.*

**1.** $\sin 13° \sin 9°$

**2.** $\cos 34° \cos 39°$

**3.** $\sin 16° \cos 20°$

**4.** $\cos 9° \sin 8°$

**5.** $\sin(5°)\cos(10°)$

**6.** $\cos(6°)\cos(8°)$

**7.** $\cos\left(\frac{\pi}{6}\right)\cos\left(\frac{\pi}{5}\right)$

**8.** $\sin\left(\frac{2\pi}{9}\right)\sin\left(\frac{3\pi}{4}\right)$

**9.** $\cos(5y^2)\cos(7y^2)$

**10.** $\cos 3t \sin 5t$

**11.** $\sin(2s - 1)\cos(s + 1)$

**12.** $\sin(3t - 1)\sin(2t + 3)$

*Find the exact value of each product. Do not use a calculator.*

**13.** $\sin(52.5°)\sin(7.5°)$

**14.** $\cos(105°)\cos(75°)$

**15.** $\sin\left(\frac{13\pi}{24}\right)\cos\left(\frac{5\pi}{24}\right)$

**16.** $\cos\left(\frac{5\pi}{24}\right)\sin\left(-\frac{\pi}{24}\right)$

*Use the sum-to-product identities to rewrite each expression.*

**17.** $\sin 12° - \sin 8°$

**18.** $\sin 7° + \sin 11°$

**19.** $\cos 80° - \cos 87°$

**20.** $\cos 44° + \cos 31°$

**21.** $\sin 3.6 - \sin 4.8$

**22.** $\sin 5.1 + \sin 6.3$

**23.** $\cos\left(\frac{\pi}{3}\right) - \cos\left(\frac{\pi}{5}\right)$

**24.** $\cos\left(\frac{1}{2}\right) + \cos\left(\frac{2}{3}\right)$

**25.** $\cos(5y - 3) - \cos(3y + 9)$

**26.** $\cos(6t^2 - 1) + \cos(4t^2 - 1)$

**27.** $\sin 5\alpha - \sin 8\alpha$

**28.** $\sin 3s + \sin 5s$

*Find the exact value of each sum. Do not use a calculator.*

**29.** $\sin(75°) + \sin(15°)$

**30.** $\sin(285°) - \sin(15°)$

**31.** $\cos\left(-\frac{\pi}{24}\right) - \cos\left(\frac{7\pi}{24}\right)$

**32.** $\cos\left(\frac{5\pi}{24}\right) + \cos\left(\frac{\pi}{24}\right)$

*Rewrite each expression in the form of $A \sin(x + C)$.*

**33.** $\sin x - \cos x$

**34.** $2 \sin x + 2 \cos x$

**35.** $-\frac{1}{2}\sin x + \frac{\sqrt{3}}{2}\cos x$

**36.** $\frac{\sqrt{2}}{2}\sin x - \frac{\sqrt{2}}{2}\cos x$

**37.** $\frac{\sqrt{3}}{2}\sin x - \frac{1}{2}\cos x$

**38.** $-\frac{\sqrt{3}}{2}\sin x - \frac{1}{2}\cos x$

*Write each function in the form $y = A\sin(x + C)$. Then graph at least one cycle and state the amplitude, period, and phase shift.*

**39.** $y = -\sin x + \cos x$

**40.** $y = \sin x + \sqrt{3}\cos x$

**41.** $y = \sqrt{2}\sin x - \sqrt{2}\cos x$

**42.** $y = 2\sin x - 2\cos x$

**43.** $y = -\sqrt{3}\sin x - \cos x$

**44.** $y = -\frac{1}{2}\sin x - \frac{\sqrt{3}}{2}\cos x$

*For each function, determine the exact amplitude and find the phase shift in radians (to the nearest tenth).*

**45.** $y = 3\sin x + 4\cos x$

**46.** $y = \sin x + 5\cos x$

**47.** $y = -6\sin x + \cos x$

**48.** $y = -\sqrt{5}\sin x + 2\cos x$

**49.** $y = -3\sin x - 5\cos x$

**50.** $y = -\sqrt{2}\sin x - \sqrt{7}\cos x$

*Prove that each equation is an identity.*

**51.** $\frac{\sin 3t - \sin t}{\cos 3t + \cos t} = \tan t$

**52.** $\frac{\sin 3x + \sin 5x}{\sin 3x - \sin 5x} = -\frac{\tan 4x}{\tan x}$

**53.** $\frac{\cos x - \cos 3x}{\cos x + \cos 3x} = \tan 2x \tan x$

**54.** $\frac{\cos 5y + \cos 3y}{\cos 5y - \cos 3y} = -\cot 4y \cot y$

**55.** $\cos^2 x - \cos^2 y = -\sin(x + y)\sin(x - y)$

**56.** $\sin^2 x - \sin^2 y = \sin(x + y)\sin(x - y)$

**57.** $\left(\sin\frac{x+y}{2} + \cos\frac{x+y}{2}\right)\left(\sin\frac{x-y}{2} + \cos\frac{x-y}{2}\right) = \sin x + \cos y$

**58.** $\sin 2A \sin 2B = \sin^2(A + B) - \sin^2(A - B)$

**59.** $\cos^2(A - B) - \cos^2(A + B) = \sin^2(A + B) - \sin^2(A - B)$

**60.** $(\sin A + \cos A)(\sin B + \cos B) = \sin(A + B) + \cos(A - B)$

*Solve each problem.*

**61.** *Motion of a Spring* A block is attached to a spring and set in motion, as in Example 7. For this block and spring, the location on the surface at any time $t$ in seconds is given in meters by $x = \sqrt{3}\sin t + \cos t$. Use the reduction formula to rewrite this function, and find the maximum distance reached by the block from the resting position.

**62.** *Motion of a Spring* A block hanging from a spring, as shown in the figure, oscillates in the same manner as the block of Example 7. If a 1-kg block attached to a certain spring is given an upward velocity of 0.3 m/sec from a point 0.5 m below its resting position, then its position at any time $t$ in seconds is given in meters by $x = -0.3\sin t + 0.5\cos t$. Use the reduction formula to find the maximum distance that the block travels from the resting position.

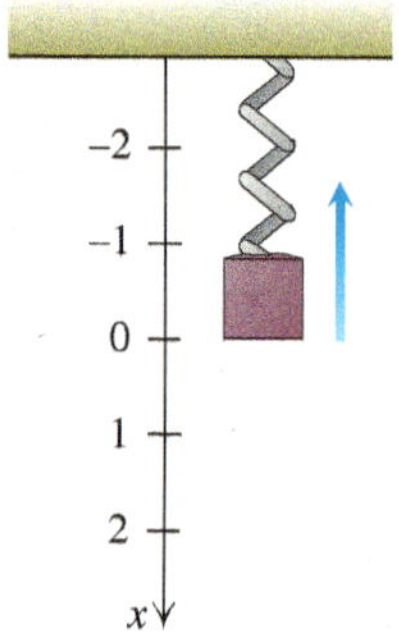


**Figure for Exercise 62**

## FOR WRITING/DISCUSSION

**63.** Derive the identity $\cos(2x) = \cos^2 x - \sin^2 x$ from a product-to-sum identity.

**64.** Derive the identity $\sin(2x) = 2\sin x\cos x$ from a product-to-sum identity.

**65.** Prove the three product-to-sum identities that were not proved in the text.

**66.** Prove the three sum-to-product identities that were not proved in the text.

## ▶ RETHINKING

**67.** Complete the sum and difference identities for tangent.

**a.** $\tan(x + y) =$ ________

**b.** $\tan(x - y) =$ ________

**68.** Find the exact value of $\sin(\alpha + \beta)$ if $\sin\alpha = 1/3$ and $\sin\beta = 1/2$ with $\alpha$ in quadrant II and $\beta$ and in quadrant I.

**69.** Find the exact value of $\tan(x/2)$ given that $\sin(x) = -\sqrt{8/9}$ and $\pi < x < 3\pi/2$.

**70.** Find the exact value of $\sin(x/2)$ given that $\cos(x) = -1/4$ and $\pi/2 < x < \pi$.

**71.** Find the exact value of $\sin(2y)$ given that $\sin(y) = -4/5$ and $270° < y < 360°$.

**72.** Find the acute angles (to the nearest tenth of a degree) for a right triangle whose sides are 5 miles, 12 miles, and 13 miles.

## THINKING OUTSIDE THE BOX XLVII & XLVIII

*Overlapping Circles* Two lawn sprinklers that each water a circular region with radius $a$ are placed $a$ feet apart. What is the total area watered by the sprinklers?

*Marching Ant* An ant starts in the center of the upper left square of an 8 by 8 chessboard. He must crawl to the lower right square of the board, passing through each square exactly once and not passing through a vertex of any square. Either find a path or explain why there is no such path.

## POP QUIZ 6.5

**1.** Use a product-to-sum identity to simplify $\frac{1}{2}[\sin(2\alpha + \beta) - \sin(2\alpha - \beta)]$.

**2.** Rewrite $\cos(2\alpha) + \cos(4\alpha)$ as a product using a sum-to-product identity.

**3.** Write $y = \sqrt{3}\sin x - \cos x$ in the form $y = A\sin(x + C)$.

**4.** Find the amplitude for $y = 3\sin x - 5\cos x$?

LINKING
## concepts... For Individual or Group Explorations

### Maximizing the Total Profit

*Profits at The Christmas Store vary periodically, with a high of* \$50,000 *in December and a low of* \$10,000 *in June, as shown in the accompanying graph. The owner of the Christmas Store also owns The Pool Store, where profits reach a high of* \$80,000 *in August and a low of* \$20,000 *in February. Assume that the profit function for each store is a sine wave.*

**a)** Write the profit function for The Christmas Store as a function of the month and sketch its graph.

**b)** Write the profit function for The Pool Store as a function of the month and sketch its graph.

**c)** Write the total profit as a function of the month and sketch its graph. What is the period?

**d)** Use the maximum feature of a graphing calculator to find the owner's maximum total profit and the month in which it occurs.

**e)** Find the owner's minimum total profit and the month in which it occurs.

**f)** We know that $y = a\sin x + b\cos x$ is a sine function. However, the sum of two arbitrary sine or cosine functions is not necessarily a sine function. Find an example in which the graph of the sum of two sine functions does not look like a sine curve.

**g)** Do you think that the sum of any two sine functions is a periodic function? Explain.

# 6.6 Conditional Trigonometric Equations

An identity is satisfied by *all* values of the variable for which both sides are defined. In this section we investigate **conditional equations,** those equations that are not identities but that have *at least one* solution. For example, the equation $\sin x = 0$ is a conditional equation. Because the trigonometric functions are periodic, conditional equations involving trigonometric functions usually have infinitely many solutions.

The equation $\sin x = 0$ is satisfied by $x = 0, \pm\pi, \pm 2\pi, \pm 3\pi$, and so on. All of these solutions are of the form $k\pi$, where $k$ is an integer. In this section we will solve conditional equations and see that identities play a fundamental role in their solution.

## Cosine Equations

The most basic conditional equation involving cosine is of the form $\cos x = a$, where $a$ is a number in the interval $[-1, 1]$. If $a$ is not in $[-1, 1]$, $\cos x = a$ has no solution. For $a$ in $[-1, 1]$, the equation $x = \cos^{-1} a$ provides one solution in the interval $[0, \pi]$. From this single solution we can determine all of the solutions because of the periodic nature of the cosine function. We must remember also that the cosine of an arc is the $x$-coordinate of its terminal point on the unit circle. So arcs that terminate at opposite ends of a vertical chord in the unit circle have the same cosine.

### EXAMPLE 1 Solving a cosine equation

Find all real numbers that satisfy each equation.

**a.** $\cos x = 1$ **b.** $\cos x = 0$ **c.** $\cos x = -1/2$

**Solution**

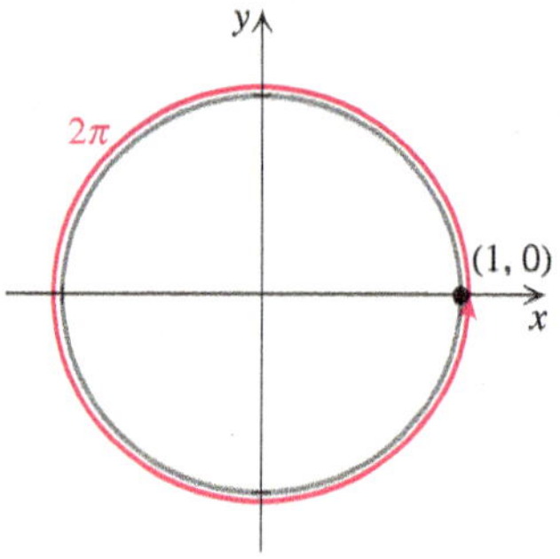

Figure 6.25

**a.** One solution to $\cos x = 1$ is

$$x = \cos^{-1}(1) = 0.$$

Since the period of cosine is $2\pi$, any integral multiple of $2\pi$ can be added to this solution to get additional solutions. So the equation is satisfied by $0, \pm 2\pi, \pm 4\pi$, and so on. Now $\cos x = 1$ is satisfied only if the arc of length $x$ on the unit circle has terminal point $(1, 0)$, as shown in Fig. 6.25. Thus there are no more solutions. The solution set is written as

$$\{x \mid x = 2k\pi\},$$

where $k$ is any integer.

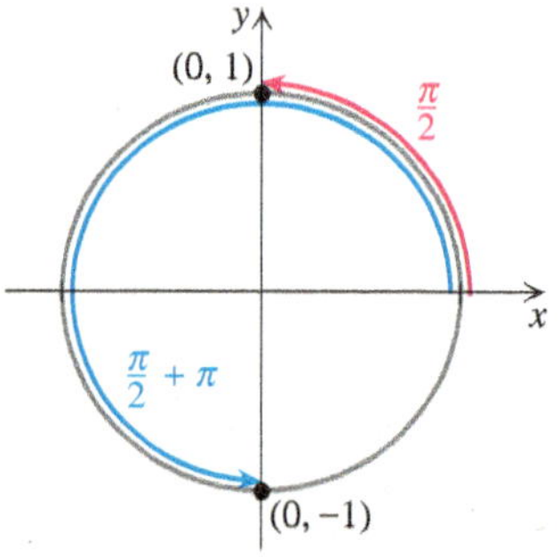

Figure 6.26

**b.** One solution to $\cos x = 0$ is

$$x = \cos^{-1}(0) = \frac{\pi}{2}.$$

The terminal point for the arc of length $\pi/2$ is $(1, 0)$. So any arc length of the form $\pi/2 + 2k\pi$ is a solution to $\cos x = 0$. However, an arc of length $x$ that terminates at $(0, -1)$ also satisfies $\cos x = 0$, as shown in Fig. 6.26, and these arcs are not included in the form $\pi/2 + 2k\pi$. Since the distance between $(1, 0)$ and $(0, -1)$ on the unit circle is $\pi$, all arcs that terminate at these points are of the form $\pi/2 + k\pi$. Thus the solution set is

$$\left\{x \,\middle|\, x = \frac{\pi}{2} + k\pi\right\},$$

where $k$ is any integer.

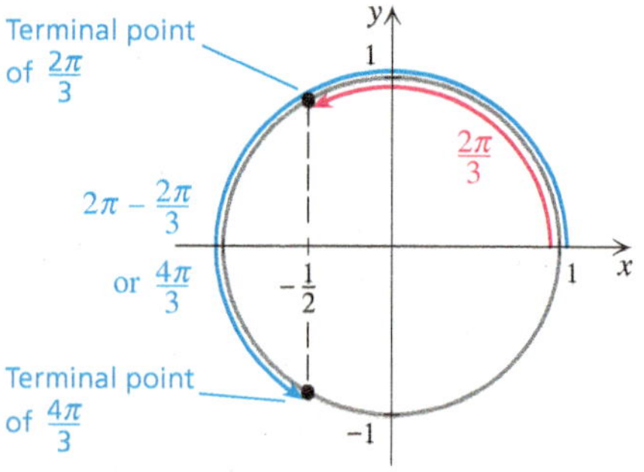

Figure 6.27

**c.** One solution to the equation $\cos x = -1/2$ is

$$x = \cos^{-1}\left(-\frac{1}{2}\right) = \frac{2\pi}{3}.$$

Since the period of cosine is $2\pi$, all arcs of length $2\pi/3 + 2k\pi$ (where $k$ is any integer) satisfy the equation. The arc of length $4\pi/3$ also terminates at a point with $x$-coordinate $-1/2$, as shown in Fig. 6.27.

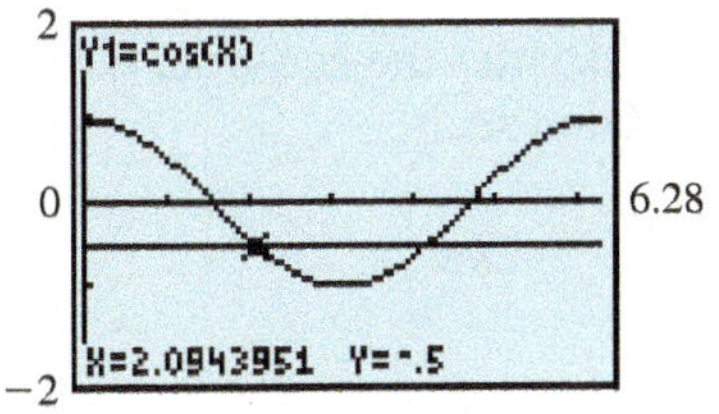


Figure 6.28

So $4\pi/3$ is also a solution to $\cos x = -1/2$, but it is not included in the form $2\pi/3 + 2k\pi$. Thus the solution set to $\cos x = -1/2$ is

$$\left\{x \,\middle|\, x = \frac{2\pi}{3} + 2k\pi \quad \text{or} \quad x = \frac{4\pi}{3} + 2k\pi\right\},$$

where $k$ is any integer. Note that the arcs of length $2\pi/3$ and $4\pi/3$ terminate at opposite ends of a vertical chord in the unit circle.

Note also that the calculator graph of $y_1 = \cos x$ in Fig. 6.28 crosses the horizontal line $y_2 = -1/2$ twice in the interval $[0, 2\pi]$, at $2\pi/3$ and $4\pi/3$.

▶**TRY THIS.** Find all real numbers that satisfy $\cos x = \sqrt{3}/2$. ■

The procedure used in Example 1(c) can be used to solve $\cos x = a$ for any nonzero $a$ between $-1$ and 1. First find two values of $x$ between 0 and $2\pi$ that satisfy the equation. (In general, $s = \cos^{-1}(a)$ and $2\pi - s$ are the values.) Next write all solutions by adding $2k\pi$ to the first two solutions. The solution sets for $a = -1, 0,$ and 1 are easier to find and remember. All of the cases for solving $\cos x = a$ are summarized next. The letter $k$ is used to represent any integer. You should not simply memorize this summary, but you should be able to solve $\cos x = a$ for any real number $a$ with the help of a unit circle.

**SUMMARY**

**Solving $\cos x = a$**

1. If $-1 < a < 1$ and $a \neq 0$, then the solution set to $\cos x = a$ is $\{x \mid x = s + 2k\pi \text{ or } x = 2\pi - s + 2k\pi\}$, where $s = \cos^{-1} a$.
2. The solution set to $\cos x = 1$ is $\{x \mid x = 2k\pi\}$.
3. The solution set to $\cos x = 0$ is $\{x \mid x = \pi/2 + k\pi\}$.
4. The solution set to $\cos x = -1$ is $\{x \mid x = \pi + 2k\pi\}$.
5. If $|a| > 1$, then $\cos x = a$ has no solution.

## Sine Equations

To solve $\sin x = a$, we use the same strategy that we used to solve $\cos x = a$. We look for the smallest nonnegative solution and then build on it to obtain all solutions. However, the "first" solution obtained from $s = \sin^{-1} a$ might be negative because the range for the function $\sin^{-1}$ is $[-\pi/2, \pi/2]$. In this case $s + 2\pi$ is positive and we build on it to find all solutions. Remember that the sine of an arc on the unit circle is the $y$-coordinate of the terminal point of the arc. So arcs that terminate at opposite ends of a horizontal chord in the unit circle have the same $y$-coordinate for the terminal point and the same sine. This fact is the key to finding all solutions of a sine equation.

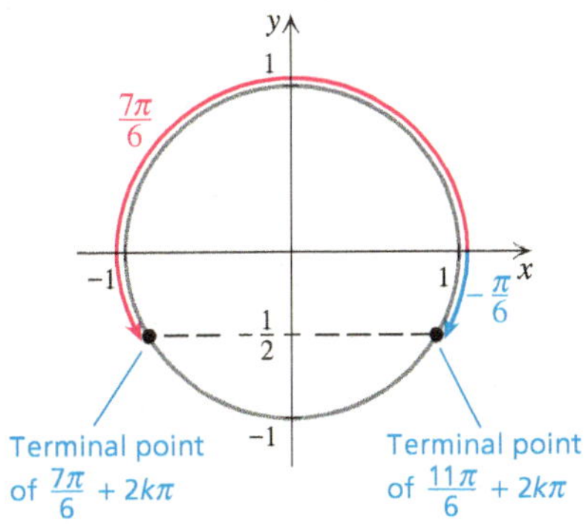


Figure 6.29

### EXAMPLE 2 Solving a sine equation

Find all real numbers that satisfy $\sin x = -1/2$.

**Solution**

One solution to $\sin x = -1/2$ is

$$x = \sin^{-1}\left(-\frac{1}{2}\right) = -\frac{\pi}{6}.$$

See Fig. 6.29. The smallest positive arc with the same terminal point as $-\pi/6$ is $-\pi/6 + 2\pi = 11\pi/6$. Since the period of the sine function is $2\pi$, all arcs of the form $11\pi/6 + 2k\pi$ have the same terminal point and satisfy $\sin x = -1/2$. The

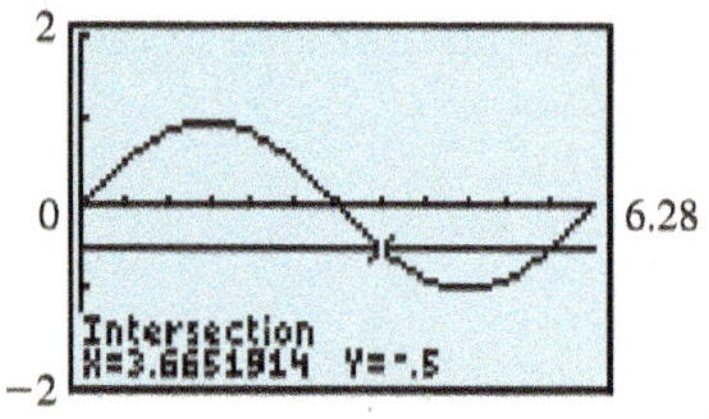


Figure 6.30

smallest positive arc that terminates at the other end of the horizontal chord shown in Fig. 6.29 is $\pi - (-\pi/6)$ or $7\pi/6$. So $7\pi/6$ also satisfies the equation, but it is not included in the form $11\pi/6 + 2k\pi$. Thus the solution set is

$$\left\{x \,\middle|\, x = \frac{7\pi}{6} + 2k\pi \quad \text{or} \quad x = \frac{11\pi}{6} + 2k\pi\right\},$$

where $k$ is any integer.

Note that the calculator graph of $y_1 = \sin x$ in Fig. 6.30 crosses the horizontal line $y_2 = -1/2$ twice in the interval $[0, 2\pi]$, at $7\pi/6$ and $11\pi/6$.

▸**TRY THIS.** Find all real numbers that satisfy $\sin x = 1/2$. ■

The procedure used in Example 2 can be used to solve $\sin x = a$ for any nonzero $a$ between $-1$ and 1. First find two values between 0 and $2\pi$ that satisfy the equation. (In general, $s = \sin^{-1} a$ and $\pi - s$ work when $s$ is positive; $s + 2\pi$ and $\pi - s$ work when $s$ is negative.) Next write all solutions by adding $2k\pi$ to the first two solutions. The equations $\sin x = 0$, $\sin x = 1$, and $\sin x = -1$ have solutions that are similar to the corresponding cosine equations. We can summarize the different solution sets to $\sin x = a$ as follows. You should not simply memorize this summary, but you should be able to solve $\sin x = a$ for any real number $a$ with the help of a unit circle.

**SUMMARY**

**Solving $\sin x = a$**

1. If $-1 < a < 1$, $a \neq 0$, and $s = \sin^{-1} a$, then the solution set to $\sin x = a$ is

$$\{x \mid x = s + 2k\pi \text{ or } x = \pi - s + 2k\pi\} \text{ for } s > 0,$$
$$\{x \mid x = s + 2\pi + 2k\pi \text{ or } x = \pi - s + 2k\pi\} \text{ for } s < 0.$$

2. The solution set to $\sin x = 1$ is $\{x \mid x = \pi/2 + 2k\pi\}$.
3. The solution set to $\sin x = 0$ is $\{x \mid x = k\pi\}$.
4. The solution set to $\sin x = -1$ is $\{x \mid x = 3\pi/2 + 2k\pi\}$.
5. If $|a| > 1$, then $\sin x = a$ has no solution.

## Tangent Equations

Tangent equations are a little simpler than sine and cosine equations, because the tangent function is one-to-one in its fundamental cycle, while sine and cosine are not one-to-one in their fundamental cycles. In the next example, we see that the solution set to $\tan x = a$ consists of any single solution plus multiples of $\pi$.

### EXAMPLE 3 Solving a tangent equation

Find all solutions, in degrees.

**a.** $\tan \alpha = 1$ **b.** $\tan \alpha = -1.34$

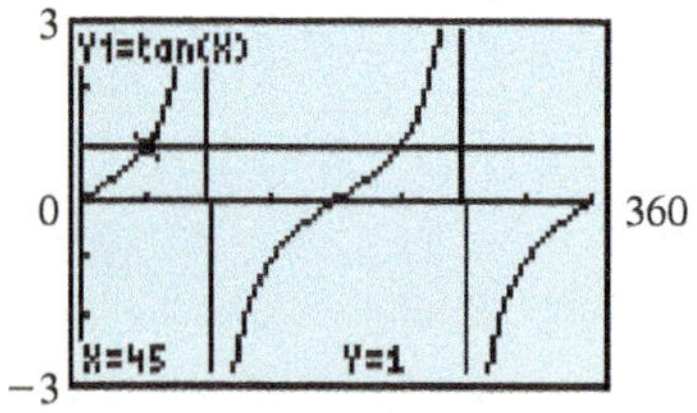


Figure 6.31

**Solution**

**a.** Since $\tan^{-1}(1) = 45°$ and the period of tangent is $180°$, all angles of the form $45° + k180°$ satisfy the equation. There are no additional angles that satisfy the equation. The solution set to $\tan \alpha = 1$ is

$$\{\alpha \mid \alpha = 45° + k180°\}.$$

The calculator graphs of $y_1 = \tan x$ and $y_2 = 1$ in Fig. 6.31 support the conclusion that the solutions to $\tan x = 1$ are $180°$ apart. □

**b.** Since $\tan^{-1}(-1.34) = -53.3°$, one solution is $\alpha = -53.3°$. Because all solutions to $\tan \alpha = -1.34$ differ by a multiple of 180°, $-53.3° + 180°$, or 126.7° is the smallest positive solution. So the solution set is

$$\{\alpha \mid \alpha = 126.7° + k180°\}.$$

▸**TRY THIS.** Find all angles $\alpha$ in degrees that satisfy $\tan \alpha = \sqrt{3}$. ■

To solve $\tan x = a$ for any real number $a$, we first find the smallest nonnegative solution. (In general, $s = \tan^{-1} a$ works if $s > 0$, and $s + \pi$ works if $s < 0$.) Next, add on all integral multiples of $\pi$. The solution to the equation $\tan x = a$ is summarized as follows.

**SUMMARY**

**Solving $\tan x = a$**

If $a$ is any real number and $s = \tan^{-1} a$, then the solution set to $\tan x = a$ is $\{x \mid x = s + k\pi\}$ for $s \geq 0$, and $\{x \mid x = s + \pi + k\pi\}$ for $s < 0$.

In the summaries for solving sine, cosine, and tangent equations, the domain of $x$ is the set of real numbers. Similar summaries can be made if the domain of $x$ is the set of degree measures. Start by finding $\sin^{-1} a$, $\cos^{-1} a$, or $\tan^{-1} a$ in degrees, then write the solution set in terms of multiples of 180° or 360°.

## Equations Involving Multiple Angles

Equations can involve expressions such as $\sin 2x$, $\cos 3\alpha$, or $\tan(s/2)$. These expressions involve a multiple of the variable rather than a single variable such as $x$, $\alpha$, or $s$. In this case we solve for the multiple just as we would solve for a single variable and then find the value of the single variable in the last step of the solution.

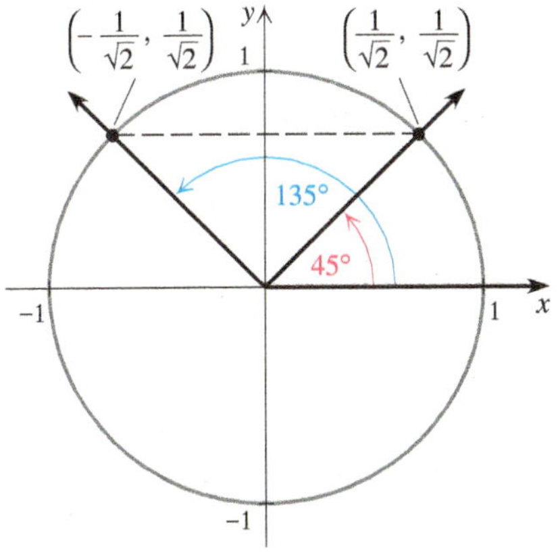

Figure 6.32

### EXAMPLE 4 A sine equation involving a double angle

Find all solutions in degrees to $\sin 2\alpha = 1/\sqrt{2}$.

**Solution**

The only values for $2\alpha$ between 0° and 360° that satisfy the equation are 45° and 135°. (Note that $\sin^{-1}(1/\sqrt{2}) = 45°$ and $135° = 180° - 45°$.) See Fig. 6.32. So proceed as follows:

$$\sin 2\alpha = \frac{1}{\sqrt{2}}$$

$$2\alpha = 45° + k360° \quad \text{or} \quad 2\alpha = 135° + k360°$$

$$\alpha = 22.5° + k180° \quad \text{or} \quad \alpha = 67.5° + k180° \qquad \text{Divide each side by 2.}$$

The solution set is $\{\alpha \mid \alpha = 22.5° + k180° \text{ or } \alpha = 67.5° + k180°\}$, where $k$ is any integer.

▸**TRY THIS.** Find all angles $\alpha$ in degrees that satisfy $\sin 2\alpha = 1$. ■

Note that in Example 4, all possible values for $2\alpha$ are found and *then* divided by 2 to get all possible values for $\alpha$. Finding $\alpha = 22.5°$ and then adding on multiples of 360° will not produce the same solutions. Observe the same procedure in the next example, where we wait until the final step to divide each side by 3.

### EXAMPLE 5 A tangent equation involving angle multiples

Find all solutions to $\tan 3x = -\sqrt{3}$ in the interval $(0, 2\pi)$.

### Solution

First find the smallest positive value for $3x$ that satisfies the equation. Then form all of the solutions by adding on multiples of $\pi$. Since $\tan^{-1}(-\sqrt{3}) = -\pi/3$, the smallest positive value for $3x$ that satisfies the equation is $-\pi/3 + \pi$, or $2\pi/3$. So proceed as follows:

$$\tan 3x = -\sqrt{3}$$

$$3x = \frac{2\pi}{3} + k\pi$$

$$x = \frac{2\pi}{9} + \frac{k\pi}{3} \quad \text{Divide each side by 3.}$$

The solutions between 0 and $2\pi$ occur if $k = 0, 1, 2, 3, 4,$ and 5. If $k = 6$, then $x$ is greater than $2\pi$. Thus the solution set is

$$\left\{\frac{2\pi}{9}, \frac{5\pi}{9}, \frac{8\pi}{9}, \frac{11\pi}{9}, \frac{14\pi}{9}, \frac{17\pi}{9}\right\}.$$

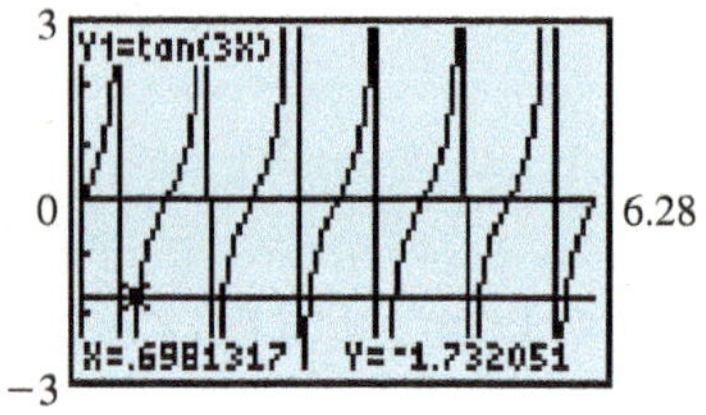


Figure 6.33

The graphs of $y_1 = \tan(3x)$ and $y_2 = -\sqrt{3}$ in Fig. 6.33 show the six solutions in the interval $(0, 2\pi)$.

▸**TRY THIS.** Find all solutions to $\tan(4x) = 1$ in the interval $(0, \pi)$. ■

## More Complicated Equations

More complicated equations involving trigonometric functions are solved by first solving for $\sin x$, $\cos x$, or $\tan x$, and then solving for $x$. In solving for the trigonometric functions, we may use trigonometric identities or properties of algebra, such as factoring or the quadratic formula. In stating formulas for solutions to equations, we will continue to use the letter $k$ to represent any integer.

### EXAMPLE 6 An equation solved by factoring

Find all solutions in the interval $[0, 2\pi)$ to the equation

$$\sin 2x = \sin x.$$

### Solution

First we use the identity $\sin 2x = 2 \sin x \cos x$ to get all of the trigonometric functions written in terms of the variable $x$ alone. Then we rearrange and factor:

$$\sin 2x = \sin x$$

$$2 \sin x \cos x = \sin x \quad \text{By the double-angle identity}$$

$$2 \sin x \cos x - \sin x = 0 \quad \text{Subtract } \sin x \text{ from each side.}$$

$$\sin x(2 \cos x - 1) = 0 \quad \text{Factor.}$$

$$\sin x = 0 \quad \text{or} \quad 2 \cos x - 1 = 0 \quad \text{Set each factor equal to 0.}$$

$$x = k\pi \quad \text{or} \quad 2 \cos x = 1$$

$$\cos x = \frac{1}{2}$$

$$x = \frac{\pi}{3} + 2k\pi \quad \text{or} \quad x = \frac{5\pi}{3} + 2k\pi$$

The only solutions in the interval $[0, 2\pi)$ are $0, \pi/3, \pi,$ and $5\pi/3$.

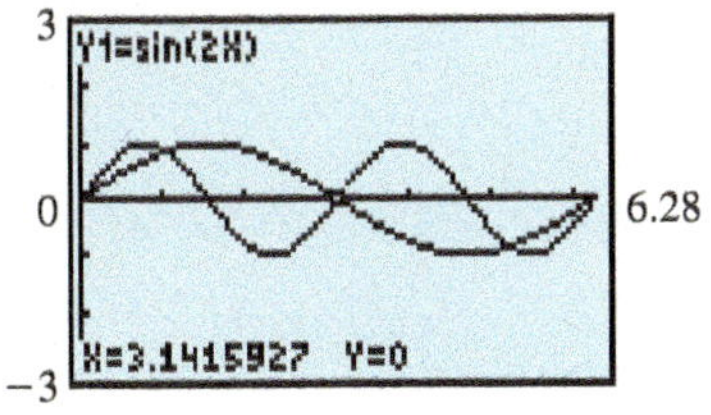


Figure 6.34

The graphs of $y_1 = \sin(2x)$ and $y_2 = \sin(x)$ in Fig. 6.34 show the four solutions to $\sin 2x = \sin x$ in the interval $[0, 2\pi)$.

▶**TRY THIS.** Find all solutions to $\sin(2x) = \cos(x)$ in the interval $(0, 2\pi)$. ■

In algebraic equations, we generally do not divide each side by any expression that involves a variable, and the same rule holds for trigonometric equations. In Example 6, we did not divide by $\sin x$ when it appeared on opposite sides of the equation. If we had, the solutions 0 and $\pi$ would have been lost.

In the next example, an equation of quadratic type is solved by factoring.

## EXAMPLE 7 An equation of quadratic type

Find all solutions to the equation

$$6\cos^2\left(\frac{x}{2}\right) - 7\cos\left(\frac{x}{2}\right) + 2 = 0$$

in the interval $[0, 2\pi)$. Round approximate answers to four decimal places.

### Solution

Let $y = \cos(x/2)$ to get a quadratic equation:

$$6y^2 - 7y + 2 = 0$$

$$(2y - 1)(3y - 2) = 0 \quad \text{Factor.}$$

$$2y - 1 = 0 \quad \text{or} \quad 3y - 2 = 0$$

$$y = \frac{1}{2} \quad \text{or} \quad y = \frac{2}{3}$$

$$\cos\frac{x}{2} = \frac{1}{2} \quad \text{or} \quad \cos\frac{x}{2} = \frac{2}{3} \quad \text{Replace } y \text{ by } \cos(x/2).$$

$$\frac{x}{2} = \frac{\pi}{3} + 2k\pi \text{ or } \frac{5\pi}{3} + 2k\pi \quad \text{or} \quad \frac{x}{2} \approx 0.8411 + 2k\pi \text{ or } 5.4421 + 2k\pi$$

Now multiply by 2 to solve for $x$:

$$x = \frac{2\pi}{3} + 4k\pi \text{ or } \frac{10\pi}{3} + 4k\pi \quad \text{or} \quad x \approx 1.6821 + 4k\pi \text{ or } 10.844 + 4k\pi$$

For all $k$, $10\pi/3 + 4k\pi$ and $10.844 + 4k\pi$ are outside the interval $[0, 2\pi)$. The solutions in the interval $[0, 2\pi)$ are $2\pi/3$ and 1.6821.

The graph of $y = 6(\cos(x/2))^2 - 7\cos(x/2) + 2$ in Fig. 6.35 appears to cross the $x$-axis at $2\pi/3$ and 1.6821.

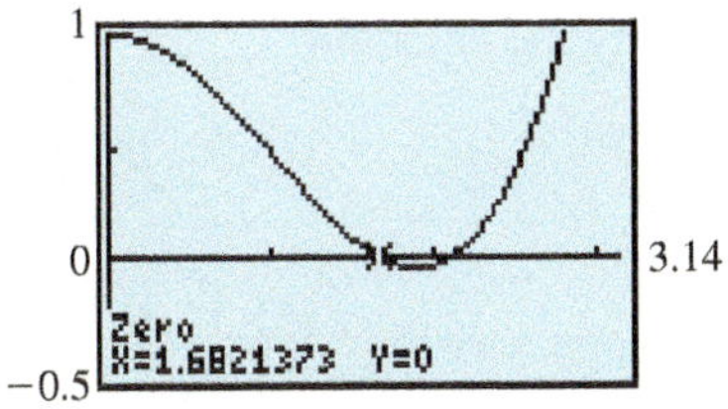


Figure 6.35

▶**TRY THIS.** Find all solutions to $2\sin^2(x) - \sin(x) - 1 = 0$ in the interval $(0, 2\pi)$. ■

For a trigonometric equation to be of quadratic type, it must be written in terms of a trigonometric function and the square of that function. In the next example, an identity is used to convert an equation involving sine and cosine to one with only sine. This equation is of quadratic type like Example 7, but it does not factor.

## EXAMPLE 8 An equation solved by the quadratic formula

Find all solutions to the equation

$$\cos^2\alpha - 0.2\sin\alpha = 0.9$$

in the interval $[0°, 360°)$. Round answers to the nearest tenth of a degree.

### Solution

$$\cos^2 \alpha - 0.2 \sin \alpha = 0.9$$

$$1 - \sin^2 \alpha - 0.2 \sin \alpha = 0.9 \qquad \text{Replace } \cos^2 \alpha \text{ with } 1 - \sin^2 \alpha.$$

$$\sin^2 \alpha + 0.2 \sin \alpha - 0.1 = 0 \qquad \text{An equation of quadratic type}$$

$$\sin \alpha = \frac{-0.2 \pm \sqrt{(0.2)^2 - 4(1)(-0.1)}}{2} \qquad \text{Use } a = 1, b = 0.2, \text{ and } c = -0.1 \text{ in the quadratic formula.}$$

$$\sin \alpha \approx 0.2317 \quad \text{or} \quad \sin \alpha \approx -0.4317$$

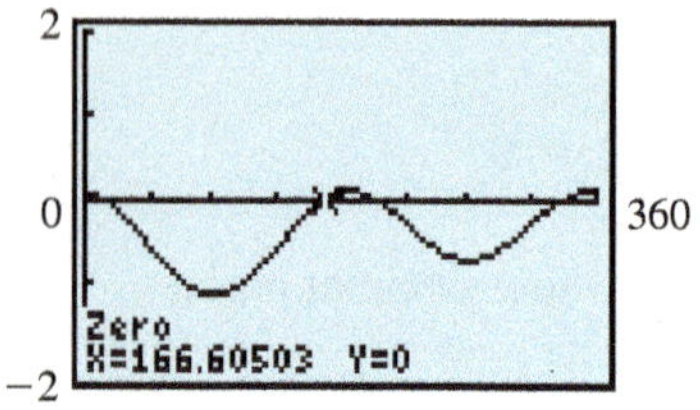

Figure 6.36

Find two positive solutions to $\sin \alpha = 0.2317$ in $[0, 360°)$ by using a calculator to get $\alpha = \sin^{-1}(0.2317) \approx 13.4°$ and $180° - \alpha \approx 166.6°$. Now find two positive solutions to $\sin \alpha = -0.4317$ in $[0, 360°)$. Using a calculator, we get $\alpha = \sin^{-1}(-0.4317) \approx -25.6°$ and $180° - \alpha \approx 205.6°$. Since $-25.6°$ is negative, we use $-25.6° + 360°$ or $334.4°$ along with $205.6°$ as the two solutions. We list all possible solutions as

$$\alpha \approx 13.4°, 166.6°, 205.6°, \text{ or } 334.4° \ (+k360° \text{ in each case}).$$

The solutions in $[0°, 360°)$ are 13.4°, 166.6°, 205.6°, and 334.4°.

The graph of $y = \cos^2 x - 0.2 \sin x - 0.9$ in Fig. 6.36 appears to cross the $x$-axis four times in the interval $[0°, 360°)$.

▶**TRY THIS.** Find all solutions to $\cos \alpha - \sin^2 \alpha = 0$ in the interval $[0°, 360°)$. ■

The Pythagorean identities $\sin^2 x = 1 - \cos^2 x$, $\csc^2 x = 1 + \cot^2 x$, and $\sec^2 x = 1 + \tan^2 x$ are frequently used to replace one function by the other. If an equation involves sine and cosine, cosecant and cotangent, or secant and tangent, we might be able to square each side and then use these identities.

## EXAMPLE 9 Square each side of the equation

Find all values of $y$ in the interval $[0, 360°)$ that satisfy the equation

$$\tan 3y + 1 = \sqrt{2} \sec 3y.$$

### Solution

$$(\tan 3y + 1)^2 = (\sqrt{2} \sec 3y)^2 \qquad \text{Square each side.}$$

$$\tan^2 3y + 2 \tan 3y + 1 = 2 \sec^2 3y$$

$$\tan^2 3y + 2 \tan 3y + 1 = 2(\tan^2 3y + 1) \qquad \text{Since } \sec^2 x = \tan^2 x + 1$$

$$-\tan^2 3y + 2 \tan 3y - 1 = 0 \qquad \text{Subtract } 2 \tan^2 3y + 2 \text{ from both sides.}$$

$$\tan^2 3y - 2 \tan 3y + 1 = 0$$

$$(\tan 3y - 1)^2 = 0$$

$$\tan 3y - 1 = 0$$

$$\tan 3y = 1$$

$$3y = 45° + k180°$$

Because we squared each side, we must check for extraneous roots. First check $3y = 45°$. If $3y = 45°$, then $\tan 3y = 1$ and $\sec 3y = \sqrt{2}$. Substituting these values into $\tan 3y + 1 = \sqrt{2} \sec 3y$ gives us

$$1 + 1 = \sqrt{2} \cdot \sqrt{2}. \qquad \text{Correct}$$

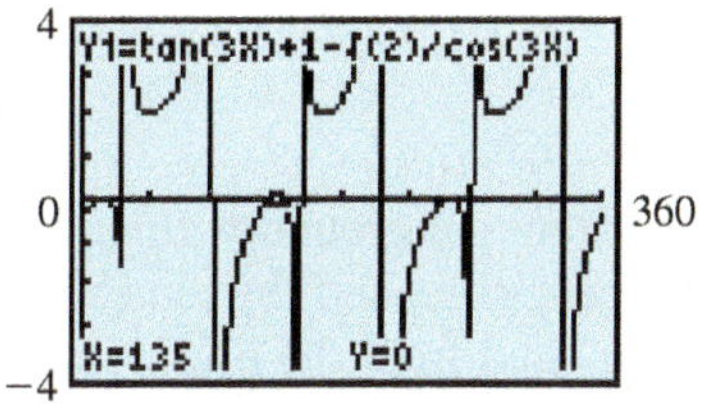


Figure 6.37

If we add any *even* multiple of 180°, or any multiple of 360°, to 45°, we get the same values for tan $3y$ and sec $3y$. So for any $k$, $3y = 45° + k360°$ satisfies the original equation. Now check 45° plus *odd* multiples of 180°. If $3y = 225°$ $(k = 1)$, then

$$\tan 3y = 1 \qquad \text{and} \qquad \sec 3y = -\sqrt{2}.$$

These values do not satisfy the original equation. Since tan $3y$ and sec $3y$ have these same values for $3y = 45° + k180°$ for any odd $k$, the only solutions are of the form $3y = 45° + k360°$, or $y = 15° + k120°$. The solutions in the interval $[0°, 360°)$ are 15°, 135°, and 255°.

The graph of $y = \tan(3x) + 1 - \sqrt{2}/\cos(3x)$ in Fig. 6.37 appears to touch the $x$-axis at these three locations.

▸**TRY THIS.** Find all $\alpha$ in $[0°, 360°)$ that satisfy $\sin\alpha - \cos\alpha = \frac{1}{\sqrt{2}}$. ■

There is no single method that applies to all trigonometric equations, but the following strategy will help you to solve them.

**STRATEGY**

**Solving Trigonometric Equations**

1. Know the solutions to $\sin x = a$, $\cos x = a$, and $\tan x = a$.
2. Solve an equation involving only multiple angles as if the equation had a single variable.
3. Simplify complicated equations by using identities. Try to get an equation involving only one trigonometric function.
4. If possible, factor to get different trigonometric functions into separate factors.
5. For equations of quadratic type, solve by factoring or the quadratic formula.
6. Square each side of the equation, if necessary, so that identities involving squares can be applied. (Check for extraneous roots.)

**HISTORICAL NOTE**

Archimedes of Syracuse (287 B.C.–212 B.C.) was a Greek mathematician, physicist, engineer, astronomer, and philosopher. Many consider him one of the greatest mathematicians in antiquity. Archimedes seems to have carried out the first theoretical calculation of $\pi$, concluding that $223/71 < \pi < 22/7$. He used a limiting process very similar to what you will do if you try the Concepts of Calculus at the end of this chapter.

## Solving a Spring Equation

Next, we consider the motion of a weight on a spring, discussed in Section 6.5.

### EXAMPLE 10 Solving a spring equation

In Example 7 of Section 6.5, a weight in motion attached to a spring had location $x$ given by $x = 2\sin t - \cos t$. For what values of $t$ is the weight at position $x = 0$?

**Solution**

To solve $2\sin t - \cos t = 0$, divide each side by $2\cos t$. We can divide by $2\cos t$ because the values of $t$ for which $2\cos t$ is zero do not satisfy the original equation.

$$2\sin t = \cos t$$

$$\frac{\sin t}{\cos t} = \frac{1}{2}$$

$$\tan t = \frac{1}{2}$$

Since the weight is set in motion at time $t = 0$, the values of $t$ are positive. Since $\tan^{-1}(1/2) = 0.46$, the weight is at position $x = 0$ for $t = 0.46 + k\pi$ for $k$ a nonnegative integer.

▸**TRY THIS.** The location $x$ for a weight on a spring is given by $x = 3\sin t - 2\cos t$ where $t$ is time in seconds. Find all $t$ for which $x = 0$. ■

## Modeling Projectile Motion

The distance $d$ (in feet) traveled by a projectile fired from the ground with an angle of elevation $\theta$ is related to the initial velocity $v_0$ (in feet per second) by the equation $v_0^2 \sin 2\theta = 32d$. If the projectile is pictured as being fired from the origin into the first quadrant, then the $x$- and $y$-coordinates (in feet) of the projectile at time $t$ (in seconds) are given by $x = v_0 t \cos\theta$ and $y = -16t^2 + v_0 t \sin\theta$.

### EXAMPLE 11 The path of a projectile

A catapult is placed 100 feet from the castle wall, which is 35 feet high. The soldier wants the burning bale of hay to clear the top of the wall and land 50 feet inside the castle wall. If the initial velocity of the bale is 70 feet per second, then at what angle should the bale of hay be launched so that it will travel 150 feet and pass over the castle wall?

#### Solution

Use the equation $v_0^2 \sin 2\theta = 32d$ with $v_0 = 70$ feet per second and $d = 150$ feet to find $\theta$:

$$70^2 \sin 2\theta = 32(150)$$

$$\sin 2\theta = \frac{32(150)}{70^2} \approx 0.97959$$

The launch angle $\theta$ must be in the interval $(0°, 90°)$, so we look for values of $2\theta$ in the interval $(0°, 180°)$. Since $\sin^{-1}(0.97959) \approx 78.4°$, both $78.4°$ and $180° - 78.4° = 101.6°$ are possible values for $2\theta$. So possible values for $\theta$ are $39.2°$ and $50.8°$. See Fig. 6.38.

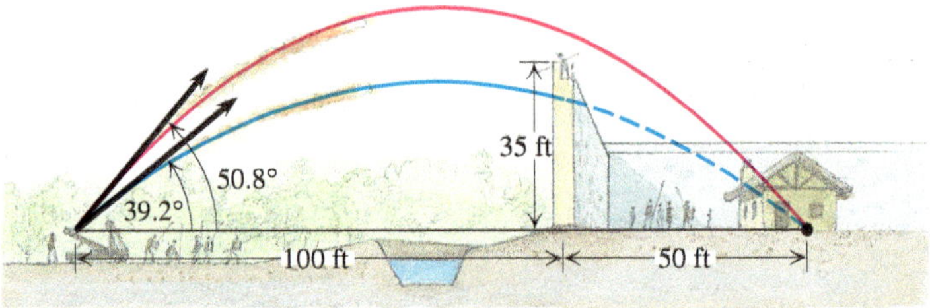


**Figure 6.38**

Use the equation $x = v_0 t \cos\theta$ to find the time at which the bale is 100 feet from the catapult (measured horizontally) by using each of the possible values for $\theta$:

$$100 = 70t\cos(39.2°) \qquad 100 = 70t\cos(50.8°)$$

$$t = \frac{100}{70\cos(39.2°)} \approx 1.84 \text{ seconds} \qquad t = \frac{100}{70\cos(50.8°)} \approx 2.26 \text{ seconds}$$

Use the equation $y = -16t^2 + v_0 t \sin\theta$ to find the altitude of the bale at time $t = 1.84$ seconds and $t = 2.26$ seconds:

$$y = -16(1.84)^2 + 70(1.84)\sin 39.2° \qquad y = -16(2.26)^2 + 70(2.26)\sin 50.8°$$

$$\approx 27.2 \text{ feet} \qquad \approx 40.9 \text{ feet}$$

If the burning bale is launched on a trajectory with an angle of $39.2°$, then it will have an altitude of only 27.2 feet when it reaches the castle wall. If it is launched

■ **Foreshadowing Calculus**

A standard calculus problem is to show that a projectile's maximum distance for a given velocity is achieved at 45°. As illustrated in Example 11, two angles can be used to achieve a distance less than the maximum, one less than 45° and one greater than 45°.

with an angle of 50.8°, then it will have an altitude of 40.9 feet when it reaches the castle wall. Since the castle wall is 35 feet tall, the 50.8° angle must be used for the bale to reach its intended target.

**TRY THIS.** Find the launch angle so that a projectile with initial velocity of 100 ft per second will travel 200 feet.

## FOR thought... True or False? Explain.

1. The only solutions to $\cos\alpha = 1/\sqrt{2}$ in $[0°, 360°)$ are 45° and 135°.
2. The only solution to $\sin x = -0.55$ in $[0, \pi)$ is $\sin^{-1}(-0.55)$.
3. $\{x \mid x = -29° + k360°\} = \{x \mid x = 331° + k360°\}$, where $k$ is any integer.
4. The solution set to $\tan x = -1$ is $\{x \mid x = \frac{7\pi}{4} + k\pi\}$, where $k$ is any integer.
5. $2\cos^2 x + \cos x - 1 = (2\cos x - 1)(\cos x + 1)$ is an identity.
6. The equation $\sin^2 x = \sin x \cos x$ is equivalent to $\sin x = \cos x$.
7. One solution to $\sec x = 2$ is $\frac{1}{\cos^{-1}(2)}$.
8. The solution set to $\cot x = 3$ for $x$ in $[0, \pi)$ is $\{\tan^{-1}(1/3)\}$.
9. The equation $\sin x = \cos x$ is equivalent to $\sin^2 x = \cos^2 x$.
10. $\{x \mid 3x = \frac{\pi}{2} + 2k\pi\} = \{x \mid x = \frac{\pi}{6} + 2k\pi\}$, where $k$ is any integer.

## EXERCISES 6.6

*Find all real numbers that satisfy each equation. Do not use a calculator. See the summaries for solving* $\cos x = a$, $\sin x = a$, *and* $\tan x = a$ *on pages 469, 470, and 471.*

1. $\cos x = -1$
2. $\cos x = 0$
3. $\sin x = 0$
4. $\sin x = 1$
5. $\sin x = -1$
6. $\cos x = 1$
7. $\cos x = 1/2$
8. $\cos x = \sqrt{2}/2$
9. $\sin x = \sqrt{2}/2$
10. $\sin x = \sqrt{3}/2$
11. $\tan x = 1$
12. $\tan x = \sqrt{3}/3$
13. $\cos x = -\sqrt{3}/2$
14. $\cos x = -\sqrt{2}/2$
15. $\sin x = -\sqrt{2}/2$
16. $\sin x = -\sqrt{3}/2$
17. $\tan x = -1$
18. $\tan x = -\sqrt{3}$

*Find all angles in degrees that satisfy each equation. Round approximate answers to the nearest tenth of a degree.*

19. $\cos\alpha = 0$
20. $\cos\alpha = -1$
21. $\sin\alpha = 1$
22. $\sin\alpha = -1$
23. $\tan\alpha = 0$
24. $\tan\alpha = -1$
25. $\cos\alpha = 0.873$
26. $\cos\alpha = -0.158$
27. $\sin\alpha = -0.244$
28. $\sin\alpha = 0.551$
29. $\tan\alpha = 5.42$
30. $\tan\alpha = -2.31$

*Find all real numbers that satisfy each equation.*

31. $\cos(x/2) = 1/2$
32. $2\cos 2x = -\sqrt{2}$
33. $\cos 3x = 1$
34. $\cos 2x = 0$
35. $2\sin(x/2) - 1 = 0$
36. $\sin 2x = 0$
37. $2\sin 2x = -\sqrt{2}$
38. $\sin(x/3) + 1 = 0$
39. $\tan 2x = \sqrt{3}$
40. $\sqrt{3}\tan(3x) + 1 = 0$
41. $\tan 4x = 0$
42. $\tan 3x = -1$
43. $\sin(\pi x) = 1/2$
44. $\tan(\pi x/4) = 1$
45. $\cos(2\pi x) = 0$
46. $\sin(3\pi x) = 1$

*Find all values of* $\alpha$ *in* $[0°, 360°)$ *that satisfy each equation.*

47. $2\sin\alpha = -\sqrt{3}$
48. $\tan\alpha = -\sqrt{3}$
49. $\sqrt{2}\cos 2\alpha - 1 = 0$
50. $\sin 6\alpha = 1$

**51.** $\sec 3\alpha = -\sqrt{2}$

**52.** $\csc(5\alpha) + 2 = 0$

**53.** $\cot(\alpha/2) = \sqrt{3}$

**54.** $\sec(\alpha/2) = \sqrt{2}$

*Find all values of α in degrees that satisfy each equation. Round approximate answers to the nearest tenth of a degree.*

**55.** $\sin 3\alpha = 0.34$

**56.** $\cos 2\alpha = -0.22$

**57.** $\sin 3\alpha = -0.6$

**58.** $\tan 4\alpha = -3.2$

**59.** $\sec 2\alpha = 4.5$

**60.** $\csc 3\alpha = -1.4$

**61.** $\csc(\alpha/2) = -2.3$

**62.** $\cot(\alpha/2) = 4.7$

*Find all real numbers in the interval $[0, 2\pi)$ that satisfy each equation. Round approximate answers to the nearest tenth. See the strategy for solving trigonometric equations on page 475.*

**63.** $3 \sin^2 x = \sin x$

**64.** $2 \tan^2 x = \tan x$

**65.** $2 \cos^2 x + 3 \cos x = -1$

**66.** $2 \sin^2 x + \sin x = 1$

**67.** $5 \sin^2 x - 2 \sin x = \cos^2 x$

**68.** $\sin^2 x - \cos^2 x = 0$

**69.** $\tan x = \sec x - \sqrt{3}$

**70.** $\csc x - \sqrt{3} = \cot x$

**71.** $\sin x + \sqrt{3} = 3\sqrt{3} \cos x$

**72.** $6 \sin^2 x - 2 \cos x = 5$

**73.** $\tan x \sin 2x = 0$

**74.** $3 \sec^2 x \tan x = 4 \tan x$

**75.** $\sin 2x - \sin x \cos x = \cos x$

**76.** $2 \cos^2 2x - 8 \sin^2 x \cos^2 x = -1$

**77.** $\sin x \cos(\pi/4) + \cos x \sin(\pi/4) = 1/2$

**78.** $\sin(\pi/6)\cos x - \cos(\pi/6)\sin x = -1/2$

**79.** $\sin 2x \cos x - \cos 2x \sin x = -1/2$

**80.** $\cos 2x \cos x - \sin 2x \sin x = 1/2$

**81.** $4 \cdot 16^{\sin^2(x)} = 64^{\sin(x)}$

**82.** $\frac{1}{2} \cdot 4^{\cos^2(x)} = 2^{\cos(x)}$

*Find all values of θ in the interval $[0°, 360°)$ that satisfy each equation. Round approximate answers to the nearest tenth of a degree.*

**83.** $\cos^2\left(\frac{\theta}{2}\right) = \sec \theta$

**84.** $2 \sin^2\left(\frac{\theta}{2}\right) = \cos \theta$

**85.** $2 \sin \theta = \cos \theta$

**86.** $3 \sin 2\theta = \cos 2\theta$

**87.** $\sin 3\theta = \csc 3\theta$

**88.** $\tan^2 \theta - \cot^2 \theta = 0$

**89.** $\tan^2 \theta - 2 \tan \theta - 1 = 0$

**90.** $\cot^2 \theta - 4 \cot \theta + 2 = 0$

**91.** $9 \sin^2 \theta + 12 \sin \theta + 4 = 0$

**92.** $12 \cos^2 \theta + \cos \theta - 6 = 0$

**93.** $\dfrac{\tan 3\theta - \tan \theta}{1 + \tan 3\theta \tan \theta} = \sqrt{3}$

**94.** $\dfrac{\tan 3\theta + \tan 2\theta}{1 - \tan 3\theta \tan 2\theta} = 1$

**95.** $8 \cos^4 \theta - 10 \cos^2 \theta + 3 = 0$

**96.** $4 \sin^4 \theta - 5 \sin^2 \theta + 1 = 0$

**97.** $\sec^4 \theta - 5 \sec^2 \theta + 4 = 0$

**98.** $\cot^4 \theta - 4 \cot^2 \theta + 3 = 0$

*Solve each equation. (These equations are types that will arise in Chapter 7.)*

**99.** $\dfrac{\sin 33.2°}{a} = \dfrac{\sin 45.6°}{13.7}$

**100.** $\dfrac{\sin 49.6°}{55.1} = \dfrac{\sin 88.2°}{b}$

**101.** $\dfrac{\sin \alpha}{23.4} = \dfrac{\sin 67.2°}{25.9}$ for $0° < \alpha < 90°$

**102.** $\dfrac{\sin 9.7°}{15.4} = \dfrac{\sin \beta}{52.9}$ for $90° < \beta < 180°$

**103.** $(3.6)^2 = (5.4)^2 + (8.2)^2 - 2(5.4)(8.2)\cos \alpha$ for $0° < \alpha < 90°$

**104.** $(6.8)^2 = (3.2)^2 + (4.6)^2 - 2(3.2)(4.6)\cos \alpha$ for $90° < \alpha < 180°$

*One way to solve an equation with a graphing calculator is to rewrite the equation with 0 on the right-hand side, then graph the function that is on the left-hand side. The x-coordinate of each x-intercept of the graph is a solution to the original equation. For each equation, find all real solutions (to the nearest tenth) in the interval $[0, 2\pi)$.*

**105.** $\sin(x/2) = \cos 3x$

**106.** $2 \sin x = \csc(x + 0.2)$

**107.** $\dfrac{x}{2} - \dfrac{\pi}{6} + \dfrac{\sqrt{3}}{2} = \sin x$

**108.** $x^2 = \sin x$

*Solve each problem.*

**109.** *Motion of a Spring* A block is attached to a spring and set in motion on a frictionless plane. Its location on the surface at any time $t$ in seconds is given in meters by $x = \sqrt{3} \sin 2t + \cos 2t$. For what values of $t$ is the block at its resting position $x = 0$?

**110.** *Motion of a Spring* A block is set in motion hanging from a spring and oscillates about its resting position $x = 0$ according to the function $x = -0.3 \sin 3t + 0.5 \cos 3t$. For what values of $t$ is the block at its resting position $x = 0$?

**111.** *Wave Action* The vertical position of a floating ball in an experimental wave tank is given by the equation $x = 2 \sin(\pi t/3)$, where $x$ is the number of feet above sea level and $t$ is the time in seconds. For what values of $t$ is the ball $\sqrt{3}$ ft above sea level?

**112.** *Periodic Sales* The number of car stereos sold by a national department store chain varies seasonally and is a function of the month of the year. The function

$$x = 6.2 + 3.1\sin\left(\frac{\pi}{6}(t - 9)\right)$$

gives the anticipated sales (in thousands of units) as a function of the number of the month ($t = 1, 2, \ldots, 12$). In what month does the store anticipate selling 9300 units?

**113.** *Firing an M-16* A soldier is accused of breaking a window 3300 ft away during target practice. If the muzzle velocity for an M-16 is 325 ft/sec, then at what angle would it have to be aimed for the bullet to travel 3300 ft? The distance $d$ (in feet) traveled by a projectile fired at an angle $\theta$ is related to the initial velocity $v_0$ (in feet per second) by the equation $v_0^2 \sin 2\theta = 32d$.

**114.** *Firing an M-16* If you were accused of firing an M-16 into the air and breaking a window 4000 ft away, what would be your defense?

**115.** *Choosing the Right Angle* Cincinnati Reds centerfielder Ken Griffey, Jr., fields a ground ball and attempts to make a 230-ft throw to home plate. Given that Griffey commonly makes long throws at 90 mph, find the two possible angles at which he can throw the ball to home plate. Find the time saved by choosing the smaller angle.

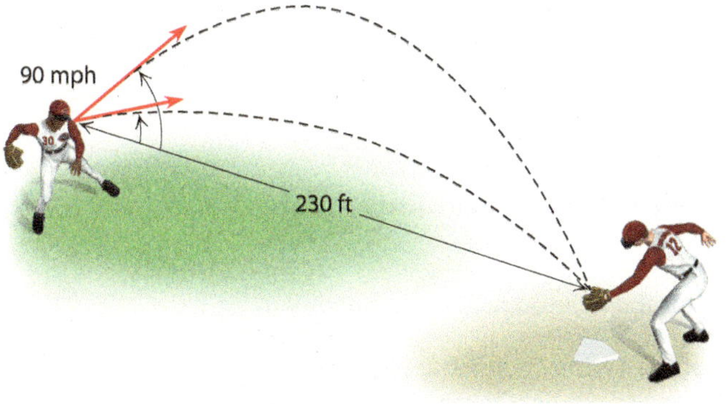


**Figure for Exercise 115**

**116.** *Muzzle Velocity* The 8-in. (diameter) howitzer on the U.S. Army's M-110 can propel a projectile a distance of 18,500 yd. If the angle of elevation of the barrel is 45°, then what muzzle velocity (in feet per second) is required to achieve this distance?

## ▶ RETHINKING

**117.** Write the function $y = \sin x - \cos x$ in the form $y = A\sin(x + C)$, then state the amplitude, period, and phase shift of the function.

**118.** Find the exact value of $\sin(\alpha + \beta)$ if $\sin\alpha = 1/3$ and $\sin\beta = 1/2$ with $\alpha$ in quadrant I and $\beta$ in quadrant II.

**119.** Find $\sin(\pi/2 - x)$, if $\cos x = 3/4$.

**120.** Find the exact value of $\tan(x/2)$ given that $\sin(x) = \sqrt{8/9}$ and $3\pi/2 < x < 2\pi$.

**121.** Find the exact value of $\sin(x/2)$ given that $\cos(x) = 1/4$ and $3\pi/2 < x < 2\pi$.

**122.** Use an identity to simplify each expression.

**a.** $\sin(3.5)\cos(2.1) + \cos(3.5)\sin(2.1)$

**b.** $\sin(2x)\cos(x) - \cos(2x)\sin(x)$

**c.** $2\sin(4.8)\cos(4.8)$

## THINKING OUTSIDE THE BOX XLIX

*Two Common Triangles* An equilateral triangle with sides of length 1 and an isosceles right triangle with legs of length 1 are positioned as shown in the accompanying diagram. Find the exact area of the shaded triangle.

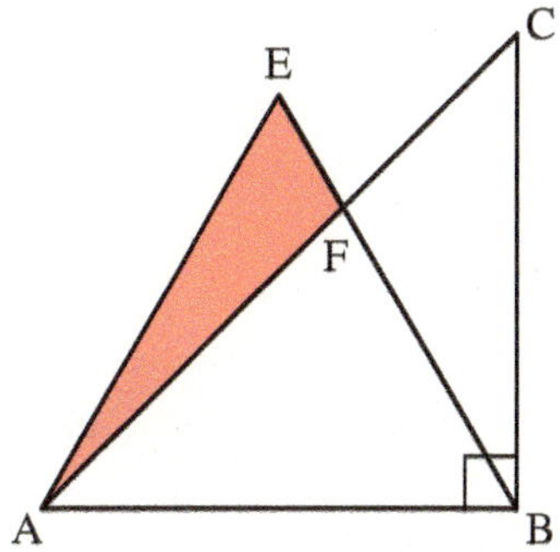


**Figure for Thinking Outside the Box XLIX**

## POP QUIZ 6.6

*Find all angles α in degrees that satisfy each equation.*

**1.** $\sin\alpha = \sqrt{2}/2$

**2.** $\cos\alpha = 1/2$

**3.** $\tan\alpha = -1$

*Find all real numbers in $[0, 2\pi]$ that satisfy each equation.*

**4.** $\sin(x/2) = 1/2$

**5.** $\cos(x) = 1$

**6.** $\tan(2x) = 1$

LINKING
# concepts... For Individual or Group Explorations

## Modeling Baseball Strategy

*An outfielder picks up a ground ball and wants to make a quick* 150-ft *throw to first base. To save time, he does not throw the ball directly to the first baseman. He skips the ball off the artificial surface at* 75 ft *so that the ball reaches first base after a bounce, as shown in the figure. Let's examine this strategy. The equations*

$$x = v_0 t \cos\theta \qquad \text{and} \qquad y = -16t^2 + v_0 t \sin\theta + h_0$$

*give the coordinates of the ball at time $t$ seconds for an initial angle of $\theta$ degrees, initial velocity $v_0$* ft/sec, *and initial height $h_0$* ft. *Assume that the ball is thrown at* 130 ft/sec *from a height of* 5 ft *and the ball is caught at a height of* 5 ft.

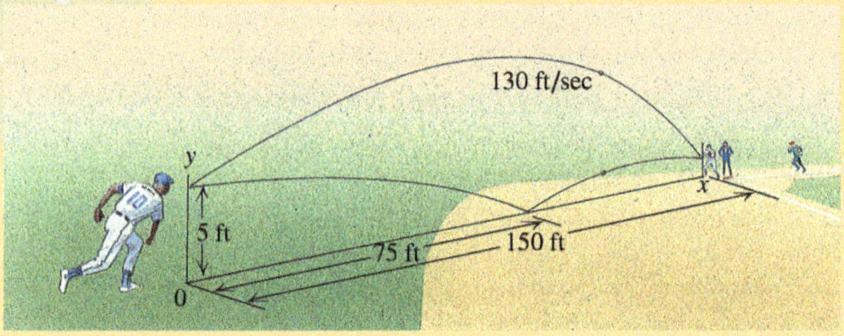


**a)** At what angle from horizontal must the ball be thrown to reach first base without a skip?

**b)** How long does it take the ball to get to first base without a skip?

**c)** At what angle from horizontal must the ball be thrown so that it strikes the ground at 75 ft?

**d)** How long does it take for the ball to reach the skip point at 75 ft?

**e)** Assume that the path of the ball is symmetric with respect to the skip point and double your answer in part (d) to get the time that it takes the ball to reach first base.

**f)** How much time is saved by skipping the ball?

**g)** How long would it take for the ball to reach first base if it could be thrown in a straight line?

**h)** Discuss the assumptions made for this model. Did we ignore anything that might affect our results?

# Highlights

## 6.1 Basic Identities

| | | |
|---|---|---|
| **Identity** | An equation that is satisfied by every number for which both sides are defined | $\sin^2 x + \cos^2 x = 1$<br>$\tan x = \sin(x)/\cos(x)$ |
| **Reciprocal Identities** | $\sin\alpha = \dfrac{1}{\csc\alpha}$ $\cos\alpha = \dfrac{1}{\sec\alpha}$ $\tan\alpha = \dfrac{1}{\cot\alpha}$<br>$\csc\alpha = \dfrac{1}{\sin\alpha}$ $\sec\alpha = \dfrac{1}{\cos\alpha}$ $\cot\alpha = \dfrac{1}{\tan\alpha}$ | |
| **Pythagorean Identities** | $\sin^2 x + \cos^2 x = 1$ $1 + \cot^2 x = \csc^2 x$ $\tan^2 x + 1 = \sec^2 x$ | |
| **Odd Identities** | $\sin(-x) = -\sin(x)$ $\csc(-x) = -\csc(x)$<br>$\tan(-x) = -\tan(x)$ $\cot(-x) = -\cot(x)$ | |
| **Even Identities** | $\cos(-x) = \cos(x)$ $\sec(-x) = \sec(x)$ | |

## 6.2 Verifying Identities

| | |
|---|---|
| **Strategy** | 1. Convert one side of the identity into the expression on the other side.<br>2. Write all functions in terms of sines and cosines, then simplify.<br>3. Multiply the numerator and denominator of a rational expression by either the numerator or denominator of the desired rational expression.<br>4. If the numerator of a rational expression is a sum or difference, separate it into two rational expressions.<br>5. Combine a sum or difference of two rational expressions into a single rational expression. |

## 6.3 Sum and Difference Identities

| | |
|---|---|
| **Cosine of a Sum or Difference** | $\cos(\alpha + \beta) = \cos\alpha\cos\beta - \sin\alpha\sin\beta$<br>$\cos(\alpha - \beta) = \cos\alpha\cos\beta + \sin\alpha\sin\beta$ |
| **Sine of a Sum or Difference** | $\sin(\alpha + \beta) = \sin\alpha\cos\beta + \cos\alpha\sin\beta$<br>$\sin(\alpha - \beta) = \sin\alpha\cos\beta - \cos\alpha\sin\beta$ |
| **Tangent of a Sum or Difference** | $\tan(\alpha \pm \beta) = \dfrac{\tan\alpha \pm \tan\beta}{1 \mp \tan\alpha\tan\beta}$ |
| **Cofunction Identities** | The value of any trigonometric function at $u$ equals the value of its cofunction at $\pi/2 - u$. |

## 6.4 Double-Angle and Half-Angle Identities

| **Double-Angle Identities** | **Half-Angle Identities** |
|---|---|
| $\sin 2x = 2\sin x\cos x$ | $\sin\dfrac{x}{2} = \pm\sqrt{\dfrac{1-\cos x}{2}}$ |
| $\cos 2x = \cos^2 x - \sin^2 x$ | $\cos\dfrac{x}{2} = \pm\sqrt{\dfrac{1+\cos x}{2}}$ |
| $= 2\cos^2 x - 1$ | $\tan\dfrac{x}{2} = \pm\sqrt{\dfrac{1-\cos x}{1+\cos x}}$ |
| $= 1 - 2\sin^2 x$ | $= \dfrac{\sin x}{1+\cos x}$ |
| $\tan 2x = \dfrac{2\tan x}{1-\tan^2 x}$ | $= \dfrac{1-\cos x}{\sin x}$ |

## 6.5 Product and Sum Identities

| **Product-to-Sum Identities** | **Sum-to-Product Identities** |
|---|---|
| $\sin A \cos B = \frac{1}{2}[\sin(A + B) + \sin(A - B)]$ | $\sin x + \sin y = 2\sin\left(\frac{x+y}{2}\right)\cos\left(\frac{x-y}{2}\right)$ |
| $\sin A \sin B = \frac{1}{2}[\cos(A - B) - \cos(A + B)]$ | $\sin x - \sin y = 2\cos\left(\frac{x+y}{2}\right)\sin\left(\frac{x-y}{2}\right)$ |
| $\cos A \sin B = \frac{1}{2}[\sin(A + B) - \sin(A - B)]$ | $\cos x + \cos y = 2\cos\left(\frac{x+y}{2}\right)\cos\left(\frac{x-y}{2}\right)$ |
| $\cos A \cos B = \frac{1}{2}[\cos(A - B) + \cos(A + B)]$ | $\cos x - \cos y = -2\sin\left(\frac{x+y}{2}\right)\sin\left(\frac{x-y}{2}\right)$ |

| | | |
|---|---|---|
| **Reduction Formula** | If $\alpha$ is an angle in standard position whose terminal side contains $(a, b)$ and $x$ is a real number, then $a\sin x + b\cos x = \sqrt{a^2 + b^2}\sin(x + \alpha)$. | $\alpha = \tan^{-1}(4/3)$<br>$3\sin x + 4\cos x$<br>$= 5\sin(x + \tan^{-1}(4/3))$ |

## 6.6 Conditional Trigonometric Equations

| | | |
|---|---|---|
| **Solving $\cos x = a$** | If $\lvert a\rvert \le 1$, find all solutions in $[0, 2\pi]$, then add all multiples of $2\pi$ to them and simplify. If $\lvert a\rvert > 1$ there are no solutions. | $\cos x = 0, x = \frac{\pi}{2} + k\pi$<br>$\cos x = \frac{1}{2}, x = \frac{\pi}{3} + 2k\pi$<br>or $x = \frac{5\pi}{3} + 2k\pi$ |
| **Solving $\sin x = a$** | If $\lvert a\rvert \le 1$, find all solutions in $[0, 2\pi]$, then add all multiples of $2\pi$ to them and simplify. If $\lvert a\rvert > 1$ there are no solutions. | $\sin x = 0, x = k\pi$<br>$\sin x = \frac{1}{2}, x = \frac{\pi}{6} + 2k\pi$<br>or $x = \frac{5\pi}{6} + 2k\pi$ |
| **Solving $\tan x = a$** | If $a$ is a real number, find all solutions in $[-\pi/2, \pi/2]$, then add multiples of $\pi$ to them and simplify. | $\tan x = 1, x = \frac{\pi}{4} + k\pi$ |

# Chapter 6 Review Exercises

*Simplify each expression.*

**1.** $(1 - \sin\alpha)(1 + \sin\alpha)$

**2.** $\csc x \tan x + \sec(-x)$

**3.** $(1 - \csc x)(1 - \csc(-x))$

**4.** $\dfrac{\cos^2 x - \sin^2 x}{\sin 2x}$

**5.** $\dfrac{1}{1 + \sin\alpha} - \dfrac{\sin(-\alpha)}{\cos^2\alpha}$

**6.** $2\sin\left(\frac{\pi}{2} - \alpha\right)\cos\left(\frac{\pi}{2} - \alpha\right)$

**7.** $\dfrac{2\tan 2s}{1 - \tan^2 2s}$

**8.** $\dfrac{\tan 2w - \tan 4w}{1 + \tan 2w \tan 4w}$

**9.** $\sin 3\theta \cos 6\theta - \cos 3\theta \sin 6\theta$

**10.** $\dfrac{\sin 2y}{1 + \cos 2y}$

**11.** $\dfrac{1 - \cos 2z}{\sin 2z}$

**12.** $\cos^2\left(\frac{x}{2}\right) - \sin^2\left(\frac{x}{2}\right)$

*Match each expression with an equivalent expression (a)–(h).*

**13.** $\sin(20°)$ **14.** $\tan(85°)$ **15.** $\cos(90°)$

**16.** $\cot(40°)$ **17.** $\sec(\pi/6)$ **18.** $\csc(\pi/8)$

**19.** $\sin(5\pi/12)$ **20.** $\cot(\pi/3)$

**a.** $\csc(\pi/3)$ **b.** $\sec(3\pi/8)$ **c.** $\sin(0°)$

**d.** $\tan(50°)$ **e.** $\cos(70°)$ **f.** $\tan(\pi/6)$

**g.** $\cos(\pi/12)$ **h.** $\cot(5°)$

*Use identities to find the exact values of the remaining five trigonometric functions at $\alpha$.*

**21.** $\cos \alpha = -5/13$ and $\pi/2 < \alpha < \pi$

**22.** $\tan \alpha = 5/12$ and $\pi < \alpha < 3\pi/2$

*Find the exact values of the six trigonometric functions at $\alpha$.*

**23.** $\sin\left(\frac{\pi}{2} - \alpha\right) = -3/5$ and $\pi < \alpha < 3\pi/2$

**24.** $\csc\left(\frac{\pi}{2} - \alpha\right) = 3$ and $0 < \alpha < \pi/2$

**25.** $\sin(\alpha/2) = 3/5$ and $3\pi/4 < \alpha/2 < \pi$

**26.** $\cos(\alpha/2) = -1/3$ and $\pi/2 < \alpha/2 < 3\pi/4$

*Determine whether each equation is an identity. Prove your answer.*

**27.** $(\sin x + \cos x)^2 = 1 + \sin 2x$

**28.** $\cos(A - B) = \cos A \cos B - \sin A \sin B$

**29.** $\csc^2 x - \cot^2 x = \tan^2 x - \sec^2 x$

**30.** $\sin^2\left(\frac{x}{2}\right) = \frac{\sin^2 x}{2 + \sin 2x \csc x}$

*Determine whether each function is odd, even, or neither.*

**31.** $f(x) = \frac{\sin x - \tan x}{\cos x}$

**32.** $f(x) = 1 + \sin^2 x$

**33.** $f(x) = \frac{\cos x - \sin x}{\sec x}$

**34.** $f(x) = \csc^3 x - \tan^3 x$

**35.** $f(x) = \frac{\sin x \tan x}{\cos x + \sec x}$

**36.** $f(x) = \sin x + \cos x$

*Match each given expression with an equivalent expression (a)–(j).*

**37.** $\cos 4$

**38.** $\sin 4$

**39.** $\sin 2 + \sin 4$

**40.** $\sin 4 - \sin 2$

**41.** $2 \sin 4 \cos 4$

**42.** $\cos^2 4 - \sin^2 4$

**43.** $\cos 4 + \cos 2$

**44.** $\cos 4 - \cos 2$

**45.** $\frac{\sin 4}{1 + \cos 4}$

**46.** $\frac{1 + \cos 4}{2}$

**a.** $2 \cos 3 \sin 1$ **b.** $\sin 8$ **c.** $\tan 2$

**d.** $\cos 8$ **e.** $2 \sin 3 \cos 1$ **f.** $\sin(\pi/2 - 4)$

**g.** $-\sin(-4)$ **h.** $2 \cos 3 \cos 1$ **i.** $-2 \sin 3 \sin 1$

**j.** $\cos^2 2$

*Prove that each of the following equations is an identity.*

**47.** $\sec 2\theta = \frac{1 + \tan^2 \theta}{1 - \tan^2 \theta}$

**48.** $\tan^2 \theta = \frac{1 - \cos 2\theta}{1 + \cos 2\theta}$

**49.** $\sin^2\left(\frac{x}{2}\right) = \frac{\csc^2 x - \cot^2 x}{2 \csc^2 x + 2 \csc x \cot x}$

**50.** $\cot(-x) = \frac{1 - \sin^2 x}{\cos(-x)\sin(-x)}$

**51.** $\cot(\alpha - 45^\circ) = \frac{1 + \tan \alpha}{\tan \alpha - 1}$

**52.** $\cos(\alpha + 45^\circ) = \frac{\cos \alpha - \sin \alpha}{\sqrt{2}}$

**53.** $\frac{\sin 2\beta}{2 \csc \beta} = \sin^2 \beta \cos \beta$

**54.** $\sin(45^\circ - \beta) = \frac{\cos 2\beta}{\sqrt{2}(\cos \beta + \sin \beta)}$

**55.** $\frac{\cot^3 y - \tan^3 y}{\sec^2 y + \cot^2 y} = 2 \cot 2y$

**56.** $\frac{\sin^3 y - \cos^3 y}{\sin y - \cos y} = \frac{2 + \sin 2y}{2}$

**57.** $\cos 4x = 8 \sin^4 x - 8 \sin^2 x + 1$

**58.** $\cos 3x = \cos x(1 - 4 \sin^2 x)$

**59.** $\sin^4 2x = 16 \sin^4 x - 32 \sin^6 x + 16 \sin^8 x$

**60.** $1 - \cos^6 x = 3 \sin^2 x - 3 \sin^4 x + \sin^6 x$

*Use an appropriate identity to find the exact value of each expression.*

**61.** $\tan(-\pi/12)$

**62.** $\sin(-\pi/8)$

**63.** $\sin(-75^\circ)$

**64.** $\cos(105^\circ)$

*Write each function in the form $y = A \sin(x + C)$, and graph the function for $-2\pi \le x \le 2\pi$. Determine the amplitude and phase shift.*

**65.** $y = 4 \sin x + 4 \cos x$

**66.** $y = \sqrt{3} \sin x + 3 \cos x$

**67.** $y = -2 \sin x + \cos x$

**68.** $y = -2 \sin x - \cos x$

*Find all real numbers that satisfy each equation.*

**69.** $2 \cos 2x + 1 = 0$

**70.** $2 \sin 2x + \sqrt{3} = 0$

**71.** $(\sqrt{3} \csc x - 2)(\csc x - 2) = 0$

**72.** $(\sec x - \sqrt{2})(\sqrt{3} \sec x + 2) = 0$

**73.** $2 \sin^2 x + 1 = 3 \sin x$

**74.** $4 \sin^2 x = \sin x + 3$

**75.** $-8\sqrt{3} \sin \frac{x}{2} = -12$

**76.** $-\cos \frac{x}{2} = \sqrt{2} + \cos \frac{x}{2}$

**77.** $\cos \frac{x}{2} - \sin x = 0$

**78.** $\sin 2x = \tan x$

**79.** $\cos 2x + \sin^2 x = 0$

**80.** $\tan \frac{x}{2} = \sin x$

**81.** $\sin x \cos x + \sin x + \cos x + 1 = 0$

**82.** $\sin 2x \cos 2x - \cos 2x + \sin 2x - 1 = 0$

*Find all angles $\alpha$ in $[0°, 360°)$ that satisfy each equation.*

**83.** $\sin\alpha\cos\alpha = \frac{1}{2}$

**84.** $\cos 2\alpha = \cos\alpha$

**85.** $\sin\alpha = \cos\alpha + 1$

**86.** $\cos\alpha\csc\alpha = \cot^2\alpha$

**87.** $\sin^2\alpha + \cos^2\alpha = \frac{1}{2}$

**88.** $\sec^2\alpha - \tan^2\alpha = 0$

**89.** $4\sin^4 2\alpha = 1$

**90.** $\sin 2\alpha = \tan\alpha$

**91.** $\tan 2\alpha = \tan\alpha$

**92.** $\tan\alpha = \cot\alpha$

**93.** $\sin 2\alpha\cos\alpha + \cos 2\alpha\sin\alpha = \cos 3\alpha$

**94.** $\cos 2\alpha\cos\alpha - \sin 2\alpha\sin\alpha = \cot 3\alpha$

*Use the sum-to-product identities to rewrite each expression as a product.*

**95.** $\cos 15° + \cos 19°$

**96.** $\cos 4° - \cos 6°$

**97.** $\sin(\pi/4) - \sin(-\pi/8)$

**98.** $\sin(-\pi/6) + \sin(\pi/12)$

*Use the product-to-sum identities to write each expression as a sum or difference.*

**99.** $2\sin 11°\cos 13°$

**100.** $2\sin 8°\sin 12°$

**101.** $2\cos(x/4)\cos(x/3)$

**102.** $2\cos s\sin 3s$

*Solve each problem.*

**103.** Find the exact value of $\sin(\alpha - \beta)$ if $\sin\alpha = \sqrt{3}/2$ and $\cos\beta = \sqrt{2}/2$ with $\alpha$ in quadrant II and $\beta$ in quadrant I.

**104.** Find the exact value of $\sin(\alpha + \beta)$ if $\sin\alpha = -\sqrt{2}/2$ and $\sin\beta = 1/2$ with $\alpha$ in quadrant IV and $\beta$ in quadrant II.

**105.** Find the exact value of $\cos(\alpha - \beta)$ if $\sin\alpha = \sqrt{3}/2$ and $\cos\beta = -\sqrt{2}/2$ with $\alpha$ in quadrant I and $\beta$ in quadrant II.

**106.** Find the exact value of $\cos(\alpha + \beta)$ if $\sin\alpha = \sqrt{2}/2$ and $\sin\beta = 1/2$ with $\alpha$ in quadrant II and $\beta$ in quadrant I.

**107.** *Motion of a Spring* A block is set in motion hanging from a spring and oscillates about its resting position $x = 0$ according to the function $x = 0.6\sin 2t + 0.4\cos 2t$, where $x$ is in centimeters and $t$ is in seconds. For what values of $t$ in the interval $[0, 3]$ is the block at its resting position $x = 0$?

**108.** *Battle of Gettysburg* The Confederates had at least one 24-lb mortar at the battle of Gettysburg in the Civil War. If the muzzle velocity of a projectile was 400 ft/sec, then at what angles could the cannon be aimed to hit the Union Army 3000 ft away? The distance $d$ (in feet) traveled by a projectile fired at an angle $\theta$ is related to the initial velocity $v_0$ (in feet per second) by the equation $v_0^2\sin 2\theta = 32d$.

### THINKING OUTSIDE THE BOX L

*Two Wrongs Make Right* In the following addition problem each letter represents a different digit from 1 through 9 and zero is not allowed. Find digits that make the addition correct.

```
  WRONG
+ WRONG
  RIGHT
```

# Chapter 6 Test

*Use identities to simplify each expression.*

**1.** $\sec x\cot x\sin 2x$

**2.** $\sin 2t\cos 5t + \cos 2t\sin 5t$

**3.** $\dfrac{1}{1-\cos y} + \dfrac{1}{1+\cos y}$

**4.** $\dfrac{\tan(\pi/5) + \tan(\pi/10)}{1 - \tan(\pi/5)\tan(\pi/10)}$

*Prove that each of the following equations is an identity.*

**5.** $\dfrac{\sin\beta\cos\beta}{\tan\beta} = 1 - \sin^2\beta$

**6.** $\dfrac{1}{\sec\theta - 1} - \dfrac{1}{\sec\theta + 1} = 2\cot^2\theta$

**7.** $\cos\left(\dfrac{\pi}{2} - x\right)\cos(-x) = \dfrac{\sin(2x)}{2}$

**8.** $\tan\dfrac{t}{2}\cos^2 t - \tan\dfrac{t}{2} = \dfrac{\sin t}{\sec t} - \sin t$

*Find all solutions to each equation.*

**9.** $\sin(-\theta) = 1$

**10.** $\cos 3s = \frac{1}{2}$

**11.** $\tan 2t = -\sqrt{3}$

**12.** $\sin 2\theta = \cos\theta$

*Find all values of $\alpha$ in $[0°, 360°)$ that satisfy each equation. Round approximate answers to the nearest tenth of a degree.*

**13.** $3\sin^2\alpha - 4\sin\alpha + 1 = 0$

**14.** $\dfrac{\tan 2\alpha - \tan 7\alpha}{1 + \tan 2\alpha\tan 7\alpha} = 1$

*Solve each problem.*

**15.** Write $y = \sin x - \sqrt{3}\cos x$ in the form $y = A\sin(x + C)$ and graph the function. Determine the period, amplitude, and phase shift.

**16.** Given that $\csc\alpha = 2$ and $\alpha$ is in quadrant II, find the exact values of $\alpha$ for the remaining five trigonometric functions.

**17.** Determine whether the function $f(x) = x \sin x$ is odd, even, or neither.

**18.** Use an appropriate identity to find the exact value of $\sin(-\pi/12)$.

**19.** Prove that $\tan x + \tan y = \tan(x + y)$ is not an identity.

**20.** A car with worn shock absorbers hits a pothole and oscillates about its normal riding position. At time $t$ (in seconds) the front bumper is distance $d$ (in inches) above or below its normal position, where $d = 2 \sin 3t - 4 \cos 3t$. For what values of $t$ (to the nearest tenth of a second) in the interval $[0, 4]$ is the front bumper at its normal position $d = 0$?

## TYING IT ALL together...

### Chapters 1–6

*Determine whether each function is even or odd.*

**1.** $f(x) = 3x^3 - 2x$ **2.** $f(x) = 2|x|$ **3.** $f(x) = x^3 + \sin x$ **4.** $f(x) = x^3 \sin x$

**5.** $f(x) = x^4 - x^2 + 1$ **6.** $f(x) = 1/x$ **7.** $f(x) = \dfrac{\sin x}{x}$ **8.** $f(x) = |\sin x|$

*Determine whether each equation is an identity. Prove your answer.*

**9.** $\sin(\alpha + \beta) = \sin \alpha + \sin \beta$ **10.** $(\alpha + \beta)^2 = \alpha^2 + \beta^2$ **11.** $3x + 5x = 8x$

**12.** $(x + 3)^2 = x^2 + 6x + 9$ **13.** $\sin^{-1}(x) = \dfrac{1}{\sin x}$ **14.** $\sin^2 x = \sin(x^2)$

*Solve each right triangle that has the given parts.*

**15.** $\alpha = 30°$ and $a = 4$ **16.** $a = \sqrt{3}$ and $b = 1$ **17.** $\cos \beta = 0.3$ and $b = 5$ **18.** $\sin \alpha = 0.6$ and $a = 2$

*In Chapter 7 we will study expressions like the following, which combine trigonometric functions and complex numbers. Write each expression in the form a + bi, where a and b are real numbers.*

**19.** $\cos(\pi/3) + i \sin(\pi/3)$ **20.** $2\left(\cos \dfrac{2\pi}{3} + i \sin \dfrac{2\pi}{3}\right)$ **21.** $(\cos 225° + i \sin 225°)^2$

**22.** $(\cos 3° + i \sin 3°)(\cos 3° - i \sin 3°)$ **23.** $(2 + i)^3$ **24.** $(\sqrt{2} - \sqrt{2}i)^4$

*Fill in the blanks.*

**25.** A degree is divided into 60 equal parts called ________.

**26.** A minute is divided into 60 equal parts called ________.

**27.** A circle with radius one is called a(n) ________ circle.

**28.** The relationship between degree and radian measure of angles is 180 degrees equals ________ radians.

**29.** The length of the arc intercepted by a central angle of $\alpha$ radians on a circle of radius $r$ is ________.

**30.** If $\omega$ is the angular velocity of a point in motion on a circle of radius $r$, then the linear velocity of the point is ________.

**31.** If $\alpha$ is an angle in standard position and $(x, y)$ is the point of intersection of the terminal side and the unit circle, then $\sin \alpha =$ ________ and $\cos \alpha =$ ________.

**32.** The identity $\sin^2 \alpha + \cos^2 \alpha = 1$ is called the ________ identity.

**33.** The value of $|A|$ is called the ________ of the function $y = A \sin x$.

**34.** For the function $y = \sin(x - C)$, $C$ is called the ________.

CONCEPTS OF

# calculus...

## Area of a circle and $\pi$

*The idea of a limiting value to a process was used long before it was formalized in calculus. Ancient mathematicians discovered the formula for the area of a circle and the value of $\pi$ using the formula for the area of a triangle $A = \frac{1}{2}bh$, trigonometry, and the idea of limits. For the following exercises imagine that you are an ancient mathematician (with a modern calculator), but you have never heard of $\pi$ or the formula $A = \pi r^2$.*

### Exercises

**1.** A regular pentagon is inscribed in a circle of radius $r$, as shown in the following figure.

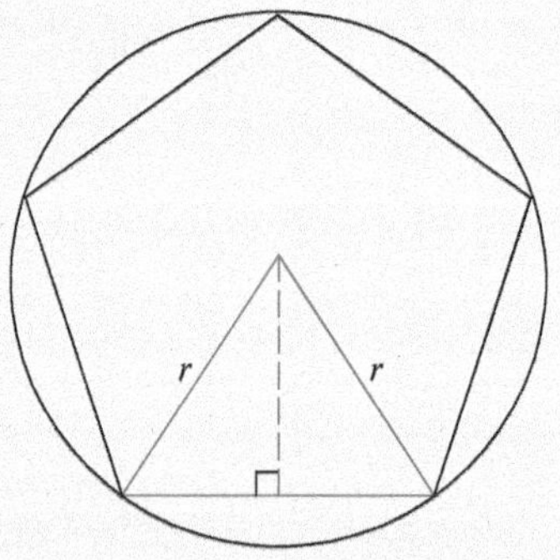


The pentagon is made up of five isosceles triangles. Show that the area of the pentagon is $\frac{5\sin(72°)}{2}r^2$.

**2.** A regular polygon of $n$ sides is inscribed in a circle of radius $r$. Write the area of the $n$-gon in terms of $r$ and $n$.

**3.** The area of a regular $n$-gon inscribed in a circle of radius $r$ is a constant multiple of $r^2$. Find the constant for a decagon, kilogon, and megagon.

**4.** What happens to the shape of the inscribed $n$-gon as $n$ increases?

**5.** What would you use as a formula for the area of a circle of radius $r$? (You have never heard of $\pi$.)

**6.** Find a formula for the perimeter of an $n$-gon inscribed in a circle of radius $r$.

**7.** Find a formula for the circumference of a circle of radius $r$. (You have never heard of $\pi$.)

**8.** Suppose that the $\pi$-key on your calculator is broken. What expression could you use to calculate $\pi$ accurate to nine decimal places?

# 7 Applications of Trigonometry

Mechanical devices that perform manipulative tasks under their own power (robots) are as old as recorded history. As early as 3000 B.C., Egyptians built waterclocks and articulated figures.

The first industrial robot joined GM's production line in 1961. In the 21st century, robotic "steel collar" workers will increasingly perform boring, repetitive, and dangerous jobs associated with a high degree of human error. They will assemble parts, mix chemicals, and work in areas that would be deadly to their human counterparts.

**WHAT YOU WILL learn...** In this chapter, we will see that trigonometry is the mathematics you need to model repetitive motion. We will learn how to use trigonometry to position a robot's arm. We'll also learn how trigonometry is used in designing machines and finding forces acting on parts.

# 7.1 The Law of Sines

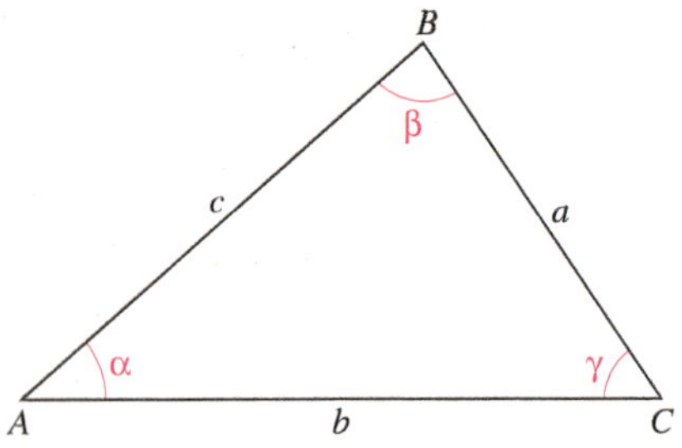


Figure 7.1

In Chapter 5 we used trigonometry to solve right triangles. In this section and the next we will learn to solve any triangle for which we have enough information to determine the shape of the triangle.

## Oblique Triangles

Any triangle without a right angle is called an **oblique triangle.** As usual, we use $\alpha$, $\beta$, and $\gamma$ for the angles of a triangle and $a$, $b$, and $c$, respectively, for the lengths of the sides opposite those angles, as shown in Fig. 7.1. The vertices at angles $\alpha$, $\beta$, and $\gamma$ are labeled $A$, $B$, and $C$, respectively. To solve an oblique triangle, we must know at least three parts of the triangle, at least one of which must be the length of a side. We can classify the different cases for the three known parts as follows:

1. One side and any two angles (ASA or AAS)
2. Two sides and a nonincluded angle (SSA)
3. Two sides and an included angle (SAS)
4. Three sides (SSS)

We can actually solve all of these cases by dividing the triangles into right triangles and using right triangle trigonometry. Since that method is quite tedious, we develop the *law of sines* and the *law of cosines.* The first two cases can be handled with the law of sines. We will discuss the last two cases in Section 7.2, when we develop the law of cosines.

## The Law of Sines

The **law of sines** says that *the ratio of the sine of an angle and the length of the side opposite the angle is the same for each angle of a triangle.*

**Theorem: The Law of Sines**

In any triangle,

$$\frac{\sin \alpha}{a} = \frac{\sin \beta}{b} = \frac{\sin \gamma}{c}.$$

**PROOF** Either the triangle is an acute triangle (all acute angles) or it is an obtuse triangle (one obtuse angle). We consider the case of the obtuse triangle here and leave the case of the acute triangle as Exercise 47. Triangle $ABC$ is shown in Fig. 7.2 with an altitude of length $h_1$ drawn from point $C$ to the opposite side and an altitude of length $h_2$ drawn from point $B$ to the extension of the opposite side.

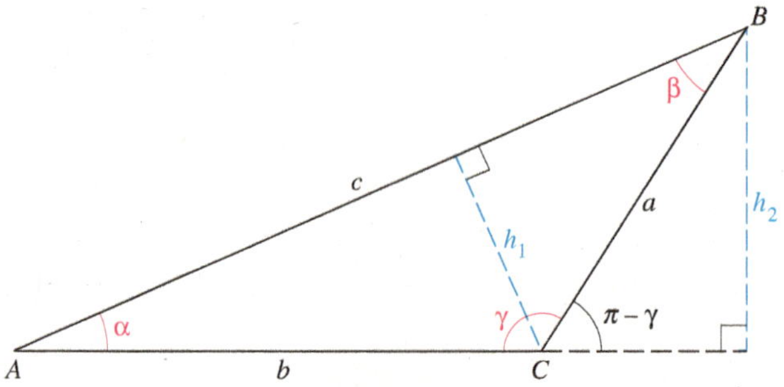


Figure 7.2

Since $\alpha$ and $\beta$ are now in right triangles, we have

$$\sin \alpha = \frac{h_1}{b} \quad \text{or} \quad h_1 = b \sin \alpha$$

and

$$\sin \beta = \frac{h_1}{a} \quad \text{or} \quad h_1 = a \sin \beta.$$

Replace $h_1$ by $b \sin \alpha$ in the equation $h_1 = a \sin \beta$ to get

$$b \sin \alpha = a \sin \beta.$$

Dividing each side of the last equation by $ab$ yields

$$\frac{\sin \alpha}{a} = \frac{\sin \beta}{b}.$$

Using the largest right triangle in Fig. 7.2, we have

$$\sin \alpha = \frac{h_2}{c} \quad \text{or} \quad h_2 = c \sin \alpha.$$

Note that the acute angle $\pi - \gamma$ in Fig. 7.2 is the reference angle for the obtuse angle $\gamma$. We know that sine is positive for both acute and obtuse angles, so $\sin \gamma = \sin(\pi - \gamma)$. Since $\pi - \gamma$ is an angle of a right triangle, we can write

$$\sin \gamma = \sin(\pi - \gamma) = \frac{h_2}{a} \quad \text{or} \quad h_2 = a \sin \gamma.$$

From $h_2 = c \sin \alpha$ and $h_2 = a \sin \gamma$ we get $c \sin \alpha = a \sin \gamma$ or

$$\frac{\sin \alpha}{a} = \frac{\sin \gamma}{c}$$

and we have proved the law of sines. ■

The law of sines can also be written in the form

$$\frac{a}{\sin \alpha} = \frac{b}{\sin \beta} = \frac{c}{\sin \gamma}.$$

In solving triangles, it is usually simplest to use the form in which the unknown quantity appears in the numerator.

In our first example we use the law of sines to solve a triangle for which we are given two angles and an included side. Remember that if the dimensions given for triangles are measurements, then any answers obtained from those measurements are only as accurate as the least accurate of the measurements. (See the discussion of significant digits in Chapter 5.)

## EXAMPLE 1 Given two angles and an included side (ASA)

Given $\beta = 34°$, $\gamma = 64°$, and $a = 5.3$, solve the triangle.

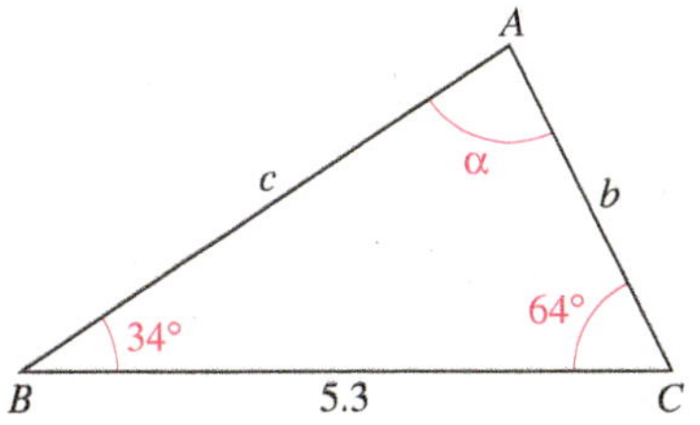

Figure 7.3

### Solution

To sketch the triangle, first draw side $a$, then draw angles of approximately 34° and 64° on opposite ends of $a$. Label all parts (see Fig. 7.3). Since the sum of the three angles of a triangle is 180°, the third angle $\alpha$ is 82°. By the law of sines,

$$\frac{5.3}{\sin 82°} = \frac{b}{\sin 34°}$$

$$b = \frac{5.3 \sin 34°}{\sin 82°} \approx 3.0.$$

Again, by the law of sines,

$$\frac{c}{\sin 64°} = \frac{5.3}{\sin 82°}$$

$$c = \frac{5.3 \sin 64°}{\sin 82°} \approx 4.8.$$

So $\alpha = 82°$, $b \approx 3.0$, and $c \approx 4.8$ solves the triangle.

▸**TRY THIS.** Given $\alpha = 28°$, $\beta = 66°$, and $c = 8.2$, solve the triangle. ■

The AAS case is similar to the ASA case. If two angles are known, then the third can be found by using the fact that the sum of all angles is 180°. If we know all angles and any side, then we can proceed to find the remaining sides with the law of sines as in Example 1.

## The Ambiguous Case (SSA)

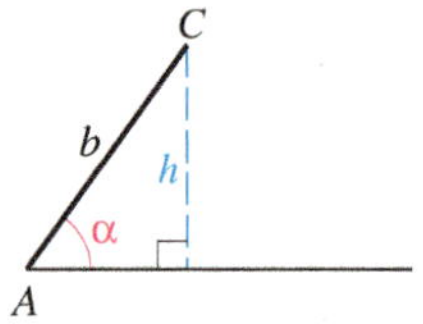


Draw α and $b$: find $h$

**Figure 7.4**

In the AAS and ASA cases we are given any two angles with positive measures and the length of any side. If the total measure of the two angles is less than 180°, then a unique triangle is determined. However, for two sides and a *nonincluded* angle (SSA), there are several possibilities. So the SSA case is called the **ambiguous case.** Drawing the diagram in the proper order will help you in understanding the ambiguous case.

Suppose we are given an acute angle $\alpha$ ($0° < \alpha < 90°$) and sides $a$ and $b$. Side $a$ is opposite the angle $\alpha$, and side $b$ is adjacent to it. Draw an angle of approximate size $\alpha$ in standard position with terminal side of length $b$, as shown in Fig. 7.4. Don't draw in side $a$ yet. Let $h$ be the distance from $C$ to the initial side of $\alpha$. Since $\sin \alpha = h/b$, we have $h = b \sin \alpha$. Now we are ready to draw in side $a$, but there are four possibilities for its location.

1. If $a < h$, then no triangle can be formed, because $a$ cannot reach from point $C$ to the initial side of $\alpha$. This situation is shown in Fig. 7.5(a).
2. If $a = h$, then exactly one right triangle is formed, as in Fig. 7.5(b).
3. If $h < a < b$, then exactly two triangles are formed, because $a$ reaches to the initial side in two places, as in Fig. 7.5(c).
4. If $a \geq b$, then only one triangle is formed, as in Fig. 7.5(d).

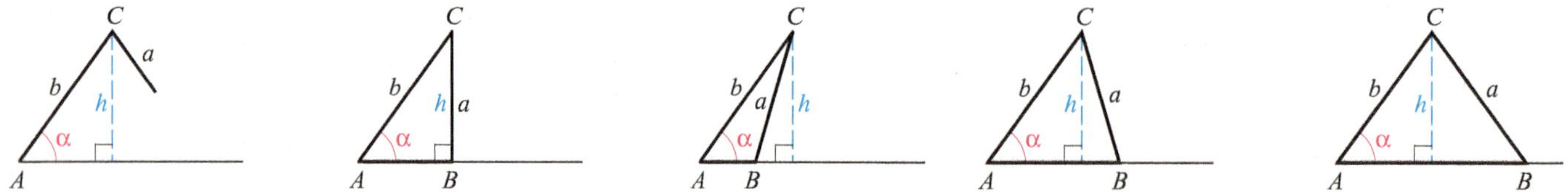


$a < h$: no triangle (a) $a = h$: one triangle (b) $h < a < b$: two triangles (c) $a \geq b$: one triangle (d)

**Figure 7.5**

If we start with $\alpha$, $a$, and $b$, where $90° \leq \alpha < 180°$, then there are only two possibilities for the number of triangles determined. If $a \leq b$, then no triangle is formed, as in Fig. 7.6(a). If $a > b$, one triangle is formed, as in Fig. 7.6(b).

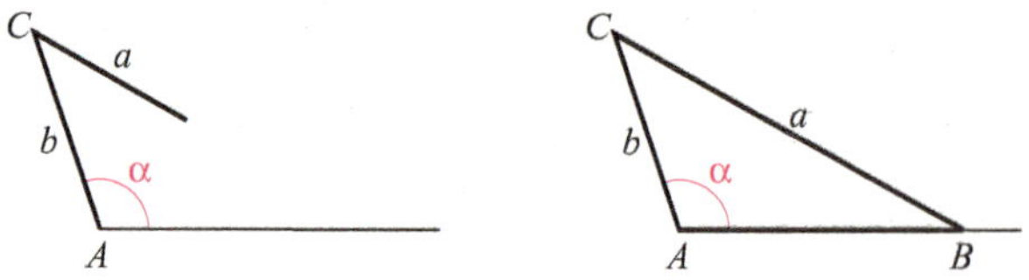


$a \leq b$: no triangle $a > b$: one triangle

**Figure 7.6**

It is not necessary to memorize all of the SSA cases shown in Figs. 7.5 and 7.6. If you draw the triangle for a given problem in the order suggested, then it will be clear how many triangles are possible with the given parts. We must decide how many triangles are possible, before we can solve the triangle(s).

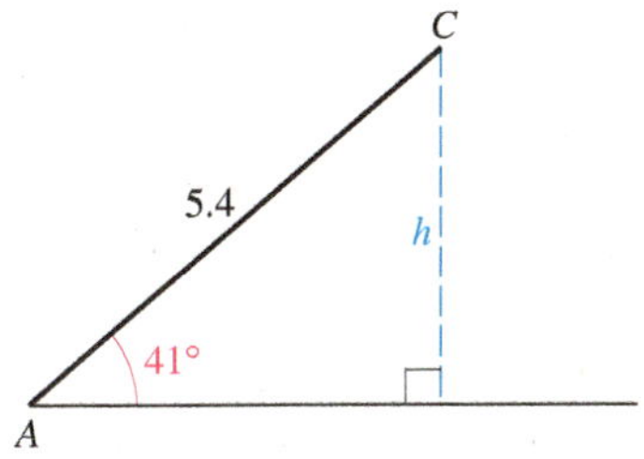


**Figure 7.7**

## EXAMPLE 2 SSA with no triangle

Given $\alpha = 41°$, $a = 3.3$, and $b = 5.4$, solve the triangle.

### Solution

Draw the 41° angle and label its terminal side 5.4, as shown in Fig. 7.7. Side $a$ must go opposite $\alpha$, but do not put it in yet. Find the length of the altitude $h$ from $C$. Using a trigonometric ratio, we get $\sin 41° = h/5.4$, or $h = 5.4 \sin 41° \approx 3.5$. Since $a = 3.3$, $a$ is *shorter* than the altitude $h$, and side $a$ will not reach from point $C$ to the initial side of $\alpha$. So there is no triangle with the given parts.

▸**TRY THIS.** Given $\beta = 38°$, $b = 2.9$, and $c = 5.9$, solve the triangle. ■

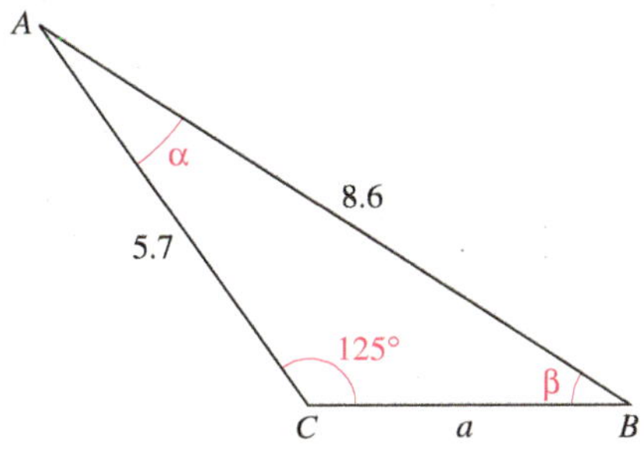


**Figure 7.8**

## EXAMPLE 3 SSA with one triangle

Given $\gamma = 125°$, $b = 5.7$, and $c = 8.6$, solve the triangle.

### Solution

Draw the 125° angle and label its terminal side 5.7. Since $8.6 > 5.7$, side $c$ will reach from point $A$ to the initial side of $\gamma$, as shown in Fig. 7.8, and form a single triangle. To solve this triangle, we use the law of sines:

$$\frac{\sin 125°}{8.6} = \frac{\sin \beta}{5.7}$$

$$\sin \beta = \frac{5.7 \sin 125°}{8.6}$$

$$\beta = \sin^{-1}\left(\frac{5.7 \sin 125°}{8.6}\right) \approx 32.9°$$

Since the sum of $\alpha$, $\beta$, and $\gamma$ is 180°, $\alpha = 22.1°$. Now use $\alpha$ and the law of sines to find $a$:

$$\frac{a}{\sin 22.1°} = \frac{8.6}{\sin 125°}$$

$$a = \frac{8.6 \sin 22.1°}{\sin 125°} \approx 3.9$$

▸**TRY THIS.** Given $\beta = 38°$, $b = 6.4$, and $c = 5.9$, solve the triangle. ■

## EXAMPLE 4 SSA with two triangles

Given $\beta = 56.3°$, $a = 8.3$, and $b = 7.6$, solve the triangle.

### Solution

Draw an angle of approximately 56.3°, and label its terminal side 8.3. Side $b$ must go opposite $\beta$, but do not put it in yet. Find the length of the altitude $h$ from point $C$ to the initial side of $\beta$, as shown in Fig. 7.9(a). Since $\sin 56.3° = h/8.3$, we get

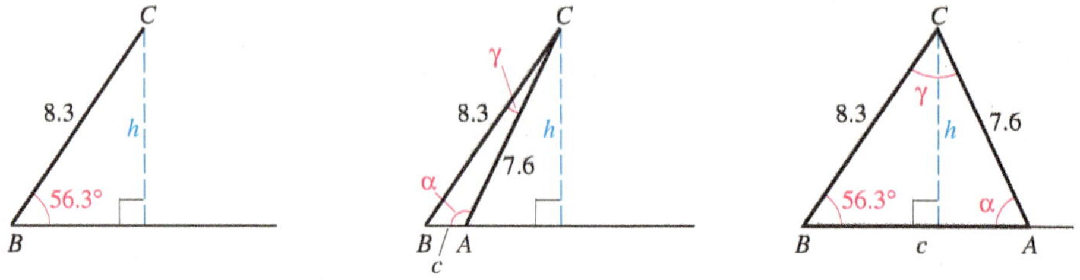


(a) (b) (c)

**Figure 7.9**

$h = 8.3 \sin 56.3° \approx 6.9$. Because $b$ is longer than $h$ but shorter than $a$, there are two triangles that satisfy the given conditions, as shown in parts (b) and (c) of Fig. 7.9. In either triangle we have

$$\frac{\sin \alpha}{8.3} = \frac{\sin 56.3°}{7.6}$$

$$\sin \alpha = 0.9086.$$

This equation has two solutions in $[0°, 180°]$. For part (c) we get $\alpha = \sin^{-1}(0.9086) = 65.3°$, and for part (b) we get $\alpha = 180° - 65.3° = 114.7°$. Using the law of sines, we get $\gamma = 9.0°$ and $c = 1.4$ for part (b), and we get $\gamma = 58.4°$ and $c = 7.8$ for part (c).

►**TRY THIS.** Given $\beta = 38°$, $b = 4.7$, and $c = 5.9$, solve the triangle. ■

The remaining cases for solving triangles are presented in the next section, which also contains a summary of how to proceed in each case. You may wish to refer to it now.

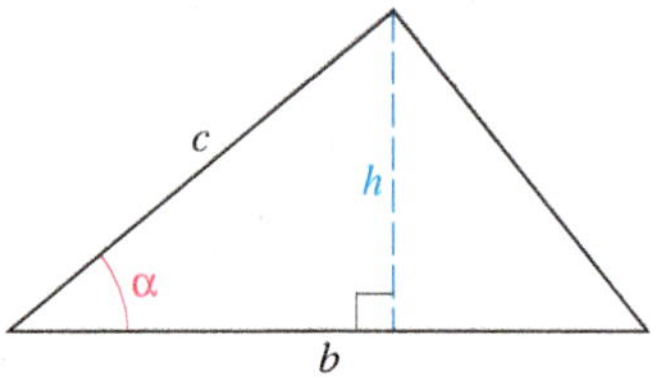


**Figure 7.10**

## Area of a Triangle

The formula $A = \frac{1}{2}bh$ gives the area of a triangle in terms of a side and the altitude to that side. We can rewrite this formula in terms of two sides and their included angle. Consider the triangle shown in Fig. 7.10, in which $\alpha$ is an acute angle and $h$ is inside the triangle. Since $\sin \alpha = h/c$, we get $h = c \sin \alpha$. Substitute $c \sin \alpha$ for $h$ in the formula to get

$$A = \frac{1}{2}bc \sin \alpha.$$

This formula is also correct in all other possible cases: $\alpha$ is acute and $h$ is the side of the triangle opposite $\alpha$, $\alpha$ is acute and $h$ lies outside the triangle, $\alpha$ is a right angle, and $\alpha$ is obtuse. (See Exercise 48.) Likewise, the formula can be written using the angles $\beta$ or $\gamma$. We have the following theorem.

**Theorem: Area of a Triangle**

The area of a triangle is given by

$$A = \frac{1}{2}bc \sin \alpha, \quad A = \frac{1}{2}ac \sin \beta, \quad \text{and} \quad A = \frac{1}{2}ab \sin \gamma.$$

Surveyors describe property in terms of the lengths of the sides and the angles between those sides. In the next example we find the area of a piece of property, using the new area formula.

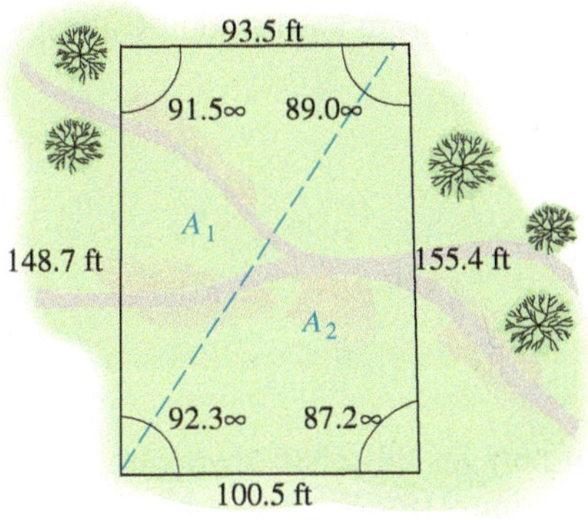


**Figure 7.11**

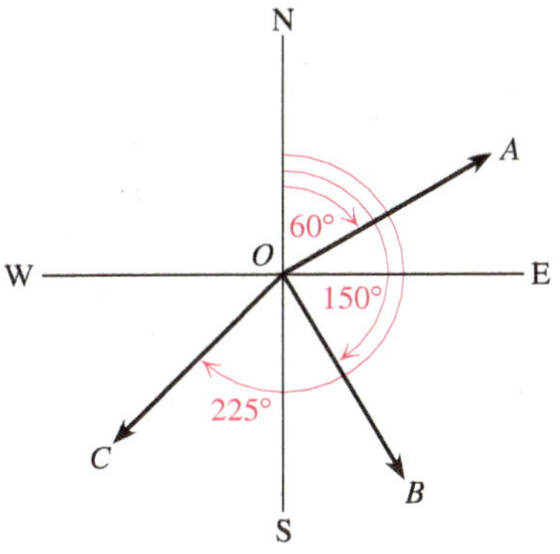


**Figure 7.12**

## EXAMPLE 5 Area of a quadrilateral

The town of Hammond is considering the purchase of a four-sided piece of property to be used for a playground. The dimensions of the property are given in Fig. 7.11. Find the area of the property in square feet.

### Solution

Divide the property into two triangles, as shown in Fig. 7.11, and find the area of each:

$$A_1 = \frac{1}{2}(148.7)(93.5) \sin 91.5° \approx 6949.3 \text{ square feet}$$

$$A_2 = \frac{1}{2}(100.5)(155.4) \sin 87.2° \approx 7799.5 \text{ square feet}$$

The total area of the property is 14,748.8 square feet.

▸**TRY THIS.** Two sides of a triangular piece of property are 244 feet and 206 feet, and the angle between these sides is 87.4°. Find the area to the nearest square foot. ■

## Bearing

The measure of an angle that describes the direction of a ray is called the **bearing** of the ray. In air navigation, bearing is given as a nonnegative angle less than 360° measured in a clockwise direction from a ray pointing due north. So in Fig. 7.12 the bearing of ray $\overrightarrow{OA}$ is 60°, the bearing of ray $\overrightarrow{OB}$ is 150°, and the bearing of $\overrightarrow{OC}$ is 225°.

## EXAMPLE 6 Using bearing in solving triangles

A bush pilot left the Fairbanks Airport in a light plane and flew 100 miles toward Fort Yukon in still air on a course with a bearing of 18°. She then flew due east (bearing 90°) for some time to drop supplies to a snowbound family. After the drop, her course to return to Fairbanks had a bearing of 225°. What was her maximum distance from Fairbanks?

### Solution

Figure 7.13 shows the course of her flight. To change course from bearing 18° to bearing 90° at point $B$, the pilot must add 72° to the bearing. So $\angle ABC$ is 108°. A bearing of 225° at point $C$ means that $\angle BCA$ is 45°. Finally, we obtain $\angle BAC = 27°$ by subtracting 18° and 45° from 90°.

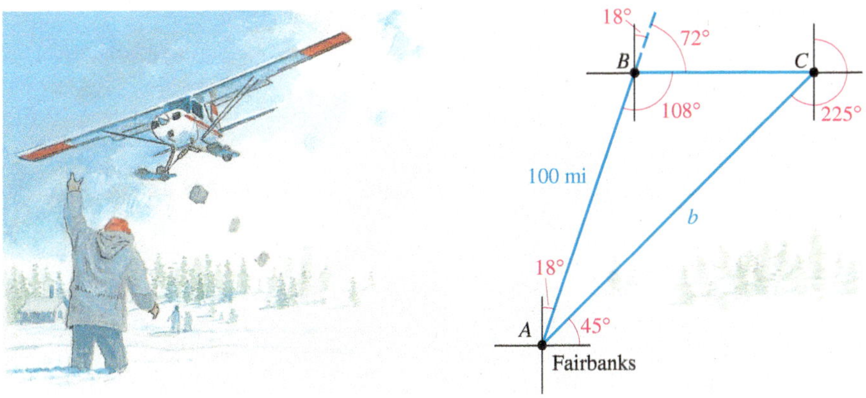


**Figure 7.13**

We can find the length of $\overline{AC}$ (the maximum distance from Fairbanks) by using the law of sines:

$$\frac{b}{\sin 108°} = \frac{100}{\sin 45°}$$

$$b = \frac{100 \cdot \sin 108°}{\sin 45°} \approx 134.5$$

So the pilot's maximum distance from Fairbanks was 134.5 miles.

▸**TRY THIS.** A pilot flew north from the airport for 88 miles, then east until he headed back to the airport on a course with bearing 210°. What was his maximum distance from the airport? ■

In marine navigation and surveying, the bearing of a ray is the acute angle the ray makes with a ray pointing due north or due south. Along with the acute angle, directions are given that indicate in which quadrant the ray lies. For example, in Fig. 7.12, $\overrightarrow{OA}$ has a bearing 60° east of north (N60°E), $\overrightarrow{OB}$ has a bearing 30° east of south (S30°E), and $\overrightarrow{OC}$ has a bearing 45° west of south (S45°W).

## One More Application

In the next example we find the height of an object from a distance. This same problem was solved in Example 7 of Section 5.6. Note how much easier the solution is here using the law of sines.

### EXAMPLE 7 Finding the height of an object from a distance

The angle of elevation of the top of a water tower from point $A$ on the ground is 19.9°. From point $B$, 50.0 feet closer to the tower, the angle of elevation is 21.8°. What is the height of the tower?

#### Solution

Let $y$ represent the height of the tower. All angles of triangle $ABC$ in Fig. 7.14 can be determined.

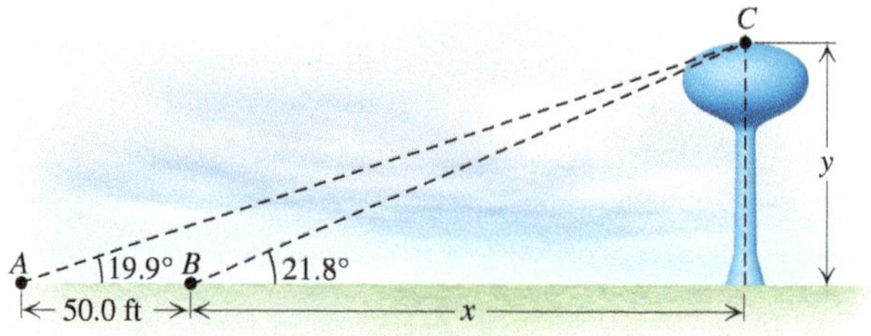


**Figure 7.14**

Apply the law of sines in triangle $ABC$:

$$\frac{a}{\sin 19.9°} = \frac{50}{\sin 1.9°}$$

$$a = \frac{50 \sin 19.9°}{\sin 1.9°} \approx 513.3 \text{ feet}$$

Now, using the smaller right triangle, we have $\sin 21.8° = y/513.3$ or $y = 513.3 \sin 21.8° \approx 191$. So the height of the tower is 191 feet.

**▶TRY THIS.** The angle of elevation of the top of a building from point $A$ on the ground is 24.2°. From point $B$, which is 44.5 feet closer, the angle of elevation is 38.1°. What is the height of the building? ■

## FOR thought... True or False? Explain.

1. If we know the measures of two angles of a triangle, then the measure of the third angle is determined.
2. If $\frac{\sin 9°}{a} = \frac{\sin 17°}{88}$, then $a = \frac{\sin 17°}{88 \sin 9°}$.
3. The equation $\frac{\sin \alpha}{5} = \frac{\sin 44°}{18}$ has exactly one solution in $[0, 180°]$.
4. One solution to $\frac{\sin \beta}{2.3} = \frac{\sin 39°}{1.6}$ is $\beta = \sin^{-1}\left(\frac{2.3 \sin 39°}{1.6}\right)$.
5. $\frac{\sin 60°}{\sqrt{3}} = \frac{\sin 30°}{1}$
6. No triangle exists with $\alpha = 60°$, $b = 10$ feet, and $a = 500$ feet.
7. There is exactly one triangle with $\beta = 30°$, $c = 20$, and $b = 10$.
8. There are two triangles with $\alpha = 135°$, $b = 17$, and $a = 19$.
9. The area of a triangle is determined by the lengths of two sides and the measure of the included angle.
10. A right triangle's area is one-half the product of the lengths of its legs.

## EXERCISES 7.1

*Fill in the blank.*

1. Any triangle without a right angle is a(n) ________ triangle.
2. According to the ________, the ratio of the sine of an angle and the length of the side opposite the angle is the same for each angle of a triangle.

*Solve each triangle.*

3.

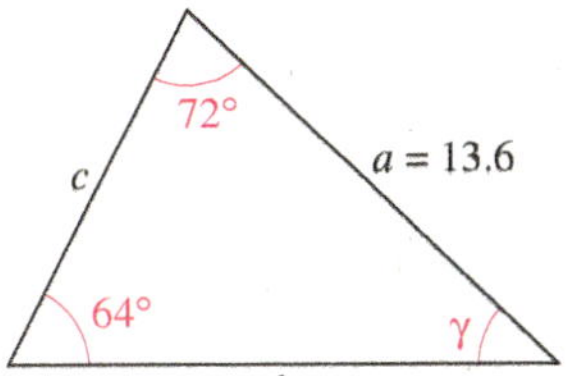


4.

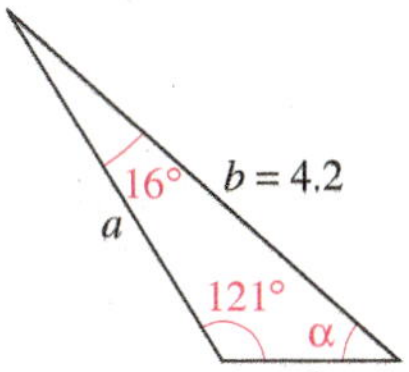


5.

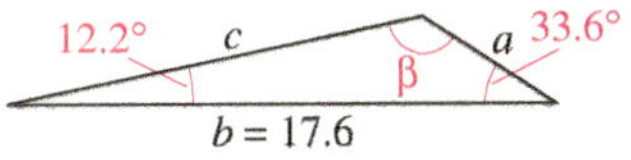


6.

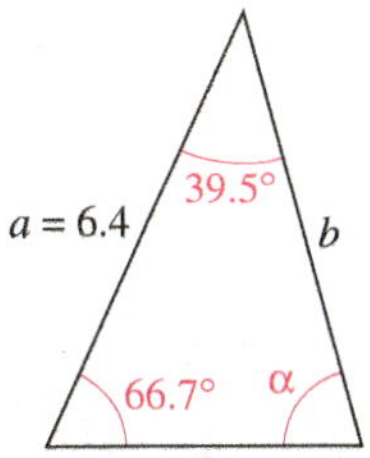


*Solve each triangle with the given parts.*

7. $\alpha = 10.3°$, $\gamma = 143.7°$, $c = 48.3$
8. $\beta = 94.7°$, $\alpha = 30.6°$, $b = 3.9$
9. $\beta = 120.7°$, $\gamma = 13.6°$, $a = 489.3$
10. $\alpha = 39.7°$, $\gamma = 91.6°$, $b = 16.4$

*Determine the number of triangles with the given parts and solve each triangle.*

11. $\alpha = 39.6°$, $c = 18.4$, $a = 3.7$
12. $\beta = 28.6°$, $a = 40.7$, $b = 52.5$
13. $\gamma = 60°$, $b = 20$, $c = 10\sqrt{3}$
14. $\alpha = 41.2°$, $a = 8.1$, $b = 10.6$
15. $\beta = 138.1°$, $c = 6.3$, $b = 15.6$
16. $\gamma = 128.6°$, $a = 9.6$, $c = 8.2$
17. $\beta = 32.7°$, $a = 37.5$, $b = 28.6$
18. $\alpha = 30°$, $c = 40$, $a = 20$
19. $\gamma = 99.6°$, $b = 10.3$, $c = 12.4$
20. $\alpha = 75.3$, $a = 12.4$, $b = 9.8$

*Find the area of each triangle with the given parts.*

21. $a = 12.9$, $b = 6.4$, $\gamma = 13.7°$
22. $b = 42.7$, $c = 64.1$, $\alpha = 74.2°$
23. $\alpha = 39.4°$, $b = 12.6$, $a = 13.7$
24. $\beta = 74.2°$, $c = 19.7$, $b = 23.5$

**25.** $\alpha = 42.3°, \beta = 62.1°, c = 14.7$

**26.** $\gamma = 98.6°, \beta = 32.4°, a = 24.2$

**27.** $\alpha = 56.3°, \beta = 41.2°, a = 9.8$

**28.** $\beta = 25.6°, \gamma = 74.3°, b = 17.3$

*Find the area of each region. Give the exact area for Exercise 29 and round to the nearest 10 square feet for Exercise 30.*

**29.**

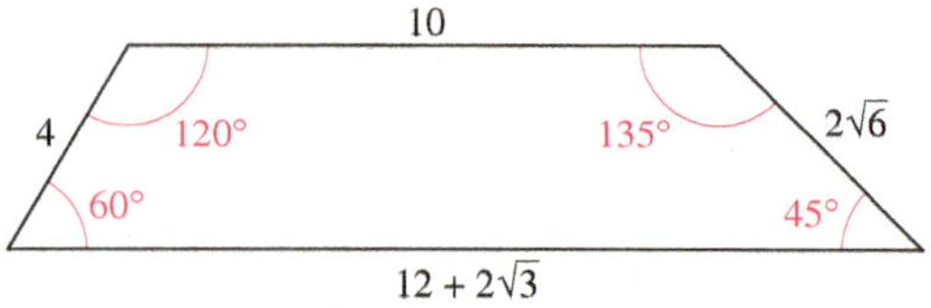


**30.**

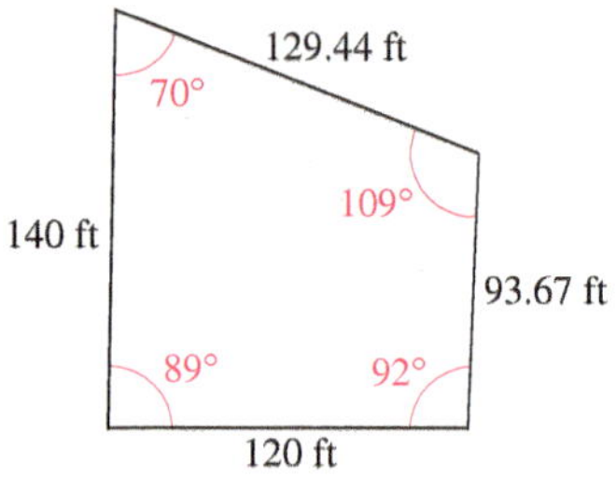


*Solve each problem.*

**31.** *Observing Traffic* A traffic report helicopter left the WKPR studios on a course with a bearing of 210°. After flying 12 mi to reach interstate highway 20, the helicopter flew due east along I-20 for some time. The helicopter headed back to WKPR on a course with a bearing of 310° and reported no accidents along I-20. For how many miles did the helicopter fly along I-20?

**32.** *Course of a Fighter Plane* During an important NATO exercise, an F-14 Tomcat left the carrier Nimitz on a course with a bearing of 34° and flew 400 mi. Then the F-14 flew for some distance on a course with a bearing of 162°. Finally, the plane flew back to its starting point on a course with a bearing of 308°. What distance did the plane fly on the final leg of the journey?

**33.** *Surveying Property* A surveyor locating the corners of a triangular piece of property started at one corner and walked 480 ft in the direction of N36°W to reach the next corner. The surveyor turned and walked S21°W to get to the next corner of the property. Finally, the surveyor walked in the direction N82°E to get back to the starting point. What is the area of the property in square feet?

**34.** *Sailing* Joe and Jill set sail from the same point, with Joe sailing in the direction of S4°E and Jill sailing in the direction S9°W. After 4 hr, Jill was 2 mi due west of Joe. How far had Jill sailed?

**35.** *Cellular One* The angle of elevation of the top of a cellular telephone tower from point $A$ on the ground is 18.1°. From point $B$, 32.5 ft closer to the tower, the angle of elevation is 19.3°. What is the height of the tower?

**36.** *Moving Back* A surveyor determines that the angle of elevation of the top of a building from a point on the ground is 30.4°. He then moves back 55.4 ft and determines that the angle of elevation is 23.2°. What is the height of the building?

**37.** *Designing an Addition* A 40-ft-wide house has a roof with a 6-12 pitch (the roof rises 6 ft for a run of 12 ft). The owner plans a 14-ft-wide addition that will have a 3-12 pitch to its roof. Find the lengths of $\overline{AB}$ and $\overline{BC}$ in the accompanying figure.

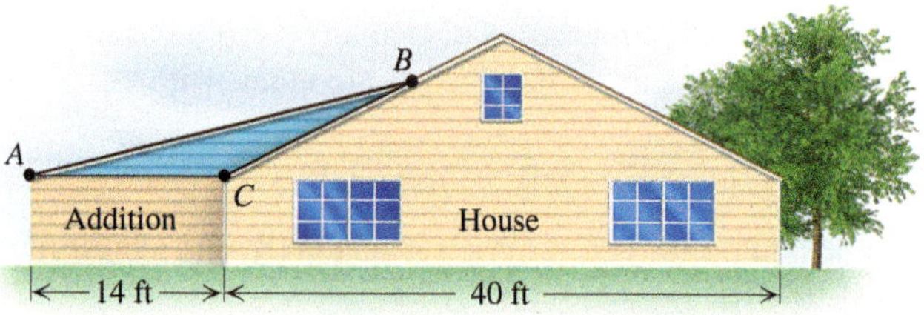


**Figure for Exercise 37**

**38.** *Saving an Endangered Hawk* A hill has an angle of inclination of 36°, as shown in the figure. A study completed by a state's highway commission showed that the placement of a highway requires that 400 ft of the hill, measured horizontally, be removed. The engineers plan to leave a slope alongside the highway with an angle of inclination of 62°, as shown in the figure. Located 750 ft up the hill measured from the base is a tree containing the nest of an endangered hawk. Will this tree be removed in the excavation?

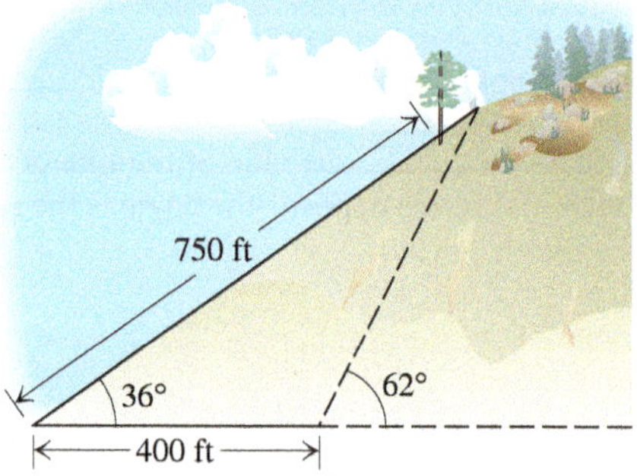


**Figure for Exercise 38**

**39.** *Making a Kite* A kite is made in the shape shown in the figure. Find the surface area of the kite in square inches.

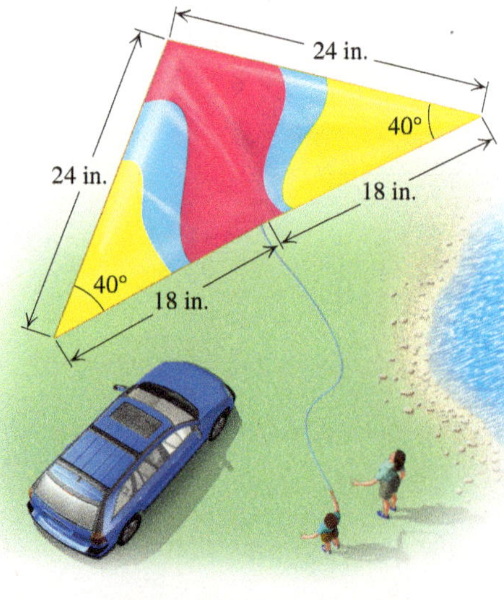


**Figure for Exercise 39**

**40.** *Area of a Wing* The F-106 Delta Dart once held a world speed record of Mach 2.3. Its sweptback triangular wings have the dimensions given in the figure. Find the area of one wing in square feet.

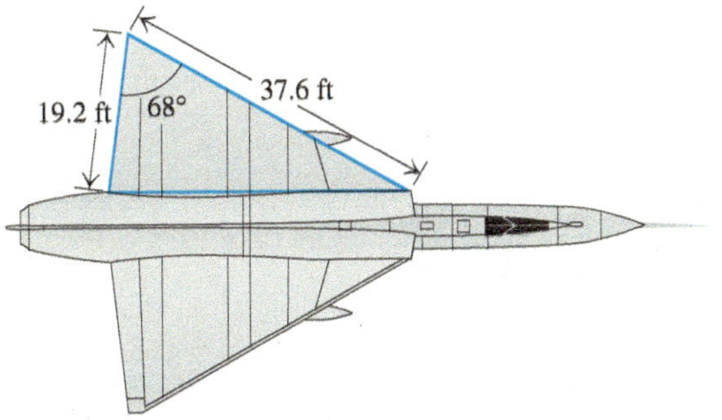


**Figure for Exercise 40**

**41.** *Filtering the Sun* When the sun is directly overhead, the sun's light is filtered by approximately 10 miles of atmosphere. As the sun sets, the intensity decreases because the light must pass a greater distance through the atmosphere, as shown in the figure.

**a.** Find the distance $d$ in the figure when the angle of elevation of the sun is 30°. Use 3950 miles as the radius of the earth.

**b.** If the sun is directly overhead at 12 noon, then at what time (to the nearest tenth of a second) is the angle of elevation of the sun 30°? The distance from the earth to the sun is 93 million miles.

**c.** Find the distance that the sunlight passes through the atmosphere at sunset.

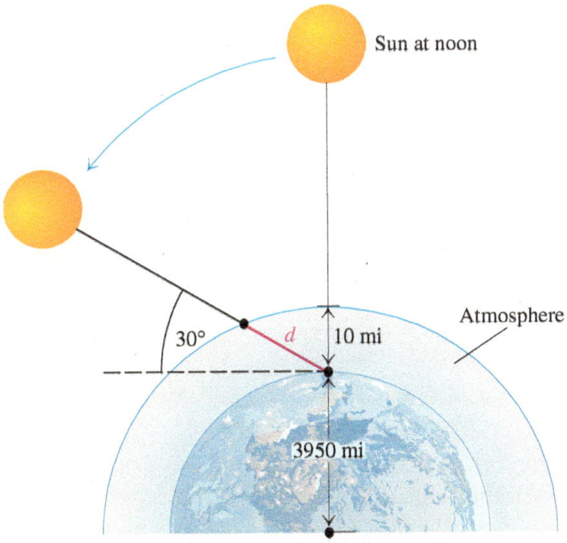


**Figure for Exercise 41**

**42.** *Flat Earth* Assume that the earth is flat with a 10-mile-thick layer of atmosphere. Find the distance that the sunlight passes through the atmosphere when the angle of elevation of the sun is 30°. Compare your answer to that obtained in part (a) of the previous exercise.

**43.** *Shot Down* A cruise missile is traveling straight across the desert at 548 mph at an altitude of 1 mile, as shown in the figure. A gunner spots the missile coming in his direction and fires a projectile at the missile when the angle of elevation of the missile is 35°. If the speed of the projectile is 688 mph, then for what angle of elevation of the gun will the projectile hit the missile?

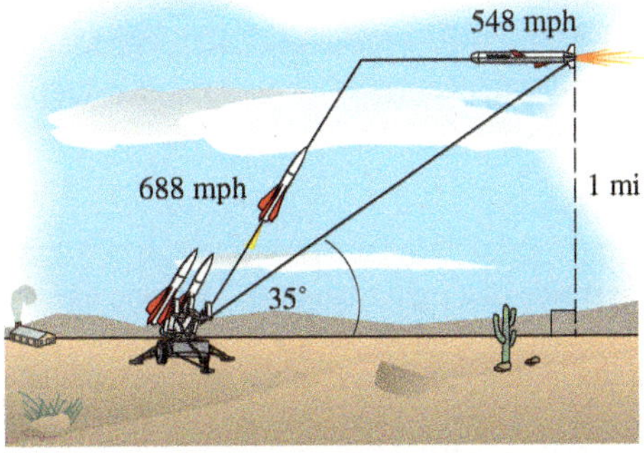


**Figure for Exercise 43**

**44.** *Angle of Completion* When the ball is snapped, Smith starts running at a 50° angle to the line of scrimmage. At the moment when Smith is at a 60° angle from Jones, Smith is running at 17 ft/sec and Jones passes the ball at 60 ft/sec to Smith. However, to complete the pass, Jones must lead Smith by the angle $\theta$ shown in the figure. Find $\theta$. (Of course, Jones finds $\theta$ in his head. Note that $\theta$ can be found without knowing any distances.)

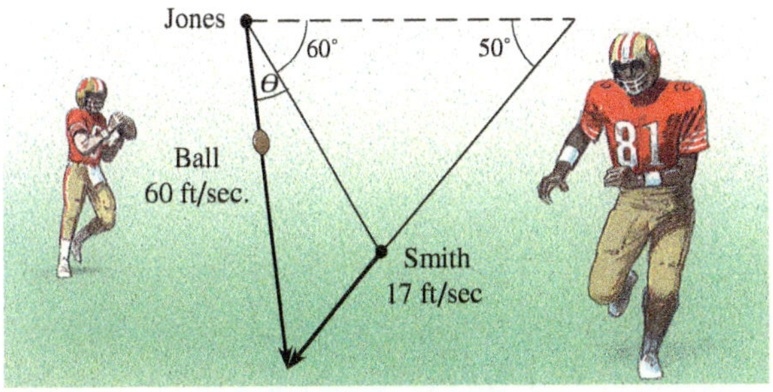


**Figure for Exercise 44**

**45.** *Rabbit and Fox* A rabbit starts running from point $A$ in a straight line in the direction 30° from the north at 3.5 ft/sec, as shown in the accompanying diagram. At the same time a fox starts running in a straight line from a position 30 ft to the west of the rabbit at 6.5 ft/sec. The fox chooses his path so that he will catch the rabbit at point $C$. In how many seconds will the fox catch the rabbit?

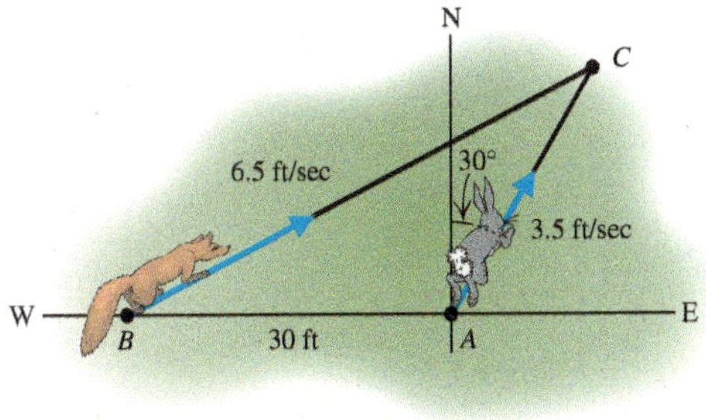


**Figure for Exercise 45**

**46.** *Head Start* How many seconds will it take for the fox in Exercise 45 to catch the rabbit if the rabbit has a 1-second head start?

## FOR WRITING/DISCUSSION

**47.** *Law of Sines* Prove the law of sines for the case in which the triangle is an acute triangle.

**48.** *Area of a Triangle* Prove the trigonometric formula for the area of a triangle in the cases mentioned in the text but not proved in the text.

## ▶ RETHINKING

**49.** The shortest side of a right triangle is 7 cm and one of the acute angles is 64°. Find the length of the hypotenuse and the length of the longer leg. Round to the nearest tenth of a centimeter.

**50.** Suppose that $\alpha$ is an angle in standard position with its terminal side in quadrant III such that $\sin \alpha = -5/6$. Find exact values for $\cos \alpha$ and $\tan \alpha$.

**51.** Determine the amplitude, period, phase shift, and range for the function $y = 5\sin(4x - \pi) - 3$.

**52.** Determine the period, asymptotes, and range for the function $y = 3\csc(\pi x - \pi) + 2$.

**53.** Simplify the expression $3 - \dfrac{3}{\csc^2(x)}$.

**54.** Simplify the expression $(\sin x + \cos x)^2 - \sin(2x)$.

## THINKING OUTSIDE THE BOX LI & LII

*Packing Beans* A can of green beans has a 3-in. diameter and 4-in. height. How many cans of beans will fit in a box that measures 24 in. long, 24 in. wide, and 4 in. high on the inside?

*Working Together* Big Red can build a brick wall by himself in 10 hr, whereas Slim would take 12 hr to build the same wall by himself. If they work together they talk sports, and each of them lays 10 fewer bricks per hour. To get the job done in a hurry, the boss assigns both men and they finish the wall in 6 hr. How many bricks are in the wall?

## POP QUIZ 7.1

**1.** If $\alpha = 8°$, $\beta = 121°$, and $c = 12$, then what is $\gamma$?

**2.** If $\alpha = 20.4°$, $\beta = 27.3°$, and $c = 38.5$, then what is $a$?

**3.** If $\alpha = 33.5°$, $a = 7.4$, and $b = 10.6$, then what is $\beta$?

**4.** Find the area of the triangle in which $a = 6$ ft, $b = 15$ ft, and $\gamma = 66.7°$.

## LINKING concepts... For Individual or Group Explorations

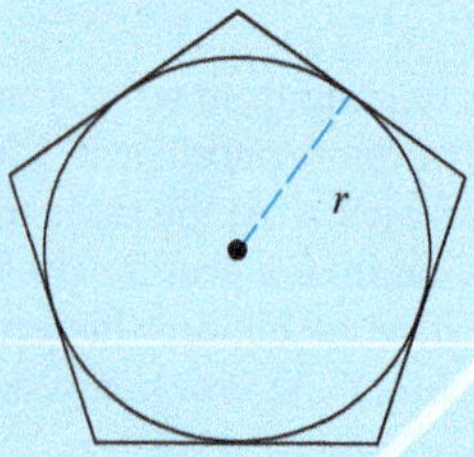


### Discovering Area and Circumference Formulas

*In these exercises you will discover the area and circumference formulas for the circle. You will work with regular polygons that circumscribe a circle of radius r, and you will use the new formula for the area of a triangle:*

$$A = \frac{1}{2}bc\sin\alpha$$

**a)** Suppose a regular pentagon circumscribes a circle of radius $r$, as shown in the figure. Show that the area of the pentagon is $5r^2\tan(36°)$.

**b)** Now suppose that a regular polygon of $n$ sides circumscribes a circle of radius $r$. Show that the area of the $n$-gon is given by $A = nr^2\tan(180°/n)$.

**c)** Express the result of part (b), using the language of variation discussed in Section 2.6.

**d)** Find the constant of proportionality for a decagon, kilogon, and megagon.

**e)** What happens to the shape of the $n$-gon as $n$ increases? So what can you conclude is the formula for the area of a circle of radius $r$?

**f)** Use trigonometry to find a formula for the perimeter of an $n$-gon that circumscribes a circle of radius $r$.

**g)** Use the result of part (f) to find a formula for the circumference of a circle of radius $r$.

**h)** Graph the function $y = x\tan(\pi/x)$, where $x$ is a real number (radian mode). Identify all vertical and horizontal asymptotes.

# 7.2 The Law of Cosines

In Section 7.1 we discussed the AAS, ASA, and SSA cases for solving triangles using the law of sines. For the SAS and SSS cases, which cannot be handled with the law of sines, we develop the law of cosines. This law is a generalization of the Pythagorean theorem and applies to any triangle. However, it is usually stated and used only for oblique triangles because it is not needed for right triangles.

## The Law of Cosines

The **law of cosines** gives a formula for the square of any side of an oblique triangle in terms of the other two sides and their included angle.

**Theorem: Law of Cosines**

If triangle $ABC$ is an oblique triangle with sides $a$, $b$, and $c$ and angles $\alpha$, $\beta$, and $\gamma$, then

$$a^2 = b^2 + c^2 - 2bc\cos\alpha,$$

$$b^2 = a^2 + c^2 - 2ac\cos\beta,$$

$$c^2 = a^2 + b^2 - 2ab\cos\gamma.$$

**PROOF** Given triangle $ABC$, position the triangle as shown in Fig. 7.15. The vertex $C$ is in the first quadrant if $\alpha$ is acute and in the second if $\alpha$ is obtuse. Both cases are shown in Fig. 7.15.

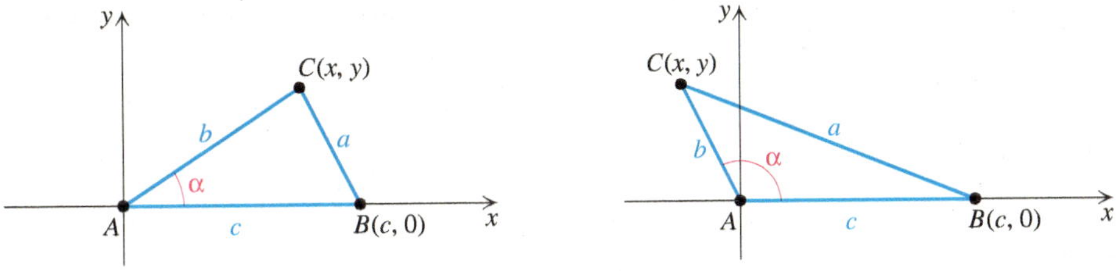


**Figure 7.15**

In either case, the $x$-coordinate of $C$ is $x = b\cos\alpha$, and the $y$-coordinate of $C$ is $y = b\sin\alpha$. The distance from $C$ to $B$ is $a$, but we can also find that distance by using the distance formula:

$$\begin{aligned} a &= \sqrt{(b\cos\alpha - c)^2 + (b\sin\alpha - 0)^2} \\ a^2 &= (b\cos\alpha - c)^2 + (b\sin\alpha)^2 \\ &= b^2\cos^2\alpha - 2bc\cos\alpha + c^2 + b^2\sin^2\alpha \\ &= b^2(\cos^2\alpha + \sin^2\alpha) + c^2 - 2bc\cos\alpha \end{aligned}$$

Using $\cos^2 x + \sin^2 x = 1$, we get the first equation of the theorem:

$$a^2 = b^2 + c^2 - 2bc \cos \alpha$$

Similar arguments with $B$ and $C$ at $(0, 0)$ produce the other two equations. ■

In any triangle with unequal sides, the largest angle is opposite the largest side, and the smallest angle is opposite the smallest side. We need this fact in the first example.

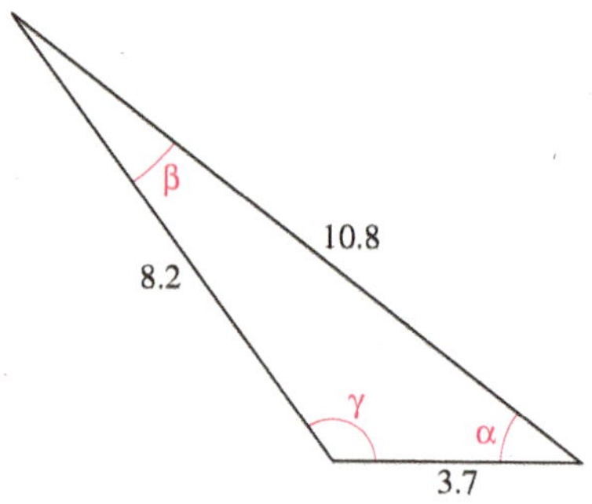


Figure 7.16

## EXAMPLE 1 Given three sides of a triangle (SSS)

Given $a = 8.2$, $b = 3.7$, and $c = 10.8$, solve the triangle.

### Solution

Draw the triangle and label it as in Fig. 7.16. Since $c$ is the longest side, use

$$c^2 = a^2 + b^2 - 2ab \cos \gamma$$

to find the largest angle $\gamma$:

$$-2ab \cos \gamma = c^2 - a^2 - b^2$$

$$\cos \gamma = \frac{c^2 - a^2 - b^2}{-2ab} = \frac{(10.8)^2 - (8.2)^2 - (3.7)^2}{-2(8.2)(3.7)} \approx -0.5885$$

$$\gamma = \cos^{-1}(-0.5885) \approx 126.1°$$

We could finish with the law of cosines, but in this case it is simpler to use the law of sines, which involves fewer computations:

$$\frac{\sin \beta}{b} = \frac{\sin \gamma}{c}$$

$$\frac{\sin \beta}{3.7} = \frac{\sin 126.1°}{10.8}$$

$$\sin \beta \approx 0.2768$$

There are two solutions to $\sin \beta = 0.2768$ in $[0°, 180°]$. However, $\beta$ must be less than 90°, because $\gamma$ is 126.1°. So $\beta = \sin^{-1}(0.2768) \approx 16.1°$. Finally, $\alpha = 180° - 126.1° - 16.1° = 37.8°$.

►**TRY THIS.** Given $a = 3.8$, $b = 9.6$, and $c = 7.7$, solve the triangle. ■

In solving the SSS case, we always *find the largest angle first,* using the law of cosines. The remaining two angles must be acute angles. So when using the law of sines to find another angle, we need only find an acute solution to the equation.

In Section 7.1 we learned that two adjacent sides and a nonincluded angle (SSA) might not determine a triangle. To determine a triangle with three given sides (SSS), the sum of the lengths of any two must be greater than the length of the third side. This fact is called the **triangle inequality.** To understand the triangle inequality, try to draw a triangle with sides of lengths 1 in., 2 in., and 5 in. Two sides and the included angle (SAS) will determine a triangle provided the angle is between 0° and 180°.

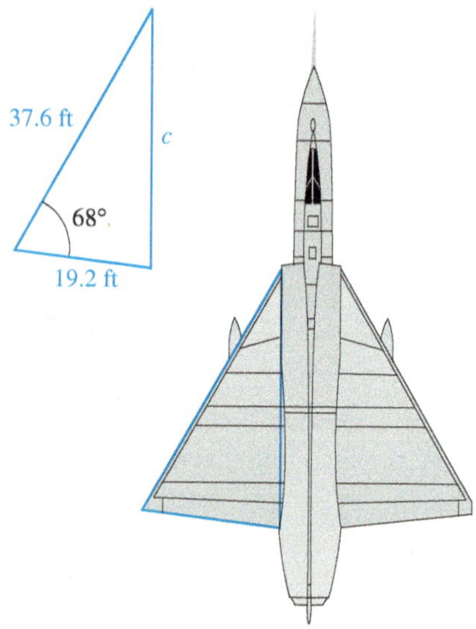


Figure 7.17

## EXAMPLE 2 Given two sides and an included angle (SAS)

The wing of the F-106 Delta Dart is triangular in shape, with the dimensions given in Fig. 7.17. Find the length of the side labeled $c$ in Fig. 7.17.

### Solution

Using the law of cosines, we find $c$ as follows:

$$c^2 = (19.2)^2 + (37.6)^2 - 2(19.2)(37.6)\cos 68^\circ \approx 1241.5$$

$$c \approx \sqrt{1241.5} \approx 35.2 \text{ feet}$$

▶**TRY THIS.** Given $b = 5.8$, $c = 3.6$, and $\alpha = 39.5^\circ$, find $a$. ■

**PROCEDURE**

**Solving Triangles (In all cases draw pictures.)**

**ASA** (For example $\alpha$, $c$, $\beta$)

1. Find $\gamma$ using $\gamma = 180^\circ - \alpha - \beta$.
2. Find $a$ and $c$ using the law of sines.

**SSA** (For example $a$, $b$, $\alpha$)

1. Find $h = b \sin \alpha$. If $h > a$, then there is no triangle.
2. If $h = a$, then there is one right triangle ($\beta = 90^\circ$ and $b$ is the hypotenuse). Solve it using right triangle trigonometry.
3. If $h < a < b$ there are two triangles, one with $\beta$ acute and one with $\beta$ obtuse. Find the acute $\beta$ using the law of sines. Subtract it from $180^\circ$ to get the obtuse $\beta$. In each of the two triangles, find $\gamma$ using $\gamma = 180^\circ - \alpha - \beta$ and $c$ using the law of sines.
4. If $a \geq b$ there is only one triangle and $\beta$ is acute ($\alpha$ or $\gamma$ might be obtuse). Find $\beta$ using the law of sines. Then find $\gamma$ using $\gamma = 180^\circ - \alpha - \beta$. Find $c$ using the law of sines.

**SSS** (For example $a$, $b$, $c$)

1. Find the largest angle using the law of cosines. The largest angle is opposite the largest side.
2. Find another angle using the law of sines.
3. Find the third angle by subtracting the first two from $180^\circ$.

**SAS** (For example $b$, $\alpha$, $c$)

1. Find $a$ using the law of cosines.
2. Use the law of sines to find $\beta$ if $b < c$ or $\gamma$ if $c < b$. If $b = c$, then find either $\beta$ or $\gamma$.
3. Find the last angle by subtracting the first two from $180^\circ$.

## Area of a Triangle by Heron's Formula

In Section 7.1 we saw the formulas $A = \frac{1}{2}bh$ and $A = \frac{1}{2}bc \sin \alpha$ for the area of a triangle. For one formula we must know a side and the altitude to that side, and for the other we need two sides and the measure of their included angle. Using the law of cosines, we can get a formula for the area of a triangle that involves only the lengths of the sides. This formula is known as **Heron's area formula,** named after Heron of Alexandria, who is believed to have discovered it in about A.D. 75.

**Heron's Area Formula**

The area of a triangle with sides $a$, $b$, and $c$ is given by the formula

$$A = \sqrt{S(S-a)(S-b)(S-c)},$$

where $S = (a + b + c)/2$.

**PROOF** First rewrite the equation $a^2 = b^2 + c^2 - 2bc\cos\alpha$ as follows:

$$2bc\cos\alpha = b^2 + c^2 - a^2$$
$$4b^2c^2\cos^2\alpha = (b^2 + c^2 - a^2)^2$$

Now write the area formula $A = \frac{1}{2}bc\sin\alpha$ in terms of $4b^2c^2\cos\alpha$:

$$A = \frac{1}{2}bc\sin\alpha$$
$$4A = 2bc\sin\alpha \qquad \text{Multiply each side by 4.}$$
$$16A^2 = 4b^2c^2\sin^2\alpha \qquad \text{Square both sides.}$$
$$16A^2 = 4b^2c^2(1 - \cos^2\alpha) \qquad \text{Pythagorean identity}$$
$$16A^2 = 4b^2c^2 - 4b^2c^2\cos^2\alpha \qquad \text{Distributive property}$$

Using the expression for $4b^2c^2\cos^2\alpha$ obtained above, we get

$$16A^2 = 4b^2c^2 - (b^2 + c^2 - a^2)^2.$$

The last equation could be solved for $A$ in terms of the lengths of the three sides, but it would not be Heron's formula. To get Heron's formula, let

$$S = \frac{(a + b + c)}{2} \quad \text{or} \quad 2S = a + b + c.$$

It can be shown (with some effort) that

$$4b^2c^2 - (b^2 + c^2 - a^2)^2 = 2S(2S - 2a)(2S - 2b)(2S - 2c).$$

Using this fact in the above equation yields Heron's formula:

$$16A^2 = 16S(S - a)(S - b)(S - c)$$
$$A^2 = S(S - a)(S - b)(S - c)$$
$$A = \sqrt{S(S - a)(S - b)(S - c)}$$

■

## EXAMPLE 3 Area of a triangle using only the sides

A piece of property in downtown Houston is advertised for sale at \$45 per square foot. If the lengths of the sides of the triangular lot are 220 feet, 234 feet, and 160 feet, then what is the asking price for the lot?

### Solution

Using Heron's formula with $S = \frac{220 + 234 + 160}{2} = 307$, we get

$$A = \sqrt{307(307 - 220)(307 - 234)(307 - 160)} \approx 16{,}929.69 \text{ square feet.}$$

At \$45 per square foot, the asking price is \$761,836.

▸**TRY THIS.** Given $a = 12$, $b = 8$, and $c = 6$, find the area of the triangle. ■

## Applications

In the next example we use the law of cosines to find a formula for the length of a chord in terms of the radius of the circle and the central angle subtended by the chord.

## EXAMPLE 4 A formula for the length of a chord

A central angle $\alpha$ in a circle of radius $r$ intercepts a chord of length $a$. Write a formula for $a$ in terms of $\alpha$ and $r$.

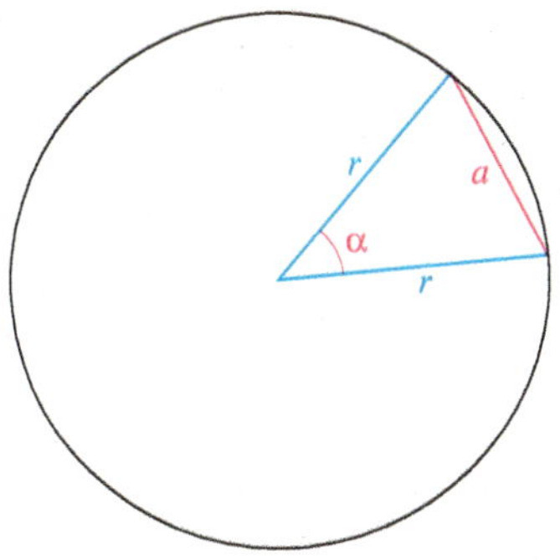


Figure 7.18

### Solution

We can apply the law of cosines to the triangle formed by the chord and the radii at the endpoints of the chord, as shown in Fig. 7.18.

$$a^2 = r^2 + r^2 - 2r^2 \cos \alpha$$
$$= r^2(2 - 2 \cos \alpha)$$
$$a = r\sqrt{2 - 2 \cos \alpha}$$

►**TRY THIS.** Find the length of the chord intercepted by a central angle of 33.8° in a circle of radius 22.4 feet.

When you reach for a light switch, your brain controls the angles at your elbow and your shoulder that put your hand at the location of the switch. An arm on a robot works in much the same manner. In the next example we see how the law of sines and the law of cosines are used to determine the proper angles at the joints so that a robot's hand can be moved to a position given in the coordinates of the workspace.

## EXAMPLE 5 Positioning a robotic arm

A robotic arm with a 0.5-meter segment and a 0.3-meter segment is attached at the origin, as shown in Fig. 7.19. The computer-controlled arm is positioned by rotating each segment through angles $\theta_1$ and $\theta_2$, as shown in Fig. 7.19. Given that we want to have the end of the arm at the point $(0.7, 0.2)$, find $\theta_1$ and $\theta_2$ to the nearest tenth of a degree.

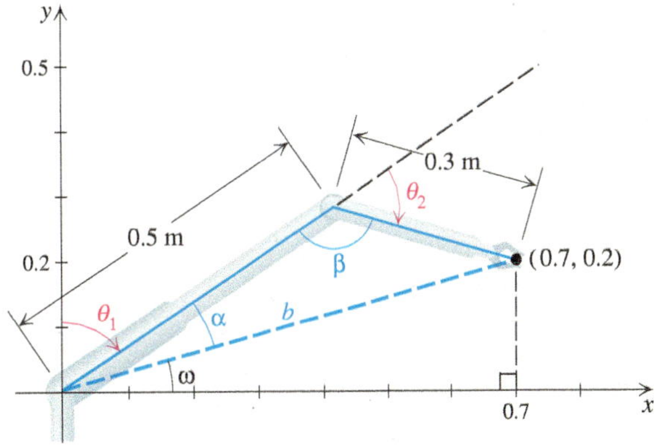


Figure 7.19

### Solution

In the right triangle shown in Fig. 7.19, we have $\tan \omega = 0.2/0.7$ and

$$\omega = \tan^{-1}\left(\frac{0.2}{0.7}\right) \approx 15.95°.$$

To get the angles to the nearest tenth, we use more accuracy along the way and round to the nearest tenth only on the final answer. Find $b$ using the Pythagorean theorem:

$$b = \sqrt{0.7^2 + 0.2^2} \approx 0.73$$

Find $\beta$ using the law of cosines:

$$(0.73)^2 = 0.5^2 + 0.3^2 - 2(0.5)(0.3) \cos \beta$$
$$\cos \beta = \frac{0.73^2 - 0.5^2 - 0.3^2}{-2(0.5)(0.3)} = -0.643$$
$$\beta = \cos^{-1}(-0.643) \approx 130.02°$$

Find $\alpha$ using the law of sines:

$$\frac{\sin \alpha}{0.3} = \frac{\sin 130.02°}{0.73}$$

$$\sin \alpha = \frac{0.3 \cdot \sin 130.02°}{0.73} = 0.3147$$

$$\alpha = \sin^{-1}(0.3147) \approx 18.34°$$

Since $\theta_1 = 90° - \alpha - \omega$,

$$\theta_1 = 90° - 18.34° - 15.95° = 55.71°.$$

Since $\beta$ and $\theta_2$ are supplementary,

$$\theta_2 = 180° - 130.02° = 49.98°.$$

So the longer segment of the arm is rotated 55.7° and the shorter segment is rotated 50.0°.

▸**TRY THIS.** Solve the triangle that has vertices $A(0, 0)$, $B(1, 5)$, and $C(8, 3)$. ■

Because the angles in Example 5 describe a clockwise rotation, the angles could be given negative signs to indicate the direction of rotation. In robotics, the direction of rotation is important, because there may be more than one way to position a robotic arm at a desired location. In fact, Example 5 has infinitely many solutions. See if you can find another one.

## FOR thought... True or False? Explain.

1. If $\gamma = 90°$ in triangle $ABC$, then $c^2 = a^2 + b^2$.
2. If $a$, $b$, and $c$ are the sides of a triangle, then $a = \sqrt{c^2 + b^2 - 2bc \cos \gamma}$.
3. If $a$, $b$, and $c$ are the sides of any triangle, then $c^2 = a^2 + b^2$.
4. The smallest angle of a triangle lies opposite the shortest side.
5. The equation $\cos \alpha = -0.3421$ has two solutions in $[0°, 180°]$.
6. If the largest angle of a triangle is obtuse, then the other two are acute.
7. The equation $\sin \beta = 0.1235$ has two solutions in $[0°, 180°]$.
8. In the SSS case of solving a triangle it is best to find the largest angle first.
9. There is no triangle with sides $a = 3.4$, $b = 4.2$, and $c = 8.1$.
10. There is no triangle with $\alpha = 179.9°$, $b = 1$ inch, and $c = 1$ mile.

## EXERCISES 7.2

*Fill in the blank.*

1. The ________ gives a formula for the square of any side of an oblique triangle in terms of the other two sides and their included angle.
2. According to the ________, the sum of the lengths of any two sides of a triangle is greater than the length of the third side.

*Solve each triangle.*

3.

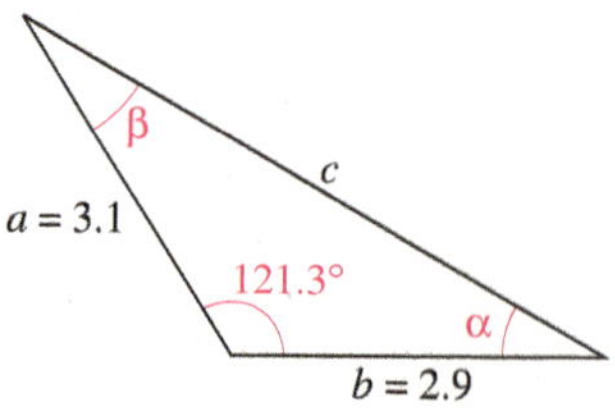


4.

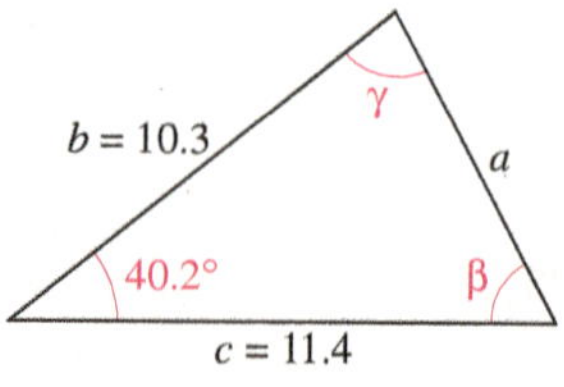

**5.**

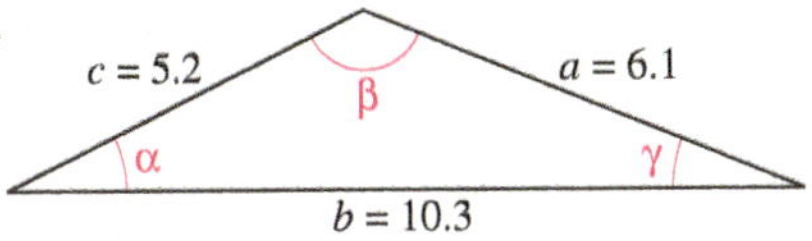


**6.**

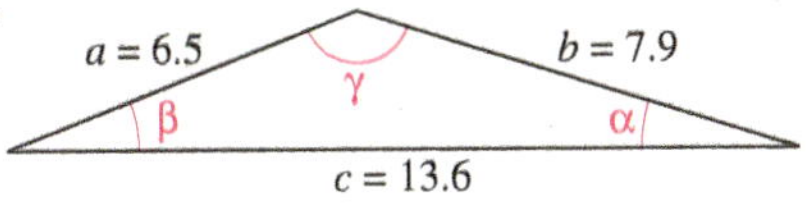


*Solve each triangle with the given information. See the procedure for solving triangles on page 501.*

**7.** $a = 6.8, c = 2.4, \beta = 10.5°$

**8.** $a = 1.3, b = 14.9, \gamma = 9.8°$

**9.** $a = 18.5, b = 12.2, c = 8.1$

**10.** $a = 30.4, b = 28.9, c = 31.6$

**11.** $b = 9.3, c = 12.2, \alpha = 30°$

**12.** $a = 10.3, c = 8.4, \beta = 88°$

**13.** $a = 6.3, b = 7.1, c = 6.8$

**14.** $a = 4.1, b = 9.8, c = 6.2$

**15.** $a = 7.2, \beta = 25°, \gamma = 35°$

**16.** $b = 12.3, \alpha = 20°, \gamma = 120°$

*Determine the number of triangles with the given parts.*

**17.** $a = 3, b = 4, c = 7$

**18.** $a = 2, b = 9, c = 5$

**19.** $a = 10, b = 5, c = 8$

**20.** $a = 3, b = 15, c = 16$

**21.** $c = 10, \alpha = 40°, \beta = 60°, \gamma = 90°$

**22.** $b = 6, \alpha = 62°, \gamma = 120°$

**23.** $b = 10, c = 1, \alpha = 179°$

**24.** $a = 10, c = 4, \beta = 2°$

**25.** $b = 8, c = 2, \gamma = 45°$

**26.** $a = \sqrt{3}/2, b = 1, \alpha = 60°$

*Find the area of each triangle using Heron's formula.*

**27.** $a = 16, b = 9, c = 10$

**28.** $a = 12, b = 8, c = 17$

**29.** $a = 3.6, b = 9.8, c = 8.1$

**30.** $a = 5.4, b = 8.2, c = 12.0$

**31.** $a = 346, b = 234, c = 422$

**32.** $a = 124.8, b = 86.4, c = 154.2$

*Use the most appropriate formula for the area of a triangle to find the area of each triangle.*

**33.**

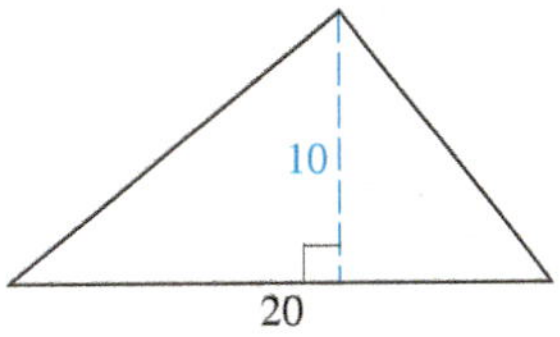


**34.**

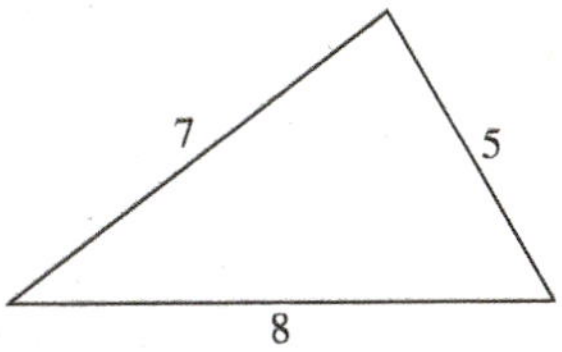


**35.**

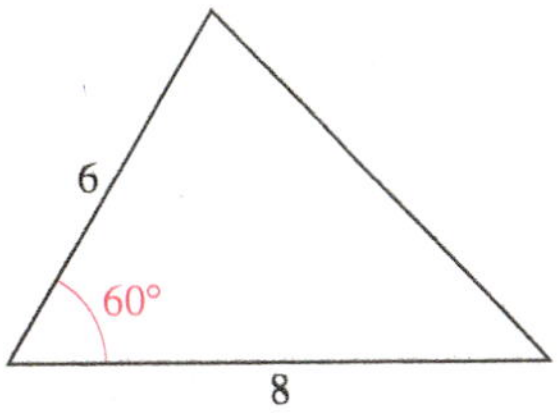


**36.**

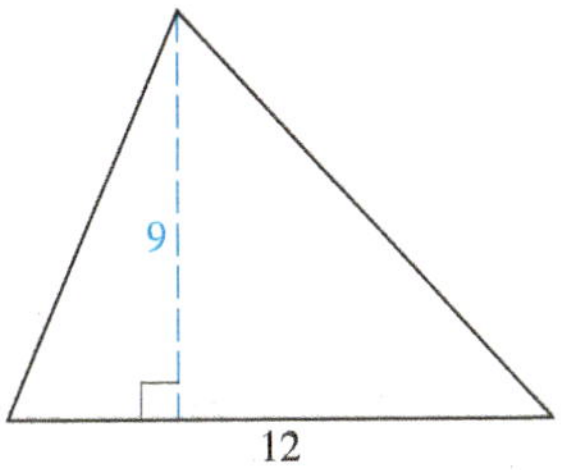


**37.**

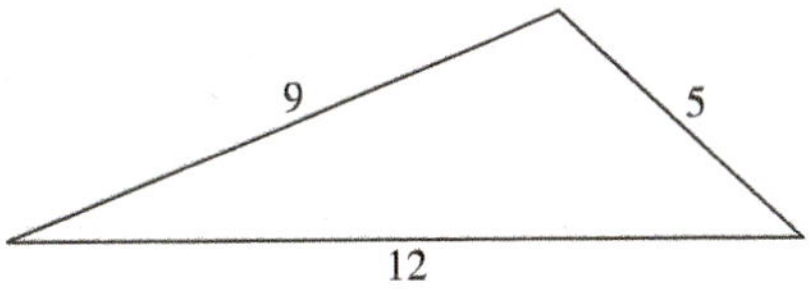


**38.**

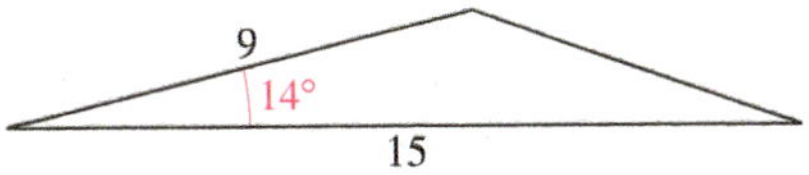


*Solve each problem.*

**39.** *Length of a Chord* What is the length of the chord intercepted by a central angle of 19° in a circle of radius 30 ft?

**40.** *Boating* The boat shown in the accompanying figure is 3 mi from both lighthouses and the angle between the line of sight to the lighthouses is 20°. Find the distance between the lighthouses. **HINT** The distance is the length of a chord.

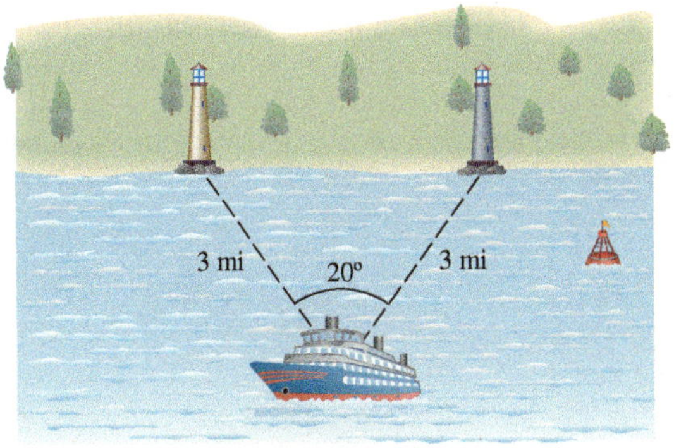


**Figure for Exercise 40**

**41.** *The Pentagon* The Pentagon in Washington D.C. is 921 ft on each side, as shown in the accompanying figure. What is the distance from a vertex to the center of the Pentagon?

**HINT** The sides of the Pentagon can be viewed as chords in a circle.

**Figure for Exercise 41**

**42.** *A Hexagon* If the length of each side of a regular hexagon is 10 ft, then what is the distance from a vertex to the center?

**43.** *Hiking* Jan and Dean started hiking from the same location at the same time. Jan hiked at 4 mph with bearing N12°E, and Dean hiked at 5 mph with bearing N31°W. How far apart were they after 6 hr?

**44.** *Flying* Andrea and Carlos left the airport at the same time. Andrea flew at 180 mph on a course with bearing 80°, and Carlos flew at 240 mph on a course with bearing 210°. How far apart were they after 3 hr?

**45.** *Positioning a Solar Panel* A solar panel with a width of 1.2 m is positioned on a flat roof, as shown in the figure. What is the angle of elevation $\alpha$ of the solar panel?

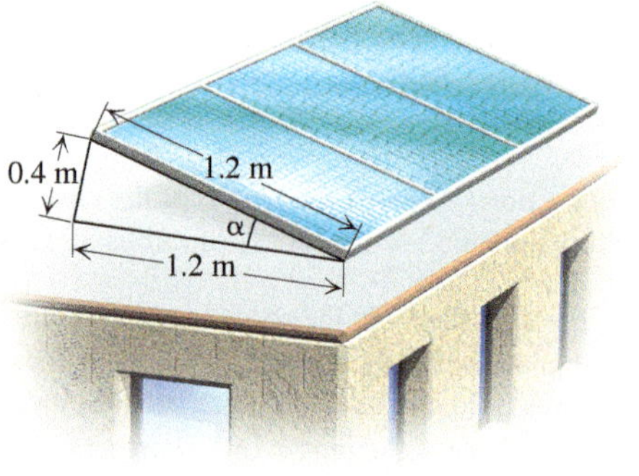


**Figure for Exercise 45**

**46.** *Installing an Antenna* A 6-ft antenna is installed at the top of a roof, as shown in the figure. A guy wire is to be attached to the top of the antenna and to a point 10 ft down the roof. If the angle of elevation of the roof is 28°, then what length guy wire is needed?

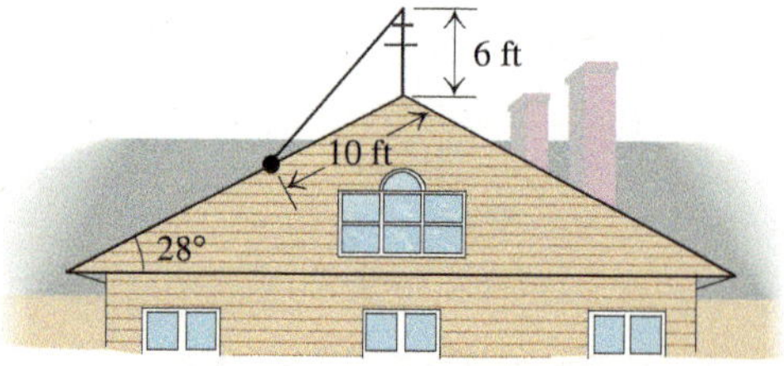


**Figure for Exercise 46**

**47.** *Adjacent Pipes* An engineer wants to position three pipes at the vertices of a triangle, as shown in the figure. If the pipes $A$, $B$, and $C$ have radii 2 in., 3 in., and 4 in., respectively, then what are the measures of the angles of the triangle $ABC$?

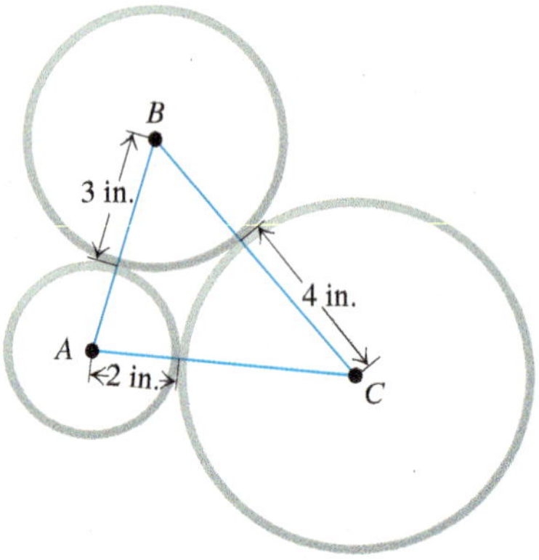


**Figure for Exercise 47**

**48.** *Firing a Torpedo* A submarine sights a moving target at a distance of 820 m. A torpedo is fired 9° ahead of the target, as shown in the drawing, and travels 924 m in a straight line to hit the target. How far has the target moved from the time the torpedo is fired to the time of the hit?

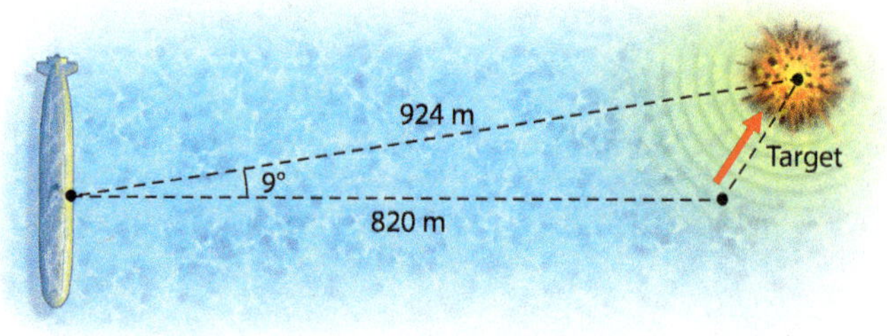


**Figure for Exercise 48**

**49.** *Planning a Tunnel* A tunnel is planned through a mountain to connect points $A$ and $B$ on two existing roads, as shown in the figure. If the angle between the roads at point $C$ is 28°, what is the distance from point $A$ to $B$? Find $\angle CBA$ and $\angle CAB$ to the nearest tenth of a degree.

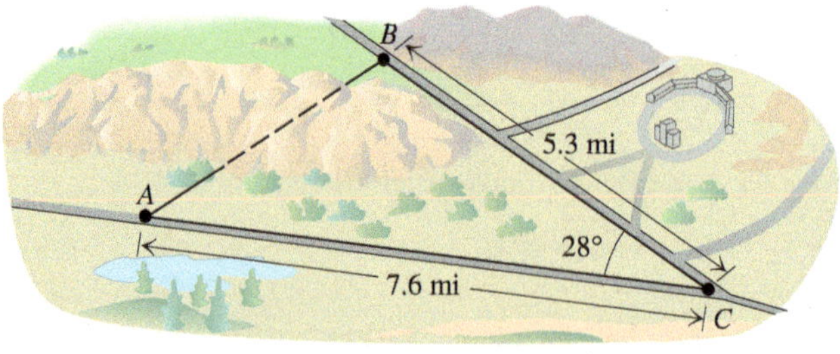


**Figure for Exercise 49**

**50.** *Scattering Angle* On June 30, 1861, Comet Tebutt, one of the greatest comets, was visible even before sunset. One of the factors that causes a comet to be extra bright is a small *scattering angle* $\theta$, shown in the accompanying figure. When Comet Tebutt was at its brightest, it was 0.133 a.u. from the earth, 0.894 a.u. from the sun, and the earth was 1.017 a.u. from the sun. Find the *phase angle* $\alpha$ and the scattering angle $\theta$ for Comet Tebutt on June 30, 1861. (One astronomical unit (a.u.) is the average distance between the earth and the sun.)

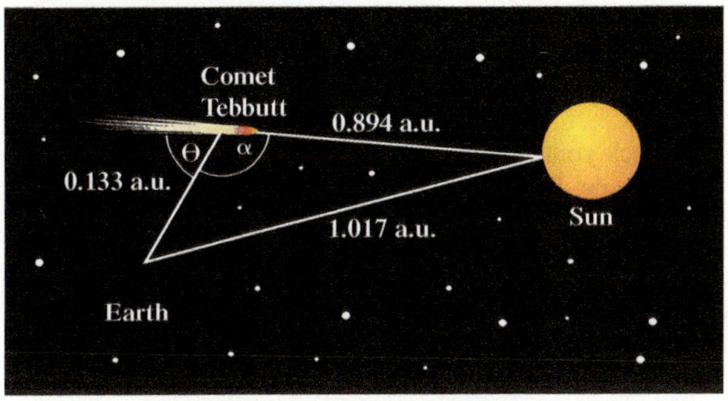


**Figure for Exercise 50**

**51.** *Side of a Pentagon* A regular pentagon is inscribed in a circle of radius 10 m. Find the length of a side of the pentagon.

**52.** *Central Angle* A central angle $\alpha$ in a circle of radius 5 m intercepts a chord of length 1 m. What is the measure of $\alpha$?

**53.** *Positioning a Human Arm* A human arm consists of an upper arm of 30 cm and a lower arm of 30 cm, as shown in the figure. To move the hand to the point $(36, 8)$, the human brain chooses angle $\theta_1$ and $\theta_2$, as shown in the figure. Find $\theta_1$ and $\theta_2$ to the nearest tenth of a degree.

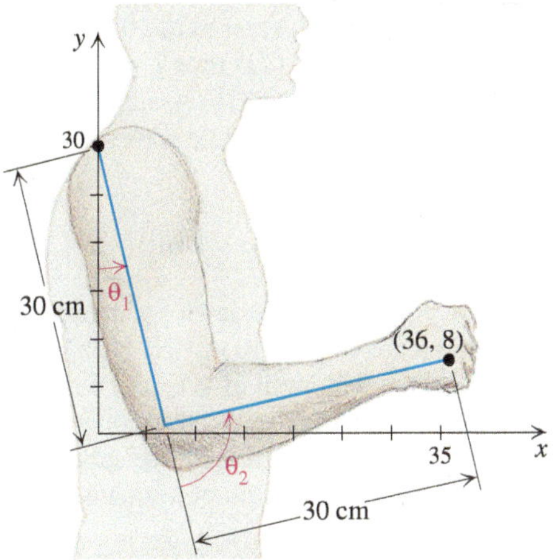


**Figure for Exercise 53**

**54.** *Making a Shaft* The end of a steel shaft for an electric motor is to be machined so that it has three flat sides of equal widths, as shown in the figure. Given that the radius of the shaft is 10 mm and the length of the arc between each pair of flat sides must be 2.5 mm, find the width $s$ of each flat side.

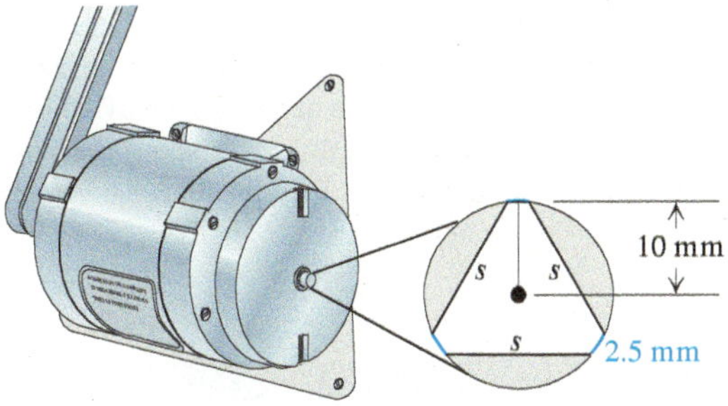


**Figure for Exercise 54**

**55.** *Total Eclipse* Astronomers refer to the angles alpha and beta ($\alpha$ and $\beta$) in the accompanying figure as the "diameters" of the sun and the moon, respectively. These angles/diameters vary as the sun, moon, and earth travel in their orbits. When the moon moves between the earth and the sun, a total eclipse occurs, provided the moon's apparent diameter is larger than that of the sun, or if beta > alpha. The actual diameters of the sun and moon are 865,000 miles and 2163 miles, respectively, and those distances do not vary.

**a.** The distance from the earth to the sun varies from 91,400,000 miles to 94,500,000 miles. Find the minimum and maximum $\alpha$ to the nearest hundredth of a degree.
HINT Use the law of cosines.

**b.** The distance from the earth to the moon varies from 225,800 miles to 252,000 miles. Find the minimum and maximum $\beta$ to the nearest hundredth of a degree.
HINT Use the law of cosines.

**c.** Is it possible for the earth, the moon, and the sun to be in perfect alignment without a total eclipse occurring?

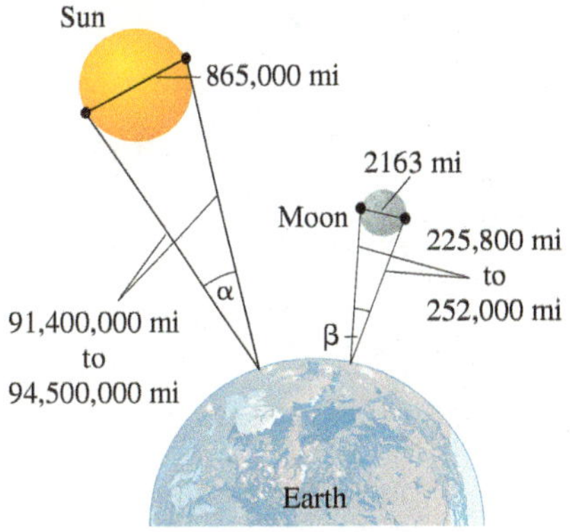


**Figure for Exercise 55**

**56.** *Total Eclipse on Jupiter* The distance from Jupiter to the sun varies from $7.406 \times 10^8$ km to $8.160 \times 10^8$ km. The diameter of the sun is $1.39 \times 10^6$ km.

**a.** Determine the minimum and maximum diameter of the sun to the nearest hundredth of a degree as seen from Jupiter.
HINT See the previous exercise and use the law of cosines.

**b.** The distance from Jupiter to its moon Callisto is $1.884 \times 10^6$ km and Callisto's diameter is 2420 km. Determine the diameter of Callisto as seen from Jupiter.
HINT Use the law of cosines.

**c.** Is it possible for Callisto to produce a total eclipse of the sun on Jupiter?

**57.** *Attack of the Grizzly* A forest ranger is 150 ft above the ground in a fire tower when she spots an angry grizzly bear east of the tower with an angle of depression of 10°, as shown in the figure. Southeast of the tower she spots a hiker with an angle of depression of 15°. Find the distance between the hiker and the angry bear.

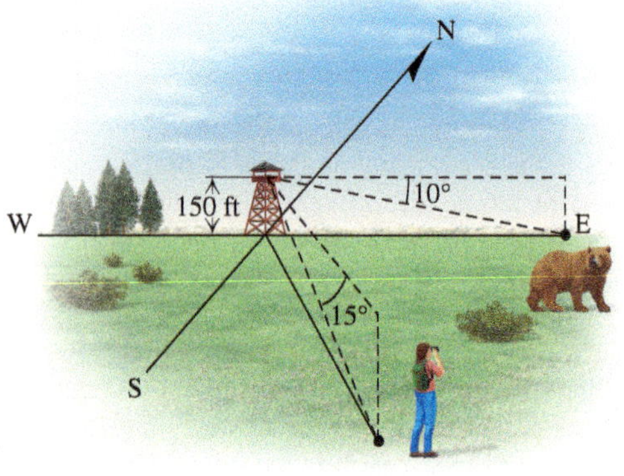


**Figure for Exercise 57**

**58.** *Smuggler's Blues* A smuggler's boat sets out at midnight at 20 mph on a course that makes a 40° angle with the shore, as shown in the figure. One hour later a DEA boat sets out also at 20 mph from a dock 80 miles up the coast to intercept the smuggler's boat. Find the angle $\theta$ (in the figure) for which the DEA boat will intercept the smuggler's boat. At what time will the interception occur?

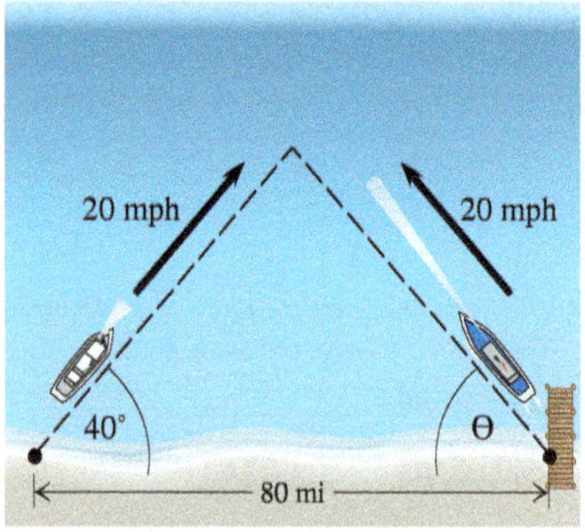


**Figure for Exercise 58**

## FOR WRITING/DISCUSSION

**59.** A central angle $\theta$ in a circle of radius $r$ intercepts a chord of length $a$, where $0° \leq \theta \leq 180°$. Show that $a = 2r\sin(\theta/2)$.

**60.** Explain why the second largest side in a triangle with unequal sides is opposite an acute angle.

**61.** Explain why the Pythagorean theorem is a special case of the law of cosines.

**62.** Find the area of the triangle with sides 37, 48, and 86, using Heron's formula. Explain your result.

**63.** Find the area of the triangle with sides 31, 87, and 56, using Heron's formula. Explain your result.

**64.** Find the area of the triangle with sides of length 6 ft, 9 ft, and 13 ft by using the formula

$$A = \frac{1}{4}\sqrt{4b^2c^2 - (b^2 + c^2 - a^2)^2},$$

and check your result using a different formula for the area of a triangle. Prove that this formula gives the area of any triangle with sides $a$, $b$, and $c$.

## RETHINKING

**65.** Solve the triangle with $\alpha = 108.1°$, $\beta = 18.6°$, and $c = 28.6$.

**66.** Determine the number of triangles with $\alpha = 22.5°$, $a = 5.1$, and $b = 12.6$.

**67.** Find all solutions to $(\sin x - 1)(\sin x + 1) = 0$ in the interval $(0, 2\pi)$.

**68.** Find all solutions to $\sin(2x)\cos(2x) = 0$ in the interval $(0, 2\pi)$.

**69.** State the domain and range for each function.

**a.** $f(x) = \sin^{-1}(x)$

**b.** $f(x) = \arccos(x)$

**c.** $f(x) = \tan^{-1}(x)$

**70.** In a controlled experiment, the temperature is 0° C at time $t = 0$. The temperature is increased to 10° C at time $t = 4$ and then decreased to $-10°$ C at time $t = 12$. The temperature returns to 0° C at time $t = 16$. Assuming the temperature on the time interval $[0, 16]$ is a sine wave, write the temperature $y$ as a function of the time $t$.

## THINKING OUTSIDE THE BOX LIII

*Watering the Lawn* Josie places her lawn sprinklers at the vertices of a triangle that has sides of 9 m, 10 m, and 11 m. The sprinklers water in circular patterns with radii of 4, 5, and 6 m. No area is watered by more than one sprinkler. What amount of area inside the triangle is not watered by any of the three sprinklers? Give your answer to the nearest thousandth of a square meter.

## POP QUIZ 7.2

**1.** If $\alpha = 12.3°$, $b = 10.4$, and $c = 8.1$, then what is $a$?

**2.** If $a = 6$, $b = 7$, and $c = 12$, then what is $\gamma$?

**3.** Find the exact area of the triangle with sides of lengths 7, 8, and 9 using Heron's formula.

LINKING concepts... For Individual or Group Explorations

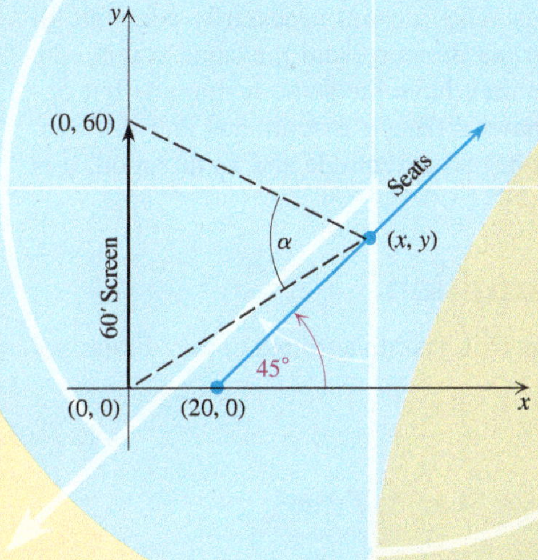


## Modeling the Best View

*A large-screen theater has a screen that is 60 feet high, as shown in the figure. The seats are placed on a 45° incline so that the seats are close to the screen.*

**a)** Find the distance to the top of the screen and to the bottom of the screen for a person sitting at coordinates (30, 10).

**b)** Find the viewing angle $\alpha$ for a person sitting at (30, 10).

**c)** Write the viewing angle $\alpha$ as a function of $x$ and graph the function using a graphing calculator.

**d)** If the good seats are those for which the viewing angle is greater than 60°, then what are the $x$-coordinates of the good seats?

**e)** If the best seat has the largest viewing angle, then what are the coordinates of the best seat?

# 7.3 Vectors

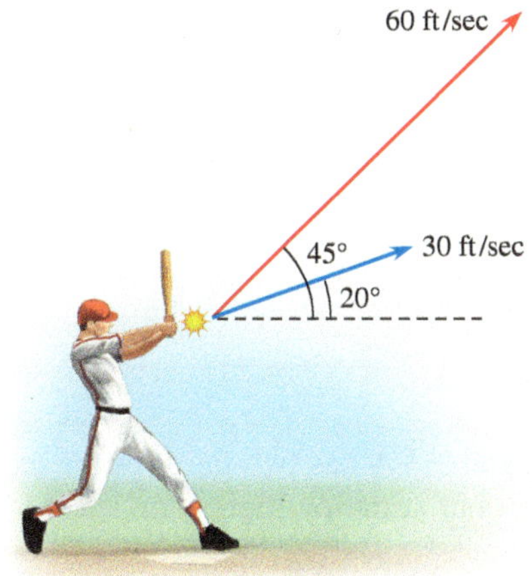


**Figure 7.20**

By using the law of sines and the law of cosines, we can find all of the missing parts of any triangle for which we have enough information to determine its shape. In this section we will use these and many other tools of trigonometry in the study of vectors.

## Definitions

Quantities such as length, area, volume, temperature, and time have magnitude only and are completely characterized by a single real number with appropriate units (such as feet, degrees, or hours). Such quantities are called **scalar quantities,** and the corresponding real numbers are **scalars.** Quantities that involve both a magnitude and a direction, such as velocity, acceleration, and force, are **vector quantities,** and they can be represented by **directed line segments.** These directed line segments are called **vectors.**

For an example of vectors, consider two baseballs hit into the air, as shown in Fig. 7.20. One is hit with an initial velocity of 30 feet per second at an angle of 20° from the horizontal, and the other is hit with an initial velocity of 60 feet per second at an angle of 45° from the horizontal. The length of a vector represents the **magnitude** of the vector quantity. So the vector representing the faster baseball is drawn twice as long as the other vector. The **direction** is indicated by the position of the vector and the arrowhead at one end.

We have used the notation $\overline{AB}$ to name a line segment with endpoints $A$ and $B$ and $\overrightarrow{AB}$ to name a ray with initial point $A$ and passing through $B$. When studying vectors, the notation $\overrightarrow{AB}$ is used to name a vector with **initial point** $A$ and **terminal point** $B$. The vector $\overrightarrow{AB}$ terminates at $B$, while the ray $\overrightarrow{AB}$ goes beyond $B$. Since the direction of the vector is the same as the order of the letters, to indicate a vector in print we will omit the arrow and use boldface type. In handwritten work, an arrow over the letter or letters indicates a vector. Thus **AB** and $\overrightarrow{AB}$ represent the same vector. The vector **BA** is a vector with initial point $B$ and terminal point $A$, and is not the

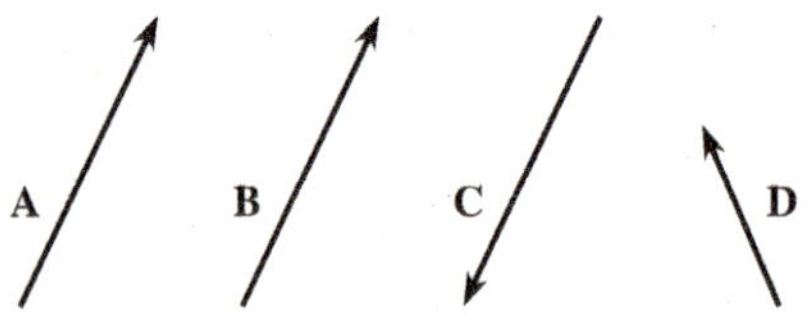


Figure 7.21

same as **AB.** A vector whose endpoints are not specified is named a single uppercase or lowercase letter. For example, **b, B,** $\vec{b}$, and $\vec{B}$ are names of vectors. The magnitude of vector **A** is written as $|\mathbf{A}|$.

Two vectors are **equal** if they have the same magnitude and the same direction. Equal vectors may be in different locations. They do not necessarily coincide. In Fig. 7.21, $\mathbf{A} = \mathbf{B}$ because they have the same direction and the same magnitude. Vector **B** is *not* equal to vector **C** because they have the same magnitude but opposite directions. $\mathbf{C} \neq \mathbf{D}$ because they have different magnitudes and different directions. The **zero vector** is a vector that has no magnitude and no direction. It is denoted by a boldface zero, **0.**

## Scalar Multiplication and Addition

Suppose two tugboats are pulling on a barge that has run aground in the Mississippi River, as shown in Fig. 7.22.

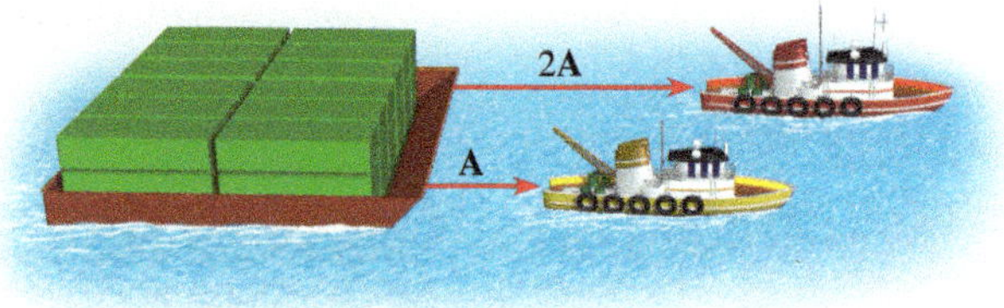


Figure 7.22

The smaller boat exerts a force of 2000 pounds in an easterly direction and its force is represented by the vector **A.** If the larger boat exerts a force of 4000 pounds in the same direction, then its force can be represented by the vector 2**A.** This example illustrates the operation of **scalar multiplication.**

**Definition: Scalar Multiplication**

For any scalar $k$ and vector **A,** $k\mathbf{A}$ is a vector with magnitude $|k|$ times the magnitude of **A.** If $k > 0$, then the direction of $k\mathbf{A}$ is the same as the direction of **A.** If $k < 0$, the direction of $k\mathbf{A}$ is opposite to the direction of **A.** If $k = 0$, then $k\mathbf{A} = \mathbf{0}$.

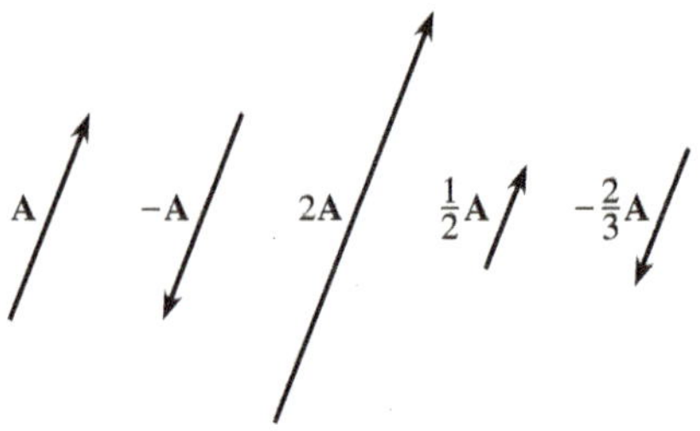


Figure 7.23

The vector 2**A** has twice the magnitude of **A,** and $-\frac{2}{3}\mathbf{A}$ has two-thirds of the magnitude. For $-1\mathbf{A}$, we write $-\mathbf{A}$. Some examples of scalar multiplication are shown in Fig. 7.23.

Suppose two draft horses are pulling on a tree stump with forces of 200 pounds and 300 pounds, as shown in Fig. 7.24, with an angle of 65° between the forces.

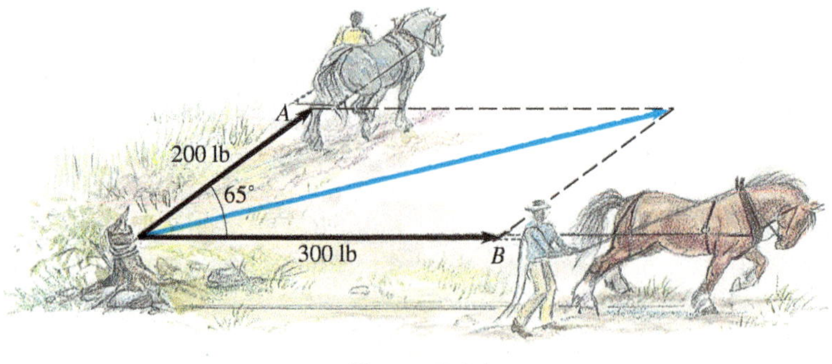


Figure 7.24

The two forces are represented by vectors **A** and **B.** If **A** and **B** had the same direction, then there would be a total force of 500 pounds acting on the stump. However, a total force of 500 pounds is not achieved because of the angle between the forces. It can be shown by experimentation that one force acting along the diagonal of the parallelogram seen in Fig. 7.24, with a magnitude equal to the length of the diagonal, has the same effect on the stump as the two forces **A** and **B.** In physics, this result is known as the **parallelogram law.** The single force **A** + **B** acting along the diagonal is called the **sum** or **resultant** of **A** and **B.** The parallelogram law motivates our definition of vector addition.

**Definition: Vector Addition**

To find the resultant or sum **A** + **B** of any vectors **A** and **B,** position **B** (without changing its magnitude or direction) so that the initial point of **B** coincides with the terminal point of **A,** as in Fig. 7.25. The vector that begins at the initial point of **A** and terminates at the terminal point of **B** is the vector **A** + **B.**

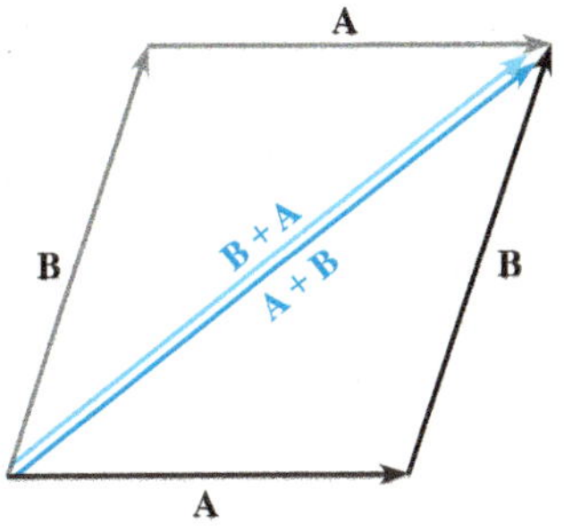

Figure 7.25

Note that the vector **A** + **B** in Fig. 7.25 coincides with the diagonal of a parallelogram whose adjacent sides are **A** and **B.** We can see from Fig. 7.25 that **A** + **B** = **B** + **A,** and vector addition is commutative. If **A** and **B** have the same direction or opposite directions, then no parallelogram is formed, but **A** + **B** can be found by the procedure given in the definition. Each vector in the sum **A** + **B** is called a **component** of the sum.

For every vector **A,** there is a vector −**A,** the **opposite** of **A,** having the same magnitude as **A** but opposite direction. The sum of a vector and its opposite is the zero vector, **A** + (−**A**) = **0.** Vector subtraction is defined like subtraction of real numbers. For any two vectors **A** and **B, A** − **B** = **A** + (−**B**). Fig. 7.26 shows **A** − **B.**

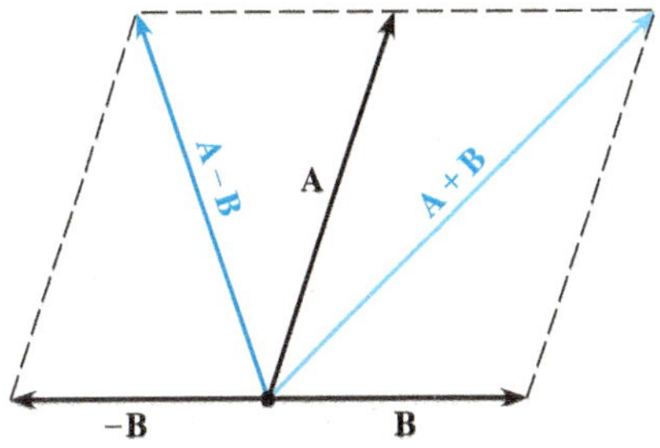

Figure 7.26

The law of sines and the law of cosines can be used to find the magnitude and direction of a resultant vector. Remember that in any parallelogram the opposite sides are equal and parallel, and adjacent angles are supplementary. The diagonals of a parallelogram do not bisect the angles of a parallelogram unless the adjacent sides of the parallelogram are equal in length.

## Horizontal and Vertical Components

Any nonzero vector **w** is the sum of a **vertical component** and a **horizontal component,** as shown in Fig. 7.27. The horizontal component is denoted $\mathbf{w}_x$ and the vertical component is denoted $\mathbf{w}_y$. The vector **w** is the diagonal of the rectangle formed by the vertical and horizontal components.

If a vector **w** is placed in a rectangular coordinate system so that its initial point is the origin (as in Fig. 7.28), then **w** is called a **position vector** or **radius vector.** A position vector is a convenient representative to focus on when considering all

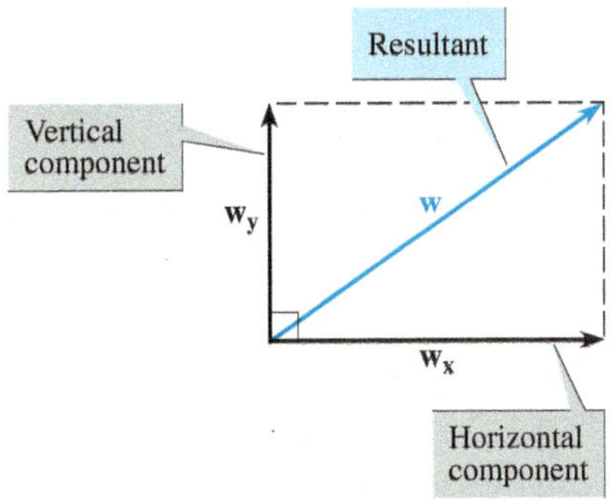

Figure 7.27

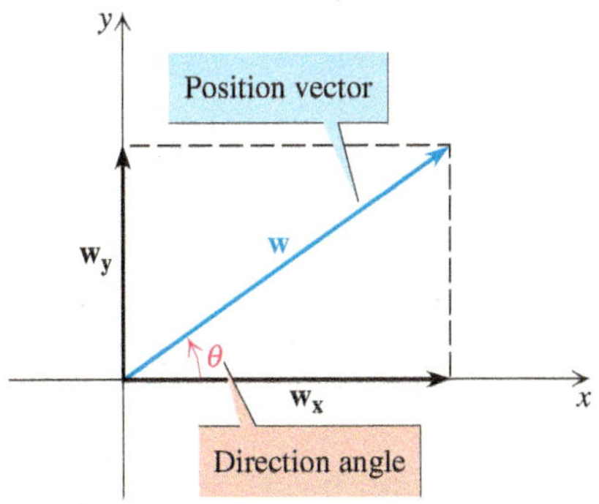

Figure 7.28

vectors that are equal to a certain vector. The angle $\theta$ ($0° \le \theta < 360°$) formed by the positive $x$-axis and a position vector is the **direction angle** for the position vector (or any other vector that is equal to the position vector).

If the vector $\mathbf{w}$ has magnitude $r$, direction angle $\theta$, horizontal component $\mathbf{w_x}$, and vertical component $\mathbf{w_y}$, as shown in Fig. 7.28, then by using trigonometric ratios we get

$$\cos\theta = \frac{|\mathbf{w_x}|}{r} \quad \text{and} \quad \sin\theta = \frac{|\mathbf{w_y}|}{r}$$

or

$$|\mathbf{w_x}| = r\cos\theta \quad \text{and} \quad |\mathbf{w_y}| = r\sin\theta.$$

If the direction of $\mathbf{w}$ is such that $\sin\theta$ or $\cos\theta$ is negative, then we can write

$$|\mathbf{w_x}| = |r\cos\theta| \quad \text{and} \quad |\mathbf{w_y}| = |r\sin\theta|.$$

### EXAMPLE 1 Finding horizontal and vertical components

Find the magnitude of the horizontal and vertical components for a vector $\mathbf{v}$ with magnitude 8.3 and direction angle 121.3°.

#### Solution

The vector $\mathbf{v}$ and its horizontal and vertical components $\mathbf{v_x}$ and $\mathbf{v_y}$ are shown in Fig. 7.29. The magnitudes of $\mathbf{v_x}$ and $\mathbf{v_y}$ are found as follows:

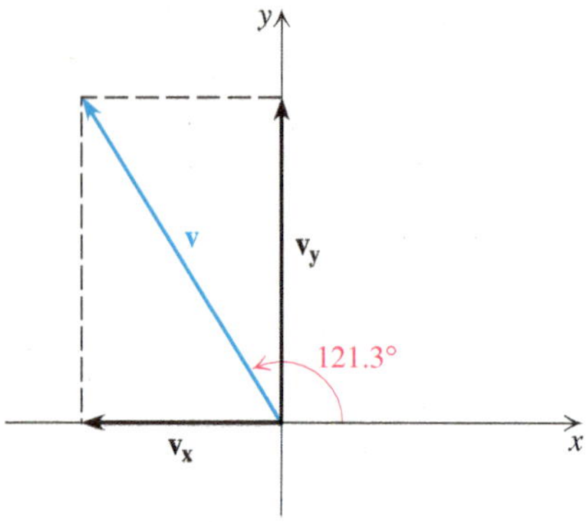


Figure 7.29

$$|\mathbf{v_x}| = |8.3\cos 121.3°| = 4.3$$
$$|\mathbf{v_y}| = |8.3\sin 121.3°| = 7.1$$

The direction angle for $\mathbf{v_x}$ is 180°, and the direction angle for $\mathbf{v_y}$ is 90°.

**▸TRY THIS.** Find the magnitude of the horizontal and vertical components for a vector $\mathbf{v}$ with magnitude 5.6 and direction angle 22°. ■

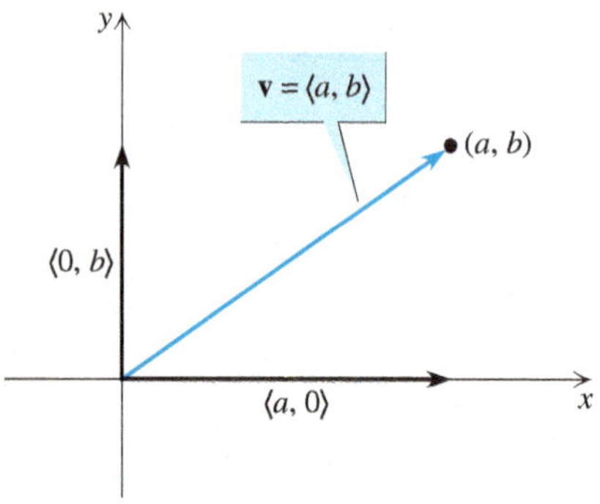


Figure 7.30

## Component Form of a Vector

Any vector is the resultant of its horizontal and vertical components. Since the horizontal and vertical components of a vector determine the vector, it is convenient to use a notation for vectors that involves them. The notation $\langle a, b\rangle$ is used for the position vector with terminal point $(a, b)$, as shown in Fig. 7.30. The form $\langle a, b\rangle$ is called **component form** because its horizontal component is $\langle a, 0\rangle$ and its vertical component is $\langle 0, b\rangle$. Since the vector $\mathbf{v} = \langle a, b\rangle$ extends from (0, 0) to $(a, b)$, its magnitude is the distance between these points:

$$|\mathbf{v}| = \sqrt{a^2 + b^2}$$

In Example 2 we find the magnitude and direction for a vector given in component form, and in Example 3 we find the component form for a vector given as a directed line segment.

### EXAMPLE 2 Finding magnitude and direction from component form

Find the magnitude and direction angle of the vector $\mathbf{v} = \langle -2, -5\rangle$.

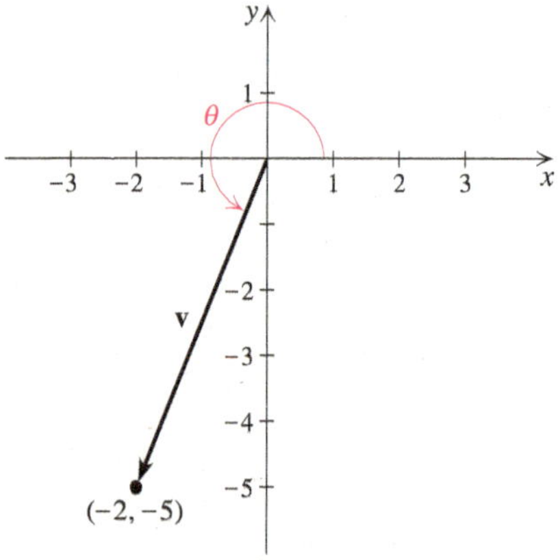


Figure 7.31

#### Solution

The vector $\mathbf{v}$ shown in Fig. 7.31 has magnitude

$$|\mathbf{v}| = \sqrt{(-2)^2 + (-5)^2} = \sqrt{29}.$$

To find the direction angle $\theta$, use trigonometric ratios to get $\sin\theta = -5/\sqrt{29}$. Since $\sin^{-1}(-5/\sqrt{29}) = -68.2°$, we have $\theta = 180° - (-68.2°) = 248.2°$.

►**TRY THIS.** Find the magnitude and direction angle for $\mathbf{v} = \langle 2, -6\rangle$. ■

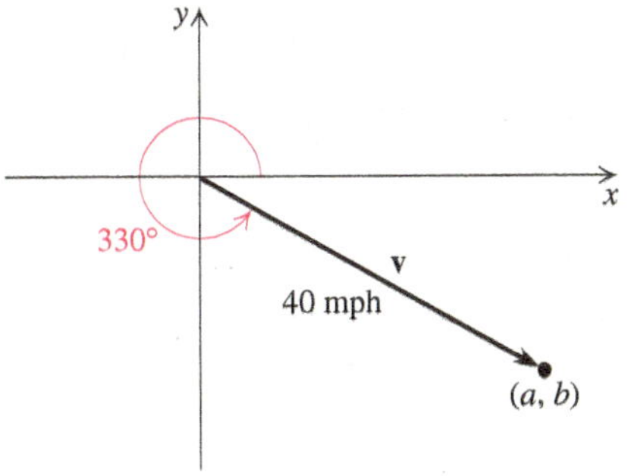


Figure 7.32

## EXAMPLE 3 Finding the component form given magnitude and direction

Find the component form for a vector of magnitude 40 mph with direction angle 330°.

### Solution

To find the component form, we need the terminal point $(a, b)$ for a vector of magnitude 40 mph positioned as shown in Fig. 7.32 Since $a = r\cos\theta$ and $b = r\sin\theta$, we have

$$a = 40\cos 330° = 40\frac{\sqrt{3}}{2} = 20\sqrt{3} \qquad \text{and}$$

$$b = 40\sin 330° = 40\left(-\frac{1}{2}\right) = -20.$$

So the component form of the vector is $\langle 20\sqrt{3}, -20\rangle$.

►**TRY THIS.** Find the component form for a vector of magnitude 50 mph with direction angle 120°. ■

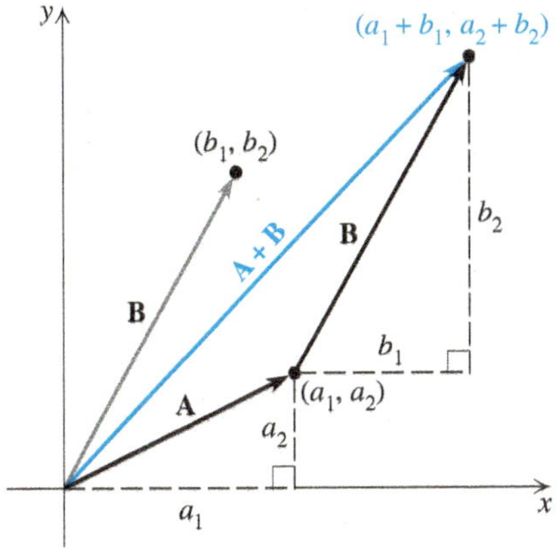


Figure 7.33

We originally defined vectors as directed line segments and performed operations with these directed line segments. When vectors are written in component form, operations with vectors are easier to perform. Figure 7.33 illustrates the addition of $\mathbf{A} = \langle a_1, a_2\rangle$ and $\mathbf{B} = \langle b_1, b_2\rangle$, where $(a_1, a_2)$ and $(b_1, b_2)$ are in the first quadrant. It is easy to see that the endpoint of $\mathbf{A} + \mathbf{B}$ is $(a_1 + b_1, a_2 + b_2)$ and so $\mathbf{A} + \mathbf{B} = \langle a_1 + b_1, a_2 + b_2\rangle$. The sum can be found in component form by adding the components instead of drawing directed line segments. The component forms for the three operations that we have already studied and a new operation called **dot product** are given in the following box.

**Rules for Scalar Product, Vector Sum, Vector Difference, and Dot Product**

If $\mathbf{A} = \langle a_1, a_2\rangle$, $\mathbf{B} = \langle b_1, b_2\rangle$, and $k$ is a scalar, then

1. $k\mathbf{A} = \langle ka_1, ka_2\rangle$ **Scalar product**
2. $\mathbf{A} + \mathbf{B} = \langle a_1 + b_1, a_2 + b_2\rangle$ **Vector sum**
3. $\mathbf{A} - \mathbf{B} = \langle a_1 - b_1, a_2 - b_2\rangle$ **Vector difference**
4. $\mathbf{A}\cdot\mathbf{B} = a_1b_1 + a_2b_2$ **Dot product**

## EXAMPLE 4 Operations with vectors in component form

Let $\mathbf{w} = \langle -3, 2\rangle$ and $\mathbf{z} = \langle 5, -1\rangle$. Perform the operations indicated.

**a.** $\mathbf{w} - \mathbf{z}$ **b.** $-8\mathbf{z}$ **c.** $3\mathbf{w} + 4\mathbf{z}$ **d.** $\mathbf{w}\cdot\mathbf{z}$

### Solution

**a.** $\mathbf{w} - \mathbf{z} = \langle -3, 2\rangle - \langle 5, -1\rangle = \langle -8, 3\rangle$
**b.** $-8\mathbf{z} = -8\langle 5, -1\rangle = \langle -40, 8\rangle$
**c.** $3\mathbf{w} + 4\mathbf{z} = 3\langle -3, 2\rangle + 4\langle 5, -1\rangle = \langle -9, 6\rangle + \langle 20, -4\rangle = \langle 11, 2\rangle$
**d.** $\mathbf{w}\cdot\mathbf{z} = \langle -3, 2\rangle\cdot\langle 5, -1\rangle = -15 + (-2) = -17$

►**TRY THIS.** Let $\mathbf{u} = \langle -1, -3\rangle$ and $\mathbf{v} = \langle 3, -4\rangle$. Find $\mathbf{u} + 3\mathbf{v}$ and $\mathbf{u}\cdot\mathbf{v}$. ■

## The Angle Between Two Vectors

If $\mathbf{A} = k\mathbf{B}$ for a nonzero scalar $k$, then **A** and **B** are **parallel** vectors. If **A** and **B** have the same direction ($k > 0$), the angle between **A** and **B** is 0°. If they have opposite directions ($k < 0$), the angle between them is 180°. If **A** and **B** are nonparallel vectors with the same initial point, then the vectors **A, B,** and $\mathbf{A} - \mathbf{B}$ form a triangle, as in Fig. 7.34. The **angle between the vectors A and B** is the angle $\alpha$ shown in Fig. 7.34. The angle between two vectors is in the interval $[0°, 180°]$. If the angle between **A** and **B** is 90°, then the vectors are **perpendicular** or **orthogonal.**

The following theorem relates the angle between two vectors, the magnitudes of the vectors, and the dot product of the vectors. The theorem is simply a vector version of the law of cosines.

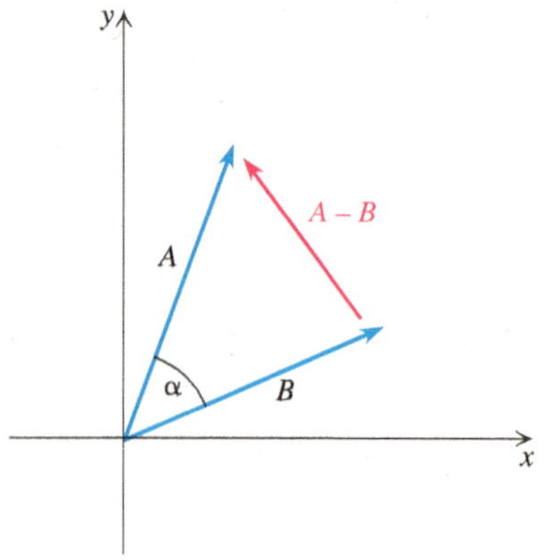


Figure 7.34

**Theorem: Dot Product**

> If **A** and **B** are nonzero vectors and $\alpha$ is the angle between them, then
> $$\cos\alpha = \frac{\mathbf{A}\cdot\mathbf{B}}{|\mathbf{A}||\mathbf{B}|}.$$

**PROOF** Let $\mathbf{A} = \langle a_1, a_2\rangle$, $\mathbf{B} = \langle b_1, b_2\rangle$, and $\mathbf{A} - \mathbf{B} = \langle a_1 - b_1, a_2 - b_2\rangle$. The vectors **A, B,** and $\mathbf{A} - \mathbf{B}$ form a triangle, as shown in Fig. 7.34. Apply the law of cosines to this triangle and simplify as follows.

$$|\mathbf{A}|^2 + |\mathbf{B}|^2 - 2|\mathbf{A}||\mathbf{B}|\cos\alpha = |\mathbf{A} - \mathbf{B}|^2$$
$$(a_1)^2 + (a_2)^2 + (b_1)^2 + (b_2)^2 - 2|\mathbf{A}||\mathbf{B}|\cos\alpha = (a_1 - b_1)^2 + (a_2 - b_2)^2$$
$$-2|\mathbf{A}||\mathbf{B}|\cos\alpha = -2a_1b_1 - 2a_2b_2$$
$$|\mathbf{A}||\mathbf{B}|\cos\alpha = a_1b_1 + a_2b_2$$
$$|\mathbf{A}||\mathbf{B}|\cos\alpha = \mathbf{A}\cdot\mathbf{B}$$
$$\cos\alpha = \frac{\mathbf{A}\cdot\mathbf{B}}{|\mathbf{A}||\mathbf{B}|}$$
■

Note that $\cos\alpha = 0$ if and only if $\mathbf{A}\cdot\mathbf{B} = 0$. So two vectors are perpendicular if and only if their dot product is zero. Two vectors are parallel if and only if $\cos\alpha = \pm 1$.

### EXAMPLE 5 Angle between two vectors

Find the smallest positive angle between each pair of vectors.

**a.** $\langle -3, 2\rangle, \langle 4, 5\rangle$ **b.** $\langle -5, 9\rangle, \langle 9, 5\rangle$

#### Solution

**a.** Use the dot product theorem to find the cosine of the angle and then a calculator to find the angle:

$$\cos\alpha = \frac{\langle -3, 2\rangle \cdot \langle 4, 5\rangle}{|\langle -3, 2\rangle||\langle 4, 5\rangle|} = \frac{-2}{\sqrt{13}\sqrt{41}}$$
$$\alpha = \cos^{-1}(-2/(\sqrt{13}\sqrt{41})) \approx 94.97°$$

**b.** Because $\langle -5, 9\rangle \cdot \langle 9, 5\rangle = 0$, the vectors are perpendicular and the angle between them is 90°.

▶**TRY THIS.** Find the smallest positive angle between $\langle 1, 3\rangle$ and $\langle 5, 2\rangle$. ■

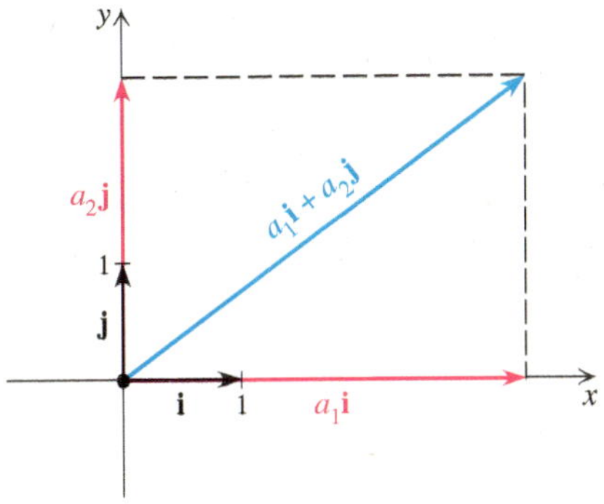


Figure 7.35

## Unit Vectors

The vectors $\mathbf{i} = \langle 1, 0 \rangle$ and $\mathbf{j} = \langle 0, 1 \rangle$ are called **unit vectors** because each has magnitude one. For any vector $\langle a_1, a_2 \rangle$, as shown in Fig. 7.35, we have

$$\langle a_1, a_2 \rangle = a_1 \langle 1, 0 \rangle + a_2 \langle 0, 1 \rangle = a_1\mathbf{i} + a_2\mathbf{j}.$$

The form $a_1\mathbf{i} + a_2\mathbf{j}$ is called a **linear combination** of the vectors **i** and **j**. These unit vectors are thought of as fundamental vectors, because any vector can be expressed as a linear combination of them.

### EXAMPLE 6 Unit vectors

Write each vector as a linear combination of the unit vectors **i** and **j**.

**a.** $\mathbf{A} = \langle -2, 6 \rangle$ **b.** $\mathbf{B} = \langle -4, -1 \rangle$

**Solution**

**a.** $\mathbf{A} = \langle -2, 6 \rangle = -2\mathbf{i} + 6\mathbf{j}$ **b.** $\mathbf{B} = \langle -4, -1 \rangle = -4\mathbf{i} - \mathbf{j}$

▸**TRY THIS.** Write $\langle -1, 7 \rangle$ as a linear combination of **i** and **j**. ■

## Applications of Vectors

When we use vectors to model forces acting on an object, we usually ignore friction and assume that all forces are acting on a single point. In the next example, we find the force that results when the two horses mentioned earlier in this section are pulling on a tree stump.

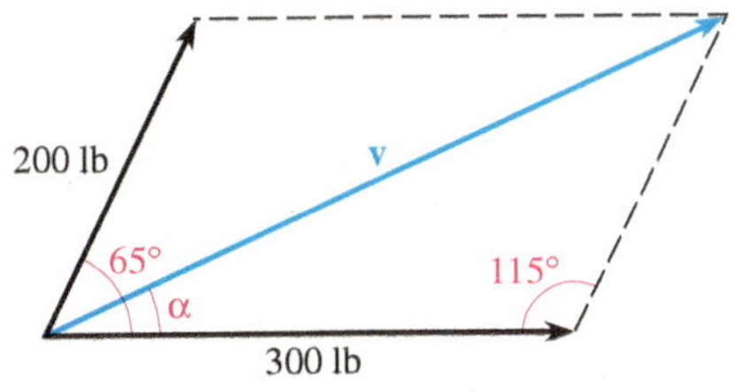


Figure 7.36

### EXAMPLE 7 Magnitude and direction of a resultant

Two draft horses are pulling on a tree stump with forces of 200 pounds and 300 pounds, as shown in Fig. 7.36. If the angle between the forces is 65°, then what is the magnitude of the resultant force? What is the angle between the resultant and the 300-pound force?

**Solution**

The resultant coincides with the diagonal of the parallelogram shown in Fig. 7.36. We can view the resultant as a side of a triangle in which the other two sides are 200 and 300, and their included angle is 115°. The law of cosines can be used to find the magnitude of the resultant vector **v**:

$$|\mathbf{v}|^2 = 300^2 + 200^2 - 2(300)(200)\cos 115° = 180{,}714.19$$

$$|\mathbf{v}| = 425.1$$

We can use the law of sines to find the angle $\alpha$ between the resultant and the 300-pound force:

$$\frac{\sin \alpha}{200} = \frac{\sin 115°}{425.1}$$

$$\sin \alpha = 0.4264$$

Since $\sin^{-1}(0.4264) = 25.2°$, both 25.2° and 154.8° are solutions to $\sin \alpha = 0.4264$. However, $\alpha$ must be smaller than 65°, so the angle between the resultant and the 300-pound force is 25.2°. So one horse pulling in the direction of the resultant with a force of 425.1 pounds would have the same effect on the stump as the two horses pulling at an angle of 65°.

▸**TRY THIS.** Forces of 100 pounds and 200 pounds are acting on a point. If the angle between the forces is 30°, then what is the magnitude of the resultant? ■

Example 7 can be solved with horizontal and vertical components. The horizontal component of the 200-pound force is 200 cos 65°, or 84.5 pounds. The vertical component is 200 sin 65°, or 181.3. The 300-pound force and the 84.5-pound force are in the same direction. So the original problem is equivalent to forces of 181.3 pounds and 384.5 pounds acting at a right angle to each other, as in Fig. 7.37. Now the resultant force is the diagonal of a rectangle and its magnitude is $\sqrt{181.3^2 + 384.5^2}$, or 425.1 pounds.

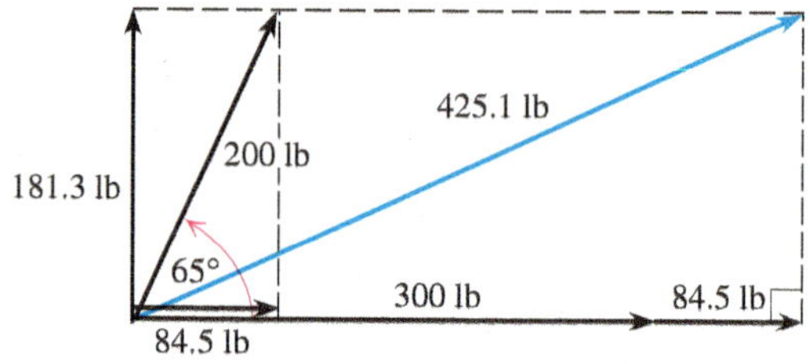


Figure 7.37

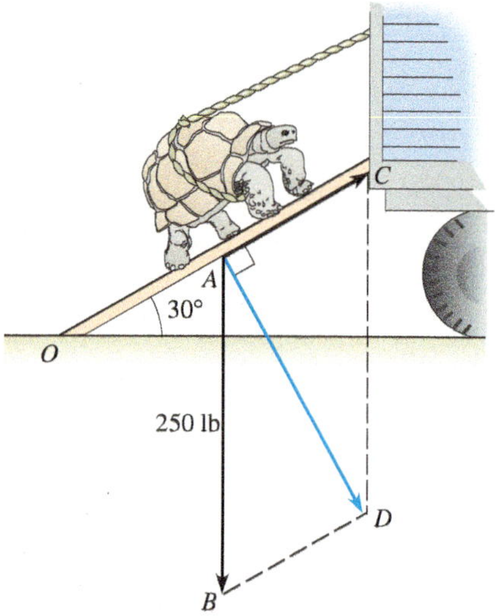


Figure 7.38

## EXAMPLE 8 Finding a force

Workers at the Audubon Zoo must move a giant tortoise to his new home. Find the amount of force required to pull a 250-pound tortoise up a ramp leading into a truck. The angle of elevation of the ramp is 30°. See Fig. 7.38.

### Solution

The weight of the tortoise is a 250-pound force in a downward direction, shown as vector **AB** in Fig. 7.38. The tortoise exerts a force against the ramp at a 90° angle with the ramp, shown as vector **AD.** The magnitude of **AD** is less than the magnitude of **AB** because of the incline of the ramp. The force required to pull the tortoise up the ramp is vector **AC** in Fig. 7.38. The force against the ramp, **AD,** is the resultant of **AB** and **AC.** Since $\angle O = 30°$, we have $\angle OAB = 60°$ and $\angle BAD = 30°$. Because $\angle CAD = 90°$, we have $\angle ADB = 90°$. We can find any of the missing parts to the right triangle $ABD$. Since

$$\sin 30° = \frac{\text{opposite}}{\text{hypotenuse}} = \frac{|\mathbf{BD}|}{|\mathbf{AB}|}$$

we have

$$|\mathbf{BD}| = |\mathbf{AB}| \sin 30° = 250 \cdot \frac{1}{2} = 125 \text{ pounds.}$$

Because the opposite sides of a parallelogram are equal, the magnitude of **AC** is also 125 pounds.

▶**TRY THIS.** Find the amount of force required to push an 800-pound block of ice up a ramp that is inclined 10°. ■

If an airplane heads directly against the wind or with the wind, the wind speed is subtracted from or added to the air speed of the plane to get the ground speed of the plane. When the wind is at some other angle to the direction of the plane, some portion of the wind speed will be subtracted from or added to the air speed to determine the ground speed. In addition, the wind causes the plane to travel on a course different from where it is headed. We can use the vector $\mathbf{v}_1$ to represent the heading and air speed of the plane, as shown in Fig. 7.39.

The vector $\mathbf{v}_2$ represents the wind direction and speed. The resultant of $\mathbf{v}_1$ and $\mathbf{v}_2$ is the vector $\mathbf{v}_3$, where $\mathbf{v}_3$ represents the course and ground speed of the plane.

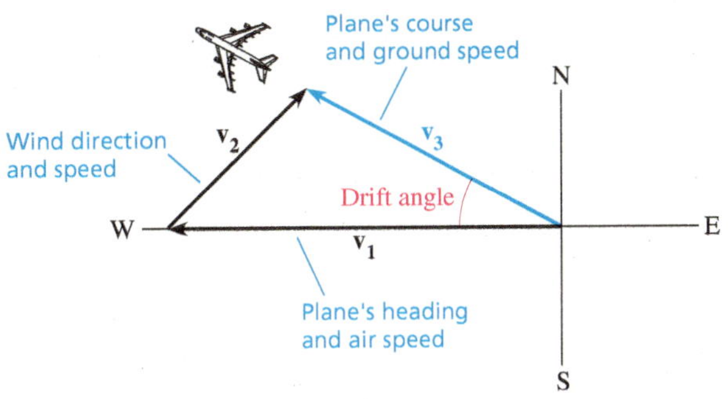


Figure 7.39

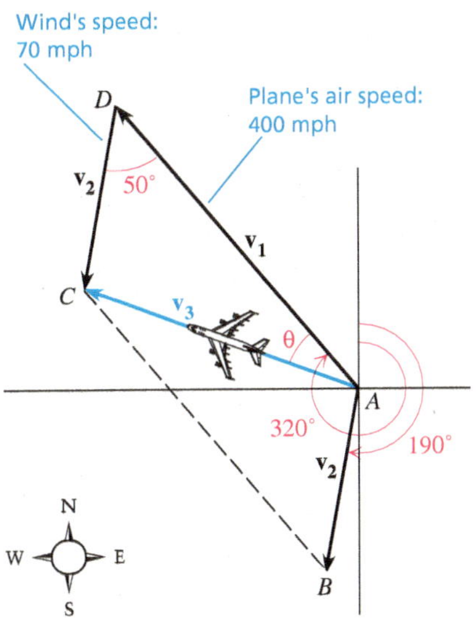


Figure 7.40

The angle between the heading and the course is the **drift angle.** Recall that the bearing of a vector used to describe direction in air navigation is a nonnegative angle smaller than 360° measured in a clockwise direction from due north. For example, the wind in Fig. 7.39 is out of the southwest and has a bearing of 45°. The vector $\mathbf{v}_1$, describing the heading of the plane, points due west and has a bearing of 270°.

## EXAMPLE 9 Finding course and ground speed of an airplane

The heading of an executive's Lear jet has a bearing of 320°. The wind is 70 mph with a bearing of 190°. Given that the air speed of the plane is 400 mph, find the drift angle, the ground speed, and the course of the airplane.

### Solution

Figure 7.40 shows vectors for the heading $\mathbf{v}_1$, the wind $\mathbf{v}_2$, and the course $\mathbf{v}_3$. Subtract the bearing of the wind from that of the heading to get $\angle DAB = 320° - 190° = 130°$. Since $ABCD$ is a parallelogram, $\angle CDA = 50°$. Apply the law of cosines to triangle $ACD$ to get

$$|\mathbf{v}_3|^2 = 70^2 + 400^2 - 2(70)(400)\cos 50° = 128{,}903.9$$
$$|\mathbf{v}_3| \approx 359.0.$$

The ground speed is approximately 359.0 mph. The drift angle $\theta$ is found by using the law of sines:

$$\frac{\sin\theta}{70} = \frac{\sin 50°}{359.0}$$
$$\sin\theta = 0.1494$$

Since $\theta$ is an acute angle, $\theta = \sin^{-1}(0.1494) = 8.6°$. The course $\mathbf{v}_3$ has a bearing of $320° - 8.6° = 311.4°$.

►**TRY THIS.** A jet is headed northwest with an air speed of 500 mph. The wind is 100 mph with a bearing of 200°. Find the drift angle, the ground speed, and the course of the jet. ■

## FOR thought... True or False? Explain.

1. The vector $2\mathbf{v}$ has the same direction as $\mathbf{v}$ but twice the magnitude of $\mathbf{v}$.

2. The magnitude of $\mathbf{A} + \mathbf{B}$ is the sum of the magnitudes of $\mathbf{A}$ and $\mathbf{B}$.

3. The magnitude of $-\mathbf{A}$ is equal to the magnitude of $\mathbf{A}$.

4. For any vector $\mathbf{A}$, $\mathbf{A} + (-\mathbf{A}) = \mathbf{0}$.

5. The parallelogram law says that the opposite sides of a parallelogram are equal in length.

**6.** In the coordinate plane, the direction angle is the angle formed by the $y$-axis and a vector with initial point at the origin.

**7.** If **v** has magnitude $r$ and direction angle $\theta$, then the horizontal component of **v** is a vector with direction angle $0°$ and magnitude $r \cos \theta$.

**8.** The magnitude of $\langle 3, -4 \rangle$ is 5.

**9.** The vectors $\langle -2, 3 \rangle$ and $\langle -6, 9 \rangle$ have the same direction angle.

**10.** The direction angle of $\langle -2, 2 \rangle$ is $\cos^{-1}(-2/\sqrt{8})$.

## EXERCISES 7.3

*Fill in the blank.*

**1.** A(n) ________ quantity involves both magnitude and direction.

**2.** Two vectors with the same magnitude and direction are ________ vectors.

**3.** The length of a vector represents the ________ of the vector quantity.

**4.** If **A** and **B** are vectors, then **A** + **B** is the ________ or ________ of **A** and **B.**

**5.** The ________ indicates that the resultant of two vectors lies along the diagonal of the parallelogram formed by the two vectors.

**6.** The angle formed by the positive $x$-axis and a position vector is the ________ angle.

**7.** The form $\langle a, b \rangle$ is the ________ form of a vector.

**8.** If the angle between two vectors is 90°, then the vectors are ________ or ________.

*For each given pair of vectors* **A** *and* **B,** *draw the vectors* **A** + **B** *and* **A** − **B.**

**9.**

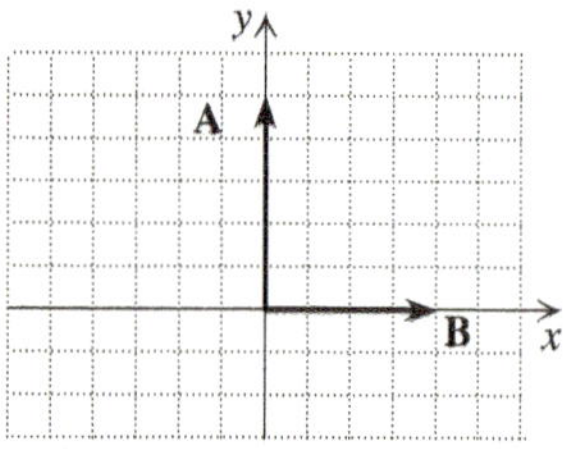


**10.**

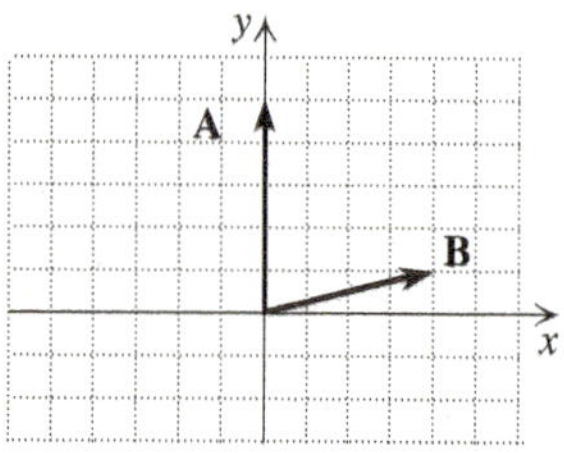


**11.**

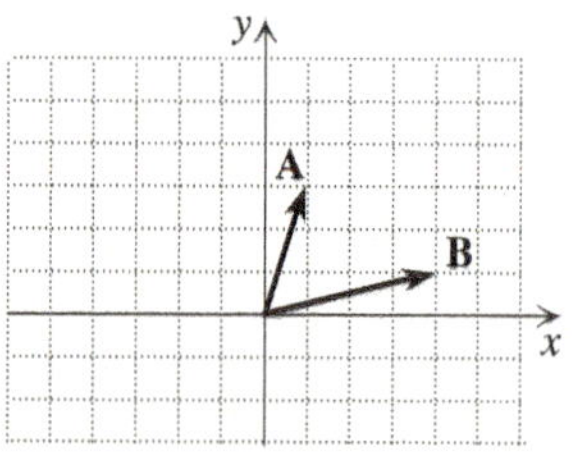


**12.**

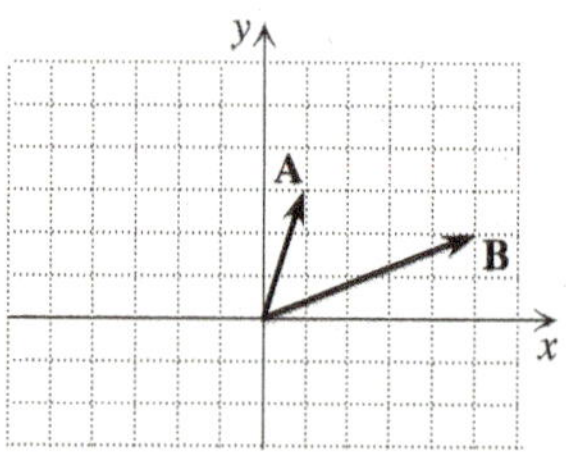


**13.**

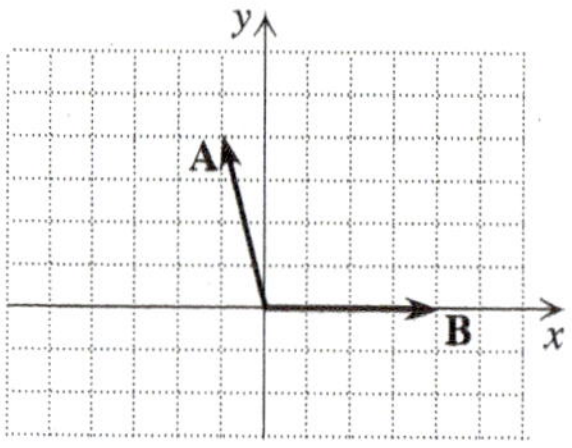


**14.**

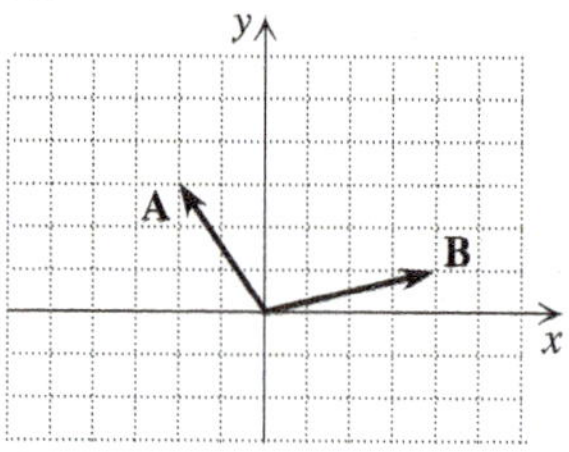


*Determine which of the position vectors shown in the accompanying figure has the given magnitude and direction angle.*

**15.** Magnitude 2 and direction angle 90°

**16.** Magnitude 3 and direction angle 180°

**17.** Magnitude $3\sqrt{2}$ and direction angle 45°

**18.** Magnitude 4 and direction angle 30°

**19.** Magnitude $3\sqrt{2}$ and direction angle 135°

**20.** Magnitude 5 and direction angle 120°

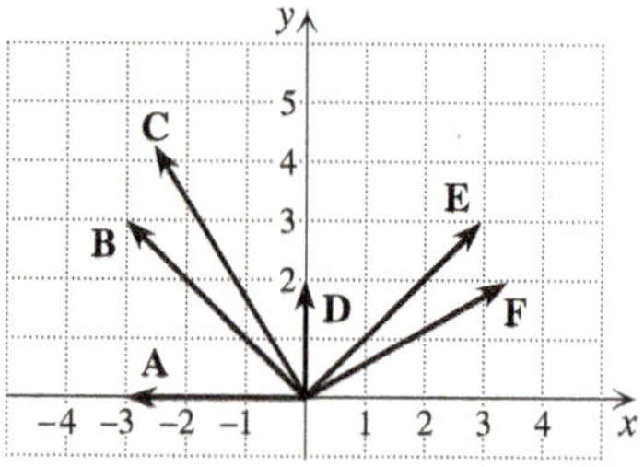


**Figure for Exercises 15–20**

*Find the magnitude of the horizontal and vertical components for each vector* **v** *with the given magnitude and given direction angle* $\theta$.

**21.** $|\mathbf{v}| = 4.5, \theta = 65.2°$

**22.** $|\mathbf{v}| = 6000, \theta = 13.1°$

**23.** $|\mathbf{v}| = 8000, \theta = 155.1°$

**24.** $|\mathbf{v}| = 445, \theta = 211.1°$

**25.** $|\mathbf{v}| = 234, \theta = 248°$

**26.** $|\mathbf{v}| = 48.3, \theta = 349°$

*Find the magnitude and direction angle of each vector.*

**27.** $\langle \sqrt{3}, 1 \rangle$

**28.** $\langle -1, \sqrt{3} \rangle$

**29.** $\langle -\sqrt{2}, \sqrt{2} \rangle$

**30.** $\langle \sqrt{2}, -\sqrt{2} \rangle$

**31.** $\langle 8, -8\sqrt{3} \rangle$

**32.** $\langle -1/2, -\sqrt{3}/2 \rangle$

**33.** $\langle 5, 0 \rangle$

**34.** $\langle 0, -6 \rangle$

**35.** $\langle -3, 2 \rangle$

**36.** $\langle -4, -2 \rangle$

**37.** $\langle 3, -1 \rangle$

**38.** $\langle 2, -6 \rangle$

*Find the component form for each vector* **v** *with the given magnitude and direction angle* $\theta$.

**39.** $|\mathbf{v}| = 8, \theta = 45°$

**40.** $|\mathbf{v}| = 12, \theta = 120°$

**41.** $|\mathbf{v}| = 290, \theta = 145°$

**42.** $|\mathbf{v}| = 5.3, \theta = 321°$

**43.** $|\mathbf{v}| = 18, \theta = 347°$

**44.** $|\mathbf{v}| = 3000, \theta = 209.1°$

*Let* $\mathbf{r} = \langle 3, -2 \rangle$, $\mathbf{s} = \langle -1, 5 \rangle$, *and* $\mathbf{t} = \langle 4, -6 \rangle$. *Perform the operations indicated. Write the vector answers in the form* $\langle a, b \rangle$.

**45.** $5\mathbf{r}$

**46.** $-4\mathbf{s}$

**47.** $2\mathbf{r} + 3\mathbf{t}$

**48.** $\mathbf{r} - \mathbf{t}$

**49.** $\mathbf{s} + 3\mathbf{t}$

**50.** $\dfrac{\mathbf{r} + \mathbf{s}}{2}$

**51.** $\mathbf{r} - (\mathbf{s} + \mathbf{t})$

**52.** $\mathbf{r} - \mathbf{s} - \mathbf{t}$

**53.** $\mathbf{r} \cdot \mathbf{s}$

**54.** $\mathbf{s} \cdot \mathbf{t}$

*Find the angle to the nearest tenth of a degree between each given pair of vectors.*

**55.** $\langle 2, 1 \rangle, \langle 3, 5 \rangle$

**56.** $\langle 2, 3 \rangle, \langle 1, 5 \rangle$

**57.** $\langle -1, 5 \rangle, \langle 2, 7 \rangle$

**58.** $\langle -2, -5 \rangle, \langle 1, -9 \rangle$

**59.** $\langle -6, 5 \rangle, \langle 5, 6 \rangle$

**60.** $\langle 2, 7 \rangle, \langle 7, -2 \rangle$

*Determine whether each pair of vectors is parallel, perpendicular, or neither.*

**61.** $\langle -2, 3 \rangle, \langle 6, 4 \rangle$

**62.** $\langle 2, 3 \rangle, \langle 8, 12 \rangle$

**63.** $\langle 1, 7 \rangle, \langle -2, -14 \rangle$

**64.** $\langle 2, -4 \rangle, \langle 2, 1 \rangle$

**65.** $\langle 5, 3 \rangle, \langle 2, 5 \rangle$

**66.** $\langle 2, 6 \rangle, \langle 6, 2 \rangle$

*Write each vector as a linear combination of the unit vectors* **i** *and* **j**.

**67.** $\langle 2, 1 \rangle$

**68.** $\langle 1, 5 \rangle$

**69.** $\langle -3, \sqrt{2} \rangle$

**70.** $\langle \sqrt{2}, -5 \rangle$

**71.** $\langle 0, -9 \rangle$

**72.** $\langle -1/2, 0 \rangle$

**73.** $\langle -7, -1 \rangle$

**74.** $\langle 1, 1 \rangle$

*Given that* $\mathbf{A} = \langle 3, 1 \rangle$ *and* $\mathbf{B} = \langle -2, 3 \rangle$, *find the magnitude and direction angle for each of the following vectors.*

**75.** $\mathbf{A} + \mathbf{B}$

**76.** $\mathbf{A} - \mathbf{B}$

**77.** $-3\mathbf{A}$

**78.** $5\mathbf{B}$

**79.** $\mathbf{B} - \mathbf{A}$

**80.** $\mathbf{B} + \mathbf{A}$

**81.** $-\mathbf{A} + \dfrac{1}{2}\mathbf{B}$

**82.** $\dfrac{1}{2}\mathbf{A} - 2\mathbf{B}$

*For each given pair of vectors, find the magnitude and direction angle of the resultant.*

**83.**

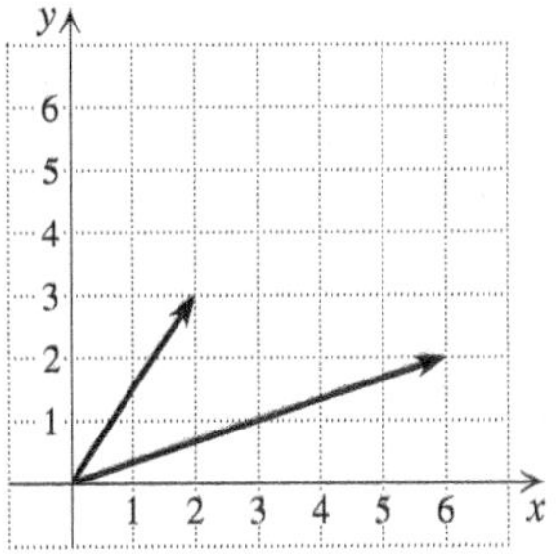

**84.**

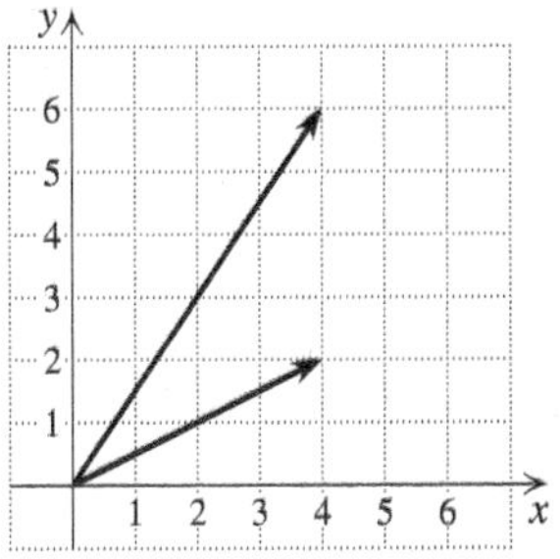

**85.**

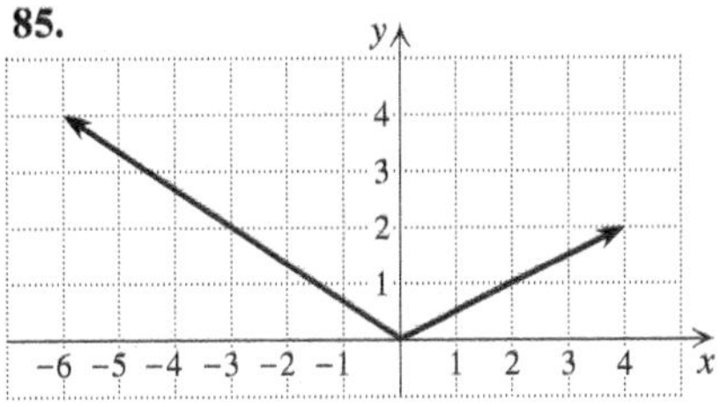

**86.**

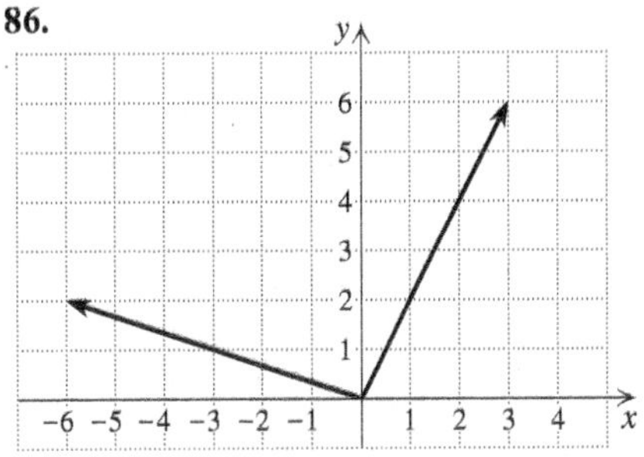

**87.**

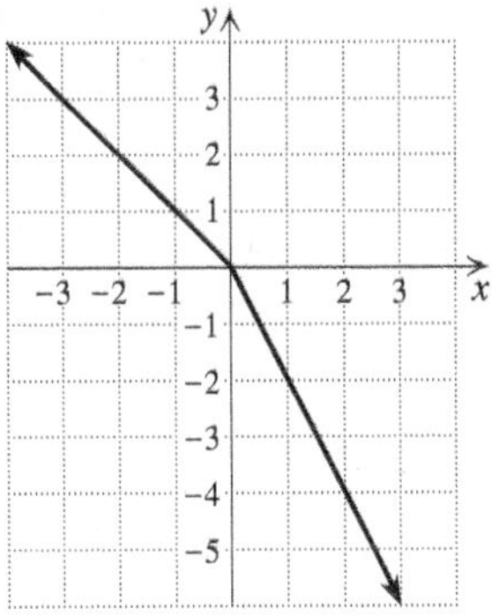

**88.**

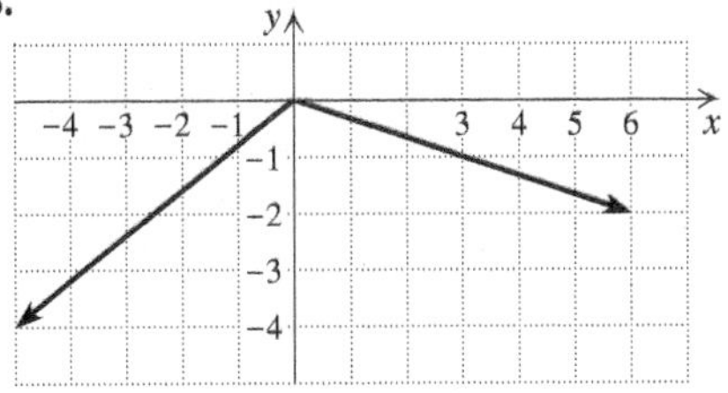

*In each case, find the magnitude of the resultant force and the angle between the resultant and each force.*

**89.** Forces of 3 lb and 8 lb act at an angle of 90° to each other.

**90.** Forces of 2 lb and 12 lb act at an angle of 60° to each other.

**91.** Forces of 4.2 newtons (a unit of force from physics) and 10.3 newtons act at an angle of 130° to each other.

**92.** Forces of 34 newtons and 23 newtons act at an angle of 100° to each other.

*Solve each problem.*

**93.** *Magnitude of a Force* The resultant of a 10-lb force and another force has a magnitude of 12.3 lb at an angle of 23.4° with the 10-lb force. Find the magnitude of the other force and the angle between the two forces.

**94.** *Magnitude of a Force* The resultant of a 15-lb force and another force has a magnitude of 9.8 lb at an angle of 31° with the 15-lb force. Find the magnitude of the other force and the angle between the other force and the resultant.

**95.** *Moving a Donkey* Two prospectors are pulling on ropes attached around the neck of a donkey that does not want to move. One prospector pulls with a force of 55 lb, and the other pulls with a force of 75 lb. If the angle between the ropes is 25°, as shown in the figure, then how much force must the donkey use in order to stay put? (The donkey knows the proper direction in which to apply his force.)

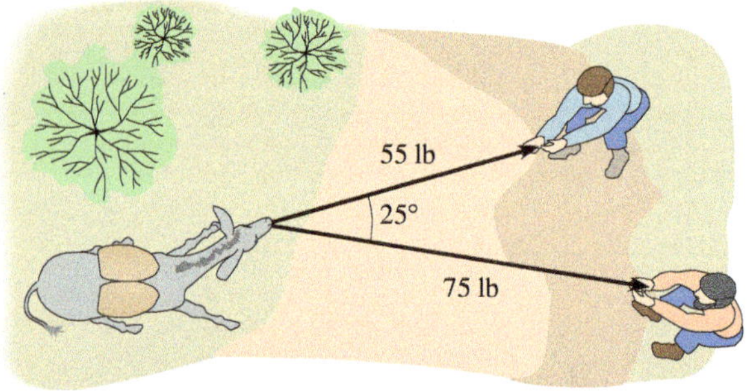


**Figure for Exercise 95**

**96.** *Pushing a Shopping Cart* Ronnie, Phyllis, and Ted are conducting a vector experiment in a Wal-Mart parking lot. Ronnie is pushing a cart containing Phyllis to the east at 5 mph while Ted is pushing it to the north at 3 mph. What is Phyllis's speed and in what direction (measured from north) is she moving?

**97.** *Gaining Altitude* An airplane with an air speed of 520 mph is climbing at an angle of 30° from the horizontal. What are the magnitudes of the horizontal and vertical components of the speed vector?

**98.** *Acceleration of a Missile* A missile is fired with an angle of elevation of 22°, with an acceleration of 30 m/sec$^2$. What are the magnitudes of the horizontal and vertical components of the acceleration vector?

**99.** *Rock and Roll* In Roman mythology, Sisyphus, king of Corinth, revealed a secret of Zeus and thus incurred the god's wrath. As punishment, Zeus banished him to Hades, where he was doomed for eternity to roll a rock uphill, only to have it roll back on him. If Sisyphus stands in front of a 4000-lb spherical rock on a 20° incline, as shown in the figure, then what force applied in the direction of the incline would keep the rock from rolling down the incline?

HINT The result of the 4000-lb force and the force of Sisyphus is perpendicular to the incline.

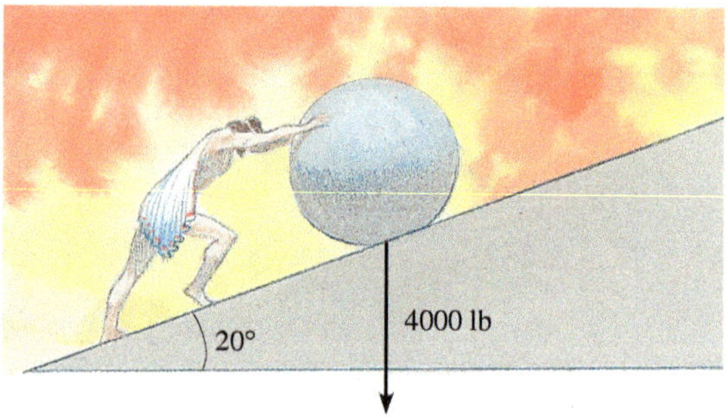


**Figure for Exercise 99**

**100.** *Weight of a Ball* A solid steel ball is placed on a 10° incline. If a force of 3.2 lb in the direction of the incline is required to keep the ball in place, then what is the weight of the ball?

**101.** *Winch Force* Find the amount of force required for a winch to pull a 3000-lb car up a ramp that is inclined at 20°.

**102.** *Ice Force* If the amount of force required to push a block of ice up an ice-covered driveway that is inclined at 25° is 100 lb, then what is the weight of the block?

**103.** *Super Force* If Superman exerts 1000 lb of force to prevent a 5000-lb boulder from rolling down a hill and crushing a bus full of children, then what is the angle of inclination of the hill?

**104.** *Sisy's Slope* If Sisyphus exerts a 500-lb force in rolling his 4000-lb spherical boulder uphill, then what is the angle of inclination of the hill?

**105.** *Due East* A plane is headed due east with an air speed of 240 mph. The wind is from the north at 30 mph. Find the bearing for the course and the ground speed of the plane.

**106.** *Due West* A plane is headed due west with an air speed of 300 mph. The wind is from the north at 80 mph. Find the bearing for the course and the ground speed of the plane.

**107.** *Ultralight* An ultralight is flying northeast at 50 mph. The wind is from the north at 20 mph. Find the bearing for the course and the ground speed of the ultralight.

**108.** *Superlight* A superlight is flying northwest at 75 mph. The wind is from the south at 40 mph. Find the bearing for the course and the ground speed of the superlight.

**109.** *Course of an Airplane* An airplane is heading on a bearing of 102° with an air speed of 480 mph. If the wind is out of the northeast (bearing 225°) at 58 mph, then what are the bearing of the course and the ground speed of the airplane?

**110.** *Course of a Helicopter* The heading of a helicopter has a bearing of 240°. If the 70-mph wind has a bearing of 185° and the air speed of the helicopter is 195 mph, then what are the bearing of the course and the ground speed of the helicopter?

**111.** *Going with the Flow* A river is 2000 ft wide and flowing at 6 mph from north to south. A woman in a canoe starts on the eastern shore and heads west at her normal paddling speed of 2 mph. In what direction (measured clockwise from north) will she actually be traveling? How far downstream from a point directly across the river will she land?

**112.** *Crossing a River* If the woman in Exercise 111 wants to go directly across the river and she paddles at 8 mph, then in what direction (measured clockwise from north) must she aim her canoe? How long will it take her to go directly across the river?

**113.** *Distance and Rate* A trigonometry student wants to cross a river that is 0.2 mi wide and has a current of 1 mph. The boat goes 3 mph in still water.

**a.** Write the distance the boat travels as a function of the angle $\beta$ shown in the figure.

**b.** Write the actual speed of the boat as a function of $\alpha$ and $\beta$.

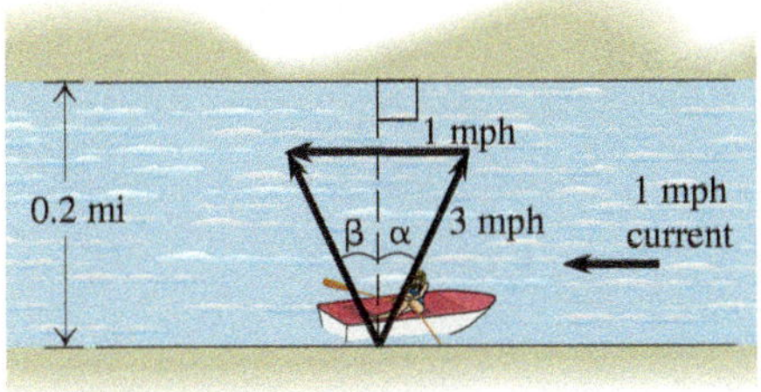


**Figure for Exercise 113**

**c.** Write the time for the trip as a function of $\alpha$. Find the angle $\alpha$ for which the student will cross the river in the shortest amount of time.

**114.** *My Three Elephants* Amal uses three elephants to pull a very large log out of the jungle. The papa elephant pulls with 800 lb of force, the mama elephant pulls with 500 lb of force, and the baby elephant pulls with 200 lb of force. The angles between the forces are shown in the figure. What is the magnitude of the resultant of all three forces? If mama is pulling due east, then in what direction will the log move?

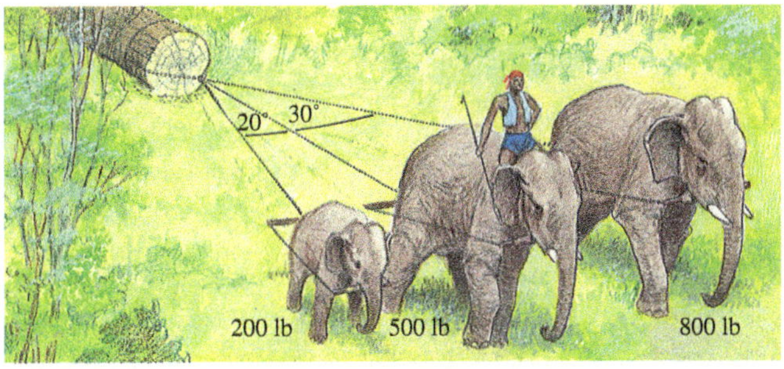


**Figure for Exercise 114**

## FOR WRITING/DISCUSSION

**115.** *Distributive* Prove that scalar multiplication is distributive over vector addition, first using the component form and then using a geometric argument.

**116.** *Associative* Prove that vector addition is associative, first using the component form and then using a geometric argument.

## RETHINKING

**117.** Solve the triangle with $\beta = 122.1°$, $a = 19.4$, and $b = 22.6$.

**118.** Solve the triangle with $\alpha = 22.1°$, $a = 19.4$, and $c = 144.2$.

**119.** Solve the triangle with $\alpha = 33.2°$, $b = 9.4$, and $c = 4.3$.

**120.** Find all solutions to the equation $2\cos(x) + 1 = 0$. Use $k$ to represent any integer.

**121.** Find all solutions to the equation $4\sin^2(3x) - 3 = 0$ in the interval $(0, \pi)$.

**122.** Find the period, asymptotes, and range for the function $y = 5\tan(\pi x + \pi)$.

## THINKING OUTSIDE THE BOX LIV

*Related Angles*

**a.** Find the triangle with the smallest perimeter and whole-number sides such that the measure of one of the angles is twice the measure of another angle.

**b.** Find general expressions that could be used to determine infinitely many such triangles.

**c.** Find the next two larger such triangles.

## POP QUIZ 7.3

**1.** Find $|\mathbf{v}_x|$ and $|\mathbf{v}_y|$ if $|\mathbf{v}| = 5.6$ and the direction angle of $\mathbf{v}$ is 33.9°.

**2.** Find the magnitude and direction angle of the vector $\mathbf{v} = \langle -2, 6 \rangle$.

**3.** Find $\mathbf{v} - \mathbf{w}$, $3\mathbf{v}$, and $\mathbf{v} \cdot \mathbf{w}$, if $\mathbf{v} = \langle -1, 3 \rangle$ and $\mathbf{w} = \langle 2, 6 \rangle$.

**4.** Find the smallest positive angle between $\langle 1, 4 \rangle$ and $\langle 2, 6 \rangle$.

**5.** An airplane is headed due east with an air speed of 300 mph. The wind is from the north at 60 mph. What is the ground speed and heading of the airplane?

## LINKING concepts... For Individual or Group Explorations

### Minimizing the Total Time

*Suk wants to go from point A to point B, as shown in the figure. Point B is 1 mile upstream and on the north side of the 0.4-mi-wide river. The speed of the current is 0.5 mph. Her boat will travel 4 mph in still water. When she gets to the north side she will travel by bicycle at 6 mph. Let $\alpha$ represent the bearing of her course and $\beta$ represent the drift angle, as shown in the figure.*

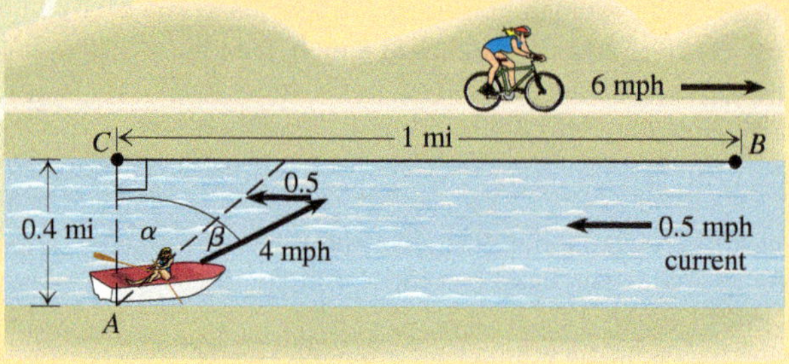


**a)** When $\alpha = 12°$, find $\beta$ and the actual speed of the boat.

**b)** Find the total time for the trip when $\alpha = 12°$.

**c)** Write the total time for the trip as a function of $\alpha$ and graph the function.

**d)** Find the approximate angle $\alpha$ that minimizes the total time for the trip and find the corresponding $\beta$.

# 7.4 Trigonometric Form of Complex Numbers

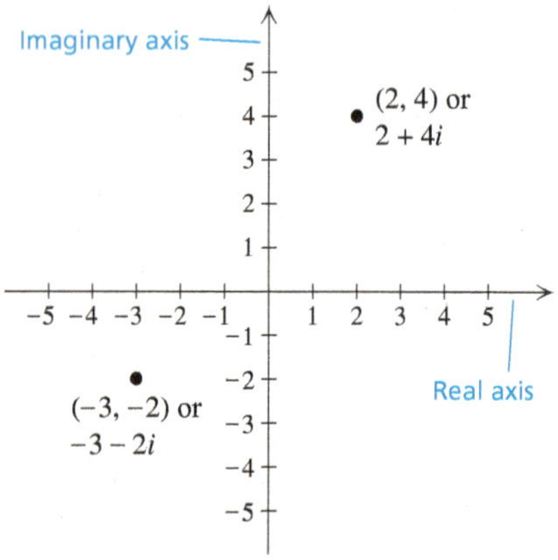


The Complex Plane

**Figure 7.41**

Until now we have concentrated on applying the trigonometric functions to real-life situations. So the trigonometric form of complex numbers, introduced in this section, may be somewhat surprising. However, with this form we can expand our knowledge about complex numbers and find powers and roots of complex numbers that we could not find without it.

## The Complex Plane

A complex number $a + bi$ can be thought of as an ordered pair $(a, b)$, where $a$ is the real part and $b$ is the imaginary part of the complex number. Ordered pairs representing complex numbers are found in a coordinate system just like ordered pairs of real numbers. The horizontal axis is called the **real axis,** and the vertical axis is called the **imaginary axis,** as shown in Fig. 7.41. This coordinate system is called the **complex plane.** The complex plane provides an order to the complex number system and allows us to treat complex numbers like vectors.

The absolute value of a real number is the distance on the number line between the number and the origin. The **absolute value of a complex number** is the distance between the complex number and the origin in the complex plane.

**Definition: Absolute Value of $a + bi$**

The absolute value of the complex number $a + bi$ is defined by

$$|a + bi| = \sqrt{a^2 + b^2}.$$

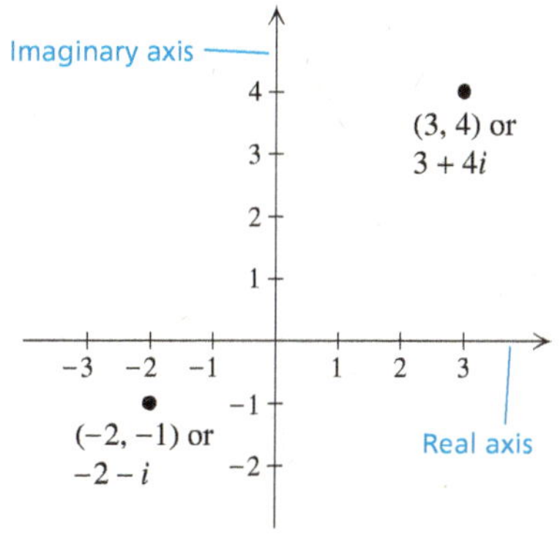


Figure 7.42

The absolute value of $a + bi$ is the same as the magnitude of the vector $\langle a, b \rangle$.

## EXAMPLE 1 Graphing complex numbers

Graph each complex number and find its absolute value.

**a.** $3 + 4i$ **b.** $-2 - i$

### Solution

**a.** The complex number $3 + 4i$ is located in the first quadrant, as shown in Fig. 7.42, three units to the right of the origin and four units upward. Using the definition of absolute value of a complex number, we have

$$|3 + 4i| = \sqrt{3^2 + 4^2} = 5.$$

**b.** The complex number $-2 - i$ is located in the third quadrant (Fig. 7.42) and

$$|-2 - i| = \sqrt{(-2)^2 + (-1)^2} = \sqrt{5}.$$

▸**TRY THIS.** Find the absolute value of $5 - i$. ■

## Trigonometric Form of a Complex Number

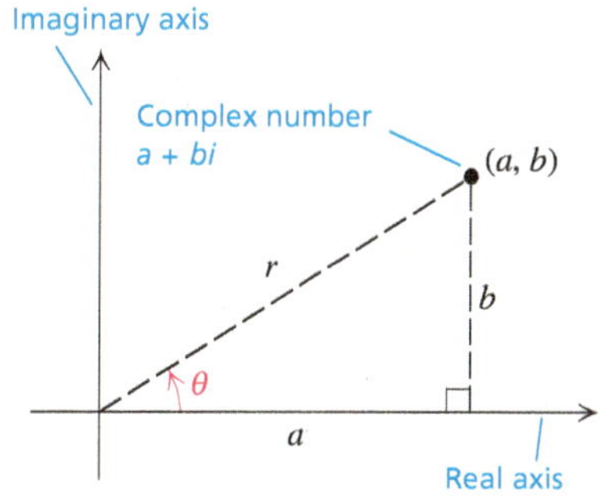


Figure 7.43

Multiplication and division of complex numbers in the standard form $a + bi$ are rather complicated, as is finding powers and roots, but in trigonometric form these operations are simpler and more natural. A complex number is written in **trigonometric form** by a process similar to finding the horizontal and vertical components of a vector.

Consider the complex number $a + bi$ shown in Fig. 7.43, where

$$r = |a + bi| = \sqrt{a^2 + b^2}.$$

Since $(a, b)$ is a point on the terminal side of $\theta$ in standard position, we have

$$a = r\cos\theta \quad \text{and} \quad b = r\sin\theta.$$

Substituting for $a$ and $b$, we get

$$a + bi = (r\cos\theta) + (r\sin\theta)i = r(\cos\theta + i\sin\theta).$$

**Definition: Trigonometric Form of a Complex Number**

If $z = a + bi$ is a complex number, then the **trigonometric form** of $z$ is

$$z = r(\cos\theta + i\sin\theta),$$

where $r = \sqrt{a^2 + b^2}$ and $\theta$ is an angle in standard position whose terminal side contains the point $(a, b)$. An abbreviation for $r(\cos\theta + i\sin\theta)$ is $r \text{ cis } \theta$.

The number $r$ is called the **modulus** of $a + bi$, and $\theta$ is called the **argument** of $a + bi$. Since there are infinitely many angles whose terminal side contains $(a, b)$, the trigonometric form of a complex number is not unique. However, we usually use the smallest nonnegative value for $\theta$ that satisfies

$$a = r\cos\theta \quad \text{or} \quad b = r\sin\theta.$$

In the next two examples, we convert from standard to trigonometric form and vice versa.

■ **Foreshadowing Calculus**

Beginning calculus courses concentrate on real-valued functions. In more advanced courses we study the calculus of complex-valued functions. The complex plane is used there to give a visual image of the complex numbers.

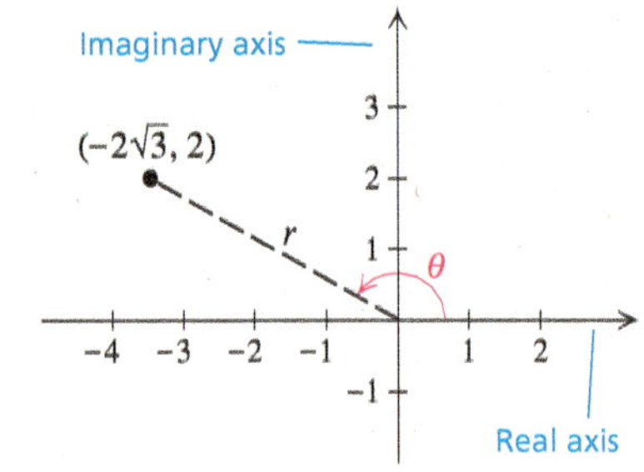


**Figure 7.44**

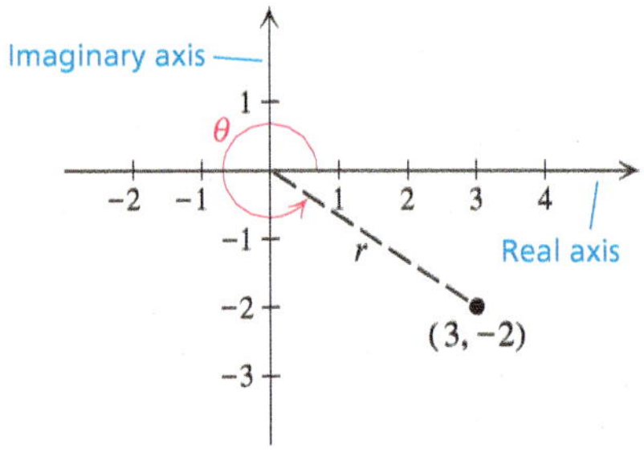


**Figure 7.45**

## EXAMPLE 2 Writing a complex number in trigonometric form

Write each complex number in trigonometric form.

**a.** $-2\sqrt{3} + 2i$ **b.** $3 - 2i$

### Solution

**a.** First we locate $-2\sqrt{3} + 2i$ in the complex plane, as shown in Fig. 7.44. Find $r$ using $a = -2\sqrt{3}$, $b = 2$, and $r = \sqrt{a^2 + b^2}$:

$$r = \sqrt{(-2\sqrt{3})^2 + 2^2} = 4$$

Since $a = r \cos \theta$, we have $-2\sqrt{3} = 4 \cos \theta$ or

$$\cos \theta = -\frac{\sqrt{3}}{2}.$$

The smallest nonnegative solution to this equation is $\theta = 5\pi/6$. Because the terminal side of $5\pi/6$ goes through $(-2\sqrt{3}, 2)$, we get

$$-2\sqrt{3} + 2i = 4\left(\cos \frac{5\pi}{6} + i \sin \frac{5\pi}{6}\right).$$

**b.** Locate $3 - 2i$ in the complex plane, as shown in Fig. 7.45. Next, find $r$ by using $a = 3$, $b = -2$, and $r = \sqrt{a^2 + b^2}$:

$$r = \sqrt{(3)^2 + (-2)^2} = \sqrt{13}$$

Since $a = r \cos \theta$, we have $3 = \sqrt{13} \cos \theta$ or

$$\cos \theta = \frac{3}{\sqrt{13}}.$$

Use a calculator to get

$$\cos^{-1}\left(\frac{3}{\sqrt{13}}\right) = 33.7°.$$

Because $\theta$ is an angle whose terminal side contains $(3, -2)$, choose $\theta = 360° - 33.7° = 326.3°$. So

$$3 - 2i = \sqrt{13}(\cos 326.3° + i \sin 326.3°).$$

▶**TRY THIS.** Write $1 + 2i$ in trigonometric form using degree measure. ■

## EXAMPLE 3 Writing a complex number in standard form

Write each complex number in the form $a + bi$.

**a.** $\sqrt{2}\left(\cos \frac{\pi}{4} + i \sin \frac{\pi}{4}\right)$ **b.** $3.6(\cos 143° + i \sin 143°)$

### Solution

**a.** Since $\cos(\pi/4) = \sqrt{2}/2$ and $\sin(\pi/4) = \sqrt{2}/2$, we have

$$\sqrt{2}\left(\cos \frac{\pi}{4} + i \sin \frac{\pi}{4}\right) = \sqrt{2}\left(\frac{\sqrt{2}}{2} + i\frac{\sqrt{2}}{2}\right)$$
$$= 1 + i.$$

**b.** Use a calculator to find cos 143° and sin 143°:

$$3.6(\cos 143° + i \sin 143°) \approx -2.875 + 2.167i$$

▶**TRY THIS.** Write $12\left(\cos \frac{\pi}{6} + i \sin \frac{\pi}{6}\right)$ in standard form. ■

## Products and Quotients in Trigonometric Form

In standard form, multiplication and division of complex numbers are rather complicated whereas addition and subtraction are simple. In trigonometric form addition and subtraction are complicated whereas multiplication and division are simple. We multiply the complex numbers

$$z_1 = r_1(\cos \theta_1 + i \sin \theta_1) \qquad \text{and} \qquad z_2 = r_2(\cos \theta_2 + i \sin \theta_2)$$

as follows:

$$\begin{aligned} z_1 z_2 &= r_1(\cos \theta_1 + i \sin \theta_1) \cdot r_2(\cos \theta_2 + i \sin \theta_2) \\ &= r_1 r_2(\cos \theta_1 + i \sin \theta_1)(\cos \theta_2 + i \sin \theta_2) \\ &= r_1 r_2[(\cos \theta_1 \cos \theta_2 - \sin \theta_1 \sin \theta_2) + i(\sin \theta_1 \cos \theta_2 + \cos \theta_1 \sin \theta_2)] \end{aligned}$$

Using the identities for cosine and sine of a sum, we get

$$z_1 z_2 = r_1 r_2[\cos(\theta_1 + \theta_2) + i \sin(\theta_1 + \theta_2)].$$

*The product of $z_1$ and $z_2$ can be found by multiplying their moduli $r_1$ and $r_2$ and adding their arguments $\theta_1$ and $\theta_2$.*

We find the quotient of $z_1$ and $z_2$ as follows:

$$\begin{aligned} \frac{z_1}{z_2} &= \frac{r_1(\cos \theta_1 + i \sin \theta_1)}{r_2(\cos \theta_2 + i \sin \theta_2)} \\ &= \frac{r_1(\cos \theta_1 + i \sin \theta_1)(\cos \theta_2 - i \sin \theta_2)}{r_2(\cos \theta_2 + i \sin \theta_2)(\cos \theta_2 - i \sin \theta_2)} \\ &= \frac{r_1}{r_2}[(\cos \theta_1 \cos \theta_2 + \sin \theta_1 \sin \theta_2) + i(\sin \theta_1 \cos \theta_2 - \cos \theta_1 \sin \theta_2)] \\ &= \frac{r_1}{r_2}[\cos(\theta_1 - \theta_2) + i \sin(\theta_1 - \theta_2)] \end{aligned}$$

*The quotient of $z_1$ and $z_2$ can be found by dividing their moduli $r_1$ and $r_2$ and subtracting their arguments $\theta_1$ and $\theta_2$.*

These two results are stated in the following theorem.

**Theorem: Product and Quotient of Complex Numbers**

If $z_1 = r_1(\cos \theta_1 + i \sin \theta_1)$ and $z_2 = r_2(\cos \theta_2 + i \sin \theta_2)$, then

$$z_1 z_2 = r_1 r_2[\cos(\theta_1 + \theta_2) + i \sin(\theta_1 + \theta_2)]$$

and

$$\frac{z_1}{z_2} = \frac{r_1}{r_2}[\cos(\theta_1 - \theta_2) + i \sin(\theta_1 - \theta_2)].$$

## EXAMPLE 4 A product and quotient using trigonometric form

Find $z_1z_2$ and $z_1/z_2$ for $z_1 = 6(\cos 2.4 + i\sin 2.4)$ and $z_2 = 2(\cos 1.8 + i\sin 1.8)$. Express the answers in the form $a + bi$.

### Solution

Find the product by multiplying the moduli and adding the arguments:

$$\begin{aligned} z_1z_2 &= 6 \cdot 2(\cos(2.4 + 1.8) + i\sin(2.4 + 1.8)) \\ &= 12(\cos 4.2 + i\sin 4.2) \\ &= 12\cos 4.2 + 12i\sin 4.2 \\ &\approx -5.88 - 10.46i \end{aligned}$$

Find the quotient by dividing the moduli and subtracting the arguments:

$$\begin{aligned} \frac{z_1}{z_2} &= \frac{6}{2}(\cos(2.4 - 1.8) + i\sin(2.4 - 1.8)) \\ &= 3(\cos(0.6) + i\sin(0.6)) \\ &= 3\cos 0.6 + 3i\sin 0.6 \\ &\approx 2.48 + 1.69i \end{aligned}$$

▶**TRY THIS.** Find $z_1z_2$ for $z_1 = 4\left(\cos\frac{\pi}{12} + i\sin\frac{\pi}{12}\right)$ and $z_2 = 8\left(\cos\frac{\pi}{12} + i\sin\frac{\pi}{12}\right)$. ■

## EXAMPLE 5 A product and quotient using trigonometric form

Use trigonometric form to find $z_1z_2$ and $z_1/z_2$, given that $z_1 = -2 + 2i\sqrt{3}$ and $z_2 = \sqrt{3} + i$.

### Solution

First convert $z_1$ and $z_2$ into trigonometric form. Since $\sqrt{(-2)^2 + (2\sqrt{3})^2} = 4$ and $120°$ is a positive angle whose terminal side contains $(-2, 2\sqrt{3})$, we have

$$z_1 = 4(\cos 120° + i\sin 120°).$$

Since $\sqrt{(\sqrt{3})^2 + 1^2} = 2$ and $30°$ is a positive angle whose terminal side contains $(\sqrt{3}, 1)$, we have

$$z_2 = 2(\cos 30° + i\sin 30°).$$

To find the product, we multiply the moduli and add the arguments:

$$\begin{aligned} z_1z_2 &= 2 \cdot 4[\cos(120° + 30°) + i\sin(120° + 30°)] \\ &= 8(\cos 150° + i\sin 150°) \end{aligned}$$

Using $\cos 150° = -\sqrt{3}/2$ and $\sin 150° = 1/2$, we get

$$z_1z_2 = 8\left(-\frac{\sqrt{3}}{2} + i\frac{1}{2}\right) = -4\sqrt{3} + 4i.$$

To find the quotient, we divide the moduli and subtract the arguments:

$$\frac{z_1}{z_2} = \frac{4}{2}[\cos(120° - 30°) + i\sin(120° - 30°)]$$
$$= 2(\cos 90° + i\sin 90°)$$
$$= 2(0 + i) = 2i$$

Check by performing the operations with $z_1$ and $z_2$ in standard form.

►**TRY THIS.** Find $z_1z_2$ using trigonometric form for $z_1 = 2\sqrt{3} + 2i$ and $z_2 = 3 + 3i\sqrt{3}$. ■

## Complex Conjugates

In standard form, $a + bi$ and $a - bi$ are complex conjugates. Their product is the real number $a^2 + b^2$. In trigonometric form, $a + bi = r(\cos\theta + i\sin\theta)$, and we could write $a - bi = r(\cos\theta - i\sin\theta)$. But trigonometric form has a plus sign in it. To multiply in trigonometric form (by multiplying the moduli and adding the arguments), the plus sign is necessary. It can be shown that $a - bi = r(\cos(-\theta) + i\sin(-\theta))$. (See Exercise 79.) Now if you multiply the moduli and add the arguments, the product is a real number. (See Exercise 80.)

### EXAMPLE 6 Complex conjugates in trigonometric form

Find the product of $2(\cos(\pi/3) + i\sin(\pi/3))$ and its conjugate using multiplication in trigonometric form.

**Solution**

The conjugate is $2(\cos(-\pi/3) + i\sin(-\pi/3))$. We find the product by multiplying the moduli and adding the arguments:

$$2\left(\cos\frac{\pi}{3} + i\sin\frac{\pi}{3}\right)\cdot 2\left(\cos\left(-\frac{\pi}{3}\right) + i\sin\left(-\frac{\pi}{3}\right)\right) = 4(\cos 0 + i\sin 0) = 4$$

►**TRY THIS.** Find the product of $6(\cos 30° + i\sin 30°)$ and its conjugate using multiplication in trigonometric form. ■

## FOR thought... True or False? Explain.

1. The complex number $3 - 4i$ lies in quadrant IV of the complex plane.
2. The absolute value of $-2 - 5i$ is $2 + 5i$.
3. If $\theta$ is an angle whose terminal side contains $(1, -3)$, then $\cos\theta = 1/\sqrt{10}$.
4. If $\theta$ is an angle whose terminal side contains $(-2, -3)$, then $\tan\theta = 2/3$.
5. $i = 1(\cos 0° + i\sin 0°)$
6. The smallest positive solution to $\sqrt{3} = 2\cos\theta$ is $\theta = 30°$.
7. The modulus of $2 - 5i$ is $\sqrt{29}$.
8. An argument for $2 - 4i$ is $\theta = 360° - \cos^{-1}(1/\sqrt{5})$.
9. $3(\cos\pi/4 + i\sin\pi/4)\cdot 2(\cos\pi/2 + i\sin\pi/2) = 6(\cos 3\pi/4 + i\sin 3\pi/4)$
10. $\dfrac{3(\cos\pi/4 + i\sin\pi/4)}{2(\cos\pi/2 + i\sin\pi/2)} = 1.5(\cos\pi/4 + i\sin\pi/4)$

## EXERCISES 7.4

*Fill in the blank.*

**1.** In the complex plane, the horizontal axis is the ________ axis and the vertical axis is the ________ axis.

**2.** The value of $\sqrt{a^2 + b^2}$ is the ________ of the complex number $a + bi$.

**3.** In the trigonometric form of a complex number $r(\cos\theta + i\sin\theta)$, $r$ is the ________ and $\theta$ is the ________.

**4.** The product of two complex numbers in trigonometric form is found by ________ their moduli and ________ their arguments.

*Graph each complex number, and find its absolute value.*

**5.** $8i$ **6.** $-3i$ **7.** $-9$

**8.** $-\sqrt{6}$ **9.** $2 - 6i$ **10.** $-1 - i$

**11.** $-2 + 2i\sqrt{3}$ **12.** $-\sqrt{3} - i$ **13.** $\frac{1}{\sqrt{2}} - \frac{i}{\sqrt{2}}$

**14.** $\frac{\sqrt{3}}{2} + \frac{i}{2}$ **15.** $3 + 3i$ **16.** $-4 - 4i$

*Write each complex number in trigonometric form, using degree measure for the argument.*

**17.** $8$ **18.** $-7$ **19.** $i\sqrt{3}$

**20.** $-5i$ **21.** $-3 + 3i$ **22.** $4 - 4i$

**23.** $-\frac{3}{\sqrt{2}} + \frac{3i}{\sqrt{2}}$ **24.** $\frac{\sqrt{3}}{6} + \frac{i}{6}$ **25.** $-\sqrt{3} + i$

**26.** $-2 - 2i\sqrt{3}$ **27.** $3 + 4i$ **28.** $-2 + i$

**29.** $-3 + 5i$ **30.** $-2 - 4i$ **31.** $3 - 6i$

**32.** $5 - 10i$

*Write each complex number in the form $a + bi$.*

**33.** $\sqrt{2}(\cos 45° + i\sin 45°)$ **34.** $6(\cos 30° + i\sin 30°)$

**35.** $\frac{\sqrt{3}}{2}(\cos 150° + i\sin 150°)$

**36.** $12(\cos(\pi/10) + i\sin(\pi/10))$

**37.** $\frac{1}{2}(\cos 3.7 + i\sin 3.7)$

**38.** $4.3(\cos(\pi/9) + i\sin(\pi/9))$

**39.** $3(\cos 90° + i\sin 90°)$ **40.** $4(\cos 180° + i\sin 180°)$

**41.** $\sqrt{3}(\cos(3\pi/2) + i\sin(3\pi/2))$

**42.** $8.1(\cos\pi + i\sin\pi)$ **43.** $\sqrt{6}(\cos 60° + i\sin 60°)$

**44.** $0.5(\cos(5\pi/6) + i\sin(5\pi/6))$

*Perform the indicated operations. Write the answer in the form $a + bi$.*

**45.** $2(\cos 150° + i\sin 150°) \cdot 3(\cos 300° + i\sin 300°)$

**46.** $\sqrt{3}(\cos 45° + i\sin 45°) \cdot \sqrt{2}(\cos 315° + i\sin 315°)$

**47.** $\sqrt{3}(\cos 10° + i\sin 10°) \cdot \sqrt{2}(\cos 20° + i\sin 20°)$

**48.** $8(\cos 100° + i\sin 100°) \cdot 3(\cos 35° + i\sin 35°)$

**49.** $[3(\cos 45° + i\sin 45°)]^2$

**50.** $[\sqrt{5}(\cos(\pi/12) + i\sin(\pi/12))]^2$

**51.** $\dfrac{4(\cos(\pi/3) + i\sin(\pi/3))}{2(\cos(\pi/6) + i\sin(\pi/6))}$

**52.** $\dfrac{9(\cos(\pi/4) + i\sin(\pi/4))}{3(\cos(5\pi/4) + i\sin(5\pi/4))}$

**53.** $\dfrac{4.1(\cos 36.7° + i\sin 36.7°)}{8.2(\cos 84.2° + i\sin 84.2°)}$

**54.** $\dfrac{18(\cos 121.9° + i\sin 121.9°)}{2(\cos 325.6° + i\sin 325.6°)}$

*Find $z_1z_2$ and $z_1/z_2$ for each pair of complex numbers, using trigonometric form. Write the answer in the form $a + bi$.*

**55.** $z_1 = 4 + 4i, z_2 = -5 - 5i$

**56.** $z_1 = -3 + 3i, z_2 = -2 - 2i$

**57.** $z_1 = \sqrt{3} + i, z_2 = 2 + 2i\sqrt{3}$

**58.** $z_1 = -\sqrt{3} + i, z_2 = 4\sqrt{3} - 4i$

**59.** $z_1 = 2 + 2i, z_2 = \sqrt{2} - i\sqrt{2}$

**60.** $z_1 = \sqrt{5} + i\sqrt{5}, z_2 = -\sqrt{6} - i\sqrt{6}$

**61.** $z_1 = 3 + 4i, z_2 = -5 - 2i$

**62.** $z_1 = 3 - 4i, z_2 = -1 + 3i$

**63.** $z_1 = 2 - 6i, z_2 = -3 - 2i$

**64.** $z_1 = 1 + 4i, z_2 = -4 - 2i$

**65.** $z_1 = 3i, z_2 = 1 + i$

**66.** $z_1 = 4, z_2 = -3 + i$

*For each given complex number, determine its complex conjugate in trigonometric form.*

**67.** $3\left(\cos\frac{\pi}{4} + i\sin\frac{\pi}{4}\right)$

**68.** $\sqrt{2}\left(\cos\frac{\pi}{8} + i\sin\frac{\pi}{8}\right)$

**69.** $2\sqrt{3}(\cos(-20°) + i\sin(-20°))$

**70.** $9(\cos 14° + i\sin 14°)$

*Find the product of the given complex number and its complex conjugate in trigonometric form.*

**71.** $3\left(\cos\frac{\pi}{6} + i\sin\frac{\pi}{6}\right)$

**72.** $5\left(\cos\frac{\pi}{7} + i\sin\frac{\pi}{7}\right)$

**73.** $2(\cos 7° + i\sin 7°)$

**74.** $6(\cos 5° + i\sin 5°)$

*Solve each problem.*

**75.** Given that $z = 4 + 4i$, find $z^2$ by writing $z$ in trigonometric form and computing $z \cdot z$.

**76.** Given that $z = -3 + 3i$, find $z^2$ by writing $z$ in trigonometric form and computing $z \cdot z$.

**77.** Given that $z = 3 + 3i$, find $z^3$ by writing $z$ in trigonometric form and computing the product $z \cdot z \cdot z$.

**78.** Given that $z = \sqrt{3} + i$, find $z^4$ by writing $z$ in trigonometric form and computing the product $z \cdot z \cdot z \cdot z$.

## FOR WRITING/DISCUSSION

**79.** Show that the complex conjugate of $z = r(\cos\theta + i\sin\theta)$ is $\bar{z} = r(\cos(-\theta) + i\sin(-\theta))$.

**80.** Find the product of $z$ and $\bar{z}$, using the trigonometric forms from Exercise 79.

**81.** Show that the reciprocal of $z = r(\cos\theta + i\sin\theta)$ is $z^{-1} = r^{-1}(\cos\theta - i\sin\theta)$, provided $r \neq 0$.

**82.** Given that $z = r(\cos\theta + i\sin\theta)$, find $z^2$ and $z^{-2}$, provided $r \neq 0$.

**83.** Find the sum of $6(\cos 9° + i\sin 9°)$ and $3(\cos 5° + i\sin 5°)$. Find the sum of $1 + 3i$ and $5 - 7i$. Is it easier to add complex numbers in trigonometric form or standard form?

**84.** Find the quotient when $6(\cos 9° + i\sin 9°)$ is divided by $3(\cos 5° + i\sin 5°)$. Find the quotient when $1 + 3i$ is divided by $5 - 7i$. Is it easier to divide complex numbers in trigonometric form or standard form?

## ▶ RETHINKING

**85.** Find the dot product of the vectors $\langle -2, 6\rangle$ and $\langle 3, 5\rangle$.

**86.** Find the smallest positive angle between the vectors $\langle -3, 5\rangle$ and $\langle 1, 6\rangle$.

**87.** Solve the triangle in which $a = 5$, $b = 7$, and $c = 10$.

**88.** Find the area of the triangle in which $\alpha = 10.6°$, $b = 5.7$ feet, and $c = 12.2$ feet.

**89.** A tall building casts a shadow of length 230 feet when the angle of elevation of the sun is 48°. Find the height of the building.

**90.** Simplify the expression $\dfrac{1 - \sin^2 x \csc^2 x + \sin^2 x}{\cos^2 x}$.

## THINKING OUTSIDE THE BOX LV

*Two Squares and Two Triangles* The two squares with areas of 25 and 36 shown in the accompanying figure are positioned so that $AB = 7$. Find the exact area of triangle $TSC$.

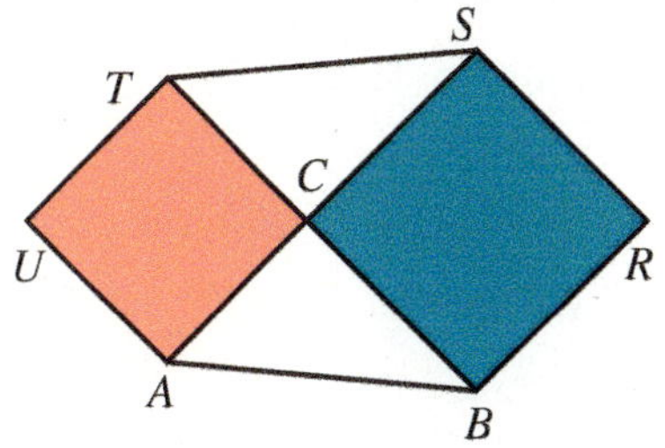


**Figure for Thinking Outside the Box LV**

## POP QUIZ 7.4

**1.** Find the absolute value of $3 - i$.

**2.** Write $5 + 5i$ in trigonometric form.

**3.** Write $2\sqrt{3}(\cos(5\pi/6) + i\sin(5\pi/6))$ in standard form.

**4.** Find $zw$ if $z = \sqrt{2}(\cos(\pi/12) + i\sin(\pi/12))$ and $w = 3\sqrt{2}(\cos(\pi/12) + i\sin(\pi/12))$.

**5.** Find the product of $\sqrt{2}(\cos(\pi/4) + i\sin(\pi/4))$ and its complex conjugate.

# 7.5 Powers and Roots of Complex Numbers

**HISTORICAL NOTE**

Abraham De Moivre (1667–1754) was a French mathematician famous for De Moivre's theorem and for his work on the normal distribution and probability theory. He was elected a Fellow of the Royal Society in 1697, and was a friend of Isaac Newton, Edmund Halley, and James Stirling. De Moivre correctly predicted the day of his own death. Noting that he was sleeping 15 minutes longer each day, De Moivre surmised that he would die on the day he would sleep for 24 hours, which yielded the date November 27, 1754. He did indeed pass away on that day.

We know that the square root of a negative number is an imaginary number, but we have not yet encountered roots of imaginary numbers. In this section we will use the trigonometric form of a complex number to find any power or root of a complex number.

## De Moivre's Theorem

The powers of a complex number can be found by repeated multiplication. For example,

$$\begin{aligned}(-\sqrt{3}+i)^3 &= (-\sqrt{3}+i)(-\sqrt{3}+i)(-\sqrt{3}+i)\\ &= (2-2i\sqrt{3})(-\sqrt{3}+i)\\ &= -2\sqrt{3}+2i+6i+2\sqrt{3}\\ &= 8i\end{aligned}$$

Using this method on a higher power such as $(-\sqrt{3}+i)^8$ would be rather tedious. However, with trigonometric form, any power of a complex number can be found quite easily.

Recall that the product of two complex numbers in trigonometric form is found by multiplying the moduli and adding the arguments. So if $z = r(\cos\theta + i\sin\theta)$, then

$$z^2 = r(\cos\theta + i\sin\theta)\cdot r(\cos\theta + i\sin\theta) = r^2(\cos 2\theta + i\sin 2\theta)$$

and

$$z^3 = r^2(\cos 2\theta + i\sin 2\theta)\cdot r(\cos\theta + i\sin\theta) = r^3(\cos 3\theta + i\sin 3\theta).$$

Of course, $z^4 = r^4(\cos 4\theta + i\sin 4\theta)$, and so on. This idea is stated in general as follows.

**De Moivre's Theorem**

> If $z = r(\cos\theta + i\sin\theta)$ is a complex number and $n$ is any positive integer, then
>
> $$z^n = r^n(\cos n\theta + i\sin n\theta).$$

De Moivre's theorem, named after the French mathematician Abraham De Moivre (1667–1754), can be proved by using mathematical induction, which is presented in Chapter 11.

### EXAMPLE 1 A power of a complex number

Use De Moivre's theorem to simplify $(-\sqrt{3}+i)^8$. Write the answer in the form $a + bi$.

#### Solution

First write $-\sqrt{3}+i$ in trigonometric form. The modulus is 2, and the argument is 150°. So

$$-\sqrt{3}+i = 2(\cos 150° + i\sin 150°).$$

Use De Moivre's theorem to find the eighth power:

$$\begin{aligned}(-\sqrt{3}+i)^8 &= [2(\cos 150° + i\sin 150°)]^8\\ &= 2^8[\cos(8\cdot 150°) + i\sin(8\cdot 150°)]\\ &= 256[\cos 1200° + i\sin 1200°]\end{aligned}$$

Since $1200° = 3\cdot 360° + 120°$,

$$\cos 1200° = \cos 120° = -\frac{1}{2} \quad\text{and}\quad \sin 1200° = \sin 120° = \frac{\sqrt{3}}{2}.$$

Use these values to simplify the trigonometric form of $(-\sqrt{3} + i)^8$:

$$(-\sqrt{3} + i)^8 = 256\left(-\frac{1}{2} + i\frac{\sqrt{3}}{2}\right) = -128 + 128i\sqrt{3}$$

▸**TRY THIS.** Simplify $(1 + i)^6$. ■

We have seen how De Moivre's theorem is used to find a positive integral power of a complex number. It is also used for finding roots of complex numbers.

## Roots of a Complex Number

In the real number system, $a$ is an $n$th root of $b$ if $a^n = b$. We have a similar definition of $n$th root in the complex number system.

**Definition: $n$th Root of a Complex Number**

> The complex number $w = a + bi$ is an $n$th root of the complex number $z$ if
> $$(a + bi)^n = z.$$

In Example 1, we saw that $(-\sqrt{3} + i)^8 = -128 + 128i\sqrt{3}$. So $-\sqrt{3} + i$ is an eighth root of $-128 + 128i\sqrt{3}$.

The definition of $n$th root does not indicate how to find an $n$th root, but a formula for all $n$th roots can be found by using trigonometric form. Suppose that the complex number $w$ is an $n$th root of the complex number $z$, where the trigonometric forms of $w$ and $z$ are

$$w = s(\cos\alpha + i\sin\alpha) \qquad \text{and} \qquad z = r(\cos\theta + i\sin\theta).$$

By De Moivre's theorem, $w^n = s^n(\cos n\alpha + i\sin n\alpha)$. Since $w^n = z$, we have

$$w^n = s^n(\cos n\alpha + i\sin n\alpha) = r(\cos\theta + i\sin\theta).$$

This last equation gives two different trigonometric forms for $w^n$. The absolute value of $w^n$ in one form is $s^n$ and in the other it is $r$. So $s^n = r$, or $s = r^{1/n}$. Because the argument of a complex number is any angle in the complex plane whose terminal side goes through the point representing the complex number, any two angles used for the argument must differ by a multiple of $2\pi$. Since the argument for $w^n$ is either $n\alpha$ or $\theta$, we have $n\alpha = \theta + 2k\pi$ or

$$\alpha = \frac{\theta + 2k\pi}{n}.$$

We can get $n$ different values for $\alpha$ from this formula by using $k = 0, 1, 2, \ldots, n - 1$. These results are summarized in the following theorem.

**Theorem: $n$th Roots of a Complex Number**

> For any positive integer $n$, the complex number $z = r(\cos\theta + i\sin\theta)$ has exactly $n$ distinct $n$th roots, given by the expression
> $$r^{1/n}\left[\cos\left(\frac{\theta + 2k\pi}{n}\right) + i\sin\left(\frac{\theta + 2k\pi}{n}\right)\right]$$
> for $k = 0, 1, 2, \ldots, n - 1$.

The $n$ distinct $n$th roots of a complex number all have the same modulus, $r^{1/n}$. What distinguishes between these roots is the $n$ different arguments determined by

$$\alpha = \frac{\theta + 2k\pi}{n}$$

for $k = 0, 1, 2, \ldots, n - 1$. If $\theta$ is measured in degrees, then the $n$ different arguments are determined by

$$\alpha = \frac{\theta + k360°}{n}$$

for $k = 0, 1, 2, \ldots, n - 1$. Note that if $k \geq n$, we get new values for $\alpha$ but no new $n$th roots, because the values of $\sin \alpha$ and $\cos \alpha$ have already occurred for some $k < n$. For example, if $k = n$, then $\alpha = \theta/n + 2\pi$, and the values for $\cos \alpha$ and $\sin \alpha$ are the same as they were for $\alpha = \theta/n$, which corresponds to $k = 0$.

## EXAMPLE 2 Finding *n*th roots

Find all of the fourth roots of the complex number $-8 + 8i\sqrt{3}$.

### Solution

First find the modulus of $-8 + 8i\sqrt{3}$:

$$\sqrt{(-8)^2 + (8\sqrt{3})^2} = 16$$

Since the terminal side of $\theta$ contains $(-8, 8\sqrt{3})$, we have $\cos \theta = -8/16 = -1/2$. Choose $\theta = 120°$. So the fourth roots are generated by the expression

$$16^{1/4}\left[\cos\left(\frac{120° + k360°}{4}\right) + i \sin\left(\frac{120° + k360°}{4}\right)\right].$$

Evaluating this expression for $k = 0, 1, 2$, and $3$ (because $n = 4$, $n - 1 = 3$), we get

$$2[\cos 30° + i \sin 30°] = \sqrt{3} + i$$
$$2[\cos 120° + i \sin 120°] = -1 + i\sqrt{3}$$
$$2[\cos 210° + i \sin 210°] = -\sqrt{3} - i$$
$$2[\cos 300° + i \sin 300°] = 1 - i\sqrt{3}.$$

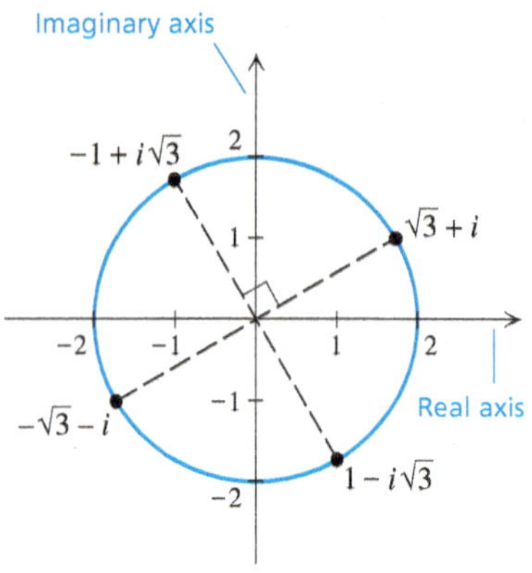


Fourth Roots of $-8 + 8i\sqrt{3}$

**Figure 7.46**

▶**TRY THIS.** Find all fourth roots of $-8 - 8i\sqrt{3}$. ■

The graph of the original complex number in Example 2 and its four fourth roots reveals an interesting pattern. Since the modulus of $-8 + 8i\sqrt{3}$ is 16, $-8 + 8i\sqrt{3}$ lies on a circle of radius 16 centered at the origin. Since the modulus of each fourth root is $16^{1/4} = 2$, the fourth roots lie on a circle of radius 2 centered at the origin, as shown in Fig. 7.46. Note how the fourth roots in Fig. 7.46 divide the circumference of the circle into four equal parts. In general, the $n$th roots of $z = r(\cos \theta + i \sin \theta)$ divide the circumference of a circle of radius $r^{1/n}$ into $n$ equal parts. This symmetric arrangement of the $n$th roots on a circle provides a simple check on whether the $n$th roots are correct.

Consider the equation $x^6 - 1 = 0$. By the fundamental theorem of algebra, this equation has six solutions in the complex number system if multiplicity is considered. Since this equation is equivalent to $x^6 = 1$, each solution is a sixth root of 1. The roots of 1 are called the **roots of unity.** The fundamental theorem of algebra does not guarantee that all of the roots are unique, but the theorem on the $n$th roots of a complex number does.

## EXAMPLE 3 The six sixth roots of unity

Solve the equation $x^6 - 1 = 0$.

### Solution

The solutions to the equation are the sixth roots of 1. Since $1 = 1(\cos 0° + i \sin 0°)$, the expression for the sixth roots is

$$1^{1/6}\left[\cos\left(\frac{0° + k360°}{6}\right) + i \sin\left(\frac{0° + k360°}{6}\right)\right].$$

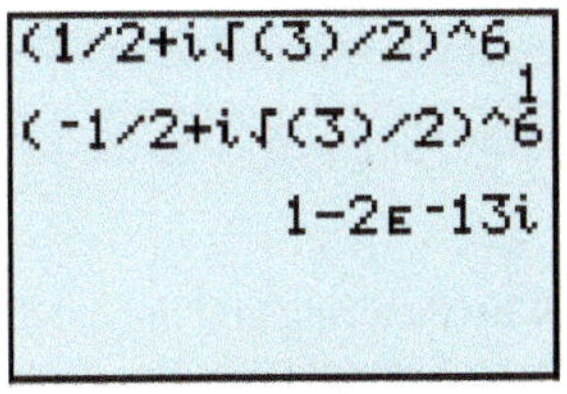


Figure 7.47

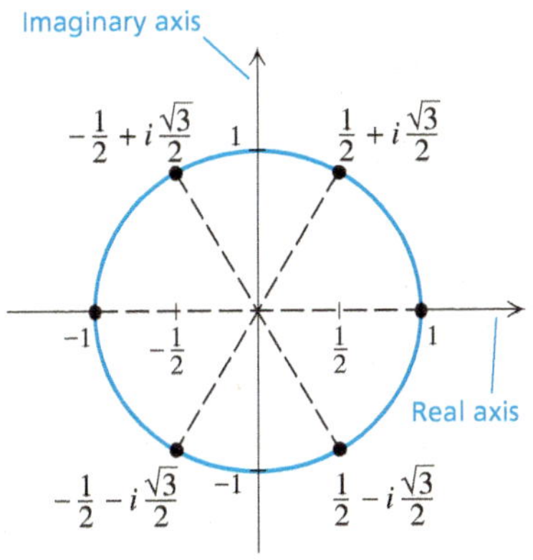


Sixth Roots of Unity

Figure 7.48

Evaluating this expression for $k = 0, 1, 2, \ldots, 5$ gives the six roots:

$$\cos 0° + i \sin 0° = 1$$

$$\cos 60° + i \sin 60° = \frac{1}{2} + i\frac{\sqrt{3}}{2}$$

$$\cos 120° + i \sin 120° = -\frac{1}{2} + i\frac{\sqrt{3}}{2}$$

$$\cos 180° + i \sin 180° = -1$$

$$\cos 240° + i \sin 240° = -\frac{1}{2} - i\frac{\sqrt{3}}{2}$$

$$\cos 300° + i \sin 300° = \frac{1}{2} - i\frac{\sqrt{3}}{2}$$

You can check the six roots using a graphing calculator that handles complex numbers, as shown in Fig. 7.47. Note that the calculator does not always get the exact answer.

▶**TRY THIS.** Find all complex solutions to $x^3 - 27 = 0$. ■

A graph of the sixth roots of unity found in Example 3 is shown in Fig. 7.48. Notice that they divide the circle of radius 1 into six equal parts. According to the conjugate pairs theorem of Chapter 3, the complex solutions to a polynomial equation with real coefficients occur in conjugate pairs. In Example 3, the complex solutions of $x^6 - 1 = 0$ did occur in conjugate pairs. This fact gives us another method of checking whether the six roots are correct. In general, the $n$th roots of any real number occur in conjugate pairs. Note that the fourth roots of the imaginary number in Example 2 did not occur in conjugate pairs.

## FOR thought... True or False? Explain.

1. $(2 + 3i)^2 = 4 + 9i^2$
2. If $z = 2(\cos 120° + i \sin 120°)$, then $z^3 = 8i$.
3. If $z$ is a complex number with modulus $r$, then the modulus of $z^4$ is $r^4$.
4. If the argument of $z$ is $\theta$, then the argument of $z^4$ is $\cos 4\theta$.
5. $[\cos(\pi/3) + i \sin(\pi/6)]^2 = \cos(2\pi/3) + i \sin(\pi/3)$
6. If $\cos \alpha = \cos \beta$, then $\alpha = \beta + 2k\pi$ for some integer $k$.
7. One of the fifth roots of 32 is $2(\cos 72° + i \sin 72°)$.
8. All solutions to $x^8 = 1$ in the complex plane lie on the unit circle.
9. All solutions to $x^8 = 1$ lie on the axes or the lines $y = \pm x$.
10. The equation $x^4 + 81 = 0$ has two real and two imaginary solutions.

## EXERCISES 7.5

*Use De Moivre's theorem to simplify each expression. Write the answer in the form $a + bi$.*

1. $[3(\cos 30° + i \sin 30°)]^3$
2. $[2(\cos 45° + i \sin 45°)]^5$
3. $[\sqrt{2}(\cos 120° + i \sin 120°)]^4$
4. $[\sqrt{3}(\cos 150° + i \sin 150°)]^3$
5. $[\cos(\pi/12) + i \sin(\pi/12)]^8$
6. $[\cos(\pi/6) + i \sin(\pi/6)]^9$
7. $[\sqrt{6}(\cos(2\pi/3) + i \sin(2\pi/3))]^4$

**8.** $[\sqrt{18}(\cos(5\pi/6) + i\sin(5\pi/6))]^3$

**9.** $[4.3(\cos 12.3° + i\sin 12.3°)]^5$

**10.** $[4.9(\cos 37.4° + i\sin 37.4°)]^6$

*Simplify each expression, by using trigonometric form and De Moivre's theorem. Write the answer in the form $a + bi$.*

**11.** $(2 + 2i)^3$ **12.** $(1 - i)^3$

**13.** $(\sqrt{3} - i)^4$ **14.** $(-2 + 2i\sqrt{3})^4$

**15.** $(-3 - 3i\sqrt{3})^5$ **16.** $(2\sqrt{3} - 2i)^5$

**17.** $(2 + 3i)^4$ **18.** $(4 - i)^5$

**19.** $(2 - i)^4$ **20.** $(-1 - 2i)^6$

**21.** $(1.2 + 3.6i)^3$ **22.** $(-2.3 - i)^3$

*Find the indicated roots. Express answers in trigonometric form.*

**23.** The square roots of $4(\cos 90° + i\sin 90°)$

**24.** The cube roots of $8(\cos 30° + i\sin 30°)$

**25.** The fourth roots of $\cos 120° + i\sin 120°$

**26.** The fifth roots of $32(\cos 300° + i\sin 300°)$

**27.** The sixth roots of $64(\cos\pi + i\sin\pi)$

**28.** The fourth roots of $16[\cos(3\pi/2) + i\sin(3\pi/2)]$

*Find the indicated roots in the form $a + bi$. Check by graphing the roots in the complex plane.*

**29.** The cube roots of 1 **30.** The cube roots of 8

**31.** The fourth roots of 16 **32.** The fourth roots of 1

**33.** The fourth roots of $-1$ **34.** The fourth roots of $-16$

**35.** The cube roots of $i$ **36.** The cube roots of $-8i$

**37.** The square roots of $-2 + 2i\sqrt{3}$

**38.** The square roots of $-4i$ **39.** The square roots of $1 + 2i$

**40.** The cube roots of $-1 + 3i$

*Solve each equation. Express answers in the form $a + bi$.*

**41.** $x^3 + 1 = 0$ **42.** $x^3 + 125 = 0$

**43.** $x^4 - 81 = 0$ **44.** $x^4 + 81 = 0$

**45.** $x^2 + 2i = 0$ **46.** $ix^2 + 3 = 0$

**47.** $x^7 - 64x = 0$ **48.** $x^9 - x = 0$

**49.** $x^5 + 5x^3 + 8x^2 + 40 = 0$

**50.** $x^5 + x^3 - 27x^2 - 27 = 0$

*Solve each equation. Express answers in trigonometric form.*

**51.** $x^5 - 2 = 0$ **52.** $x^5 + 3 = 0$

**53.** $x^4 + 3 - i = 0$ **54.** $ix^3 + 2 - i = 0$

*Solve each problem.*

**55.** Write the expression $[\cos(\pi/3) + i\sin(\pi/6)]^3$ in the form $a + bi$.

**56.** Solve the equation $x^6 - 1 = 0$ by factoring and the quadratic formula. Compare your answers to Example 3 of this section.

**57.** Solve $x^2 + (-1 + i)x - i = 0$.
**HINT** Use the quadratic formula and De Moivre's theorem.

**58.** Solve $x^2 + (-1 - 3i)x + (-2 + 2i) = 0$.
**HINT** Use the quadratic formula and De Moivre's theorem.

## FOR WRITING/DISCUSSION

**59.** Explain why $x^6 - 2x^3 + 1 = 0$ has three distinct solutions, $x^6 - 2x^3 = 0$ has four distinct solutions, and $x^6 - 2x = 0$ has six distinct solutions.

**60.** Find all real numbers $a$ and $b$ for which it is true that $\sqrt{a + bi} = \sqrt{a} + i\sqrt{b}$.

## ▶ RETHINKING

**61.** Find the absolute value of the complex number $3 + 5i$.

**62.** Write $4 - 4i$ in trigonometric form using degree measure for the argument.

**63.** Perform the indicated operation:

$$2(\cos(\pi/6) + i\sin(\pi/6)) \cdot 3(\cos(\pi/3) + i\sin(\pi/3))$$

Write the result in $a + bi$ form.

**64.** Find the amount of force required to push a 600-pound motorcycle up a ramp that is inclined at 20°.

**65.** Find the magnitude and direction angle for the vector $\langle -3, -9 \rangle$.

**66.** Solve the triangle in which $a = 5$, $b = 9$, and $c = 12$.

## THINKING OUTSIDE THE BOX LVI

*Double-Boxed* A rectangular box contains a delicate statue. The shipping department places the box containing the statue inside a 3 ft by 4 ft rectangular box as shown from above in the accompanying figure. If the box containing the statue is 1 ft wide, then what is its length? Find a four-decimal place approximation.

**Figure for Thinking Outside the Box LVI**

## POP QUIZ 7.5

1. Use De Moivre's theorem to simplify $(1 + i)^8$.

2. Find all fourth roots of $-16$.

# 7.6 Polar Equations

The Cartesian coordinate system is a system for describing the location of points in a plane. It is not the only coordinate system available. In this section we will study the **polar coordinate system.** In this coordinate system, a point is located by using a *directed distance* and an *angle* in a manner that will remind you of the magnitude and direction angle of a vector and the modulus and argument for a complex number.

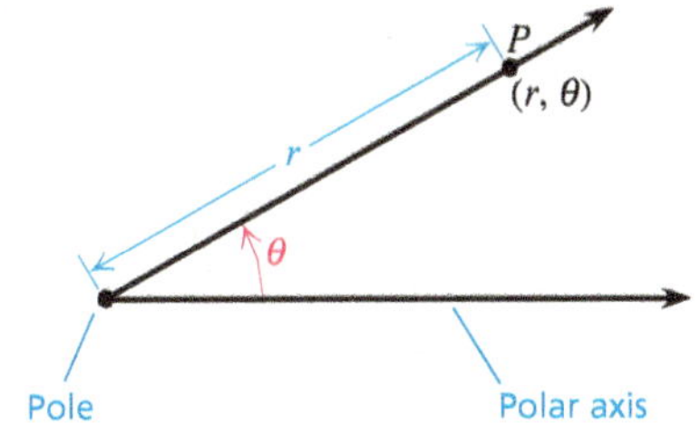


Figure 7.49

## Polar Coordinates

In the rectangular coordinate system, points are named according to their position with respect to an $x$-axis and a $y$-axis. In the polar coordinate system, we have a fixed point called the **pole** and a fixed ray called the **polar axis.** A point $P$ has coordinates $(r, \theta)$, where $r$ is the directed distance from the pole to $P$ and $\theta$ is an angle whose initial side is the polar axis and whose terminal side contains the point. See Fig. 7.49.

Since we are so familiar with rectangular coordinates, we retain the $x$- and $y$-axes when using polar coordinates. The pole is placed at the origin, and the polar axis is placed along the positive $x$-axis. The angle $\theta$ is any angle (in degrees or radians) in standard position whose terminal side contains the point. As usual, $\theta$ is positive for a counterclockwise rotation and negative for a clockwise rotation. In polar coordinates, $r$ can be any real number. For example, the ordered pair $(2, \pi/4)$ represents the point that lies two units from the origin on the terminal side of the angle $\pi/4$. The point $(0, \pi/4)$ is at the origin. The point $(-2, \pi/4)$ lies two units from the origin on the line through the terminal side of $\pi/4$ but in the direction opposite to $(2, \pi/4)$. See Fig. 7.50. Any ordered pair in polar coordinates names a single point, but the coordinates of a point in polar coordinates are not unique. For example, $(-2, \pi/4)$, $(2, 5\pi/4)$, and $(2, -3\pi/4)$ all name the same point.

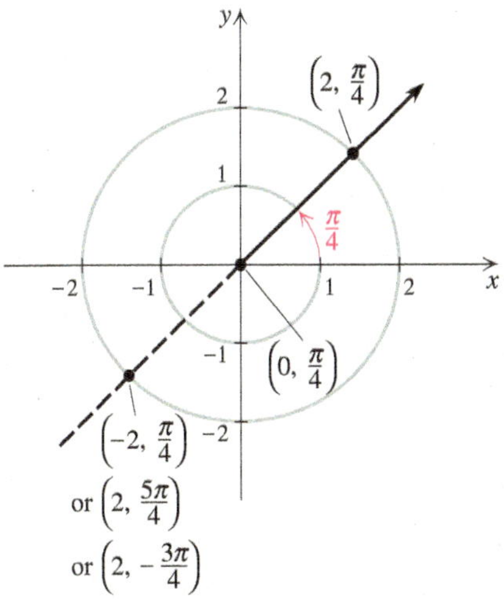


Figure 7.50

### EXAMPLE 1 Plotting points in polar coordinates

Plot the points whose polar coordinates are $(2, 5\pi/6)$, $(-3, \pi)$, $(1, -\pi/2)$, and $(-1, 450°)$.

#### Solution

The terminal side of $5\pi/6$ lies in the second quadrant, so $(2, 5\pi/6)$ is two units from the origin along this ray. The terminal side of the angle $\pi$ points in the direction of the negative $x$-axis, but since the first coordinate of $(-3, \pi)$ is negative, the point is located three units in the direction opposite to the direction of the ray. So $(-3, \pi)$ lies three units from the origin on the positive $x$-axis. The point $(-3, \pi)$ is the same point as $(3, 0)$. The terminal side of $-\pi/2$ lies on the negative $y$-axis. Thus $(1, -\pi/2)$ lies one unit from the origin on the negative $y$-axis. The terminal side of $450°$ lies on the positive $y$-axis. Since the first coordinate of $(-1, 450°)$ is negative, the point is located one unit in the opposite direction. The point $(-1, 450°)$ is the same as $(1, -\pi/2)$. All points are shown in Fig. 7.51.

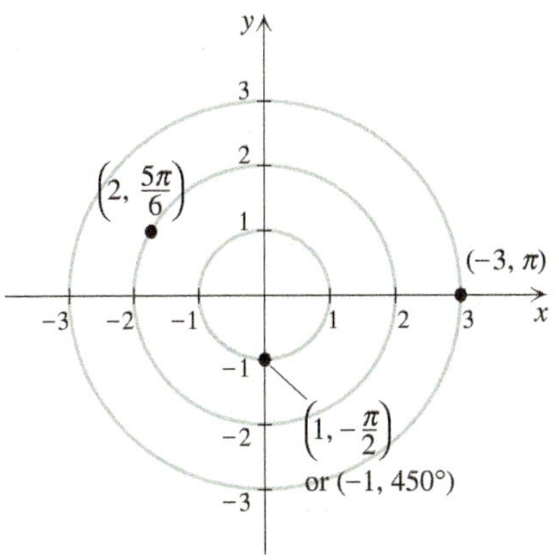


Figure 7.51

**TRY THIS.** Plot $(2, 3\pi/4)$ and $(-1, \pi/2)$ in polar coordinates. ■

If you search for *polar coordinate graph paper* on the Internet, you will find sites that have free grids that you can print like the one in Fig. 7.52. Using grids like this will make your graphing in polar coordinates easier and more accurate. Or, make copies of this page and use the grid given here.

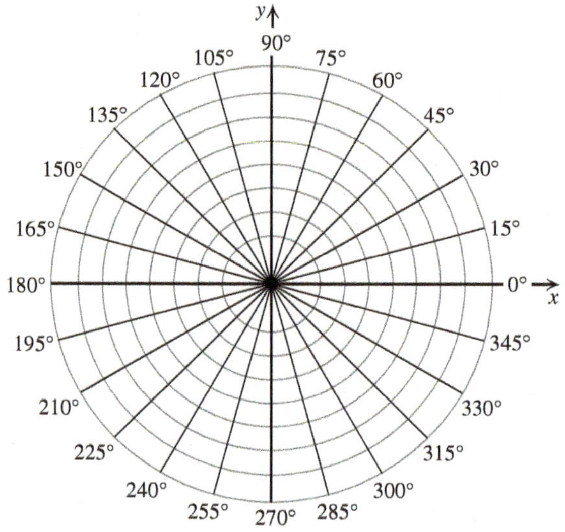


Figure 7.52

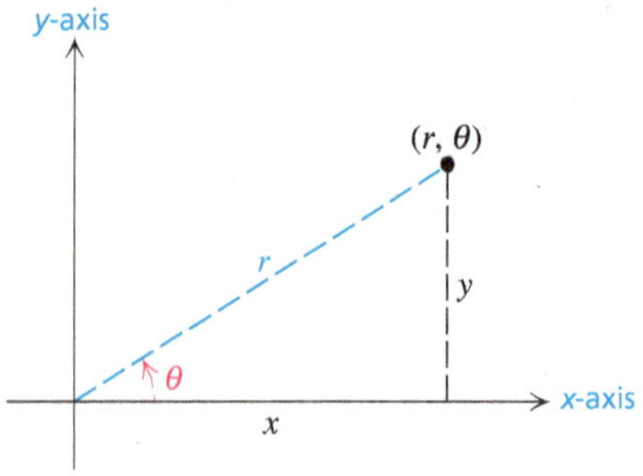


Figure 7.53

## Polar-Rectangular Conversions

To graph equations in polar coordinates, we must be able to switch the coordinates of a point in either system to the other. Suppose that $(r, \theta)$ is a point in the first quadrant with $r > 0$ and $\theta$ acute, as shown in Fig. 7.53. If $(x, y)$ is the same point in rectangular coordinates, then $x$ and $y$ are the lengths of the legs of the right triangle shown in Fig. 7.53. Using the Pythagorean theorem and trigonometric ratios, we have

$$x^2 + y^2 = r^2, \qquad x = r\cos\theta, \qquad y = r\sin\theta, \qquad \text{and} \qquad \tan\theta = \frac{y}{x}.$$

It can be shown that these equations hold for any point $(r, \theta)$, except that $\tan\theta$ is undefined if $x = 0$. The following rules for converting from one system to the other follow from these relationships.

**Polar-Rectangular Conversion Rules**

To convert $(r, \theta)$ to rectangular coordinates $(x, y)$, use

$$x = r\cos\theta \qquad \text{and} \qquad y = r\sin\theta.$$

To convert $(x, y)$ to polar coordinates $(r, \theta)$, use

$$r = \sqrt{x^2 + y^2}$$

and any angle $\theta$ in standard position whose terminal side contains $(x, y)$.

There are several ways to find an angle $\theta$ whose terminal side contains $(x, y)$. One way to find $\theta$ (provided $x \neq 0$) is to find an angle that satisfies $\tan\theta = y/x$ and goes through $(x, y)$. Remember that the angle $\tan^{-1}(y/x)$ is between $-\pi/2$ and $\pi/2$ and will go through $(x, y)$ only if $x > 0$. If $x = 0$, then the point $(x, y)$ is on the $y$-axis, and you can use $\theta = \pi/2$ or $\theta = -\pi/2$ as appropriate.

## EXAMPLE 2 Polar-rectangular conversion

**a.** Convert $(6, 210°)$ to rectangular coordinates.
**b.** Convert the rectangular coordinates $(-3, 6)$ to polar coordinates.

### Solution

**a.** Use $r = 6$, $\cos 210° = -\sqrt{3}/2$, and $\sin 210° = -1/2$ in the formulas $x = r\cos\theta$ and $y = r\sin\theta$:

$$x = 6\left(-\frac{\sqrt{3}}{2}\right) = -3\sqrt{3} \quad \text{and} \quad y = 6\left(-\frac{1}{2}\right) = -3$$

Figure 7.54

So $(6, 210°)$ in polar coordinates is $(-3\sqrt{3}, -3)$ in rectangular coordinates.
You can check this result with a graphing calculator, as shown in Fig. 7.54. □

**b.** To convert $(-3, 6)$ to polar coordinates, find $r$:

$$r = \sqrt{(-3)^2 + 6^2} = 3\sqrt{5}$$

Figure 7.55

Use a calculator to find $\tan^{-1}(-2) \approx -63.4°$. To get an angle whose terminal side contains $(-3, 6)$, use $\theta = 180° - 63.4° = 116.6°$. So $(-3, 6)$ in polar coordinates is $(3\sqrt{5}, 116.6°)$. Since there are infinitely many representations for any point in polar coordinates, this answer is not unique. In fact, another possibility is $(-3\sqrt{5}, -63.4°)$.
You can check this result with a graphing calculator, as shown in Fig. 7.55.

▸**TRY THIS.** Convert $(3, 45°)$ to rectangular coordinates and convert $(-2, 2\sqrt{3})$ to polar coordinates. ■

## Polar Equations

An equation in two variables (typically $x$ and $y$) that is graphed in the rectangular coordinate system is called a **rectangular** or **Cartesian equation.** An equation in two variables (typically $r$ and $\theta$) that is graphed in the polar coordinate system is called a **polar equation.** Certain polar equations are easier to graph than the equivalent Cartesian equations. We can graph a polar equation in the same way that we graph a rectangular equation; that is, we can simply plot enough points to get the shape of the graph. However, since most of our polar equations involve trigonometric functions, finding points on these curves can be tedious. A graphing calculator can be used to great advantage here.

## EXAMPLE 3 Graphing a polar equation

Sketch the graph of the polar equation $r = 2\cos\theta$.

### Solution

If $\theta = 0°$, then $r = 2\cos 0° = 2$. So, the ordered pair $(2, 0°)$ is on the graph. If $\theta = 30°$, then $r = 2\cos 30° = \sqrt{3}$, and $(\sqrt{3}, 30°)$ is on the graph. These ordered pairs and several others that satisfy the equation are listed in the following table:

| $\theta$ | 0° | 30° | 45° | 60° | 90° | 120° | 135° | 150° | 180° |
|---|---|---|---|---|---|---|---|---|---|
| $r$ | 2 | $\sqrt{3}$ | $\sqrt{2}$ | 1 | 0 | −1 | $-\sqrt{2}$ | $-\sqrt{3}$ | −2 |

Plot these points and draw a smooth curve through them to get the graph shown in Fig. 7.56. If $\theta$ is larger than 180° or smaller than 0°, we get different ordered pairs, but they are all located on the curve drawn in Fig. 7.56. For example, $(-\sqrt{3}, 210°)$ satisfies $r = 2 \cos \theta$, but it has the same location as $(\sqrt{3}, 30°)$.

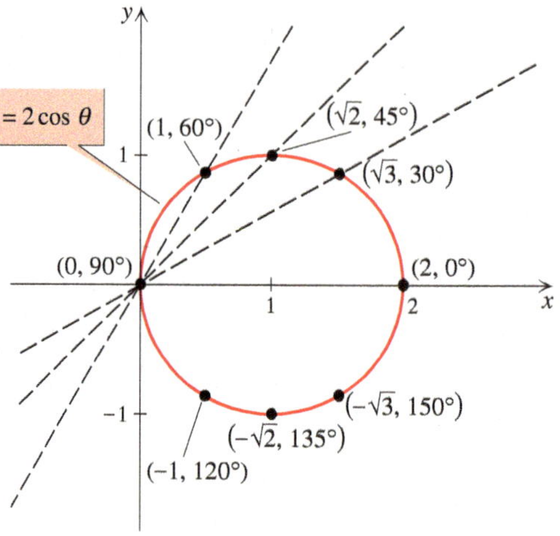


Figure 7.56

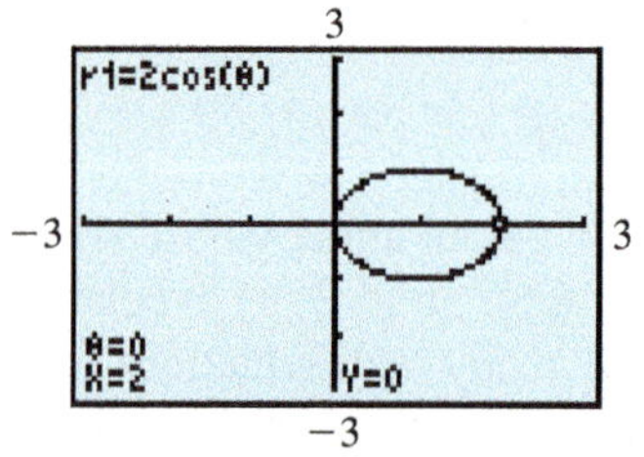


Figure 7.57

To check this graph with a calculator, set the mode to polar and enter $r = 2 \cos \theta$. The graph in Fig. 7.57 supports the graph in Fig. 7.56.

▶**TRY THIS.** Graph $r = 4 \sin \theta$ in polar coordinates. ■

The graph $r = 2 \cos \theta$ in Fig. 7.56 looks like a circle. To verify that it is a circle, we can convert the polar equation to an equivalent rectangular equation because we know the form of the equation of a circle in rectangular coordinates. This conversion is done in Example 6.

In the next example, a simple polar equation produces a curve that is not usually graphed when studying rectangular equations because the equivalent rectangular equation is quite complicated.

## EXAMPLE 4 Graphing a polar equation

Sketch the graph of the polar equation $r = 3 \sin 2\theta$.

### Solution

The ordered pairs in the following table satisfy the equation $r = 3 \sin 2\theta$. The values of $r$ are rounded to the nearest tenth.

| $\theta$ | 0° | 15° | 30° | 45° | 60° | 90° | 135° | 180° | 225° | 270° | 315° | 360° |
|---|---|---|---|---|---|---|---|---|---|---|---|---|
| $r$ | 0 | 1.5 | 2.6 | 3 | 2.6 | 0 | −3 | 0 | 3 | 0 | −3 | 0 |

As $\theta$ varies from 0° to 90°, the value of $r$ goes from 0 to 3, then back to 0, creating a loop in quadrant I. As $\theta$ varies from 90° to 180°, the value of $r$ goes from 0 to −3 then back to 0, creating a loop in quadrant IV. As $\theta$ varies from 180° to 270°, a loop in quadrant III is formed; and as $\theta$ varies from 270° to 360°, a loop in quadrant II is formed. If $\theta$ is chosen greater than 360° or less than 0°, we get the same points over and over because the sine function is periodic. The graph of $r = 3 \sin 2\theta$ is shown in Fig. 7.58. The graph is called a **four-leaf rose.**

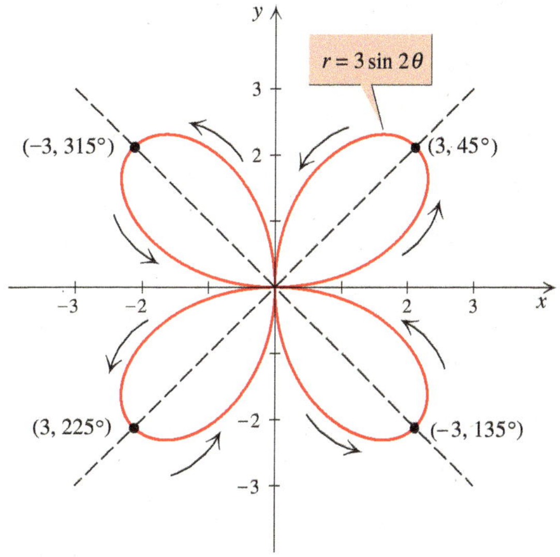


Figure 7.58

With a graphing calculator you can make a table of ordered pairs, as shown in Fig. 7.59. Make this table yourself and scroll through the table to see how the radius oscillates between 3 and −3 as the angle varies. The calculator graph of $r = 3\sin(2\theta)$ in polar mode with $\theta$ between 0° and 360° is shown in Fig. 7.60. This graph supports the graph shown in Fig. 7.58.

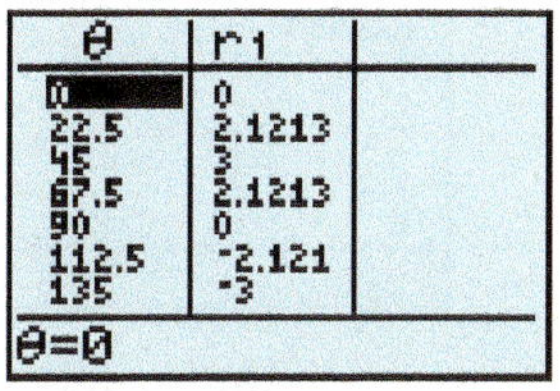


Figure 7.59

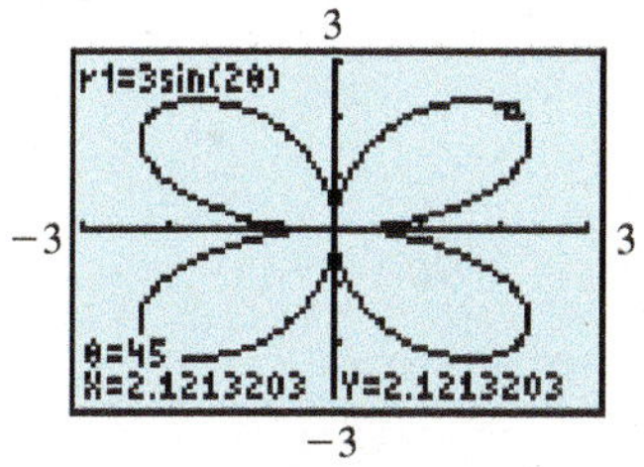


Figure 7.60

▶**TRY THIS.** Graph $r = \cos 2\theta$ in polar coordinates. ■

## EXAMPLE 5 Graphing a polar equation

Sketch the graph of the polar equation $r = \theta$, where $\theta$ is in radians and $\theta \geq 0$.

### Solution

The ordered pairs in the following table satisfy $r = \theta$. The values of $r$ are rounded to the nearest tenth.

| $\theta$ | 0 | $\pi/4$ | $\pi/2$ | $3\pi/4$ | $\pi$ | $3\pi/2$ | $2\pi$ | $3\pi$ |
|---|---|---|---|---|---|---|---|---|
| $r$ | 0 | 0.8 | 1.6 | 2.4 | 3.1 | 4.7 | 6.3 | 9.4 |

The graph of $r = \theta$, called **the spiral of Archimedes,** is shown as a smooth curve in Fig. 7.61 on the next page. As the value of $\theta$ increases, the value of $r$ increases, causing the graph to spiral out from the pole. There is no repetition of points as there was in Examples 3 and 4, because no periodic function is involved.

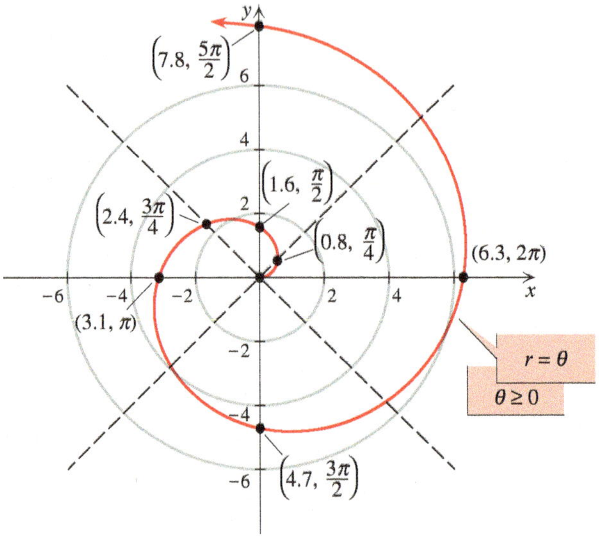


**Figure 7.61**

With a graphing calculator you can make a table of ordered pairs for $r = \theta$, as shown in Fig. 7.62. The calculator graph of $r = \theta$ in polar coordinates for $\theta$ ranging from 0 through 40 radians is shown in Fig. 7.63. This graph shows much more of the spiral than Fig. 7.61. It supports the conclusion that the graph of $r = \theta$ is a spiral.

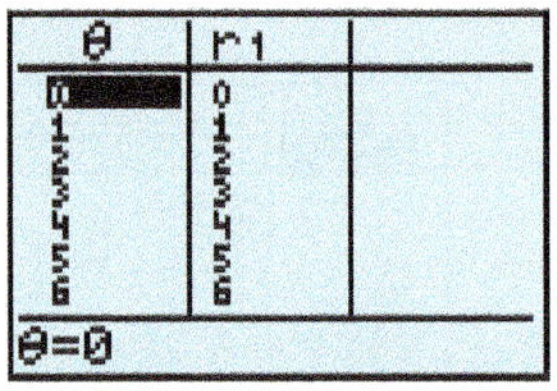


**Figure 7.62**

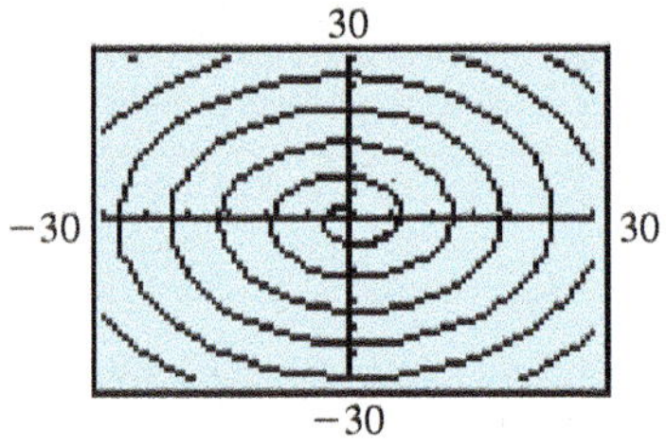


**Figure 7.63**

▶**TRY THIS.** Graph $r = -\theta$ for $\theta$ in radians and $\theta \geq 0$. ■

The Function Gallery on page 542 shows the graphs of several types of polar equations.

## Converting Equations

We know that certain types of rectangular equations have graphs that are particular geometric shapes, such as lines, circles, and parabolas. We can use our knowledge of equations in rectangular coordinates with equations in polar coordinates (and vice versa) by converting the equations from one system to the other. For example, we can determine whether the graph of $r = 2 \cos \theta$ in Example 3 is a circle by finding the equivalent Cartesian equation and deciding whether it is the equation of a circle. When converting from one system to another, we use the relationships

$$x^2 + y^2 = r^2, \qquad x = r \cos \theta, \qquad \text{and} \qquad y = r \sin \theta.$$

### EXAMPLE 6 Converting a polar equation to a rectangular equation

Write an equivalent rectangular equation for the polar equation $r = 2 \cos \theta$.

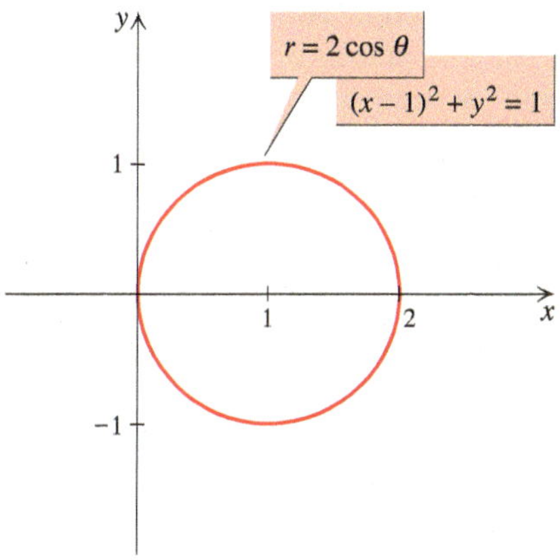


Figure 7.64

### Solution

First multiply each side of $r = 2\cos\theta$ by $r$ to get $r^2 = 2r\cos\theta$. Now eliminate $r$ and $\theta$ by making substitutions using $r^2 = x^2 + y^2$ and $x = r\cos\theta$:

$$r^2 = 2r\cos\theta$$
$$x^2 + y^2 = 2x$$

From Section 1.3, the standard equation of a circle with center $(h, k)$ and radius $r$ is $(x - h)^2 + (y - k)^2 = r^2$. Complete the square to get $x^2 + y^2 = 2x$ into the standard form of the equation of a circle:

$$x^2 - 2x + y^2 = 0$$
$$x^2 - 2x + 1 + y^2 = 0 + 1$$
$$(x - 1)^2 + y^2 = 1$$

Since we recognize the rectangular equation as the equation of a circle centered at $(1, 0)$ with radius 1, the graph of $r = 2\cos\theta$ shown in Fig. 7.64 is a circle centered at $(1, 0)$ with radius 1.

▶**TRY THIS.** Convert $r = 3\sin\theta$ to a rectangular equation. ■

In the next example we convert the rectangular equation of a line and circle into polar coordinates.

## EXAMPLE 7 Converting a rectangular equation to a polar equation

For each rectangular equation, write an equivalent polar equation.

**a.** $y = 3x - 2$ **b.** $x^2 + y^2 = 9$

### Solution

**a.** Substitute $x = r\cos\theta$ and $y = r\sin\theta$, and then solve for $r$:

$$y = 3x - 2$$
$$r\sin\theta = 3r\cos\theta - 2$$
$$r\sin\theta - 3r\cos\theta = -2$$
$$r(\sin\theta - 3\cos\theta) = -2$$
$$r = \frac{-2}{\sin\theta - 3\cos\theta}$$

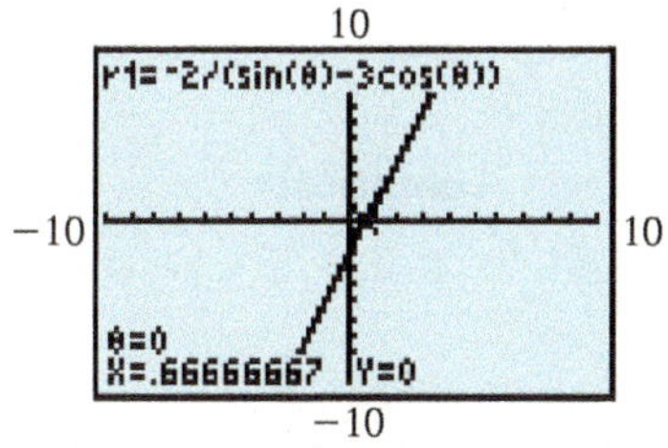


Figure 7.65

The graph in Fig. 7.65 supports the conclusion that we have found the polar coordinate form for the line $y = 3x - 2$. □

**b.** Substitute $r^2 = x^2 + y^2$ to get polar coordinates:

$$x^2 + y^2 = 9$$
$$r^2 = 9$$
$$r = \pm 3$$

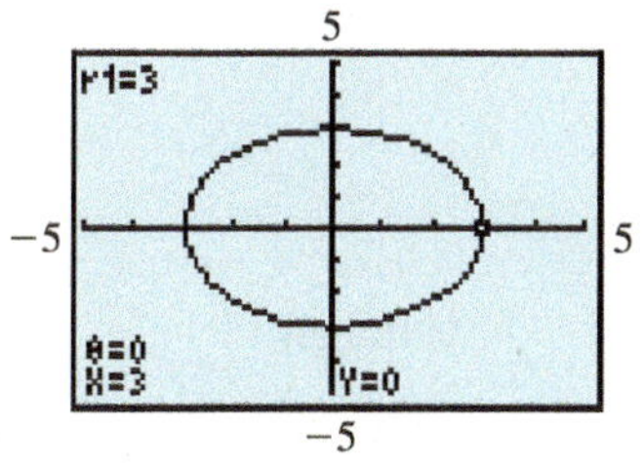


Figure 7.66

A polar equation for a circle of radius 3 centered at the origin is $r = 3$. The equation $r = -3$ is also a polar equation for the same circle.

The graph in Fig. 7.66 supports the conclusion that $r = 3$ is a circle in polar coordinates. Graph $r = 3$ on your calculator, using a viewing window that makes the circle look round.

▶**TRY THIS.** Convert $y = -2x + 5$ into a polar equation. ■

In Example 7 we saw that a straight line has a rather simple equation in rectangular coordinates but a more complicated equation in polar coordinates. A circle centered at the origin has a very simple equation in polar coordinates but a more complicated equation in rectangular coordinates. The graphs of simple polar equations are typically circular or somehow "centered" at the origin.

## FUNCTION gallery... Functions in Polar Coordinates

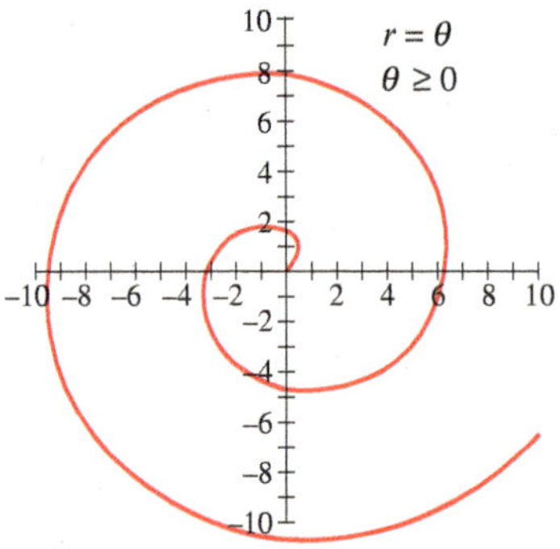


**Spiral**

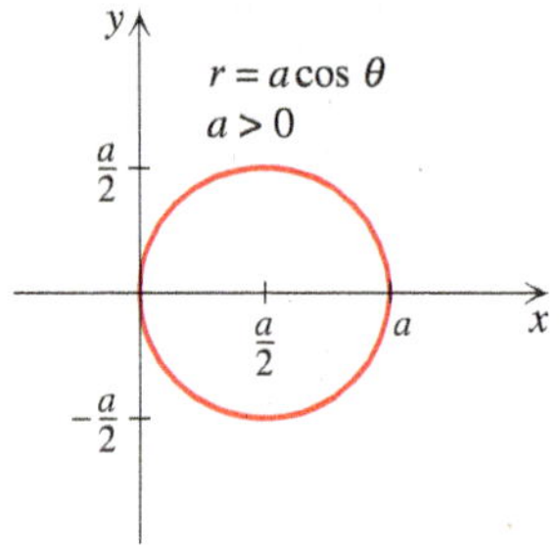


**Circle**

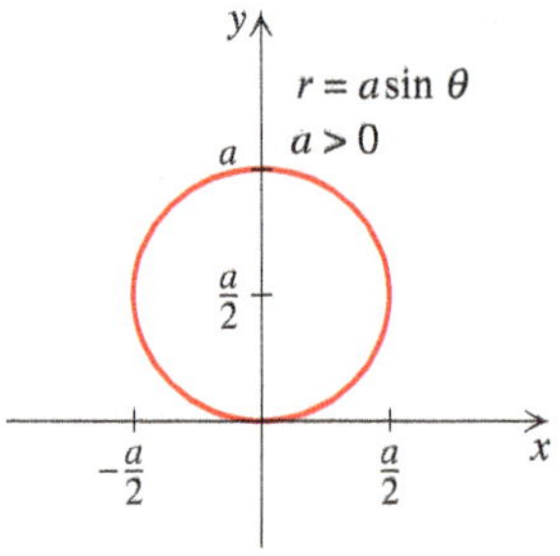


**Circle**

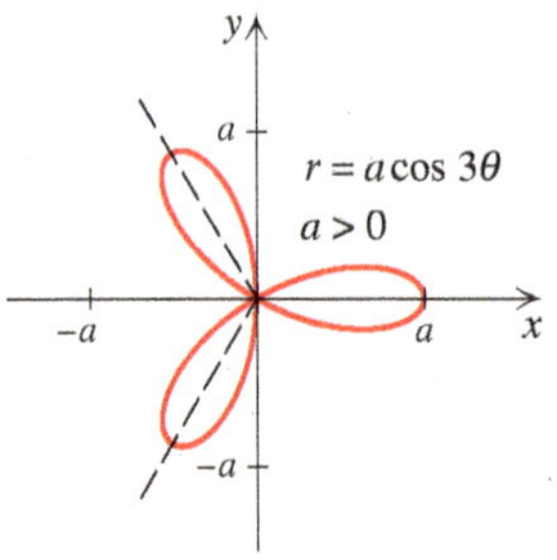


**Three-Leaf Rose**

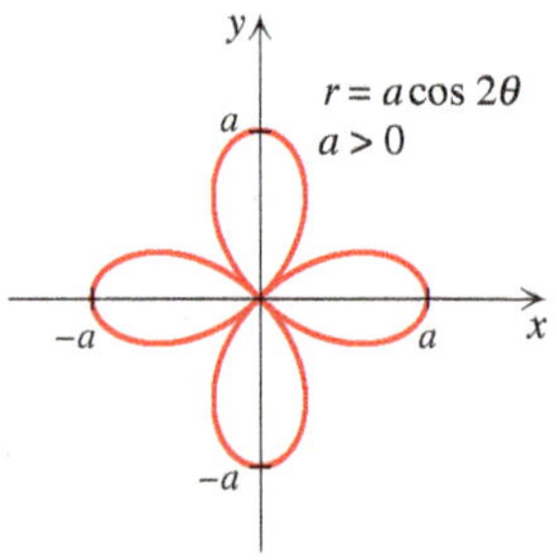


**Four-Leaf Rose**

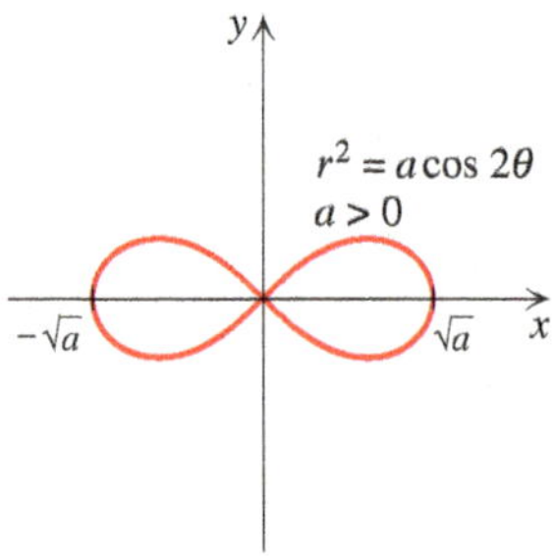


**Lemniscate**

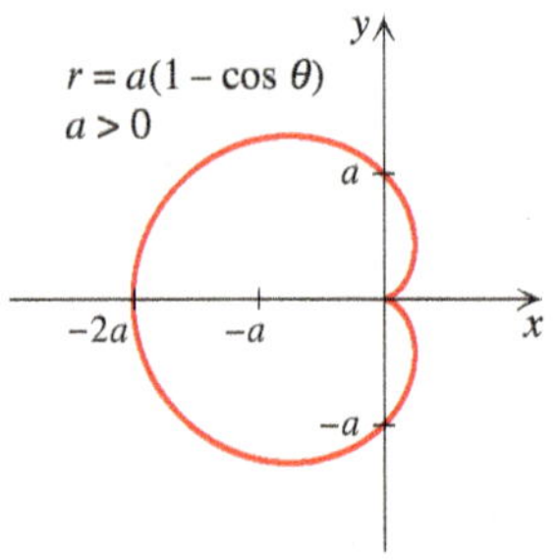


**Cardioid**

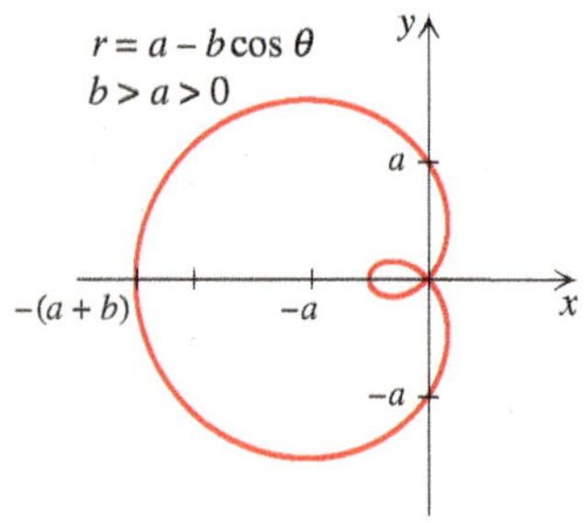


**Limaçon**

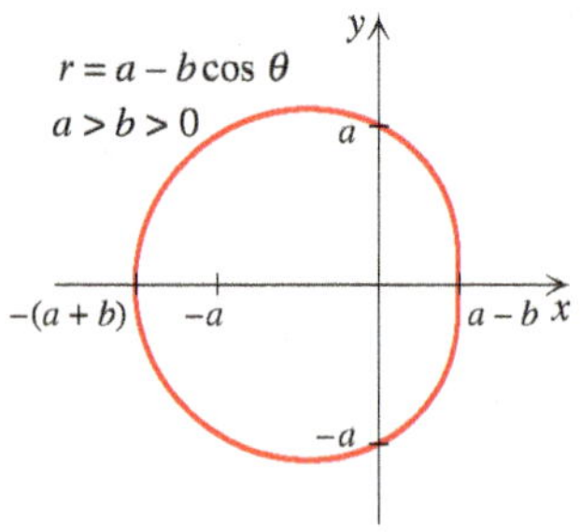


**Limaçon**

## FOR thought... True or False? Explain.

**1.** The distance of the point $(r, \theta)$ from the origin depends only on $r$.

**2.** The distance of the point $(r, \theta)$ from the origin is $r$.

**3.** The ordered pairs $(2, \pi/4)$, $(2, -3\pi/4)$, and $(-2, 5\pi/4)$ all represent the same point in polar coordinates.

**4.** The equations relating rectangular and polar coordinates are $x = r\sin\theta$, $y = r\cos\theta$, and $x^2 + y^2 = r^2$.

**5.** The point $(-4, 225°)$ in polar coordinates is $(2\sqrt{2}, 2\sqrt{2})$ in rectangular coordinates.

**6.** The graph of $0 \cdot r + \theta = \pi/4$ in polar coordinates is a straight line.

**7.** The graphs of $r = 5$ and $r = -5$ are identical.

**8.** The ordered pairs $(-\sqrt{2}/2, \pi/3)$ and $(\sqrt{2}/2, \pi/3)$ satisfy $r^2 = \cos 2\theta$.

**9.** The graph of $r = 1/\sin\theta$ is a vertical line.

**10.** The graphs of $r = \theta$ and $r = -\theta$ are identical.

## EXERCISES 7.6

*Fill in the blank.*

**1.** In the polar coordinate system, the ________ axis is placed along the positive $x$-axis.

**2.** In the polar coordinate system, the ________ is placed at the origin.

*Find polar coordinates for each given point, using radian measure for the angle.*

**3.**

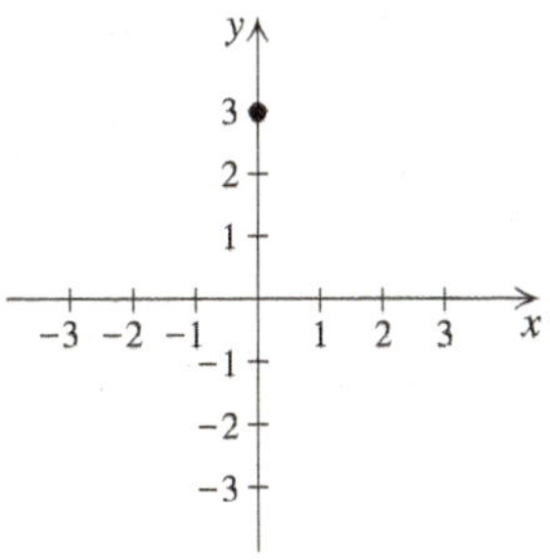

**4.**

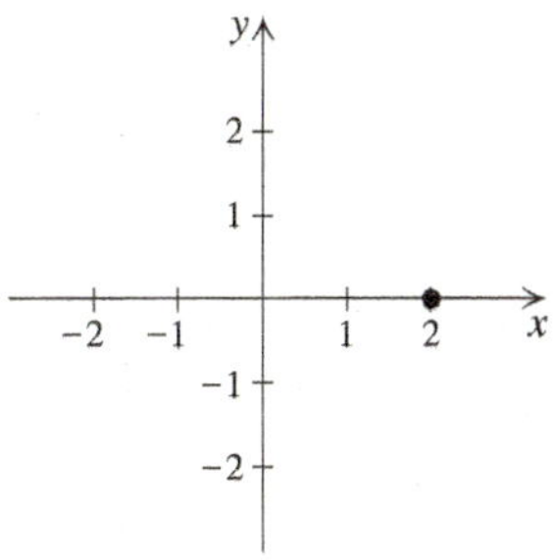

**5.**

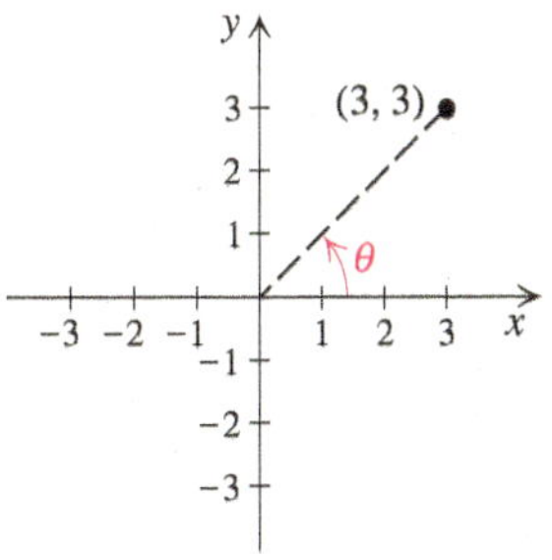


**6.**

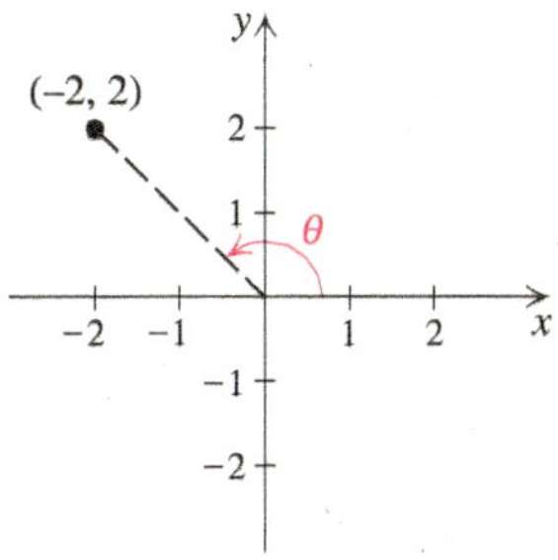


*Plot the points whose polar coordinates are given.*

**7.** $(2, 0°)$ **8.** $(-3, 0°)$ **9.** $(0, 35°)$

**10.** $(0, 90°)$ **11.** $(3, \pi/6)$ **12.** $(2, \pi/4)$

**13.** $(-2, 2\pi/3)$ **14.** $(-1, \pi/6)$ **15.** $(2, -\pi/4)$

**16.** $(1, -2\pi/3)$ **17.** $(3, -225°)$ **18.** $(2, -180°)$

**19.** $(-2, 45°)$ **20.** $(-3, 30°)$ **21.** $(4, 390°)$

**22.** $(3, 13\pi/6)$

*Convert the polar coordinates of each point to rectangular coordinates.*

**23.** $(4, 0°)$ **24.** $(-5, 0°)$

**25.** $(0, \pi/4)$ **26.** $(0, -\pi)$

**27.** $(1, \pi/6)$ **28.** $(2, \pi/4)$

**29.** $(-3, 3\pi/2)$ **30.** $(-2, 2\pi)$

**31.** $(\sqrt{2}, 135°)$ **32.** $(\sqrt{3}, 150°)$

**33.** $(-\sqrt{6}, -60°)$ **34.** $(-\sqrt{2}/2, -45°)$

*Convert the rectangular coordinates of each point to polar coordinates. Use degrees for $\theta$.*

**35.** $(\sqrt{3}, 3)$ **36.** $(4, 4)$ **37.** $(-2, 2)$

**38.** $(-2, 2\sqrt{3})$ **39.** $(0, 2)$ **40.** $(-2, 0)$

**41.** $(-3, -3)$ **42.** $(2, -2)$ **43.** $(1, 4)$

**44.** $(-2, 3)$ **45.** $(\sqrt{2}, -2)$ **46.** $(-2, -\sqrt{3})$

*Convert the polar coordinates of each point to rectangular coordinates rounded to the nearest hundredth.*

**47.** $(4, 26°)$ **48.** $(-5, 33°)$ **49.** $(2, \pi/7)$

**50.** $(3, 2\pi/9)$ **51.** $(-2, 1.1)$ **52.** $(6, 2.3)$

*Convert the rectangular coordinates of each point to polar coordinates. Round r to the nearest tenth and θ to the nearest tenth of a degree.*

**53.** $(4, 5)$

**54.** $(-5, 3)$

**55.** $(-2, -7)$,

**56.** $(3, -8)$

*Sketch the graph of each polar equation.*

**57.** $r = 2 \sin \theta$

**58.** $r = 3 \cos \theta$

**59.** $r = 3 \cos 2\theta$

**60.** $r = -2 \sin 2\theta$

**61.** $r = 2\theta$ for $\theta$ in radians

**62.** $r = \theta$ for $\theta \leq 0$ and $\theta$ in radians

**63.** $r = 1 + \cos \theta$ (cardioid)

**64.** $r = 1 - \cos \theta$ (cardioid)

**65.** $r^2 = 9 \cos 2\theta$ (lemniscate)

**66.** $r^2 = 4 \sin 2\theta$ (lemniscate)

**67.** $r = 4 \cos 2\theta$ (four-leaf rose)

**68.** $r = 3 \sin 2\theta$ (four-leaf rose)

**69.** $r = 2 \sin 3\theta$ (three-leaf rose)

**70.** $r = 4 \cos 3\theta$ (three-leaf rose)

**71.** $r = 1 + 2 \cos \theta$ (limaçon)

**72.** $r = 2 + \cos \theta$ (limaçon)

**73.** $r = 3.5$

**74.** $r = -5$

**75.** $\theta = 30°$

**76.** $\theta = 3\pi/4$

*For each polar equation, write an equivalent rectangular equation.*

**77.** $r = 4 \cos \theta$

**78.** $r = 2 \sin \theta$

**79.** $r = \dfrac{3}{\sin \theta}$

**80.** $r = \dfrac{-2}{\cos \theta}$

**81.** $r = 3 \sec \theta$

**82.** $r = 2 \csc \theta$

**83.** $r = 5$

**84.** $r = -3$

**85.** $\theta = \dfrac{\pi}{4}$

**86.** $\theta = 0$

**87.** $r = \dfrac{2}{1 - \sin \theta}$

**88.** $r = \dfrac{3}{1 + \cos \theta}$

*For each rectangular equation, write an equivalent polar equation.*

**89.** $x = 4$

**90.** $y = -6$

**91.** $y = -x$

**92.** $y = x\sqrt{3}$

**93.** $x^2 = 4y$

**94.** $y^2 = 2x$

**95.** $x^2 + y^2 = 4$

**96.** $2x^2 + y^2 = 1$

**97.** $y = 2x - 1$

**98.** $y = -3x + 5$

**99.** $x^2 + (y - 1)^2 = 1$

**100.** $(x + 1)^2 + y^2 = 4$

*Graph each pair of polar equations on the same screen of your calculator and use the trace feature to estimate the polar coordinates of all points of intersection of the curves. Check your calculator manual to see how to graph polar equations on your calculator.*

**101.** $r = 1, r = 2 \sin 3\theta$

**102.** $r = \sin \theta, r = \sin 2\theta$

**103.** $r = 3 \sin 2\theta, r = 1 - \cos \theta$

**104.** $r = 3 \sin 4\theta, r = 2$

## FOR WRITING/DISCUSSION

**105.** Explain why $\theta$ must be radian measure for the equation $r = 2\theta$, but $\theta$ can be in radians or degrees for the equation $r = 2 \cos \theta$.

**106.** Show that a polar equation for the straight line $y = mx$ is $\theta = \tan^{-1} m$.

## ▶ RETHINKING

**107.** Use De Moivre's theorem to simplify $(1 + i)^{12}$.

**108.** Find all fourth roots of $-8 - 8i\sqrt{3}$.

**109.** Find all solutions in radians to $\sin x = 0.88$. Round to the nearest hundredth.

**110.** Simplify $\dfrac{\cos(2x)}{\sin^2(x)} - \csc^2(x)$.

**111.** Find a function of the form $y = A\cos(B(x - C)) + D$ that has the same graph as $y = \sin(2x - \pi/4)$.

**112.** The five key points on one cycle of a sine wave are $(\pi/3, -1)$, $(2\pi/3, -2)$, $(\pi, -3)$, $(4\pi/3, -2)$, and $(5\pi/3, -1)$. Find the equation of the wave in the form $y = A\cos(B(x - C)) + D$.

## THINKING OUTSIDE THE BOX LVII

*Laying Pipe* A circular pipe with radius 1 is placed in a V-shaped trench whose sides form an angle of $\theta$ radians. In the cross section shown here, the pipe touches the sides of the trench at points $A$ and $B$.

**a.** Find the area inside the circle and below the line segment $AB$ in terms of $\theta$ (the blue area).

**b.** Find the area below the circle and inside the trench in terms of $\theta$ (the red area).

**c.** If $\theta$ is nearly equal to $\pi$, then the blue area and the red area are both very small. If $\theta$ is equal to $\pi$ both areas are zero. As $\theta$ approaches $\pi$ there is a limit to the ratio of the blue area to the red area. Use a calculator to determine this limit.

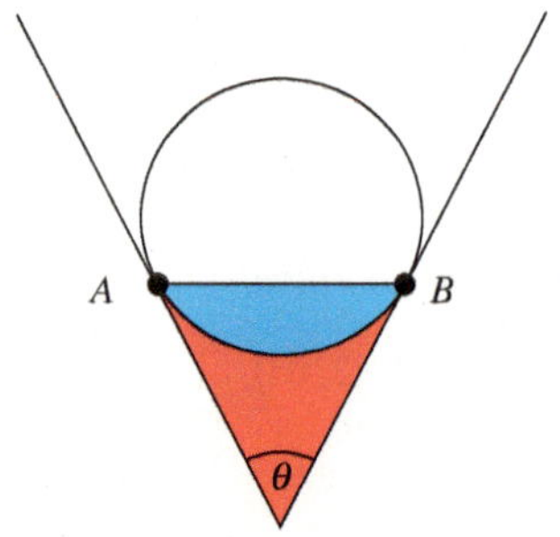


**Figure for Thinking Outside the Box LVII**

## POP QUIZ 7.6

1. If the polar coordinates of a point are $(-1, 15\pi/4)$, then in which quadrant does the point lie?
2. Convert $(4, 150°)$ to rectangular coordinates.
3. Convert the rectangular coordinates $(-2, 2)$ to polar coordinates using radians for the angle.
4. Convert the polar equation $r = 4\cos\theta$ into a rectangular equation.
5. Find the center and radius of the circle $r = -16\sin\theta$.
6. Convert $y = x + 1$ into polar coordinates.

# 7.7 Parametric Equations

We know how to graph points in the plane using the rectangular coordinate system and using polar coordinates. Points in the plane can also be located using parametric equations. We have already used parametric equations when we gave the $x$- and $y$-coordinates of a projectile as functions of time in Chapter 6. In this section we will study parametric equations in more detail.

## Graphs of Parametric Equations

If $f(t)$ and $g(t)$ are functions of $t$, where $t$ is in some interval of real numbers, then the equations $x = f(t)$ and $y = g(t)$ are called **parametric equations.** The variable $t$ is called the **parameter** and the graph of the parametric equations is said to be defined **parametrically.** If the parameter is thought of as time, then we know when each point of the graph is plotted. If no interval is specified for $t$, then $t$ is assumed to be any real number for which both $f(t)$ and $g(t)$ are defined.

### EXAMPLE 1 Graphing a line segment

Graph the parametric equations $x = 3t - 2$ and $y = t + 1$ for $t$ in the interval $[0, 3]$. Determine the domain (the set of $x$-coordinates) and the range (the set of $y$-coordinates) for the function or relation that you graphed.

**Solution**

Make a table of ordered pairs corresponding to values of $t$ between 0 and 3:

| $t$ | $x$ | $y$ |
|---|---|---|
| 0 | −2 | 1 |
| 1 | 1 | 2 |
| 2 | 4 | 3 |
| 3 | 7 | 4 |

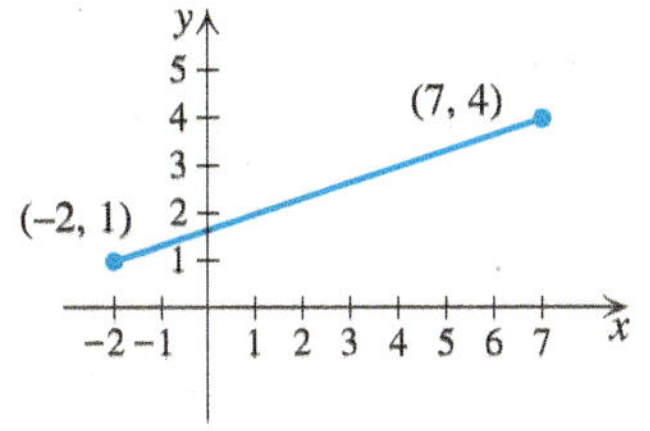

Figure 7.67

Since $x = 3t - 2$, we have $t = \frac{x+2}{3}$. Since $t$ is in the interval $[0, 3]$, we have $0 \le \frac{x+2}{3} \le 3$. Solving for $x$, we get $-2 \le x \le 7$. So $x$ is in the interval $[-2, 7]$. The graph is not the whole line. It is the line segment with endpoints $(-2, 1)$ and $(7, 4)$ shown in Fig. 7.67. We use solid dots at the end points to show that they are included in the graph. The domain is the interval $[-2, 7]$ and the range is the interval $[1, 4]$.

To check Example 1 with a graphing calculator, set your calculator to parametric mode and enter the parametric equations, as shown in Fig. 7.68(a). Set the limits on the viewing window and the parameter as in Fig. 7.68(b). The graph is shown in Fig. 7.68(c).

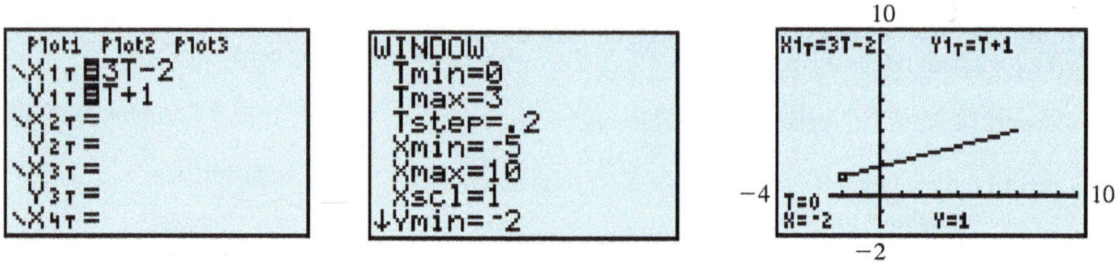


(a) (b) (c)

**Figure 7.68**

▸**TRY THIS.** Graph $x = t + 5$ and $y = 2t - 1$ for $t$ in $[0, 5]$. ■

## Eliminating the Parameter

In rectangular coordinates, we know that $y = mx + b$ is a line, $(x - h)^2 + (y - k)^2 = r^2$ is a circle, and $y = ax^2 + bx + c$ is a parabola. Since we have little experience with parametric equations, it may not be obvious when a system of parametric equations has a familiar graph. However, it is often possible to identify a graph by eliminating the parameter and writing an equation involving only $x$ and $y$.

### EXAMPLE 2 Eliminating the parameter

Eliminate the parameter and identify the graph of the parametric equations. Determine the domain (the set of $x$-coordinates) and the range (the set of $y$-coordinates).

**a.** $x = 3t - 2, y = t + 1, -\infty < t < \infty$
**b.** $x = 7 \sin t, y = 7 \cos t, -\infty < t < \infty$

#### Solution

**a.** Solve $y = t + 1$ for $t$ to get $t = y - 1$. Now replace $t$ in the other equation by $y - 1$:

$$x = 3(y - 1) - 2$$
$$x = 3y - 5$$
$$3y = x + 5$$
$$y = \frac{1}{3}x + \frac{5}{3}$$

After eliminating the parameter, we get $y = \frac{1}{3}x + \frac{5}{3}$, which is the equation of a line with slope $1/3$ and $y$-intercept $(0, 5/3)$. Because $-\infty < t < \infty$ and both $x$ and $y$ are linear functions of $t$, the doman of $y = \frac{1}{3}x + \frac{5}{3}$ is $(-\infty, \infty)$ and its range is $(-\infty, \infty)$. So the graph is the entire line. Note that in Example 1 these same parametric equations with a different interval for $t$ determined a line segment.

**b.** The simplest way to eliminate the parameter in this case is to use the trigonometric identity $\sin^2(\theta) + \cos^2(\theta) = 1$. Because $\sin t = x/7$ and $\cos t = y/7$, we have $(x/7)^2 + (y/7)^2 = 1$ or $x^2 + y^2 = 49$. So the graph is a circle centered at the origin with radius 7. The domain of this relation is $[-7, 7]$ and the range is $[-7, 7]$. You can check this conclusion with a graphing calculator, as shown in Fig. 7.69.

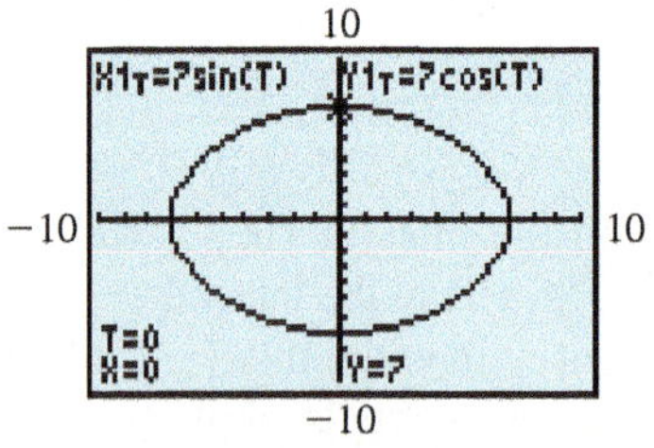


**Figure 7.69**

▸**TRY THIS.** Eliminate the parameter and identify the graph $x = 4t - 9$ and $y = -t + 1$ for $t$ in $(-\infty, \infty)$. Determine the domain and range. ■

## Writing Parametric Equations

Because a nonvertical straight line has a unique slope and $y$-intercept, it has a unique equation in slope-intercept form. However, a polar equation for a curve is not unique and neither are parametric equations for a curve. For example, consider the line $y = 2x + 1$. For parametric equations we could let $x = t$ and $y = 2t + 1$. We could also let $x = 4t$ and $y = 8t + 1$. We could even write $x = t^3 + 7$ and $y = 2t^3 + 15$. Each of these pairs of parametric equations produces the line $y = 2x + 1$.

### EXAMPLE 3 Writing parametric equations for a line segment

Write parametric equations for the line segment between $(1, 3)$ and $(5, 8)$ for $t$ in the interval $[0, 2]$, with $t = 0$ corresponding to $(1, 3)$ and $t = 2$ corresponding to $(5, 8)$.

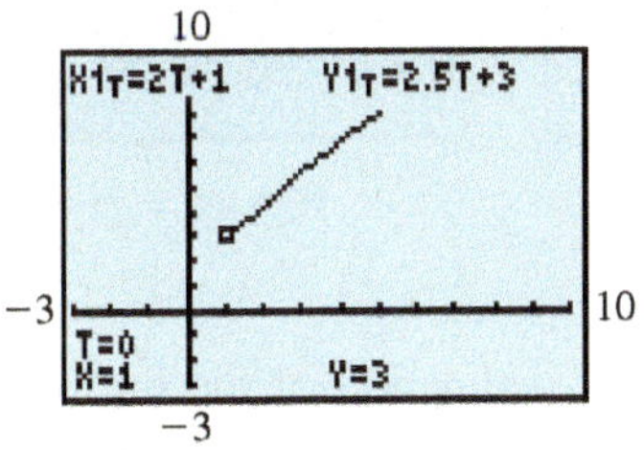

Figure 7.70

**Solution**

We can make both parametric equations linear functions of $t$. If $x = mt + b$ and $t = 0$ corresponds to $x = 1$, then $1 = m \cdot 0 + b$ and $b = 1$. So $x = mt + 1$. If $t = 2$ corresponds to $x = 5$, then $5 = m \cdot 2 + 1$ or $m = 2$. So we have $x = 2t + 1$. Using similar reasoning for the $y$-coordinates we get $y = 2.5t + 3$. You can use a graphing calculator to check, as shown in Fig. 7.70.

**▸TRY THIS.** Write parametric equations for the line segment between $(1, 2)$ and $(8, 10)$ with $t$ in the interval $[0, 1]$, where $t = 0$ corresponds to $(1, 2)$ and $t = 1$ corresponds to $(8, 10)$. ■

While it may not seem obvious how to write parametric equations for a particular rectangular equation, there is a simple way to find parametric equations for a polar equation $r = f(\theta)$. Because $x = r\cos\theta$ and $y = r\sin\theta$, we can substitute $f(\theta)$ for $r$ and write $x = f(\theta)\cos\theta$ and $y = f(\theta)\sin\theta$. In this case the parameter is $\theta$ and we have parametric equations for the polar curve. The parametric equations for a polar curve can be used to graph a polar curve on a calculator that is capable of handling parametric equations but not polar equations.

### EXAMPLE 4 Converting a polar equation to parametric equations

Write parametric equations for the polar equation $r = 1 - \cos\theta$.

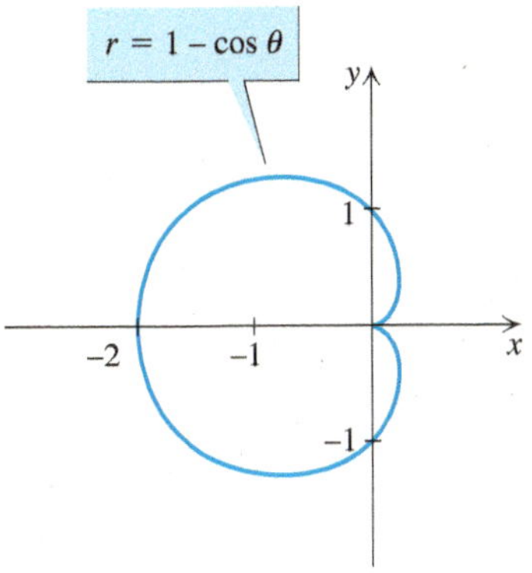

Figure 7.71

**Solution**

Replace $r$ by $1 - \cos\theta$ in the equations $x = r\cos\theta$ and $y = r\sin\theta$ to get $x = (1 - \cos\theta)\cos\theta$ and $y = (1 - \cos\theta)\sin\theta$. We know from Section 7.6 that the graph of $r = 1 - \cos\theta$ is the cardioid shown in Fig. 7.71. So the graph of these parametric equations is the same cardioid.

The calculator graph of the parametric equations in Fig. 7.72 appears to be a cardioid and supports the graph shown in Fig. 7.71.

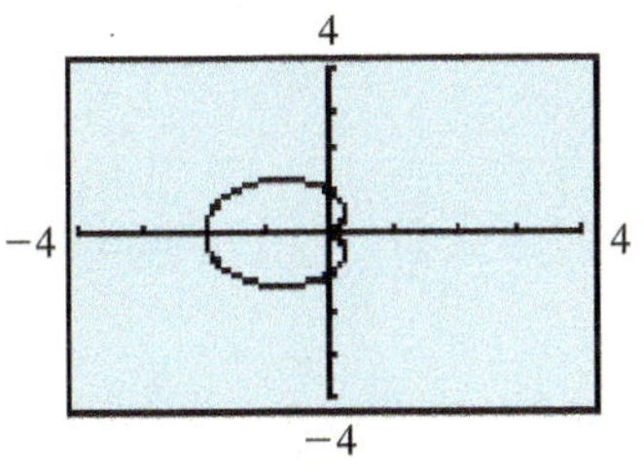

Figure 7.72

**▸TRY THIS.** Write parametric equations for the polar equation $r = 3\cos\theta$. ■

## Shooting Baskets with Parametric Equations

With $x$ and $y$ in feet and $t$ in seconds, the parametric equations for the path of a projectile are

$$x = v_0(\cos\theta)t \quad \text{and} \quad y = -16t^2 + v_0(\sin\theta)t + h_0$$

where $\theta$ is the angle of inclination in degrees, $v_0$ is the initial velocity in feet per second, and $h_0$ is the initial height in feet. With these equations and a graphing calculator you can illustrate the flight of a basketball.

To enter these equations into a graphing calculator, first set the MODE as in Fig. 7.73. We will throw the ball from the location (0, 6) to a basket at (20, 10). To get these locations to appear on the screen we use STAT PLOT. Use STAT EDIT to enter these ordered pairs in L1 and L2 as shown in Fig. 7.74. Then use STAT PLOT to turn on the feature that plots a square mark at these locations as in Fig. 7.75.

**Figure 7.73**

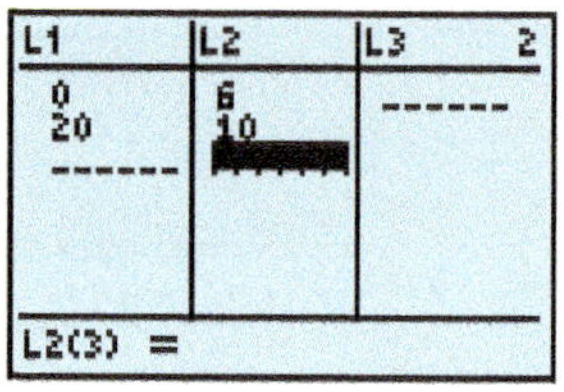

**Figure 7.74**

**Figure 7.75**

Using $A$ for the initial velocity, $B$ for the angle of inclination of the throw, and $h_0 = 6$ feet, enter the parametric equations using $Y =$ as in Fig. 7.76. To get the "ball," move the cursor to the left of $X_{1T}$ and press enter until a small circle appears to the left of $X_{1T}$ as in Fig. 7.76. Set the window so that the time $T$ ranges from 0 to 2 seconds in steps of 0.1 second, as in Fig. 7.77. Set the window so that $-2 \leq x \leq 25$ and $0 \leq y \leq 20$. Press QUIT to enter the initial values for $A$ and $B$ on the home screen using the STO button as in Figure 7.78.

**Figure 7.76**

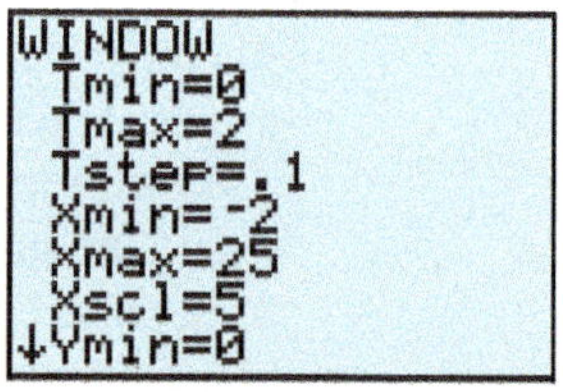

**Figure 7.77**

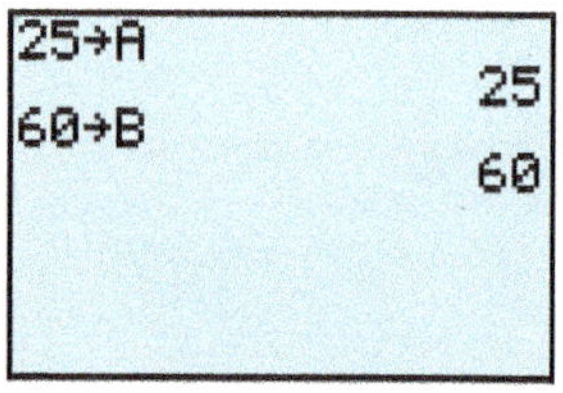

**Figure 7.78**

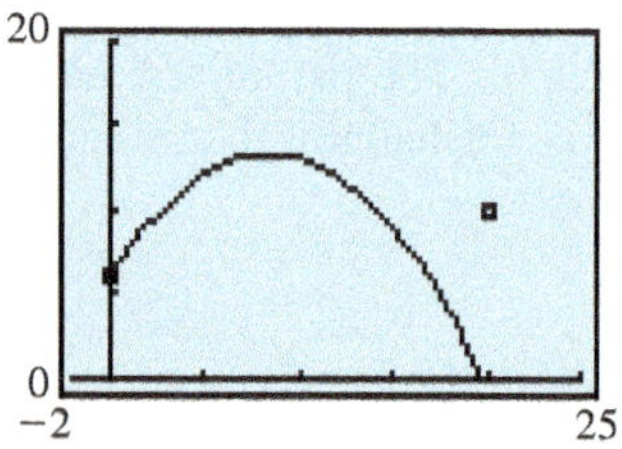

**Figure 7.79**

Now press GRAPH to see the ball tossed toward the basket. After the toss, the path of the ball will appear as in Fig. 7.79.

With the initial velocity of 25 feet per second and the angle 60°, the ball falls short of the basket as seen in Fig. 7.79. Enter a new angle and velocity on the home screen until you find a combination for which the path of the ball goes through the basket at (20, 10). Before pressing GRAPH, press DRAW and ClrDraw to clear the old path from the graph screen.

## FOR thought... True or False? Explain.

**1.** If $x = 3t + 1$ and $y = 4t - 2$, then $t$ is the variable and $x$ and $y$ are the parameters.

**2.** Parametric equations are graphed in the rectangular coordinate system.

**3.** The graph of $x = 0.5t$ and $y = 2t + 1$ is a straight line with slope 4.

**4.** The graph of $x = \cos t$ and $y = \sin t$ is a sine wave.

5. The graph of $x = 3t + 1$ and $y = 6t - 1$ for $0 \le t \le 3$ includes the point $(2, 1)$.

6. The graph of $x = w^2 - 3$ and $y = w + 5$ for $-2 < w < 2$ includes the point $(1, 7)$.

7. The graph of $x = e^t$ and $y = e^t$ lies entirely within the first quadrant.

8. The graph of $x = -\sin t$ and $y = \cos t$ for $0 < t < \pi/2$ lies entirely within the second quadrant.

9. The parametric equations $x = e^t$ and $y = e^t$ have the same graph as $x = \ln t$ and $y = \ln t$.

10. The polar equation $r = \cos\theta$ can be graphed using the parametric equations $x = \cos^2\theta$ and $y = \cos\theta\sin\theta$.

## EXERCISES 7.7

*Fill in the blank.*

1. If both $x$ and $y$ are functions of $t$, then $x = f(t)$ and $y = g(t)$ are ________ equations.

2. If $x = f(t)$ and $y = g(t)$, then $t$ is the ________.

*Complete the table that accompanies each pair of parametric equations.*

3. $x = 4t + 1, y = t - 2$, for $0 \le t \le 3$

| $t$ | $x$ | $y$ |
|---|---|---|
| 0 | | |
| 1 | | |
| | 7 | |
| | | 1 |

4. $x = 3 - t, y = 2t + 5$, for $2 \le t \le 7$

| $t$ | $x$ | $y$ |
|---|---|---|
| 2 | | |
| 3 | | |
| | −2 | |
| | | 19 |

5. $x = t^2, y = 3t - 1$, for $1 \le t \le 5$

| $t$ | $x$ | $y$ |
|---|---|---|
| 1 | | |
| 2.5 | | |
| | 5 | |
| | | 11 |
| | 25 | |

6. $x = \sqrt{t}, y = t + 4$, for $0 \le t \le 9$

| $t$ | $x$ | $y$ |
|---|---|---|
| 0 | | |
| 2 | | |
| 4 | | |
| | | 12 |
| | 3 | |

*Graph each pair of parametric equations in the rectangular coordinate system. Determine the domain (the set of x-coordinates) and the range (the set of y-coordinates).*

7. $x = 3t - 2, y = t + 3$, for $0 \le t \le 4$

8. $x = 4 - 3t, y = 3 - t$, for $1 \le t \le 3$

9. $x = t - 1, y = t^2$, for $t$ in $(-\infty, \infty)$

10. $x = t - 3, y = 1/t$, for $t$ in $(-\infty, \infty)$

11. $x = \sqrt{w}, y = \sqrt{1 - w}$, for $0 < w < 1$

12. $x = \ln t, y = t + 3$, for $-2 < t < 2$

13. $x = \cos t, y = \sin t$

14. $x = 0.5t, y = \sin t$

*Eliminate the parameter and identify the graph of each pair of parametric equations. Determine the domain (the set of x-coordinates) and the range (the set of y-coordinates).*

15. $x = 4t - 5, y = 3 - 4t$

16. $x = 5t - 1, y = 4t + 6$

17. $x = -4\sin 3t, y = 4\cos 3t$

18. $x = 2\sin t\cos t, y = 3\sin 2t$

19. $x = t/4, y = e^t$

20. $x = t - 5, y = t^2 - 10t + 25$

21. $x = \tan t, y = 2\tan t + 3$

22. $x = \tan t, y = -\tan^2 t + 3$

*Write a pair of parametric equations that will produce the indicated graph. Answers may vary.*

23. The line segment starting at $(2, 3)$ with $t = 0$ and ending at $(5, 9)$ with $t = 2$

24. The line segment starting at $(-2, 4)$ with $t = 3$ and ending at $(5, -9)$ with $t = 7$

25. That portion of the circle $x^2 + y^2 = 4$ that lies in the third quadrant

26. That portion of the circle $x^2 + y^2 = 9$ that lies below the $x$-axis

27. The vertical line through $(3, 1)$

28. The horizontal line through $(5, 2)$

29. The circle whose polar equation is $r = 2\sin\theta$

30. The four-leaf rose whose polar equation is $r = 5\sin(2\theta)$

*Graph the following pairs of parametric equations with the aid of a graphing calculator. These are uncommon curves that would be difficult to describe in rectangular or polar coordinates.*

31. $x = \cos 3t, y = \sin t$

32. $x = \sin t, y = t^2$

33. $x = t - \sin t, y = 1 - \cos t$ (cycloid)

**34.** $x = t - \sin t, y = -1 + \cos t$ (inverted cycloid)

**35.** $x = 4\cos t - \cos 4t, y = 4\sin t - \sin 4t$ (epicycloid)

**36.** $x = \sin^3 t, y = \cos^3 t$ (hypocycloid)

*The following problems involve the parametric equations for the path of a projectile*

$$x = v_0(\cos\theta)t \qquad \text{and} \qquad y = -16t^2 + v_0(\sin\theta)t + h_0,$$

*where $\theta$ is the angle of inclination of the projectile at the launch, $v_0$ is the initial velocity of the projectile in feet per second, and $h_0$ is the initial height of the projectile in feet.*

**37.** An archer shoots an arrow from a height of 5 ft at an angle of inclination of 30° with a velocity of 300 ft/sec. Write the parametric equations for the path of the projectile and sketch the graph of the parametric equations.

**38.** If the arrow of Exercise 37 strikes a target at a height of 5 ft, then how far is the target from the archer?

**39.** For how many seconds is the arrow of Exercise 37 in flight?

**40.** What is the maximum height reached by the arrow in Exercise 37?

## ▶ RETHINKING

**41.** Write a rectangular equation that is equivalent to the polar equation $r = 8\cos\theta$.

**42.** Find all complex solutions to the equation $x^4 + 1 = 0$.

**43.** Find the trigonometric form for the complex number $3 - 3i\sqrt{3}$. Use radian measure for the argument.

**44.** Solve the equation $w = \frac{1}{3}\cos(4b)$ for $b$ where $0 \le b \le \frac{\pi}{4}$.

**45.** Find all solutions to $2\cos^2(x) + \cos(x) - 1 = 0$ on the interval $[0, 2\pi]$.

**46.** Find the exact value of $\tan\alpha$ if $\sin\alpha = -7/8$ and $\pi < \alpha < 3\pi/2$.

## THINKING OUTSIDE THE BOX LVIII

*Lakefront Property* A man-made lake in the shape of a triangle is bounded on each of its sides by a square lot as shown in the figure. The square lots are 8, 13, and 17 acres, respectively. What is the area of the lake in square feet?

**HINT** One acre is 43,560 square feet.

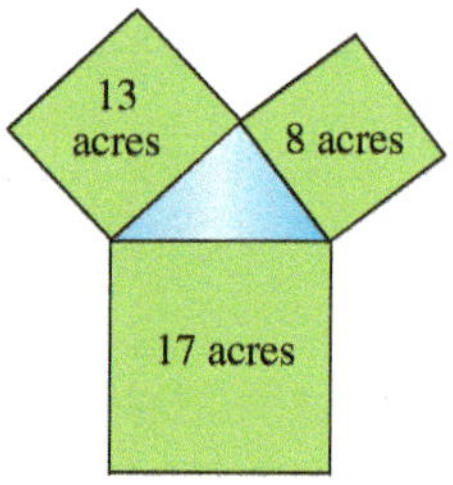


**Figure for Thinking Outside the Box LVIII**

## POP QUIZ 7.7

1. The graph of $x = 2t + 5$ and $y = 3t - 7$ for $t$ in $[3, 5]$ is a line segment. What are the endpoints?

2. Eliminate the parameter and identify the graph of $x = 3\cos t$ and $y = 3\sin t$ for $-\infty < t < \infty$.

3. Write parametric equations for the line segment between $(0, 1)$ and $(3, 5)$, where $t = 0$ corresponds to $(0, 1)$ and $t = 4$ corresponds to $(3, 5)$.

## LINKING concepts... For Individual or Group Explorations

### Distance

*We know that the distance $d$ between $(x_1, y_1)$ and $(x_2, y_2)$ in rectangular coordinates is given by the formula*

$$d = \sqrt{(x_2 - x_1)^2 + (y_2 - y_1)^2}.$$

*In the following exercises you will investigate distance in other coordinate systems.*

**a)** Show that the distance between $(r_1, \theta_1)$ and $(r_2, \theta_2)$ in polar coordinates is given by

$$d = \sqrt{r_1^2 + r_2^2 - 2r_1r_2\cos(\theta_2 - \theta_1)}$$

by converting the distance formula in rectangular coordinates into polar coordinates.

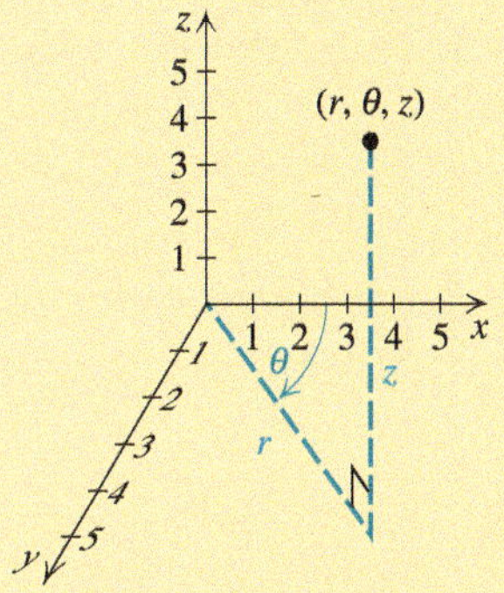


**Cylindrical coordinates**

**b)** Find the distance formula given in part (a) without using the distance formula in rectangular coordinates.

**c)** The *cylindrical coordinate system* for three-dimensional space is $(r, \theta, z)$ where $r$ and $\theta$ are polar coordinates in the $xy$-plane and $z$ is the altitude above or below the $xy$-plane, as shown in the accompanying figure. Show that the distance $d$ between $(r_1, \theta_1, z_1)$ and $(r_2, \theta_2, z_2)$ is given by

$$d = \sqrt{r_1^2 + r_2^2 - 2r_1r_2\cos(\theta_2 - \theta_1) + (z_1 - z_2)^2}.$$

**d)** Points on the surface of the earth (a sphere of radius $r$) are located by two coordinates $(\alpha, \beta)$, where $\alpha$ is longitude and $\beta$ is latitude. The distance $S$ between two points on the surface of the earth is the length of an arc through the two points. Show that the distance $S$ between $(\alpha_1, \beta_1)$ and $(\alpha_2, \beta_2)$ is given by

$$S = r \cdot \cos^{-1}(\sin\beta_1 \sin\beta_2 + \cos\beta_1 \cos\beta_2 \cos(\alpha_2 - \alpha_1)),$$

assuming that $\alpha$ is the angle west of the prime meridian and $\beta$ is the angle from the equator (positive in the northern hemisphere and negative in the southern hemisphere).

**e)** Find the longitude and latitude for Paris, France, and Chicago, Illinois, and use the formula in part (d) to calculate the distance between them. Using the radius of the earth as 3963 miles, you should get a distance of 4140 miles. Explain any discrepancies.

# Highlights

## 7.1 The Law of Sines

| | | |
|---|---|---|
| **Law of Sines** | In any triangle $\frac{\sin\alpha}{a} = \frac{\sin\beta}{b} = \frac{\sin\gamma}{c}$. | $a = \sqrt{3}, b = 1, c = 2$<br>$\alpha = \pi/3, \beta = \pi/6, \gamma = \pi/2$<br>$\frac{\sin(\frac{\pi}{3})}{\sqrt{3}} = \frac{\sin(\frac{\pi}{6})}{1} = \frac{\sin(\frac{\pi}{2})}{2}$ |
| **Area of a Triangle** | The area of a triangle is one-half the product of any two sides and the sine of the angle between them: $A = \frac{1}{2}bc\sin\alpha$. | Equilateral triangle with sides of length 4:<br>$A = \frac{1}{2}\cdot 4\cdot 4\cdot\sin(60°) = 4\sqrt{3}$ |

## 7.2 The Law of Cosines

| | | |
|---|---|---|
| **Law of Cosines** | If $a$ is the side opposite angle $\alpha$ in any triangle, then $a^2 = b^2 + c^2 - 2bc\cos\alpha$. | $a = 4, b = 5, c = 6$<br>$4^2 = 5^2 + 6^2 - 60\cos\alpha$ |
| **Heron's Area Formula** | If $a$, $b$, and $c$ are the sides of a triangle and $S = (a + b + c)/2$, then $A = \sqrt{S(S - a)(S - b)(S - c)}$. | $S = (4 + 6 + 8)/2 = 9$<br>$A = \sqrt{9\cdot 5\cdot 3\cdot 1} = \sqrt{135}$ |

## 7.3 Vectors

| | | |
|---|---|---|
| **Component Form** | The vector with initial point $(0, 0)$ and terminal point $(a, b)$ is denoted $\langle a, b\rangle$. | $\langle 2, 3\rangle$ has initial point $(0, 0)$ and terminal point $(2, 3)$. |
| **Magnitude** | $\lvert\langle a, b\rangle\rvert = \sqrt{a^2 + b^2}$ | $\lvert\langle 2, 3\rangle\rvert = \sqrt{13}$ |

| | | |
|---|---|---|
| **Scalar Product** | If $k$ is a scalar and **A** is a vector, then $\|k\mathbf{A}\| = \|k\| \cdot \|\mathbf{A}\|$. If $k > 0$, $k\mathbf{A}$ has the same direction as **A.** If $k < 0$, $k\mathbf{A}$ has the opposite direction as **A.** | $4\langle 1, 2\rangle = \langle 4, 8\rangle$<br>$-4\langle 1, 2\rangle = \langle -4, -8\rangle$ |
| **Resultant of A and B** | If **B** is placed so that its initial point coincides with the terminal point of **A,** then **A** + **B** is the vector from the initial point of **A** to the terminal point of **B.** | $\mathbf{A} = \langle 1, 4\rangle, \mathbf{B} = \langle 2, 5\rangle$<br>$\mathbf{A} + \mathbf{B} = \langle 3, 9\rangle$ |
| **Dot Product** | $\langle a, b\rangle \cdot \langle c, d\rangle = ac + bd$ | $\langle 1, 4\rangle \cdot \langle 2, 5\rangle = 22$ |
| **Angle Between Two Vectors** | If $\alpha$ is the angle between nozero vectors **A** and **B,** then $\cos\alpha = \dfrac{\mathbf{A}\cdot\mathbf{B}}{\|\mathbf{A}\|\|\mathbf{B}\|}$. | $\mathbf{A} = \langle 1, 4\rangle, \mathbf{B} = \langle 2, 5\rangle$<br>$\cos\alpha = \dfrac{22}{\sqrt{17}\sqrt{29}}$ |

## 7.4 Trigonometric Form of Complex Numbers

| | | |
|---|---|---|
| **Absolute Value** | $\|a + bi\| = \sqrt{a^2 + b^2}$ | $\|2 + 3i\| = \sqrt{13}$ |
| **Trigonometric Form** | $z = a + bi = r(\cos\theta + i\sin\theta)$, where $a = r\cos\theta, b = r\sin\theta, r = \sqrt{a^2 + b^2}$ and $\theta$ is an angle in standard position whose terminal side contains $(a, b)$. | $z = \sqrt{3} + i$<br>$= 2\left(\cos\frac{\pi}{6} + i\sin\frac{\pi}{6}\right)$ |
| **Multiplying and Dividing** | $z_1 = r_1(\cos\theta_1 + i\sin\theta_1)$<br>$z_2 = r_2(\cos\theta_2 + i\sin\theta_2)$<br>$z_1z_2 = r_1r_2[\cos(\theta_1 + \theta_2) + i\sin(\theta_1 + \theta_2)]$<br>$\dfrac{z_1}{z_2} = \dfrac{r_1}{r_2}[\cos(\theta_1 - \theta_2) + i\sin(\theta_1 - \theta_2)]$ | $z_1 = 8\left(\cos\frac{\pi}{3} + i\sin\frac{\pi}{3}\right)$<br>$z_2 = 2\left(\cos\frac{\pi}{6} + i\sin\frac{\pi}{6}\right)$<br>$z_1z_2 = 16\left(\cos\frac{\pi}{2} + i\sin\frac{\pi}{2}\right)$<br>$\dfrac{z_1}{z_2} = 4\left(\cos\frac{\pi}{6} + i\sin\frac{\pi}{6}\right)$ |

## 7.5 Powers and Roots of Complex Numbers

| | | |
|---|---|---|
| **De Moivre's Theorem** | If $z = r(\cos\theta + i\sin\theta)$ and $n$ is a positive integer, then $z^n = r^n(\cos n\theta + i\sin n\theta)$. | $z = 2\left(\cos\frac{\pi}{6} + i\sin\frac{\pi}{6}\right)$<br>$z^3 = 8\left(\cos\frac{\pi}{2} + i\sin\frac{\pi}{2}\right)$ |
| **Roots** | The $n$ distinct $n$th roots of $r(\cos\theta + i\sin\theta)$ are $r^{1/n}\left[\cos\left(\frac{\theta + 2k\pi}{n}\right) + i\sin\left(\frac{\theta + 2k\pi}{n}\right)\right]$ for $k = 0, 1, 2, \ldots, n - 1$. | Square roots of $z$:<br>$\sqrt{2}\left(\cos\frac{\pi}{12} + i\sin\frac{\pi}{12}\right)$<br>$\sqrt{2}\left(\cos\frac{13\pi}{12} + i\sin\frac{13\pi}{12}\right)$ |

## 7.6 Polar Equations

| | | |
|---|---|---|
| **Polar Coordinates** | If $r > 0$, then $(r, \theta)$ is $r$ units from the origin on the terminal side of $\theta$ in standard position. If $r < 0$, then $(r, \theta)$ is $\|r\|$ units from the origin on the extension of the terminal side of $\theta$. | $(2, \pi/4)$<br>$(-2, 5\pi/4)$ |
| **Converting** | Polar to rectangular: $x = r\cos\theta, y = r\sin\theta$.<br>Rectangular to polar: $r = \sqrt{x^2 + y^2}$ and the terminal side of $\theta$ contains $(x, y)$. | Polar: $(2, \pi/4)$<br>Rectangular: $(\sqrt{2}, \sqrt{2})$ |

## 7.7 Parametric Equations

| | | |
|---|---|---|
| **Parametric Equations** | $x = f(t)$ and $y = g(t)$ where $t$ is a parameter in some interval of real numbers. | $x = 2t, y = t^2$ for $t$ in $(0, 5)$ |
| **Converting to Rectangular** | Eliminate the parameter. | $y = (x/2)^2$ for $x$ in $(0, 10)$ |

# Chapter 7 Review Exercises

*Solve each triangle that exists with the given parts. If there is more than one triangle with the given parts, then solve each one.*

**1.** $\gamma = 48°, a = 3.4, b = 2.6$

**2.** $a = 6, b = 8, c = 10$

**3.** $\alpha = 13°, \beta = 64°, c = 20$

**4.** $\alpha = 50°, a = 3.2, b = 8.4$

**5.** $a = 3.6, b = 10.2, c = 5.9$

**6.** $\beta = 36.2°, \gamma = 48.1°, a = 10.6$

**7.** $a = 30.6, b = 12.9, c = 24.1$

**8.** $\alpha = 30°, a = \sqrt{3}, b = 2\sqrt{3}$

**9.** $\beta = 22°, c = 4.9, b = 2.5$

**10.** $\beta = 121°, a = 5.2, c = 7.1$

*Find the area of each triangle.*

**11.**

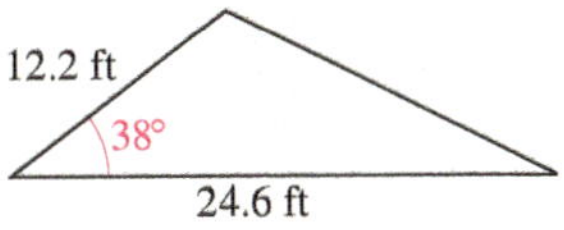


**12.**

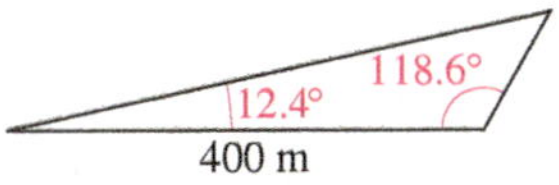


**13.**

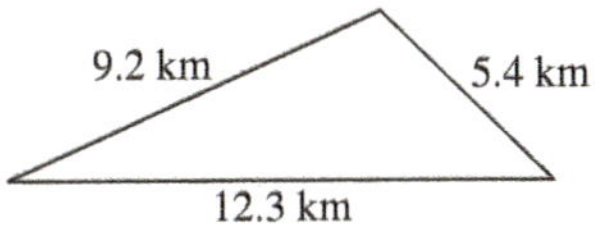


**14.**

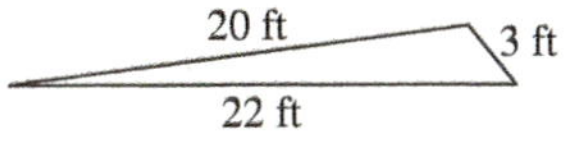


*Find the magnitude of the horizontal and vertical components for each vector* **v** *with the given magnitude and given direction angle* $\theta$.

**15.** $|\mathbf{v}| = 6, \theta = 23.3°$

**16.** $|\mathbf{v}| = 4.5, \theta = 156°$

**17.** $|\mathbf{v}| = 3.2, \theta = 231.4°$

**18.** $|\mathbf{v}| = 7.3, \theta = 344°$

*Find the magnitude and direction for each vector.*

**19.** $\langle 2, 3 \rangle$

**20.** $\langle -4, 3 \rangle$

**21.** $\langle -3.2, -5.1 \rangle$

**22.** $\langle 2.1, -3.8 \rangle$

*Find the component form for each vector* **v** *with the given magnitude and direction angle.*

**23.** $|\mathbf{v}| = \sqrt{2}, \theta = 45°$

**24.** $|\mathbf{v}| = 6, \theta = 60°$

**25.** $|\mathbf{v}| = 9.1, \theta = 109.3°$

**26.** $|\mathbf{v}| = 5.5, \theta = 344.6°$

*Perform the vector operations. Write your answer in the form* $\langle a, b \rangle$ *if the answer is a vector.*

**27.** $2\langle -3, 4 \rangle$

**28.** $-3\langle 4, -1 \rangle$

**29.** $\langle 2, -5 \rangle - 2\langle 1, 6 \rangle$

**30.** $3\langle 1, 2 \rangle + 4\langle -1, -2 \rangle$

**31.** $\langle -1, 5 \rangle \cdot \langle 4, 2 \rangle$

**32.** $\langle -4, 7 \rangle \cdot \langle 7, 4 \rangle$

*Rewrite each vector* **v** *in the form* $a_1\mathbf{i} + a_2\mathbf{j}$, *where* $\mathbf{i} = \langle 1, 0 \rangle$ *and* $\mathbf{j} = \langle 0, 1 \rangle$.

**33.** In component form, $\mathbf{v} = \langle -4, 8 \rangle$.

**34.** In component form, $\mathbf{v} = \langle 3.2, -4.1 \rangle$.

**35.** The direction angle for **v** is 30° and its magnitude is 7.2.

**36.** The magnitude of **v** is 6 and it has the same direction as the vector $\langle 2, 5 \rangle$.

*Find the absolute value of each complex number.*

**37.** $3 - 5i$

**38.** $3.6 + 4.8i$

**39.** $\sqrt{5} + i\sqrt{3}$

**40.** $-2\sqrt{2} + 3i\sqrt{5}$

*Write each complex number in trigonometric form, using degree measure for the argument.*

**41.** $-4.2 + 4.2i$

**42.** $3 - i\sqrt{3}$

**43.** $-2.3 - 7.2i$

**44.** $4 + 9.2i$

*Write each complex number in the form* $a + bi$.

**45.** $\sqrt{3}(\cos 150° + i \sin 150°)$

**46.** $\sqrt{2}(\cos 225° + i \sin 225°)$

**47.** $6.5(\cos 33.1° + i \sin 33.1°)$

**48.** $14.9(\cos 289.4° + i \sin 289.4°)$

*Find the product and quotient of each pair of complex numbers, using trigonometric form.*

**49.** $z_1 = 2.5 + 2.5i, z_2 = -3 - 3i$

**50.** $z_1 = -\sqrt{3} + i, z_2 = -2 - 2i\sqrt{3}$

**51.** $z_1 = 2 + i, z_2 = 3 - 2i$

**52.** $z_1 = -3 + i, z_2 = 2 - i$

*Use De Moivre's theorem to simplify each expression. Write the answer in the form $a + bi$.*

**53.** $[2(\cos 45° + i \sin 45°)]^3$

**54.** $[\sqrt{3}(\cos 210° + i \sin 210°)]^4$

**55.** $(4 + 4i)^3$

**56.** $(1 - i\sqrt{3})^4$

*Find the indicated roots. Express answers in the form $a + bi$.*

**57.** The square roots of $i$

**58.** The cube roots of $-i$

**59.** The cube roots of $\sqrt{3} + i$

**60.** The square roots of $3 + 3i$

**61.** The cube roots of $2 + i$

**62.** The cube roots of $3 - i$

**63.** The fourth roots of $625i$

**64.** The fourth roots of $-625i$

*Convert the polar coordinates of each point to rectangular coordinates.*

**65.** $(5, 60°)$

**66.** $(-4, 30°)$

**67.** $(\sqrt{3}, 100°)$

**68.** $(\sqrt{5}, 230°)$

*Convert the rectangular coordinates of each point to polar coordinates. Use radians for $\theta$.*

**69.** $(-2, -2\sqrt{3})$

**70.** $(-3\sqrt{2}, 3\sqrt{2})$

**71.** $(2, -3)$

**72.** $(-4, -5)$

*Sketch the graph of each polar equation.*

**73.** $r = -2 \sin \theta$

**74.** $r = 5 \sin 3\theta$

**75.** $r = 2 \cos 2\theta$

**76.** $r = 1.1 - \cos \theta$

**77.** $r = 500 + \cos \theta$

**78.** $r = 500$

**79.** $r = \dfrac{1}{\sin \theta}$

**80.** $r = \dfrac{-2}{\cos \theta}$

*For each polar equation, write an equivalent rectangular equation.*

**81.** $r = \dfrac{1}{\sin \theta + \cos \theta}$

**82.** $r = -6 \cos \theta$

**83.** $r = -5$

**84.** $r = \dfrac{1}{1 + \sin \theta}$

*For each rectangular equation, write an equivalent polar equation.*

**85.** $y = 3$

**86.** $x^2 + (y + 1)^2 = 1$

**87.** $x^2 + y^2 = 49$

**88.** $2x + 3y = 6$

*Sketch the graph of each pair of parametric equations.*

**89.** $x = 3t, y = 3 - t$, for $t$ in $(0, 1)$

**90.** $x = t - 3, y = t^2$, for $t$ in $(-\infty, \infty)$

**91.** $x = -\sin t, y = -\cos t$, for $t$ in $[0, \pi/2]$

**92.** $x = -\cos t, y = \sin t$, for $t$ in $[0, \pi]$

*Solve each problem.*

**93.** *Resultant Force* Forces of 12 lb and 7 lb act at a 30° angle to each other. Find the magnitude of the resultant force and the angle that the resultant makes with each force.

**94.** *Course of a Cessna* A twin-engine Cessna is heading on a bearing of 35° with an air speed of 180 mph. If the wind is out of the west (bearing 90°) at 40 mph, then what are the bearing of the course and the ground speed of the airplane?

**95.** *Dividing Property* Mrs. White Eagle gave each of her children approximately half of her four-sided lot in Gallup by dividing it on a diagonal. If Susan's piece is 482 ft by 364 ft by 241 ft and Seth's piece is 482 ft by 369 ft by 238 ft, then which child got the larger piece?

**96.** *Area, Area, Area* A surveyor found that two sides of a triangular lot were 135.4 ft and 164.1 ft, with an included angle of 86.4°. Find the area of this lot, using each of the three area formulas.

**97.** *Pipeline Detour* A pipeline was planned to go from $A$ to $B$, as shown in the figure. However, Mr. Smith would not give permission for the pipeline to cross his property. The pipeline was laid 431 ft from $A$ to $C$ and then 562 ft from $C$ to $B$. If $\angle C$ is 122° and the cost of the pipeline was \$21.60/ft, then how much extra was spent to go around Mr. Smith's property?

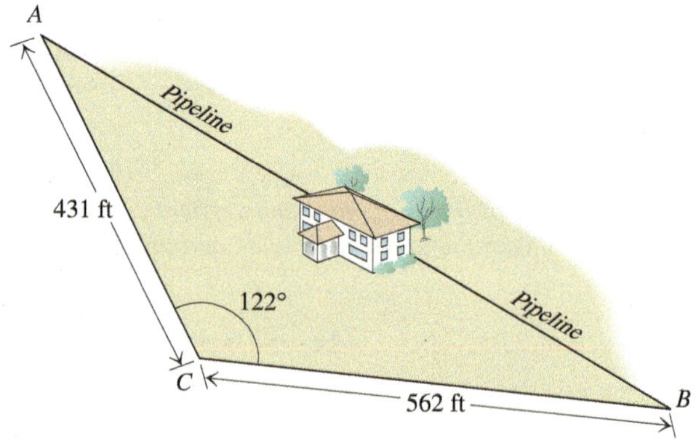


**Figure for Exercise 97**

**98.** *In the Wrong Place* In a lawsuit filed against a crane operator, a pedestrian of average height claims that he was struck by a wrecking ball. At the time of the accident, the operator had the ball extended 40 ft from the end of the 60-ft boom as shown in the figure, and the angle of elevation of the boom was 53°. How far from the crane would the pedestrian have to stand to be struck by this wrecking ball?

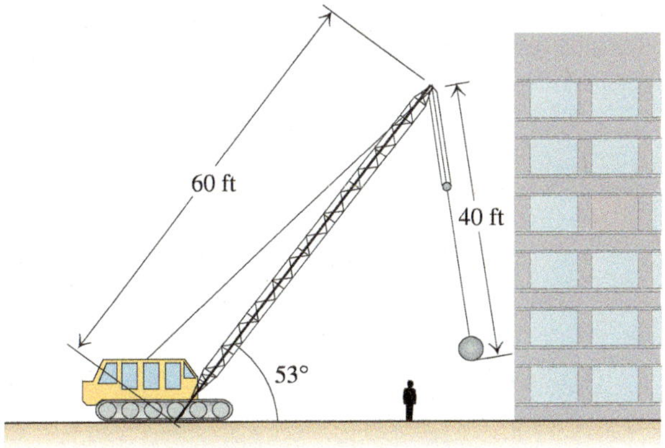


Figure for Exercise 98

**99.** *Detroit Pistons* The pistons in a gasoline engine are connected to a crankshaft, as shown in the figure. If the length of the connecting rod is $c$ and the radius of revolution around the center of the crankshaft is $r$, then the distance from the center of the crankshaft to the center of the piston, $a$, varies from $c + r$ to $c - r$.

**a.** Show that $a = \sqrt{c^2 - r^2 \sin^2 \theta} + r \cos \theta$.

**b.** Suppose that the crankshaft is rotating at 426 rpm and time $t = 0$ min corresponds to $\theta = 0$ and $a = c + r$. If $c = 12$ in., $r = 2$ in., and $t = 0.1$ min, then what is $a$?

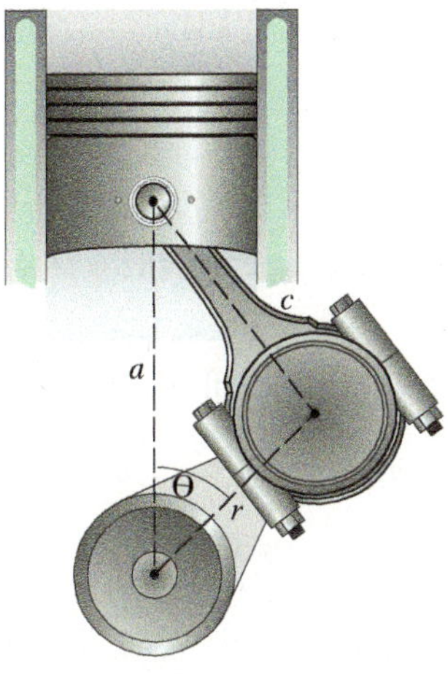


Figure for Exercise 99

### THINKING OUTSIDE THE BOX LIX

*Tiling a Room* Tile Mart is selling new T-shaped ceramic tiles as shown in the figure. Each tile is 3 ft long and 2 ft wide, and covers four square feet.

**a.** Is it possible to tile completely an 8 ft by 8 ft room with these T-shaped tiles? No cutting, breaking, or overlapping of the tiles is allowed. Explain.

**b.** Is it possible to tile a 6 ft by 6 ft room with these tiles?

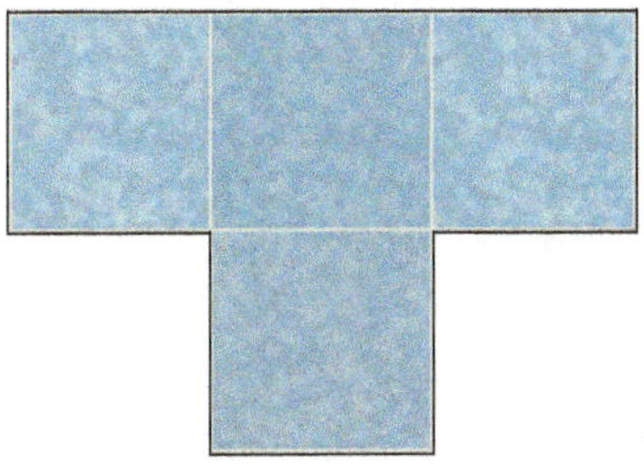

Figure for Thinking Outside the Box LIX

# Chapter 7 Test

*Determine the number of triangles with the given parts and solve each triangle.*

**1.** $\alpha = 30°, b = 4, a = 2$

**2.** $\alpha = 60°, b = 4.2, a = 3.9$

**3.** $a = 3.6, \alpha = 20.3°, \beta = 14.1°$

**4.** $a = 2.8, b = 3.9, \gamma = 17°$

**5.** $a = 4.1, b = 8.6, c = 7.3$

*Given* $\mathbf{A} = \langle -3, 2 \rangle$ *and* $\mathbf{B} = \langle 1, 4 \rangle$, *find the magnitude and direction angle for each of the following vectors.*

**6.** $\mathbf{A} + \mathbf{B}$ **7.** $\mathbf{A} - \mathbf{B}$ **8.** $3\mathbf{B}$

*Write each complex number in trigonometric form, using degree measure for the argument.*

**9.** $3 + 3i$ **10.** $-1 + i\sqrt{3}$ **11.** $-4 - 2i$

*Perform the indicated operations. Write the answer in the form $a + bi$.*

**12.** $3(\cos 20° + i \sin 20°) \cdot 2(\cos 25° + i \sin 25°)$

**13.** $[2(\cos 10° + i \sin 10°)]^9$ **14.** $\dfrac{3(\cos 63° + i \sin 63°)}{2(\cos 18° + i \sin 18°)}$

*Give the rectangular coordinates for each of the following points in the polar coordinate system.*

**15.** $(5, 30°)$ **16.** $(-3, -\pi/4)$ **17.** $(33, 217°)$

*Sketch the graph of each equation in polar coordinates.*

**18.** $r = 5 \cos \theta$

**19.** $r = 3 \cos 2\theta$

*Solve each problem.*

**20.** Find the area of the triangle in which $a = 4.1$ m, $b = 6.8$ m, and $c = 9.5$ m.

**21.** A vector $\mathbf{v}$ in the coordinate plane has direction angle $\theta = 37.2°$ and $|\mathbf{v}| = 4.6$. Find real numbers $a_1$ and $a_2$ such that $\mathbf{v} = a_1\mathbf{i} + a_2\mathbf{j}$, where $\mathbf{i} = \langle 1, 0 \rangle$ and $\mathbf{j} = \langle 0, 1 \rangle$.

**22.** Find all of the fourth roots of $-81$.

**23.** Write an equation equivalent to $x^2 + y^2 + 5y = 0$ in polar coordinates.

**24.** Write an equation equivalent to $r = 5 \sin 2\theta$ in rectangular coordinates.

**25.** Find a pair of parametric equations whose graph is the line segment joining the points $(-2, -3)$ and $(4, 5)$.

**26.** The bearing of an airplane is 40° with an air speed of 240 mph. If the wind is out of the northwest (bearing 135°) at 30 mph, then what are the bearing of the course and the ground speed of the airplane?

## FUNCTION gallery... Some Basic Functions of Trigonometry

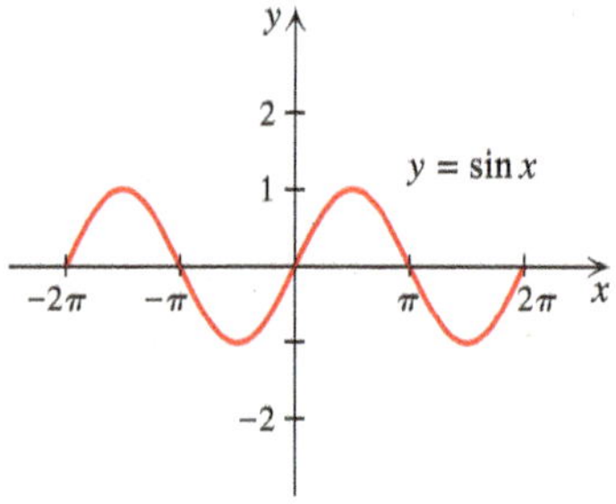


**Period $2\pi$**
**Amplitude 1**

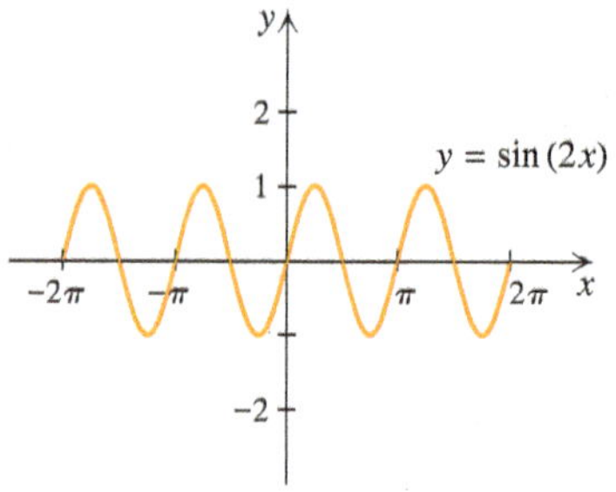


**Period $\pi$**
**Amplitude 1**

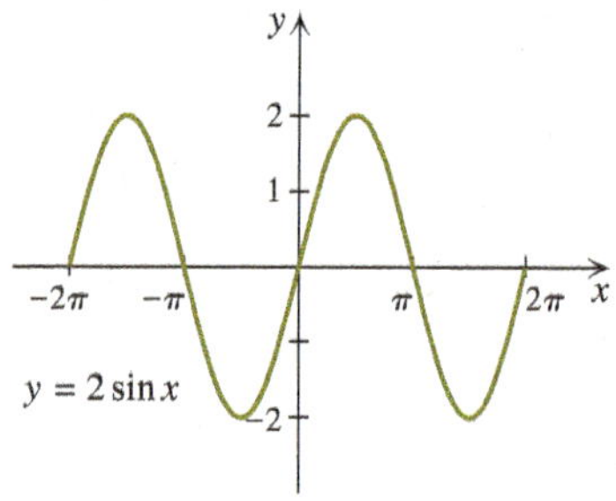


**Period $2\pi$**
**Amplitude 2**

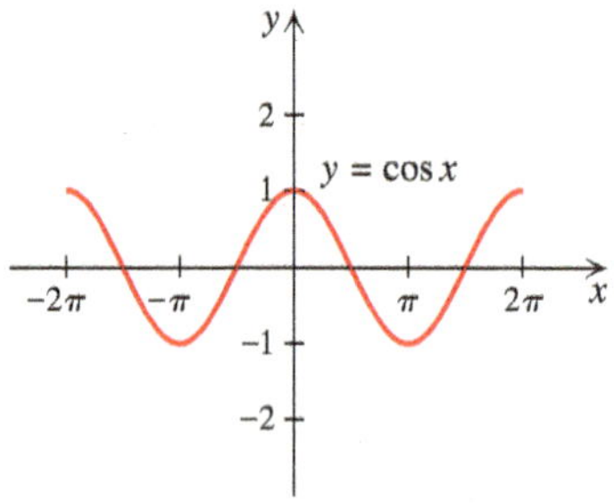


**Period $2\pi$**
**Amplitude 1**

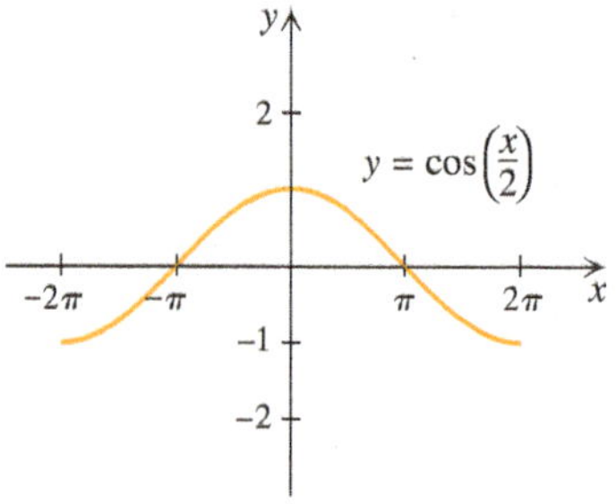


**Period $4\pi$**
**Amplitude 1**

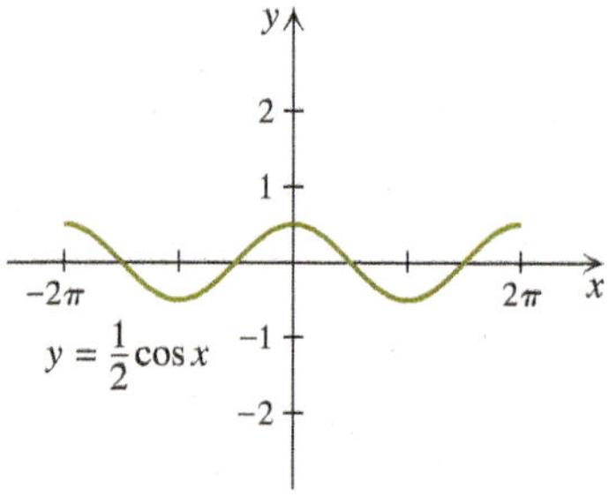


**Period $2\pi$**
**Amplitude $\frac{1}{2}$**

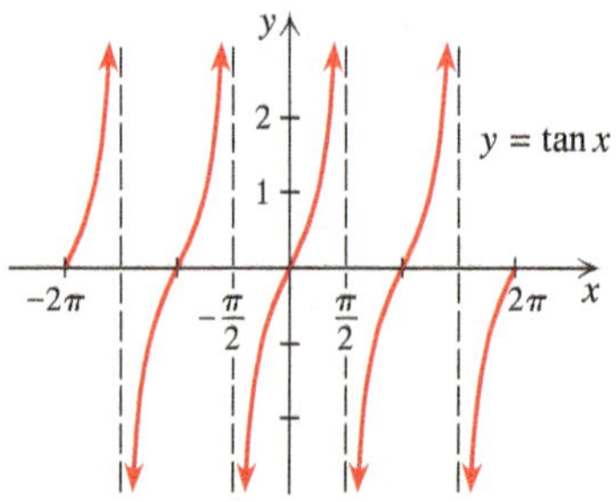


**Period $\pi$**

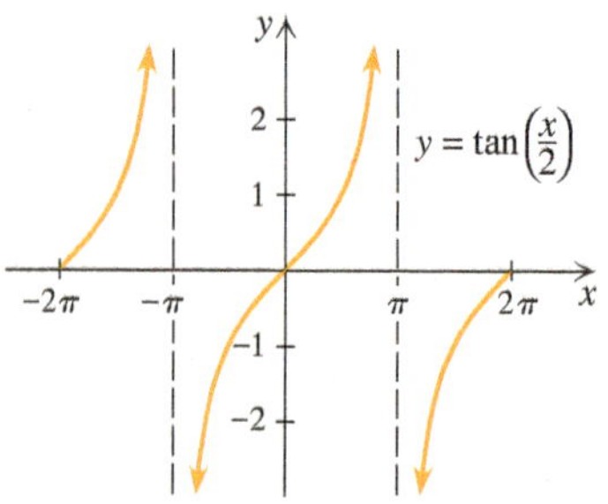


**Period $2\pi$**

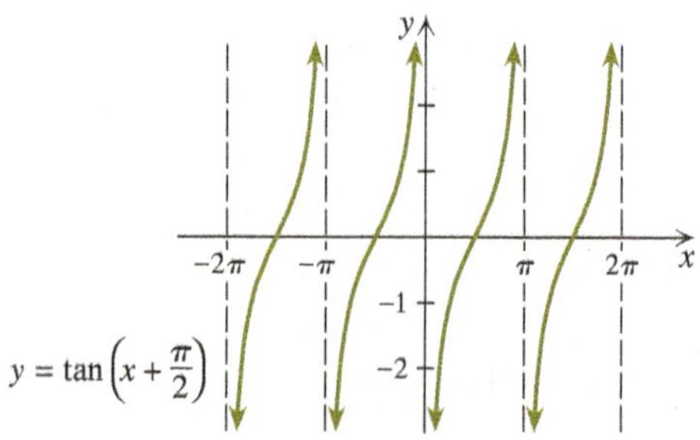


**Period $\pi$**

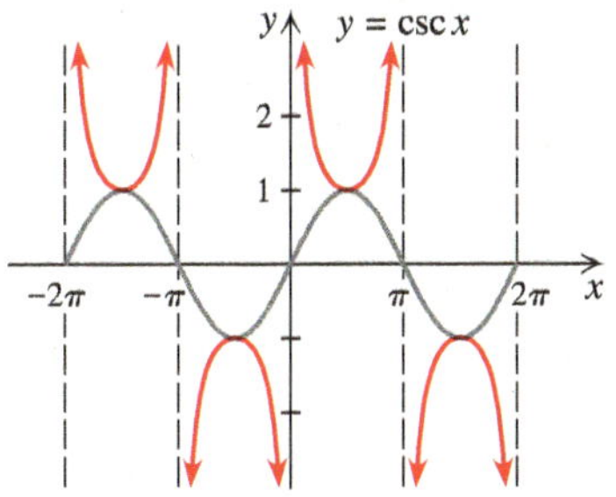


**Period 2π**

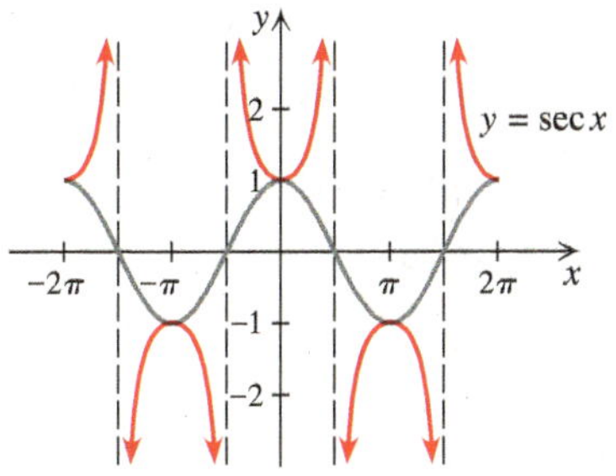


**Period 2π**

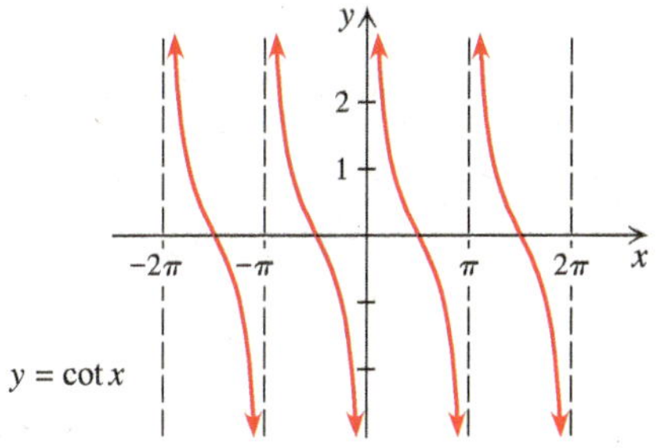


**Period π**

## TYING IT ALL together...

### Chapters 1–7

*Find all real and imaginary solutions to each equation.*

**1.** $x^4 - x = 0$
**2.** $x^3 - 2x^2 - 5x + 6 = 0$
**3.** $x^6 + 2x^5 - x - 2 = 0$
**4.** $x^7 - x^4 + 2x^3 - 2 = 0$
**5.** $2\sin 2x - 2\cos x + 2\sin x = 1$
**6.** $4x\sin x + 2\sin x - 2x = 1$
**7.** $e^{\sin x} = 1$
**8.** $\sin(e^x) = 1/2$
**9.** $2^{2x-3} = 32$
**10.** $\log(x - 1) - \log(x + 2) = 2$

*Sketch the graph of each function using rectangular or polar coordinates as appropriate.*

**11.** $y = \sin x$
**12.** $y = e^x$
**13.** $r = \sin\theta$
**14.** $r = \theta$
**15.** $y = \sqrt{\sin x}$
**16.** $y = \ln(\sin x)$
**17.** $r = \sin(\pi/3)$
**18.** $y = x^{1/3}$

*Evaluate each expression.*

**19.** $\log(\sin(\pi/2))$
**20.** $\sin(\log(1))$
**21.** $\cos(\ln(e^\pi))$
**22.** $\ln(\cos(2\pi))$
**23.** $\sin^{-1}(\log_2(\sqrt{2}))$
**24.** $\cos^{-1}(\ln(\sqrt{e}))$
**25.** $\tan^{-1}(\log(0.1))$
**26.** $\tan^{-1}(\ln(e))$

*Perform the indicated operations and simplify.*

**27.** $\dfrac{1}{x+2} + \dfrac{1}{x-2}$
**28.** $\dfrac{1}{1-\sin x} + \dfrac{1}{1+\sin x}$
**29.** $\dfrac{1}{2+\sqrt{3}} + \dfrac{1}{2-\sqrt{3}}$
**30.** $\log\left(\dfrac{1}{x+2}\right) + \log\left(\dfrac{1}{x-2}\right)$
**31.** $\dfrac{x^2-9}{2x-6}\cdot\dfrac{4x+12}{x^2+6x+9}$
**32.** $\dfrac{1-\sin^2(x)}{2\cos^2(x)-\sin(2x)}\cdot\dfrac{\cos^2(x)-\sin^2(x)}{2\cos^2(x)+\sin(2x)}$
**33.** $\dfrac{\sqrt{8}}{2-\sqrt{3}}\cdot\dfrac{\sqrt{2}}{4+\sqrt{12}}$
**34.** $\log\left(\dfrac{x^2+3x+2}{x^2+5x+6}\right) - \log\left(\dfrac{x+1}{x+3}\right)$
**35.** $\dfrac{\dfrac{1}{x-2}+\dfrac{1}{x+2}}{\dfrac{1}{x^2-4}-\dfrac{1}{x+2}}$
**36.** $\dfrac{\dfrac{1}{\sin(2x)}-\tan(x)}{\dfrac{\cot(x)}{2}-\dfrac{\tan(x)}{2}}$

*(Continued on next page)*

*Fill in the blanks.*

**37.** If the period of a sine wave is $P$, then $1/P$ is called the ________ of the wave.

**38.** The line $x = \pi/2$ is called a(n) ________ for the graph of $y = \tan x$.

**39.** If $f(-x) = f(x)$ then $f$ is called a(n) ________ function.

**40.** If $f(-x) = -f(x)$ then $f$ is called a(n) ________ function.

**41.** If $\alpha$ is an acute angle in a right triangle, then $\sin(\alpha)$ is the length of the side ________ $\alpha$ divided by the length of the ________.

**42.** If $\alpha$ is an acute angle in a right triangle, then $\cos(\alpha)$ is the length of the side ________ to $\alpha$ divided by the length of the ________.

**43.** A central angle of one radian in a unit circle intercepts an arc of length ________.

**44.** The ________ for the function $y = \sin(B(x - C))$ is $2\pi/B$.

**45.** The identities $\sin^2 x + \cos^2 x = 1$, $1 + \cot^2 x = \csc^2 x$, and $\tan^2 x + 1 = \sec^2 x$ are called the ________ identities.

**46.** Cosine and secant are ________ functions, whereas the other four trigonometric functions are ________.

# CONCEPTS OF calculus...

## Limits and asymptotes

*Rational, exponential, logarithmic, and trigonometric functions can have asymptotes. An asymptote is basically a "boundary line" for the graph of a function. An asymptote indicates the behavior of the graph in the long run, and in calculus the idea of limits is used to describe this behavior. Consider the graph of* $f(x) = \frac{1}{x}$ *shown in the following figure. The figure includes the appropriate limit notation for each asymptote.*

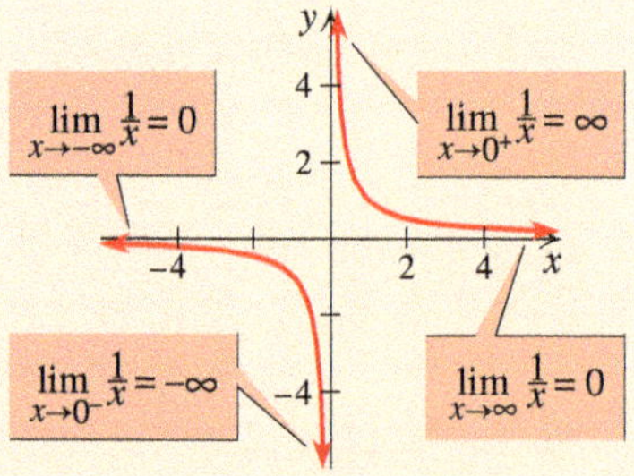


*Note that* $\lim_{x\to0^+} \frac{1}{x} = \infty$ *means that as x approaches* 0 *through real numbers larger than* 0*, the graph is approaching the vertical asymptote* $x = 0$ *(the y-axis) going upward. The notation* $\lim_{x\to0^-} \frac{1}{x} = -\infty$ *means that as x approaches* 0 *through real numbers smaller than* 0*, the graph is approaching the vertical asymptote* $x = 0$ *going downward.*

## Exercises

**1.** Determine whether each of these limits is $\infty$ or $-\infty$.

**a.** $\lim_{x\to0^+} \frac{-1}{x}$ **b.** $\lim_{x\to0^-} \frac{-1}{x}$

**c.** $\lim_{x\to1^+} \frac{x-2}{x-1}$ **d.** $\lim_{x\to1^-} \frac{x-2}{x-1}$

**e.** $\lim_{x\to\frac{\pi}{2}^+} \tan(x)$ **f.** $\lim_{x\to\frac{\pi}{2}^-} \tan(x)$

**g.** $\lim_{x\to\frac{\pi}{2}^+} \sec(x)$ **h.** $\lim_{x\to\pi^-} \csc(x)$

**i.** $\lim_{x\to0^+} \ln(x)$ **j.** $\lim_{x\to0^-} \ln(-x)$

**2.** Find each limit.

**a.** $\lim_{x\to\infty} \frac{-1}{x}$ **b.** $\lim_{x\to-\infty} \frac{-1}{x}$

**c.** $\lim_{x\to\infty} \frac{x-2}{x-1}$ **d.** $\lim_{x\to-\infty} \frac{x-2}{x-1}$

**e.** $\lim_{x\to\infty} \frac{3x-4}{9x-2}$ **f.** $\lim_{x\to-\infty} \frac{3x-4}{9x-2}$

**g.** $\lim_{x\to\infty} \tan^{-1}(x)$ **h.** $\lim_{x\to-\infty} \tan^{-1}(x)$

**i.** $\lim_{x\to-\infty} e^x$ **j.** $\lim_{x\to\infty} e^{-x^2}$

# Appendix A

## Solutions to Try This Exercises

### Chapter 1

**Section 1**

**1.1.1** $5(3x - 2) = 5 - 7(x - 1)$
$$15x - 10 = 5 - 7x + 7$$
$$22x = 22$$
$$x = 1$$
Check: $5(3 \cdot 1 - 2) = 5 - 7(1 - 1)$
$$5 = 5$$
The solution set is $\{1\}$.

**1.1.2** $x(x - 1) - 6 = (x - 3)(x + 2)$
$$x^2 - x - 6 = x^2 - x - 6$$
Since both sides are identical, it is an identity.

**1.1.3**
$$\frac{2}{x-3} - \frac{3}{x+3} = \frac{4}{x^2 - 9}$$
$$(x-3)(x+3)\left(\frac{2}{x-3} - \frac{3}{x+3}\right) = (x^2 - 9)\frac{4}{x^2-9}$$
$$2(x + 3) - 3(x - 3) = 4$$
$$-x + 15 = 4$$
$$-x = -11$$
$$x = 11$$
Check: $\frac{2}{11-3} - \frac{3}{11+3} = \frac{1}{28}, \frac{4}{11^2 - 9} = \frac{1}{28}$
The solution set to this conditional equation is $\{11\}$.

**1.1.4** $\frac{2}{3.4x} - \frac{1}{8.9} = \frac{4}{4.7}$
$$\frac{2}{3.4x} = \frac{4}{4.7} + \frac{1}{8.9}$$
$$\frac{2}{3.4} = x\left(\frac{4}{4.7} + \frac{1}{8.9}\right)$$
$$\frac{\frac{2}{3.4}}{\frac{4}{4.7} + \frac{1}{8.9}} = x$$
$$x \approx 0.611$$
Check: $\frac{2}{3.4(0.611)} - \frac{1}{8.9} \approx 0.851, \frac{4}{4.7} \approx 0.851$
The solution set is $\{0.611\}$.

**1.1.5** $|2x - 3| = 5$
$$2x - 3 = 5 \quad \text{or} \quad 2x - 3 = -5$$
$$2x = 8 \quad \text{or} \quad 2x = -2$$
$$x = 4 \quad \text{or} \quad x = -1$$
Check: $|2 \cdot 4 - 3| = 5$ and $|2 \cdot (-1) - 3| = 5$
The solution set is $\{-1, 4\}$.

**1.1.6** Solve the following equation.
$$25{,}000 = 355.9x + 11{,}075.3$$
$$13{,}924.7 = 355.9x$$
$$x \approx 39$$
So 39 years after 1990, or in 2029 the median income will reach \$25,000.

**Section 2**

**1.2.1** $A = \frac{1}{2}hb_1 + \frac{1}{2}hb_2$
$$2A = hb_1 + hb_2$$
$$2A = h(b_1 + b_2)$$
$$h = \frac{2A}{b_1 + b_2}$$

**1.2.2** $A = \frac{1}{2}h(b_1 + b_2)$
$$20 = \frac{1}{2} \cdot 2(b_1 + 3)$$
$$20 = b_1 + 3$$
$$17 = b_1$$

**1.2.3** Let $x$ represent the price of the computer and $0.05x$ the amount of tax.
$$x + 0.05x = 1506.75$$
$$1.05x = 1506.75$$
$$x = 1435$$
$$0.05x = 71.75$$
The amount of tax was \$71.75.

**1.2.4** Let $x$ represent the width and $5x - 20$ represent the length.
$$2W + 2L = P$$
$$2x + 2(5x - 20) = 800$$
$$12x - 40 = 800$$
$$12x = 840$$
$$x = 70$$
$$5x - 20 = 330$$
So the length is 330 cm.

**1.2.5** Let $x$ represent her rate uphill and $x + 2$ her rate downhill. The distance is $6x$ or $3(x + 2)$.
$$6x = 3(x + 2)$$
$$6x = 3x + 6$$
$$3x = 6$$
$$x = 2$$
$$6x = 12$$
The total distance hiked is 24 miles.

**1.2.6** Let $x$ represent her speed on the return trip. Her time to work is $1/3$ hr and her time for the return trip is $20/x$. Her average speed of 50 mph is the total distance divided by the total time:
$$\frac{20 + 20}{\frac{1}{3} + \frac{20}{x}} = 50$$
$$40 = 50\left(\frac{1}{3} + \frac{20}{x}\right)$$
$$\frac{4}{5} = \frac{1}{3} + \frac{20}{x}$$
$$15x \cdot \frac{4}{5} = 15x\left(\frac{1}{3} + \frac{20}{x}\right)$$
$$12x = 5x + 300$$
$$7x = 300$$
$$x = 42\frac{6}{7}$$
Her average speed on the return trip was $42\frac{6}{7}$ mph.

**1.2.7** Let $x$ represent the number of gallons of 40% acid solution.
$$0.40x + 0.20(30) = 0.35(x + 30)$$
$$0.40x + 6 = 0.35x + 10.5$$
$$0.05x = 4.5$$
$$x = 90$$
Use 90 gallons of 40% acid solution.

**1.2.8** Let $x$ represent the number of hours the small pipe is used and $x - 2$ represent the number of hours the large pipe is used.

$$\frac{1}{12}x + \frac{1}{8}(x - 2) = 1$$
$$\frac{1}{12}x + \frac{1}{8}x - \frac{1}{4} = 1$$
$$24\left(\frac{1}{12}x + \frac{1}{8}x - \frac{1}{4}\right) = 24 \cdot 1$$
$$2x + 3x - 6 = 24$$
$$5x = 30$$
$$x = 6$$

It will take 6 hours to fill the tank.

## Section 3

**1.3.1**

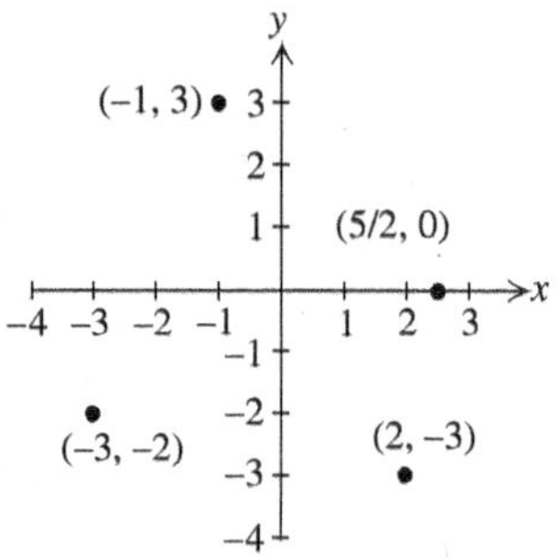


**1.3.2** $\sqrt{(-3 - (-1))^2 + (-2 - 4)^2} = \sqrt{4 + 36} = \sqrt{40} = 2\sqrt{10}$

**1.3.3** Add the corresponding coordinates and divide by 2:

$$\left(\frac{4 + 3}{2}, \frac{-1 + 1/2}{2}\right) = \left(\frac{7}{2}, \frac{-1/2}{2}\right) = \left(\frac{7}{2}, -\frac{1}{4}\right)$$

**1.3.4** The midpoint of the diagonal with endpoints $(0, 0)$ and $(3, 5)$ is $(3/2, 5/2)$. The midpoint of the diagonal with endpoints $(4, 1)$ and $(-1, 4)$ is also $(3/2, 5/2)$. So the diagonals bisect each other.

**1.3.5** The circle has center $(-2, 4)$ and radius 5.

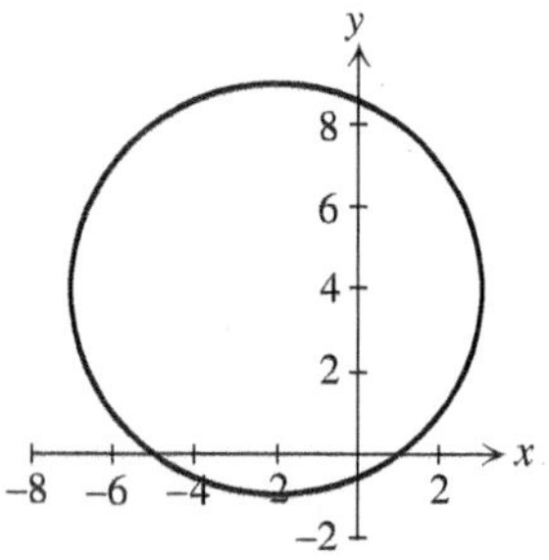

$(x + 2)^2 + (y - 4)^2 = 25$

**1.3.6** The radius is the distance from the center $(2, -1)$ to $(3, 6)$:

$$\sqrt{(2 - 3)^2 + (-1 - 6)^2} = \sqrt{50}$$

The equation is $(x - 2)^2 + (y + 1)^2 = 50$.

**1.3.7** Complete the squares:

$$x^2 + 3x + \frac{9}{4} + y^2 - 2y + 1 = 0 + \frac{9}{4} + 1$$
$$\left(x + \frac{3}{2}\right)^2 + (y - 1)^2 = \frac{13}{4}$$

The graph is a circle with center $(-3/2, 1)$ and radius $\sqrt{13}/2$ or about 1.8.

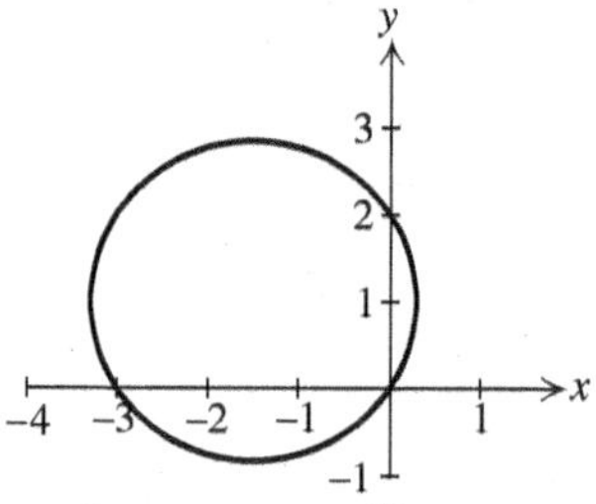

$(x + \frac{3}{2})^2 + (y - 1)^2 = \frac{13}{4}$

**1.3.8** If $x = 0$, then $5y = 10$ or $y = 2$. If $y = 0$, then $2x = 10$ or $x = 5$. Draw a line through the intercepts $(0, 2)$ and $(5, 0)$.

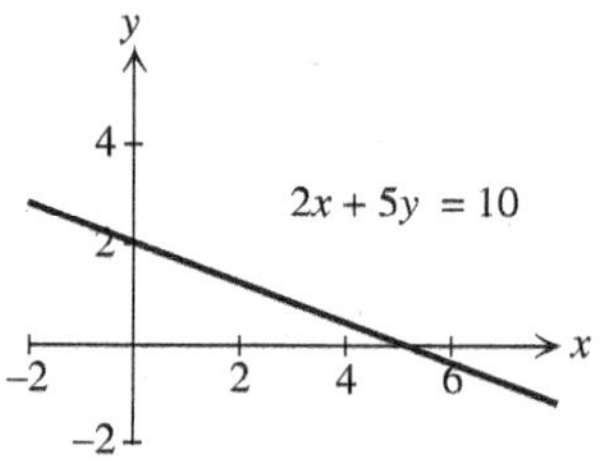


**1.3.9** The ordered pairs $(-2, 5)$, $(0, 5)$, and $(2, 5)$ satisfy $y = 5$. So the graph is a horizontal line.

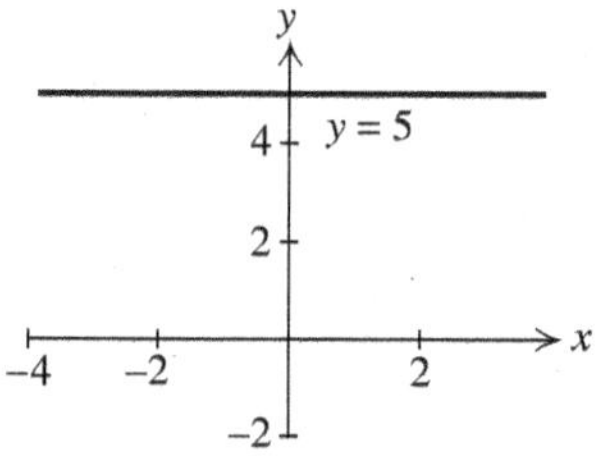


**1.3.10** Graph $y = 0.34(x - 2.3) + 4.5$ and find the $x$-intercept. The $x$-intercept is $(-10.93529, 0)$ and the solution to the equation is approximately $-10.93529$.

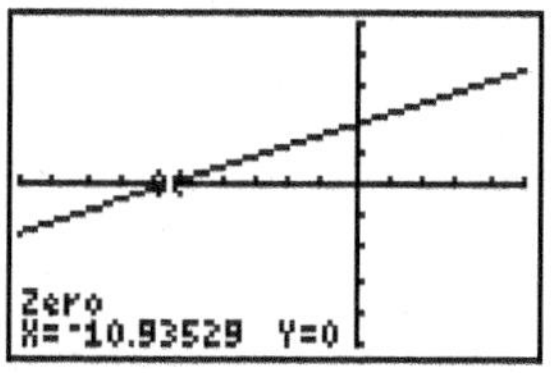


## Section 4

**1.4.1** slope $= \frac{-3 - 5}{-1 - (-2)} = \frac{-8}{1} = -8$

**1.4.2** Use the slope $-8$ from the preceding answer and the point $(-2, 5)$ in point-slope form.

$$y - 5 = -8(x - (-2))$$
$$y - 5 = -8x - 16$$
$$y = -8x - 11$$

**1.4.3** Solve the equation for $y$:

$$3x + 5y = 15$$
$$5y = -3x + 15$$
$$y = -\frac{3}{5}x + 3$$

The slope is $-\frac{3}{5}$ and the $y$-intercept is $(0, 3)$.

**1.4.4** Use $m = 1/4$ and $(x, y) = (4, -6)$ in $y = mx + b$:

$$-6 = \frac{1}{4} \cdot 4 + b$$
$$-6 = 1 + b$$
$$-7 = b$$

So the equation is $y = \frac{1}{4}x - 7$.

**1.4.5** Start at the $y$-intercept $(0, 1)$ and use the slope $-3/2$ to locate a second point $(2, -2)$ which is down 3 and 2 to the right from $(0, 1)$:

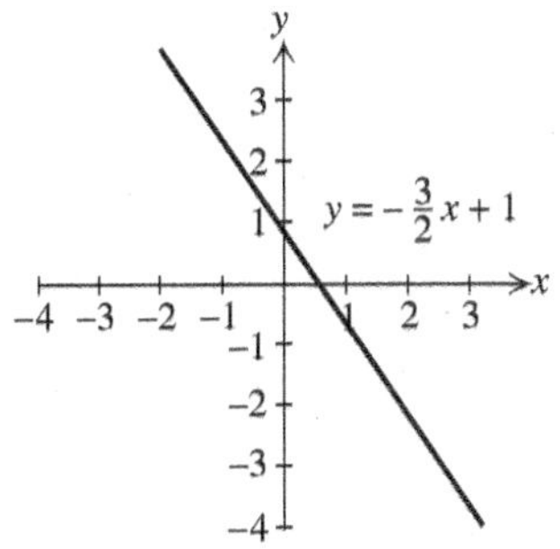


**1.4.6** Use slope-intercept form:

$$y = \frac{3}{4}x + \frac{1}{2}$$
$$4y = 3x + 2$$
$$-3x + 4y = 2$$
$$3x - 4y = -2$$

**1.4.7** The slope of any line parallel to $y = -\frac{1}{2}x + 9$ is $-\frac{1}{2}$. Use slope $-\frac{1}{2}$ and $(2, 4)$ in the point-slope form:

$$y - 4 = -\frac{1}{2}(x - 2)$$
$$y - 4 = -\frac{1}{2}x + 1$$
$$y = -\frac{1}{2}x + 5$$

**1.4.8** Write $2x - y = 8$ as $y = 2x - 8$. The slope of any line perpendicular to $y = 2x - 8$ is $-\frac{1}{2}$. Use $-\frac{1}{2}$ and $(-2, 1)$ in point-slope form:

$$y - 1 = -\frac{1}{2}(x - (-2))$$
$$y - 1 = -\frac{1}{2}x - 1$$
$$y = -\frac{1}{2}x$$

**1.4.9** The slope of the side with endpoints $(0, 0)$ and $(5, 2)$ is $\frac{2-0}{5-0}$ or $\frac{2}{5}$. The slope of the side with endpoints $(5, 2)$ and $(1, 12)$ is $\frac{12-2}{1-5}$ or $-\frac{5}{2}$. So the sides are perpendicular and the triangle is a right triangle.

**1.4.10** Find the equation of the line through $(100, 70)$ and $(200, 90)$. The slope is $\frac{90-70}{200-100}$ or 0.20. Use the point-slope formula with $C$ for cost in dollars and $m$ for miles:

$$C - 70 = 0.20(m - 100)$$
$$C - 70 = 0.20m - 20$$
$$C = 0.20m + 50$$

**1.4.11** Let $F$ be the number of file cabinets and $B$ be the number of bookshelves. Solve $100F + 150B = 3000$ for $F$:

$$100F = -150B + 3000$$
$$F = -1.5B + 30$$

The slope $-1.5$ means that increasing $B$ by 1 causes $F$ to decrease by 1.5.

## Section 5

**1.5.1 a.** Since the points appear to be approximately in line, the relationship is linear.
**b.** Since the points appear to be in an approximate parabolic shape, the relationship is nonlinear.

**1.5.2** Graph the data and draw a line that approximately fits the data. At 2011 the second coordinate should be about 4.5. So the cost in 2011 should be about \$4.50.

**1.5.3** Enter the data into your calculator and find the regression equation $C = 1.926x + 12.093$ where $x$ is the number of years after 1980. $C(30) \approx \$69.87$.

## Section 6

**1.6.1** The complex number $i - 5$ is imaginary because the coefficient of $i$ is nonzero. In standard form it is written as $-5 + i$.

**1.6.2** $(4 - 3i)(1 + 2i) = 4 - 3i + 8i - 6i^2$
$= 4 + 5i - 6(-1) = 10 + 5i$

**1.6.3** $i^{35} = (i^4)^8 i^3 = 1^8(-i) = -i$

**1.6.4** $(3 - 5i)(3 + 5i) = 9 - 25i^2 = 34$

**1.6.5** $\frac{4}{1+i} = \frac{4(1-i)}{(1+i)(1-i)}$
$= \frac{4-4i}{2} = 2 - 2i$

**1.6.6** $\frac{2 - \sqrt{-12}}{2} = \frac{2 - 2i\sqrt{3}}{2} = 1 - i\sqrt{3}$

## Section 7

**1.7.1**
$$x^2 - 7x - 18 = 0$$
$$(x - 9)(x + 2) = 0$$
$$x - 9 = 0 \text{ or } x + 2 = 0$$
$$x = 9 \text{ or } \quad x = -2$$

The solution set is $\{-2, 9\}$.

**1.7.2**
$$(x - 3)^2 = 16$$
$$x - 3 = \pm\sqrt{16}$$
$$x = 3 \pm 4$$
$$x = 7 \text{ or } -1$$

The solution set is $\{-1, 7\}$.

**1.7.3**
$$2x^2 - 4x - 1 = 0$$
$$x^2 - 2x - \frac{1}{2} = 0$$
$$x^2 - 2x + 1 = \frac{1}{2} + 1$$
$$(x - 1)^2 = \frac{3}{2}$$
$$x - 1 = \pm\sqrt{\frac{3}{2}}$$
$$x = 1 \pm \frac{\sqrt{6}}{2} = \frac{2 \pm \sqrt{6}}{2}$$

The solution set is $\left\{\frac{2-\sqrt{6}}{2}, \frac{2+\sqrt{6}}{2}\right\}$.

**1.7.4**
$$2x^2 - 3x - 2 = 0$$
$$x = \frac{3 \pm \sqrt{(-3)^2 - 4(2)(-2)}}{2(2)}$$
$$= \frac{3 \pm \sqrt{25}}{4} = \frac{3 \pm 5}{4}$$

The solution set is $\left\{-\frac{1}{2}, 2\right\}$.

**1.7.5** $b^2 - 4ac = (-7)^2 - 4(5)(9) = -131$
Since the discriminant is negative, the equation has no real solutions.

**1.7.6** Let $x$ be Josh's average speed and $x + 5$ be Bree's average speed. Their times differ by one-half hour:

$$\frac{100}{x} - \frac{90}{x+5} = \frac{1}{2}$$
$$2x(x + 5)\left(\frac{100}{x} - \frac{90}{x+5}\right) = 2x(x + 5)\frac{1}{2}$$
$$200x + 1000 - 180x = x^2 + 5x$$
$$-x^2 + 15x + 1000 = 0$$
$$x^2 - 15x - 1000 = 0$$
$$(x - 40)(x + 25) = 0$$
$$x = 40 \text{ or } x = -25$$

Josh averaged 40 mph and Bree 45 mph.

**1.7.7** The ball is back on the earth when $h = 0$:

$$-16t^2 + 40t + 6 = 0$$
$$t = \frac{-40 \pm \sqrt{(40)^2 - 4(-16)(6)}}{2(-16)}$$
$$t = \frac{-40 \pm \sqrt{1984}}{-32} = \frac{-40 \pm 8\sqrt{31}}{-32}$$
$$t = \frac{5 \pm \sqrt{31}}{4} \approx 2.64 \quad \text{or} \quad -0.14$$

The ball is in the air for $(5 + \sqrt{31})/4$ or about 2.64 seconds.

**1.7.8** Let $x$ and $x + 2$ represent the lengths of the legs.

$$x^2 + (x + 2)^2 = 6^2$$
$$2x^2 + 4x + 4 = 36$$
$$x^2 + 2x - 16 = 0$$
$$x = \frac{-2 \pm \sqrt{2^2 - 4(1)(-16)}}{2(1)} = \frac{-2 \pm \sqrt{68}}{2}$$
$$= \frac{-2 \pm 2\sqrt{17}}{2} = -1 \pm \sqrt{17}$$

The short leg is $-1 + \sqrt{17}$ or about 3.1 ft and the long leg is $1 + \sqrt{17}$ or about 5.1 ft.

**1.7.9** Let $x$ be the number of years after 1980. Quadratic regression yields

$$C = 0.0495x^2 - 0.973x + 11.850.$$

So $C(30) \approx \$27.21$.

## Section 8

**1.8.1** The interval $(-\infty, 5]$ consists of all real numbers that are less than or equal to 5, and that is the solution set to $x \leq 5$.

**1.8.2** $2 - 5x \leq 7$

$$-5x \leq 5$$
$$x \geq -1$$

The solution set is $[-1, \infty)$ and it is graphed as follows.

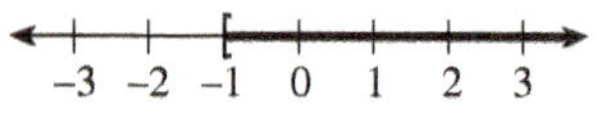


**1.8.3**

$$\frac{1}{2}x + \frac{1}{3} \leq \frac{1}{3}x + 1$$
$$6\left(\frac{1}{2}x + \frac{1}{3}\right) \leq 6\left(\frac{1}{3}x + 1\right)$$
$$3x + 2 \leq 2x + 6$$
$$x \leq 4$$

The solution set is $(-\infty, 4]$ and it is graphed as follows.

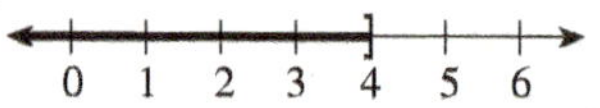


**1.8.4** $A \cup B$ consists of all real numbers between 1 and 9. So $A \cup B = (1, 9)$. $A \cap B$ consists of all real numbers that belong to both $A$ and $B$. So $A \cap B = [4, 6)$.

**1.8.5** $2x > -4$ and $4 - x \geq 0$

$x > -2$ and $-x \geq -4$

$x > -2$ and $x \leq 4$

The solution set is $(-2, 4]$.

**1.8.6** $3x + 2 > -1$ and $5 < -3 - 4x$

$3x > -3$ and $4x < -8$

$x > -1$ and $x < -2$

Since $(-\infty, -2) \cap (-1, \infty) = \varnothing$, the solution set is the empty set, $\varnothing$.

**1.8.7** $|x - 6| - 3 \leq -2$

$$|x - 6| \leq 1$$
$$-1 \leq x - 6 \leq 1$$
$$5 \leq x \leq 7$$

The solution set is $[5, 7]$.

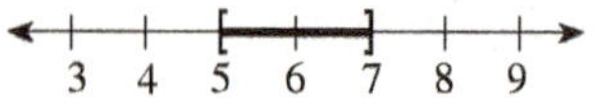


**1.8.8** Let $x$ be the third commission.

$$100 < \frac{80 + 90 + x}{3} < 110$$
$$300 < 170 + x < 330$$
$$130 < x < 160$$

The third commission must be in the interval $(130, 160)$ or between \$130 and \$160.

**1.8.9**

$$\frac{|x - 10|}{10} < 0.01$$
$$|x - 10| < 0.1$$
$$-0.1 < x - 10 < 0.1$$
$$9.9 < x < 10.1$$

If the actual amount dispensed is between 9.9 and 10.1 gallons, then the pump is certified as accurate.

# Chapter 2

## Section 1

**2.1.1** Tax is determined by finding 5% of $p$ and rounding to the nearest cent. So $t$ is a function of $p$. One cannot determine $p$ from $t$ because 10 cents in tax is paid on an item that costs \$2 or \$2.01. So $p$ is not a function of $t$.

**2.1.2** Since no vertical line can be drawn so that it crosses this graph more than once, $y$ is a function of $x$.

**2.1.3 a.** Since $(5, 5)$ and $(5, 7)$ have the same first coordinate and different second coordinates, the relation is not a function.

**b.** Since no ordered pair in the table has the same first coordinate and different second coordinates, the relation is a function.

**2.1.4** Since $(-1, 1)$ and $(-1, -1)$ both satisfy $x^3 + y^2 = 0$, the equation does not define $y$ as a function of $x$.

**2.1.5** Since $\sqrt{x + 3}$ is a real number only if $x + 3 \geq 0$ or $x \geq -3$, the domain is $[-3, \infty)$. Since $\sqrt{x + 3} \geq 0$, the range is $[0, \infty)$. Since $y$ is uniquely determined by $y = \sqrt{x + 3}$, this relation is a function.

**2.1.6 a.** $f(4) = 4 - 3 = 1$

**b.** If $f(x) = 9$, then $x - 3 = 9$ and $x = 12$.

**2.1.7** Replace $x$ with $x + 2$ to get

$$f(x + 2) = (x + 2)^2 - 4$$
$$= x^2 + 4x + 4 - 4$$
$$= x^2 + 4x$$

**2.1.8**

$$\frac{28{,}645 - 13{,}837}{2003 - 2009} = -2468$$

The average rate of change was $-\$2468$ per year.

**2.1.9**

$$\frac{f(x + h) - f(x)}{h} = \frac{(x + h)^2 - (x + h) - (x^2 - x)}{h}$$
$$= \frac{x^2 + 2xh + h^2 - x - h - x^2 + x}{h}$$
$$= \frac{2xh + h^2 - h}{h} = 2x + h - 1$$

**2.1.10** For a square, $P = 4s$. So $s = P/4$ expresses the side as a function of the perimeter.

## Section 2

**2.2.1** Since no vertical line crosses the graph more than once, $y$ is a function of $x$.

| $x$ | $-4$ | $-2$ | $0$ | $2$ | $4$ |
|---|---|---|---|---|---|
| $y = \frac{1}{2}x^2$ | 8 | 2 | 0 | 2 | 8 |

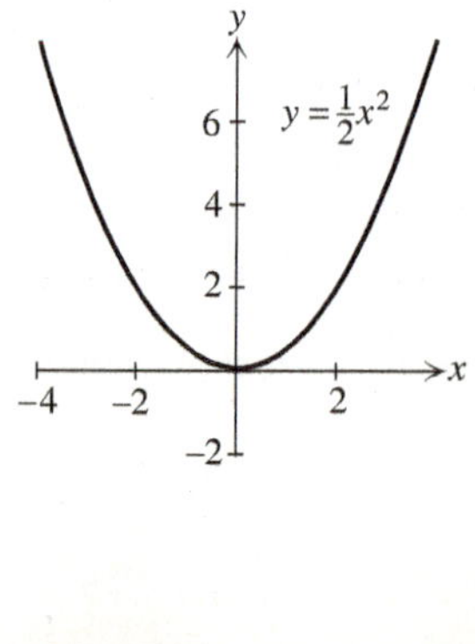

Since any real number can be used for $x$, the domain is $(-\infty, \infty)$. Since $y \geq 0$, the range is $[0, \infty)$.

**2.2.2** Since no vertical line crosses the graph more than once, $y$ is a function of $x$.

| $x$ | 1 | 0 | $-3$ | $-8$ |
|---|---|---|---|---|
| $y = \sqrt{1-x}$ | 0 | 1 | 2 | 3 |

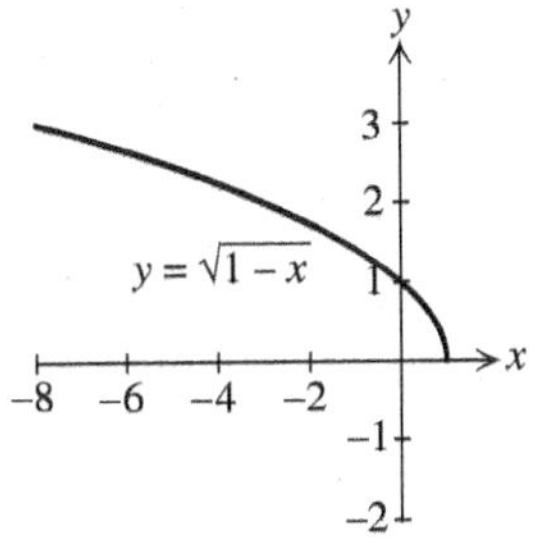


Since $1 - x \geq 0$ or $x \leq 1$ the domain is $(-\infty, 1]$. Since $y \geq 0$, the range is $[0, \infty)$.

**2.2.3** Since $(-1, 1)$ and $(-1, -1)$ both satisfy $x = -y^2$, $y$ is not uniquely determined by $x$ and the equation is not a function. Also, $y$ is not a function of $x$ by the vertical line test.

| $x = -y^2$ | $-4$ | $-1$ | 0 | $-1$ | $-4$ |
|---|---|---|---|---|---|
| $y$ | $-2$ | $-1$ | 0 | 1 | 2 |

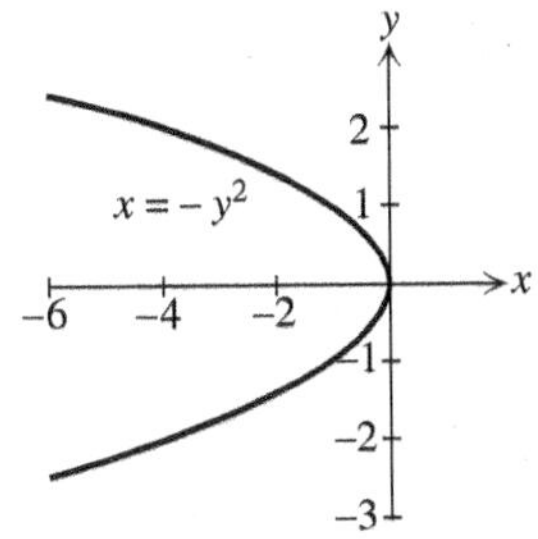


Since any real number can be used for $y$, the range is $(-\infty, \infty)$. Since $x \leq 0$, the domain is $(-\infty, 0]$.

**2.2.4** Since no vertical line crosses the graph more than once, $y$ is a function of $x$.

| $x$ | 0 | 1 | 8 | 27 |
|---|---|---|---|---|
| $y = -\sqrt[3]{x}$ | 0 | $-1$ | $-2$ | $-3$ |

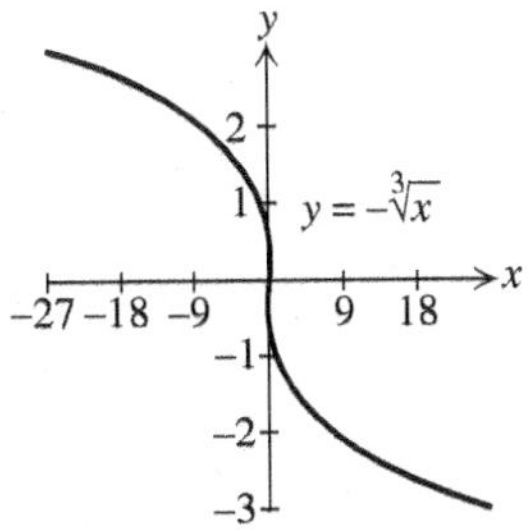


Since any real number can be used for $x$, the domain is $(-\infty, \infty)$. Since any real number can occur for $y$, the range is $(-\infty, \infty)$.

**2.2.5**

| $x$ | $-3$ | 0 | 3 |
|---|---|---|---|
| $y = -\sqrt{9 - x^2}$ | 0 | $-3$ | 0 |

The graph is a semicircle.

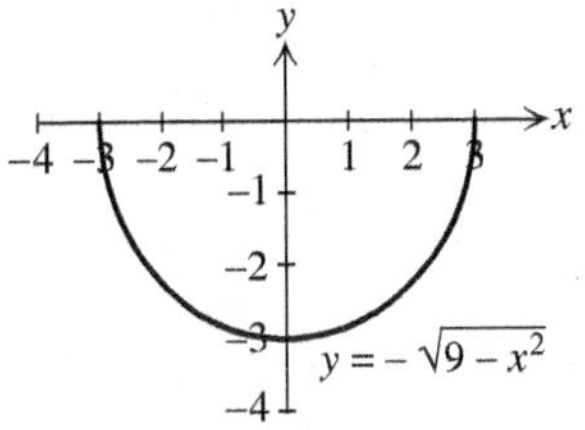


Since $x$ must be between $-3$ and 3 inclusive, the domain is $[-3, 3]$. Since $y$ is between $-3$ and 0 inclusive, the range is $[-3, 0]$.

**2.2.6** Since no vertical line crosses the graph more than once, $y$ is a function of $x$.

| $x$ | 0 | $\pm 1$ | $\pm 2$ |
|---|---|---|---|
| $y = \lvert x\rvert + 2$ | 2 | 3 | 4 |

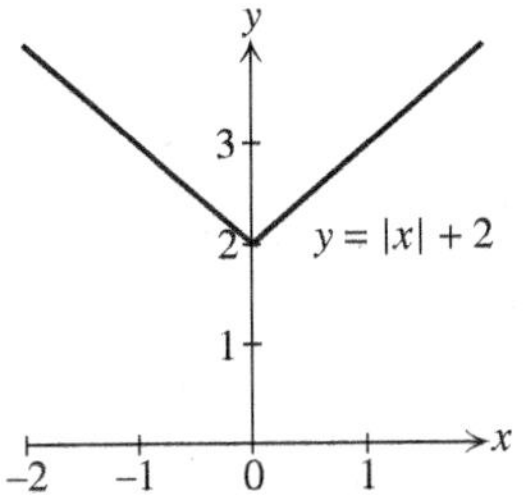


Since $x$ can be any real number, the domain is $(-\infty, \infty)$. Since $y \geq 2$, the range is $[2, \infty)$.

**2.2.7** For $x \geq 0$ the graph is a line with slope 1 starting at the origin. For $x < 0$ the graph is a line with slope $-2$.

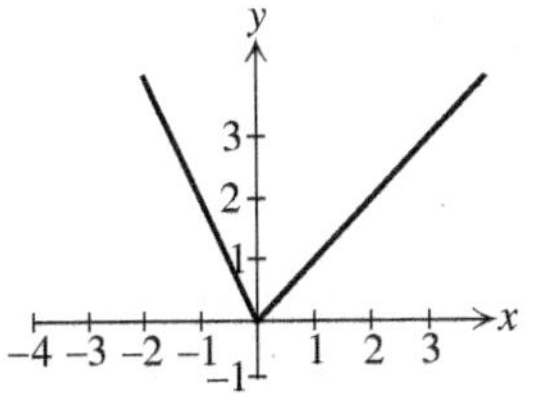


Since $x$ can be any real number, the domain is $(-\infty, \infty)$. Since $y \geq 0$, the range is $[0, \infty)$.

**2.2.8**

| $x$ | $[0, 1)$ | $[1, 2)$ | $[2, 3)$ |
|---|---|---|---|
| $y = -[\![x]\!]$ | 0 | $-1$ | $-2$ |

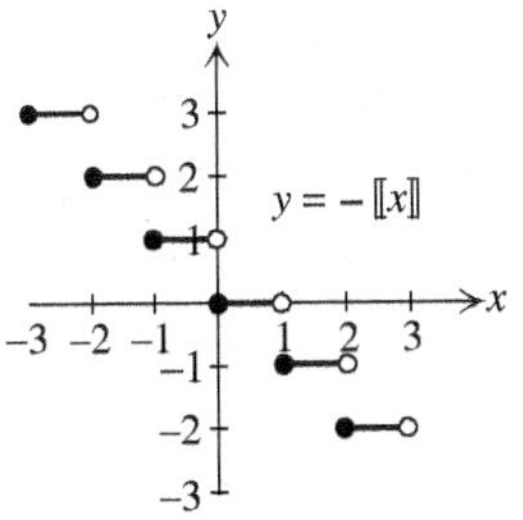


Since any real number can be used for $x$, the domain is $(-\infty, \infty)$. The range is the set of integers.

**2.2.9**

| $x$ | $[0, 1)$ | $[1, 2)$ | $[2, 3)$ |
|---|---|---|---|
| $y = [\![x + 2]\!]$ | 2 | 3 | 4 |

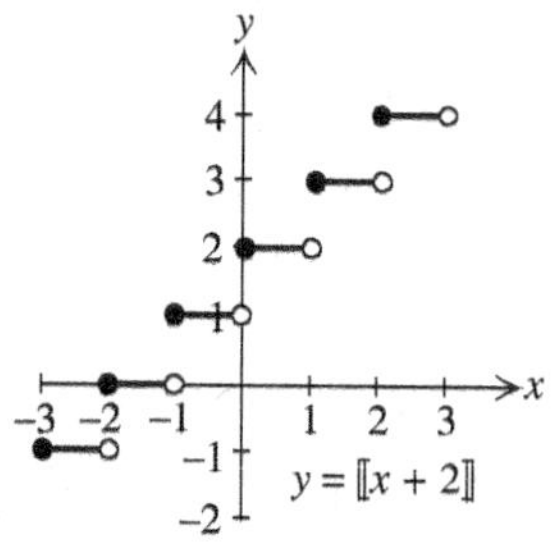


Since any real number can be used for $x$, the domain is $(-\infty, \infty)$. The range is the set of integers.

**2.2.10** Graph $f(x) = -3x$ as follows.

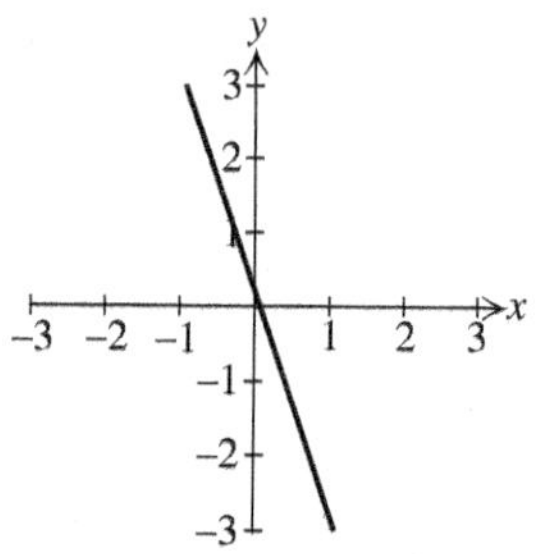


Since the $y$-coordinates are decreasing as we move from left to right on the graph, the function is decreasing.

**2.2.11** Graph $f(x)$ as follows.

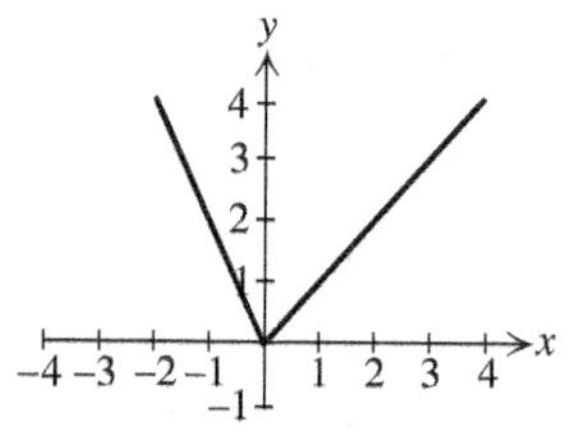


The function is decreasing on $(-\infty, 0]$ and increasing on $[0, \infty)$.

## Section 3

**2.3.1** Note that $g$ lies two units to the right of $f$ and $h$ lies one unit to the left of $f$.

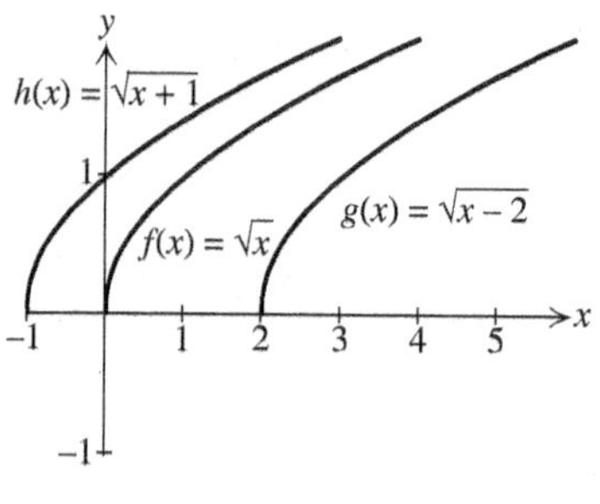


**2.3.2** Note that $f(x) = (x - 2)^2$ goes through $(1, 1)$, $(2, 0)$, and $(3, 1)$ and lies two units to the right of $y = x^2$.

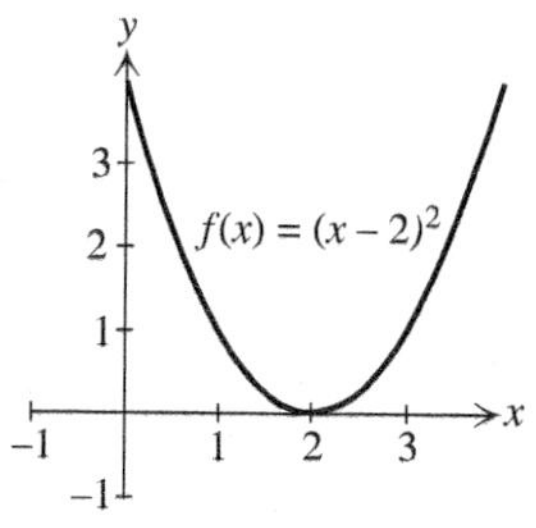


**2.3.3** The graph of $g(x) = -\sqrt{x}$ is a reflection of the graph of $f(x) = \sqrt{x}$. Note that $g$ lies below the $x$-axis and $f$ lies above the $x$-axis.

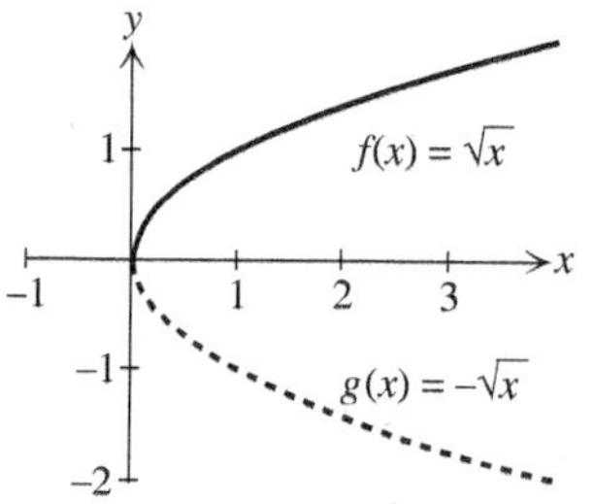


**2.3.4** The graph of $g(x) = 3|x|$ is obtained by stretching the graph of $f(x) = |x|$. The graph of $h(x) = \frac{1}{3}|x|$ is obtained by shrinking the graph of $f(x)$.

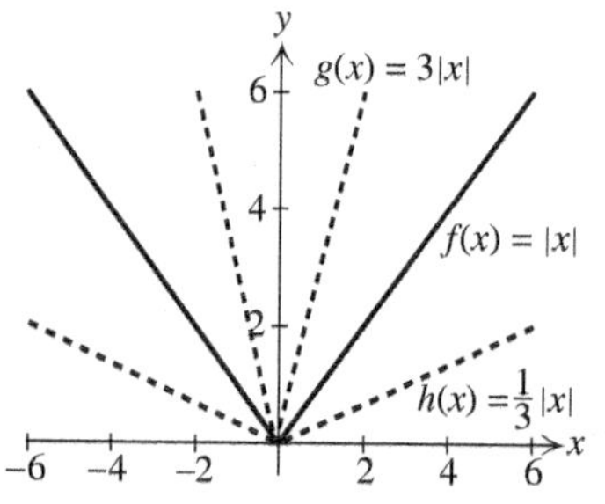


**2.3.5** Note that $g$ lies one unit above $f$ and $h$ lies two units below $f$.

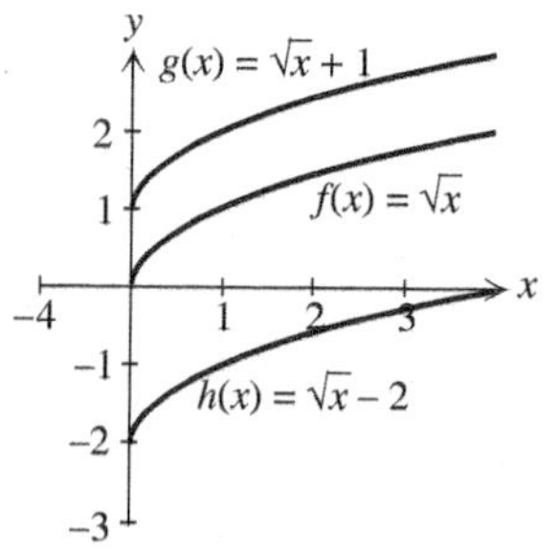

**2.3.6** The graph of $y = |x|$ is translated one unit to the left, stretched by a factor of 2, reflected in the $x$-axis, and finally translated four units upward to obtain the graph of $y = 4 - 2|x + 1|$. The graph is v-shaped and extends downward from $(-1, 4)$.

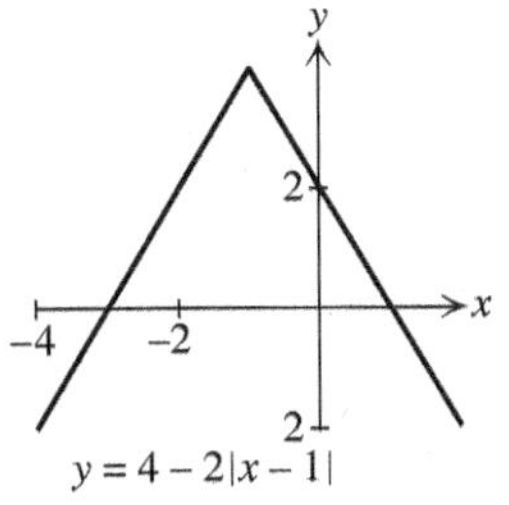


**2.3.7** The graph of $y = x$ is stretched by a factor of 2, reflected in the $x$-axis, and translated five units upward to obtain the graph of $y = -2x + 5$.

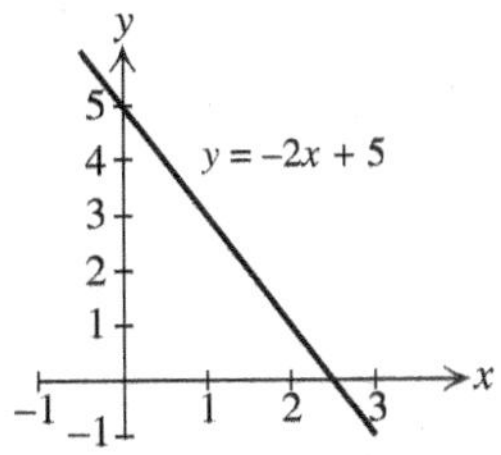


**2.3.8** Since $f(-x) = -2(-x)^2 + 5 = -2x^2 + 5$, we have $f(-x) = f(x)$. So the graph is symmetric about the $y$-axis.
**2.3.9** The graph of $y = 2 - |x - 1|$ is shown here.

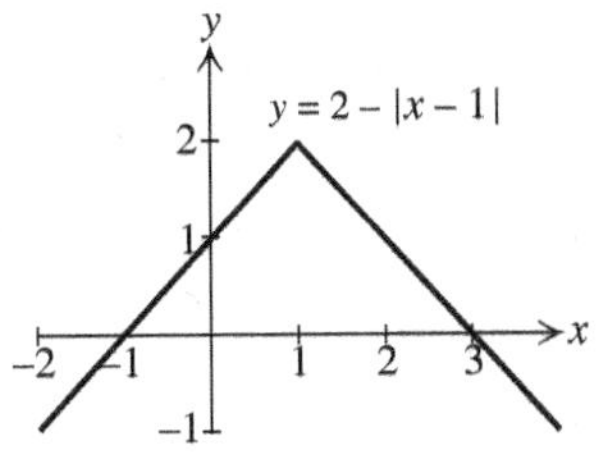


Since the graph is below the $x$-axis when $x \le -1$ or when $x \ge 3$, the solution set to the inequality is $(-\infty, -1] \cup [3, \infty)$.

## Section 4

**2.4.1** $(h + j)(5) = h(5) + j(5) = 5^2 + 3 \cdot 5 = 40$,
$(h \cdot j)(2) = h(2) \cdot j(2) = 2^2 \cdot 3(2) = 24$,
$(h/j)(a) = h(a)/j(a) = a^2/(3a) = a/3$
**2.4.2** Note that $(h + j)(2) = h(2) + j(2) = 10 + 5 = 15$, $(h + j)(6) = h(6) + j(6) = 8 + 0 = 8$, and $(h + j)(4)$ is undefined because 4 is not in the domain of $j$. So $h + j = \{(2, 15), (6, 8)\}$ and the domain is $\{2, 6\}$. Note that $(h/j)(2) = h(2)/j(2) = 10/5 = 2$, $(h/j)(4)$ is undefined because 4 is not in the domain of $j$, and $(h/j)(6)$ is undefined because $j(6) = 0$. So $h/j = \{(2, 2)\}$ and the domain is $\{2\}$.
**2.4.3** The domain of $h + j$ is the intersection of the domains of $h$ and $j$. So $(h + j)(x) = \sqrt{x} + x$ and the domain is $[0, \infty)$. Since $j(0) = 0$, the domain of $h/j$ is $(0, \infty)$ and $(h/j)(x) = \sqrt{x}/x$.
**2.4.4** Note that $(j \circ h)(2) = j(h(2)) = j(0) = 7$, $(j \circ h)(4) = j(h(4)) = j(0) = 7$, and $(j \circ h)(6)$ is undefined. So $j \circ h = \{(2, 7), (4, 7)\}$ and the domain is $\{2, 4\}$.
**2.4.5** Note that $h(5) = \sqrt{5 - 1} = 2$ and $j(5) = 2 \cdot 5 = 10$. So $(h \circ j)(5) = h(j(5)) = h(10) = \sqrt{10 - 1} = 3$ and $(j \circ h)(5) = j(h(5)) = j(2) = 2(2) = 4$.
**2.4.6** Since $(h \circ j)(x) = h(j(x)) = h(3x) = \sqrt{3x + 3}$ we must have $3x + 3 \ge 0$ or $x \ge -1$. So the domain is $[-1, \infty)$.
**2.4.7** The function $K$ is the composition of subtracting 3, taking the square root, and then multiplying by 2, in that order. So $K = h \circ f \circ g$. Check: $h(f(g(x))) = h(f(x - 3)) = h(\sqrt{x - 3}) = 2\sqrt{x - 3}$
**2.4.8** Substitute $r = C/(2\pi)$ into $d = 2r$ to get $d = 2 \cdot \dfrac{C}{2\pi}$ or $d = \dfrac{C}{\pi}$.
**2.4.9** In $C(x) = 0.10x$, $x$ is revenue. Since $R(x) = 80x$, replace $x$ with $80x$ to get $C(x) = 0.10(80x) = 8x$ where $x$ is the number of books.

## Section 5

**2.5.1** This function is one-to-one because no two ordered pairs have different first coordinates and the same second coordinate. Note that $\pi$, 3.14, and 22/7 are three different numbers.
**2.5.2** Since no horizontal line can cross the graph in (a) more than once, the function is one-to-one. Since the horizontal line $y = -2$ crosses the graph in (b) more than once, the function is not one-to-one.
**2.5.3** If $h(x_1) = h(x_2)$, then $5x_1^2 = 5x_2^2$ or $x_1^2 = x_2^2$. But $x_1^2 = x_2^2$ does not imply that $x_1 = x_2$, because $2^2 = (-2)^2$. So the function is not one-to-one.
**2.5.4** Since there are no ordered pairs with the same second coordinate and different first coordinates, the function is invertible and $h^{-1} = \{(1, 2), (4, 3), (0, 4)\}$.
**2.5.5** Since $h^{-1} = \{(1, 2), (4, 3), (0, 4)\}$, $h(3) = 4$, $h^{-1}(4) = 3$, and $(h \circ h^{-1})(1) = h(h^{-1}(1)) = h(2) = 1$.
**2.5.6** Since $f$ is a composition of multiplying $x$ by 2/3. and then adding 6, $f^{-1}$ is a composition of subtracting 6 and then dividing by 2/3 (or multiplying by 3/2) so $f^{-1}(x) = \frac{3}{2}(x - 6) = \frac{3}{2}x - 9$.
**2.5.7** The domain of $f$ is $[-2, \infty)$ and the range is $[0, \infty)$. So $f^{-1}(x) = x^2 - 2$ for $x \ge 0$.

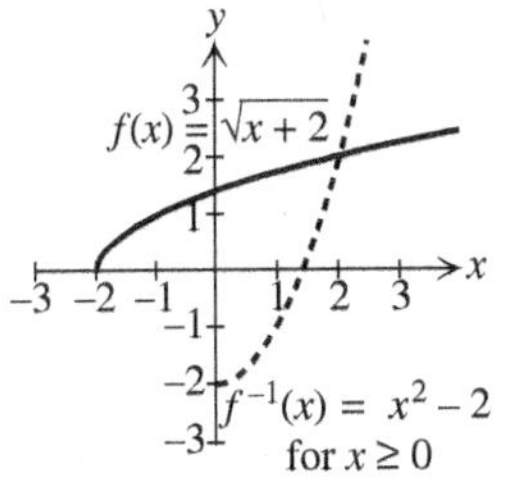


**2.5.8** First switch $x$ and $y$ in $y = x^3 - 5$ then solve for $y$:

$$\begin{aligned} x &= y^3 - 5 \\ y^3 &= x + 5 \\ y &= \sqrt[3]{x + 5} \\ h^{-1}(x) &= \sqrt[3]{x + 5} \end{aligned}$$

**2.5.9** Find $f(g(x))$ and $g(f(x))$:

$$f(g(x)) = 2\left(\frac{x - 1}{2}\right) + 1 = x - 1 + 1 = x$$

$$g(f(x)) = \frac{(2x + 1) - 1}{2} = \frac{2x}{2} = x$$

Since these equations are correct for any real number $x$, the functions are inverses of each other.

## Section 6

**2.6.1** Since $c = ks$ and $3.60 = k(12)$, we have $k = 0.30$. So if $s = 16$, then $c = 0.30(16) = 4.80$. The cost of a 16-ounce smoothie is \$4.80.
**2.6.2** Since $t = k/n$ and $12 = k/4$, we have $k = 48$. So if $n = 6$, then $t = 48/6 = 8$. So it takes 6 rakers 8 hours to complete the job.
**2.6.3** Since $c = kLH$ and $3000 = k(200 \cdot 5)$, we have $k = 3$. So if $L = 250$ and $H = 6$, then $c = 3 \cdot 250 \cdot 6 = 4500$. So the cost of a 6-foot fence that is 250 feet long is \$4500.
**2.6.4** Since the variation is direct, we use multiplication: $M = kwz^3$.
**2.6.5** Use $W = 480$, $m = 5$, and $t = 4$ in $W = km/t^2$ to find $k$:

$$\begin{aligned} 480 &= \frac{k(5)}{4^2} \\ k &= 1536 \end{aligned}$$

Now use $k = 1536$, $m = 3$, and $t = 6$ to find $W$:

$$W = \frac{1536(3)}{6^2} = 128$$

## Chapter 3

### Section 1

**3.1.1** $f(x) = 2x^2 - 8x + 9$
$f(x) = 2(x^2 - 4x) + 9$
$f(x) = 2(x^2 - 4x + 4 - 4) + 9$
$f(x) = 2(x^2 - 4x + 4) - 8 + 9$
$f(x) = 2(x - 2)^2 + 1$

**3.1.2** $f(x) = -2x^2 - 4x + 1$
$f(x) = -2(x^2 + 2x) + 1$
$f(x) = -2(x^2 + 2x + 1 - 1) + 1$
$f(x) = -2(x^2 + 2x + 1) + 2 + 1$
$f(x) = -2(x + 1)^2 + 3$

The graph is a parabola opening downward from $(-1, 3)$. The graph goes through $(0, 1)$ and $(-2, 1)$.

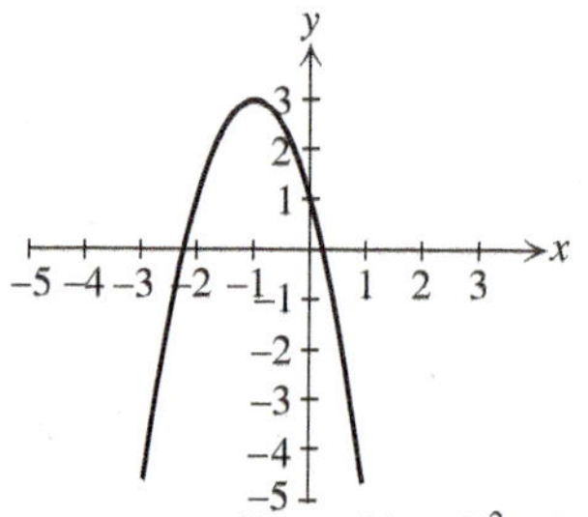


$f(x) = -2(x + 1)^2 + 3$

**3.1.3** $x = \frac{-b}{2a} = \frac{-(-12)}{2(3)} = 2$

$f(2) = 3(2^2) - 12(2) + 5 = -7$

The vertex is $(2, -7)$.

**3.1.4** The vertex is $(5, -4)$ and the axis of symmetry is $x = 5$. Since the parabola opens upward from the vertex, the range is $[-4, \infty)$ and $-4$ is the minimum value of the function. The function is decreasing on $(-\infty, 5]$ and increasing on $[5, \infty)$.

**3.1.5** If $x = 0$, then $y = 3(0 - 5)^2 - 4 = 71$. If $y = 0$, then $3(x - 5)^2 - 4 = 0$ or $x = 5 \pm \sqrt{\frac{4}{3}} = 5 \pm \frac{2\sqrt{3}}{3} = \frac{15 \pm 2\sqrt{3}}{3}$.

The intercepts are $(0, 71)$ and $\left(\frac{15 \pm 2\sqrt{3}}{3}, 0\right)$.

$y = 3(x - 5)^2 - 4$

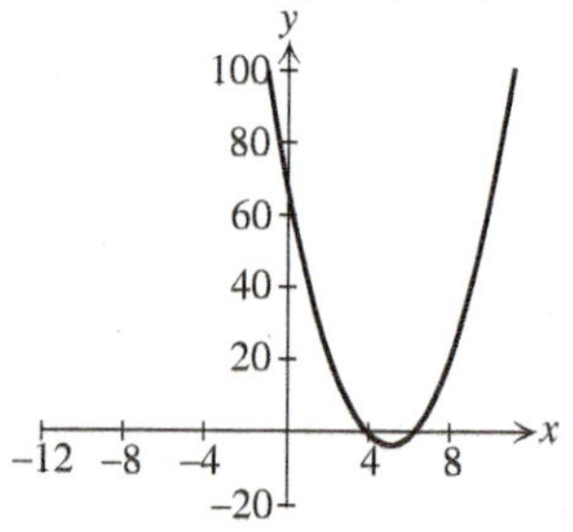


**3.1.6** Find the $x$-intercepts for $y = x^2 + 6x + 8$:

$$x^2 + 6x + 8 = 0$$
$$(x + 2)(x + 4) = 0$$
$$x = -2 \quad \text{or} \quad x = -4$$

The $x$-intercepts are $(-4, 0)$ and $(-2, 0)$. Since the parabola $y = x^2 + 6x + 8$ opens upward, the $y$-coordinates are positive outside the $x$-intercepts and zero at the $x$-intercepts. So the solution set to $x^2 + 6x + 8 \ge 0$ is $(-\infty, -4] \cup [-2, \infty)$.

**3.1.7** Solve $x^2 - 2x - 4 = 0$:

$$x = \frac{2 \pm \sqrt{4 - 4(1)(-4)}}{2(1)} = 1 \pm \sqrt{5}$$

The numbers $1 - \sqrt{5}$ and $1 + \sqrt{5}$ divide the number line into three intervals, $(-\infty, 1 - \sqrt{5})$, $(1 - \sqrt{5}, 1 + \sqrt{5})$, and $(1 + \sqrt{5}, \infty)$. Test one point from each interval in $x^2 - 2x - 4 < 0$ to see that the solution set is $(1 - \sqrt{5}, 1 + \sqrt{5})$.

**3.1.8** For $h(t) = -16t^2 + 96t + 40$ we have $a = -16$, $b = 96$, and $-b/(2a) = -96/(-32) = 3$. Since $h(3) = -16(3^2) + 96(3) + 40 = 184$, the vertex is $(3, 184)$. Since the parabola opens downward, 184 is the maximum value of the function and the maximum height above the ground is 184 feet.

**3.1.9** Since $L + W = 25$, $W = 25 - L$.

$$A = L(25 - L) = -L^2 + 25L$$
$$L = \frac{-b}{2a} = \frac{-25}{2(-1)} = 12.5$$

If $L = 12.5$, then $W = 12.5$ and the rectangle with the largest area is a 12.5-m by 12.5-m square.

### Section 2

**3.2.1**

$$\begin{array}{r} x - 6 \\ x - 1\overline{)x^2 - 7x + 9} \\ \underline{x^2 - \phantom{7}x} \\ -6x + 9 \\ \underline{-6x + 6} \\ 3 \end{array}$$

Since the remainder is 3, $P(1) = 3$.

**3.2.2**

| | | | | |
|---|---|---|---|---|
| −2 | 1 | 0 | −7 | 5 |
| | | −2 | 4 | 6 |
| | 1 | −2 | −3 | 11 |

The quotient is $x^2 - 2x - 3$ and the remainder is 11.

**3.2.3**

| | | | | |
|---|---|---|---|---|
| 3 | 1 | 1 | 0 | −9 |
| | | 3 | 12 | 36 |
| | 1 | 4 | 12 | 27 |

$P(3) = 27$

**3.2.4**

| | | | | |
|---|---|---|---|---|
| 1 | 1 | 0 | −3 | 2 |
| | | 1 | 1 | −2 |
| | 1 | 1 | −2 | 0 |

Since the remainder is 0, $x - 1$ is a factor.

$$P(x) = (x - 1)(x^2 + x - 2)$$
$$= (x - 1)^2(x + 2)$$

**3.2.5** The factors of 2 are 1 and 2. The factors of 3 are 1 and 3. All possible factors of 3 over factors of 2 are $\pm 1$, $\pm 3$, $\pm\frac{1}{2}$, and $\pm\frac{3}{2}$.

**3.2.6** Try the possible rational zeros.

| | | | | |
|---|---|---|---|---|
| −1 | 2 | 1 | −4 | −3 |
| | | −2 | 1 | 3 |
| | 2 | −1 | −3 | 0 |

$$h(x) = (x + 1)(2x^2 - x - 3)$$
$$= (x + 1)(x + 1)(2x - 3)$$

The zeros are $-1$ and $3/2$.

**3.2.7** The possible rational zeros come from factors of 8 over factors of 3:

$$\pm\left(1, 2, 4, 8, \frac{1}{3}, \frac{2}{3}, \frac{4}{3}, \frac{8}{3}\right)$$

Use synthetic division to find a rational zero:

| | | | | |
|---|---|---|---|---|
| $\frac{2}{3}$ | 3 | −2 | 12 | −8 |
| | | 2 | 0 | 8 |
| | 3 | 0 | 12 | 0 |

Solve $3x^2 + 12 = 0$:

$$3x^2 = -12$$
$$x^2 = -4$$
$$x = \pm\sqrt{-4} = \pm 2i$$

The three zeros are $2/3$ and $\pm 2i$.

## Section 3

**3.3.1** Set each factor equal to zero to get 0 as a root with multiplicity 3, $-2$ as a root with multiplicity 2, and $5/2$ as a root.

**3.3.2** If the coefficients are to be real, then both $-i$ and its conjugate $i$ must be roots.

$$(x - 3)(x - i)(x + i) = 0$$
$$(x - 3)(x^2 + 1) = 0$$
$$x^3 - 3x^2 + x - 3 = 0$$

**3.3.3** The number of variations in sign for $P(x) = x^3 - 5x^2 + 4x + 3$ is 2. So there are either 0 or 2 positive roots. The number of variations in sign for $P(-x) = -x^3 - 5x^2 - 4x + 3$ is 1. So there is exactly 1 negative root. Since there must be a total of 3 roots, there is one negative root and 2 positive roots or one negative root and 2 imaginary roots.

**3.3.4** The number of variations in sign for $P(x) = x^4 - 6x^2 + 10$ is 2 and for $P(-x) = x^4 - 6x^2 + 10$ it is also 2. So there are either 0 or 2 positive roots and 0 or 2 negative roots. So the possibilities are 0 positive, 0 negative, 4 imaginary; 0 positive, 2 negative, 2 imaginary; 2 positive, 0 negative, 2 imaginary; 2 positive, 2 negative, 0 imaginary.

**3.3.5** The first positive integer for which all terms of the bottom row of synthetic division are nonnegative is 3. So 3 is the best upper bound for the roots. The first negative integer for which the terms of the bottom row alternate in sign is $-5$. So $-5$ is the best lower bound for the roots by the theorem on bounds.

**3.3.6** Try possible rational roots with synthetic division:

$$\begin{array}{c|cccc} \frac{1}{2} & 2 & -5 & -8 & 5 \\ & & 1 & -2 & -5 \\ \hline & 2 & -4 & -10 & 0 \end{array}$$

$$2x^2 - 4x - 10 = 0$$
$$x^2 - 2x - 5 = 0$$
$$x = \frac{2 \pm \sqrt{4 - 4(1)(-5)}}{2(1)} = 1 \pm \sqrt{6}$$

The roots are $1/2$, $1 - \sqrt{6}$, and $1 + \sqrt{6}$.

## Section 4

**3.4.1** $x^3 - 2x^2 + 5x - 10 = 0$

$$x^2(x - 2) + 5(x - 2) = 0$$
$$(x^2 + 5)(x - 2) = 0$$
$$x^2 + 5 = 0 \quad \text{or} \quad x - 2 = 0$$
$$x = \pm i\sqrt{5} \quad \text{or} \quad x = 2$$

The solution set is $\{2, -i\sqrt{5}, i\sqrt{5}\}$.

**3.4.2**

$$x^5 = 27x^2$$
$$x^5 - 27x^2 = 0$$
$$x^2(x - 3)(x^2 + 3x + 9) = 0$$
$$x^2 = 0 \quad \text{or} \quad x = 3 \quad \text{or} \quad x = \frac{-3 \pm \sqrt{-27}}{2}$$
$$= -\frac{3}{2} \pm \frac{3}{2}i\sqrt{3}$$

The solution set is $\{0, 3, -\frac{3}{2} \pm \frac{3}{2}i\sqrt{3}\}$.

**3.4.3** $\sqrt{x} + 12 = x$

$$\sqrt{x} = x - 12$$
$$x = x^2 - 24x + 144$$
$$0 = x^2 - 25x + 144$$
$$0 = (x - 9)(x - 16)$$
$$x = 9 \quad \text{or} \quad x = 16$$

Since 9 does not satisfy $\sqrt{x} + 12 = x$ the solution set is $\{16\}$.

**3.4.4** $\sqrt{3x - 2} - \sqrt{x} = 2$

$$\sqrt{3x - 2} = 2 + \sqrt{x}$$
$$3x - 2 = 4 + 4\sqrt{x} + x$$
$$2x - 6 = 4\sqrt{x}$$
$$x - 3 = 2\sqrt{x}$$
$$x^2 - 6x + 9 = 4x$$
$$x^2 - 10x + 9 = 0$$
$$(x - 9)(x - 1) = 0$$
$$x = 9 \quad \text{or} \quad x = 1$$

Since $\sqrt{3x - 2} - \sqrt{x} = 2$ is not satisfied by 1, the solution set is $\{9\}$.

**3.4.5** $x^{-4/5} = 16$

$$(x^{-4/5})^{-5/4} = \pm 16^{-5/4}$$
$$x = \pm\frac{1}{32}$$

The solution set is $\{\pm 1/32\}$.

**3.4.6** $x^4 - 9x^2 + 20 = 0$

$$(x^2 - 5)(x^2 - 4) = 0$$
$$x^2 - 5 = 0 \quad \text{or} \quad x^2 - 4 = 0$$
$$x = \pm\sqrt{5} \quad \text{or} \quad x = \pm 2$$

The solution set is $\{\pm\sqrt{5}, \pm 2\}$.

**3.4.7** $(x^2 + x)^2 - 8(x^2 + x) + 12 = 0$

$$u^2 - 8u + 12 = 0$$
$$(u - 6)(u - 2) = 0$$
$$u - 6 = 0 \quad \text{or} \quad u - 2 = 0$$
$$x^2 + x - 6 = 0 \quad \text{or} \quad x^2 + x - 2 = 0$$
$$(x + 3)(x - 2) = 0 \quad \text{or} \quad (x + 2)(x - 1) = 0$$
$$x = -3 \quad \text{or} \quad x = 2 \quad \text{or} \quad x = -2 \quad \text{or} \quad x = 1$$

The solution set is $\{-3, -2, 1, 2\}$.

**3.4.8** $x^{2/3} - x^{1/3} - 6 = 0$

$$(x^{1/3} - 3)(x^{1/3} + 2) = 0$$
$$x^{1/3} = 3 \quad \text{or} \quad x^{1/3} = -2$$
$$x = 27 \quad \text{or} \quad x = -8$$

The solution set is $\{-8, 27\}$

**3.4.9** $|x^2 - x - 4| = 2$

$$x^2 - x - 4 = 2 \quad \text{or} \quad x^2 - x - 4 = -2$$
$$x^2 - x - 6 = 0 \quad \text{or} \quad x^2 - x - 2 = 0$$
$$(x - 3)(x + 2) = 0 \quad \text{or} \quad (x - 2)(x + 1) = 0$$
$$x = 3 \quad \text{or} \quad x = -2 \quad \text{or} \quad x = 2 \quad \text{or} \quad x = -1$$

The solution set is $\{-2, -1, 2, 3\}$.

**3.4.10** $|x| = |x - 1|$

$$x = x - 1 \quad \text{or} \quad x = -(x - 1)$$
$$0 = -1 \quad \text{or} \quad 2x = 1$$
$$\text{or} \quad x = \frac{1}{2}$$

The solution set is $\{1/2\}$.

**3.4.11** Let $D = 10$ and solve for $P$:

$$10 = \sqrt{P} - \sqrt{P - 400}$$
$$\sqrt{P - 400} = \sqrt{P} - 10$$
$$P - 400 = P - 20\sqrt{P} + 100$$
$$-500 = -20\sqrt{P}$$
$$25 = \sqrt{P}$$
$$625 = P$$

At \$625 the demand is 10 suits per month.

## Section 5

**3.5.1** Since $f(-x) = x^3 + 4x$, we have $f(-x) = -f(x)$ and the graph is symmetric about the origin.

**3.5.2** Since $x - 1$ occurs with an odd power, the graph crosses the $x$-axis at $(1, 0)$. Since $x + 5$ occurs with an even power, the graph does not cross the $x$-axis at $(-5, 0)$.

**3.5.3** Since the leading coefficient is negative and the degree is odd, as $x \to \infty, y \to -\infty$. As $x \to -\infty, y \to \infty$.

**3.5.4** Since $f(-x) = f(x)$ the graph is symmetric about the $y$-axis. Since $f(x) = x^2(x - 2)(x + 2)$ the graph crosses the $x$-axis at $(2, 0)$ and $(-2, 0)$ and touches but does not cross at $(0, 0)$. As $x \to \pm\infty, y \to \infty$.

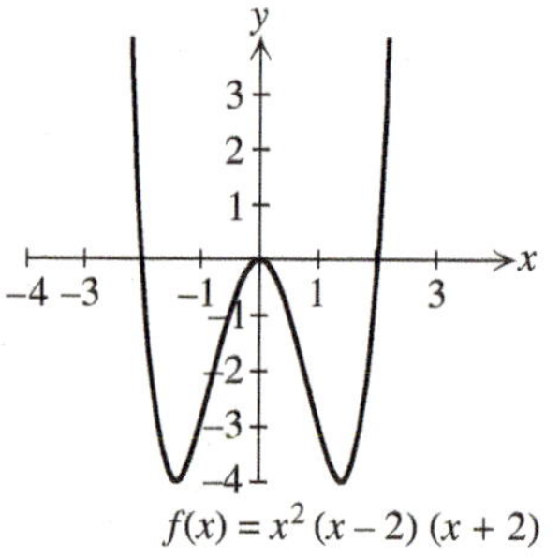


**3.5.5** The solutions to $x^4 - 4x^2 = 0$ are $-2$, 0, and 2. They divide the number line into four intervals $(-\infty, -2)$, $(-2, 0)$, $(0, 2)$, and $(2, \infty)$. Testing one number from each interval in $x^4 - 4x^2 < 0$ reveals that the solution set to the inequality is $(-2, 0) \cup (0, 2)$.

## Section 6

**3.6.1** Since $x^2 - 9 = 0$ only if $x = \pm 3$, the domain is $(-\infty, -3) \cup (-3, 3) \cup (3, \infty)$.

**3.6.2** The vertical lines $x = -3$ and $x = 3$ are vertical asymptotes. Since the degree of the denominator is larger than the degree of the numerator, the $x$-axis is a horizontal asymptote.

**3.6.3**

$$\begin{array}{r} 3x + 3 \phantom{)} \\ x - 1\overline{)3x^2 + 0x - 4} \\ \underline{3x^2 - 3x} \phantom{- 4} \\ 3x - 4 \\ \underline{3x - 3} \\ -1 \end{array}$$

$$\frac{3x^2 - 4}{x - 1} = 3x + 3 + \frac{-1}{x - 1}$$

The line $y = 3x + 3$ is an oblique asymptote and the line $x = 1$ is a vertical asymptote.

**3.6.4** The vertical asymptotes are $x = \pm 3$ and the horizontal asymptote is $y = 0$. The $x$-intercept is $(1, 0)$.

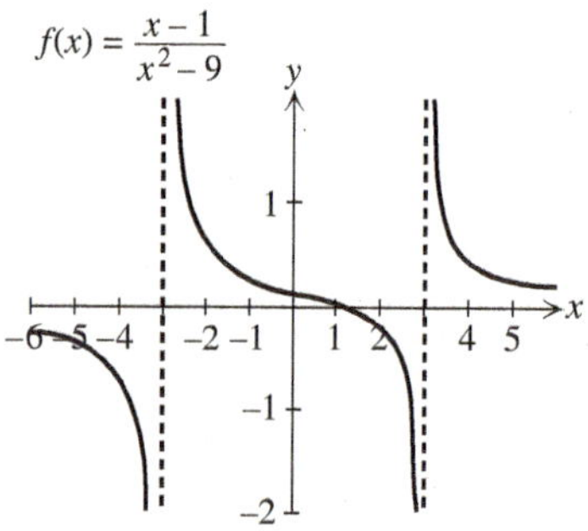


**3.6.5** The line $x = 2$ is a vertical asymptote and the $x$-axis is a horizontal asymptote. The $x$-intercept is $(0, 0)$.

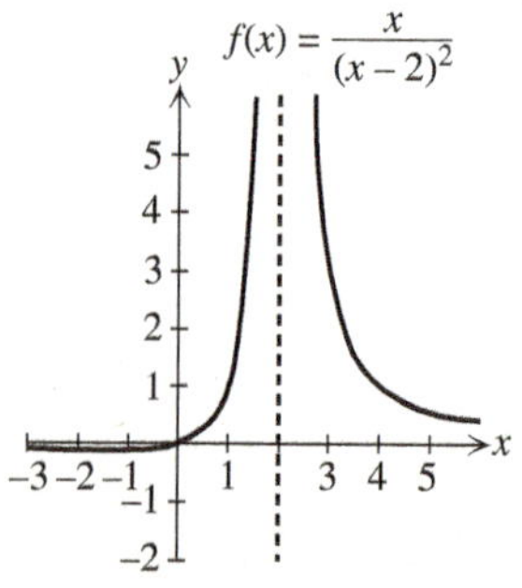


**3.6.6** From 3.6.3, the oblique asymptote is $y = 3x + 3$ and the vertical asymptote is $x = 1$.

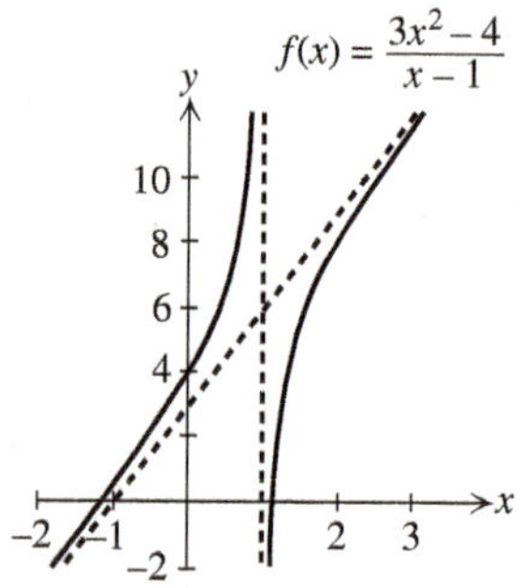


**3.6.7** If $x \neq 1$ and $x \neq -1$, then $f(x) = \frac{1}{x + 1}$. So there is a vertical asymptote at $x = -1$ and a hole at $(1, 1/2)$.

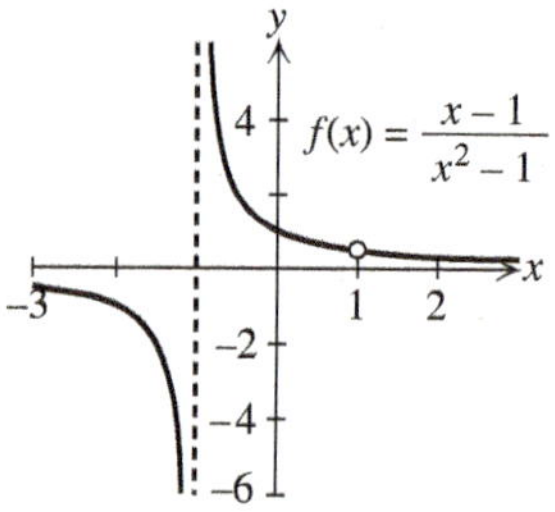


**3.6.8**

$$\frac{1}{x - 1} \geq 1$$

$$\frac{1}{x - 1} - 1 \geq 0$$

$$\frac{1}{x - 1} - \frac{x - 1}{x - 1} \geq 0$$

$$\frac{2 - x}{x - 1} \geq 0$$

The rational expression is undefined if $x = 1$ and has value 0 if $x = 2$. These critical numbers divide the number line into three intervals, $(-\infty, 1)$, $(1, 2)$, and $(2, \infty)$. Test one point in each interval to determine that $\frac{2 - x}{x - 1}$ is positive only on the interval $(1, 2)$. Since the value of $\frac{2 - x}{x - 1}$ is zero if $x = 2$, the solution set is $(1, 2]$.

**3.6.9** The expression is undefined if $x = 1$ and has a value of 0 if $x = -3$. These numbers divide the number line into three intervals, $(-\infty, -3)$, $(-3, 1)$, and $(1, \infty)$. Test one number in each interval to see that $\frac{x + 3}{x - 1} \geq 0$ is satisfied for $x$ in $(-\infty, -3] \cup (1, \infty)$.

**3.6.10** The average cost per mile is given by $C = \frac{0.80x + 100}{x}$ where $x$ is the number of miles. As $x$ increases, $C$ approaches the horizontal asymptote, which is \$0.80 per mile.

# Chapter 4

## Section 1

**4.1.1** If $f(x) = 9^x$, then $f(-2) = 9^{-2} = 1/81$ and $f(1/2) = 9^{1/2} = 3$.

**4.1.2** The ordered pairs $(-2, 1/81)$, $(0, 1)$, and $(2, 81)$ are on the graph. The domain is $(-\infty, \infty)$ and the range is $(0, \infty)$.

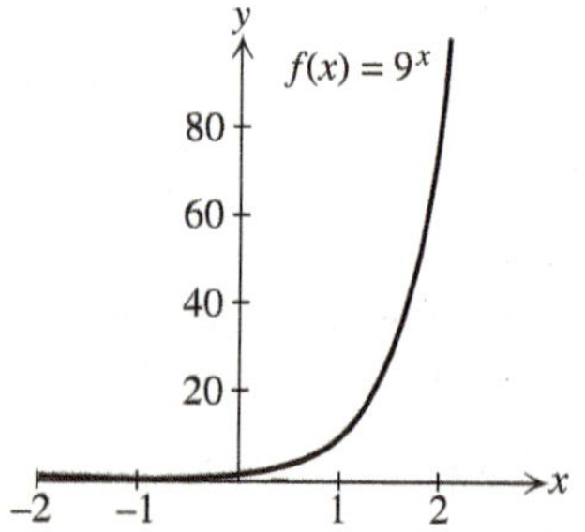

**4.1.3** The ordered pairs $(-2, 81)$, $(0, 1)$, and $(2, 1/81)$ are on the graph. The domain is $(-\infty, \infty)$ and the range is $(0, \infty)$.

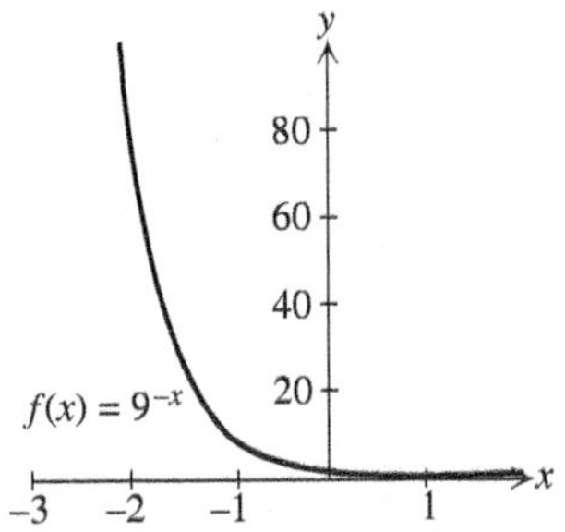


**4.1.4** The ordered pairs $(1, 3/2)$, $(2, 2)$, and $(3, 3)$ are on the graph. The domain is $(-\infty, \infty)$ and the range is $(1, \infty)$.

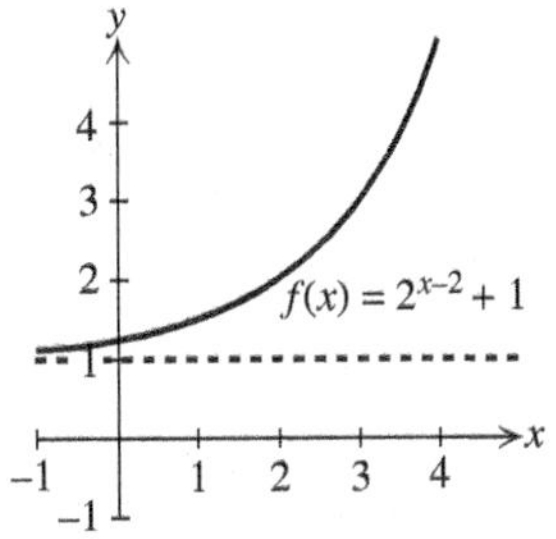


**4.1.5** The ordered pairs $(-1, -4)$, $(0, -1)$, and $(1, -1/4)$ are on the graph. The domain is $(-\infty, \infty)$ and the range is $(-\infty, 0)$.

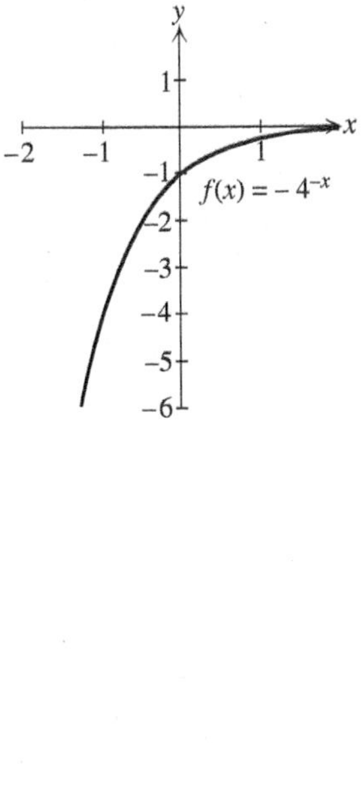


**4.1.6** $2^{-x} = \frac{1}{8}$

$$
\begin{aligned}
2^{-x} &= 2^{-3}\\
-x &= -3\\
x &= 3
\end{aligned}
$$

**4.1.7** $3^{4-x} = \frac{1}{9}$

$$
\begin{aligned}
3^{4-x} &= 3^{-2}\\
4 - x &= -2\\
-x &= -6\\
x &= 6
\end{aligned}
$$

**4.1.8** $A = 9000\left(1 + \frac{0.054}{12}\right)^{12 \cdot 6}$

$\approx \$12{,}434.78$

**4.1.9** $A = 8000e^{0.063(7.25)}$

$\approx \$12{,}631.47$

**4.1.10** $A = 2.9e^{-1.21\times10^{-4}(0)} = 2.9$

The initial amount is 2.9 grams.

$A = 2.9e^{-1.21\times10^{-4}(6500)} \approx 1.3$ grams

## Section 2

**4.2.1** Since $4^{-2} = 1/16$, $\log_4(1/16) = -2$.

**4.2.2** Since $10^2 = 100$, $\log(100) = 2$.

**4.2.3** The ordered pairs $(1/6, -1)$, $(1, 0)$, $(6, 1)$, and $(36, 2)$ are on the graph. The domain is $(0, \infty)$ and the range is $(-\infty, \infty)$.

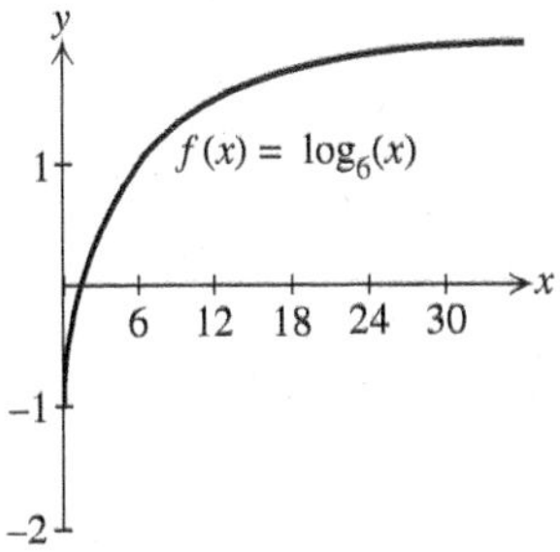


**4.2.4** The ordered pairs $(36, -2)$, $(6, -1)$, $(1, 0)$, $(1/6, 1)$, and $(1/36, 2)$ are on the graph. The domain is $(0, \infty)$ and the range is $(-\infty, \infty)$.

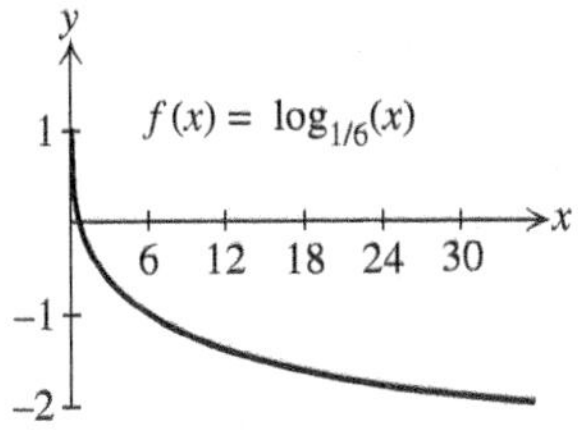


**4.2.5** The ordered pairs $(0, 0)$, $(1, -1)$, $(3, -2)$, $(7, -3)$, and $(-1/2, 1)$ are on the graph. The domain is $(-1, \infty)$ and the range is $(-\infty, \infty)$.

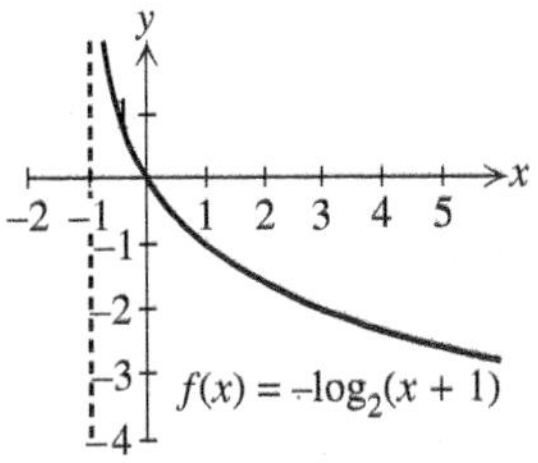


**4.2.6** Since $y = a^x$ is equivalent to $\log_a(y) = x$, $\log_4(w) = 7$ is equivalent to $w = 4^7$.

**4.2.7**

$$
\begin{aligned}
f(x) &= 3 \cdot e^{5x-1}\\
y &= 3 \cdot e^{5x-1}\\
x &= 3 \cdot e^{5y-1}\\
\frac{x}{3} &= e^{5y-1}\\
\ln\left(\frac{x}{3}\right) &= 5y - 1\\
5y &= \ln\left(\frac{x}{3}\right) + 1\\
y &= \frac{1}{5}\ln\left(\frac{x}{3}\right) + \frac{1}{5}\\
f^{-1}(x) &= \frac{1}{5}\ln\left(\frac{x}{3}\right) + \frac{1}{5}
\end{aligned}
$$

**4.2.8** $\log_5(2x) = -3$

$$
\begin{aligned}
2x &= 5^{-3}\\
2x &= \frac{1}{125}\\
x &= \frac{1}{250}
\end{aligned}
$$

**4.2.9** $10^{3x} = 70$

$$
\begin{aligned}
3x &= \log(70)\\
x &= \frac{\log(70)}{3} \approx 0.6150
\end{aligned}
$$

**4.2.10** $$\begin{aligned} 12 &= 4 \cdot e^{0.05t} \\ 3 &= e^{0.05t} \\ 0.05t &= \ln(3) \\ t &= \frac{\ln(3)}{0.05} \approx 21.9722 \end{aligned}$$

The time is 21.9722 years or 21 years and 355 days (to the nearest day).

### Section 3

**4.3.1** $10^{\log(5w)} = 5w$, $\log(10^p) = p$

**4.3.2** $\log(x) + \log(y) = \log(xy)$

**4.3.3** $$\begin{aligned} \log(5x) - \log(5) &= \log(5x/5) \\ &= \log(x) \end{aligned}$$

**4.3.4** $\ln(27) = \ln(3^3) = 3 \cdot \ln(3)$

**4.3.5** $$\begin{aligned} \ln(45) &= \ln(3^2 \cdot 5) \\ &= 2 \cdot \ln(3) + \ln(5) \end{aligned}$$

**4.3.6** $$\begin{aligned} \log\left(\frac{x^2}{5a}\right) &= \log(x^2) - \log(5a) \\ &= 2 \cdot \log(x) - [\log(5) + \log(a)] \\ &= 2 \cdot \log(x) - \log(5) - \log(a) \end{aligned}$$

**4.3.7** $$\begin{aligned} &\ln(x) - \ln(y) - 2 \cdot \ln(z) \\ &= \ln\left(\frac{x}{y}\right) - \ln(z^2) \\ &= \ln\left(\frac{x}{yz^2}\right) \end{aligned}$$

**4.3.8** $$\begin{aligned} 5.44^x &= 2.3 \\ \log(5.44^x) &= \log(2.3) \\ x \cdot \log(5.44) &= \log(2.3) \\ x &= \frac{\log(2.3)}{\log(5.44)} \\ x &\approx 0.4917 \end{aligned}$$

**4.3.9** $$\begin{aligned} 400 &= 100\left(1 + \frac{0.05}{365}\right)^{365t} \\ 4 &= \left(1 + \frac{0.05}{365}\right)^{365t} \\ 4 &\approx 1.000136986^{365t} \\ 365t &\approx \log_{1.000136986}(4) \\ t &\approx \frac{1}{365} \cdot \frac{\ln(4)}{\ln(1.000136986)} \\ &\approx 27.72778621 \text{ yr} \\ &\approx 27 \text{ years } 266 \text{ days} \end{aligned}$$

**4.3.10** $$\begin{aligned} 300 &= 100\left(1 + \frac{r}{365}\right)^{365(20)} \\ 3 &= \left(1 + \frac{r}{365}\right)^{7300} \\ 1 + \frac{r}{365} &= 3^{1/7300} \\ r &= 365(3^{1/7300} - 1) \\ r &\approx 5.5\% \end{aligned}$$

### Section 4

**4.4.1** $$\begin{aligned} \log(2x + 1) &= 3 \\ 2x + 1 &= 10^3 \\ 2x &= 999 \\ x &= 499.5 \end{aligned}$$

**4.4.2** $$\begin{aligned} \log(x - 1) + \log(3) &= \log(x) - \log(4) \\ \log(3x - 3) &= \log(x/4) \\ 3x - 3 &= \frac{x}{4} \\ 12x - 12 &= x \\ 11x &= 12 \\ x &= \frac{12}{11} \end{aligned}$$

**4.4.3** $$\begin{aligned} (1.05)^{3t} &= 8 \\ 3t &= \log_{1.05}(8) \\ t &= \frac{1}{3} \cdot \frac{\ln(8)}{\ln(1.05)} \\ t &\approx 14.2067 \end{aligned}$$

**4.4.4** $$\begin{aligned} 3^{x-1} &= 2^x \\ \ln(3^{x-1}) &= \ln(2^x) \\ (x - 1)\ln(3) &= x \cdot \ln(2) \\ x \cdot \ln(3) - \ln(3) &= x \cdot \ln(2) \\ x \cdot \ln(3) - x \cdot \ln(2) &= \ln(3) \\ x(\ln(3) - \ln(2)) &= \ln(3) \\ x \cdot \ln(3/2) &= \ln(3) \\ x &= \frac{\ln(3)}{\ln(3/2)} \approx 2.7095 \end{aligned}$$

**4.4.5** First find $r$:

$$\begin{aligned} 1 &= 2e^{1\times10^6 r} \\ 0.5 &= e^{1\times10^6 r} \\ 1 \times 10^6 r &= \ln(0.5) \\ r &= \frac{\ln(0.5)}{1 \times 10^6} \approx -6.93 \times 10^{-7} \end{aligned}$$

Next find $t$:

$$\begin{aligned} 0.40 &= 1e^{-6.93\times10^{-7} t} \\ -6.93 \times 10^{-7} t &= \ln(0.40) \\ t &= \frac{\ln(0.40)}{-6.93 \times 10^{-7}} \\ &\approx 1.32 \times 10^6 \text{ years} \end{aligned}$$

**4.4.6** $$\begin{aligned} \frac{1}{2}A_0 &= A_0 e^{4r} \\ \frac{1}{2} &= e^{4r} \\ 4r &= \ln(1/2) \\ r &= \frac{\ln(1/2)}{4} \approx -0.1732868 \end{aligned}$$

If 99% is eliminated, 1% remains. Now find the time:

$$\begin{aligned} 0.01A_0 &= A_0 e^{-0.1732868t} \\ 0.01 &= e^{-0.1732868t} \\ -0.1732868t &= \ln(0.01) \\ t &= \frac{\ln(0.01)}{-0.1732868} \approx 26.6 \end{aligned}$$

So 99% will be eliminated in 26.6 days.

**4.4.7** The difference in temperature goes from 200° to 90° in 4 minutes.

$$\begin{aligned} 90 &= 200e^{4r} \\ e^{4r} &= 0.45 \\ 4r &= \ln(0.45) \\ r &= \frac{\ln(0.45)}{4} \approx -0.1996 \end{aligned}$$

Let $t$ be the time required for the difference to go from 90° to 70°.

$$\begin{aligned} 70 &= 90e^{-0.1996t} \\ e^{-0.1996t} &= 7/9 \\ -0.1996t &= \ln(7/9) \\ t &= \frac{\ln(7/9)}{-0.1996} \\ &\approx 1.2589 \text{ minutes} \end{aligned}$$

**4.4.8** $$\begin{aligned} 2000 &= 120{,}000 \frac{0.075/12}{1 - (1 + 0.075/12)^{-12t}} \\ 1 - (1 + 0.075/12)^{-12t} &= 0.375 \\ 0.625 &= (1 + 0.075/12)^{-12t} \\ -12t \cdot \ln(1 + 0.075/12) &= \ln(0.625) \\ t &= \frac{\ln(0.625)}{-12 \cdot \ln(1 + 0.075/12)} \\ &\approx 6.2863 \text{ yr} \\ &\approx 6 \text{ years } 3 \text{ mo} \end{aligned}$$

## Chapter 5

### Section 1

**5.1.1** $$\begin{aligned} 10° + 1 \cdot 360° &= 370° \\ 10° + 2 \cdot 360° &= 730° \\ 10° + (-1) \cdot 360° &= -350° \\ 10° + (-2) \cdot 360° &= -710° \end{aligned}$$

**5.1.2** $-690 + 360k = 390$

$$360k = 1080$$
$$k = 3$$

Since $-690°$ and $390°$ differ by a multiple of $360°$, they are coterminal.

**5.1.3** $-890° + 3(360°) = 190°$

Since $-890°$ is coterminal with $190°$ (which is in quadrant III), $-890°$ lies in quadrant III.

**5.1.4** $15' = 15 \text{ min} \cdot \dfrac{1 \text{ deg}}{60 \text{ min}} = \dfrac{1}{4} \text{deg}$

$$12'' = 12 \text{ sec} \cdot \frac{1 \text{ deg}}{3600 \text{ sec}} = \frac{1}{300} \text{deg}$$
$$35°\,15'\,12'' = \left(35 + \frac{1}{4} + \frac{1}{300}\right) \text{deg}$$
$$\approx 35.2533°$$

**5.1.5** $0.321 \text{ deg} \cdot \dfrac{60 \text{ min}}{1 \text{ deg}} = 19.26 \text{ min}$

$$0.26 \text{ min} \cdot \frac{60 \text{ sec}}{1 \text{ min}} = 15.6 \text{ sec}$$
$$56.321° = 56°\,19'\,15.6''$$

**5.1.6** $210° = 210 \text{ deg} \cdot \dfrac{\pi \text{ rad}}{180 \text{ deg}}$

$$= \frac{7\pi}{6} \text{ rad}$$

**5.1.7** $\dfrac{5\pi}{3} \text{ rad} = \dfrac{5\pi}{3} \text{ rad} \cdot \dfrac{180 \text{ deg}}{\pi \text{ rad}}$

$$= 300°$$

**5.1.8**

$$-\frac{\pi}{3} + 1 \cdot 2\pi = -\frac{\pi}{3} + \frac{6\pi}{3} = \frac{5\pi}{3}$$
$$-\frac{\pi}{3} + 2 \cdot 2\pi = -\frac{\pi}{3} + \frac{12\pi}{3} = \frac{11\pi}{3}$$
$$-\frac{\pi}{3} + (-1) \cdot 2\pi = -\frac{\pi}{3} - \frac{6\pi}{3} = -\frac{7\pi}{3}$$
$$-\frac{\pi}{3} + (-2) \cdot 2\pi = -\frac{\pi}{3} - \frac{12\pi}{3} = -\frac{13\pi}{3}$$

**5.1.9** Since $45° = \pi/4$ radians,

$$s = \alpha r = \frac{\pi}{4} \cdot 8 \text{ in.} = 2\pi \text{ in.} \approx 6.28 \text{ in.}$$

**5.1.10** Since $\alpha = s/r$, $\alpha = 3/50$ rad.

So $\alpha = \dfrac{3}{50} \text{ rad} \cdot \dfrac{180 \text{ deg}}{\pi \text{ rad}} \approx 3.4°$.

**5.1.11** $\omega = \dfrac{5 \text{ rev}}{\text{sec}} \cdot \dfrac{2\pi \text{ rad}}{\text{rev}} = 10\pi \text{ rad/sec}$

$$\approx 31.416 \text{ rad/sec}$$
$$v = r\omega$$
$$= 10 \text{ ft} \cdot \frac{10\pi \text{ rad}}{\text{sec}} \cdot \frac{1 \text{ mi}}{5280 \text{ ft}} \cdot \frac{3600 \text{ sec}}{1 \text{ hr}}$$
$$\approx 214.199 \text{ mi/hr}$$

**5.1.12** $v = \omega r$

$$= \frac{2\pi \text{ rad}}{24 \text{ hr}} \cdot 3950 \text{ mi} \cdot \frac{5280 \text{ ft}}{1 \text{ mi}} \cdot \frac{1 \text{ hr}}{3600 \text{ sec}}$$
$$\approx 1516.691 \text{ ft/sec}$$

## Section 2

**5.2.1** The terminal side of $-\pi/2$ intersects the unit circle at $(0, -1)$. So $\sin(-\pi/2) = -1$ and $\cos(-\pi/2) = 0$.

**5.2.2** The terminal side of $-5\pi/4$ intersects the unit circle at $(-\sqrt{2}/2, \sqrt{2}/2)$. So $\sin(-5\pi/4) = \sqrt{2}/2$ and $\cos(-5\pi/4) = -\sqrt{2}/2$.

**5.2.3** The terminal side of $-\pi/6$ intersects the unit circle at $(\sqrt{3}/2, -1/2)$. So $\sin(-\pi/6) = -1/2$ and $\cos(-\pi/6) = \sqrt{3}/2$.

**5.2.4** Since the terminal side of $135°$ is in quadrant II, the reference angle is $180° - 135°$ or $45°$. In radians it is $\pi/4$.

**5.2.5** The reference angle for $135°$ is $45°$. So $\sin(135°) = \sin(45°) = \sqrt{2}/2$ and $\cos(135°) = -\cos(45°) = -\sqrt{2}/2$.

**5.2.6** With a calculator in radian mode, $\sin(55.6) \approx -0.8126$. With a calculator in degree mode, $\cos(34.2°) \approx 0.8271$.

**5.2.7** $\sin^2\alpha + \left(\dfrac{1}{4}\right)^2 = 1$

$$\sin^2\alpha = \frac{15}{16}$$
$$\sin\alpha = \pm\sqrt{\frac{15}{16}} = \pm\frac{\sqrt{15}}{4}$$

Since $\sin\alpha < 0$ in quadrant IV, $\sin\alpha = -\sqrt{15}/4$.

**5.2.8** Let $t = 3$ to get

$$x = -2\sin(2 \cdot 3) + 3\cos(2 \cdot 3)$$
$$\approx 3.4$$

So the weight is 3.4 cm below equilibrium.

## Section 3

**5.3.1** The curve goes through $(0, 0)$, $(\pi/2, 4)$, $(\pi, 0)$, $(3\pi/2, -4)$, and $(2\pi, 0)$.

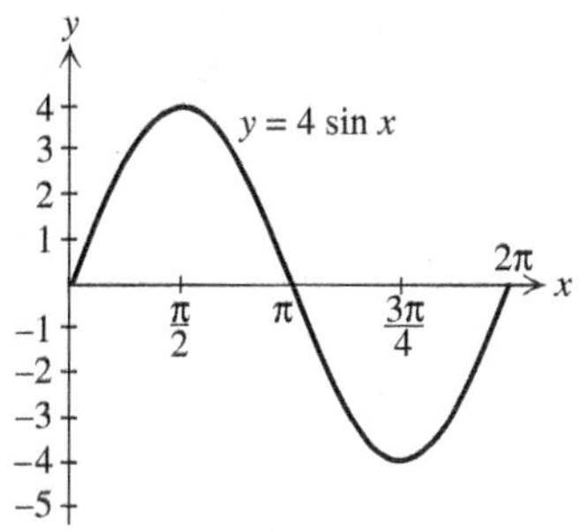


**5.3.2** The maximum $y$-coordinate is 4 and the minimum is $-4$. The amplitude is $\left|\dfrac{1}{2}(4 - (-4))\right|$ or 4.

**5.3.3** The curve goes through $(0, 4)$, $(\pi/2, 0)$, $(\pi, -4)$, $(3\pi/2, 0)$, and $(2\pi, 4)$. The amplitude is 4.

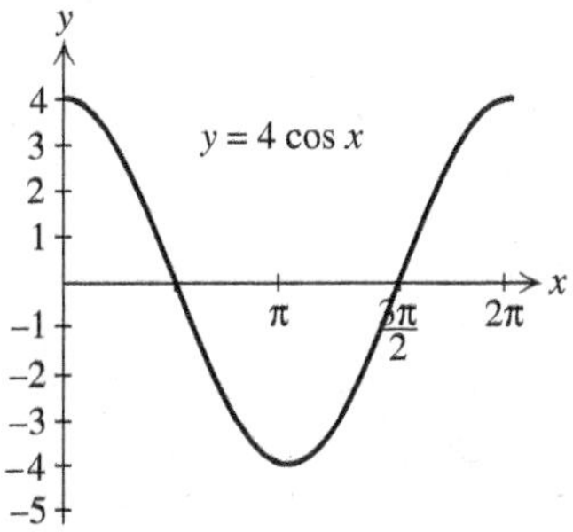


**5.3.4** The curve goes through $(\pi/6, 1)$, $(2\pi/3, 0)$, $(7\pi/6, -1)$, $(5\pi/3, 0)$, and $(13\pi/6, 1)$. The phase shift is $\pi/6$.

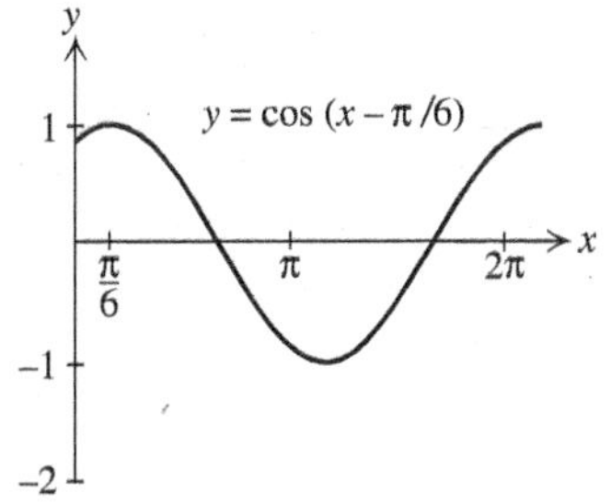

**5.3.5** The curve goes through $(-\pi/4, 1)$, $(\pi/4, 2)$, $(3\pi/4, 1)$, $(5\pi/4, 0)$, and $(7\pi/4, 1)$. The phase shift is $-\pi/4$.

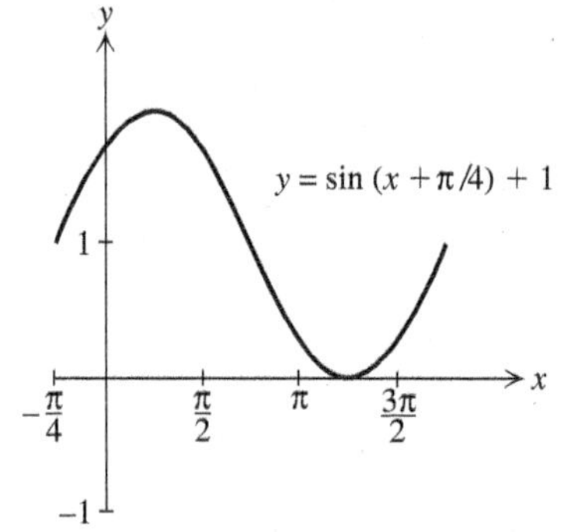


**5.3.6** The period is $\pi$. So two cycles occur in the interval $[0, 2\pi]$. The curve goes through $(0, 1)$, $(\pi/4, 0)$, $(\pi/2, -1)$, $(3\pi/4, 0)$, and $(\pi, 1)$.

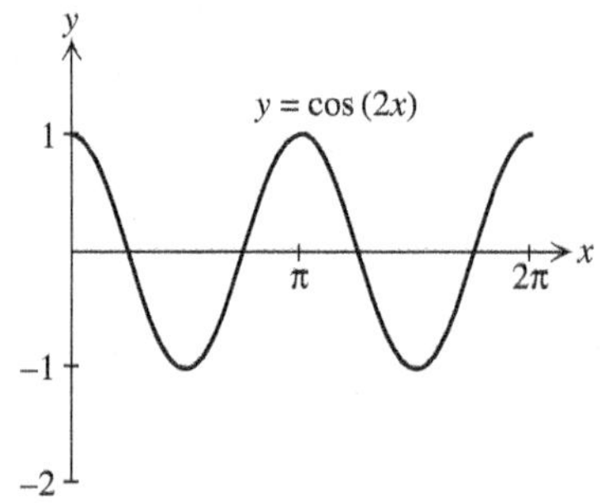


**5.3.7** The period is $2\pi/\pi$ or 2. So two cycles occur in the interval $[0, 4]$. The curve goes through $(0, 1)$, $(1/2, 0)$, $(1, -1)$, $(3/2, 0)$, and $(2, 1)$.

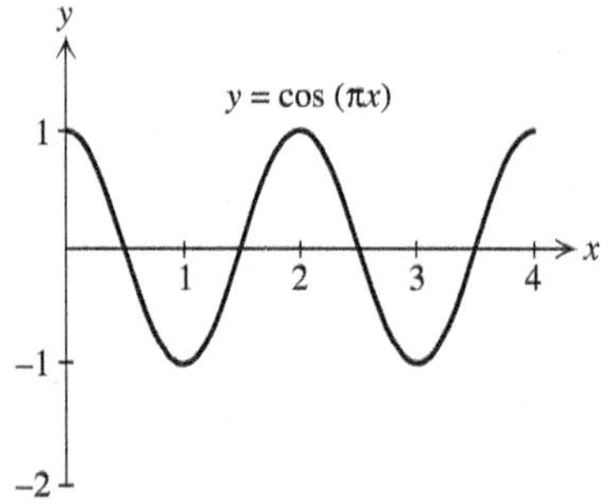


**5.3.8** Rewrite the equation as $y = 3\sin\left(2\left(x - \frac{\pi}{2}\right)\right) - 1$. The period is $2\pi/2$ or $\pi$. The amplitude is 3 and the phase shift is $\pi/2$. So one cycle occurs in the interval $[-\pi/2, \pi/2]$. The curve goes through $(-\pi/2, -1)$, $(-\pi/4, 2)$, $(0, -1)$, $(\pi/4, -4)$, and $(\pi/2, -1)$.

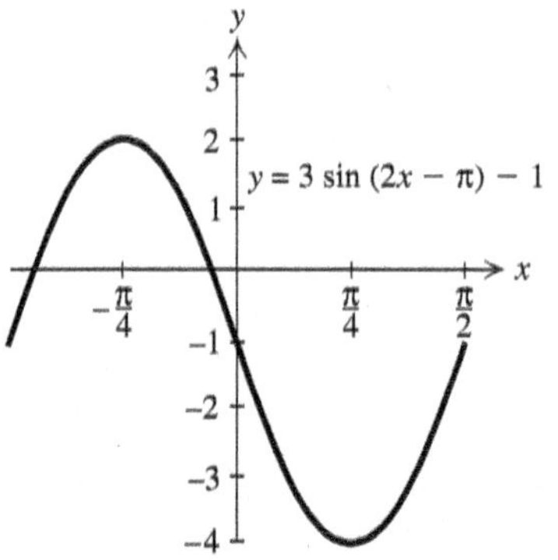


**5.3.9** Rewrite the equation as $y = -2\cos\left(2\left(x + \frac{\pi}{2}\right)\right) + 1$. The period is $2\pi/2$ or $\pi$. The amplitude is 2 and the phase shift is $-\pi/2$. So one cycle occurs in the interval $[-\pi/2, \pi/2]$. The curve goes through $(-\pi/2, -1)$, $(-\pi/4, 1)$, $(0, 3)$, $(\pi/4, 1)$, and $(\pi/2, -1)$.

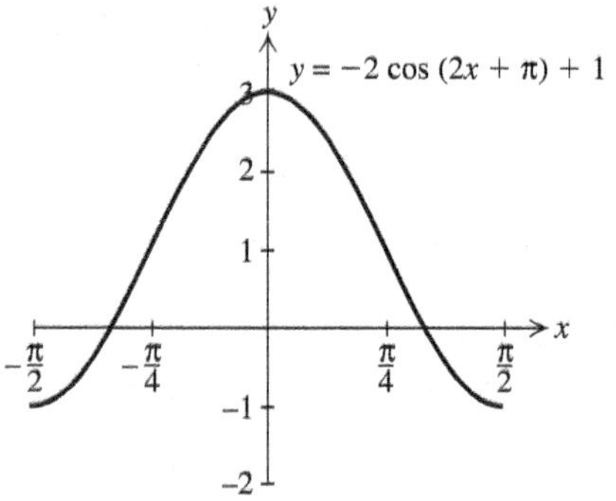


**5.3.10** The period is $2\pi/(100\pi)$ or $1/50$. Since $F = 1/P$, $F = 50$.
**5.3.11** Since the range is $[-1, 5]$ the amplitude is 3. Since one cycle occurs on $[1, 5]$ the period is 4 and the phase shift is 1.

So $y = 3\sin\left(\frac{\pi}{2}(x - 1)\right) + 2$.

**5.3.12** Enter the five points into your calculator and use sinsusoidal regression to get $y = 3.003\sin(1.562x - 1.497) + 1.972$.

## Section 4

**5.4.1** In quadrant IV, only cosine and secant are positive. So
$\sin(-\pi/4) = -\sqrt{2}/2$,
$\cos(-\pi/4) = \sqrt{2}/2$,
$\tan(-\pi/4) = -1$,
$\sec(-\pi/4) = 1/\cos(-\pi/4) = \sqrt{2}$,
$\csc(-\pi/4) = 1/\sin(-\pi/4) = -\sqrt{2}$, and
$\cot(-\pi/4) = 1/\tan(-\pi/4) = -1$.
**5.4.2** $\sec(\pi/22) = 1/\cos(\pi/22) \approx 1.0103$
$\csc(-14.5) = 1/\sin(-14.5) \approx -1.0696$
**5.4.3** The period is $\pi/(2\pi)$ or $1/2$. So two cycles occur on the interval $[-1/4, 3/4]$.

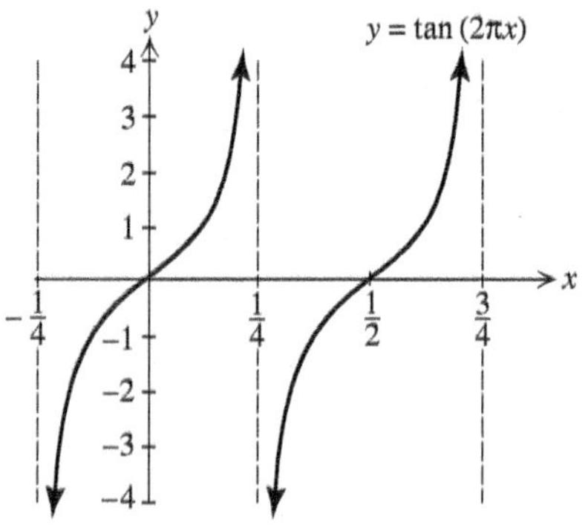

**5.4.4** Rewrite the equation as $y = 3\cot\left(2\left(x + \frac{\pi}{2}\right)\right)$. The period is $\pi/2$. So two cycles occur on the interval $[-\pi/2, \pi/2]$. The vertical asymptotes are at 0 and $\pm\pi/2$.

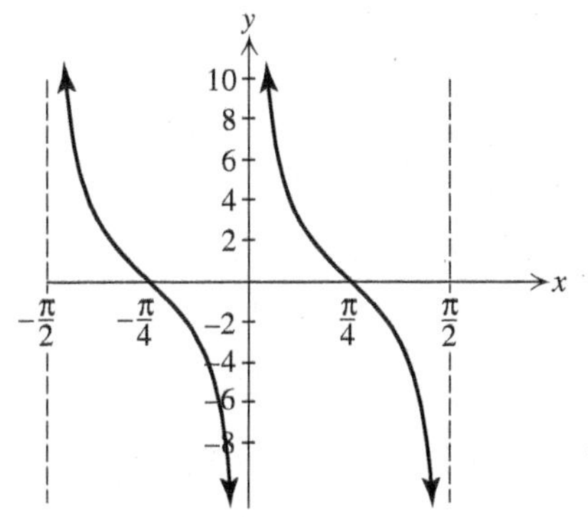


$y = 3\cot(2x + \pi)$

**5.4.5** The period is $2\pi/2$ or $\pi$ and the range is $(-\infty, -3] \cup [3, \infty)$. First graph one cycle of $y = 3\cos(2x)$ on $[0, \pi]$ as a dashed curve.

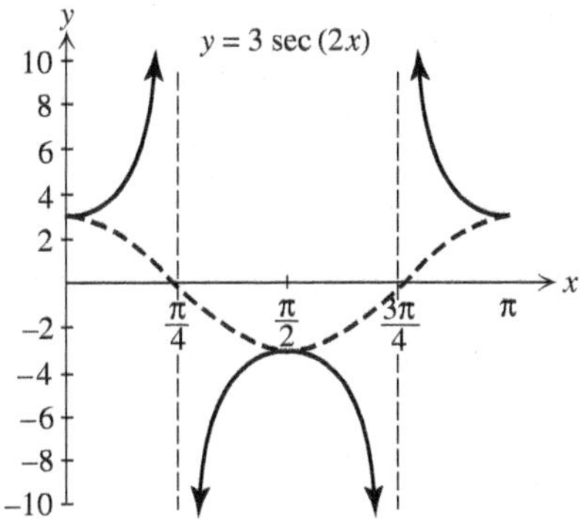


**5.4.6** The period is $2\pi/2$ or $\pi$ and the range is $(-\infty, -1] \cup [1, \infty)$. First graph one cycle of $y = \sin(2x - \pi)$ on $[0, \pi]$ as a dashed curve.

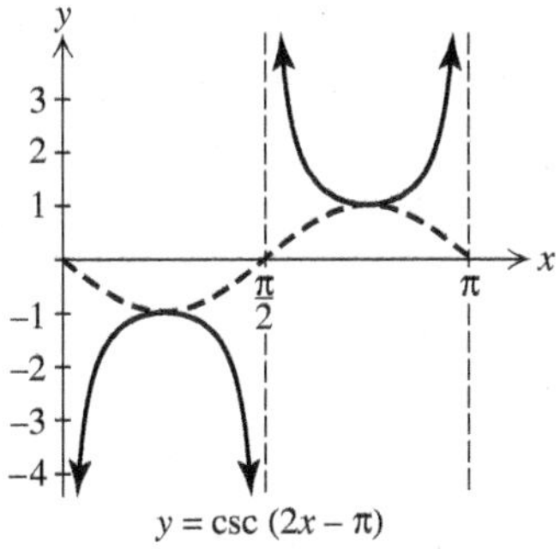


$y = \csc(2x - \pi)$

## Section 5

**5.5.1** Since $-\pi/4$ is in the interval $[-\pi/2, \pi/2]$ and $\sin(-\pi/4) = -\sqrt{2}/2$, $\sin^{-1}(-\sqrt{2}/2) = -\pi/4$.
**5.5.2** Since $-60°$ is in the interval $[-90°, 90°]$ and $\sin(-60°) = -\sqrt{3}/2$, $\sin^{-1}(-\sqrt{3}/2) = -60°$.
**5.5.3** Use a calculator in degree mode to get $\alpha = \sin^{-1}(0.3) \approx 17.5°$.
**5.5.4** Since $5\pi/6$ is in the interval $[0, \pi]$ and $\cos(5\pi/6) = -\sqrt{3}/2$, $\cos^{-1}(-\sqrt{3}/2) = 5\pi/6$.
**5.5.5** Use a calculator in degree mode to get $\alpha = \cos^{-1}(-0.4) \approx 113.6°$.
**5.5.6** Since $\pi/3$ is in the interval $[-\pi/2, \pi/2]$ and $\tan(\pi/3) = \sqrt{3}$, $\arctan(\sqrt{3}) = \pi/3$.
**5.5.7** $\cot^{-1}(1.3) = \tan^{-1}(1/1.3) \approx 0.6557$
**5.5.8** Note that $\sin(5\pi/6) = \sin(\pi/6) = 1/2$. So $\sin^{-1}(\sin(5\pi/6)) = \sin^{-1}(1/2) = \pi/6$ because the range of $\sin^{-1}$ is $[-\pi/2, \pi/2]$.
**5.5.9**

$$\begin{aligned} f(x) &= 2\cos(3x) - 1 \\ y &= 2\cos(3x) - 1 \\ x &= 2\cos(3y) - 1 \\ \cos(3y) &= \frac{x+1}{2} \\ 3y &= \cos^{-1}\left(\frac{x+1}{2}\right) \\ y &= \frac{1}{3}\cos^{-1}\left(\frac{x+1}{2}\right) \\ f^{-1}(x) &= \frac{1}{3}\cos^{-1}\left(\frac{x+1}{2}\right) \end{aligned}$$

Since $-1 \le \cos(3x) \le 1$, the range of $f$ is the interval $[-3, 1]$ and the domain of $f^{-1}$ is $[-3, 1]$.

## Section 6

**5.6.1** Since $r = \sqrt{2^2 + 3^2} = \sqrt{13}$, $\sin\alpha = 3/\sqrt{13} = 3\sqrt{13}/13$, $\cos\alpha = 2/\sqrt{13} = 2\sqrt{13}/13$, and $\tan\alpha = 3/2$.
**5.6.2**

$$\begin{aligned} c &= \sqrt{2^2 + 4^2} = 2\sqrt{5} \\ \sin\alpha &= 4/(2\sqrt{5}) = 2\sqrt{5}/5 \\ \cos\alpha &= 2/(2\sqrt{5}) = \sqrt{5}/5 \\ \tan\alpha &= 4/2 = 2 \end{aligned}$$

**5.6.3** Since $\alpha = 60°$, $\beta = 30°$ and $\gamma = 90°$. Since $c = 2$, $\sin(60°) = a/2$ and $a = 2 \cdot \sqrt{3}/2 = \sqrt{3}$. Since $c = 2$, $\sin(30°) = b/2$ and $b = 2 \cdot 1/2 = 1$. So the angles are 30°, 60°, and 90°, and the sides opposite those angles are 1, $\sqrt{3}$, and 2, respectively.
**5.6.4**

$$\begin{aligned} c &= \sqrt{2^2 + 5^2} = \sqrt{29} \\ \alpha &= \sin^{-1}(2/\sqrt{29}) \approx 21.8° \\ \beta &= \sin^{-1}(5/\sqrt{29}) \approx 68.2° \\ \gamma &= 90° \end{aligned}$$

**5.6.5** The least number of significant digits in 24 and 5.461 is 2. So $24\sin(5.461°) \approx 2.3$.
**5.6.6** The height of the tower $a$ is the side opposite 38.2° in a right triangle in which the other leg $b$ is 344 feet. So $\tan(38.2°) = a/344$ or $a = 344\tan(38.2°) \approx 271$ feet.
**5.6.7** Let $y$ be the height of the antenna and $x$ be the distance from the second location to the antenna. So $\tan 63.1° = y/x$ and $\tan 44.2° = y/(x + 100)$. Substitute.

$$\begin{aligned} \tan 44.2° &= \frac{y}{\dfrac{y}{\tan 63.1°} + 100} \\ \tan 44.2°\left(\frac{y}{\tan 63.1°} + 100\right) &= y \\ \left(\frac{\tan 44.2°}{\tan 63.1°} - 1\right)y &= -100 \cdot \tan 44.2° \\ y &\approx 192 \text{ feet} \end{aligned}$$

**5.6.8** The line of sight to the horizon is perpendicular to the radius at the horizon. So the acute angle $\alpha$ at the center of the earth in this right triangle is $\cos^{-1}(3950/3957)$. The width of the path is the length of the arc intercepted by a central angle of $2\alpha$ in a circle of radius 3950 miles. So using the arc-length formula and radian mode, the width is $2 \cdot \cos^{-1}(3950/3957) \cdot 3950$ or about 470 miles.
**5.6.9** Draw a circle for the earth. The plane, San Antonio, and the center of the earth are the vertices of a triangle with a right angle at San Antonio. One leg of the right triangle is $r$ miles (the radius of the earth) and the hypotenuse is $r + 5$ miles. The angle at the center of the earth between these two sides is $\alpha$ radians. Since $\alpha$ intercepts an arc of length 200 miles, we have $200 = \alpha r$ or $\alpha = 200/r$. Now use right triangle trigonometry to get $\cos(200/r) = r/(r + 5)$. Graph $y = \cos(200/r) - r/(r + 5)$ and find the $r$ intercept to solve this equation. You should get about 4004 miles for $r$.

## Chapter 6

### Section 1

**6.1.1** $\dfrac{\tan x \csc x}{\sec x} = \dfrac{\dfrac{\sin x}{\cos x}\cdot\dfrac{1}{\sin x}}{\dfrac{1}{\cos x}}$

$= \dfrac{\sin x}{\cos x}\cdot\dfrac{1}{\sin x}\cdot\dfrac{\cos x}{1} = 1$

**6.1.2** $1 + \cot^2 x = \csc^2 x$

$\cot^2 x = \csc^2 x - 1$

$\cot x = \pm\sqrt{\dfrac{1}{\sin^2 x} - 1}$

**6.1.3** $\csc^2 x = 1 + \cot^2 x$

$\csc x = \pm\sqrt{1 + \cot^2 x}$

$\sin x = \pm\dfrac{1}{\sqrt{1 + \cot^2 x}}$

In quadrant II sine is positive and cosine is negative.

$\sin\alpha = \dfrac{1}{\sqrt{1 + (-1/3)^2}}$

$= \dfrac{1}{\sqrt{10/9}} = \dfrac{3\sqrt{10}}{10}$

$\cos\alpha = -\sqrt{1 - \sin^2\alpha}$

$= -\sqrt{1 - 9/10} = -\sqrt{10}/10$

**6.1.4** Let $\theta =$ arccos $x$. Since $\cos\theta = x$, $\sec(\arccos(x)) = \sec(\theta) = 1/\cos\theta = 1/x$.

**6.1.5** $\csc(-x)\tan(-x) = (-\csc x)(-\tan x)$

$= \csc(x)\tan(x) = \dfrac{1}{\sin x}\cdot\dfrac{\sin x}{\cos x}$

$= \sec x$

**6.1.6** $f(-x) = \csc(-x) + \tan(-x)$

$= -\csc(x) - \tan(x)$

$= -(\csc x + \tan x) = -f(x)$

Since $f(-x) = -f(x)$ the function is an odd function.

**6.1.7** If $t = \pi/3$, then

$\cos(3t) = \cos(\pi) = -1$ and

$3\cos(t) = 3\cos(\pi/3) = 3/2$.

So the equation is not an identity.

### Section 2

**6.2.1** $1 - \sec(x)\csc(x)\tan(x)$

$= 1 - \dfrac{1}{\cos x}\dfrac{1}{\sin x}\dfrac{\sin x}{\cos x}$

$= 1 - \sec^2 x$

$= -\tan^2 x$

**6.2.2** $(1 - 2\sin x)(1 + 2\sin x)$

$= 1 - 4\sin^2 x$

**6.2.3** $\sin^2 x + 4\sin x + 4 = (\sin x + 2)^2$

**6.2.4** $\dfrac{\csc x - 1}{\cot x} = \dfrac{(\csc x - 1)(\csc x + 1)}{(\cot x)(\csc x + 1)}$

$= \dfrac{\csc^2 x - 1}{(\cot x)(\csc x + 1)}$

$= \dfrac{\cot^2 x}{(\cot x)(\csc x + 1)} = \dfrac{\cot x}{\csc x + 1}$

**6.2.5** $\dfrac{\sec x - \cos x}{\cos x} = \dfrac{\sec x}{\cos x} - \dfrac{\cos x}{\cos x}$

$= \sec^2 x - 1 = \tan^2 x$

**6.2.6** $\dfrac{1}{1 - \sec x} + \dfrac{1}{1 + \sec x}$

$= \dfrac{1(1 + \sec x) + 1(1 - \sec x)}{(1 - \sec x)(1 + \sec x)}$

$= \dfrac{2}{1 - \sec^2 x} = \dfrac{2}{-\tan^2 x}$

$= -2\cot^2 x$

**6.2.7** $\dfrac{1 - \cos^2(-t)}{\sin(-t)} = \dfrac{1 - \cos^2(t)}{-\sin(t)}$

$= \dfrac{\sin^2(t)}{-\sin(t)} = -\sin t$

$\tan(-t)\cos(-t) = -\tan(t)\cos(t)$

$= -\dfrac{\sin t}{\cos t}\cos t = -\sin t$

Since both sides simplify to the same expression, the equation is an identity.

### Section 3

**6.3.1** $\cos(105°) = \cos(60° + 45°)$

$= \cos 60°\cos 45° - \sin 60°\sin 45°$

$= \dfrac{1}{2}\cdot\dfrac{\sqrt{2}}{2} - \dfrac{\sqrt{3}}{2}\cdot\dfrac{\sqrt{2}}{2} = \dfrac{\sqrt{2} - \sqrt{6}}{4}$

**6.3.2** $\cos(75°) = \cos(135° - 60°)$

$= \cos 135°\cos 60° + \sin 135°\sin 60°$

$= -\dfrac{\sqrt{2}}{2}\cdot\dfrac{1}{2} + \dfrac{\sqrt{2}}{2}\cdot\dfrac{\sqrt{3}}{2} = \dfrac{\sqrt{6} - \sqrt{2}}{4}$

**6.3.3** Use the result of 6.3.2:

$\sin(15°) = \cos(90° - 15°)$

$= \cos(75°) = \dfrac{\sqrt{6} - \sqrt{2}}{4}$

**6.3.4** $\sin(75°) = \sin(30° + 45°)$

$= \sin 30°\cos 45° + \cos 30°\sin 45°$

$= \dfrac{1}{2}\cdot\dfrac{\sqrt{2}}{2} + \dfrac{\sqrt{3}}{2}\cdot\dfrac{\sqrt{2}}{2} = \dfrac{\sqrt{2} + \sqrt{6}}{4}$

**6.3.5** $\tan(75°) = \tan(30° + 45°)$

$= \dfrac{\tan 30° + \tan 45°}{1 - \tan 30°\tan 45°}$

$= \dfrac{\dfrac{\sqrt{3}}{3} + 1}{1 - \dfrac{\sqrt{3}}{3}\cdot 1} = \dfrac{\sqrt{3} + 3}{3 - \sqrt{3}}$

$= \dfrac{12 + 6\sqrt{3}}{6} = 2 + \sqrt{3}$

**6.3.6** $\sin\alpha\cos 3\alpha + \cos\alpha\sin 3\alpha$

$= \sin(\alpha + 3\alpha) = \sin(4\alpha)$

**6.3.7** Since $\sin\alpha = 1/3$,

$\left(\dfrac{1}{3}\right)^2 + \cos^2\alpha = 1$ or $\cos\alpha = -\dfrac{2\sqrt{2}}{3}$.

Since $\cos\beta = -1/3$,

$\sin^2\beta + \left(-\dfrac{1}{3}\right)^2 = 1$ or $\sin\beta = \dfrac{2\sqrt{2}}{3}$.

$\cos(\alpha - \beta) = \cos\alpha\cos\beta + \sin\alpha\sin\beta$

$= -\dfrac{2\sqrt{2}}{3}\left(-\dfrac{1}{3}\right) + \dfrac{1}{3}\cdot\dfrac{2\sqrt{2}}{3}$

$= \dfrac{4\sqrt{2}}{9}$

### Section 4

**6.4.1** $\sin(60°) = 2\sin 30°\cos 30°$

$= 2\cdot\dfrac{1}{2}\cdot\dfrac{\sqrt{3}}{2} = \dfrac{\sqrt{3}}{2}$

**6.4.2** $\sin(-22.5°) = -\sqrt{\dfrac{1 - \cos(-45°)}{2}}$

$= -\sqrt{\dfrac{1 - \dfrac{\sqrt{2}}{2}}{2}}$

$= -\dfrac{\sqrt{2 - \sqrt{2}}}{2}$

**6.4.3** $\cos 75^\circ = \cos\frac{150^\circ}{2}$

$$= \sqrt{\frac{1+\cos 150^\circ}{2}}$$

$$= \sqrt{\frac{1+\left(-\frac{\sqrt{3}}{2}\right)}{2}}$$

$$= \sqrt{\frac{2-\sqrt{3}}{4}} = \frac{\sqrt{2-\sqrt{3}}}{2}$$

**6.4.4** $\tan(-22.5^\circ) = \frac{\sin(-45^\circ)}{1+\cos(-45^\circ)}$

$$= \frac{-\frac{\sqrt{2}}{2}}{1+\frac{\sqrt{2}}{2}} = -\frac{\sqrt{2}}{2+\sqrt{2}}$$

$$= -\frac{2\sqrt{2}-2}{2} = 1-\sqrt{2}$$

**6.4.5** Cosine is negative in quadrant II and positive in quadrant I. Since $\sin\alpha = 1/3$, $\left(\frac{1}{3}\right)^2 + \cos^2\alpha = 1$ or $\cos\alpha = -\frac{2\sqrt{2}}{3}$.

$$\cos\frac{\alpha}{2} = \sqrt{\frac{1+\cos\alpha}{2}} = \sqrt{\frac{1-\frac{2\sqrt{2}}{3}}{2}}$$

$$= \sqrt{\frac{3-2\sqrt{2}}{6}}$$

**6.4.6** $\cos(3x) = \cos(x+2x)$
$= \cos x\cos 2x - \sin x\sin 2x$
$= \cos x(\cos^2 x - \sin^2 x) - \sin x(2\sin x\cos x)$
$= \cos^3 x - \cos x\sin^2 x - 2\sin^2 x\cos x$
$= \cos^3 x - 3\cos x\sin^2 x$

**6.4.7** $\sin^2\left(\frac{x}{2}\right)\cos^2\left(\frac{x}{2}\right)$

$$= \frac{1-\cos x}{2}\cdot\frac{1+\cos x}{2}$$

$$= \frac{1-\cos^2 x}{4} = \frac{\sin^2 x}{4}$$

## Section 5

**6.5.1** $\sin(4x)\cos(3x) = \frac{1}{2}[\sin(7x)+\sin(x)]$

**6.5.2** $\sin(52.5^\circ)\cos(7.5^\circ)$

$$= \frac{1}{2}[\sin(60^\circ)+\sin(45^\circ)]$$

$$= \frac{1}{2}\left(\frac{\sqrt{3}}{2}+\frac{\sqrt{2}}{2}\right) = \frac{\sqrt{3}+\sqrt{2}}{4}$$

**6.5.3** $\cos(4x)+\cos(2x) = 2\cos 3x\cos x$

**6.5.4** $\sin(105^\circ)+\sin(15^\circ)$
$= 2\sin(60^\circ)\cos(45^\circ)$
$= 2\cdot\frac{\sqrt{3}}{2}\cdot\frac{\sqrt{2}}{2} = \frac{\sqrt{6}}{2}$

**6.5.5** The terminal side of $3\pi/4$ goes through $(-2, 2)$. So
$-2\sin x + 2\cos x$
$= \sqrt{(-2)^2+2^2}\sin(x+3\pi/4)$
$= 2\sqrt{2}\sin(x+3\pi/4)$

**6.5.6** The terminal side of $-\pi/4$ contains $(1,-1)$.
So $y = \sin x - \cos x = \sqrt{2}\sin(x-\pi/4)$. The amplitude is $\sqrt{2}$ and the phase shift is $\pi/4$.

**6.5.7** The amplitude is $\sqrt{a^2+b^2}$ or $\sqrt{10}$. So the maximum distance that the block travels from rest is $\sqrt{10}$ cm.

## Section 6

**6.6.1** One solution is $x = \cos^{-1}(\sqrt{3}/2) = \pi/6$.
Also, $\cos(11\pi/6) = \sqrt{3}/2$. So the solution set is $\{x \mid x = \pi/6 + 2k\pi \text{ or } x = 11\pi/6 + 2k\pi\}$, where $k$ is any integer.

**6.6.2** One solution is $x = \sin^{-1}(1/2) = \pi/6$. Also, $\sin(5\pi/6) = 1/2$. So the solution set is $\{x \mid x = \pi/6 + 2k\pi \text{ or } x = 5\pi/6 + 2k\pi\}$, where $k$ is any integer.

**6.6.3** Since $\alpha = \tan^{-1}(\sqrt{3}) = 60^\circ$, the solution set is $\{x \mid x = 60^\circ + k180^\circ\}$, where $k$ is any integer.

**6.6.4** Since $2\alpha = \sin^{-1}(1) = 90^\circ$, we have $2\alpha = 90^\circ + k360^\circ$. So the solution set is $\{\alpha \mid \alpha = 45^\circ + k180^\circ\}$, where $k$ is any integer.

**6.6.5** Since $4x = \tan^{-1}(1) = \pi/4$, $4x = \pi/4 + k\pi$ and $x = \pi/16 + k\pi/4$. The solutions between 0 and $\pi$ occur if $k = 0, 1, 2, 3$. So the solution set is $\left\{\frac{\pi}{16}, \frac{5\pi}{16}, \frac{9\pi}{16}, \frac{13\pi}{16}\right\}$.

**6.6.6** $\sin(2x) = \cos(x)$
$2\sin x\cos x - \cos x = 0$
$\cos x(2\sin x - 1) = 0$

$$\cos x = 0 \text{ or } \sin x = \frac{1}{2}$$

$$x = \frac{\pi}{2} + k\pi \text{ or } x = \frac{\pi}{6} + 2k\pi \text{ or } x = \frac{5\pi}{6} + 2k\pi$$

The only solutions in the interval $(0, 2\pi)$ are $\pi/6$, $\pi/2$, $5\pi/6$, and $3\pi/2$.

**6.6.7** $2\sin^2(x) - \sin(x) - 1 = 0$
$(\sin x - 1)(2\sin x + 1) = 0$

$$\sin x = 1 \text{ or } \sin x = -\frac{1}{2}$$

$$x = \frac{\pi}{2} + 2k\pi \text{ or } x = \frac{7\pi}{6} + 2k\pi \text{ or } x = \frac{11\pi}{6} + 2k\pi$$

The only solutions in the interval $(0, 2\pi)$ are $\pi/2$, $7\pi/6$, and $11\pi/6$.

**6.6.8** $\cos\alpha - \sin^2\alpha = 0$
$\cos\alpha - (1-\cos^2\alpha) = 0$
$\cos^2\alpha + \cos\alpha - 1 = 0$

Apply the quadratic formula: $\cos\alpha = \frac{-1\pm\sqrt{5}}{2} \approx 0.6180$ or $-1.6180$

Note that $\cos\alpha = -1.6180$ has no solutions. So $\alpha \approx 51.8^\circ + k360^\circ$ or $308.2^\circ + k360^\circ$. The only solutions in the interval $[0^\circ, 360^\circ)$ are $51.8^\circ$ and $308.2^\circ$.

**6.6.9**

$$\sin\alpha - \cos\alpha = \frac{1}{\sqrt{2}}$$

$$\sin^2\alpha - 2\sin\alpha\cos\alpha + \cos^2\alpha = \frac{1}{2}$$

$$1 - 2\sin\alpha\cos\alpha = \frac{1}{2}$$

$$2\sin\alpha\cos\alpha = \frac{1}{2}$$

$$\sin 2\alpha = \frac{1}{2}$$

$2\alpha = 30^\circ + k360^\circ$ or $150^\circ + k360^\circ$
$\alpha = 15^\circ + k180^\circ$ or $75^\circ + k180^\circ$

Only $15^\circ$, $75^\circ$, $195^\circ$, and $255^\circ$ are in the interval $[0^\circ, 360^\circ)$. Checking reveals that only $75^\circ$ and $195^\circ$ satisfy the original equation.

**6.6.10** $3\sin t - 2\cos t = 0$

$$\frac{\sin t}{\cos t} = \frac{2}{3}$$

$$\tan t = \frac{2}{3}$$

$$t = \tan^{-1}\left(\frac{2}{3}\right) + k\pi$$

So $x = 0$ when $t \approx 0.588 + k\pi$, where $k$ is a nonnegative integer.

**6.6.11** *Use* $v_0^2\sin 2\theta = 32d$
$100^2\sin 2\theta = 32(200)$
$\sin 2\theta = 0.64$

Since $\sin^{-1}(0.64) \approx 39.8^\circ$, $2\theta \approx 39.8^\circ + k360^\circ$ or $140.2^\circ + k360^\circ$ and $\theta \approx 19.9^\circ + k180^\circ$ or $70.1^\circ + k180^\circ$. Since the launch angle must be acute, the only possibilities are $19.9^\circ$ and $70.1^\circ$.

# Chapter 7

## Section 1

**7.1.1** Note that $c$ is the side between $\alpha$ and $\beta$ and that $\gamma = 86°$.

$$\frac{a}{\sin 28°} = \frac{b}{\sin 66°} = \frac{8.2}{\sin 86°}$$
$$a = \sin 28° \cdot \frac{8.2}{\sin 86°} \approx 3.9$$
$$b = \sin 66° \cdot \frac{8.2}{\sin 86°} \approx 7.5$$

**7.1.2** First draw angle $\beta$ with vertex $B$ and mark off side $c$ on one ray of the angle. The other endpoint of $c$ is vertex $A$. Next find the altitude from $A$ to the opposite side.

$$h = 5.9 \sin 38° \approx 3.6$$

Since $b = 2.9$ and that is less than 3.6, side $b$ cannot reach from $A$ to the opposite side. So there is no triangle with these parts.

**7.1.3** First draw angle $\beta$ with vertex $B$ and mark off side $c$ on one ray of the angle. The other endpoint of $c$ is vertex $A$. Since $b$ is larger than $c$, there is one obtuse triangle that is formed.

$$\frac{a}{\sin \alpha} = \frac{6.4}{\sin 38°} = \frac{5.9}{\sin \gamma}$$
$$\sin \gamma = \sin 38° \cdot \frac{5.9}{6.4}$$
$$\gamma = \sin^{-1}\left(\sin 38° \cdot \frac{5.9}{6.4}\right) \approx 34.6°$$

Now $\alpha \approx 107.4°$.

$$a = \sin 107.4° \cdot \frac{6.4}{\sin 38°} \approx 9.9$$

**7.1.4** First draw angle $\beta$ with vertex $B$ and mark off side $c$ on one ray of the angle. The other endpoint of $c$ is vertex $A$. Next find the altitude from $A$ to the opposite side.

$$h = 5.9 \sin 38° \approx 3.6$$

Since $b = 4.7$ and that is larger than 3.6 and smaller than 5.9, it will reach from $A$ to the opposite side in two places. So there are two triangles.

$$\frac{a}{\sin \alpha} = \frac{4.7}{\sin 38°} = \frac{5.9}{\sin \gamma}$$
$$\sin \gamma = \sin 38° \cdot \frac{5.9}{4.7}$$

Since $\sin^{-1}\left(\sin 38° \cdot \frac{5.9}{4.7}\right) \approx 50.6°$, $\gamma$ is 50.6° or 129.4°. Now if $\gamma = 50.6°$, then $\alpha \approx 91.4°$ and

$$a = \sin 91.4° \cdot \frac{4.7}{\sin 38°} \approx 7.6.$$

For the obtuse triangle in which $\gamma = 129.4°$, we have $\alpha \approx 12.6°$ and

$$a = \sin 12.6° \cdot \frac{4.7}{\sin 38°} \approx 1.7.$$

**7.1.5** $A = \frac{1}{2} \cdot 244 \cdot 206 \cdot \sin 87.4°$
$\approx 25{,}106$ square feet

**7.1.6** The triangle is a 30-60-90 triangle in which the side opposite 60° is 88. The maximum distance from the airport is side $c$, the hypotenuse. Using right triangle ratios, $\sin 60° = 88/c$ or $c \approx 101.6$ mi.

**7.1.7** Let $C$ be the point at the top of the building and $D$ be the point on the ground directly below $C$. In $\Delta ABC$, $m\angle A = 24.2°$, $m\angle B = 141.9°$, and $m\angle C = 13.9°$. Find $a$:

$$\frac{a}{\sin 24.2°} = \frac{44.5}{\sin 13.9°}$$
$$a \approx 75.9$$

In right triangle $BCD$, $\sin 38.1° = CD/a$ or $CD = a \sin 38.1 \approx 46.9$ feet.

## Section 2

**7.2.1** Use the law of cosines to find the largest angle:

$$9.6^2 = 3.8^2 + 7.7^2 - 2(3.8)(7.7)\cos\beta$$
$$\beta = \cos^{-1}\left(\frac{9.6^2 - 3.8^2 - 7.7^2}{-2(3.8)(7.7)}\right) \approx 108.4°$$

Use the law of sines to find the next angle:

$$\frac{3.8}{\sin \alpha} = \frac{9.6}{\sin 108.4°}$$
$$\alpha = \sin^{-1}\left(\frac{3.8 \sin 108.4°}{9.6}\right) \approx 22.1°$$

Now subtract $\alpha$ and $\beta$ from 180° to get $\gamma \approx 49.5°$.

**7.2.2** Use the law of cosines to find $a$:

$$a^2 = 5.8^2 + 3.6^2 - 2(5.8)(3.6)\cos 39.5°$$
$$a \approx 3.8$$

**7.2.3** $S = \frac{12 + 8 + 6}{2} = 13$

$$A = \sqrt{13(13-12)(13-8)(13-6)} = \sqrt{455} \approx 21.3$$

**7.2.4** $a = r\sqrt{2 - 2\cos\alpha}$
$= 22.4\sqrt{2 - 2\cos 33.8°}$
$\approx 13.0$ feet

**7.2.5** Use the distance formula to find the lengths of the sides: $AB = \sqrt{26}$, $BC = \sqrt{53}$, and $AC = \sqrt{73}$.
Use the law of cosines to find the largest angle:

$$73 = 53 + 26 - 2\sqrt{53}\sqrt{26}\cos\beta$$
$$\beta \approx 85.4°$$

Use the law of sines to find $\alpha$:

$$\frac{\sqrt{53}}{\sin \alpha} = \frac{\sqrt{73}}{\sin 85.4°}$$
$$\alpha \approx 58.1°$$
$$\gamma \approx 36.5°$$

## Section 3

**7.3.1** $|\mathbf{v}_x| = |5.6 \cos 22°| \approx 5.2$
$|\mathbf{v}_y| = |5.6 \sin 22°| \approx 2.1$

**7.3.2** First find the magnitude:

$$|\mathbf{v}| = \sqrt{2^2 + (-6)^2} = 2\sqrt{10}$$

Use trigonometric ratios to find the direction angle $\theta$:

$$\sin^{-1}\left(\frac{-6}{2\sqrt{10}}\right) \approx -71.6°$$

Since $\theta$ must be in $[0°, 360°)$, $\theta \approx -71.6° + 360° = 288.4°$

**7.3.3** For terminal point $(a, b)$

$$a = r\cos\theta = 50\cos(120°) = 50(-1/2) = -25$$
$$b = r\sin\theta = 50\sin(120°) = 50(\sqrt{3}/2) = 25\sqrt{3}.$$

So the component form is $\langle -25, 25\sqrt{3}\rangle$.

**7.3.4** $\mathbf{u} + 3\mathbf{v} = \langle -1, -3\rangle + 3\langle 3, -4\rangle = \langle 8, -15\rangle$
$\mathbf{u} \cdot \mathbf{v} = \langle -1, -3\rangle \cdot \langle 3, -4\rangle = -3 + 12 = 9$

**7.3.5** $\cos\alpha = \frac{\langle 1, 3\rangle \cdot \langle 5, 2\rangle}{|\langle 1, 3\rangle| \cdot |\langle 5, 2\rangle|}$
$= \frac{11}{\sqrt{10}\sqrt{29}}$
$\alpha \approx 49.8°$

**7.3.6** $\langle -1, 7\rangle = -1\langle 1, 0\rangle + 7\langle 0, 1\rangle$
$= -1\mathbf{i} + 7\mathbf{j}$

**7.3.7** Let $\mathbf{v}$ be the resultant.

$$|\mathbf{v}|^2 = 100^2 + 200^2 - 2\cdot 100\cdot 200\cos 150^\circ$$
$$|\mathbf{v}| = 290.9 \text{ pounds}$$

**7.3.8** Draw a figure and label it as in Example 8.

$$|\mathbf{BD}| = |\mathbf{AB}|\sin 10^\circ = 800\sin 10^\circ \approx 138.9 \text{ pounds}$$

**7.3.9** Draw a figure and label it as in Example 9. Heading northwest is 315°. The vector $\mathbf{v}_3$ is the course. The angle between $\mathbf{v}_1$ and $\mathbf{v}_2$ is $315^\circ - 200^\circ$ or 115°. The angle at point $D$ is $180^\circ - 115^\circ$ or 65°.

$$|\mathbf{v}_3|^2 = 100^2 + 500^2 - 2\cdot 100\cdot 500\cos 65^\circ$$
$$|\mathbf{v}_3| = 466.6 \text{ mph (the ground speed)}$$

If $\theta$ is the drift angle, then $\dfrac{\sin\theta}{100} = \dfrac{\sin 65^\circ}{466.6}$ or $\theta \approx 11.2^\circ$. So the bearing of the course is $315^\circ - 11.2^\circ$ or 303.8°.

## Section 4

**7.4.1** $|5 - i| = \sqrt{5^2 + (-1)^2} = \sqrt{26}$

**7.4.2** For $z = 1 + 2i$, $r = \sqrt{1^2 + 2^2} = \sqrt{5}$. From $a = r\cos\theta$ we get $1 = \sqrt{5}\cos\theta$ or $\theta \approx 63.4^\circ$. Since the terminal side of 63.4° goes through $(1, 2)$, $z \approx \sqrt{5}(\cos 63.4^\circ + i\sin 63.4^\circ)$.

**7.4.3** $12\left(\cos\dfrac{\pi}{6} + i\sin\dfrac{\pi}{6}\right)$

$$= 12\left(\frac{\sqrt{3}}{2} + i\frac{1}{2}\right) = 6\sqrt{3} + 6i$$

**7.4.4** Multiply the moduli and add the arguments:

$$z_1z_2 = 32\left(\cos\frac{\pi}{6} + i\sin\frac{\pi}{6}\right)$$
$$= 32\left(\frac{\sqrt{3}}{2} + i\frac{1}{2}\right) = 16\sqrt{3} + 16i$$

**7.4.5** For $z_1 = 2\sqrt{3} + 2i$, $r = \sqrt{12 + 4} = 4$. So $a = r\cos\theta$ yields $\cos\theta = \sqrt{3}/2$ and $\theta = 30^\circ$. For $z_2 = 3 + 3i\sqrt{3}$, $r = \sqrt{9 + 27} = 6$. So $a = r\cos\theta$ yields $\cos\theta = 1/2$ and $\theta = 60^\circ$. So

$$z_1 = 4(\cos 30^\circ + i\sin 30^\circ),$$
$$z_2 = 6(\cos 60^\circ + i\sin 60^\circ),$$

and

$$z_1z_2 = 24(\cos 90^\circ + i\sin 90^\circ)$$
$$= 24(0 + i) = 24i$$

**7.4.6** The conjugate of $6(\cos 30^\circ + i\sin 30^\circ)$ is $6(\cos(-30^\circ) + i\sin(-30^\circ))$. The product is $36(\cos 0^\circ + i\sin 0^\circ)$ or $36(1 + 0i) = 36$.

## Section 5

**7.5.1**
$$1 + i = \sqrt{2}(\cos 45^\circ + i\sin 45^\circ)$$
$$(1 + i)^6 = (\sqrt{2})^6(\cos(270^\circ) + i\sin(270^\circ))$$
$$= 8(0 + (-1)i) = -8i$$

**7.5.2** Write $-8 - 8i\sqrt{3}$ in trigonometric form as $16(\cos 240^\circ + i\sin 240^\circ)$. So

$$2\left(\cos\left(\frac{240^\circ + k360^\circ}{4}\right) + i\sin\left(\frac{240^\circ + k360^\circ}{4}\right)\right)$$

generates the fourth roots for $k = 0, 1, 2, 3$:

$$2(\cos(60^\circ) + i\sin(60^\circ)) = 1 + i\sqrt{3}$$
$$2(\cos(150^\circ) + i\sin(150^\circ)) = -\sqrt{3} + i$$
$$2(\cos(240^\circ) + i\sin(240^\circ)) = -1 - i\sqrt{3}$$
$$2(\cos(330^\circ) + i\sin(330^\circ)) = \sqrt{3} - i$$

**7.5.3** The solutions are the three cube roots of 27, where $27 = 27(\cos 0^\circ + i\sin 0^\circ)$. The roots are given by

$$27^{1/3}\left(\cos\left(\frac{0^\circ + k360^\circ}{3}\right) + i\sin\left(\frac{0^\circ + k360^\circ}{3}\right)\right)$$

where $k = 0, 1, 2$:

$$3(\cos 0^\circ + i\sin 0^\circ) = 3$$
$$3(\cos 120^\circ + i\sin 120^\circ) = -\frac{3}{2} + \frac{3\sqrt{3}}{2}i$$
$$3(\cos 240^\circ + i\sin 240^\circ) = -\frac{3}{2} - \frac{3\sqrt{3}}{2}i$$

## Section 6

**7.6.1**

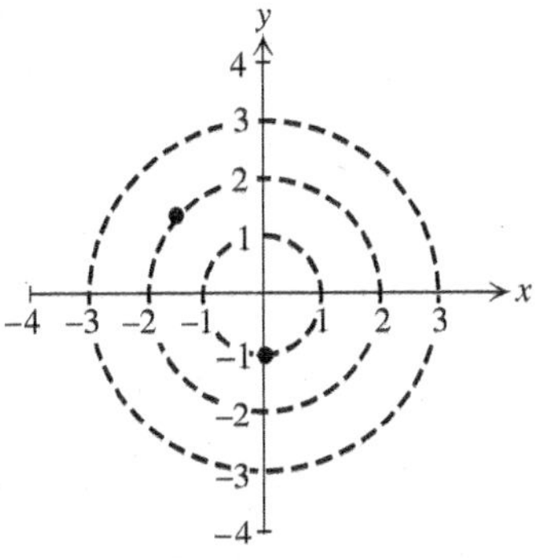


**7.6.2** $x = 3\cos 45^\circ = 3\sqrt{2}/2$

$y = 3\sin 45^\circ = 3\sqrt{2}/2$

So $(3, 45^\circ)$ in rectangular coordinates is $(3\sqrt{2}/2, 3\sqrt{2}/2)$.

For $(-2, 2\sqrt{3})$, $r = \sqrt{4 + 12} = 4$ and $\theta = \cos^{-1}(-2/4) = 120^\circ$.

So $(-2, 2\sqrt{3})$ in polar coordinates is $(4, 120^\circ)$.

**7.6.3** Some pairs that satisfy $r = 4\sin\theta$ are $(0^\circ, 0)$, $(45^\circ, 2\sqrt{2})$, $(90^\circ, 4)$, $(135^\circ, 2\sqrt{2})$, and $(180^\circ, 0)$.

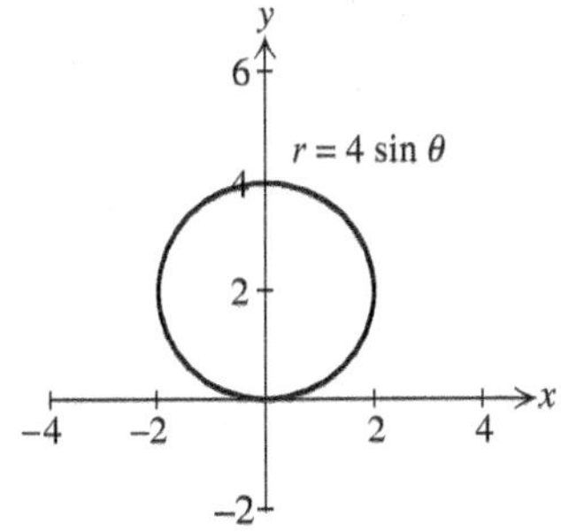


**7.6.4** Some pairs that satisfy $r = \cos 2\theta$ are $(0^\circ, 1)$, $(22.5^\circ, \sqrt{2}/2)$, $(45^\circ, 0)$, $(67.5^\circ, -\sqrt{2}/2)$, and $(90^\circ, -1)$.

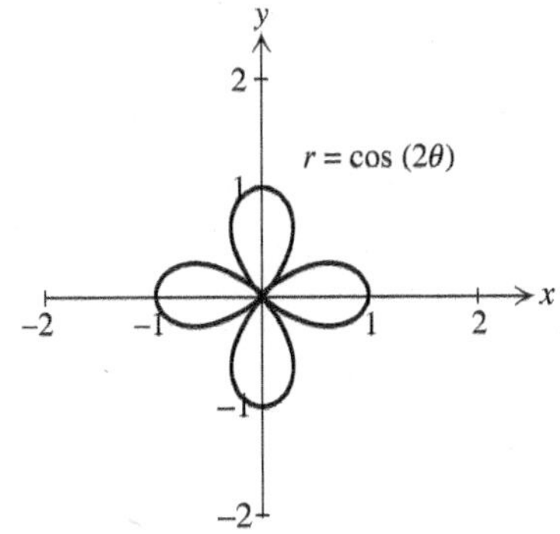


**7.6.5** Some pairs that satisfy $r = -\theta$ are $(0, 0)$, $(\pi/4, -\pi/4)$, $(\pi/2, -\pi/2)$, $(3\pi/4, -3\pi/4)$, and $(\pi, -\pi)$.

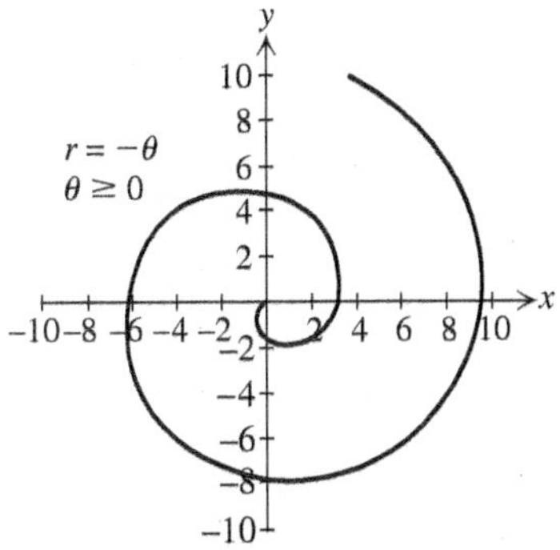

**7.6.6**

$$r = 3\sin\theta$$
$$r^2 = 3r\sin\theta$$
$$x^2 + y^2 = 3y$$

This rectangular equation can be rewritten in the standard form for a circle:

$$x^2 + y^2 - 3y = 0$$
$$x^2 + (y - 3/2)^2 = 9/4$$

**7.6.7**

$$y = -2x + 5$$
$$r\sin\theta = -2r\cos\theta + 5$$
$$r\sin\theta + 2r\cos\theta = 5$$
$$r(\sin\theta + 2\cos\theta) = 5$$
$$r = \frac{5}{\sin\theta + 2\cos\theta}$$

## Section 7

**7.7.1** The graph is a line segment with endpoints $(5, -1)$ corresponding to $t = 0$ and $(10, 9)$ corresponding to $t = 5$.

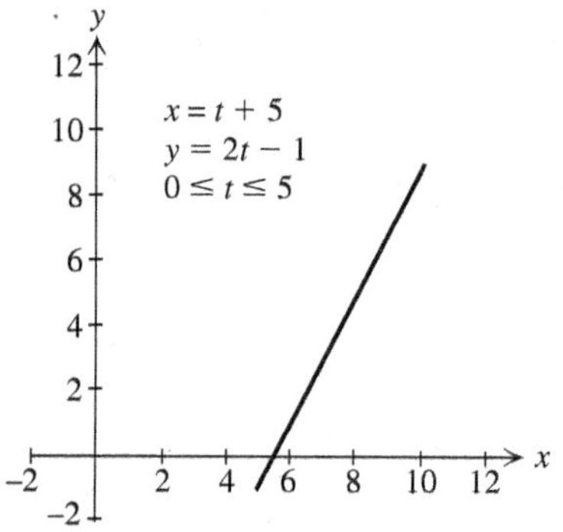


**7.7.2** If $y = -t + 1$, then $t = 1 - y$. Substitute $t = 1 - y$ into $x = 4t - 9$ to get $x = 4(1 - y) - 9$ or $x + 4y = -5$, which is a straight line. The domain is $(-\infty, \infty)$ and the range is $(-\infty, \infty)$.

**7.7.3** If $x = mt + b$, then $1 = m \cdot 0 + b$ and $8 = m \cdot 1 + b$ yields $b = 1$ and $m = 7$. So $x = 7t + 1$. If $y = mt + b$, then $2 = m \cdot 0 + b$ and $10 = m \cdot 1 + b$ yields $b = 2$ and $m = 8$.
So $y = 8t + 2$.

**7.7.4** Let $r = 3\cos\theta$ in $x = r\cos\theta$ and $y = r\sin\theta$ to get $x = 3\cos^2\theta$ and $y = 3\cos\theta\sin\theta$. Graph the parametric equations to see that the graph is a circle with diameter 3.

# Appendix B

## Basic Algebra Review

# B.1 Real Numbers and Their Properties

The real numbers and their properties form the foundation for the study of algebra.

### The Real Numbers

The set of **real numbers** is the set of numbers that correspond to points on the number line. There are several important subsets of the real numbers.

**Subsets of the Real Numbers**

| | |
|---|---|
| **Counting or natural numbers** | $\{1, 2, 3, \ldots\}$ |
| **Whole numbers** | $\{0, 1, 2, 3, \ldots\}$ |
| **Integers** | $\{\ldots, -3, -2, -1, 0, 1, 2, 3, \ldots\}$ |
| **Rational numbers** | Numbers of the form $\frac{a}{b}$, where $a$ and $b$ are integers with $b \neq 0$ |
| **Irrational numbers** | Real numbers that are not rational |

Using decimal notation, the rational numbers are the numbers that are repeating or terminating decimals, and the irrational numbers are the nonrepeating nonterminating decimals. For example, the number $0.595959\ldots$ is a rational number because the three dots after the 9 mean that the pair 59 repeats indefinitely. In the number $5.010010001\ldots$, the three dots mean that the pattern continues indefinitely. So each group of zeros contains one more zero than the previous group. Because no group of digits repeats, $5.010010001\ldots$ is an irrational number.

The square root of any positive integer that is not a perfect square is irrational. So $\sqrt{2}$, $\sqrt{3}$, and $\sqrt{5}$ are irrational. The number $\pi$ is irrational also. However, it is difficult to see that numbers like $\sqrt{2}$ and $\pi$ are irrational because their decimal representations are not apparent. Since a calculator operates with a fixed number of decimal places, it gives us only a *rational approximation* for an irrational number such as $\sqrt{2}$ or $\pi$.

**EXAMPLE 1** Classifying numbers

Determine whether each statement is true or false and explain.

**a.** The number 0 is a real number.
**b.** The number $\pi$ is a rational number.
**c.** Every real number is rational.
**d.** No rational number is irrational.

**Solution**

**a.** True, because 0 is a whole number and the whole numbers are real.
**b.** False, because $\pi$ is irrational.
**c.** False, because the irrational numbers are real numbers.
**d.** True, because the irrational numbers and the rational numbers have no numbers in common.

## Properties of the Real Numbers

The operations of addition and multiplication on the real numbers have the following properties.

**Properties of the Real Numbers**

For any real numbers $a$, $b$, and $c$:

| | |
|---|---|
| $a + b$ and $ab$ are real numbers | **Closure Property** |
| $a + b = b + a$ and $ab = ba$ | **Commutative Properties** |
| $a + (b + c) = (a + b) + c$ and $a(bc) = (ab)c$ | **Associative Properties** |
| $a(b + c) = ab + ac$ | **Distributive Property** |
| $0 + a = a$ and $1 \cdot a = a$ | **Identity Properties** |
| $0 \cdot a = 0$ | **Multiplication Property of Zero** |
| For each real number $a$, there is a unique real number $-a$ such that $a + (-a) = 0$. | **Additive Inverse Property** |
| For each nonzero number $a$, there is a unique real number $1/a$ such that $a \cdot 1/a = 1$. | **Multiplicative Inverse Property** |

Zero is the **additive identity** and 1 is the **multiplicative identity.** The number $-a$ is the **additive inverse** or **opposite** of $a$, and $1/a$ is the **multiplicative inverse** or **reciprocal** of $a$.

Note that the properties stated here involve only addition and multiplication. We define subtraction as $a - b = a + (-b)$ and division as $a \div b = a \cdot 1/b$ for $b \neq 0$. Note that $a - b$ is called the **difference** of $a$ and $b$ and $a \div b$ is called the **quotient** of $a$ and $b$.

### EXAMPLE 2 Using the properties

Complete each statement using the property named.

**a.** $a7 =$ ________, commutative **b.** $2x + 4 =$ ________, distributive

**c.** $8($________$) = 1$, multiplicative inverse **d.** $\frac{1}{3}(3x) =$ ________, associative

**Solution**

**a.** $a7 = 7a$ **b.** $2x + 4 = 2(x + 2)$

**c.** $8\left(\frac{1}{8}\right) = 1$ **d.** $\frac{1}{3}(3x) = \left(\frac{1}{3} \cdot 3\right)x$ ■

## Additive Inverses

We read $-7$ as negative 7, but we do not read $-a$ as negative $a$. The symbol $-a$ is read as the additive inverse or opposite of $a$. If $a$ is positive, then $-a$ is negative. If $a$ is negative, then $-a$ is positive. Some useful properties of additive inverses or opposites follow.

**Properties of Opposites**

For any real numbers $a$ and $b$:

| | |
|---|---|
| **1.** $-1 \cdot a = -a$ | The product of $-1$ and $a$ is the opposite of $a$. |
| **2.** $-(-a) = a$ | The opposite of the opposite of $a$ is $a$. |
| **3.** $-(a - b) = b - a$ | The opposite of $a - b$ is $b - a$. |

### EXAMPLE 3 Using properties of opposites

Use the properties of opposites to complete each equation.

**a.** $-(-\pi) =$ ______ **b.** $-1(-2) =$ ______
**c.** $-1(x - h) =$ ______

**Solution**

**a.** $-(-\pi) = \pi$
**b.** $-1(-2) = -(-2) = 2$
**c.** $-1(x - h) = -(x - h) = h - x$ ■

## Absolute Value

The **absolute value** of $a$ (in symbols $|a|$) is the distance from $a$ to 0 on a number line. Since both 3 and $-3$ are three units from 0 on the number line, $|3| = 3$ and $|-3| = 3$. A symbolic definition of absolute value follows.

**Definition: Absolute Value**

For any real number $a$, $|a| = \begin{cases} a & \text{if } a \geq 0 \\ -a & \text{if } a < 0. \end{cases}$

Note that if $a < 0$, then $-a$ is positive. So $|a| \geq 0$ for any real number $a$.

### EXAMPLE 4 Finding the absolute value of a number

Use the symbolic definition of absolute value to simplify each expression.

**a.** $|5.6|$ **b.** $|0|$ **c.** $|-3|$

**Solution**

**a.** Since $5.6 \geq 0$, we use the equation $|a| = a$ to get $|5.6| = 5.6$.
**b.** Since $0 \geq 0$, we use the equation $|a| = a$ to get $|0| = 0$.
**c.** Since $-3 < 0$, we use $|a| = -a$ to get $|-3| = -(-3) = 3$. ■

Absolute value has the following properties.

**Properties of Absolute Value**

For any real numbers $a$ and $b$:

1. $|a| \geq 0$ — The absolute value of any number is nonnegative.
2. $|-a| = |a|$ — Additive inverses have the same absolute value.
3. $|a \cdot b| = |a| \cdot |b|$ — The absolute value of a product is the product of the absolute values.
4. $\left|\dfrac{a}{b}\right| = \dfrac{|a|}{|b|}, b \neq 0$ — The absolute value of a quotient is the quotient of the absolute values.
5. The distance between $a$ and $b$ on a number line is $|a - b|$.

## Exponential Expressions

We use positive integral exponents to indicate the number of times a number occurs in a product. For example, $2 \cdot 2 \cdot 2 \cdot 2$ is written as $2^4$. We read $2^4$ as "the fourth power of 2" or "2 to the fourth power."

**Definition: Positive Integral Exponents**

For any positive integer $n$:

$$a^n = \underbrace{a \cdot a \cdot a \cdot \cdots \cdot a}_{n \text{ factors of } a}$$

We call a the **base,** $n$ the **exponent** or **power,** and $a^n$ an **exponential expression.**

We read $a^n$ as "$a$ to the $n$th power." For $a^1$ we usually omit the exponent and just write $a$. We refer to the exponents 2 and 3 as squares and cubes. For example, $3^2$ is read "3 squared," $2^3$ is read "2 cubed," $x^4$ is read "$x$ to the fourth," $b^5$ is read "$b$ to the fifth," and so on. To evaluate an expression such as $-3^2$, we square 3 first, then take the opposite. So $-3^2 = -9$ and $(-3)^2 = (-3)(-3) = 9$.

### EXAMPLE 5 Evaluating exponential expressions

Evaluate.

**a.** $4^3$ **b.** $(-2)^4$ **c.** $-2^4$

**Solution**

**a.** $4^3 = 4 \cdot 4 \cdot 4 = 16 \cdot 4 = 64$
**b.** $(-2)^4 = (-2)(-2)(-2)(-2) = 16$
**c.** $-2^4 = -(2 \cdot 2 \cdot 2 \cdot 2) = -16$

## Arithmetic Expressions

The result of writing numbers in a meaningful combination with the ordinary operations of arithmetic is called an **arithmetic expression** or simply an **expression.** The **value** of an arithmetic expression is the real number obtained when all operations are performed. Symbols such as parentheses, brackets, braces, absolute value bars, and fraction bars are called **grouping symbols.** Operations within grouping symbols are performed first.

### EXAMPLE 6 Evaluating an arithmetic expression with grouping symbols

Evaluate each expression.

**a.** $(-7 \cdot 3) + (5 \cdot 8)$ **b.** $\dfrac{3 - 9}{-2 - (-5)}$ **c.** $3 - |5 - (2 \cdot 9)|$

**Solution**

**a.** Perform the operations within the parentheses first and remove the parentheses:

$$(-7 \cdot 3) + (5 \cdot 8) = -21 + 40 = 19$$

**b.** Since the fraction bar acts as a grouping symbol, we evaluate the numerator and denominator before dividing.

$$\frac{3 - 9}{-2 - (-5)} = \frac{-6}{3} = -2$$

**c.** First evaluate within the innermost grouping symbols:

$$3 - |5 - (2 \cdot 9)| = 3 - |5 - 18| \quad \text{Innermost grouping symbols}$$
$$= 3 - |-13| \quad \text{Innermost grouping symbols}$$
$$= 3 - 13 \quad \text{Evaluate the absolute value.}$$
$$= -10 \quad \text{Subtract.}$$

■

## The Order of Operations

When some or all grouping symbols are omitted in an expression, we evaluate the expression using the following order of operations.

**Order of Operations**

1. Evaluate exponential expressions.
2. Perform multiplication and division in order from left to right.
3. Perform addition and subtraction in order from left to right.

### EXAMPLE 7 Using the order of operations to evaluate an expression

Evaluate each expression.

**a.** $3 - 4 \cdot 2^3$ **b.** $5 \cdot 8 \div 4 \cdot 2$ **c.** $3 - 4 + 9 - 2$

**Solution**

**a.** By the order of operations evaluate $2^3$, then multiply, and then subtract:

$$3 - 4 \cdot 2^3 = 3 - 4 \cdot 8 = 3 - 32 = -29$$

**b.** In an expression with only multiplication and division, the operations are performed from left to right:

$$5 \cdot 8 \div 4 \cdot 2 = 40 \div 4 \cdot 2 = 10 \cdot 2 = 20$$

**c.** In an expression with only addition and subtraction, the operations are performed from left to right:

$$3 - 4 + 9 - 2 = -1 + 9 - 2 = 8 - 2 = 6$$

■

## Algebraic Expressions

When we write numbers and one or more variables in a meaningful combination with the ordinary operations of arithmetic, the result is called an **algebraic expression,** or simply an expression. The **value of an algebraic expression** is the value of the arithmetic expression that is obtained when the variables are replaced by real numbers.

### EXAMPLE 8 Evaluating an algebraic expression

Find the value of $b^2 - 4ac$ when $a = -1$, $b = -2$, and $c = 3$.

**Solution**

Replace the variables by the appropriate numbers:

$$b^2 - 4ac = (-2)^2 - 4(-1)(3) = 16$$

■

The **domain** of an algebraic expression in one variable is the set of all real numbers that can be used for the variable. For example, the domain of $1/x$ is the set of nonzero real numbers, because division by 0 is undefined. Two algebraic expressions in one variable are **equivalent** if they have the same domain and if they have the same value for each member of the domain. The expressions $1/x$ and $x/x^2$ are equivalent.

An expression that is the product of a number and one or more variables raised to powers is called a **term.** Expressions such as $3x$, $2kab^3$, and $\pi r^2$ are terms. In the term $3x$, 3 and $x$ are called **factors.** The **coefficient** of any variable part of a term is the product of the remaining factors in the term. For example, the coefficient of $x$ in $3x$ is 3. The coefficient of $ab^3$ in $2kab^3$ is $2k$ and the coefficient of $b^3$ is $2ka$. If two terms contain the same variables with the same exponents, then they are called **like terms.** The distributive property allows us to **combine like terms:** $3x + 2x = (3 + 2)x = 5x$.

To **simplify** an expression means to find a simpler-looking equivalent expression. The properties of the real numbers are used to simplify expressions.

### EXAMPLE 9 Using properties to simplify an expression

Simplify each expression.

**a.** $-4x - (6 - 7x)$ **b.** $\frac{1}{2}x - \frac{3}{4}x$ **c.** $-6(x - 3) - 3(5 - 7x)$

**Solution**

**a.**
$$\begin{aligned} -4x - (6 - 7x) &= -4x + [-(6 + (-7x))] && \text{Definition of subtraction} \\ &= -4x + [-1(6 + (-7x))] && \text{First property of opposites} \\ &= -4x + [(-6) + 7x] && \text{Distributive property} \\ &= [-4x + 7x] + (-6) && \text{Commutative and associative properties} \\ &= 3x - 6 && \text{Combine like terms.} \end{aligned}$$

**b.**
$$\begin{aligned} \frac{1}{2}x - \frac{3}{4}x &= \frac{2}{4}x - \frac{3}{4}x && \text{Write } \tfrac{1}{2} \text{ as } \tfrac{2}{4} \text{ to obtain a common denominator.} \\ &= -\frac{1}{4}x && \text{Combine like terms.} \end{aligned}$$

**c.**
$$\begin{aligned} -6(x - 3) - 3(5 - 7x) &= -6x + 18 - 15 + 21x && \text{Distributive property} \\ &= 15x + 3 && \text{Combine like terms.} \end{aligned}$$

## EXERCISES B.1

*Determine which numbers in the set* $\{-3.5, -\sqrt{2}, -1, 0, 1, \sqrt{3}, 3.14, \pi, 4.3535\ldots, 5.090090009\ldots\}$ *are members of the following sets.*

**1.** Real numbers

**2.** Rational numbers

**3.** Irrational numbers

**4.** Integers

**5.** Whole numbers

**6.** Natural numbers

*Complete each statement using the property named.*

**7.** $7 + x =$ ________, commutative

**8.** $5(4y) =$ ________, associative

**9.** $5(x + 3) =$ ________, distributive

**10.** $-3(x - 4) =$ ________, distributive

**11.** $\frac{1}{2}x + \frac{1}{2} =$ ________, distributive

**12.** $-5x + 10 =$ ________, distributive

**13.** $-13 + (4 + x) =$ __________, associative

**14.** $yx =$ __________, commutative

**15.** $0.125($__________$) = 1$, multiplicative inverse

**16.** $-3 + ($__________$) = 0$, additive inverse

*Use the properties of opposites to complete each equation.*

**17.** $-(-\sqrt{3}) =$ **18.** $-1(-6.4) =$

**19.** $-1(x^2 - y^2) =$ **20.** $-(1 - a^2) =$

*Use the symbolic definition of absolute value to simplify each expression.*

**21.** $|7.2|$ **22.** $|0/3|$ **23.** $|-\sqrt{5}|$ **24.** $|-3/4|$

*Evaluate each expression.*

**25.** $4^3$ **26.** $3^4$ **27.** $-7^2$

**28.** $-9^2$ **29.** $(-4)^2$ **30.** $(-10)^4$

*Evaluate each expression.*

**31.** $(2 \cdot 5) - (3 \cdot 6)$ **32.** $(5 - 3)(2 - 6)$

**33.** $|3 - (4 \cdot 5)| - 5$ **34.** $5 - |4 - (2 \cdot 3)|$

**35.** $|-4 \cdot 3| - |-3 \cdot 5|$ **36.** $(-8 \cdot 3) - |-3 \cdot 7|$

**37.** $\dfrac{-2 - (-6)}{-5 - (-9)}$ **38.** $\dfrac{4 - (-3)}{-3 - (-1)}$

*Use the order of operations to evaluate each expression.*

**39.** $4 - 5 \cdot 3^2$ **40.** $4 + 2(-6)^2$

**41.** $3 - 4 + 5 - 7 - 4$ **42.** $4 - 3 + 2 - 5 + 6$

**43.** $3 \cdot 6 + 2 \cdot 4$ **44.** $-2 \cdot 9 + 3 \cdot 5$

**45.** $26 \cdot \frac{1}{5} \div \frac{1}{2} \cdot 5$ **46.** $\frac{4}{3} \cdot 50(0.75) \div 2$

**47.** $(3 \cdot 4 - 1)(1 + 2 \cdot 4)$ **48.** $-2 - 3(5 - 2 \cdot 8)$

**49.** $2 - 3\,|3 - 4 \cdot 6|$ **50.** $1 - (3 - |1 - 2 \cdot 3|)$

**51.** $7^2 - 2(-3)(-6)$ **52.** $(-3)^2 - 4(-2)(-5)$

**53.** $(-2 - 3)^2 - (4 - (-1))^2$

**54.** $-(5 - (-2))^2 + (6 - 7)^2$

*Evaluate each expression if $a = -2$, $b = 3$, and $c = 4$.*

**55.** $b^2 - 4ac$ **56.** $(b - 4ac)^2$

**57.** $\dfrac{a - c}{b - c}$ **58.** $\dfrac{a^2 - c}{b^3 + c^4}$

*Use the properties of the real numbers to simplify each expression.*

**59.** $-5x + 3x$ **60.** $-5x - (-8x)$

**61.** $x - 0.15x$ **62.** $x + 3 - 0.9x$

**63.** $-3(2xy)$ **64.** $\frac{1}{2}(8wz)$

**65.** $\frac{1}{2}(6 - 4x)$ **66.** $\frac{1}{4}(8x - 4)$

**67.** $\dfrac{6x - 2y}{2}$ **68.** $\dfrac{-9 - 6x}{-3}$

**69.** $(3 - 4x) + (x - 9)$ **70.** $(9x - 3) + (4 - 6x)$

**71.** $x - 0.03(x + 200)$ **72.** $y - 0.9(y - 3000)$

**73.** $-2(4 - x) - 3(3 - 3x)$ **74.** $5(4 - 2x) - 2(x - 5)$

# B.2 Exponents and Radicals

We reviewed positive integral exponents in the last section. In this section we review negative integral exponents, rational exponents, and radicals.

## Negative Integral Exponents

We use a negative sign in an exponent to represent multiplicative inverses or reciprocals. For negative exponents we do not allow the base to be zero because zero does not have a reciprocal.

**Definition: Negative Integral Exponents**

If $a$ is a nonzero real number and $n$ is a positive integer, then $a^{-n} = \frac{1}{a^n}$.

## EXAMPLE 1 Evaluating expressions that have negative exponents

Simplify each expression without using a calculator, then check with a calculator.

**a.** $4 \cdot 2^{-3}$ **b.** $\left(\frac{2}{3}\right)^{-3}$ **c.** $\frac{6^{-2}}{2^{-3}}$

### Solution

**a.** $4 \cdot 2^{-3} = 4 \cdot \frac{1}{2^3} = 4 \cdot \frac{1}{8} = \frac{1}{2}$

**b.** $\left(\frac{2}{3}\right)^{-3} = \frac{1}{\left(\frac{2}{3}\right)^3} = \frac{1}{\frac{2}{3} \cdot \frac{2}{3} \cdot \frac{2}{3}} = \frac{1}{\frac{8}{27}} = \frac{27}{8}$ Note that $\left(\frac{2}{3}\right)^{-3} = \left(\frac{3}{2}\right)^3$.

**c.** $\frac{6^{-2}}{2^{-3}} = \frac{\frac{1}{6^2}}{\frac{1}{2^3}} = \frac{1}{6^2} \cdot \frac{2^3}{1} = \frac{8}{36} = \frac{2}{9}$ Note that $\frac{6^{-2}}{2^{-3}} = \frac{2^3}{6^2}$. ■

Example 1(b) illustrates the fact that a fractional base can be inverted, if the sign of the exponent is changed. Example 1(c) illustrates the fact that a factor of the numerator or denominator can be moved from the numerator to the denominator or vice versa as long as we change the sign of the exponent.

**Rules for Negative Exponents and Fractions**

If $a$ and $b$ are nonzero real numbers and $m$ and $n$ are integers, then

$$\left(\frac{a}{b}\right)^{-m} = \left(\frac{b}{a}\right)^m \quad \text{and} \quad \frac{a^{-m}}{b^{-n}} = \frac{b^n}{a^m}.$$

Be careful with the rules of exponents when sums or differences are involved. For example, $\frac{2^{-1} - 1}{3^{-1}} \neq \frac{3^1 - 1}{2^1}$ because

$$\frac{2^{-1} - 1}{3^{-1}} = \frac{\frac{1}{2} - 1}{\frac{1}{3}} = -\frac{3}{2} \quad \text{and} \quad \frac{3^1 - 1}{2^1} = 1.$$

## Rules of Exponents

So far we have defined positive and negative integral exponents. Zero as an exponent is defined as follows.

**Definition: Zero Exponent**

If $a$ is a nonzero real number, then $a^0 = 1$.

Using the definitions of positive, negative, and zero exponents, we can show that the following rules hold for any integral exponents.

**Rules for Integral Exponents**

If $a$ and $b$ are nonzero real numbers and $m$ and $n$ are integers, then

1. $a^m a^n = a^{m+n}$ **Product Rule**
2. $\frac{a^m}{a^n} = a^{m-n}$ **Quotient Rule**
3. $(a^m)^n = a^{mn}$ **Power of a Power Rule**
4. $(ab)^n = a^n b^n$ **Power of a Product Rule**
5. $\left(\frac{a}{b}\right)^n = \frac{a^n}{b^n}$ **Power of a Quotient Rule**

The rules for integral exponents are used to simplify expressions in the next example.

**EXAMPLE 2** Simplifying expressions with integral exponents

Simplify each expression. Write your answer without negative exponents. Assume that all variables represent nonzero real numbers.

**a.** $(3x^2)^3(-2x^{-2})$ **b.** $\left(\dfrac{a^5b^{-1}}{a^7}\right)^4$

**Solution**

**a.**
$$\begin{aligned}(3x^2)^3(-2x^{-2}) &= 3^3(x^2)^3(-2x^{-2}) && \text{Power of a product rule}\\ &= 27x^6(-2x^{-2}) && \text{Power of a power rule}\\ &= -54x^4 && \text{Product rule}\end{aligned}$$

**b.**
$$\begin{aligned}\left(\frac{a^5b^{-1}}{a^7}\right)^4 &= \frac{(a^5b^{-1})^4}{(a^7)^4} && \text{Power of a quotient rule}\\ &= \frac{(a^5)^4(b^{-1})^4}{(a^7)^4} && \text{Power of a product rule}\\ &= \frac{a^{20}b^{-4}}{a^{28}} && \text{Power of a power rule}\\ &= a^{-8}b^{-4} && \text{Quotient rule}\\ &= \frac{1}{a^8b^4} && \text{Definition of negative exponent}\end{aligned}$$

■

## Roots

Since $2^4 = 16$ and $(-2)^4 = 16$, both 2 and $-2$ are fourth roots of 16. The *n*th root of a number is defined in terms of the *n*th power.

**Definition: *n*th Roots**

> If $n$ is a positive integer and $a^n = b$, then $a$ is called an ***n*th root** of $b$.
> If $a^2 = b$, then $a$ is a **square root** of $b$. If $a^3 = b$, then $a$ is the **cube root** of $b$.

If $n$ is even and $a$ is an $n$th root of $b$, then $a$ is called an **even root** of $b$. If $n$ is odd and $a$ is an $n$th root of $b$, then $a$ is called an **odd root** of $b$. Every positive real number has *two* real even roots, a positive root and a negative root. For example, both 5 and $-5$ are square roots of 25 because $5^2 = 25$ and $(-5)^2 = 25$. Moreover, every real number has exactly *one* real odd root. For example, because $2^3 = 8$ and 3 is odd, 2 is the only real cube root of 8. Because $(-2)^3 = -8$ and 3 is odd, $-2$ is the only real cube root of $-8$.

Finding an $n$th root is the reverse of finding an $n$th power, so we use the notation $a^{1/n}$ for the $n$th root of $a$. For example, since the positive square root of 25 is 5, we write $25^{1/2} = 5$.

**Definition: Exponent 1/*n***

> If $n$ is a positive even integer and $a$ is positive, then $a^{1/n}$ denotes the **positive real *n*th root of *a*** and is called the **principal *n*th root of *a*.**
>
> If $n$ is a positive odd integer and $a$ is any real number, then $a^{1/n}$ denotes the real $n$th root of $a$.
>
> If $n$ is a positive integer, then $0^{1/n} = 0$.

### EXAMPLE 3 Evaluating expressions involving exponent $1/n$

Evaluate each expression.

**a.** $4^{1/2}$ **b.** $(-8)^{1/3}$ **c.** $(-4)^{1/2}$

**Solution**

**a.** Because the positive real square root of 4 is 2, $4^{1/2} = 2$.
**b.** Because the real cube root of $-8$ is $-2$, $(-8)^{1/3} = -2$.
**c.** Since the definition of $n$th root does not include an even root of a negative number, $(-4)^{1/2}$ has not yet been defined. Even roots of negative numbers do exist in the complex number system, but an even root of a negative number is not a real number. ■

## Rational Exponents

A rational exponent indicates both a root and a power. The expression $a^{m/n}$ is defined as the $m$th power of the $n$th root of $a$.

**Definition: Rational Exponents**

If $m$ and $n$ are positive integers, then

$$a^{m/n} = (a^{1/n})^m,$$

provided that $a^{1/n}$ is a real number.

Note that $a^{1/n}$ is not real when $a$ is negative and $n$ is even. According to the definition of rational exponents, expressions such as $(-25)^{-3/2}$, $(-43)^{1/4}$, and $(-1)^{2/2}$ are not defined because each of them involves an even root of a negative number. Note that some authors define $a^{m/n}$ only for $m/n$ in lowest terms. In that case the fourth power of the square root of three could *not* be written as $3^{4/2}$. This author prefers the more general definition given above.

A negative rational exponent indicates a reciprocal just as a negative integral exponent does. So $7^{-2/3} = 1/7^{2/3}$. The root or the power indicated by a rational exponent can be evaluated in either order. That is, $(a^{1/n})^m = (a^m)^{1/n}$, provided $a^{1/n}$ is real. However, for mental evaluation the following order is best.

**Procedure: Evaluating $a^{-m/n}$**

To evaluate $a^{-m/n}$ mentally,

**1.** find the $n$th root of $a$, **2.** raise it to the $m$ power, **3.** find the reciprocal.

### EXAMPLE 4 Evaluating expressions with rational exponents

Evaluate each expression.

**a.** $(-8)^{2/3}$ **b.** $27^{-2/3}$ **c.** $100^{6/4}$

**Solution**

**a.** Mentally, the cube root of $-8$ is $-2$ and the square of $-2$ is 4. In symbols:

$$(-8)^{2/3} = ((-8)^{1/3})^2 = (-2)^2 = 4$$

**b.** Mentally, the cube root of 27 is 3, the square of 3 is 9, and the reciprocal of 9 is $\frac{1}{9}$. In symbols:

$$27^{-2/3} = \frac{1}{(27^{1/3})^2} = \frac{1}{3^2} = \frac{1}{9}$$

**c.** $100^{6/4} = 100^{3/2} = 10^3 = 1000$ ■

Note how we reduced the exponent in Example 4(c). However, exponents can be reduced only on expressions that are defined. For example, $(-1)^{2/2} \neq (-1)^1$ because $(-1)^{2/2}$ is an undefined expression.

The rules for integral exponents stated previously in this section also hold for rational exponents. Note that the power of a power rule can fail if the base is negative. For example, $(x^2)^{1/2} = x$ according to the rule. However, if $x$ is a negative number the left side of this equation is positive and the right side is negative. So $(x^2)^{1/2} = x$ is only correct if $x$ is nonnegative. The equation $(x^2)^{1/2} = |x|$ is correct for any $x$.

### EXAMPLE 5 Simplifying expressions with rational exponents

Use the rules of exponents to simplify each expression. Assume that the variables represent positive real numbers. Write answers without negative exponents.

**a.** $x^{2/3}x^{4/3}$ **b.** $(x^4y^{1/2})^{1/4}$ **c.** $\left(\frac{a^{3/2}b^{2/3}}{a^2}\right)^3$

**Solution**

**a.** $x^{2/3}x^{4/3} = x^{6/3}$ Product rule

$= x^2$ Simplify the exponent.

**b.** $(x^4y^{1/2})^{1/4} = (x^4)^{1/4}(y^{1/2})^{1/4}$ Power of a product rule

$= xy^{1/8}$ Power of a power rule

**c.** $\left(\frac{a^{3/2}b^{2/3}}{a^2}\right)^3 = \frac{(a^{3/2})^3(b^{2/3})^3}{(a^2)^3}$ Power of a quotient rule

$= \frac{a^{9/2}b^2}{a^6}$ Power of power rule

$= a^{-3/2}b^2$ Quotient rule $\left(\frac{9}{2} - 6 = -\frac{3}{2}\right)$

$= \frac{b^2}{a^{3/2}}$ Definition of negative exponents ■

## Radical Notation

The exponent $1/n$ and the **radical sign** $\sqrt[n]{\ }$ are both used to indicate the $n$th root.

**Definition: Radical Notation**

> If $n$ is a positive integer and $a$ is a number for which $a^{1/n}$ is defined, then the expression $\sqrt[n]{a}$ is called a **radical,** and
>
> $$\sqrt[n]{a} = a^{1/n}.$$
>
> If $n = 2$, we write $\sqrt{a}$ rather than $\sqrt[2]{a}$.

The number $a$ is called the **radicand** and $n$ is the **index** of the radical. Expressions such as $\sqrt{-3}$, $\sqrt[4]{-81}$, and $\sqrt[6]{-1}$ do not represent real numbers because each is an even root of a negative number.

## EXAMPLE 6 Evaluating radicals

Evaluate each expression.

**a.** $\sqrt{49}$ **b.** $\sqrt[3]{-1000}$ **c.** $\sqrt[4]{\frac{16}{81}}$ **d.** $\sqrt[3]{125^2}$

### Solution

**a.** The symbol $\sqrt{49}$ indicates the positive square root of 49. So $\sqrt{49} = 49^{1/2} = 7$. Writing $\sqrt{49} = \pm 7$ is incorrect.

**b.** $\sqrt[3]{-1000} = (-1000)^{1/3} = -10$ Check that $(-10)^3 = -1000$.

**c.** $\sqrt[4]{\frac{16}{81}} = \left(\frac{16}{81}\right)^{1/4} = \frac{2}{3}$ Check that $\left(\frac{2}{3}\right)^4 = \frac{16}{81}$.

**d.** $\sqrt[3]{125^2} = (125^2)^{1/3} = 125^{2/3} = 5^2 = 25$ ■

Since $a^{1/n} = \sqrt[n]{a}$, expressions involving rational exponents can be written with radicals.

**Rule: Converting $a^{m/n}$ to Radical Notation**

If $a$ is a real number and $m$ and $n$ are integers for which $\sqrt[n]{a}$ is real, then

$$a^{m/n} = (\sqrt[n]{a})^m = \sqrt[n]{a^m}.$$

## EXAMPLE 7 Converting rational exponents to radicals

Write each expression in radical notation. Assume that all variables represent positive real numbers. Simplify the radicand if possible.

**a.** $2^{2/3}$ **b.** $(3x)^{3/4}$ **c.** $2(x^2 + 3)^{-1/2}$

### Solution

**a.** $2^{2/3} = \sqrt[3]{2^2} = \sqrt[3]{4}$

**b.** $(3x)^{3/4} = \sqrt[4]{(3x)^3} = \sqrt[4]{27x^3}$

**c.** $2(x^2 + 3)^{-1/2} = 2 \cdot \frac{1}{(x^2 + 3)^{1/2}} = \frac{2}{\sqrt{x^2 + 3}}$ ■

## The Product and Quotient Rules for Radicals

Using the power of a product and the power of a quotient rules for rational exponents, we can write

$$(ab)^{1/n} = a^{1/n}b^{1/n} \quad \text{and} \quad \left(\frac{a}{b}\right)^{1/n} = \frac{a^{1/n}}{b^{1/n}}.$$

These rules are expressed in radical notation as follows.

**Rules: Product and Quotient Rules for Radicals**

For any positive integer $n$ and real numbers $a$ and $b$ $(b \neq 0)$,

1. $\sqrt[n]{ab} = \sqrt[n]{a} \cdot \sqrt[n]{b}$ **Product Rule for Radicals**
2. $\sqrt[n]{\frac{a}{b}} = \frac{\sqrt[n]{a}}{\sqrt[n]{b}}$ **Quotient Rule for Radicals**

provided that all of the roots are real.

In words, the $n$th root of a product (or quotient) is the product (or quotient) of the $n$th roots.

An expression that is the square of a term that is free of radicals is called a **perfect square.** For example, $9x^6$ is a perfect square because $9x^6 = (3x^3)^2$. Likewise, $27y^{12}$ is a **perfect cube.** In general, an expression that is the $n$th power of an expression free of radicals is a **perfect *n*th power.** In the next example, the product and quotient rules for radicals are used to simplify radicals containing perfect squares, cubes, and so on.

### EXAMPLE 8 Using the product and quotient rules for radicals

Simplify each radical expression. Assume that all variables represent positive real numbers.

**a.** $\sqrt[3]{125a^6}$ **b.** $\sqrt{\dfrac{3}{16x^2}}$

**Solution**

**a.** Both 125 and $a^6$ are perfect cubes. So use the product rule to simplify:

$$\sqrt[3]{125a^6} = \sqrt[3]{125} \cdot \sqrt[3]{a^6} = 5a^2 \quad \text{Since } \sqrt[3]{a^6} = a^{6/3} = a^2$$

**b.** Since 16 is a perfect square, use the quotient rule to simplify the radical:

$$\sqrt{\frac{3}{16x^2}} = \frac{\sqrt{3}}{\sqrt{16x^2}} = \frac{\sqrt{3}}{4x}$$

■

## Simplified Form and Rationalizing the Denominator

We have been simplifying radical expressions by just making them look simpler. However, a radical expression is in *simplified form* only if it satisfies the following three specific conditions.

**Definition: Simplified Form for Radicals of Index $n$**

A radical of index $n$ in **simplified form** has

1. ***no*** perfect $n$th powers as factors of the radicand,
2. ***no*** fractions inside the radical, and
3. ***no*** radicals in a denominator.

The process of removing radicals from a denominator is called **rationalizing the denominator.**

### EXAMPLE 9 Simplified form of a radical expression

Write each radical expression in simplified form. Assume that all variables represent positive real numbers.

**a.** $\sqrt{20x^8y^9}$ **b.** $\dfrac{9}{\sqrt{3}}$ **c.** $\sqrt[3]{\dfrac{3}{5a^4}}$

**Solution**

**a.** Use the product rule to factor the radical, putting all perfect squares in the first factor:

$$\begin{aligned}\sqrt{20x^8y^9} &= \sqrt{4x^8y^8}\sqrt{5y} && \text{Product rule}\\ &= 2x^4y^4\sqrt{5y} && \text{Simplify the first radical.}\end{aligned}$$

**b.** Since $\sqrt{3}$ appears in the denominator, we multiply the numerator and denominator by $\sqrt{3}$ to rationalize the denominator:

$$\frac{9}{\sqrt{3}} = \frac{9\sqrt{3}}{\sqrt{3}\sqrt{3}} = \frac{9\sqrt{3}}{3} = 3\sqrt{3}$$

**c.** To rationalize this denominator, we must get a perfect cube in the denominator. The radicand $5a^4$ can be made into the perfect cube $125a^6$ by multiplying by $25a^2$.

$$\sqrt[3]{\frac{3}{5a^4}} = \frac{\sqrt[3]{3}}{\sqrt[3]{5a^4}} \quad \text{Quotient rule for radicals}$$

$$= \frac{\sqrt[3]{3}\sqrt[3]{25a^2}}{\sqrt[3]{5a^4}\sqrt[3]{25a^2}} \quad \text{Multiply numerator and denominator by } \sqrt[3]{25a^2}.$$

$$= \frac{\sqrt[3]{75a^2}}{\sqrt[3]{125a^6}} \quad \text{Product rule for radicals}$$

$$= \frac{\sqrt[3]{75a^2}}{5a^2} \quad \text{Since } (5a^2)^3 = 125a^6$$

■

## Operations with Radical Expressions

We can use the properties of radicals to add, subtract, multiply, and divide radical expressions with the same index.

### EXAMPLE 10 Operations with radicals of the same index

Perform each operation and write the answer in simplified form for radicals. Assume that each variable represents a positive real number.

**a.** $\sqrt{20} + \sqrt{5}$ **b.** $\sqrt[4]{4y^3}\sqrt[4]{12y^2}$ **c.** $\sqrt{40} \div \sqrt{5}$

**Solution**

**a.**
$$\sqrt{20} + \sqrt{5} = \sqrt{4}\sqrt{5} + \sqrt{5} \quad \text{Product rule for radicals}$$
$$= 2\sqrt{5} + \sqrt{5} \quad \text{Simplify.}$$
$$= (2 + 1)\sqrt{5} \quad \text{Distributive property}$$
$$= 3\sqrt{5} \quad \text{Simplify.}$$

**b.**
$$\sqrt[4]{4y^3}\sqrt[4]{12y^2} = \sqrt[4]{48y^5} \quad \text{Product rule for radicals}$$
$$= \sqrt[4]{16y^4}\sqrt[4]{3y} \quad \text{Factor out the perfect fourth powers.}$$
$$= 2y\sqrt[4]{3y} \quad \text{Simplify.}$$

**c.**
$$\sqrt{40} \div \sqrt{5} = \sqrt{\frac{40}{5}} \quad \text{Quotient rule for radicals}$$
$$= \sqrt{8} \quad \text{Divide.}$$
$$= \sqrt{4}\sqrt{2} \quad \text{Product rule for radicals}$$
$$= 2\sqrt{2} \quad \text{Simplify.}$$

■

Radical expressions that can be added or subtracted using the distributive property as in Example 10(a) are called **like radicals.** Note that it is not necessary to write out the distributive step. We can simply write $2\sqrt{5} + \sqrt{5} = 3\sqrt{5}$ just like we write $2x + x = 3x$.

## EXERCISES B.2

*Evaluate each expression.*

1. $3^{-4}$
2. $\dfrac{1}{2^{-3}}$
3. $6^{-1} + 5^{-1}$
4. $2^0 + 2^{-1}$
5. $\dfrac{3^{-2}}{6^{-3}}$
6. $\dfrac{3^{-1}}{2^3}$
7. $\left(\dfrac{1}{2}\right)^{-3}$
8. $\left(-\dfrac{1}{10}\right)^{-4}$

*Simplify each expression. Write answers without negative exponents. Assume that all variables represent nonzero real numbers.*

9. $(-3x^2y^3)(2x^9y^8)$
10. $(-6a^7b^4)(3a^3b^5)$
11. $y^3y^2 + 2y^4y$
12. $x^2x^5 + x^3x^4$
13. $-1(2x^3)^2$
14. $(-3y^{-1})^{-1}$
15. $\left(\dfrac{-2x^2}{3}\right)^3$
16. $\left(\dfrac{-1}{2a}\right)^{-2}$
17. $\dfrac{6x^7}{2x^3}$
18. $\dfrac{-9x^2y}{3xy^2}$
19. $\left(\dfrac{y^2}{5}\right)^{-2}$
20. $\left(-\dfrac{y^2}{2a}\right)^4$
21. $\left(\dfrac{1}{2}x^{-4}y^3\right)\left(\dfrac{1}{3}x^4y^{-6}\right)$
22. $\left(\dfrac{1}{3}a^{-5}b\right)(a^4b^{-1})$
23. $\left(\dfrac{-3m^{-1}n}{-6m^{-1}n^{-1}}\right)^2$
24. $\left(\dfrac{-p^{-1}q^{-1}}{-3pq^{-3}}\right)^{-2}$

*Evaluate each expression.*

25. $-9^{1/2}$
26. $27^{1/3}$
27. $64^{1/2}$
28. $-144^{1/2}$
29. $(-64)^{1/3}$
30. $81^{1/4}$
31. $(-27)^{4/3}$
32. $125^{-2/3}$
33. $8^{-4/3}$
34. $4^{-3/2}$

*Simplify each expression. Assume that all variables represent positive real numbers. Write your answers without negative exponents.*

35. $(x^4y)^{1/2}$
36. $(a^{1/2}b^{1/3})^2$
37. $(2a^{1/2})(3a)$
38. $(-3y^{1/3})(-2y^{1/2})$
39. $\dfrac{6a^{1/2}}{2a^{1/3}}$
40. $\dfrac{-4y}{2y^{2/3}}$
41. $(a^2b^{1/2})(a^{1/3}b^{1/2})$
42. $(4^{3/4}a^2b^3)(4^{3/4}a^{-2}b^{-5})$
43. $\left(\dfrac{x^6y^3}{z^9}\right)^{1/3}$
44. $\left(\dfrac{x^{1/2}y}{y^{1/2}}\right)^3$

*Evaluate each radical expression.*

45. $\sqrt{900}$
46. $\sqrt{400}$
47. $\sqrt[3]{-8}$
48. $\sqrt[3]{64}$
49. $\sqrt[3]{-\dfrac{8}{1000}}$
50. $\sqrt[4]{\dfrac{1}{625}}$
51. $\sqrt[4]{16^3}$
52. $\sqrt[3]{8^5}$

*Write each expression involving rational exponents in radical notation, and each expression involving radicals in exponential notation.*

53. $10^{2/3}$
54. $-2^{3/4}$
55. $3y^{-3/5}$
56. $a(b^4 + 1)^{-1/2}$
57. $\dfrac{1}{\sqrt{x}}$
58. $-4\sqrt{x^3}$
59. $\sqrt[5]{x^3}$
60. $\sqrt[3]{x^3 + y^3}$

*Simplify each radical expression. Assume that all variables represent positive real numbers.*

61. $\sqrt{16x^2}$
62. $\sqrt{121y^4}$
63. $\sqrt[3]{8y^9}$
64. $\sqrt[3]{125x^{18}}$
65. $\sqrt{\dfrac{xy}{100}}$
66. $\sqrt{\dfrac{t}{81}}$
67. $\sqrt[3]{\dfrac{-8a^3}{b^{15}}}$
68. $\sqrt[4]{\dfrac{16t^4}{y^8}}$

*Write each radical expression in simplified form. Assume that all variables represent positive real numbers.*

69. $\sqrt{28}$
70. $\sqrt{50}$
71. $\dfrac{1}{\sqrt{5}}$
72. $\dfrac{7}{\sqrt{7}}$
73. $\sqrt{\dfrac{x}{8}}$
74. $\sqrt{\dfrac{3y}{20}}$
75. $\sqrt[3]{40}$
76. $\sqrt[3]{54}$
77. $\sqrt[3]{-250x^4}$
78. $\sqrt[3]{-24a^5}$
79. $\sqrt[3]{\dfrac{1}{2}}$
80. $\sqrt[3]{\dfrac{3x}{25}}$

*Perform the indicated operations and simplify your answer. Assume that all variables represent positive real numbers.*

81. $3\sqrt{6} + 9 - 5\sqrt{6}$
82. $3\sqrt{2} + 8 - 5\sqrt{2}$
83. $\sqrt{8} + \sqrt{20} - \sqrt{12}$
84. $\sqrt{18} - \sqrt{50} + \sqrt{12} - \sqrt{75}$
85. $(-2\sqrt{3})(5\sqrt{6})$
86. $(-3\sqrt{2})(-2\sqrt{3})$
87. $(3\sqrt{5a})(4\sqrt{5a})$
88. $(-2\sqrt{6})(3\sqrt{6})$
89. $\sqrt{18a} \div \sqrt{2a^4}$
90. $\sqrt{21x^7} \div \sqrt{3x^2}$
91. $\sqrt{20x^3} + \sqrt{45x^3}$
92. $\sqrt[3]{16a^4} + \sqrt[3]{54a^4}$

# B.3 Polynomials

In this section we will review some basic facts about polynomials.

## Definitions

A term was defined in Section B.1 as the product of a number and one or more variables raised to powers. A **polynomial** is simply a single term or a finite sum of terms in which the powers of the variables are whole numbers. A polynomial in one variable is defined as follows.

**Definition: Polynomial in One Variable $x$**

If $n$ is a nonnegative integer and $a_0, a_1, a_2, \ldots, a_n$ are real numbers, then

$$a_nx^n + a_{n-1}x^{n-1} + a_{n-2}x^{n-2} + \cdots + a_1x + a_0$$

is a **polynomial** in one variable $x$.

In algebra a single number is often referred to as a **constant.** The last term $a_0$ is called the **constant term.**

Polynomials with one, two, and three terms are called **monomials, binomials,** and **trinomials,** respectively. We usually write the terms of a polynomial in a single variable so that the exponents are in descending order from left to right. When a polynomial is written in this manner, the coefficient of the first term is the **leading coefficient.** The **degree** of a polynomial in one variable is the highest power of the variable in the polynomial. A constant such as 5 is a monomial with zero degree because $5 = 5x^0$. First-, second-, and third-degree polynomials are called **linear, quadratic,** and **cubic polynomials,** respectively.

### EXAMPLE 1 Using the definitions

Find the degree and leading coefficient of each polynomial and determine whether the polynomial is a monomial, binomial, or trinomial.

**a.** $\dfrac{x^3}{2} - \dfrac{1}{8}$ **b.** $5y^2 + y - 9$ **c.** $3w$

**Solution**

Polynomial (a) is a third-degree binomial, (b) is a second-degree trinomial, and (c) is a first-degree monomial. The leading coefficients are 1/2, 5, and 3, respectively. We can also describe (a) as a cubic polynomial, (b) as a quadratic polynomial, and (c) as a linear polynomial. ■

## Naming and Evaluating Polynomials

Polynomials are often used to model quantities such as profit, revenue, and cost. A profit polynomial might be named $P$. For example, if the expression $3x - 10$ is a profit polynomial, we write $P = 3x - 10$ or $P(x) = 3x - 10$. $P$ and $P(x)$ (read "$P$ of $x$") both represent the profit when $x$ units are sold. If $x = 6$, then the value of the polynomial is $3 \cdot 6 - 10$ or 8. If $x = 7$ the value is 11. Using the $P(x)$ notation we write $P(6) = 8$ and $P(7) = 11$. With the $P(x)$ notation it is clear that the profit for six units is 8 and the profit for seven units is 11. The $P(x)$ notation is called **function notation.** Functions are discussed in great detail in Chapter 2.

### EXAMPLE 2 Evaluating a polynomial

Let $P(x) = x^2 - 5$ and $C(x) = -x^3 + 5x - 3$. Find the following.

**a.** $P(3)$ **b.** $C(10)$

#### Solution

**a.** $P(3) = 3^2 - 5 = 4$
**b.** $C(10) = -10^3 + 5(10) - 3 = -1000 + 50 - 3 = -953$ ■

## Addition and Subtraction of Polynomials

We add or subtract polynomials by adding or subtracting the like terms. You can arrange the work horizontally, as in Example 3, or vertically, as in Example 4.

### EXAMPLE 3 Adding and subtracting polynomials horizontally

Find each sum or difference.

**a.** $(3x^3 - x + 5) + (-8x^3 + 3x - 9)$ **b.** $(x^2 - 5x) - (3x^2 - 4x - 1)$

#### Solution

**a.** We use the commutative and associative properties of addition to rearrange the terms.

$$(3x^3 - x + 5) + (-8x^3 + 3x - 9) = (3x^3 - 8x^3) + (-x + 3x) + (5 - 9)$$
$$= -5x^3 + 2x - 4 \quad \text{Combine like terms.}$$

**b.** The first step is to distribute the multiplication by $-1$ over the three terms of the second polynomial, changing the sign of every term.

$$(x^2 - 5x) - (3x^2 - 4x - 1) = x^2 - 5x - 3x^2 + 4x + 1 \quad \text{Distributive property}$$
$$= -2x^2 - x + 1 \quad \text{Combine like terms.}$$

■

### EXAMPLE 4 Adding and subtracting polynomials vertically

Find each sum or difference.

**a.** $(3x^3 - x + 5) + (-8x^3 + 3x - 9)$ **b.** $(x^2 - 5x) - (3x^2 - 4x - 1)$

#### Solution

**a.** Add:
$$\begin{array}{r} 3x^3 - \phantom{3}x + 5 \\ -8x^3 + 3x - 9 \\ \hline -5x^3 + 2x - 4 \end{array}$$

**b.** Subtract:
$$\begin{array}{r} x^2 - 5x \phantom{{}-1} \\ 3x^2 - 4x - 1 \\ \hline -2x^2 - \phantom{4}x + 1 \end{array}$$
■

## Multiplication of Polynomials

To multiply two polynomials, we use the distributive property.

### EXAMPLE 5 Multiplying polynomials

Use the distributive property to find each product.

**a.** $-3x(2x - 3)$ **b.** $(x^2 - 3x + 4)(2x - 3)$ **c.** $(x + 5)(2x - 3)$

### Solution

**a.** $-3x(2x - 3) = -6x^2 + 9x$ Distributive property

**b.** $(x^2 - 3x + 4)(2x - 3) = x^2(2x - 3) - 3x(2x - 3) + 4(2x - 3)$ Distributive property

$= 2x^3 - 3x^2 - 6x^2 + 9x + 8x - 12$ Distributive property

$= 2x^3 - 9x^2 + 17x - 12$ Combine like terms.

**c.** $(x + 5)(2x - 3) = x(2x - 3) + 5(2x - 3)$ Distributive property

$= 2x^2 - 3x + 10x - 15$ Distributive property

$= 2x^2 + 7x - 15$ Combine like terms. ■

## Using FOIL

The product of two binomials (as in Example 5c) results in four terms:

the product of the First term of each,
the product of the Outer terms,
the product of the Inner terms, and
the product of the Last term of each.

We use FOIL as a memory aid for these four products. The FOIL method allows us to quickly obtain the product of two binomials.

### EXAMPLE 6 Multiplying binomials using FOIL

Find each product using FOIL.

**a.** $(x + 4)(2x - 3)$

**b.** $(a^2 - 3)(2a - 3)$

### Solution

**a.** $(x + 4)(2x - 3) = \overbrace{2x^2}^{F} - \overbrace{3x}^{O} + \overbrace{8x}^{I} - \overbrace{12}^{L} = 2x^2 + 5x - 12$

**b.** $(a^2 - 3)(2a - 3) = 2a^3 - 3a^2 - 6a + 9$ ■

## Special Products

The products $(a + b)^2$, $(a - b)^2$, and $(a + b)(a - b)$ are called the **special products.** We could use FOIL to find these products, but it is better to memorize the following rules for finding these products quickly.

**The Special Products**

| | |
|---|---|
| $(a + b)^2 = a^2 + 2ab + b^2$ | **The Square of a Sum** |
| $(a - b)^2 = a^2 - 2ab + b^2$ | **The Square of a Difference** |
| $(a + b)(a - b) = a^2 - b^2$ | **The Product of a Sum and a Difference** |

### EXAMPLE 7 Finding special products

Find each product by using the special product rules.

**a.** $(2x + 3)^2$ **b.** $(x^3 - 9)^2$ **c.** $(3x + 5)(3x - 5)$

**Solution**

**a.** To find $(2x + 3)^2$ substitute $2x$ for $a$ and 3 for $b$ in $(a + b)^2 = a^2 + 2ab + b^2$:

$$(2x + 3)^2 = (2x)^2 + 2(2x)(3) + 3^2$$
$$= 4x^2 + 12x + 9$$

**b.** To find $(x^3 - 9)^2$ substitute $x^3$ for $a$ and 9 for $b$ in $(a - b)^2 = a^2 - 2ab + b^2$:

$$(x^3 - 9)^2 = (x^3)^2 - 2(x^3)(9) + 9^2$$
$$= x^6 - 18x^3 + 81$$

**c.** $(3x + 5)(3x - 5) = (3x)^2 - 5^2 = 9x^2 - 25$ ■

The expressions $3 - \sqrt{6}$ and $3 + \sqrt{6}$ are called **conjugates.** Their product is a rational number because of the rule $(a + b)(a - b) = a^2 - b^2$. This fact is used to rationalize a denominator in the next example.

### EXAMPLE 8 Using conjugates to rationalize a denominator

Simplify the expression $\frac{\sqrt{3}}{3 - \sqrt{6}}$.

**Solution**

Multiply the numerator and denominator by $3 + \sqrt{6}$, the conjugate of $3 - \sqrt{6}$:

$$\frac{\sqrt{3}}{3 - \sqrt{6}} = \frac{\sqrt{3}(3 + \sqrt{6})}{(3 - \sqrt{6})(3 + \sqrt{6})} = \frac{3\sqrt{3} + \sqrt{18}}{9 - 6}$$
$$= \frac{3\sqrt{3} + 3\sqrt{2}}{3} = \sqrt{3} + \sqrt{2}$$ ■

## Division of Polynomials

If 20 is divided by 3, the quotient is 6 and the remainder is 2. So we can write $20 = 6 \cdot 3 + 2$. If the **dividend** $P(x)$ and the **divisor** $D(x)$ are polynomials such that $D(x) \neq 0$ and the degree of $P(x)$ is greater than or equal to the degree of $D(x)$, then there exist unique polynomials, the **quotient** $Q(x)$ and the **remainder** $R(x)$, such that

$$P(x) = Q(x)D(x) + R(x),$$

where $R(x) = 0$ or the degree of $R(x)$ is less than the degree of $D(x)$. To find the quotient and remainder polynomials, we use an algorithm that is similar to the well-known algorithm for dividing whole numbers.

### EXAMPLE 9 Using the division algorithm to divide polynomials

Find the quotient and remainder when $x^3 - 8$ is divided by $x - 2$.

**Solution**

To keep the division organized, insert $0x^2$ and $0x$ for the missing $x^2$ and $x$ terms.

$$\begin{array}{rll}
 & x^2 + 2x + 4 & \\
x - 2 \,\big) & x^3 + 0x^2 + 0x - 8 & x^3 \div x = x^2 \\
 & \underline{x^3 - 2x^2} & x^2(x - 2) = x^3 - 2x^2 \\
 & 2x^2 + 0x & 0x^2 - (-2x^2) = 2x^2 \\
 & \underline{2x^2 - 4x} & 2x(x - 2) = 2x^2 - 4x \\
 & 4x - 8 & 4(x - 2) = 4x - 8 \\
 & \underline{4x - 8} & \\
 & 0 &
\end{array}$$

The quotient is $x^2 + 2x + 4$ and the remainder is 0. To check, find the product $(x^2 + 2x + 4)(x - 2)$.

The dividend is equal to the quotient times the divisor plus the remainder:

$$\text{dividend} = (\text{quotient})(\text{divisor}) + \text{remainder}$$

We can also express this relationship as follows:

$$\frac{\text{dividend}}{\text{divisor}} = \text{quotient} + \frac{\text{remainder}}{\text{divisor}}$$

## EXERCISES B.3

*Find the degree and leading coefficient of each polynomial. Determine whether the polynomial is a monomial, binomial, or trinomial.*

**1.** $x^3 - 4x^2 + \sqrt{5}$ **2.** $-x^7 - 6x^4$

**3.** $x - 3x^2$ **4.** $x + 5 + x^2$

**5.** 79 **6.** $-\dfrac{x}{\sqrt{2}}$

*Let* $P(x) = x^2 - 3x + 2$ *and* $M(x) = -x^3 + 5x^2 - x + 2$. *Find the following.*

**7.** $P(-2)$ **8.** $P(-1)$ **9.** $M(-3)$ **10.** $M(50)$

*Find each sum or difference.*

**11.** $(3x^2 - 4x) + (5x^2 + 7x - 1)$

**12.** $(-3x^2 - 4x + 2) + (5x^2 - 8x - 7)$

**13.** $(4x^2 - 3x) - (9x^2 - 4x + 3)$

**14.** $(x^2 + 2x + 4) - (x^2 + 4x + 4)$

**15.** $(4ax^3 - a^2x) - (5a^2x^3 - 3a^2x + 3)$

**16.** $(x^2y^2 - 3xy + 2x) - (6x^2y^2 + 4y - 6x)$

*Perform the indicated operation.*

**17.** Add: $\begin{array}{r} 3x - 4 \\ \underline{-x + 3} \end{array}$

**18.** Add: $\begin{array}{r} -2x^2 - 5 \\ \underline{3x^2 - 6} \end{array}$

**19.** Subtract: $\begin{array}{r} x^2 \quad\quad - 8 \\ \underline{-2x^2 + 3x - 2} \end{array}$

**20.** Subtract: $\begin{array}{r} -2x^2 - 5x + 9 \\ \underline{4x^2 - 7x \quad\quad} \end{array}$

*Use the distributive property to find each product.*

**21.** $-3a^3(6a^2 - 5a + 2)$ **22.** $-2m(m^2 - 3m + 9)$

**23.** $(3b^2 - 5b + 2)(b - 3)$

**24.** $(-w^2 - 5w + 6)(w + 5)$

**25.** $(2x - 1)(4x^2 + 2x + 1)$ **26.** $(3x - 2)(9x^2 + 6x + 4)$

**27.** $(x + 5)(x^2 - 5x + 25)$ **28.** $(a + 3)(a^2 - 3a + 9)$

**29.** $(x - 4)(z + 3)$ **30.** $(a - 3)(b + c)$

**31.** $(a - b)(a^2 + ab + b^2)$ **32.** $(a + b)(a^2 - ab + b^2)$

*Find each product using FOIL.*

**33.** $(a + 9)(a - 2)$ **34.** $(z - 3)(z - 4)$

**35.** $(2y - 3)(y + 9)$ **36.** $(2y - 1)(3y + 4)$

**37.** $(2x - 9)(2x + 9)$ **38.** $(4x - 6y)(4x + 6y)$

**39.** $(2x^2 + 4)(3x^2 + 5)$

**40.** $(3x^3 - 2)(5x^3 + 6)$

**41.** $(2x + 5)^2$

**42.** $(5x - 3)^2$

*Find each product using the special product rules.*

**43.** $(3x + 5)^2$

**44.** $(x^3 - 2)^2$

**45.** $(x^2 - 3)(x^2 + 3)$ **46.** $(2z^3 + 1)(2z^3 - 1)$

**47.** $(\sqrt{2} - 5)(\sqrt{2} + 5)$ **48.** $(6 - \sqrt{3})(6 + \sqrt{3})$

**49.** $(3x^3 - 4)^2$ **50.** $(2x^2y^3 + 1)^2$

**51.** $(2xy - 5)^2$ **52.** $(-3x^4 - 2)^2$

*Simplify each expression by rationalizing the denominator.*

**53.** $\dfrac{\sqrt{10}}{\sqrt{5} - 2}$ **54.** $\dfrac{\sqrt{3}}{\sqrt{2} - \sqrt{3}}$

**55.** $\dfrac{\sqrt{6}}{6 + \sqrt{3}}$ **56.** $\dfrac{\sqrt{2}}{\sqrt{8} + \sqrt{3}}$

*Find the quotient and remainder when the first polynomial is divided by the second.*

**57.** $x^2 + 6x + 9,\ x + 3$

**58.** $x^2 - 3x - 54,\ x - 9$

**59.** $a^3 - 1,\ a - 1$

**60.** $b^6 + 8,\ b^2 + 2$

**61.** $x^2 + 3x + 3,\ x - 2$

**62.** $3x^2 - x + 4,\ x + 2$

**63.** $2x^2 - 5,\ x + 3$

**64.** $-4x^2 + 1,\ x - 1$

**65.** $x^3 - 2x^2 - 2x - 3,\ x - 3$

**66.** $x^3 + 3x^2 - 3x + 4,\ x + 4$

**67.** $6x^2 - 7x + 2,\ 2x + 1$

**68.** $-3x^2 + 4x + 9,\ 2 - x$

**69.** $2x^3 + 3x^2 - 7x - 12,\ x^2 - 4$

**70.** $x^3 + 2x^2 + x - 3,\ x^2 - 1$

**71.** $x^3 + 2x^2 - 7x - 6,\ x^2 - x - 2$

**72.** $2x^3 - 3x^2 - 19x - 8,\ x^2 - 2x - 8$

*Perform the indicated operations mentally. Write down only the answer.*

**73.** $(x - 4)(x + 6)$

**74.** $(z^4 + 5)(z^4 - 4)$

**75.** $(2a^5 - 9)(a^5 + 3)$

**76.** $(3b^2 + 1)(2b^2 + 5)$

**77.** $(y - 3) - (2y + 6)$

**78.** $(a^2 + 3) - (a^2 - 6)$

**79.** $(w + 4)^2$

**80.** $(t - 2)^2$

**81.** $3y^2(y^3 - 3x)$

**82.** $a^4b(a^2b^2 + 1)$

**83.** $(6b^3 - 3b^2) \div (3b^2)$

**84.** $(3w - 8) \div (8 - 3w)$

**85.** $(3w^2 - 2n)^2$

**86.** $(7y^2 - 3x)^2$

# B.4 Factoring Polynomials

In this section we will factor polynomials. **Factoring** "reverses" multiplication.

## Factoring Out the Greatest Common Factor

To factor $6x^2 - 3x$, notice that $3x$ is a monomial that can be divided evenly into each term. We can use the distributive property to write

$$6x^2 - 3x = 3x(2x - 1).$$

We call this process **factoring out** $3x$. Both $3x$ and $2x - 1$ are **factors** of $6x^2 - 3x$. Since 3 is a factor of $6x^2$ and $3x$, 3 is a **common factor** of the terms of the polynomial. The **greatest common factor** (GCF) is a monomial that includes every number or variable that is a factor of all terms of the polynomial. The monomial $3x$ is the greatest common factor of $6x^2 - 3x$. Usually the common factor has a positive coefficient, but at times it is useful to factor out a common factor with a negative coefficient.

### EXAMPLE 1 Factoring out the greatest common factor

Factor out the greatest common factor from each polynomial, first using the GCF with a positive coefficient and then using a negative coefficient.

**a.** $9x^4 - 6x^3 + 12x^2$ **b.** $x^2y + 10xy + 25y$

**Solution**

**a.** $9x^4 - 6x^3 + 12x^2 = 3x^2(3x^2 - 2x + 4)$
$= -3x^2(-3x^2 + 2x - 4)$

**b.** $x^2y + 10xy + 25y = y(x^2 + 10x + 25)$
$= -y(-x^2 - 10x - 25)$ ■

## Factoring by Grouping

Some four-term polynomials can be factored by **grouping** the terms in pairs and factoring out a common factor from each pair of terms.

### EXAMPLE 2 Factoring four-term polynomials by grouping

Factor each polynomial by grouping.

**a.** $x^3 + x^2 + 3x + 3$ **b.** $aw + bc - bw - ac$

### Solution

**a.** Factor the common factor $x^2$ out of the first two terms and the common factor 3 out of the last two terms:

$$x^3 + x^2 + 3x + 3 = x^2(x + 1) + 3(x + 1) \quad \text{Factor out common factors.}$$
$$= (x + 1)(x^2 + 3) \quad \text{Factor out the common factor } (x + 1).$$

**b.** We must first arrange the polynomial so that the first group of two terms has a common factor and the last group of two terms also has a common factor.

$$aw + bc - bw - ac = aw - bw - ac + bc \quad \text{Rearrange.}$$
$$= w(a - b) - c(a - b) \quad \text{Factor out common factors.}$$
$$= (w - c)(a - b) \quad \text{Factor out } a - b.$$

■

## Factoring $ax^2 + bx + c$

Factoring by grouping can be used also to factor a trinomial that is the product of two binomials.

### EXAMPLE 3 Factoring $ax^2 + bx + c$ with $a = 1$

Factor each trinomial.

**a.** $x^2 - 5x - 14$ **b.** $x^2 + 4x - 21$

### Solution

**a.** First find two numbers that have a product of $-14$ and a sum of $-5$. The numbers are $-7$ and 2. To get four terms that we can factor by grouping, we replace $-5x$ with $-7x + 2x$.

$$x^2 - 5x - 14 = x^2 - 7x + 2x - 14 \quad \text{Replace } -5x \text{ with } -7x + 2x.$$
$$= (x - 7)x + (x - 7)2 \quad \text{Factor out common factors.}$$
$$= (x - 7)(x + 2) \quad \text{Factor out } (x - 7).$$

Check by using FOIL. Note that once we have found $-7$ and 2, we can skip the grouping step and simply write the answer $(x - 7)(x + 2)$.

**b.** Two numbers that have a product of $-21$ and a sum of 4 are 7 and $-3$.

$$x^2 + 4x - 21 = (x + 7)(x - 3)$$

Check by using FOIL. ■

To factor $ax^2 + bx + c$ with $a \neq 1$, we can use the ***ac*-method.**

**Procedure: The *ac*-Method for Factoring $ax^2 + bx + c$ with $a \neq 1$**

To factor $ax^2 + bx + c$ with $a \neq 1$:

1. Find two numbers whose sum is $b$ and whose product is $ac$.
2. Replace $b$ with the sum of these two numbers.
3. Factor the resulting four-term polynomial by grouping.

## EXAMPLE 4 Factoring $ax^2 + bx + c$ with $a \neq 1$

Factor each trinomial using the *ac*-method.

**a.** $2x^2 + 5x + 2$ **b.** $6x^2 - x - 12$

### Solution

**a.** Since $ac = 2 \cdot 2 = 4$ and $b = 5$, we need two numbers that have a product of 4 and a sum of 5. The numbers are 4 and 1.

$$2x^2 + 5x + 2 = 2x^2 + 4x + x + 2 \quad \text{Replace } 5x \text{ by } 4x + x.$$
$$= (x + 2)2x + (x + 2)1 \quad \text{Factor by grouping.}$$
$$= (x + 2)(2x + 1) \quad \text{Check using FOIL.}$$

**b.** Since $ac = 6(-12) = -72$ and $b = -1$, we need two numbers that have a product of $-72$ and a sum of $-1$. The numbers are 8 and $-9$.

$$6x^2 - x - 12 = 6x^2 - 9x + 8x - 12 \quad \text{Replace } -x \text{ by } -9x + 8x.$$
$$= (2x - 3)3x + (2x - 3)4 \quad \text{Factor by grouping.}$$
$$= (2x - 3)(3x + 4) \quad \text{Check using FOIL.}$$

## Factoring the Special Products

Trinomials of the form $a^2 + 2ab + b^2$ and $a^2 - 2ab + b^2$ are called **perfect square trinomials.** We can factor the perfect square trinomials and the difference of two squares by using the same rules that we used to obtain these special products.

**Factoring the Special Products**

| | |
|---|---|
| $a^2 - b^2 = (a + b)(a - b)$ | **Difference of Two Squares** |
| $a^2 + 2ab + b^2 = (a + b)^2$ | **Perfect Square Trinomial** |
| $a^2 - 2ab + b^2 = (a - b)^2$ | **Perfect Square Trinomial** |

## EXAMPLE 5 Factoring the special products

Factor each polynomial.

**a.** $4x^2 - 1$ **b.** $x^2 - 6x + 9$ **c.** $9y^2 + 30y + 25$

### Solution

**a.** $4x^2 - 1 = (2x)^2 - 1^2$ Recognize the difference of two squares.

$$= (2x + 1)(2x - 1)$$

**b.** $x^2 - 6x + 9 = x^2 - 2 \cdot 3 \cdot x + 3^2$ Recognize the perfect square trinomial.

$$= (x - 3)^2$$

**c.** $9y^2 + 30y + 25 = (3y)^2 + 2(3y)(5) + 5^2$ Recognize the perfect square trinomial.

$$= (3y + 5)^2$$

## Factoring the Difference and Sum of Two Cubes

The following formulas are used to factor the difference of two cubes and the sum of two cubes. You should verify these formulas using multiplication.

**Factoring the Difference and Sum of Two Cubes**

$a^3 - b^3 = (a - b)(a^2 + ab + b^2)$ **Difference of Two Cubes**

$a^3 + b^3 = (a + b)(a^2 - ab + b^2)$ **Sum of Two Cubes**

### EXAMPLE 6 Factoring differences and sums of two cubes

Factor each polynomial.

**a.** $x^3 - 27$ **b.** $8w^6 + 125z^3$

**Solution**

**a.** Since $x^3 - 27 = x^3 - 3^3$, we use $a = x$ and $b = 3$ in the formula for factoring the difference of two cubes.

$$x^3 - 27 = (x - 3)(x^2 + 3x + 9)$$

**b.** Since $8w^6 + 125z^3 = (2w^2)^3 + (5z)^3$, we use $a = 2w^2$ and $b = 5z$ in the formula for factoring the sum of two cubes.

$$8w^6 + 125z^3 = (2w^2 + 5z)(4w^4 - 10w^2z + 25z^2)$$ ■

## Factoring Completely

Polynomials that cannot be factored using integral coefficients are called **prime** or **irreducible over the integers.** For example, $a^2 + 1$, $b^2 + b + 1$, and $x + 5$ are prime polynomials because they cannot be expressed as a product (in a nontrivial manner). A polynomial is **factored completely** when it is written as a product of prime polynomials. When factoring polynomials, we usually do not factor integers that are common factors. For example, $4x^2(2x - 3)$ is factored completely even though the coefficient 4 could be factored.

### EXAMPLE 7 Factoring completely

Factor each polynomial completely.

**a.** $2w^4 - 32$ **b.** $-6x^7 + 6x$

**Solution**

**a.** $2w^4 - 32 = 2(w^4 - 16)$ Factor out the greatest common factor.

$= 2(w^2 - 4)(w^2 + 4)$ Difference of two squares

$= 2(w - 2)(w + 2)(w^2 + 4)$ Difference of two squares

The polynomial is now factored completely because $w^2 + 4$ is prime.

**b.** $-6x^7 + 6x = -6x(x^6 - 1)$ Greatest common factor

$= -6x(x^3 - 1)(x^3 + 1)$ Difference of two squares

$= -6x(x - 1)(x^2 + x + 1)(x + 1)(x^2 - x + 1)$ Difference of two cubes; sum of two cubes

The polynomial is factored completely because all of the factors are prime. ■

## EXERCISES B.4

*Factor out the greatest common factor from each polynomial, first using a positive coefficient on the GCF and then using a negative coefficient.*

**1.** $6x^3 - 12x^2$

**2.** $12x^2 + 18x^3$

**3.** $-ax^3 + 5ax^2 - 6ax$

**4.** $-sa^3 + sb^3 - sb$

**5.** $m - n$ **6.** $y - x$

*Factor each polynomial by grouping.*

**7.** $x^3 + 2x^2 + 5x + 10$ **8.** $2w^3 - 2w^2 + 3w - 3$

**9.** $y^3 - y^2 - 3y + 3$ **10.** $x^3 + x^2 - 7x - 7$

**11.** $ady - w + d - awy$ **12.** $xy + ab + by + ax$

**13.** $x^2y^2 + ab - ay^2 - bx^2$ **14.** $6yz - 3y - 10z + 5$

*Factor each trinomial.*

**15.** $x^2 + 10x + 16$ **16.** $x^2 + 7x + 12$

**17.** $x^2 - 4x - 12$ **18.** $y^2 + 3y - 18$

**19.** $m^2 - 12m + 20$ **20.** $n^2 - 8n + 7$

**21.** $t^2 + 5t - 84$ **22.** $s^2 - 6s - 27$

**23.** $2x^2 - 7x - 4$ **24.** $3x^2 + 5x - 2$

**25.** $8x^2 - 10x - 3$ **26.** $18x^2 - 15x + 2$

**27.** $6y^2 + 7y - 5$ **28.** $15x^2 - 14x - 8$

**29.** $12b^2 + 17b + 6$ **30.** $8h^2 + 22h + 9$

*Factor each special product.*

**31.** $t^2 - u^2$ **32.** $9t^2 - v^2$

**33.** $t^2 + 2t + 1$ **34.** $m^2 + 10m + 25$

**35.** $4w^2 - 4w + 1$ **36.** $9x^2 - 12xy + 4y^2$

**37.** $9z^2x^2 + 24zx + 16$ **38.** $25t^2 - 20tw^3 + 4w^6$

*Factor each sum or difference of two cubes.*

**39.** $t^3 - u^3$ **40.** $m^3 + n^3$

**41.** $a^3 - 8$ **42.** $b^3 + 1$

**43.** $27y^3 + 8$ **44.** $1 - 8a^6$

**45.** $27x^3y^6 - 8z^9$

**46.** $8t^3h^3 + n^9$

*Factor each polynomial completely.*

**47.** $-3x^3 + 27x$ **48.** $a^4b^2 - 16b^2$

**49.** $16t^4 + 54w^3t$ **50.** $8a^6 - a^3b^3$

**51.** $a^3 + a^2 - 4a - 4$

**52.** $2b^3 + 3b^2 - 18b - 27$

**53.** $x^4 - 2x^3 - 8x + 16$ **54.** $a^4 - a^3 + a - 1$

**55.** $-36x^3 + 18x^2 + 4x$ **56.** $-6a^4 - a^3 + 15a^2$

**57.** $a^7 - a^6 - 64a + 64$

**58.** $a^5 - 4a^4 - 4a + 16$

**59.** $-6x^2 - x + 15$ **60.** $-6x^2 - 9x + 42$

# B.5 Rational Expressions

In this section we will review the basic operations with rational expressions.

## Reducing

A **rational expression** is a ratio of two polynomials in which the denominator is not the zero polynomial. The **domain** of a rational expression is the set of all real numbers that can be used in place of the variable.

## EXAMPLE 1 Domain of a rational expression

Find the domain of each rational expression.

**a.** $\dfrac{2x - 1}{x + 3}$ **b.** $\dfrac{2x + 4}{(x + 2)(x + 3)}$ **c.** $\dfrac{1}{x^2 + 8}$

### Solution

**a.** The domain is the set of all real numbers except those that cause $x + 3$ to have a value of 0. So $-3$ is excluded from the domain, because $x + 3$ has a value of 0 for $x = -3$. We write the domain in set notation as $\{x \mid x \neq -3\}$.

**b.** The domain is the set of all real numbers except $-2$ and $-3$, because replacing $x$ by either of these numbers would cause the denominator to be 0. The domain is written in set notation as $\{x \mid x \neq -2 \text{ and } x \neq -3\}$.

**c.** The value of $x^2 + 8$ is positive for any real number $x$. So the domain is the set of all real numbers, $R$. ■

In arithmetic we learned that each rational number has infinitely many equivalent forms. This fact is due to the **basic principle of rational numbers.**

**Basic Principle of Rational Numbers**

If $a$, $b$, and $c$ are integers with $b \neq 0$ and $c \neq 0$, then

$$\frac{ac}{bc} = \frac{a}{b}.$$

The basic principle holds true when $a$, $b$, and $c$ are real numbers as well as when they are integers, but we use it most often with integers, when we reduce fractions. For example, we reduce 3/6 as follows:

$$\frac{3}{6} = \frac{1 \cdot 3}{2 \cdot 3} = \frac{1}{2}$$

Note that since 3 and 6 have a common factor of 3, we divide both the numerator and denominator by 3 to reduce the fraction. We reduce rational expressions in the same manner. Factor the numerator and denominator completely, then *divide out* the common factors. A rational expression is in *lowest terms* when all common factors have been divided out.

## EXAMPLE 2 Reducing to lowest terms

Reduce each rational expression to lowest terms.

**a.** $\dfrac{2x + 4}{x^2 + 5x + 6}$

**b.** $\dfrac{b - a}{a^3 - b^3}$

**c.** $\dfrac{x^2z^3}{x^5z}$

### Solution

**a.** $\dfrac{2x + 4}{x^2 + 5x + 6} = \dfrac{2(x + 2)}{(x + 2)(x + 3)}$ Factor the numerator and denominator.

$= \dfrac{2}{x + 3}$ Divide out the common factor $x + 2$.

**b.** $\dfrac{b-a}{a^3-b^3} = \dfrac{-1(a-b)}{(a-b)(a^2+ab+b^2)}$ Factor $-1$ out of $b - a$.

$= \dfrac{-1}{a^2+ab+b^2}$

**c.** $\dfrac{x^2z^3}{x^5z} = \dfrac{(x^2z)(z^2)}{(x^2z)(x^3)} = \dfrac{z^2}{x^3}$ (The quotient rule yields the same result.) ■

Be careful when reducing. The only way to reduce rational expressions is to factor and divide out the common *factors*. Identical terms that are not factors cannot be eliminated from a rational expression. For example,

$$\frac{x+3}{3} \neq x$$

for all real numbers, because 3 is not a factor of the numerator.

## Multiplication and Division

We multiply two rational numbers by multiplying their numerators and their denominators. For example, $\frac{2}{3}\cdot\frac{5}{7} = \frac{10}{21}$.

**Definition: Multiplication of Rational Numbers**

If $a/b$ and $c/d$ are rational numbers, then

$$\frac{a}{b}\cdot\frac{c}{d} = \frac{ac}{bd}.$$

We multiply rational expressions in the same manner as rational numbers. Of course, any common factor can be divided out as we do when reducing rational expressions.

### EXAMPLE 3 Multiplying rational expressions

Find each product.

**a.** $\dfrac{2a-2b}{6}\cdot\dfrac{9a}{a^2-b^2}$ **b.** $\dfrac{x-1}{x^2+4x+4}\cdot\dfrac{x+2}{x^2+2x-3}$

**Solution**

**a.** $\dfrac{2a-2b}{6}\cdot\dfrac{9a}{a^2-b^2} = \dfrac{2(a-b)}{2\cdot 3}\cdot\dfrac{3\cdot 3a}{(a-b)(a+b)}$ Factor completely.

$= \dfrac{3a}{a+b}$ Divide out common factors.

**b.** $\dfrac{x-1}{x^2+4x+4}\cdot\dfrac{x+2}{x^2+2x-3} = \dfrac{x-1}{(x+2)^2}\cdot\dfrac{x+2}{(x+3)(x-1)}$ Factor completely.

$= \dfrac{1}{x^2+5x+6}$ Divide out common factors. ■

We divide rational numbers by multiplying by the reciprocal of the divisor, or *invert and multiply*. For example, $6 \div \frac{1}{2} = 6\cdot 2 = 12$.

**Definition: Division of Rational Numbers**

If $a/b$ and $c/d$ are rational numbers with $c \neq 0$, then

$$\frac{a}{b} \div \frac{c}{d} = \frac{a}{b}\cdot\frac{d}{c}.$$

Rational expressions are divided in the same manner as rational numbers.

### EXAMPLE 4 Dividing rational expressions

Perform the indicated operations.

**a.** $\dfrac{9}{2x} \div \dfrac{3}{x}$ **b.** $\dfrac{4 - x^2}{6} \div \dfrac{x - 2}{2}$

**Solution**

**a.** $\dfrac{9}{2x} \div \dfrac{3}{x} = \dfrac{9}{2x} \cdot \dfrac{x}{3}$ Invert and multiply.

$= \dfrac{3 \cdot 3}{2x} \cdot \dfrac{x}{3}$ Factor completely.

$= \dfrac{3}{2}$ Divide out common factors.

**b.** $\dfrac{4 - x^2}{6} \div \dfrac{x - 2}{2} = \dfrac{-1(x - 2)(x + 2)}{2 \cdot 3} \cdot \dfrac{2}{x - 2}$ Invert and multiply.

$= \dfrac{-x - 2}{3}$ Divide out common factors. ■

Note that the division in Example 4(a) is valid only if $x \neq 0$. The division in Example 4(b) is valid only if $x \neq 2$ because $x - 2$ appears in the denominator after the rational expression is inverted.

## Building Up the Denominator

The addition of fractions can be carried out only when their denominators are identical. To get a required denominator, we may **build up** the denominator of a fraction. To build up the denominator, we use the basic principle of rational numbers in the reverse of the way we use it for reducing. We multiply the numerator and denominator of a fraction by the same nonzero number to get an equivalent fraction.

### EXAMPLE 5 Writing equivalent rational expressions

Convert the first rational expression into an equivalent one that has the indicated denominator.

**a.** $\dfrac{3}{2a}, \dfrac{?}{6ab}$ **b.** $\dfrac{x - 1}{x + 2}, \dfrac{?}{x^2 + 6x + 8}$ **c.** $\dfrac{a}{3b - a}, \dfrac{?}{a^2 - 9b^2}$

**Solution**

**a.** Compare the two denominators. Since $6ab = 2a(3b)$, we multiply the numerator and denominator of the first expression by $3b$:

$$\frac{3}{2a} = \frac{3 \cdot 3b}{2a \cdot 3b} = \frac{9b}{6ab}$$

**b.** Factor the second denominator and compare it to the first. Since $x^2 + 6x + 8 = (x + 2)(x + 4)$, we multiply the numerator and denominator by $x + 4$:

$$\frac{x - 1}{x + 2} = \frac{(x - 1)(x + 4)}{(x + 2)(x + 4)} = \frac{x^2 + 3x - 4}{x^2 + 6x + 8}$$

**c.** Factor the second denominator as

$$a^2 - 9b^2 = (a - 3b)(a + 3b) = -1(3b - a)(a + 3b).$$

Since $3b - a$ is a factor of $a^2 - 9b^2$, we multiply the numerator and denominator by $-1(a + 3b)$:

$$\frac{a}{3b - a} = \frac{a(-1)(a + 3b)}{(3b - a)(-1)(a + 3b)} = \frac{-a^2 - 3ab}{a^2 - 9b^2}$$

## Addition and Subtraction

Fractions can be added or subtracted only if their denominators are identical. For example, $\frac{1}{3} + \frac{1}{3} = \frac{2}{3}$ and $\frac{7}{12} - \frac{2}{12} = \frac{5}{12}$.

**Definition: Addition and Subtraction of Rational Numbers**

If $a/b$ and $c/d$ are rational numbers, then

$$\frac{a}{b} + \frac{c}{b} = \frac{a + c}{b} \quad \text{and} \quad \frac{a}{b} - \frac{c}{b} = \frac{a - c}{b}.$$

For fractions with different denominators, we build up one or both denominators to get denominators that are equal to the least common multiple (LCM) of the denominators. The **least common denominator (LCD)** is the smallest number that is a multiple of all of the denominators. Use the following steps to find the LCD.

**Procedure: Finding the LCD**

1. Factor each denominator completely.
2. Write a product using each factor that appears in a denominator.
3. For each factor, use the highest power of that factor that occurs in the denominators.

For example, to find the LCD for 10 and 12, we write $10 = 2 \cdot 5$ and $12 = 2^2 \cdot 3$. The LCD contains the factors 2, 3, and 5. Using the highest power of each we get $2^2 \cdot 3 \cdot 5 = 60$ for the LCD. So, to add fractions with denominators of 10 and 12, we build up each fraction to a denominator of 60:

$$\frac{1}{10} + \frac{1}{12} = \frac{1 \cdot 6}{10 \cdot 6} + \frac{1 \cdot 5}{12 \cdot 5} = \frac{6}{60} + \frac{5}{60} = \frac{11}{60}$$

We use the same method to add or subtract rational expressions.

### EXAMPLE 6 Adding and subtracting rational expressions

Perform the indicated operations.

**a.** $\frac{x}{x - 1} + \frac{2x + 3}{x^2 - 1}$ **b.** $\frac{x}{x^2 + 6x + 9} - \frac{x - 3}{x^2 + 5x + 6}$

**Solution**

**a.**
$$\frac{x}{x - 1} + \frac{2x + 3}{x^2 - 1} = \frac{x}{x - 1} + \frac{2x + 3}{(x - 1)(x + 1)}$$ Factor denominators completely

$$= \frac{x(x + 1)}{(x - 1)(x + 1)} + \frac{2x + 3}{(x - 1)(x + 1)}$$ Build up using the LCD $(x - 1)(x + 1)$.

$$= \frac{x^2 + x + 2x + 3}{(x - 1)(x + 1)}$$ Add the fractions.

$$= \frac{x^2 + 3x + 3}{(x - 1)(x + 1)}$$ Simplify the numerator.

**b.** $$\frac{x}{x^2+6x+9} - \frac{x-3}{x^2+5x+6} = \frac{x}{(x+3)^2} - \frac{x-3}{(x+2)(x+3)}$$
$$= \frac{x(x+2)}{(x+3)^2(x+2)} - \frac{(x-3)(x+3)}{(x+2)(x+3)(x+3)}$$
$$= \frac{x^2+2x}{(x+3)^2(x+2)} - \frac{x^2-9}{(x+3)^2(x+2)}$$
$$= \frac{2x+9}{(x+3)^2(x+2)}$$

## EXERCISES B.5

*Find the domain of each rational expression.*

**1.** $\frac{x-3}{x+2}$

**2.** $\frac{x^2-1}{x-5}$

**3.** $\frac{x^2-9}{(x-4)(x+2)}$

**4.** $\frac{2x-3}{(x+1)(x-3)}$

**5.** $\frac{x+1}{(x-3)(x+3)}$

**6.** $\frac{x+2}{(x+2)(x+1)}$

**7.** $\frac{3x^2-2x+1}{x^2+3}$

**8.** $\frac{-2x^2-7}{3x^2+8}$

*Reduce each rational expression to lowest terms.*

**9.** $\frac{3x-9}{x^2-x-6}$

**10.** $\frac{-2x-4}{x^2-3x-10}$

**11.** $\frac{10a-8b}{12b-15a}$

**12.** $\frac{a^2-b^2}{b-a}$

**13.** $\frac{a^3b^6}{a^2b^3-a^4b^2}$

**14.** $\frac{18u^6v^5+24u^3v^3}{42u^2v^5}$

**15.** $\frac{x^4y^5z^2}{x^7y^3z}$

**16.** $\frac{t^3u^7}{-t^8u^5}$

**17.** $\frac{a^3-b^3}{a^2-b^2}$

**18.** $\frac{a^3+b^3}{a^2+b^2}$

**19.** $\frac{ab+3a-by-3y}{a^2-2ay+y^2}$

**20.** $\frac{x^4-16}{x^4+8x^2+16}$

*Find the products or quotients.*

**21.** $\frac{2a}{3b^2}\cdot\frac{9b}{14a^2}$

**22.** $\frac{14w}{51y}\cdot\frac{3w}{7y}$

**23.** $\frac{12a}{7}\div\frac{2a^3}{49}$

**24.** $\frac{20x}{y^3}\div\frac{30}{y^5}$

**25.** $\frac{a^2-9}{3a-6}\cdot\frac{a^2-4}{a^2-a-6}$

**26.** $\frac{6x^2+x-1}{6x+3}\cdot\frac{15}{9x^2-1}$

**27.** $\frac{x^2-y^2}{9}\div\frac{x^2+2xy+y^2}{18}$

**28.** $\frac{a^3-b^3}{a^2-2ab+b^2}\div\frac{2a^2+2ab+2b^2}{9a^2-9b^2}$

**29.** $\frac{x^2-y^2}{-3xy}\cdot\frac{6x^2y^3}{2y-2x}$

**30.** $\frac{a^2-a-2}{2}\cdot\frac{1}{4-a^2}$

**31.** $\frac{wx-x}{x^2}\div\frac{1-w^2}{2}$

**32.** $\frac{2+x}{2}\div\frac{x^2-4}{4}$

*Convert the first rational expression into an equivalent one that has the indicated denominator.*

**33.** $\frac{4}{3a}, \frac{?}{12a^2}$

**34.** $\frac{a+2}{4a^2}, \frac{?}{20a^3b}$

**35.** $\frac{x-5}{x+3}, \frac{?}{x^2-9}$

**36.** $\frac{x+2}{x-8}, \frac{?}{16-2x}$

**37.** $\frac{x}{x+5}, \frac{?}{x^2+6x+5}$

**38.** $\frac{3a-b}{a+b}, \frac{?}{9b^2-9a^2}$

**39.** $\frac{t}{2t+2}, \frac{?}{2t^2+4t+2}$

**40.** $\frac{x-1}{2x+4}, \frac{?}{4x^2-16x-48}$

*Find the least common denominator (LCD) for each given pair of rational expressions.*

**41.** $\frac{1}{4ab^2}, \frac{7}{6a^2b^3}$

**42.** $\frac{3}{2x^2y}, \frac{a}{5xy}$

**43.** $\frac{-7a}{3a+3b}, \frac{5b}{2a+2b}$

**44.** $\frac{1}{3a-3b}, \frac{2}{a^2-b^2}$

**45.** $\frac{2x}{x^2+5x+6}, \frac{3x}{x^2-x-6}$

**46.** $\frac{x+7}{2x^2+7x-15}, \frac{x-5}{2x^2-5x+3}$

*Perform the indicated operations.*

47. $\dfrac{3}{2x} + \dfrac{1}{6}$

48. $\dfrac{-7}{3a^2b} + \dfrac{4}{6ab^2}$

49. $\dfrac{x+3}{x-1} - \dfrac{x+4}{x+1}$

50. $\dfrac{x+2}{x-3} - \dfrac{x^2+3x-2}{x^2-9}$

51. $3 + \dfrac{1}{a}$

52. $-1 - \dfrac{3}{c}$

53. $t - 1 - \dfrac{1}{t+1}$

54. $w + \dfrac{1}{w-1}$

55. $\dfrac{x}{x^2+3x+2} + \dfrac{x-1}{x^2+5x+6}$

56. $\dfrac{x-1}{x^2+x-6} - \dfrac{x-2}{x^2+4x+3}$

57. $\dfrac{1}{x-3} - \dfrac{5}{6-2x}$

58. $\dfrac{5}{4-x^2} - \dfrac{2x}{x-2}$

59. $\dfrac{y^2}{x^3-y^3} + \dfrac{x+y}{x^2+xy+y^2}$

60. $\dfrac{ab}{a^3+b^3} + \dfrac{a}{2a^2-2ab+2b^2}$

61. $\dfrac{x-2}{2x^2+7x-15} - \dfrac{x+1}{2x^2-5x+3}$

62. $\dfrac{x-1}{2x^2-5x-3} - \dfrac{x}{4x^2-1}$

# Answers to Exercises

## Chapter 1

### Section 1.1

***For Thought:*** **1.** T **2.** T **3.** F **4.** T **5.** F **6.** T **7.** F **8.** F **9.** F **10.** F

***Exercises:***

**1.** equation **3.** equivalent **5.** identity **7.** conditional equation **9.** No **11.** Yes **13.** $\{5/3\}$ **15.** $\{-2\}$ **17.** $\{1/2\}$ **19.** $\{11\}$ **21.** $\{-24\}$ **23.** $\{-6\}$ **25.** $\{-2/5\}$ **27.** $R$, identity **29.** $\{0\}$, conditional **31.** $\{9\}$, conditional **33.** $\varnothing$, inconsistent **35.** $\{x \mid x \neq 0\}$, identity **37.** $\{w \mid w \neq 1\}$, identity **39.** $\{x \mid x \neq 0\}$, identity **41.** $\{9/8\}$, conditional **43.** $\{x \mid x \neq 3 \text{ and } x \neq -3\}$, identity **45.** $\varnothing$, inconsistent **47.** $\{-2\}$, conditional **49.** $\{-19.952\}$ **51.** $\{36.28\}$ **53.** $\{-2.562\}$ **55.** $\{1/3\}$ **57.** $\{0.199\}$ **59.** $\{0.425\}$ **61.** $\{-0.380\}$ **63.** $\{-8, 8\}$ **65.** $\{-4, 12\}$ **67.** $\{6\}$ **69.** $\varnothing$ **71.** $\{-2, 5\}$ **73.** $\{-23, 41\}$ **75.** $\{-10, 0\}$ **77.** $\{2/3\}$ **79.** $\varnothing$ **81.** $\{200\}$ **83.** $\{4\}$ **85.** $\{0\}$ **87.** $\{-1/2\}$ **89.** $\{-10\}$ **91.** $\{5\}$ **93.** $\{-8, -4\}$ **95.** $\{3/2\}$ **97.** $\varnothing$ **99.** $\{x \mid x \neq 2 \text{ and } x \neq -2\}$, identity **101.** $\{x \mid x \neq -3 \text{ and } x \neq 2\}$, identity **103.** $\varnothing$, inconsistent **105. a.** About 1995 **b.** Increasing **c.** 2015 **107.** \$18,260.87 **109.** 250,000 **111.** $(\sqrt{3} - 1)/2$

### Section 1.2

***For Thought:*** **1.** F **2.** F **3.** F **4.** T **5.** T **6.** F **7.** F **8.** T **9.** F **10.** F

***Exercises:***

**1.** formula **3.** uniform **5.** $r = \dfrac{I}{Pt}$ **7.** $C = \dfrac{5}{9}(F - 32)$ **9.** $b = \dfrac{2A}{h}$ **11.** $y = \dfrac{C - Ax}{B}$ **13.** $R_1 = \dfrac{RR_2R_3}{R_2R_3 - RR_3 - RR_2}$ **15.** $n = \dfrac{a_n - a_1 + d}{d}$ **17.** $a_1 = \dfrac{S(1 - r)}{1 - r^n}$ **19.** $D = \dfrac{5.688 - L + F\sqrt{S}}{2}$ **21.** $R = D/T$ **23.** $W = A/L$ **25.** $r = d/2$ **27.** 5.4% **29.** 2.5 hr **31.** $-5°$C **33.** \$37,250 **35.** 33.9 in. **37.** \$46,000 **39.** \$60,000 **41.** 16 ft, 7 ft, 7 ft **43.** 6400 ft$^2$ **45.** 112.5 mph **47.** 48 mph **49.** \$106,000 **51.** Northside 600, Southside 900 **53.** 28.8 hr **55.** 2 P.M. **57.** 64.85 acres **59.** 225 ft **61.** 4.15 ft **63.** 1.998 hectares **65.** \$102,039 **67.** 8/3 liters **69.** 1250 gallons **71.** 4 lb apples, 16 lb apricots **73.** 3 dimes, 5 nickels **75.** 120 milliliters **77.** 2.5 gal **79.** 3 hr 5 min **81.** 8 gal of 15%, 12 gal of 10% **83. a.** About 1992 **b.** 1992 **85.** 1 hr, 3 min, 6 sec, no **87.** $\{11/2\}$ **89.** $\{10{,}000\}$ **91.** $\varnothing$

### Section 1.3

***For Thought:*** **1.** F **2.** F **3.** F **4.** F **5.** T **6.** F **7.** T **8.** T **9.** T **10.** F

***Exercises:***

**1.** ordered **3.** Cartesian **5.** circle **7.** linear equation **9.** (4, 1), I **11.** (1, 0), $x$-axis **13.** (5, −1), IV **15.** (−4, −2), III **17.** (−2, 4), II **19.** 5, (2.5, 5) **21.** $2\sqrt{2}$, (0, −1) **23.** 25, (17/2, 1) **25.** $6, \left(\dfrac{-2 + 3\sqrt{3}}{2}, \dfrac{5}{2}\right)$ **27.** $\sqrt{74}$, (−1.3, 1.3) **29.** $|a - b|, \left(\dfrac{a + b}{2}, 0\right)$ **31.** $\dfrac{\sqrt{\pi^2 + 4}}{2}, \left(\dfrac{3\pi}{4}, \dfrac{1}{2}\right)$

**33.** (0, 0), 4

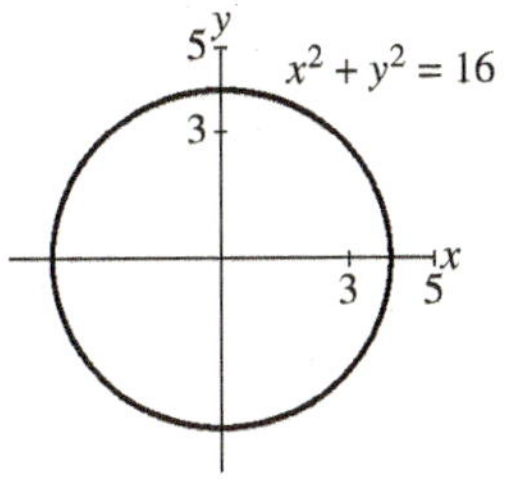


**35.** (−6, 0), 6

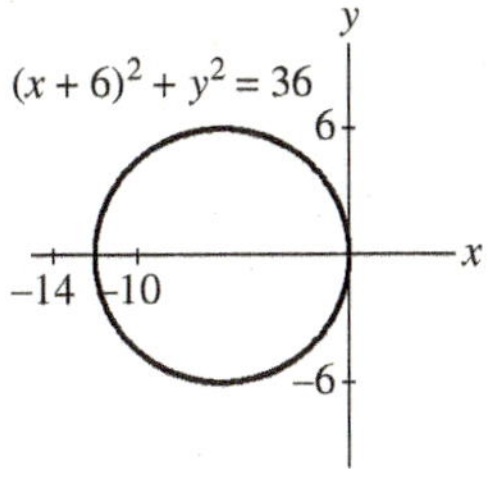


**37.** (−1, 0), 5

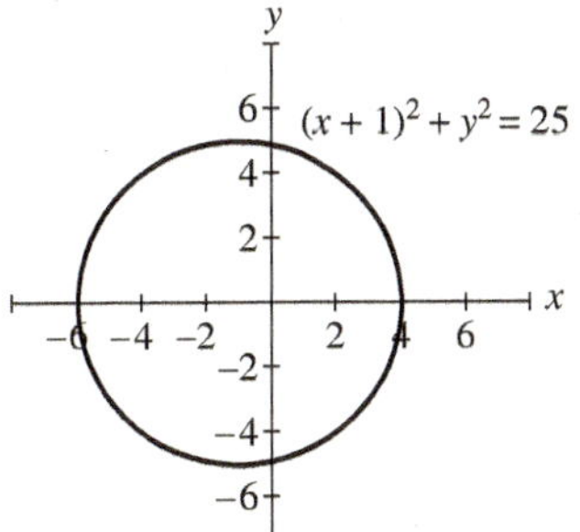


**39.** (2, −2), $2\sqrt{2}$

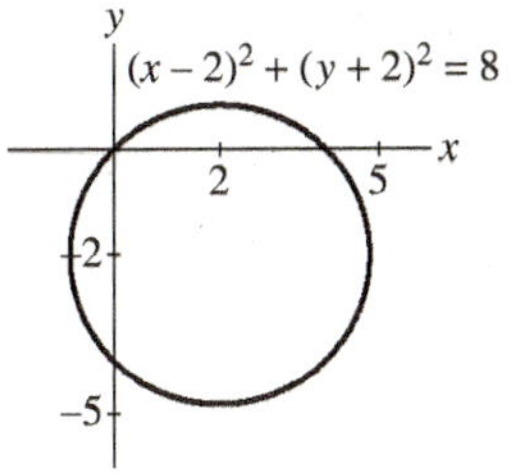


**41.** $x^2 + y^2 = 49$ **43.** $(x + 2)^2 + (y - 5)^2 = 1/4$ **45.** $(x - 3)^2 + (y - 5)^2 = 34$ **47.** $(x - 5)^2 + (y + 1)^2 = 32$ **49.** (0, 0), 3 **51.** (0, −3), 3

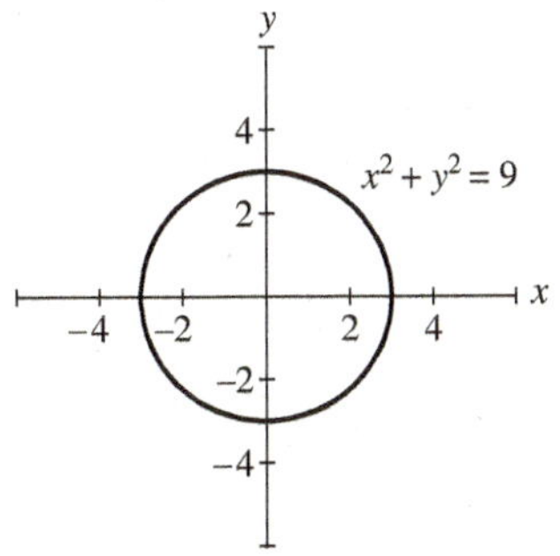


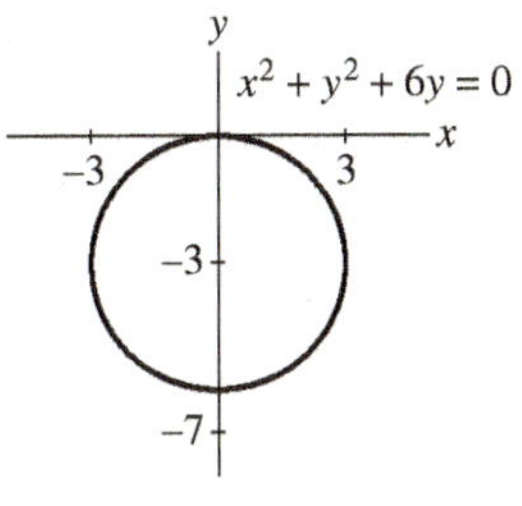


**53.** (−3, −4), 5 **55.** (3/2, −1), 2

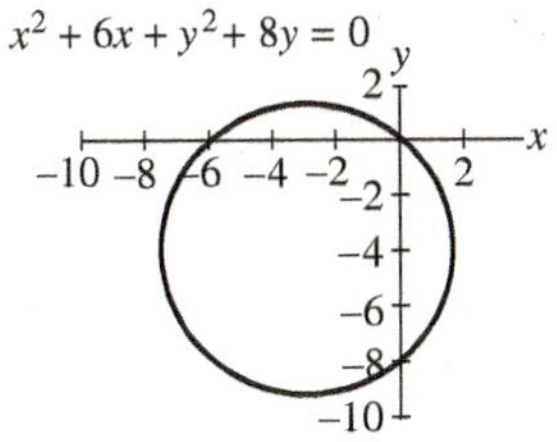


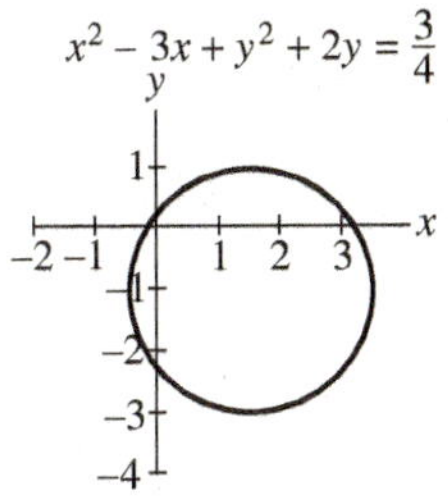


**57.** (3, 4), 5 **59.** (2, 3/2), 5/2

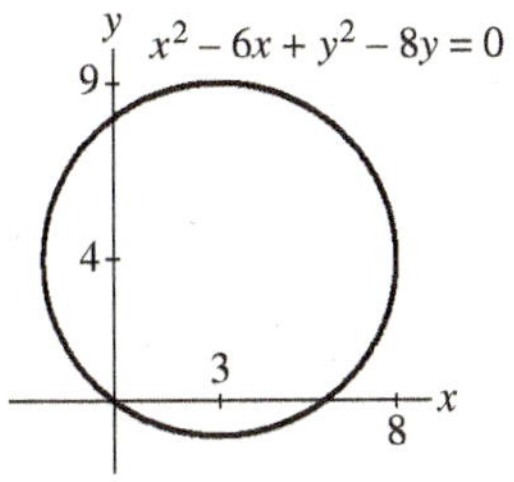


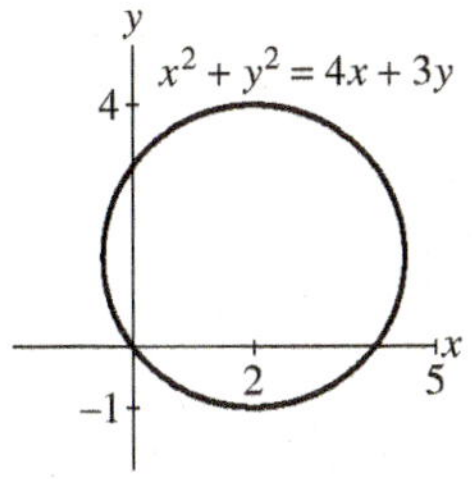

**61.** (1/4, −1/6), 1/6

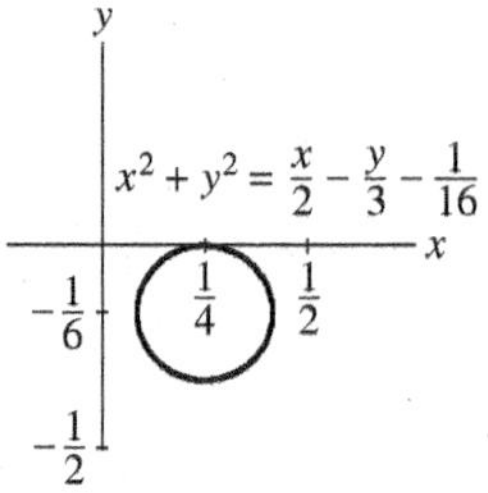


**83.**

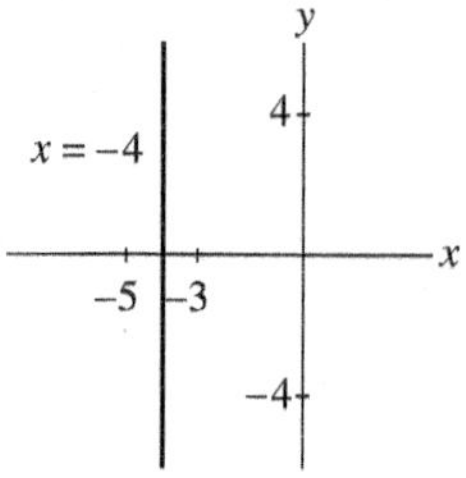


**85.**

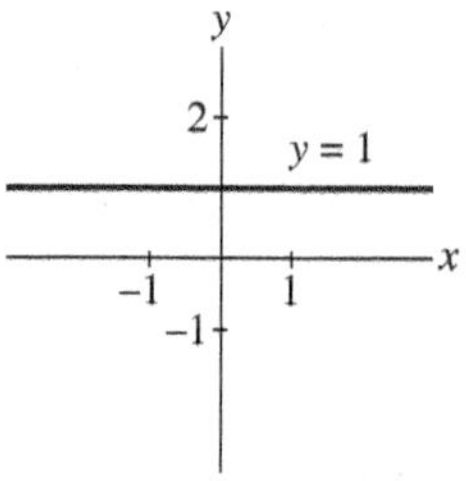


**63. a.** $x^2 + y^2 = 49$ **b.** $(x - 1)^2 + y^2 = 20$
**c.** $(x - 1)^2 + (y - 2)^2 = 13$
**65. a.** $(x - 2)^2 + (y + 3)^2 = 4$ **b.** $(x + 2)^2 + (y - 1)^2 = 1$
**c.** $(x - 3)^2 + (y + 1)^2 = 9$ **d.** $x^2 + y^2 = 1$
**67.** (0, −4), (4/3, 0) **69.** (0, −6), (2, 0)

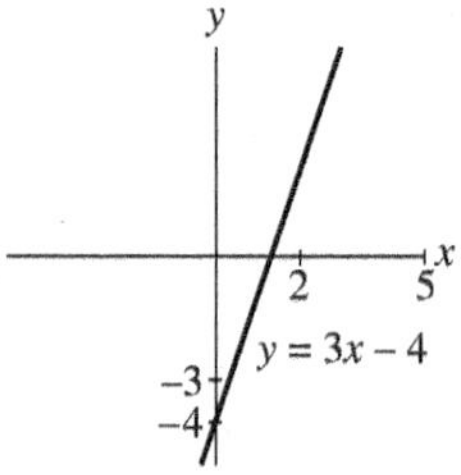


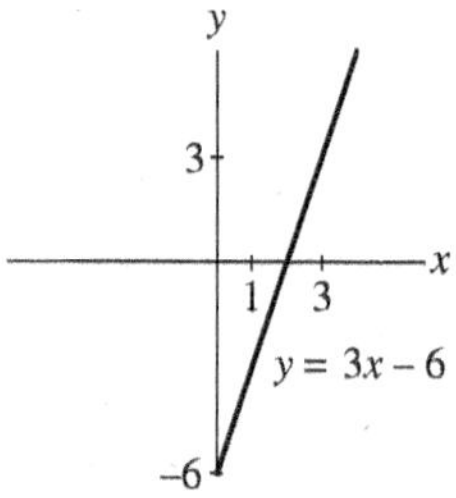


**71.** (0, 30), (−90, 0) **73.** (0, 600), (−800, 0)

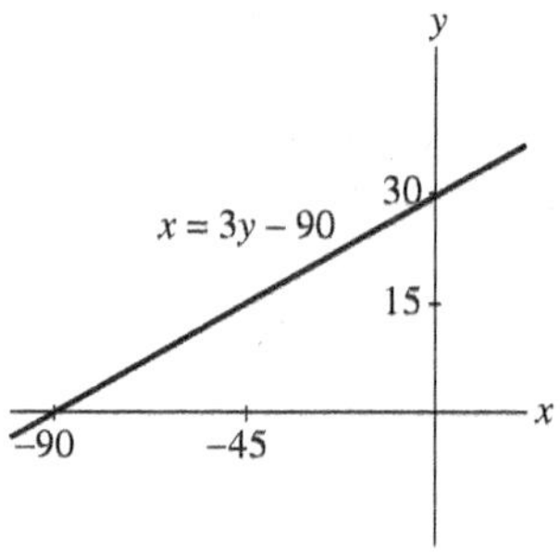


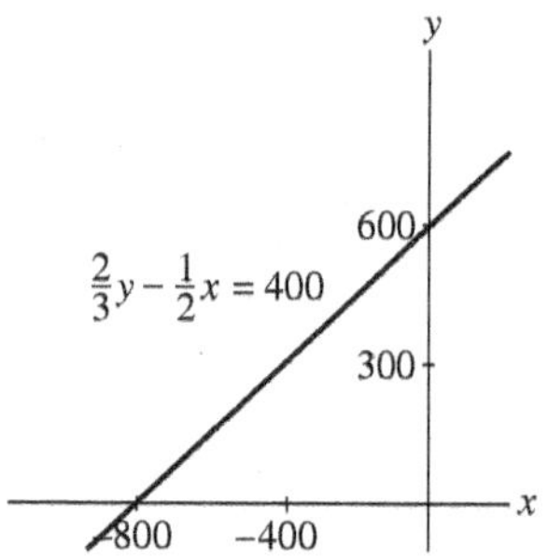


**75.** (0, 0.0025), (0.005, 0) **77.** (0, 2500), (5000, 0)

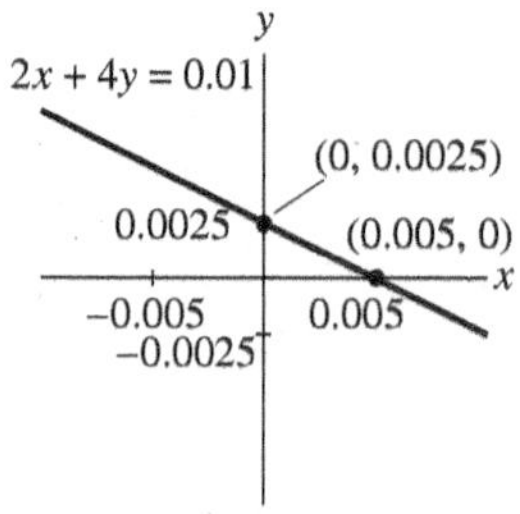


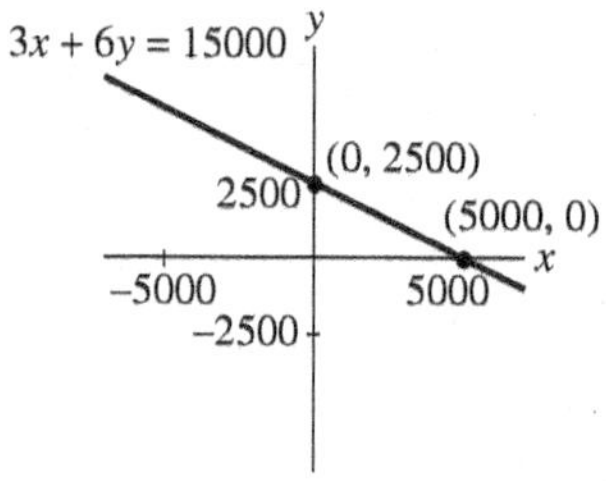


**79.**

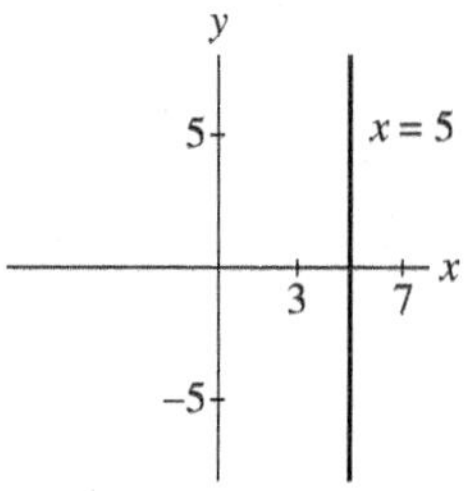


**81.**

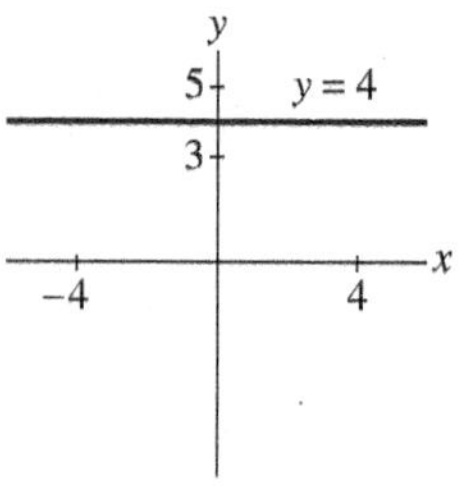


**87.** {3.6} **89.** {14} **91.** {−2.83} **93.** {558.54}
**95.** {116,566.67} **97.** {4.91}
**99. a.** (15, 22.95) The median age at first marriage in 1985 was 22.95.
**b.** 30.3; because of the units, distance is meaningless.
**101.** $C = 1.8$

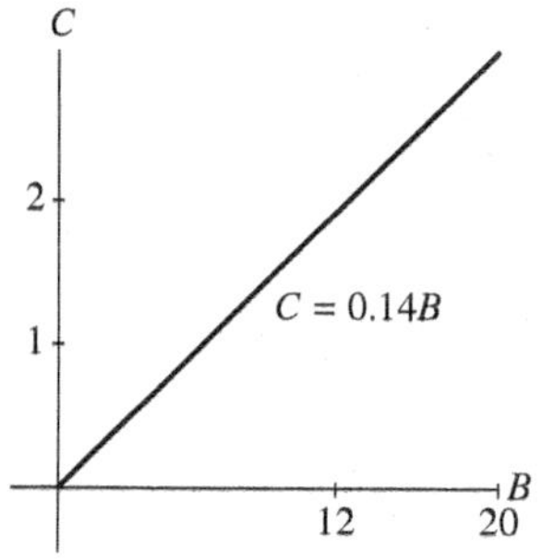


**109. a.** Conditional equation **b.** Identity **c.** Inconsistent equation
**111.** ∅ **113.** $x = \frac{d - b}{a - c}$

## Section 1.4

***For Thought:*** **1.** F **2.** F **3.** F **4.** F **5.** T **6.** F
**7.** F **8.** T **9.** F **10.** T

***Exercises:***
**1.** rise **3.** slope **5.** slope-intercept **7.** perpendicular
**9.** $\frac{1}{3}$ **11.** −4 **13.** 0 **15.** 2 **17.** No slope
**19.** $y = \frac{5}{4}x + \frac{1}{4}$ **21.** $y = -\frac{7}{6}x + \frac{11}{3}$ **23.** $y = 5$
**25.** $x = 4$ **27.** $y = \frac{2}{3}x - 1$ **29.** $y = \frac{5}{2}x + \frac{3}{2}$
**31.** $y = -2x + 4$ **33.** $y = \frac{3}{2}x + \frac{5}{2}$ **35.** $y = \frac{3}{5}x - 2, \frac{3}{5}, (0, -2)$
**37.** $y = 2x - 5, 2, (0, -5)$ **39.** $y = \frac{1}{2}x + \frac{1}{2}, \frac{1}{2}, \left(0, \frac{1}{2}\right)$
**41.** $y = 4, 0, (0, 4)$ **43.** $y = \frac{1}{4}x + 7$ **45.** $y = -\frac{1}{2}x - \frac{7}{2}$

**47.**

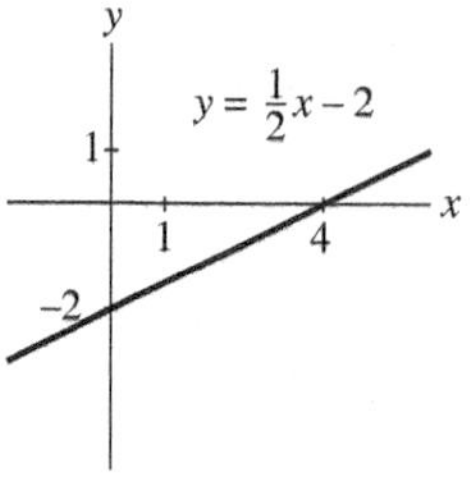


**49.**

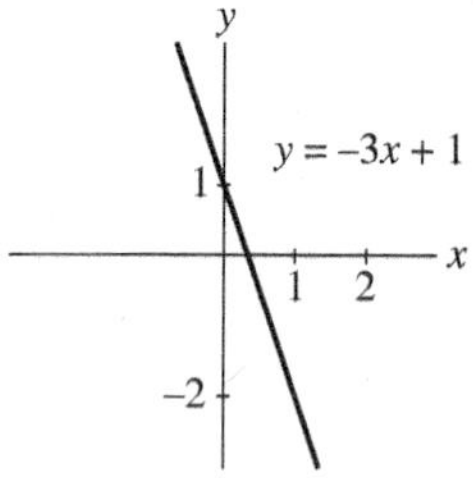

**51.**

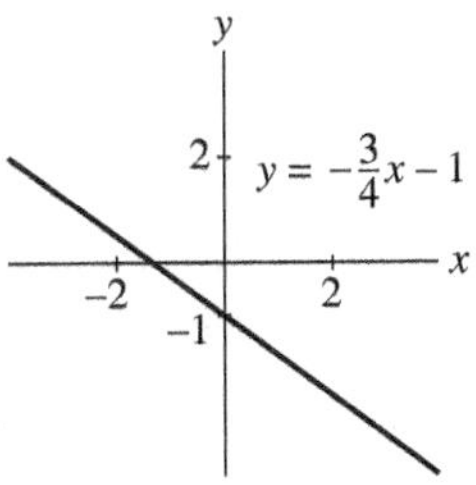


**53.**

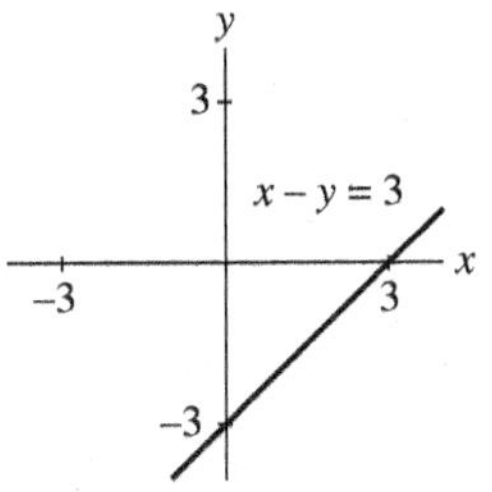


**55.**

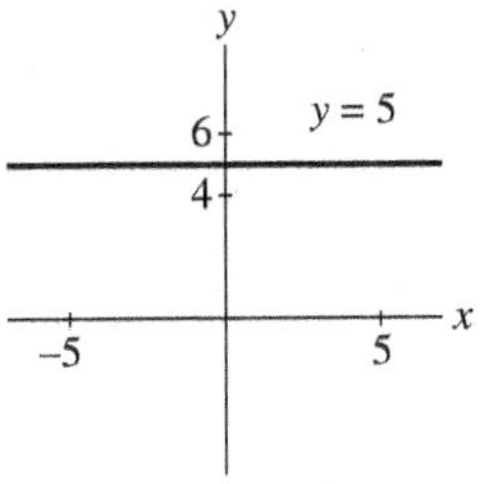


**57.** $4x - 3y = 12$ **59.** $4x - 5y = -7$ **61.** $x = -4$
**63.** $16x - 15y = 22$ **65.** $3x - 50y = -11$ **67.** 0.5 **69.** $-1$
**71.** 0 **73.** $2x - y = 4$ **75.** $3x + y = 7$ **77.** $5x - 7y = 23$
**79.** $3x + 2y = 0$ **81.** $2x + y = -5$ **83.** $y = 5$ **85.** $-5$
**87.** 8 **89.** T **91.** T **93.** F **95.** Yes, no
**97.** $y = x - 2, x^3 - 8 = (x - 2)(x^2 + 2x + 4)$
**99.** $F = \frac{9}{5}C + 32$, 302°F **101.** $c = 50 - n$, \$400
**103.** $S = -0.005D + 95$
**105.** $c = -\frac{3}{4}p + 30, -\frac{3}{4}$; if $p$ increases by 4 then $c$ decreases by 3.
**107.** $(-4, 11), (1, 8), (6, 5), (11, 2), (16, -1)$ **109.** 85/13
**111.** $\sqrt{26}/13$ **115.** $\{17/5, 23/5\}$ **117.** $(x - 2)^2 + (y - 6)^2 = 10$
**119.** $\{4\}$

## Section 1.5

***For Thought:*** **1.** T **2.** T **3.** F **4.** T **5.** T **6.** T
**7.** F **8.** F **9.** F **10.** F

***Exercises:***
**1.** scatter diagram **3.** Linear **5.** No relationship **7.** Nonlinear
**9.** Linear **11.** Linear **13.** Linear **15.** No relationship
**17.** (3.3, 160), (4.0, 193) **19.** (132, 34), (148, 25)
**21. a.** $y = 13,450x + 956,753$ **b.** 1,091,253 thousand tons
**c.** Coal use for electricity increases by 13,450 thousand tons per year.
**23. a.** $y = 44.8x + 1075.3$ **b.** 2021 **c.** \$851 billion
**25.** $p = -0.069A + 0.403$, 12.7% **27.** $\{-1\}$ **29.** $(-4, 5), \sqrt{41}$
**31.** 3/5

## Section 1.6

***For Thought:*** **1.** T **2.** T **3.** F **4.** T **5.** F **6.** F
**7.** T **8.** T **9.** T **10.** F

***Exercises:***
**1.** complex numbers **3.** imaginary number **5.** Imaginary, $0 + 6i$
**7.** Imaginary, $\frac{1}{3} + \frac{1}{3}i$ **9.** Real, $\sqrt{7} + 0i$ **11.** Real, $\frac{\pi}{2} + 0i$
**13.** $7 + 2i$ **15.** $-2 - 3i$ **17.** $4 + i\sqrt{2}$ **19.** $\frac{9}{2} + \frac{5}{6}i$
**21.** $-12 - 18i$ **23.** 26 **25.** $34 - 22i$
**27.** 29 **29.** 4 **31.** $-7 + 24i$ **33.** $1 - 4i\sqrt{5}$ **35.** $i$
**37.** $-1$ **39.** $-i$ **41.** $i$ **43.** $-i$ **45.** $-1$ **47.** 90
**49.** 17/4 **51.** 1 **53.** 12 **55.** $\frac{2}{5} + \frac{1}{5}i$ **57.** $\frac{3}{2} - \frac{3}{2}i$
**59.** $3 + 3i$ **61.** $\frac{1}{13} - \frac{5}{13}i$ **63.** $\frac{1}{34} - \frac{13}{34}i$ **65.** $-i$
**67.** $-4 + 2i$ **69.** $-6$ **71.** $-10$ **73.** $-1 + i\sqrt{5}$
**75.** $-3 + i\sqrt{11}$ **77.** $-4 + 8i$ **79.** $-1 + 2i$ **81.** $\frac{-2 + i\sqrt{2}}{2}$
**83.** $-3 - 2i\sqrt{2}$ **85.** $\frac{3 + \sqrt{21}}{2}$ **87.** 34 **89.** 6
**91.** $-\frac{8}{17} - \frac{15}{17}i$ **93.** $-1 + i$ **95.** $3 - i$
**103.** $\{-21\}$ **105.** $W = \frac{P - 2L}{2}$ **107.** 6.25 gal ethanol

## Section 1.7

***For Thought:*** **1.** F **2.** F **3.** F **4.** F **5.** F **6.** T
**7.** F **8.** F **9.** T **10.** T

***Exercises:***
**1.** quadratic **3.** discriminant **5.** $\{-4, 5\}$ **7.** $\{-2, -1\}$
**9.** $\left\{-\frac{1}{2}, 3\right\}$ **11.** $\left\{\frac{2}{3}, \frac{1}{2}\right\}$ **13.** $\{-7, 6\}$ **15.** $\{\pm\sqrt{5}\}$
**17.** $\left\{\pm i\frac{\sqrt{6}}{3}\right\}$ **19.** $\{0, 6\}$ **21.** $\{1/3\}$ **23.** $\{-2, 3\}$
**25.** $\{-2 \pm 2i\}$ **27.** $\left\{0, \frac{4}{3}\right\}$ **29.** $x^2 - 12x + 36$
**31.** $r^2 + 3r + \frac{9}{4}$ **33.** $w^2 + \frac{1}{2}w + \frac{1}{16}$ **35.** $\{-3 \pm 2\sqrt{2}\}$
**37.** $\{1 \pm \sqrt{2}\}$ **39.** $\left\{\frac{-3 \pm \sqrt{13}}{2}\right\}$ **41.** $\left\{-4, \frac{3}{2}\right\}$
**43.** $\left\{\frac{-1 \pm i\sqrt{2}}{3}\right\}$ **45.** $\{-4, 1\}$ **47.** $\left\{-\frac{1}{2}, 3\right\}$
**49.** $\left\{-\frac{1}{3}\right\}$ **51.** $\left\{\pm\frac{\sqrt{6}}{2}\right\}$ **53.** $\{2 \pm i\}$ **55.** $\{1 \pm i\sqrt{3}\}$
**57.** $\left\{\frac{1}{2} \pm \frac{3}{2}i\right\}$ **59.** $\left\{1 \pm \frac{\sqrt{3}}{2}i\right\}$ **61.** $\{-3.24, 0.87\}$
**63.** $\{-1.99, 3.40\}$ **65.** 0, 1 **67.** $-4, 0$ **69.** 172, 2
**71.** $\left\{-\frac{2}{3}, \frac{1}{2}\right\}$ **73.** $\{-3, 5\}$ **75.** 1 **77.** 0 **79.** 2
**81.** $\left\{-\frac{1}{3}, \frac{5}{3}\right\}$ **83.** $\{\pm\sqrt[4]{2}\}$ **85.** $\left\{-\frac{\sqrt{6}}{6}, \frac{\sqrt{6}}{12}\right\}$
**87.** $\{-12, 6\}$ **89.** $\left\{\frac{1 \pm \sqrt{5}}{2}\right\}$ **91.** $\{2 \pm \sqrt{3}\}$ **93.** $\{-6, 8\}$
**95.** $\varnothing$ **97.** $\{1/2\}$ **99.** $r = \pm\sqrt{\frac{A}{\pi}}$
**101.** $x = -k \pm \sqrt{k^2 - 3}$ **103.** $y = x\left(-1 \pm \frac{\sqrt{6}}{2}\right)$
**105.** 5000 or 35,000 **107.** 2.5 sec **109.** 340 ft
**111.** $2\sqrt{205} \approx 28.6$ yd **113.** 18,503.4 pounds **115.** $6 - \sqrt{10}$ ft
**117.** 10 ft/hr **119.** 4.58 m/sec, 0.93 sec
**121. a.** $y = -0.067x^2 + 1.26x + 51.14$ where $x$ is years since 1980
**b.** 2019
**123.** $\frac{9 + \sqrt{53}}{2} \approx 8.14$ days **125.** 40 lb or 20 lb
**127.** $4x - 5y = -38$ **129.** \$9000 **131.** $-2 - 3i$

## Section 1.8

***For Thought:*** **1.** T **2.** F **3.** F **4.** T **5.** F **6.** F
**7.** F **8.** F **9.** F **10.** T

***Exercises:***
**1.** interval **3.** closed **5.** compound **7.** $x < 12$
**9.** $x \geq -7$ **11.** $[-8, \infty)$ **13.** $(-\infty, \pi/2)$
**15.** $(5, \infty)$ **17.** $[2, \infty)$

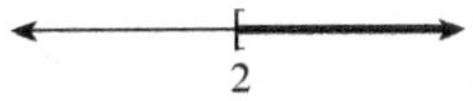

**19.** $(-\infty, 54)$

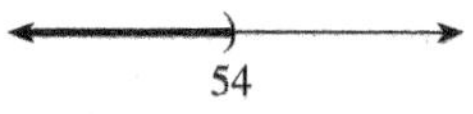


**21.** $(-\infty, 13/3]$

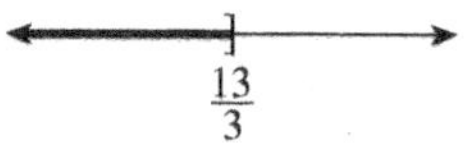


**23.** $(-\infty, 3/2]$

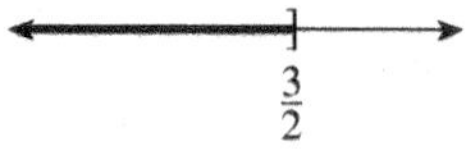


**25.** $(-\infty, 0]$

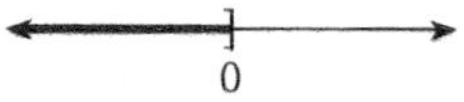


**27.** $(-\infty, -3.5)$ **29.** $(-\infty, 1.4]$ **31.** $(4, \infty)$ **33.** $(-\infty, -2]$
**35.** $(-3, \infty)$ **37.** $(-3, \infty)$ **39.** $(-5, -2)$ **41.** $\varnothing$
**43.** $(-\infty, 5]$
**45.** $(3, 6)$

**47.** $(1/2, \infty)$

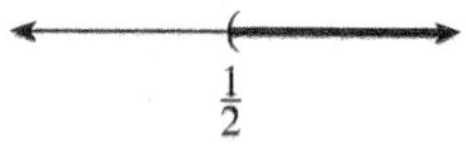


**49.** $(-3, \infty)$

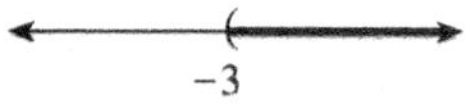


**51.** $(-\infty, \infty)$

**53.** $\varnothing$
**55.** $(2, 4)$

**57.** $(-3, 1]$

**59.** $(-1/3, 1)$

**61.** $[1, 3/2]$

**63.** $(-\infty, 0] \cup [2, \infty)$

**65.** $(-\infty, 2) \cup (8, \infty)$

**67.** $[-1, 9]$

**69.** $\varnothing$ **71.** $\varnothing$

**73.** $(-\infty, 1) \cup (3, \infty)$

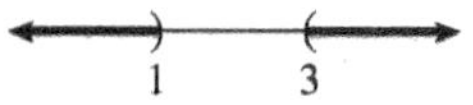


**75.** $(-\infty, 1) \cup (5, \infty)$

**77.** $|x| < 5$ **79.** $|x| > 3$ **81.** $|x - 6| < 2$ **83.** $|x - 4| > 1$
**85.** $|x| \geq 9$ **87.** $|x - 7| \leq 4$ **89.** $|x - 5| > 2$ **91.** $[2, \infty)$
**93.** $(-\infty, 2)$ **95.** $(-\infty, -3] \cup [3, \infty)$ **97.** [\$0, \$7000]
**99.** [93, 100] **101.** (86, 100] **103.** (0 in., 15 in.]
**105.** 96, 79, 68, 56, 47, yes **107. a.** $|x - 130{,}645| > 10{,}000$
**b.** $x > 140{,}645$ or $x < 120{,}645$
**109.** $\frac{|x - 35|}{35} < 0.01$, (34.65°, 35.35°) **111.** [2.26 cm, 2.32 cm]
**113. a.** Colorado, Vermont
**b.** Alabama, Georgia, Maryland, New Jersey, S. Carolina
**115.** $\{-2, 0\}$ **117.** $x + 2y = -5$ **119.** $y = \frac{w + 9}{3 - a}$

## Chapter 1 Review Exercises

**1.** $\{2/3\}$ **3.** $\left\{\frac{32}{15}\right\}$ **5.** $\{-2\}$ **7.** $\left\{-\frac{1}{3}\right\}$

**9.** $\sqrt{146}, (-1/2, -1/2)$ **11.** $\frac{\sqrt{73}}{12}, \left(\frac{3}{8}, \frac{2}{3}\right)$

**13.** $(0, 0), 5$

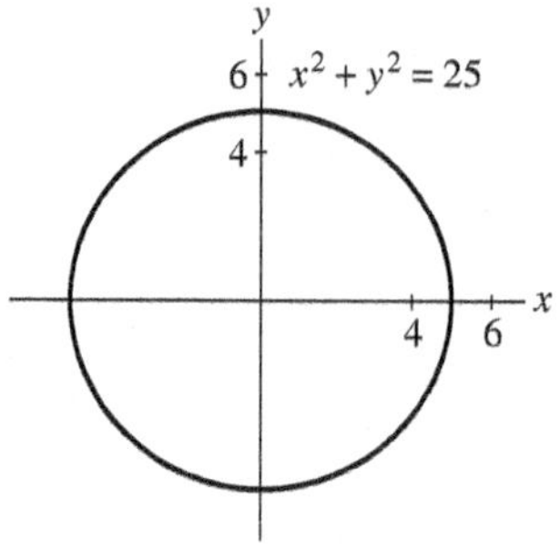


**15.** $(-2, 0), 2$

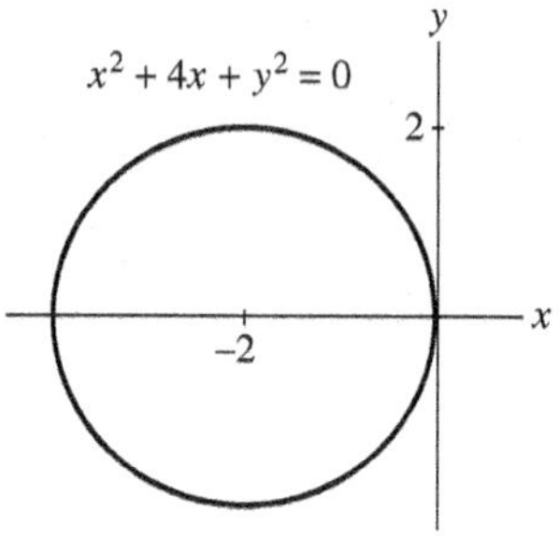


**17.** $(25, 0), (0, 25)$

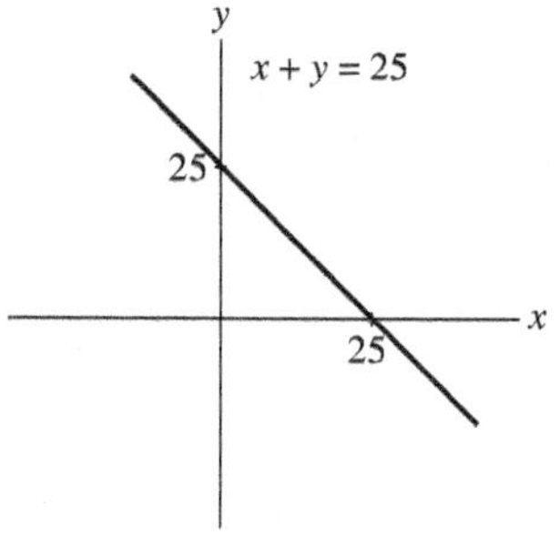


**19.** $(4/3, 0), (0, -4)$

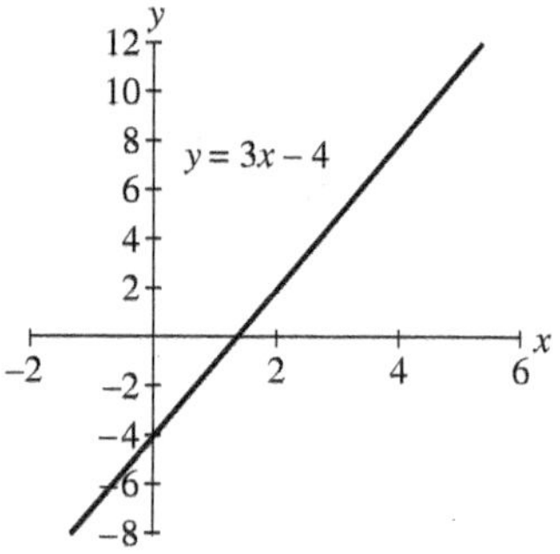


**21.** $(5, 0)$

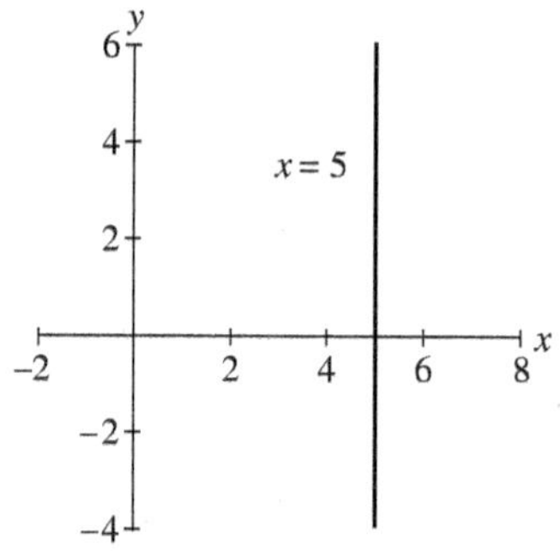


**23.** $(x + 3)^2 + (y - 5)^2 = 3$ **25.** $(4, 0), (0, -3)$
**27.** $-2$ **29.** $y = -\frac{4}{7}x + \frac{13}{7}$ **31.** $x - 3y = 14$
**33.** $y = \frac{2}{3}x - 2$ **35.** $y = \frac{1}{x - 3}$ **37.** $y = -\frac{a}{b}x + \frac{c}{b}$
**39.** 8, two real **41.** 0, one real **43.** $-1 - i$ **45.** $-9 - 40i$
**47.** 20 **49.** $-3 - 2i$ **51.** $\frac{1}{5} - \frac{3}{5}i$ **53.** $-\frac{1}{13} + \frac{5}{13}i$
**55.** $3 + i\sqrt{2}$ **57.** $\frac{3}{4} - \frac{\sqrt{5}}{4}i$ **59.** $-1 - i$ **61.** $\{\pm\sqrt{5}\}$
**63.** $\{\pm 2i\sqrt{2}\}$ **65.** $\left\{\pm i\frac{\sqrt{2}}{2}\right\}$ **67.** $\{2 \pm \sqrt{17}\}$
**69.** $\{-3, 4\}$ **71.** $\{3 \pm i\}$ **73.** $\{2 \pm \sqrt{3}\}$
**75.** $\left\{\frac{1 \pm \sqrt{6}}{2}\right\}$ **77.** $\{1 \pm i\}$ **79.** $\left\{\frac{1}{3}, 2\right\}$
**81.** $\left\{\frac{2}{3}, 2\right\}$ **83.** $\{3/2\}$ **85.** No solutions
**87.** $(3, \infty)$

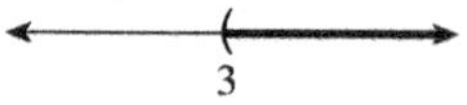


**89.** $(-\infty, 4)$

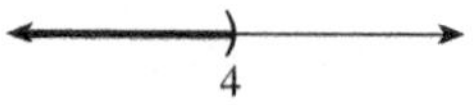

**91.** $(-\infty, -14/3)$

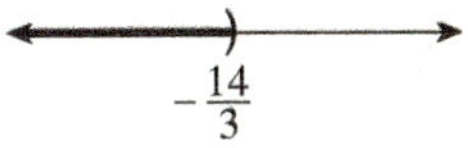


**93.** $(-1, 13]$

**95.** $(1/2, 1)$

**97.** $(-4, \infty)$

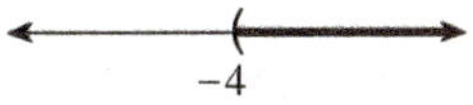


**99.** $(-\infty, 1) \cup (5, \infty)$

**101.** $\{7/2\}$

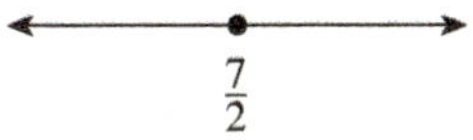


**103.** $(-\infty, \infty)$



**105.** $\{10\}$

**107.** $(-\infty, 8)$ **109.** $\frac{19 - \sqrt{209}}{4} \approx 1.14$ in. **111.** 136.4 mi

**113.** 1600 **115.** 20 mi **117.** (0, \$12.50) **119.** (7 in., 11.5 in.)

**121.** $2.15 \times 10^9$ gal, 31.8 mpg

**123. a.** $y = 17.08x - 34{,}034$ **b.** 331 million

**125.** $p = 0.008a + 0.07$, 59%

**127. a.** $y = 3.67x + 47.11$ **b.** \$102.16

**129.** $(x - 1)^2 + (y - 1)^2 = 1$,
$(x - 1 - 2\sqrt{2})^2 + (y - 2)^2 = 4$,
$(x - 5 + 2\sqrt{2})^2 + (y - 6 + 4\sqrt{2})^2 = (6 - 4\sqrt{2})^2$

### Chapter 1 Test

**1.** $\{-7\}$ **2.** $\{1\}$ **3.** $\left\{\pm\frac{\sqrt{6}}{3}\right\}$ **4.** $\{3 \pm 2\sqrt{2}\}$

**5.** $\{2, 7\}$ **6.** $\{0\}$ **7.** $\{1 \pm 2i\}$ **8.** $\{\pm i\}$

**9.**

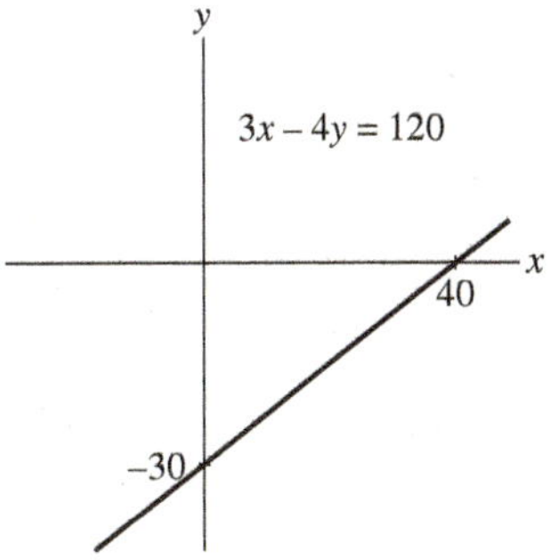


**10.**

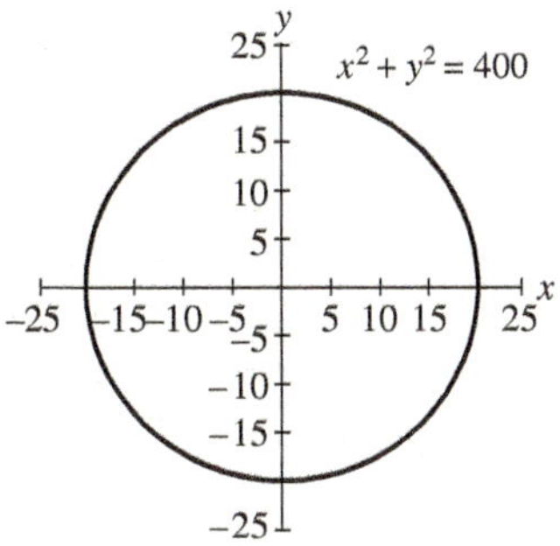


**11.**

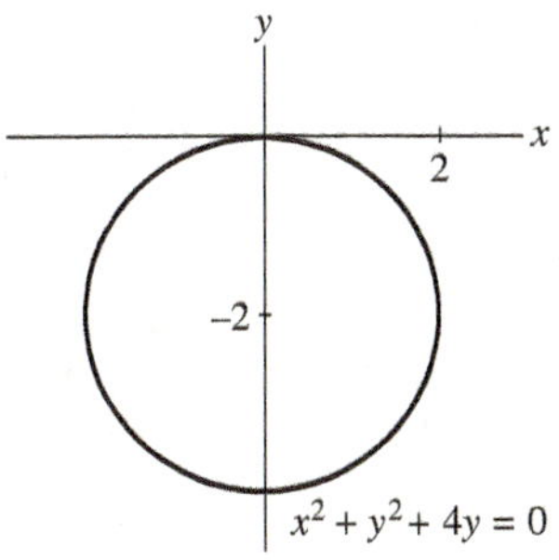


**12.**

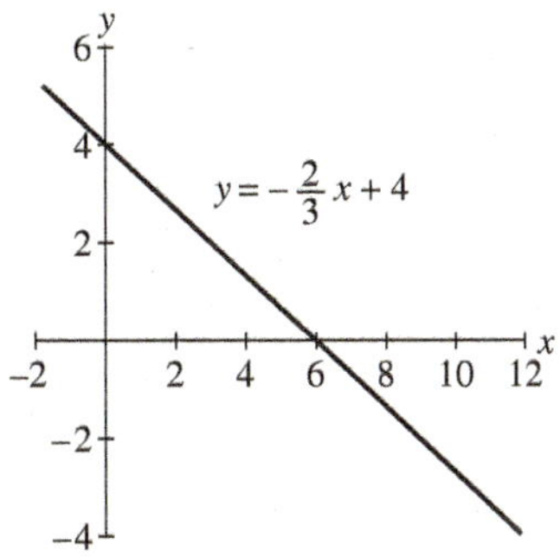


**13.**

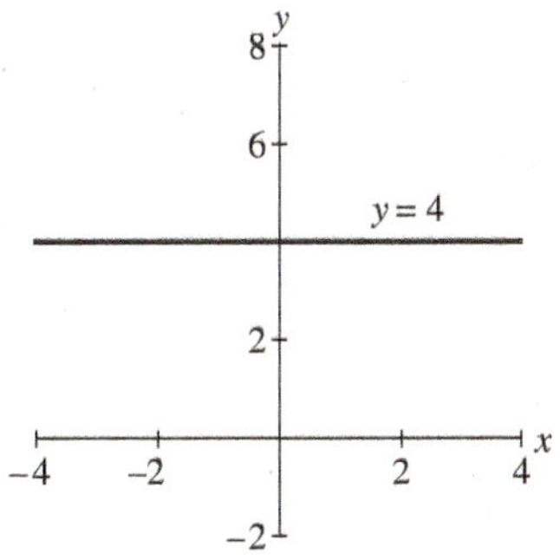


**14.**

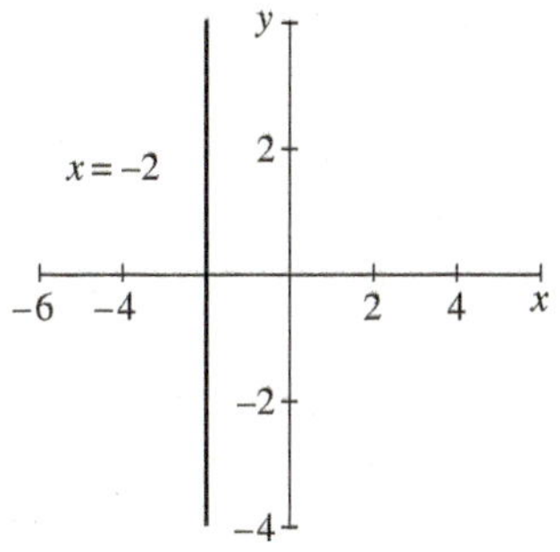


**15.** $7 - 24i$ **16.** $\frac{1}{2} - \frac{1}{2}i$ **17.** $-1 + i$ **18.** $-4 + 4i\sqrt{3}$

**19.** 3/5 **20.** −5/4 **21.** $y = -\frac{3}{2}x - \frac{1}{2}$ **22.** $y = -x - 1$

**23.** $\sqrt{34}$ **24.** $\left(0, \frac{1}{2}\right)$ **25.** −11, none **26.** $y = \frac{1}{3x + 2}$

**27.** $(-\infty, -2)$

**28.** $(6, \infty)$

**29.** $[-1, 2]$

**30.** $(-\infty, 1) \cup (5, \infty)$

**31.** 289 ft$^2$ **32.** 20 gal

**33. a.** $y = 18.4x + 311$ where $x$ is the number of years since 1997 and $y$ is in thousands of dollars. **b.** \$642,000

**34. a.** $y = -11.39x + 2159.14$,
$y = 0.30x^2 - 13.77x + 2162.71$
**b.** 2,022,000, 2,041,000

## Chapter 2

### Section 2.1

***For Thought:*** **1.** F **2.** F **3.** T **4.** F **5.** F **6.** F **7.** T **8.** T **9.** T **10.** F

***Exercises:***

**1.** relation **3.** independent, dependent **5.** difference quotient
**7.** Both **9.** $a$ is a function of $b$ **11.** $b$ is a function of $a$
**13.** Neither **15.** Both **17.** No **19.** Yes **21.** Yes
**23.** Yes **25.** No **27.** No **29.** Yes **31.** Yes
**33.** Yes **35.** No **37.** Yes **39.** Yes **41.** No
**43.** $\{-3, 4, 5\}$, $\{1, 2, 6\}$ **45.** $(-\infty, \infty)$, $\{4\}$ **47.** $(-\infty, \infty)$, $[5, \infty)$
**49.** $[-3, \infty)$, $(-\infty, \infty)$ **51.** $[4, \infty)$, $[0, \infty)$ **53.** $(-\infty, 0]$, $(-\infty, \infty)$
**55.** 6 **57.** 11 **59.** 3 **61.** 7 **63.** 22 **65.** $3a^2 - a$
**67.** $4a + 6$ **69.** $3x^2 + 5x + 2$ **71.** $4x + 4h - 2$
**73.** $6x + 2$ **75.** $6xh + 3h^2 - h$ **77.** −\$2400 per yr
**79.** −32, −48, −62.4, −63.84, and −63.984 ft/sec
**81.** −10.1 million hectares per yr **83.** 4 **85.** 3
**87.** $2x + h + 1$ **89.** $-2x - h + 1$ **91.** $\frac{3}{\sqrt{x + h} + \sqrt{x}}$
**93.** $\frac{1}{\sqrt{x + h + 2} + \sqrt{x + 2}}$ **95.** $\frac{-1}{x(x + h)}$
**97.** $\frac{-3}{(x + 2)(x + h + 2)}$
**99. a.** $A = s^2$ **b.** $s = \sqrt{A}$ **c.** $s = \frac{d\sqrt{2}}{2}$ **d.** $d = s\sqrt{2}$
**e.** $P = 4s$ **f.** $s = P/4$ **g.** $A = \frac{P^2}{16}$ **h.** $d = \sqrt{2A}$

**101.** $C = 50 + 35n$
**103. a.** Amount spent in 2004, \$9.6 billion **b.** 2010
**105.** $h = (2\sqrt{3} + 2)a$
**107.** At \$18/ticket revenue is increasing at \$1950 per dollar change in ticket price. At \$22/ticket revenue is decreasing at \$2050 per dollar change in ticket price.
**111.** $\{-9/17\}$ **113.** $2\sqrt{10}$ **115.** $\{-6, 7\}$

## Section 2.2

***For Thought:*** **1.** T **2.** F **3.** T **4.** T **5.** F **6.** T **7.** T **8.** T **9.** F **10.** T

***Exercises:***

**1.** parabola

**3.** $(-\infty, \infty), (-\infty, \infty)$, yes

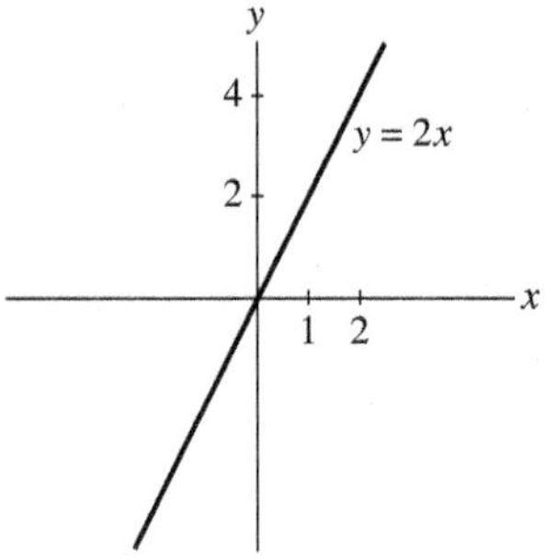


**5.** $(-\infty, \infty), (-\infty, \infty)$, yes

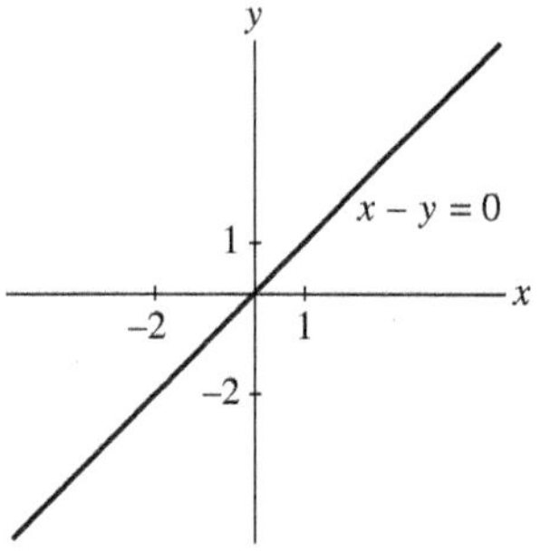


**7.** $(-\infty, \infty), \{5\}$, yes

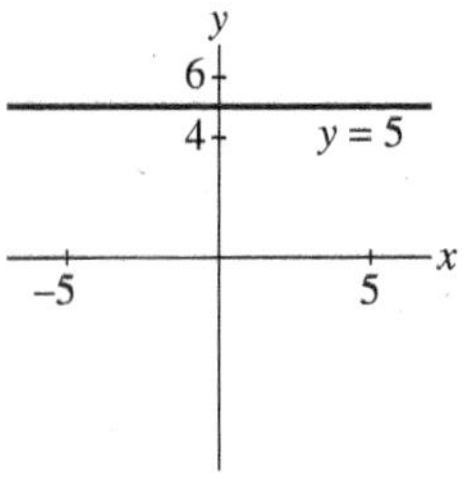


**9.** $(-\infty, \infty), [0, \infty)$, yes

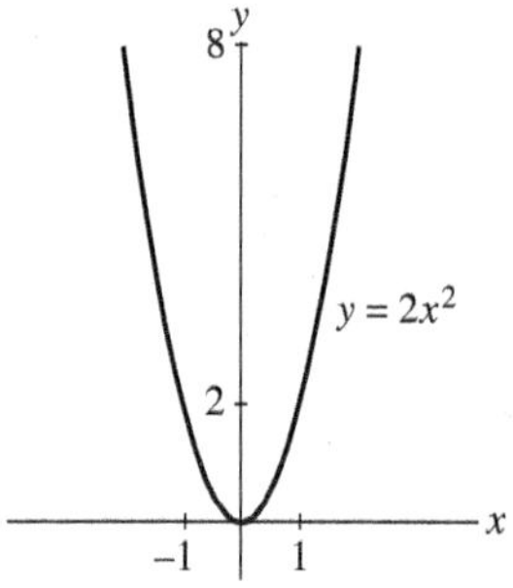


**11.** $(-\infty, \infty), (-\infty, 1]$, yes

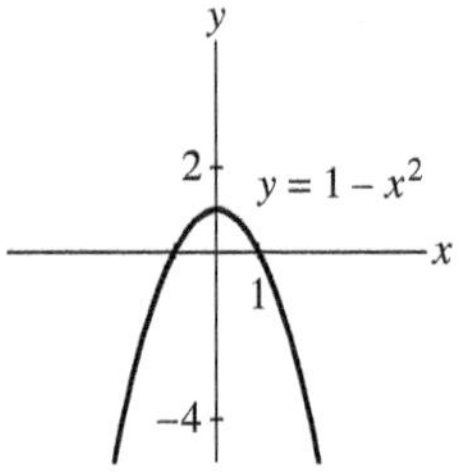


**13.** $[0, \infty), [1, \infty)$, yes

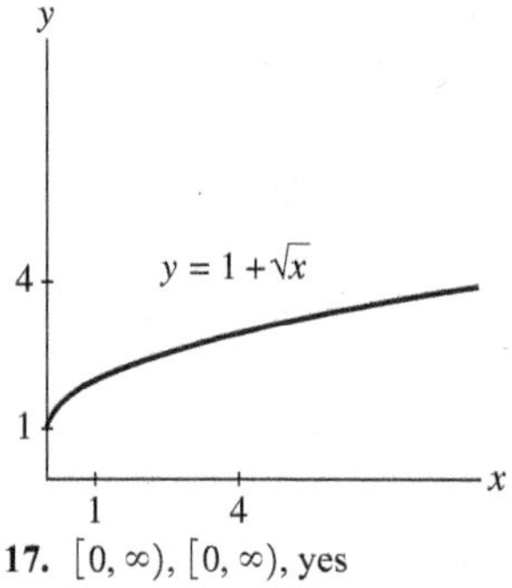


**15.** $[1, \infty), (-\infty, \infty)$, no

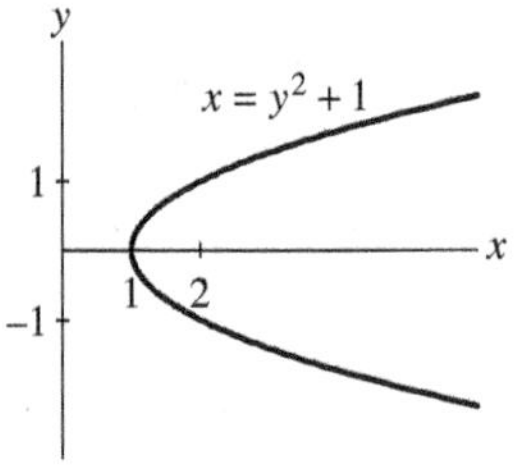


**17.** $[0, \infty), [0, \infty)$, yes

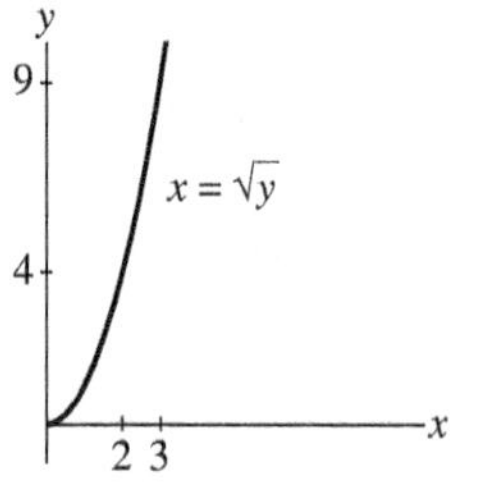


**19.** $(-\infty, \infty), (-\infty, \infty)$, yes

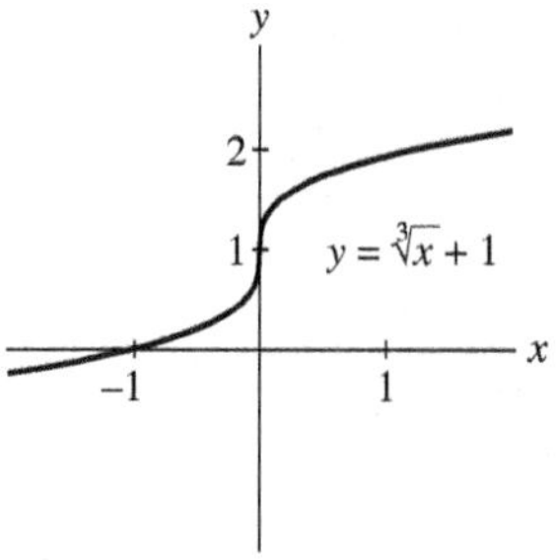


**21.** $(-\infty, \infty), (-\infty, \infty)$, yes

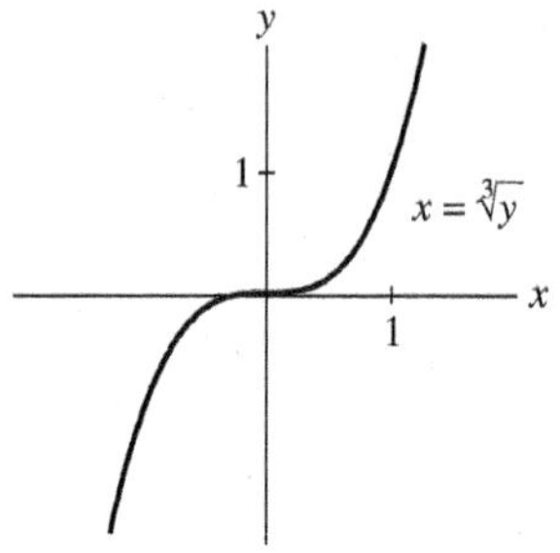


**23.** $[-1, 1], [-1, 1]$, no

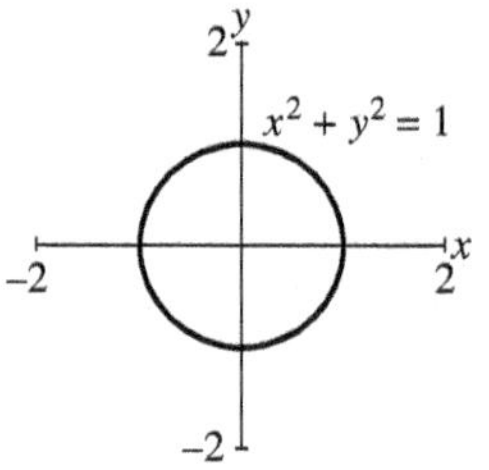


**25.** $[-1, 1], [0, 1]$, yes

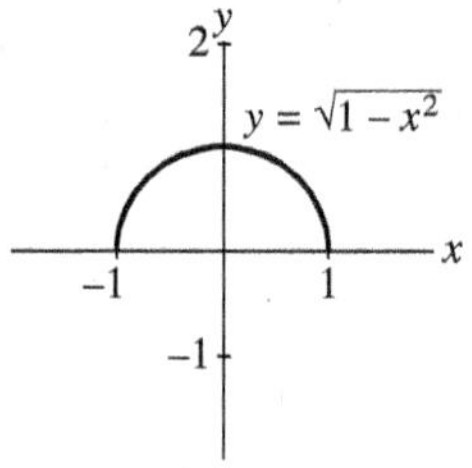


**27.** $(-\infty, \infty), (-\infty, \infty)$, yes

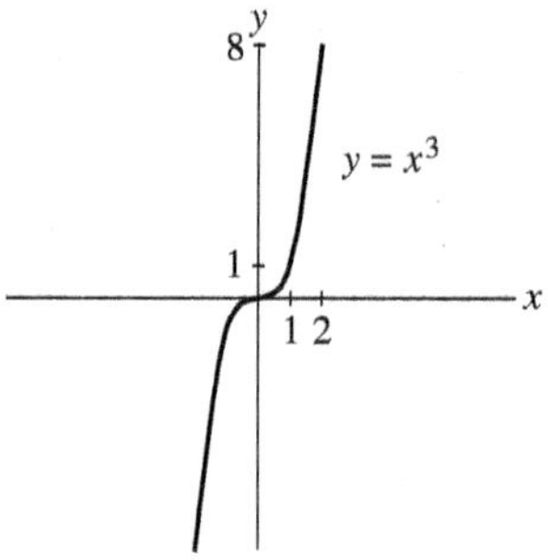


**29.** $(-\infty, \infty), [0, \infty)$, yes

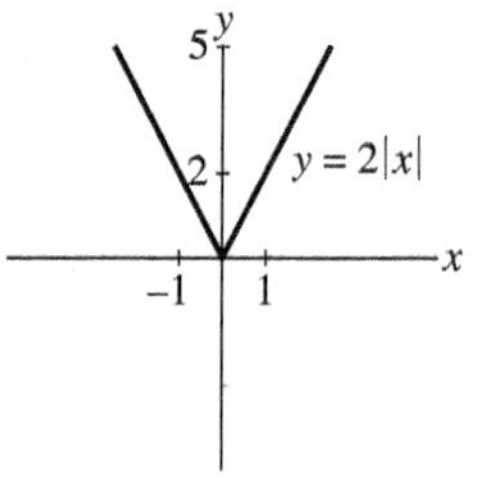


**31.** $(-\infty, \infty), (-\infty, 0]$, yes

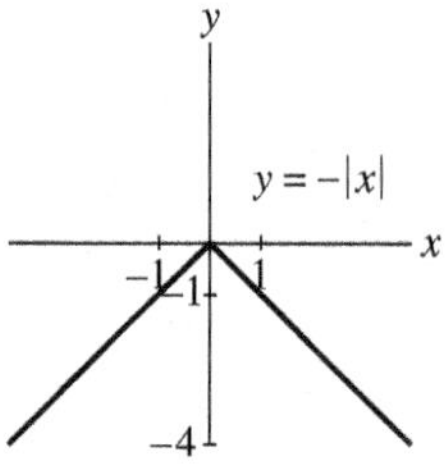


**33.** $[0, \infty), (-\infty, \infty)$, no

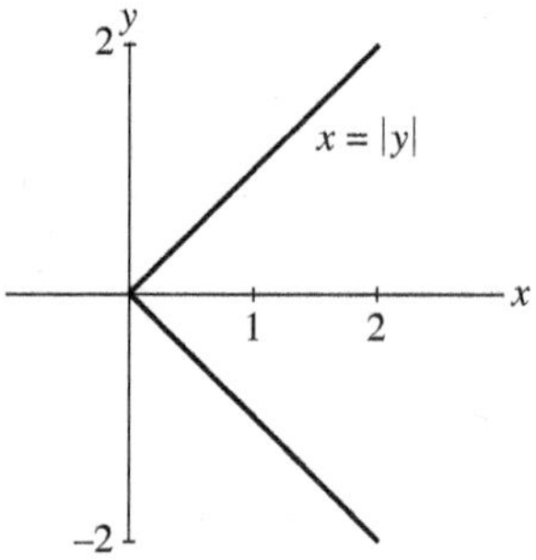


**35.** $(-\infty, \infty), \{-2, 2\}$

$$f(x) = \begin{cases} 2 & x < -1 \\ -2 & x \geq -1 \end{cases}$$

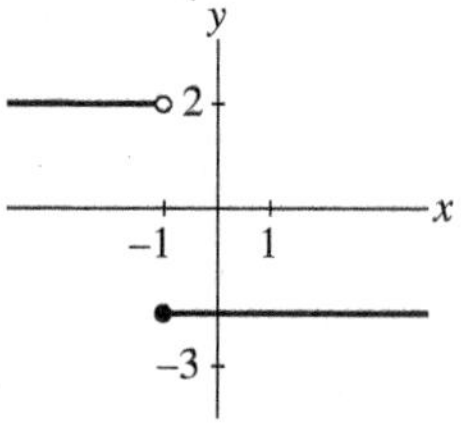


**37.** $(-\infty, \infty), (-\infty, -2] \cup (2, \infty)$

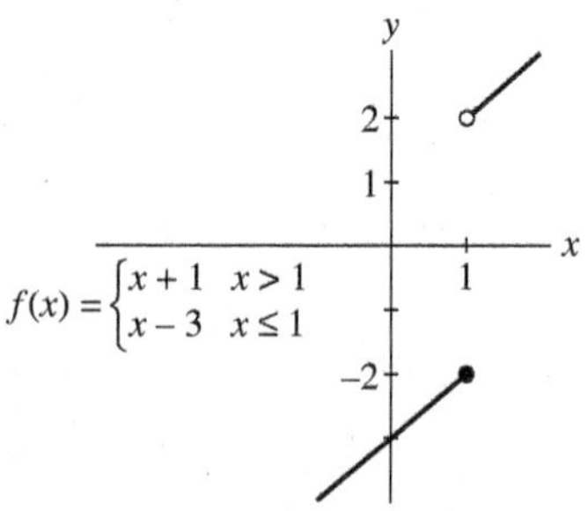

**39.** $[-2, \infty), (-\infty, 2]$

$$f(x) = \begin{cases} \sqrt{x+2} & -2 \le x \le 2 \\ 4 - x & x > 2 \end{cases}$$

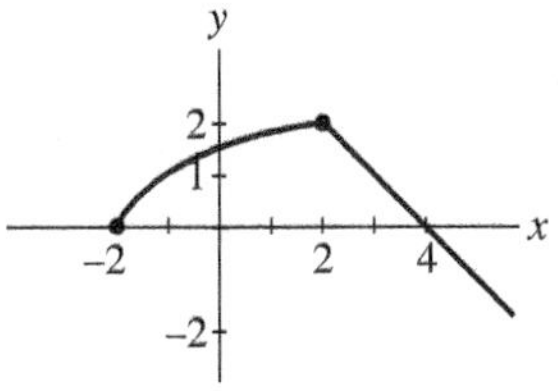


**41.** $(-\infty, \infty), [0, \infty)$

$$f(x) = \begin{cases} \sqrt{-x} & x < 0 \\ \sqrt{x} & x \ge 0 \end{cases}$$

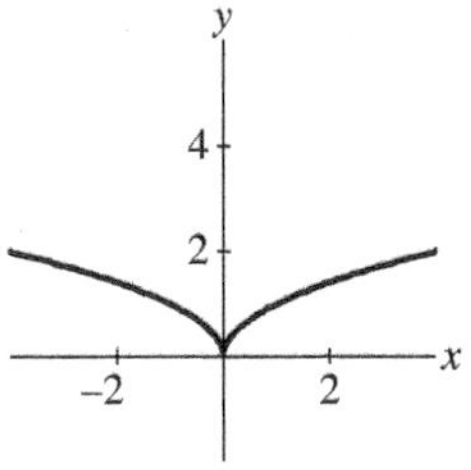


**61.** $(-\infty, 0) \cup (0, \infty), \{-1, 1\}$, constant $(-\infty, 0), (0, \infty)$

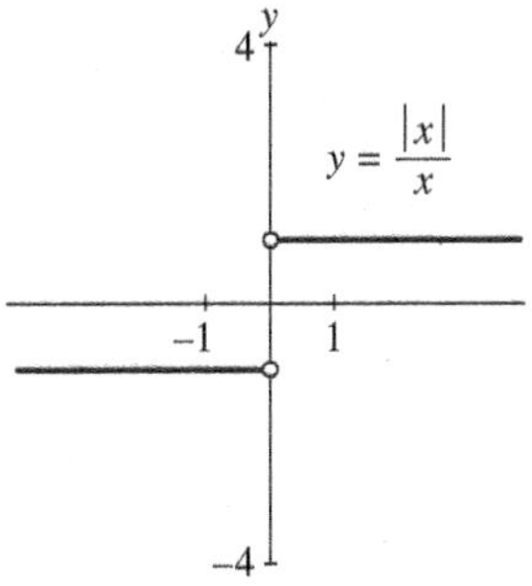


**63.** $[-3, 3], [0, 3]$, inc $[-3, 0]$, dec $[0, 3]$

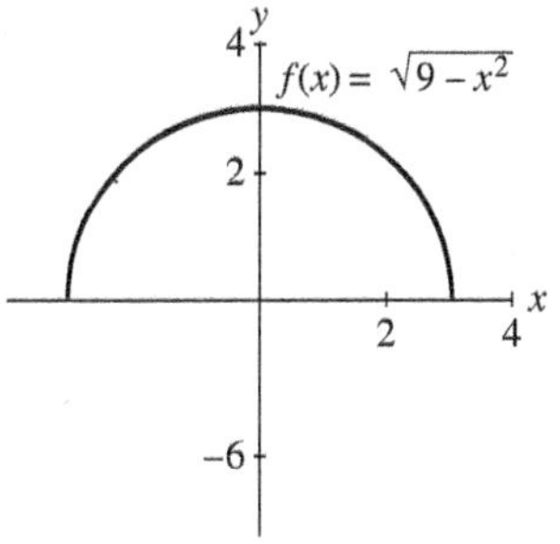


**43.** $(-\infty, \infty), (-\infty, \infty)$

$$f(x) = \begin{cases} x^2 & x < -1 \\ -x & x \ge -1 \end{cases}$$

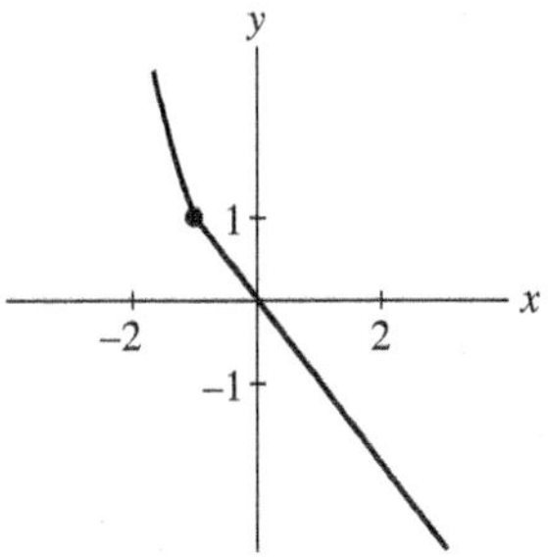


**45.** $(-\infty, \infty)$, integers

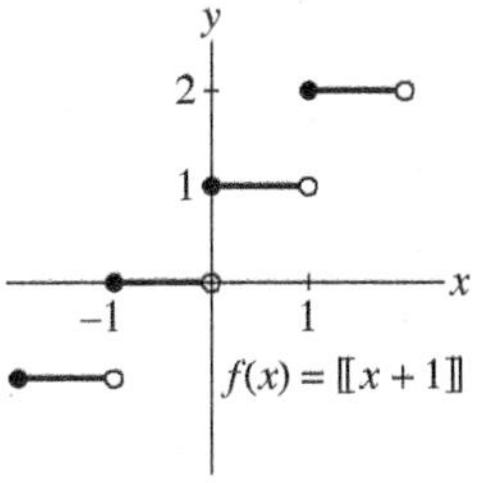


**65.** $(-\infty, \infty), (-\infty, \infty)$, inc $[-\infty, 3), [3, \infty)$

$$f(x) = \begin{cases} x + 1 & x \ge 3 \\ x + 2 & x < 3 \end{cases}$$

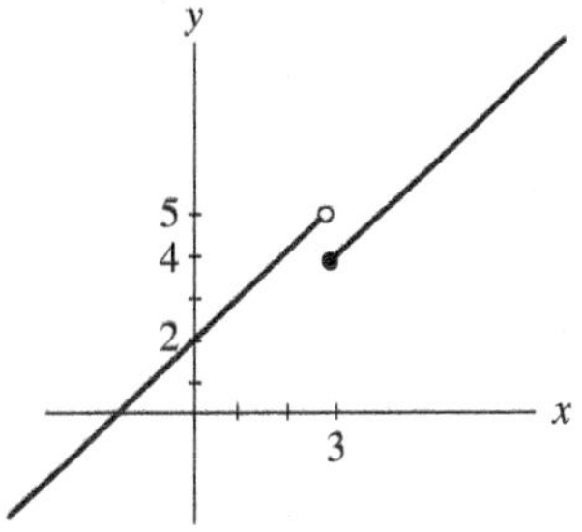


**47.** $[0, 4), \{2, 3, 4, 5\}$

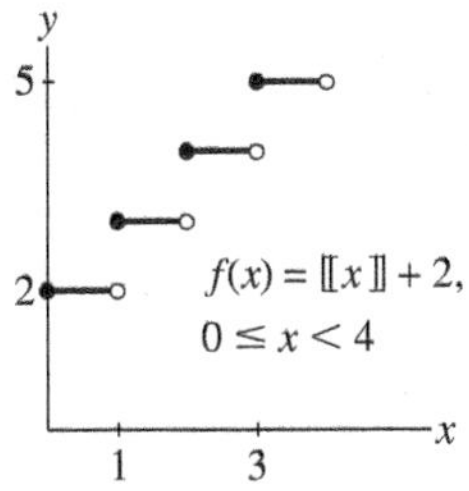


**67.** $(-\infty, \infty), (-\infty, 2]$, inc $(-\infty, -2], (-2, 0]$ dec $[0, 2), [2, \infty)$

$$f(x) = \begin{cases} x + 3 & x \le -2 \\ \sqrt{4 - x^2} & -2 < x < 2 \\ -x + 3 & x \ge 2 \end{cases}$$

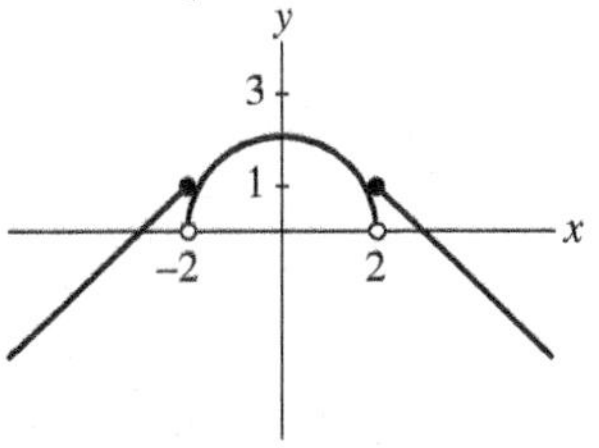


**49. a.** $D(-\infty, \infty), R(-\infty, \infty)$, dec $(-\infty, \infty)$
**b.** $D(-\infty, \infty), R(-\infty, 4]$, inc $(-\infty, 0]$, dec $[0, \infty)$
**51. a.** $D[-2, 6], R[3, 7]$, inc $[-2, 2]$, dec $[2, 6]$
**b.** $D(-\infty, 2], R(-\infty, 3]$, inc $(-\infty, -2]$, constant $[-2, 2]$
**53. a.** $D(-\infty, \infty), R[0, \infty)$, dec $(-\infty, 0]$, inc $[0, \infty)$
**b.** $D(-\infty, \infty), R(-\infty, \infty)$, dec$(-\infty, -2]$ and $[-2/3, \infty)$, inc $[-2, -2/3]$
**55. a.** $D(-\infty, \infty), R(-\infty, \infty)$, inc $(-\infty, \infty)$
**b.** $D[-2, 5], R[1, 4]$, dec $[-2, 1]$, inc $[1, 2]$, constant $[2, 5]$

**57.** $(-\infty, \infty), (-\infty, \infty)$, inc $(-\infty, \infty)$

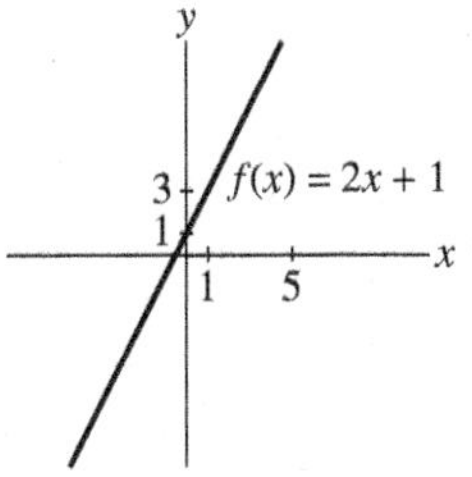


**59.** $(-\infty, \infty), [0, \infty)$, dec $(-\infty, 1]$, inc $[1, \infty)$

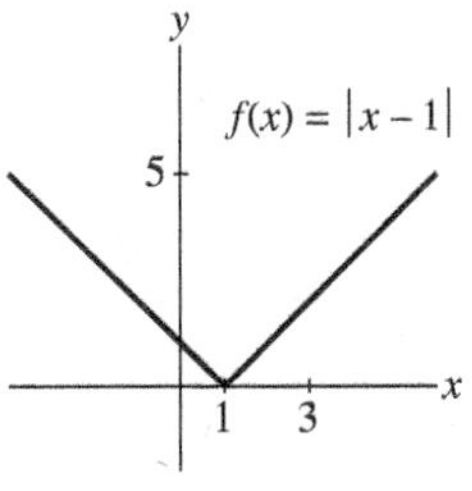


**69.** $f(x) = \begin{cases} 2 & \text{for } x > -1 \\ -1 & \text{for } x \le -1 \end{cases}$

**71.** $f(x) = \begin{cases} x - 1 & \text{for } x \ge -1 \\ -x & \text{for } x < -1 \end{cases}$

**73.** $f(x) = \begin{cases} 2x - 2 & \text{for } x \ge 0 \\ -x - 2 & \text{for } x < 0 \end{cases}$

**75.** Dec $(-\infty, 0.83]$, inc $[0.83, \infty)$
**77.** Inc $(-\infty, -1], [1, \infty)$, dec $[-1, 1]$
**79.** Dec $(-\infty, -1.73], [0, 1.73]$, inc $[-1.73, 0], [1.73, \infty)$
**81.** Inc $[30, 50], [70, \infty)$, dec $(-\infty, 30], [50, 70]$
**83.** c **85.** d
**87.** Inc $[0, 3], [6, 15]$, dec $[3, 6], [30, 39]$, constant $[15, 30]$

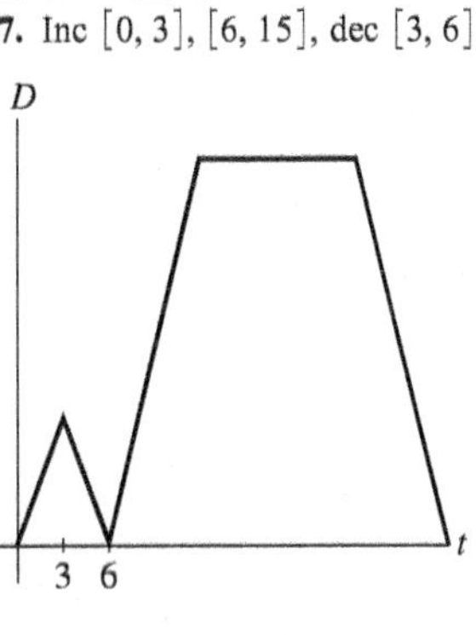

**89.** Inc [0, 2], dec [2.5, 4.5], constant [2, 2.5]

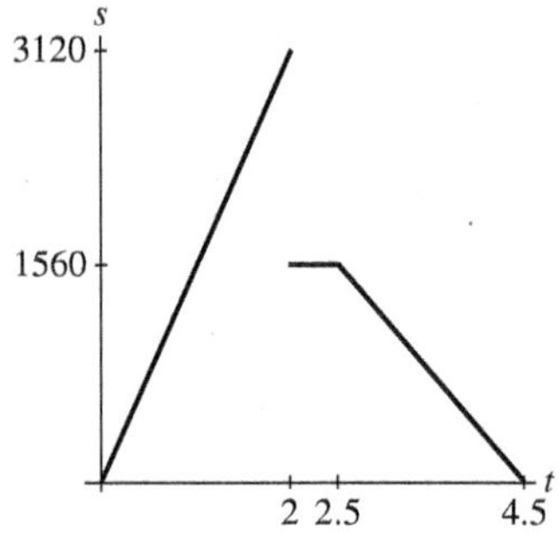


**91.** 565 million, 800 million, 14.5 million/yr
**93.** $[0, 10^4]$, $[10^4, \infty)$
**95.** $[5, \infty)$

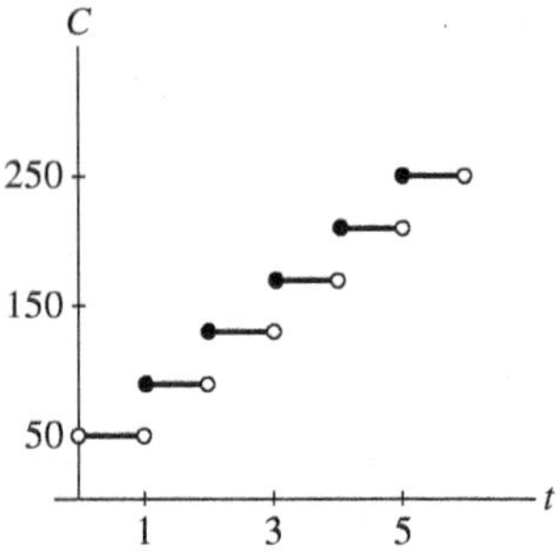


**97.** $f(x) = \begin{cases} 150 & 0 < x < 3 \\ 50x & 3 \le x \le 10 \end{cases}$

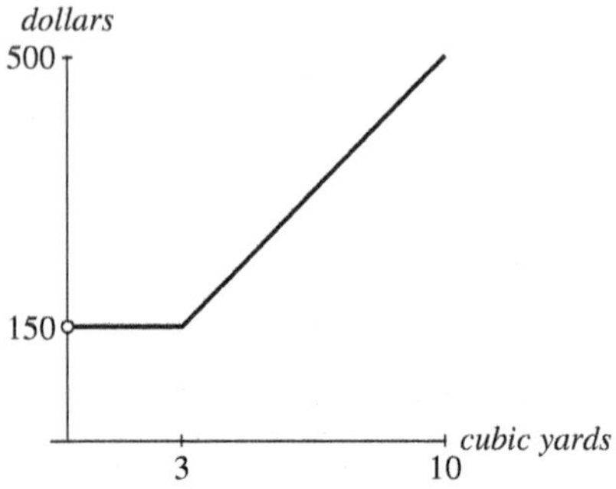


**101.** $[2, \infty)$, $[3, \infty)$ **103.** $\{55/13\}$ **105.** $\{5/2\}$

## Section 2.3

***For Thought:*** **1.** F **2.** T **3.** F **4.** T **5.** T **6.** F **7.** T **8.** F **9.** T **10.** T

***Exercises:***
**1.** rigid **3.** parabola **5.** reflection **7.** linear **9.** odd

**11.**

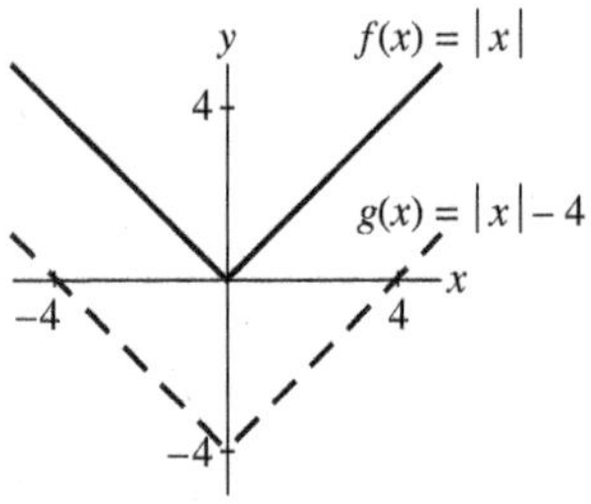


**13.**

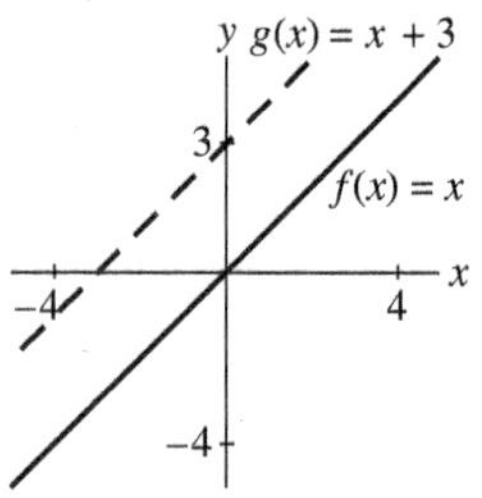


**15.**

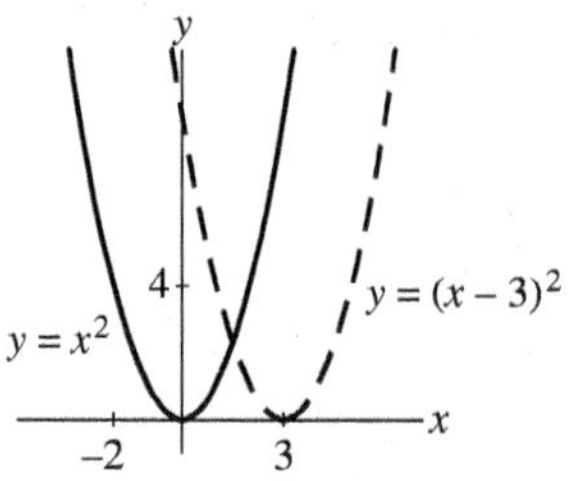


**17.**

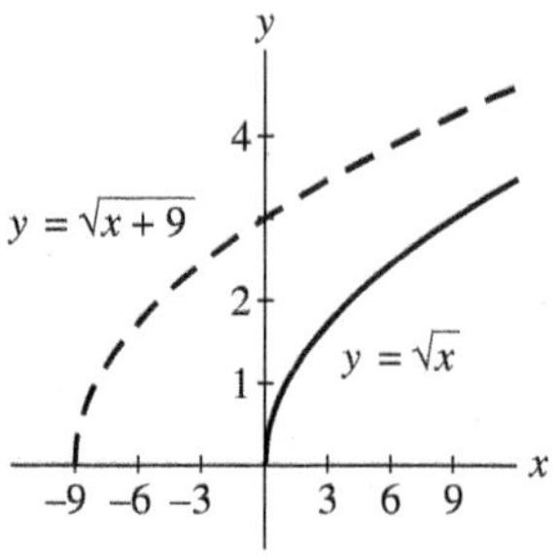


**19.**

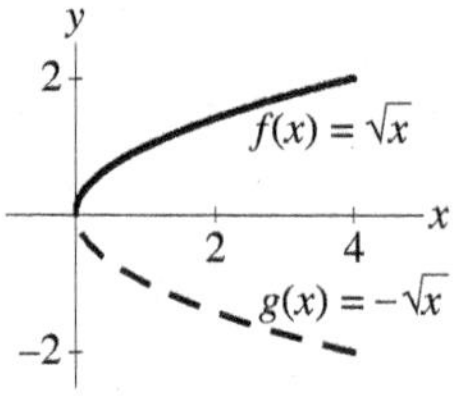


**21.**

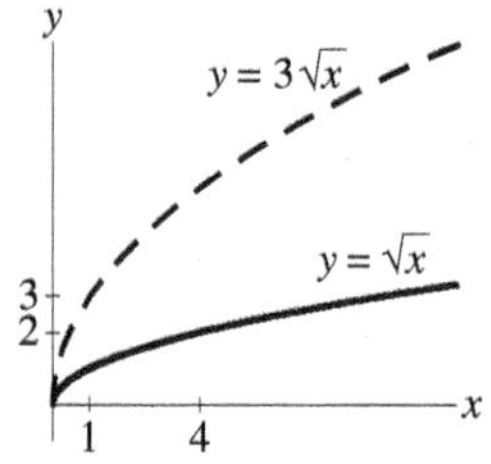


**23.**

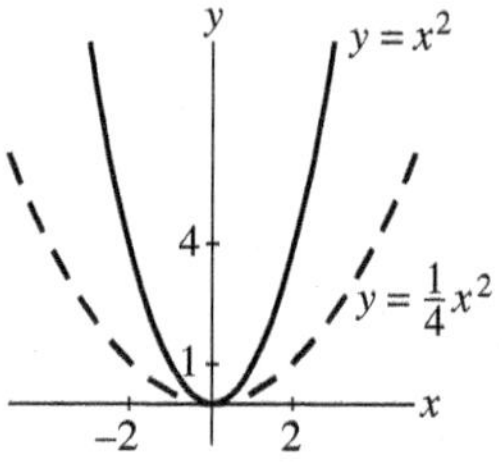


**25.**

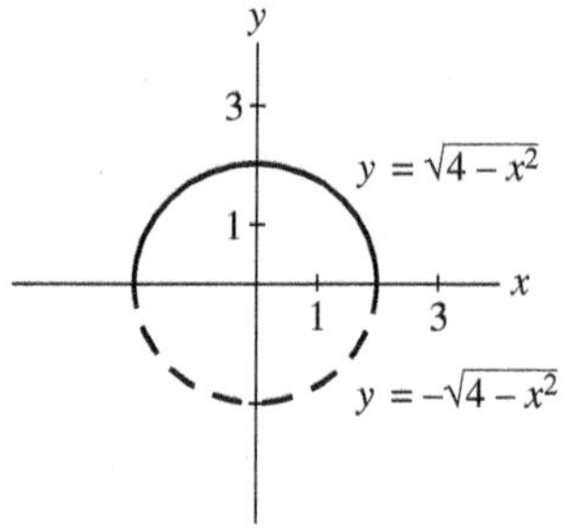


**27.** g **29.** b **31.** c **33.** f
**35.** $y = \sqrt{x} + 2$ **37.** $y = (x - 5)^2$ **39.** $y = (x - 10)^2 + 4$
**41.** $y = -3\sqrt{x} - 5$ **43.** $y = -3|x - 7| + 9$
**45.** $(-\infty, \infty)$, $[2, \infty)$

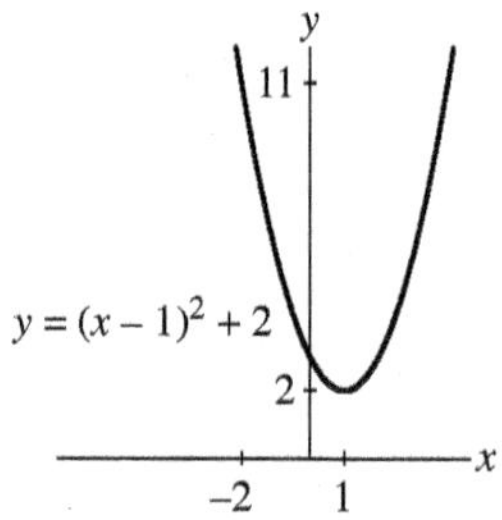


**47.** $(-\infty, \infty)$, $[3, \infty)$

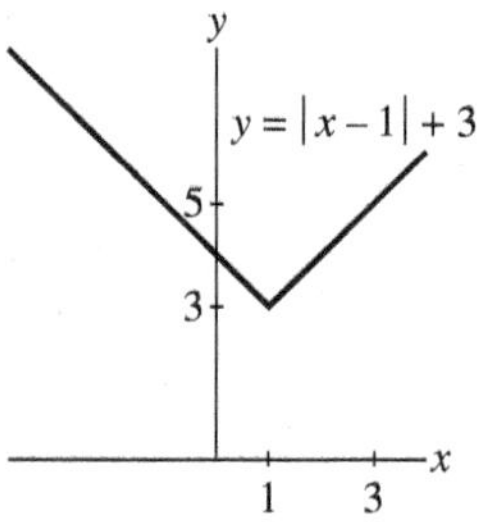


**49.** $(-\infty, \infty)$, $(-\infty, \infty)$

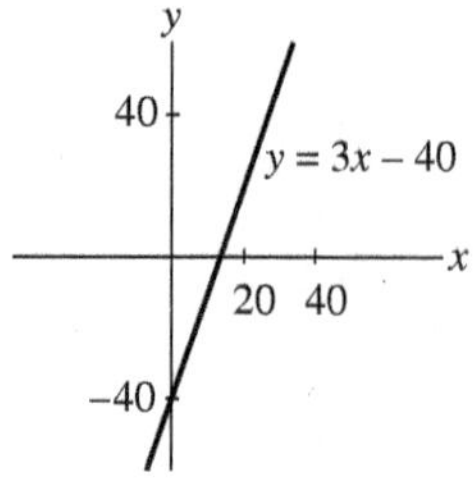


**51.** $(-\infty, \infty)$, $(-\infty, \infty)$

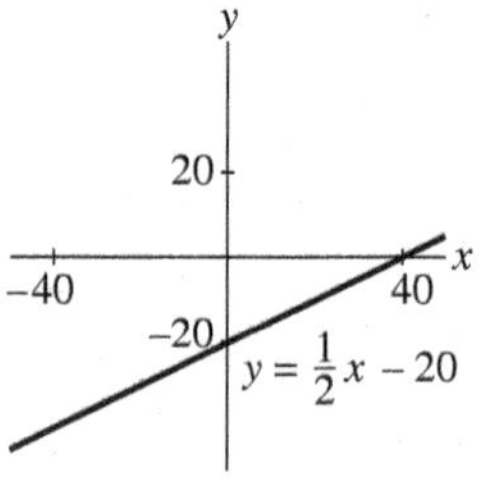

**53.** $(-\infty, \infty), (-\infty, 40]$

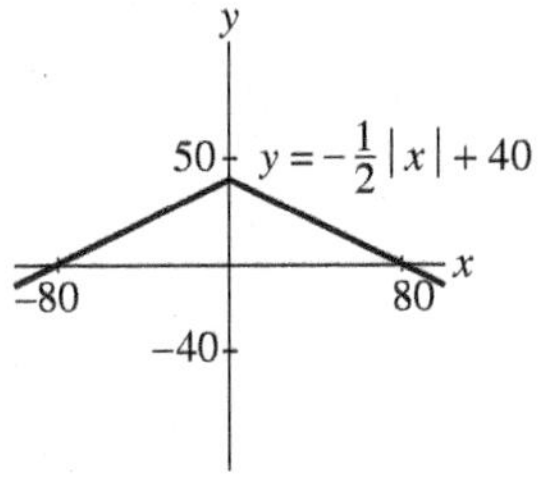


**55.** $(-\infty, \infty), (-\infty, 0]$

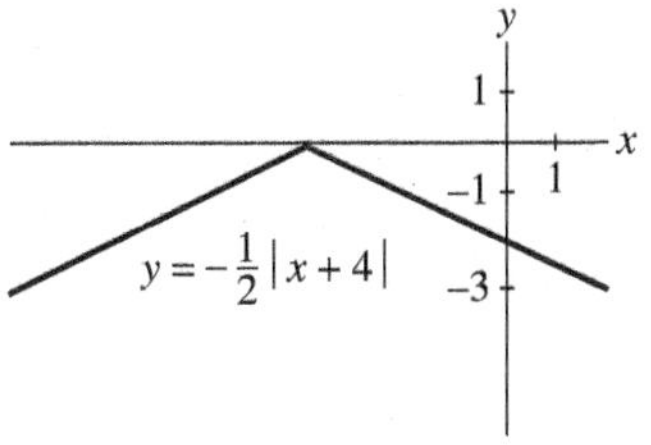


**57.** $[3, \infty), (-\infty, 1]$

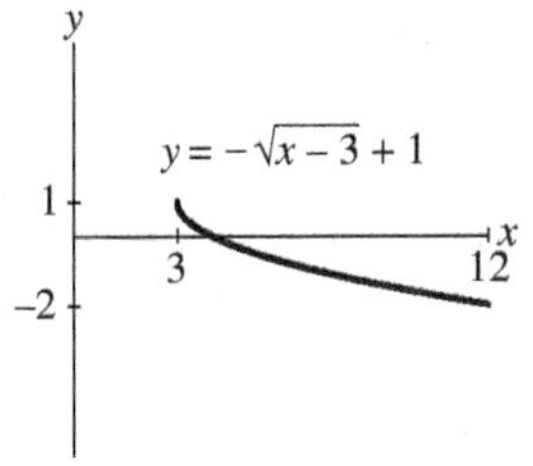


**59.** $[-3, \infty), (-\infty, 2]$

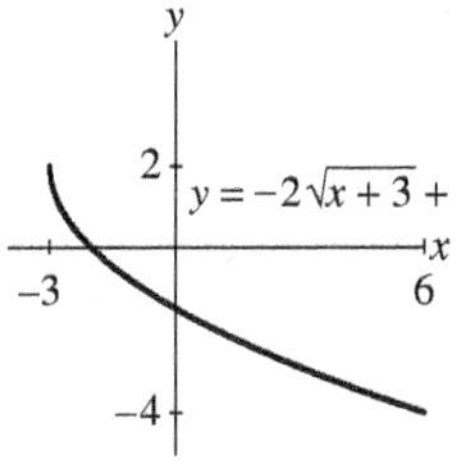


**61.** $y$-axis, even **63.** No symmetry, neither **65.** $x = -3$, neither **67.** $x = 2$, neither **69.** Origin, odd **71.** No symmetry, neither **73.** No symmetry, neither **75.** $y$-axis, even **77.** No symmetry, neither **79.** $y$-axis, even **81.** e **83.** g **85.** b **87.** c **89.** $(-\infty, -1] \cup [1, \infty)$ **91.** $(-\infty, -1) \cup (5, \infty)$ **93.** $(-2, 4)$ **95.** $[0, 25]$ **97.** $(-\infty, 2 - \sqrt{3}) \cup (2 + \sqrt{3}, \infty)$ **99.** $(-5, 5)$ **101.** $(-3.36, 1.55)$

**103. a.**

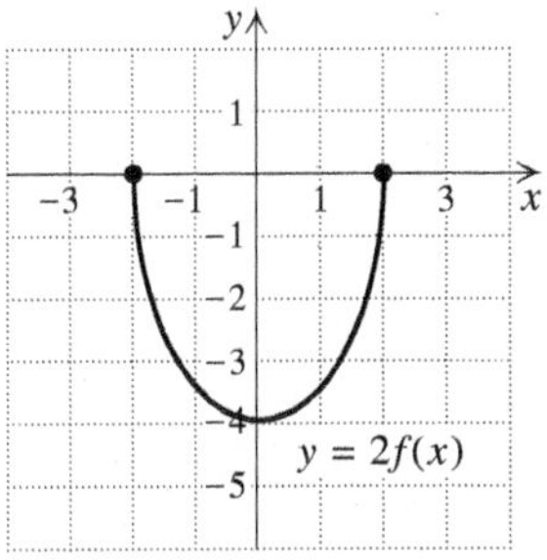


**b.**

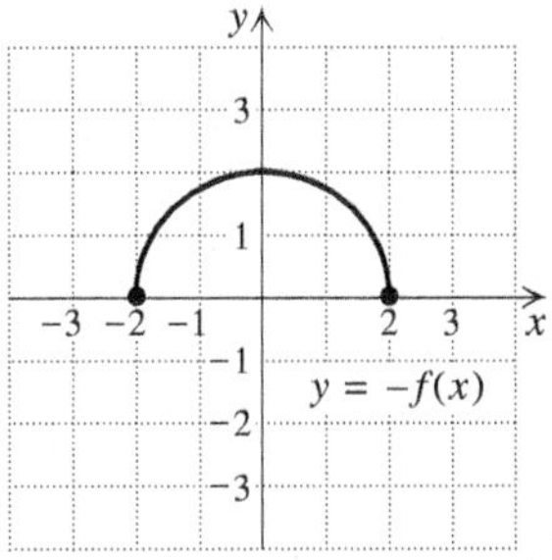


**c.**

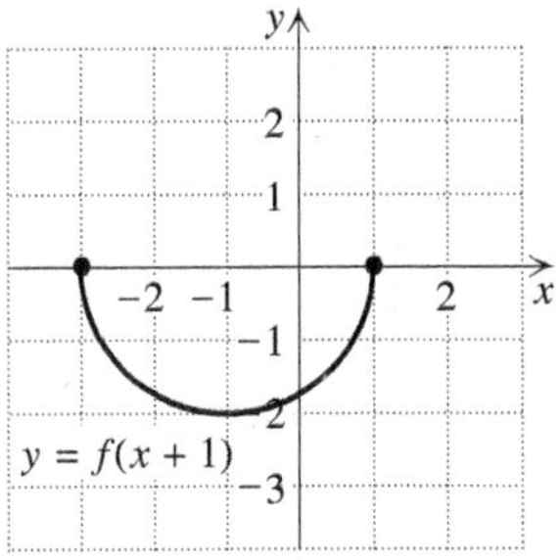


**d.**

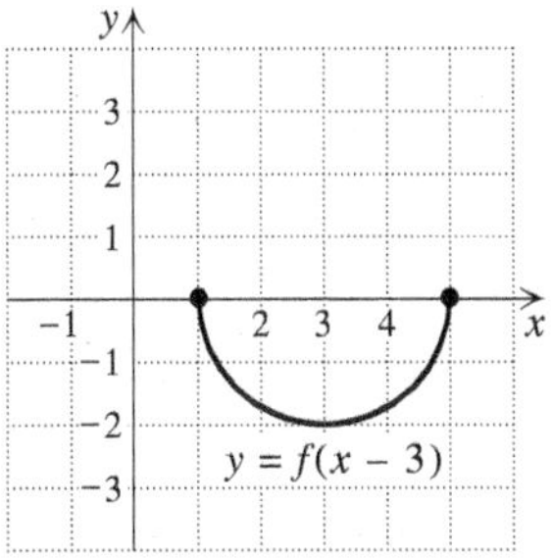


**e.**

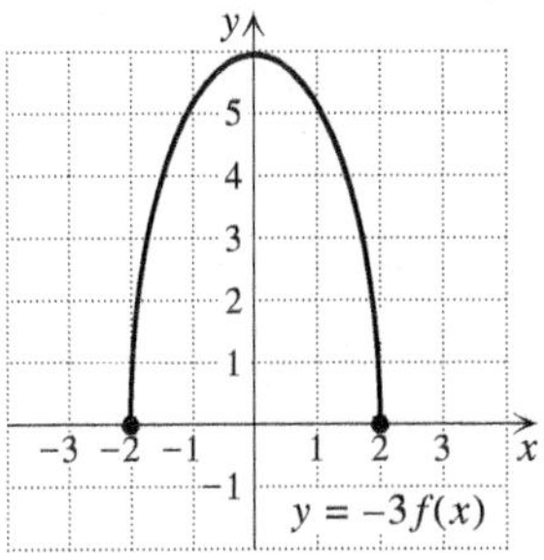


**f.**

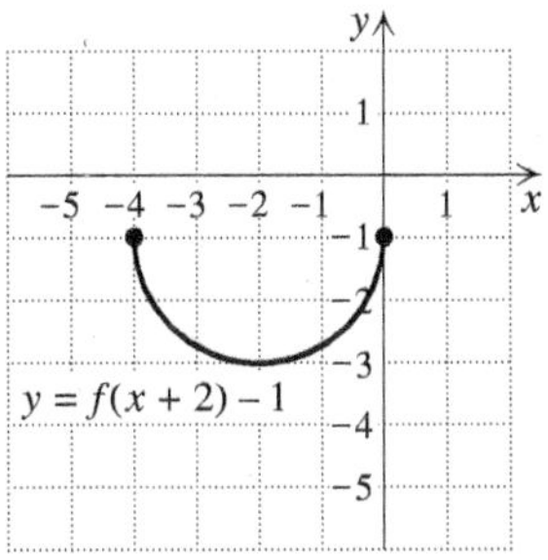


**g.**

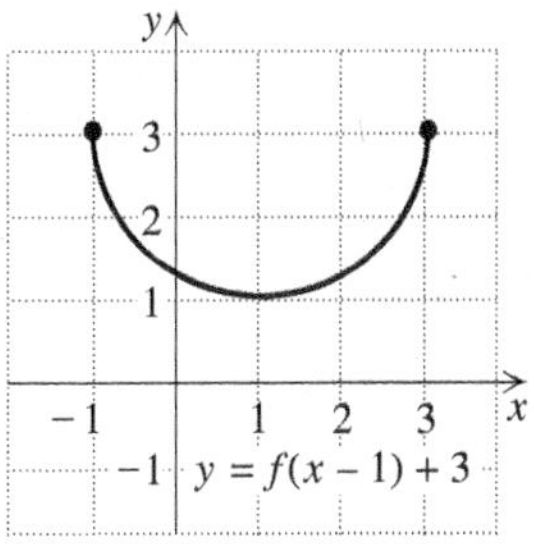


**h.**

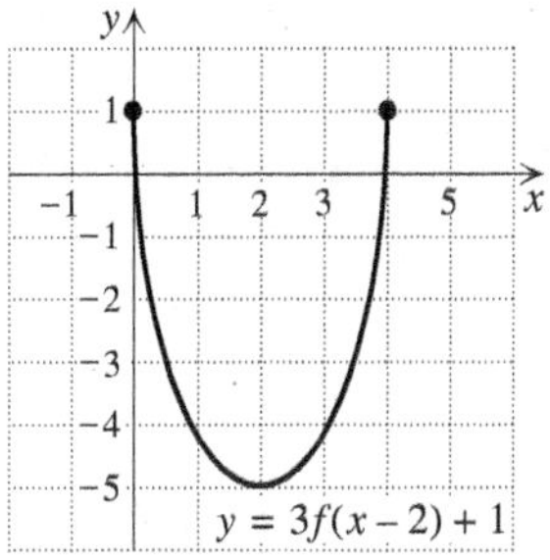


**105.** $N(x) = x + 2000$ **107.** $x > 25\%$ **109. a.** Even function **b.** Odd function **c.** Translation one unit to left **d.** Translation two units to right and three units up **111.** $4a + 4$ **113.** $(-\infty, -2/3] \cup [2/3, \infty)$ **115.** $y = \dfrac{-2}{a + b - c}$

## Section 2.4

***For Thought:*** **1.** F **2.** T **3.** T **4.** T **5.** T **6.** T **7.** F **8.** T **9.** F **10.** T

***Exercises:***

**1.** sum **3.** 1 **5.** −11 **7.** −8 **9.** 1/12 **11.** $a^2 - 3$ **13.** $a^3 - 4a^2 + 3a$ **15.** $\{(-3, 3), (2, 6)\}, \{-3, 2\}$ **17.** $\{(-3, -1), (2, -6)\}, \{-3, 2\}$ **19.** $\{(-3, 2), (2, 0)\}, \{-3, 2\}$ **21.** $\{(-3, 2)\}, \{-3\}$ **23.** $(f + g)(x) = \sqrt{x} + x - 4, [0, \infty)$ **25.** $(f - h)(x) = \sqrt{x} - \dfrac{1}{x - 2}, [0, 2) \cup (2, \infty)$ **27.** $(g \cdot h)(x) = \dfrac{x - 4}{x - 2}, (-\infty, 2) \cup (2, \infty)$ **29.** $(g/f)(x) = \dfrac{x - 4}{\sqrt{x}}, (0, \infty)$ **31.** $\{(-3, 0), (1, 0), (4, 4)\}$ **33.** $\{(1, 4)\}$ **35.** $\{(-3, 4), (1, 4)\}$ **37.** 5 **39.** 5 **41.** 59.8163 **43.** 5 **45.** 5 **47.** $a$ **49.** $3t^2 + 2$ **51.** $(f \circ g)(x) = \sqrt{x} - 2, [0, \infty)$ **53.** $(f \circ h)(x) = \dfrac{1}{x} - 2, (-\infty, 0) \cup (0, \infty)$ **55.** $(h \circ g)(x) = \dfrac{1}{\sqrt{x}}, (0, \infty)$ **57.** $(f \circ f)(x) = x - 4, (-\infty, \infty)$ **59.** $(h \circ g \circ f)(x) = \dfrac{1}{\sqrt{x - 2}}, (2, \infty)$ **61.** $(h \circ f \circ g)(x) = \dfrac{1}{\sqrt{x} - 2}, (0, 4) \cup (4, \infty)$ **63.** $F = g \circ h$ **65.** $H = h \circ g$ **67.** $N = h \circ g \circ f$ **69.** $P = g \circ f \circ g$ **71.** $S = g \circ g$ **73.** $g(x) = x^3$ and $h(x) = x - 2$ **75.** $g(x) = x + 5$ and $h(x) = \sqrt{x}$ **77.** $g(x) = 3x - 1$ and $h(x) = \sqrt{x}$, $g(x) = 3x$ and $h(x) = \sqrt{x - 1}$ **79.** $g(x) = |x|$ and $h(x) = 4x + 5$, $g(x) = 4|x|$ and $h(x) = x + 5$ **81.** $y = 6x - 1$ **83.** $y = x^2 + 6x + 7$ **85.** $y = x$ **87.** $y = (n - 4)^2$ **89.** $y = \sqrt{x + 16/8}$ **91.** $y = -x$, no **93.** $[-1, \infty), [-7, \infty)$ **95.** $[1, \infty), [0, \infty)$ **97.** $[0, \infty), [4, \infty)$ **99.** $P(x) = 28x - 200, x \geq 8$ **101.** $A = d^2/2$ **103.** $(f \circ f)(x) = 0.899x$, $(f \circ f \circ f)(x) = 0.852x$ **105.** $T(x) = 1.26x$ **107.** $D = \dfrac{1.16 \times 10^7}{L^3}$ **109.** $W = \dfrac{(8 + \pi)s^2}{8}$ **111.** $s = \dfrac{d\sqrt{2}}{2}$ **113.** No, composition **115.** $\dfrac{-3}{x(x + h)}$ **117.** $[3, \infty), (-\infty, 2]$ **119.** $(-1/5, \infty)$

## Section 2.5

***For Thought:*** **1.** F **2.** F **3.** F **4.** T **5.** F **6.** F **7.** F **8.** F **9.** F **10.** T

***Exercises:***

**1.** one-to-one **3.** inverse **5.** Yes **7.** No **9.** No

**11.** Not one-to-one **13.** One-to-one **15.** Not one-to-one
**17.** One-to-one **19.** One-to-one **21.** Not one-to-one
**23.** Not one-to-one **25.** One-to-one **27.** Invertible, $\{(3, 9), (2, 2)\}$
**29.** Not invertible **31.** Invertible, $\{(3, 3), (2, 2), (4, 4), (7, 7)\}$
**33.** Not invertible **35.** Not invertible **37.** Invertible
**39.** Invertible **41.** $\{(1, 2), (5, 3)\}$, 3, 2
**43.** $\{(-3, -3), (5, 0), (-7, 2)\}$, 0, 2 **45.** Not invertible
**47.** Not invertible
**49.** **a.** $f^{-1}(x) = \frac{x - 1}{5}$ **b.** $f^{-1}(x) = (x + 88)/3$
**c.** $f^{-1}(x) = \frac{x + 7}{3}$ **d.** $f^{-1}(x) = \frac{x - 4}{-3}$ **e.** $f^{-1}(x) = 2x + 18$
**f.** $f^{-1}(x) = -x$ **g.** $f^{-1}(x) = (x + 9)^3$ **h.** $f^{-1}(x) = \sqrt[3]{\frac{x + 7}{3}}$
**i.** $f^{-1}(x) = (x - 5)^3 + 1$ **j.** $f^{-1}(x) = \left(\frac{x}{2}\right)^3 + 7$
**51.** No **53.** Yes
**55.**

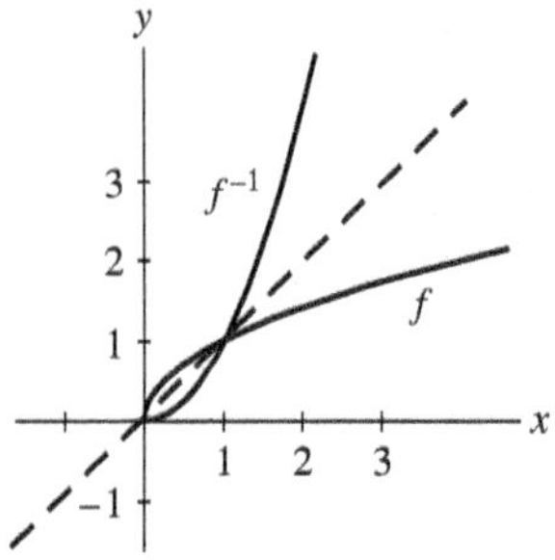


**57.**

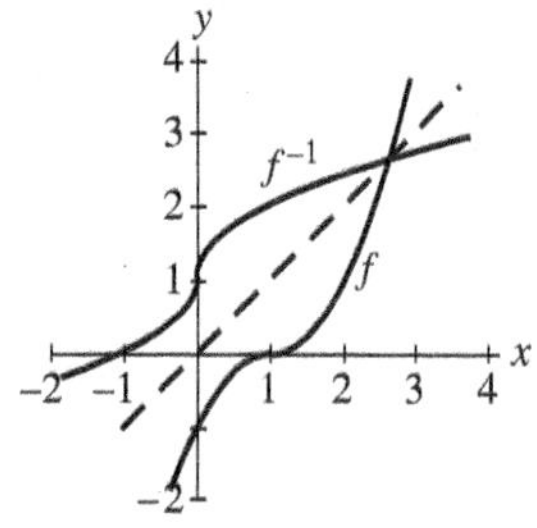


**59.** $f^{-1}(x) = \frac{x - 2}{3}$

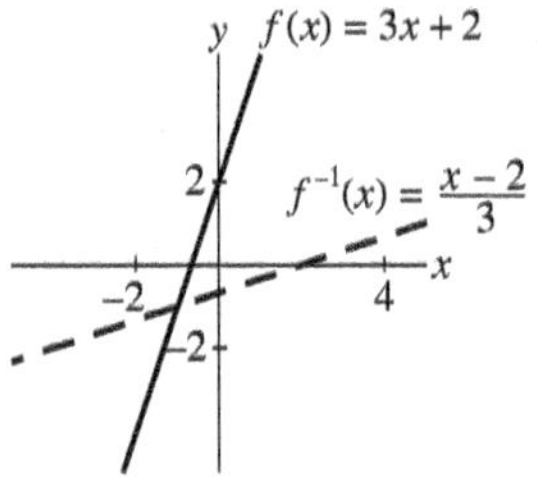


**61.** $f^{-1}(x) = \sqrt{x + 4}$

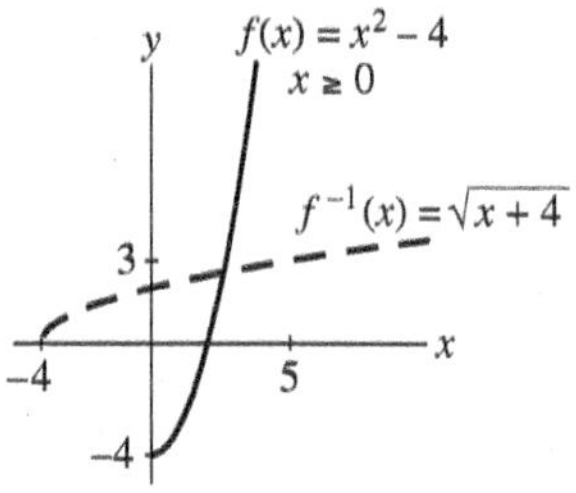


**63.** $f^{-1}(x) = \sqrt[3]{x}$

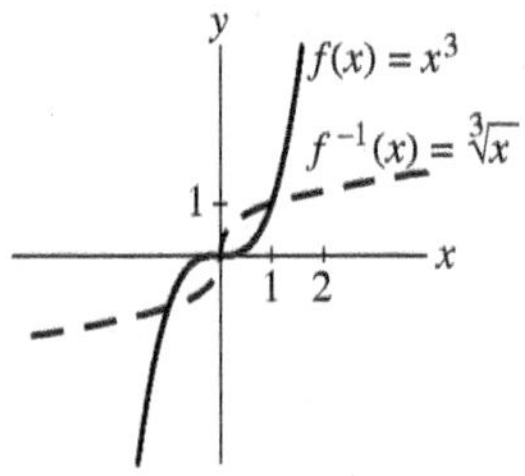


**65.** $f^{-1}(x) = (x + 3)^2$ for $x \geq -3$

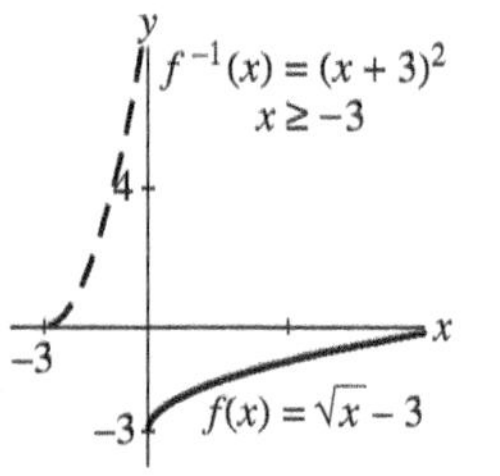


**67.** $f^{-1}(x) = \frac{x + 7}{3}$ **69.** $f^{-1}(x) = (x - 2)^2 + 3$ for $x \geq 2$
**71.** $f^{-1}(x) = -x - 9$ **73.** $f^{-1}(x) = \frac{5x + 3}{x - 1}$
**75.** $f^{-1}(x) = -\frac{1}{x}$ **77.** $f^{-1}(x) = (x - 5)^3 + 9$
**79.** $f^{-1}(x) = \sqrt{x} + 2$ **81.** $f(g(x)) = x, g(f(x)) = x$, yes
**83.** $f(g(x)) = x, g(f(x)) = |x|$, no
**85.** $f(g(x)) = x, g(f(x)) = x$, yes
**87.** $f(g(x)) = x, g(f(x)) = x$, yes
**89.** The functions $y_1$ and $y_2$ are inverses.
**91.** $C = 1.08P, P = \frac{C}{1.08}$ **93.** Yes, $r = \frac{7.89 - t}{0.39}$, 6

**95.** $w = \frac{V^2}{1.496}$, 8840 lb **97.** **a.** 10.9% **b.** $V = 50{,}000(1 - r)^5$
**103.** $1 + \frac{-5}{x + 2}$ **105.** 1, 7/5 **107.** $[-3, 3], [-3, 0], [0, 3]$
**109.** $\{15.5\}$

## Section 2.6

***For Thought:*** **1.** F **2.** F **3.** T **4.** T **5.** T **6.** F
**7.** T **8.** T **9.** T **10.** F

***Exercises:***
**1.** varies directly **3.** varies inversely **5.** $G = kn$
**7.** $V = \frac{k}{P}$ **9.** $C = khr$ **11.** $Y = \frac{kx}{\sqrt{z}}$
**13.** $A$ varies directly as the square of $r$.
**15.** The variable $y$ varies inversely as $x$.
**17.** No variation **19.** The variable $a$ varies jointly as $z$ and $w$.
**21.** $H$ varies directly as the square root of $t$ and inversely as $s$.
**23.** $D$ varies jointly as $L$ and $J$, and inversely as $W$.
**25.** $y = \frac{5}{9}x$ **27.** $T = \frac{-150}{y}$ **29.** $m = 3t^2$
**31.** $y = \frac{1.37x}{\sqrt{z}}$ **33.** $-27/2$ **35.** 1 **37.** $\sqrt{6}$ **39.** 7/4
**41.** Direct, $L_l = 12L_f$ **43.** Inverse, $P = 20/n$
**45.** Direct, $S_m = 0.6S_k$ **47.** Neither **49.** Direct, $A = 30W$
**51.** Inverse, $n = \frac{50}{p}$ **53.** 2604 lb/in.$^2$ **55.** 12.8 hr
**57.** \$50.70 **59.** \$19.84 **61.** 18.125 oz **63.** 8 ft/yr
**65.** No **67.** 38 **69.** 35.96 ft/sec
**73.** $f^{-1}(x) = (x - 1)^3 + 9$ **75.** 50 mph **77.** $x - 2y = -8$

## Chapter 2 Review Exercises

**1.** $\{-2, 0, 1\}, \{-2, 0, 1\}$, yes **3.** $(-\infty, \infty), (-\infty, \infty)$, yes

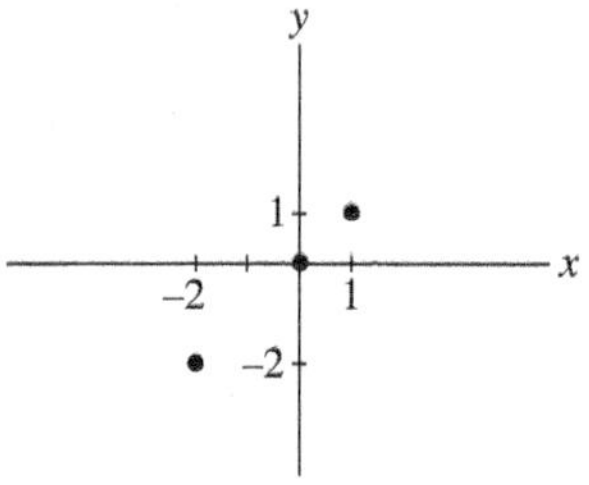


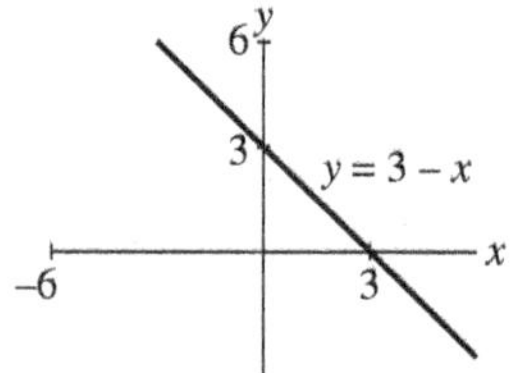


**5.** $\{2\}, (-\infty, \infty)$, no **7.** $[-0.1, 0.1], [-0.1, 0.1]$, no

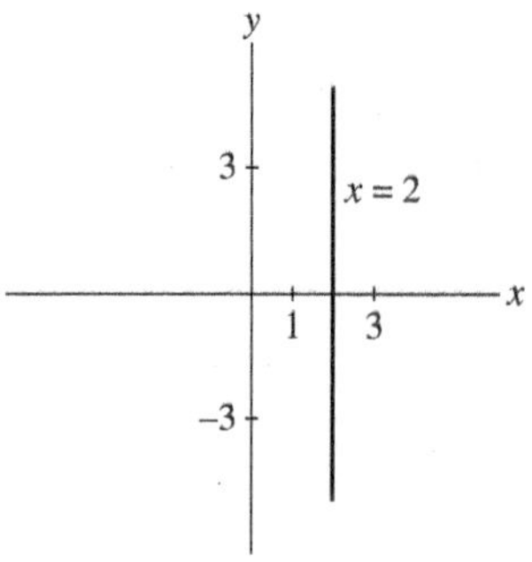


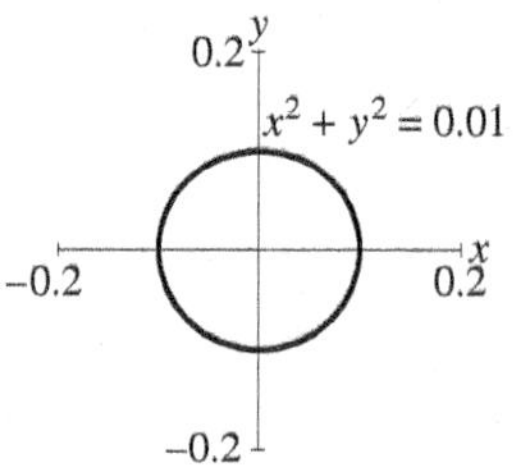


**9.** $[1, \infty), (-\infty, \infty)$, no **11.** $[0, \infty), [-3, \infty)$, yes

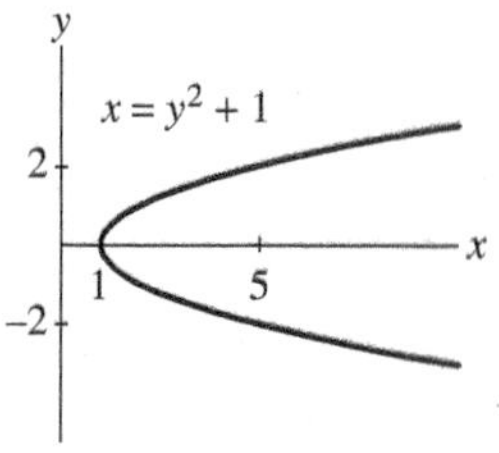


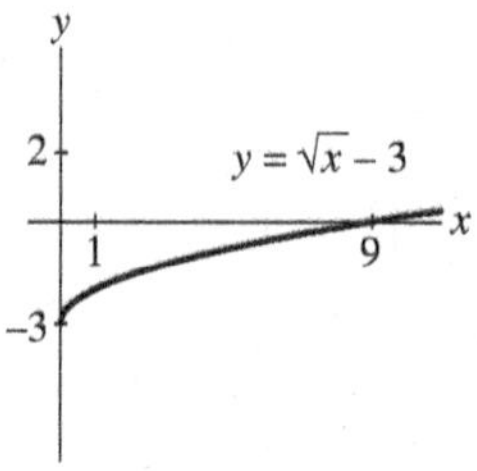

**13.** 12 **15.** 17 **17.** $\pm 4$ **19.** 17 **21.** 4 **23.** $-36$
**25.** 12 **27.** $4x^2 - 28x + 52$ **29.** $x^4 + 6x^2 + 12$
**31.** $a^2 + 2a + 4$ **33.** $6 + h$ **35.** $2x + h$ **37.** $x$
**39.** $\dfrac{x+7}{2}$

**41.**

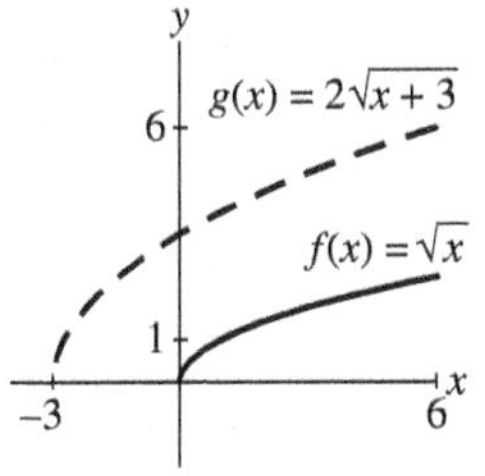


**43.**

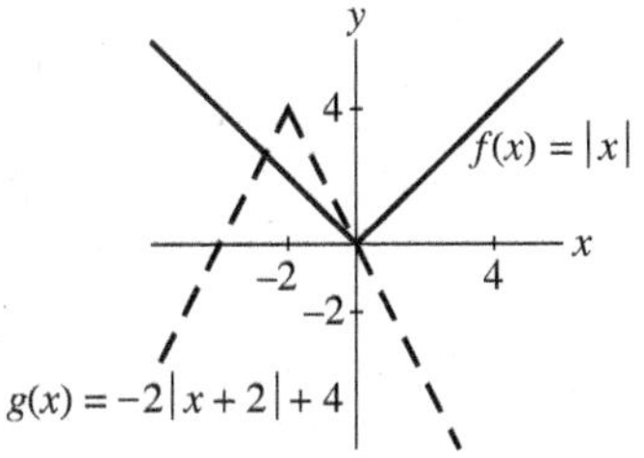


**45.**

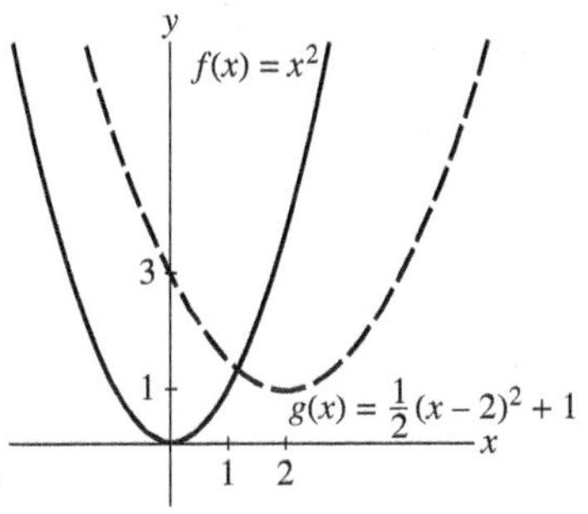


**47.** $(f \circ g)(x) = \sqrt{x} - 4$, $[0, \infty)$
**49.** $(f \circ h)(x) = x^2 - 4$, $(-\infty, \infty)$
**51.** $(g \circ f \circ h)(x) = \sqrt{x^2 - 4}$, $(-\infty, -2] \cup [2, \infty)$

**53.**

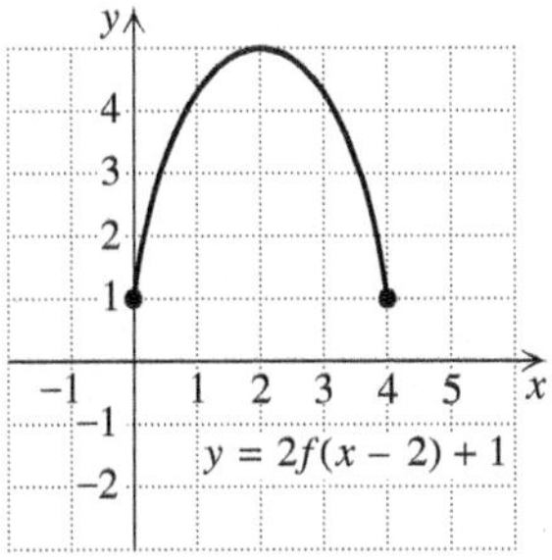


**55.**

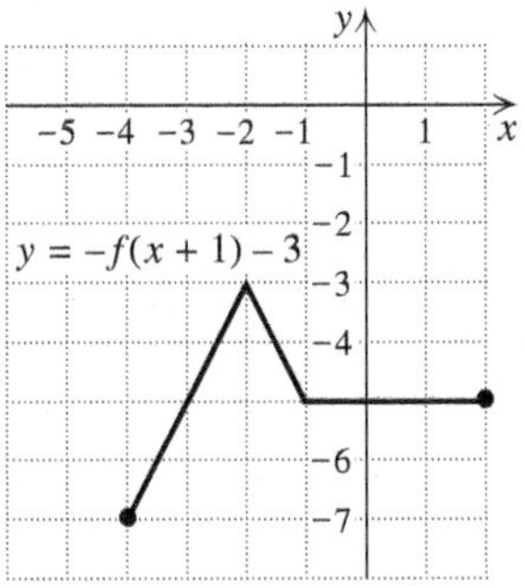


**57.**

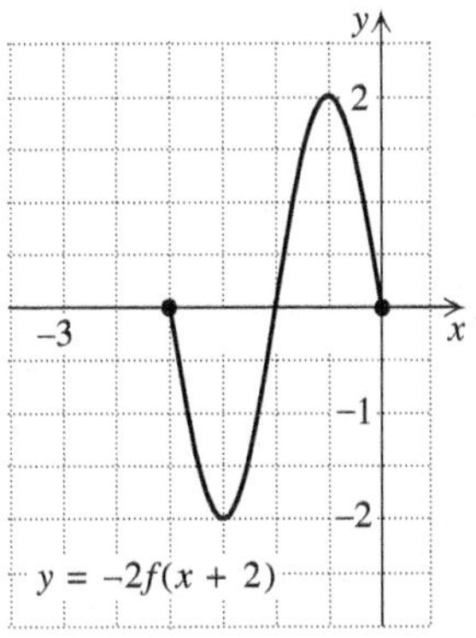


**59.**

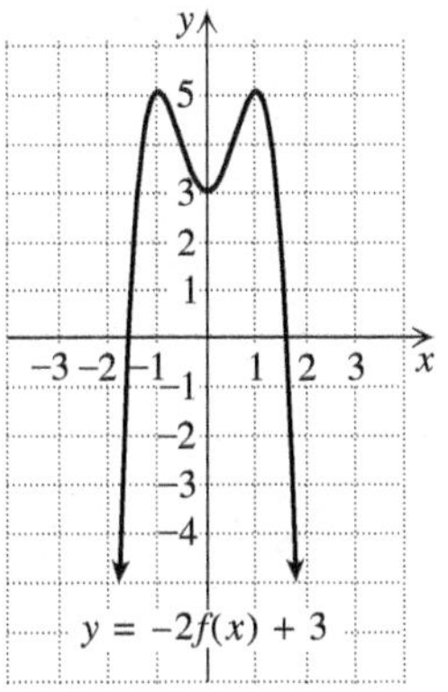


**61.** $F = f \circ g$ **63.** $H = f \circ h \circ g \circ j$
**65.** $N = h \circ f \circ j$ or $N = h \circ j \circ f$ **67.** $R = g \circ h \circ j$
**69.** $-5$ **71.** $\dfrac{-1}{2x(x+h)}$

**73.** $[-10, 10]$, $[0, 10]$, inc $[-10, 0]$, dec $[0, 10]$

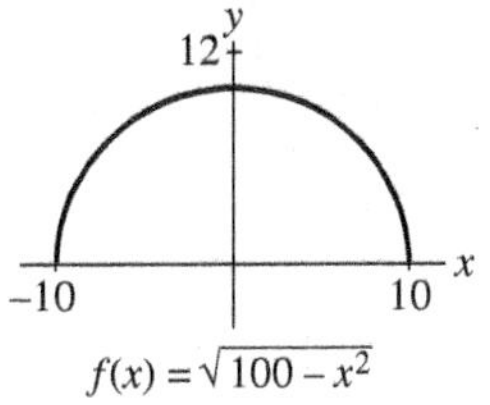


**75.** $(-\infty, \infty)$, $(-\infty, \infty)$, inc $(-\infty, \infty)$

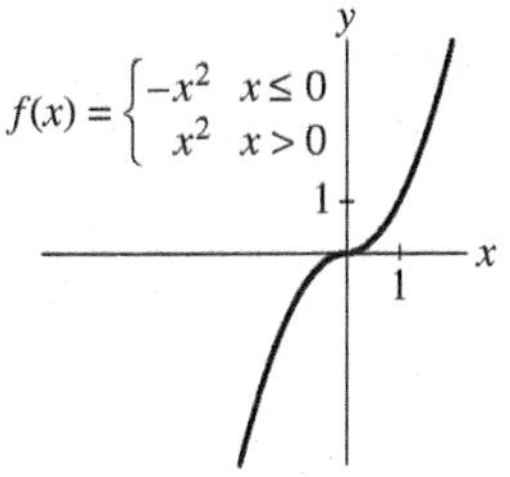


**77.** $(-\infty, \infty)$, $[-2, \infty)$, inc $[-2, 0]$, $[2, \infty)$, dec $(-\infty, -2]$ and $[0, 2]$

$$f(x) = \begin{cases} -x - 4 & x \le -2 \\ -|x| & -2 < x < 2 \\ x - 4 & x \ge 2 \end{cases}$$

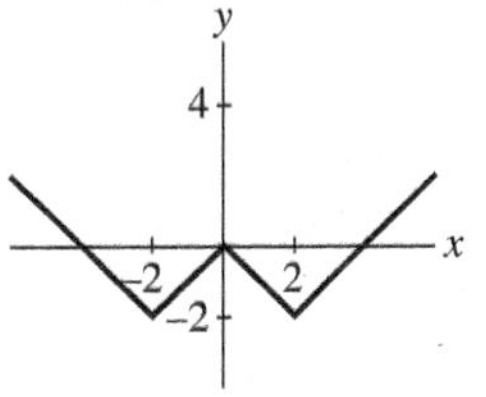


**79.** $y = |x| - 3$, $(-\infty, \infty)$, $[-3, \infty)$
**81.** $y = -2|x| + 4$, $(-\infty, \infty)$, $(-\infty, 4]$
**83.** $y = |x + 2| + 1$, $(-\infty, \infty)$, $[1, \infty)$
**85.** $y$-axis **87.** Origin **89.** Neither symmetry **91.** $y$-axis
**93.** $(-\infty, 2] \cup [4, \infty)$ **95.** $(-\sqrt{2}, \sqrt{2})$ **97.** $\varnothing$
**99.** Inverse functions **101.** Inverse functions

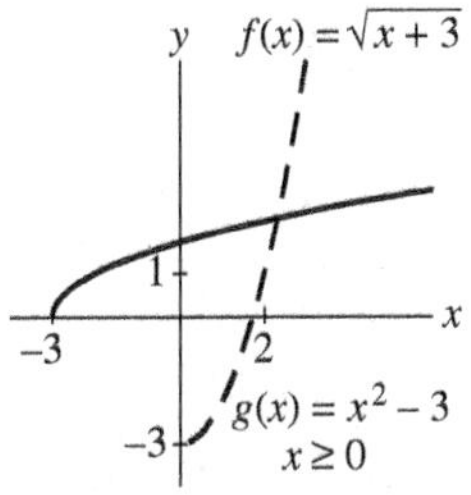


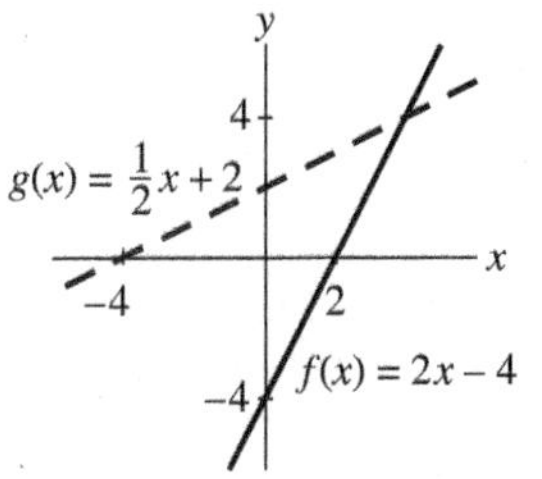


**103.** Not invertible **105.** $f^{-1}(x) = \dfrac{x+21}{3}$, $(-\infty, \infty)$, $(-\infty, \infty)$
**107.** Not invertible **109.** $f^{-1}(x) = x^2 + 9$ for $x \ge 0$, $[0, \infty)$, $[9, \infty)$
**111.** $f^{-1}(x) = \dfrac{5x+7}{1-x}$, $(-\infty, 1) \cup (1, \infty)$, $(-\infty, -5) \cup (-5, \infty)$
**113.** $f^{-1}(x) = -\sqrt{x-1}$, $[1, \infty)$, $(-\infty, 0]$
**115.** $C(x) = 1.20x + 40$, $R(x) = 2x$, $P(x) = 0.80x - 40$ where $x$ is the number of roses, 51 or more roses
**117.** $t = \dfrac{\sqrt{64-h}}{4}$, domain $[0, 64]$ **119.** $d = 2\sqrt{A/\pi}$
**121.** 0.5 in./lb **123.** 36 **125.** 61 km/hr **127.** \$30.72

## Chapter 2 Test

**1.** No **2.** Yes **3.** No **4.** Yes **5.** $\{2, 5\}$, $\{-3, -4, 7\}$
**6.** $[9, \infty)$, $[0, \infty)$ **7.** $[0, \infty)$, $(-\infty, \infty)$
**8.**

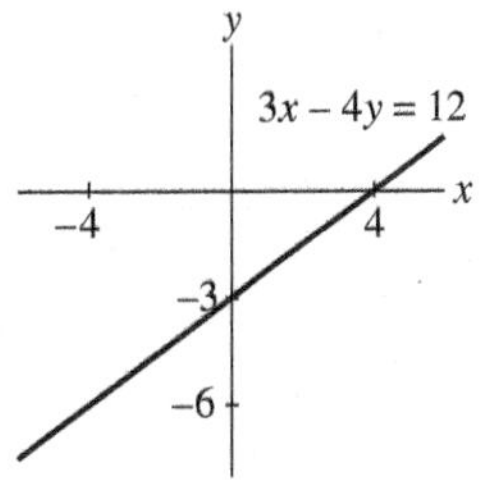


**9.**

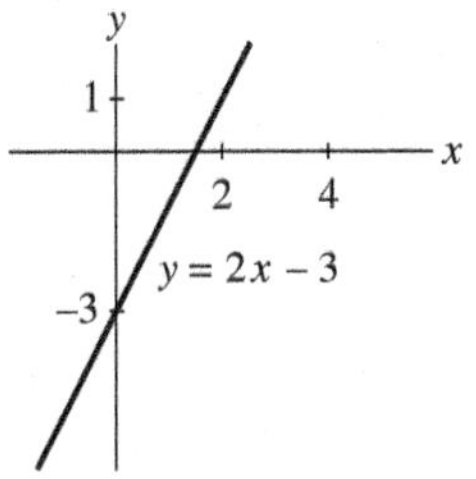

**10.**

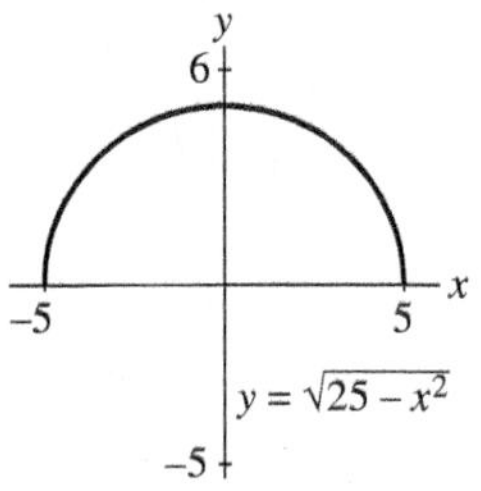


**11.**

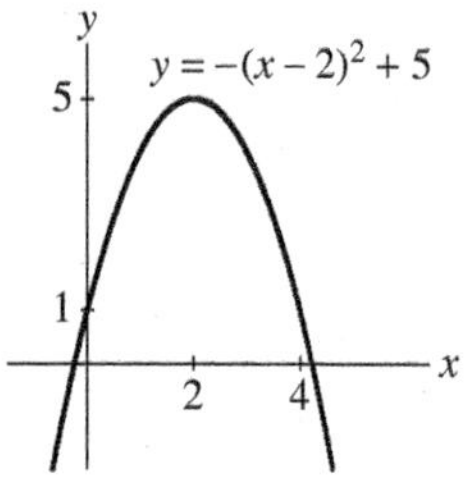


**12.** $(-\infty, \infty), (-\infty, \infty), (3, 0)$

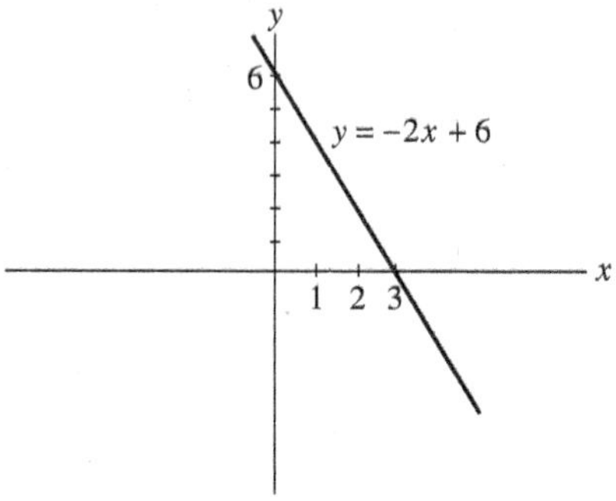


**12.**

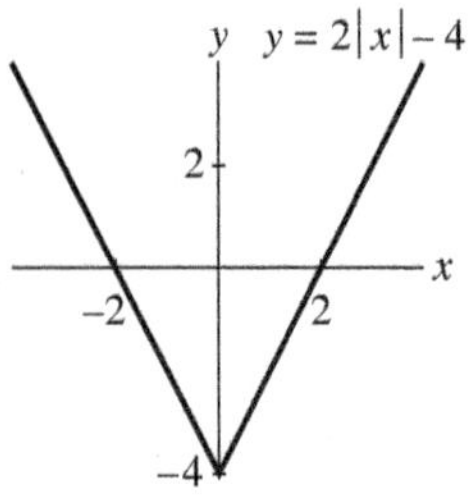


**13.**

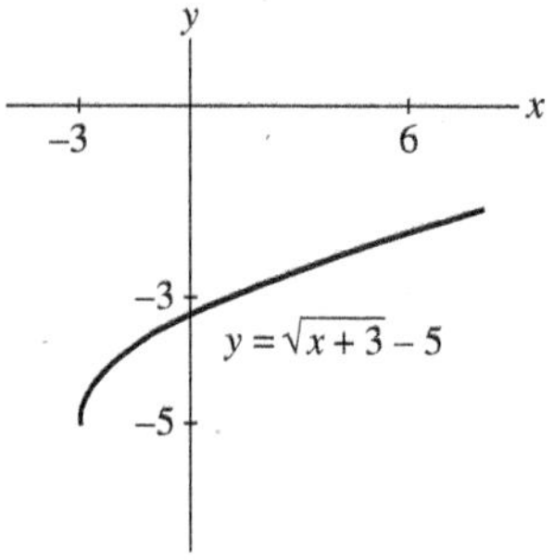


**13.** $(-\infty, \infty), [-100, \infty), (\pm 100, 0)$

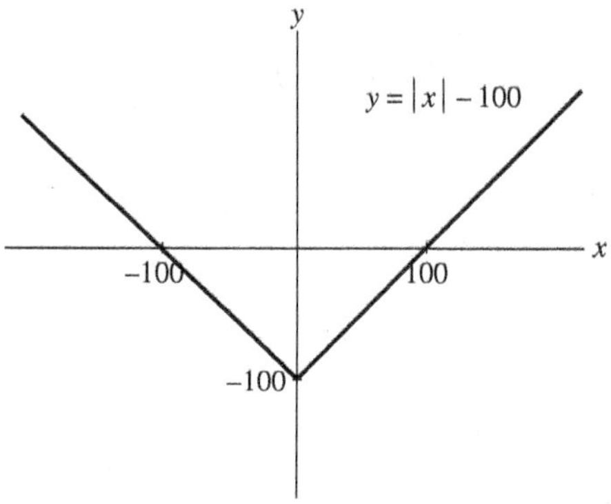


**14.**

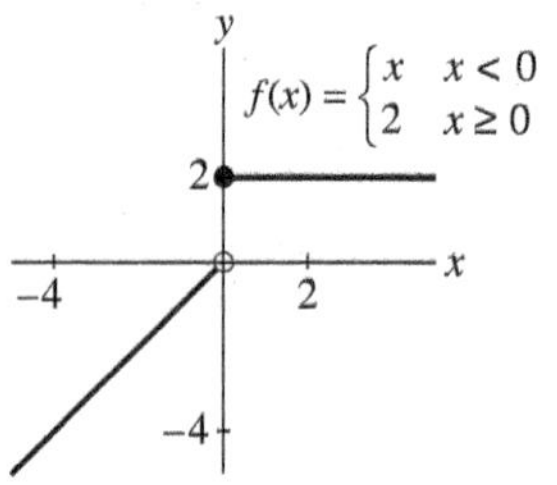


**14.** $(-\infty, \infty), (-\infty, 1], (-89.5, 0), (-90.5, 0)$

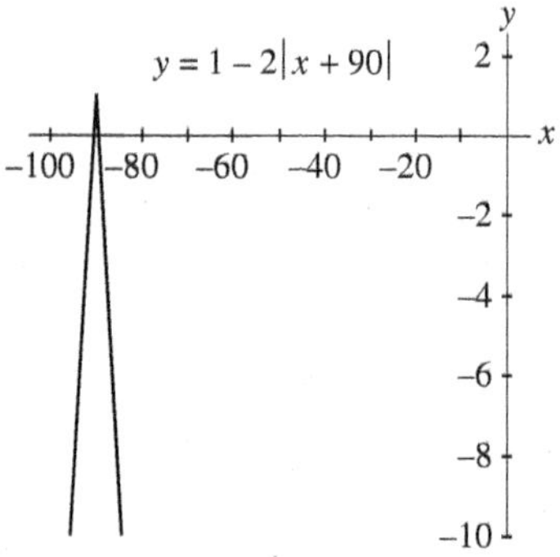


**15.** 3 **16.** $\sqrt{7}$ **17.** $\sqrt{3x + 1}$ **18.** $\dfrac{x + 1}{3}$ **19.** 45
**20.** 3 **21.** Inc $[3, \infty)$, dec $(-\infty, 3]$ **22.** $y$-axis
**23.** $(-2, 4)$ **24.** $g^{-1}(x) = (x - 3)^3 + 2$
**25.** $f^{-1}(x) = x^2 + 5$ for $x \geq 0$ **26.** \$0.125 per envelope
**27.** 12 candlepower **28.** $V = \dfrac{\sqrt{2}d^3}{4}$

## Tying It All Together Chapters 1–2

**1.** $\{3/2\}$ **2.** $\{3\}$ **3.** $\{\pm 100\}$ **4.** $\{-89.5, -90.5\}$
**5.** $\{-27.75\}$ **6.** $\varnothing$ **7.** $\left\{\dfrac{4 \pm \sqrt{2}}{2}\right\}$ **8.** $\{-5/2, -3/2\}$
**9.** $\{\pm\sqrt{5}\}$ **10.** $\varnothing$
**11.** $(-\infty, \infty), (-\infty, \infty), (3/2, 0)$

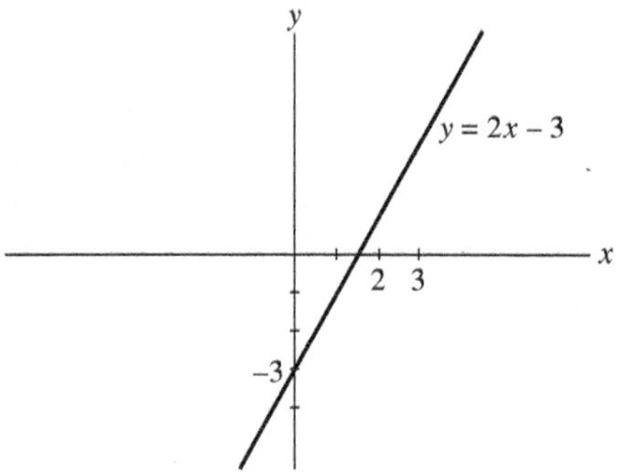


**15.** $[-30, \infty), (-\infty, 3], (-27.75, 0)$

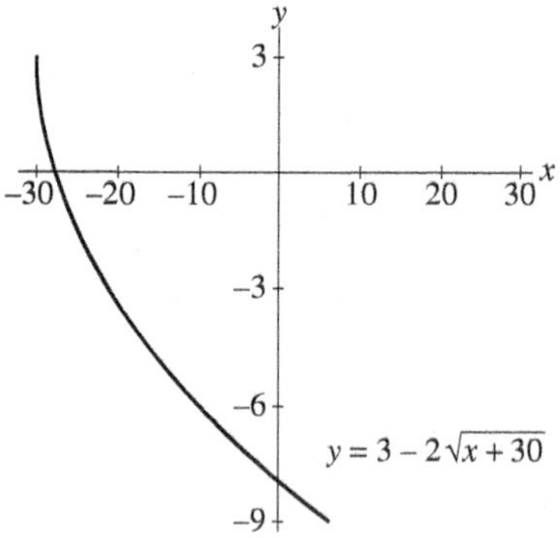

**16.** $[3, \infty)$, $[15, \infty)$, no $x$-intercepts

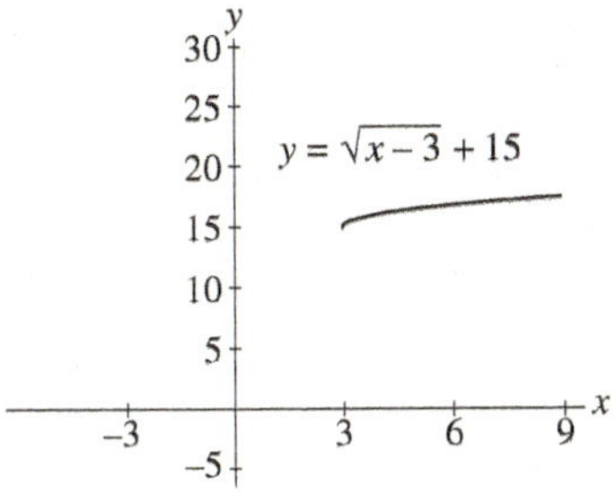


**17.** $(-\infty, \infty)$, $(-\infty, 1]$, $\left(\frac{4 \pm \sqrt{2}}{2}, 0\right)$

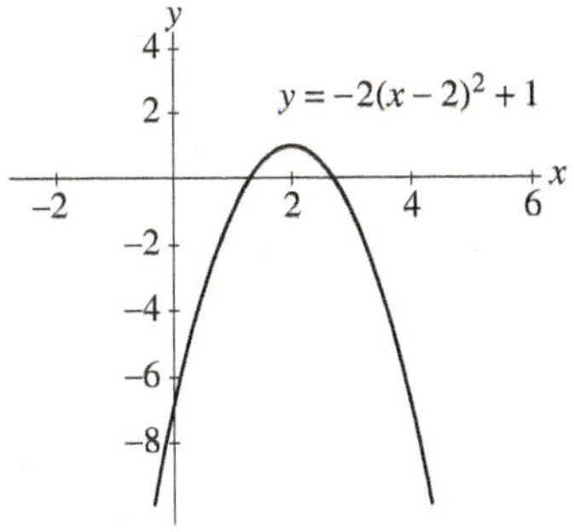


**18.** $(-\infty, \infty)$, $[-1, \infty)$, $(-5/2, 0)$, $(-3/2, 0)$

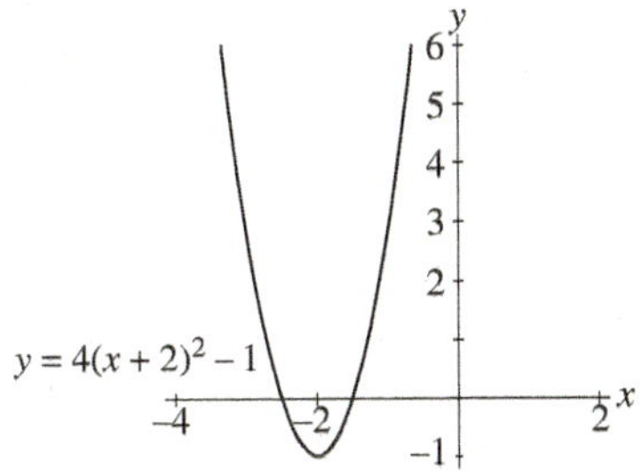


**19.** $[-3, 3]$, $[-2, 1]$, $(\pm\sqrt{5}, 0)$

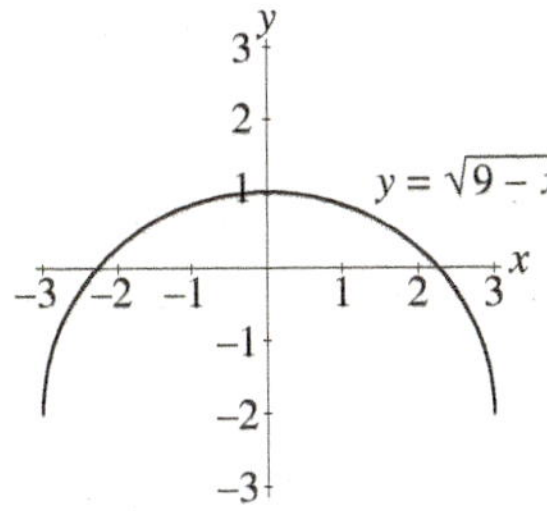


**20.** $[-7, 7]$, $[3, 10]$, no $x$-intercepts

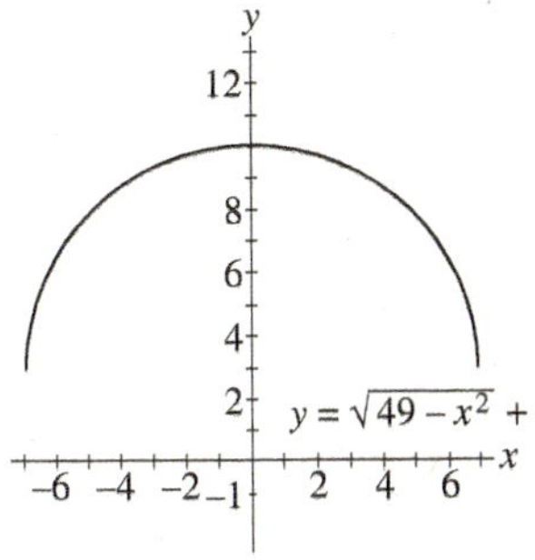


**21.** $(3/2, \infty)$ **22.** $[3, \infty)$ **23.** $(-\infty, -100] \cup [100, \infty)$
**24.** $(-90.5, -89.5)$ **25.** $[-27.75, \infty)$ **26.** $(-\infty, \infty)$
**27.** $\left(-\infty, \frac{4 - \sqrt{2}}{2}\right) \cup \left(\frac{4 + \sqrt{2}}{2}, \infty\right)$
**28.** $(-\infty, -5/2] \cup [-3/2, \infty)$ **29.** $[-\sqrt{5}, \sqrt{5}]$ **30.** $\varnothing$ **31.** 57
**32.**

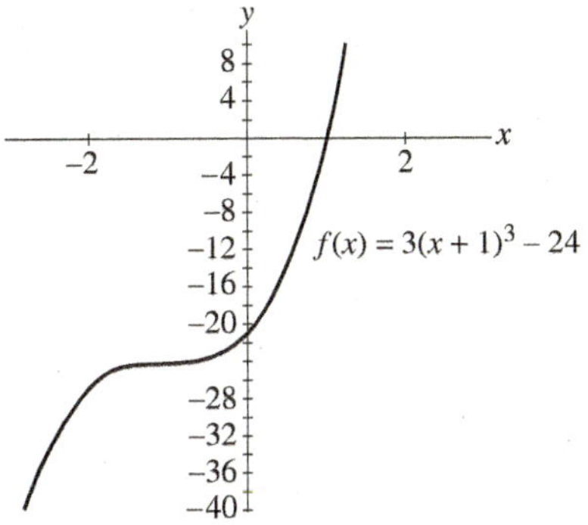


**33.** $f = K \circ F \circ H \circ G$ **34.** $\{1\}$ **35.** $[1, \infty)$

**36.** $x = \sqrt[3]{\frac{y + 24}{3}} - 1$ **37.** $f^{-1}(x) = \sqrt[3]{\frac{x + 24}{3}} - 1$

**38.**

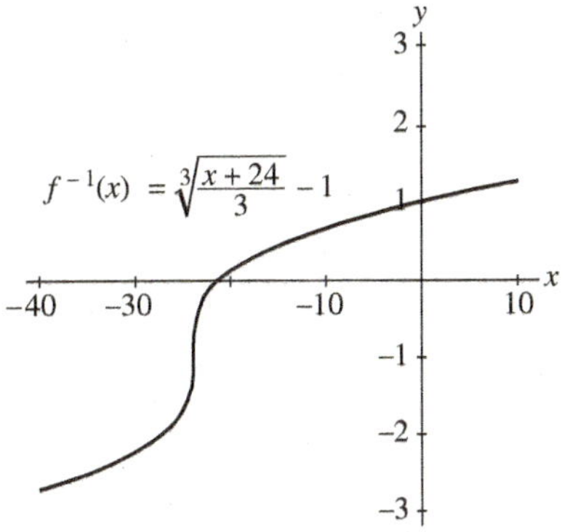


**39.** $(-21, \infty)$ **40.** $f^{-1} = G^{-1} \circ H^{-1} \circ F^{-1} \circ K^{-1}$

## Chapter 3

### Section 3.1

***For Thought:*** **1.** F **2.** F **3.** T **4.** T **5.** T **6.** T
**7.** T **8.** T **9.** T **10.** F

***Exercises:***

**1.** upward **3.** vertex **5.** minimum **7.** axis of symmetry

**9.** $y = (x + 2)^2 - 4$ **11.** $y = \left(x - \frac{3}{2}\right)^2 - \frac{9}{4}$

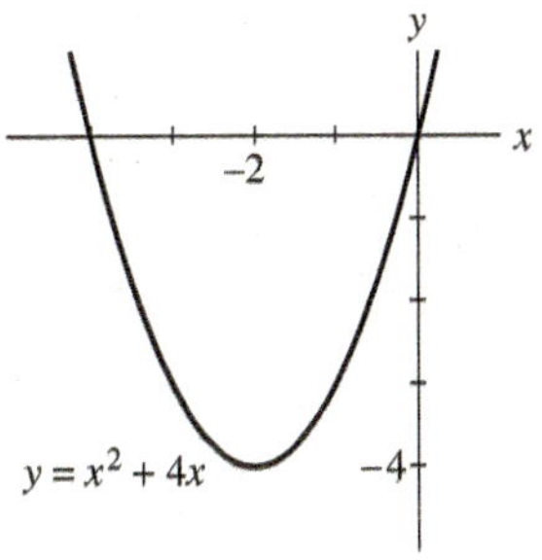


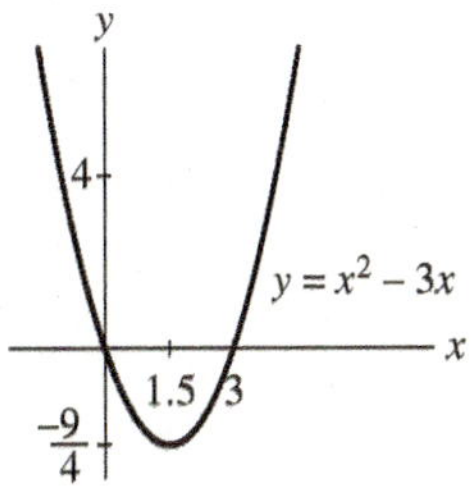

**13.** $y = 2(x-3)^2 + 4$

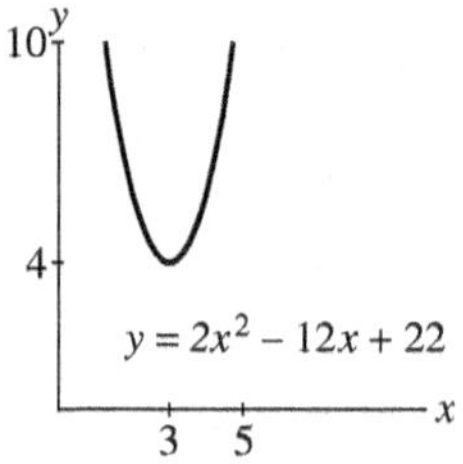


**15.** $y = -3(x-1)^2$

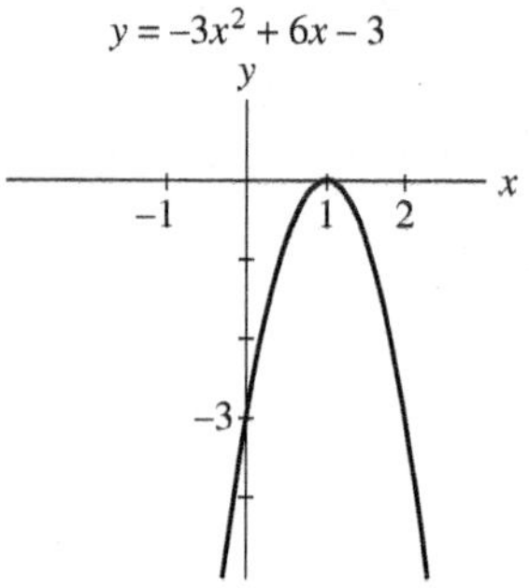


**49.** $(2, 12), x = 2, (0, 0),$ $(4, 0)$, down

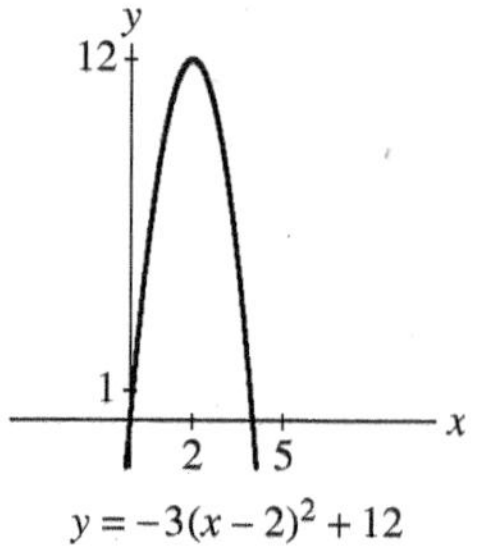


**51.** $(1, 3), x = 1, (0, 1),$ $\left(1 \pm \frac{\sqrt{6}}{2}, 0\right)$, down

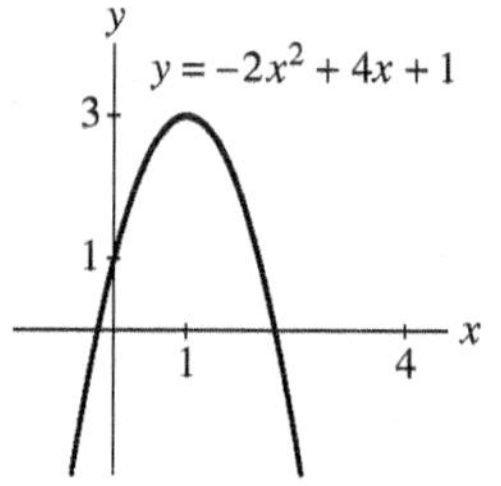


**17.** $y = \left(x + \frac{3}{2}\right)^2 + \frac{1}{4}$

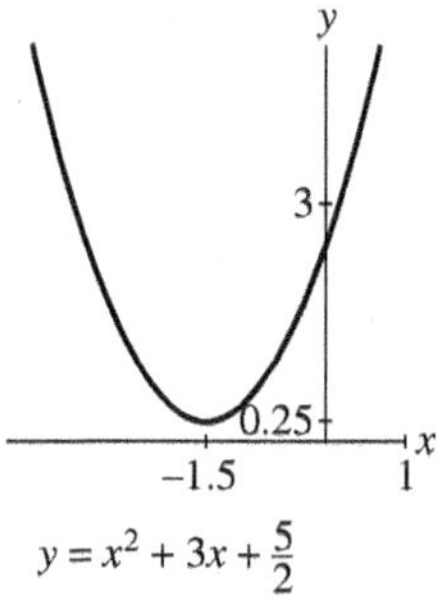


**19.** $y = -2\left(x - \frac{3}{4}\right)^2 + \frac{1}{8}$

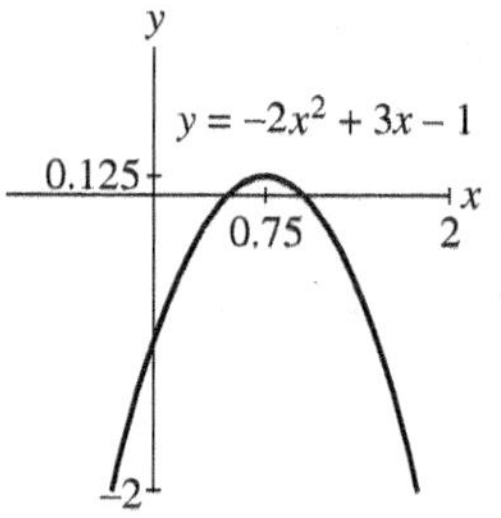


**21.** $(2, -11)$ **23.** $(4, 1)$ **25.** $(-1/3, 1/18)$
**27.** Up, $(1, -4), x = 1, [-4, \infty)$, min $-4$, dec $(-\infty, 1]$, inc $[1, \infty)$
**29.** $(-\infty, 3]$, max 3, inc $(-\infty, 0]$, dec $[0, \infty)$
**31.** $[-1, \infty)$, min $-1$, dec $(-\infty, 1]$, inc $[1, \infty)$
**33.** $[-18, \infty)$, min value $-18$, dec $(-\infty, -4]$, inc $[-4, \infty)$
**35.** $[4, \infty)$, min value 4, dec $(-\infty, 3]$, inc $[3, \infty)$
**37.** $(-\infty, 27/2]$, max $27/2$, inc $(-\infty, 3/2]$, dec $[3/2, \infty)$
**39.** $(-\infty, 9]$, max 9, inc $(-\infty, 1/2]$, dec $[1/2, \infty)$

**41.** $(0, -3), x = 0, (0, -3),$ $(\pm\sqrt{3}, 0)$, up

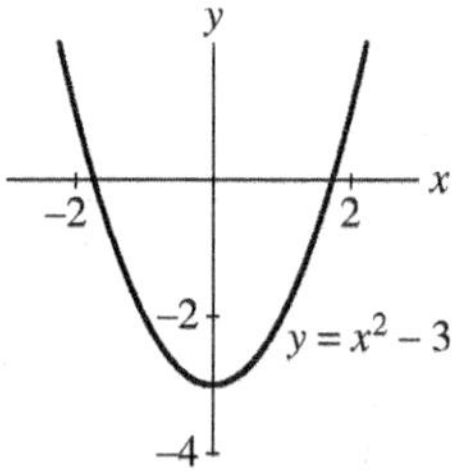


**43.** $\left(\frac{1}{2}, -\frac{1}{4}\right), x = \frac{1}{2},$ $(0, 0), (1, 0)$, up

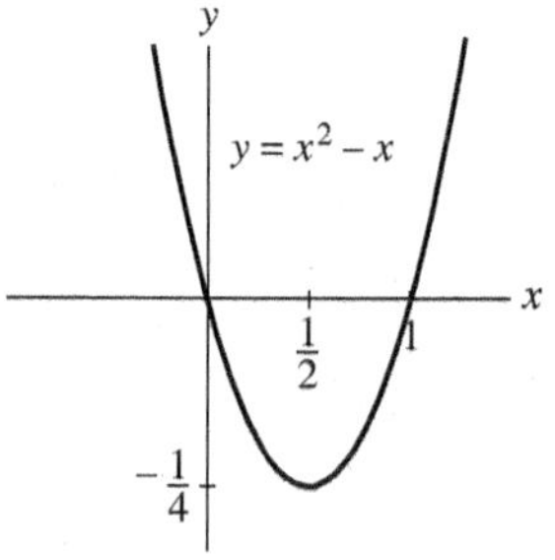


**45.** $(-3, 0), x = -3, (0, 9),$ $(-3, 0)$, up

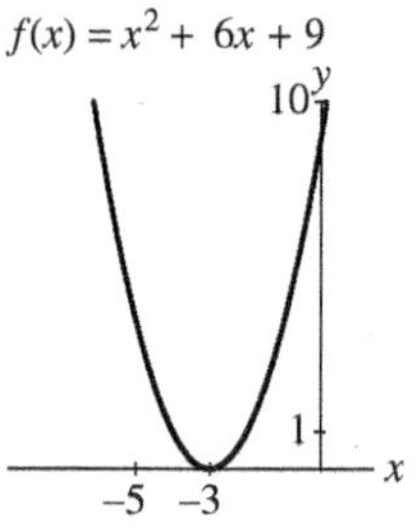


**47.** $(3, -4), x = 3, (0, 5), (1, 0),$ $(5, 0)$, up

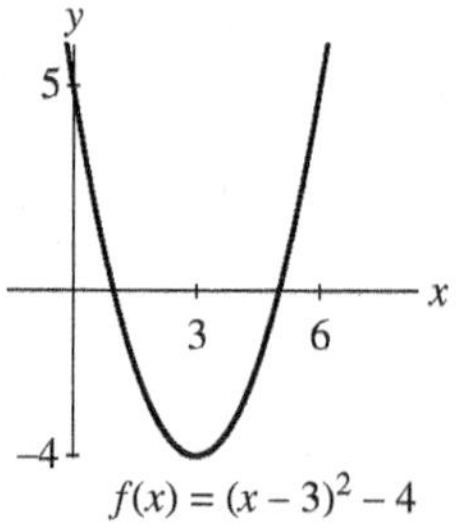


**53.** $(-\infty, -1) \cup (3, \infty)$ **55.** $(2 - \sqrt{3}, 2 + \sqrt{3})$
**57.** $(-\infty, -1/3) \cup (1/2, \infty)$ **59.** $(-\infty, -1] \cup [3, \infty)$
**61.** $(-3, 1)$ **63.** $[-3, 1]$
**65.** $(2 - \sqrt{2}, 2 + \sqrt{2})$

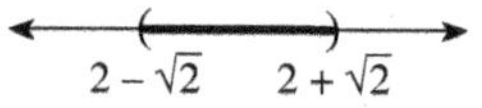


**67.** $(-\infty, -\sqrt{10}) \cup (\sqrt{10}, \infty)$

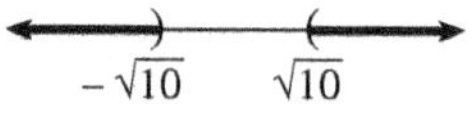


**69.** $(-\infty, 5 - \sqrt{7}) \cup [5 + \sqrt{7}, \infty]$

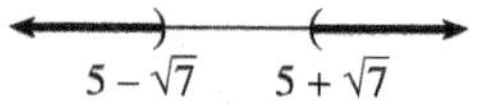


**71.** $(-\infty, \infty)$



**73.** No solution

**75.** $(-\infty, \infty)$



**77.** $(-1, 3/2)$

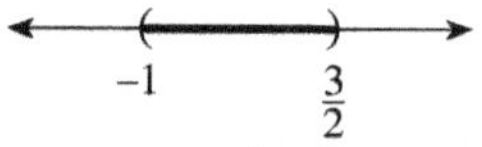


**79.** $(-\infty, -3) \cup (5, \infty)$

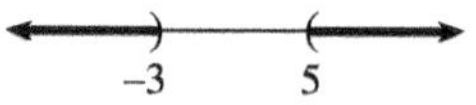


**81.** $(-\infty, -2] \cup [6, \infty)$

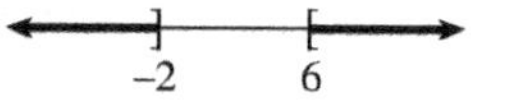


**83.** $[-4, 4]$

**85.** $\{-3\}$

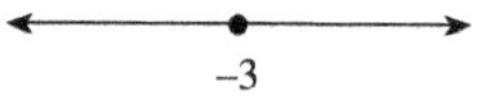


**87.** $(-\infty, 3/2) \cup (3/2, \infty)$

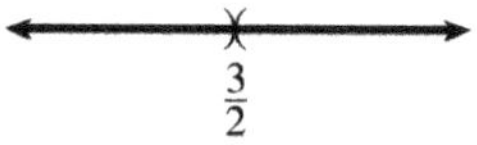


**89. a.** $\{-2, 5\}$ **b.** $\{0, 3\}$ **c.** $(-\infty, -2) \cup (5, \infty)$ **d.** $[-2, 5]$
**e.** $f(x) = \left(x - \frac{3}{2}\right)^2 - \frac{49}{4}$; Move $y = x^2$ to the right $\frac{3}{2}$ and down $\frac{49}{4}$ to obtain $f$.
**f.** $(-\infty, \infty)$, $\left[-\frac{49}{4}, \infty\right)$, minimum $-\frac{49}{4}$

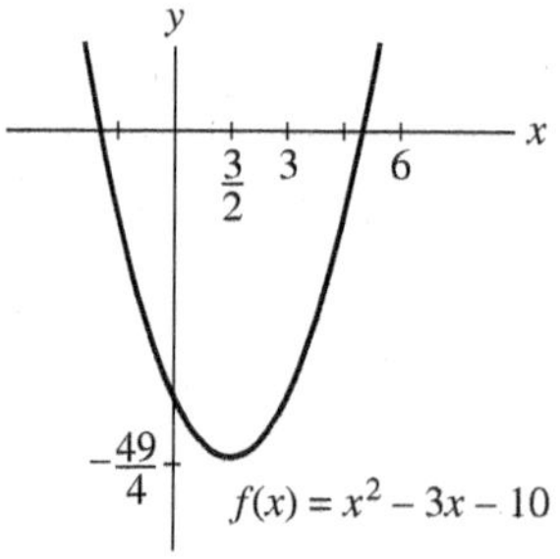


**g.** The graph of $f$ is above the $x$-axis when $x$ is in $(-\infty, -2) \cup (5, \infty)$ and on or below the $x$-axis when $x$ is in $[-2, 5]$.
**h.** $(-2, 0), (5, 0), (0, -10), x = \frac{3}{2}, \left(\frac{3}{2}, -\frac{49}{4}\right)$, opens upward, dec on $\left(-\infty, \frac{3}{2}\right]$ and inc on $\left[\frac{3}{2}, \infty\right)$
**91.** 261 ft **93. a.** 408 ft **b.** $(10 + \sqrt{102})/2 \approx 10.05$ sec
**95. a.** Approximately 100 mph **b.** 97.24 mph **c.** 13.2 gal/hr
**97.** 50 yd by 50 yd **99.** 20 ft by 30 ft **101.** 7.5 ft by 15 ft
**103.** 5 in. wide, 2.5 in. high
**105. a.** $p = 50 - n$ **b.** $R = 50n - n^2$ **c.** \$625 **107.** 1/2

**109. a.**

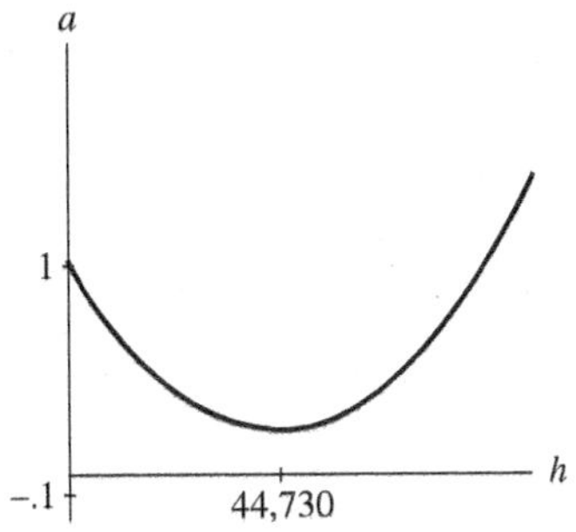


$a = 3.89 \times 10^{-10}h^2 - 3.48 \times 10^{-5}h + 1$

**b.** Decreasing **c.** Dec $[0, 44{,}730]$, inc $[44{,}730, \infty)$ **d.** No
**e.** Approximately (0, 30,000)
**111. a.** $y = -2405.7x + 36{,}397.8$, $y = 156.4x^2 - 3969.9x + 39{,}265.6$
**b.** Quadratic

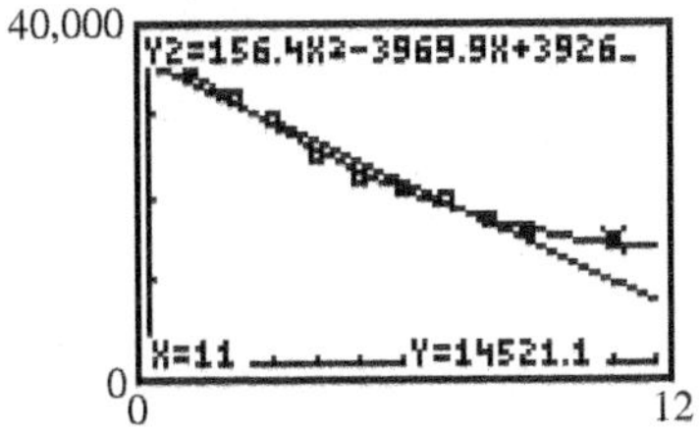


**c.** \$9935, \$14,521
**113.** 9/10 **115.** 12 **117.** $\{-3, 1, 3, 4\}$, $\{2, 5\}$

## Section 3.2

***For Thought:*** **1.** F **2.** T **3.** T **4.** T **5.** F **6.** F **7.** F **8.** T **9.** T **10.** F
***Exercises:***
**1.** zero **3.** zero, factor **5.** $x - 3, 1$ **7.** $-2x^2 + 6x - 14, 33$
**9.** $s^2 + 2, 16$ **11.** $x + 6, 13$ **13.** $-x^2 + 4x - 16, 57$
**15.** $4x^2 + 2x - 4, 0$ **17.** $2a^2 - 4a + 6, 0$
**19.** $x^3 + x^2 + x + 1, -2$ **21.** $x^4 + 2x^3 - 2x^2 - 4x - 4, -13$
**23.** 0 **25.** −33 **27.** 5 **29.** $\frac{55}{8}$ **31.** 0 **33.** 8
**35.** $(x + 3)(x + 2)(x - 1)$ **37.** $(x - 4)(x + 3)(x + 5)$
**39.** Yes **41.** No **43.** Yes **45.** No
**47.** $\pm(1, 2, 3, 4, 6, 8, 12, 24)$ **49.** $\pm(1, 3, 5, 15)$
**51.** $\pm\left(1, 3, 5, 15, \frac{1}{2}, \frac{1}{4}, \frac{1}{8}, \frac{3}{2}, \frac{3}{4}, \frac{3}{8}, \frac{5}{2}, \frac{5}{4}, \frac{5}{8}, \frac{15}{2}, \frac{15}{4}, \frac{15}{8}\right)$
**53.** $\pm\left(1, 2, \frac{1}{2}, \frac{1}{3}, \frac{2}{3}, \frac{1}{6}, \frac{1}{9}, \frac{2}{9}, \frac{1}{18}\right)$ **55.** 2, 3, 4 **57.** $-3, 2 \pm i$
**59.** $\frac{1}{2}, \frac{3}{2}, \frac{5}{2}$ **61.** $\frac{1}{2}, \frac{1 \pm i}{3}$ **63.** $-6, 3 \pm i$ **65.** $\pm i, 1, -2$
**67.** $-1, \pm\sqrt{2}$ **69.** $\frac{1}{4}, \frac{1}{3}, \frac{1}{2}$ **71.** $\frac{1}{16}, 1 \pm 2i$ **73.** $-\frac{6}{7}, \frac{7}{3}, \pm i$
**75.** $-5, -2, 1, \pm 3i$ **77.** $1, 3, 5, 2 \pm \sqrt{3}$ **79.** $2 + \frac{5}{x - 2}$
**81.** $a + \frac{5}{a - 3}$ **83.** $1 + \frac{-3c}{c^2 - 4}$ **85.** $2 + \frac{-7}{2t + 1}$
**87. a.** 6 hr **b.** ≈120 ppm **c.** ≈3 hr **d.** ≈4 hr
**89.** 5 in. by 9 in. by 14 in. **93.** $f(x) = 2\left(x - \frac{3}{4}\right)^2 - \frac{1}{8}$ **95.** $(-\infty, \infty)$
**97. a.** $6a(4a - 5)(a + 2)$ **b.** $x(x^2 + 4)(x - 2)(x + 2)$

## Section 3.3

***For Thought:*** **1.** F **2.** T **3.** T **4.** F **5.** F **6.** T **7.** F **8.** F **9.** T **10.** T
***Exercises:***
**1.** multiplicity **3.** $a - bi$ **5.** Degree 2, 5 with multiplicity 2
**7.** Degree 5, $\pm 3$, 0 with multiplicity 3
**9.** Degree 4, 0, 1 each with multiplicity 2
**11.** Degree 4, $-\frac{4}{3}, \frac{3}{2}$ each with multiplicity 2
**13.** Degree 3, $0, 2 \pm \sqrt{10}$ **15.** $x^2 + 9$ **17.** $x^2 - 2x - 1$
**19.** $x^2 - 6x + 13$ **21.** $x^3 - 8x^2 + 37x - 50$
**23.** $x^2 - 2x - 15 = 0$ **25.** $x^2 + 16 = 0$
**27.** $x^2 - 6x + 10 = 0$ **29.** $x^3 + 2x^2 + x + 2 = 0$
**31.** $x^3 + 3x = 0$ **33.** $x^3 - 5x^2 + 8x - 6 = 0$
**35.** $x^3 - 6x^2 + 11x - 6 = 0$ **37.** $x^3 - 5x^2 + 17x - 13 = 0$
**39.** $24x^3 - 26x^2 + 9x - 1 = 0$ **41.** $x^4 - 2x^3 + 3x^2 - 2x + 2 = 0$
**43.** 3 neg; 1 neg, 2 imag **45.** 1 pos, 2 neg; 1 pos, 2 imag
**47.** 4 imag **49.** 4 pos; 2 pos, 2 imag; 4 imag **51.** 4 imag and 0
**53.** $-1 < x < 3$ **55.** $-3 < x < 2$ **57.** $-1 < w < 5$
**59.** $-1 < x < 3$ **61.** $-2, 1, 5$ **63.** $-3, \frac{3 \pm \sqrt{13}}{2}$
**65.** $\pm i, 2, -4$ **67.** $-5, \frac{1}{3}, \frac{1}{2}$ **69.** 1, −2 each with multiplicity 2
**71.** 0, 2 with multiplicity 3 **73.** $0, 1, \pm 2, \pm i\sqrt{3}$
**75.** −2, −1, 1/4, 1, 3/2 **77.** $-5 < x < 6, -5 < x < 6$
**79.** $-6 < x < 6, -5 < x < 5$ **81.** $-1 < x < 23, -1 < x < 23$
**83.** 4 hr and 5 hr **85.** 3 in. **91.** $f(x) = -\frac{1}{2}x^3 + 3x^2 - \frac{11}{2}x + 3$
**93.** $\pm 1, \pm 2, \pm 3, \pm 6, \pm 1/2, \pm 3/2$ **95.** $b$ is a function of $a$.
**97.** $(f \circ g)(x) = 8x^2 - 32x + 23$

## Section 3.4

***For Thought:*** **1.** F **2.** F **3.** F **4.** F **5.** T **6.** F **7.** F **8.** T **9.** T **10.** F
***Exercises:***
**1.** $\{\pm 2, -3\}$ **3.** $\left\{-500, \pm\frac{\sqrt{2}}{2}\right\}$ **5.** $\left\{0, \frac{15 \pm \sqrt{205}}{2}\right\}$
**7.** $\{0, \pm 2\}$ **9.** $\{\pm 2, \pm 2i\}$ **11.** $\{8\}$ **13.** $\{25\}$
**15.** $\left\{\frac{1}{4}\right\}$ **17.** $\left\{\frac{2 + \sqrt{13}}{9}\right\}$ **19.** $\{-4, 6\}$ **21.** $\{9\}$
**23.** $\{5\}$ **25.** $\{10\}$ **27.** $\{\pm 2\sqrt{2}\}$ **29.** $\left\{\pm\frac{1}{8}\right\}$
**31.** $\left\{\frac{1}{49}\right\}$ **33.** $\left\{\frac{5}{4}\right\}$ **35.** $\{\pm 3, \pm\sqrt{3}\}$ **37.** $\{\pm i\sqrt{7}, \pm 1\}$
**39.** $\{\pm 3i, \pm 3\}$ **41.** $\left\{-\frac{17}{2}, \frac{13}{2}\right\}$ **43.** $\left\{\frac{3}{20}, \frac{4}{15}\right\}$
**45.** $\{-2, -1, 5, 6\}$ **47.** $\{1, 9\}$ **49.** $\{9, 16\}$ **51.** $\{8, 125\}$
**53.** $\{\pm\sqrt{7}, \pm 1\}$ **55.** $\{0, 8\}$ **57.** $\{-3, 0, 1, 4\}$
**59.** $\{-2, 4\}$ **61.** $\{1/2\}$ **63.** $\{3\}$ **65.** $\left\{\frac{1}{2}\right\}$
**67.** $\{\pm 2, -1 \pm i\sqrt{3}, 1 \pm i\sqrt{3}\}$ **69.** $\{\sqrt{3}, 2\}$
**71.** $\{-2, \pm 1\}$ **73.** $\{5 \pm 9i\}$ **75.** $\left\{\frac{1 \pm 4\sqrt{2}}{3}\right\}$
**77.** $\{\pm 2\sqrt{6}, \pm\sqrt{35}\}$ **79.** $\{\pm 3, 2\}$ **81.** $\{-11\}$ **83.** $\{2\}$
**85.** 279.56 m² **87.** 23 **89.** $\frac{25}{4}$ and $\frac{49}{4}$ **91.** 5 in.
**93.** 1600 ft² **95.** 17,419.3 lb **97.** 462.89 in.³ **99.** 1 P.M.
**101. a.** \$478 million **b.** \$572 million **103.** 27.4 m
**105.** $6 - \sqrt{10}$ ft **107.** $x^3 + 5x^2 + 4x + 20 = 0$
**109.** $(-2, 4)$ **111.** $y = -\frac{2}{3}x + 10$

## Section 3.5

***For Thought:*** **1.** F **2.** T **3.** T **4.** F **5.** T **6.** T **7.** F **8.** F **9.** T **10.** F
***Exercises:***
**1.** $y$-axis **3.** $-b/(2a)$ **5.** Symmetric about $y$-axis
**7.** Symmetric about $x = 3/2$ **9.** Neither symmetry
**11.** Symmetric about origin **13.** Symmetric about $x = 5$
**15.** Symmetric about origin **17.** Does not cross at (4, 0)

**19.** Crosses at $(1/2, 0)$ **21.** Crosses at $(1/4, 0)$
**23.** No $x$-intercepts **25.** Does not cross at $(0, 0)$, crosses at $(3, 0)$
**27.** Crosses at $(1/2, 0)$, does not cross at $(1, 0)$
**29.** Does not cross at $(-3, 0)$, crosses at $(2, 0)$ **31.** $y \to \infty$
**33.** $y \to -\infty$ **35.** $y \to -\infty$ **37.** $y \to \infty$ **39.** $y \to \infty$
**41.** Neither symmetry; crosses at $(-2, 0)$; does not cross at $(1, 0)$; $y \to \infty$ as $x \to \infty$; $y \to -\infty$ as $x \to -\infty$
**43.** Symmetric about $y$-axis; no $x$-intercepts; $y \to \infty$ as $x \to \infty$; $y \to \infty$ as $x \to -\infty$ **45.** $\infty$ **47.** $-\infty$ **49.** $\infty$ **51.** $-\infty$

**53.**

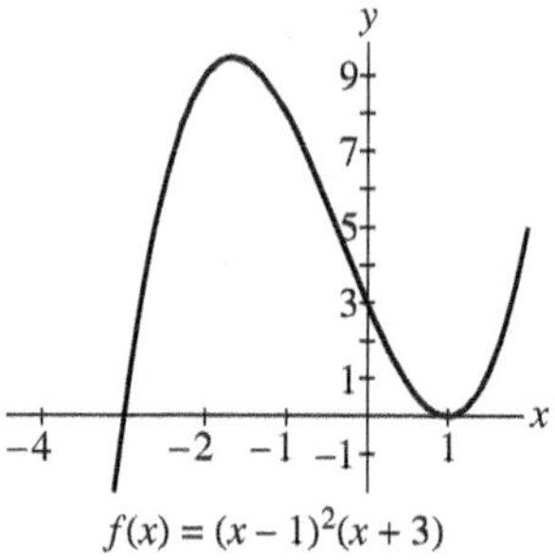

$f(x) = (x-1)^2(x+3)$

**55.**

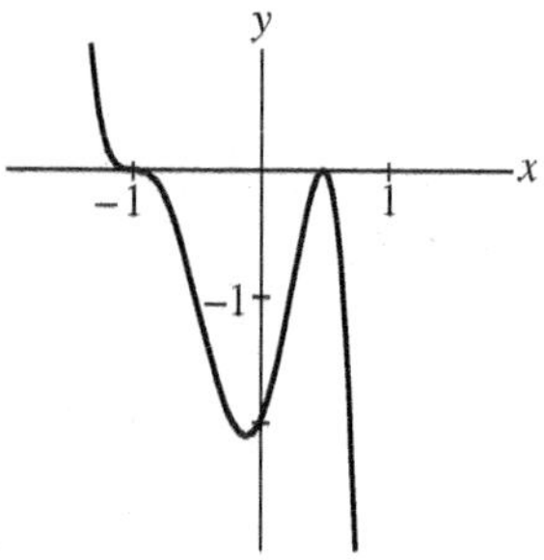

$f(x) = -2(2x-1)^2(x+1)^3$

**57.** (e) **59.** (g) **61.** (b) **63.** (c)

**65.**

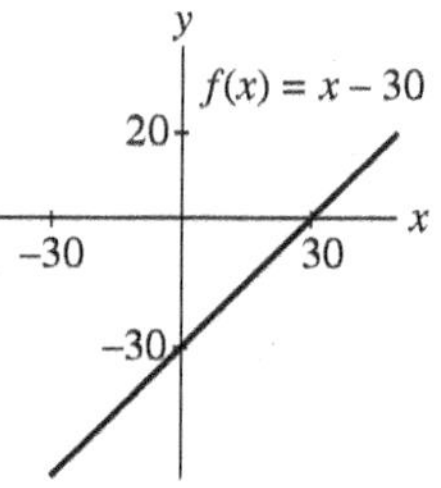


**67.**

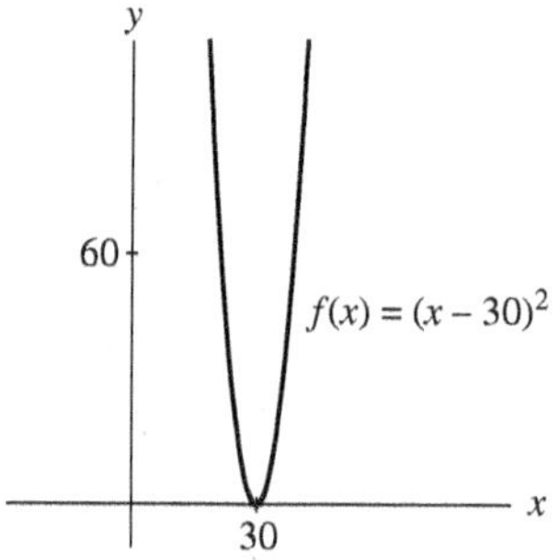


**69.**

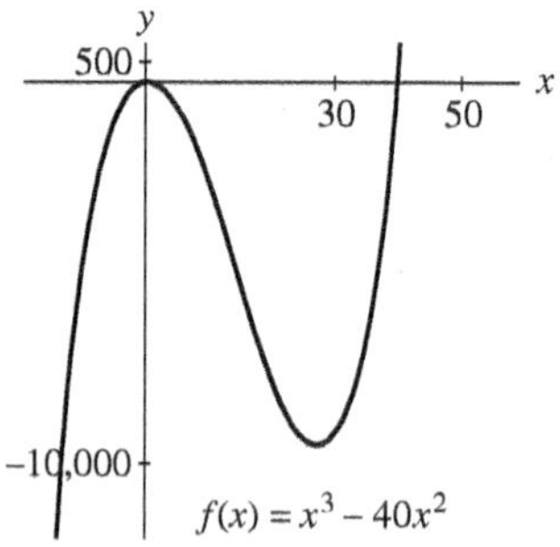


**71.**

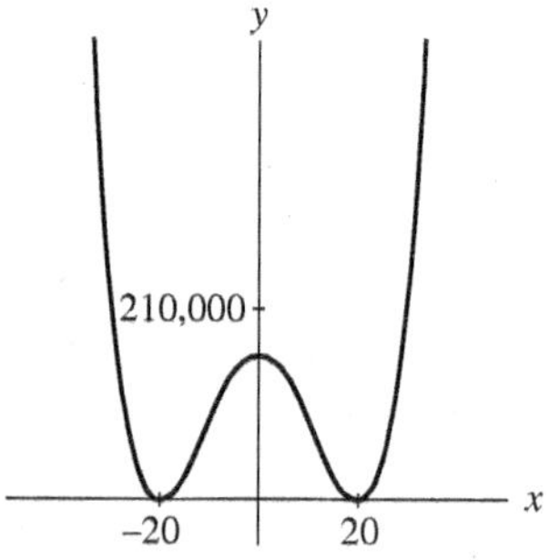

$f(x) = (x-20)^2(x+20)^2$

**73.**

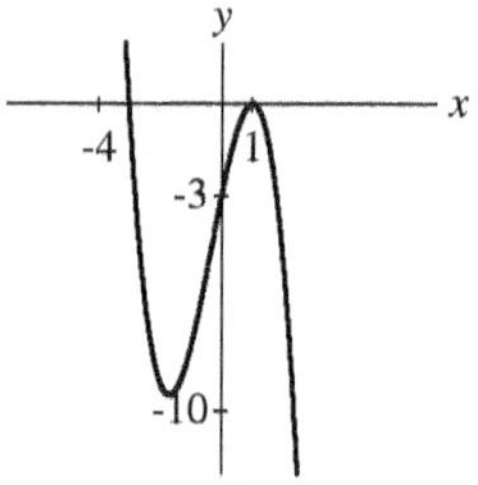

$f(x) = -x^3 - x^2 + 5x - 3$

**75.**

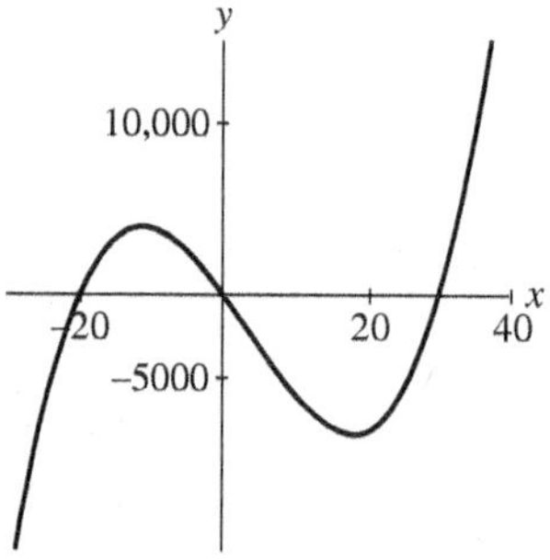

$f(x) = x^3 - 10x^2 - 600x$

**77.**

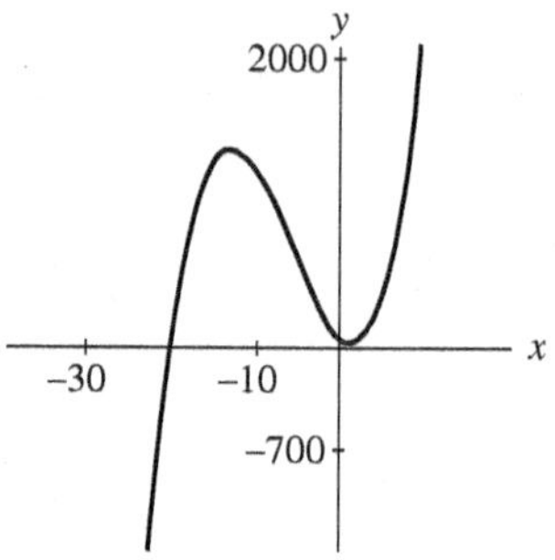

$f(x) = x^3 + 18x^2 - 37x + 60$

**79.**

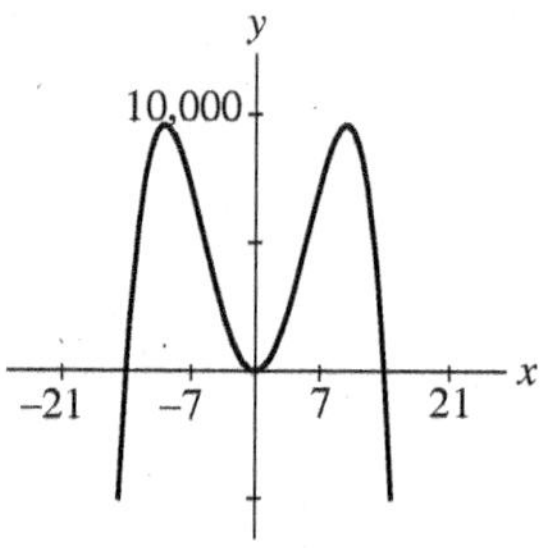

$f(x) = -x^4 + 196x^2$

**81.**

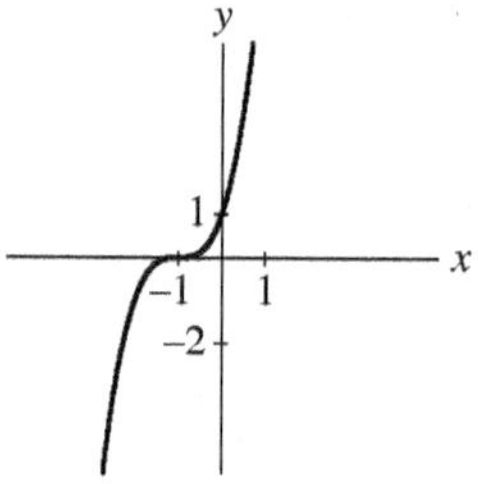

$f(x) = x^3 + 3x^2 + 3x + 1$

**83.**

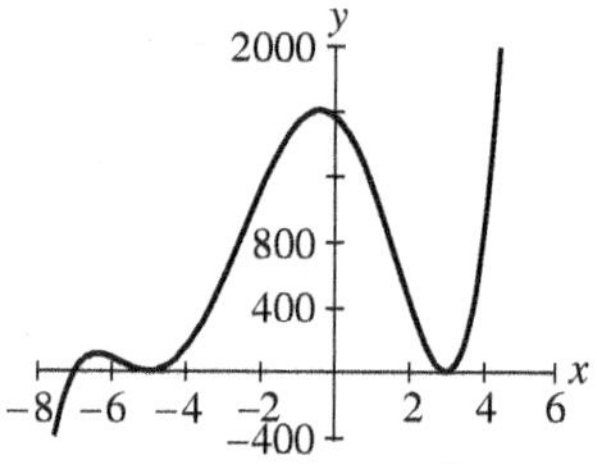

$f(x) = (x-3)^2(x+5)^2(x+7)$

**85.** $(-\sqrt{3}, 0) \cup (\sqrt{3}, \infty)$ **87.** $(-\infty, -\sqrt{2}] \cup \{0\} \cup [\sqrt{2}, \infty)$
**89.** $(-4, -1) \cup (1, \infty)$ **91.** $[-4, 2] \cup [6, \infty)$ **93.** $(-\infty, 1)$
**95.** $[-\sqrt{10}, -3] \cup [3, \sqrt{10}]$ **97.** $(-2, 0) \cup (2, \infty)$
**99.** $[-2, 0] \cup [2, \infty)$ **101.** $(-2, -1) \cup (1, 2)$
**103.** $(-\infty, -2] \cup [-1, 1] \cup [2, \infty)$ **105.** d

**107.**

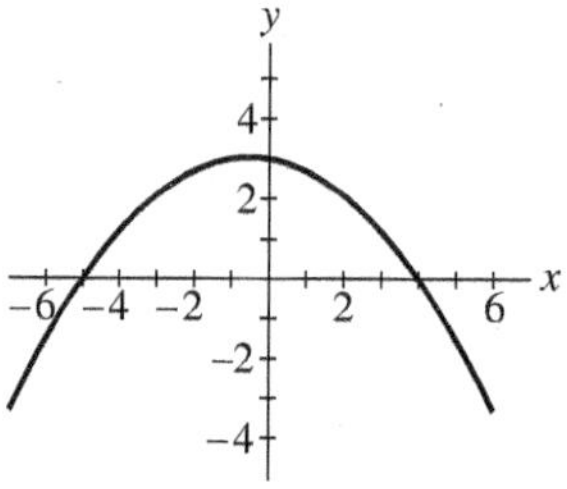

$y = -\frac{3}{20}(x+5)(x-4)$

**109.**

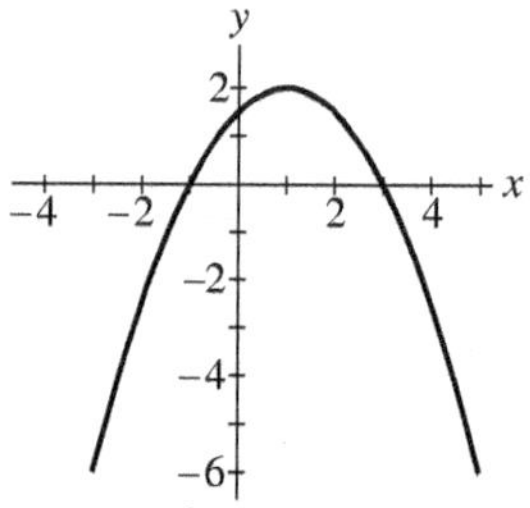

$y = -\frac{1}{2}(x+1)(x-3)$

**111.**

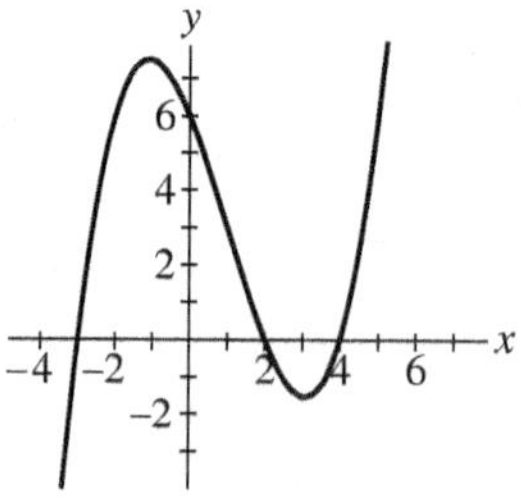

$y = \frac{1}{4}(x-2)(x+3)(x-4)$

**113.**

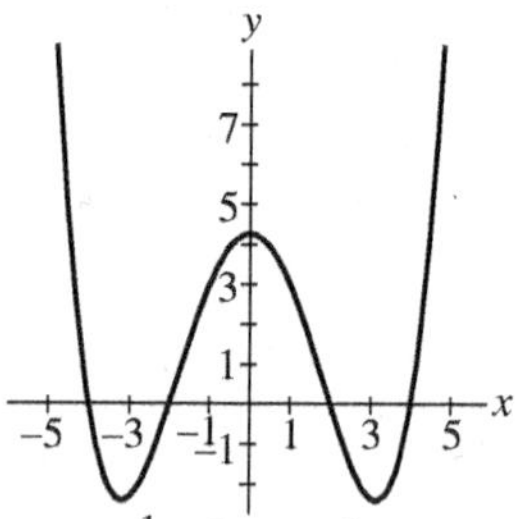

$y = \frac{1}{15}(x^2-4)(x^2-16)$

**115.** \$3,400, \$2,600, $x > \$2,200$
**117.** Decreases to 0 at 30 stores, then increases
**119.** $V = 3x^3 - 24x^2 + 48x$, 4/3 in. by 4 in. by 16/3 in.
**121.** All of the paint coats the inside of can. $r = 6.67 \times 10^{-4}$ ft and $h = 11{,}924.69$ ft, $r = 2.82$ ft and $h = 6.67 \times 10^{-4}$ ft
**123.** $\{13\}$ **125.** $x^3 + x^2 - 11x + 24, -53$
**127.** $[0, \infty), (-\infty, \infty)$

## Section 3.6

***For Thought:*** **1.** F **2.** F **3.** F **4.** F **5.** T **6.** F **7.** T **8.** F **9.** T **10.** T

***Exercises:***

**1.** rational **3.** vertical asymptote **5.** $(-\infty, -2) \cup (-2, \infty)$
**7.** $(-\infty, -2) \cup (-2, 2) \cup (2, \infty)$ **9.** $(-\infty, 3) \cup (3, \infty)$
**11.** $(-\infty, 0) \cup (0, \infty)$ **13.** $(-\infty, -1) \cup (-1, 0) \cup (0, 1) \cup (1, \infty)$
**15.** $(-\infty, -3) \cup (-3, -2) \cup (-2, \infty)$
**17.** $(-\infty, 2) \cup (2, \infty), y = 0, x = 2$
**19.** $(-\infty, 0) \cup (0, \infty), y = x, x = 0$
**21.** $x = 2, y = 0$ **23.** $x = \pm 3, y = 0$ **25.** $x = 1, y = 2$
**27.** $x = 0, y = x - 2$ **29.** $x = -1, y = 3x - 3$
**31.** $x = -2, y = -x + 6$
**33.** $x = 0, y = 0$

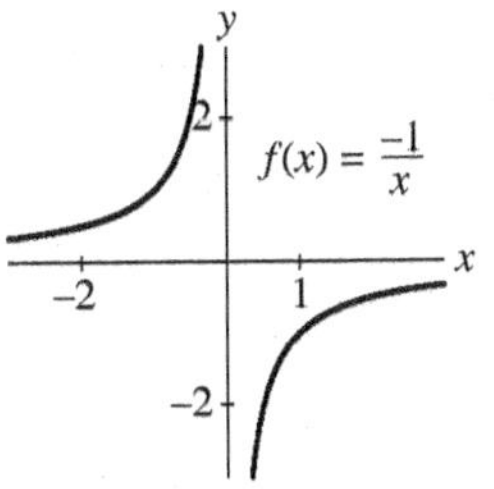

$f(x) = \frac{-1}{x}$

**35.** $x = 2, y = 0, (0, -1/2)$

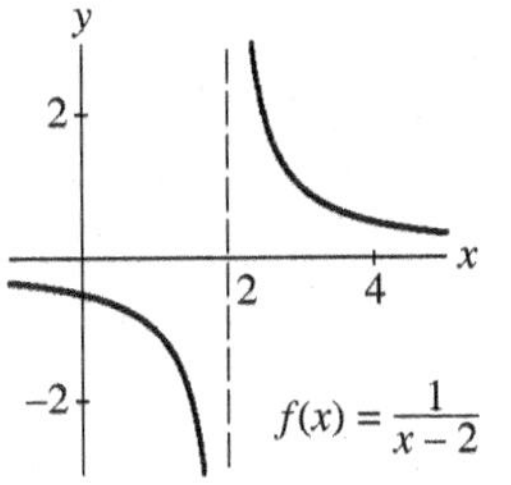

$f(x) = \frac{1}{x - 2}$

**37.** $x = \pm 2, y = 0, (0, -1/4)$

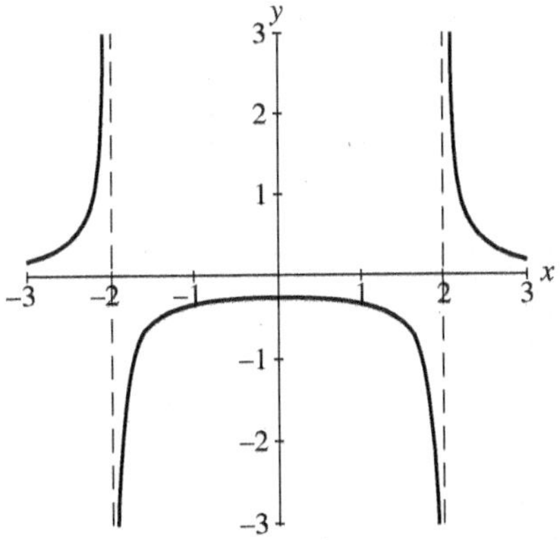

$y = \frac{1}{x^2 - 4}$

**39.** $x = -1, y = 0, (0, -1)$

$f(x) = \frac{-1}{(x + 1)^2}$

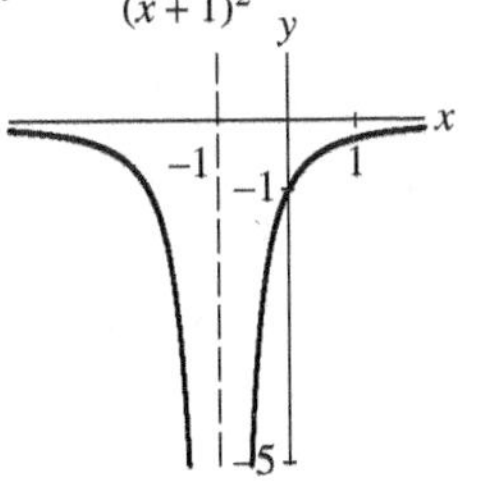

**41.** $x = 1, y = 2, (0, -1), (-1/2, 0)$

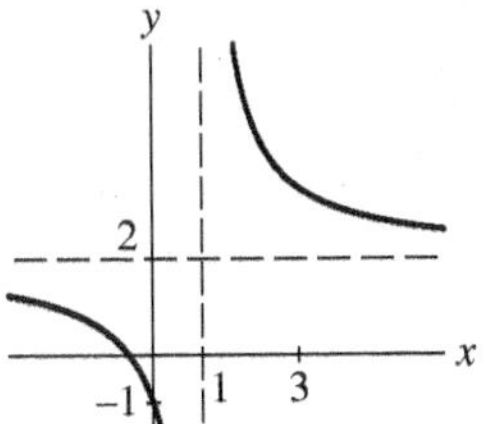

$f(x) = \frac{2x + 1}{x - 1}$

**43.** $x = -2, y = 1, (3, 0), (0, -3/2)$

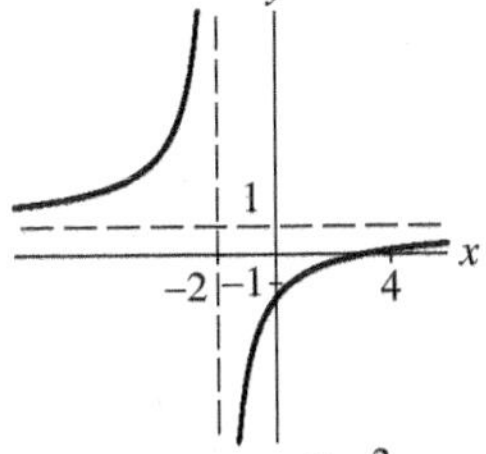

$f(x) = \frac{x - 3}{x + 2}$

**45.** $x = \pm 1, y = 0, (0, 0)$

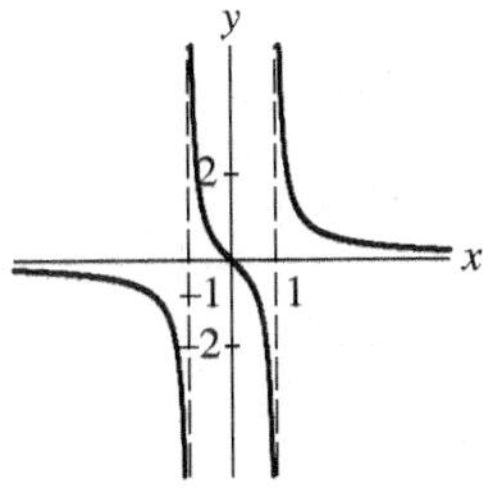

$f(x) = \frac{x}{x^2 - 1}$

**47.** $x = 1, y = 0, (0, 0)$

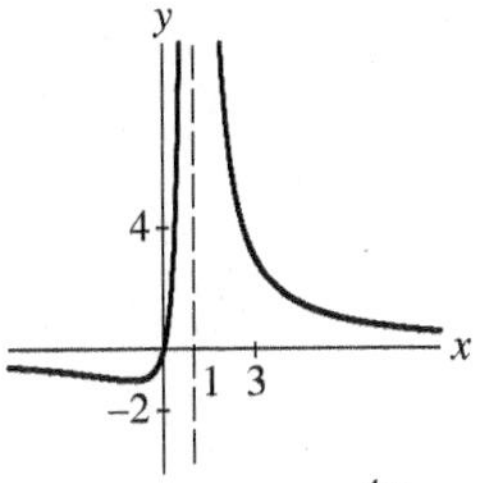

$f(x) = \frac{4x}{x^2 - 2x + 1}$

**49.** $x = \pm 3, y = -1, (0, -8/9), (\pm\sqrt{8}, 0)$

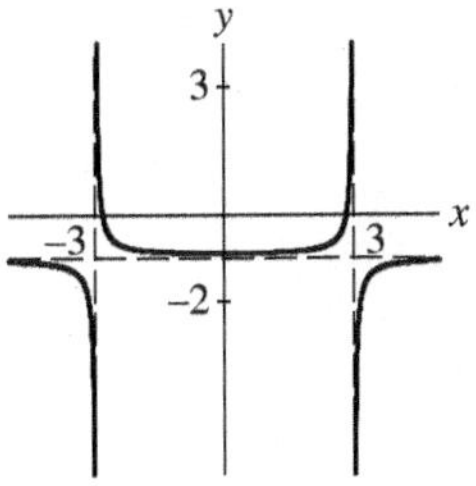

$f(x) = \frac{8 - x^2}{x^2 - 9}$

**51.** $x = -1, y = 2, (0, 2), (-2 \pm \sqrt{3}, 0)$

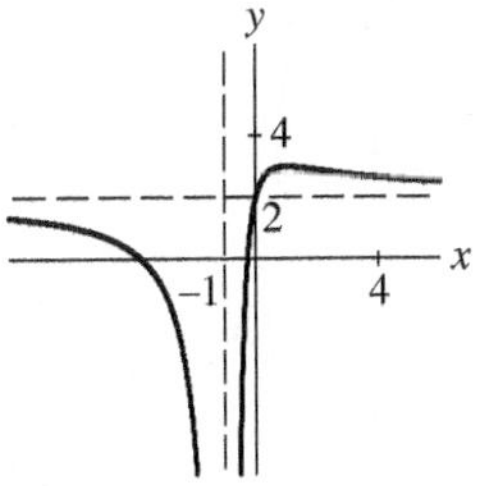

$f(x) = \frac{2x^2 + 8x + 2}{x^2 + 2x + 1}$

**53.** 0 **55.** 2 **57.** $\infty$ **59.** $\infty$

**61.** $y = x$

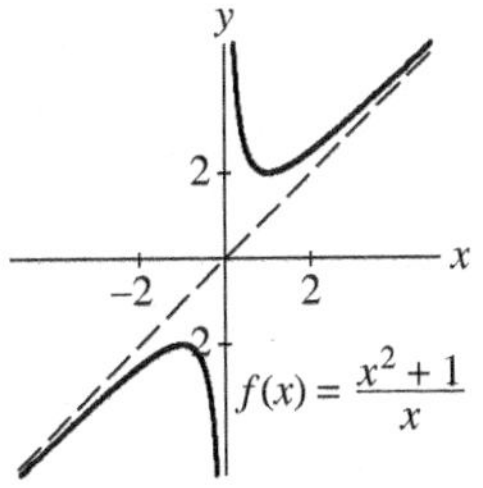

$f(x) = \frac{x^2 + 1}{x}$

**63.** $y = x$

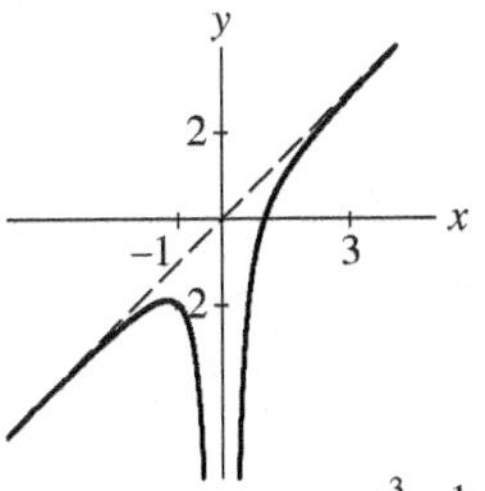

$f(x) = \frac{x^3 - 1}{x^2}$

**65.** $y = x - 1$

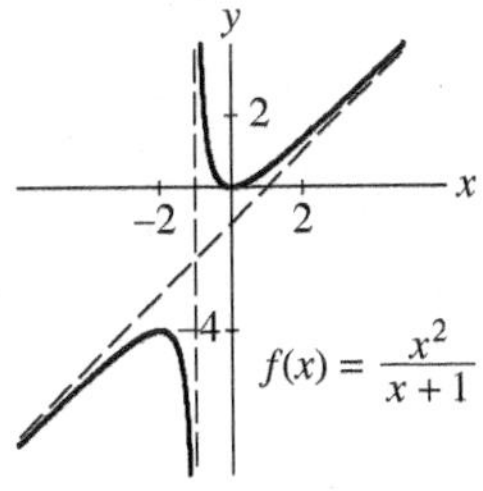

$f(x) = \frac{x^2}{x + 1}$

**67.** $y = 2x + 1$

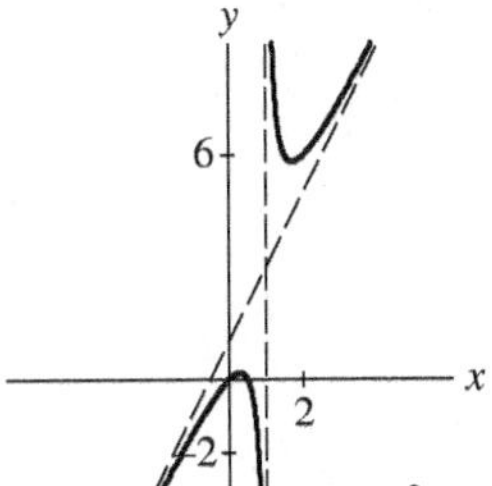

$f(x) = \frac{2x^2 - x}{x - 1}$

**69.** $y = x - 1$

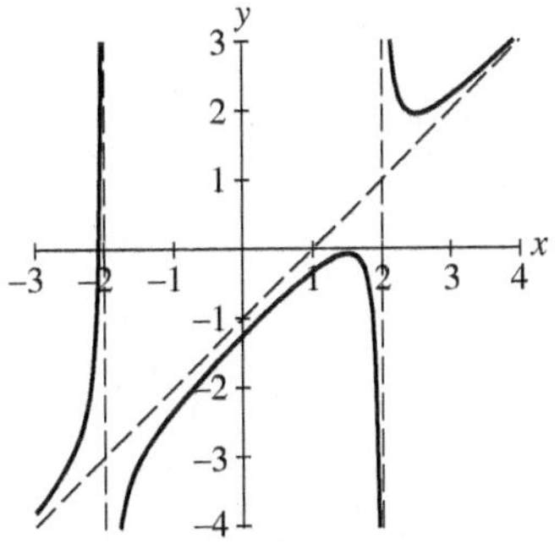

$f(x) = \frac{x^3 - x^2 - 4x + 5}{x^2 - 4}$

**71.** $y = -x + 2$

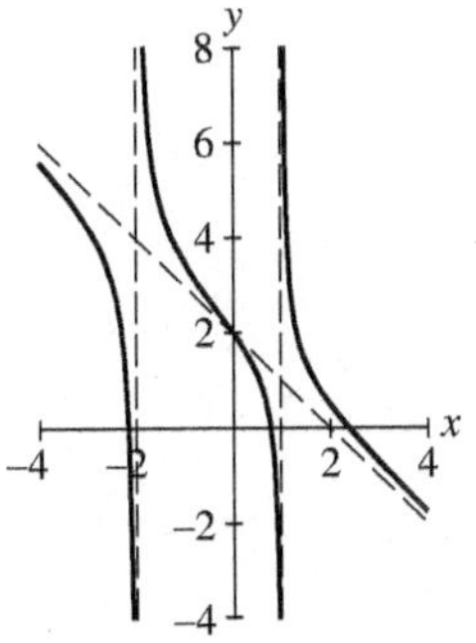

$f(x) = \frac{-x^3 + x^2 + 5x - 4}{x^2 + x - 2}$

**73.** (e) **75.** (a) **77.** (b) **79.** (c)

**81.** $x \neq \pm 1$

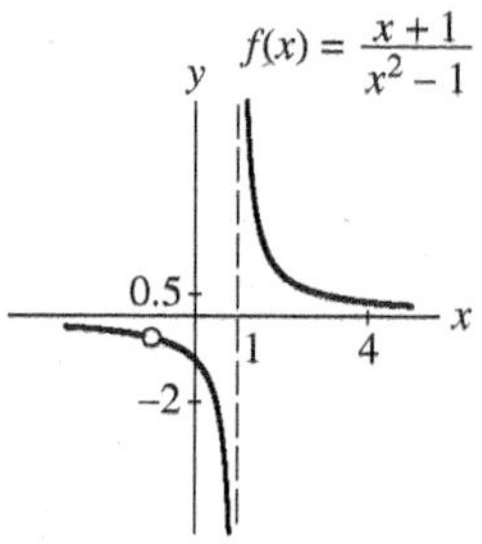


**83.** $x \neq 1$

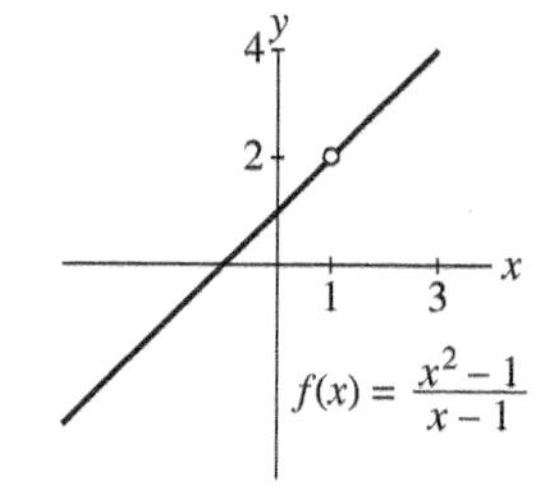


**85.** $x \neq -1, 1$

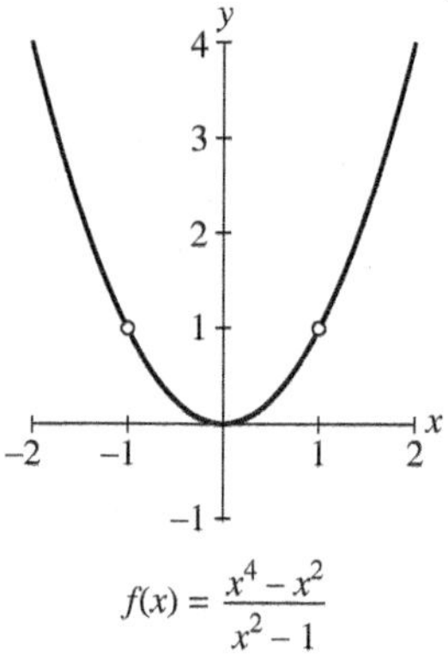


**87.** $x \neq -2, 2, 0$

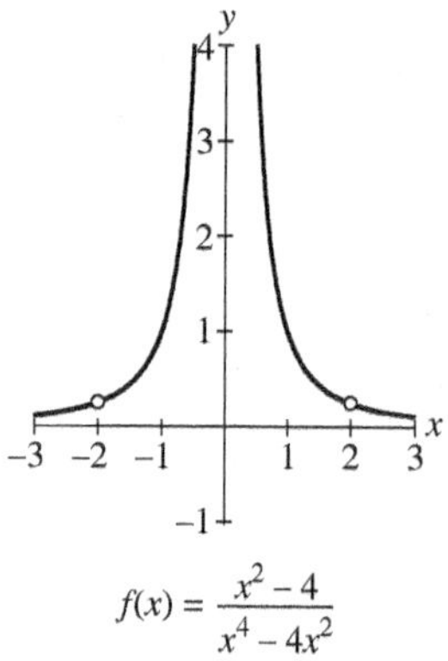


**89.**

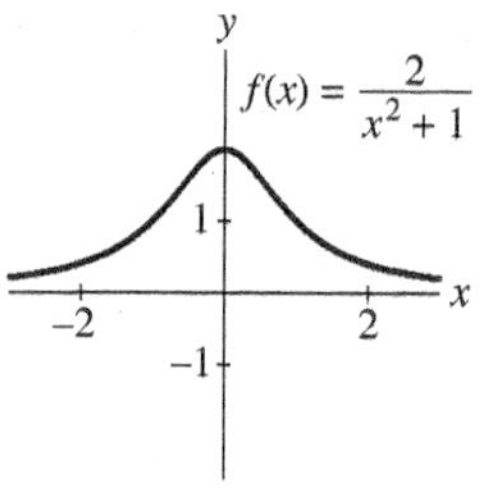


**91.**

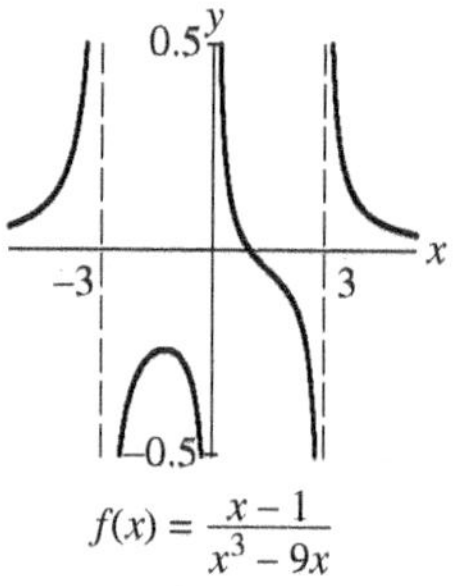


**93.**

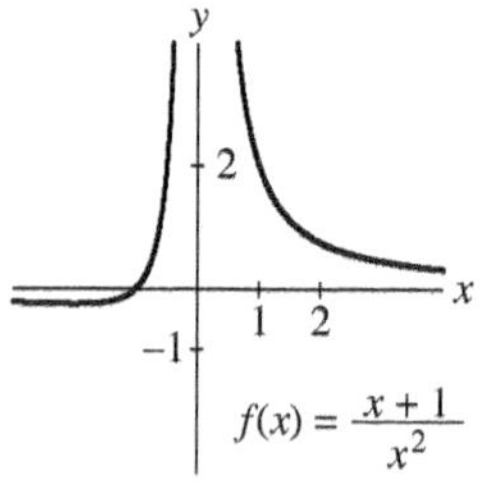


**95.** $(-2, 4]$ **97.** $(-\infty, -8) \cup (-3, \infty)$ **99.** $[-2, 3] \cup (6, \infty)$
**101.** $(-2, 3)$ **103.** $(-\infty, -2) \cup (4, 5)$ **105.** $[-1, 3] \cup (5, \infty)$
**107.** $(-\infty, -\sqrt{7}] \cup (-\sqrt{2}, \sqrt{2}) \cup [\sqrt{7}, \infty)$
**109.** $(-\infty, -3) \cup \{-1\} \cup (5, \infty)$ **111.** $(1, \infty)$ **113.** $(3, \infty)$
**115.** $(-\infty, 1) \cup (3, \infty)$ **117.** $(-\infty, 1] \cup (3, \infty)$
**119.** $(-\infty, -2) \cup (0, 2)$ **121.** $(-2, 0] \cup (2, \infty)$

**123.**

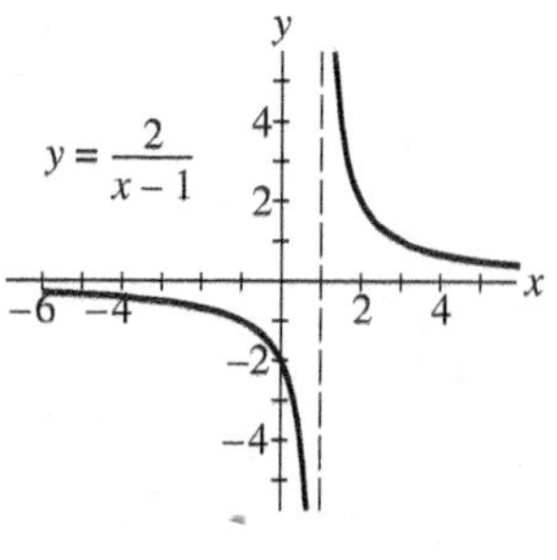


**125.**

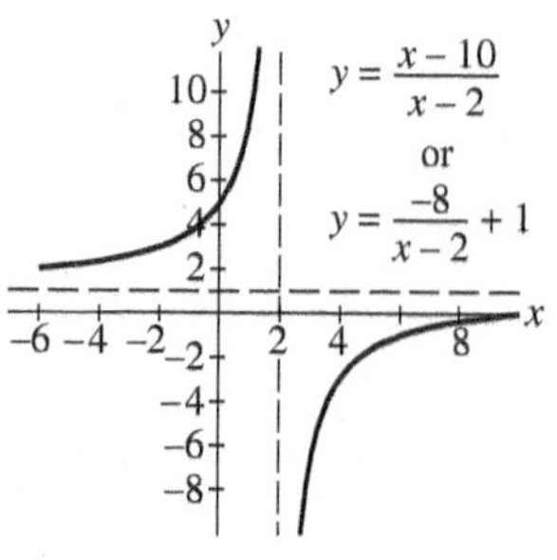


**127.**

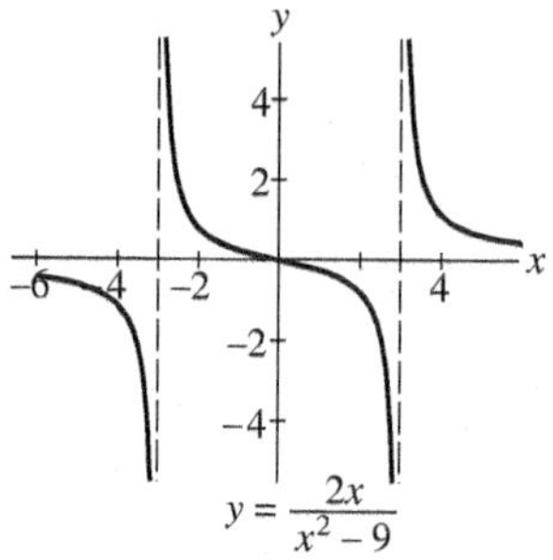


**129.**

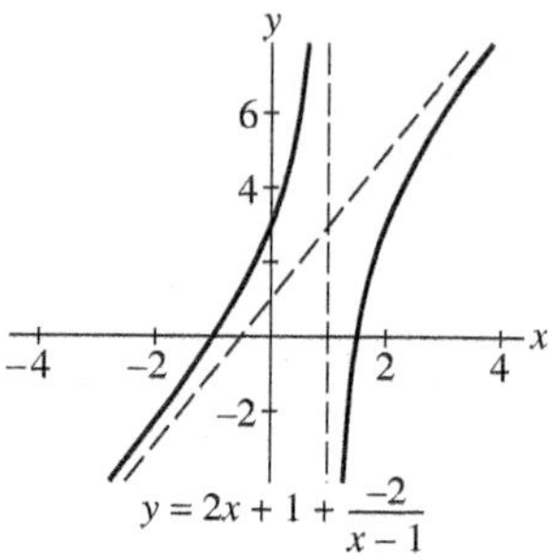


**131.** $C = \frac{100 + x}{x}$, \$2, $C \to \$1$ **133.** $S = \frac{100}{4 - x}$, $S \to \infty$ as $x \to 4$

**135. a.** $\approx 15$ min **b.** $\approx 220$ PPM
**c.** $t = 0$, PPM $= 0$; Low concentration for long time or high concentration for short time will give permanent brain damage.

**137. a.** $h = 500/(\pi r^2)$
**b.** $S = 2\pi r^2 + 1000/r$

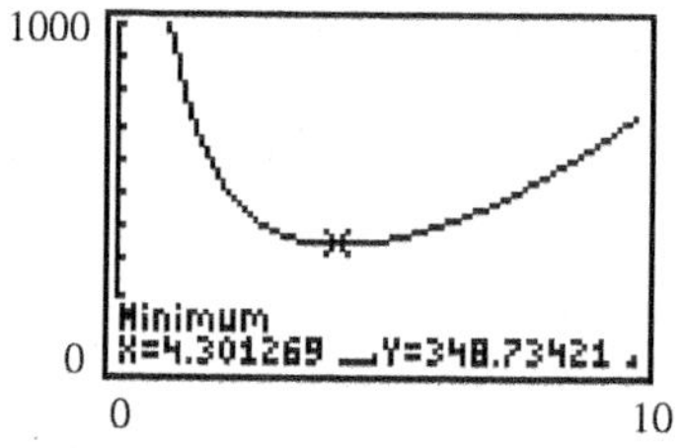


**c.** 4.3 ft **d.** \$2789.87
**141.** $(-5, 1) \cup (1, \infty)$ **143.** $\{0, \pm 2i, \pm 3i\}$
**145.** $(3/2, -31/4)$, $(-\infty, 3/2]$

## Chapter 3 Review Exercises

**1.** $f(x) = 3\left(x - \frac{1}{3}\right)^2 + \frac{2}{3}$

**3.** $(1, -3)$, $x = 1$, $\left(\frac{2 \pm \sqrt{6}}{2}, 0\right)$, $(0, -1)$

**5.** $y = -2x^2 + 4x + 6$ **7.** $1/3$ **9.** $\pm 2\sqrt{2}$

**11.** $\frac{1}{2}, \frac{-1 \pm i\sqrt{3}}{4}$ **13.** $\pm\sqrt{10}, \pm i\sqrt{10}$

**15.** $-\frac{1}{2}, \frac{1}{2}$ with multiplicity 2 **17.** $0, -1 \pm \sqrt{7}$ **19.** 83

**21.** 5 **23.** $\pm\left(1, 2, \frac{1}{3}, \frac{2}{3}\right)$ **25.** $\pm\left(1, 3, \frac{1}{2}, \frac{1}{3}, \frac{1}{6}, \frac{3}{2}\right)$

**27.** $2x^2 - 5x - 3 = 0$ **29.** $x^2 - 6x + 13 = 0$
**31.** $x^3 - 4x^2 + 9x - 10 = 0$ **33.** $x^2 - 4x + 1 = 0$
**35.** 0 with multiplicity 2, 6 imag **37.** 1 pos, 2 imag; 3 pos
**39.** 3 neg; 1 neg, 2 imag **41.** $-4 < x < 3$
**43.** $-1 < x < 8$ **45.** $-1 < x < 1$ **47.** 1, 2, 3

**49.** $\frac{1}{2}, \frac{1}{3}, \pm i$ **51.** $3, 3 \pm i$ **53.** $2, 1 \pm i\sqrt{2}$

**55.** $0, \frac{1}{2}, 1 \pm \sqrt{3}$ **57.** $\{1/5\}$ **59.** $\{\pm\sqrt{2}\}$ **61.** $\{30\}$

**63.** $\{16\}$ **65.** $\{\pm 2\}$ **67.** $\{-7, 9\}$ **69.** No solution
**71.** $\{11/4\}$ **73.** $x = 3/4$ **75.** $y$-axis **77.** Origin
**79.** $(-\infty, -2.5) \cup (-2.5, \infty)$ **81.** $(-\infty, \infty)$

**83.** $(-1, 0), (2, 0), (0, -2)$

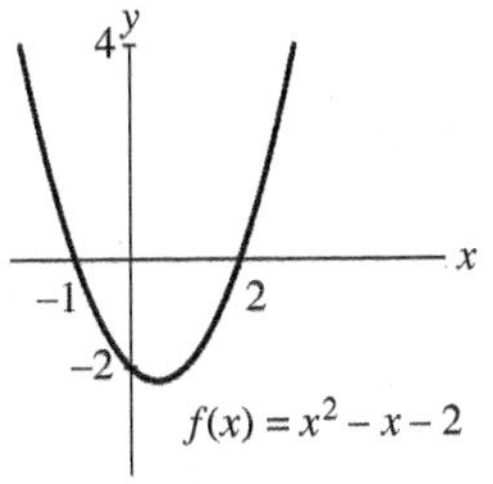


**85.** $(-1, 0), (2, 0), (0, -2)$

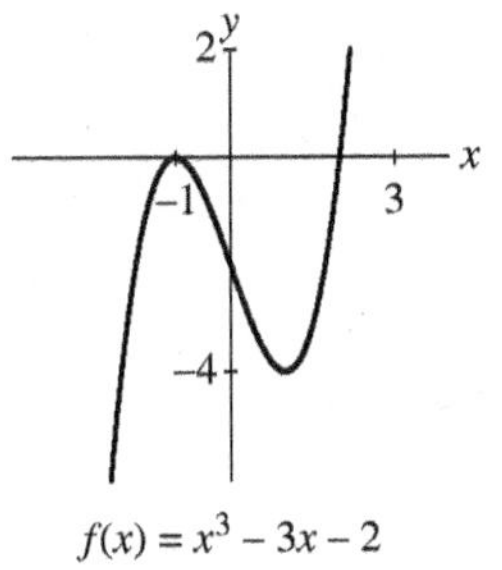


**87.** $(\pm 2, 0), (1, 0), (0, 2)$

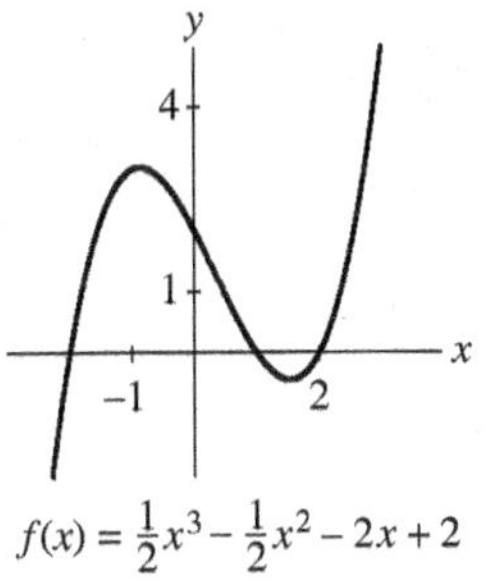


**89.** $(\pm 2, 0), (0, 4)$

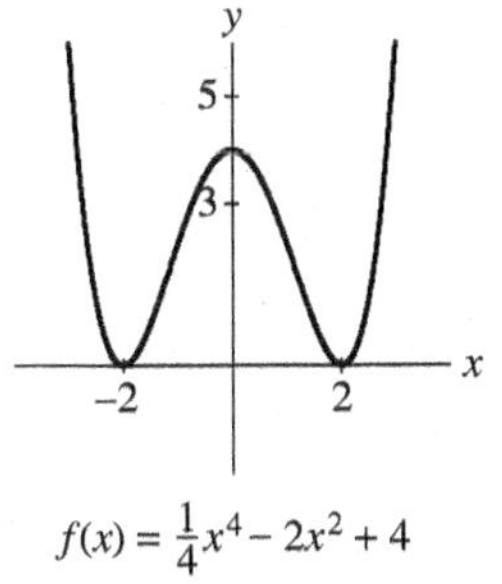


**91.** $(0, 2/3), x = -3, y = 0$

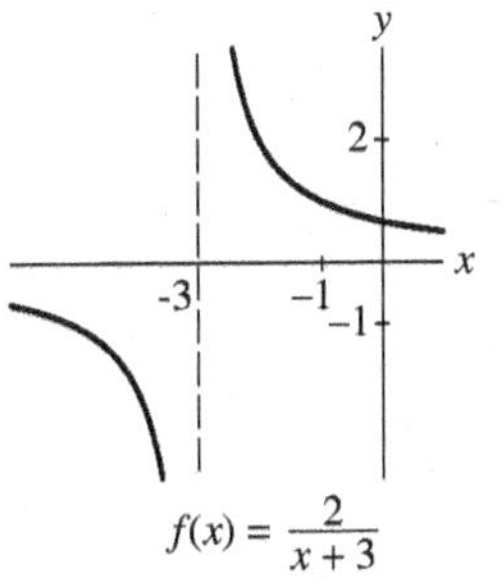


**93.** $(0, 0), x = \pm 2, y = 0$

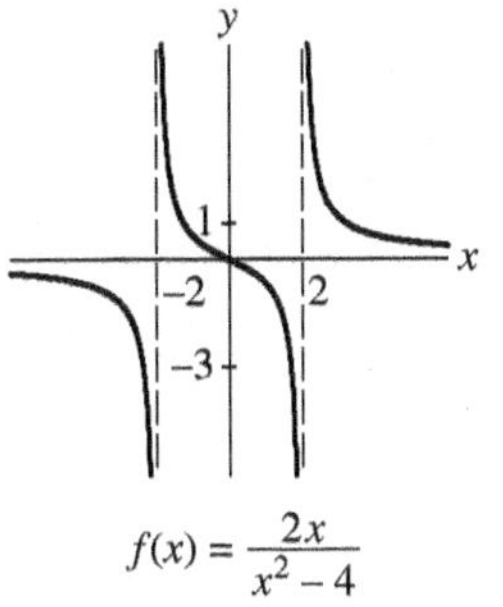


**95.** $(1, 0), \left(0, -\frac{1}{2}\right), x = 2,$ $y = x$

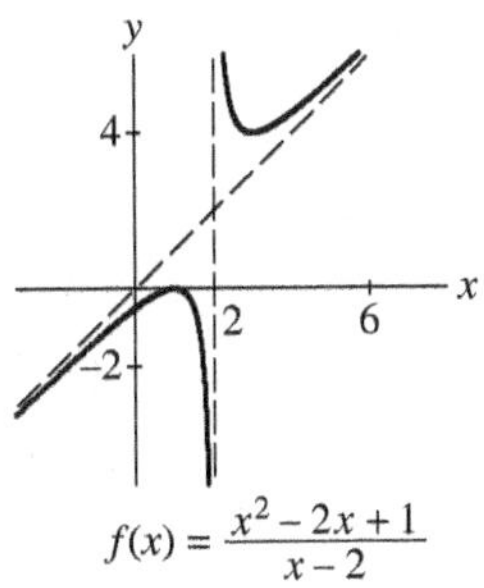


**97.** $\left(\frac{1}{2}, 0\right), \left(0, -\frac{1}{2}\right), x = 2,$ $y = -2$

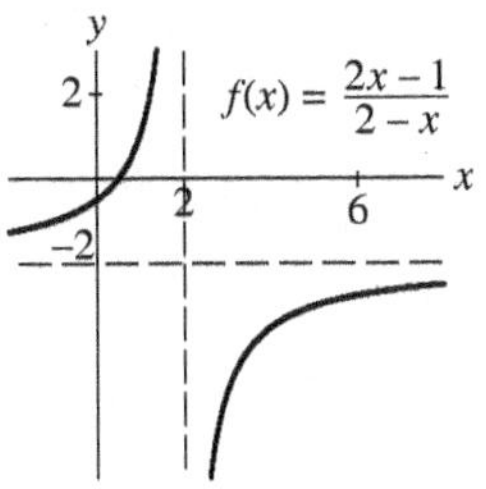


**99.** $(-2, 0), (0, 2)$

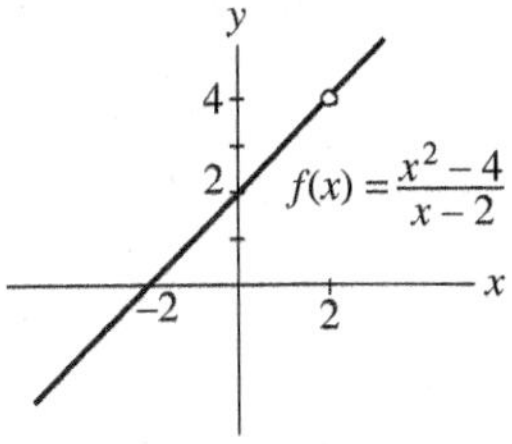


**101.** $(1/4, 1/2)$ **103.** $[-5, 3]$ **105.** $[-1/2, 1/2] \cup [100, \infty)$
**107.** $(-\infty, -2) \cup (0, \infty)$ **109.** $(-\infty, 0) \cup (0, 3) \cup (4, \infty)$
**111.** $(-\infty, 1] \cup [2, 3) \cup (4, \infty)$ **113.** $x^2 - 3x, -15$
**115.** 380.25 ft **117.** 24 ft wide, 7 ft high
**119. a.** 172.4 ft/sec **b.** $V = 200$ **c.** 200 ft/sec

## Chapter 3 Test

**1.** $y = 3(x - 2)^2 - 11$
**2.** $(2, -11), x = 2, (0, 1), \left(\frac{6 \pm \sqrt{33}}{3}, 0\right), [-11, \infty)$ **3.** $-11$
**4.** $2x^2 - 6x + 14, -37$ **5.** $-14$ **6.** $\pm\left(1, 2, 3, 6, \frac{1}{3}, \frac{2}{3}\right)$
**7.** $x^3 + 3x^2 + 16x + 48 = 0$ **8.** 2 pos, 1 neg; 1 neg, 2 imag
**9.** 256 ft **10.** $\pm 3$ **11.** $\pm 2, \pm 2i$ **12.** $1 \pm \sqrt{6}, 2$
**13.** $\pm i$ each with multiplicity 2
**14.** 2, 0 with multiplicity 2, $-\frac{3}{2}$ with multiplicity 3 **15.** $2 \pm i, \frac{1}{2}$

**16.**

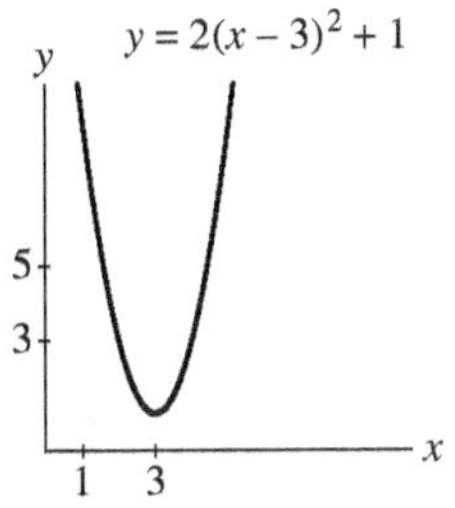


**17.**

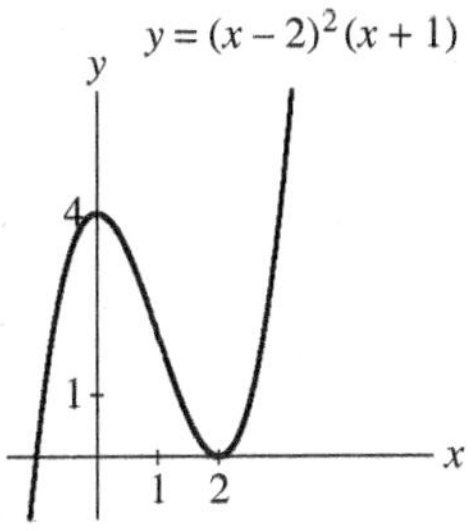


**18.**

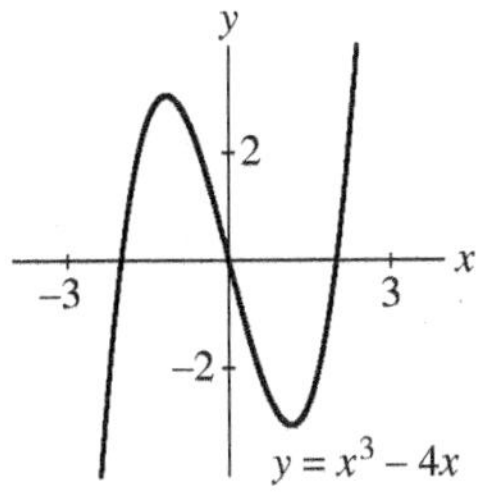


**19.** $x = 2, y = 0$

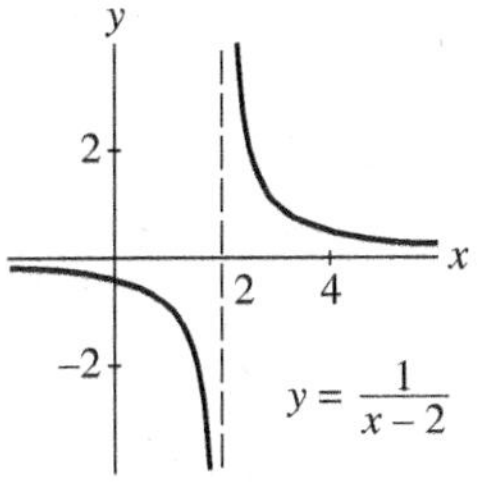


**20.** $x = 2, y = 2$

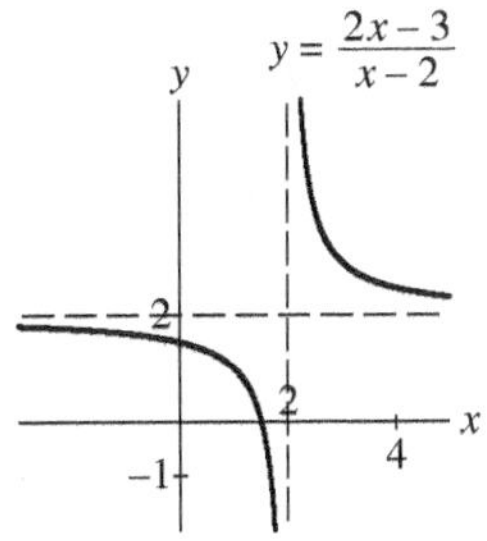


**21.** $x = 0, y = x$

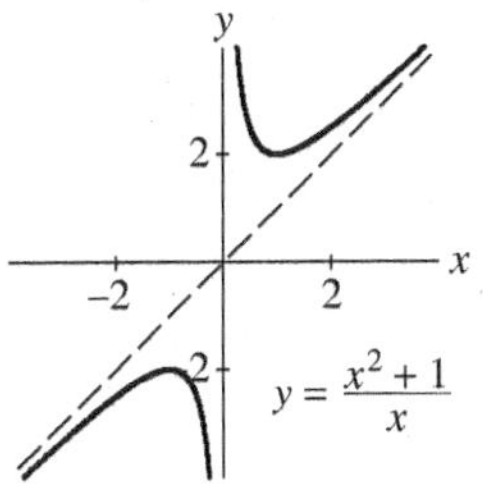


**22.** $x = \pm 2, y = 0$

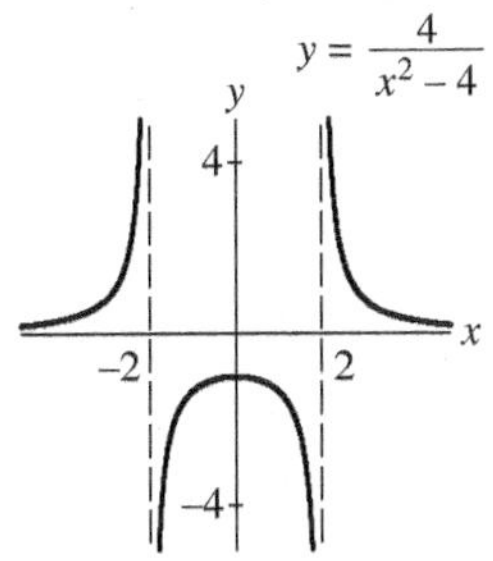


**23.**

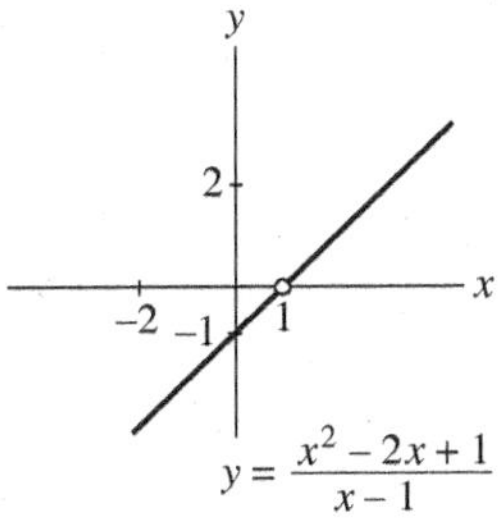


**24.** $(-2, 4)$ **25.** $(-\infty, 1/2) \cup (3, \infty)$ **26.** $(-\infty, -3] \cup (-1, 4)$
**27.** $(-\sqrt{7}, 0) \cup (\sqrt{7}, \infty)$ **28.** $\{3 \pm 3\sqrt{3}\}$ **29.** $\{16\}$

## Tying It All Together Chapters 1–3

**1.** $3/2$ **2.** $0, 3/2$ **3.** $-1$ **4.** $1$ **5.** $\pm 27$ **6.** $\pm i$
**7.** $(2, 0); (0, -4); (-\infty, \infty); -; (-\infty, \infty); (-\infty, \infty); \{2\}; (2, \infty); (-\infty, 2]; -; -; -$
**8.** $(-1, 0), (3, 0); (0, 3); (-\infty, 1]; [1, \infty); (-\infty, \infty); (-\infty, 4]; \{-1, 3\}; (-1, 3); (-\infty, -1] \cup [3, \infty); (1, 4); -; -$
**9.** $(-3, 0), (3, 0); (0, 27); (-\infty, -1]; [3, \infty); [-1, 3]; (-\infty, \infty); (-\infty, \infty); \{-3, 3\}; (-3, 3) \cup (3, \infty); (-\infty, -3] \cup \{3\}; -; -; -$
**10.** $(-1, 0), (1, 0); (0, 1/4); (-\infty, -2), (-2, 0]; [0, 2), (2, \infty); (-\infty, -2) \cup (-2, 2) \cup (2, \infty); (-\infty, 1/4] \cup (1, \infty); \{-1, 1\}; (-\infty, -2) \cup (-1, 1) \cup (2, \infty); (-2, -1] \cup [1, 2); -; x = -2, x = 2; y = 1$
**11.** $\left(-\infty, \frac{3}{2}\right)$ **12.** $\left(0, \frac{3}{2}\right)$ **13.** $(-1, \infty)$
**14.** $(-\infty, 1) \cup (2, \infty)$ **15.** $(-27, 27)$ **16.** $\varnothing$
**17.** $f^{-1}(x) = \dfrac{x + 3}{2}$ **18.** Not invertible **19.** $f^{-1}(x) = \sqrt[3]{x} - 1$
**20.** $f^{-1}(x) = \dfrac{2x - 1}{x - 1}$ **21.** Not invertible **22.** Not invertible
**23.** trinomial **24.** binomial **25.** conjugates
**26.** dividend, quotient, division, remainder
**27.** greatest common factor **28.** prime
**29.** real part, imaginary part
**30.** formula **31.** ordered pair, interval
**32.** Cartesian coordinate system

# Chapter 4

## Section 4.1

*For Thought:* **1.** F **2.** T **3.** T **4.** T **5.** T **6.** T
**7.** T **8.** F **9.** T **10.** T

*Exercises:*
**1.** algebraic **3.** exponential **5.** increasing, decreasing **7.** range
**9.** 27 **11.** $-1$ **13.** $1/8$ **15.** 16 **17.** 4 **19.** $-1/27$
**21.** 9 **23.** $1/9$ **25.** $1/2$ **27.** 8 **29.** 4 **31.** 2
**33.** $(-\infty, \infty), (0, \infty)$, inc **35.** $(-\infty, \infty), (0, \infty)$, dec

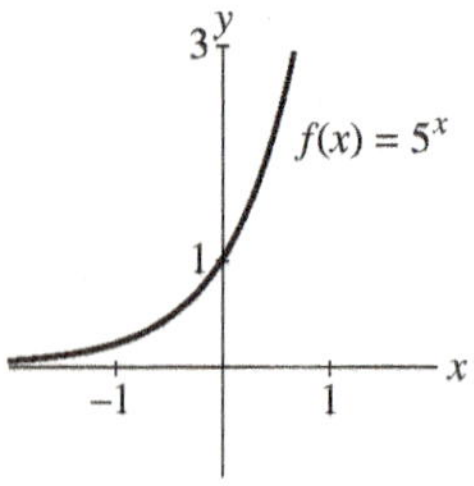


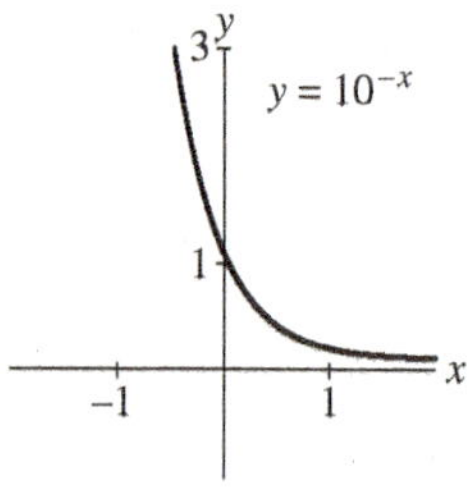


**37.** $(-\infty, \infty), (0, \infty)$, dec **39.** $\infty$ **41.** 0 **43.** 0 **45.** $\infty$

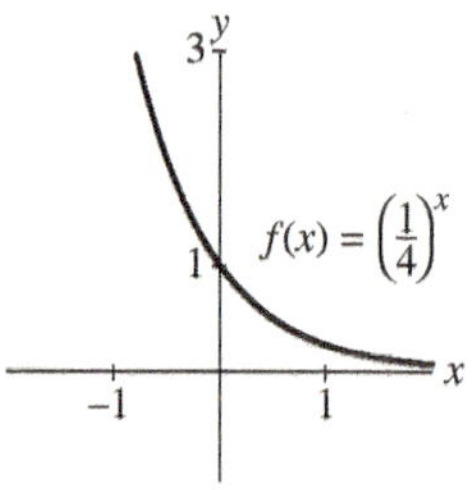


**47.** $(-\infty, \infty), (-3, \infty)$, $y = -3$, increasing

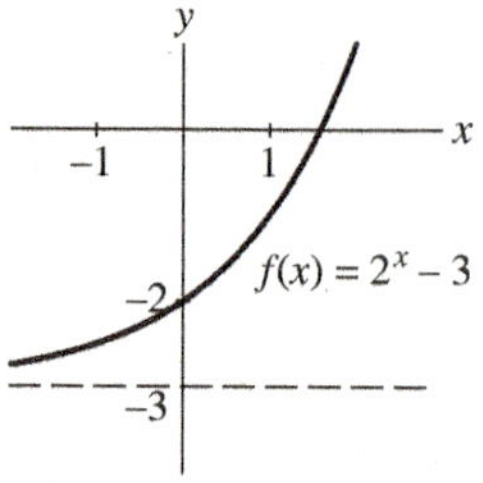


**49.** $(-\infty, \infty), (-5, \infty)$, $y = -5$, increasing

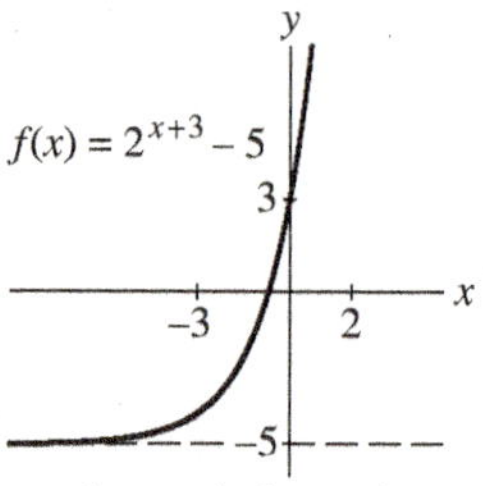


**51.** $(-\infty, \infty), (-\infty, 0)$, $y = 0$, increasing

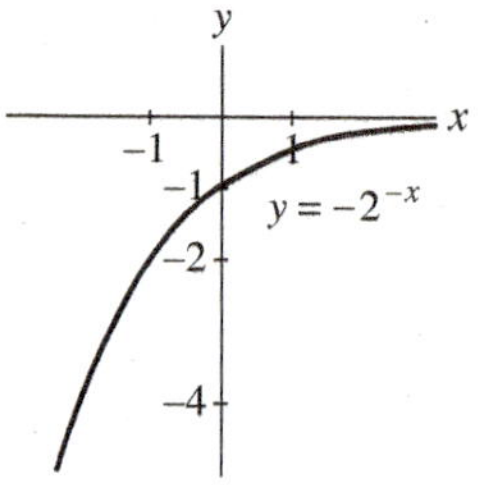


**53.** $(-\infty, \infty), (-\infty, 1)$, $y = 1$, decreasing

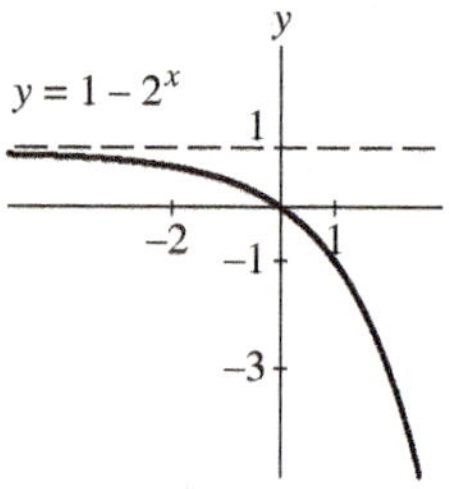


**55.** $(-\infty, \infty), (0, \infty)$, $y = 0$, increasing

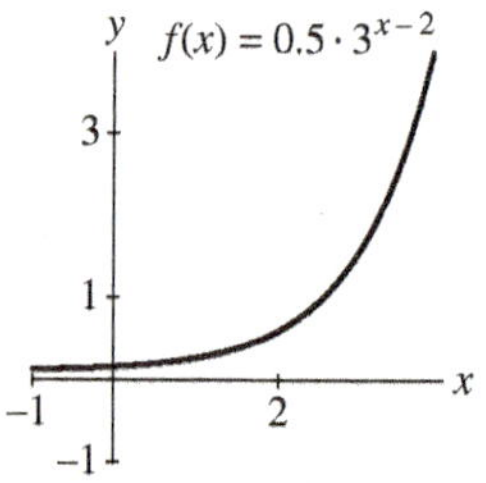


**57.** $(-\infty, \infty), (0, \infty)$, $y = 0$, decreasing

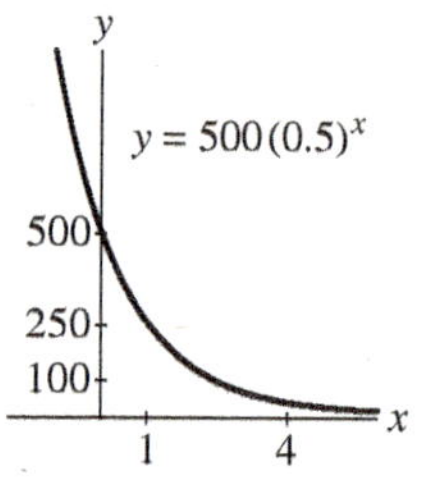


**59.** $y = 2^{x-5} - 2$ **61.** $y = -(1/4)^{x-1} - 2$ **63.** $\{6\}$
**65.** $\{-1\}$ **67.** $\{3\}$ **69.** $\{-2\}$ **71.** $\{1/3\}$ **73.** $\{-2\}$
**75.** $\{-3\}$ **77.** $\{-1\}$ **79.** $\{3\}$ **81.** 2 **83.** $-1$
**85.** 0 **87.** $-3$ **89.** 3 **91.** $-1$ **93.** 1 **95.** $-1$
**97.** $9, 1, 1/3, -2$ **99.** $1, -2, 5, 1$ **101.** $-16, -2, -1/2, 5$
**103.** **a.** \$7934.37, \$2934.37 **b.** \$8042.19, \$3042.19
**c.** \$8067.51, \$3067.51 **d.** \$8079.95, \$3079.95
**105.** **a.** \$8080.37 **b.** \$9671.31 **c.** \$7694.93 **d.** \$26,570.30
**107.** \$2121.82 **109.** \$3934.30
**111.** **a.** \$6.85 **b.** \$6.87 **113.** 200 g, 121.3 g
**115.** **a.** 400,000 **b.** 600 million
**117.** **a.** $y = 8.44(1.23)^x$ **b.** Yes **c.** 530 million
**119.** $P = 10\left(\frac{1}{2}\right)^n$ **121.** 6 **125.** $y = 2, x = -7$
**127.** $(-\infty, -1) \cup (6, \infty)$ **129.** $x = \dfrac{y + 5}{2}$

## Section 4.2

*For Thought:* **1.** T **2.** F **3.** T **4.** T **5.** F **6.** T
**7.** T **8.** F **9.** T **10.** T

*Exercises:*
**1.** logarithmic **3.** natural **5.** vertical asymptote
**7.** logarithmic family **9.** 6 **11.** $-4$ **13.** $1/4$ **15.** $-3$
**17.** 6 **19.** $-4$ **21.** $1/4$ **23.** $-3$ **25.** $-1$ **27.** 0
**29.** 1 **31.** $-5$

**33.** $(0, \infty), (-\infty, \infty)$

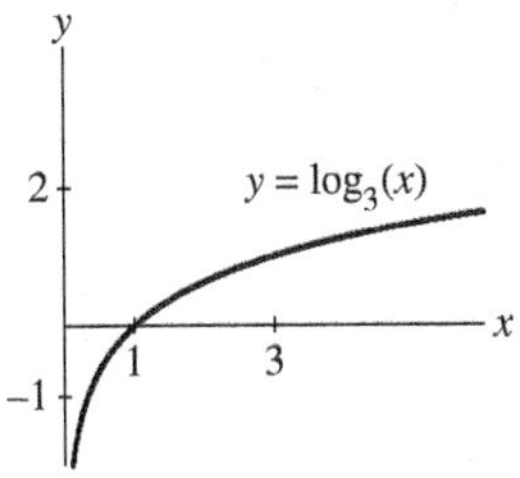


**35.** $(0, \infty), (-\infty, \infty)$

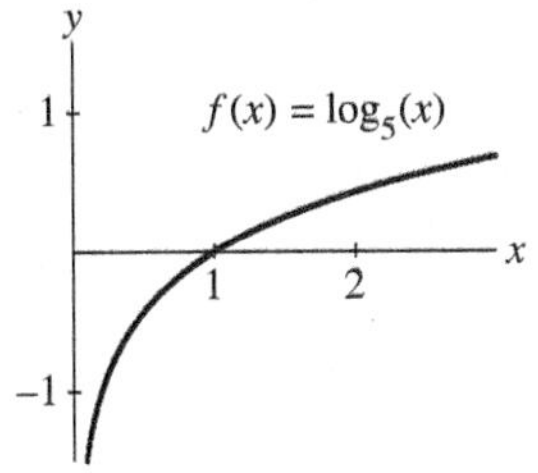


**37.** $(0, \infty), (-\infty, \infty)$

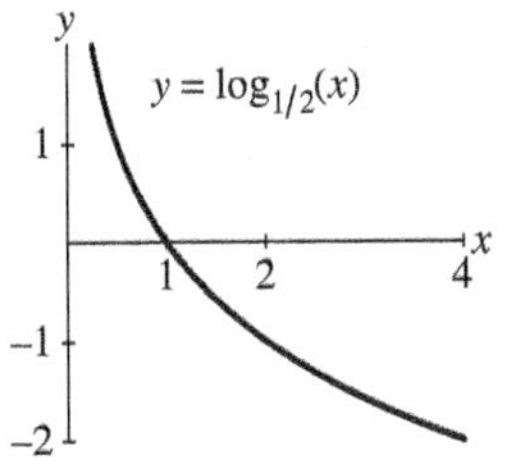


**39.** $(0, \infty), (-\infty, \infty)$

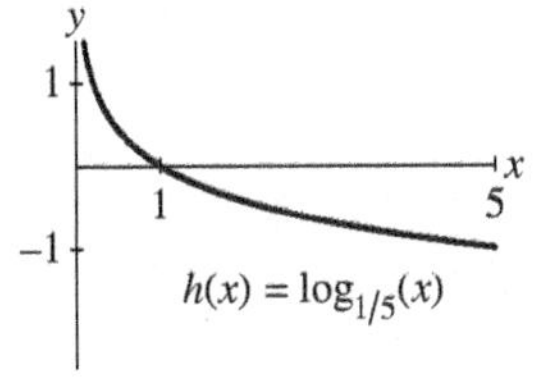


**41.** $(1, \infty), (-\infty, \infty)$

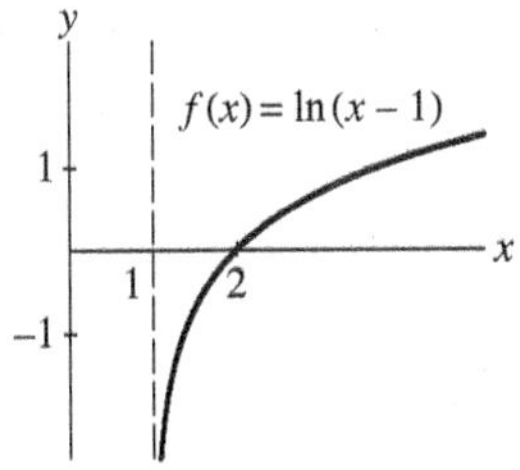


**43.** $(-2, \infty), (-\infty, \infty)$

$f(x) = -3 + \log(x + 2)$

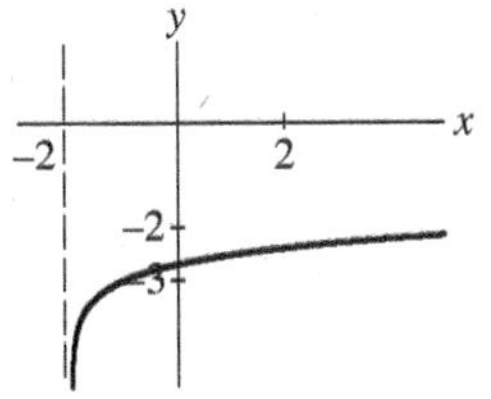


**45.** $(1, \infty), (-\infty, \infty)$

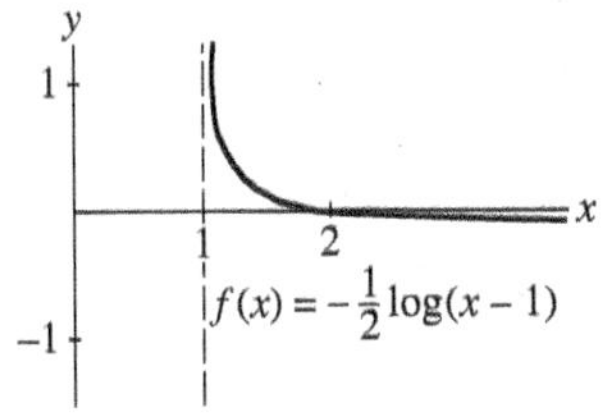


**47.** $\infty$ **49.** $\infty$ **51.** $-\infty$ **53.** $\infty$ **55.** $y = \ln(x - 3) - 4$
**57.** $y = -\log_2(x - 5) - 1$ **59.** $2^5 = 32$ **61.** $5^y = x$
**63.** $10^z = 1000$ **65.** $e^x = 5$ **67.** $x = a^m$ **69.** $\log_5(125) = 3$
**71.** $\ln(y) = 3$ **73.** $\log(y) = m$ **75.** $z = \log_a(y)$
**77.** $\log_a(n) = x - 1$ **79.** $f^{-1}(x) = \log_2(x)$ **81.** $f^{-1}(x) = 7^x$
**83.** $f^{-1}(x) = e^x + 1$ **85.** $f^{-1}(x) = \log_3(x) - 2$
**87.** $f^{-1}(x) = \log(2x - 10) + 1$ **89.** 256 **91.** $\sqrt{3}$ **93.** 4
**95.** $\log_3(77)$ **97.** 6 **99.** $3\sqrt{2}$ **101.** $-1 + \log_3(7)$
**103.** 2 **105.** 27 **107.** 1/4 **109.** 6/5 **111.** 64
**113.** 1.3979 **115.** 0.5493 **117.** −0.2231 **119.** −0.3010
**121.** **a.** 34.7 yr **b.** 17.3 yr **c.** 8.7 yr **d.** 4.3 yr
**123.** **a.** 22.0% **b.** 11.0% **c.** 5.5% **d.** 2.7%
**125.** 49 yr 125 days **127.** $r = \ln(A/P)/t$, 23.1%
**129.** **a.** 6.9 yr **131.** 3.5% **133.** 9.8 yr **135.** 1.87%, 7.3 billion
**137.** **a.** 133,412 acres **b.** 12 days
**139.** **a.** $p = -45 \cdot \log(I) + 190$ **b.** 0%
**141.** 4.1 **143.** 3.7 **145.** $c = \ln(b)$, 3.54%
**149.** $(-\infty, \infty), (-\infty, 7)$ **151.** $2 \times 10^{-19}$ **153.** $\{0, 2 \pm 3i\}$

## Section 4.3

*For Thought:* **1.** F **2.** T **3.** T **4.** T **5.** F **6.** F
**7.** F **8.** T **9.** F **10.** F
*Exercises:*
**1.** sum **3.** power **5.** $\sqrt{y}$ **7.** $y + 1$ **9.** 999
**11.** $\log(15)$ **13.** $\log_2(x^2 - x)$ **15.** $\log_4(6)$ **17.** $\ln(x^5)$
**19.** $\log_2(3) + \log_2(x)$ **21.** $\log(x) - \log(2)$
**23.** $\log(x - 1) + \log(x + 1)$ **25.** $\ln(x - 1) - \ln(x)$
**27.** $3 \cdot \log_a(5)$ **29.** $\frac{1}{2} \cdot \log_a(5)$ **31.** $-1 \cdot \log_a(5)$
**33.** $\log_a(2) + \log_a(5)$ **35.** $\log_a(5) - \log_a(2)$
**37.** $\log_a(2) + \frac{1}{2} \cdot \log_a(5)$ **39.** $2 \cdot \log_a(2) - 2 \cdot \log_a(5)$
**41.** $\log_3(5) + \log_3(x)$ **43.** $\log_2(5) - \log_2(2) - \log_2(y)$
**45.** $\log(3) + \frac{1}{2}\log(x)$ **47.** $\log(3) + (x - 1)\log(2)$
**49.** $\frac{1}{3}\ln(x) + \frac{1}{3}\ln(y) - \frac{4}{3}\ln(t)$
**51.** $\ln(6) + \frac{1}{2}\ln(x - 1) - \ln(5) - 3 \cdot \ln(x)$
**53.** $\log_2(5x^3)$ **55.** $\log_7(x^{-3})$ **57.** $\log\left(\frac{2xy}{z}\right)$ **59.** $\log\left(\frac{z\sqrt{x}}{y\sqrt[3]{w}}\right)$
**61.** $\log_4(x^{20})$ **63.** 3.1699 **65.** −3.5864 **67.** 11.8957
**69.** 2.2025 **71.** 1.5850 **73.** 0.3772 **75.** −3.5850
**77.** 13.8695 **79.** 34.3240 **81.** 0.3200 **83.** 0.0479, −24.0479
**85.** 2.0172 **87.** 1.5928 **89.** 11 yr 166 days
**91.** 44 quarters **93.** 5.7% **95.** 3.58% **97.** 4.4%
**99.** $\log(I/I_0)$, 3 **101.** $t = \frac{1}{r}\ln(P) - \frac{1}{r}\ln(P_0)$
**103.** **a.** Decreasing **b.** $n \le 4{,}892{,}961$
**105.** $MR(x) = \log\left[\left(\frac{x + 2}{x + 1}\right)^{500}\right]$, $MR(x) \to 0$
**107.** **a.** $y = 217.9(1.084)^x$ **b.** $y = 217.9e^{0.0807x}$ **c.** 8.07%
**d.** 2014 **e.** No
**109.** No, $(-\infty, 0) \cup (0, \infty), (0, \infty)$

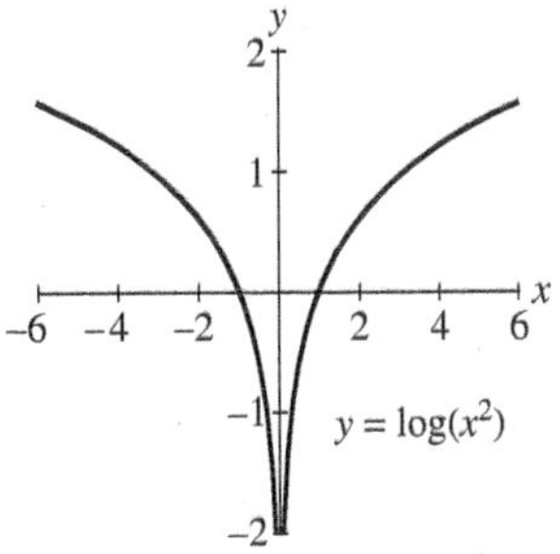


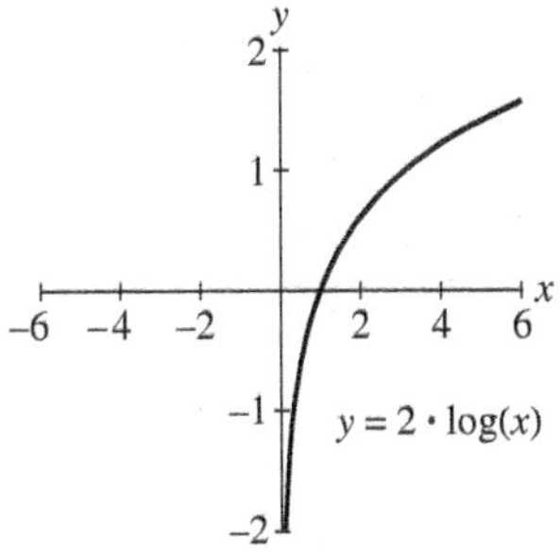


**113.** $(1, \infty), (-\infty, \infty)$ **115.** \$12,532.62 **117.** $\{-65/27\}$

## Section 4.4

*For Thought:* **1.** T **2.** T **3.** F **4.** T **5.** F **6.** F
**7.** T **8.** T **9.** T **10.** T
*Exercises:*
**1.** 8 **3.** 80 **5.** $\pm 5$ **7.** 3 **9.** 1/2 **11.** $\sqrt[3]{10}$
**13.** 1/4 **15.** 6 **17.** 5 **19.** 1/999 **21.** $\varnothing$ **23.** 20
**25.** 2 **27.** $\frac{\sqrt{5}}{2}$ **29.** $\frac{1}{2}$ **31.** $\frac{2}{\ln(6)}$ **33.** 3.8074
**35.** 3.5502 **37.** −3.0959 **39.** No solution **41.** 1.5850
**43.** 0.7677 **45.** −0.2 **47.** 2.5850 **49.** 1/3 **51.** 1, 100
**53.** 29.4872 **55.** 2 **57.** −5/4 **59.** 0.194, 2.70
**61.** −49.73 **63.** −0.767, 2, 4 **65.** $-6.93 \times 10^{-5}$
**67.** 19,035 yr ago **69.** 1507 yr **71.** 24,850 yr
**73.** 30.5% **75.** a.d. 34
**77.** **a.** −18.8% per year **b.** \$8200 **79.** 1596 million
**81.** 1 hr 11 min, forever **83.** 5:20 a.m. **85.** 10 yr 3 mo
**87.** 19,328 yr 307 days
**89.** **a.** 12,300 **b.** ≈15 yr **c.** 13.9 yr
**91.** **a.** 2211 **b.** 99% **93.** 1.32 parsecs
**95.** **a.** $P = 10^{-0.1826x+4.5}$ **b.** \$718.79 **c.** 2010
**97.** $10^{-3}$ watts/m² **99.** \$6791.91
**101.** **a.** $y = 114.8 - 12.8 \cdot \ln(x)$ in year 1960 + $x$ **b.** 65%
**c.** 2121 **d.** Quadratic or exponential

**103.** 1.105170833, 1.105170918 **105.** 2 **107.** $\{2.5883\}$
**109.** $(g \circ f)(x) = x$

## Chapter 4 Review Exercises

**1.** 64 **3.** 6 **5.** 0 **7.** 17 **9.** 6 **11.** 32 **13.** 3
**15.** 9 **17.** 3 **19.** $-3$ **21.** 3 **23.** $\log(x^2 - 3x)$
**25.** $\ln(3x^2y)$ **27.** $\log(3) + 4 \cdot \log(x)$
**29.** $\log_3(5) + \frac{1}{2}\log_3(x) - 4 \cdot \log_3(y)$ **31.** $\ln(2) + \ln(5)$
**33.** $\ln(2) + 2 \cdot \ln(5)$ **35.** $10^{10}$ **37.** 3 **39.** $-5$ **41.** $-4$
**43.** $2 + \ln(9)$ **45.** $-2$ **47.** $100\sqrt{5}$ **49.** 8 **51.** 6 **53.** 3
**55.** $3, -3, \sqrt{3}, 0$ **57.** (c) **59.** (b) **61.** (d) **63.** (e)
**65.** $(-\infty, \infty)$, $(0, \infty)$, inc, $y = 0$ **67.** $(-\infty, \infty)$, $(0, \infty)$, dec, $y = 0$

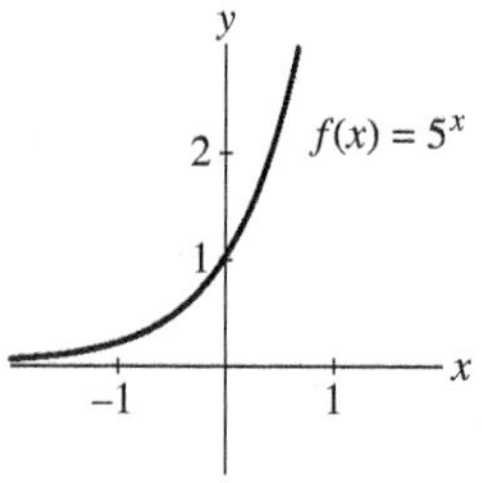


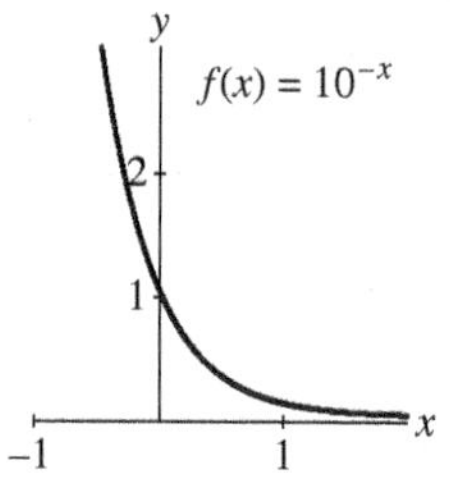


**69.** $(0, \infty)$, $(-\infty, \infty)$, inc, $x = 0$ **71.** $(-3, \infty)$, $(-\infty, \infty)$, inc, $x = -3$

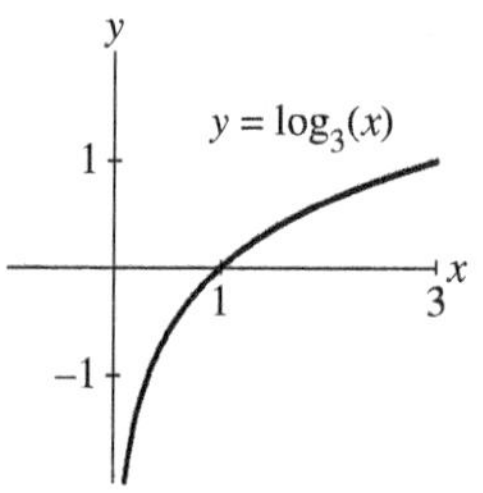


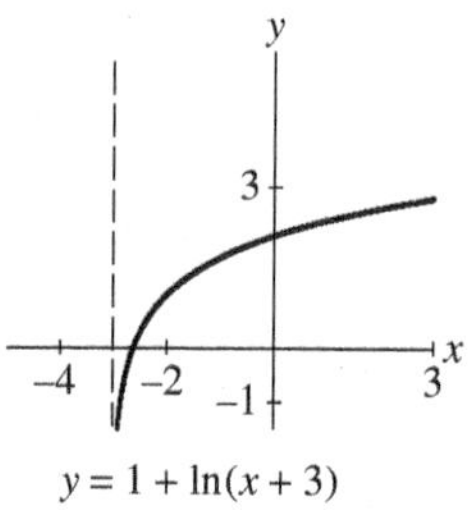


**73.** $(-\infty, \infty)$, $(1, \infty)$, inc, $y = 1$ **75.** $(-\infty, 2)$, $(-\infty, \infty)$, dec, $x = 2$

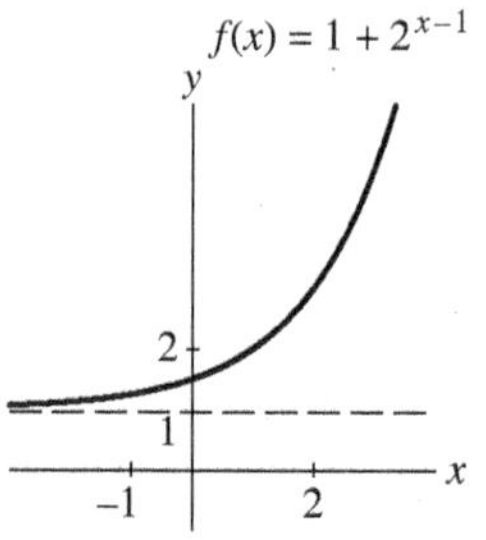


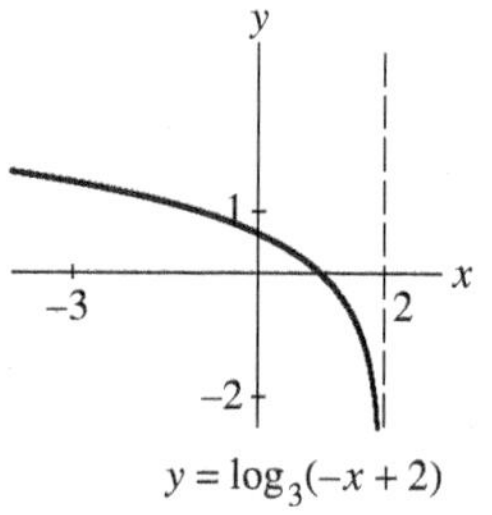


**77.** $f^{-1}(x) = \log_7(x)$ **79.** $f^{-1}(x) = 5^x$ **81.** $f^{-1}(x) = 10^{x/3} + 1$
**83.** $f^{-1}(x) = -2 + \ln(x + 3)$ **85.** 2.0959 **87.** 7.8538
**89.** $-4.4243$ **91.** T **93.** F **95.** T **97.** F **99.** F
**101.** F **103.** T **105.** The pH of A is one less than the pH of B.
**107.** \$122,296.01 **109.** 56 quarters **111.** 25 g, 18.15 g, 2166 yr
**113.** 2877 hr **115.** $-2/3, -3, -1/3$

## Chapter 4 Test

**1.** 3 **2.** $-2$ **3.** 6.47 **4.** $\sqrt{2}$ **5.** $f^{-1}(x) = e^x$
**6.** $f^{-1}(x) = -1 + \log_8(x + 3)$ **7.** $\log(xy^3)$
**8.** $\ln\left(\frac{\sqrt{x} - 1}{33}\right)$ **9.** $2 \cdot \log_a(2) + \log_a(7)$
**10.** $\log_a(7) - \log_a(2)$ **11.** 4 **12.** 18
**13.** $\frac{\ln(5)}{\ln(5) - \ln(3)} \approx 3.1507$ **14.** $1 + 3^{5.46} \approx 403.7931$
**15.** $(-\infty, \infty)$, $(1, \infty)$, inc, $y = 1$ **16.** $(1, \infty)$, $(-\infty, \infty)$, dec, $x = 1$

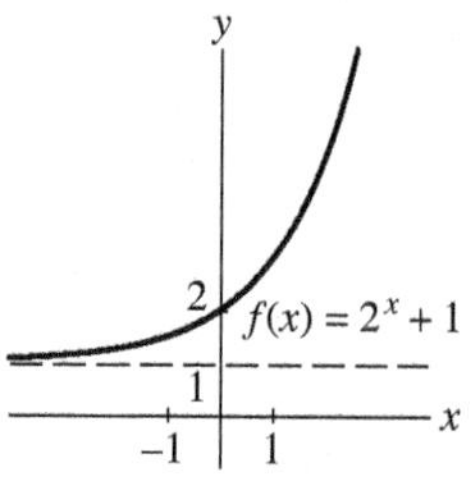


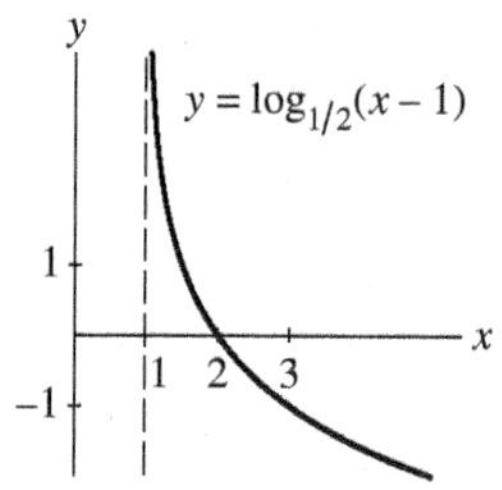


**17.** $(1, 0)$ **18.** \$9750.88, \$9906.06
**19.** 22.5 watts, 173.3 days, 428.7 days **20.** 61.5 quarters
**21.** $p = 0.86$, no

## Tying It All Together Chapters 1–4

**1.** 5, 1 **2.** 5 **3.** 19 **4.** 5 **5.** 19 **6.** $7, -1$ **7.** $2 \pm \sqrt{2}$
**8.** $-3$ **9.** 130 **10.** $\log_2(3)$ **11.** 4 **12.** $-1, 2, 3$

**13.**

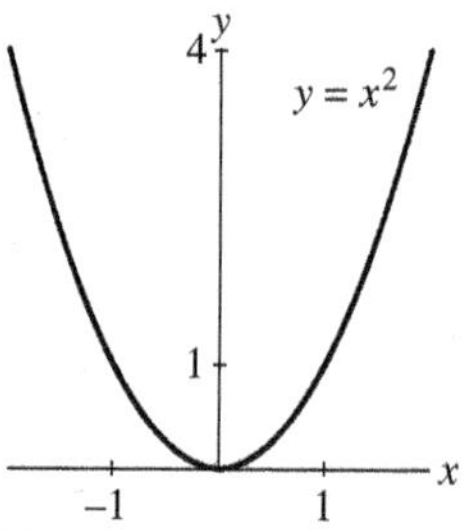


**14.**

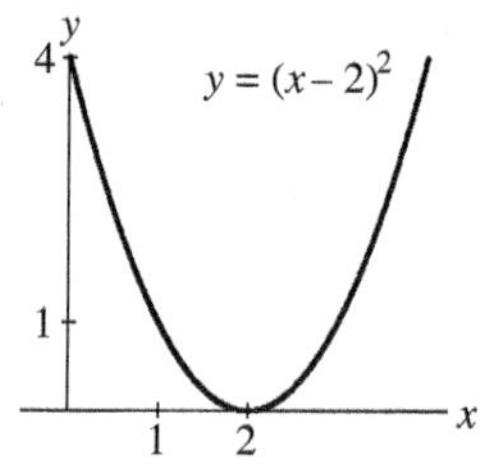


**15.**

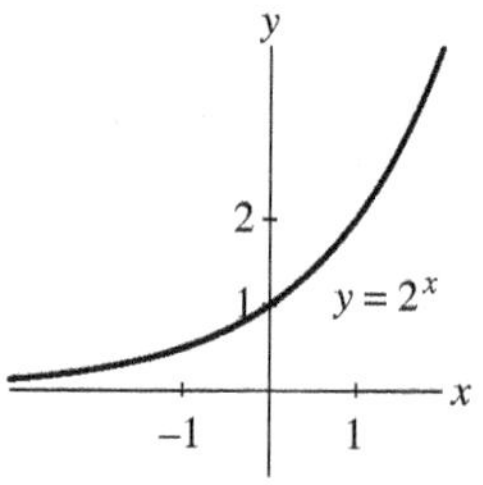


**16.**

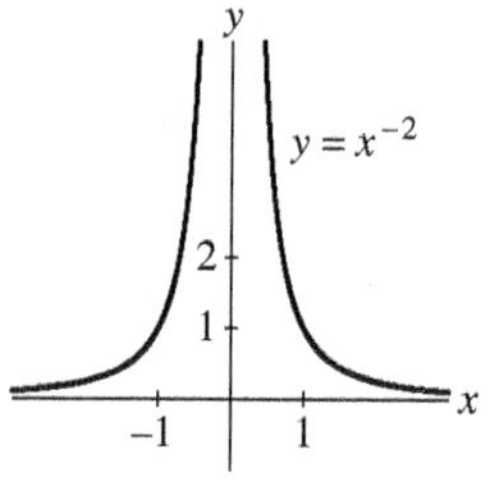


**17.**

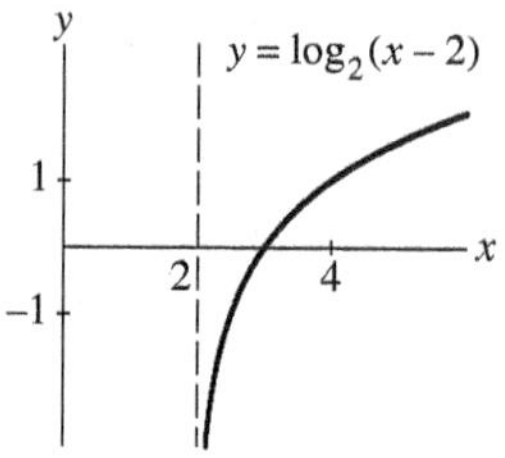


**18.**

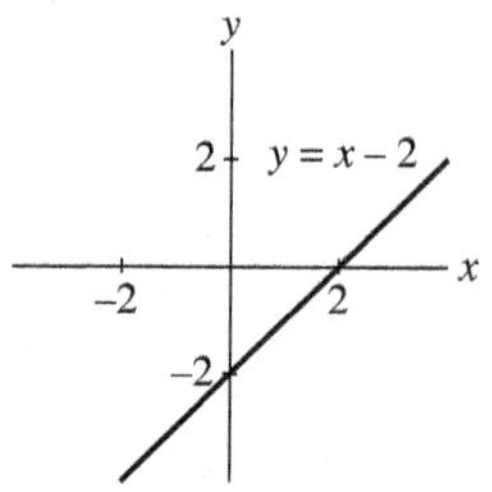


**19.**

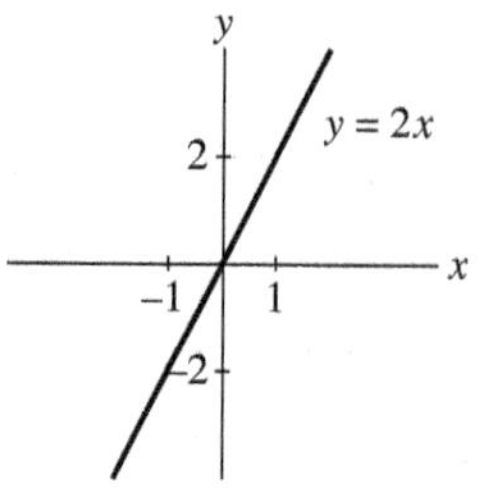


**20.**

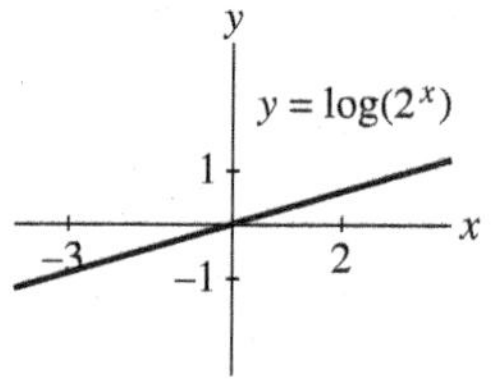

**21.**

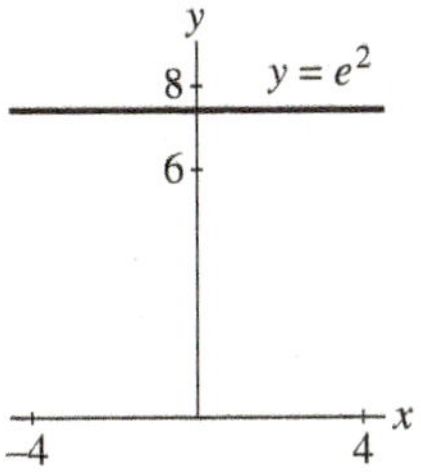


**22.**

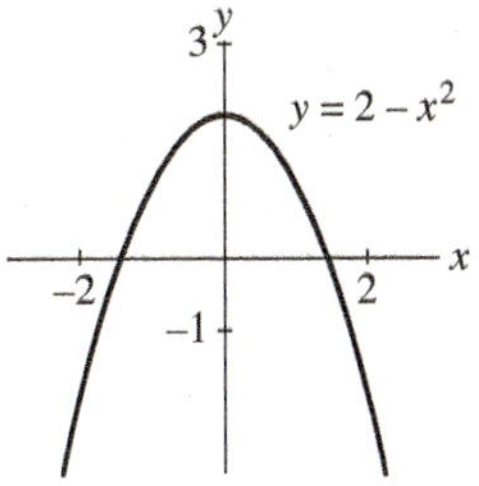


**23.**

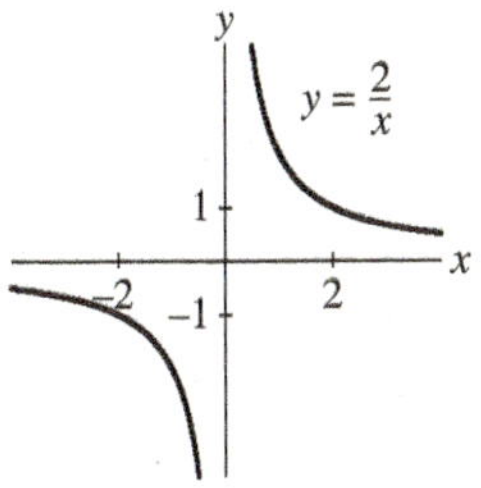


**24.**

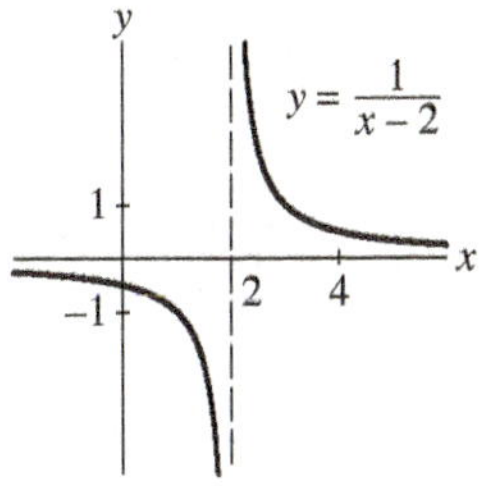


**25.** $f^{-1}(x) = 3x$ **26.** $f^{-1}(x) = -\log_3(x)$
**27.** $f^{-1}(x) = x^2 + 2$ for $x \geq 0$ **28.** $f^{-1}(x) = \sqrt[3]{x-2} + 5$
**29.** $f^{-1}(x) = (10^x + 3)^2$ **30.** $f^{-1} = \{(1, 3), (4, 5)\}$
**31.** $f^{-1}(x) = \dfrac{1}{x-3} + 5$ **32.** $f^{-1}(x) = (\ln(3 - x))^2$ for $x \leq 2$
**33.** $(p \circ m)(x) = e^{x+5}$, $(-\infty, \infty)$, $(0, \infty)$
**34.** $(p \circ q)(x) = e^{\sqrt{x}}$, $[0, \infty)$, $[1, \infty)$
**35.** $(q \circ p \circ m)(x) = \sqrt{e^{x+5}}$, $(-\infty, \infty)$, $(0, \infty)$
**36.** $(m \circ r \circ q)(x) = \ln(\sqrt{x}) + 5$, $(0, \infty)$, $(-\infty, \infty)$
**37.** $(p \circ r \circ m)(x) = e^{\ln(x+5)}$, $(-5, \infty)$, $(0, \infty)$
**38.** $(r \circ q \circ p)(x) = \ln(\sqrt{e^x})$, $(-\infty, \infty)$, $(-\infty, \infty)$
**39.** $F = f \circ g \circ h$ **40.** $H = g \circ h \circ f$ **41.** $G = h \circ g \circ f$
**42.** $M = h \circ f \circ g$ **43.** origin **44.** circle **45.** rise, run
**46.** point-slope form **47.** slope-intercept form
**48.** perpendicular **49.** parallel **50.** quadratic **51.** quadratic
**52.** equivalent

## Chapter 5

### Section 5.1

***For Thought:*** **1.** T **2.** F **3.** F **4.** F **5.** T **6.** F
**7.** F **8.** T **9.** F **10.** T

***Exercises:***
**1.** angle **3.** standard position **5.** obtuse **7.** coterminal
**9.** minute **11.** unit **13.** 420°, 780°, −300°, −660°
**15.** 344°, 704°, −376°, −736° **17.** Yes **19.** No **21.** I
**23.** III **25.** IV **27.** I **29.** 45° **31.** 60° **33.** 120°
**35.** 40° **37.** 20° **39.** 340° **41.** 13.2° **43.** −8.505°
**45.** 28.0858° **47.** 75°30′ **49.** −17°19′48″ **51.** 18°7′23″
**53.** $\pi/6$ **55.** $\pi/10$ **57.** $-3\pi/8$ **59.** $7\pi/2$ **61.** 0.653
**63.** −0.241 **65.** −0.936 **67.** 75° **69.** 315° **71.** −1080°
**73.** 136.937° **75.** $7\pi/3$, $13\pi/3$, $-5\pi/3$, $-11\pi/3$
**77.** $11\pi/6$, $23\pi/6$, $-13\pi/6$, $-25\pi/6$ **79.** $\pi$ **81.** $\pi/2$
**83.** $\pi/3$ **85.** $5\pi/3$ **87.** 2.04 **89.** No **91.** Yes **93.** I
**95.** III **97.** IV **99.** IV
**101.** $30° = \pi/6$, $45° = \pi/4$, $60° = \pi/3$, $90° = \pi/2$, $120° = 2\pi/3$, $135° = 3\pi/4$, $150° = 5\pi/6$, $180° = \pi$, $210° = 7\pi/6$, $225° = 5\pi/4$, $240° = 4\pi/3$, $270° = 3\pi/2$, $300° = 5\pi/3$, $315° = 7\pi/4$, $330° = 11\pi/6$, $360° = 2\pi$ **103.** $3\pi$ ft **105.** 209.4 mi
**107.** 1 mi **109.** 3.18 km **111.** 3102 mi **113.** 29.0°
**115.** 390,185.8 cm/min **117.** 166.6 mph **119.** 20.5 mph
**121.** 0.26 mph **123.** 41,143 km, 40,074 km **125.** 33.5 in.$^2$
**127.** 10 ft/sec and 24 rad/sec, 10 ft/sec and 40 rad/sec
**129.** 12:16:22 and 12:49:05
**131. a.** 22.5 in.$^3$ **b.** $V = \dfrac{\pi(360 - \alpha)^2 \sqrt{720\alpha - \alpha^2}}{2{,}187{,}000}$
**133.** $\{-5, 2\}$ **135.** $\{3/7\}$ **137.** $(-\infty, -3) \cup [4, \infty)$

### Section 5.2

***For Thought:*** **1.** F **2.** F **3.** T **4.** F **5.** F **6.** T
**7.** F **8.** F **9.** F **10.** T

***Exercises:***
**1.** $\sin(\alpha)$, $\cos(\alpha)$
**3.** $(1, 0)$, $(\sqrt{2}/2, \sqrt{2}/2)$, $(0, 1)$, $(-\sqrt{2}/2, \sqrt{2}/2)$, $(-1, 0)$, $(-\sqrt{2}/2, -\sqrt{2}/2)$, $(0, -1)$, $(\sqrt{2}/2, -\sqrt{2}/2)$
**5.** 0 **7.** 0 **9.** 0 **11.** 0 **13.** $\sqrt{2}/2$ **15.** $-\sqrt{2}/2$
**17.** 1/2 **19.** 1/2 **21.** $-\sqrt{3}/2$ **23.** $\sqrt{3}/2$ **25.** 1/2
**27.** 1/2 **29.** $\sqrt{3}/2$ **31.** 30°, $\pi/6$ **33.** 60°, $\pi/3$
**35.** 60°, $\pi/3$ **37.** 30°, $\pi/6$ **39.** 45°, $\pi/4$ **41.** 45°, $\pi/4$
**43.** + **45.** + **47.** − **49.** − **51.** $\sqrt{2}/2$ **53.** 1/2
**55.** $-\sqrt{2}/2$ **57.** $-\sqrt{3}/2$ **59.** $-\sqrt{2}/2$ **61.** −1/2
**63.** $\sqrt{3}/3$ **65.** −1 **67.** 1 **69.** $2 + \sqrt{3}$ **71.** $\sqrt{2}$
**73.** 0.9999 **75.** 0.4035 **77.** −0.7438 **79.** 1.0000
**81.** −0.2588 **83.** 1 **85.** 1/2 **87.** $\sqrt{2}/2$ **89.** $\sqrt{3}/2$
**91.** −12/13 **93.** −4/5 **95.** $2\sqrt{2}/3$
**97.** $x = 4 \sin t - 3 \cos t$, 3.53 **99.** 1.708 in., 1.714 in.
**101.** $\cos \alpha = \sqrt{1 - \sin^2 \alpha}$ for $\alpha$ in quadrants I or IV, $\cos \alpha = -\sqrt{1 - \sin^2 \alpha}$ for $\alpha$ in quadrants II or III **103.** 48°13′48″
**105.** $\pm(1, 2, 3, 6, 1/4, 1/2, 3/2, 3/4)$ **107.** $[3, \infty)$, $(-\infty, 1]$, function

### Section 5.3

***For Thought:*** **1.** F **2.** F **3.** F **4.** T **5.** T **6.** T
**7.** F **8.** T **9.** T **10.** T

***Exercises:***
**1.** sine wave **3.** amplitude **5.** period **7.** $y = -2 \sin(x)$, 2
**9.** $y = 3 \cos(x)$, 3 **11.** 2, $2\pi$, 0 **13.** 1, $2\pi$, $\pi/2$ **15.** 2, $2\pi$, $-\pi/3$
**17.** 1, 0, $(0, 0)$, $(\pi/2, -1)$, $(\pi, 0)$, $(3\pi/2, 1)$, $(2\pi, 0)$
**19.** 3, 0, $(0, 0)$, $(\pi/2, -3)$, $(\pi, 0)$, $(3\pi/2, 3)$, $(2\pi, 0)$

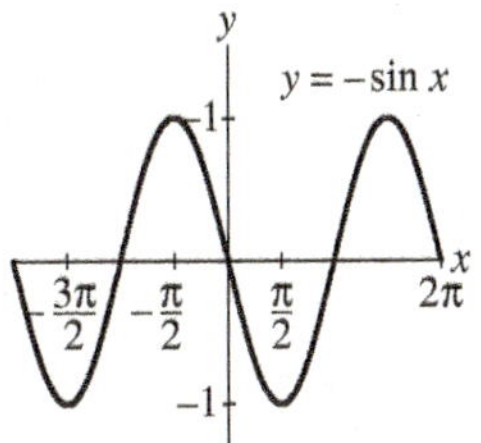


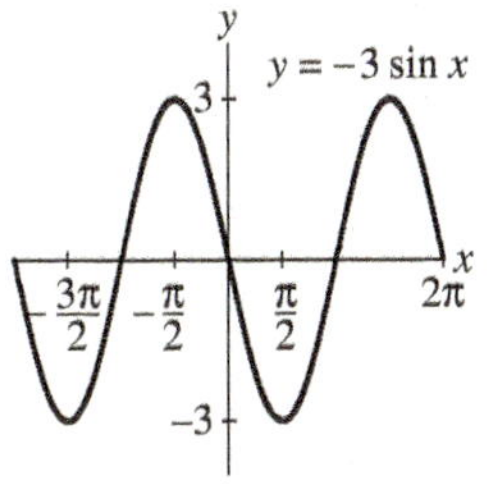


**21.** 1/2, 0, $(0, 1/2)$, $(\pi/2, 0)$, $(\pi, -1/2)$, $(3\pi/2, 0)$, $(2\pi, 1/2)$
**23.** 1, $-\pi$, $(0, 0)$, $(\pi/2, -1)$, $(\pi, 0)$, $(3\pi/2, 1)$, $(2\pi, 0)$

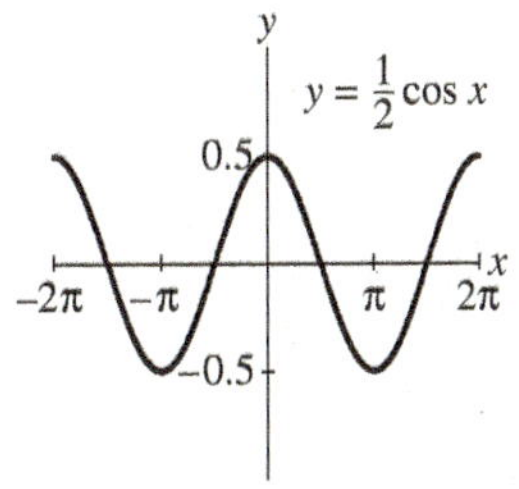


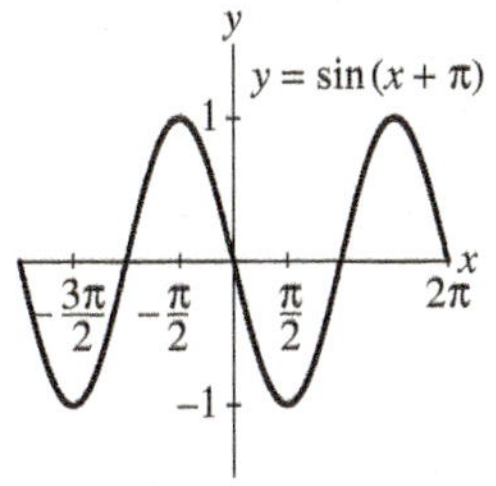

**25.** $1, \pi/3, (-2\pi/3, -1), (-\pi/6, 0), (\pi/3, 1), (5\pi/6, 0), (4\pi/3, -1)$

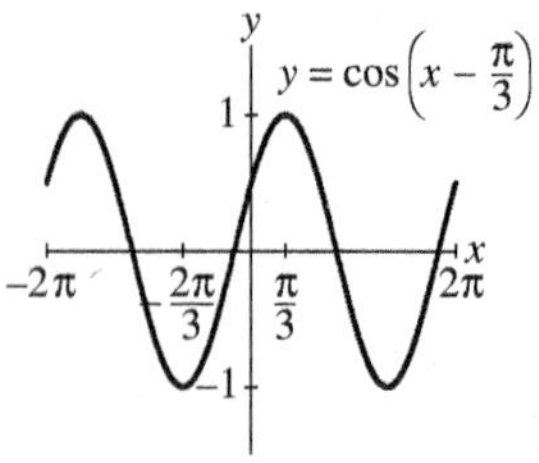


**27.** $1, 0, (0, 3), (\pi/2, 2), (\pi, 1), (3\pi/2, 2), (2\pi, 3)$

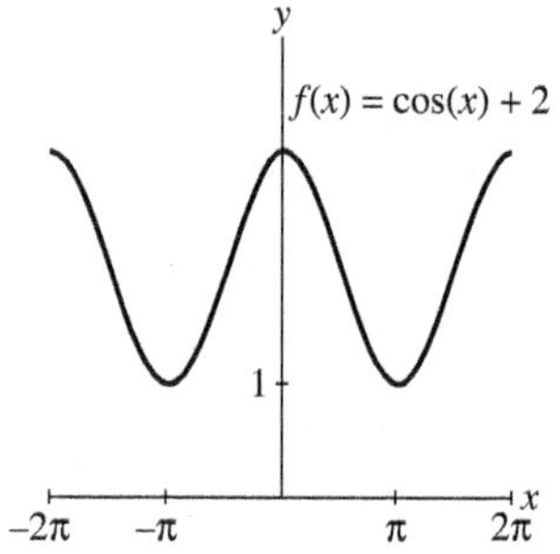


**29.** $1, 0, (0, -1), (\pi/2, -2), (\pi, -1), (3\pi/2, 0), (2\pi, -1)$

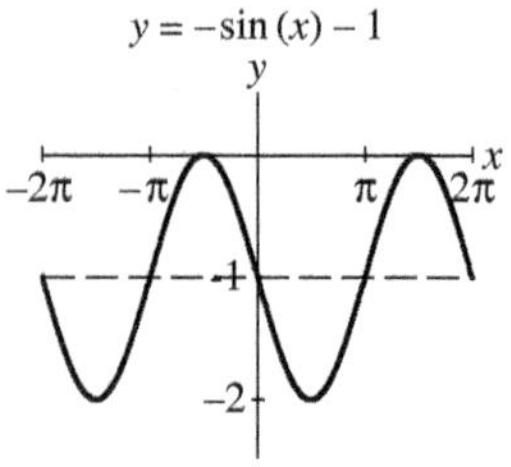


**31.** $1, -\pi/4, (-\pi/4, 2), (\pi/4, 3), (3\pi/4, 2), (5\pi/4, 1), (7\pi/4, 2)$

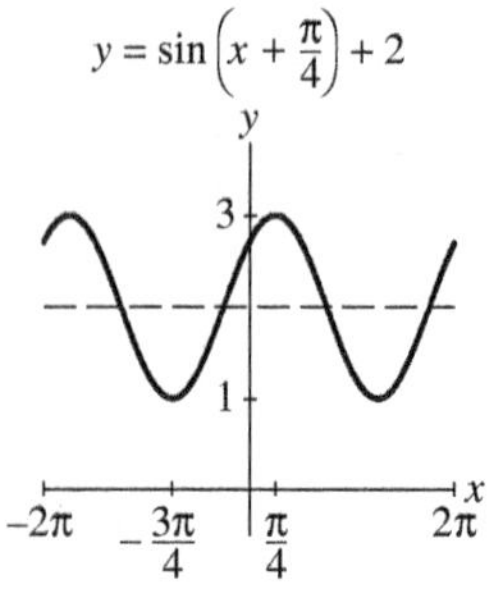


**33.** $2, -\pi/6, (-\pi/6, 3), (\pi/3, 1), (5\pi/6, -1), (4\pi/3, 1), (11\pi/6, 3)$

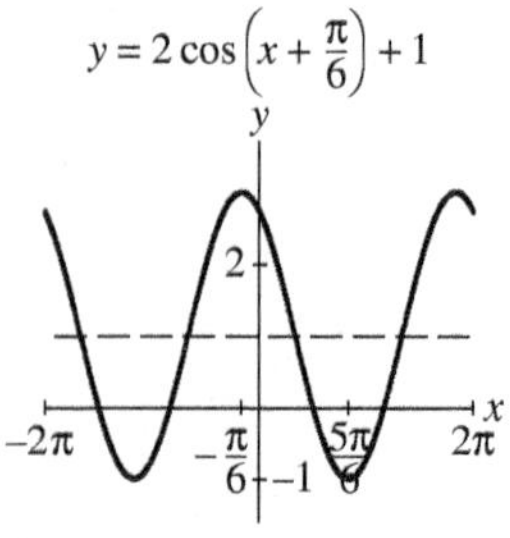


**35.** $2, \pi/3, (-\pi/6, 3), (\pi/3, 1), (5\pi/6, -1), (4\pi/3, 1), (11\pi/6, 3)$

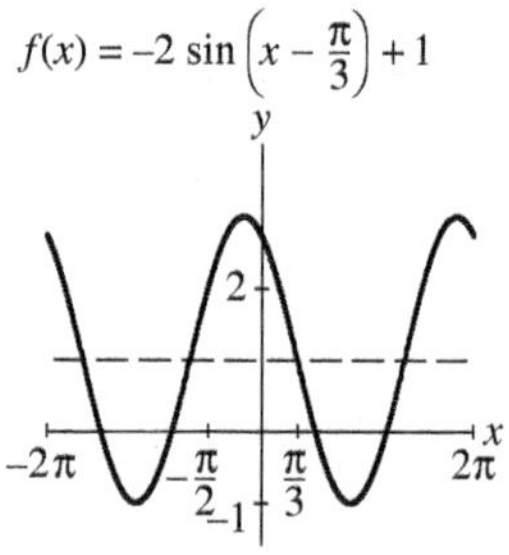


**37.** $3, \pi/2, 0$ **39.** $1, 4\pi, 0$ **41.** $2, 2\pi, \pi$ **43.** $2, \pi, -\pi/4$
**45.** $2, 4, -2$ **47.** $y = 2\sin[2(x + \pi/2)] + 5$
**49.** $y = 5\sin[\pi(x - 2)] + 4$ **51.** $y = 6\sin[4\pi(x + \pi)] - 3$
**53.** $y = -\sin(x - \pi/4) + 1$ **55.** $y = -3\cos(x - \pi) + 2$
**57.** $F(x) = \sin(3x - \pi/4)$ **59.** $F(x) = \sin(3x - 3\pi/4)$

**61.** $2\pi/3, 0, [-1, 1], (0, 0), (\pi/6, 1), (\pi/3, 0), (\pi/2, -1), (2\pi/3, 0)$

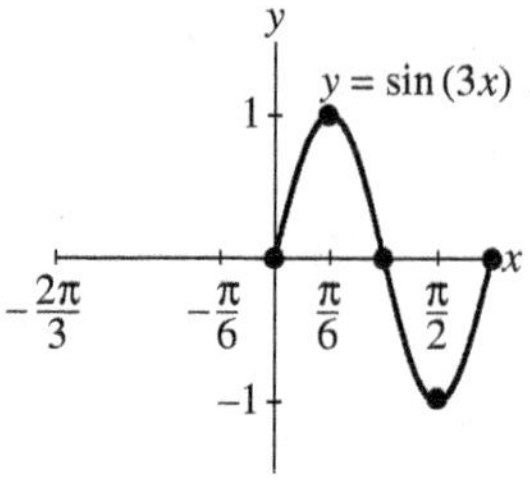


**63.** $\pi, 0, [-1, 1], (0, 0), (\pi/4, -1), (\pi/2, 0), (3\pi/4, 1), (\pi, 0)$

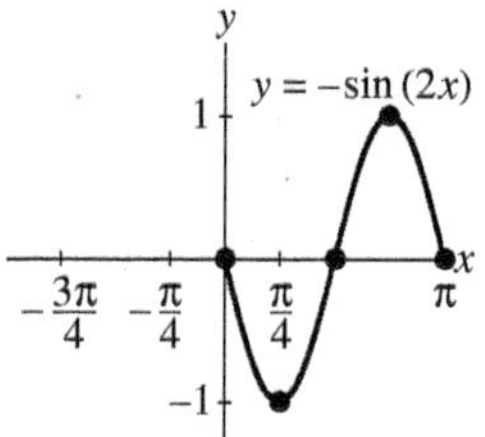


**65.** $\pi/2, 0, [1, 3], (0, 3), (\pi/8, 2), (\pi/4, 1), (3\pi/8, 2), (\pi/2, 3)$

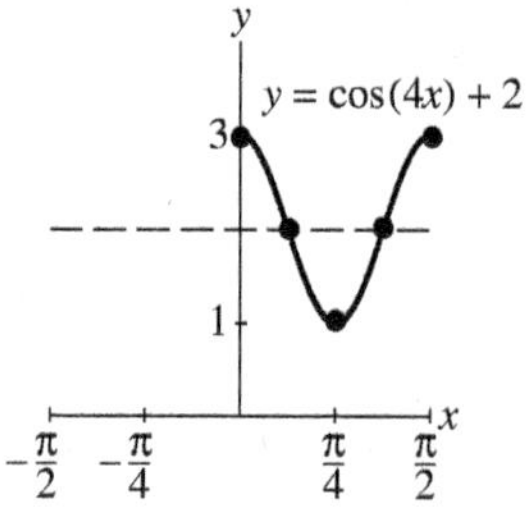


**67.** $8\pi, 0, [1, 3], (0, 2), (2\pi, 1), (4\pi, 2), (6\pi, 3), (8\pi, 2)$

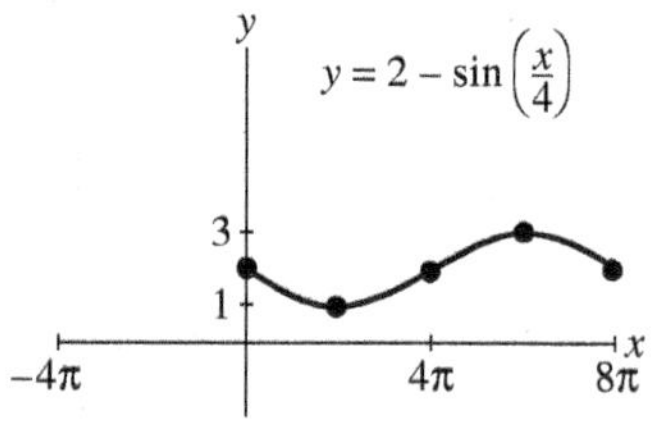


**69.** $6, 0, [-1, 1], (0, 0), (1.5, 1), (3, 0), (4.5, -1), (6, 0)$

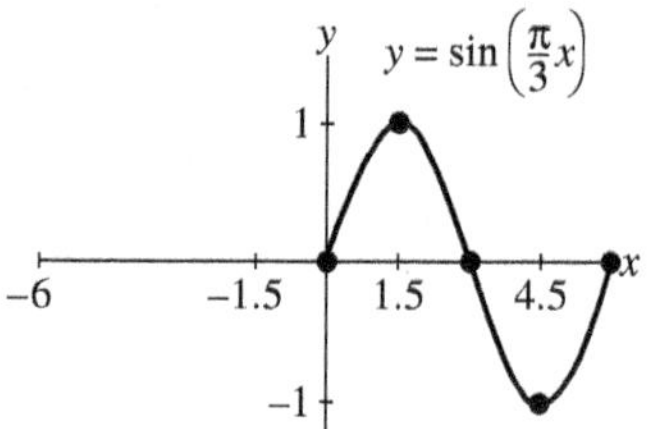


**71.** $\pi, \pi/2, [-1, 1], (\pi/2, 0), (3\pi/4, 1), (\pi, 0), (5\pi/4, -1), (3\pi/2, 0)$

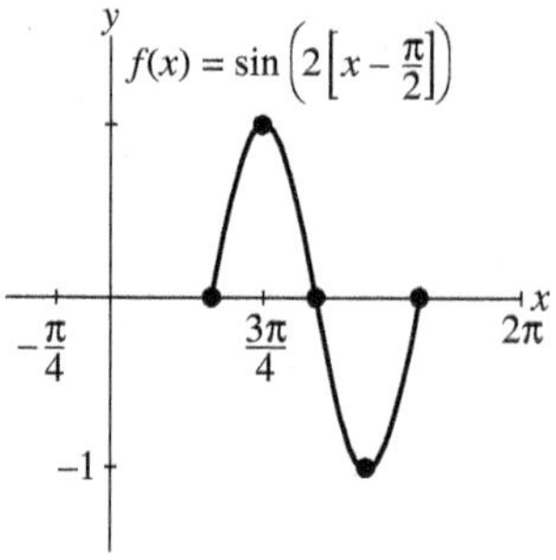


**73.** $4, -3, [-1, 1], (-3, 0), (-2, 1), (-1, 0), (0, -1), (1, 0)$

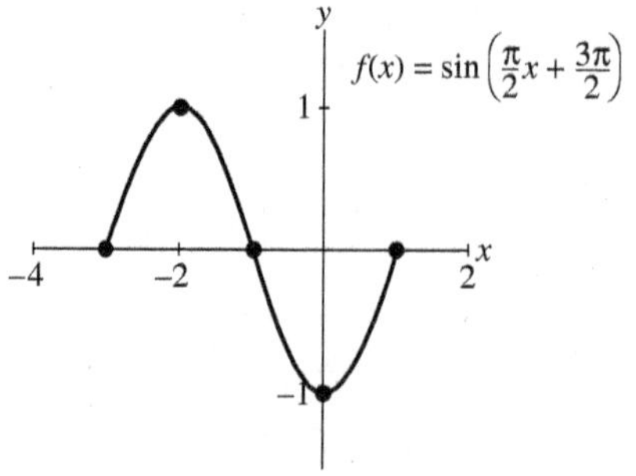

**75.** $\pi, -\pi/6, [-1, 3], (-\pi/6, 3), (\pi/12, 1), (\pi/3, -1), (7\pi/12, 1), (5\pi/6, 3)$

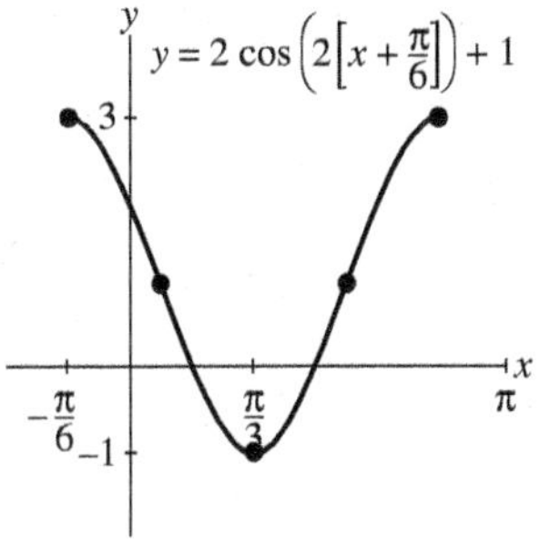


**77.** $2\pi/3, \pi/6, [-3/2, -1/2], (\pi/6, -1), (\pi/3, -3/2), (\pi/2, -1), (2\pi/3, -1/2), (5\pi/6, -1)$

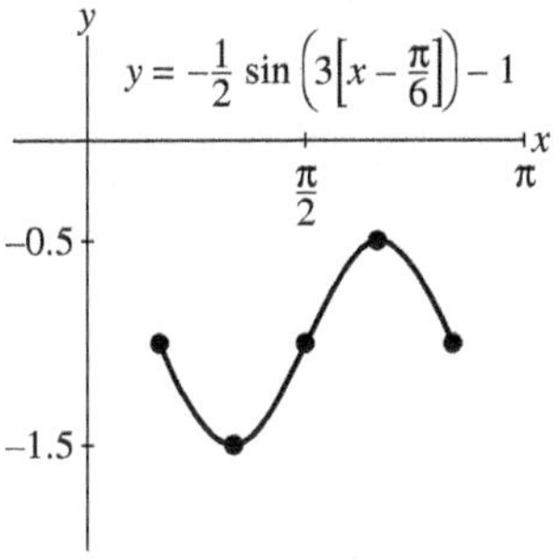


**79.** $y = 2\sin\left(2\left[x - \frac{\pi}{4}\right]\right)$ **81.** $y = 3\sin\left(\frac{3}{2}\left[x + \frac{\pi}{3}\right]\right) + 3$

**83.** 100 cycles/sec **85.** 40 cycles/hr

**87.** $x = 3\sin(2t), 3, \pi$

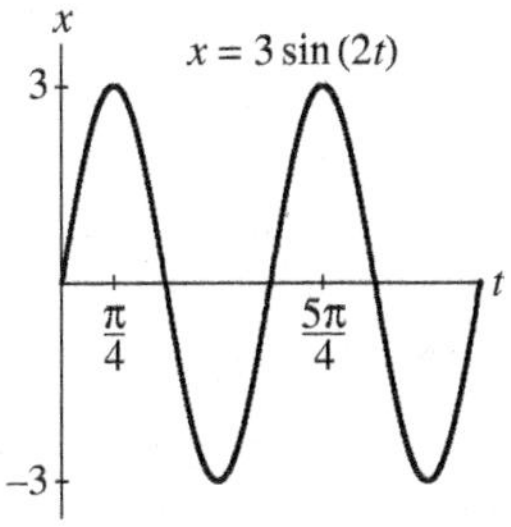


**89.** 11 yr **91. a.** 1300 cc, 500 cc **b.** 30

**93.** 12, 15,000, −3, 25,000, $y = 15{,}000\sin(\pi x/6 + \pi/2) + 25{,}000$, \$17,500

**95. a.** 40, 65, $y = 65\sin(\pi x/20)$ **b.** 40 days **c.** −38.2 m/sec **d.** Planet is between earth and Rho.

**97.** $y = \sin(\pi x/10) + 1$

**99.** $y = 50.0\sin(0.21x - 0.62) + 48.8$, 29.4 days, 99%

**103.** $5\pi/4$ **105.** 178.5 mi/hr

**107. a.** $\sqrt{2}/2$ **b.** $\sqrt{2}/2$

## Section 5.4

***For Thought:*** **1.** T **2.** F **3.** T **4.** F **5.** F **6.** T **7.** F **8.** T **9.** T **10.** T

***Exercises:***

**1.** $\tan(\alpha), \sec(\alpha)$

**3.** $\tan(0) = 0, \tan(\pi/4) = 1, \tan(\pi/2)$ undefined, $\tan(3\pi/4) = -1, \tan(\pi) = 0, \tan(5\pi/4) = 1, \tan(3\pi/2)$ undefined, $\tan(7\pi/4) = -1$

**5.** $\sqrt{3}$ **7.** −1 **9.** 0 **11.** $-\sqrt{3}/3$ **13.** $2\sqrt{3}/3$

**15.** Undefined **17.** Undefined **19.** $\sqrt{2}$ **21.** −1 **23.** $\sqrt{3}$

**25.** −2 **27.** $-\sqrt{2}$ **29.** 0 **31.** 48.0785 **33.** −2.8413

**35.** 500.0003 **37.** 1.0353 **39.** 636.6192 **41.** −1.4318

**43.** 71.6221 **45.** −0.9861 **47.** 4 **49.** $\sqrt{3}/3$ **51.** $-\sqrt{2}$

**53.** $\pi/3$

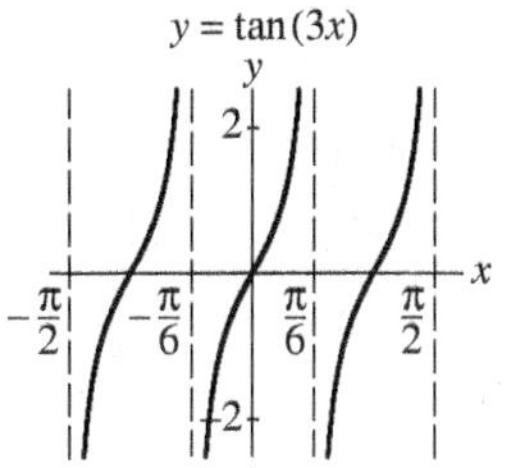


**55.** $\pi$

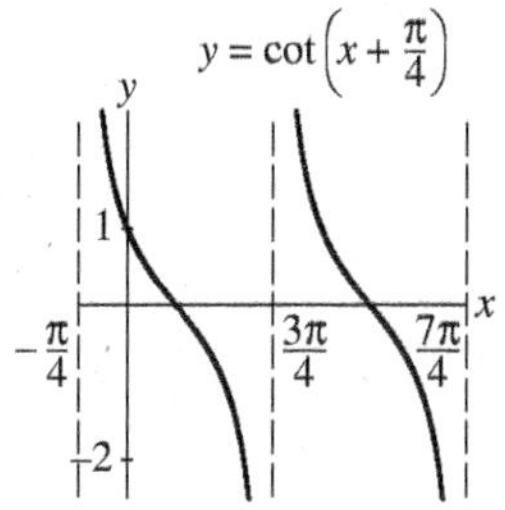


**57.** $2\pi$

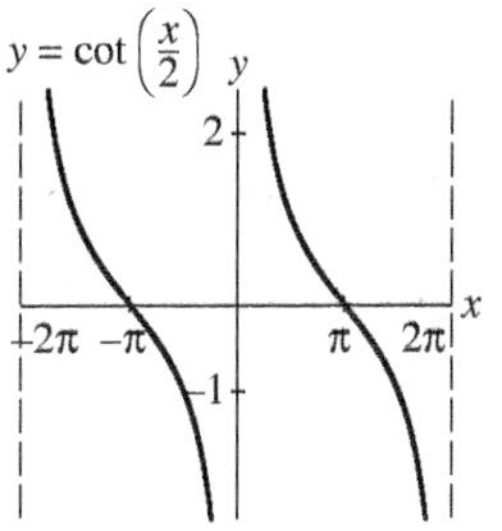


**59.** 1

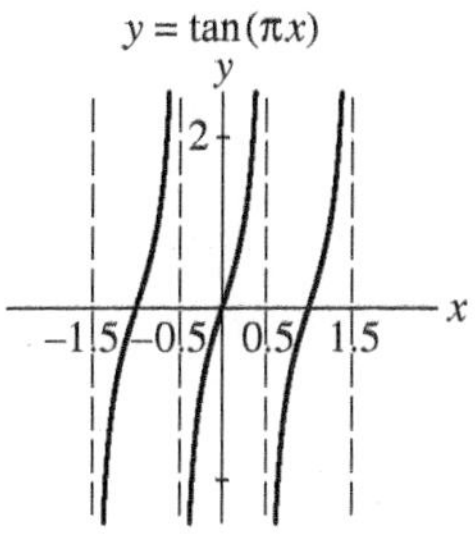


**61.** $\pi$

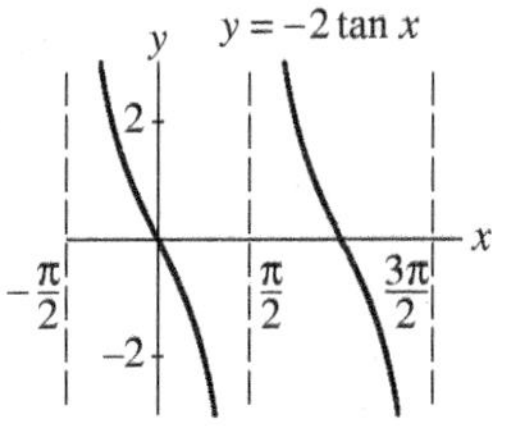


**63.** $\pi$

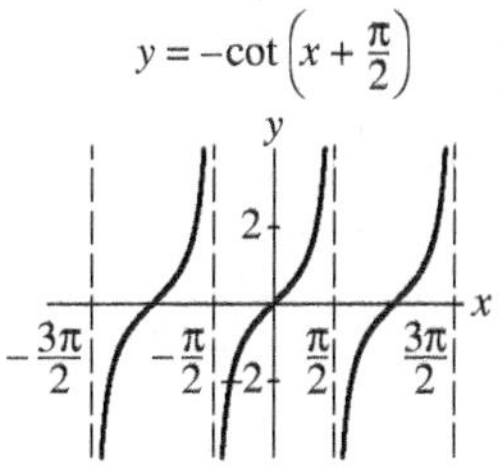


**65.** $\pi/2$

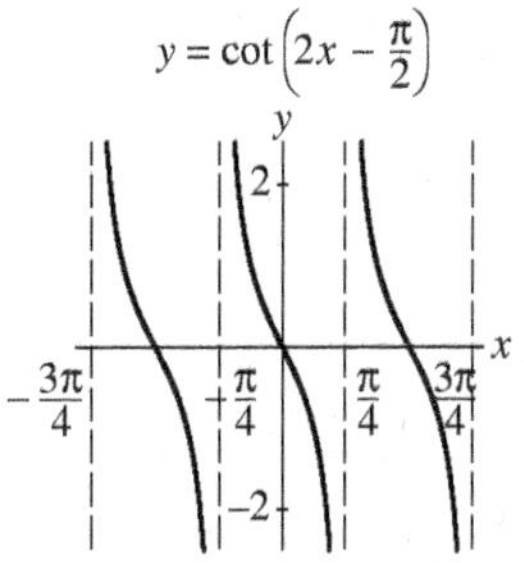


**67.** 2

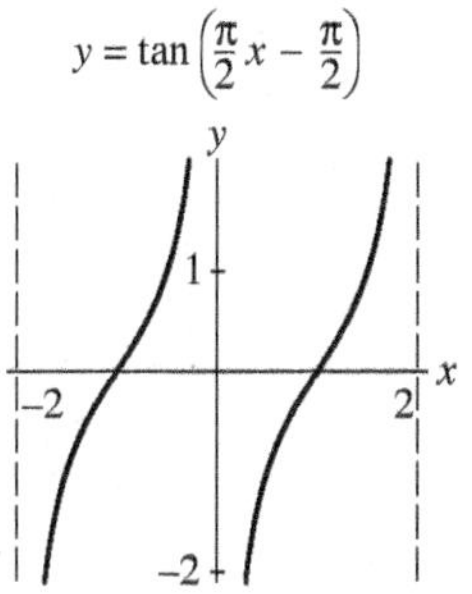


**69.** $\pi, (-\infty, -1] \cup [1, \infty)$

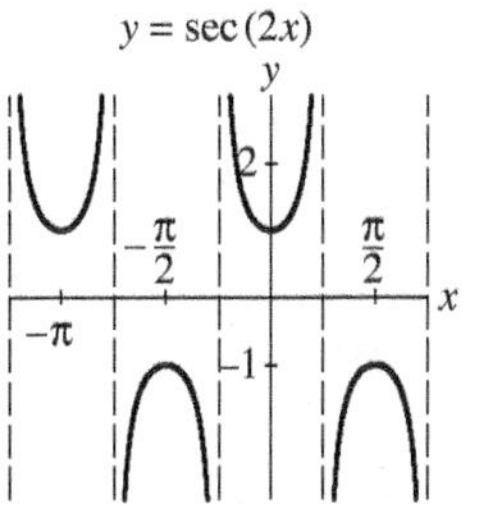


**71.** $2\pi, (-\infty, -1] \cup [1, \infty)$

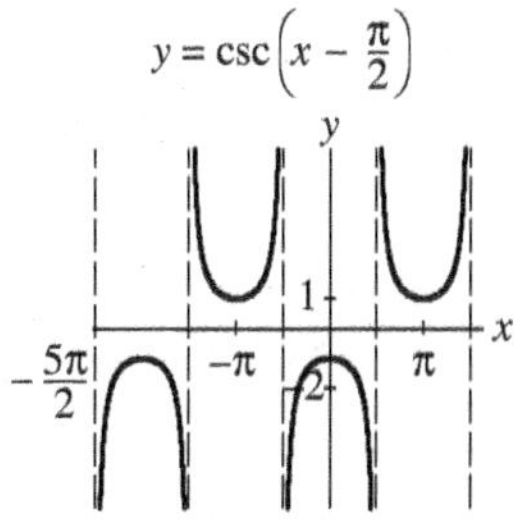

**73.** $4\pi$, $(-\infty, -1] \cup [1, \infty)$

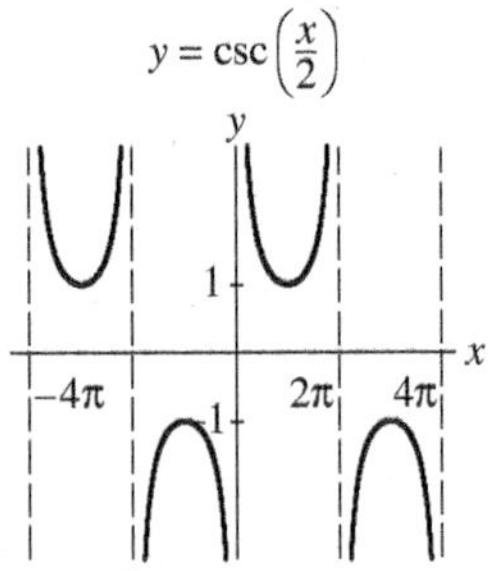


**75.** 4, $(-\infty, -1] \cup [1, \infty)$

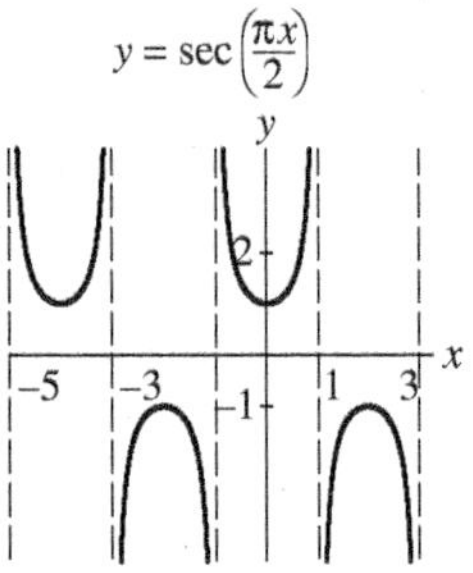


**77.** $2\pi$, $(-\infty, -2] \cup [2, \infty)$

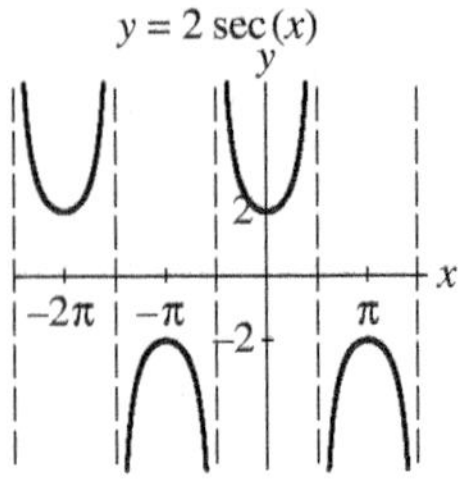


**79.** $\pi$, $(-\infty, -1] \cup [1, \infty)$

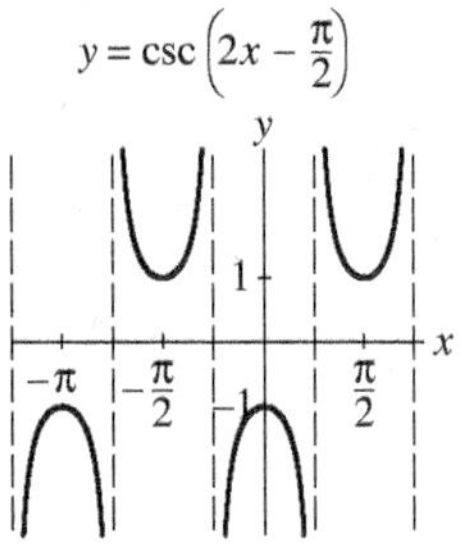


**81.** 4, $(-\infty, -1] \cup [1, \infty)$

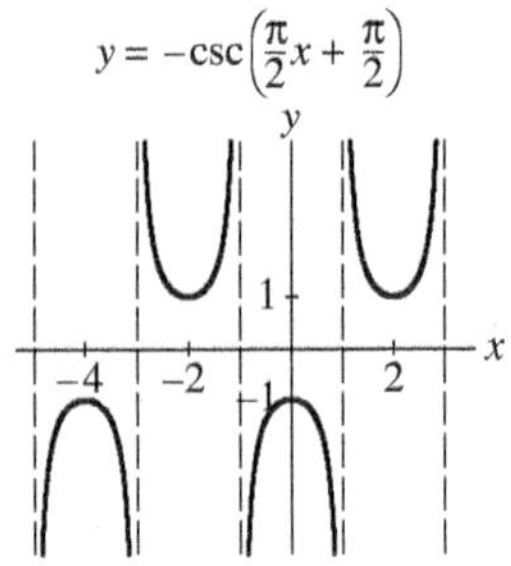


**83.** $\pi$, $(-\infty, 0] \cup [4, \infty)$

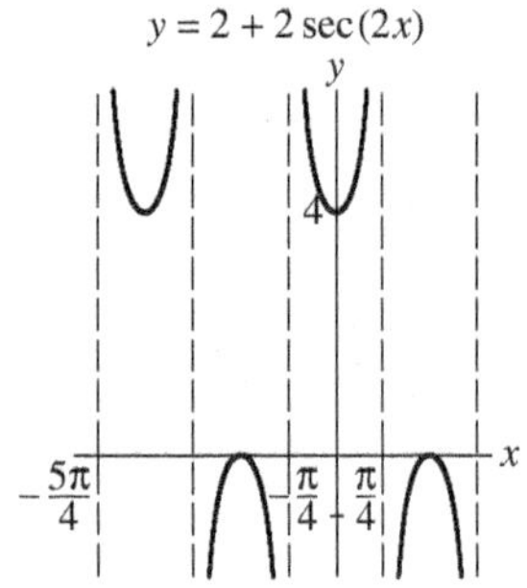


**85.** $\pi/2$, $(-\infty, \infty)$ **87.** $4\pi$, $(-\infty, -3] \cup [1, \infty)$
**89.** $\pi$, $(-\infty, -7] \cup [-1, \infty)$ **91.** $y = 3\tan(x - \pi/4) + 2$
**93.** $y = -\sec(x + \pi) + 2$
**95.**

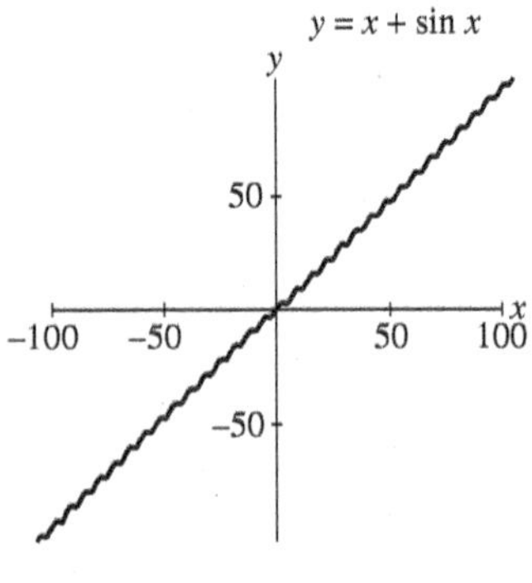


**97. a.** $y = \cos x$ **b.** $-\sin(x)$, $e^x$, $1/x$, $2x$
**99.** $-210°$ **101.** 1517 ft/sec **103.** $y = -\cos(x + \pi) + 2$

## Section 5.5

***For Thought:*** **1.** T **2.** T **3.** F **4.** F **5.** F **6.** T **7.** T **8.** T **9.** F **10.** F

***Exercises:***
**1.** domain **3.** domain **5.** $-\pi/6$ **7.** $\pi/6$ **9.** $\pi/4$ **11.** $-45°$ **13.** $30°$ **15.** $0°$ **17.** $-19.5°$ **19.** $34.6°$ **21.** $3\pi/4$ **23.** $\pi/3$ **25.** $\pi$ **27.** $135°$ **29.** $180°$ **31.** $120°$ **33.** $173.2°$ **35.** $89.9°$ **37.** $-\pi/4$ **39.** $\pi/3$ **41.** $\pi/4$ **43.** $-\pi/6$ **45.** 0 **47.** $\pi/2$ **49.** $3\pi/4$ **51.** $2\pi/3$ **53.** 0.60 **55.** 3.02 **57.** $-0.14$ **59.** 1.87 **61.** 1.15 **63.** $-0.36$ **65.** 3.06 **67.** 0.06 **69.** $\sqrt{3}$ **71.** $-\pi/6$ **73.** $\pi/6$ **75.** $\pi/4$ **77.** 1 **79.** $\pi/2$ **81.** 0 **83.** $\pi/2$ **85.** 0.8930 **87.** Undefined **89.** $-0.9802$ **91.** $-0.4082$ **93.** 3.4583 **95.** 1.0183
**97.** $f^{-1}(x) = 0.5\sin^{-1}(x)$, $[-1, 1]$
**99.** $f^{-1}(x) = \dfrac{1}{\pi}\tan^{-1}(x - 3)$, $(-\infty, \infty)$
**101.** $f^{-1}(x) = 2\sin(x - 3)$, $[3 - \pi/2, 3 + \pi/2]$ **103.** $67.1°$
**109. a.** $-1$ **b.** $\sqrt{2}/2$ **c.** $\sqrt{3}$ **d.** Undefined **e.** $-2$ **f.** Undefined **g.** $-\sqrt{3}$ **h.** $-\sqrt{2}/2$ **111.** $\pi/6$
**113.** $\pi$, $x = \pi/4 + k\pi/2$ for any integer $k$, $(-\infty, -3] \cup [3, \infty)$

## Section 5.6

***For Thought:*** **1.** F **2.** T **3.** F **4.** T **5.** T **6.** F **7.** F **8.** T **9.** T **10.** F

***Exercises:***
**1.** opposite side, hypotenuse **3.** opposite side, adjacent side
**5.** 4/5, 3/5, 4/3, 5/4, 5/3, 3/4
**7.** $3\sqrt{10}/10$, $-\sqrt{10}/10$, $-3$, $\sqrt{10}/3$, $-\sqrt{10}$, $-1/3$
**9.** $-\sqrt{3}/3$, $-\sqrt{6}/3$, $\sqrt{2}/2$, $-\sqrt{3}$, $-\sqrt{6}/2$, $\sqrt{2}$
**11.** $-1/2$, $\sqrt{3}/2$, $-\sqrt{3}/3$, $-2$, $2\sqrt{3}/3$, $-\sqrt{3}$
**13.** $\sqrt{5}/5$, $2\sqrt{5}/5$, $1/2$, $2\sqrt{5}/5$, $\sqrt{5}/5$, 2
**15.** $3\sqrt{34}/34$, $5\sqrt{34}/34$, 3/5, $5\sqrt{34}/34$, $3\sqrt{34}/34$, 5/3
**17.** 4/5, 3/5, 4/3, 3/5, 4/5, 3/4 **19.** $80.5°$ **21.** $60°$ **23.** 1.0
**25.** 0.4 **27.** $\beta = 30°$, $a = 10\sqrt{3}$, $b = 10$
**29.** $c = 10$, $\alpha = 36.9°$, $\beta = 53.1°$ **31.** $a = 5.7$, $\alpha = 43.7°$, $\beta = 46.3°$
**33.** $\beta = 74°$, $a = 5.5$, $b = 19.2$ **35.** $\beta = 50°51'$, $b = 11.1$, $c = 14.3$
**37.** 25 **39.** 0.831 **41.** 18.8 **43.** $-289$ **45.** 50 ft
**47.** 1.7 mi **49.** 43.2 m **51.** 25.1 ft **53.** 22 m, $57.6°$
**55.** No, speed $\approx$ 11.2 mph **57.** 153.1 m **59.** 4.5 km
**61.** $1.9 \times 10^{13}$ mi, $\approx$ 3.3 yr **63.** 4391 mi **65.** $75°$
**67. a.** $72(1 + 2\sqrt{2} - \pi)$ m$^2$ **b.** 88.2% **69.** 41.60 ft by 90.14 ft
**71.** 78.1 ft **73.** 5.987 ft **75.** 2.768 ft
**79. a.** $-30°$ **b.** $120°$ **c.** $-45°$
**81.** 3, 4, 1, $[4, 10]$ **83.** 2, $(-\infty, -5] \cup [5, \infty)$

## Chapter 5 Review Exercises

**1.** $28°$ **3.** $206°45'33''$ **5.** $180°$ **7.** $108°$ **9.** $300°$
**11.** $270°$ **13.** $11\pi/6$ **15.** $-5\pi/3$
**17.–18**

| $\theta$ deg | 0 | 30 | 45 | 60 | 90 | 120 | 135 | 150 | 180 |
|---|---|---|---|---|---|---|---|---|---|
| $\theta$ rad | 0 | $\frac{\pi}{6}$ | $\frac{\pi}{4}$ | $\frac{\pi}{3}$ | $\frac{\pi}{2}$ | $\frac{2\pi}{3}$ | $\frac{3\pi}{4}$ | $\frac{5\pi}{6}$ | $\pi$ |
| $\sin\theta$ | 0 | $\frac{1}{2}$ | $\frac{\sqrt{2}}{2}$ | $\frac{\sqrt{3}}{2}$ | 1 | $\frac{\sqrt{3}}{2}$ | $\frac{\sqrt{2}}{2}$ | $\frac{1}{2}$ | 0 |
| $\cos\theta$ | 1 | $\frac{\sqrt{3}}{2}$ | $\frac{\sqrt{2}}{2}$ | $\frac{1}{2}$ | 0 | $-\frac{1}{2}$ | $-\frac{\sqrt{2}}{2}$ | $-\frac{\sqrt{3}}{2}$ | $-1$ |
| $\tan\theta$ | 0 | $\frac{\sqrt{3}}{3}$ | 1 | $\sqrt{3}$ | | $-\sqrt{3}$ | $-1$ | $-\frac{\sqrt{3}}{3}$ | 0 |

**19.** $-\sqrt{2}/2$ **21.** $\sqrt{3}$ **23.** $-2\sqrt{3}/3$ **25.** 0 **27.** 0
**29.** $-1$ **31.** $\sqrt{3}/3$ **33.** $-\sqrt{2}/2$ **35.** $-2$ **37.** $-\sqrt{3}/3$
**39.** 5/13, 12/13, 5/12, 13/5, 13/12, 12/5 **41.** 0.6947 **43.** $-0.0923$
**45.** 0.1869 **47.** 1.0356 **49.** $-\pi/6$ **51.** $-\pi/4$ **53.** $\pi/4$
**55.** $\pi/6$ **57.** $90°$ **59.** $135°$ **61.** $30°$ **63.** $90°$
**65.** $c = \sqrt{13}$, $\alpha = 33.7°$, $\beta = 56.3°$ **67.** $\beta = 68.7°$, $c = 8.8$, $b = 8.2$

**69.** $2\pi/3$, $[-2, 2]$

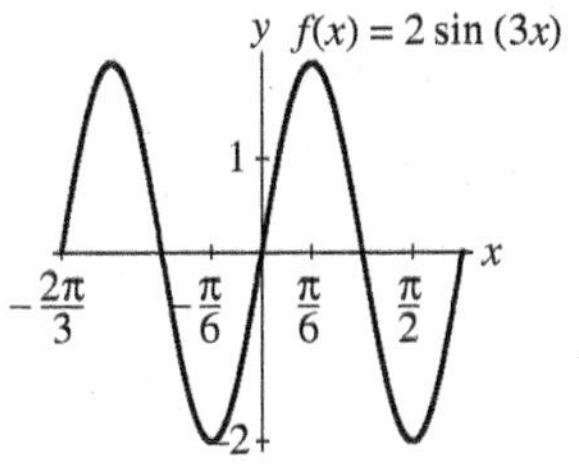


**71.** $\pi/2$, $(-\infty, \infty)$

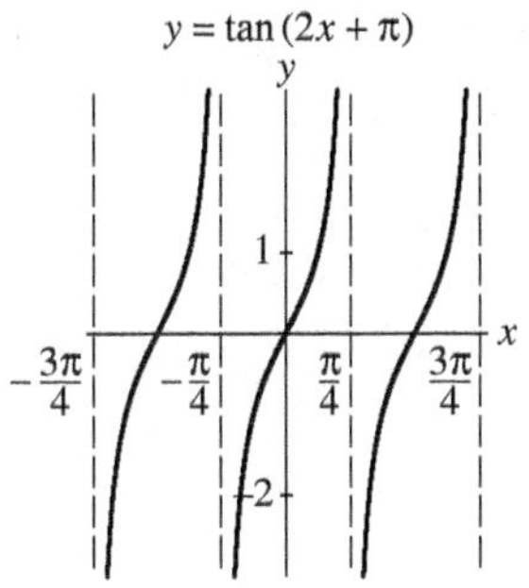


**73.** $4\pi$, $(-\infty, -1] \cup [1, \infty)$

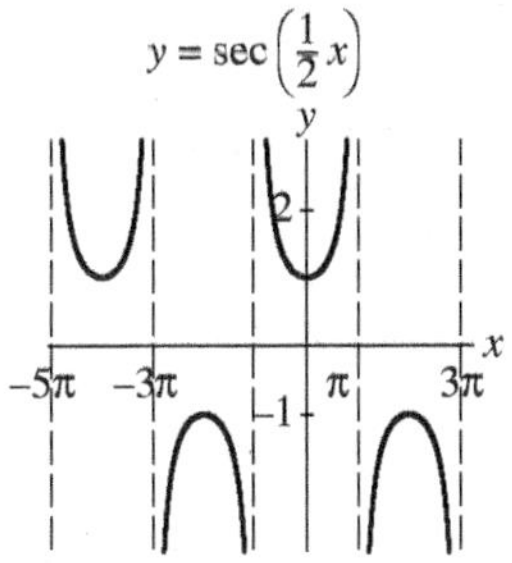


**75.** $\pi$, $[-1/2, 1/2]$

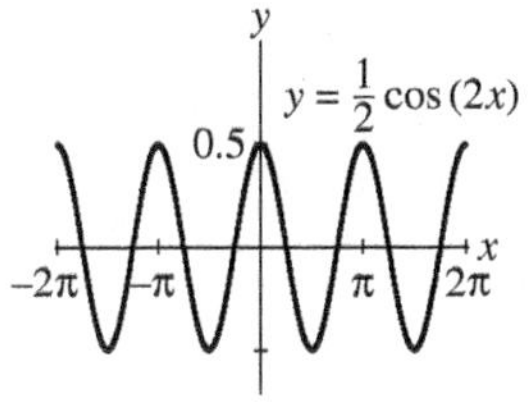


**77.** $\pi/2$, $(-\infty, \infty)$

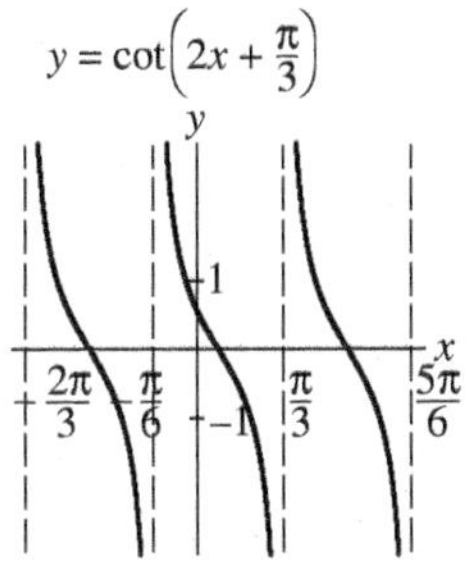


**79.** $\pi$, $(-\infty, -1/3] \cup [1/3, \infty)$

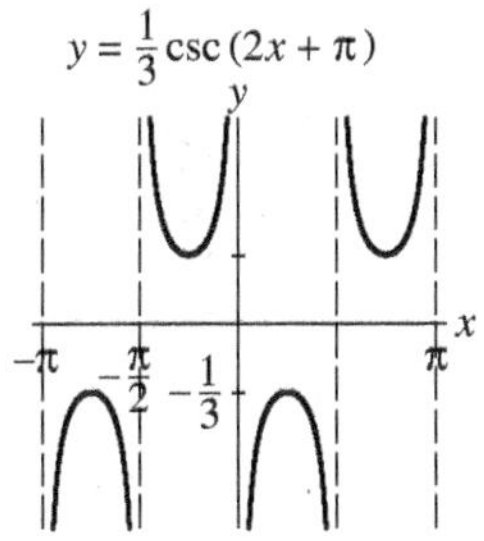


**81.** $y = 2\sin\left(\frac{\pi}{2}[x - 2]\right)$ **83.** $y = 20\sin(x) + 40$

**85.** 150° **87.** $-2\sqrt{6}/5$ **89.** 6.6 ft **91.** 53.1°

**93.** $1.08 \times 10^{-8}$ sec **95.** 94.2 ft/hr **97.** 6.9813 ft

**99.**

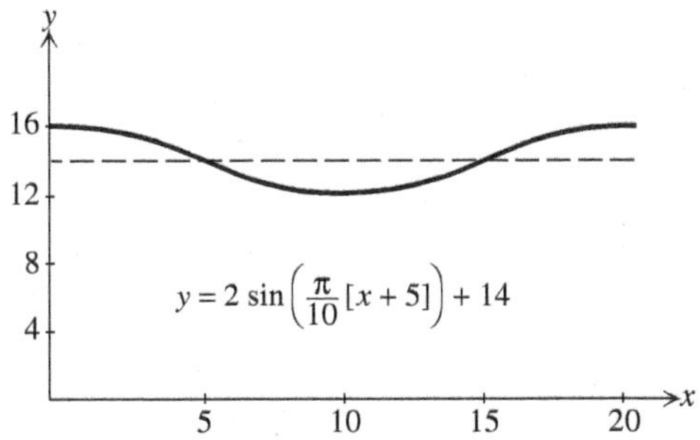


**101.** No **103.** 0.33 rad/sec **105.** 11.6 mi **107.** 56.1 in.

## Chapter 5 Test

**1.** 1/2 **2.** −1/2 **3.** −1 **4.** 2 **5.** −2 **6.** $\sqrt{3}/3$

**7.** $-\pi/6$ **8.** $2\pi/3$ **9.** $-\pi/4$ **10.** Undefined

**11.** Undefined **12.** $2\sqrt{2}/3$

**13.** $2\pi/3$, $[-3, -1]$, 1

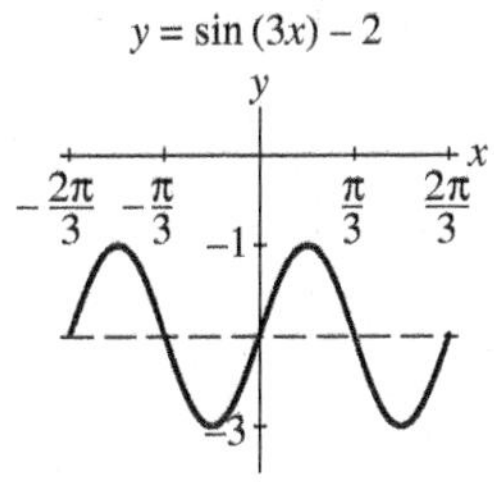


**14.** $2\pi$, $[-1, 1]$, 1

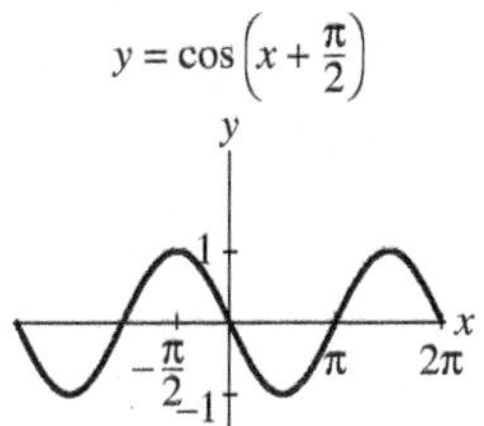


**15.** 2, $(-\infty, \infty)$

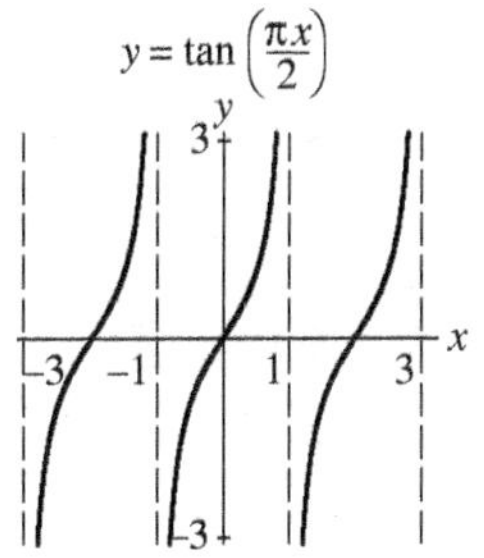


**16.** $\pi$, $[-2, 2]$, 2

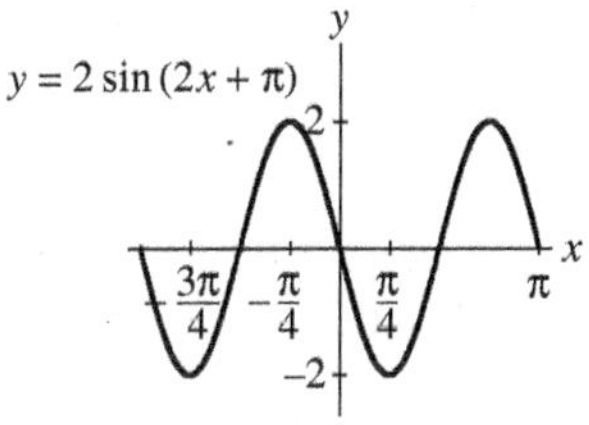


**17.** $2\pi$, $(-\infty, -2] \cup [2, \infty)$

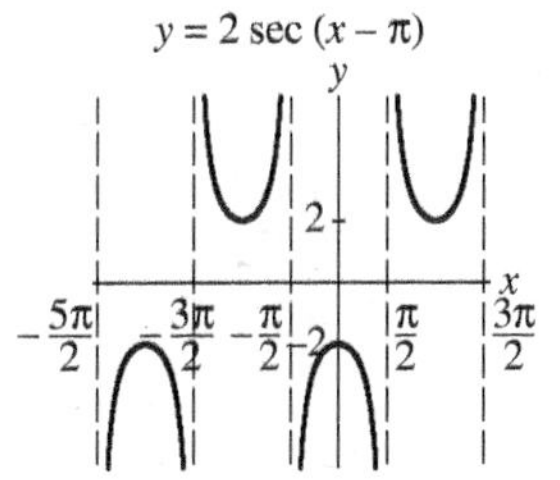


**18.** $2\pi$, $(-\infty, -1] \cup [1, \infty)$

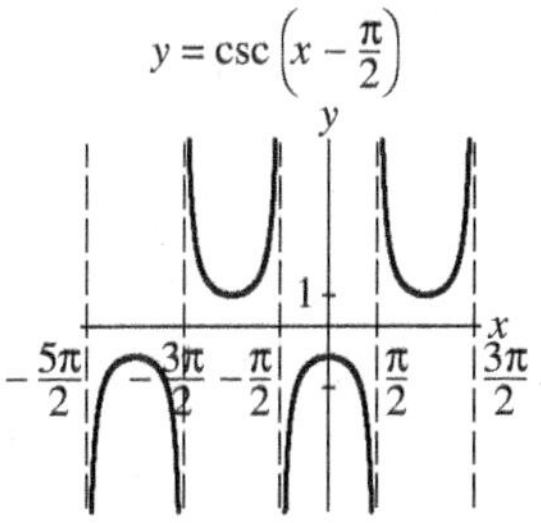


**19.** $\pi/2$, $(-\infty, \infty)$

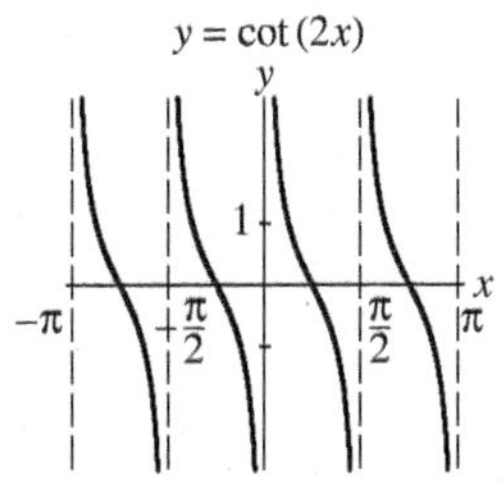


**20.** $2\pi$, $[-1, 1]$, 1

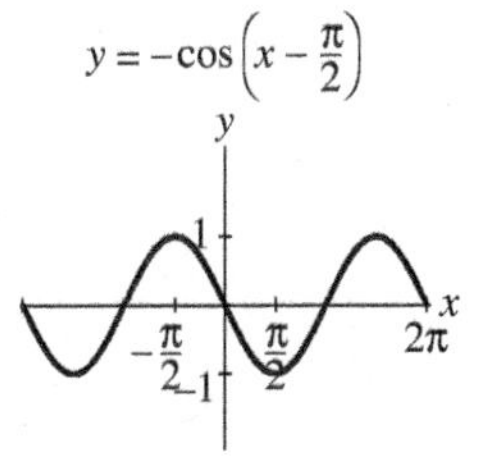


**21.** 28.85 m **22.** 134.1° **23.** $-\sqrt{15}/4$

**24.** $\sin\alpha = -2/\sqrt{29}$, $\cos\alpha = 5/\sqrt{29}$, $\tan\alpha = -2/5$, $\csc\alpha = -\sqrt{29}/2$, $\sec\alpha = \sqrt{29}/5$, $\cot\alpha = -5/2$

**25.** 647.2 rad/min **26.** 7.97 mph **27.** 12.2 m **28.** 1117 ft

**29.**

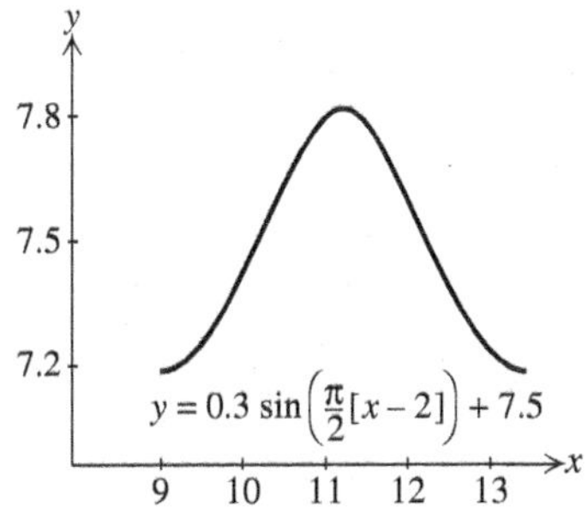

### Tying It All Together Chapters 1–5

**1.** $(-\infty, \infty)$, $(2, \infty)$

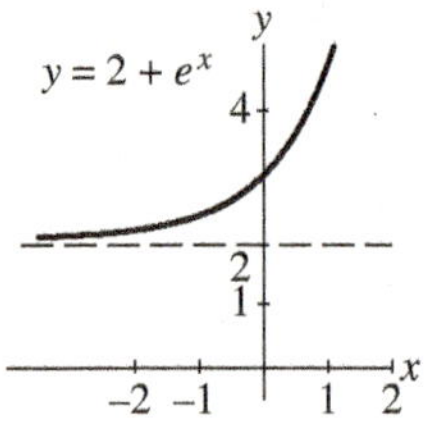


**2.** $(-\infty, \infty)$, $[2, \infty)$

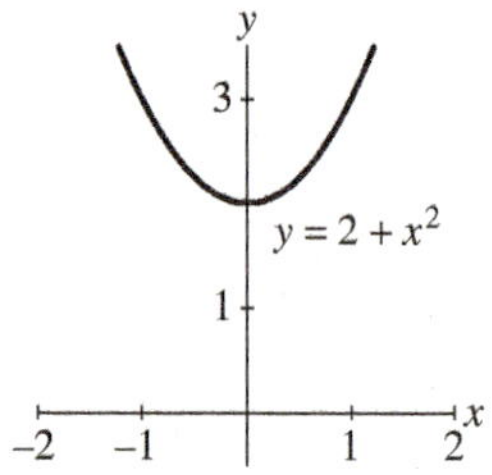


**3.** $(-\infty, \infty)$, $[1, 3]$

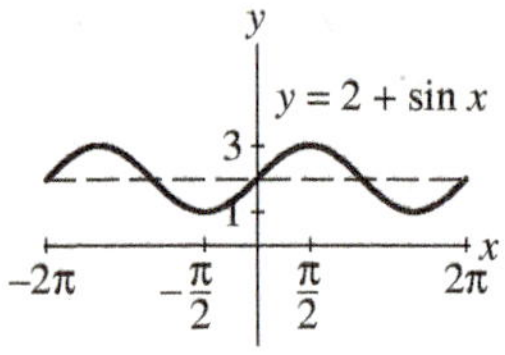


**4.** $(0, \infty)$, $(-\infty, \infty)$

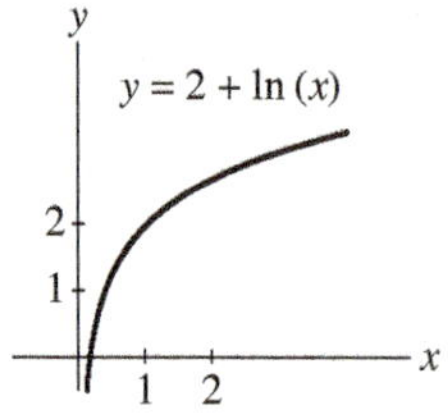


**5.** $(\pi/4, \infty)$, $(-\infty, \infty)$

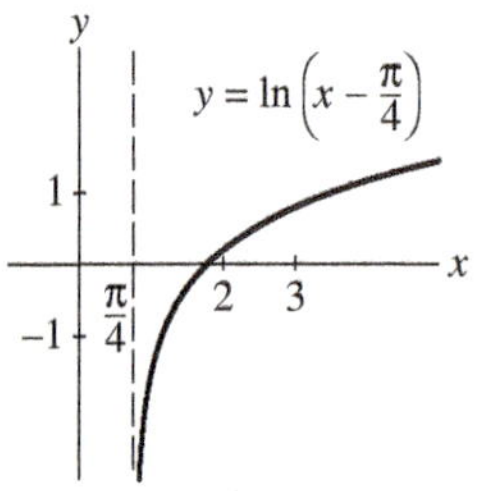


**6.** $(-\infty, \infty)$, $[-1, 1]$

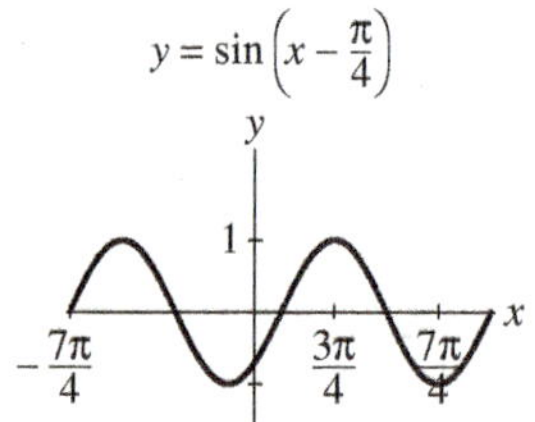


**7.** $(0, \infty)$, $(-\infty, \infty)$

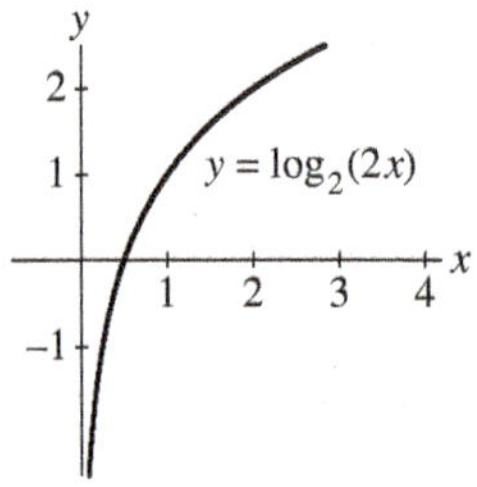


**8.** $(-\infty, \infty)$, $[-1, 1]$

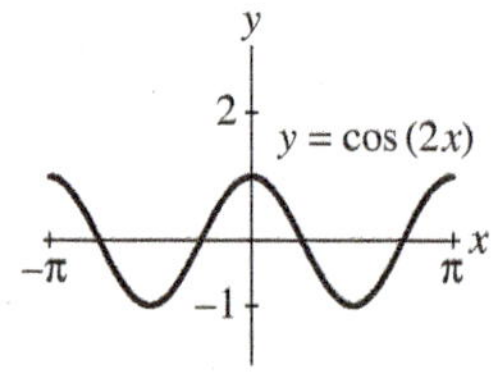


**9.** Neither **10.** Odd **11.** Even **12.** Even **13.** Odd **14.** Even **15.** Neither **16.** Odd **17.** Increasing **18.** Increasing **19.** Increasing **20.** Decreasing **21.** Decreasing **22.** Increasing **23.** perfect square **24.** exponential **25.** logarithmic **26.** common **27.** natural **28.** angle **29.** central **30.** acute **31.** quadrantal **32.** coterminal

## Chapter 6

### Section 6.1

***For Thought:*** **1.** F **2.** T **3.** T **4.** F **5.** T **6.** F **7.** F **8.** T **9.** F **10.** T

***Exercises:***

**1.** even **3.** Pythagorean **5.** odd **7.** $\sin x$ **9.** 1 **11.** $\csc x$ **13.** 1 **15.** 1 **17.** $\cos^2 \alpha$ **19.** $-\cos^2 \beta$ **21.** $2 \sin \alpha$

**23.** $\cot(x) = \pm\sqrt{\csc^2(x) - 1}$ **25.** $\sin(x) = \dfrac{\pm 1}{\sqrt{1 + \cot^2(x)}}$

**27.** $\tan(x) = \dfrac{\pm 1}{\sqrt{\csc^2(x) - 1}}$

**29.** $\sin \alpha = 1/\sqrt{5}$, $\cos \alpha = 2/\sqrt{5}$, $\csc \alpha = \sqrt{5}$, $\sec \alpha = \sqrt{5}/2$, $\cot \alpha = 2$

**31.** $\sin \alpha = -\sqrt{22}/5$, $\tan \alpha = \sqrt{22}/\sqrt{3}$, $\csc \alpha = -5/\sqrt{22}$, $\sec \alpha = -5/\sqrt{3}$, $\cot \alpha = \sqrt{3}/\sqrt{22}$

**33.** $\sin \alpha = -3/\sqrt{10}$, $\cos \alpha = 1/\sqrt{10}$, $\tan \alpha = -3$, $\csc \alpha = -\sqrt{10}/3$, $\sec \alpha = \sqrt{10}$

**35.** $\sqrt{1 - x^2}$ **37.** $1/\sqrt{1 + x^2}$ **39.** $x/\sqrt{1 - x^2}$ **41.** $\sqrt{1 + x^2}$ **43.** $\cos x$ **45.** 0 **47.** 0 **49.** $\cos^2 \alpha$ **51.** $-\cos \beta$ **53.** Odd **55.** Neither **57.** Even **59.** Even **61.** Even **63.** Odd **65.** h **67.** n **69.** m **71.** k **73.** 1 **75.** g **77.** b **79.** f **81.** d **93.** $-\tan^2 x$ **95.** $-1$ **97.** 1 **99.** $\tan^2 x$ **101.** $\csc x$ **103.** $\sin^2 x - \cos^2 x$ **105.** $\pm 2\sqrt{2}/3$ **107.** $\sqrt{1 - u^2}$

**109.** $x = \dfrac{\pi}{2} + k\pi$, where $k$ is any integer **113.** 1155 feet

**115.** 5, $\pi$, $\pi/2$, $[-2, 8]$ **117.** 0

### Section 6.2

***For Thought:*** **1.** T **2.** F **3.** T **4.** T **5.** T **6.** F **7.** T **8.** T **9.** F **10.** F

***Exercises:***

**1.** D **3.** A **5.** B **7.** H **9.** G **11.** $2\cos^2 \beta - \cos \beta - 1$ **13.** $\csc^2(x) + 2 + \sin^2(x)$ **15.** $4\sin^2(\theta) - 1$ **17.** $9\sin^2 \theta + 12 \sin \theta + 4$ **19.** $4\sin^4 y - 4 + \csc^4 y$ **21.** $\cos^2 \alpha$ **23.** $\cot^2 \alpha$ **25.** $-1$ **27.** $(2 \sin \gamma + 1)(\sin \gamma - 3)$ **29.** $(\tan \alpha - 4)(\tan \alpha - 2)$ **31.** $(2 \sec \beta + 1)^2$ **33.** $(\tan \alpha - \sec \beta)(\tan \alpha + \sec \beta)$ **35.** $(\cos \beta)(\sin \beta + 2)(\sin \beta - 1)$

**37.** $(2\sec^2 x - 1)^2$ **39.** $(\cos \alpha + 1)(\sin \alpha + 1)$ **41.** $\dfrac{\sin^2 x}{a}$

**43.** $\dfrac{3 \sin(2x)}{2}$ **45.** $\dfrac{5 \tan x}{6}$ **47.** $1 - \sin x$ **49.** $\sin x + \cos x$

**51.** $\dfrac{\sin x + 1}{\sin x + 2}$ **53.** $\sin x$ **95.** Identity **97.** Not identity

**99.** Identity **101.** Not identity **103.** Identity

**107.** $\sin^2 x + \cos^2 x = 1$, $1 + \cot^2 x = \csc^2 x$, $\tan^2 x + 1 = \sec^2 x$

**109.** $\cot x$ **111.** 4, 3, 1/2 **113.** 57.1 mph

### Section 6.3

***For Thought:*** **1.** F **2.** T **3.** T **4.** F **5.** T **6.** F **7.** T **8.** T **9.** F **10.** T

***Exercises:***

**1.** cosine **3.** $\dfrac{\sqrt{2} - \sqrt{6}}{4}$ **5.** $\dfrac{\sqrt{2} + \sqrt{6}}{4}$ **7.** 70°

**9.** $\pi/3$ **11.** 6° **13.** $\dfrac{\sqrt{2} + \sqrt{6}}{4}$ **15.** $\dfrac{\sqrt{6} - \sqrt{2}}{4}$

**17.** $2 - \sqrt{3}$ **19.** $\sqrt{3} - 2$ **21.** $\dfrac{7\pi}{12}$ **23.** $\dfrac{13\pi}{12}$

**25.** 1 **27.** 0 **29.** 1 **31.** $\sin(3k)$ **33.** 30° + 45°

**35.** 120° + 45° **37.** $\dfrac{\sqrt{6} - \sqrt{2}}{4}$ **39.** $\dfrac{\sqrt{6} + \sqrt{2}}{4}$

**41.** $2 + \sqrt{3}$ **43.** $\dfrac{\sqrt{2} - \sqrt{6}}{4}$ **45.** $\dfrac{-\sqrt{6} - \sqrt{2}}{4}$ **47.** $-2 + \sqrt{3}$

**49.** 1 **51.** $\cos(3\pi/10)$ **53.** $\tan(13\pi/42)$ **55.** sin 49° **57.** G **59.** H **61.** F **63.** A **65.** 16/65

**67.** $\dfrac{2 - \sqrt{15}}{6}$ **69.** 297/425 **71.** $-416/425$ **73.** $-\sin \alpha$

**75.** $-\cos \alpha$ **77.** $-\sin \alpha$ **79.** $\cos \alpha$ **105.** $\cos^2 \alpha$

**107.** $\pi/2$, $x = k\pi/2$ for any integer $k$

**109.** $\cos \alpha = -\sqrt{15}/4$, $\tan \alpha = -\sqrt{15}/15$, $\csc \alpha = 4$, $\sec \alpha = -4\sqrt{15}/15$, $\cot \alpha = -\sqrt{15}$

### Section 6.4

***For Thought:*** **1.** T **2.** T **3.** F **4.** T **5.** F **6.** T **7.** F **8.** F **9.** T **10.** T

***Exercises:***

**1.** 1 **3.** $\sqrt{3}$ **5.** $-1$ **7.** $\sqrt{3}$ **9.** $\dfrac{\sqrt{2 + \sqrt{3}}}{2}$

**11.** $\dfrac{\sqrt{2 - \sqrt{3}}}{2}$ **13.** $2 - \sqrt{3}$ **15.** $\dfrac{\sqrt{2 - \sqrt{2}}}{2}$

**17.** Positive **19.** Negative **21.** Negative **23.** sin 26° **25.** $\sqrt{2}/2$ **27.** $\sqrt{3}/6$ **29.** tan 6° **31.** $-\sin(7\pi/9)$ **33.** $\cos(2\pi/9)$ **35.** c **37.** g **39.** a **41.** h **43.** f

**45.** $1/\sqrt{5}, 2/\sqrt{5}, 1/2, \sqrt{5}, \sqrt{5}/2, 2$
**47.** $\sqrt{26}/26, 5\sqrt{26}/26, 1/5, \sqrt{26}, \sqrt{26}/5, 5$
**49.** $-\sqrt{15}/8, -7/8, \sqrt{15}/7, -8/\sqrt{15}, -8/7, 7/\sqrt{15}$
**51.** $-24/25, -7/25, 24/7, -25/24, -25/7, 7/24$
**67.** Not identity **69.** Not identity **71.** Not identity
**73.** Identity **75.** $-24/25$ **77.** $161/289$ **79.** $\dfrac{5\sqrt{34}-25}{3}$
**81.** $A = \dfrac{d^2}{2}\sin(2\alpha)$ **85.** $2\sec^2(x)$
**87. a.** $\cos x \cos y - \sin x \sin y$ **b.** $\cos x \cos y + \sin x \sin y$
**89. a.** $1/2$ **b.** $-1$ **c.** Undefined **d.** 2 **e.** $-1$ **f.** 1

## Section 6.5

***For Thought:*** **1.** T **2.** F **3.** T **4.** F **5.** T **6.** F **7.** T **8.** T **9.** T **10.** T

***Exercises:***

**1.** $0.5(\cos 4° - \cos 22°)$ **3.** $0.5(\sin 36° - \sin 4°)$
**5.** $0.5(\sin 15° - \sin 5°)$ **7.** $0.5(\cos(\pi/30) + \cos(11\pi/30))$
**9.** $0.5(\cos 2y^2 + \cos 12y^2)$ **11.** $0.5(\sin(3s) + \sin(s-2))$
**13.** $\dfrac{\sqrt{2}-1}{4}$ **15.** $\dfrac{\sqrt{2}+\sqrt{3}}{4}$ **17.** $2\cos 10° \sin 2°$
**19.** $2\sin 83.5° \sin 3.5°$ **21.** $-2\cos(4.2)\sin(0.6)$
**23.** $-2\sin(4\pi/15)\sin(\pi/15)$ **25.** $-2\sin(4y+3)\sin(y-6)$
**27.** $-2\cos(6.5\alpha)\sin(1.5\alpha)$ **29.** $\sqrt{6}/2$ **31.** $\dfrac{\sqrt{2-\sqrt{2}}}{2}$
**33.** $\sqrt{2}\sin(x-\pi/4)$ **35.** $\sin(x+2\pi/3)$ **37.** $\sin(x-\pi/6)$
**39.** $y = \sqrt{2}\sin(x+3\pi/4), \sqrt{2}, 2\pi, -3\pi/4$

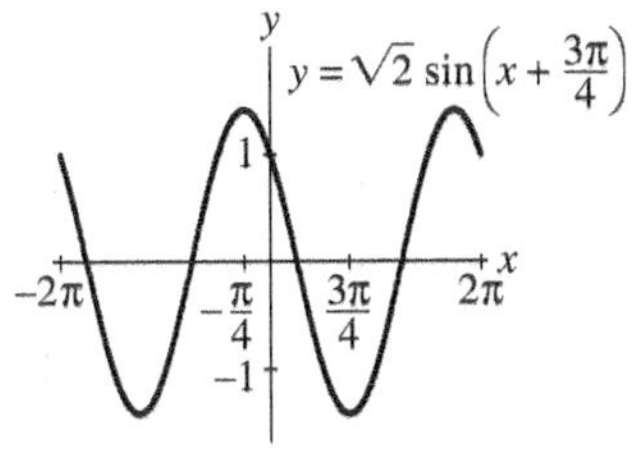


**41.** $y = 2\sin(x-\pi/4), 2, 2\pi, \pi/4$

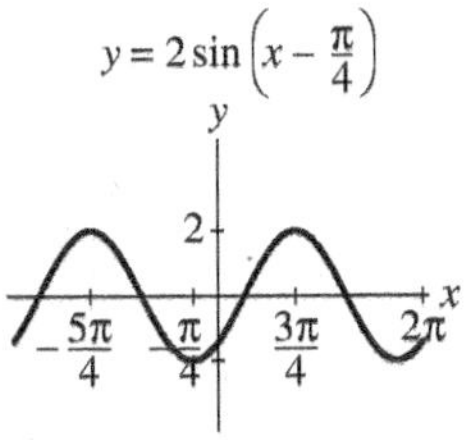


**43.** $y = 2\sin(x+7\pi/6), 2, 2\pi, -7\pi/6$

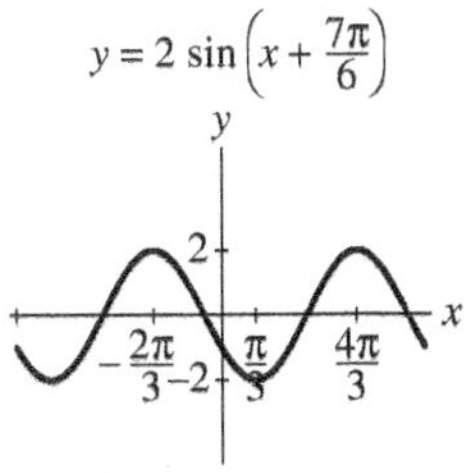


**45.** $5, -0.9$ **47.** $\sqrt{37}, -3.0$ **49.** $\sqrt{34}, -4.2$
**61.** $x = 2\sin(t+\pi/6)$, 2 m
**67. a.** $\dfrac{\tan x + \tan y}{1 - \tan x \tan y}$ **b.** $\dfrac{\tan x - \tan y}{1 + \tan x \tan y}$
**69.** $-\sqrt{2}$ **71.** $-24/25$

## Section 6.6

***For Thought:*** **1.** F **2.** F **3.** T **4.** T **5.** T **6.** F **7.** F **8.** T **9.** F **10.** F

***Exercises:***

**1.** $\{x \mid x = \pi + 2k\pi\}$ **3.** $\{x \mid x = k\pi\}$ **5.** $\left\{x \mid x = \dfrac{3\pi}{2} + 2k\pi\right\}$
**7.** $\left\{x \mid x = \dfrac{\pi}{3} + 2k\pi \text{ or } x = \dfrac{5\pi}{3} + 2k\pi\right\}$
**9.** $\left\{x \mid x = \dfrac{\pi}{4} + 2k\pi \text{ or } x = \dfrac{3\pi}{4} + 2k\pi\right\}$
**11.** $\left\{x \mid x = \dfrac{\pi}{4} + k\pi\right\}$ **13.** $\left\{x \mid x = \dfrac{5\pi}{6} + 2k\pi \text{ or } x = \dfrac{7\pi}{6} + 2k\pi\right\}$
**15.** $\left\{x \mid x = \dfrac{5\pi}{4} + 2k\pi \text{ or } x = \dfrac{7\pi}{4} + 2k\pi\right\}$
**17.** $\left\{x \mid x = \dfrac{3\pi}{4} + k\pi\right\}$ **19.** $\{x \mid x = 90° + k180°\}$
**21.** $\{x \mid x = 90° + k360°\}$ **23.** $\{x \mid x = k180°\}$
**25.** $\{x \mid x = 29.2° + k360° \text{ or } x = 330.8° + k360°\}$
**27.** $\{x \mid x = 345.9° + k360° \text{ or } x = 194.1° + k360°\}$
**29.** $\{x \mid x = 79.5° + k180°\}$
**31.** $\left\{x \mid x = \dfrac{2\pi}{3} + 4k\pi \text{ or } x = \dfrac{10\pi}{3} + 4k\pi\right\}$
**33.** $\left\{x \mid x = \dfrac{2k\pi}{3}\right\}$ **35.** $\left\{x \mid x = \dfrac{\pi}{3} + 4k\pi \text{ or } x = \dfrac{5\pi}{3} + 4k\pi\right\}$
**37.** $\left\{x \mid x = \dfrac{5\pi}{8} + k\pi \text{ or } x = \dfrac{7\pi}{8} + k\pi\right\}$ **39.** $\left\{x \mid x = \dfrac{\pi}{6} + \dfrac{k\pi}{2}\right\}$
**41.** $\left\{x \mid x = \dfrac{k\pi}{4}\right\}$ **43.** $\left\{x \mid x = \dfrac{1}{6} + 2k \text{ or } x = \dfrac{5}{6} + 2k\right\}$
**45.** $\left\{x \mid x = \dfrac{1}{4} + \dfrac{k}{2}\right\}$ **47.** $\{240°, 300°\}$
**49.** $\{22.5°, 157.5°, 202.5°, 337.5°\}$
**51.** $\{45°, 75°, 165°, 195°, 285°, 315°\}$ **53.** $\{60°\}$
**55.** $\{\alpha \mid \alpha = 6.6° + k120° \text{ or } \alpha = 53.4° + k120°\}$
**57.** $\{\alpha \mid \alpha = 72.3° + k120° \text{ or } \alpha = 107.7° + k120°\}$
**59.** $\{\alpha \mid \alpha = 38.6° + k180° \text{ or } \alpha = 141.4° + k180°\}$
**61.** $\{\alpha \mid \alpha = 668.5° + k720° \text{ or } \alpha = 411.5° + k720°\}$
**63.** $\{0, 0.3, 2.8, \pi\}$ **65.** $\{\pi, 2\pi/3, 4\pi/3\}$
**67.** $\{0.7, 2.5, 3.4, 6.0\}$ **69.** $\{11\pi/6\}$ **71.** $\{\pi/3, 4.9\}$
**73.** $\{0, \pi\}$ **75.** $\{\pi/2, 3\pi/2\}$ **77.** $\{7\pi/12, 23\pi/12\}$
**79.** $\{7\pi/6, 11\pi/6\}$ **81.** $\{\pi/6, \pi/2, 5\pi/6\}$ **83.** $\{0°\}$
**85.** $\{26.6°, 206.6°\}$ **87.** $\{30°, 90°, 150°, 210°, 270°, 330°\}$
**89.** $\{67.5°, 157.5°, 247.5°, 337.5°\}$ **91.** $\{221.8°, 318.2°\}$
**93.** $\{120°, 300°\}$ **95.** $\{30°, 45°, 135°, 150°, 210°, 225°, 315°, 330°\}$
**97.** $\{0°, 60°, 120°, 180°, 240°, 300°\}$ **99.** 10.5 **101.** 56.4°
**103.** 19.6° **105.** $\{0.4, 1.9, 2.2, 4.0, 4.4, 5.8\}$ **107.** $\{1.0\}$
**109.** $\dfrac{5\pi}{12} + \dfrac{k\pi}{2}$ for $k$ a nonnegative integer
**111.** $1 + 6k$ and $2 + 6k$ for $k$ a nonnegative integer
**113.** 44.4° or 45.6° **115.** 12.5° or 77.5°, 6.3 sec
**117.** $y = \sqrt{2}\sin(x + 7\pi/4), \sqrt{2}, 2\pi, 7\pi/4$ **119.** 3/4 **121.** $\sqrt{6}/4$

## Chapter 6 Review Exercises

**1.** $\cos^2\alpha$ **3.** $-\cot^2 x$ **5.** $\sec^2\alpha$ **7.** $\tan 4s$ **9.** $-\sin 3\theta$
**11.** $\tan z$ **13.** e **15.** c **17.** a **19.** g
**21.** $\sin\alpha = 12/13, \tan\alpha = -12/5, \csc\alpha = 13/12, \sec\alpha = -13/5, \cot\alpha = -5/12$
**23.** $\sin\alpha = -4/5, \cos\alpha = -3/5, \tan\alpha = 4/3, \csc\alpha = -5/4, \sec\alpha = -5/3, \cot\alpha = 3/4$
**25.** $\sin\alpha = -24/25, \cos\alpha = 7/25, \tan\alpha = -24/7, \csc\alpha = -25/24, \sec\alpha = 25/7, \cot\alpha = -7/24$
**27.** Identity **29.** Not identity **31.** Odd **33.** Neither
**35.** Even **37.** f **39.** e **41.** b **43.** h **45.** c
**61.** $\sqrt{3} - 2$ **63.** $-\dfrac{\sqrt{2+\sqrt{3}}}{2}$

**65.** $y = 4\sqrt{2}\sin(x + \pi/4)$, $4\sqrt{2}$, $-\pi/4$

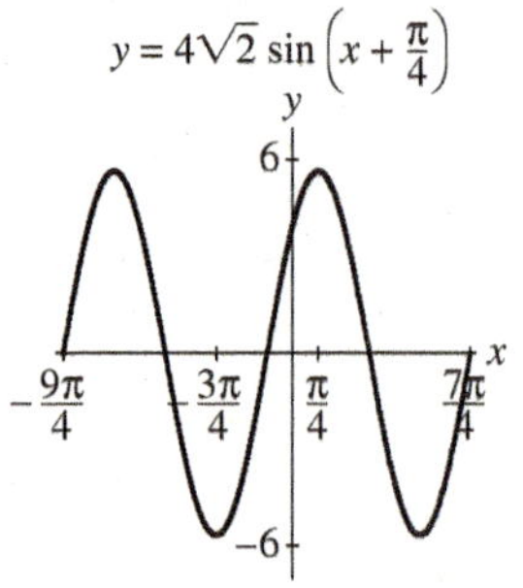


**67.** $y = \sqrt{5}\sin(x + 2.68)$, $\sqrt{5}$, $-2.68$

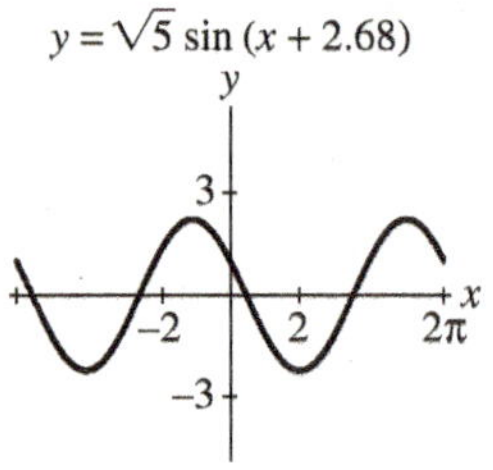


**69.** $\left\{x \middle| x = \frac{\pi}{3} + k\pi \text{ or } x = \frac{2\pi}{3} + k\pi\right\}$

**71.** $\left\{x \middle| x = \frac{\pi}{3} + 2k\pi, \frac{2\pi}{3} + 2k\pi, \frac{\pi}{6} + 2k\pi, \frac{5\pi}{6} + 2k\pi\right\}$

**73.** $\left\{x \middle| x = \frac{\pi}{6} + 2k\pi, \frac{5\pi}{6} + 2k\pi, \frac{\pi}{2} + 2k\pi\right\}$

**75.** $\left\{x \middle| x = \frac{2\pi}{3} + 4k\pi \text{ or } x = \frac{4\pi}{3} + 4k\pi\right\}$

**77.** $\left\{x \middle| x = \pi + 2k\pi, \frac{\pi}{3} + 4k\pi, \frac{5\pi}{3} + 4k\pi\right\}$

**79.** $\left\{x \middle| x = \frac{\pi}{2} + k\pi\right\}$

**81.** $\left\{x \middle| x = \pi + 2k\pi \text{ or } x = \frac{3\pi}{2} + 2k\pi\right\}$

**83.** $\{45°, 225°\}$ **85.** $\{90°, 180°\}$ **87.** $\varnothing$

**89.** $\{22.5°, 67.5°, 112.5°, 157.5°, 202.5°, 247.5°, 292.5°, 337.5°\}$

**91.** $\{0°, 180°\}$

**93.** $\{15°, 75°, 135°, 195°, 255°, 315°\}$ **95.** $2\cos 17° \cos 2°$

**97.** $2\cos\frac{\pi}{16}\sin\frac{3\pi}{16}$ **99.** $\sin 24° - \sin 2°$ **101.** $\cos\frac{x}{12} + \cos\frac{7x}{12}$

**103.** $\frac{\sqrt{6} + \sqrt{2}}{4}$ **105.** $\frac{\sqrt{6} - \sqrt{2}}{4}$ **107.** 1.28, 2.85

## Chapter 6 Test

**1.** $2\cos x$ **2.** $\sin 7t$ **3.** $2\csc^2 y$ **4.** $\tan(3\pi/10)$

**9.** $\left\{\theta \middle| \theta = \frac{3\pi}{2} + 2k\pi\right\}$

**10.** $\left\{s \middle| s = \frac{\pi}{9} + \frac{2k\pi}{3} \text{ or } s = \frac{5\pi}{9} + \frac{2k\pi}{3}\right\}$

**11.** $\left\{t \middle| t = \frac{\pi}{3} + \frac{k\pi}{2}\right\}$

**12.** $\left\{\theta \middle| \theta = \frac{\pi}{2} + k\pi \text{ or } \theta = \frac{\pi}{6} + 2k\pi \text{ or } \frac{5\pi}{6} + 2k\pi\right\}$

**13.** $\{19.5°, 90°, 160.5°\}$

**14.** $\{27°, 63°, 99°, 171°, 207°, 243°, 279°, 351°\}$

**15.** $y = 2\sin(x + 5\pi/3)$, $2\pi$, 2, $-5\pi/3$

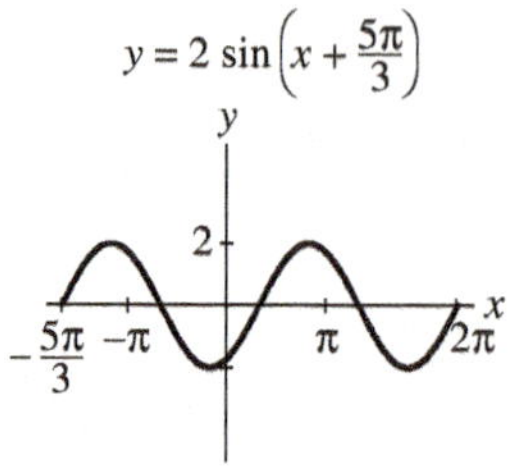


**16.** $\sin\alpha = 1/2$, $\cos\alpha = -\sqrt{3}/2$, $\tan\alpha = -1/\sqrt{3}$, $\sec\alpha = -2/\sqrt{3}$, $\cot\alpha = -\sqrt{3}$ **17.** Even

**18.** $-\frac{\sqrt{2 - \sqrt{3}}}{2}$ **20.** 0.4 sec, 1.4 sec, 2.5 sec, 3.5 sec

## Tying It All Together Chapters 1–6

**1.** Odd **2.** Even **3.** Odd **4.** Even **5.** Even
**6.** Odd **7.** Even **8.** Even **9.** Not identity
**10.** Not identity **11.** Identity **12.** Identity
**13.** Not identity **14.** Not identity
**15.** $\beta = 60°$, $b = 4\sqrt{3}$, $c = 8$
**16.** $\alpha = 60°$, $\beta = 30°$, $c = 2$
**17.** $\alpha = 17.5°$, $\beta = 72.5°$, $a = 1.6$, $c = 5.2$
**18.** $\alpha = 36.9°$, $\beta = 53.1°$, $b = 2.7$, $c = 3.3$
**19.** $\frac{1}{2} + \frac{\sqrt{3}}{2}i$ **20.** $-1 + \sqrt{3}i$ **21.** $i$ **22.** 1
**23.** $2 + 11i$ **24.** $-16$ **25.** minutes **26.** seconds
**27.** unit **28.** $\pi$ **29.** $\alpha r$ **30.** $r\omega$ **31.** $y, x$
**32.** fundamental **33.** amplitude **34.** phase shift

# Chapter 7

## Section 7.1

***For Thought:*** **1.** T **2.** F **3.** F **4.** T **5.** T **6.** F **7.** T **8.** F **9.** T **10.** T

***Exercises:***

**1.** oblique **3.** $\gamma = 44°$, $b = 14.4$, $c = 10.5$
**5.** $\beta = 134.2°$, $a = 5.2$, $c = 13.6$ **7.** $\beta = 26°$, $a = 14.6$, $b = 35.8$
**9.** $\alpha = 45.7°$, $b = 587.9$, $c = 160.8$ **11.** None
**13.** One: $\alpha = 30°$, $\beta = 90°$, $a = 10$
**15.** One: $\alpha = 26.3°$, $\gamma = 15.6°$, $a = 10.3$
**17.** Two: $\alpha_1 = 134.9°$, $\gamma_1 = 12.4°$, $c_1 = 11.4$; or $\alpha_2 = 45.1°$, $\gamma_2 = 102.2°$, $c_2 = 51.7$
**19.** One: $\alpha = 25.4°$, $\beta = 55.0°$, $a = 5.4$ **21.** 9.8 **23.** 83.4
**25.** 66.3 **27.** 37.7 **29.** $22\sqrt{3} + 6$ **31.** 18.4 mi
**33.** 97,535 $\text{ft}^2$ **35.** 159.4 ft **37.** 28.9 ft, 15.7 ft
**39.** 277.7 $\text{in.}^2$
**41. a.** 19.9 mi **b.** 1.0 sec before 4 P.M. **c.** 281 mi
**43.** 62.2° **45.** 7.5 sec **49.** 16.0 cm, 14.4 cm
**51.** 5, $\pi/2$, $\pi/4$, $[-8, 2]$ **53.** $3\cos^2(x)$

## Section 7.2

***For Thought:*** **1.** T **2.** F **3.** F **4.** T **5.** F **6.** T **7.** T **8.** T **9.** T **10.** F

***Exercises:***

**1.** law of cosines **3.** $\alpha = 30.4°$, $\beta = 28.3°$, $c = 5.2$
**5.** $\alpha = 26.4°$, $\beta = 131.3°$, $\gamma = 22.3°$
**7.** $\alpha = 163.9°$, $\gamma = 5.6°$, $b = 4.5$
**9.** $\alpha = 130.3°$, $\beta = 30.2°$, $\gamma = 19.5°$
**11.** $\beta = 48.3°$, $\gamma = 101.7°$, $a = 6.2$
**13.** $\alpha = 53.9°$, $\beta = 65.5°$, $\gamma = 60.6°$
**15.** $\alpha = 120°$, $b = 3.5$, $c = 4.8$ **17.** 0 **19.** 1
**21.** 0 **23.** 1 **25.** 0 **27.** 40.9 **29.** 13.9
**31.** 40,471.9 **33.** 100 **35.** 20.8 **37.** 20.4 **39.** 9.90 ft

**41.** 783.45 ft **43.** 20.6 mi **45.** 19.2°
**47.** $\angle A = 78.5°$, $\angle B = 57.1°$, $\angle C = 44.4°$
**49.** 3.8 mi, 111.6°, 40.4° **51.** 11.76 m **53.** $\theta_1 = 13.3°$, $\theta_2 = 90.6°$
**55. a.** 0.52° to 0.54° **b.** 0.49° to 0.55° **c.** yes
**57.** 603 ft **65.** $\gamma = 53.3°$, $a = 33.9$, $b = 11.4$ **67.** $\pi/2, 3\pi/2$
**69. a.** D: $[-1, 1]$, R: $[-\pi/2, \pi/2]$ **b.** D: $[-1, 1]$, R: $[0, \pi]$
**c.** D: $(-\infty, \infty)$, R: $(-\pi/2, \pi/2)$

## Section 7.3

***For Thought:*** **1.** T **2.** F **3.** T **4.** T **5.** F **6.** F
**7.** T **8.** T **9.** T **10.** T
***Exercises:***
**1.** vector **3.** magnitude **5.** parallelogram law **7.** component
**9.**

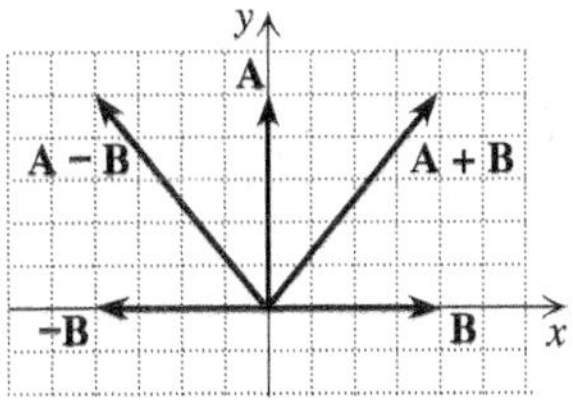


**11.**

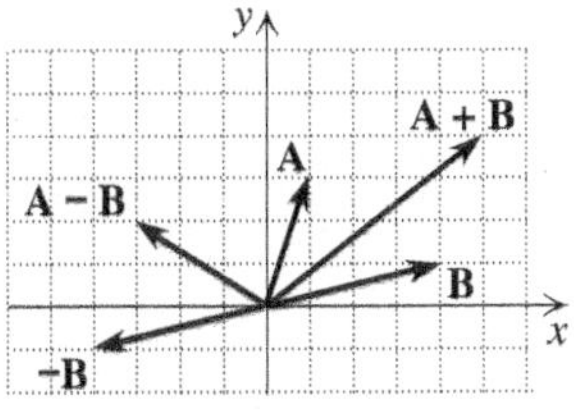


**13.**

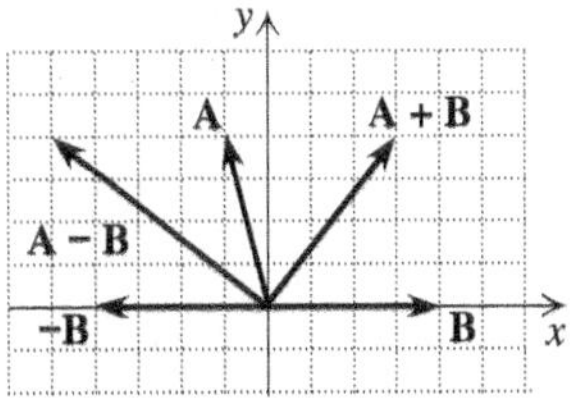


**15.** D **17.** E **19.** B **21.** $|\mathbf{v}_x| = 1.9$, $|\mathbf{v}_y| = 4.1$
**23.** $|\mathbf{v}_x| = 7256.4$, $|\mathbf{v}_y| = 3368.3$ **25.** $|\mathbf{v}_x| = 87.7$, $|\mathbf{v}_y| = 217.0$
**27.** 2, 30° **29.** 2, 135° **31.** 16, 300° **33.** 5, 0°
**35.** $\sqrt{13}$, 146.3° **37.** $\sqrt{10}$, 341.6° **39.** $\langle 4\sqrt{2}, 4\sqrt{2} \rangle$
**41.** $\langle -237.6, 166.3 \rangle$ **43.** $\langle 17.5, -4.0 \rangle$ **45.** $\langle 15, -10 \rangle$
**47.** $\langle 18, -22 \rangle$ **49.** $\langle 11, -13 \rangle$ **51.** $\langle 0, -1 \rangle$ **53.** $-13$
**55.** 32.5° **57.** 27.3° **59.** 90° **61.** Perpendicular
**63.** Parallel **65.** Neither **67.** $2\mathbf{i} + \mathbf{j}$ **69.** $-3\mathbf{i} + \sqrt{2}\mathbf{j}$
**71.** $-9\mathbf{j}$ **73.** $-7\mathbf{i} - \mathbf{j}$ **75.** $\sqrt{17}$, 76.0° **77.** $3\sqrt{10}$, 198.4°
**79.** $\sqrt{29}$, 158.2° **81.** $\sqrt{65}/2$, 172.9° **83.** $\sqrt{89}$, 32.0°
**85.** $2\sqrt{10}$, 108.4° **87.** $\sqrt{5}$, 243.4° **89.** $\sqrt{73}$ lb, 69.4°, 20.6°
**91.** 8.25 newtons, 22.9°, 107.1° **93.** 5.1 lb, 75.2° **95.** 127.0 lb
**97.** 450.3 mph, 260 mph **99.** 1368.1 lb **101.** 1026.1 lb
**103.** 11.5° **105.** 97.1°, 241.9 mph **107.** 66.5°, 38.5 mph
**109.** 108.2°, 451.0 mph **111.** 198.4°, 6000 ft
**113. a.** $d = 0.2|\sec\beta|$ **b.** $r = 3\cos(\alpha)|\sec(\beta)|$
**c.** $t = \frac{1}{15}\sec(\alpha)$, $\alpha = 0°$
**117.** $\alpha = 46.7°$, $\gamma = 11.2°$, $c = 5.2$
**119.** $a = 6.3$, $\beta = 124.7°$, $\gamma = 22.1°$
**121.** $\pi/9, 2\pi/9, 4\pi/9, 5\pi/9, 7\pi/9, 8\pi/9$

## Section 7.4

***For Thought:*** **1.** T **2.** F **3.** T **4.** F **5.** F **6.** T
**7.** T **8.** T **9.** T **10.** F
***Exercises:***
**1.** real, imaginary **3.** modulus, argument **5.** 8 **7.** 9
**9.** $2\sqrt{10}$ **11.** 4 **13.** 1 **15.** $3\sqrt{2}$
**17.** $8(\cos 0° + i\sin 0°)$ **19.** $\sqrt{3}(\cos 90° + i\sin 90°)$
**21.** $3\sqrt{2}(\cos 135° + i\sin 135°)$ **23.** $3(\cos 135° + i\sin 135°)$
**25.** $2(\cos 150° + i\sin 150°)$ **27.** $5(\cos 53.1° + i\sin 53.1°)$
**29.** $\sqrt{34}(\cos 121.0° + i\sin 121.0°)$
**31.** $3\sqrt{5}(\cos 296.6° + i\sin 296.6°)$ **33.** $1 + i$
**35.** $-\frac{3}{4} + \frac{\sqrt{3}}{4}i$ **37.** $-0.42 - 0.26i$ **39.** $3i$ **41.** $-i\sqrt{3}$
**43.** $\frac{\sqrt{6}}{2} + \frac{3\sqrt{2}}{2}i$ **45.** $6i$ **47.** $\frac{3\sqrt{2}}{2} + \frac{\sqrt{6}}{2}i$ **49.** $9i$
**51.** $\sqrt{3} + i$ **53.** $0.34 - 0.37i$ **55.** $-40i, -0.8$
**57.** $8i, \frac{\sqrt{3}}{4} - \frac{1}{4}i$ **59.** $4\sqrt{2}, i\sqrt{2}$ **61.** $-7 - 26i, -\frac{23}{29} - \frac{14}{29}i$
**63.** $-18 + 14i, \frac{6}{13} + \frac{22}{13}i$ **65.** $-3 + 3i, 1.5 + 1.5i$
**67.** $3\left(\cos\left(-\frac{\pi}{4}\right) + i\sin\left(-\frac{\pi}{4}\right)\right)$ **69.** $2\sqrt{3}(\cos 20° + i\sin 20°)$
**71.** 9 **73.** 4 **75.** $32i$ **77.** $-54 + 54i$
**83.** $6\cos(9°) + 3\cos(5°) + i(6\sin(9°) + 3\sin(5°))$, $6 - 4i$, standard form **85.** 24 **87.** $\alpha = 27.7°$, $\beta = 40.5°$, $\gamma = 111.8°$ **89.** 255.4 ft

## Section 7.5

***For Thought:*** **1.** F **2.** F **3.** T **4.** F **5.** F **6.** F
**7.** T **8.** T **9.** T **10.** F
***Exercises:***
**1.** $27i$ **3.** $-2 + 2i\sqrt{3}$ **5.** $-\frac{1}{2} + \frac{\sqrt{3}}{2}i$ **7.** $-18 + 18i\sqrt{3}$
**9.** $701.5 + 1291.9i$ **11.** $-16 + 16i$ **13.** $-8 - 8i\sqrt{3}$
**15.** $-3888 + 3888i\sqrt{3}$ **17.** $-119 - 120i$ **19.** $-7 - 24i$
**21.** $-44.928 - 31.104i$
**23.** $2(\cos 45° + i\sin 45°)$, $2(\cos 225° + i\sin 225°)$
**25.** $\cos\alpha + i\sin\alpha$ for $\alpha = 30°, 120°, 210°, 300°$
**27.** $2(\cos\alpha + i\sin\alpha)$ for $\alpha = \frac{\pi}{6}, \frac{\pi}{2}, \frac{5\pi}{6}, \frac{7\pi}{6}, \frac{3\pi}{2}, \frac{11\pi}{6}$
**29.** $1, -\frac{1}{2} \pm \frac{\sqrt{3}}{2}i$ **31.** $\pm 2, \pm 2i$
**33.** $\frac{\sqrt{2}}{2} \pm \frac{\sqrt{2}}{2}i, -\frac{\sqrt{2}}{2} \pm \frac{\sqrt{2}}{2}i$ **35.** $\pm\frac{\sqrt{3}}{2} + \frac{1}{2}i, -i$
**37.** $\pm(1 + i\sqrt{3})$ **39.** $1.272 + 0.786i, -1.272 - 0.786i$
**41.** $\frac{1}{2} \pm \frac{\sqrt{3}}{2}i, -1$ **43.** $\pm 3, \pm 3i$ **45.** $-1 + i, 1 - i$
**47.** $0, \pm 2, 1 \pm i\sqrt{3}, -1 \pm i\sqrt{3}$ **49.** $-2, \pm i\sqrt{5}, 1 \pm i\sqrt{3}$
**51.** $2^{1/5}(\cos\alpha + i\sin\alpha)$ for $\alpha = 0°, 72°, 144°, 216°, 288°$
**53.** $10^{1/8}(\cos\alpha + i\sin\alpha)$ for $\alpha = 40.4°, 130.4°, 220.4°, 310.4°$
**55.** $-\frac{1}{4} + \frac{1}{4}i$ **57.** $1, -i$ **61.** $\sqrt{34}$ **63.** $6i$
**65.** $3\sqrt{10}$, 251.6°

## Section 7.6

***For Thought:*** **1.** T **2.** F **3.** F **4.** F **5.** T **6.** T
**7.** T **8.** F **9.** F **10.** F
***Exercises:***
**1.** polar **3.** $(3, \pi/2)$ **5.** $(3\sqrt{2}, \pi/4)$
**7–21 odd**

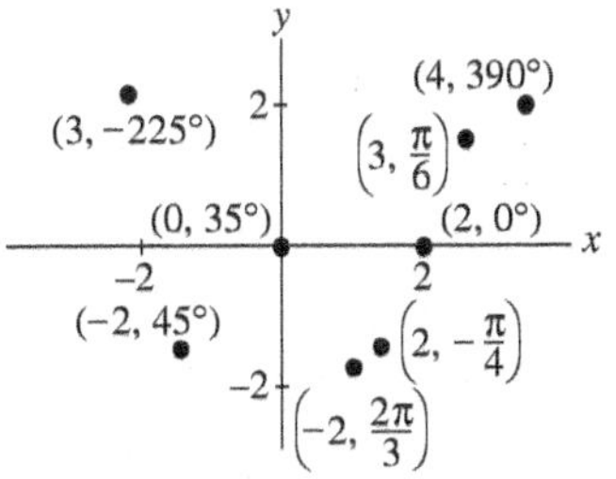


**23.** $(4, 0)$ **25.** $(0, 0)$ **27.** $\left(\frac{\sqrt{3}}{2}, \frac{1}{2}\right)$ **29.** $(0, 3)$
**31.** $(-1, 1)$ **33.** $\left(-\frac{\sqrt{6}}{2}, \frac{3\sqrt{2}}{2}\right)$ **35.** $(2\sqrt{3}, 60°)$

**37.** $(2\sqrt{2}, 135°)$ **39.** $(2, 90°)$ **41.** $(3\sqrt{2}, 225°)$
**43.** $(\sqrt{17}, 75.96°)$ **45.** $(\sqrt{6}, -54.7°)$ **47.** $(3.60, 1.75)$
**49.** $(1.80, 0.87)$ **51.** $(-0.91, -1.78)$ **53.** $(6.4, 51.3°)$
**55.** $(7.3, 254.1°)$
**57.**

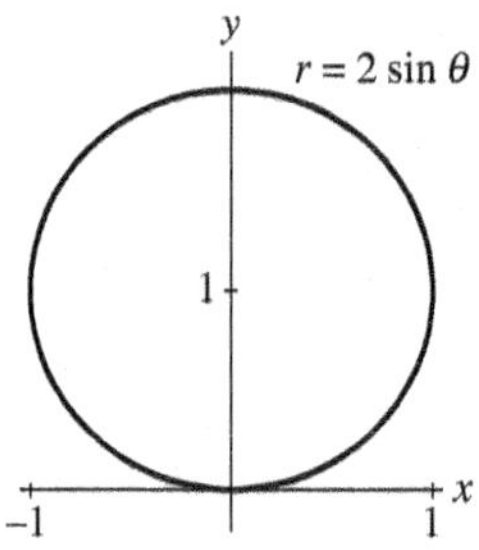


**59.**

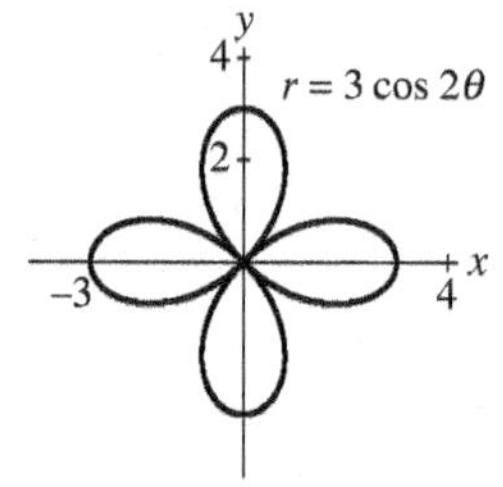


**61.**

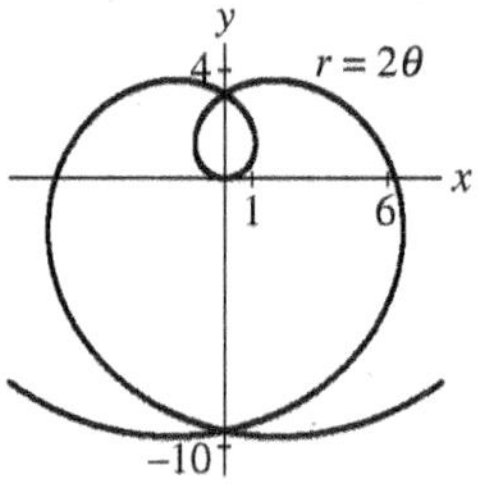


**63.**

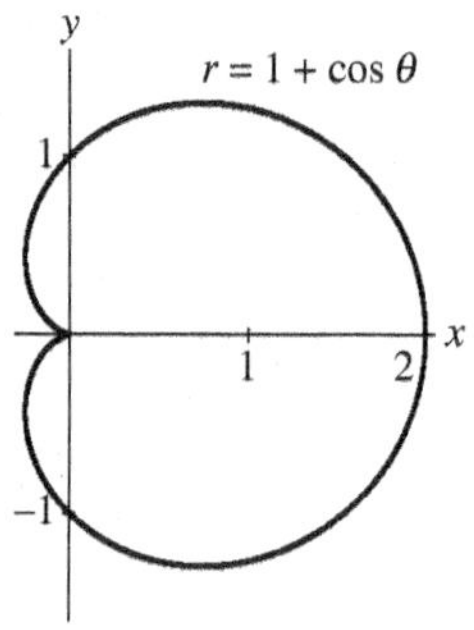


**65.**

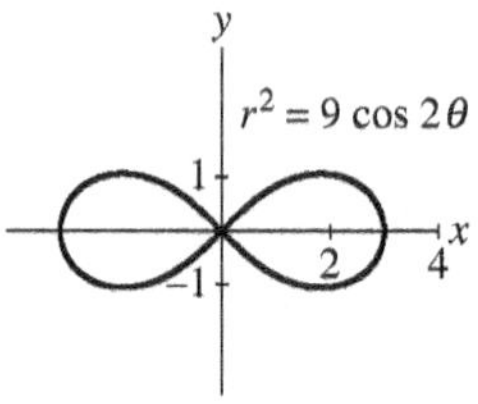


**67.**

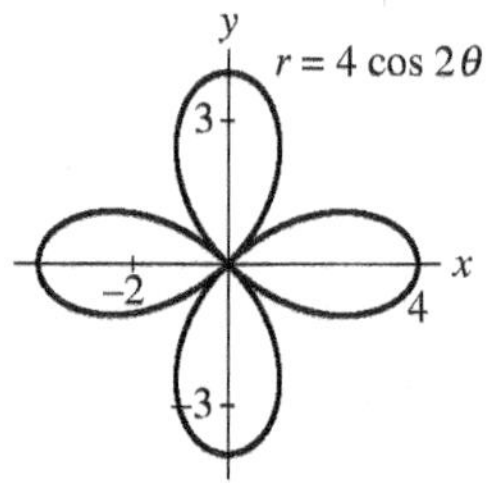


**69.**

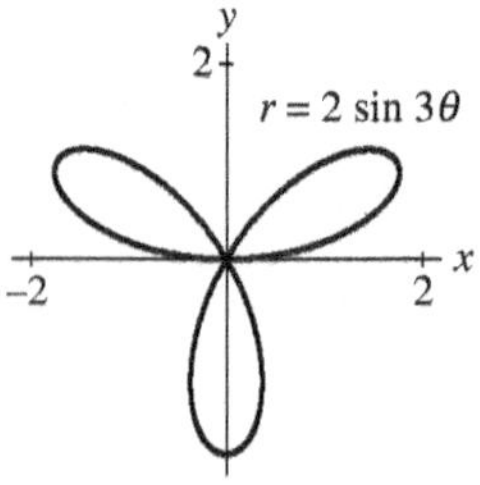


**71.**

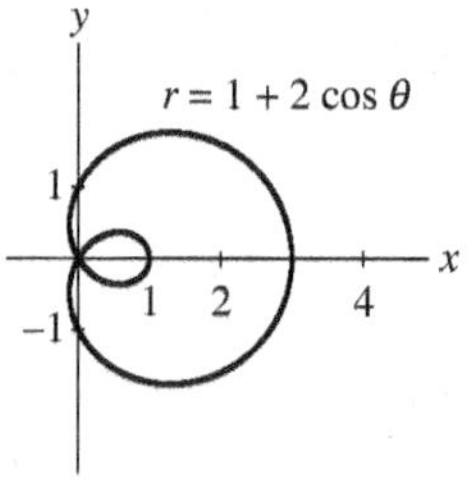


**73.**

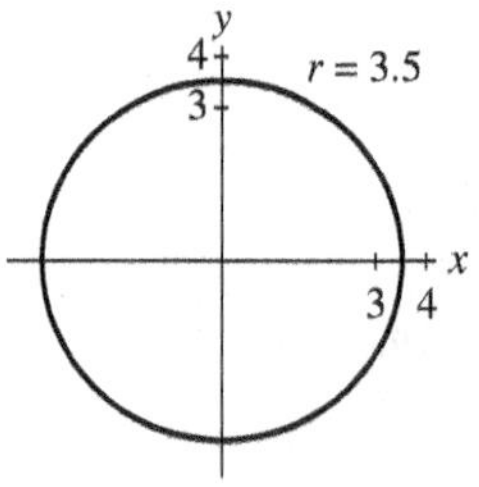


**75.**

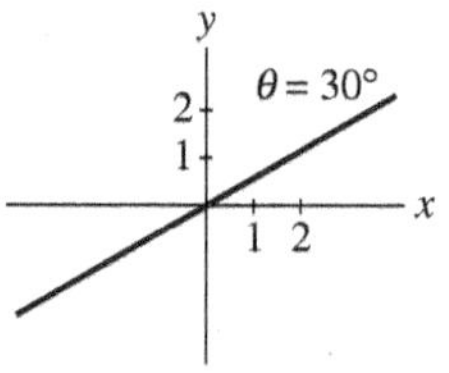


**77.** $x^2 - 4x + y^2 = 0$ **79.** $y = 3$ **81.** $x = 3$
**83.** $x^2 + y^2 = 25$ **85.** $y = x$ **87.** $x^2 - 4y = 4$

**89.** $r \cos \theta = 4$ **91.** $\theta = -\pi/4$ **93.** $r = 4 \tan \theta \sec \theta$
**95.** $r = 2$ **97.** $r = \dfrac{1}{2 \cos \theta - \sin \theta}$ **99.** $r = 2 \sin \theta$
**101.** $(1, 0.17), (1, 0.87), (1, 2.27), (1, 2.97), (1, 4.36), (1, 5.06)$
**103.** $(0, 0), (0.9, 1.4), (1.2, 1.8), (1.9, 2.8), (1.9, 3.5), (1.2, 4.5), (0.8, 4.9)$
**107.** $-64$ **109.** $1.08 + 2k\pi$ or $2.07 + 2k\pi$
**111.** $y = \cos(2(x - 3\pi/8))$

## Section 7.7

***For Thought:*** **1.** F **2.** T **3.** T **4.** F **5.** T **6.** F
**7.** T **8.** T **9.** F **10.** T

***Exercises:***
**1.** parametric
**3.**

| $t$ | $x$ | $y$ |
|---|---|---|
| 0 | 1 | −2 |
| 1 | 5 | −1 |
| 1.5 | 7 | −0.5 |
| 3 | 13 | 1 |

**5.**

| $t$ | $x$ | $y$ |
|---|---|---|
| 1 | 1 | 2 |
| 2.5 | 6.25 | 6.5 |
| $\sqrt{5}$ | 5 | $3\sqrt{5} - 1$ |
| 4 | 16 | 11 |
| 5 | 25 | 14 |

**7.** $[-2, 10], [3, 7]$

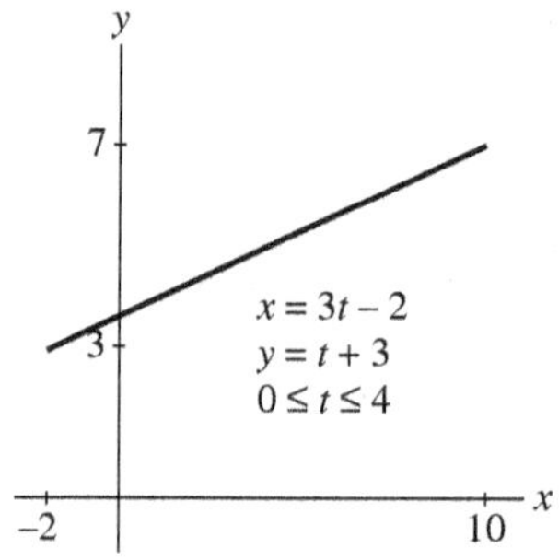


**9.** $(-\infty, \infty), [0, \infty)$

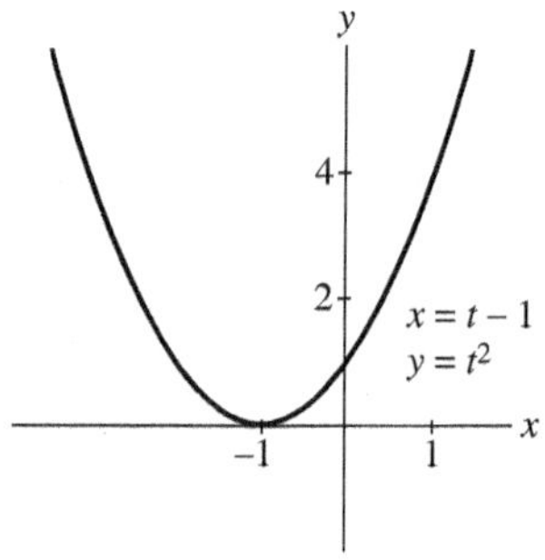


**11.** $(0, 1), (0, 1)$

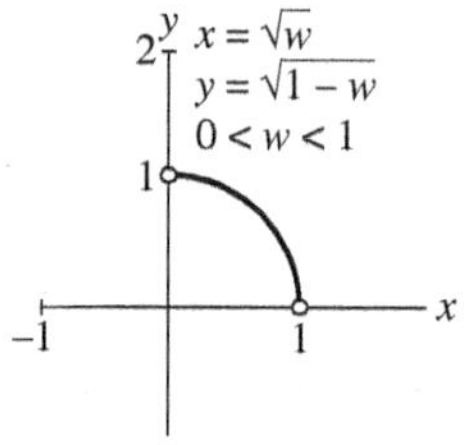


**13.** $[-1, 1], [-1, 1]$

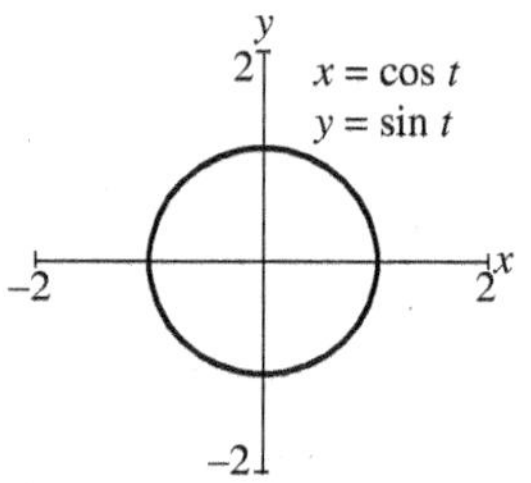


**15.** $x + y = -2$, line, $(-\infty, \infty), (-\infty, \infty)$
**17.** $x^2 + y^2 = 16$, circle, $[-4, 4], [-4, 4]$
**19.** $y = e^{4x}$, exponential, $(-\infty, \infty), (0, \infty)$
**21.** $y = 2x + 3$, line, $(-\infty, \infty), (-\infty, \infty)$
**23.** $x = \frac{3}{2}t + 2, y = 3t + 3, 0 \le t \le 2$
**25.** $x = 2 \cos t, y = 2 \sin t, \pi < t < 3\pi/2$
**27.** $x = 3, y = t, -\infty < t < \infty$ **29.** $x = \sin(2t), y = 2 \sin^2(t)$
**31.**

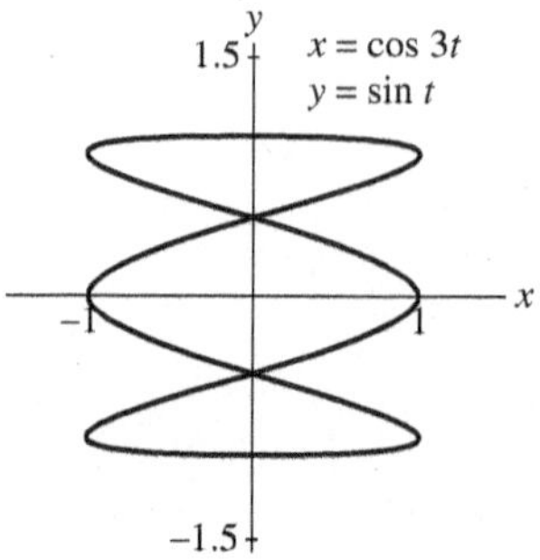


**33.**

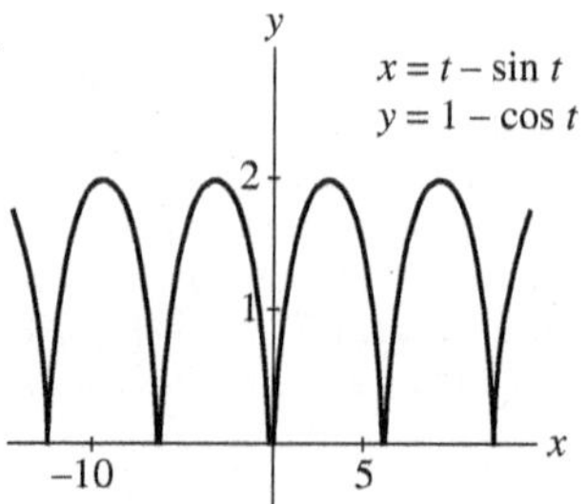

**35.**

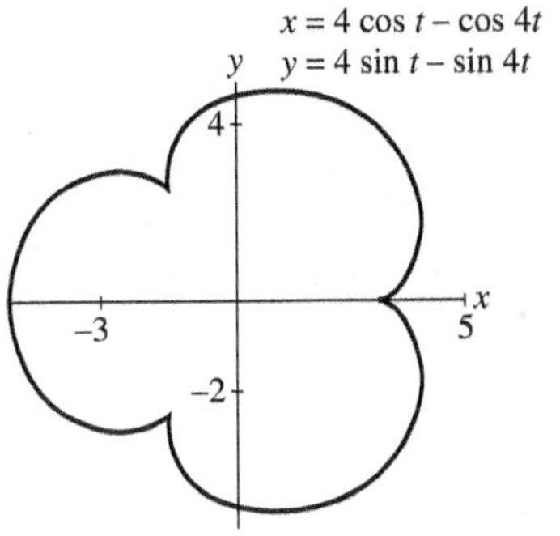


**37.**

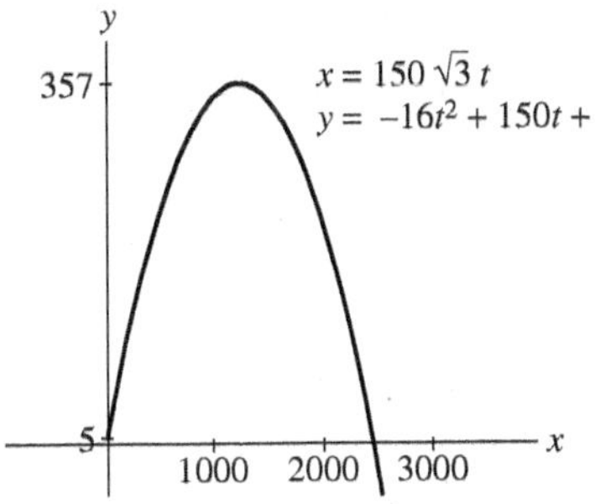


**39.** 9.4 sec **41.** $x^2 - 8x + y^2 = 0$
**43.** $6(\cos(5\pi/3) + i\sin(5\pi/3))$ **45.** $\pi/3, \pi, 5\pi/3$

## Chapter 7 Review Exercises

**1.** $\alpha = 82.7°, \beta = 49.3°, c = 2.5$
**3.** $\gamma = 103°, a = 4.6, b = 18.4$
**5.** No triangle **7.** $\alpha = 107.7°, \beta = 23.7°, \gamma = 48.6°$
**9.** $\alpha_1 = 110.8°, \gamma_1 = 47.2°, a_1 = 6.2; \alpha_2 = 25.2°, \gamma_2 = 132.8°, a_2 = 2.8$
**11.** 92.4 $\text{ft}^2$ **13.** 23.0 $\text{km}^2$ **15.** 5.5, 2.4 **17.** 2.0, 2.5
**19.** $\sqrt{13}$, 56.3° **21.** 6.0, 237.9° **23.** $\langle 1, 1\rangle$ **25.** $\langle -3.0, 8.6\rangle$
**27.** $\langle -6, 8\rangle$ **29.** $\langle 0, -17\rangle$ **31.** 6 **33.** $-4\mathbf{i} + 8\mathbf{j}$
**35.** $3.6\sqrt{3}\mathbf{i} + 3.6\mathbf{j}$ **37.** $\sqrt{34}$ **39.** $2\sqrt{2}$
**41.** $5.94(\cos 135° + i\sin 135°)$ **43.** $7.6(\cos 252.3° + i\sin 252.3°)$
**45.** $-\frac{3}{2} + \frac{\sqrt{3}}{2}i$ **47.** $5.4 + 3.5i$ **49.** $-15i, -\frac{5}{6}$
**51.** $8 - i, \frac{4}{13} + \frac{7}{13}i$ **53.** $-4\sqrt{2} + 4i\sqrt{2}$ **55.** $-128 + 128i$
**57.** $\frac{\sqrt{2}}{2} + \frac{\sqrt{2}}{2}i, -\frac{\sqrt{2}}{2} - \frac{\sqrt{2}}{2}i$
**59.** $2^{1/3}(\cos\alpha + i\sin\alpha)$ for $\alpha = 10°, 130°, 250°$
**61.** $5^{1/6}(\cos\alpha + i\sin\alpha)$ for $\alpha = 8.9°, 128.9°, 248.9°$
**63.** $5(\cos\alpha + i\sin\alpha)$ for $\alpha = 22.5°, 112.5°, 202.5°, 292.5°$
**65.** $(2.5, 5\sqrt{3}/2)$ **67.** $(-0.3, 1.7)$ **69.** $(4, 4\pi/3)$
**71.** $(\sqrt{13}, -0.98)$
**73.**

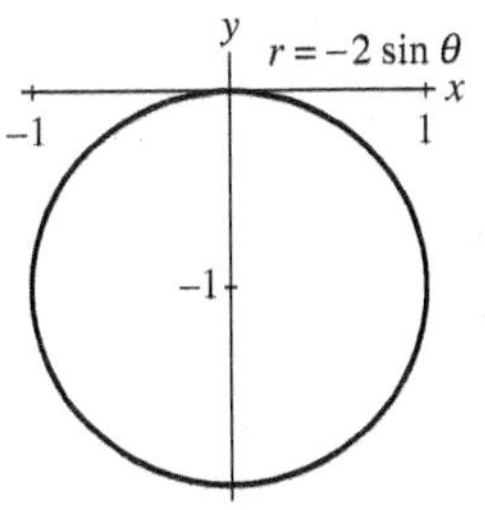


**75.**

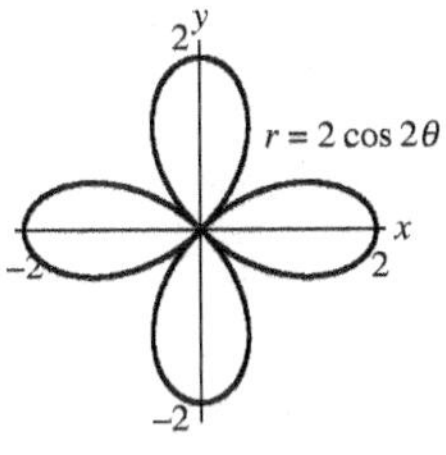


**77.**

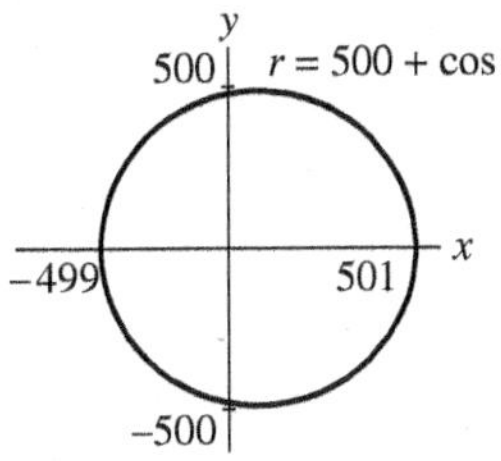


**79.**

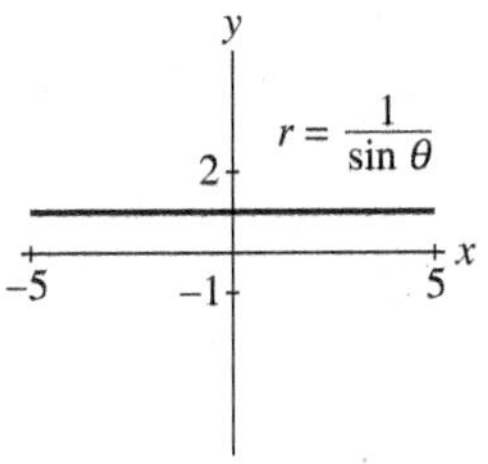


**81.** $x + y = 1$ **83.** $x^2 + y^2 = 25$ **85.** $r = \frac{3}{\sin\theta}$ **87.** $r = 7$

**89.**

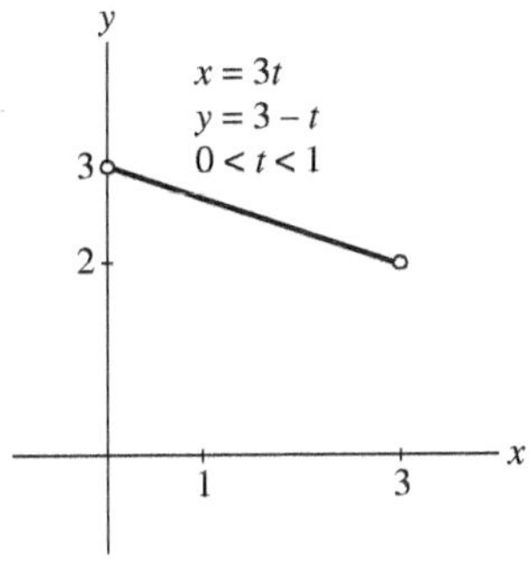


**91.**

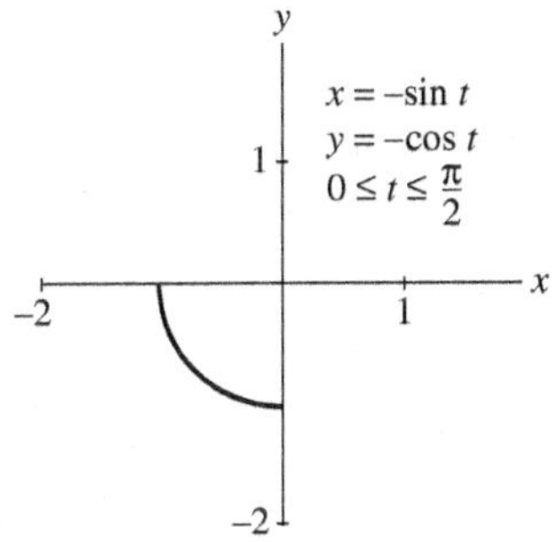


**93.** 18.4 lb, 11.0°, 19.0° **95.** Seth, 42,785 $\text{ft}^2$
**97.** $2639 **99. b.** 10.3 in.

## Chapter 7 Test

**1.** One: $\beta = 90°, \gamma = 60°, c = 2\sqrt{3}$
**2.** Two: $\beta_1 = 111.1°, \gamma_1 = 8.9°, c_1 = 0.7; \beta_2 = 68.9°, \gamma_2 = 51.1°, c_2 = 3.5$
**3.** One: $\gamma = 145.6°, b = 2.5, c = 5.9$
**4.** One: $\alpha = 33.8°, \beta = 129.2°, c = 1.5$
**5.** One: $\alpha = 28.4°, \beta = 93.7°, \gamma = 57.9°$
**6.** $2\sqrt{10}$, 108.4° **7.** $2\sqrt{5}$, 206.6° **8.** $3\sqrt{17}$, 76.0°
**9.** $3\sqrt{2}(\cos 45° + i\sin 45°)$ **10.**
**11.** $2\sqrt{5}(\cos 206.6° + i\sin 206.6°)$ **12.** $3\sqrt{2} + 3i\sqrt{2}$
**13.** $512i$ **14.** $\frac{3\sqrt{2}}{4} + \frac{3\sqrt{2}}{4}i$ **15.** $\left(\frac{5\sqrt{3}}{2}, \frac{5}{2}\right)$
**16.** $\left(-\frac{3\sqrt{2}}{2}, \frac{3\sqrt{2}}{2}\right)$ **17.** $(-26.4, -19.9)$

**18.**

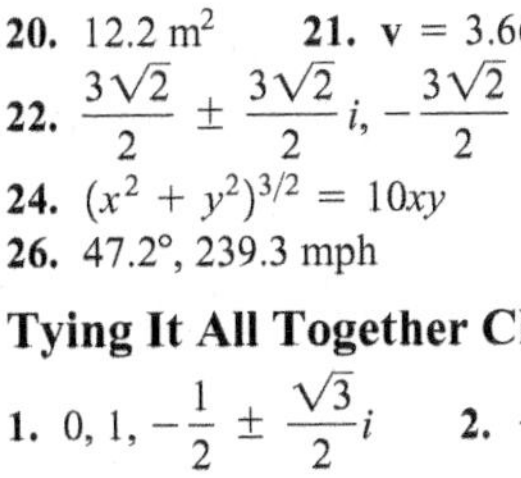


**19.**

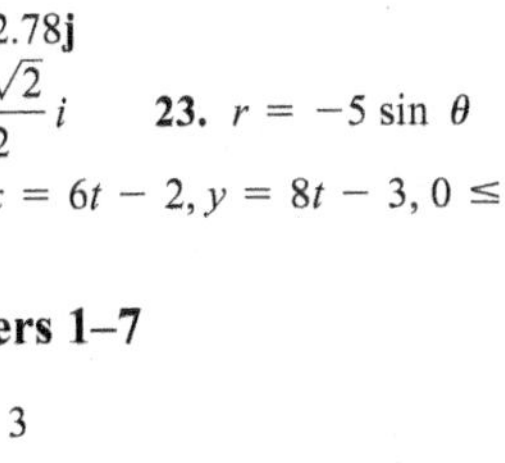


**20.** 12.2 $\text{m}^2$ **21.** $\mathbf{v} = 3.66\mathbf{i} + 2.78\mathbf{j}$
**22.** $\frac{3\sqrt{2}}{2} \pm \frac{3\sqrt{2}}{2}i, -\frac{3\sqrt{2}}{2} \pm \frac{3\sqrt{2}}{2}i$ **23.** $r = -5\sin\theta$
**24.** $(x^2 + y^2)^{3/2} = 10xy$ **25.** $x = 6t - 2, y = 8t - 3, 0 \le t \le 1$
**26.** 47.2°, 239.3 mph

## Tying It All Together Chapters 1–7

**1.** $0, 1, -\frac{1}{2} \pm \frac{\sqrt{3}}{2}i$ **2.** $-2, 1, 3$
**3.** $-2, 1, \cos\theta + i\sin\theta$ for $\theta = 72°, 144°, 216°, 288°$
**4.** $2^{-1/4} \pm i2^{-1/4}, -2^{-1/4} \pm i2^{-1/4}, 1, -\frac{1}{2} \pm \frac{\sqrt{3}}{2}i$

**5.** $\frac{\pi}{6} + 2k\pi, \frac{5\pi}{6} + 2k\pi, \frac{2\pi}{3} + 2k\pi, \frac{4\pi}{3} + 2k\pi$

**6.** $-\frac{1}{2}, \frac{\pi}{6} + 2k\pi, \frac{5\pi}{6} + 2k\pi$ **7.** $k\pi$

**8.** $\ln\left(\frac{\pi}{6} + 2k\pi\right), \ln\left(\frac{5\pi}{6} + 2k\pi\right)$ for $k$ a nonnegative integer

**9.** 4 **10.** $\varnothing$

**11.**

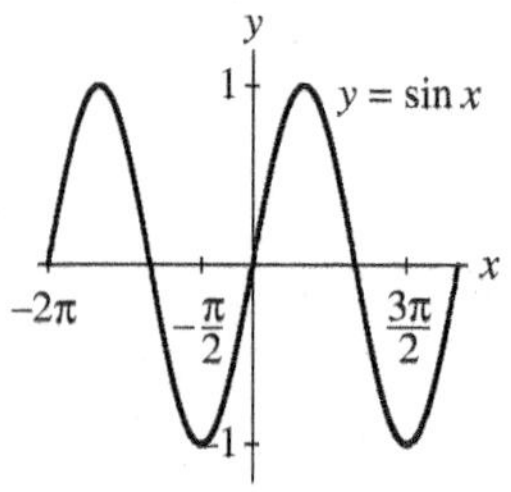


**12.**

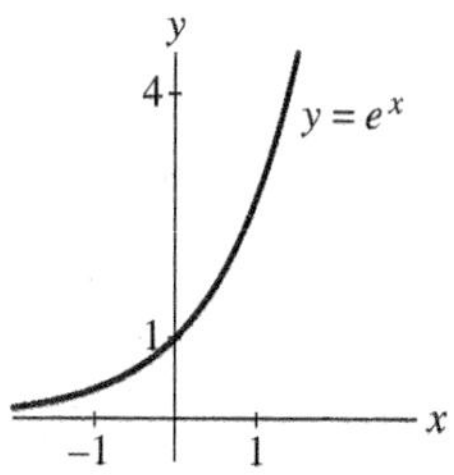


**13.**

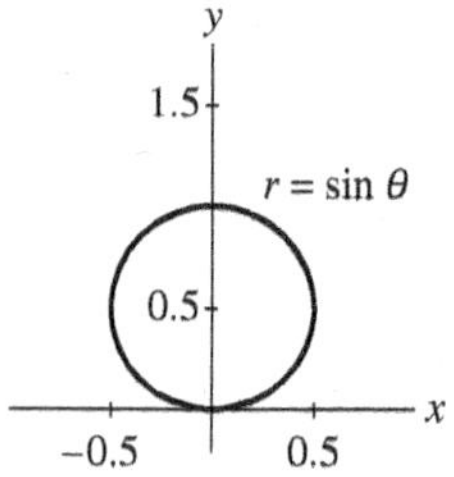


**14.**

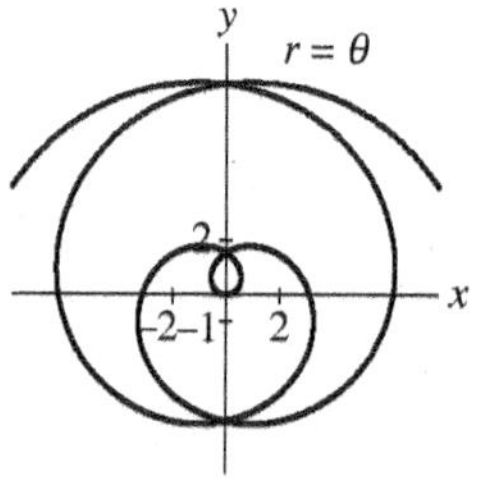


**15.**

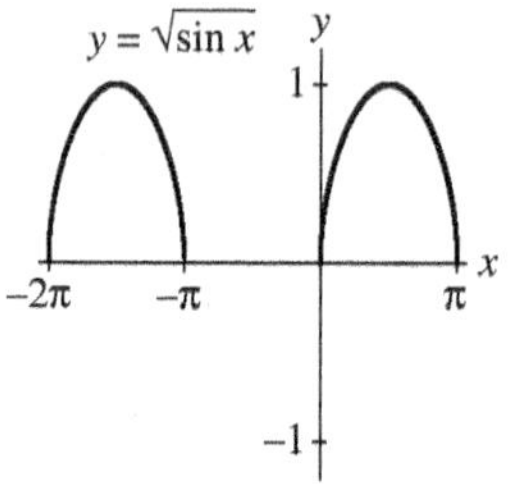


**16.**

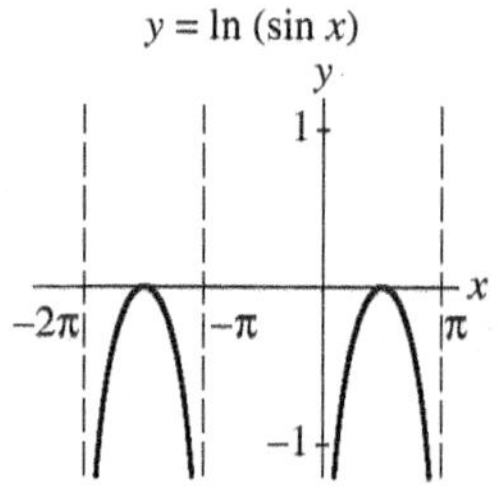


**17.**

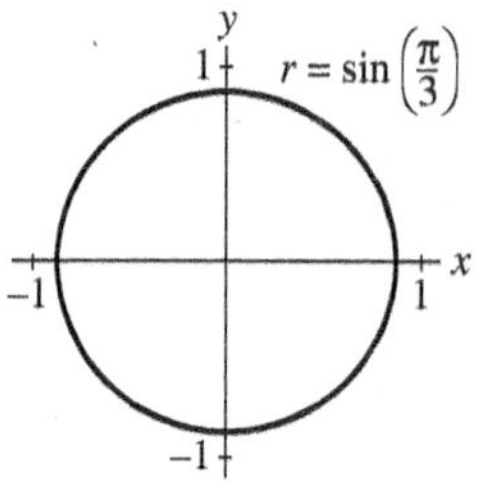


**18.**

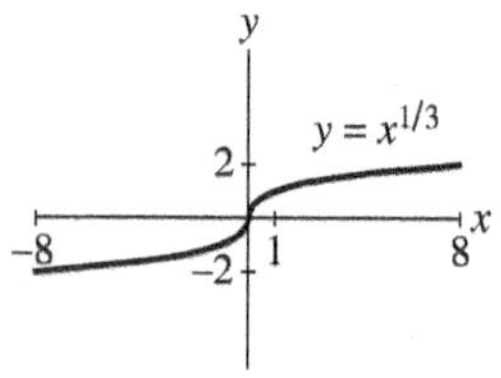


**19.** 0 **20.** 0 **21.** −1 **22.** 0 **23.** $\pi/6$ **24.** $\pi/3$

**25.** $-\pi/4$ **26.** $\pi/4$ **27.** $\frac{2x}{x^2 - 4}$ **28.** $2\sec^2(x)$ **29.** 4

**30.** $-\log(x^2 - 4)$ **31.** 2 **32.** 1/4 **33.** 2 **34.** 0

**35.** $\frac{-2x}{x - 3}$ **36.** 1 **37.** frequency **38.** vertical asymptote

**39.** even **40.** odd **41.** opposite, hypotenuse

**42.** adjacent, hypotenuse **43.** one **44.** period

**45.** Pythagorean **46.** even, odd

# Index of Applications

## Archeology



## Astronomy



## Biology/Health/Life Sciences



## Business

## Chemistry



## Construction



## Consumer



## Design

## Environment



## Geometry



## Investment



## Miscellaneous

## Navigation



## Number



## Physics



## Probability

## Production



## Science



## Sports



## Statistics/Demographics

# Index

## H



## I

## T

# Geometry

### Rectangle

Area = $LW$

Perimeter = $2L + 2W$

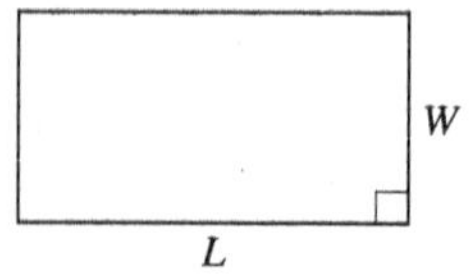


### Square

Area = $s^2$

Perimeter = $4s$

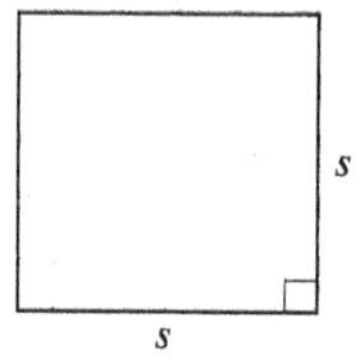


### Triangle

Area = $\frac{1}{2}bh$

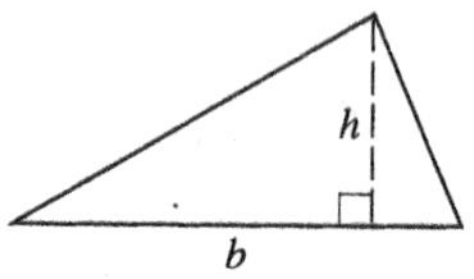


### Right Triangle

Area = $\frac{1}{2}ab$

Pythagorean theorem:
$c^2 = a^2 + b^2$

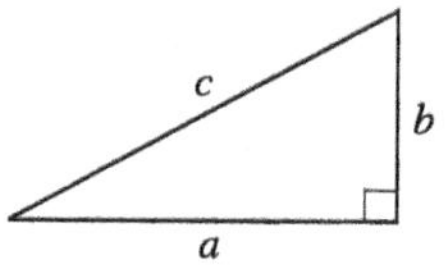


### Parallelogram

Area = $bh$

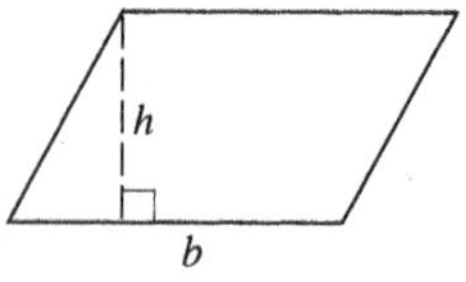


### Trapezoid

Area = $\frac{1}{2}h(b_1 + b_2)$

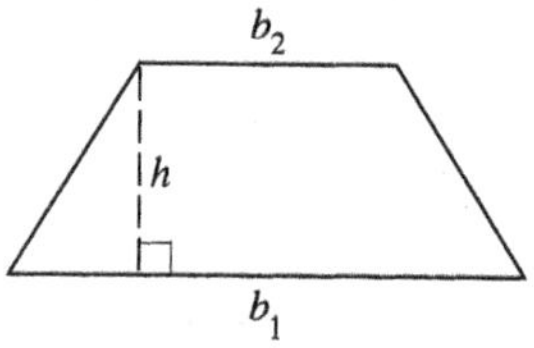


### Circle

Area = $\pi r^2$

Circumference = $2\pi r$

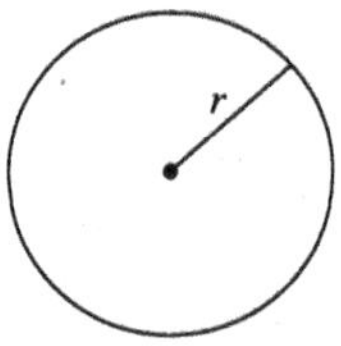


### Right Circular Cone

Volume = $\frac{1}{3}\pi r^2 h$

Lateral surface area = $\pi r\sqrt{r^2 + h^2}$

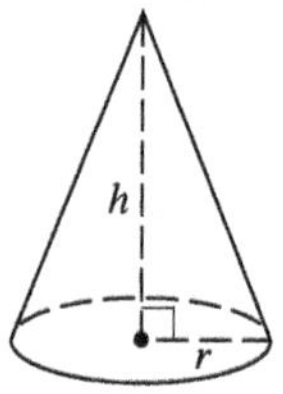


### Right Circular Cylinder

Volume = $\pi r^2 h$

Lateral surface area = $2\pi rh$

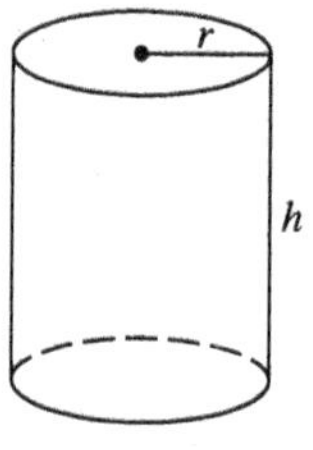


### Sphere

Volume = $\frac{4}{3}\pi r^3$

Surface area = $4\pi r^2$

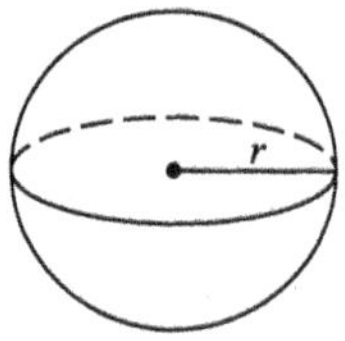


## Metric Abbreviations

| Length | | Volume | | Weight | |
|---|---|---|---|---|---|
| mm | millimeter | mL | milliliter | mg | milligram |
| cm | centimeter | cL | centiliter | cg | centigram |
| dm | decimeter | dL | deciliter | dg | decigram |
| m | meter | L | liter | g | gram |
| dam | dekameter | daL | dekaliter | dag | dekagram |
| hm | hectometer | hL | hectoliter | hg | hectogram |
| km | kilometer | kL | kiloliter | kg | kilogram |

## English-Metric Conversion

| Length | Volume (U.S.) | Weight |
|---|---|---|
| 1 in. = 2.540 cm | 1 pt = 0.4732 L | 1 oz = 28.35 g |
| 1 ft = 30.48 cm | 1 qt = 0.9464 L | 1 lb = 453.6 g |
| 1 yd = 0.9144 m | 1 gal = 3.785 L | 1 lb = 0.4536 kg |
| 1 mi = 1.609 km | | |

| Length | Volume (U.S.) | Weight |
|---|---|---|
| 1 cm = 0.3937 in. | 1 L = 2.2233 pt | 1 g = 0.0353 oz |
| 1 cm = 0.03281 ft | 1 L = 1.0567 qt | 1 g = 0.002205 lb |
| 1 m = 1.0936 yd | 1 L = 0.2642 gal | 1 kg = 2.205 lb |
| 1 km = 0.6215 mi | | |

# Algebra

## Subsets of the Real Numbers

Natural numbers = $\{1, 2, 3, \ldots\}$
Whole numbers = $\{0, 1, 2, 3, \ldots\}$
Integers = $\{\ldots -3, -2, -1, 0, 1, 2, 3, \ldots\}$
Rational = $\left\{\frac{a}{b} \middle| a \text{ and } b \text{ are integers with } b \neq 0\right\}$
Irrational = $\{x|x \text{ is not rational}\}$

## Properties of the Real Numbers

For all real numbers $a$, $b$, and $c$

| | | |
|---|---|---|
| $a + b$ and $ab$ are real numbers. | | Closure |
| $a + b = b + a$; $a \cdot b = b \cdot a$ | | Commutative |
| $(a + b) + c = a + (b + c)$; $(ab)c = a(bc)$ | | Associative |
| $a(b + c) = ab + ac$; $a(b - c) = ab - ac$ | | Distributive |
| $a + 0 = a$; $1 \cdot a = a$ | | Identity |
| $a + (-a) = 0$; $a \cdot \frac{1}{a} = 1$ | $(a \neq 0)$ | Inverse |
| $a \cdot 0 = 0$ | | Multiplication property of 0 |

## Absolute Value

$$|a| = \begin{cases} a & \text{for } a \geq 0 \\ -a & \text{for } a < 0 \end{cases}$$

$\sqrt{x^2} = |x|$ for any real $x$

$|x| = k \Leftrightarrow x = k$ or $x = -k \quad (k > 0)$

$|x| < k \Leftrightarrow -k < x < k \quad (k > 0)$

$|x| > k \Leftrightarrow x < -k$ or $x > k \quad (k > 0)$

(The symbol $\Leftrightarrow$ means "if and only if.")

## Interval Notation

$(a, b) = \{x|a < x < b\}$ $\qquad$ $[a, b] = \{x|a \leq x \leq b\}$

$(a, b] = \{x|a < x \leq b\}$ $\qquad$ $[a, b) = \{x|a \leq x < b\}$

$(-\infty, a) = \{x|x < a\}$ $\qquad$ $(a, \infty) = \{x|x > a\}$

$(-\infty, a] = \{x|x \leq a\}$ $\qquad$ $[a, \infty) = \{x|x \geq a\}$

## Exponents

$a^n = a \cdot a \cdot \cdots \cdot a$ ($n$ factors of $a$)

$a^0 = 1 \qquad a^{-n} = \frac{1}{a^n}$

$a^r a^s = a^{r+s} \qquad \frac{a^r}{a^s} = a^{r-s}$

$(a^r)^s = a^{rs} \qquad (ab)^r = a^r b^r$

$\left(\frac{a}{b}\right)^r = \frac{a^r}{b^r} \qquad \left(\frac{a}{b}\right)^{-r} = \left(\frac{b}{a}\right)^r$

## Radicals

$a^{1/n} = \sqrt[n]{a} \qquad a^{m/n} = (\sqrt[n]{a})^m = \sqrt[n]{a^m}$

$\sqrt[n]{ab} = \sqrt[n]{a} \cdot \sqrt[n]{b} \qquad \sqrt[n]{\frac{a}{b}} = \frac{\sqrt[n]{a}}{\sqrt[n]{b}}$

## Factoring

$a^2 + 2ab + b^2 = (a + b)^2$

$a^2 - 2ab + b^2 = (a - b)^2$

$a^2 - b^2 = (a + b)(a - b)$

$a^3 - b^3 = (a - b)(a^2 + ab + b^2)$

$a^3 + b^3 = (a + b)(a^2 - ab + b^2)$

## Rational Expressions

$\frac{ac}{bc} = \frac{a}{b} \qquad \frac{a}{b} + \frac{c}{d} = \frac{ad + bc}{bd}$

$\frac{a}{b} \cdot \frac{c}{d} = \frac{ac}{bd} \qquad \frac{a}{b} \div \frac{c}{d} = \frac{a}{b} \cdot \frac{d}{c}$

## Quadratic Formula

The solutions to $ax^2 + bx + c = 0$ with $a \neq 0$ are

$$x = \frac{-b \pm \sqrt{b^2 - 4ac}}{2a}.$$

## Distance Formula

The distance from $(x_1, y_1)$ to $(x_2, y_2)$, is

$$\sqrt{(x_2 - x_1)^2 + (y_2 - y_1)^2}.$$

# Algebra

### Midpoint Formula

The midpoint of the line segment with endpoints $(x_1, y_1)$ and $(x_2, y_2)$ is

$$\left(\frac{x_1 + x_2}{2}, \frac{y_1 + y_2}{2}\right).$$

### Slope Formula

The slope of the line through $(x_1, y_1)$ and $(x_2, y_2)$ is

$$\frac{y_2 - y_1}{x_2 - x_1} \quad \text{(for } x_1 \neq x_2\text{)}.$$

### Linear Function

$f(x) = mx + b$ with $m \neq 0$

Graph is a line with slope $m$.

### Quadratic Function

$f(x) = ax^2 + bx + c$ with $a \neq 0$

Graph is a parabola.

### Polynomial Function

$f(x) = a_n x^n + a_{n-1}x^{n-1} + \cdots + a_1 x + a_0$ for $n$ a nonnegative integer

### Rational Function

$f(x) = \dfrac{p(x)}{q(x)}$, where $p$ and $q$ are polynomial functions with $q(x) \neq 0$

### Exponential and Logarithmic Functions

$f(x) = a^x$ for $a > 0$ and $a \neq 1$

$f(x) = \log_a(x)$ for $a > 0$ and $a \neq 1$

### Properties of Logarithms

Base-$a$ logarithm: $y = \log_a(x) \Leftrightarrow a^y = x$

Natural logarithm: $y = \ln(x) \Leftrightarrow e^y = x$

Common logarithm: $y = \log(x) \Leftrightarrow 10^y = x$

One-to-one: $a^{x_1} = a^{x_2} \Leftrightarrow x_1 = x_2$

$\log_a(x_1) = \log_a(x_2) \Leftrightarrow x_1 = x_2$

$\log_a(a) = 1$ $\quad \log_a(1) = 0$

$\log_a(a^x) = x$ $\quad a^{\log_a(N)} = N$

$\log_a(MN) = \log_a(M) + \log_a(N)$

$\log_a(M/N) = \log_a(M) - \log_a(N)$

$\log_a(M^x) = x \cdot \log_a(M)$

$\log_a(1/N) = -\log_a(N)$

$$\log_a(M) = \frac{\log_b(M)}{\log_b(a)} = \frac{\ln(M)}{\ln(a)} = \frac{\log(M)}{\log(a)}$$

### Compound Interest

$P$ = principal, $t$ = time in years, $r$ = annual interest rate, and $A$ = amount:

$$A = P\left(1 + \frac{r}{n}\right)^{nt} \text{ (compounded } n \text{ times/year)}$$

$$A = Pe^{rt} \text{ (compounded continuously)}$$

### Variation

Direct: $y = kx$ $\quad (k \neq 0)$

Inverse: $y = k/x$ $\quad (k \neq 0)$

Joint: $y = kxz$ $\quad (k \neq 0)$

### Straight Line

Slope-intercept form: $y = mx + b$

Slope: $m$ $\quad$ $y$-intercept: $(0, b)$

Point-slope form: $y - y_1 = m(x - x_1)$

Standard form: $Ax + By = C$

Horizontal: $y = k$ $\quad$ Vertical: $x = k$

# Algebra

## Parabola

$y = a(x - h)^2 + k \qquad (a \neq 0)$

Vertex: $(h, k)$

Axis of symmetry: $x = h$

Focus: $(h, k + p)$, where $a = \dfrac{1}{4p}$

Directrix: $y = k - p$

## Circle

$(x - h)^2 + (y - k)^2 = r^2 \qquad (r > 0)$

Center: $(h, k)$ Radius: $r$

$x^2 + y^2 = r^2$

Center $(0, 0)$ Radius: $r$

## Ellipse

$\dfrac{x^2}{a^2} + \dfrac{y^2}{b^2} = 1 \qquad (a > b > 0)$

Center: $(0, 0)$ Major axis: horizontal

Foci: $(\pm c, 0)$, where $c^2 = a^2 - b^2$

$\dfrac{x^2}{b^2} + \dfrac{y^2}{a^2} = 1 \qquad (a > b > 0)$

Center: $(0, 0)$ Major axis: vertical

Foci: $(0, \pm c)$, where $c^2 = a^2 - b^2$

## Hyperbola

$\dfrac{x^2}{a^2} - \dfrac{y^2}{b^2} = 1$

Center: $(0, 0)$ Vertices: $(\pm a, 0)$

Foci: $(\pm c, 0)$, where $c^2 = a^2 + b^2$

Asymptotes: $y = \pm\dfrac{b}{a}x$

$\dfrac{y^2}{a^2} - \dfrac{x^2}{b^2} = 1$

Center: $(0, 0)$ Vertices: $(0, \pm a)$

Foci: $(0, \pm c)$, where $c^2 = a^2 + b^2$

Asymptotes: $y = \pm\dfrac{a}{b}x$

## Arithmetic Sequence

$a_1, a_1 + d, a_1 + 2d, a_1 + 3d, \ldots$

Formula for $n$th term: $a_n = a_1 + (n - 1)d$

Sum of $n$ terms:

$$S_n = \sum_{i=1}^{n} [a_1 + (i - 1)d] = \frac{n}{2}(a_1 + a_n)$$

## Geometric Sequence

$a_1, a_1r, a_1r^2, a_1r^3, \ldots$

Formula for $n$th term: $a_n = a_1r^{n-1}$

Sum of $n$ terms when $r \neq 1$:

$$S_n = \sum_{i=1}^{n} a_1r^{i-1} = \frac{a_1 - a_1r^n}{1 - r}$$

Sum of all terms when $|r| < 1$:

$$S = \sum_{i=1}^{\infty} a_1r^{i-1} = \frac{a_1}{1 - r}$$

## Counting Formulas

Factorial notation: $n! = 1 \cdot 2 \cdot 3 \cdot \cdots \cdot (n - 1) \cdot n$

Permutation: $P(n, r) = \dfrac{n!}{(n - r)!}$ for $0 \leq r \leq n$

Combination: $C(n, r) = \dbinom{n}{r} = \dfrac{n!}{(n - r)!r!}$ for $0 \leq r \leq n$

## Binomial Expansion

$(a + b)^2 = a^2 + 2ab + b^2$

$(a + b)^3 = a^3 + 3a^2b + 3ab^2 + b^3$

$(a + b)^4 = a^4 + 4a^3b + 6a^2b^2 + 4ab^3 + b^4$

$(a + b)^n = \sum_{r=0}^{n} \dbinom{n}{r} a^{n-r}b^r$, where $\dbinom{n}{r} = \dfrac{n!}{(n - r)!r!}$

# Trigonometry

## Trigonometric Functions

If the angle $\alpha$ (in standard position) intersects the unit circle at $(x, y)$, then

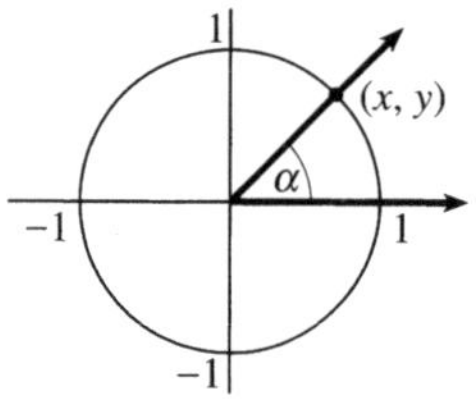


$\sin \alpha = y$ $\qquad \cos \alpha = x$ $\qquad \tan \alpha = \dfrac{y}{x}$

$\csc \alpha = \dfrac{1}{y}$ $\qquad \sec \alpha = \dfrac{1}{x}$ $\qquad \cot \alpha = \dfrac{x}{y}$

## Trigonometric Ratios

If $(x, y)$ is any point other than the origin on the terminal side of $\alpha$ and $r = \sqrt{x^2 + y^2}$, then

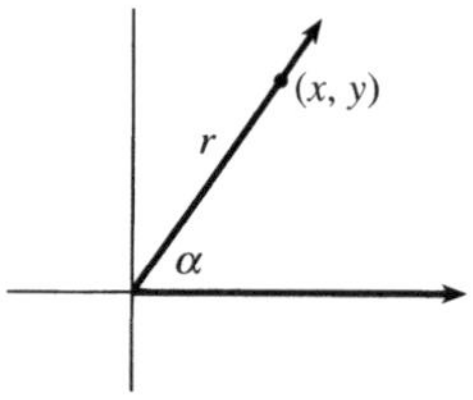


$\sin \alpha = \dfrac{y}{r}$ $\qquad \cos \alpha = \dfrac{x}{r}$ $\qquad \tan \alpha = \dfrac{y}{x}$

$\csc \alpha = \dfrac{r}{y}$ $\qquad \sec \alpha = \dfrac{r}{x}$ $\qquad \cot \alpha = \dfrac{x}{y}$

## Right Triangle Trigonometry

If $\alpha$ is an acute angle of a right triangle, then

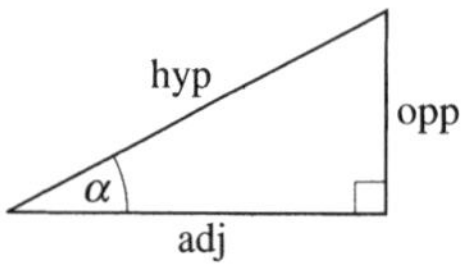


$\sin \alpha = \dfrac{\text{opp}}{\text{hyp}}$ $\qquad \cos \alpha = \dfrac{\text{adj}}{\text{hyp}}$ $\qquad \tan \alpha = \dfrac{\text{opp}}{\text{adj}}$

$\csc \alpha = \dfrac{\text{hyp}}{\text{opp}}$ $\qquad \sec \alpha = \dfrac{\text{hyp}}{\text{adj}}$ $\qquad \cot \alpha = \dfrac{\text{adj}}{\text{opp}}$

## Special Right Triangles

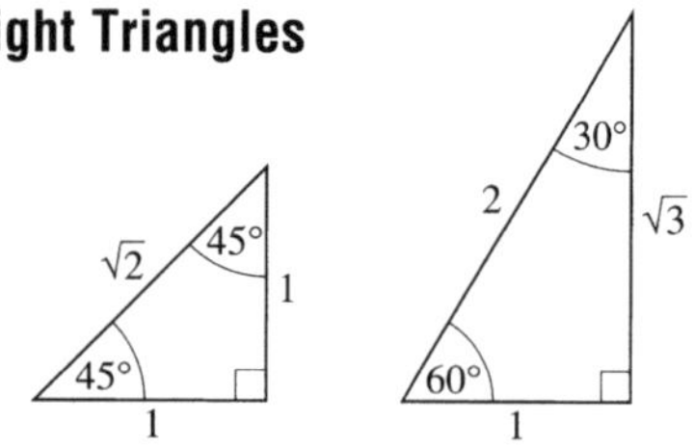


## Exact Values of Trigonometric Functions

| $x$ degrees | $x$ radians | $\sin x$ | $\cos x$ | $\tan x$ |
|---|---|---|---|---|
| 0° | 0 | 0 | 1 | 0 |
| 30° | $\frac{\pi}{6}$ | $\frac{1}{2}$ | $\frac{\sqrt{3}}{2}$ | $\frac{\sqrt{3}}{3}$ |
| 45° | $\frac{\pi}{4}$ | $\frac{\sqrt{2}}{2}$ | $\frac{\sqrt{2}}{2}$ | 1 |
| 60° | $\frac{\pi}{3}$ | $\frac{\sqrt{3}}{2}$ | $\frac{1}{2}$ | $\sqrt{3}$ |
| 90° | $\frac{\pi}{2}$ | 1 | 0 | — |

## Basic Identities

$\tan x = \dfrac{\sin x}{\cos x} = \dfrac{1}{\cot x}$ $\qquad \cot x = \dfrac{\cos x}{\sin x} = \dfrac{1}{\tan x}$

$\sin x = \dfrac{1}{\csc x}$ $\qquad \csc x = \dfrac{1}{\sin x}$

$\cos x = \dfrac{1}{\sec x}$ $\qquad \sec x = \dfrac{1}{\cos x}$

## Pythagorean Identities

$\sin^2 x + \cos^2 x = 1$ $\qquad 1 + \cot^2 x = \csc^2 x$

$\tan^2 x + 1 = \sec^2 x$

## Odd Identities

$\sin(-x) = -\sin(x)$ $\qquad \csc(-x) = -\csc(x)$

$\tan(-x) = -\tan(x)$ $\qquad \cot(-x) = -\cot(x)$

## Even Identities

$\cos(-x) = \cos(x)$ $\qquad \sec(-x) = \sec(x)$